Coral Reef Fishes

Coral Reef Fishes

Dynamics and Diversity in a Complex Ecosystem

Edited by

Peter F. Sale

Department of Biological Sciences and
Great Lakes Institute for Environmental Research
University of Windsor
Windsor, Ontario, Canada

ELSEVIER

Amsterdam Boston Heidelberg London New York Oxford
Paris San Diego San Francisco Singapore Sydney Tokyo
Academic Press is an imprint of Elsevier

Academic Press is an imprint of Elsevier
30 Corporate Drive, Suite 400, Burlington, MA 01803, USA
525 B Street, Suite 1900, San Diego, California 92101-4495, USA
84 Theobald's Road, London WC1X 8RR, UK

This book is printed on acid-free paper. ♾

Library of Congress Cataloging-in-Publication Data
Application Submitted

British Library Cataloguing-in-Publication Data
A catalogue record for this book is available from the British Library.

ISBN 13: 978-0-12-373609-3
ISBN 10: 0-12-373609-9

For information on all Academic Press publications
visit our Web site at www.books.elsevier.com

Printed in the United States of America
06 07 08 09 10 9 8 7 6 5 4 3 2 1

Chris

Here briefly, learning, one with nature.
Memories swim ever gently.

Contents

Contributors

Numbers in parentheses indicate the pages on which the authors' contributions begin.

David R. Bellwood *(5, 33), Centre for Coral Reef Biodiversity, School of Marine Biology and Aquaculture, James Cook University, Townsville, Queensland 4811, Australia*

M. Julian Caley *(81), School of Marine Biology and Aquaculture, James Cook University, Townsville, Queensland 4811, Australia*

J. Howard Choat *(57), School of Marine Biology and Aquaculture, James Cook University, Townsville, Queensland 4811, Australia*

Robert K. Cowen *(149), Rosensteil School of Marine and Atmospheric Science, University of Miami, Miami, Florida 33149*

Peter J. Doherty *(327), Australian Institute of Marine Science, Cape Ferguson, Townsville, Queensland 4810, Australia*

Graham E. Forrester *(275), Department of Biological Sciences, University of Rhode lsland, Kingston, Rhode Island 02881*

Lee A. Fuiman *(123), Department of Marine Science, University of Texas at Austin, Marine Science Institute, Port Aransas, Texas 78373*

Churchill B. Grimes *(377), National Marine Fisheries Service, Southwest Fisheries Science Center, Santa Cruz Laboratory, Santa Cruz, California 95060*

Jonathan A. Hare *(243), National Oceanic and Atmospheric Administration, National Ocean Service, National Centers for Coastal Ocean Science, Center for Coastal Fisheries and Habitat Research, Beaufort, North Carolina 28516*

Mireille L. Harmelin-Vivien *(265), Centre d'Océanologie de Marseille, CNRS UMR 6540, Université de la Méditerranée-Station Marine d'Endoume 13007 Marseille, France*

Mark A. Hixon *(303), Department of Zoology, Oregon State University, Corvallis, Oregon 97331*

Geoffrey P. Jones *(81, 221), Centre for Coral Reef Biodiversity, School of Marine Biology and Aquaculture, James Cook University, Townsville, Queensland 4811, Australia*

Jeffrey M. Lcis *(171), Ichthyology and Center for Biodiversity and Conservation Research, Australian Museum, Sydney NSW 2010, Australia*

Phillip S. Levin *(377), National Marine Fisheries Service, Seattle, Washington 98112*

Mark I. McCormick *(171, 221), School of Marine Biology and Aquaculture, James Cook University, Townsville, Queensland 4811, Australia*

Philip L. Munday *(81), Centre for Coral Reef Biodiversity, School of Marine Biology and Aquaculture, James Cook University, Townsville, Queensland 4811, Australia*

Arthur A. Myrberg, Jr. *(123), Rosensteil School of Marine and Atmospheric Science, University of Miami, Miami, Florida 33149*

Christopher W. Petersen *(103), College of the Atlantic, Bar Harbor, Maine 04609*

Serge Planes *(201), Centre de Biologie et d'Écologie Tropicale et Méditerranée, EPHE ESA 8046 CNRS, Université de Perpignan, Perpignan 66860, France*

D. Ross Robertson *(57), Smithsonian Tropical Research Institute, Balboa, Panama*

Garry R. Russ *(421), School of Marine Biology and Aquaculture, James Cook University, Townsville, Queensland 4811, Australia*

Yvonne J. Sadovy *(391), Department of Ecology & Biodiversity, The University of Hong Kong, Hong Kong*

Peter F. Sale *(361), Department of Biological Sciences and Great Lakes Institute for Environmental Research, University of Windsor, Windsor, Ontario, Canada N9B 3P4*

Mark A. Steele *(275), Department of Biological Sciences, University of Rhode Island, Kingston, Rhode Island 02881*

Simon R. Thorrold *(243), Biology Department, Woods Hole Oceanographic Institution, Woods Hole, Massachusetts 02543*

Richard R. Vance *(275), Department of Organismic Biology, Ecology and Evolution, University of California at Los Angeles, Los Angeles, California 90095*

Amanda C. J. Vincent *(391), Project Seahorse, Department of Biology, McGill University, Penfield, Montreal, Quebec H3A 1B1, Canada*

Peter C. Wainwright *(5, 33), Center for Population Biology, University of California at Davis, Davis, California 95616*

Robert R. Warner *(103), Department of Ecology, Evolution, and Marine Biology, University of California, Santa Barbara, Santa Barbara, California 93106*

Michael S. Webster *(303), Department of Zoology, Oregon State University, Corvallis, Oregon 97331*

Preface

The impetus to produce this book came in a brief phone call in 1998. Chuck Crumly, of Academic Press, called with an invitation and a deadline. Either *The Ecology of Fishes on Coral Reefs,* published in 1991, would be allowed to lapse into out-of-print status, or I would agree to produce a second edition. Looking back on all the work, I suspect it might have been wiser to say, "Let her lapse." But I didn't. During my deliberations, I thought about whether a new edition was worthwhile, whether further books on the topic were justified, and what my colleagues would say if I came seeking chapter authors. I told Academic Press that a second edition was unrealistic, but that an entirely new book, that would visit some of the same topics, was a definite possibility. Then began the search for willing contributors.

My intention from the beginning was to produce a book that would speak to graduate students, to scientists in the field, to reef managers and others interested in coral reefs, and to the wider ecological and scientific community. I am confident that this book will do so, and will open new doors that attract new people to become direct participants in this exciting field. The 30 contributors (including myself) include 15 based in the United States, 9 in Australia, two in Canada, two in France, one in Panama, and one in Hong Kong. (Lest my American colleagues read this as a sign of their preeminence, 11 of 19 chapters include authors with significant Australian experience while just 5 have exclusively American parentage. And, to keep my Aussie friends under control—some of us have left your shores, mates.) The chapters provide comprehensive coverage of the major fields of ecology of reef fishes currently being investigated, essential reviews in several cognate areas, and four chapters devoted to science of management issues. As they arrived over the last 18 months, and I had a chance to read them, their quality provided the spur to ensure I did the things I had to do to get the book to press. There is some excellent work here, and I thank each of the contributors for working hard to produce a quality product, for putting up with my demands, and for fulfilling my requests, usually in a timely manner.

The book is divided into four sections with a brief introduction to each. While the sections group together chapters with thematic similarities, there are many instances where chapters in one section make points of relevance to chapters in other sections. Nevertheless, a sequence from Chapter 1 to Chapter 19 makes reasonable sense, and, if I used it in a graduate seminar, that's the sequence I would follow.

I knew that growth in this field had been substantial, but in finalizing the bibliography for this book, I realized just how great it had been. When Paul Ehrlich (1975) reviewed the population biology of reef fishes, he did a thorough job in 36 pages and cited 313 references going back to 1908, including a handful or two from prior to 1950. In Sale (1980), I reviewed the field in 54 pages, citing 318 references, nearly all of which were from the 1960s and 1970s. *The Ecology of Fishes on Coral Reefs* required 754 pages, of which 87 pages comprised a bibliography of about 1690 citations, mostly from the 1970s and 1980s. The present book contains over 2580 citations, of which more than 60% are from 1990 or later, while just 14% are from the 1970s or earlier. Further, the present book is less comprehensive than the former, and whole fields of ecology are omitted to keep the book to manageable size. There are a lot more people doing reef fish ecology now than there were as little as 10 years ago.

The other change in this field has been the growing awareness by reef fish ecologists that our study animals are not only wonderful, but valuable, rare, and becoming rarer. I hope that this book will encourage still more ecologists to explore reef fishes as model organisms with which to ask important and fundamental ecological questions, and to this end, most chapters close with questions for the future. But I hope, even more, that this book will encourage ecologists to use their science to contribute to much more effective management of our impact on reef fish and the other components of coral reef systems. There is good, intellectually stimulating science that is desperately needed if we are to manage these systems sustainably in the future. I want somebody to write a new book on reef fish ecology in 10 years and to be able to keep it in the present tense.

I have already thanked the contributors. In putting the final manuscript together, I was helped by two undergraduates in turn: Nick Kamenos, who worked in my lab through the fall of 2000, and Allison Pratt, who worked there through the spring and summer of 2001. Each provided the careful attention that allowed me to assemble a pooled bibliography with minimal mistakes, and they did the work cheerfully. I thank Caroline Lekic who came to my aid at a critical point as we compiled the index. Finally, I cannot adequately thank two special people, Donna and Darian, who make my life worthwhile, while somehow understanding that I sometimes neglect them, only because I do love what I do.

Peter F. Sale

SECTION I

Reef Fishes

A Diversity of Adaptations and Specializations

Introduction

Ecology is a holistic science that seeks a broad understanding of the relationships among organisms, their environment, and, increasingly, humans. Though the science of ecology has its own specialties, ecologists must remain well versed in the methods, the goals, and the knowledge accumulated in other fields. In the past decade or so, rapid developments in instrumentation, in techniques and analytical approaches, and in general knowledge in fields as remote as molecular biology and physical oceanography have made keeping up with progress in these allied fields ever more difficult. Yet, more than ever, the challenges facing ecologists can be answered only by using the knowledge gained by workers in other fields, and by building collaborations that go beyond the boundaries of ecology. Several of these other fields are discussed in later parts of this book; however, this first part deliberately attempts to provoke awareness and generate new questions.

The five chapters that make up Section I all focus on ecological science. Yet each chapter also looks at an edge of ecological exploration, and will help fill in gaps in knowledge of some cognate fields. I believe that each chapter has the capacity to stimulate new questions and new approaches by those who will do the future work in reef fish ecology.

In Chapter 1, Dave Bellwood and Peter Wainwright begin at the beginning, with a review of the origins of reef fishes and reef fish faunas, reminding us of how advanced reef fish assemblages are, and how relatively recently derived are those species that now dominate coral reef systems. (That they refer to this 50-million-year-ago history as a long one confirms they are, at heart, ecologists rather than paleontologists!)

Bellwood and Wainwright review this history in a way that facilitates understanding of the biogeographic features of coral reef systems. Given that ecological studies are increasingly being done on larger spatial scales, their discussion of how reef fish assemblages differ from place to place is particularly helpful. Their question of whether reef fishes have played an important role in facilitating the development of coral reefs is particularly provocative.

Wainwright and Bellwood, in Chapter 2, shift from phylogeny to functional morphology, developing a picture of the feeding ecology of reef fishes as driven by morphological possibilities and constraints. For those readers brought up with classical ichthyology, this chapter will be a refreshing update, but not a surprise. But too many students of ecology now manage to bypass "old-fashioned" courses, and, for them, this chapter may open a new door to appreciating that ecology is the result of interactions of real organisms that have physical limitations and possibilities.

In Chapter 3, Howard Choat and Ross Robertson steer a more narrowly defined ecological path, but set out a strong argument for radically revising the way reef fish ecologists and fisheries biologists have approached demographic questions. They argue, convincingly, that it is possible to age coral reef species, and demonstrate that the results of so doing are going to cause some significant revisions in the "conventional wisdom" concerning longevity and growth rates in these animals. Given that so much fisheries science depends on knowledge of age structures and growth rates, their argument has importance for management as well as for ecology.

Geoff Jones, Julian Caley, and Philip Munday use Chapter 4 to raise a difficult ecological question—how to account for the existence of rarity. Although reef fish assemblages are noted for their high diversity relative to most other assemblages of fish, relative abundance of species is typically log-normally distributed and there are always many species that are locally rare. Many of these locally rare species also are regionally, if not universally, rare. How do we account for the successful persistence of species that are so uncommon? Chapter 4 stimulates thinking on a vexing problem, and is a reminder that knowing still more about the commonest species is not going to provide all the answers.

In Chapter 5, Chris Petersen and Bob Warner turn attention to another of the boundaries of ecology and address the behavioral adaptations of reef fish reproduction. There was a time 25–30 years ago when there was more behavioral than ecological research done on reef fishes. For reasons that are not entirely clear, the quantity of behavioral research in this system has not grown along with that of ecological research; however, studies on a few topics (by a few particular investigators) continue to demonstrate that the reef fish system is very manageable for sophisticated explorations of behavioral questions in field settings. The question of evolution of behavioral processes, particularly with respect to reproductive and parental activities, has been a fruitful area for research, and this chapter provides an introduction to this topic from two of the leaders. Their section on applied behavioral ecology should convince readers that behavioral science remains "relevant," and that there are potentially important consequences if we ignore behavioral science in managing our impacts on reef fishes.

CHAPTER 1

The History and Biogeography of Fishes on Coral Reefs

David R. Bellwood
Centre for Coral Reef Biodiversity
School of Marine Biology and Aquaculture
James Cook University
Townsville, Queensland 4811, Australia

Peter C. Wainwright
Center for Population Biology
University of California at Davis
Davis, California 95616

I. Introduction

Coral reefs have been around since the Ordovician (Wood, 1999), and throughout their 450-million-year history they have shared the oceans with fishes. Modern scleractinian-dominated coral reefs and their associated fish faunas represent only the latest manifestation of a reefal ecosystem. It is almost self-evident that history is important to coral reefs, as the reefs build on the skeletons of past generations. But what of the associated fauna? Today, fishes form an integral part of reef communities, modifying benthic community structure and forming a major conduit for the movement of energy and material. Like the reefs, reef fish faunas have been shaped by history, but this historical influence may not be as apparent. Although it is becoming increasingly clear that history plays an important role in structuring local communities (Rickleffs and Schluter, 1993a), its influence on the ecology and biogeography of fishes on coral reefs remains largely unknown.

Most studies of reef systems have addressed the question of how biogeographic and ecological patterns are maintained; relatively few consider how these patterns arose or their consequences. However, it is the combination of these two factors, origins and maintenance, that offers the clearest understanding of the nature of biogeographic patterns in reef organisms. Studies of the history of coral reefs have been largely restricted to documenting the history of the reef builders, which have left an outstanding fossil record (Wood, 1999). The history of associated faunas, and fish in particular, is less clear. However, this is changing, primarily as a result of phylogenetic analyses of reef fishes and from a reappraisal of the fossil record.

Until recently, historical considerations of reef fishes were restricted largely to studies by museum workers (e.g., Allen, Randall, Springer, Winterbottom) who examined the taxonomy, systematics, and biogeography of extant reef fishes. Paleontological information has likewise been confined to the works of specialists in museums. Workers such as Blot, Sorbini, and Tyler have provided a sound basis for the evaluation of the fossil record of reef fishes. The broader application of these findings to present-day ecology, community structure, and ecosystem function has only recently begun to be considered. Ecologists are looking increasingly at data from large temporal and spatial scales to provide a framework within which to interpret local patterns and small-scale experimental results. It is from this integration of systematics, biogeography, ecology, and paleontology that a new understanding of the nature of reef fishes is arising.

In this chapter we summarize our knowledge of the phylogenetics, paleontology, and biogeography of fishes on coral reefs and examine how these data, along with geological evidence, can aid our understanding of the role of historical factors in shaping modern coral reef fish faunas and their ecological attributes. In particular, we wish to address several specific questions:

1. What are coral reef fishes, when did they appear, and where did they come from?
2. Are Caribbean and Indo-Pacific reef fish assemblages comparable, and how do we explain major differences in reef fish assemblages across the Indo-Pacific?
3. How tight is the reef fish–coral reef association, and how do we evaluate the interaction between fishes and coral reefs?
4. What role have fishes played in the evolution of coral reefs, and is there any evidence of a change in this role over time?

II. Reef Fishes: Definitions and Distribution Patterns

Reef fishes are often seen as a distinctive and easily characterized group of fishes. However, though numerous texts and papers refer to "reef fishes," the uniting characteristics of these assemblages are rarely defined. Although there have been several attempts to characterize the essence of a reef fish, none of these descriptions has proved to be diagnostic. Bellwood (1988a) provided a classification based on the degree of ecological association between the fish and reef, in terms of the reef's role in providing food and/or shelter. A broader overview was given in Choat and Bellwood (1991), who described the ecological and taxonomic characteristics of reef fishes. In this scheme, they noted the abundance of small-gaped deep-bodied fishes on reefs, and the numerical dominance of a few families, including labrids, pomacentrids, chaetodontids, and acanthurids. Later Bellwood (1996a) established a more specific "consensus list" of reef fish families. This list comprised all families that one would find on a coral reef irrespective of its biogeographic location (i.e., Acanthuridae, Apogonidae, Blenniidae, Carangidae, Chaetodontidae, Holocentridae, Labridae, Mullidae, Pomacentridae, and Scaridae). These 10 families were regarded as characteristic reef fish families, the essence of a reef fish fauna; all are abundant and speciose on coral reefs (Fig. 1, but see Section VI below).

However, these studies have all looked at the similarities among reef fish faunas. They provide only a description of a reef fish fauna and are not diagnostic (Bellwood, 1998). Further examination of reef and nonreef areas has found that many of the characteristics of reef fish faunas may apply equally well to nonreefal fish faunas (Bellwood, 1998; Robertson, 1998b). In this chapter therefore, the term "reef fish" refers to those taxa that are found on, and are characteristic of, coral reefs (i.e., the consensus list plus taxa characteristic of reefs in specific areas).

An understanding of the nature of the differences among reef fish faunas is critical to our understanding of the evolution of reef fishes and the role of history in determining the structure of modern reef fish assemblages. The dissimilarity between reef fish faunas can be seen in Fig. 1, which contrasts the species richness in a number of fish families at four biogeographically distinct reefal locations. Several features are immediately apparent:

1. Despite a more than threefold decrease in species numbers between the Great Barrier Reef (GBR) and the Red Sea, the basic pattern remains broadly comparable. The Red Sea reef fish fauna appears to be a random subset of a comparable high-diversity Indo-Pacific system such as the GBR. Indeed, there is no significant difference between the two faunas in terms of the distribution of species in families (χ^2 16.9; p, 0.46; df, 17).

2. Although overall the data for species/familial diversity are similar in the Caribbean and Red Sea (277/45 and 281/40, respectively), the familial composition and patterns of familial species richness vary markedly. In the Caribbean, the Lethrinidae, Pseudochromidae, Siganidae, Nemipteridae, and Caesionidae (Caesioninae) are absent. Together these families comprise approximately 7% of the species in the GBR fish fauna. However, several families are relatively well represented in the Caribbean, including the Serranidae, Haemulidae, and Sparidae and the regional (East Pacific/Caribbean) endemics, the Chaenopsidae and Labrisomidae.

3. Many of the characteristic reef fish families (e.g., Labridae, Pomacentridae) are present and abundant in New Zealand, a temperate region devoid of coral reefs. A similar pattern is seen in South Africa, South Australia, and western North America. Thus, although we readily recognize them as coral reef fish families, most of these characteristic reef fish families do not disappear when coral reefs stop. These taxa are characteristic of, but not restricted to, coral reefs.

If comparable data sets collected from a range of reefal and subtropical/temperate locations are examined using a Principal Component Analysis [PCA; modified after Bellwood (1997)] clear regional groupings are apparent (Fig. 2A, C), with high-, medium-, and low-diversity, low-latitude Indo-Pacific sites laying along the first axis. The decreasing diversity at these sites generally tracks a longitudinal shift away from

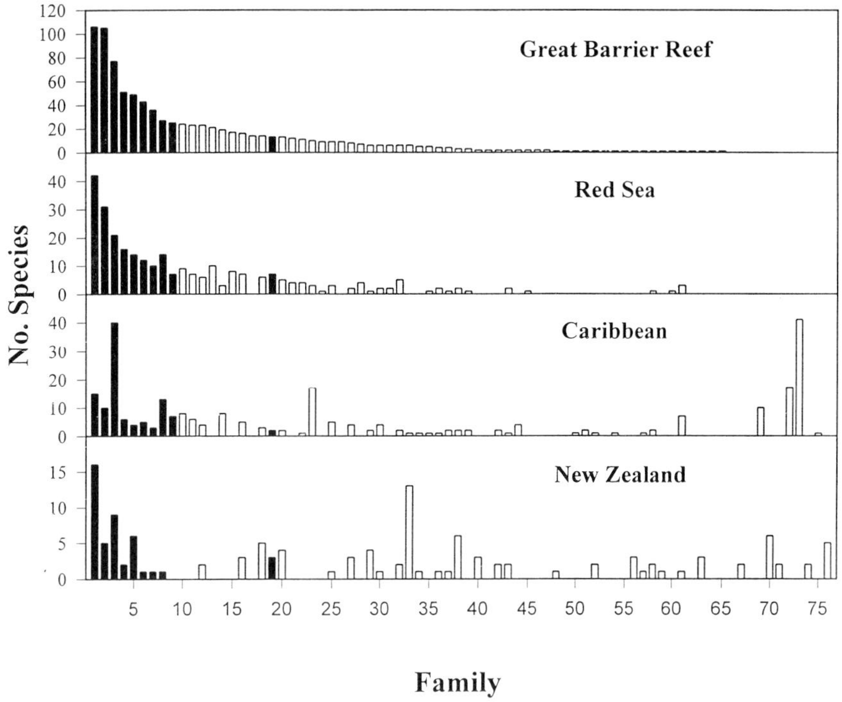

FIGURE 1 Species richness, by family, at four sites. The ranking of families at each site follows the Great Barrier Reef. Characteristic reef fish families are indicated by solid bars (after Bellwood, 1996a, 1997). Families: 1, Labridae; 2, Pomacentridae; 3, Serranidae; 4, Blenniidae; 5, Apogonidae; 6, Chaetodontidae; 7, Acanthuridae; 8, Scaridae; 9, Holocentridae; 10, Lutjanidae; 11, Pomacanthidae; 12, Scorpaenidae; 13, Lethrinidae; 14, Monacanthidae; 15, Pseudochromidae; 16, Balistidae; 17, Microdesmidae; 18, Tetraodontidae; 19, Mullidae; 20, Syngnathidae; 21, Siganidae; 22, Cirrhitidae; 23, Haemulidae; 24, Nemipteridae; 25, Ostraciidae; 26, Pinguipedidae; 27, Synodontidae; 28, Caesionidae; 29, Antennariidae; 30, Diodontidae; 31, Plesiopidae; 32, Sphyraenidae; 33, Tripterygiidae; 34, Callionymidae; 35, Ephippidae; 36, Malacanthidae; 37, Pempheridae; 38, Kyphosidae; 39, Priacanthidae; 40, Bythitidae; 41, Caracanthidae; 42, Gobiesocidae; 43, Mugilidae; 44, Opistognathidae; 45, Plotosidae; 46, Solenostomidae; 47, Trichonotidae; 48, Acanthoclinidae; 49, Aploactinidae; 50, Aulostomidae; 51, Batrachoididae; 52, Carapidae; 53, Centriscidae; 54, Centropomidae; 55, Chandidae; 56, Creediidae; 57, Dactylopteridae; 58, Echeinidae; 59, Eleotridae; 60, Fistulariidae; 61, Sparidae; 62, Teraponidae; 63, Uranoscopidae; 64, Xenisthmidae; 65, Zanclidae; 66, Albulidae; 67, Aplodactylidae; 68, Berycidae; 69, Chaenopsidae; 70, Cheilodactylidae; 71, Clinidae; 72, Cynoglossidae; 73, Labrisomidae; 74, Odacidae; 75, Ogcocephalidae; 76, Pentacerotidae.

the Indo-Australian Archipelago. Examination of the family-vectors (Fig. 2B) suggests that the first axis is associated primarily with total species richness. However, principal component 1 (PC1) does not just measure species richness. The scores reflect similar numbers of species in those families exhibiting greatest variation in the data set. The strong correlation with total species richness reflects the congruence among families in the decrease in familial species richness. This pattern is seen in the relatively uniform orientation of family-vectors around PC1, which also suggests that differences between high- and low-diversity sites are a result of the absence of taxa at low-diversity sites, i.e., there is no replacement. Low-diversity, low-latitude sites merely contain a lower number of species in the families found at high-diversity sites (as in Fig. 1). There are no "new" families that are characteristic of low-diversity sites (cf. Bellwood and Hughes, 2001).

The second axis explains only 12.3% of the variation but it appears to reflect changes in the relative composition of the assemblages in terms of temperate vs. tropical taxa (Fig. 2B). This axis separates high-latitude

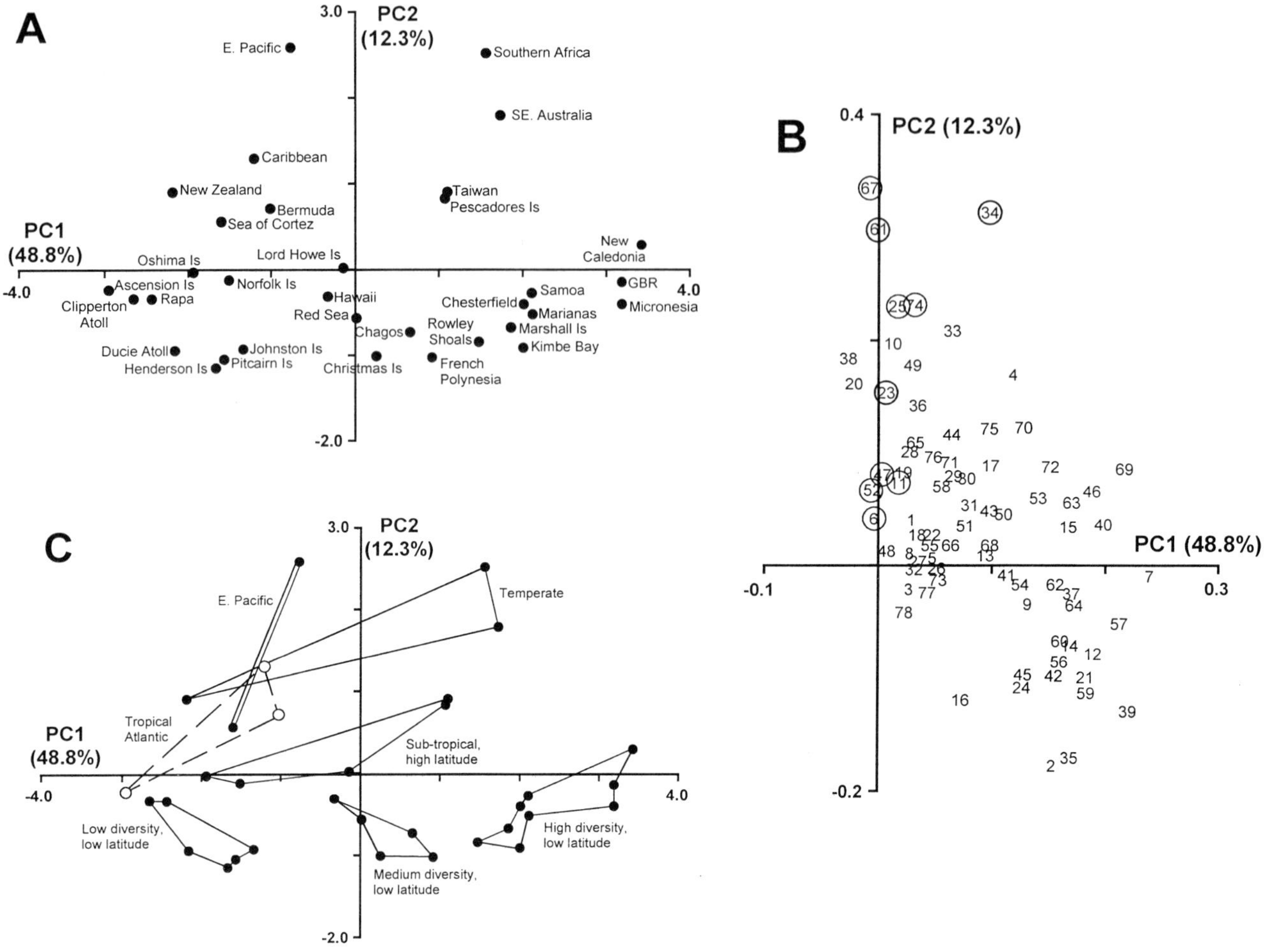

FIGURE 2 Principal Component Analysis of reef and nonreef sites. (A) Plot of sites on the first two axes. (B) Family-vectors, with families as listed below (temperate families are encircled). (C) Plot of sites on the first two axes with tropical, subtropical, and temperate sites delineated. Solid dots and solid lines indicate Indo-Pacific sites; open dots and dashed lines indicate Atlantic sites. Families: 1, Acanthoclinidae; 2, Acanthuridae; 3, Albulidae; 4, Antennariidae; 5, Aplodactylidae; 6, Aploacinidae; 7, Apogonidae; 8, Aulostomidae; 9, Balistidae; 10, Batrachoididae; 11, Berycidae; 12, Blenniidae; 13, Bythitidae; 14, Caesionidae; 15, Callionymidae; 16, Caracanthidae; 17, Carapidae; 18, Centriscidae; 19, Centropomidae; 20, Chaenopsidae; 21, Chaetodontidae; 22, Chandidae; 23, Cheilodactylidae; 24, Cirrhitidae; 25, Clinidae; 26, Creediidae; 27, Dactylopteridae; 28, Diodontidae; 29, Echeinidae; 30, Eleotridae; 31, Ephippidae; 32, Fistulariidae; 33, Gobiesocidae; 34, Haemulidae; 35, Holocentridae; 36, Kyphosidae; 37, Labridae; 38, Labrisomidae; 39, Lethrinidae; 40, Lutjanidae; 41, Malacanthidae; 42, Microdesmidae; 43, Monacanthidae; 44, Mugilidae; 45, Mullidae; 46, Nemipteridae; 47, Odacidae; 48, Ogcocephalidae; 49, Opistognathidae; 50, Ostraciidae; 51, Pempheridae; 52, Pentacerotidae; 53, Pinguipedidae; 54, Plesiopidae; 55, Plotosidae; 56, Pomacanthidae; 57, Pomacentridae; 58, Priacanthidae; 59, Pseudochromidae; 60, Scaridae; 61, Sciaenidae; 62, Scorpaenidae; 63, Serranidae; 64, Siganidae; 65, Sillaginidae; 66, Solenostomidae; 67, Sparidae; 68, Sphyranidae; 69, Syngnathidae; 70, Synodontidae; 71, Teraponidae; 72, Tetraodontidae; 73, Trichonotidae; 74, Triglidae; 75, Tripterygiidae; 76, Uranoscopidae; 77, Xenisthmidae; 78, Zanclidae.

vs. low-latitude low-diversity assemblages in the Indo-Pacific. As one moves away from the center of diversity in the Indo-Australian Archipelago, total species diversity decreases steadily with changes in both latitude and longitude. In both cases, characteristic reef fish families remain consistently well represented, whereas less speciose families are progressively lost. However, the latitudinal and longitudinal changes are not the same; high-latitude sites have a marked temperate influence.

This temperate influence is even clearer in the tropical Atlantic and tropical East Pacific sites. These sites are united by the presence of endemic families

(Chaenopsidae, Labrisomidae), the absence of several speciose Indo-Pacific families (e.g., Lethrinidae, Nemipteridae, Siganidae), and an increase in the diversity of other families (Haemulidae), including some with a strong representation in temperate waters (e.g., Sparidae, Monacanthidae). This similarity probably reflects a common history of the two areas prior to the closure of the Isthmus of Panama and a shared period of faunal loss (see Sections IV and V). The analyses suggest that the Caribbean, despite being a low-latitude tropical region with strong coral reef development, has a reef fish fauna that is more similar to those of high latitude or temperate Indo-Pacific sites than to tropical Indo-Pacific sites. The Caribbean reef fish fauna has a distinct temperate component.

The similarity between the patterns described in reef fishes and corals are striking (Bellwood and Hughes, 2001). The two groups have markedly different life histories, approaching the extremes seen in marine benthic faunas. If the biogeographic patterns seen in fish and corals reflect a common mechanism, then the processes may be operating at the regional or ecosystem level and at large temporal scales. If this is the case, then one may expect to see congruent patterns in other benthic marine taxa.

III. The Origins of Reef Fishes

A. Major Lineages

Fishes and corals both have a long tenure in the fossil record. However, at what point in the past did events begin to have a direct bearing on the ecology and distribution patterns of modern reef fish taxa? Devonian fish certainly have a legacy that passes through to modern times, but when did the history of modern reef fishes begin? The answer, it seems, is that these groups were already in place by the early Tertiary [50 million years (Ma) ago], with origins spreading back to at least the late Cretaceous (70 Ma), and possibly even to the early Cretaceous (100–130 Ma).

Most reef fish families have been placed in the order Perciformes. This order contains approximately 9293 species, and represents about 63% of all marine fish species (Nelson, 1994). The order encompasses about 75% of the fish species found on coral reefs (Randall *et al.,* 1990), including all of the characteristic reef fish families (Fig. 1). Unfortunately this order is probably paraphyletic (Johnson and Patterson, 1993). However, the Perciformes along with the Scorpaeniformes, Pleuronectiformes, and Tetraodontiformes may form a monophyletic group, the Percomorpha (*sensu* Johnson and Patterson, 1993).

Estimates of the ages of major fish groups are based on fossils or inferences from cladograms and biogeographic patterns. Fossil evidence ranges from isolated fragments, predominantly otoliths, to complete, fully articulated skeletons. Age estimates based on otoliths are consistently older than those based on complete skeletons (cf. Patterson, 1993), possibly reflecting the abundance of otoliths in the fossil record and the fact that otoliths do not require the exceptional conditions necessary for preservation of the complete fish skeleton. Identifying a fish taxon based on otoliths can be difficult because they have a limited range of characters, often of unknown phylogenetic significance. Furthermore, fossil otoliths are often worn, and considerable subjectivity may arise in character-state designations. The taxonomic utility of otoliths also varies widely between taxa (Nolf, 1985). In contrast, complete skeletons often permit fossil taxa to be incorporated into existing cladograms, providing estimates of the minimum age of specific lineages along with a great deal of information on changes in functional capabilities through time. However, complete fossil skeletons of reef fishes are rare and minimum ages based on complete skeletons are likely to underestimate the actual age of the group.

The biogeographic patterns of reef fishes observed today are the result of a long and complex history, which has probably involved a number of vicariance, dispersal, and extinction events (Fig. 3). When trying to disentangle this convoluted history, fossils provide a unique series of reference points. The utility of fossils in the study of phylogeny and biogeography has been critically appraised by Patterson (1981) and Humphreys and Parenti (1986). Fossils provide neither ancestors nor absolute ages of taxa. However, accurately dated fossils, when combined with phylogenies, can provide the minimum age of a lineage, its sister group, and all of the more basal lineages. Given this age one may be able to identify the vicariance events (i.e., environmental changes leading to the separation of populations) that were associated with the origin and subsequent diversification of lineages. Fossils also pinpoint a taxon in a location at a given time. This is particularly valuable when this location lies outside the geographic range of living forms.

The earliest record of the Perciformes is based on otoliths from the late Cretaceous (Cenomanian, 97.0–90.4 Ma) (Patterson, 1993), with the first full skeleton, *Nardoichthys*, being recorded from the upper Campanian/lower Maastrichtian (c. 74 Ma) of southern Italy (Sorbini and Bannikov, 1991). Of the remaining percomorph groups the oldest fossil, to date, is a tetraodontid *Plectocretacicus* (Sorbini, 1979) from the late

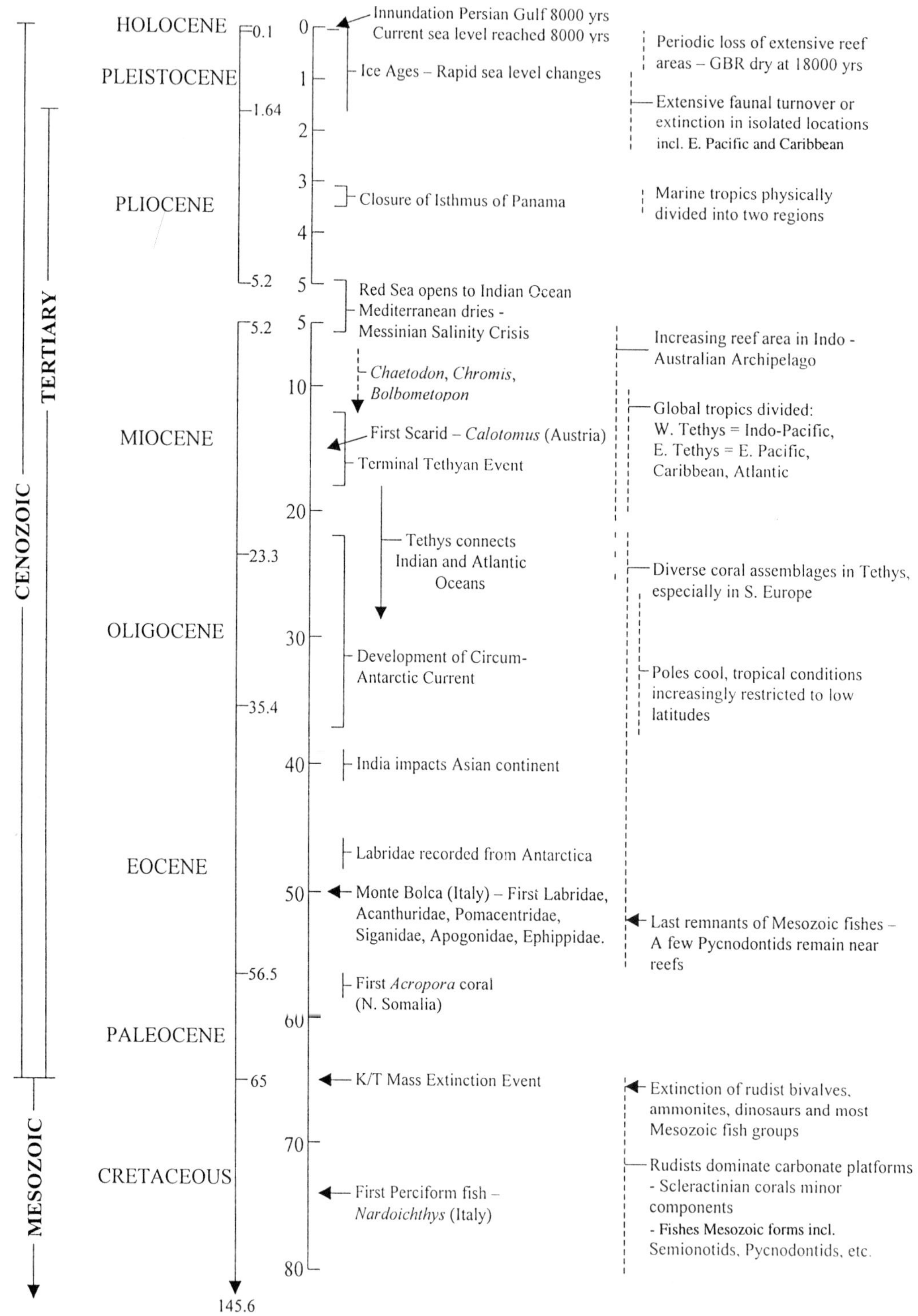

FIGURE 3 Major events in the history of fishes on coral reefs, showing the relationship between the appearance or loss of fish groups, changes in the status of coral reefs, and major biogeographic events. Earliest records of fish groups refer to identifications based on complete skeletal remains. Ages given in Ma. See text for details.

Cretaceous (Cenomanian, 97.0–90.4 Ma). The Pleuronectiformes and Scorpaeniformes are first recorded from otoliths in the Eocene (Ypresian, 56.5–50 Ma), with whole skeletons first recorded in the Lutetian (50 Ma) and Langhian (14.4 Ma), respectively (Patterson, 1993; Schultz, 1993). The oldest records of most of the characteristic reef fish families are equivocal or based on otoliths. The oldest records of the Acanthuridae, Labridae, and Pomacentridae (and the Zanclidae, Siganidae, Ephippidae, and Sparidae) based on skeletal material are Eocene (50 Ma) (Bellwood, 1996a) (Fig. 3).

The second line of evidence that may provide some indication of the age of reef fish lineages is historical biogeography, incorporating geological, biogeographic, and phylogenetic evidence. Of all perciform groups the clearest pattern of historical division is seen in the freshwater cichlids (a group that may be closely allied to the marine Pomacentridae). Today, cichlids are found in rivers and lakes throughout South America, Africa, Madagascar, and the southern extremity of the Indian subcontinent. All of these land masses are Gondwanan fragments. The break up of Gondwana extended over a period of about 80 Ma from an initial separation about 135 Ma to the final separation of South America and Africa in the North Atlantic at about 84 Ma (Lundberg, 1993). The presence of cichlids on all three continents (Africa, South America, and Asia) presents a dilemma. Either they were associated with the land masses prior to separation, cichlid origins being at least 100 Ma, or cichlids maintained contact between land masses through extensive marine connections. Lundberg (1993), in a thorough review of African–South American fish relationships, favored the latter option. This interpretation is consistent with the fossil record. The oldest cichlid fossils are from the Eocene of Africa, at about 46 Ma (Murray, 2001). This date is consistent with the oldest record of the other labroids, the Pomacentridae and Labridae, at 50 Ma (Bellwood and Sorbini, 1996). At 100 Ma the origins of the Cichlidae, based on a biogeographic model, would predate the earliest fossil of the family and that of any other extant perciform family by over 40 Ma.

Although there are brackish or fully marine dwelling cichlids on all occupied continents, Cenozoic trans-Atlantic dispersal seems unlikely given the clear regionalization seen in the major cichlid clades (cf. Stiassny, 1991). A much older origin for the Cichlidae and therefore the Perciformes must remain a possibility, with the Cichlidae being widespread across Gondwana prior to fragmentation. This would require the origin of the group to be in the order of 100 Ma. Comparable early origins have been proposed based on biogeographic evidence for congrogadins (Pseudochromidae) (Winterbottom, 1986) and teraponids (Vari, 1978). Recent biogeographic interpretations based on molecular phylogenies are consistent with these early origins. In both the Cichlidae and the aplochelioid cyprinodontiforms the molecular phylogenies strongly suggest that the distributions of taxa within the two respective groups are the result of Gondwanan fragmentation (Murphy and Collier, 1997; Farias et al, 1999). If this were the case, then the cladogram of Streelman and Karl (1997) would suggest that several reef fish lineages (labrids, pomacentrids, acanthurids, and pomacanthids) were already established prior to Gondwanan fragmentation (i.e., over 125 Ma). The biogeographic model therefore suggests that major reef fish lineages may have early Cretaceous origins.

It is almost axiomatic that the only surprise that the fossil record holds, in terms of the age of taxa, is that taxa are older than previously thought. Care must be taken, therefore, in ascribing minimum ages. Phylogenies and biogeographic patterns present interesting possibilities, otoliths provide tentative oldest recorded minimum ages, whereas the most conservative estimates are based on whole specimens. The discrepancy between historical biogeography (perciform origins 100–125 Ma) and the fossil record (74–100 Ma) is yet to be resolved. Molecular techniques that shed light on older relationships appear to be a particularly promising source of new information (cf. Streelman and Karl, 1997).

B. Reef Fish Families

So far we have examined the origins of the major lineages. In terms of the fossil record, this represents a handful of specimens in three or four species covering the whole of the Percomorpha. What about the families and genera of fishes found on reefs today? How did the diversity of fishes seen on modern reefs arise? Was it the steady accumulation of more and more complex forms, a progressive series of faunal replacements with major periods of diversification and loss, as in terrestrial mammal faunas?

Some of the answers to these questions lie in the exceptional fossil fishes collected from Monte Bolca in the foothills of the Italian Alps. Here, Eocene marine deposits have yielded a large quantity of fossil fishes of excellent preservation quality (Fig. 4). The fossil deposits of Monte Bolca have yielded over 250 species in 82 families. These specimens include the first representatives of almost all fish families found on coral reefs today (Patterson, 1993; Bellwood, 1996a). In terms of reef fish families, the Bolca fish fauna is relatively

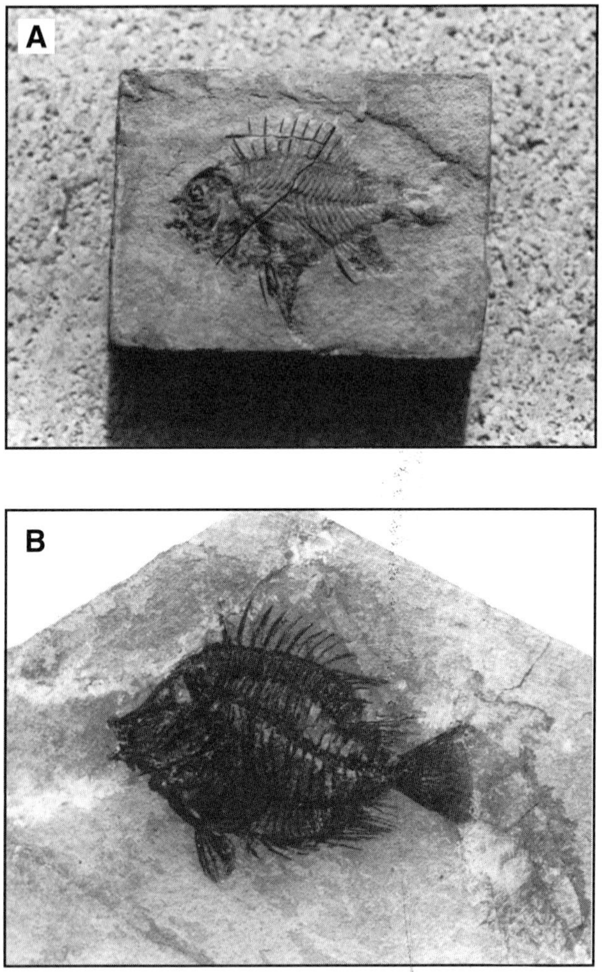

FIGURE 4 (A) *Palaeopomacentrus orphae* Bellwood & Sorbini, a fossil pomacentrid from the Eocene (50 Ma) deposits of Monte Bolca, Italy; 29.5 mm standard length. This is the first of two pomacentrid species recovered from these deposits. (B) *Lehmanichthys lessiniensis* Blot & Tyler, a fossil acanthurid from Monte Bolca; 71 mm standard length. Acanthurids were particularly well represented in Monte Bolca, with over 10 genera recorded. Reprinted with permission from Museo Civico di Storia Naturale, Verona, Italy.

complete. Thus, by the late lower Eocene (50 Ma), almost all modern reef fish families are present in a single biogeographic location. Furthermore, the benthic component of this fauna is dominated by perciform fishes. This may not appear surprising; however, only 15 Ma earlier the Perciformes is represented in the fossil record by only a few specimens of one species. Bolca also marks a period of transition, with the last remnants of ancient Mesozoic forms (i.e., pycnodontids) persisting along with modern reef fish families.

In this chapter families are frequently used to examine historical patterns. It should be noted, however, that this does not imply equivalent status to these groups (families, genera, and, to a lesser extent, species are relatively arbitrary groupings). Families are often identified by traits that are evident in fossils, providing a common taxonomic basis for comparing living and ancient assemblages (Bellwood, 1996a). Given the paucity of information on relationships among taxa, families have no stronger status than any other taxonomic level—they merely represent major lineages with sufficiently distinct body plans to suggest monophyly. The status of almost all family groupings is in need of reappraisal.

In addition to the strong links between Bolca and modern reef fish assemblages in terms of fish families, several extant perciform genera have also been recorded from the deposits, including *Acropoma*, *Pristigenys*, *Mene*, *Scatophagus*, and *Seriola* (Blot, 1980). The latter may be regarded as a member of a "characteristic" reef fish family (Carangidae), although it is a more open-water genus. Besides this example, the oldest records of extant "reef fish" genera appear to be in the Miocene with *Chaetodon* (Arambourg, 1927), *Chromis* (Bellwood and Sorbini, 1996), and *Bolbometopon* and *Calotomus* (Bellwood and Schultz, 1991).

The morphology of species in reef fish families recovered from Monte Bolca is almost indistinguishable from that of living representatives. These were not "primitive" precursors of modern forms. Their structural features and implied functional and ecological characteristics are comparable to those of modern reef fishes. Indeed, the level of preservation is such that in some cases pigment patterns can be seen, with striking similarities to living forms. A juvenile *Scatophagus* from Bolca has pigment bands on the body that are almost identical to those found on living forms, and the two earliest pomacentrids, *Palaeopomacentrus orphae* and *Lorenzichthys olihan*, both have an ocellus on the dorsal fin comparable to those seen on juvenile pomacentrids today (Bellwood, 1999; Bellwood and Sorbini, 1996). On a dive along the coast of the Tethys Sea in the Bolca region 50 million years ago one would see a fish fauna little different from that in the tropics today. Most reef fish families would be represented, complete with "modern" morphological attributes. Modern underwater fish identification sheets would suffice to identify many of the fish families.

Bolca marks the starting point in the known evolution of most reef fish families. Their presence in one location 50 million years ago highlights the stability of the taxonomic and morphological characteristics of tropical benthic marine fish faunas throughout the Cenozoic. This suggests that the broad similarities in the familial composition of modern reef fish faunas may reflect an old shared history rather than recent colonization, and that familial differences between reef regions may be explained by subsequent events. It is this post-Eocene history in which the differences between major reef regions probably arose.

IV. Barriers and Vicariance Events in the Evolution and Biogeography of Reef Fishes

Throughout the Cretaceous the Tethys Sea was the dominant tropical marine seaway. During most of this period, there was widespread reef construction. These reefs were dominated by rudist bivalves, although hermatypic corals and algae were present (Kauffman and Fagerstrom, 1993). Scleractinian corals were a minor structural component (Kauffman and Sohl, 1974; Wood, 1999). Toward the end of the Upper Cretaceous the rudist reefs disappeared, to be replaced sometime later by scleractinian coral reefs. By the Eocene, some 20 Ma after the loss of the rudists, the Tethys Sea had an essentially modern tropical fauna. Reefs were dominated by scleractinian corals, and modern reef fish families were abundant in the vicinity. From these reef fishes we can trace a continuous history through to those fishes living on reefs today.

Throughout their history the tropical seas in which we find reef fishes have been repeatedly divided, with each fragment having a different history through to the present. In some areas, the occupants were completely eliminated; in other areas they prospered, their individual fates being dependent on a complex series of interacting factors, including geographic location, regional connectivity, size, bathymetry, and the timing of the separation.

Barriers separating marine populations vary widely, from complete physical barriers such as land bridges, to partial barriers resulting from distance, currents, or ecology. The barriers may be permanent or intermittent. In several areas barriers can be clearly identified, but in other areas barriers are only inferred, being marked by faunal breaks with no clear geological or biological explanation (Fig. 5). Barriers have been widely implicated in the regional increase in reef fish species, through vicariance (Woodland, 1983; McManus, 1985; Springer, 1988; but see Springer and Williams, 1990), with isolation followed by perturbation being a common theme.

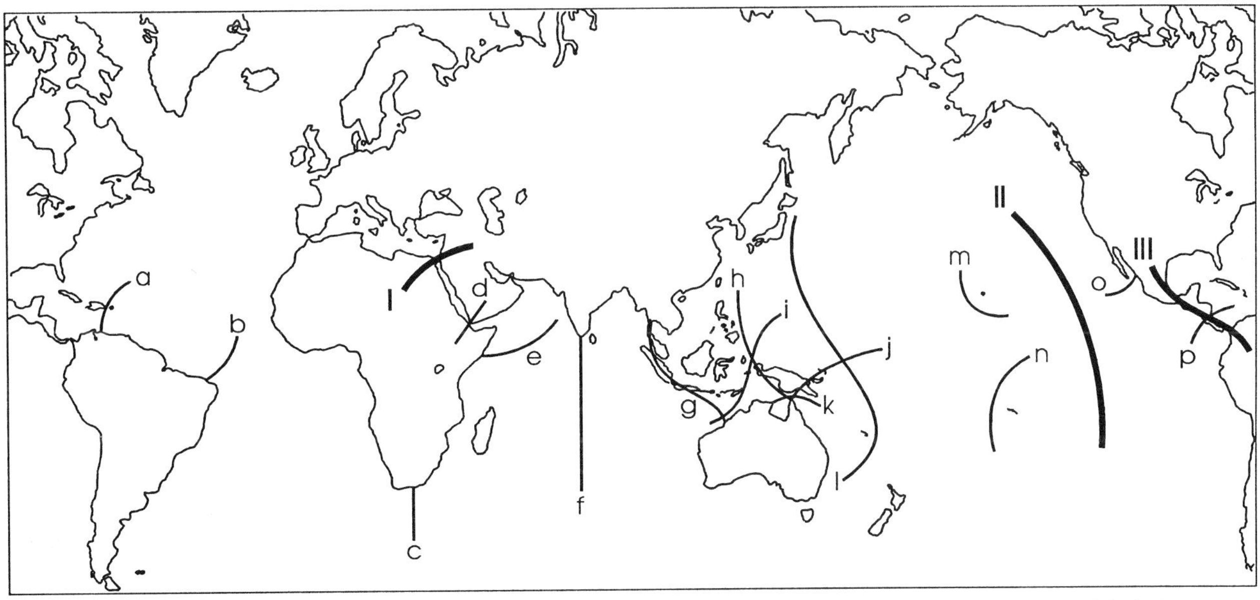

FIGURE 5 Faunal boundaries and proposed locations of vicariance events affecting reef fish taxa (modified after Blum, 1989; with kind permission from Kluwer Academic Publishers). (I) Terminal Tethyan Event (Red Sea land bridge) 12–18 Ma. (II) East Pacific Barrier 0–65 Ma. (III) Isthmus of Panama 3.1–3.5 Ma. Letters denote specific references to reef fishes: a, Hastings and Springer (1994); b and c, Blum (1989); d, Springer and Williams (1994); e, Blum (1989) and Springer and Williams (1994); f, Winterbottom (1986), Springer (1988) Blum (1989), and Springer and Williams (1994); g, h, and i, Indo-Australian Archipelago (see text); j and k, Springer (1988) and Blum (1989); l, Springer (1982); m and n, Blum (1989); o and p, Hastings and Springer (1994) and Lessios *et al.* (1995).

The evolution of modern reef fish families has been largely confined to the past 90 Ma. For most of this period the Tethys was the dominant tropical sea. It provided a broad marine seaway connecting the Atlantic and Indian oceans until the Miocene. This connectivity was reflected by a considerable degree of faunal overlap, with numerous cosmopolitan species (Adams and Ager, 1967; Dilley 1973). Within the Tethys, however, regional faunal differentiation has been recorded during almost every major time period. The evidence suggests that this is the result of a series of temporally distinct vicariance events dividing successive populations, often in the same location. The major events that have shaped tropical marine fish faunas are outlined below, with a summary in Fig. 3.

A. Cooling at High Latitudes

During the late Cretaceous and early Cenozoic there was the potential for extensive connectivity between the oceans, both longitudinally and latitudinally. Latitudinal temperature gradients were not as strong as today. In the Eocene, for example, Antarctica had a temperate climate and a fish fauna that included families that may be found on coral reefs today [e.g., Labridae (Long, 1992); Oplegnathidae (Cione *et al.*, 1994)]. The complete separation of Gondwanan fragments, with the opening of the Australian–Antarctic seaway and the Drake Passage between Antarctica and South America, permitted the formation of the circum-Antarctic current. This effectively isolated Antarctica and was associated with the formation of a steep thermal gradient between the tropics and the South Pole (Veevers and Ettriem, 1988). By 37 Ma, at the end of the Eocene, the circum-Antarctic current was in place and the poles had permanent ice sheets. With increased separation of the continents, the circum-Antarctic current progressively increased in magnitude and the water cooled further. It is likely that this effectively locked in the tropics, preventing significant movement of tropical forms between the oceans at high latitudes. Tropical interoceanic connectivity was probably largely restricted to the Tethys and it is here where further division is seen.

B. The Terminal Tethyan Event

The terminal Tethyan event (TTE) has been regarded as one of the most important events in marine biogeography. Contact between the African and Eurasian continental plates raised a land bridge in the Middle East. This marked the end of a tropical marine connection between the Indian and Atlantic oceans, and the end of the Tethys Sea. Estimates of the timing of the TTE vary, although the final closure is usually placed between 12 and 18 Ma (Adams *et al.*, 1983; Rögl and Steininger, 1983). The TTE has been associated with the division between Caribbean and Indo-Pacific sister taxa (e.g., Blum, 1989). Estimates of the impact of the TTE on marine faunas vary. The TTE provides a firm minimum age for a split in Tethyan populations. However, this is a minimum age. Much earlier divergences are possible and, in many cases, probable. By the Miocene, the Tethys had been reduced to a narrow channel with seas in peripheral basins (Paratethys). Connectivity between the western (Atlantic) province and the eastern (Indian Ocean) province was probably minimal. The TTE probably represented only the final stage of a progressive division between these two provinces. Fossil evidence suggests that the TTE may have been important for dividing some coral and echinoid species, but most genera were separated prior to the TTE (Rosen and Smith, 1988). Major divisions between Atlantic and Indo-Pacific taxa are likely to reflect an earlier separation. In terms of reef fish, the TTE may have been of limited significance. It provides a minimum age for the possible divergence of some lineages, but there is considerable evidence to suggest that for genera, at least, most divisions occurred prior to the TTE.

C. Tethyan Provinciality Prior to the TTE

For reef fishes, the clearest picture of the relative importance of the TTE is provided by the parrot fishes, for which we have a fully resolved cladogram of genera and a clear, albeit limited, fossil record (Bellwood, 1994). (*Note*: Although the parrot fishes are currently placed in the Scaridae, they almost certainly represent a derived clade within the Labridae.) Today, the scarids are represented by four genera in the remnants of the west Tethys (including *Nicholsina* and *Cryptotomus*), and seven genera in the Indo-Pacific (including *Calotomus*). Separation of these taxa prior to the TTE is strongly suggested by the topology of the cladogram of the family (Fig. 6A), the resultant area cladogram (Fig. 6B), and the record of a fossil *Calotomus* in mid-Miocene deposits (14 Ma) in Austria at about the time of the TTE (Bellwood and Schultz, 1991). Together, these data suggest that the two west Tethyan genera, *Nicholsina* and *Cryptotomus*, and the east Tethyan *Calotomus* are all at least 14 Ma old and were established in their respective provinces prior to the TTE.

The distribution of the remaining scarid genera, with *Sparisoma* restricted to the Caribbean and

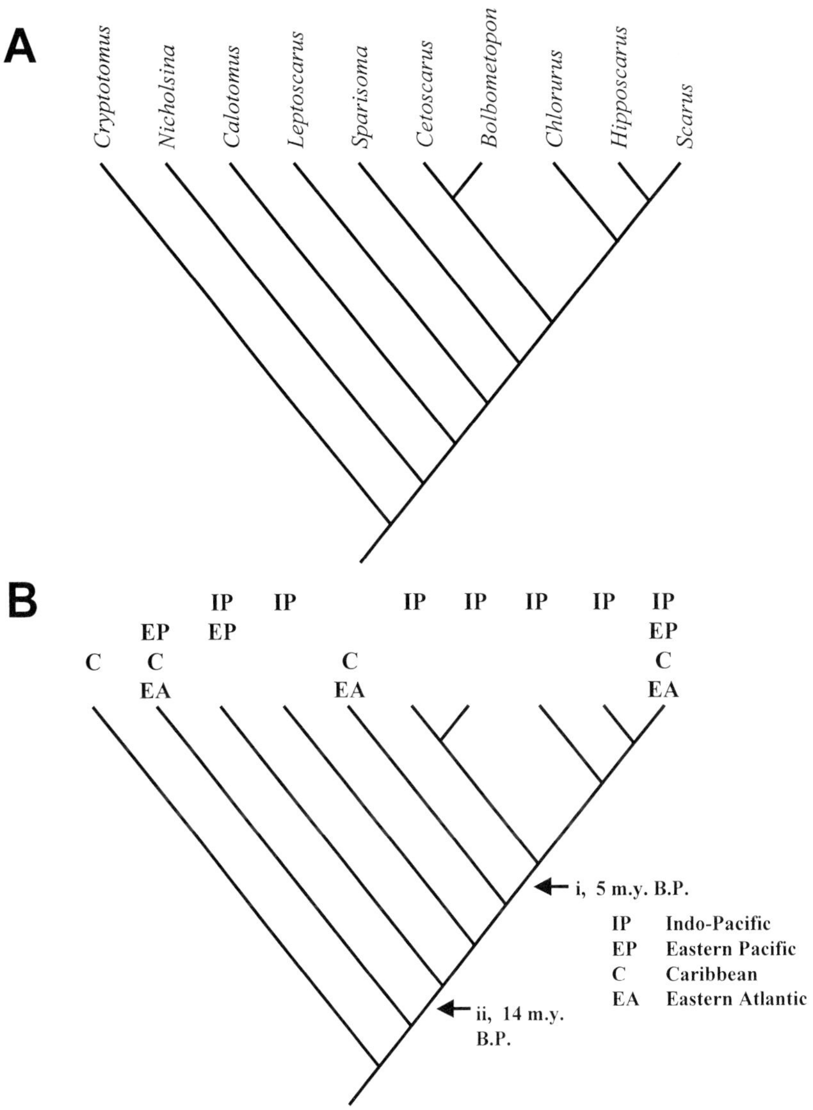

FIGURE 6 (A) Cladogram of scarid genera. (B) Area-cladogram of scarids; i and ii indicate the minimum age (million years before present) of divergence based on fossil evidence (with permission after Bellwood, 1994).

Leptoscarus, *Bolbometopon*, *Cetoscarus*, *Chlorurus*, and *Hipposcarus* restricted to the Indo-Pacific, suggests that these lineages may also have been present and regionally segregated prior to the TTE. The alternative explanation of mutual reciprocal extinctions is less parsimonious. Thus, in this family at least, although it appears that the major faunal divisions fall on either side of the location of the TTE in the Middle East, the actual division may have predated the final closure of the Tethys Sea. Comparable pre-TTE provinciality has been suggested based on fossil data for corals and echinoids (Rosen and Smith, 1988).

Given that these faunal divisions may predate the TTE, are there any clear indications of the principal vicariance events associated with these divisions? Unfortunately, the fossil record for marine taxa does not, at present, permit detailed resolution of the various vicariance events in the Middle East region. Adams (1981) has raised the possibility of a land bridge in the Middle East region of the Tethys during both the Paleocene and Oligocene. It is also possible that shallow seas in the Middle East region would have provided an effective barrier to marine dispersal for some time prior to the formation of a land bridge, by restricting current flow and increasing susceptibility to rapid salinity changes and periodic habitat loss. Furthermore, given the loss of Mediterranean taxa during the Messinian salinity crisis (5–6 Ma), observations based on living taxa may not be able to resolve the relationships between Tethyan fragments beyond comparisons between the tropical Atlantic (Caribbean) and Indo-Pacific. At this level of resolution one cannot separate TTE events from divergences as far back as the Cretaceous (see below). At this point, the only consistent indication is that some coral reef fish genera were probably present, with some regional differentiation, prior to the TTE. After the Messinian crisis the Mediterranean probably no longer represented a Tethyan relict, but rather an offshoot of the post-Pliocene Atlantic (but see Jaume and Boxshall, 1996). An alternative explanation for a Tethyan division is provided for ostracids (Tetraodontiformes) by Klassen (1995), who suggests that the two ostracine lineages were separated by raised landmasses between the Americas in the late Cretaceous (100–80 Ma).

Extant reef fish genera common to both the Atlantic and Indo-Pacific oceans may be the result of either a widespread pre-TTE distribution or subsequent dispersal. There is evidence supporting both hypotheses. Several reef fish genera have fossil records from the Miocene, with some extending back to the Eocene (Section III), indicating pre-TTE origins. In two examples (*Naso* and *Oplegnathus*), fossil evidence suggests that these taxa were present in both the Atlantic and Indo-Pacific prior to the TTE (Section II). However, post-TTE origins in the Indo-Pacific with subsequent dispersal into the Caribbean prior to the closure of the Isthmus of Panama have been suggested for two reef fish genera, *Scarus* (Bellwood, 1994) and *Bodianus* (Gomon, 1997). Detailed species-level phylogenies will be required to evaluate the relative contribution of post-TTE dispersal to the Caribbean fish fauna.

D. Cretaceous Provinciality: Division from the Beginning

During the Cretaceous and early Cenozoic there was pantropical marine connectivity through the Tethys seaway (Barron and Peterson, 1989). However, this connectivity did not preclude biogeographic differentiation between regions. Although tropical marine faunas of the Lower Cretaceous were relatively cosmopolitan, in the Upper Cretaceous the Caribbean and Mediterranean became increasingly distinct (Hallam, 1973). Coates (1973), for example, records the first signs of the Caribbean region as a distinct marine biogeographic province in the Aptian–Albian (97–124.5 Ma), with the appearance of endemic genera of corals, rudists, and nonrudist bivalves. The Caribbean remains distinct throughout the remainder of the Cretaceous, with the greatest degree of endemicity in all three taxa during the Cretaceous being recorded in the Maastrichtian (65–74 Ma). A similar pattern is reported in the larger foraminifera (Dilley, 1973). The proposed vicariance event, which resulted in this initial separation of the Caribbean and Mediterranean regions of the Tethys, is the spreading of the Atlantic ridge system and the expansion of the proto-Atlantic. Restriction of water movement between the Americas would have reinforced the extent of isolation by reducing trans-Pacific colonization (cf. Klassen, 1995). Thus, when we look at the origins of reef fishes, the major lineages probably arose in a system marked by some degree of provinciality and in which the principal barriers were already in place and were becoming increasingly effective, i.e., the spreading Atlantic and narrowing Tethyan seaway.

E. Isthmus of Panama

The Pliocene raising of the Isthmus of Panama (IOP) marks the final closure of the tropical seas into two discrete regions. Estimates of the timing of this division vary from 1.8 (Keller *et al.*, 1989) to 3.5 Ma (Coates *et al.*, 1992), with recent estimates of the first complete closure of the IOP around 3.5–3.1 Ma (Coates and Obando, 1996). For fish, the observed impact is predominantly at the species level. There remain about a dozen species of shore fish that span the isthmus with little or no morphological differentiation, and several closely related species pairs. Reef fish examples include mullids (Stepien *et al.*, 1994) and blennies (Hastings, 1990). Along with the divisions on either side of the IOP, there is also consistent evidence of divisions along the East Pacific coast. In one group of pomacentrids, this north–south division appears to predate the final closure of the isthmus, with evidence of a more recent link between the southern East Pacific and the Caribbean than between northern and southern East Pacific forms (Lessios *et al.*, 1995). In contrast, Hastings and Springer (1994) suggest that, for some blennioid fishes, comparable East Pacific divisions occurred after the closure of the IOP.

Within the Caribbean, there is a suggestion of a broad division of the region into northern and southern biogeographic provinces. There are several reef fish species pairs with broadly overlapping northeast–southwest distributions, e.g., *Pomacanthus arcuatus*/*P. paru*, *Holacanthus bermudensis*/*H. ciliaris*, and *Centropyge argi*/*C. aurantonotus*. Hastings and Springer (1994) suggest that there is more overlap in the distributions of Caribbean species than in closely related species in the East Pacific, possibly reflecting more discrete patches of suitable habitat in the latter region. A north–south division within the Caribbean with further subdivisions has been suggested by Domeier (1994) based on *Hypoplectrus*, a reef-associated serranid. A comparable division into north–south faunal provinces has been proposed for Pliocene mollusc faunas (Petuch, 1982; Vermeij and Petuch, 1986), although the northern province was restricted to the Florida region. A more distinct division between the Caribbean and southwest Atlantic fish assemblages has been described (Rosa and Moura, 1997; Floeter and Gasparini, 2000), with several sister species occurring on either side of the mouth of the Amazon. Although there is some evidence of continuity of benthic marine faunas (Moura *et al.*, 1999), the river appears to present a significant barrier for marine species.

The direct impact of the IOP may have been of limited significance for reef fish taxa. The East Pacific barrier appears to have been a relatively effective barrier and would have limited the effect of the IOP in the Pacific to the isolation of populations along the East Pacific coastline. Here, the greatest effect was probably mediated through the combined effects of isolation and subsequent faunal loss [as in other taxa; cf. J. B. C. Jackson *et al.*, (1993)]. The IOP prevented recolonization of the East Pacific by Caribbean taxa. The IOP marks the latest land bridge in this region. However, there may have been earlier land connections in this location in the Paleogene 30–60 Ma (White, 1986) and Cretaceous 100–80 Ma (Smith *et al.*, 1981). Klassen (1995) cites the latter event as an alternative explanation for an east–west Tethyan division in ostracids (Tetraodontiformes).

The IOP stands as a good example of the nature of land barriers. Although studied in considerable detail, the final date of closure remains uncertain. It appears that the isthmus was completely closed around 3.5–3.1 Ma, with a possible breakdown and marine passage between 2.4 and 2.0 Ma (Coates and Obando, 1996; Cronin and Dowsett, 1996). Furthermore, there is increasing evidence that the shallow waters formed by the rising isthmus represented a significant ecological barrier between Caribbean and Pacific marine systems, with the possibility of speciation on either side of the isthmus since the late Miocene (Jackson *et al.*, 1996; Vermeij, 1997).

F. Gondwanan Fragmentation

Gondwanan fragmentation is widely believed to have been one of the major geological events that has influenced the distribution patterns of plants and animals on the world's continents. In marine systems, it may also have had a direct impact on the temperate marine faunas of the southern continents, with clear links between the temperate fishes of South Africa, South America, and Australia. Although these continents all possess numerous regional endemics, they also share a number of fish taxa that may reflect Gondwanan associations. These taxa include the Aplodactylidae, Latridae, Congiopodidae, and genera or species in the Cheilodactylidae, Labridae, Sciaenidae, and Sparidae (Wilson and Allen, 1987). The relative importance of vicariance associated with Gondwanan fragmentation and dispersal via the west wind drift remains to be determined, but Gondwanan fragmentation appears to have been a significant factor in the biogeography of temperate fish taxa (cf. Wilson and Allen, 1987).

Evidence of an impact of Gondwanan fragmentation on reef fishes is limited, although three studies are noteworthy. Vari (1978) and Winterbottom (1986) identified Gondwanan fragmentation as the most likely explanation for the observed distribution patterns in teraponids and congragadids, respectively, and Springer (1988) identified the northern movement of India and its collision with Eurasia at about 40 Ma as a major vicariance event dividing the common ancestor of two species groups of reef-associated blennies (*Ecsenius*). Chao (1986) even suggested a late Jurassic origin for the Sciaenidae, with associated Gondwanan links. The main problem with the Gondwanan vicariance scenario, as noted by Winterbottom (1986), Springer (1988), and Briggs (1989), is that the inferred events require that the common ancestral species be extremely old, from about 40–100 Ma, which clearly conflicts with the fossil record (see Section III above).

An alternative explanation for some apparent Gondwanan links is provided by Woodland (1986), who proposed a founder-principle scenario to explain the observed patterns of siganids, with colonization of Australia from Asia as the continent moved into the tropics. In this scenario, the date of colonization is more "reasonable" given that tropical conditions were reestablished in northern Australia by about 15 Ma (Davies, 1988). It may be noteworthy in this context that significant reef growth in the Indo-Australian Archipelago was not recorded until the Miocene (Wilson and Rosen, 1998). Molecular data may help to resolve this dilemma, because the two scenarios have markedly different inferred ages for species divisions and different divergence patterns.

G. East Pacific Barrier

Today, the East Pacific Barrier (Ekman, 1953; Briggs, 1961) separates the Indo-Pacific and East Pacific faunas by an expanse of deep open ocean approximately 5000 km wide. The East Pacific Barrier has almost certainly been in effect since the early Miocene and probably throughout the Cenozoic (Rosen and Smith, 1988). As such, it acted with the Terminal

Tethyan Event to effectively divide the world's tropical seas in two and, after the closure of the Isthmus of Panama, to isolate the East Pacific tropical fauna.

It has been suggested that during most of the Cretaceous, passage of shallow-water benthos across the Pacific was restricted by wide expanses of water similar to the East Pacific Barrier with the exception of a short period during the Campanian/Maastrichtian (83–65 Ma), when a series of volcanic "stepping-stones" is proposed to explain the apparent spread of shallow-water taxa from the Caribbean to the West Pacific (Skelton, 1988). One of the fish groups that may have crossed the Pacific from east to west is the Embiotocidae, which has 20 species off California and three off Japan. However, as livebearers with no pelagic stage, the Embiotocidae are poor candidates for oceanic dispersal across island chains, and movement around the Pacific rim remains a more likely option. Furthermore, as noted above, any Cretaceous connections would require the taxa to be considerably older than the fossil record would suggest.

The East Pacific Barrier is one of the few widely accepted barriers that does not require a "hard" physical separation of marine populations, e.g., land bridges. An interesting issue that arises from the consideration of such "soft" barriers is the reliance on interpretations based on present-day bathymetry and ocean currents. There is increasing evidence that past ocean circulation patterns were markedly different from today and that even on relatively recent time scales they could have a marked impact on gene flow. In both reef bivalves (Benzie and Williams, 1997) and fishes (Doherty *et al.*, 1995), genetic studies of West Pacific populations have identified barriers that are not apparent based on existing patterns of marine connectivity. Understanding the role of past currents in shaping patterns of connectivity between reef systems is a difficult but significant goal in historical biogeography.

H. Indo-Australian Archipelago: Center of Origin or a Refuge?

One of the most enduring representations in texts of marine biogeography is the "bullseye" pattern of species/generic diversity, with the center of diversity in the Indo-Australian Archipelago (IAA) and a decline in numbers as one moves latitudinally or longitudinally into the Indian Ocean or across the Pacific (Fig. 7). This pattern is found in numerous marine groups, from corals and echinoids to reef fishes. That such patterns are repeated in numerous marine taxa suggests that there may be a general explanation, although a unifying explanation has remained elusive. Explaining these plots has been the focus of numerous works (Wallace, 1997). These revolve around three basic models that describe the center as (1) the center of origin, (2) a region

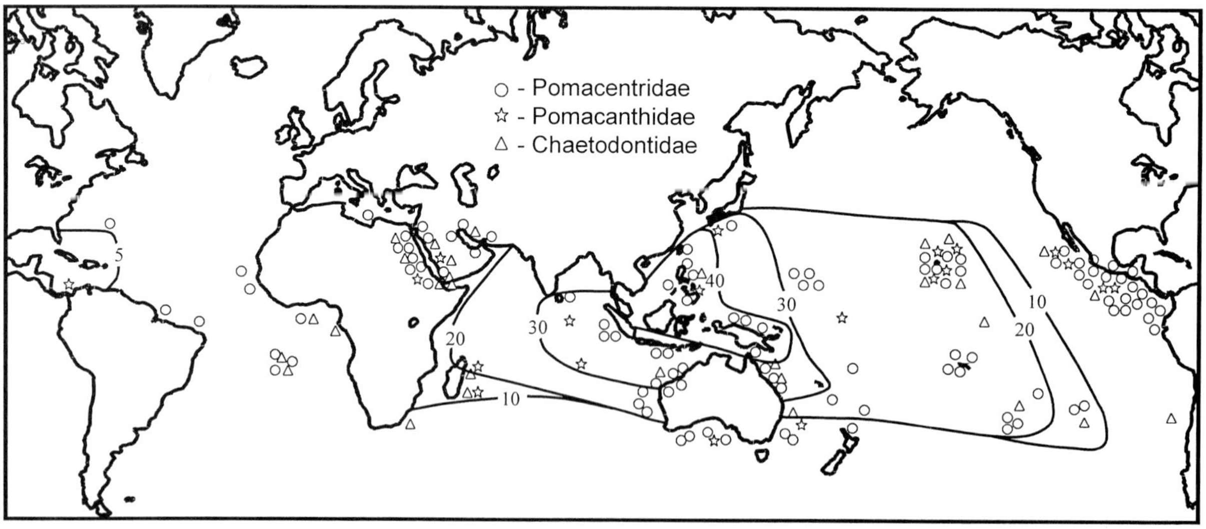

FIGURE 7 Patterns of species richness and endemicity in three reef fish families (Chaetodontidae, Pomacanthidae, Pomacentridae). Lines indicate the pattern of species richness in the Chaetodontidae; the patterns for the other two families were comparable. Positions of endemics (open symbols) are approximate and refer only to those species recorded from a single locality [based on data from Allen (1979, 1991)]. Using values for the Chaetodontidae and Pomacanthidae, respectively, the most conspicuous centers of endemicity are the Red Sea (28%, 46%), Hawaii (60%, 20%), and East Pacific (100%, 75%).

of overlap, or (3) a refuge. Many early explanations were based on center-of-origin theories, which assume that the center of diversity is also the center of origin, with each species dispersing from this center by its own means. This has been a particularly well-favored explanation, apparently supported by the clear propensity for marine taxa to disperse during their pelagic larval stage. It has been applied to reef fish on several occasions (Allen, 1975; McManus, 1985; Myers, 1989).

Center-of-origin theories, like many others, are often based on the unique features of the area: extensive shallow-water geological complexity, and contacts with two major biogeographic regions. Shallow basins may promote speciation within the region at low sea levels (McManus, 1985; Springer and Williams, 1994), with the area acting as a true center of origin. Alternatively, the extensive shallow habitats may reduce faunal losses, e.g. as a result of habitat reduction during sea level changes (cf. Potts, 1985; Myers, 1989, Paulay, 1996), thus acting as a refuge (Rosen, 1984; Wilson and Rosen 1998) or centre of accumulation (Palumbi, 1997; Bellwood and Hughes, 2001). The tectonic complexity of the area and its position between two major biogeographic realms also increase the number of potential sources of new taxa (a region of overlap). Thus, as Parenti (1991) notes, "a continent is part of the biogeographic regions of all the oceans it contacts." The same applies to archipelagos.

The high diversity in the region may also be due, at least in part, to faunal overlap. The area includes several representatives that are otherwise restricted to the Indian Ocean or West Pacific biogeographic regions. Woodland (1983) described the region as a "zone of overlap" for siganids. This pattern of overlap is consistent with the data of Donaldson (1986) and Blum (1989) for cirrhitids and chaetodontids, respectively. It also appears to hold true for some corals (Wallace, 1997), although for both fish and corals, the total species numbers in the IAA is boosted by a number of regional endemics (but see Bellwood and Hughes, 2001).

Despite the different scenarios proposed to explain the high species richness, most workers seem to agree on the underlying mechanism: vicariance at various sites in or around the Indo-Australian Archipelago during Pleistocene sea level changes leading to speciation. Examples of reef fish taxa displaying apparent patterns of vicariance in this region include *Myripristis* spp. (Greenfield, 1968), *Amphiprion* (Allen, 1972), *Siganus* spp. (Woodland, 1983), *Congragadus subductens* subpopulations (Winterbottom *et al.,* 1984), *Chaetodon* spp. (Blum, 1989), and several blenny species groups (Springer, 1988; Springer and Williams, 1994). Randall (1998) lists further examples from 15 fish families that may include geminate species pairs. In all these examples, the inferred age of the species or their immediate common ancestors is less than 2 Ma. Such recent species divisions are supported by molecular analyses (McMillan and Palumbi, 1995).

However, there may be other factors involved. The same "bullseye" center of diversity pattern is seen in genera as well as species. This raises the question: Are the factors underlying generic and species diversity patterns similar? One line of evidence suggests that they are, but that it is the role of the IAA as a refuge, not its role as a location for vicariance events, that is common to both species and genera. Of the 31 chaetodontid genera, subgenera, and species groups considered by Blum (1989), most have both Indian Ocean and Pacific Ocean representatives (27 of 31). The remaining four groups, *Amphichaetodon*, *Chelmonops*, *Johnrandalia*, and a *Hemitaurichthys* species subgroup, are all peripheral Pacific Ocean endemics with sister taxa in the Indian Ocean. There are no endemic chaetodontid genera in the Indo-Australian Archipelago.

For chaetodontid species, the relatively high diversity in the IAA appears to be largely a result of (1) overlap of species from adjacent biogeographic regions and (2) low species richness (= loss of species?) in peripheral locations. For genera, there is no evidence of overlap by adjacent groups (although extensive dispersal may mask earlier divisions). For genera and the nonpaired species (only 8 of the 49 chaetodontids in the IAA are species pairs), the role of the IAA as a refuge may be the most important consideration (Bellwood and Hughes, 2001). Sea level changes may split populations and foster speciation, but for genera the most important effect may be the loss of peripheral species, the overall effect being one of range reduction rather than vicariance.

Although the IAA has been regarded as a key location for Plio-Pleistocene vicariance, much earlier vicariance events are also possible. Woodland (1986) identified divisions either side of Wallace's line in several marine taxa, including two genera of ovoviviparous reef sharks (family Hemiscyllidae). While most authors propose Plio-Pleistocene vicariance events for such divisions, Woodland (1986) notes that divisions could date back to the early Miocene, 20–25 Ma, coinciding with the northern movement of Australia. He also points out that these alternatives (Pleistocene sea level changes and movement of Australia) are not mutually exclusive. Furthermore, Springer and Williams (1994) discuss the possibilities of earlier divisions in the IAA ca. 8–16 Ma as a result of Indonesian region tectonic activity that changed surface circulation patterns in the Indian and Pacific oceans. Modern geological evidence has highlighted the tectonic complexity of this area (R. Hall, 1998) and its potential role in the development of Neogene reefal systems (Wilson and Rosen, 1998). Overall, the IAA is clearly an important location for marine vicariance events; however, the timing, cause, nature, and significance of these events remain to be determined.

If one examines endemics as potential indicators of speciation events there is a strongly congruent pattern within the Chaetodontidae, Pomacanthidae, and Pomacentridae. In all three families the endemics are largely peripheral, all laying outside the center of diversity, with the Red Sea, Hawaii, and the East Pacific being conspicuous centers of endemicity (Fig. 7). A similar pattern has been described in the siganid subgenus *Lo* (Woodland, 1986), and in two gastropod groups, Cypraeidae (Kay, 1990) and *Conus* (Kohn, 1985). The mechanism, however, is unclear. Peripheral areas may be marked by relatively high rates of origination vs. extinction. Alternatively, species in peripheral areas may be more likely to remain isolated and thus recorded as endemics (peripheral relicts are considered unlikely). In the IAA endemics may appear to be lost as a result of rapid range extension. Apparent endemicity based on presence/absence data can also result from limited sampling and recent descriptions. Given the data currently available it is not possible to resolve these alternatives, although fossil evidence in other taxa with short generation times [e.g., *Conus* (Kohn, 1985); *Cyprea* (Kay, 1990)] lends some support to the suggestion of extensive peripheral speciation in reef-associated taxa (but see Palumbi *et al.,* 1997).

It should be noted that there is a high probability of all the above factors working in concert, the IAA being both a source of vicariance (a center of origin and center of overlap) and a refuge (a center of accumulation), with the more peripheral areas being marked by endemism and extensive faunal loss. In such considerations it is critical to distinguish theories concerning the origins of species from those concerned with the maintenance of species. The two may not necessarily occur in the same location. For reef fishes, a resolution of the relative importance of these factors in explaining high diversity in the IAA remains elusive. However, with a more detailed description of species distributions, robust species-level cladograms, and molecular data, this issue is likely to be resolved. One of the immediate challenges is the selection of appropriate taxonomic units. So far, most barriers have been identified based on the distribution patterns of species pairs, with most species identified based on color patterns. In fishes, color patterns may not reflect genetic separation (McMillan and Palumbi, 1995), although they may change rapidly and provide a basis for maintenance of discrete morphs (cf. Domeier, 1994). The problem of separating discrete species or subspecies is even greater in corals (cf. Willis *et al.,* 1997). For reef fish, at least, a resolution may be possible.

I. Conclusions

As more data become available it is becoming increasingly clear that congruent divisions in distribution patterns may not reflect a single vicariance event. Congruent patterns at different taxonomic levels in several key locations suggest that either (1) different taxa were affected at different times or (2) a single event affected taxa in markedly different ways. At present these alternatives cannot be resolved. Biogeography based on analyses of distribution patterns, even with cladograms, can only identify the possible location(s) and sequence of vicariance events. Congruence emphasizes the relative importance of locations. However, given the possibility of several temporally separate vicariance events in several key locations, another set of information is needed to provide details of the timing of events. Geology provides the timing of some events but their biological significance can only be inferred. The two most promising sources of information are the fossil record and molecular data. Fossil data are excellent because they provide information on both the minimum age and past locations of taxa. However, fossils are unavailable for many reef taxa. Molecular data are not restricted in this respect and may provide useful age estimates. Indeed, phylogeographic hypotheses and a knowledge of inter- and intraspecific relationships promise to yield invaluable information on historical patterns of connectivity and the origins of lineages. The combination of fossil evidence, molecular systematics, and vicariance biogeography (cf. Reid *et al.,* 1996; Bernardi *et al.,* 2000) offers an exciting avenue for future research in reef fish biology.

For reef fishes, we are beginning to identify the location of major vicariance events. The challenge is to decipher the timing and nature of these events. It is becoming increasingly apparent that there is a need to critically reevaluate the nature of marine barriers. In the past a great deal of work has revolved around hard barriers, e.g., land bridges, where there is a clear physical separation of populations. However, marine taxa appear to respond to a wide range of soft barriers. Of these, the East Pacific Barrier is well documented. Similar barriers probably operate at smaller scales. Even around hard barriers, such as the Isthmus of Panama and the Terminal Tethyan Event, there is increasing evidence of ecological barriers to marine taxa prior to land bridge formation. Perhaps the best example of the importance of soft barriers is Springer's (1982) classic study of Pacific plate biogeography. Here, fish and non-fish taxa appear to be closely linked to a specific continental plate. The nature of the barrier is unclear. Why do so many taxa with widespread or oceanic distributions and planktotrophic larvae not cross the plate margins? It is as if there is an invisible barrier in midocean. Indeed, given the recent advances in our understanding of the biology of fish larvae, we may be able to begin to understand the nature of such barriers (see Chapters 6 to 9). As with the pioneering work of Leon Croizat (Croizat *et al.*, 1974), it may be the simplest of patterns that provides the foundation for a quantum leap in our understanding of the nature of barriers in marine biogeography.

V. Postvicariance Survival Patterns: Fate after Isolation

In the previous section, barriers that isolated regional fish faunas were identified. The subsequent fate of these faunas, however, may vary widely depending on the component taxa and regional characteristics. The extent and nature of subsequent diversification or loss may have a profound effect on the composition, ecology, and functional attributes of surviving faunas. In reef fishes, the available evidence does not permit detailed analyses of the fate of faunas in various regions. However, based on the data in Section II the most marked difference between reef fish faunas is seen between the Caribbean–eastern Pacific and the Indo-Pacific regions, the remnants of the east and west Tethyan provinces. In the following sections, therefore, we restrict comparisons to these two major biogeographical realms.

Today, Caribbean reefs support only about 22% of the number of fish species found on Indo-Pacific reefs and about 80% of the families. The data in Section II suggested that the difference between these areas is primarily a result of a lack of taxa in the Caribbean, in that there is little evidence of faunal replacement. With the exception of the Chaenopsidae, Labrisomidae, and Inermiidae, Caribbean reefs merely possess a subset of the families found in the Indo-Pacific. There are two possible scenarios: the missing taxa were either never present, or they were present but have been subsequently lost. The best way to evaluate these alternatives

TABLE 1 Species Richness of Extant Perciform Reef Fish Families Recorded from Monte Bolca, Italy, and Recent Biogeographic Distributions

Family	Species at Bolca	Occurrence in biogeographic region	
		Atlantic	Indo-Pacific
Acanthuridae	13	X	X
Apogonidae	2	X	X
Blenniidae	1	X	X
Carangidae	4	X	X
Centropomidae	2	X	X
Enoplosidae	1		X
Ephippidae	2	X	X
Gobiidae	1	X	X
Haemulidae	1	X	X
Labridae	2	X	X
Monodactylidae	1		X
Mullidae	1	X	X
Pomacentridae	2	X	X
Scatophagidae	1		X
Scombridae	5	X	X
Serranidae	2	X	X
Siganidae	2		X
Sparidae	10	X	X
Sphyraenidae	1	X	X
Zanclidae	1		X

is to examine the fossil record. Of all extant perciform families represented in Monte Bolca (see section IIIB.), most occur today in both the Caribbean and the Indo-Pacific (Table 1). However, although all 20 are represented today in the Indo-Pacific, only 15 are recorded from the Caribbean. There are no families recorded in Monte Bolca that are found solely in the Caribbean. The Caribbean today appears to possess only a portion of an old and possibly widespread fish fauna, which was largely retained in the Indo-Pacific. The proximity of Bolca to the Atlantic (5000 km upstream, along a coastline) would suggest that the missing families were present in at least the East Atlantic.

Given that the Caribbean has been faunistically distinct since the Cretaceous (Section III), there remains the possibility that these families were never present in the Caribbean. Unfortunately, the fossil record of fishes in this region during the Cenozoic is poor, and direct evidence for the loss of fish taxa in the Caribbean is sparse. There are only two records that support the suggestion that the region is characterized by loss rather than absence. An Eocene (?) acanthurid fossil from Antigua, West Indies (previously identified as *Naso*) has been placed in the genus *Eonaso*, as an extinct putative sister taxon to *Naso* (Tyler, 1997). Today, no member of the *Naso–Eonaso* lineage remains in the Atlantic. Similarly, the Oplegnathidae (Cione *et al.*, 1994) has been recorded from the Miocene of both North America and Europe. Today, this family is found only in the Indo-Pacific.

Faunal loss in the Caribbean and East Pacific has been recorded in numerous other marine taxa (Vermeij and Petuch, 1986; J.B.C. Jackson *et al.*, 1993; Edinger and Risk, 1994), with a major period of faunal turnover in the Plio-Pleistocene (Jackson *et al.*, 1996). There is a strong likelihood that this period also marked a period of change in fish faunas. These studies have identified a number of factors that may have been implicated in the differential loss of taxa from the East Pacific and tropical Atlantic including changes in turbidity, productivity, temperature, and circulation patterns. Cooling of the oceans during the Plio-Pleistocene, in particular, has been closely linked with regional losses (e.g., Stanley, 1984; Jackson, 1994), although temperature alone appears unlikely to explain the observed faunal loss (Jackson *et al.*, 1996). This applies equally well to reef fish families. Their presence in subtropical waters (Section II) suggests that they would be relatively insensitive to temperature changes. Furthermore, shallow reefs persisted in the Caribbean throughout the Neogene (Johnson *et al.*, 1995; Budd *et al.*, 1996).

There are several factors associated with low sea levels, which may have been detrimental to reefs and fishes on reefs. During low sea levels there would be a marked decrease in the area available for shallow-water taxa. Given that reefs and reef fishes are largely restricted to the top 50 m, a drop in sea levels of 180 m would result in reefs being relocated off continental shelves. Taking the area enclosed by the 0- to 50-m isobath and comparing it to the area enclosed by the 150- to 200-m isobaths, as a proxy for comparable shallow waters during Pleistocene lows, the area of shallow water in the Caribbean was reduced by about 89% (Fig. 8). Paulay (1990) provided a comparable explanation for the loss of bivalves on oceanic islands in the West Pacific but highlighted the role of key regions (e.g., Australia's northwest shelf and Fiji) where a shallow sloping bathymetry would provide refuges for shallow-water taxa during low sea stands. Lower sea levels may decrease not only the area of shallow habitats but also the nature of these habitats, with an increase in the proportion of benthic habitats between 0 and 50 m being restricted to relatively steep escarpments along continental margins. In the Caribbean, this resulted in an estimated change in the mean slope of coastal shallow waters from 1.4 to 10.1 m km^{-1}. Furthermore, there would probably be extensive loss of shallow banks, lagoons, and sediment aprons that are characteristic of the shallow continental shelves (cf. Potts, 1985; Myers, 1989; Paulay, 1990; Domeier, 1994). The two effects are quite different. The former is just a proportional loss of area; the latter may result in the total loss of a given habitat from a whole region. The former (area changes) alone have been linked with speciation in reef fishes (Domeier, 1994), whereas habitat loss has been identified as a significant factor in the loss of taxa during the Pleistocene in crustacea (Dall, 1991), corals (Potts, 1985), bivalves (Paulay, 1990, 1996), and reef fishes (Myers, 1989).

Restriction of shallow-water habitats to the edge of continental shelves and more exposed locations on isolated land masses may increase the impact of other potential disruptive factors, including cool upwellings, turbidity, hyposalinity, and storm damage. Fleminger (1986) presents evidence for enhanced cool upwelling in the Indo-Australian Archipelago during Pleistocene periods of low sea level, arguing that this cooler water may have acted as a thermal barrier effectively separating stenothermal populations. Furthermore, he suggests that mean wind speeds and upwellings were enhanced during these cool low-water periods. In addition, coastal freshwater runoff and silt loads may be increased, because both are dependent on rainfall and land area (Schopf, 1980). As shallow waters are restricted, the detrimental effect of runoff may be increased because the runoff is concentrated near the narrow reefs along the shelf break rather than being diluted inshore over broad continental shelves. Springer and Willams (1990, 1994) argue that these effects, when combined, may be responsible for the loss of reef fish taxa in the Indo-Australian Archipelago. Edinger and Risk (1994) describe a comparable scenario for coral losses in the Caribbean during the Oligocene-Miocene.

Overall, it appears that shallow-water faunas were subject to a wide range of potential detrimental effects during low sea stands. The Pleistocene sea level changes are the best documented, but similar effects would be expected during any of the Cenozoic marine regressions, including those in the Oligocene and Miocene. For reef fishes, many of these factors have been implicated in the loss of species, but they offer little in terms of an explanation for the loss of families from the whole of the Caribbean, particularly because many families have broad habitat associations (Section II). These broad habitat associations are noteworthy when examining the families that appear to be missing from the Caribbean. Of those taxa that were present in Monte Bolca but are absent

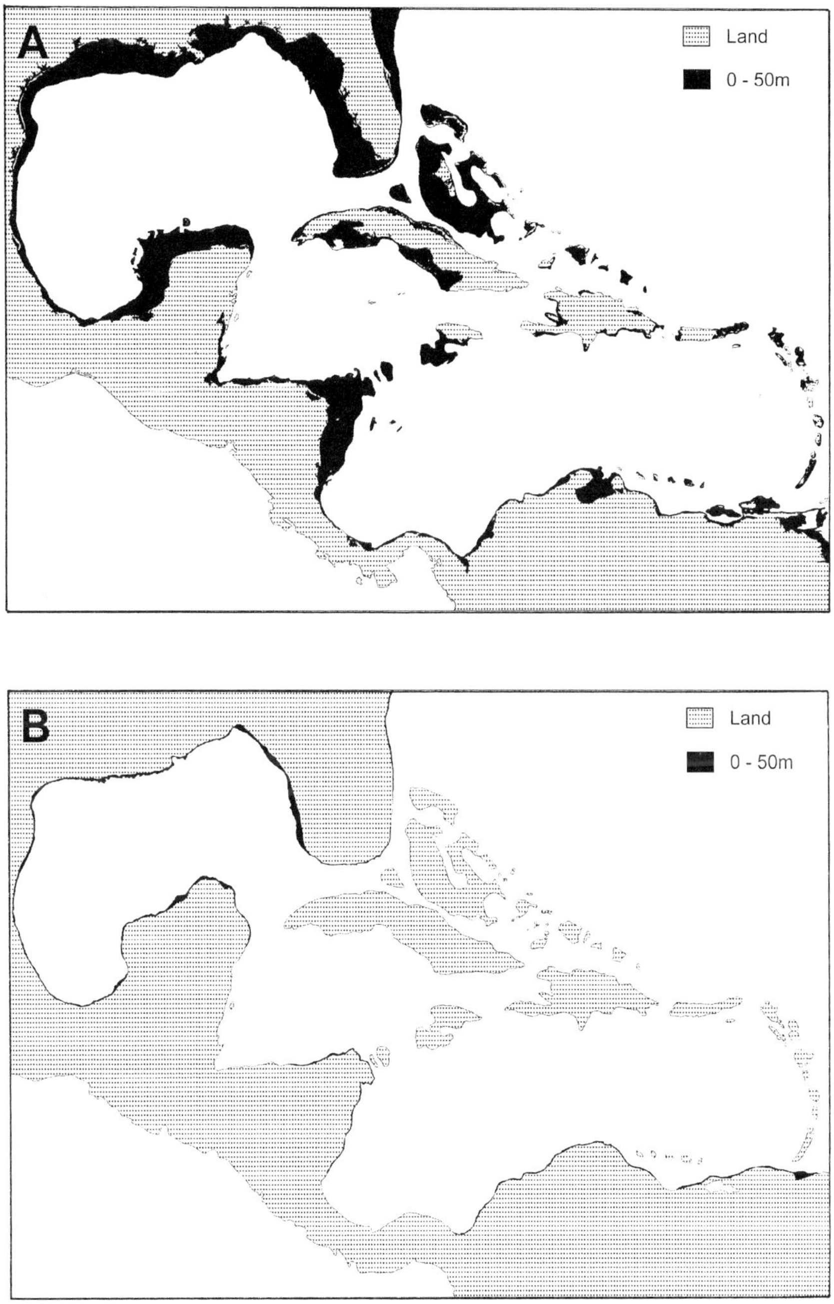

FIGURE 8 Estimated changes in shallow-water habitat availability in the Caribbean. (A) Area of shallow water enclosed by the 0- to 50-m isobaths today. (B) Area enclosed by the 150- to 200-m isobaths (coastal margins indicated only), taken as a proxy for shallow areas during glacial low sea stands, with sea levels 150 m below present. During such low sea stands, the area of coastal shallow-water habitat was only approximately 11% of that in the region today and the mean slope of coastal benthic habitats increased from 1.4 to 10 m km^{-1}.

from the Atlantic today, only two have strong reef associations (Zanclidae, Siganidae), and of these, one (Siganidae) contains several nonreef species. The remaining families are either estuarine (Monodactylidae, Scatophagidae) or temperate coastal forms (Enoplosidae). Furthermore, of the families that are absent from the Caribbean but present in the Indo-Pacific, many have associations with nonreef habitats, including seagrass and soft sediments (Lethrinidae, Nemipteridae) and estuaries (Plotosidae, Teraponidae, Aploactinidae). Only the Pseudochromidae is strongly reef associated. The absence of these families from the Caribbean strongly suggests that the loss of fish taxa was associated with changes that impacted a wide range of coastal and shallow-water habitats, not just coral reefs. In this context, it is interesting to note that during the Plio-Pleistocene the loss of reef corals was most marked in seagrass communities; diverse coral-rich communities appeared to fare relatively well (Budd *et al.*, 1996).

VI. The History and Nature of the Reef–Fish Relationship

Today, many fish species are intimately associated with coral reefs, and it is on reefs that numerous fish families reach their greatest species diversity and abundance. Documenting diversity patterns and examining the factors responsible for producing and maintaining species diversity are major goals in ecology. In this respect coral reefs offer an exciting challenge, with over 1200 fish species on the Great Barrier Reef alone (Randall *et al.*, 1990) and over 250 species on a single reef (Russell, 1983). Many studies have highlighted the close relationship between fish species and various reef attributes, such as habitat complexity (e.g., McCormick, 1995; Chabernet *et al.*, 1997). However, to what extent can we assume that coral reefs have been the arena in which these reef fishes evolved? Today, coral reefs are sites of high diversity, but do they also represent sites of origin? Given the diversity of fishes on reefs and the long tenure of reefs in the fossil record, it appears logical that the fish probably evolved on reefs. Conditions on the reef all appear to be "favorable," with numerous niches, abundant food, high productivity, structural complexity, and habitat continuity through time. Yet, as was noted in Section II, few of the characteristic reef fish families are restricted to coral reefs. The focus of this section, therefore, is to examine the nature of this reef–fish relationship in an evolutionary context.

Direct examination of the fossil record offers little assistance in evaluating this relationship (Bellwood, 1998). However, phylogenetic data provide an alternative line of evidence that may give a clearer indication of the history of the reef fish relationship. Phylogenetic studies provide a basis for examining not only relationships between taxa but also the evolution of various character states. These characters may include behavioral, trophic, and ecological traits, including habitat associations (Brooks and McLennan, 1991; Winterbottom and McLennan, 1993).

There are two possible scenarios:

1. *Coral reefs as the site of origin of reef fish lineages.* Today, coral reefs support a vast array of fish lineages. Coral reefs were present in some form prior to the origins of these lineages. Did they therefore provide the environment within which these fish lineages arose? In the fossil record, modern scleractinian-dominated coral reefs and modern reef fishes first appear and then diversify at approximately the same time. In the early Cenozoic, coral reefs may have filled an ecological vacuum (cf. Boucot, 1983) and provided a habitat within which basal percomorphs could rapidly diversify. Coral reefs may therefore represent the site of origin and the site for the maintenance of reef fish faunas.

2. *Coral reefs as a benign sanctuary.* Coral reefs may merely provide a habitat capable of supporting a diverse fish fauna. Reefs may have acquired lineages from existing nonreef faunas, acting as a sanctuary for the maintenance of diversity with no specific role in the origins of this diversity.

The relative importance of these two alternatives can be assessed using cladograms, by mapping then optimizing habitat details (sensu Winterbottom and McLennan, 1993). If the first scenario (coral reefs as the site of origin of reef fish lineages) is correct, then one would expect to find basal taxa living on reefs and that this is the inferred habitat of the hypothetical ancestral taxon. This would suggest that these lineages have lived on coral reefs from their earliest beginnings. If the second (benign sanctuary) scenario is correct, then the reef dwellers should be derived and the basal taxa and hypothetical ancestral taxon would occupy nonreef habitats. This would suggest that the lineages evolved in nonreef areas followed by a movement onto coral reefs. Furthermore, if fossil data are incorporated into the cladograms, the timing of these inferred events can be estimated.

Cladograms of higher taxa and habitat utilization patterns are available for four reef fish lineages: hypsigenyine labrids, scarids, the Acanthuroidei, and the Chaetodontidae (Fig. 9). Today all of these lineages are closely associated with reefs, and include many of the "characteristic" reef fish families (Fig. 1). Mapping and optimizing the principal habitat utilization patterns of these taxa reveal an interesting pattern, with clear links to nonreef habitats:

1. In the hypsigenyine labrids [Fig. 9A; cladogram from Gomon (1997)], the basal divisions all incorporate temperate or deep-water lineages. It appears that a reef-dwelling mode arose at least twice within this clade, in both of the two main lineages. In one lineage, the reef-dwelling genus *Choerodon* is derived from a lineage that lives in deep (40–240 m), soft-sediment habitats. In the second major lineage, the reef-dwellers *Clepticus* and *Bodianus* both have sister taxa living on temperate rocky coasts. The inferred habitat of the hypothetical ancestor of the Hypsigenyini is equivocal. However, there are strong links with both temperate waters and deep soft-sediment habitats. The age of these hypsigenyine lineages is unknown, although the first record of a putative hypsigenyine labrid is from Monte Bolca (Bellwood, 1990). It is interesting to

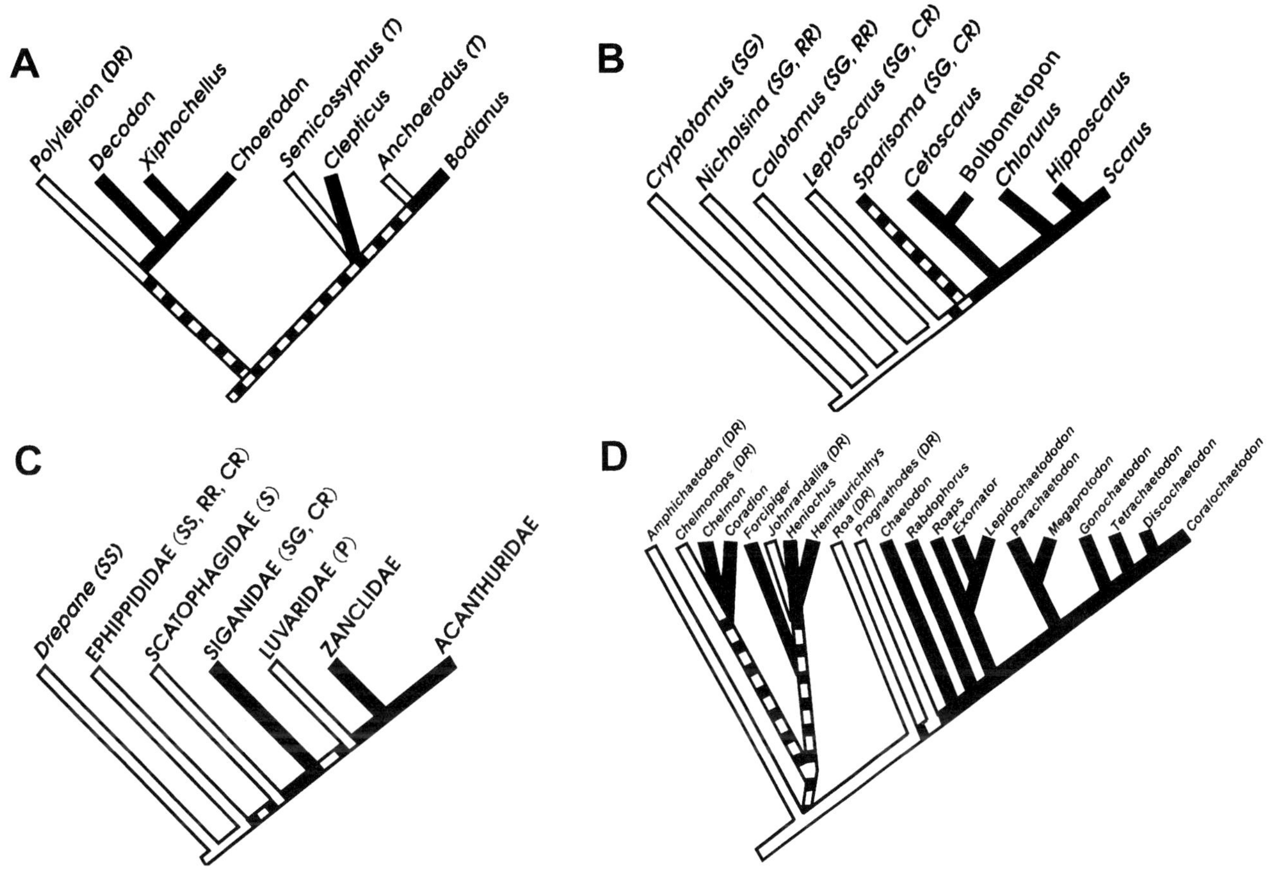

FIGURE 9 Patterns of habitat use in four "reef fish" lineages. (A) Hypsigenyine labrid genera. (B) Scarid genera. (C) Acanthuroid families and lineages. (D) Chaetodontid species groups. Open bars, nonreef habitats; solid bars, coral reef habitats; dashed bars, equivocal. Abbreviations: T, temperate coastlines; SG, seagrass; RR, rocky reef; CR, coral reef; SS, soft sediments; P, pelagic.

note that a slightly younger Eocene hypsigenyine labrid has been recorded from Antarctica (Long, 1992), at a time when Antarctica had a temperate coastal margin. Furthermore, the Antarctic specimen is similar in structure to extant taxa living on temperate rocky shores.

2. The scarids (Fig. 9B) are a derived clade within the Labridae. Today, they are a conspicuous and abundant component of reef fish assemblages. However, if habitat associations are examined based on a cladogram of the family, there is a clear indication that the lineage lived initially in seagrasses and that the early evolution of the group was predominantly off-reef (Bellwood, 1994). Fossil evidence indicates that the seagrass-dwelling forms are at least 15 Ma old and that the move onto reefs occurred at least 5 million years ago (Bellwood and Schultz, 1991; Bellwood, 1994).

3. In the Acanthuroidei [Fig. 9C; after Tyler *et al.* (1989)] all of the basal taxa (and *Drepane*) have strong associations with coastal soft-substratum nonreef habitats. In the Ephippidae a few species occur on reefs as adults. However, the juveniles of these forms are most frequently reported from estuaries and coastal mangroves (Kuiter, 1996), suggesting that reef dwelling was secondarily derived. The Ephippidae, Scatophagidae, Siganidae, Zanclidae, and Acanthuridae have all been recorded from Monte Bolca (Bellwood, 1996a). This suggests that some of these taxa may have already moved onto reefs by the Eocene (50 Ma).

4. Finally, the Chaetodontidae (Fig. 9D), one of the most conspicuous and brightly colored of all reef fish families, also appears to have nonreef origins. Using the cladogram of chaetodontid genera and subgenera of Ferry-Graham *et al.* (2001b) and the habitat and depth data of Allen (1979), an interesting pattern emerges. The basal taxa live predominantly in deep water (at least 20 m, most below 50 m, maximum 200 m), usually on drop-offs or over rock (cf. Pyle and Chave, 1994). Many of the records are from temperate or marginal

reef areas. Overall, the data suggest that chaetodontids arose as a deep-water lineage, living on rocky substrata and feeding on benthic invertebrates, possibly in a subtropical region. Furthermore, it appears from the cladogram that the chaetodontids may have colonized coral reefs on at least two separate occasions. It is noteworthy that, as in many groups, the coral reef lineages are particularly speciose when compared to nonreef lineages. A coral-reef-dwelling habit may not have been implicated in the origins of reef fish lineages, but it may have played an instrumental role in the diversification of these lineages (e.g., *Chaetodon*, *Scarus*, and *Halichoeres*).

Other families with strong reef associations may also have nonreef origins. The Pomacanthidae, like the Chaetodontidae, have several genera or species groups living in deep water, often over rock on steep slopes or drop-offs (e.g., *Aplomichthys*, *Genicanthus*, and *Centropyge*). Given the provisional phylogeny in Allen (1979) there appears to be the possibility of a basal deep-water, nonreef association. The Holocentridae may also be derived from deep-water taxa. Many beryciform taxa, including "basal" groups, inhabit deep waters (cf. Johnson and Patterson, 1993; Nelson, 1994). In both families, further analyses require a clear resolution of the systematic relationships of the component taxa.

Given the likelihood of nonreef origins for several lineages of fishes that are conspicuous on reefs today, the next question arises: What was the fate of these lineages on the reef? Did they proliferate without further input from nonreef habitats? It is apparent from Section IV that many reef fish species appear to have arisen as a result of vicariance events, and that the increasing diversity may be at least partially the result of repeated cycles of division, speciation, and intermixing. However, relatively few studies of reef fishes have considered the impact of vicariance events on nonreef marine habitats. For reef fishes it may be assumed that major vicariance events isolate both reef and nonreef habitats. However, the possible effect of nonreef habitat utilization patterns may be important. For example, in the Indo-Australian Archipelago, the presence of cool or nonreefal environments may not represent a barrier to the basal taxa of many reef fish lineages. The historical biogeography of coral reef fishes may incorporate habitats other than coral reefs (Bellwood, 1997).

Nonreefal habitats may continue to be a source of new species. The location of endemics in three reef fish families, the Pomacentridae, Pomacanthidae, and Chaetodontidae (Fig. 7), are broadly congruent, with numerous peripheral endemics, i.e., away from the center of diversity in the Indo-Australian Archipelago (see Section IV). Many of these endemics are in nonreefal habitats. The presence of numerous endemics in peripheral locations, including many marginal or nonreef areas, points to the possibility of a continuing role for nonreef areas in the origins of evolutionary novelty and the generation of fish diversity on coral reefs. Data from a comparable high-diversity system (rain forests) have emphasized the importance of noncore habitats in the generation of diversity (Smith *et al.*, 1997).

These nonreef associations have implications for our view of reef fishes. First, historical links with nonreef habitats may help explain current ecological links with nonreef habitats, e.g., reef fish taxa settling in seagrass beds or mangroves (e.g., Shulman and Ogden, 1987; Bellwood and Choat, 1989; Parrish, 1989). Second, these observations may help to explain the functional, morphological, and ecological characteristics of extant reef fishes, particularly in considering exaptations (sensu Gould and Vrba, 1982). For example, the highly modified jaws and pharyngeal apparatus of scarids probably evolved in taxa that fed on seagrasses and seagrass epiphytes. Only later were these structures used by taxa feeding on coral reefs (Bellwood, 1994). Finally, nonreef associations will have a major bearing on biogeographic studies. Distributions of reef fish taxa may not be delineated by coral reefs. The history of reef fishes may depend on the distribution, continuity, and connectivity of habitats other than coral reefs. Nonreef areas may act as refugia for reef fish in both ecological and evolutionary contexts.

VII. Functional Aspects of the Reef–Fish Association

Fishes play a variety of functional roles on coral reefs, with direct involvement in numerous reef processes (i.e., interacting with the movement of energy or material). The role of fishes in reef processes, however, is not dependent solely on the number of species or individuals present, but on the identity of the species and the nature of their functional role. In terms of functional roles, all fishes are not equal. At a functional level the role may transcend taxonomic boundaries, as in the changes associated with the mass mortality of *Diadema* in the Caribbean (Hughes, 1994), where the echinoid may have replaced fishes as the dominant grazer prior to the mass mortality event. In this case the functional role passed from fishes to a single invertebrate species.

When considering the implications of historical factors in the ecology of modern reef fish faunas, the immediate challenge is in interpreting the available data. The structures of modern reef fish faunas are often described based on species richness or relative abundance (e.g., Williams, 1982; Thresher, 1991), although a few have considered functional categories (e.g., Williams and Hatcher, 1983). There has been a recent surge of interest in the functional group approach, effectively extending the "guild" concept. It permits interactions and patterns to be examined without the need for detailed species-level observations (cf. Hay, 1994; Steneck and Dethier, 1994). This approach offers a means of gathering broad-scale data on community composition and trophic interactions. It may therefore be useful to consider the information base necessary to identify meaningful functional groups among reef fishes and the possible role of history in determining the nature of these groups.

A. Functional Groups on Coral Reefs: Piscine Bioerosion as a Case Study

Bioerosion refers to the removal of material by living organisms. On coral reefs there are numerous taxa that bore or excavate carbonate substrata (Hutchings, 1986). Of these taxa, the parrotfishes (scarids) and the echinoids are the dominant external bioeroders. Both scarids and echinoids erode the substratum by removing calcareous material as they graze the epilithic algal matrix (sensu Wilson and Bellwood, 1997). Scarids have long been recognized as a distinctive group of bioeroders, with numerous studies documenting their erosional activities (Hutchings, 1986). This designation was based largely on their feeding mode, which often leaves bite marks or scars, and gut contents that invariably include a large proportion of finely ground carbonate. Scarids have been long regarded as a relatively uniform group of grazing or scraping herbivores that erode the substratum during feeding (reviews by Hutchings, 1986; Horn, 1989; Steneck, 1988).

However, based on observations of the functional morphology, ecology, and feeding behavior of scarids, Bellwood and Choat (1990) suggested that only a minority of Indo-Pacific scarid species are significant bioeroders. A comparable pattern is seen in the Caribbean (Bruggemann *et al.*, 1996). Thus, in terms of bioerosion, the functional group of bioeroders is composed of a small number of species not scarids per se (Fig. 10). As species composition and species densities can vary widely among sites, comparisons of total erosion between biogeographic regions are tenuous. Nevertheless,

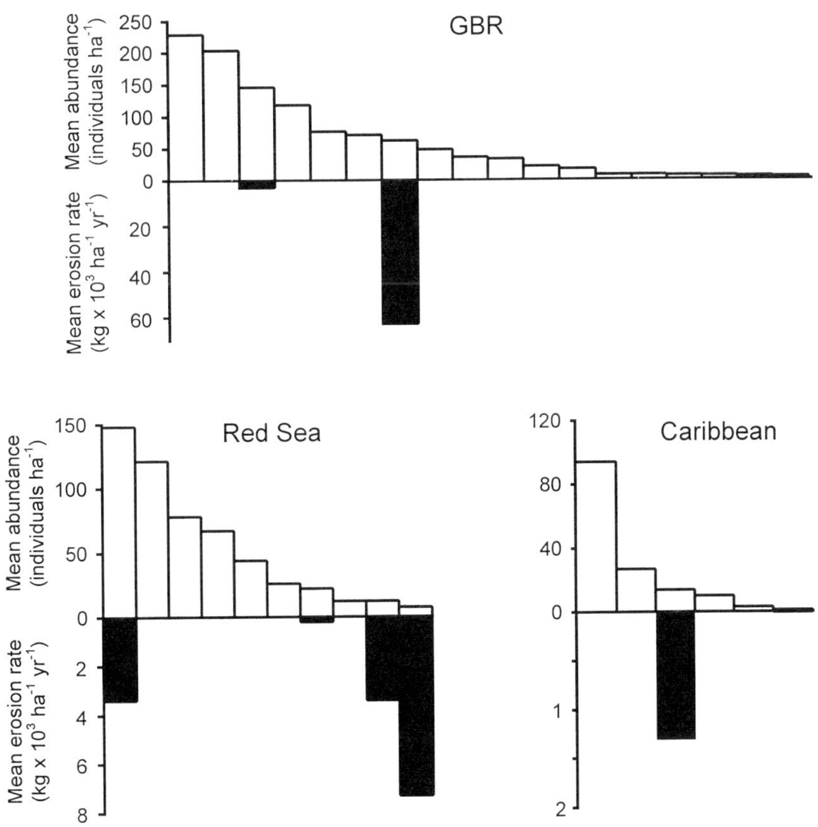

FIGURE 10 Patterns of scarid abundance and respective bioerosion rates at sites in the Indo-Pacific [Red Sea and Great Barrier Reef (GBR)] and Caribbean. [Caribbean values based on Lewis and Wainwright (1985) and Bruggemann *et al.* (1996); GBR values based on Bellwood (1996b); Red Sea values from C. M. Roberts (unpublished data)]. The following species are found in each region, in order of decreasing abundance. GBR: *Scarus rivulatus, Chlorurus sordidus, S. globiceps, S. frenatus, S. schlegeli, C. microrhinos, S. spinus, S. niger, S. psittacus, S. altipinnis, S. oviceps, S. rubroviolaceus, C. bleekeri, S. ghobban, Cetoscarus bicolor.* Red Sea: *Chlorurus sordidus, S. ferrugineus, Hipposcarus harid, S. niger, S. frenatus, S. psittacus, S. ghobban, S. fuscopurpureus, Cetoscarus bicolor, C. gibbus.* Caribbean: *S. croicensis, Sparisoma aurofrenatum, Sp. viride, Sp. chrysopterum, Sp. rubripinne, S. vetula.*

two features are noteworthy in the patterns of erosion in the two regions. First, species richness and species densities have little impact on erosion rates because, in both locations, bioerosional activity is restricted to a few, often relatively rare, species. Despite the fact that the Caribbean is characterized by fewer species and genera, the functional abilities of the fish community may be unaffected. In terms of scarid bioerosion, neither the number of species or genera nor even total numbers of individuals can provide a sound basis for estimating erosion rates; this requires estimates of the abundance of eroding species only. Second, there is no overlap in the taxa responsible for erosion in the two regions, with *Sparisoma* in the Caribbean and *Chlorurus*, *Bolbometopon*, and *Cetoscarus* in the Indo-Pacific. This difference appears to be historical (Section IV) and may have direct implications in terms of the nature of erosion in the two regions.

In both the Caribbean and the Indo-Pacific, scarid erosion is dependent on a few species, with local patterns dependent on the abundance, size, and behavior of these key species. In the Indo-Pacific, one of the primary eroders, *Chlorurus microrhinos* (formerly *Chlorurus gibbus*), has very specific feeding patterns, with most erosion being from convex epilithic algal-covered substrata in shallow water, near the reef edge (Bellwood, 1995b). Sediment is preferentially released in deeper water, with fish swimming up to 60 m away from the feeding sites. In one area an estimated 67 kg of sediment was deposited in two small "defecation sites" over a 4-day period (Bellwood, 1995b). Comparable species-specific behavior is seen in *Sparisoma viride*, the primary eroding species in the Caribbean (Bruggemann *et al.*, 1996). In both regions these species are removing material at a rate that is comparable to local calcification rates (Bellwood, 1995a; Bruggemann *et al.*, 1996).

Given that one can identify the principal taxa involved in bioerosion (and the morphological features associated with an excavating feeding mode), one can focus efforts looking at the biology of eroding species. It is thus possible to provide a detailed picture of the role of these species on the reef in terms of erosion, habitat modification, and sediment production (e.g., Bellwood, 1995b, 1996b; Bruggemann *et al.*, 1996). Furthermore, the close relationship between jaw morphology and excavation potential provides a basis for examining the fossil record for evidence of the history of bioerosion by scarids. A Miocene fossil *Bolbometopon* jaw fragment from Sri Lanka (Bellwood and Schultz, 1991), for example, indicates that excavating scarids have been present in the Indo-Pacific for at least 5.2 Ma. The cladogram of genera and current biogeographic distributions (Section IV), however, suggest that excavating forms may be considerably older.

Using the cladogram and data on feeding mode in scarids it is possible to track the evolution of the various feeding modes. For example, eroders appear to have arisen from seagrass-dwelling browsing forms, Whereas the excavating (eroding) feeding mode arose independently as a result of hypermorphosis in both *Sparisoma* in the Caribbean and its sister taxon, the *Bolbometopon–Cetoscarus–Chlorurus–Hipposcarus–Scarus* lineage, in the Indo-Pacific (Bellwood, 1994; Bernardi *et al.*, 2000). Given that the Caribbean and Indo-Pacific probably had different scarid taxa prior to the Neogene, and that the two areas have had markedly different histories, it is striking that scarid erosion in the two systems is so similar, with most erosion restricted to one or two large species. The disparate histories of the Caribbean and Indo-Pacific offer an exciting format for examining broader questions of ecosystem function, assembly rules, trophic networks, and the relationship between species diversity, cladogenesis, trophic diversity, and functional redundancy. Central to these studies is the recognition and definition of appropriate functional groups (see Chapter 2).

B. Functional Groups in the History of Coral Reefs

In the evolution of marine systems major faunal change has often been marked by mass extinction events followed by rapid diversification (Boucot, 1983). The role of biotic interactions in evolution has been the subject of considerable debate. Of all examples, the Mesozoic Marine Revolution (MMR) (Vermeij, 1977) stands out as a clear example of historical changes in the nature of benthic communities that have been explained by changes in ecological interactions. In the MMR, changes in the composition of benthic faunas, and the form of component species, are interpreted as a response to an increase in predation intensity. This change in predation pressure is based on the appearance and diversification of numerous predator groups in the Mesozoic. It has been argued that these changes accounted for the appearance of many antipredator adaptations, including increased external ornamentation in gastropods (Vermeij, 1977), and infaunalization, boring, and cementation in bivalves and other sessile taxa (Stanley, 1977; Harper, 1991; Wood, 1999).

A second ecologically driven revolution may have coincided with the massive radiation in marine fishes during the late Mesozoic and early Cenozoic: a Cenozoic Marine Grazing Revolution. At this time, the appearance of several herbivorous fish groups may mark the origins of fish-based herbivory on coral reefs, with benthic algae forming a key food source for members of the relatively new and rapidly expanding group, the perciform fishes. Today, perciform fishes dominate modern benthic marine systems. The combination of the MMR and Cenozoic changes has been interpreted as two parts of a long-term trend. Jackson and McKinney (1990) argue that the observed trends (i.e., changes in distribution, prey shape, and infaunalization) can be readily explained as part of a progressive escalation in the defenses employed by prey to counter increasing predation intensity, and an increase in the diversity of forms preying and grazing on the marine benthos.

During the Cenozoic, coral reefs represented a critical arena in which these changes in the nature of trophic interactions may be seen, in that they represent one of the most complex assemblages of marine benthic invertebrates and one of the primary habitats that may be occupied by the rapidly expanding perciform fish faunas. Following a global hiatus of 7–10 Ma in the late Cretaceous–Paleocene, when there was no significant deposition of shallow tropical carbonate platforms (Kauffman and Fagerstrom, 1993), there arose a new and widespread coral reef system that, by the Eocene, appears to have been characterized by the taxa and possibly processes observed on coral reefs today.

The presence of modern reef taxa in the Eocene is readily established by examination of the fossil record. This period marks the first record of most fish families found on modern reefs (Bellwood, 1996a) and many modern scleractinian genera (Veron, 1995). It is harder to determine if processes were similar. However, there are several lines of evidence to suggest that the new reefs of the early Cenozoic were playing by "modern" rules.

Predators capable of excavating or crushing calcareous skeletons have been around since at least the Jurassic. Among the neopterygiian fishes, the pycnodonts were a major group of durophages throughout the Mesozoic (Nursall, 1996). However, the Cenozoic appears to mark a major change in the nature of fish-based predation, from the more carnivorous Mesozoic fish faunas, with relatively few durophagous taxa (cf. Viohl, 1990; Maisey, 1994; Tintori, 1995), to Cenozoic faunas with numerous durophagous taxa and a suite of grazing or browsing herbivores. There appears to have been a major shift in the relationship between fishes and benthic substrata, probably including coral reefs. In the Mesozoic, reef–fish interactions were probably indirect, with reefs providing structure and shelter for potential prey. In the Cenozoic, numerous fish groups appeared that were capable of feeding directly on the reef surface. These fishes probably included epilithic algal grazers and algal browsers, with the presence of families including the Acanthuridae, Siganidae, and Pomacentridae (Bellwood, 1996a; Bellwood and Sorbini, 1996). Today, these fish groups are major determinants of benthic community structures on coral reefs (Steneck, 1988).

On modern reefs, most reef substrata are dominated by corals, coralline algae, and short epilithic algal communities. The maintenance of these groups appears to be dependent on grazing pressure. Removal or reduction of grazing activity may result in an increase in algae and a decline in corals. This scenario has been documented over many scales, from sites over months to reefs over years, with increasing evidence that many components of the benthic community are dependent on intensive grazing pressure (Carpenter, 1986; Hughes, 1994; Steneck, 1997).

C. Grazing through History

Throughout the Phanerozoic there have been shallow-water reef structures with associated grazers. Steneck (1983) documented the changes in the relative intensity of excavating grazing on hard substrata, with a progressive increase in the number of taxa and grazing depth. The most marked change, particularly in excavation depth, was in the late Mesozoic and Cenozoic corresponding with the increase in piscine herbivory. It appears that the "new" Cenozoic reefs were different from their Mesozoic precursors in that the fish fauna contained a diverse array of percomorph fishes that were able to play a key role in shaping benthic communities. Moving onto reefs from other habitats, these fishes would be in position to stamp their impression on evolving coral reefs.

Mesozoic grazers were predominantly invertebrates, which were relatively slow moving and left only shallow scars and minimal structural damage. Only in the late Mesozoic were deep-grazing limpets and urchins recorded. Among the fishes, the pycnodontids are widely regarded as the dominant durophagous group feeding on benthic hard substrata (Viohl, 1990; Maisey, 1991, 1994; Tintori, 1995; Nursall, 1996). At this time there were few, if any, clearly recognizable grazing or browsing fish. Piscine herbivory may date from the Devonian, but there is little evidence to suggest that an herbivorous feeding mode arose prior to the Cenozoic. In the Cenozoic, in addition to the invertebrate grazers, vertebrate grazers appeared in significant numbers with several different fish groups represented (Bellwood, 1996a). These new grazing forms were all characterized by high mobility and in some taxa the potential for deep scars and significant structural damage. These new herbivores could have radically changed the nature of grazing, combining intensive grazing with the potential for selectivity and mobility.

Marine vertebrate grazing may not have been restricted to fishes. Chelonioid turtles have been recorded since the upper Jurassic (Carroll, 1988), although the origins of herbivory within this group are unclear. Today only two species are predominantly herbivorous, *Chelonia midas* and *Chelonia agassisi*, the former occasionally grazing algae on reefs. The earliest record of the family to which these species belong is in the early Cenozoic (Legler and Georges, 1993). The other major group of vertebrate herbivores, the Sirenians (dugongs and manatees), likewise dates back to the early Cenozoic, with the first records from the Lower Eocene (Carroll, 1988).

The impact of these changes in grazing pressure is hard to assess. The fossil record of tropical fleshy algae is poor. However, several lines of evidence suggest that grazing pressure increased during the late Cretaceous and Cenozoic and that there was an associated change in the structure of benthic communities. This is most clearly seen in the coralline algae and corals. Historically there have been two major groups of coralline algae, the Solenoporaceae (Cambrian–Miocene) and the Corallinaceae (Carboniferous–recent). The two forms differ markedly in their inferred resistance to grazing, with the former being much more susceptible to damage. Steneck (1983) related changes in the relative abundance of the two algal forms to changes in the levels of herbivory, arguing that the decline in solenopores was correlated with the appearance of deep grazing urchins, limpets, and fishes. A comparable decrease in delicate (herbivore-susceptible) branching corallines was recorded in the tropics after the Eocene (Johnson, 1961). The inferred scenario is one of increased herbivory, both in intensity and excavation depth, with a concomitant decrease in the abundance of macrophytes. This results in a system dominated by close-cropped turfs, where decreased competition between algae and corals permits coralline algae and corals to proliferate.

Scleractinian corals have been a significant component of reefs since the mid–late Triassic. From the late Triassic onward, coral assemblages included taxa with large colonies and with a strong probability of a well-established zooxanthellate association (Wood, 1995). Coates and Jackson (1985) argue that the risk of mortality is the primary selective force in shaping the geometry of epibenthic marine invertebrates. In addition, they suggest that the nature and extent of coloniality and module integration are directly related to this mortality risk. Today reef corals are predominantly multiserial (erect or encrusting) hermatypic (i.e., containing zooxanthellae) with small corallites and high levels of integration (Coates and Jackson, 1985). This growth form confers not only the ability to grow to large sizes and great flexibility in growth morphology, it also compartmentalizes damage, with the ability to regenerate from fragments or to rapidly repair damage (Jackson and Hughes, 1985; Meesters *et al.*, 1994). This growth form is also strongly linked with an association with zooxanthellae (Coates and Jackson, 1985), an association that brings additional benefits to corals with the potential for faster calcification and rapid growth (Barnes and Chalker, 1990). In many respects therefore, modern scleractinian corals possess a suite of traits that may limit the impact of predation. Although driven largely by a decline in one family, an increase in the degree of colony integration during the Cenozoic may be indicative of changes in the nature of mortality risks during this period (Coates and Jackson, 1985; Wood, 1995).

There are several other taxa in which changes in taxonomic composition, distribution, or morphology have been linked to increased predation, bioerosion, or grazing pressure by fishes during the late Mesozoic and Cenozoic. This includes the loss of shallow-water stalked crinoids (Meyer and Macurda, 1977) and articulate brachiopods (Stanley, 1974; Logan, 1977), and morphological changes in bryozoans (Jackson and McKinney, 1990). Changes in the structure of non-calcareous algal assemblages are difficult to determine because there are few well-preserved specimens and numerous unreliable records (Dawson, 1966). Changes in the algal community can only be inferred; however, the patterns seen in corals, calcareous algae, and other benthic taxa suggest that the Cenozoic would be marked by the rise of grazing-resistant, integrated, clonal algae, possibly with burrowing or endolithic components—in all, the characteristics of a modern turf or epilithic algal community.

Today, the structure of the jaws and multidenticulate dentition of many acanthurids are closely associated with the browsing or grazing of short, filamentous, epilithic algae (Jones, 1968; Purcell and Bellwood, 1993). The presence in Monte Bolca of several acanthurid genera with comparable dentition (Blot and Tyler, 1990) suggests that these fishes would have been able to graze short-turfing algae. With over eight genera and representing over 4% of the nonclupeid specimens recorded from Monte Bolca (Bellwood, 1996a), these early acanthurids may have been a significant group of epilithic algal grazers. Other potential grazing groups include the Siganidae, and the

Pomacentridae, and, to a lesser extent, the Zanclidae, Scatophagidae, and Ephippidae.

Given the evolutionary history of reef fishes and modern scleractinian coral reefs it appears that the appearance and proliferation of modern scleractinian-dominated coral reefs in the late Eocene/Eocene were not the catalyst in the origin and the early diversification of reef fish lineages. These lineages were almost certainly established in nonreefal habitats prior to the widespread appearance of modern coral reefs. However, the converse may not be true: the presence of reef fishes may have been instrumental in the formation of modern coral reefs and their successful invasion of oligotrophic tropical waters. Although the dominant constructional elements of modern reefs—scleractinian corals and coralline algae—have been around since the Triassic, the evolution and development of modern reefs may have been dependent on the presence of modern fish groups, particularly the herbivores.

Most Phanerozoic reefs were calcified algal or soft-substrate heterotrophic communities that were trophically unlike modern coral–algal reefs (Wood, 1993, 1995). An increase in the intensity and nature of grazing and predation provided a dramatically different environment for reef builders in the late Cretaceous and early Cenozoic. Suppression of macroalgae would give corals and grazing-resistant coralline algae a competitive opportunity, and may have underpinned their successful proliferation and invasion of oligotrophic environments. Reefs have been present throughout the Phanerozoic. It is unlikely that reef development per se required fishes, but it is likely that the nature of reef development and the composition of benthic communities were shaped by fishes, the result being a system dominated by grazing-resistant algae, leading to modern benthic communities that are strongly influenced by, and possibly dependent on, intensive predation and grazing.

VIII. Discussion and Conclusions

It is clear that knowledge of history can change the way we view living reef fish faunas. The study of marine evolution and biogeography is at an exciting stage. Phylogenetic analyses have provided the stimulus for a reevaluation of historical information and fostered a synergism between disciplines that is increasingly evident in biogeography today. Numerous workers have recognized the extent to which local patterns are dependent on regional factors, which in turn are shaped by historical and biogeographic events (cf. Rickleffs and Schluter, 1993a). The global decline in biodiversity has highlighted the need to understand the processes operating at regional and large biogeographic scales.

Reef fishes represent one of the most diverse vertebrate assemblages. With a 50-million-year history and biogeographic ranges that can span the entire tropical Indo-Pacific, from Africa to the Gulf of California, they offer a unique insight into the role of history and biogeography in marine systems. The fishes seen on reefs today are the product of a long and complex history, with superimposed patterns of origin and loss extending over at least 50 million years. Indeed, regional fish faunas may have been shaped by events as far back as the Cretaceous. The Indo-Pacific and Caribbean probably share a common faunal history, yet today the species richness in the Caribbean is only about a fifth of that in the Indo-Pacific, with strong temperate influences. The two regions have been shaped by their subsequent period of isolation, with the Caribbean being characterized by a history of decline and loss.

For reef fishes, the inclusion of historical evidence in our analyses has already challenged us to reappraise our understanding of the term "reef fishes" and the nature of interactions between fishes and the reef. Reef fishes defy simple definition, with nonreef habitats playing a critical role in the origin and maintenance of reef fish taxa. Reef fish assemblages appear to be a collection of lucky survivors. Those taxa that survived the vagaries of division and regional extinctions form the basis of Recent reef fish assemblages in major biogeographic realms (e.g., Indo-Pacific, Atlantic, East Pacific). Within these realms subsequent vicariance events may lead to speciation and regional subdivision. However, many fish assemblages appear to be relatively constrained, as marked by the similarity in the familial composition and relative species richness of families in the Red Sea and on the Great Barrier Reef.

Reef fish faunas also represent a collection of ancient body forms. The staggering diversity of species on reefs today, belies this ancient history. Much of this diversity may be only skin deep, with diversity being based largely on differences in color patterns; the body plans have changed little over millions of years. Most families, and their associated structural features, were established 50 Ma ago and some lineages may even extend back into the Cretaceous.

This long and complex history challenges us to look at the way that the fishes and the reef interact. It appears that many fish families do not need reefs. In both evolutionary and ecological terms, coral reefs represent only one of a range of suitable habitats. Reef associations may be largely facultative. In contrast, the presence of fishes appears to have been of critical importance in the evolution of modern coral reefs.

There are four major components in biogeographic studies: (1) the description of patterns, (2) elucidating the origins of the patterns, (3) examining the mechanisms maintaining patterns, and (4) determining the ecological consequences of these patterns. In reef fish studies, progress has been made in all four areas. For reef fishes, our knowledge will continue to grow, not only from the steady accumulation of more fossils and phylogenies, but also from the application of new methods. Fossils continue to provide new data, but further analyses of both form (cf. Foote, 1996) and function (cf. Valkenburgh, 1994; Jernavall *et al.*, 1996) offer promising avenues of investigation in order to expand our understanding of the nature and impact of changes in form through time. Furthermore, the full impact of molecular studies has yet to be felt in our analyses of reef fish biogeography, although the initial results offer exciting insights, with ancient origins for the major lineages and extensive turnover and intermixing, at a species level, in relatively recent times.

Reliable species lists are now available for most major reef areas and interesting patterns are emerging, with the identification of the location of possible vicariance events. These locations are identified largely on discontinuities in the distributions of color morphs or species. However, the nature and timing of these vicariance events remain to be determined. Many other questions remain: What are the primary determinants of species richness on coral reefs and to what extent does congruence between fishes and nonfish reef taxa (e.g. Bellwood and Hughes, 2001) reflect general biogeographic processes? The maintenance of assemblages and biogeographic patterns is still an area of active investigation although regional comparisons of processes maintaining assemblages (cf. Hughes *et al.*, 1999) are limited.

Of all components, the last is arguably the most important: the ecological consequences of historical events and biogeographic patterns. Reef fish assemblages are not uniform. Their differences are the product of both historical and ecological factors. The consequences for management of reef resources are far reaching. How can the Red Sea thrive with only a quarter of the fish species on the Great Barrier Reef? Do reef fish assemblages follow assembly rules? And to what extent would such rules apply to functional groups? To what extent is ecosystem function shaped by species richness and regional history? At what point will declining species richness compromise ecosystem function? And to what extent is the answer dependent on biogeographic location: are peripheral sites more vulnerable than those close to high-diversity locations? Clearly, we must look beyond species numbers and even biogeographic patterns in species abundances. A knowledge of the functional abilities and role of fishes on reefs appears to hold the key to our understanding of the significance of biodiversity in reef ecosystems.

Preliminary data have identified patterns that will serve as a basis for future research into the nature and significance of biodiversity. Within the Indo-Pacific there are strong gradations in species richness that permit comparisons of faunas of differing diversity within a biogeographic region, whereas comparisons between the Caribbean and Indo-Pacific permit comparisons between regions with markedly different histories. Such analyses will provide a rich source of comparative data to examine the role of history in shaping faunas and the role of species richness in the function of reef ecosystems.

Acknowledgments

The ideas in this chapter have been shaped by numerous colleagues who have patiently explained the nuances of their research so that we could explore the links between paleontology, functional morphology, and ecology. Their tolerance of our eclectic questions is gratefully acknowledged. In particular we thank C. Patterson, L. Sorbini, J. Tyler, P. Forey, R. Wood, J. Long, B. Rosen, P. Ahlberg, S. Turner, M. Westneat, C. Wallace, G. Paulay, D. Goujet, O. Schultz, G. Young, M. Coates, and T. Flannery. Thanks also to C. Roberts for unpublished data and O. Bellwood, A. Hoey, M. Marnane, S. Purcell, and E. Vytopil for help with the figures. This chapter was improved as a result of helpful discussions with J. H. Choat, T. P. Hughes, J. M. Leis, D. R. Robertson, and R. Steneck. This work was supported by the Australian Research Council and National Science Foundation.

CHAPTER 2

Ecomorphology of Feeding in Coral Reef Fishes

Peter C. Wainwright
Center for Population Biology
University of California at Davis
Davis, California 95616

David R. Bellwood
Centre for Coral Reef Biodiversity
School of Marine Biology and Aquaculture
James Cook University
Townsville, Queensland
4811, Australia

I. Introduction

Once an observer gets past the stunning coloration, surely no feature inspires wonder in coral reef fishes so much as their morphological diversity. From large-mouthed groupers, to beaked parrotfish, barbeled goatfish, long-snouted trumpet fish, snaggle-toothed tusk fish, tube-mouthed planktivores, and fat-lipped sweet lips, coral reef fishes display a dazzling array of feeding structures. And, even the most casual fish watcher quickly gets a sense that this diversity means something, that fish form is related to what the animals eat. Clearly there is something to this impression, but just how are head and body morphology connected to prey choice? Are we really able to predict what a reef fish eats from studying its jaws and teeth? What are the major axes of diversification that are seen in reef fishes? Which morphological and ecological trophic types show the most common convergence? What ecological insights into reef processes have been gained from consideration of the functional design of fish feeding mechanisms?

In this chapter we explore the relationship between fish anatomy and feeding habits. Our overall goal is to show how an understanding of the functional morphology of fish feeding mechanisms can be a powerful tool when considering several ecological issues in coral reef fish biology. We have attempted to identify generalities, the major patterns that seem to cut across phylogenetic and geographic boundaries. We begin by constructing a rationale for how functional morphology can be used to enhance our insight into some long-standing ecological questions. We then review the fundamental mechanical issues associated with feeding in fishes, and the basic design features of the head that are involved in prey capture and prey processing. This sets the stage for a discussion of how the mechanical properties of fish feeding systems have been modified during reef fish diversification. With this background, we consider some of the major conclusions that have been drawn from studies of reef fish feeding ecomorphology. Because of space constraints we discuss only briefly the role of sensory modalities—vision, olfaction, electroreception, and hearing—but these are also significant and diverse elements of the feeding arsenal of coral reef fishes and entire review chapters could be written on each one.

II. How Does Morphology Influence Ecology?

There is a strong intuitive awareness of the link between the design of organisms and their natural history. We instinctively see the connection between the large mouth and sharp raptorial teeth of the coral trout, *Plectropomus leopordus,* and the observation that it preys largely on other fishes. Similarly, the massive platelike jaws and the bulky jaw muscles of the porcupine fish, *Diodon hystrix,* seem well suited to this fish's diet of hard-shelled molluscs. Why do these associations between gross morphology and feeding habits appeal to our sense of order? We see in these examples morphological attributes that we expect to enhance the ability of these fishes to perform the key tasks that are involved in feeding on their prey. Our experience with other fishes suggests that *Diodon* has unusually large jaw muscles and bones to enable an exceptionally powerful bite capable of crushing the hard shells of gastropods and bivalves. The key here is that there is a causal connection between morphology and performance, or the ability of the fish to perform key tasks. The connections between morphology and performance on the one hand, and between performance and ecological patterns on the other hand, are the focus of ecomorphology, the study of the functional and ecological consequences of organismal diversity (van der Klaauw, 1948; Barel, 1983; Motta and Kotrschal, 1992; Wainwright and Reilly, 1994; Koehl, 1996).

A. Linking Morphology and Performance

The design of physiological systems determines their performance properties. Just as the component parts of man-made machines can be modified to produce engines with differing performance qualities, so too are the elements of biological machines modified during evolution to produce diversity in performance of the overall system. So, in principle, understanding how physiological and biomechanical systems function and knowing the role of the various component parts provides the foundation for interpreting their ecological role. Research fields such as functional morphology, physiology, and biomechanics aim to learn how biological systems work, providing insights into design and thus a framework for interpreting diversity. These disciplines establish our best understanding of the causal connection between design and performance.

Biological function involves more than gross anatomy, and significant modifications to functional systems can occur at many levels of organization. Thus, one might consider the consequences of changes in the mechanical advantage of the jaw muscle as it crosses the jaw joint, the consequences of different myosin isoforms for muscle contraction speed, the spectral properties of different visual pigments, or the digestive properties of various enzymes. But, whether the problem is one of mechanics or physiology, there are two valuable consequences of accomplishing a thorough analysis of how a functional system works. First, a complete study of a system provides a rigorous framework for understanding why differences in design have the specific consequences for performance that they have. It becomes possible to interpret morphological diversity in a performance landscape. Second, if one understands how a system functions, one can make well-informed predictions about the performance of similar systems in which performance has not been measured. The significance of this point lies in the common practice of using correlations between morphological and ecological variables as a basis for generalizing to unstudied taxa. Such an approach will be strongest when thorough functional analyses underlie the selection of key morphological traits for study. Spurious correlations between morphological and ecological traits can come about for many reasons, such as underlying phylogenetic patterns, or the fact that often many morphological features are integrated so tightly that correlations will appear even when there is no causal connection. For example, the size of almost every structure in the body changes with growth in concert with overall body size, so that many structures will show a correlation with an ontogenetic change in feeding habits even if they are not related to feeding behavior.

B. The Role of Behavioral Performance in Shaping Ecological Patterns

In general, the performance capacity of individual animals is a central factor in determining patterns of resource use, such as space and prey, and ultimately their fitness. This link may be starkly direct, as in the consequences of escape locomotor performance for avoiding predation, but, direct or indirect, the ability to perform various tasks enters into the equations that determine usage of virtually all resources that an animal seeks during its life. Here we are concerned with understanding the factors that shape the feeding habits of reef fishes, in particular the connection between feeding performance and prey consumption.

The first and most obvious effect of performance on feeding habits is that limits on performance place constraints on the range of prey that potentially can be utilized. A consequence of this observation is that all individuals will be functionally incapable of successfully feeding on most of the possible prey in their environment. Many reef fishes use suction feeding to capture prey that they swallow whole. In suction-feeding predators, the size of the oral aperture directly limits the size of prey that can be eaten (Werner, 1977; Wainwright and Richard, 1995) [in some taxa it is the size of the pharynx (Lawrence, 1957)]. Given the tremendous variety of possible prey items available to a fish on a coral reef, its feeding performance places discrete boundaries on what it *can* eat. Thus, the limits of feeding performance define a potential feeding niche.

There is considerable room for further refinement of this potential niche and it is usually the case that fishes do not feed randomly on the prey that they are minimally capable of eating (Wainwright, 1988; Huckins, 1997). Here again, feeding abilities are a central factor in determining the ultimate shape of prey-use curves. Consider the forms of foraging models that estimate net energy return as a function of three primary factors: rates of encounter with the prey, the probability that the predator will attack an encountered prey, and the probability that the attack is successful (O'Brien, 1979; Greene, 1983; Osenberg and Mittelbach, 1989). For each of these factors the behavioral abilities of the predator to perform relevant tasks may play a major role in shaping the function. For example, encounter rates can be sharply limited by the sensory abilities of the predator. Visual acuity sets the minimum size of prey that planktivores can see, thus having the effect of drastically reducing encounter rates of planktivorous predators with extremely small prey (Li *et al.*, 1985).

Handling-time curves are shaped by the abilities of the predator. A typical pattern is one in which handling time, or energy expended per prey item, increases exponentially as prey size increases (Werner, 1977; Mittelbach, 1984; Stein *et al.*, 1984; Osenberg and Mittelbach, 1989; Huckins, 1997). Because the energy gained per prey item also increases with prey size, there is a prey size that maximizes net energy return for any combination of predator and prey type. The shape of the handling-time curve can differ for different predators, so that for each the prey size, or type, that maximizes net energy return will change (e.g., Werner, 1977; Huckins, 1997). Regardless of whether predators do indeed forage to optimize energy gain, the effect of behavioral performance on these components of foraging models illustrates the basis of how functional diversity can lead to specific, quantitative expectations of the differences among fish species in their feeding habits (Lavin and McPhail, 1986; Wainwright, 1988; Norton, 1991; Sanderson, 1991). Thus, there is a causal link between functional morphology of fish feeding mechanisms and the prey they eat, and the link is provided by the relationships between the phenotype and feeding performance on the one hand, and role of the latter in shaping patterns of prey use on the other hand.

The impact of diversity of feeding mechanisms reaches into other important components of fish ecology. Resource-use patterns may play a causal role in shaping patterns of abundance and habitat distributions in fishes (Mittelbach and Osenberg, 1994; Wainwright, 1996). This connection has been successfully made in some temperate, freshwater fish communities (Keast and Webb, 1966; Werner and Hall, 1974; Mittelbach, 1984), but has not been extended into coral reef fishes as yet. Competitive interactions between bluegill and pumpkinseed sunfish in North American lakes result in bluegill emerging superior in utilizing zooplankton, the preferred prey of both species (Mittelbach, 1984). However, pumpkinseed sunfish gain a competitive refuge by feeding on snails, a less preferred prey resource, but one that pumpkinseed are notably better than bluegill at utilizing (Mittelbach, 1984; Osenberg and Mittelbach, 1989). Zooplankton are abundant in open water and snails are abundant in the vegetated littoral zone, and the abundance of bluegill and pumpkinseed sunfish is proportional to the relative amount of these two habitats in different lakes (Mittelbach, 1984).

It is clear that to understand the foraging decisions that a fish makes one wants to know the shape of the handling-time curve, and other measures of feeding ability and how they impact on foraging decisions. But, what is the significance of understanding the functional basis of feeding ability? Is it not enough to measure the behavioral ability? Why should the ecologist be concerned with the details of biomechanics and musculoskeletal functional morphology that underlie prey capture and handling skills in fishes? Perhaps the major connection here is that understanding the causal basis for differences among species in feeding ability greatly enhances our ability to make predictions in new systems for which we possess limited information. If we understand why the shapes of the prey handling curves, and hence prey-use patterns, differ for two species, then we have a more solid foundation for making predictions about the shape of these curves in taxa for which few ecological data exist. As future work draws connections between prey-use patterns and the abundance and distribution of coral reef species, our understanding of the functional basis of feeding habits will enhance our ability to predict patterns of fish abundance and distribution in coral reefs altered by human fishing pressure and changes in global weather patterns. Furthermore, given that coral reef fish communities are so species rich, the potential ability to use carefully chosen morphological traits as indicators of feeding ability offers an efficient route toward gaining some understanding of the performance capacity of these exceptionally diverse fish communities.

III. The Biomechanical Basis of Feeding Performance

The past three decades have seen tremendous advances in our understanding of the functional morphology of fish feeding mechanisms and the mechanical basis of feeding performance. Several methods have figured prominently in this field, including anatomical and mechanical studies of the muscle–skeleton systems involved in feeding (Tchernavin, 1953; Alexander, 1967; Gosline, 1971; Barel, 1983; Westneat, 1990; Motta and Wilga, 1999), high-speed kinematic and video analyses (Grobecker and Pietsch, 1979; Richard and Wainwright, 1995; Wainwright and Shaw, 1999; Viladiu *et al.,* 1999), endoscopic video recording of particle motion inside the buccal cavity (reviewed in Sanderson and Wassersug, 1993), electromyography to study the use of muscles during feeding behaviors (Lauder, 1980; Sibbing *et al.,* 1986; Wainwright, 1989a; Wilga and Motta, 2000), catheter-tipped transducers to measure suction pressures (Lauder, 1983a; Grubich and Wainwright, 1997; Nemeth, 1997), and the use of particle flow velocimetry to visualize the water movement created by feeding fishes (Ferry-Graham and Lauder, 2001). In this section we describe general principles of the functional morphology of feeding in teleost fishes. We emphasize key design themes of the skull that have been shown to be particularly influential in shaping fish feeding performance and the morphological variables that can be used to estimate their mechanical properties.

A. Prey Capture

1. Methods of Prey Capture

There are generally two stages to feeding in teleost fishes, prey capture and processing; these may employ different functional systems of the skull. Fishes employ one or more of their arsenal of sensory systems to locate prey before capturing them with their oral jaw system (Fig. 1A). Capture is followed by one or more processing behaviors. The prey may be separated from unwanted material, such as sand or other debris, by winnowing behaviors, or a protective armor may be cracked to access the flesh of the prey, as in the case of a molluscivore. During these processing behaviors teleost fishes usually employ a second set of jaws, the pharyngeal jaw apparatus (Fig. 2), a musculoskeletal system formed from toothed gill arch bones and muscles that is mechanically independent of the oral jaw apparatus. Both the oral jaws and the pharyngeal jaw apparatus of coral reef fishes exhibit extensive functional diversity. This functional diversity corresponds to ecological diversity, because nearly all conceivable prey are used by one reef fish or another.

There are three broad methods of prey capture in fishes (Liem, 1980): suction feeding, ram feeding, and manipulation. When suction feeding, the fish rapidly expands the buccal cavity, creating a pressure gradient between this space and the area around the head (Figs. 3 and 4). Water comes rushing into the buccal cavity through the mouth to eliminate this pressure gradient (Muller *et al.,* 1982; Muller and Osse, 1984; Lauder, 1985). Prey that are in the region of the oral aperture are entrained in this water flow and accelerated into the mouth. Suction feeding is believed to be the primitive method of prey capture in teleost fishes (Lauder, 1985) and is seen to some degree in almost all species of coral reef fishes. It is the most common method of prey capture in predators of mobile animal prey.

Suction feeding (Muller *et al.,* 1982; Lauder, 1983a; Muller and Osse, 1984) is frequently contrasted with ram feeding (Norton and Brainerd, 1993; Van Damme and Aerts, 1997), in which the predator overtakes the prey with forward movement of the body or protruding jaws. Extreme examples of ram feeding exist on coral reefs in animals such as *Manta* and the whale shark, which sometimes capture prey in an opened mouth while they swim through dense swarms of planktonic prey (Sanderson and Wassersug, 1993; Sanderson *et al.,* 1996) [note, however, that there are also other mechanisms of suspension feeding (Goodrich *et al.,* 2000)]. However, pure ram and suction feeding appear to be relatively rare, and most feeding events employ a combination of the two (Norton, 1991, 1995; Norton and Brainerd, 1993). Ram feeding can be distinguished in practice from suction feeding based on the relative role of forward mouth movement (ram) and water flow toward the mouth (suction) in moving the prey into the oral cavity. Ram feeders face an additional challenge in having to eliminate the bow wave that may be formed if they are forcing an opened mouth through the water (Van Damme and Aerts, 1997). This problem can be relieved somewhat if the fish opens the opercular space, allowing water to flow freely through the oral cavity, but this only partly eliminates the formation of the bow wave. The solution for most predators that use considerable ram (e.g., barracuda and many serranids) is to overtake the prey with body movements, and finish the attack with an explosive strike of suction feeding (Figs. 4 and 5). At the very least, some compensatory suction is needed to eliminate the bow wave (Summers *et al.,* 1998).

Ram may be accomplished by forward motion of the entire body, or in combination with, or separately by, protrusion of the jaws, termed "jaw-ram" feeding, as seen in many zooplanktivores who use jaw protrusion to move their mouth quickly toward the prey before using a burst of suction to draw the prey into the buccal cavity (Motta, 1984a; Coughlin, 1994). This is also illustrated by the feeding method used by the sling-jaw wrasse, *Epibulus insidiator,* a predator of elusive prey that possesses the most protrusible jaws known in

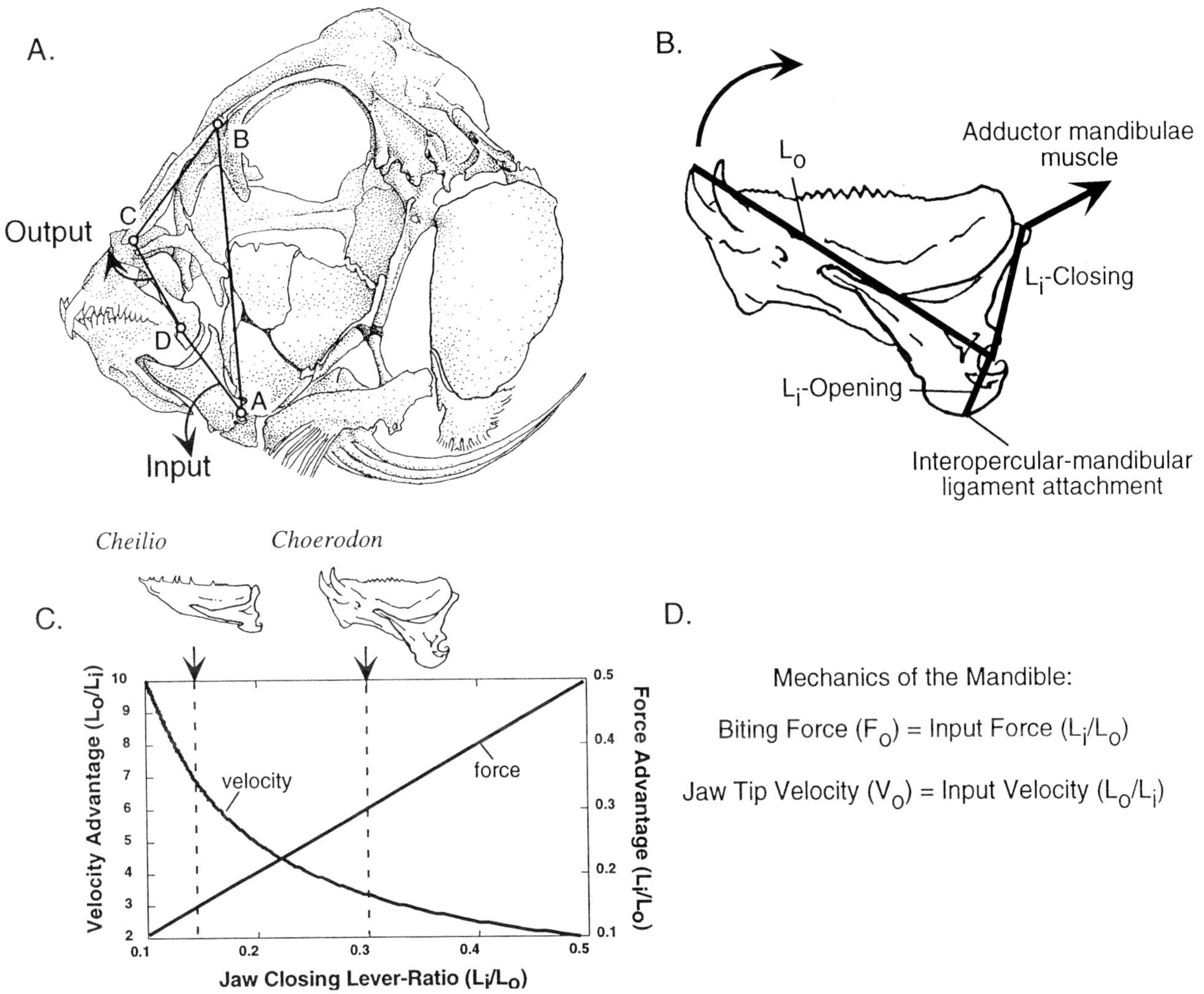

FIGURE 1 Lever mechanics of the oral jaws in fishes. (A) Illustration of the anterior jaws four-bar linkage described by Westneat (1991) for labrid fishes, based on the Caribbean wrasse *Xyrichtys martinicensis*. This planar four-bar system consists of four skeletal links connected by joints A–D. Change in any angle must be accommodated by changes in all other angles in the linkage. In this system, lower jaw depression opens the angle at joint A. This motion results in opening of the angle of joint C, an action that involves anterior rotation of the maxilla, and thus protrusion of the upper jaw. See text for additional discussion. (B) Lever system of the fish mandible. The mandible rotates about the articular-quadrate joint, and the distance of attachment of the jaw adductor muscle on the mandible from the joint is the mechanical advantage of this muscle, or the inlever for jaw closing. Similarly, the mechanical advantage of jaw depression is the distance of the interopercular-mandibular ligament from the jaw joint. (C and D) In the mechanical system of the mandible, force and velocity of jaw motion trade-off as a function of the jaw-closing lever-ratio, as given in the equations. The mandible of *Choerodon venustus* transmits adductor muscle force better than that of *Cheilio inermis,* but the *Cheilio* jaw has greater velocity advantage. *Choerodon* is a durophagous molluscivore, whereas *Cheilio* feeds on elusive prey, including other fishes.

teleost fishes (Fig. 3) (Westneat and Wainwright, 1989; Westneat, 1991). Some combination of ram and suction feeding is the dominant mode of prey capture in species of many coral reef fish families, including the Apogonidae, Anteneriidae, Aulostomidae, Carangidae, Haemulidae, Holocentridae, Lethrinidae, Lutjanidae, Nemipteridae, Pomacentridae, Scorpaenidae, and Serranidae, and many others.

The third method of prey capture is manipulation, in which the jaws are directly applied to the prey and used to remove it from the substratum (Liem, 1980; Wainwright and Turingan, 1993; Alfaro and

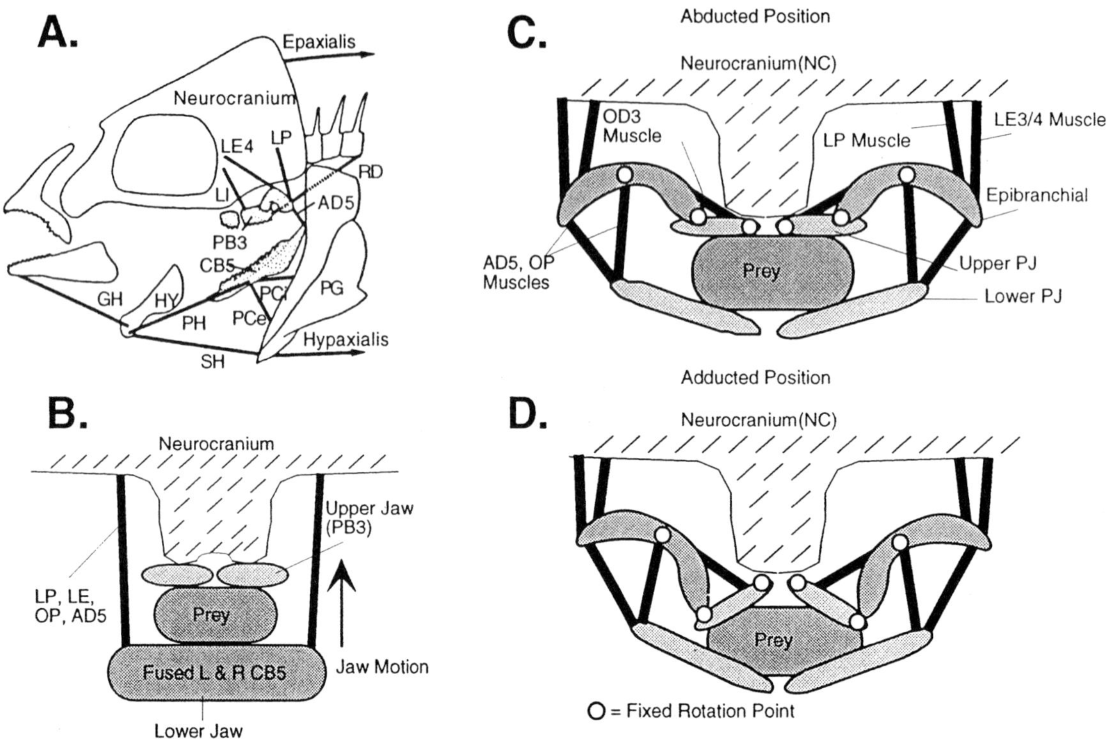

FIGURE 2 Diagram of the pharyngeal jaw (PJ) apparatus of teleost fishes. (A) Lateral view schematic showing the location of the PJ apparatus in the posterior region of the branchial chamber, with several of the major muscles indicated by thick black lines. Prey are captured by the oral jaws, and processed mostly in the PJ apparatus. (B) Diagram of the condition of the PJ apparatus in labrid fishes, in which the primitively paired lower jaw bones are fused into a single lower jaw and there is a direct muscular connection from the neurocranium to the lower jaw by the levator posterior muscle. Compare with C and D, which show the PJ apparatus of a generalized perciform (lutjanid, serranid, lethrinid, haemulid, etc.), illustrating the mechanism of biting. The epibranchial bone is rotated by action of several muscles, pressing down against the dorsal surface of the toothed upper jaw.

Westneat, 1999). Across teleosts in general this feeding method is found less frequently compared to suction and ram feeding (Liem, 1980), but this category characterizes several of the major lineages of coral reef fishes, and within this behavior there has been marked evolution in various groups of these fishes. Most reef herbivores use manipulation, including the Scaridae, Acanthuridae, Siganidae, some Ostraciidae, and the Blenniidae. Furthermore, other dominant coral reef groups use manipulation to take animal prey, including many Labridae, Chaetodontidae, Pomacanthidae, Zanclidae, Balistidae, Tetraodontidae, and Diodontidae. Perhaps no single feature characterizes the trophic nature of coral reef fish assemblages as well as the extensive use of manipulation as a dominant method of prey capture.

Suction, ram, and manipulation feeding emphasize different elements of performance. Suction and ram feeding highlight speed of motion, whereas manipulation is usually enhanced by a stronger biting action (Wainwright and Richard, 1995). Interestingly, in mechanical systems, speed and force trade off, and herein lies a dominant theme of the mechanical design of reef fish feeding systems: the contrast between systems designed for speed and those built for strength. Below we describe the functional morphology of the basic mechanism of prey capture in teleost fishes and the key design parameters that can be modified to produce jaws with greater speed or strength.

2. Functional Morphology of Prey Capture in Teleost Fishes

Whether capturing prey by ram, suction, or manipulation all teleost fishes use a common feeding apparatus constructed of a homologous network of muscles, bones, and soft connective tissue. A dominant feature of the fish feeding mechanism that sets it apart from other vertebrate groups is the large number of moving elements. More than 20 major skeletal components are put into motion by about 40 muscles. Fortunately, it is possible to focus on a relatively small number of elements to embody the major features of feeding mechanics.

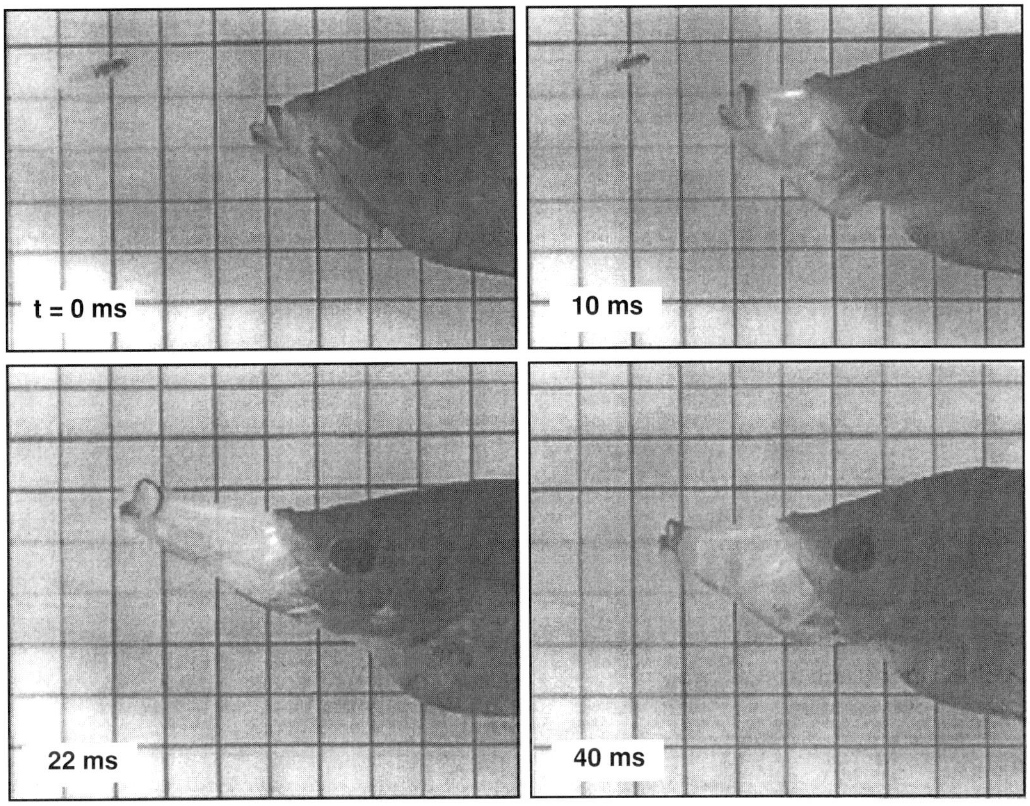

FIGURE 3 Four frames from a high-speed video recording of prey capture in the sling-jaw wrasse, *Epibulus insidiator*. This species has a highly modified jaw mechanism (Westneat and Wainwright, 1989) that permits extensive and rapid lower jaw protrusion (note the short time required for jaw extension). Associated with the origin of this mechanism, *Epibulus* also exhibits a change to feeding on elusive prey such as fishes and small prawns (Westneat, 1995). Video courtesy of L. Ferry-Graham and P. Wainwright.

Ram and suction feeding involve the rapid expansion of the buccal cavity by nearly simultaneous elevation of the neurocranium, lateral expansion of the cheek bones (suspensorium), ventral depression of the floor of the buccal cavity by depression of hyoid elements, and frequently some anterior expansion by depression of the lower jaw and protrusion of the upper jaw (Figs. 3 and 4). The major muscles that participate in these actions include the epaxialis, which dorsally rotates the neurocranium on the vertebral column; the sternohyoideus, which retracts the hyoid bar; the hypaxialis, which retracts and stabilizes the pectoral girdle; and the levator arcus palatini, which laterally rotates the suspensorium (Fig. 6) (Liem, 1970; Sanderson, 1988; Wainwright and Turingan, 1993). Muscles that produce the reverse actions (adduction of the jaws and suspensorium) include the adductor mandibulae and adductor arcus palatini muscles (Fig. 6) (Ballintijn *et al.*, 1972; Friel and Wainwright, 1999). During suction feeding buccal expansion may be coupled with strong adduction of the gill bars to prevent communication between the buccal and opercular cavities (Lauder 1980, 1983a). Adduction of the jaws during manipulation behaviors involves the actions of the adductor mandibulae complex in a variety of rasping, nipping, scraping, and forceful biting actions (Wainwright and Turingan, 1993; Ralston and Wainwright, 1997; Alfaro and Westneat, 1999).

Teeth function differently during ram/suction feeding than they do during manipulation feeding. For ram/suction feeding the teeth mainly act as a friction device, preventing captured prey from escaping back out of the mouth. Large teeth in ram/suction feeders are typically raptorial and associated with capture of particularly large and elusive prey, such as other fishes. Teeth are more diverse in manipulators, often reflecting the method used by the fish to extract prey from the substratum. In some taxa, such as many wrasses, the teeth are relatively large and recurved, though not sharp, as seen in piscivores, and are used in gripping relatively large invertebrate prey from within the reef. Manipulating predators (e.g., the Chaetodontidae) of smaller

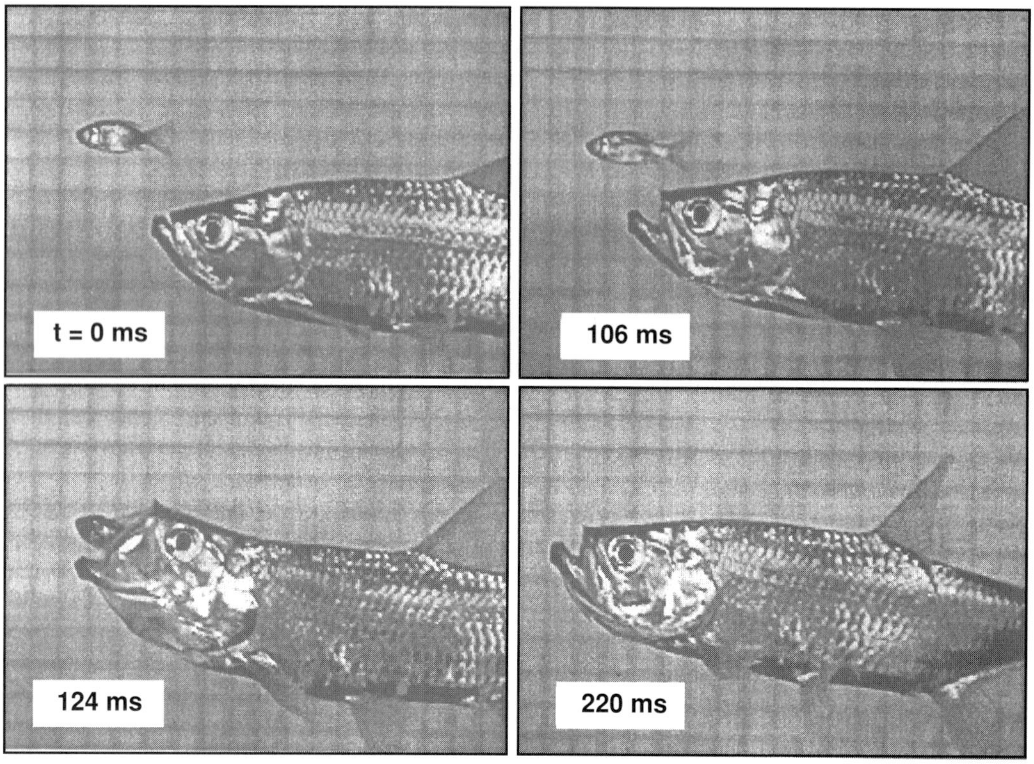

FIGURE 4 Sequence from a high-speed video of a tarpon, *Megalops atlanticus*, capturing a fish. This species illustrates the explosive expansion of the head, together with head elevation and hyoid depression, that characterizes prey capture in coral reef teleosts that use suction and ram feeding to take prey. Video courtesy of J. Grubich.

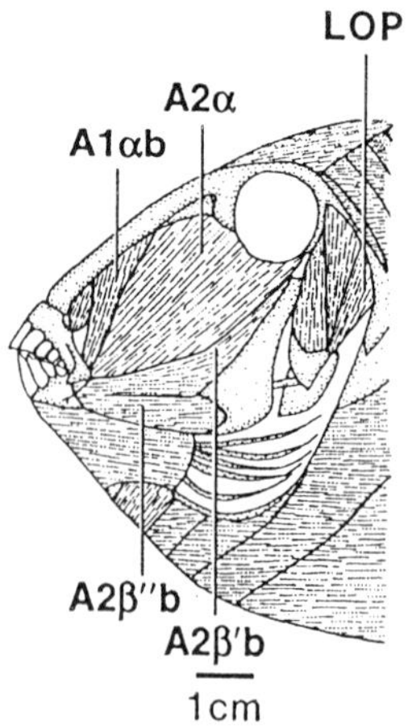

FIGURE 5 Diagram of the head of the gray triggerfish, *Balistes capriscus*, illustrating the superficial muscles of the feeding mechanism. This species feeds on infaunal invertebrates such as sand dollars, molluscs, brachyuran crabs, and polychaetes. All muscles labeled with terms beginning with an "A" are sections of the adductor mandibulae muscle; LOP, the levator operculi muscle (Wainwright and Friel, 2000; reprinted by permission of Wiley Liss, Inc., a subsidiary of John Wiley & Sons, Inc.).

invertebrate prey possess smaller teeth, sometimes arranged in pads. Herbivores show variation from the beaklike structures of the parrotfish to single rows of complex crowned teeth as seen in the surgeonfish. Teeth of manipulators frequently have significant iron deposits (Motta, 1984b, 1987; Suga *et al.*, 1989) that appear to enhance tooth strength.

Aspects of the functional morphology of feeding have been studied in representatives of many of the major coral reef groups, including the Blenniidae (Goldschmid and Kotrschal, 1985; Kotrschal, 1988, 1989a,b), the Labridae (Rognes, 1973; van Hasselt, 1978, 1979a,b, 1980; Tedman, 1980a,b; Sanderson, 1988, 1990, 1991; Wainwright, 1988; Westneat and Wainwright, 1989; Westneat, 1990, 1991, 1994, 1995; Clifton and Motta, 1998), the Scaridae (Monod, 1951; Board, 1956; Tedman, 1980a,b; Clements and Bellwood, 1988; Gobalet, 1989; Bellwood and Choat, 1990; Bellwood, 1994; Bullock and Monod, 1997; Alfaro and Westneat, 1999), the Chaetodontidae (Motta, 1982, 1984b, 1985, 1987, 1988, 1989; Sano, 1989), the Serranidae (Mullaney and Gale, 1996; Viladiu *et al.*, 1999), the Mullidae

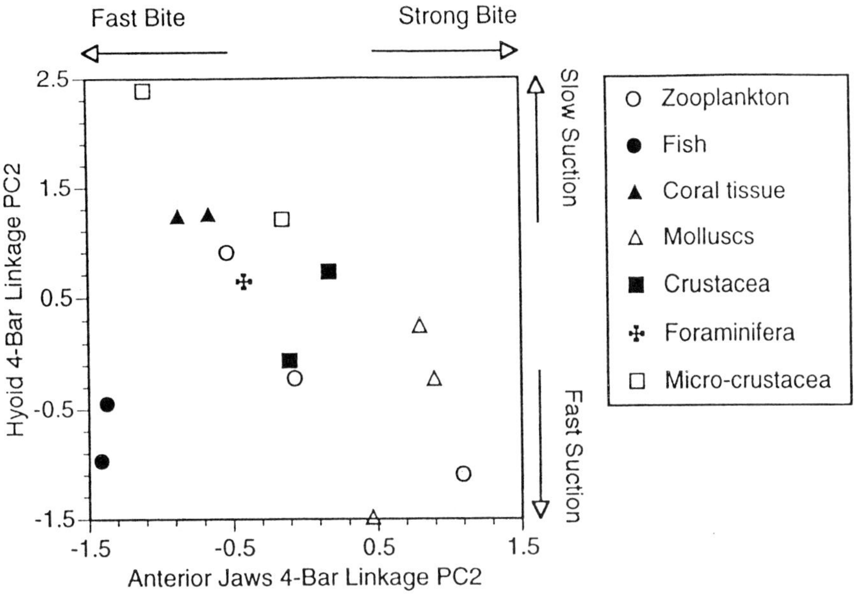

FIGURE 6 Plots of scores of 16 Great Barrier Reef labrid species in the mechanical space of the four-bar linkage systems of the oral jaws and the hyoid apparatus. Separate Principal Component Analyses were run on each linkage system from a data set of 228 specimens from 81 wrasse species of the GBR region. In each analysis the first principal component (PC) was a size factor and the second PC was the major shape axis. Mean PC2 scores were calculated for each species and examples of several major trophic types are shown. In most (but not all) cases, the species shown for each trophic group are thought to represent independent origins of that trophic habit. Note that most trophic groups tend to occupy specific regions of four-bar space. The one exception is the zooplanktivores, which are represented by considerable mechanical diversity in both four-bar systems. In order of increasing anterior jaws PC2 score, the species shown are *Cheilio inermis, Oxycheilinus digrammus, Anampses neuguinaicus, Labropsis australis, Labrichthys unilineatus, Pseudojuloides cerasinus, Macropharyngodon meleagris, Hemigymnus melapterus, Halichoeres ornatissimus, Leptojulis cyanopleura, Thalassoma jansenii, Choerodon jordani, Cheilinus fasciatus, Bodianus loxozonus,* and *Cirrhilabrus punctatus*. Data are from Wainwright, *et al.* (2002b). With permission from Wainwright and Friel, 2000, Wiley-Liss, Inc., a subsidiary of John Wiley & Sons, Inc.

(Gosline, 1984; McCormick, 1993, 1995; McCormick and Shand, 1992), the Acanthuridae (Jones, 1968; Purcell and Bellwood, 1993), Tetraodontiformes (Sarkar, 1960; Turingan and Wainwright, 1993; Wainwright and Turingan, 1993, 1997; Turingan, 1994; Turingan *et al.,* 1995; Friel and Wainwright, 1997, 1998, 1999; Ralston and Wainwright, 1997; Wainwright and Friel, 2000), the Pleuronectiforms (Gibb, 1995, 1996), and several elasmobranchs (Motta and Wilga, 1999; Wilga and Motta, 2000).

An interesting aspect of the ram/suction vs. manipulation categorization of prey capture mechanisms is that the performance attributes that are expected to enhance each feeding mode are different. In general, ram/suction feeding is expected to emphasize speed and power of jaw and head motion. Manipulation highlights forcefulness of movements, and fine motor control of the jaws in the case of taxa that pick at individual items. As we shall see in the next section, force and speed of motion trade off in the mechanical systems that underlie motion in the fish prey capture apparatus. This design trade-off reveals a major dimension of fish skull diversity, and has considerable ecological consequences.

3. Linkage Mechanics of Prey Capture

The head movements associated with prey capture in fishes have been modeled in a variety of constructs that simplify the anatomy of the system and attempt to highlight the major mechanical relationships among the muscles and bones involved. The most widely discussed of these are the lever systems of the lower jaw associated with mouth opening and closing (Barel, 1983; Wainwright and Richard, 1995; Wainwright and Shaw,

1999), and several four-bar linkage systems that have been proposed to govern the mechanics of upper jaw protrusion (Westneat, 1990), hyoid depression (Muller, 1987, 1989), and lower jaw depression via opercular rotation (Aerts and Verraes, 1974; Anker, 1974). Four-bar linkages are complex lever systems and as models they provide a framework for understanding the consequences that specific morphological changes will have for skull motion during feeding behavior. A major value of these mechanical models is that they provide insight into the functional basis of major aspects of feeding performance in fishes, such as the speed and forcefulness of jaw opening and closing, the rate of oral expansion, and the rate of jaw protrusion. Studies have shown that various groups of coral reef fishes have radiated considerably within the context of these mechanical models (Wainwright and Richard, 1995; Westneat, 1995).

The four-bar linkages that have been developed reduce interconnecting networks of skull bones into a planar series of four stiff links of specific length, connected in a series (Fig. 1). At any given time the length of the diagonal will determine the conformation of the four-bar linkage. The basis of the linkage as a mechanism of skeletal motion is that a change in the joint angle between any two links results in compensatory changes in the other three angles. For example, in any four-bar linkage, the value of the angle between two links determines all other angles in the system (Fig. 1). Thus, with the linkage drawn in Fig. 1, a muscle that increases the angle at joint A will cause changes in the other angles, in particular, joint C. Notice that in this case, the linkage conformation allows the muscle to depress the lower jaw (link A–D), and simultaneously effect anterior–dorsal rotation of the maxilla (link C–D). This illustrates one of the interesting and important features of four-bar linkages: they can transform input motion into output motion in very different directions. Much like a simple lever acting across a fulcrum, four-bar linkages result in a specific mechanical advantage for the input muscle. Thus, evolutionary modifications of the relative lengths of the links result in hyoid depression, jaw protrusion, and mandible depression systems with differing mechanical properties.

The opercular linkage (Anker, 1974; Aerts and Verraes, 1974) transmits opercular rotation, via contraction of the levator operculi muscle, into posterior tension on the interopercular–mandibular ligament, resulting in jaw depression. The hyoid linkage (Muller, 1987) uses cranial elevation as the input motion, and ventral depression of the intersection point of the left and right hyoid elements is the output. The anterior jaws linkage (Westneat, 1990) uses lower jaw depression as the input to produce upper jaw protrusion by rotation of the maxilla (Fig. 1). These three linkages have been modeled only as planar systems, so their applicability to fish taxa that exhibit considerable lateral motion of the skull during feeding is unclear. However, Westneat (1990, 1991) has tested these three mechanisms with kinematic data from films of prey capture sequences in several species of wrasses from the Great Barrier Reef. He found both the hyoid and anterior jaws linkages performed quite well in accounting for skeletal motions, but that the opercular linkage was inadequate as a mechanism of lower jaw depression. The latter finding was not entirely unexpected because lower jaw depression is also known to be linked to hyoid depression by a ligament between the posterior end of the ceratohyal and the mandible (the hyomandibular ligament) that is found in virtually all fishes (Lauder, 1982; Wilga *et al.*, 2000). The levator operculi muscle that provides input to the opercular linkage is small and weak relative to the sternohyoideus muscle, which provides input to the hyoid linkage, and thus the action of the hyoid linkage in depressing the jaws may dominate any action of the opercular linkage. The opercular linkage may function during feeding in nonlabrid taxa, however, because this four-bar system has been successfully associated with respiratory movements of some other taxa (Barel *et al.*, 1977; Anker, 1974). The opercular linkage appears to function strongly in larval fishes and may represent a largely vestigial linkage in the adults of many lineages.

The mandible of the lower jaw can also be treated as a mechanical lever (Fig. 1), rotating about the joint between the quadrate and articular bones (Barel, 1983; Westneat, 1994; Wainwright and Richard, 1995). The important actions of jaw opening and closing are directly tied to levers on the mandible. The jaw-closing adductor mandibulae muscle attaches on the mandible near the coronoid process, and has a moment arm equal to the distance from the jaw joint to the insertion of the muscle. Jaw opening is effected by tension on the interopercular mandibular ligament acting on the posterior–ventral tip of the mandible, with a moment arm equal to the distance between insertion of the ligament and the jaw joint. Confirmation of the role of these two levers in determining the speed (Wainwright and Shaw, 1999; Wainwright *et al.*, 2000) and force (Hernandez and Motta, 1997) of jaw movement has met with considerable success.

B. Prey Processing

1. Pharyngeal Jaw Functional Morphology

Prey often require considerable processing before they can be swallowed and digested. For example, it may be necessary to separate the prey from a mouthful of sand or to overcome the structural defenses of a mollusc. Many of these behaviors involve the actions of the pharyngeal jaw apparatus, a set of modified gill arch muscles and bones located at the posterior end of the pharynx (Fig. 2). Many fish taxa use a variety of pharyngeal behaviors to process prey beyond simple intraoral transport, including winnowing (Drucker and Jensen, 1991), crushing (Lauder, 1983b,c; Wainwright, 1987, 1988, 1989a; Grubich, 2000), grinding/triturating (Gobalet, 1989), and chewing (Lauder, 1983b; Wainwright, 1989a; Vandewalle *et al.*, 1992). Other taxa may accomplish the same tasks with the oral jaws, including prey crushing (Ralston and Wainwright, 1997; Turingan, 1994) and winnowing.

The basic organization of the pharyngeal apparatus as seen in the majority of coral reef fish families involves paired lower pharyngeal jaw elements, paired upper jaw elements, and a number of muscles that connect these structures to the neurocranium, pectoral girdle, and other branchial bones (Fig. 2A). The mechanisms of action of this apparatus were discussed by Lauder (1983c) and Vandewalle *et al.* (1992) and a more formal model of the mechanics of biting was developed by Wainwright (1989b) and Galis and Drucker (1996). Forceful biting is produced by contraction of dorsal levator muscles that rotate the epibranchial bones, causing them to press on the dorsal surface of the upper-jaw bones (Figs. 2C and 2D). A number of reef fishes with this generalized pharyngeal apparatus are durophagous, specializing on hard prey such as bivalve and gastropod molluscs. Examples include several members of the Haemulidae (Wainwright, 1989a,b), Sciaenidae (Grubich, 2000), Gerriidae (Randall, 1967), and Carangidae (Grubich, 2000). These taxa all exhibit hypertrophication of pharyngeal jaw bones and muscles, frequently with fused or more robust teeth, and in the cases that have been studied, molluscivores exhibit a derived pattern of muscle activity when crushing prey (Lauder, 1983b; Grubich, 2000).

A derived condition of the pharyngeal apparatus is found in labrid fishes (Yamaoka, 1978; Liem and Sanderson, 1986; Stiassny and Jensen, 1987; Wainwright, 1987). Generalized labrids have a single lower pharyngeal jaw element that is suspended from the neurocranium by modified branchial muscles (Fig. 2B). The upper jaws have well-developed and buttressed joints with the underside of the neurocranium. This configuration results in a strong pharyngeal bite, because the biting action is produced more directly by muscle action than in the more generalized condition (Fig. 2B). Many species of wrasses are durophagous (Hiatt and Strasburg, 1960; Randall, 1967; Hobson, 1974; Wainwright, 1987, 1988; Clifton and Motta, 1998). Many other species of wrasses are not durophagous and the diversity of functional morphology of the pharyngeal apparatus in this group is generally poorly known.

Within the Labridae, the pharyngeal jaws are modified further in the parrotfishes, a group that uses beaklike jaws to nip, scrape, and gouge algae and associated material from hard reef substrata (Bellwood and Choat, 1990; Bellwood, 1994; Bruggemann *et al.*, 1994a,b,c, 1996). The mixture of living and dead material is ground to a fine slurry in the pharyngeal jaws, which show modifications from the generalized labrid condition and permit extensive anterior–posterior motion of the lower jaws on the upper jaws (Clements and Bellwood, 1988; Gobalet, 1989; Bellwood, 1994).

2. Performance of Pharyngeal Jaw Function

Attempts to assess the morphological basis of prey processing performance of reef fishes mostly have focused on limits of crushing strength, a capacity that generally has a relatively clear basis in the size and organization of the muscles and jaw bones involved in generating and delivering a forceful bite (Wainwright, 1987, 1988; Hernandez and Motta, 1997; Ralston and Wainwright, 1997). There is some indication of a performance trade-off in the pharyngeal jaws, because a strong bite may be acquired at the cost of a large jaw gape, a trend that would limit the size of prey that can be processed by strong species (Wainwright, 1991). Estimates of crushing strength for individual fishes based on muscle morphology and the lever mechanics of the jaws have proved to be remarkably accurate in predicting actual strength of feeding in animals. In most cases this has been done by comparing estimates of biting strength in the fishes to their ability to crush molluscan prey of known resistance (Wainwright, 1987; Osenberg and Mittelbach, 1989; Hernandez and Motta, 1997; Huckins, 1997). Except in molluscivory, little attention has been paid to the performance of prey processing systems and their possible role in shaping feeding performance and patterns of prey use in reef fishes.

IV. Ecological Consequences of Functional Morphology

No habitat on Earth has been the focus of as much ecomorphological insight into fish trophic ecology as have coral reefs. The trophic biology of reef fishes is intimately intertwined with the mechanics of their feeding systems, and the inspiring morphological diversity seen in reef fishes has profound implications for their ecological diversity. In this section we identify a series of major themes that have emerged from research on reef fish trophic ecomorphology. This is not meant to be an exhaustive list of the lessons that have been learned in this area, but we hope to identify the dominant patterns.

Morphology is not a perfect predictor of prey-use patterns, of course. In part, this is because different lineages have different ancestral body forms, and so although evolutionary changes in feeding habits may generally be correlated with evolutionary changes in feeding morphology (Westneat, 1995; Ferry-Graham *et al.*, 2001b), it is not the case that lineages that converge on a common prey type would be expected to exhibit identical jaw and body form. Thus, midwater zooplanktivory has evolved numerous times in reef fishes, including members of the Labridae, Pomacentridae, Haemulidae, Lutjanidae, Serranidae, Acanthuridae, Chaetodontidae, Pomacanthidae, Balistidae, and others. In an absolute sense, zooplanktivorous members of these groups usually resemble their nonzooplankton-feeding close relatives more than they do other zooplanktivores. However, in every case, the zooplanktivores do appear to show consistent changes in their feeding functional morphology, when compared to their close relatives that do not eat zooplankton: reduced mouth size, enhanced jaw protrusion, reduced adductor mandibulae muscles and pharyngeal structures, enlarged eyes, and higher sustained swimming abilities.

A. Reef Fishes Are Models of Trophic Radiation

The exceptional species richness of reef fish communities alone makes them prime candidates for studies of trophic radiation. Several clades of teleosts have radiated primarily and extensively on coral reefs, and are represented today by large numbers of species. Although none of these families can be considered exclusively reef dwelling, among the most prominent reef radiations are the Gobiidae (>1500 species), Labridae [now defined to include both wrasses and parrotfishes (Gomon, 1997); >450 species], the Pomacentridae (>300 species), Chaetodontidae plus Pomacanthidae (>200 species combined), and the Acanthuridae (>70 species). In terms of ecology and morphology, these groups vary considerably in their diversity. Labrids are the most disparate, representing feeding habits as diverse as piscivores, zooplanktivores, ectoparasite predators, molluscivores, polychaete predators, decapod crab predators, coral predators, coral mucous feeders, herbivores, amphipod predators, and predators of various echinoderms. This diversity in patterns of prey use is matched by the morphological diversity of labrids. Species vary in body size from a few grams (e.g., *Labroides*), up to 100 kg or more (e.g., *Cheilinus undulatus*). Skull and body shape vary tremendously as well, from the elongate *Cheilio* to the deep-bodied species of *Cheilinus, Lachnolaimus,* and *Choerodon.*

Trophic diversity in labrids is strongly reflected by interspecific variation in the structures of the feeding apparatus. In an analysis of the anterior jaws and hyoid four-bar morphospace in a sample of 228 specimens from 81 wrasse species from the Great Barrier Reef, the major shape axes from Principal Component Analyses of each four-bar system reflected trade-offs between strength and speed (Fig. 6). Species that share feeding habits tend to occupy similar regions of this four-bar space. Two species representing lineages that have independently evolved piscivory, *Oxycheilinus digrammus* and *Cheilio inermis,* cluster in the lower left corner of this plot, indicating that they both possess an oral jaw apparatus modified for speed of motion and a hyoid apparatus modified for strength. Molluscivores, such as *Bodianus loxozonus, Choerodon jordani,* and *Cheilinus fasciatus,* typically have strong oral jaws (Fig. 6).

Further insights into the ecomorphology of this group of wrasses is gained from an inspection of muscle mass residuals (from regressions on body mass). Muscle mass can be used as a proxy for muscle strength (Calow and Alexander, 1973; Wainwright, 1987). The adductor mandibulae (the major jaw-closing muscle that adducts the anterior jaws four-bar linkage) and the sternohyoideus (the muscle most responsible for hyoid depression, and therefore generation of suction during the expansion of the hyoid four-bar linkage) both range in size among species by about 10-fold, after correcting for body size (Fig. 7). Even more variable is the size of the levator posterior muscle (Fig. 2B; the major pharyngeal jaw-closing muscle), which differs 500-fold in mass between the mollusc-crushing *Cheilinus fasciatus* and the coral mucous-feeding *Labrichthys unilineatus* (Fig. 7). Ecomorphological relationships are revealed by these key muscles and the size of the mouth opening. Coral mucous feeders have exceptionally small levator posterior muscles and small mouths, whereas piscivores have strong adductor mandibulae muscles (that power a linkage system built for speed), average sized levator posterior muscles, and a large mouth (Fig. 7). Molluscivores have strong pharyngeal jaws (large levator posterior muscles) and a large mouth. Zooplanktivores have the smallest adductor mandibulae muscles, weak pharyngeal jaws, smaller than average mouths, and sternohyoideus muscles.

The Pomacentridae offer a notable contrast to labrid diversity. With about two-thirds as many species,

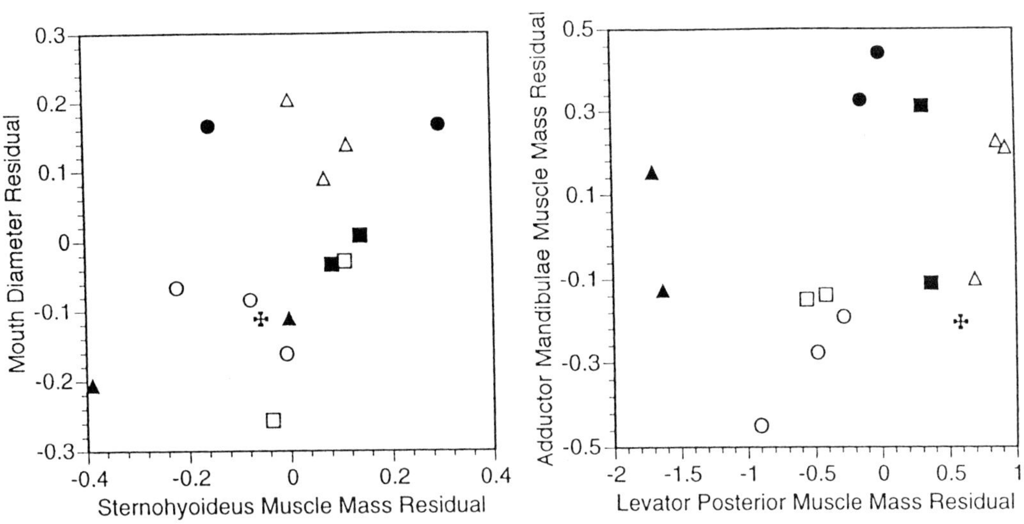

FIGURE 7 Plots of mean residuals (from regressions on Log body mass) of Log-transformed values of mouth diameter, and the masses of three feeding muscles from the heads of 16 species of wrasses from the Great Barrier Reef. Residuals are based on regressions of 228 specimens from 81 species. The sternohyoideus muscle retracts and depresses the hyoid apparatus and is believed to be the major creator of suction pressure during suction. The adductor mandibulae is the oral jaw-closing muscle, and the levator posterior muscle is the major jaw-closing muscle of the pharyngeal jaw apparatus. Species are coded by trophic category. Note how variable these muscles are in different species (e.g., there is a range of almost three orders of magnitude in size of the levator posterior muscle), and that species tend to cluster by trophic habit: molluscivores have strong pharyngeal jaws and large mouths, but average size sternohyoideus and adductor mandibulae muscles. In order of increasing sternohyoideus muscle mass residual, the species shown are *Labrichthys unilineatus, Leptojulis cyanopleura, Cheilio inermis, Cirrhilabrus punctatus, Macropharyngodon meleagris, Anampses neuguinaicus, Pseudojuloides cerasinus, Labropsis australis, Choerodon jordani, Bodianus loxozonus, Halichoeres ornatissimus, Hemigymnus melapterus, Cheilinus fasciatus, Thalassoma jansenii,* and *Oxycheilinus digrammus.* Data are from Wainwright, *et al.* (in press). ○ Zooplankton ▲ Coral tissue ■ Crustacea □ Micro-crustacea ● Fish △ Molluscs ✠ Foraminifera.

damselfishes show little of the ecological diversity found in labrids. Damselfishes are herbivorous, zooplantivorous, and some feed on small benthic invertebrates. None are known to be durophagous, specialized piscivores or many of the other specialized feeding habits seen in labrids. This relatively restricted ecological diversity is associated with what seems to be low morphological variation (Emery, 1973). Although there has never been any attempt to make such a comparison (Gluckman and Vandewalle, 1998), the general appearance is that pomacentrids are not as morphologically disparate as labrids.

The Chaetodontidae plus Pomacanthidae (butterfly fishes and angelfishes, respectively) are believed to be a monophyletic group (Blum, 1988; Ferry-Graham *et al.*, 2001b). Within these groups species can be classified as predators of benthic prey, which the fishes bite from the substratum, or open-water predators, usually of zooplankton. The former category includes predators on a range of prey taxa, including sponges, algae, zoanthids, coral, polychaetes, and other invertebrates, but in all cases the fishes directly bite the substrate to capture the prey. Within this group there is considerable diversity in jaw shape and especially in dentition (Motta, 1985, 1987, 1988, 1989). Suction and ram feeding are used by the midwater predators (Motta, 1982, 1988; Ferry-Graham *et al.*, 2001a,b). Motta (1988) analyzed the opercular four-bar linkage that contributes to jaw depression in several *Chaetodon* species and found a relationship between the form of this linkage and jaw movement times.

Several butterflyfishes have exceptionally elongate jaws, and studies of these species have mostly revealed that they are biting predators, often using their long jaws to reach invertebrates that are taking refuge inside the complex habitat of the reef substratum. However, the species with the longest jaws, *Forcipiger longirostris*, is known to feed exclusively on small, elusive caridean shrimps (Hobson, 1974; Harmelin-Vivien and Bouchon-Navarro, 1983). This species has a modified jaw apparatus that results in novel movement patterns during the strike, permitting extreme protrusion of the upper and lower jaws (Ferry-Graham *et al.*, 2001a,b). Surprisingly, this species does not appear to be an exceptional suction feeder, but rather uses rapid jaw protrusion (i.e., ram feeding) to overtake wary, elusive prey. Comparative studies of linkage mechanics in angel fishes have not been conducted, but like butterflyfishes the pomacanthids are either benthic biters or zooplankton feeders. It is therefore likely that many of the same patterns seen in butterflyfishes will apply to this group.

The acanthuroid fishes are a monophyletic group that includes surgeonfishes (Acanthuridae), rabbitfishes (Siganidae), the moorish idol (*Zanclus*), and the pelagic *Luvarus* (Winterbottom and McLennan, 1993; Tang *et al.*, 1999). Like the Chaetodontidae plus Pomacanthidae, acanthuroids mostly feed by biting attached prey, although zooplanktivory has evolved at least three times within the group (Winterbottom and McLennan, 1993). Most surgeonfishes and rabbitfishes feed on attached algae and/or detritus (Jones, 1968; Bryan, 1975; Woodland, 1990; Purcell and Bellwood, 1993), taking their food from the benthos, and a wide variety of morphological and physiological modifications of the jaws, pharyngeal apparatus, and intestines, underlie a considerable radiation within this general feeding pattern (Jones, 1968; Purcell and Bellwood, 1993). Some of these modifications include changes in the mechanics of the jaws, oral jaw dentition, variation in gill raker form and pharyngeal tooth structure, a muscular gizzard, and variation in the size and shape of the intestines (Jones, 1968; Mok, 1977; Purcell and Bellwood, 1993; Choat and Clements, 1998). In addition, some surgeonfishes harbor in their intestines specialized bacterial symbionts that appear to aid in the digestion of structural carbohydrates (Montgomery and Pollak, 1988; Clements *et al.*, 1989).

There are many other interesting and important radiations of fishes on reefs not discussed above. Among the better studied from the standpoint of the ecomorphological diversity are the Tetraodontiformes (Sarkar, 1960; Turingan, 1994; Turingan *et al.*, 1995; Ralston and Wainwright, 1997), the blennioid fishes (Kotrschal, 1988, 1989a,b), and the Serranidae (Wainwright and Richard, 1995; Mullaney and Gale, 1996; St. John, 1999; Viladiu *et al.*, 1999).

B. Feeding Mechanics Constrain Patterns of Prey Use

The ability of fishes to find, capture, and handle various prey plays a central role in shaping patterns of prey use in natural populations. These abilities have their basis in the mechanical design of the feeding apparatus, in various sensory systems, and in behavioral modifications. This link between the morphology or physiology of fishes and a major ecological feature is strongly intuitive, but it is important to ask how important these connections really are in shaping the ecology of reef fishes. Do differences among species in the construction of the feeding apparatus and in feeding abilities adequately account for diversity in patterns of prey use? What are the major functional properties of prey along which fish feeding mechanisms have radiated? In this section we first review the major ecomorphological axes seen in the feeding mechanisms of reef fishes, and then we discuss the specific evidence for how functional morphology of the feeding apparatus shapes patterns of prey use.

1. Major Axes of Ecomorphological Radiation in Reef Fishes

The taxonomic range of prey items eaten by reef fishes is as broad as the diversity of life on coral reefs (Randall, 1967; Hiatt and Strasburg, 1960; Hobson, 1974). Virtually all organisms on reefs are prone to predation by reef fishes at some point in their life history. This fact can make the task of searching for generalities in ecomorphological associations overwhelming. One approach to summarizing the diversity of feeding habits in reef fishes is to consider the key functional properties of the prey that predators must overcome to feed on them successfully. These attributes may include elements of predator avoidance, such as elusiveness, or where the prey are located and what must be accomplished to extract the prey from the environment. Constructing "prey functional groups" makes it possible to place phylogenetically diverse prey organisms into a manageable few categories defined by some mechanical or physical variable that can then be related to the construction and performance of the feeding mechanisms used to obtain them. This approach can be used to generate predictions for the specific performance properties that are expected to be enhanced in species that take prey from a particular prey functional group. Here we identify major "prey functional groups" of reef fishes and discuss observations on the functional morphology of feeding in the various fish groups that have evolved feeding habits in these areas.

a. Hard-Shelled Prey Many prey are protected by a hard outer covering, carapace, or shell that must be cracked and disassembled if the digestive juices of the predator's intestines are to gain access to the digestible parts of the body. Examples of hard-shelled prey include gastropod and bivalve molluscs, echinoids, some decapod crustaceans, some shell-bearing anellids, foraminifera, and organisms that bore into the hard, calcareous substrate of dead coral. Hard-shelled prey are often little challenge to capture once they have been detected, but the challenge to the predator comes in handling them.

Most teleost fishes that feed largely on hard-shelled prey crush them in either the pharyngeal jaw apparatus (Labridae, some Carangidae, some Haemulidae, some Sciaenidae) or the oral jaws (Diodontidae, Tetraodontidae, Balistidae, Sparidae). The functional basis of feeding performance in these predators lies in the ability of the fish to deliver large forces to the prey. Modifications of the pharyngeal and oral jaws that facilitate the delivery of a forceful bite include enlarged biting musculature, which is often pinnate in organization, robust bones that can withstand the forces generated by the muscles, and a favorable mechanical advantage in the biting mechanism that maximizes, or amplifies, delivery of muscular tension to the protective shell of the prey (Wainwright, 1987, 1988; Turingan, 1994; Turingan *et al.*, 1995; Hernandez and Motta, 1997).

Two trade-offs have been noted in the construction of jaw systems that deliver large forces. First, because lever mechanics dictate that a high mechanical advantage must occur at the direct cost of velocity transfer, jaw systems of hard-shelled predators are expected to be slow moving (Wainwright and Richard, 1995). Second, one cost in increasing angles of pinnation in muscles is that this will limit the extensibility of the muscle compared to a similar muscle with a parallel fibered arrangement. A series of labrid fishes from the Caribbean were found to show a marked trade-off in strength of the pharyngeal jaw apparatus and the extent to which the jaws could open (Wainwright, 1991). Strong species were limited to smaller prey.

b. Large, Firmly Attached Prey Many prey gain protection by having a tenacious grip of the substrate, often combined with locating themselves in hard-to-reach interstices of the reef. Included here are many gastropod and bivalve molluscs, including some that do not have to be crushed once captured, such as limpets and chitons, echinoids, and some decapod crabs. Predators of these prey must be able to grasp them firmly in their oral jaws and to pull them out of their protective resting location. Examples of these types of predators include many Labridae (e.g., *Bodianus,* many *Thalassoma,* and *Semicossyphus*) and Balistidae (Randall, 1967; Turingan *et al.,* 1995; Westneat, 1991, 1994, 1995). Predators feeding in this category have been found to have a strong oral jaw bite, with a large adductor mandibulae muscle (Turingan, 1994; Turingan *et al.,* 1995) and a high mechanical advantage in the jaws (Westneat, 1994; Wainwright and Richard, 1995). These trends are most apparent when the fishes are compared to close relatives that feed on other prey (Turingan *et al.,* 1995; Westneat, 1995).

c. Small, Firmly Attached Prey Many other firmly attached prey are much smaller relative to the predator and are collected in groups by predatory fishes that scrape them from the substratum. These prey include epilithic algae, many sponges, and scleractinian coral polyps. Fishes that feed on these prey include many species in the Pomacanthidae, Chaetodontidae, Ostraciidae, Monacanthidae, Scaridae, Pomacentridae, Blenniidae, Acanthuridae, and Siganidae. These fishes often have small mouths with high mechanical advantage of the adductor muscles (Wainwright and Richard, 1995). The teeth either form a single cutting edge, as in scarids, acanthurids, kyphosids, ostraciids, monacanthids, and some pomacentrids, or they form a broad pad, many tooth rows deep, as seen in pomacanthids, chaetodontids, and some pomacentrids. Tooth form is often a key factor associated with trophic diversity within this "small-attached" category (Jones, 1968; Motta, 1984b, 1987, 1989). Scarids that gouge deep troughs in the calcareous substratum have larger adductor muscles than do species that graze the surface of similar substrata (Bellwood and Choat, 1990), but data on adductor muscle design in most taxa in this category have not yet been collected.

d. Zooplankton Many reef fishes feed individually on the small organisms of the plankton that are carried onto the reef by currents. Included in this prey category are copepods, siphonophores, pteropods, and the eggs and larvae of fishes, crustaceans, and other invertebrates. Predators of the zooplankton are often seen in schools hovering over the upstream end of the reef, feeding on individual plankters. The primary challenge for these fishes appears to be in seeing the prey items, many of which are protected by being partially translucent. Zooplanktivorous fishes typically have smaller mouths compared to their close relatives, and some have moderately protrusible jaws that form a distinctly circular aperture when opened. A zooplanktivorous habit has evolved independently in many groups of reef fishes, often repeatedly. Among the families that have given rise to zooplanktivorous taxa are the Pomacentridae (Emery, 1973), Labridae, Pomacanthidae (*Genicanthus*; Randall *et al.,* 1997), Chaetodontidae (Motta, 1982; Harmelin-Vivien and Bouchon-Navarro, 1983), Lutjanidae (Carpenter, 1987), Serranidae, Inermiidae, Congridae, Balistidae (Turingan, 1994), and Acanthuridae (Winterbottom and McLennan, 1993) [see Randall (1967) for diet data on planktivores of all of these families except Pomacanthidae]. Prey-type-related diversity in the feeding mechanism within this trophic group has not been described, although on any given reef there are clearly differences in body size, mouth size, and finer detail of branchial and jaw anatomy among species of zooplanktivores. Foraging location in the water column has been proposed to be related to swimming ability, in that it may be related to escape ability of these fish in the face of their own predators. Hobson and Chess (1978) observed that fish species with body forms that appear to confer greater swimming abilities feed in locations farthest from the reef. To date, no comprehensive study of the relationship between body form and swimming ability in these fishes has been conducted. It also appears that zooplanktivores as a group are faster swimmers than are their close relatives.

e. Large, Elusive Prey Other fishes, cephalopods, and several types of crustaceans present predatory fishes with a special challenge. These prey are relatively large and alert and possess well-developed escape responses. Fishes that specialize in this prey functional category typically have larger mouths compared to their close relatives, raptorial teeth, and large adductor mandibulae (jaw-closing) muscles, and the mouth-opening and -closing mechanical systems are modified toward a high-velocity transfer (Wainwright and Richard, 1995; Westneat, 1995). These fishes typically employ a stealth tactic to approach their prey, striking with an explosive lunge that involves considerable ram and suction (Fig. 4). The large mouth allows them to take large prey, the large adductor muscles enhance the strength and power of the jaw grip, and the mechanical system of the jaw translates muscular contractions into rapid motion. Included in this category are representatives of the Serranidae (Wainwright and Richard, 1995; Viladiu *et al.,* 1999), Labridae (Westneat, 1991, 1994, 1995), Antennariidae (Grobecker and Pietsch, 1979, 1987), Scorpaenidae (Grobecker, 1983), Lutjanidae, Aulostomidae, Scorpaenidae, Muraenidae, Carangidae, Sphyraenidae, and a number of other groups. Some of these families are exclusively in this category (e.g., Sphyraenidae and Antennariidae), but in other groups this mode has evolved as part of a trophic radiation within the family (e.g., Serranidae, Labridae, and Lutjanidae).

The families of fishes that are characterized by predation on the "firmly attached prey" category often give rise to zooplanktivores (e.g., Acanthuridae, Chaetodontidae, Pomacentridae, and Balistidae). However, only one group, the Labridae, includes several lineages of predators of "firmly attached prey" that have given rise to predators of "large, elusive prey." It is possible that the functional morphology of a predator of firmly attached prey is simpler to modify into an effective zooplanktivore than into a predator of large, elusive prey. Several groups that are primarily made up of predators of large, elusive prey have given rise to highly successful lineages of zooplanktivores. Examples include the Serranidae, which includes the zooplanktivorous Anthiinae; the Lutjanidae, which gave rise (perhaps separately) to the caesionids (Carpenter, 1987, 1990, 1993), *Pinjalo* (Randall *et al.,* 1997), and *Ocyurus* (Randall, 1967); and the Haemulidae, which apparently gave rise to the Inermiidae (Johnson, 1981). Thus, based on the weight of evidence from modern reef fish assemblages, zooplanktivory has evolved frequently from both benthic biting predators and predators of large, elusive prey.

f. Mobile Benthic Prey Many reef fishes feed on relatively small invertebrate and fish prey that live in very close association with the substratum. These elusive prey are usually found within the matrix of the reef. Common microhabitats that are exploited include the interstices of the reef: coral, dead coral, algal turf, and sand. Here the predators apparently use suction feeding, and sometimes employ biting with the strike. Included in this category of predators are many members of the Labridae, Lutjanidae, Haemulidae, Lethrinids, Mullidae, Nemipteridae, Serranidae, Gobiidae, Chaetodontidae, and several other groups. The morphology of these predators is often very similar to that of species that feed on large, elusive prey, except that body size or the structures of the feeding apparatus are scaled down to a smaller size.

A subset of this group includes fishes that are familiar components of the coral reef, but which actually do not feed on the reef, instead only using it as a refuge during the day when they are inactive. Most Haemulidae, Lutjanidae, Lethrinidae, Sciaenidae, and Nemipteridae actually feed in nearby seagrass beds or in sandy areas that lie adjacent to reefs (Randall, 1967; Hiatt and Strasburg, 1960). These fishes feed mostly at night on mobile benthic prey and sometimes on zooplankton.

g. Sand-Dwelling Prey Prey that live buried in sand present two types of challenges to predatory fishes. They can be both difficult to locate and difficult to separate from the mouthful of sand that accompanies them when captured. A variety of behaviors are used to locate infaunal prey that live in sand. Jets of water blown out of the mouth are used to excavate buried prey in numerous species of the Balistidae (Fricke, 1971, 1975; Frazer *et al.,* 1991; Wainwright and Turingan, 1997), Monacanthidae, Ostraciidae, Labridae, and Haemulidae. Some haemulids and sciaenids appear to locate buried prey through the use of mechanosensory pits located in the lower jaw and other structures of the skull. Goatfishes protrude their snout into the sand and use a pair of mobile barbels mounted on the end of the hyoid apparatus to locate buried prey mechanically (Gosline, 1984).

Labrids use a particularly diverse battery of behaviors to locate buried prey. Several species will grip and move large rocks or coral with their oral jaws in order to search the space under the rock (e.g., *Novaculichthys,* some *Choerodon,* and some *Halichoeres*). Buried prey will be located by probing the substrate with the snout, digging with the snout, blowing jets of water, and even fanning the sand with the pectoral fin (P. C. Wainwright and D. R. Bellwood, unpublished observations on *Choerodon schoenlinii*). These behaviors are generally poorly documented and their phylogenetic distribution within the Labridae has not been explored.

Infaunal prey are typically captured with a mouthful of sand and a winnowing behavior is typically then used to separate the prey and sand. Recorded in some haemulids, labrids, mullids, and gobies, among others, the morphological basis of winnowing is not well understood, but it appears usually to involve a combination of water motion inside the buccal cavity and fine manipulation by the pharyngeal jaw apparatus (Liem, 1986; Drucker and Jensen, 1991).

h. Nocturnal vs. Diurnal Feeding The basis for resource partitioning along the 24-hour temporal axis is based largely on visual and other sensory modalities. These systems are beyond the scope of this review, but we mention this major axis of ecological radiation in reef fishes, in part because the implications of feeding in low light at night, with respect to the functional morphology of the prey capture apparatus, have not been explored. We note that nocturnal reef fish tend to be either predators of large, elusive prey, mobile benthic prey, or zooplankton. Very few nocturnal fish predators feed on hard-shelled prey (e.g., *Diodon*) or firmly attached prey.

2. The Mechanical Basis of Feeding Patterns in Fishes

In the previous section we described a context for organizing the bewildering diversity of reef fish feeding habits. The identification of functional groups of prey allows one to organize trophic diversity along lines that may be more readily viewed from the standpoint of functional morphology of the feeding mechanism. What are the chief performance features that we expect to be important for the predators in each category, and how might the feeding mechanism be modified to enhance these features? Although very general correlations can be described between key morphological features and patterns of prey use in reef fishes, in the vast majority of cases there is a lack of strong experimental data that would address thoroughly the causal basis of feeding performance.

One focus of work on the ecomorphology of reef fishes has been on the relative importance of, and functional basis of strength and speed in, the feeding mechanism. Recall from Section III, A, 3 that the mechanics of linkage systems and muscles create a trade-off between force transmission and velocity transfer. The mechanical advantage of the adductor mandibulae (jaw-closing muscle) on the mandible can be modified to enhance the forcefulness of the bite, or to enhance the speed of jaw-closing, but both cannot be enhanced simultaneously. Indeed, interspecific analyses of fishes have shown that speed of motion of the mandible during jaw closing has a strong negative correlation with the mechanical advantage of the adductor mandibulae muscle (Wainwright and Shaw, 1999; Wainwright *et al.,* 2000).

An analogous situation exists in the four-bar linkage mechanics that have been described to govern hyoid depression, and upper and lower jaw motion in the fish skull (Westneat, 1991; Muller, 1996). Westneat (1994, 1995), in elegant and thorough analyses that corrected for phylogenetic relatedness among taxa, has shown that the evolution of mechanical properties in the skull linkage systems is strongly correlated with changes in feeding habits in cheiline wrasses. Historical changes to taxa that feed more heavily on elusive prey are significantly correlated with changes in skull linkage mechanics that enhance velocity transfer. Similarly, changes to feeding on prey that require a forceful bite are associated with linkage changes that enhance force transmission in the jaws.

Attempts to identify the morphological features that are good indicators of the ability to feed on hard-shelled prey have met with considerable success. Mollusc crushing performance appears to be directly related to biting force, a parameter that has a direct morphological basis in the size of biting muscles. Working with the pharyngeal jaw apparatus of labrid fishes in the Caribbean, Wainwright (1987, 1988) showed that the physiological cross-sectional area of one dominant muscle (the levator posterior muscle) accurately reflected the crushing strength of individual fishes feeding on gastropods. Maximal crushing strength of fishes was shown to be an active constraint on the ranges of prey that were eaten. Because labrid species differed in their crushing strength, among species this led to differences in feeding habits that could be accounted for by difference in crushing ability (Wainwright, 1988).

Performance in oral jaw crushing has also been found to be predicted accurately by jaw adductor muscle morphology and lever mechanics of the lower jaw (Hernandez and Motta, 1997).

C. Scaling of Feeding Mechanics Results in Strong Ontogenetic Effects

One of the most striking generalities about reef fishes, and fishes in general, is that patterns of prey use change ontogenetically in virtually every species. These changes are typically drastic, usually taking species from being zooplanktivores when they are larvae and newly metamorphosed, through a series of distinct stages in which one or another prey type dominates the diet. Well-documented examples of ontogenetic diet changes in reef fishes include members of the Serranidae (Randall, 1965; Mullaney and Gale, 1996; St John, 1999), Labridae (Wainwright, 1988; Kanashiro, 1998), parrotfishes (Bellwood, 1988b), Mullidae (Lukosek and McCormick, 2002), and others. Ontogenetic diet shifts reflect the influence of life stage and body size on habitat use and feeding capabilities. Almost all reef fishes begin as planktonic larvae (when they feed mostly on small planktonic animals) and most move through substantial or subtle habitat shifts after they settle onto the reef, then grow and mature. To some extent, feeding habits must be constrained by what is available in the habitat.

As fishes grow the mechanical properties of their feeding mechanism change. Even in the simplest scaling case, wherein shape is maintained during growth, the larger feeding apparatus will exhibit its most effective feeding performance on larger prey (Werner, 1977; Wainwright, 1987). Changes in optimal or preferred prey size often result in changes in the taxonomic composition of the diet.

An example of this phenomenon is provided in an analysis of feeding habits of several Caribbean species of the Serranidae (Wainwright and Richard, 1995). Serranids make an interesting case because there are relatively few shape differences between species, and ontogeny produces large animals that are similar in shape to small animals. The major morphological axis of the serranid radiation is body size. Several small species may be only a few centimeters long when fully grown, but the large species of *Epinephalus* may exceed 2 m and weigh over 250 kg (Randall, 1983). However, the shape of those species that vary so much in size is strikingly similar. In the 18 species analyzed from the Caribbean there was a strong match between body size and the dominant prey in the diet (Fig. 8). The smallest species feed on copepods and similar small crustaceans; somewhat larger species feed on various free-moving prawns and at about 200 mm body size the dominant prey is decapod crabs, and above that most species feed on fishes and cephalopod molluscs. Detailed ontogenetic data for one species from the Western Atlantic, *Mycteroperca microlepis,* reveal that this species matches this body size/diet pattern during its ontogeny. Some exceptions to this trend exist, most notably in taxa that diverge significantly from the generalized serranid body plan and show distinct trophic specializations. For example, *Paranthias furcifer* is a midwater species with a relatively small mouth and gracile feeding structures. At 200 mm this species is a zooplanktivore, rather than a crab or fish predator like most groupers of this size (Fig. 8).

Studies of the scaling of feeding mechanisms in other reef fish groups indicate that strength often scales more rapidly than expected under models of isometry (Wainwright, 1988; Westneat, 1994, 1995; Hernandez and Motta, 1997). In Caribbean labrids, diet breadth increased during ontogeny, as fishes are increasingly strong-jawed, but once a biting strength of 5 Newtons was achieved, individuals became increasingly specialized on a taxonomically narrow diet of molluscs (Wainwright, 1987, 1988; Clifton and Motta, 1998). An interesting area for future work will be to develop mechanical models of the scaling of suction feeding performance, and to explore the consequences of this relationship on prey-use patterns in suction-feeding predators.

D. Behavioral Differences between Species Can Affect Prey Use

Feeding functional morphology provides one level at which the interaction between predator and prey can be modified during evolution to produce differences in patterns of prey use. However, mechanics of the feeding apparatus are brought to bear on prey through the filter of animal behavior. Species with quite similar feeding morphology may have different patterns of prey use because they feed in different habitats, or they may feed in exactly the same habitat but use subtle variations of feeding kinematics or strategies to obtain prey. Particularly striking examples of both are found within the Acanthuridae and the Scaridae.

The surgeonfishes *Ctenochaetus striatus* and *Acanthurus nigrofuscus* of the Great Barrier Reef are morphologically similar species that appear to be completely sympatric, even to the point of feeding on the same turf-algae-covered rock (Purcell and Bellwood,

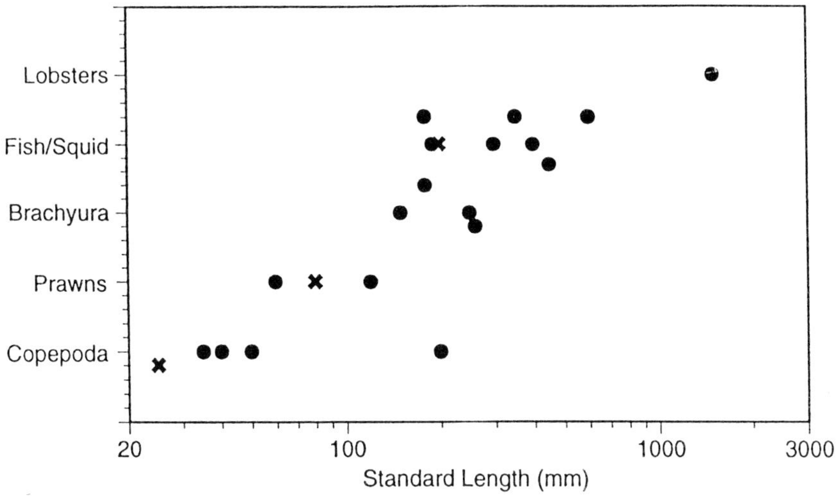

FIGURE 8 Plots of major prey eaten vs. body size for 18 species of groupers (Serranidae) from the Caribbean. The effects of body size are seen in this fish group, which varies mostly in size rather than shape. The three "x" symbols represent data from three size classes of *Mycteroperca microlepis* (Mullaney and Gale, 1996). Data are from Randall (1967); plot redrawn from Wainwright and Richard (1995), with kind permission from Kluwer Academic Publishers.

1993). However, they use different prey: *A. nigrofuscus* grazes on turf algae, whereas *C. striatus* feeds on the particulate and epiphytic material that occurs on the surface of the turf-algae blades. Subtle differences in jaw morphology and the kinematics of biting behavior reveal the basis of these different feeding habits. The teeth of *C. striatus* are finer and form a brushlike surface, but the major factor that accounts for the difference in prey use is biting behavior. Algae are sheared or torn by the teeth of *A. nigrofuscus* while the jaws are simultaneously adducted and the head is quickly rotated laterally. The mouth of *C. striatus* is opened wider and the surface of the turf algae is brushed by the fine teeth during a quick adduction of the lower jaw (Purcell and Bellwood, 1993). Similar subtleties in biting behavior have been used to account for interspecific differences in prey-use patterns of chaetodontids (Motta, 1988, 1989).

Other species of surgeonfishes appear to share similar jaw morphology and patterns of prey use, and yet segregate spatially. *Acanthurus nigrofuscus* and *Acanthurus lineatus* have very similar tooth form and feeding behavior, and both nip and tear turf algae from hard substrata (Choat 1991). However, *A. lineatus* is territorial and excludes other species, including *A. nigrofuscus*, from its shallow-water feeding areas (Choat and Bellwood, 1985; Robertson and Gaines, 1986). Here, niche separation occurs along a spatial axis due to behavioral interference in two species that feed on similar prey with similar feeding methods and morphology.

Parrotfishes provide a second example of how behavioral differences can underlie marked ecological variation among close relatives. Two functional groups of parrotfishes were identified by Bellwood and Choat (1990), who found Great Barrier Reef scarids either excavated deep bites into rocky substratum or scraped the surface of the same hard surfaces. In this case, substantial morphological differences were subsequently discovered in the two groups; representative excavating species had heavier jaw musculature, heavier jaw bones, and stronger ligamentous connections in the jaws compared to the scraping species (Bellwood and Choat, 1990). In this example, the differences produced by these patterns of feeding kinematics result in profound consequences for the impact of the two groups on coral reef ecology, in that the excavating species probably account for the majority of reef bioerosion by fishes (Kiene, 1988; Bellwood, 1995a,b).

As in the acanthurids, however, parrotfishes can also show habitat segregation within an ecomorph. On the Great Barrier Reef, *Scarus frenatus* and *Scarus niger* both exhibit the scraping morphology and behavior (Bellwood and Choat, 1990). However, these species segregate spatially, with *S. frenatus* usually being found in shallower locations (Russ, 1984). *Scarus frenatus* may defend the shallower feeding sites in behavioral interactions with *S. niger* (D. R. Bellwood, unpublished data).

There is a lesson in the behavioral differences between species: although analyzing the functional design of a feeding mechanism allows the researcher to identify potential feeding niches and make inferences about relative feeding abilities in different species, a number of other factors interact with the inherent abilities of a predator, all of which ultimately determine usage patterns of prey and other resources. Interactions between species on the reef are one factor that can further shape resource-use patterns, as do the distribution and value of the prey resource. Being able to infer relative feeding performance from morphological observations does not necessarily also provide direct insight into patterns of resource use.

E. Feeding Mechanics and Prey Use Can Vary among Populations

One understudied aspect of reef fish feeding ecomorphology that is relevant to the process of diversification is variation within species. Freshwater fishes, particularly lake-dwelling taxa, have provided numerous cases of intraspecific polymorphisms associated with divergent forms within species that make use of different prey resources (Robinson and Wilson, 1994). These systems are often characterized by being relatively species poor, so that trophic niches are unfilled by existing species (Robinson and Wilson, 1994; Schluter, 1996; Smith and Skulason, 1996; Bell and Andrews, 1997; Robinson and Schluter, 1999). Similar patterns of polymorphism or descriptions of potentially adaptive intraspecific variation in general are lacking for marine systems. Although the prevailing views of the existence of polymorphisms in lakes very much focus on the isolated and depauperate nature of these faunas, there are very few empirical data on marine taxa. Thus, it is unclear to what extent marine forms exhibit polymorphisms or extensive trophic variation within species.

Several studies have examined patterns of prey use and trophic morphology within species of reef fishes (Turingan *et al.*, 1995; Cutwa and Turingan, 2000; Durie and Turingan, 2001). In each of these analyses, fish populations were found to differ markedly in feeding habits and in trophic morphology. Examples include several triggerfishes (Turingan *et al.*, 1995; Durie and Turingan, 2001), a sparid (Cutwa and Turingan, 2000), and a haemulid (Turingan and Roth, 2001). In some cases, different populations appear to emphasize the use of different prey capture modes, as in the case of *Xanthichtys ringens,* which feeds predominantly on zooplankton at Mona Island, but grazes on benthic sponges in Puerto Rico (Turingan *et al.*, 1995). Among populations variation in trophic morphology can be extensive, but in none of these cases is it clear whether the differences observed are the result of developmental plasticity or genetic variation among the populations. We should note that although genetic polymorphisms are thought to be common in the temperate lake systems, an exclusive role of phenotypic plasticity has only been experimentally refuted in a few cases (Schluter and McPhail, 1993; Smith and Skulason, 1996). Adaptive developmental plasticity may be expected in reef fishes, and in marine fishes in general, because offspring may typically recruit away from their natal reefs, to locations where prey availability and community structure can differ considerably from the habitat of their parents. This uncertainty in the environment of the offspring may lead to some adaptive flexibility in the development of the phenotype.

Other issues concerning intraspecific variation in reef fishes remain unexplored. Are trophically generalized species sometimes composed of populations of specialized individuals? Do the complex life histories of labrids, serranids, and other lineages result in sex-specific patterns of prey use? What are the patterns of natural selection on reef fish feeding mechanisms, and how do they vary across broad biogeographic ranges? These and other questions await future researchers of reef fish ecomorphology.

F. Other Areas of Progress in Reef Fish Ecomorphology

In this chapter we have focused on the functional morphology of prey capture and prey processing in juvenile and adult reef fishes, but there are a number of related areas in which an ecomorphological framework has also proved insightful. Here, we briefly mention several and indicate some potential areas of future research.

Work on the ecomorphology of locomotion has shown patterns of association between habitat use and functional design of the swimming apparatus in reef fishes. Like most dominant reef fish groups, labrids swim predominantly by use of their pectoral fins. However, within the family there is diversity in the mechanical mechanisms that are used by labrids to generate thrust with their fins, with extremes being represented by either a paddling mechanism or a general flapping behavior (Westneat, 1996; Walker and Westneat, 2000). Theoretical results suggest that a paddling mechanism is best suited to a rounded fin morphology, whereas effectiveness in the flapping behavior is maximized by an elongate, high-aspect-ratio fin (Walker and Westneat, 2000). Furthermore, the paddling behavior is thought to be most efficient at slow swimming speeds, and the flapping behavior to be most efficient at high swimming speeds (Vogel, 1994; Walker and Westneat, 2000). Laboratory observations in flow tank swimming trials with four wrasse species indicate that swimming speed is positively correlated with pectoral fin aspect ratio. Field data on 41 labrid species on the Great Barrier Reef further support this pattern, because swimming speeds of undisturbed fishes on the reef were positively correlated with pectoral fin aspect ratio (Wainwright *et al.*, in press). Fin shape, and hence swimming mechanism and performance, are also correlated with patterns of habitat use in labrids. A strong pattern was seen within reef locations, and across the shelf of the Great Barrier Reef, in which shallow, high-energy, high-water-flow areas were dominated by labrids with high-aspect-ratio fins (Bellwood and Wainwright, 2001; Fulton *et al.*, 2001).

Gut morphology is extremely diverse in reef fishes and may be strongly correlated with diet (Motta, 1988; Horn, 1989). Ecomorphological relationships between the anatomy of the digestive tract of reef fishes and the prey type have been known since the pioneering work of Al-Hussaini (1947), who showed that fishes that depend on poor-quality diets, such as herbivores and coral predators, tend to have longer intestines than do carnivores, whose diets are nutritionally denser. Comparisons across families have supported this observation, with comparable relationships between gut length and diet in the Labridae, Pomacentridae, and Chaetodontidae (Elliott and Bellwood, 2002). Although based on soft anatomy, the digestive tract is likely to be subject to a set of constraints and trade-offs comparable to those seen in the mechanical operations of the oral and pharyngeal jaws. Initial observations, for example, suggest that particle size of the ingesta may be strongly correlated with intestinal diameter (J. K. Elliott and D. R. Bellwood, unpublished data).

Although ecomorphology has primarily been applied to the study of juvenile and adult fishes, its application to the biology of the enigmatic larval stage of reef fishes has provided some particularly useful insights. Reef fish larvae are morphologically diverse and in many cases this diversity has functional consequences. In the past decade a number of studies have examined the relationship between larval reef fish morphology, performance, and ecology. This includes studies of sensory systems (McCormick and Shand, 1992; Job and Bellwood, 1996, 2000; Shand, 1997) and locomotor systems (Stobutzki and Bellwood, 1994, 1997, 1998; Fisher *et al.*, 2000; Bellwood and Fisher, 2001), and changes during the transition at settlement (McCormick, 1993). These studies have highlighted the fact that, contrary to earlier assumptions, reef fish larvae do not drift as passive particles in ocean currents, and like their reef-based counterparts, they have a rich and varied suite of behaviors and abilities that shape their ecology. Reef fish larvae include species that are capable of swimming for short periods at over 40 body lengths a second (Fisher *et al.*, 2000), or for longer periods, covering the equivalent of over 100 km without food or rest (Stobutzki and Bellwood, 1997). Studies of the visual capabilities of larval fishes indicate remarkable sensitivity to light and, although light attenuation limits the ability of larvae to feed at depth, some species are able to feed successfully at depths of 250 m in open ocean conditions (Job and Bellwood, 1996). Species differ considerably in light sensitivity and swimming ability and this may underlie considerable variation among fishes in the depth range over which they can feed and their potential to shape dispersal by active locomotion (Job and Bellwood, 2000). Further ecomorphological analyses are likely to yield additional surprises about the early life history stages of reef fishes.

G. Reef Fish Ecomorphology in Evolutionary Biology

Our ability to understand patterns of ecomorphological diversification in coral reef fish groups depends on our ability to infer the historical sequences of these radiation events. One of the most powerful approaches to gaining historical insights involves the use of phylogenetic hypotheses of the group in question to infer a transformation series of particular functional systems and to remove phylogenetic effects from comparative analyses of trait correlations, which form the basis of many ecomorphological studies.

A phylogenetic hypothesis of tetraodontiform fishes (puffers, triggerfishes, filefishes, boxfishes, and their relatives) (Winterbottom, 1974; Tyler, 1980) was used to infer the evolutionary history of the inflation mechanism found in pufferfishes (Wainwright and Turingan, 1997). A key finding in this study was that major elements of the inflation mechanism (the pattern of muscle activation used in inflation behavior) appeared to have evolved earlier than inflation behavior in tetraodontiform phylogeny. Changes to complex functional characters may be acquired gradually in the evolution of a group until some breakthrough form is achieved that has major implications for the subsequent evolution of the clade possessing the feature. Phylogenies provide a powerful tool for investigating the sequences of changes in functional systems that lead to major functional innovations (Stiassny and Jensen, 1987; Lauder, 1990).

Phylogenetic hypotheses have also provided insights into the correlation between the evolution of functional morphological traits and ecological traits. In an analysis of the cheiline labrids from the Indo-Pacific, Westneat (1995) tested the hypothesis that evolutionary changes in the four-bar linkage mechanics of the skull were associated with evolutionary changes in prey-use patterns. The work revealed a strong association between the two, such that historical changes in skull functional morphology were predictably associated with changes in feeding habits. Patterns of historical association between morphological evolution and feeding habits were also explored by Winterbottom and McLennan (1993) in a study of acanthuroid fishes. Use of the phylogeny allowed these authors to conclude that zooplanktivory had evolved several times within this group of reef fishes, rather than just once. Phylogenies likewise suggest that an excavating bite evolved more than once in parrot fishes (Bellwood, 1994; Bernardi *et al.,* 2000). Another major roll of phylogenetic hypotheses involves their use during interpretations of historical biogeogrpahic patterns (e.g., McMillan and Palumbi, 1995). This area was explored in Chapter 1.

V. Prospectus

Coral reef fishes have taught us a tremendous amount about how the functional design of organisms relates to their ecology, and promise many exciting new areas of inquiry in the future. At least two practical features have helped propel coral reef fishes into the forefront of ecomorphological research. First, coral reefs house the most species-rich vertebrate communities on Earth, and, thus, there is a tremendous amount of diversity to attract the attention of researchers. Second, perhaps no major aquatic system on Earth lends itself so well to observational work as do coral reefs, where divers can work comfortably while submerged in clear water. This combination of spectacular diversity and unparalleled accessibility has inspired about 40 years of research on the ecological consequences of the design of the reef fish feeding apparatus and a far larger body of published research based on underwater observations than is found for any other group of fishes. In this chapter we have attempted to summarize the most prominent findings of research in reef fish ecomorphology, but where is this field headed? What major questions will become the focus of future research and where might the greatest promise lie in this field? How will further refinements in our ability to infer feeding abilities of fishes from their morphology help us gain insights into deeper ecological and evolutionary questions?

To a large extent the fish faunas of the world's reefs are now described. We expect that one of the biggest developments that will impact ecomorphological research on coral reef fishes in the next 25 years is the exponentially increasing rate at which phylogenetic hypotheses are being developed for the world's organisms (Pagel, 1997). Modern molecular sequencing methods have improved drastically and largely underlie the rapid growth rate of systematic hypotheses. With the increase in the number of phylogenies will come the ability to test many historical hypotheses with greater rigor than has been possible in the past (Martins, 2000). Do labrids actually represent a radiation of unparalleled magnitude within reef fish communities? Do pomacentrids represent a radiation of species that involved relatively minor morphological evolution? If we can calibrate the amount of time that lineages have been in existence, it will be possible to use phylogenies as the basis for comparisons of the magnitude of functional evolution within major reef fish clades. We expect that in the future such questions will be addressed with some quantitative and statistical rigor. Ecomorphological methods already give us the ability to interpret morphology quantitatively in a mechanical or functional context.

Other major questions about the history of fish feeding ecomorphology may be approached initially without explicit phylogenetic information before being integrated into formal comparative studies. Are there mechanical hotspots in skull morphospace toward which reef fishes have repeatedly evolved, or does mechanical design of the skull show a continuous distribution in morphospace? Have reef fish assemblages with different histories (i.e., the Indo-Pacific vs. the Caribbean) evolved to occupy similar ecomorphological space? Reef fish lineages, because they are diverse and accessible, will continue to be a rich system for documenting patterns of diversification and functional evolution.

By focusing attention on the causal link between functional design, performance, and patterns of resource use, ecomorphology will continue to take a prominent position in our attempts to understand not just the remarkable number of fish species on coral reefs, but also the implications of this diversity of form for ecological issues ranging from population and local community structure up to biogeographic patterns and species distributions. One key to understanding how so many species coexist on reefs is to recognize how functional design limits patterns of prey use in species, and thus permits species to partition the resource base. It remains to be seen how important this process is in structuring reef fish communities. One of the obstacles to approaching the issue is the magnitude of the diversity. It has been difficult to characterize the pairwise interactions between so many potentially interacting species and combine this information into a coherent picture of these communities. One approach to this that has yet to be developed would be to use an ecomorphological framework and assume that similarity in functional

morphological traits implies similarity in feeding ability and ecological role. Such an approach obviously involves major assumptions about how morphology and patterns of resource use map onto one another, but our understanding of feeding biomechanics is such that it is possible now to infer general trends in feeding ability from anatomy. Carefully selected morphological traits could be used as surrogates for species-specific feeding abilities, and thus related to detailed habitat distribution data to address questions concerning whether taxa with similar functional abilities tend to be negatively associated with each other in microhabitat use, whether reef zones typically have communities of fishes that occupy a similar range of ecomorphs, and whether taxa with only certain trophic abilities are able to make use of certain microhabitats.

Although reef fishes have been major contributors to phylogenetically based studies of evolution, one area that has received remarkably little attention is speciation in reef fishes. Considerable insights have been gained into speciation in several freshwater fish systems, including cichlids in Central America and Africa (Markert *et al.*, 1999; Wilson *et al.*, 2000), and the low-diversity lakes of north-temperate systems (Robinson *et al.*, 2000; Rundle *et al.*, 2000). Coral reef fishes, with their planktonic larvae that offer the potential for long-distance dispersal, and their exceptionally high species richness, offer some major contrasts with these freshwater systems. The north-temperate lakes are extremely low diversity (sometimes involving only one or two species), and in general, freshwater lake systems are more prone to spatial segregation of populations, because the lakes are separated and may remain so in periods between glaciations. How does fish speciation occur on coral reefs? Are species typically generated in geographically and reproductively isolated locations before spreading out across wider geographic areas? Or, can speciation occur at the center of species richness, driven by some other means of premating isolation? The presence of endemics in peripheral, isolated areas such as Hawaii suggests that areas like this can be important generators of species. But is this the major mode of speciation in coral reef fishes? And what is the influence of trophic adaptation? It will eventually be possible to develop a clear idea about how most reef fish species are formed and specifically whether trophic specialization and the diversity of feeding ecomorphology are key players in the process or instead represent the consequences of secondary contact of species spread from distant areas of origination.

Ecomorphology is all about the integration of functional morphology and ecology. It is about understanding how organisms get to be built the way they are built, and the consequences of their design for patterns of resource use, interactions with other species, and patterns of evolution. The groundwork has been laid in coral reef fishes. Our understanding of the major design features of fish feeding systems is solid, although not entirely complete. The new generation of reef fish ecomorphologists will be able to use this foundation to examine the role of trophic biology in structuring communities, the evolution of fish feeding biomechanics, and the roll of trophic ecomorphology in reef fish speciation. These are indeed exciting times.

Acknowledgments

Many people have contributed to the rich research history of reef fish ecomorphology. We have benefited in particular from the work of, and our interactions with, J.H. Choat, L. Ferry-Graham, P. Motta, J. Randall, R. Turingan, and M. Westneat. Valuable, detailed comments on drafts of this chapter were provided by C. Peterson, P. Motta, G. Bergmann, and C. Durie. We thank the Australian Research Council and the National Science Foundation for financial support of our research.

CHAPTER 3

Age-Based Studies

J. Howard Choat
School of Marine Biology and Aquaculture
James Cook University
Townsville, Queensland 4811, Australia

D. Ross Robertson
Smithsonian Tropical Research Institute
Balboa, Panama

I. Introduction

There exists a vast literature on the age and growth of fishes (Hilborne and Walters, 1992; Secor *et al.*, 1995a). Much of this reflects a century of research on the major stocks of exploited fishes, primarily those of temperate water, open-ocean, and deep-water habitats. In contrast, there is relatively little age-based demographic information available on coral reef fishes. There are a number of reasons for advocating a more comprehensive understanding of the age-based demography of coral reef fishes. The primary reason relates to the increasing exploitation of reef fish stocks by both artisanal and commercial fisheries, some of the latter (e.g., the live fish trade) being internationally based (Jennings and Polunin, 1996; Birkeland, 1997a; Lee and Sadovy, 1998). Demographic information will provide a better basis for reef fish conservation and management.

Assembling age-based data for coral reef fishes will be a complex task. Most species have wide distributions spanning significant geographic gradients (Myers, 1999). Within localized areas a species will occur in a variety of habitats. As poikilotherms, fish are sensitive to prevailing environmental conditions (Atkinson, 1994). Growth and life history patterns of reef fishes can be expected to vary, at geographic, latitudinal, and habitat scales. An emerging picture of the demography of reef fishes suggests three things. Many taxa will be relatively long-lived (Munro and Polunin, 1997), with life-spans exceeding 15 years; they will have highly distinctive patterns of growth, and age-based demographic features will show a strong phylogenetic structure (Hart and Russ, 1996; Newman *et al.*, 1996; Choat and Axe, 1996; Choat *et al.*, 1996). Most importantly, it is becoming evident that size and age in many taxa of reef fishes may be decoupled (Hart and Russ, 1996; Newman *et al.*, 1996, 2000b). Although reef fishes occur in a wide range of sizes, large size may not correlate with longevity. In addition, many species of coral reef fishes may have complex sexual ontogenies, which can in turn influence size and sex-specific growth patterns (Sadovy and Shapiro, 1987; Choat *et al.*, 1996).

The arguments for developing an age-based demography of coral reef fishes are compelling. Estimates of age-specific variation in growth, reproductive, and mortality rates are the foundation of population biology (Cole, 1954; Stearns, 1992). Knowledge of age structure and mean and maximum life-spans is an important element of fisheries science (Beverton and Holt, 1957; Beverton, 1992; Roff, 1984). Warner and Chesson (1985) suggest that under conditions of fluctuating recruitment rates extended life-spans may be expected. If populations of reef fishes do display extended life-spans, with individuals accumulating in older age classes, then these will be especially vulnerable to exploitation (Musick, 1999a). In deep-sea fishes otolith increments have revealed a picture of long life, slow growth, and populations dominated by older individuals (Smith *et al.*, 1995; Stewart *et al.*, 1995; McFarlane and Beamish, 1995). Similar conditions may well occur in many coral reef fish taxa.

As an introduction to the major sections of this chapter we identify three issues that are important for a better understanding of age-based dynamics for coral reef fishes: (1) the methods used in age-based approaches, (2) tropical/temperate comparisons of fish faunas and the distinctive features of coral reef fishes, and (3) the issue of phylogenetic influences on patterns

of growth and longevity among the perciform lineages of coral reefs.

A. Development of Age-Based Approaches

Estimation of fish age is best achieved through analysis of incremental structures that occur in otoliths. The most useful information is provided by counts of annuli obtained from sections of sagittal otoliths. The choice of sagittal otoliths follows from the demonstration that they continue to accrete calcareous material in a linear fashion throughout the life of the fish, regardless of the pattern of somatic growth (Secor *et al.*, 1995a). This also provides a means of obtaining estimates of age from otolith weight data (Worthington *et al.*, 1995). The patterns of spatial variation in fish growth indicate, however, that it will be necessary to calibrate the relationship between age and otolith weight for each sampling locality.

Other structures such as scales, fin spines, and vertebral centra may also record annual increments, but there are drawbacks to their use for tropical fishes, especially if species are long-lived. The most conveniently sampled structures are scales. However, there may be difficulties in detecting annuli near the margins of scales in long-lived species (Secor *et al.*, 1995a). Some species shed scales easily and hence possess numerous replacement scales that do not record the entire growth history, and many taxa such as acanthurids do not possess scales suitable for reading. Analysis of otolith structure, composition, and weights remains the most effective method for aging of fishes. For long-lived species the most unbiased results are obtained from sectioned rather than whole otoliths (Secor *et al.*, 1995a).

Despite the capacity of otoliths to provide a basis for age estimation, there have been relatively few attempts to use this approach for coral reef fishes. This reflects in part the initially discouraging results of otolith examination in tropical fishes (Munro, 1983). Moreover, many biologists based in the tropics have not had access to appropriate laboratory facilities. Length-based analyses of growth (Pauly, 1998) and reproduction (Sadovy, 1996) have been important in developing a demographic data base for reef fishes. However, if the relationship between size and age is decoupled then length-based analyses may be misleading, especially for long-lived species (Hilborne and Walters, 1992).

Daily otolith increments have proved useful for the analysis of growth patterns in larval, juvenile and short-lived fishes (Chambers and Miller, 1995). However, there may be difficulties associated with the reading of daily rings in fishes living longer than 10 years. For these, annual increments are more appropriate. Fowler (1995) examined sagittal otoliths in 28 species of coral reef fishes and detected interpretable structures suggesting annual increments in 25 of these. Fowler also reviewed information on the validation of annual increments in coral reef fishes. Of 35 instances in which validation has been attempted, only twice was validation not achieved because no periodicity in increment structure was detected. In four examples validation was recorded as "preliminary" only.

Since Fowler's review a number of additional validations of annuli have been achieved for coral reef fishes (Hart and Russ, 1996; Newman *et al.*, 1996; Choat and Axe, 1996; Choat *et al.*, 1996; Manickchand-Heileman and Phillip, 2000; Cappo *et al.*, 2000; Hernamen *et al.*, 2000; Pilling *et al.*, 2000). The most comprehensive studies have been on the family Lutjanidae, for which validation was achieved for 11 coral reef species (Cappo *et al.*, 2000; Newman *et al.*, 2000a,b). Combined with previous estimates of age and growth in reef-associated lutjanid fishes (Loubens, 1978; Manooch, 1987) and work on East Pacific species (Rocha-Olivares, 1998), these studies provide an informative picture of one group of coral reef fishes. Lifespans in excess of 30 years were recorded in a number of species.

The most cost-effective approaches to validation of increments as annual check marks are through chemical markers (Hart and Russ, 1996; Newman *et al.*, 1996; Choat and Axe, 1996; Choat *et al.*, 1996; Cappo *et al.*, 2000; Hernamen *et al.*, 2000) or marginal increment analysis (Manickchand-Heileman and Phillip, 2000; Pilling *et al.*, 2000). There are three types of difficulties in using these approaches for coral reef fishes. First, although examples of sagittal otoliths with clearly defined annuli can be obtained from coral reef fish sampled at low latitudes, the consistency with which rings are displayed may be poor relative to fish from temperate waters (Fowler, 1995). Second, many coral reef fish species are globally abundant but are rare on a local scale (Munday and Jones, 1998). This limits the opportunity for tagging and retrieving chemically marked fishes for validation studies. Third, much coral reef research is still expeditionary in nature. Year-round access is often difficult, a fact that limits the opportunity for marginal increment analysis as a validation tool. However, the last of these difficulties is rapidly vanishing as tropical fisheries authorities realize the value of such information.

A number of additional validation options are available for coral reef fishes. Major environmental

signals at known dates, such as those generated by El Niño events, may modify growth schedules of fishes and be incorporated into otolith structures (Meekan *et al.*, 1999). Age determinations based on elemental or isotopic concentrations in otoliths provide a technically demanding but potentially valuable approach to the issue of validating the temporal basis of otolith increments (Campana, 1999). Annual periodicity in Sr:Ca ratios derived from the analysis of sagittal otoliths has been reported for one coral reef species (Sadovy and Severin, 1992). Variation in oxygen isotope cycles reflecting annual patterns of temperature provides a potential avenue for validation (Campana, 1999). However, their utility for fishes that occur in environments with limited annual temperature ranges has yet to be tested. Chemical techniques require sophisticated laboratory facilities. For many tropical countries marginal increment analysis may be the most cost-effective means of age validation.

The chemical dating of otoliths based on the decay of radioisotopes provides an accurate method for determining age in fishes, especially longer lived species (Campana, 1999). To date, the main focus for such age determinations has been on long-lived temperate water and deep-water fishes. Work on temperate sciaenid fishes (Pereira *et al.*, 1995; Jones and Wells, 1998) based on radiometric validations has provided estimates of life-spans in the order of 60–110 years. Kalish (1993) suggested the analysis of radiocarbon derived from nuclear testing and incorporated into the otolith matrix as an objective and accurate age validation procedure. The study of Kalish *et al.* (1996) is particularly significant in that it provided age estimations of *Thunnus maccoyi* in excess of 30 years, 50% longer than estimations that had been incorporated into management models for this species. However, age groups of fishes most sensitive to nuclear bomb radiocarbon are those born between 1958 and 1965 (Campana, 1999). Even long-lived fishes are now starting to pass beyond this window of opportunity.

Future validation of age estimates in coral reef fishes must also investigate the use of stable isotopes and radioisotopes, ideally based on international collaboration with respect to analytical facilities. Despite the complexity of these techniques and their interpretation (Campana, 1999), they hold the greatest promise of precision and objectivity of age determinations, especially when used in association with more traditional approaches. Problems associated with viewing of increment structures in individual fishes from low latitudes may be resolved by using validation techniques based on a combination of chemical markers and otolith growth models (Cappo *et al.*, 2000).

B. Tropical–Temperate Comparisons

The majority of our information on fish growth patterns and age-based population dynamics comes from the study of the great temperate water fish stocks, including clupeoids, gadids, and pleuronectids (Beverton and Holt, 1957). There is now a comprehensive information base on these groups, a reflection of their prior abundance and economic importance, but relatively little information on shallow-water perciform fishes (Beverton, 1992; He, 2001). Are studies of temperate groups an appropriate foundation for the development of an age-based approach to the demography of coral reef fishes dominated by perciformes?

Coral reef fishes and their habitats display highly distinctive characteristics that differ substantially from temperate groups listed above. A number of reef fish taxa, including acanthurids and lutjanids, display distinctive age-specific growth patterns (Newman *et al.*, 1996; Choat and Axe, 1996). Young fish grow rapidly until they reach maturity then abruptly reduce somatic growth. This results in a highly characteristic "square" growth curve with numerous age classes accumulating in a few size classes. In contrast, tropical serranid fishes of the genus *Plectropomus* show relatively short life-spans and growth curves more consistent with those displayed by clupeoid, gadid, and pleuronectid fishes (Beverton, 1992). Both types of growth curves are illustrated in Fig. 1. The "square" growth curves are characteristic of a number of coral reef fishes, including lutjanids and acanthurids, and may generate age-specific properties very different from those seen in the more characteristic growth curves of temperate species.

The physical structure and hydrodynamic environment of coral reefs are more complex (Boehlert, 1996) than are those of higher latitudes. Fisheries biologists in temperate waters can access a substantial body of information on the gyres and current systems that influence movement patterns and larval transport of fishes (Sinclair, 1988). Fisheries research in higher latitudes has built on networks of fisheries laboratories and workers with collaborative research agendas. The establishment of age-based approaches has been one of these. The result has been a common approach to methodologies and an integrated perspective on fish population dynamics. For example, knowledge of the hydrodynamic environment of the coastal seas and of the seasonal patterns of spawning in temperate species has played an important part in the recognition of fish stocks and their distinctive demographic characteristics (Beverton and Holt, 1957; Harden-Jones, 1968; Cushing, 1976; Isles and Sinclair, 1982).

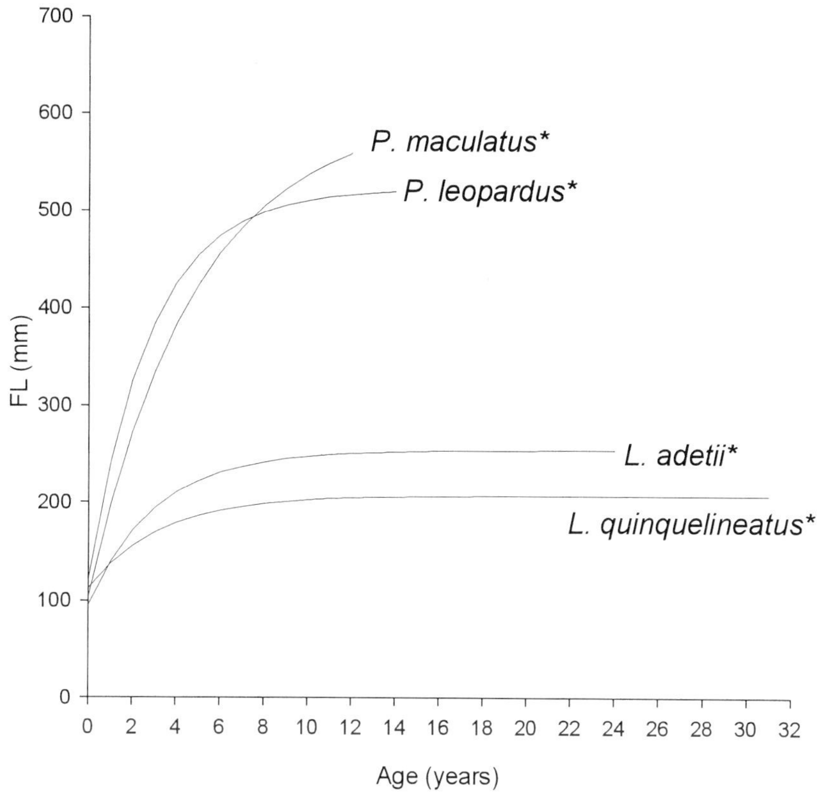

FIGURE 1 von Bertalanffy growth curves fitted to size-at-age data for two species of *Plectropomus* and two species of *Lutjanus* to show contrasting growth and longevity patterns. *Plectropomus* data from Ferreira and Russ (1992, 1994); *Lutjanus* data from Newman *et al.* (1996). * Age increments validated.

In contrast, fisheries for coral reef species are often localized, artisanal, and targeted to multispecies assemblages (Dalzell and Adams, 1997; Jennings and Lock, 1996). This reflects the widely distributed, circumtropical nature of coral reefs and their accessibility through a variety of fishing methods. There have been limited opportunities for the development of common methodologies, or to focus research on a few representative taxa. The complexity of the hydrodynamic environments of coral reefs has made analysis of dispersal patterns difficult (Leis, 1991a; Boehlert, 1996), although work by Jones *et al.* (1999) has revealed a surprising degree of recruitment back to natal reefs. Studies of genetic structuring (Planes, 1993; Shulman and Bermingham, 1995) are providing a perspective on levels of gene flow and connectivity in reef fish populations. However, there is still little information on the spatial variation in demographic structure of reef fish populations or on the degree of concordance between demographic and genetic characteristics.

In the absence of a comprehensive understanding of age-based dynamics the use of length-based approaches has provided a substantial data base and the opportunity to compare the life histories of temperate and tropical fishes (Pauly, 1998). Such a comparison was provided by Pauly (1994), who argued that tropical fishes have smaller asymptotic sizes, shorter life-spans, higher fecundities, and higher natural mortalities.

The argument that tropical fishes possess smaller mean sizes, shorter life-spans, and higher mortality rates compared to their temperate counterparts has considerable significance for coral reef fisheries. However, it is unclear that the exploited fishes of coral reefs have a smaller average asymptotic size compared to their temperate water counterparts. Coral reefs harbor numerous taxa of small fishes, especially in the speciose families Gobiidae, Blennidae, Apogonidae, and Pomacentridae. In these families the maximum size rarely exceeds 100 mm standard length (SL). However, in a number of abundant but heavily exploited taxa (Acanthuridae, Haemulidae, Labridae, Lethrinidae, Lutjanidae, Scaridae, and Serranidae) the mean length for species is close to or greater than 300 mm SL. Moreover, the maximum size achieved by species ranges from 100 to 1000 mm SL in many families (Munday and Jones, 1998; Randall *et al.*, 1990; Böhlke and Chaplin, 1993). Although tropical fish faunas harbor many small species, the distinctive feature is the range of sizes achieved within a taxon, rather than the mean size.

Second, is there a consistent relationship between longevity and size? In a preliminary analysis of 38 taxa, Thresher (1991) identified a positive relationship between maximum size and maximum age. However, additional aging has revealed long life-spans in small species (Newman *et al.*, 1996; Choat and Axe, 1996). A phylogenetic component in the distribution of life-spans would introduce additional variability into size/longevity relationships. For example, small species of lutjanids may have slower growth rates and longer lives compared to large serranids of the genus *Plectropomus* (Ferreira and Russ, 1992, 1994; Newman *et al.*, 1996).

Finally, are tropical reef fishes more fecund than their temperate equivalents? Comparisons between tropical and temperate species are difficult to obtain. Intuitively it seems that tropical fishes should produce more eggs per year compared to temperate fishes, because the latter may be constrained to seasonal breeding episodes. For example, Davis and West (1993) suggest a monthly frequency of up to 22 spawning episodes for a tropical lutjanid. Robertson *et al.* (1999) monitored population spawning behavior of *Thalassoma bifasciatum* and found that an average of 75% of mature females spawn on a given day but that the clutch size among such females averaged only approximately one-third of the size-specific maximum.

What is clear is that tropical fishes display a wide range of spawning periodicities (Robertson, 1991a; Sadovy, 1996; Robertson and Kaufman, 1998) and that it will be difficult to make general comparisons with temperate species. Two useful comparative measures can be identified: (1) annual fecundity as a percentage of body weight—this may be much higher in species that spawn frequently throughout the year compared to those that spawn less frequently, because clutch weight does not exceed 35% of total weight, even among species that spawn only once per year (Burt *et al.*, 1988; Robertson, 1990); and (2) lifetime fecundity (Sadovy, 1996; Jones and Wells, 1998). However, there are few estimates of this for any tropical fishes and particularly for the larger species. Warner and colleagues have provided robust estimates of lifetime fecundities for *Thalassoma bifaciatum* (Warner, 1984a). Doherty (1983b) has also estimated lifetime fecundity for a pomacentrid and Kuwamura *et al.* (1996) for gobies. There are few estimates available for the larger species of coral reef fishes. Reproductive patterns in coral reef fishes are complex, with protogyny the dominant mode in many abundant species. With the exception of *Thalassoma bifacsiatum* the age distribution of reproductive effort in protogynous species is not well understood in any tropical fish.

C. Comparison of Age and Growth Patterns among Lineages

Tropical reef fish faunas include several distinctive evolutionary lineages (Tyler *et al.*, 1989; Helfman *et al.*, 1997) with different morphologies and reproductive modes. The majority of species within this group are perciformes with dispersive larvae (Sale, 1991a). Beyond this it is not possible to generalize life history characteristics across lineages and the great diversity precludes detailed demographic analysis by taxon. However, estimates of longevity among the different lineages can provide a basis for identifying different life histories in reef species and their capacity to respond to exploitation.

To illustrate the diversity of life histories in reef fishes we will compare data on size and age among two ecologically similar but phylogenetically distinct lineages of reef fishes. The lineages chosen, surgeonfishes (Acanthuridae) and parrot fishes (Scaridae), are speciose groups of percomorph fishes, representatives of the suborders Acanthuroidei and Labroidei, respectively. The Acanthuridae are a monophyletic and highly distinctive lineage within the acanthuroid fishes (Tyler *et al.*, 1989; Winterbottom and McLennan, 1993; Tang *et al.*, 1999). The scarid and labrid fishes in association with the odacids display a number of structural characteristics of the pharyngeal apparatus that support the argument of the existence of a single major lineage of labroid fishes (Kaufman and Liem, 1982). However, evidence for monophyly, especially molecular information, remains a key issue in the analysis of labroid phylogeny (Johnson, 1993). Species in both lineages share a number of similar ecological features (Choat, 1991). The size ranges and numbers of species in each family are almost identical (Munday and Jones, 1998), although the Acanthuridae have a greater mean size. Both are circumtropical in distribution with paleontological evidence of occurrence in tropical reef environments from the Eocene (Acanthuridae) and the Miocene (Scaridae) (Bellwood, 1994, 1996a,b). Although all scarids are basically benthic-feeding-grazers, acanthurids show greater ecological diversity. A number of species, especially in the genera *Naso* and *Acanthurus*, are midwater plankton feeders as opposed to grazers of shallow reef substrata (Jones, 1968).

Each lineage has a distinctive set of morphological and structural characteristics. Scarid fishes exhibit typical labroid features of a fusiform body with a total length to greatest body depth ratio of ~2.2 and a substantial skeletal investment in a pharyngeal apparatus with a grinding mode of action. Acanthurid fishes are laterally compressed, with a total length to

greatest body depth ratio of ~2.9, have a muscular gizzard rather than a grinding pharyngeal apparatus, and have a highly distinctive set of single knives or multiple fixed blades on the tail base. Further details of the structural characteristics and evolutionary relationships within each group may be found in a variety of publications (Bellwood, 1994; Winterbottom and McLennan, 1993). Despite differences in structural and morphological features the most abundant taxa in each lineage display similar feeding patterns (Clements and Choat, 1995; Choat and Clements, 1998).

Collectively these two taxa represent a substantial proportion of reef fish biomass, with individual taxa, especially those in the genera *Scarus, Acanthurus*, and *Ctenochaetus*, being the most abundant fishes in the size range of 200–400 mm SL. Both groups are of major ecological importance in reef dynamics, with grazing activities implicated in removal of algal biomass and bioerosion (Bellwood, 1995a; Bruggemann *et al.*, 1994a; Steneck, 1988). The pattern of ecological similarities and structural and reproductive contrasts makes these abundant reef fishes good candidates for a comparative investigation of life history and demographic features. The key variables are size distributions, growth rates, patterns of growth, age structures, and maximum life-span in populations of representatives of each taxon.

We also refer to published data available for other groups of reef fishes, especially lutjanids and serranids (Manooch, 1987; Fowler, 1995; Sadovy *et al.*, 1992), a reflection of their traditional importance in reef fisheries. However, the expanding scale of reef fishing (Dalzell and Adams, 1997; Jennings and Lock, 1996) has resulted in an increased pressure on other groups of large reef fishes for which we have relatively little demographic information. Those abundant fishes, usually identified as coral reef herbivores, now bear the brunt of increasing fishing on many reef systems.

In summary, age-based approaches will provide a sound foundation for the analysis of reef fish population dynamics. However, the resultant picture is likely to be highly complex. This reflects both the nature of the reef environment and the diverse perciform assemblages that inhabit coral reefs and sustain multispecies fisheries. Attempts to compare tropical and temperate fish faunas will be confounded by this complexity, due to decoupling of age and size and the diverse reproductive behaviors and ontogenies of reef fishes. Phylogenetic factors are likely to contribute strongly to the diversity of life histories and reproductive modes seen in reef fishes.

II. Hypotheses for Testing and Refinement

Our intention for the remainder of this chapter is to present the issues as hypotheses for further testing and refinement. The most desirable outcome would be more attempts to estimate ages and the validation of age-based life history features in coral reef fish. Our approach is guided by three observations: (1) there are relatively few examples of size-at-age data for coral reef fishes, although the published information suggests a variety of growth patterns and maximum life-spans. (Manooch, 1987); (2) lineages of reef fishes differ considerably with respect to mean life-span and reproductive mode (Munday and Jones, 1998); and (3) many coral reef fishes are protogynous, with populations composed of large males and smaller females, which intuitively one would expect to be older and younger, respectively (Helfman *et al.*, 1997). However, for most protogynous species we lack information on age-specific growth rates, age structures, and life-spans that would allow a test of this expectation.

Further discussion will follow the sequence of the five questions listed below, using acanthurids and scarids as examples.

- What is the distribution of life-spans in the chosen groups and are they consistent or do they vary among regions?
- Is there a predictable relationship between body size and maximum age and is any such relationship consistent among taxa?
- Are there taxon-specific patterns in growth curves generated by size at age data among the different groups?
- Are maximum age estimates and growth parameters consistent over geographic and environmental gradients in the same taxa?
- In protogynous species are size and age distributions concordant?

A. What Is the Distribution of Life-Spans among Acanthurids and Scarids: Are They Consistent or Do They Vary among Regions?

To investigate this question we have compared species of acanthurids and scarids sampled from the western Pacific and the tropical Atlantic regions. Four questions are posed. (1) Are there taxon-specific

differences in longevity among these groups? (2) Are taxon-specific differences consistent between regions? (3) Are there overall differences in longevity among regions? (4) Are these differences reflected in otolith dimensions in each taxon and region?

In order to compare life-spans we used an estimate of maximum age based on the mean of the upper 25% of the annuli counts for each species. Given the relatively small sample sizes for some species this provided a more robust and conservative estimate than an estimate of T_{max} based on the oldest individual in the sample (Beverton, 1992). To examine the relationship between otolith dimensions and age we used the ratio of the distance from the otolith core to the edge of the proximal margin of the sagittal section (height) to the greatest distance across the dorsal to ventral surfaces of the section (width). The orientation is shown in Fig. 2. The ratio of height to width was used because substantial differences in sagittae size and morphology existed between the two taxa. We made two assumptions about otolith growth. First, sectioned sagittal otoliths will show an increase in thickness (defined as height in Fig. 2) throughout the life of the fish as increments are added through time. Second, the rate of thickening will decrease as distance between increments is reduced in older fishes. The material for the analysis was collected from coral and rocky reefs in the West Pacific Great Barrier Reef (GBR) and the western and central regions of the tropical Atlantic. The taxa sampled are representative of the abundant species at each locality (Meekan and Choat, 1997).

A plot of relative otolith dimensions vs. age revealed a clear structure in the data set (Fig. 2A). The relative height of sagittal otoliths increases in a predictable manner with increasing age even among distinct taxa. The ratio of otolith sectional height to width increased as a power of the mean maximum age, showing that the rate of accretion of otolith width declines as a function of maximum life-span. The specific points from this data set were as follows:

1. Acanthurids had consistently greater maximum ages compared to scarids both in the tropical Atlantic and the West Pacific. With the exception of *Bolbometopon muricatum* (the largest scarid), scarids had maximum ages of <20 years, with most being considerably less than that. Most acanthurids are long-lived and a number of species exceed 30 years.
2. Scarids of the genus *Sparisoma* (tropical Atlantic) had low maximum ages relative to members of the genus *Scarus* from the West Pacific. This could represent both locality and phylogenetic sources of variation.
3. Acanthurids in the West Pacifc achieve greater maximum ages than do tropical Atlantic species. Atlantic acanthurids are relatively short-lived, especially those from the Caribbean proper.
4. There are differences in the pattern of otolith growth relative to age in tropical Atlantic vs. West Pacific fishes (Fig. 2A). The relationships are adequately represented by power curves. The patterns were driven by species of the genus *Acanthurus* (Fig. 2B). Atlantic species have relatively thicker otoliths compared to their West Pacific counterparts.

The maximum ages recorded for acanthurids approach those seen in other shallow-water long-lived fishes such as sciaenids (Jones and Wells, 1998), although they are less than those of deep-sea fishes. Atlantic and West Pacific members of the same taxa appear to have shorter life-spans and different otolith growth allometries. A more direct interregional comparison of life-spans and otolith dimensions for scarids must await increased sampling of Atlantic species of the genus *Scarus*.

B. Is There a Predictable Relationship between Body Size and Maximum Age?

Is size a useful proxy for age? The estimates of maximum age provide an opportunity to examine the relationship between size and age in each group (Fig. 3). Estimates of T_{max} (Beverton, 1992) and the von Bertalanffy parameters for the acanthurid and scarid populations used in Fig. 3 are shown in Table 1. For scarids there is a predictable pattern, with 64% of the variance in maximum age explained by increase in size. Increased sampling will likely tighten this relationship. For example, more comprehensive collections of *Scarus rubroviolaceus* from the Seychelles indicated maximum ages of 18–20 years for fish 400 mm SL. (E. Grandcourt, personal communication). This would make this species consistent with other large scarids (*Chlorurus microrhinos, Cetoscarus bicolor*) in terms of size/age relationships. We would also expect the *B. muricatum* age estimates to shift to the right because the preliminary collections did not sample the largest individuals that occur on the Great Barrier Reef. However, it should be pointed out that some smaller species, most notably *Scarus frenatus*, are older than "expected" for their maximum size.

In contrast, no relationship between age and size is discernible among the species of acanthurids. Very small species such as *Zebrasoma scopas* and *Acanthurus nigricans* displayed greater ages than did the

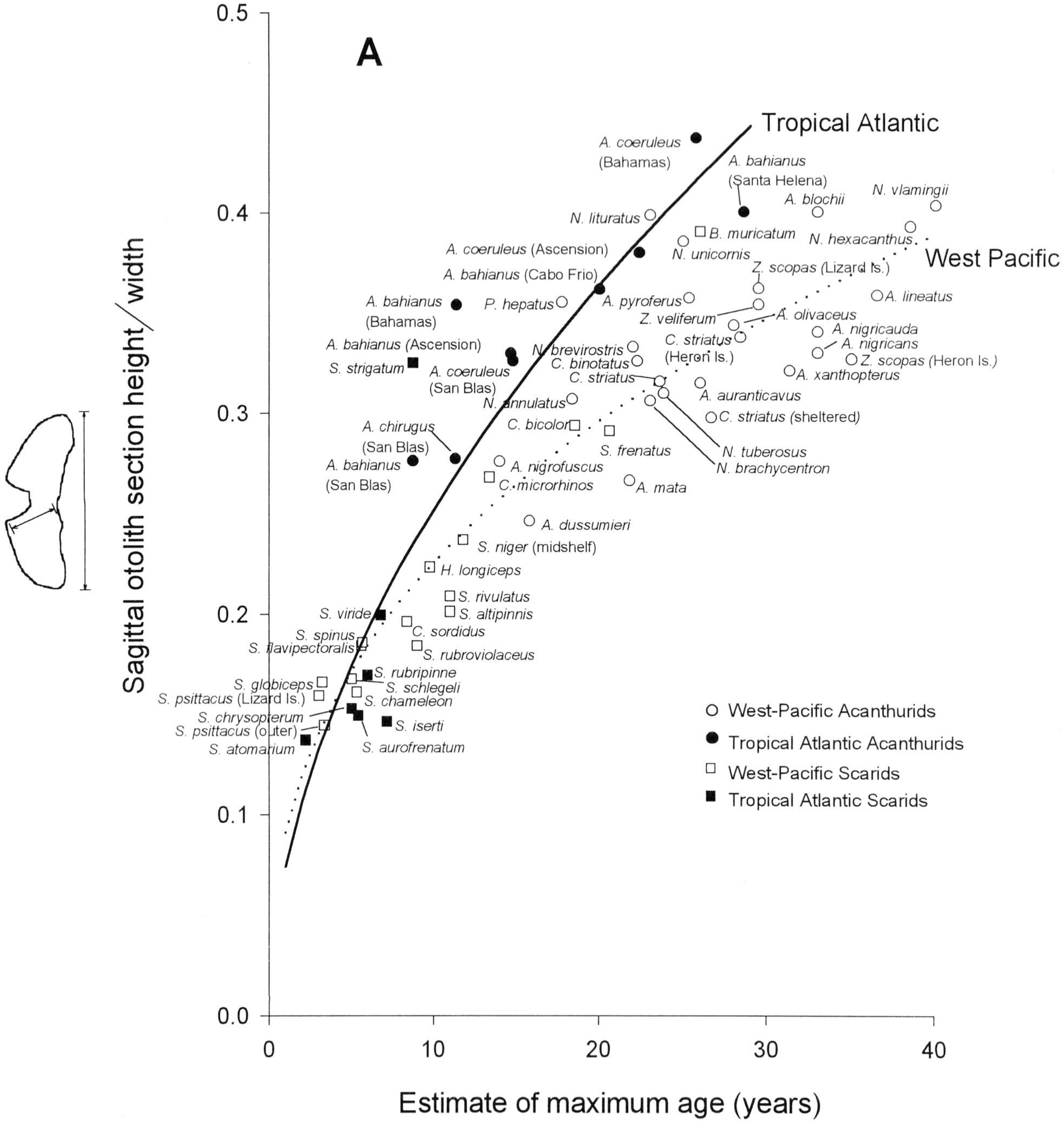

FIGURE 2 (A) The relationship between estimated maximum age and the ratio of sectioned otolith height to width in 50 taxa and populations of acanthurid and scarid fishes sampled from the West Pacific and tropical Atlantic. Maximum age is estimated as the mean of the upper 25% of the sampled age distribution for each taxon. West Pacific: $y = 0.0911x^{0.394}$, $r^2 = 0.791$. Tropical Atlantic: $y = 0.0743x^{0.531}$, $r^2 = 0.774$. (B) The relationship between estimated maximum age and the ratio of sectioned otolith height to width in species of the genus *Acanthurus* from the tropical Atlantic and the West Pacific. For Atlantic and Caribbean data five populations of *Acanthurus bahianus* and three populations of *Acanthurus coeruleus* were included. West Pacific *Acanthurus:* $y = 0.0879x^{0.397}$, $r^2 = 0.551$. Tropical Atlantic *Acanthurus:* $y = 0.0128x^{0.353}$, $r^2 = 0.541$.

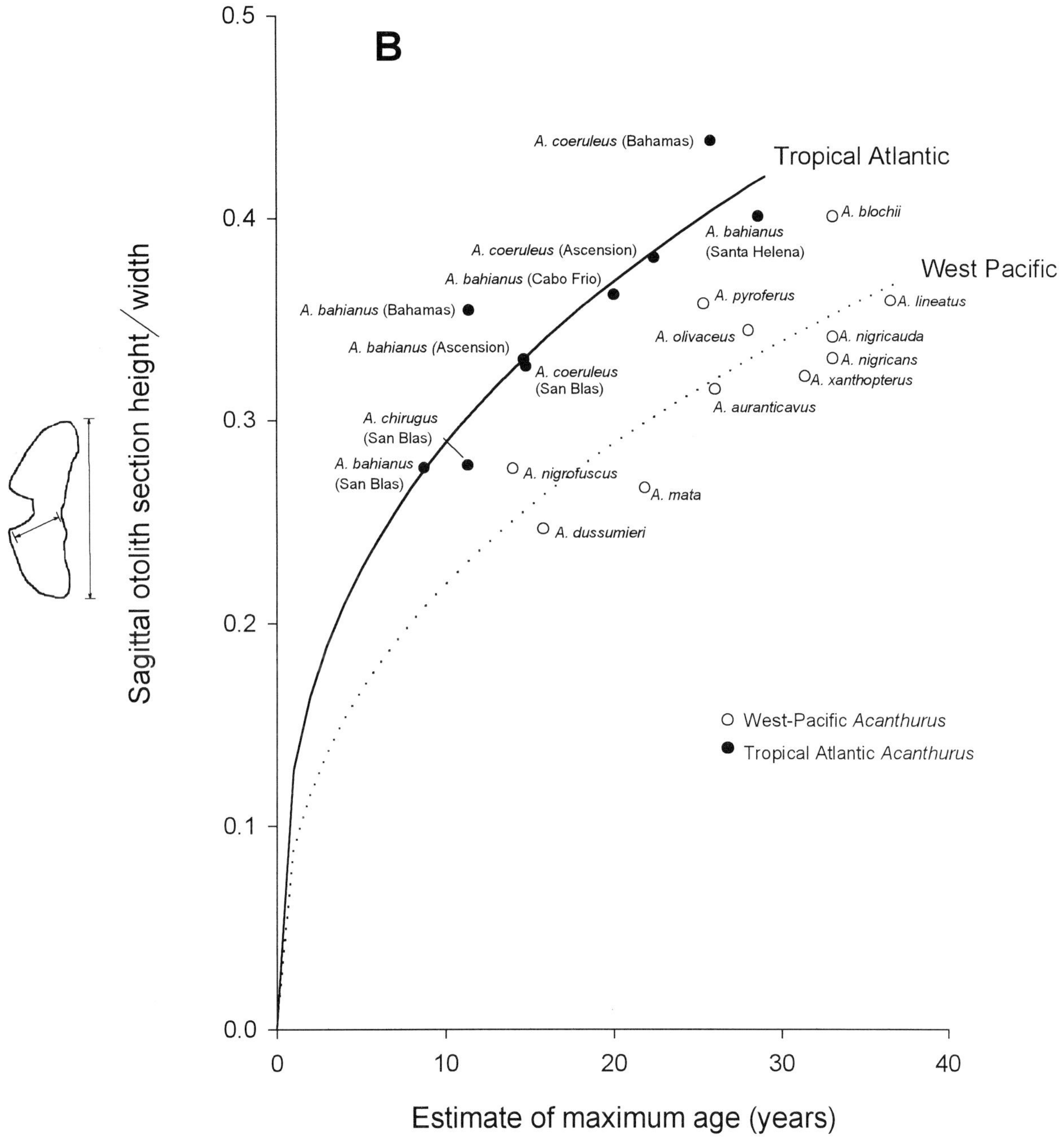

FIGURE 2 (*continued*)

majority of members of the genus *Naso,* a genus that includes the largest acanthurids. Nor is there any consistent pattern evident within any of the five genera. The largest members of the genus *Acanthurus* sampled in this series, *A. xanthopterus, A. mata,* and *A. dussumieri,* were shorter-lived than several of the smaller species, *A. nigricauda, A. lineatus,* and *A. nigricans.* A similar lack of patterns occurred in the other speciose genus *Naso*. In the Acanthuridae size and age are effectively decoupled.

C. Are There Taxon-Specific Patterns in Growth Curves?

We examined patterns of somatic growth using size-at-age data for a selection of species. This

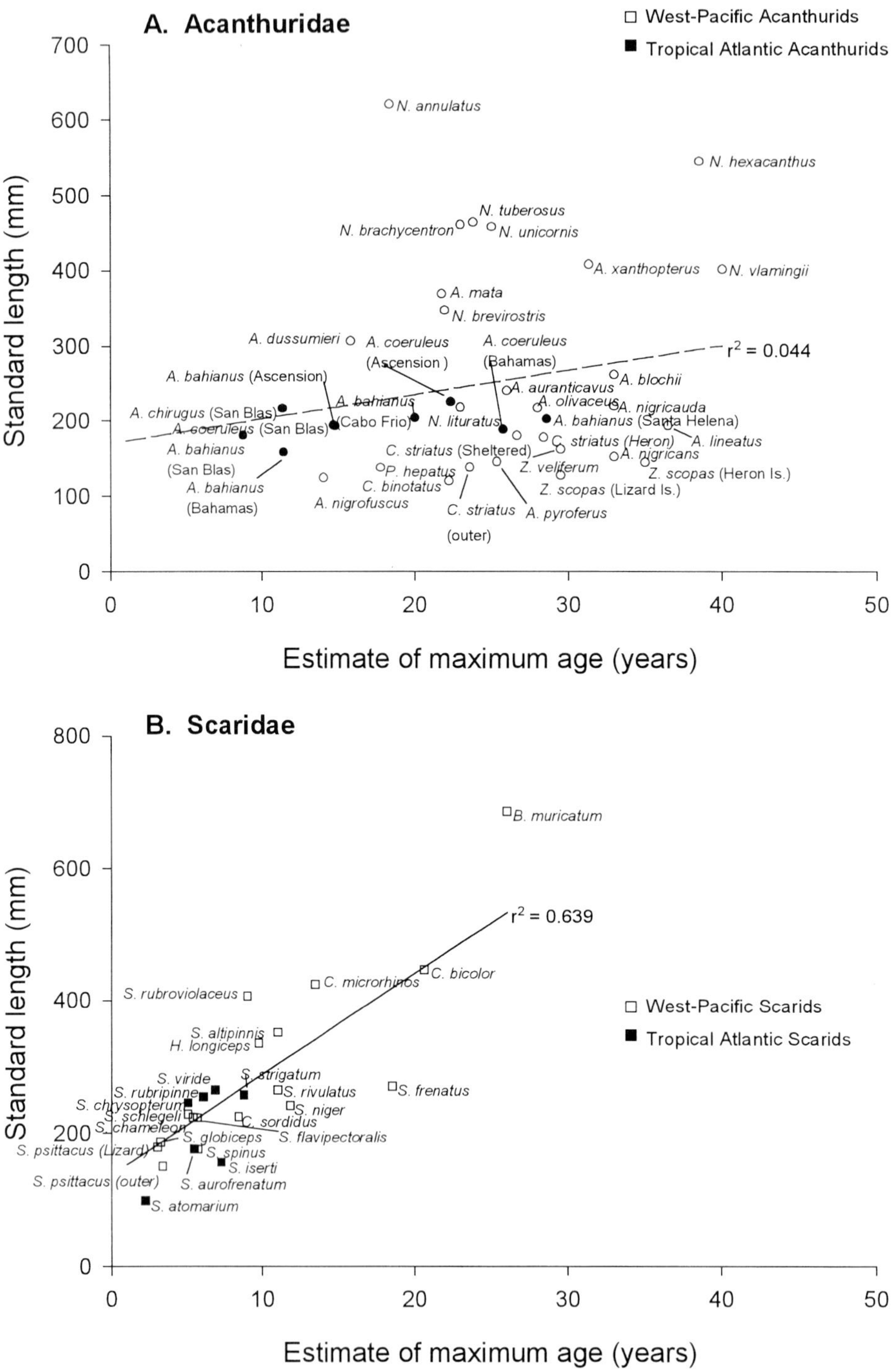

FIGURE 3 The relationship between L_∞ (expressed as SL) and maximum age (see Figure 2) in 50 taxa of acanthurid and scarid fishes. Lines are least-squares regressions. (A) West Pacific and tropical Atlantic Acanthuridae. (B) West Pacific and tropical Atlantic Scaridae.

TABLE 1 von Bertalanffy Parameters and T_{max} (*sensu* Beverton, 1992) for Acanthurid and Scarid Populations used in Fig. 3

	Parameter/estimate				Parameter/estimate		
Family/species	K	L_∞	T_{max}	Family/species	K	L_∞	T_{max}
Acanthuridae				Acanthuridae (*cont.*)			
Acanthurus auranticavus	0.676	235	30	*Naso unicornis*	0.489	386	30
Acanthurus bahianus (St. Helena Is.)	1.241	189	31	*Naso vlamingii*	0.264	379	45
Acanthurus bahianus (Ascension Is.)	1.061	176	19	*Zebrasoma scopas*	0.425	133	33
Acanthurus bahianus (Cabo Frio)	1.201	183	22	*Zebrasoma veliferum*	0.282	208	27
Acanthurus bahianus (San Blas)	1.584	148	10	Scaridae			
Acanthurus bahianus	0.985	156	13	*Bolbometopon muricatum*	0.136	694	33
(Lee Stocking Is.)				*Cetoscarus bicolor*	0.255	421	21
Acanthurus blochii	0.250	276	35	*Chlorurus microrhinos*	0.301	430	15
Acanthurus chirugus (San Blas)	0.705	200	13	*Chlorurus sordidus* (Heron Is.)	0.493	229	8
Acanthurus coeruleus (Ascension Is.)	0.403	193	37	*Chlorurus sordidus* (midshelf)	1.083	193	9
Acanthurus coeruleus (San Blas)	0.477	183	16	*Chlorurus sordidus* (outer)	1.215	158	8
Acanthurus coeruleus	0.817	174	27	*Hipposcarus longiceps*	0.278	350	12
(Lee Stocking Is.)				*Scarus altipinnis*	0.253	377	13
Acanthurus dussumieri	0.296	308	28	*Scarus chameleon*	0.716	231	6
Acanthurus lineatus	0.462	183	42	*Scarus frenatus* (midshelf)	0.844	232	19
Acanthurus mata	0.219	383	23	*Scarus frenatus* (outer)	1.251	195	7
Acanthurus nigricans	0.280	143	34	*Scarus iserti* (San Blas)	1.235	117	8
Acanthurus nigrofuscus	0.783	120	16	*Scarus niger* (midshelf)	0.736	238	16
Acanthurus olivaceus	1.066	210	33	*Scarus niger* (outer)	0.989	176	23
Acanthurus pyroferus	0.422	143	28	*Scarus psittacus* (midshelf)	1.190	172	5
Acanthurus xanthopterus	0.287	426	34	*Scarus psittacus* (outer)	1.505	175	4
Ctenochaetus striatus (exposed)	0.974	154	32	*Scarus rivulatus* (midshelf)	0.220	309	8
Ctenochaetus striatus (sheltered)	0.803	175	36	*Scarus schlegeli*	0.403	239	8
Ctenochaetus striatus (Heron Is.)	0.802	105	28	*Sparisoma atomarium* (San Blas)	1.828	101	3
Naso annulatus	0.213	626	23	*Sparisoma aurofrenatum* (San Blas)	0.669	178	7
Naso brachycentron	0.429	434	31	*Sparisoma chrysopterum* (San Blas)	0.627	258	5
Naso brevirostris	0.402	304	25	*Sparisoma rubripinne* (San Blas)	0.811	238	7
Naso hexacanthus	0.221	527	44	*Sparisoma strigatum* (St. Helena Is.)	0.993	252	10
Naso lituratus	0.755	211	39	*Sparisoma viride* (Lee Stocking Is.)	0.458	357	9
Naso tuberosus	0.428	459	25	*Sparisoma viride* (San Blas)	0.559	280	7

selection was made to facilitate comparison within and between each taxonomic group and between fishes from different biogeographic regions. Figure 3 shows that the relationship between maximum size and maximum age differed between the two groups. Were there also differences in patterns of somatic growth in acanthurids and scarids? The analyses are based on estimates of the von Bertalanffy growth function with parameters L_t (length at time t), L_∞ (mean asymptotic length), K (the growth coefficient that defines the growth rate to L_∞), and t_0 (the hypothetical age at which a fish would have length 0). To facilitate comparisons between the groups the intercepts of the growth curves generated by the model were constrained to approximate sizes for recruitment from the pelagic to benthic populations in each taxon (Leis and Rennis, 1983). These were 25 mm SL for West Pacific acanthurids, 27 mm for Atlantic acanthurids (Robertson, 1991b), and 10 mm SL for scarids from both regions.

Information from other groups of reef fishes [for example, the snapper genus *Lutjanus* (Newman *et al.*, 1996) and the serranid genus *Plectropomus* (Ferreira and Russ, 1992, 1994)] shows that species in different genera have different growth patterns and life-spans. In the former there is rapid initial growth to an asymptotic size that remains constant over the reproductive life of the fish, with 60–80% of linear growth being accomplished within the first 15% of the life-span. The species illustrated in Fig. 1 are small but long-lived. This effectively decouples size and age. In contrast, although the species of *Plectropomus* achieve substantially greater sizes compared to the lutjanids, they are relatively short-lived, with somatic growth occurring over a greater part of the size range, resulting in a discernible relationship between age and size. Although one obvious interpretation of these differences is that shorter life-spans of *Plectropomus* reflect the effects of fishing for these sought-after species, this is not the case (Ferreira and Russ, 1995).

Size-at-age data for 15 species of scarid and 17 species of acanthurid were fitted to the von Bertalanffy growth function (Figs. 4 and 5; Table 2). The growth curves confirmed the relative distribution of ages seen in Fig. 1, with the mean maximum age of acanthurids being 29 years and that of scarids 12 years. For both groups the Atlantic and Caribbean taxa had lower maximum ages than did equivalent-sized West Pacific species. However, the von Bertalanffy parameters K and L_∞ for both groups had similar mean values, although K for scarids was higher than it was for acanthurids (0.733 vs. 0.566). Mean L_∞ values were 288 mm for scarids and 295 for acanthurids. The two groups generated different growth curves.

The genera *Acanthurus, Ctenochaetus,* and *Zebrasoma* showed rapid initial growth, achieving a consistent L_∞ value and resulting in the characteristic square growth curve similar to that seen in small lutjanids (Newman *et al.*, 1996). This was most obvious in small species of the genus *Acanthurus,* such as *A. olivaceus* and *A. lineatus.* Atlantic acanthurids displayed the same type of square growth curves seen in West Pacific congeners despite having life-spans only half that of equivalent West Pacific species. These small acanthurids generated high K values. Members of the genus *Naso* were long-lived. Most displayed flat growth curves although *N. annulatus* and *N. tuberosus* were exceptions. Despite greater sizes of the *Naso* species the distribution of life-spans was similar to that seen in *Acanthurus, Ctenochaetus,* and *Zebrasoma.* For scarids the growth curves for the selected species also reflected the distribution of life-spans shown in Fig. 3. For the West Pacific species the relationship between size and age can be clearly seen in the distribution of life-spans from the smallest (*Scarus psittacus*) to the largest (*Bolbometopon muricatum*) species. The greatest age recorded was 33 years for *B. muricatum,* although this will be a conservative estimate, because the samples did not include the largest individuals observed. The Caribbean species again showed shorter life-spans compared to the West Pacific species. When examined on the same temporal scale as the West Pacific species the Caribbean scarids of the genus *Sparisoma* show very short life-spans characterized by continuous and rapid somatic growth.

The von Bertalanffy parameters K and L_∞ can be plotted to provide summaries of growth performance in a variety of taxa. Pauly (1998) has presented such a summary based on 4826 sets of parameter estimates of the von Bertalanffy growth function. The negative relationship generated by plotting K on L_∞ identifies the "growth space" used by fishes. We have provided a similar plot for the 47 species sampled (Fig. 6). This generated the familiar negative relationship, confirming the conclusions of Pauly (1994, 1998) that fishes either tend to remain small and have high K values or become large and have low K values (Fig. 6A).

The value K represents only one aspect of the fish growth regime, that rate at which L_∞ is approached. Other growth-related features that may be important to management are not reflected in such plots. For example, despite major differences in longevity between acanthurids and scarids these taxa were not obviously partitioned in Fig. 6. Thus although the parameters K and L_∞ allow comparison of growth patterns they do not capture the important life history features that arise from differences in life-span. Couplets of species closely associated in Fig. 6 as a result of their similar K and L_∞ values have life-spans varying by a factor of 2–4, e.g., *Chlorurus microrhinos* ($K = 0.307$, $L_\infty = 428$, $T_{max} = 15$) vs. *Acanthurus xanthopterus* ($K = 0.298$, $L_\infty = 425$, $T_{max} = 33$); *Chlorurus sordidus* ($K = 1.117$, $L_\infty = 193$, $T_{max} = 9$) vs. *Acanthurus olivaceus* ($K = 1.144$, $L_\infty = 209$, $T_{max} = 33$); and *Hipposcarus longiceps* ($K = 0.282$, $L_\infty = 350$, $T_{max} = 12$) vs. *Naso vlamingii* ($K = 0.282$, $L_\infty = 376$, $T_{max} = 45$).

If K is plotted on T_{max} (Fig. 6B) a negative trend is present but with greater scatter than the relationship between K and L_∞. Moreover, the taxa predictably segregate in terms of T_{max}. This can be seen in estimates of size at sexual maturity as a proportion of L_∞, and age at sexual maturity as a proportion of T_{max} (Table 3). Four groups of fish are considered: West Pacific species of *Acanthurus* and *Naso* and West Pacific scarids of the genera *Chlorurus* and *Scarus* (Table 3). For *Acanthurus* and *Naso,* respectively, sexual maturity is achieved at 6 and 15% of T_{max}. For the scarids the estimate is 19%. For the acanthurids 86 and 85% of size at L_∞ is achieved by sexual maturity. For scarids the estimate is 68%. The value 6% of T_{max} for *Acanthurus* is lower than the value for *Chlorurus* and *Scarus* by a factor of three and is indicative of a life history pattern with an extended reproductive span.

D. Within a Taxon, Are Maximum Age Estimates and Growth Parameters Consistent over Geographic and Environmental Gradients?

Most species of reef fishes occur in a number of distinctive habitats within a reef and also have broad

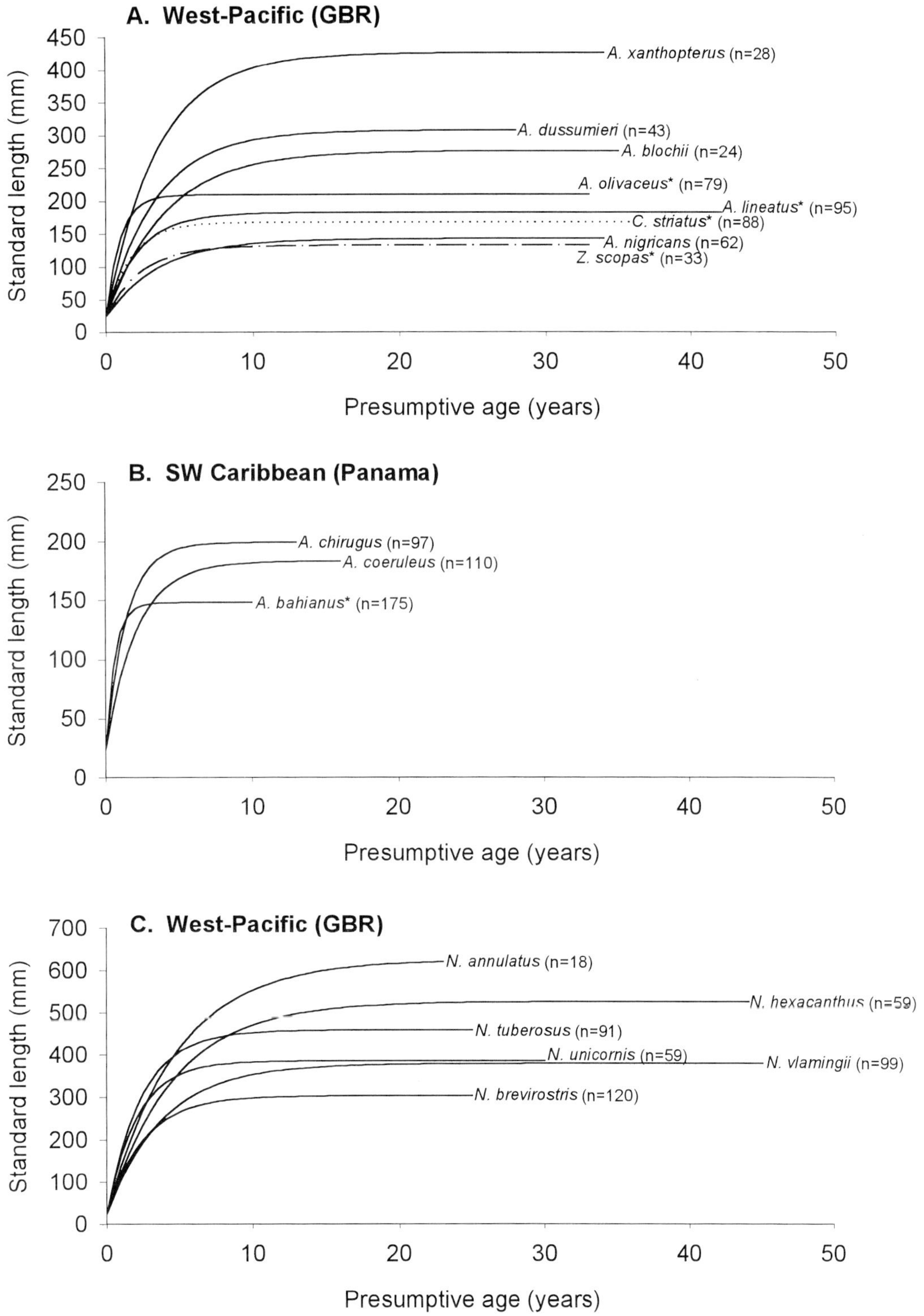

FIGURE 4 von Bertalanffy growth curves fitted to size-at-age data for 17 acanthurid taxa (von Bertalanffy parameters in Table 1; size-at-age data available from J. H. Choat). (A) West Pacific (northern GBR) taxa of *Acanthurus*, *Zebrasoma*, *and Ctenochaetus*. (B) Caribbean taxa of *Acanthurus* from the San Blas sampling locality. (C) West Pacific (northern GBR) taxa of *Naso*. *Validation of annual increments achieved.

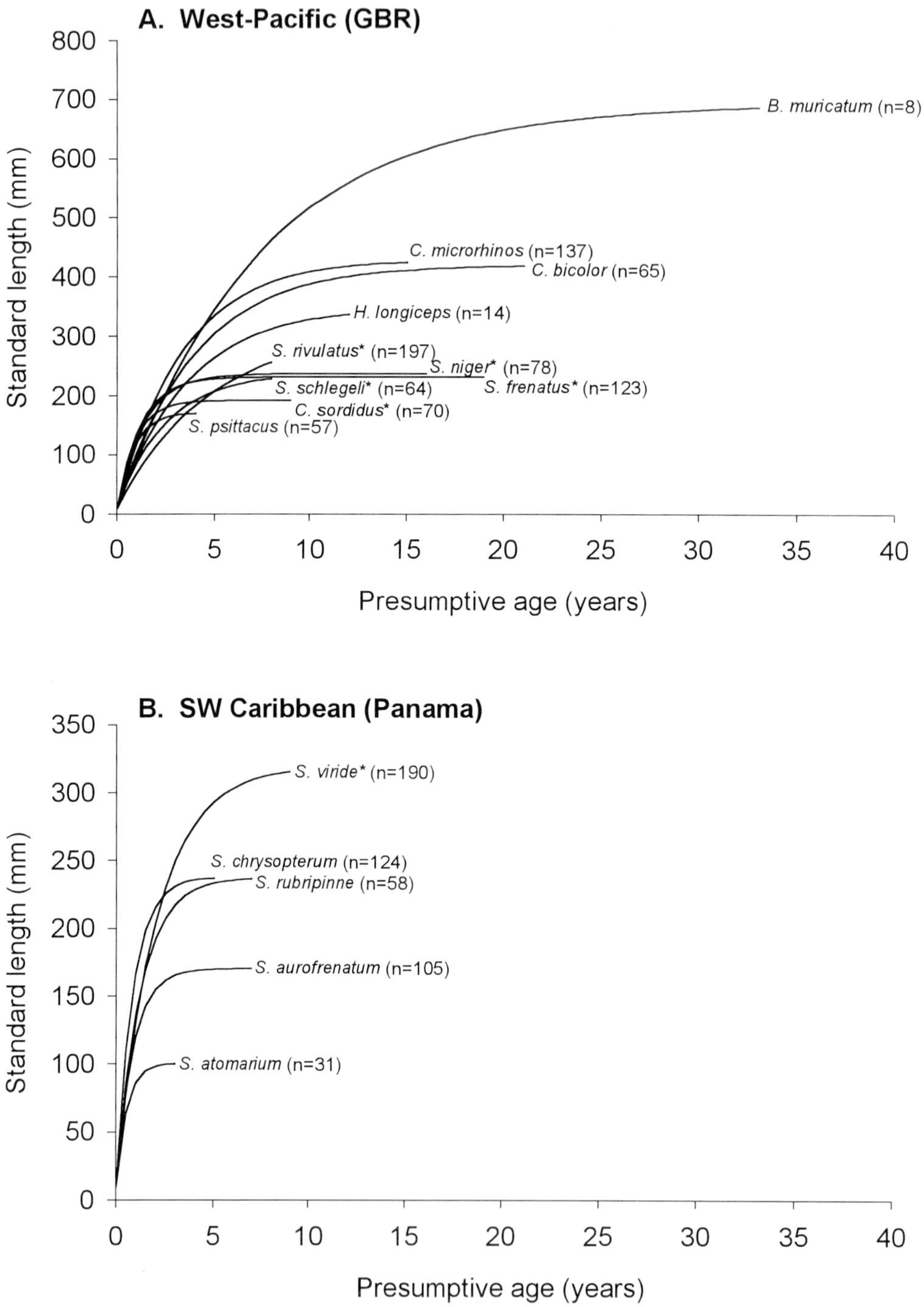

FIGURE 5 von Bertalanffy growth curves fitted to size-at-age data for 15 scarid taxa (von Bertalanffy parameters in Table 1; size-at-age data available from J. H. Choat). (A) West Pacific (northern GBR) taxa of *Bolbometopon*, *Chlorurus*, *Hipposcarus*, and *Scarus*. (B) Caribbean taxa of *Sparisoma* from the San Blas sampling locality. *Validation of annual increments achieved.

TABLE 2 von Bertalanffy Parameters: Sample Size and Size Range for Acanthurids (Fig. 4) and Scarids (Fig. 5)

	Parameter/estimate				
Family/species	L_∞	K	t_0	n	Size range
Acanthuridae					
Acanthurus bahianus	148.32	1.584	−0.12	175	31–198
Acanthurus blochii	276.07	0.250	−0.38	24	129–328
Acanthurus chirugus	199.61	0.705	−0.19	97	55–230
Acanthurus coeruleus	183.48	0.477	−0.31	110	30–206
Acanthurus dussumieri	307.67	0.296	−0.29	43	140–334
Acanthurus lineatus	182.74	0.462	−0.32	95	92–206
Acanthurus nigricans	142.86	0.280	−0.69	62	84–160
Acanthurus olivaceus	209.74	1.066	−0.12	79	95–248
Acanthurus xanthopterus	426.25	0.287	−0.21	28	111–473
Ctenochaetus striatus	168.00	0.591	−0.27	88	103–197
Naso annulatus	625.76	0.213	−0.19	18	179–630
Naso brevirostris	303.75	0.402	−0.21	120	86–364
Naso hexacanthus	527.18	0.221	−0.22	59	148–577
Naso tuberosus	459.39	0.428	−0.13	91	144–517
Naso unicornis	385.77	0.489	−0.14	59	176–494
Naso vlamingii	379.10	0.264	−0.26	99	136–459
Zebrasoma scopas	132.82	0.425	−0.49	33	76–157
Scaridae					
Bolbometopon muricatum	693.81	0.136	−0.11	8	231–710
Cetoscarus bicolor	420.56	0.255	−0.09	65	155–461
Chlorurus microrhinos	429.89	0.301	−0.08	137	105–499
Chlorurus sordidus	192.95	1.083	−0.05	70	98–245
Hipposcarus longiceps	350.07	0.278	−0.10	14	191–374
Scarus frenatus	232.36	0.844	−0.05	123	109–295
Scarus niger	238.01	0.736	−0.06	78	121–278
Scarus psitticus	172.10	1.190	−0.05	57	104–215
Scarus rivulatus	308.50	0.220	−0.15	197	88–290
Scarus schlegeli	238.81	0.403	−0.11	64	89–249
Sparisoma atomarium	101.08	1.828	−0.06	31	50–101
Sparisoma aurofrenatum	170.89	1.163	−0.05	105	91–234
Sparisoma chrysopterum	237.97	1.176	−0.04	124	74–278
Sparisoma rubripinne	237.57	0.811	−0.05	58	112–272
Sparisoma viride	318.96	0.498	−0.06	190	32–379

geographical distributions. Plasticity in growth and reproductive parameters in poikilotherms in response to environmental variation is well known (Atkinson, 1994). The issues of geographic variation and plasticity in growth and life history parameters are critical to our understanding of reef fish population dynamics. The capacity to age fishes over major parts of their geographic range provides a means of assessing systematic trends in life history parameters in response to both latitudinal and local gradients in environmental conditions. This approach requires estimation of ages of fishes by examining annuli in sectioned otoliths in fishes collected adjacent to the equator. Such sampling is still in its infancy, although sectioned otoliths from the Seychelles, 4° south latitude, show consistent and regular increment structures (E. Grandcourt, personal communication). The examples presented here are taken from size-at-age data collected from two species of widely distributed acanthurid, *Acanthurus bahianus* in the West and central Atlantic and *Ctenochaetus striatus* from southern and central western Pacific reefs (Fig. 7).

The two sampling gradients had the following features. For *A. bahianus* the gradient in the tropical Atlantic extended from Cabo Frio (southwestern Atlantic) to the Bahamas (northwestern Atlantic), covering in absolute terms 16° of latitude and 6° Celsius range in mean annual temperature. For *C. striatus*,

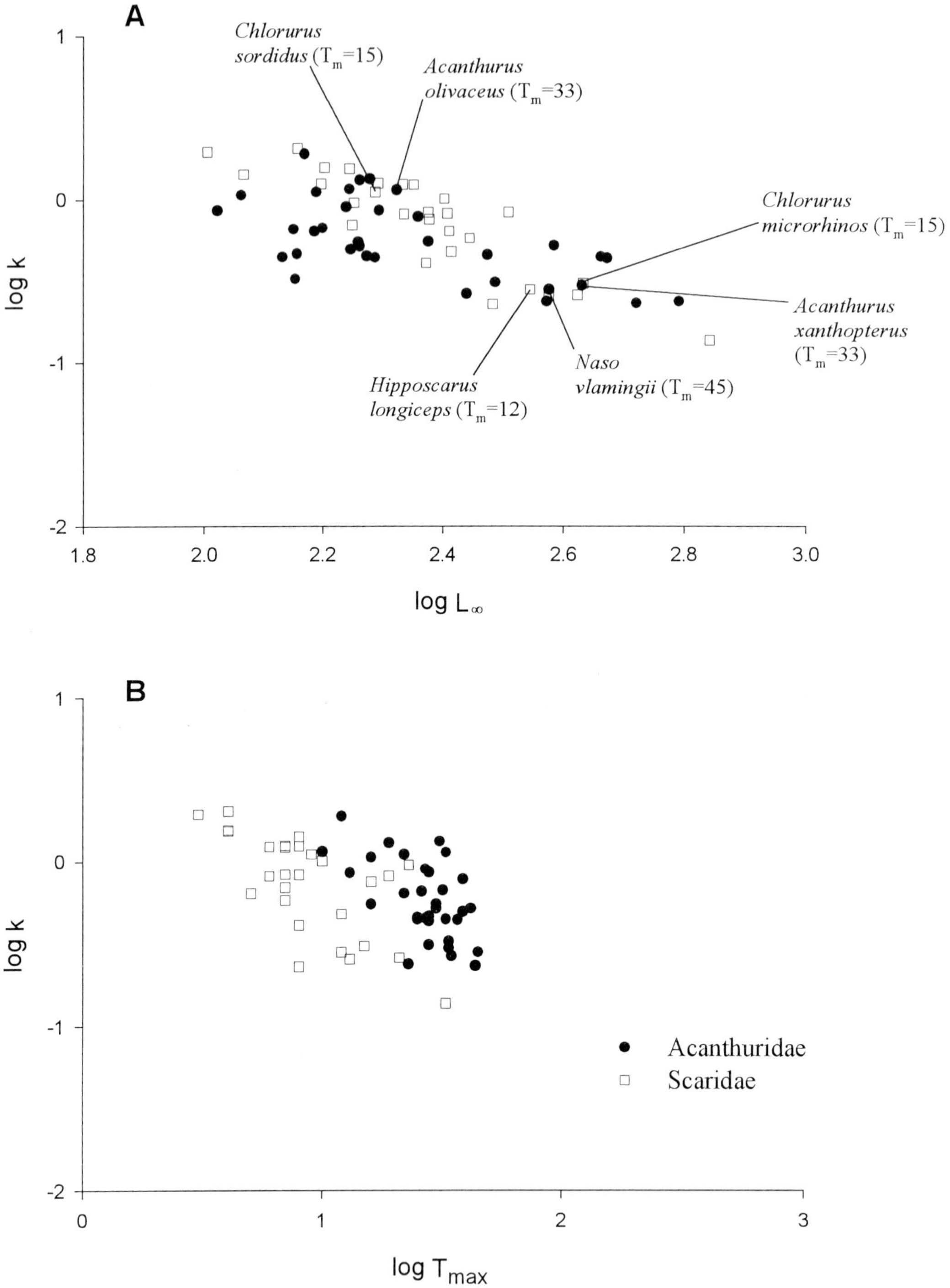

FIGURE 6 (A) Relationship between K and L_{∞} for 34 acanthurid populations (25 taxa) and 28 samples of scarid populations (22 taxa). The three couplets of species show pairs of acanthurids and scarids with similar combinations of K and L_{∞} values but with distinctive T_{max} values (*sensu* Beverton, 1992). (B) Relationship between K and T_{max} for the same data set.

TABLE 3 Estimates of K and Values for L_∞, T_{max}, (*sensu* Beverton, 1992) T_m (Age at Maturity), and L_m (Size at Maturity) for Six Species of Acanthurid and Six Species of Scarid

	Parameter/estimate						
Family/species	L_∞	K	T_{max}	T_m	L_m	A[a]	B[b]
Acanthuridae							
Acanthurus olivaceus	210	1.066	33	2	180	86%	6%
Acanthurus lineatus	183	0.462	42	3	160	87%	7%
Ctenochaetus striatus	168	0.591	36	2	135	80%	6%
Naso brevirostris	304	0.402	25	4	250	82%	16%
Naso hexacanthus	527	0.221	44	7	450	85%	16%
Naso tuberosus	459	0.428	25	4	350	76%	16%
Scaridae							
Chlorurus microrhinos	430	0.301	15	4	300	70%	27%
Cetoscarus bicolor	421	0.255	21	5	300	71%	24%
Chlorurus sordidus	193	1.083	9	1.5	100	52%	17%
Scarus frenatus	232	0.844	19	2	180	78%	11%
Scarus niger	238	0.736	16	2	175	74%	13%
Scarus psittacus	172	1.190	4	1	110	64%	25%

[a] $A = \%L_\infty$, the proportion of L_∞ achieved at sexual maturity.
[b] $B = \%T_{max}$, the proportion of T_{max} achieved at sexual maturity.

the gradient in the southern and central Indo-Pacific extended from Heron Island (southern GBR) to Kimbe Bay (New Britain), covering 18° of latitude and 4° mean annual temperature. Acanthurids were sampled in as close as possible to similar habitats at the following localities. For *A. bahianus* the localities were St. Helena and Ascension Island (south central Atlantic), Cabo Frio (southern Brazil), San Blas Point (southwestern Caribbean), and Lee Stocking Island (Bahamas, northwestern Atlantic). Over this range *A. bahianus* occurs on both rocky and coral reefs. For *C. striatus* the localities were Heron Island (southern Great Barrier Reef), Lizard Island (northern Great Barrier Reef), and Kimbe Bay (New Britain). For the West Pacific, gradient samples at different sites in the vicinity of Lizard Island are included (Choat and Axe, 1996). Size-at-age data from sagittal otoliths were used to construct growth curves and to estimate K, L_∞, and T_{max} from samples from each locality. At Heron Island insufficient specimens were obtained to construct a growth curve, although samples at the upper end of the size range were available to provide an estimate of T_{max}.

Plotting estimates of T_{max} against latitude revealed a positive relationship between maximum age estimates and latitude for the West Pacific but not for the Atlantic locations. In the South Atlantic, latitude and temperature are not closely coupled due to the influences of cold oceanic currents (Santa Helena) and upwelling (Cabo Frio). A plot of T_{max} on mean annual sea temperature revealed a negative relationship between age and temperature in both the West Pacific and the Atlantic localities. Sampling at Lizard Island localities incorporating a range of habitat types confirmed the presence of localized variation in life history features.

The size-at-age data from each locality were fitted to the von Bertalanffy equation to display comparative growth patterns. The West Pacific material (*C. striatus*) showed a reduction in asymptotic size and shorter life-spans at Kimbe Bay site (5° south latitude) (Fig. 8A). The samples from Kimbe Bay showed an L_∞ 41% lower compared to the sheltered reef samples from Lizard Island. A clearer case for latitudinal variation in growth and life history parameters is seen in the analysis for *A. bahianus* (Fig. 8B). The characteristic square growth curve occurred at each locality, but the maximum sizes and ages achieved varied systematically with location. The L_∞ for the San Blas samples was 20% smaller than that for the St. Helena samples. The 95% ellipsoidal confidence regions around the von Bertalanffy parameters were used to compare the growth functions among populations (Kimura, 1980) (Fig. 8C). Statistically significant variation among locations occurred in the life history parameters. The tropical Atlantic populations of Santa Helena, Cabo Frio, and Ascension showed significantly higher L_∞ compared to the Caribbean populations from Bahamas and San Blas.

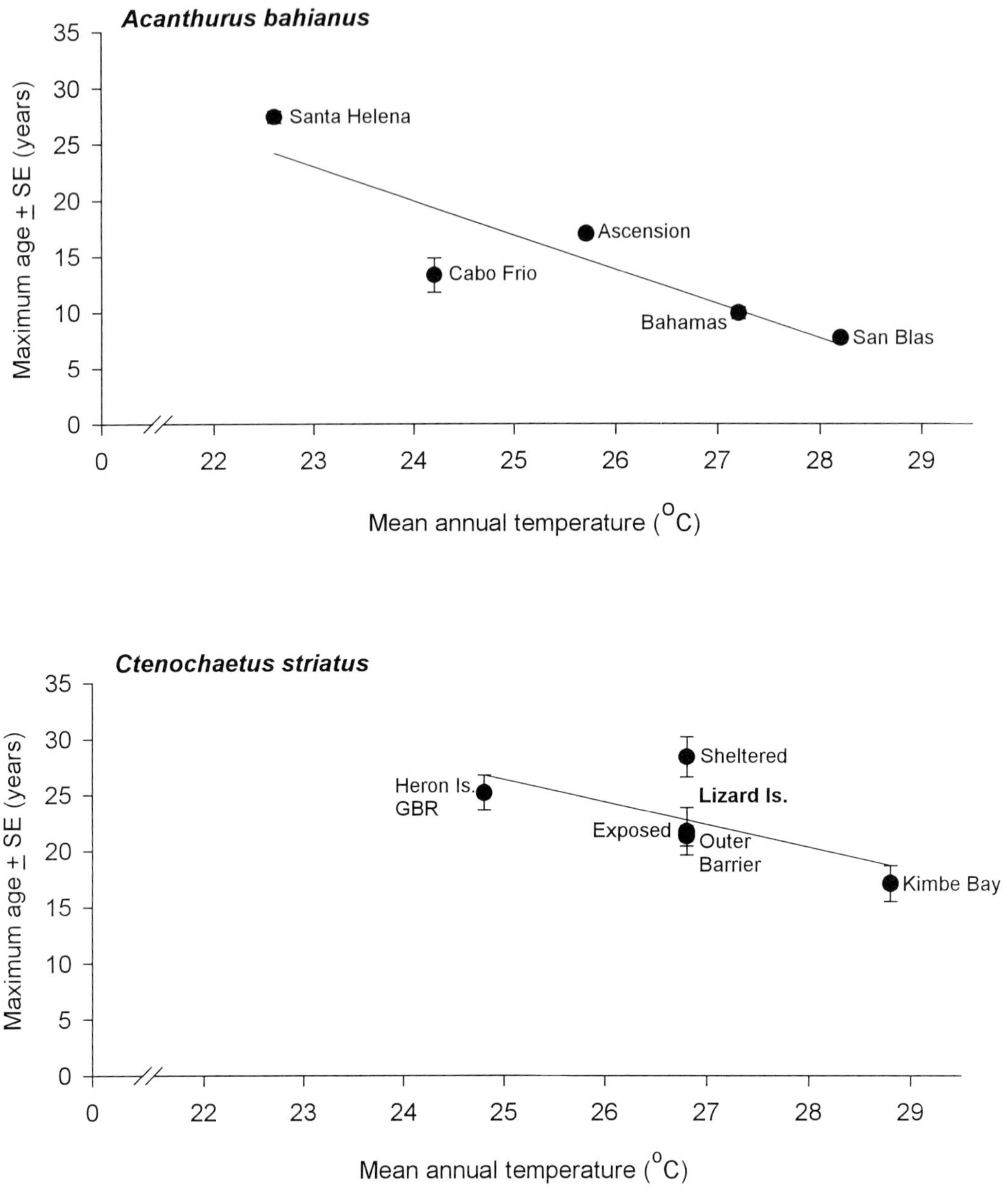

FIGURE 7 Relationship between estimates of maximum age and average annual sea surface temperature for populations of *Acanthurus bahianus* sampled at five geographic localities from the tropical Atlantic and Caribbean and five populations of *Ctenochaetus striatus* samples from three localities on the Great Barrier Reef and New Britain. The Lizard Island locality was subsampled by habitat. Temperature data from IGOSS (1998) and D'Croz and Robertson (1997). Localities are at the following latitudes: for the Indo-Pacific—Heron Island, 24°S; Lizard Island, 14°S; Kimbe Bay, 5°S; for the tropical Atlantic and Caribbean—Santa Helena, 16°S; Ascension, 8°S; Cabo Frio, 23°S; San Blas, 9°N, Bahamas, 24°N.

Each species maintained similar patterns of variation in longevity and growth over a wide range of environmental conditions but showed systematic changes in K, L_∞, and T_{max} correlated with variation in temperature. This included a threefold increase in T_{max} in *A. bahianus* from the San Blas Archipelago to St. Helena.

E. In Protogynous Species Are Size and Age Distributions Concordant?

A large number of coral reef fishes are protogynous hermaphrodites (Sadovy, 1996). Labroid fishes are the most abundant protogynous hermaphrodites on

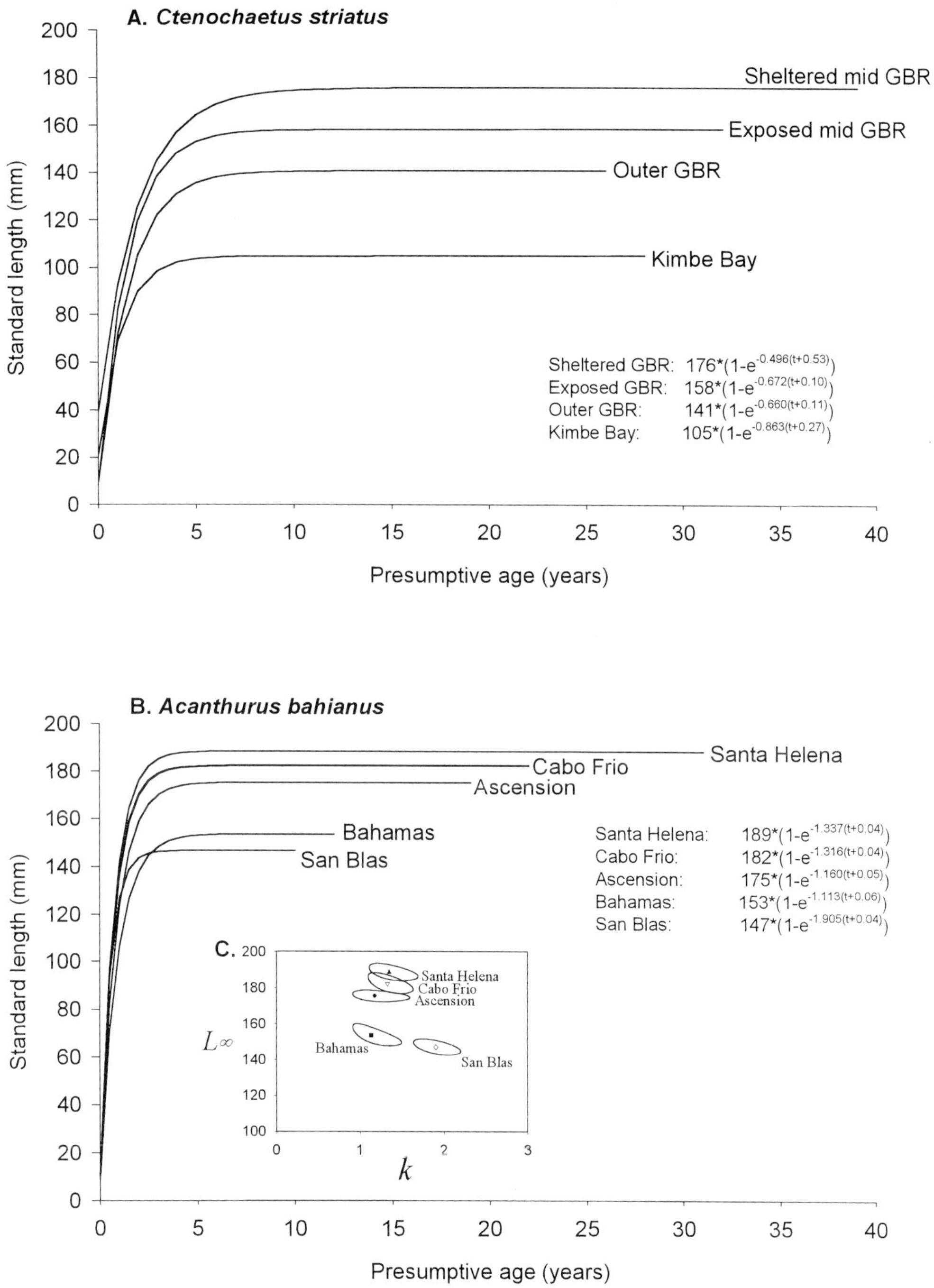

FIGURE 8 von Bertalanffy growth curves fitted to size-at-age data for four populations of *Ctenochaetus striatus* and five populations of *Acanthurus bahianus* (size-at-age data available from J. H. Choat). (A) *Ctenochaetus striatus* growth curves and von Bertalanffy equations for four populations. (B) *Acanthurus bahianus* growth curves and von Bertalanffy equations for five populations. (C) Comparison of five von Bertalanffy growth curves for populations of *Acanthurus bahianus* showing cross-sections of 95% confidence regions around least-squares estimates of K and L_∞ following Kimura (1980). *Validation of annual increments achieved.

coral reefs, with scarids being the numerically dominant taxon of larger protogynous fishes. Scarids well illustrate the need for age-based approaches to the analysis of protogynous life history patterns. Sexually dichromatic scarids present a picture of large terminal males and smaller initial-phase females. Intuitively one would expect that the population should also be partitioned by age. Large terminal males represent the oldest members of the population, whereas females are smaller and younger and are recruited into the terminal male population by sex reversal, following the loss (removal or departure) of large terminal males. Females are young, abundant, and have a capacity to replenish the large-male population by sex reversal. Representative size frequency distributions of three species of scarids illustrate these points (Fig. 9). However, when size/age relations are considered a more complicated picture emerges. Size-at-age plots for protogynous species such as these are often messy, with a considerable range of sizes in different age groups. Partitioning plots by sex reduces the variation, especially in intermediate age groups (Choat *et al.*, 1996). At a given size, secondary (sex-changed) males are larger than females. However, small mature females may represent some of the oldest members of the population. Comparison of age and size distributions (Fig. 9) for the three scarids illustrates these points. In the case of *Scarus frenatus* and *C. microrhinos* age plots revealed a number of females that are as old as, but substantially smaller than, the largest terminal-phase males.

In summary, a review of growth parameters and age structures in two ecologically similar groups of reef fishes confirms that analysis of the population dynamics of reef fishes is likely to be complex. There will be marked variation in the patterns of growth, in the relationships between size and age, and in the distribution of maximum ages among different lineages. Protogynous patterns of sexual development and sex-specific differences in growth rate will be a further complicating factor.

Reliable estimates of maximum age and of age structure are of critical importance in evaluating the population dynamics of reef fishes and their capacity to respond to altered recruitment and mortality rates.

III. Discussion

Age-based studies of tropical reef fishes have been relatively slow moving. In some instances good descriptions of age-specific demography have been obtained from the otoliths of fishes from low-latitude reef environments (Sadovy *et al.*, 1992). However, for many others, including species under increasing fishing pressure, we lack such information. This is partly due to the perception that otoliths of fishes from low-latitude reef systems do not show consistent formation of annual rings. Detection and analysis of periodic rings in otoliths sampled from fishes in low-latitude coral reef systems is a challenge. It is our experience that consistent patterns can be demonstrated, although it is necessary to examine large numbers of otoliths to achieve this. There are relatively few instances in which the first otolith examined reveals incremental structures of the clarity seen in temperate reef fishes.

In size-at-age plots presented here we have used the term "presumptive age." In many taxa the annual periodicity of check marks has not been validated. For most reef species validation has been achieved through chemical marking of calcareous structures or marginal increment analysis. However, validation that satisfies protocols of independent methodologies and application over the full range of the age distribution are difficult to achieve in coral reef fishes due to their diversity, relative rareness of individual species, and, in many instances, extended life-spans.

Much of the debate concerning age estimation in fishes has been focused appropriately on the accuracy and precision by which otolith increments are recognized, validated, and counted. This can be problematical in species with numerous increments crowded near the margins of otolith sections. Moreover, the processes controlling increment formation in tropical fishes are poorly understood (Fowler, 1995). Understanding the interplay between physiological and environmental factors in the formation of internal otolith structures will facilitate a more accurate approach to the analysis of age-based dynamics. The fact that some temperate species form clearly defined increments in their sagittae that are not referable to an annual schedule of deposition (Morales-Nin *et al.*, 1998) confirms the need for validations. However, an emerging consensus of age determination studies is that consistent increments observed in sectioned sagittal otoliths in both tropical and temperate species may be reliably interpreted as having an annual periodicity. There may be exceptions to this interpretation. For example, Milton *et al.* (1995) suggested that rings observed in the otoliths of lutjanids may not be formed annually (but see Cappo *et al.*, 2000). Despite such issues it is important to provide age estimations as hypotheses for further testing and to help prioritize future work.

The potential problems faced in coral reef fisheries are similar to those encountered in fisheries for deep-water species. The inability of management procedures to prevent the serious overexploitation of deep-water

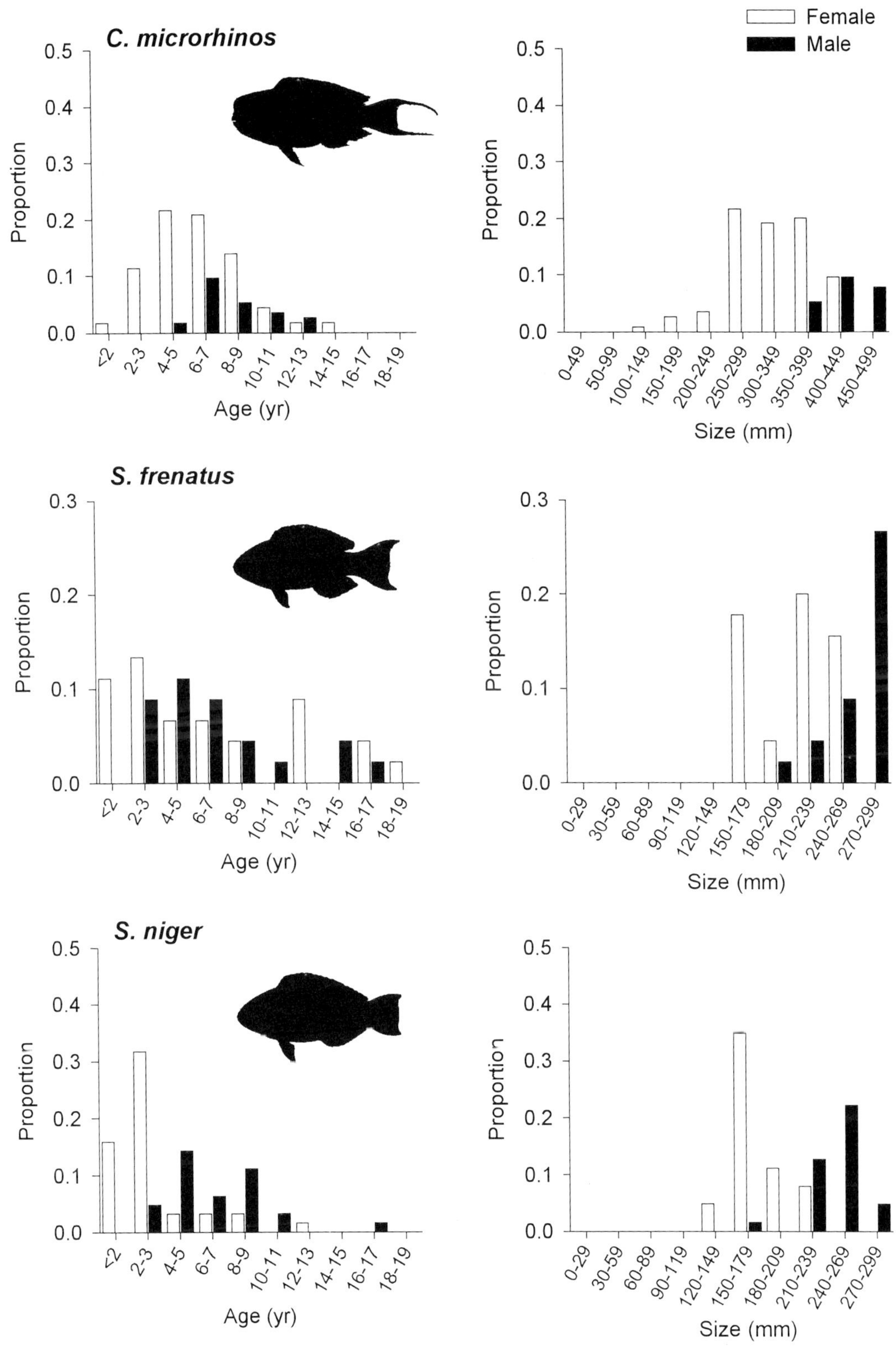

FIGURE 9 Age and size distributions by sex for three species of protogynous scarids: *Chlorurus microrhinos*, *Scarus frenatus*, and *Scarus niger* sampled from the West Pacific (GBR).

fishes, especially the orange roughy *Hoplostethus atlanticus* (Tracey and Horn, 1999), can be traced in part to the lack of appreciation that the fishery was targeting long-lived and slow-growing fishes. The basic issue for fisheries and life history studies is not one of determining whether the particular otolith section reveals either 6 or 10 presumed annual check marks, but whether the fish lives 10 years or 40 years. Underestimation of ages may lead to unrealistically high estimates of growth and natural mortality rates, an issue that has been especially problematical in, but not confined to, deep-water fishes.

There is increasing recognition of the need to conserve reef fish stocks and to gather the information necessary to identify levels of vulnerability based on life history features. Roberts and Hawkins (1999) identified extended life-spans, slow growth rates, and low natural mortality rates as characteristics that will render marine species vulnerable to extinction through human activities, including overexploitation. Roberts (1997a) also identified the difficulties facing managers of multispecies fisheries targeting taxa with a range of life history features. An additional dimension to this problem is seen in the taxa used by Roberts (1997a) to identify contrasting life history tactics. Acanthurids were provided as examples of *r* selected species and serranids were provided as examples of *K* selected species, i.e., rapidly growing, short-lived vs. slow-growing, longer lived, respectively. A number of workers (see Musick, 1999a, for a summary) have pointed out that *r* selected species may have higher maximum sustainable yields compared to *K* selected species. Our data indicate that, in contrast to the previous conclusions, in many reef environments acanthurids rather than serranids may have life history features more characteristic of a *K* selection regime. The estimates of growth rates and longevity from size-at-age data that provide the core of this chapter are an attempt to provide a more inclusive framework for the analysis of reef fish population phenomena. They are designed as hypotheses to be challenged or confirmed by additional studies on the age–structure of tropical fishes and as alternatives to length-based approaches. Given the variation in relationships between size and age discussed here, the most appropriate outcome would be an increased emphasis on age estimations and the validation of presumed annual check marks in the otoliths of coral reef fishes.

A recurring theme in studies of coral reef fishes concerns the extent to which we may generalize about ecological and life history attributes. The analysis of life-spans in acanthurid and scarid fishes shows that ecologically similar species can differ substantially in demographic characteristics. The literature suggests that this type of partitioning among different lineages of reef fishes is common. For example, some pomacentrids with a total length of 60 mm may have maximum ages of approximately 16–18 years (Fowler and Doherty, 1992; Doherty and Fowler, 1994a), similar to ages achieved by the serranids *Plectropomus maculatus* and *Plectropomus leopardus* for fork lengths of 680 and 500 mm, respectively (Ferreira and Russ, 1992, 1994), and *Epinephelus guttatus* for a 458-mm fork length) (Sadovy *et al.*, 1992). Very large serranids such as *Epinephelus itajara* (2065-mm fork length) may be only 33 years old (Bullock *et al.*, 1992), equivalent to many acanthurids that are 160–210 mm in fork length, or a tenth of the length and a thousandth of the serranid weight. Age estimates for lutjanids demonstrate that relatively small species may be substantially longer lived compared to large serranids (Newman *et al.*, 1996). Such decoupling of size and age and differences in vital rates between and within lineages of reef fishes mean that it will be difficult to generalize about responses to environmental change or human impacts without a considerable data base on life history characteristics.

The growth patterns generated by the size-at-age data revealed different patterns of growth among our two chosen taxa. Acanthurids displayed square growth curves, partitioned into a short ascending arm where rapid growth to achieve asymptotic size and sexual maturity occurs followed by an extended life-span where reproductive output occurs with minimal somatic growth. Similar patterns have been recorded for small species of lutjanids (Newman *et al.*, 1996). These growth patterns show certain similarities to temperate sciaenids (Jones and Wells, 1998) and deep-water oreosomatids (Stewart *et al.*, 1995) and hoplostethids (Smith *et al.*, 1995), in which growth to the start of the horizontal arm of the growth curve occurs in approximately the first 18–25% of the life-span. The scarids generate more familiar growth curves in which size increases gradually throughout life, although there are exceptions (Choat *et al.*, 1996). The growth patterns observed in large scarids such as *C. microrhinos* are similar to those of a number of serranids (Ferreira and Russ, 1994; Sadovy *et al.*, 1992) and gadid, clupeoid, and pleuronectiform fishes (Beverton, 1992).

Analytical treatments of fish growth and life history features (Beverton, 1992; Pauly, 1998) rely heavily on the von Bertalanffy parameters, K and L_∞. Making useful comparisons based only on these parameters is difficult. First, fishes with very different life history features may have similar values of K and L_∞ (Fig. 6). These parameters do not capture differences in longevity that can generate different life history features, notably the reproductive lifetime. Second, von Bertalanffy growth curves are frequently fitted to truncated size-at-age data that do not adequately span the age range. A number of studies have recognized the consequences of exclusion of particular age classes from estimates of growth parameters and the difficulties this causes for the comparison of growth among species or populations (Mulligan and Leaman, 1992). The estimated values of K and t_0 are sensitive to the age composition of the sample, with exclusion of the younger age classes in particular resulting in the underestimation of K (Ferreira and Russ, 1994). This effect will be particularly problematical in species with square growth curves (Craig, 1999). Some standardization of the structure of data sets used to estimate K and L_∞ is required. In comparisons within a taxon we constrained the growth curve to the estimated size at settlement.

Sensitivity of K to the age composition of the sample causes a number of difficulties in the estimation of growth parameters for coral reef fish populations. First, coral reef fishes are subject to a wide range of fisheries practices and the age of "recruitment" to a fishery may vary accordingly. Very small individuals are selectively targeted in some species, e.g., newly settled siganids (Bryan and Madraisau, 1977). Second, there is relatively little information on age and size at reproductive maturity, and third, there is evidence of very different patterns of growth among species with similar size ranges. Comparisons of K for different regional populations will be confounded if there is no control for the effects of variation in the distribution of ages used in different samples.

Beverton (1992) has provided a number of analytical protocols for comparing life history parameters in fishes, including the concept of the growth–maturity–longevity (GML) plot. This captures important features of life-span and reproductive output in different groups of fishes. Plotting of the dimensionless ratios K/T_{max} and L_m/L_∞, where L_m is the length at which 50% of a year-class reach maturity, provides a demonstration of the distinctive differences between the demographic patterns of long-lived cold-water fishes of the genus *Sebastes* (average $T_{max} = 54$) when compared to the shorter lived Gadiformes (average $T_{max} = 14$) and Pleuronectiformes (average $T_{max} = 23$). Such an approach may be informative for reef fishes, provided estimates of K are standardized and more extensive data sets for T_{max} are available.

Growth parameters and life-spans within a species vary over environmental gradients, although the pattern of growth is preserved. For the acanthurids examined both size and longevity varied substantially over the average annual temperature gradients examined. There are likely to be a number of explanations for these patterns. For example, habitats in lower latitudes will support a greater diversity and density of fishes (Meekan and Choat, 1997). This in turn may result in decreased asymptotic size through competitively driven reductions in growth rate or increased mortality due to predation. Decreasing temperature may reduce growth rate (Pauly, 1998), which in turn seems to be associated with increased life-spans. A more potent source of variation in life history features may relate to latitudinal variation in reproductive output. Robertson (1991a) examined the relationship between seasonal cycles and reproductive outputs and concluded that relatively benign tropical conditions would permit year-round spawning. Greater amplitude of seasonal cycling would have the capacity to reduce spawning duration and reproductive outputs. Reduced outputs and more variable recruitment expected in tropical species near the latitudinal limits of their range may select for extended life-spans.

A potential source of confounding in the comparisons of longevity is the unmeasured effect of fishing mortality. Fish populations at localities within the Caribbean with intense artisanal fisheries are more likely to be impacted by fishing compared to populations in the peripheral or isolated localities. It could be argued that the short life-spans of fishes from the San Blas Archipelago and Lee Stocking Island might reflect fishing pressure. However, although the San Blas Kuna peoples do harvest the marine resources of the sampled reefs, small herbivorous reef fishes such as *Acanthurus bahianus* are not a targeted catch and the method (traps) most likely to yield such species is not widely used (D. R. Robertson, personal observation). At Lee Stocking Island there is no intensive trap fishery (J. H. Choat and D. R. Robertson, personal observations) for small reef species as occurs in Jamaica and St. Thomas. This issue should be resolved by comparing T_{max} and age structure in conspecific populations from fished and unfished sites with similar environmental profiles, preferably at the same location

An alternative explanation to variation in fishing pressure is a trade-off between life-span and reproduction. Individuals at the periphery of their species range may reproduce and recruit less frequently and maintain longer lived populations compared to those at the center of the range. The differences between West Pacific and Caribbean acanthurids and scarids in terms of maximum age achieved are striking. However, we acknowledge that these comparisons may need reinterpretation once the largest Caribbean scarids, *Scarus gaucamia, S. coelestinus,* and *S. coeruleus,* are included. Moreover, a more expanded data base is required to ensure that the comparisons between Indo-Pacific and Caribbean scarids are not confounded by phylogenetic factors. It is also important that studies comparing growth rates, life-spans, and recruitment rates (as a proxy for reproductive outputs) be carried out under conditions that control for fishing mortality and locality effects in each ocean system. Marine reserves offer the best opportunity for such tests that could be accomplished without significant damage to populations of abundant species in either the short or the long term.

The analysis of growth and age structure in protogynous species provides further evidence of distinctive growth patterns in different groups of reef fishes. For scarids there is strong evidence of sex-specific growth differentials resulting in males being larger than females at a given size (van Rooij *et al.,* 1995; Choat *et al.,* 1996). One result is the presence of large, young males in the population. Another is that some females may live as long as males. Sex-specific growth patterns in protogynous species have been attributed to a growth spurt in males at the time of or following sexual transition (Charnov, 1982). Evidence from scarids supports this interpretation (Choat *et al.,* 1996). Some serranid fishes—for example, *Epinephelus guttatus* (Sadovy *et al.,* 1992)—do not, however, show evidence of either long-lived females relative to males or evidence of sex-specific growth patterns. Not only will protogynous species and gonochoristic species require different approaches to harvesting (Sadovy, 1996), but also different lineages of protogynous fishes may have distinct demographies that also require different approaches to management.

Growth spurts in secondary male fishes are arguably associated with the change in energy investment from eggs to sperm. The characteristically small testes of large, fast-growing secondary male scarids suggest a trade-off between somatic and reproductive growth. Francis and Barlow (1993) provide an alternative interpretation for differential growth rates in males and females. Faster male growth rates in Midas cichlids were shown experimentally to be a result of growth trajectories established early in life by a combination of developmental conditions and social interactions. For protogynous species, faster growing members of the population may differentiate as secondary males. Hence faster growth rates may be a predisposing condition for sex reversal, rather than a consequence. Analysis of growth histories of individual fishes through back-calculation of otolith increments would help establish the pattern of growth in secondary males and resolve the issue of resultant growth spurts vs. preestablished growth trajectories. Comparison between protogynous serranid and labroid fishes with a focus on growth chronologies and sex-specific growth patterns would be particularly useful.

Regardless of the detail in interpreting demographic events in protogynous fishes it is clear that age-estimation studies are critical for determining the most appropriate management options. As Sadovy (1996) has indicated, mechanisms governing sex change in reef fishes are sensitive to a number of factors—physiological, behavioral and ecological. Estimates of fecundity for different size and age classes of the female population are critical for both management and life history studies. Estimations of fecundity are complicated by wide variation in mating systems, spawning behaviors, and population and individual spawning periodicities in reef fishes (Robertson, 1991a; Robertson *et al.,* 1999).

Throughout this chapter we have discussed growth phenomena in some species in which validation of annual growth increments has not been achieved. To reiterate, we are in full agreement with the importance of validation, but argue that given the diversity of reef fishes and the manifest and rapidly accelerating influence of fishing pressure, a preliminary analysis of growth patterns is important in determining future research priorities. The possibility that many reef species may be long-lived and display growth patterns that do not sit comfortably with traditional analytical protocols must be noted. The degree of plasticity in growth parameters and vital statistics within species that have wide distributions is also a key issue. Additional studies and validations are required to confirm or challenge the patterns we have described. Increased continuity of access to coral reefs will enhance the prospects for marginal increment analysis. In many situations and especially in equatorial areas validations based on chemical marking of sagittae should be expanded to incorporate otolith growth models (Cappo *et al.,* 2000) and the possibilities of radiochemical dating of otoliths further investigated.

Acknowledgments

We thank Garry Russ, John Ackerman, Lynda Axe, David Bellwood, Kendall Clements, Jake Kritzer, and Phil Munday for the discussion of ideas in this paper, and John Ackerman, Jake Kritzer, Will Robbins, Ernesto Pena, and Ileana Hernandez for assistance with preparation of otoliths, data analysis, and field and laboratory assistance. Work in the Pacific was carried out on the GBR and supported by the Lizard Island Research Station. We thank the Administrators of Ascension and St.Helena, the U.S. Air force, for logistical support; Carlos Ferreira and the Institudo de Estudos do Mar for logistical support at Cabo Frio; the Kuna Congress for permission to work in San Blas, and James Cook University; and the Australian Research Council and the Smithsonian Tropical Research Institution for funding.

CHAPTER 4

Rarity in Coral Reef Fish Communities

Geoffrey P. Jones
Centre for Coral Reef Biodiversity
School of Marine Biology and Aquaculture
James Cook University
Townsville, Queensland 4811 Australia

Philip L. Munday
Centre for Coral Reef Biodiversity
School of Marine Biology and Aquaculture
James Cook University
Townsville, Queensland 4811 Australia

M. Julian Caley
School of Marine Biology and Aquaculture
James Cook University
Townsville, Queensland 4811 Australia

I. Introduction

One of the greatest challenges in ecology is to explain the inequality of abundance of taxonomic groups, and, in fact, why a majority of the species in any ecosystem are relatively rare (Preston, 1948; Gaston and Lawton, 1990a; Gaston, 1994a; Kunin and Gaston, 1997). Unfortunately, rare species are rarely studied, so relatively little information is available on the processes constraining their distributions or limiting their numbers (Kunin and Gaston, 1993). Although the ecological processes influencing rare species may not be fundamentally different from those affecting common species, this cannot be assumed. Direct comparisons of the ecology and life histories of common and rare species, in a range of different taxa, are only now beginning to reveal the essential causes of rarity (Gaston, 1994a), but almost all the progress in this field has come from studies of terrestrial organisms. Gaston's (1994a) book entitled "Rarity," one of the first attempts to review this subject comprehensively, contained almost no marine examples.

Many factors may explain why the issue of rarity has been avoided by marine ecologists. The taxonomic status of rare species in many marine taxa is uncertain (Ray and Grassle, 1991). It is always difficult to sample rare species and this may be particularly true in aquatic environments. Published accounts of the geographic distributions of rare marine species are not always reliable because rare species may not be detected by the sampling method. Estimates of the abundances of rare species are almost always unreliable, because typical sample unit sizes and replication are invariably appropriate only for common species (Andrew and Mapstone, 1987). However, there is no fundamental reason why studies on rare marine species cannot make a major contribution to our general understanding of the phenomenon of rarity. Indeed, we cannot assume that the processes leading to rarity in terrestrial organisms apply to their marine counterparts. The prevalence of species with dispersive larval stages and consequently an "open" population structure may require new explanations of rarity. Therefore, comprehensive studies of marine taxa are necessary before a general synthesis of this fundamental ecological issue can be attained.

The general scarcity of studies on rarity in marine organisms extends to coral reef fish communities, although this topic has received some recent attention (McAllister *et al.*, 1994; Ormond and Roberts, 1997; Hawkins *et al.*, 2000; Morris *et al.*, 2000). There are many reasons why focusing on coral reef fishes may enable us to reach a greater understanding of this general ecological issue. Reef fishes exhibit the highest diversity of all vertebrate communities on Earth and are therefore likely to contain a greater number of rare species as compared to their terrestrial counterparts. Among marine organisms, the taxonomic status and patterns of distribution and abundance of coral reef fishes are among the best known. Also, there is sufficient published information available to examine the patterns and potential causes of rarity in a number of reef fish families. This review combines a comprehensive analysis of disparate information on patterns and correlates of rarity in reef fish communities with an evaluation of how they fit into the existing theoretical framework.

In spite of their potential to contribute to our understanding of rarity issues, community-level studies of reef fishes have historically been more concerned with species richness and coexistence, than with explaining differences in relative abundance (see Sale, 1977, 1980). In the majority of community-oriented studies of reef fishes, it is often difficult to extract information on the abundance of rarer species, and in many cases rarer species are completely ignored. The extent to which patterns in species diversity reported in these studies are artifacts of differences in sampling effort, which will determine how many rare species are detected, is very often uncertain. Because of these limitations we presently know very little about why or how some communities of reef fishes support more rare species than others.

Population-level studies of reef fishes have also provided little information relevant to the understanding of rarity issues. Numerous studies have been carried out on common species in a few well-studied reef fish families (Jones, 1991). These studies have documented the predictable spatial patterns and substantial temporal variability that characterize reef fish populations (Doherty and Williams, 1988; Sale, 1991b; Williams, 1991). Much of this temporal variability appears to be the result of fluctuations in larval mortality and dispersal (Williams, 1991; Doherty and Fowler, 1994a; Caley *et al.*, 1996). Despite this variation, for particular geographic regions and habitats, common species tend to stay common and rare species tend to stay rare. These observations represent an enigma! In a system in which recruitment variability is the rule, we are lacking a robust explanation of why certain species consistently recruit in greater numbers than others and maintain larger populations. In a system in which the potential for widespread dispersal is extreme, we do not have a satisfactory explanation of why reef fishes do not continue to expand their geographic ranges. Although there is also ample evidence that both resource availability and predation can limit reef fish populations (Forrester, 1990; Jones, 1991; Hixon, 1991; Caley, 1993, 1995a; Hixon and Beets, 1993; Caley and St. John, 1996; Hixon and Carr, 1997; Munday *et al.*, 1997, 2001), the potential roles of these processes in causing rarity remain obscure. Are rare species simply those that have specialized on a few scarce resources or have few refuges from predation? Or do we need to focus on fundamentally different processes in order to explain consistently low abundance?

With the increasing concern over the loss of biodiversity through human-induced extinction, the developing field of conservation biology has provided a new impetus for the study of rare species (Soule and Wilcox, 1980; Simberloff, 1988). Although many species are naturally rare and are not necessarily endangered, it is also probably true that those most threatened with extinction will be among those that are rare. An understanding of patterns in rarity (Which species are rare? Where are they located? What is their population size?) and the processes that cause particular species to persist in a rare state (Why are species rare?) forms the necessary biological basis for the management of species that may be threatened by virtue of their limited distribution or abundance. This review focuses on these questions, because without this information, the long-term environmental consequences of rarity will remain poorly understood.

Perhaps the explanation as to why so few studies have focused on rare marine organisms is their apparent resistance to extinction (Estes *et al.*, 1989; Jones and Kaly, 1995). There have been only a handful of extinctions of such animals in the past 200 years (e.g., the eelgrass limpet, *Lottia alveus*) (Carlton *et al.*, 1999). No recent extinctions of rare coral reef fishes have been recorded, although Hawkins *et al.* (2000) have drawn attention to some suspected cases. However, a number of large, heavily exploited species are considered threatened, particularly those with restricted geographic ranges or restricted spawning aggregations (Jones and Kaly, 1995; Morris *et al.*, 2000). The Nassau grouper (*Epinephelus striatus*), for example, has been overfished to the extent that one third of the documented traditional spawning aggregations can no longer be located (Sadovy, 1993). There is reason to believe that the threat of extinction for coral reef fishes is increasing as the scale and intensity of human impact expand (Hawkins *et al.*, 2000). The increasing stresses being placed on coral reefs highlight the overdue concern for rare species in this environment (Roberts, 1993; Roberts and Hawkins, 1999). Marine protected areas have been proposed as a means of maintaining biodiversity, but these areas will serve only this purpose if the size of the protected area is scaled to encompass sustainable breeding populations of rare species (Jones *et al.*, 1993). Without knowledge of the patterns of rarity in these species, such scaling cannot proceed in an informed manner.

A. What Is a Rare Species?

Our discussion above assumes an intuitive understanding of what constitutes a rare species. It is now widely accepted that there are two fundamental elements to rarity: low abundance and restricted geographic range (Harper, 1981; Gaston, 1994a, 1998). However, rarity has many other connotations that are prevalent in the literature. In attempting to encompass these broader meanings, Rabinowitz (1981) included a third element of rarity: species found in a few specialized habitats. A combination of some or all of these characteristics leads to seven potential forms of rarity (and one form of commonness). Some attempt has been made to classify different taxa according to these types of rarity. For example, in sand pine scrubs in Florida, plants are often rare in terms of geographic distribution, and amphibians and reptiles are rare in terms of habitat specificity (McCoy and Mushinsky, 1992). Under these three definitions, patterns of rarity in coral reef fishes remain uncertain. Although there are many examples of reef fish species living in highly specialized habitats (e.g., anemone fishes, *Amphiprion*; coral gobies, *Gobiodon*), many of these would not be considered rare. One can find examples of coral reef fishes with extremely small geographic ranges [e.g., the Banggai Island cardinal fish, *Pterapogon kauderni*, restricted to a few small islands in the Indonesian archipelago; see Allen (2000)], but much larger ranges are certainly the norm (McAllister *et al.*, 1994). It may well be that coral reef fish are most often rare in terms of low local abundance, but patterns of abundance throughout a species' geographic range have generally not been described. The possibility that all species are common somewhere in their range is an attractive notion, but establishing whether this is true could be a daunting task. To date there has been no comprehensive analysis of patterns in range size, abundance, and habitat specificity for any groups of coral reef fishes.

In this review, we use the term *rarity* in the same sense as Gaston (1994a), to distinguish species with low local abundance and/or small geographic range. Classifying species as rare under Rabinowitz' (1981) definition potentially confounds these two widely accepted forms of rarity with another potential cause of rarity: habitat specialization. Habitat specialization need not necessarily equate with low numerical abundance. For example, a habitat-specialized species may have low local abundance but be sparsely distributed over a large geographic range. Certainly, specialists may be locally scarce if their preferred habitat is uncommon and they may also be restricted in their regional distribution (Glazier, 1980; Hodgson, 1986). However, Hanski *et al.* (1993) found no evidence that habitat specialists are generally less abundant or widespread than are habitat-generalist species. Both habitat specialization and the availability of resources provided by the habitat may play a causal role in determining patterns of distribution and abundance (Brown, 1984; Gaston *et al.*, 1997). Therefore, throughout this review we restrict ourselves to considering rarity as low abundance and/or small geographic range, and consider resource specialization along with other potential causes of rarity.

It has been argued that a positive relationship between distribution and abundance is a fundamental law in ecology (Lawton, 1999). Species conforming to such a pattern are subject to a "double jeopardy," because either their restricted distribution or low abundance may increase their extinction risks (Gaston, 1998). Among terrestrial animals it is common to find a positive relationship between geographic range and population density, suggesting that these parameters could be linked by a common cause (Hanski, 1982; Brown, 1984; Gaston and Lawton, 1988a; Gaston, 1996; Johnson, 1998). Therefore, although low abundance and restricted geographic range are two forms of rarity, they cannot *a priori* be assumed to be independent. To date, the relationship between range size and abundance for coral reef fishes remains unexplored. This will be examined for the first time in this review.

Although we assess rarity on two axes, abundance and geographic range, what constitutes a rare species along these axes? Past analyses have viewed rarity both as a discrete and as a continuous variable. Clearly, there is no universal or distinct boundary that separates rare from common species, either in terms of local abundance or geographic range. However, it may be useful in some situations to delineate rare from common. For example, it may be useful for pragmatic reasons to employ a rarity cutoff when dealing with conservation issues or when unique qualities of rare species are to be emphasized (Gaston, 1994a). This pragmatism motivated Gaston's (1994a) suggestion that the first quartile of abundance or geographic range be adopted across rarity studies to provide uniformity of definition. Such a uniform definition, however, constrains all communities, irrespective of their total diversity, or habitat and taxonomic affiliations, to contain a similar proportion of rare species (Holt, 1997). Our goal here is to explore relationships between abundance and geographic range of coral reef fish species with a variety of other biological parameters. In order to explore such relationships the continuum from common to rare species is of particular interest. Therefore, in this review we avoid routinely applying any conventional cutoffs to define rarity.

B. What Causes Rarity?

A large number of factors have been implicated as causes of rarity in terrestrial environments, both in terms of geographic range and abundance, but no one factor predominates among studies (Gaston, 1994a). Because the evidence for causes of rarity comes mainly from interspecific comparisons, our understanding in this area is based almost entirely on correlational information. Yet, many of the same trends emerge within and among a range of unrelated taxa, suggesting in some cases a common underlying cause. Body size in animals, for example, is frequently negatively correlated with abundance (e.g., Damuth, 1981; Peters, 1983; Gaston and Lawton, 1988b; Currie, 1993; Gaston and Blackburn, 1996). Even when the relationship is not clear-cut, large species tend to be rare, but small species can exhibit a range of densities (e.g., Cotgreave, 1993; Blackburn *et al.*, 1993; Blackburn and Gaston, 1997). Range size appears to be an increasing function of body size, with large species nearly always having large ranges (e.g., Gaston and Lawton, 1988b; Brown and Maurer, 1989; Arita *et al.*, 1990, Brown and Nicoletto, 1991). Again, small species can exhibit great variability in range size and it may be that body size just sets the possible minimum range size (Gaston, 1994a). Given that local abundance is positively correlated with range size, some of these patterns seem contradictory and worthy of more detailed analysis.

A number of other life history and ecological characteristics also correlate with rarity. Species with asexual reproduction (e.g., Hedderson, 1992), lower reproductive effort (e.g., Glazier, 1980; Paine, 1990), and poor dispersal ability (Juliano, 1983; Holldobler and Wilson, 1990; McDonald *et al.*, 1995) are often rare. In addition, species whose ranges are centered at higher latitudes often exhibit greater latitudinal ranges, a pattern now referred to as Rapoport's rule (Stevens, 1989). This is thought to be caused by increasing environmental tolerance of species at these latitudes (Stevens, 1989). In some terrestrial animals, geographic range is correlated with altitudinal distribution, further supporting the notion that tolerance to physical conditions may be an important determinant of geographic range (Glazier, 1980). Other ecological factors that correlate with rarity include specialized habitat requirements, resource availability (Brown *et al.*, 1995), and a poor ability to establish in new areas and marginal habitats (Rabinowitz, 1978; Glazier, 1980). The sheer range of correlated factors that may or may not be direct causes of rarity suggests that distinguishing among them will take considerable time and ingenuity.

The factors correlating with rarity are only just beginning to be examined for marine organisms (e.g., Reaka, 1980; Marquet *et al.*, 1990; Munday and Jones, 1998; Hawkins *et al.*, 2000). At this stage we can only speculate on their relative importance in marine environments. Although we are unable at this point to predict how rarity may be related in a general sense to any of the variables previously identified for terrestrial organisms, many groups of marine animals and their terrestrial counterparts exhibit variation in the same parameters. For example, differences in body size of several orders of magnitude are exhibited by some marine taxa, but body size–abundance–range relationships in these groups have not been widely addressed. A few studies confirm patterns for terrestrial species. For example, Marquet *et al.* (1990) found that population densities in rocky intertidal communities were inversely correlated with body size, and Reaka (1980) found range size to be an increasing function of body size in coral-dwelling mantis shrimps. Munday and Jones (1998) showed that coral reef fish abundance was only loosely related to body size, with large species seldom being very abundant. In terms of environmental constraints, species associated with small seas separated by land barriers and species associated with islands that are isolated by great distances in oceanic areas may exhibit a tendency to have small geographic ranges, but this has not been examined. Latitudinal patterns in the distribution of marine fish do not appear to conform to Rapoport's rule (Rohde *et al.*, 1993), although coral reef fishes have not been analyzed separately. Depth ranges of some marine organisms appear to increase for deeper water species (Stevens, 1996), which may predispose them toward larger geographic ranges. Marine species with high fecundity and dispersing larvae often occupy greater geographic ranges compared to species with direct development (Jablonski, 1986). Marine organisms vary in the length of their larval life, but this has not been correlated with patterns of numerical commonness and rarity [but see Thresher (1991) for geographic range]. Although it has been suggested that ecological processes such as habitat or dietary selection and predation pressure (Lubchenco, 1978; Estes *et al.*, 1982) may cause rarity, these have not been fully investigated. Clearly, for marine organisms there is a long overdue need to examine the strength of associations between rarity and this range of potentially important characteristics, both in terms of abundance and geographic range.

C. Coral Reef Fishes: Scope of Review

Here we use published descriptions of species' geographic ranges and abundances to examine patterns of commonness and rarity among coral reef fish species. We then use these patterns to test predictions generated by hypotheses of the causes of rarity. In so doing we accomplish three things: (1) we exclude for reef fishes some hypotheses regarding the causes of rarity that have been supported in the past by studies of terrestrial taxa, (2) we provide evidence that some patterns of rarity are consistent between marine and terrestrial taxa, and (3) we identify areas where, in our opinions, further research is warranted.

We have confined our analyses to five families of reef fishes for which we have reliable and accessible information on the geographic extent of occurrence (*sensu* Gaston, 1994b) for most species. These are the Chaetodontidae, Lutjanidae, Pomacentridae, Pomacanthidae, and Serranidae. Reliable estimates of abundance of species from a series of geographic locations are much less readily available. Therefore, we have confined our comparisons of abundance to the family Chaetodontidae, the only family of reef fishes for which sufficient abundance data are available for a large number of species at a range of geographic locations. When possible, we supplement these data on abundance with information on species from other families. We also confine most of our abundance analyses to comparisons of rank abundances in order to avoid many of the difficulties associated with comparing estimates of density over different sampling intensities and spatial, temporal, and taxonomic scales (McArdle *et al.*, 1990).

Understanding the reasons for observed patterns of rarity will ultimately require the applications of phylogenetically controlled comparisons (Harvey *et al.*, 1995). Because there are currently few detailed reef fish phylogenies, we have restricted our discussion here in most cases to exploration of rarity patterns, not their causes. In the absence of phylogenies, however, it can still be informative to explore patterns of rarity. For example, if extinction risk is related to abundance, predictors of abundance such as body size might be useful for predicting extinction risk irrespective of how body size–abundance relationships evolved. Furthermore, comparisons of reef fish rarity either among families or between reef fishes and other taxa provide sufficient phylogenetic control to begin to postulate possible reasons. More sophisticated, phylogenetically controlled tests will need to await the production of appropriate phylogenies.

The specific questions we have addressed are as follows: (1) What are the global patterns in the geographic ranges of coral reef fishes? (2) How do geographic ranges relate to location, latitude, body size, dispersal ability, and/or habitat specialization? (3) Is there a positive relationship between geographic range and abundance in coral reef fishes? (4) Does local abundance relate to body size or habitat specialization? (5) Are there consistent patterns in the relative abundance of common and rare species at different locations, or at the same location at different times?

II. Geographic Range

A. General Patterns

In general, marine organisms have larger geographic ranges, as compared to terrestrial organisms (Rapoport, 1994). This pattern is consistent with the world's oceans being larger than most landmasses, with few geographical barriers, and with many marine species having a dispersive larval stage. This general pattern for marine species holds also for species of coral reef fishes (McAllister *et al.*, 1994). The geographic ranges of many coral reef fishes are larger than the largest landmasses, although the actual areas of occupancy within ranges are smaller, due to the limited area of reef development (Hawkins *et al.*, 2000). In spite of the paucity of land barriers, some constraints on the geographic range size of reef fishes are still apparent. For example, in our analysis of five taxa, Indo-Pacific and Atlantic reef fishes exhibit similar latitudinal ranges, but the maximum longitudinal range of Indo-Pacific fishes can be twice that of Atlantic fishes (Fig. 1; maximum longitudinal range: Indo-Pacific fishes, 250°; Atlantic fishes, 127°). Consequently, the typical range shape of Indo-Pacific reef fishes is elongate on an east–west axis whereas Atlantic reef fishes tend to exhibit more symmetrically shaped ranges (Fig. 1). These patterns reflect the spatial arrangements of these ocean basins and the distribution of reef habitats.

Although coral reef fishes tend on average to have larger geographic ranges than most terrestrial organisms, the frequency distributions of geographic range sizes are very similar in shape (e.g., Brown *et al.*, 1996; Gaston, 1998). Within the five families of coral reef fishes examined, most species have relatively small geographic ranges, although a few species have very large ones (Fig. 2) (see also Ormond and Roberts, 1997;

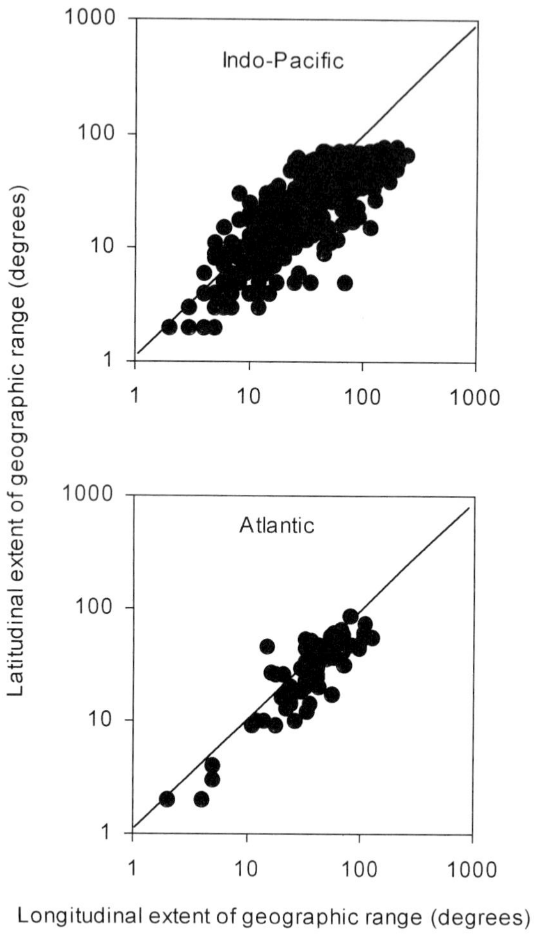

FIGURE 1 Latitudinal versus longitudinal extent of geographic ranges for Indo-Pacific and Atlantic coral reef fishes. Range limits estimated from Allen (1985) (Lutjanidae), Heemstra and Randall (1993) (Epinephelinae), Allen (1994) (Pomacentridae), and Allen *et al.* (1998) (Chaetodontidae and Pomacanthidae); $N = 670$ species from the Indo-Pacific ocean and 113 species from the Atlantic Ocean.

Hawkins *et al.*, 2000). The causes of these patterns are presently unknown, but must result from some combination of the effects of speciation, extinction, and rate of population expansion (Gaston, 1998). If geographic range size was not a function of speciation and extinction, approximately log-normal distributions of species range sizes would be expected (Gaston, 1998). Most frequency distributions of species' geographic range sizes (Gaston, 1998; Lawton, 1999), however, are left-skewed on a log geographic-range axis, including those of coral reef fishes presented here. This suggests that patterns of speciation and/or extinction might be biased in relation to geographic range size. Although this interpretation of the importance of speciation and extinction processes is consistent with observations of high rates of extinction in both marine and terrestrial taxa (Jablonski, 1987; Buzas and Culver, 1991; Caley, 1997; Miller, 1997), it has not been addressed specifically for coral reef fishes.

There are no particular geographic regions that exhibit a concentration of coral reef fish species that are relatively rare in terms of their geographic range size (Fig. 3). That is, species that have a geographic range less than 5% of the maximum range size recorded for a family can be found throughout tropical and subtropical seas (see also Hawkins *et al.*, 2000). Although the greatest diversity of coral reef fishes is found in the Philippines–Indonesian region, this is not associated with a greater number of species with a small range size at this location. This suggests that greater diversity is not explained by the presence of a greater number of rare species in terms of range size, a pattern consistent with a diverse array of mostly terrestrial taxa (Caley, 1997). Greater proportions of the species that are present on isolated islands in the three oceans appear to have restricted geographic ranges. Subtropical and temperate species of the five reef fish families discussed in this chapter also appear to have smaller geographic ranges (Fig. 3). Endemicity is particularly high at subtropical islands at the periphery of Oceania (Randall, 1998). Also, many species found in enclosed bodies of water, such as the Red Sea, have relatively small geographic ranges. Such species may have a limited ability to expand their geographic ranges. Of the five families examined here, the Pomacentridae appears to contain a greater proportion of species with restricted geographic ranges.

B. Latitude

The geographical ranges of organisms living at lower latitudes often have smaller latitudinal extent compared to ranges of organisms living at higher latitudes (Stevens, 1989), although this pattern is not universal (Gaston, 1999). The underlying mechanism proposed to explain this pattern (Rapoport's latitudinal rule) is the increased tolerance to seasonal variation in environmental conditions by species occupying higher latitudes. Species in the tropics are thought to have narrower tolerance compared to those at higher latitudes because seasonal changes are less extreme (Stevens, 1989, 1996). Consequently, species in the tropics, as compared to those at higher latitudes, might exploit a narrower band of latitudes. If Rapoport's rule holds for coral reef fishes, then species living near the equator should have smaller latitudinal ranges than do species living in more southern or northern latitudes. Rohde *et al.* (1993) and our extension of their analyses (Fig. 4)

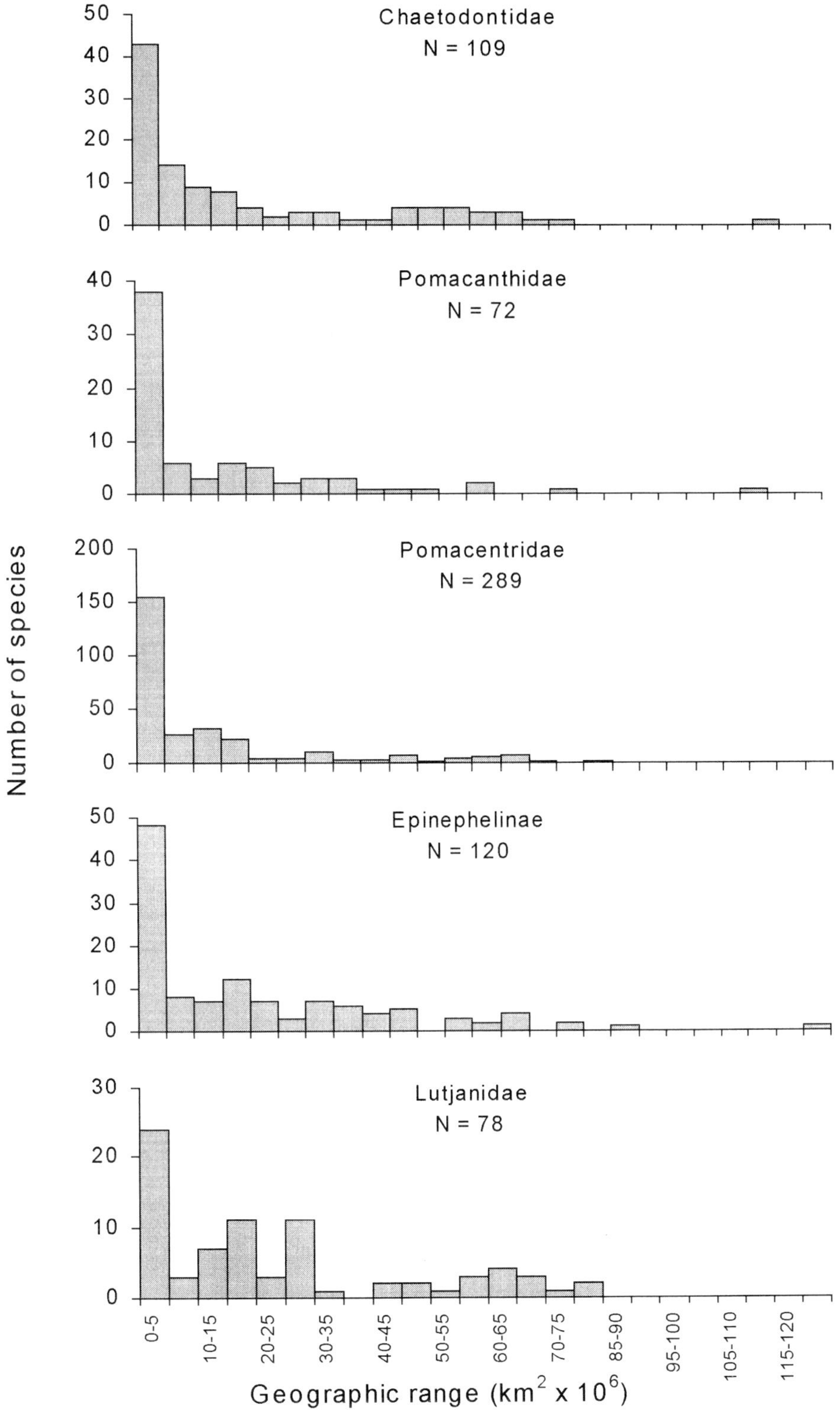

FIGURE 2 Geographic range of occurrence (km^2) for species of coral reef fishes from the Indo-Pacific ocean (N = number of species). Geographic range sizes were estimated from equal-area maps [see Gaston (1994b), for discussion on measuring geographic range size] following the range limits given in Allen (1985, 1994), Heemstra and Randall (1993), and Allen *et al.* (1998).

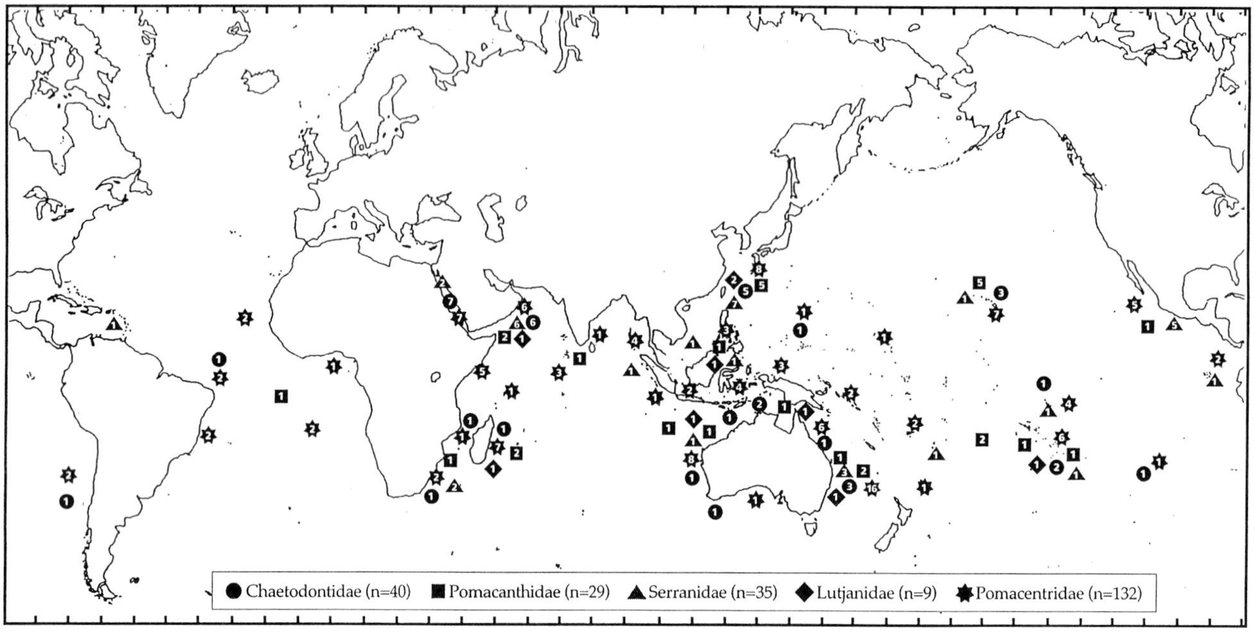

FIGURE 3 Map showing the location of the species with the smallest geographical ranges in the families Chaetodontidae, Lutjanidae, Pomacanthidae, Pomacentridae, and Serranidae (subfamily Epinephelinae). Within each family, species with small geographic ranges were defined as those species with less than 5% of the average range size of the 5% of species with the largest ranges. The number in each symbol is the number of species with small ranges, and the location is the approximate range midpoint for these species. Data from same sources as Fig. 1.

strongly suggest that Rapoport's rule does not hold for tropical reef fishes. In fact, from these analyses the opposite pattern emerges. Species with the smallest geographic ranges are most likely to be those from higher latitudes. This is true for a range of families, including obligate reef dwellers such as the Chaetodontidae and predatory fishes such as the Epinephelinae (Fig. 4). Stevens (1996) has argued that latitudinal gradients in range size consistent with Rapoport's rule might not be found in marine fishes unless a clear distinction is made between shallow- and deep-water fishes. As a consequence of their habitat association, coral reef fishes used in these analyses are largely restricted to shallow waters and, therefore, should provide a reliable test of latitude–range size relationships.

Why reef fishes should exhibit a latitude–range size relationship opposite to Rapoport's rule is unclear. However, both physiological tolerances and the potential for range extension may be important determinants of this pattern. Coral reef fishes living near the equator have the potential to expand their ranges either north or south without immediately encountering environments outside their physiological tolerances. In contrast, the geographic ranges of coral reef fishes in high latitudes appear to be constrained at their temperate limits by their physiological tolerance to cold water and the latitudinal extent of coral reefs. Larvae of some coral reef fishes are frequently transported to higher latitudes, where reefs are absent, but are unable to survive winter sea temperatures (McBride and Able, 1998). The asymptote on the plot of range shapes of Indo-Pacific fishes (Fig. 1) is consistent with the notion that habitat distribution and/or environmental conditions limit the geographic range of coral reef fishes. Indo-Pacific fishes are apparently constrained in the extent of their latitudinal ranges more so than the extent of their longitudinal ranges. Furthermore, the maximum latitudinal range of both Indo-Pacific and Atlantic fishes is approximately 70° (Fig. 1), indicating that a uniform set of factors might control the latitudinal range limits of coral reef fishes. Why the more temperate species cannot expand their ranges at their tropical limits is uncertain.

C. Body Size

Geographic range size of animals is often positively associated with body size (Brown *et al.*, 1996). These range size–body size relationships are usually triangular, with small-bodied species exhibiting the largest variance in geographic range sizes and the minimum

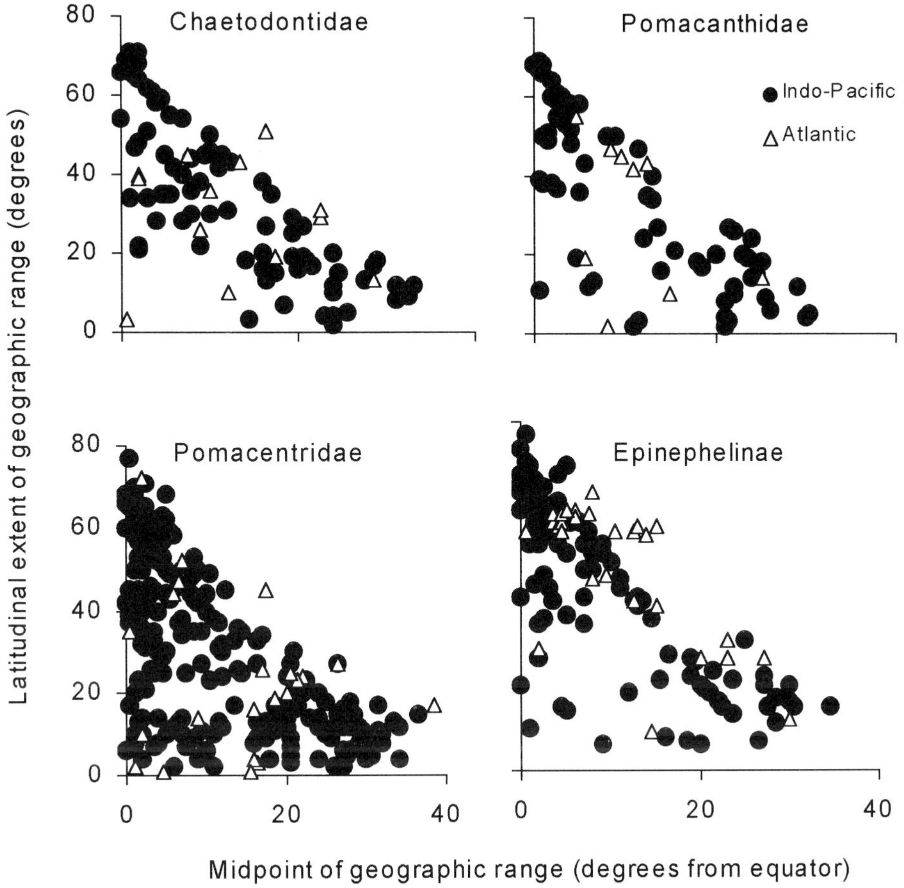

FIGURE 4 Latitudinal extent of geographic range versus midpoint of geographic range for Indo-Pacific and Atlantic coral reef fish. Data sources as per Fig. 1. Rohde *et al.* (1993) have conducted a similar analysis for the family Lutjanidae and other groups of marine fishes not shown here. Their analyses identified similar triangular relationships between midpoints of geographic ranges and their latitudinal extent.

range size increasing with body size (Gaston, 1994a). Consequently, small species are more likely than large species to have small geographic ranges. Previously published range size–body size relationships have used body mass as an estimate of body size. Body mass estimates are not readily available for a wide range of coral reef fish species, so we have used length in our analyses. Because body mass should scale as (body length)^3, estimating body size in this way should not affect the essence of the relationship between range size and body size, if such a relationship exists.

The range size–body size relationship for coral reef fishes (Fig. 5) is broadly consistent with range size–body size relationships reported for other taxa [e.g., neotropical forest mammals (Gaston, 1994a); North American land birds and land mammals (Brown, 1995)]. Small-bodied species exhibit geographic range sizes from the smallest to the largest, but large-bodied species do not exhibit small ranges. This relationship for coral reef fishes, however, differs in one important aspect from previously reported relationships for terrestrial taxa. The range sizes of the largest bodied species of coral reef fishes vary over approximately 2.5 orders of magnitude, whereas range sizes for the largest land bird and land mammal species only vary over approximately 1 order of magnitude (Gaston, 1994a; Brown, 1995). This greater variation of range size in coral reef fishes may be related to their greater dispersal capabilities. If so, range size–body size relationships for other marine taxa should also show greater variation of range sizes for larger bodied species.

It has been suggested that for coral reef fishes, small species might tend to have small geographic ranges because they are mostly demersal spawners, whereas larger species are mostly pelagic spawners (Barlow, 1981; Thresher, 1991; Munday and Jones, 1998). The

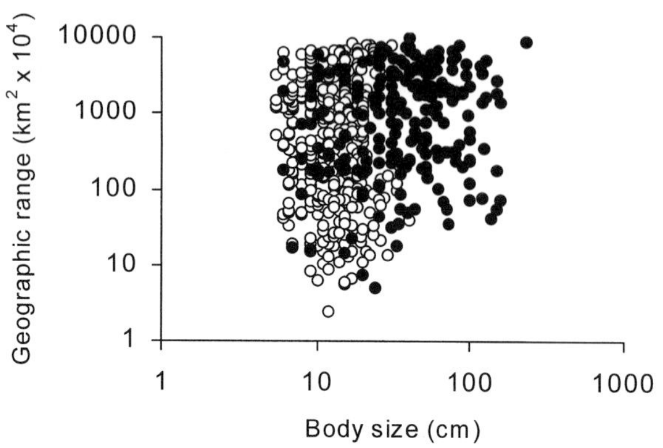

FIGURE 5 Geographic range of occurrence (km^2) versus body size (total length) for species of coral reef fishes from the Indo-Pacific ocean. Open circles are demersal spawners (Pomacentridae, 287 species), filled circles are pelagic spawners (Chaetodontidae, 108 species; Epinephelinae, 117 species; Lutjanidae, 77 species; Pomacanthidae, 73 species). Data sources and techniques as per Fig. 2.

eggs of demersal spawners are retained at the natal reef until hatching, whereas the eggs of pelagic spawners are immediately subject to hydrodynamic forces that could aid long-distance dispersal. Consequently, demersal spawners might tend to have lower rates of widespread dispersal. However, Fig. 6 shows no clear distinction between the range sizes of the Pomacentridae (demersal spawners) and all other species in our analysis (pelagic spawners). Therefore, based on these data, range size does not appear to be limited by spawning mode.

The relationship between range size and body size observed across all reef fish taxa does not hold within families of reef fishes (Fig. 6). At this taxonomic scale, both small- and large-bodied species are equally likely to have small geographic ranges. Why range size–body size relationships do not hold within families of coral reef fishes is unclear, but evidence suggests that there is a phylogenetic component to these relationships that may be related to similar body size and dispersal characteristics of related species. Using taxonomically matched data sets, Hawkins *et al.* (2000) found that species with extremely restricted ranges ($<50{,}000\ km^2$) were smaller in size than those with very large ranges. Clearly, although body size explains some of the variation in geographic range, it may not be the most important factor.

D. Dispersal Ability

The geographic range of a species should be related, at least in part, to its dispersal capabilities (e.g., McDonald *et al.*, 1995). Nearly all coral reef fishes have a planktonic larval phase that is likely to aid in their spread among patches of reef separated by deep water barriers. All else being equal, the duration of its planktonic phase should contribute to a species' dispersal capabilities. To date, however, support for larval duration as a determinant of geographic range is equivocal. No relationship between pelagic larval duration and geographical range has been detected for angelfishes (Thresher and Brothers, 1985), damselfishes (Wellington and Victor, 1989a; Brothers and Thresher, 1985; Victor and Wellington, 2000), or wrasses (Victor and Wellington, 2000). There was no clear relationship between pelagic larval duration and range size in this study with families pooled (Fig. 7) except that the species with the longest pelagic duration also had the largest geographic ranges. Pelagic larval duration explained only 8% of the variation in range size with all species pooled. Most of the range size variation that was explained, seems attributable to a weak and positive relationship for butterfly fishes ($r^2 = 0.3$, $p = 0.03$, $n = 16$). With butterfly fishes removed, only 1% of the range size variation was attributable to pelagic duration.

Dispersal ability might be reflected in the degree of genetic relatedness of widely separated populations (Shulman, 1998). For example, Doherty *et al.* (1995) found a negative correlation between pelagic larval duration and genetic heterogeneity between populations of damselfishes on the Great Barrier Reef. In contrast, Shulman and Bermingham (1995) found a very poor relationship between genetic structure and length of larval life among populations of Caribbean reef fishes. Perhaps the Caribbean may be sufficiently small to generate panmixia across all available habitats, regardless of the length of pelagic larval development.

Reproductive mode has been linked to dispersal characteristics by a number of authors (Barlow, 1981; Thresher, 1991; Munday and Jones, 1998) and the potential importance of larval behavior in the dispersal of reef fishes has been highlighted in a number of papers (Stobutzki and Bellwood, 1997; Leis *et al.*, 1996; Cowen and Sponaugle, 1997). It seems likely that the relationship between geographic range and dispersal characteristics will involve a combination of factors, including larval duration, hydrodynamic processes, and the behavior of larvae. Further studies suggest that offspring of species with pelagic larvae do not necessarily disperse away from natal reefs, regardless of whether they have narrow or broad geographic ranges (Jones *et al.*, 1999; Swearer *et al.*, 1999). Even species with no larval dispersal can exhibit the extremes in terms of geographic range size.

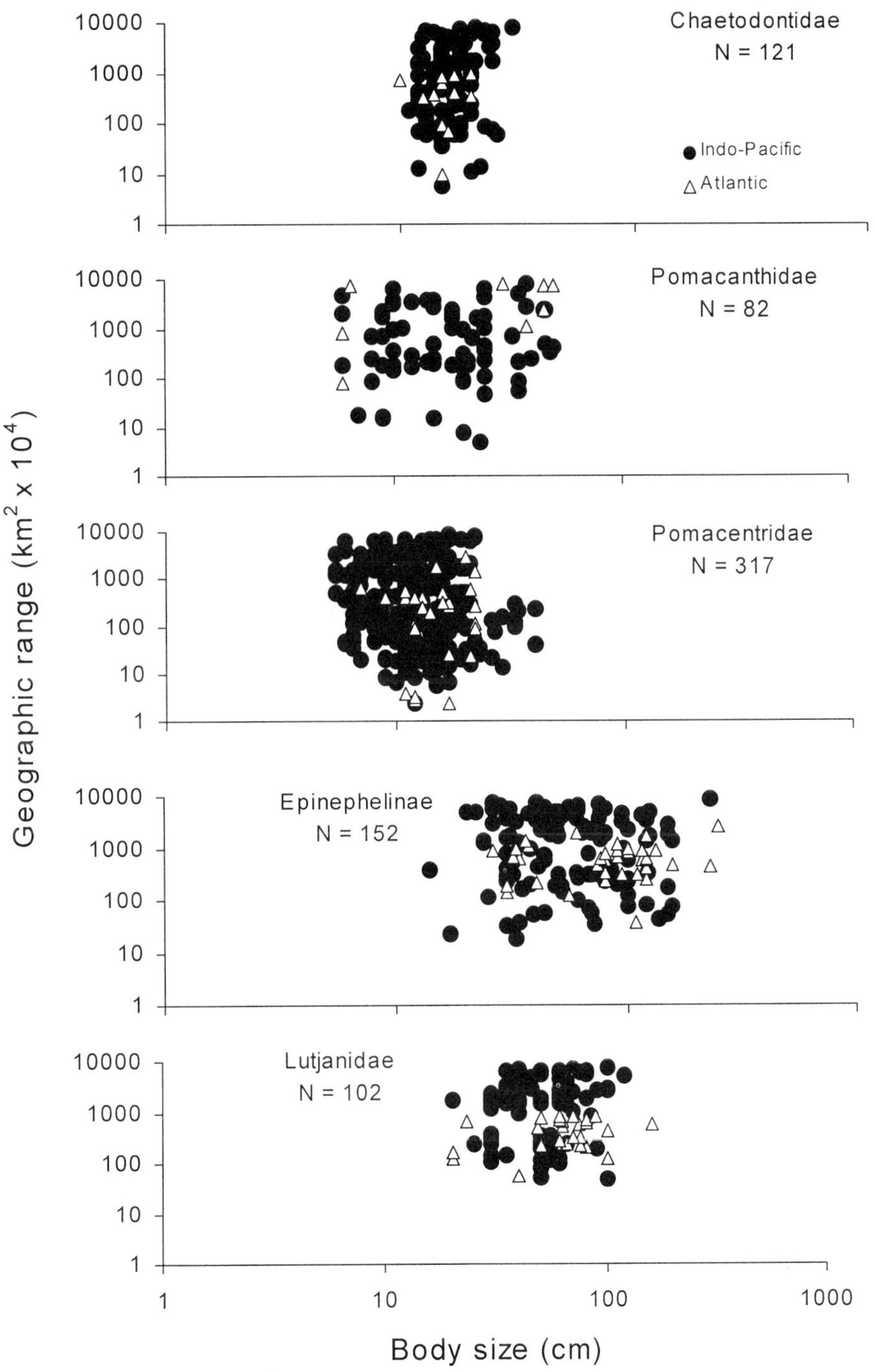

FIGURE 6 Family-level analysis of geographic range of occurrence (km^2) versus body size (total length) for species of coral reef fishes from the Indo-Pacific and Atlantic oceans (N = number of species). Data sources and techniques as per Fig. 2.

The only damselfish lacking a dispersive larval stage, *Acanthochromis polyacanthus*, has a broad Indo-Pacific distribution greater than most confamilials with pelagic larvae (Allen, 1994). In contrast, the direct-developing Banggai Island cardinal fish, *Pterapogon kauderni*, is restricted to a few small islands in the Indonesian archipelago. Clearly, factors affecting movement or ability to colonize new locations during the reef-associated stage of reef fish life histories may also be important in determining geographic ranges.

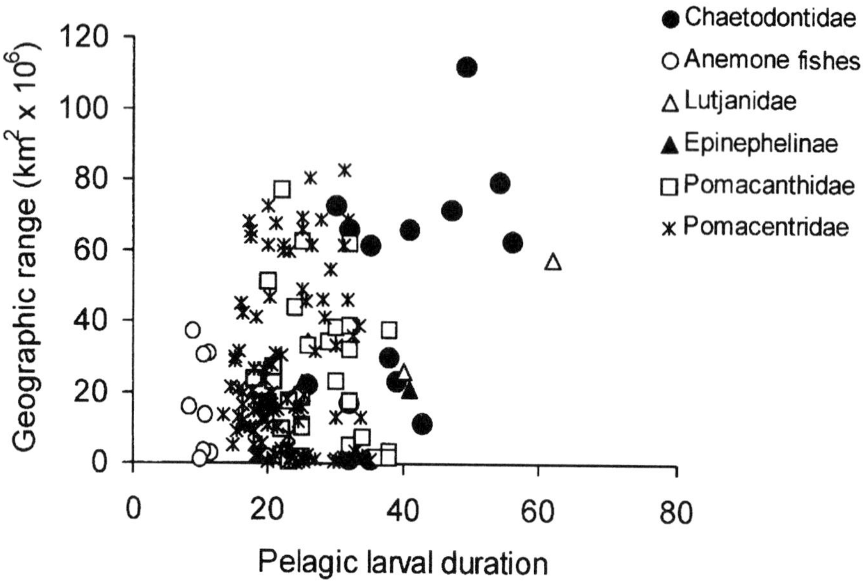

FIGURE 7 Geographic range of occurrence (km^2) versus mean pelagic larval duration for species of coral reef fishes from the Indo-Pacific ocean. Anemone fishes are illustrated separately from other members of the family Pomacentridae because of their short pelagic larval durations. *Acanthochromis polyacanthus*, which has no free-swimming larval stage but has a large geographic range (14.6 million km^2), is not included in the analysis. Data sources and techniques for estimating geographic range size as per Fig. 2. Estimates of mean pelagic duration from Brothers *et al.* (1983), Brothers and Thresher (1985), Leis (1989), Thresher *et al.* (1989), Wellington and Victor (1989a), Doherty *et al.* (1994b), and Wilson and McCormick (1999).

E. Specialization

Widespread dispersal does not guarantee widespread establishment. For example, the geographic range of resource specialists may be limited by the geographic range of their resources (Brown, 1984). Specialist species are expected to have smaller geographic ranges because the spatial extent of suitable resources is likely to be more restricted for them than for species that can use a wide variety of resources. For reef fishes, the only support for this theory is a study showing that species with smaller depth ranges also have smaller geographic ranges (Hawkins *et al.*, 2000).

Specialization in butterflyfishes and anemone fishes is apparently uncorrelated with geographic range size. Using the dietary classifications for butterfly fishes in Allen *et al.* (1998), range size and dietary specialization are unrelated (Fig. 8). Similarly, there is no clear relationship between the number of anemone species inhabited and geographic range for anemone fishes, except that the species with the smallest ranges inhabit only one species of anemone and the least specialized has species the greatest geographic range (Fig. 9). Generally, the geographic ranges of anemone fishes are far smaller than the ranges of their anemone hosts (Fautin and Allen, 1992) and there is no relationship between the geographic ranges of anemone fishes and the ranges of all the anemones they inhabit ($r^2 = 0.08$, $p = 0.15$). From these analyses it seems unlikely that, in these two groups of fishes, resource distributions limit the geographic ranges of specialist species any more than they do generalist ones. With so few examples, the role of specialization requires further investigation.

III. Abundance

Species abundances are often correlated with geographic range size (Hanski, 1982; Brown, 1984; Gaston and Lawton, 1988a; Gaston, 1996; Johnson, 1998), body size (Peters, 1983; Gaston and Lawton, 1988b; Arita *et al.*, 1990; Currie, 1993), and the degree of resource specialization (Brown, 1984; Hodgson, 1986; McCoy and Mushinsky, 1992) in a wide variety of taxa. Below we examine the association between species abundances of coral reef fishes and each of these variables. Although some species in all communities are likely to have low abundance, little is currently known

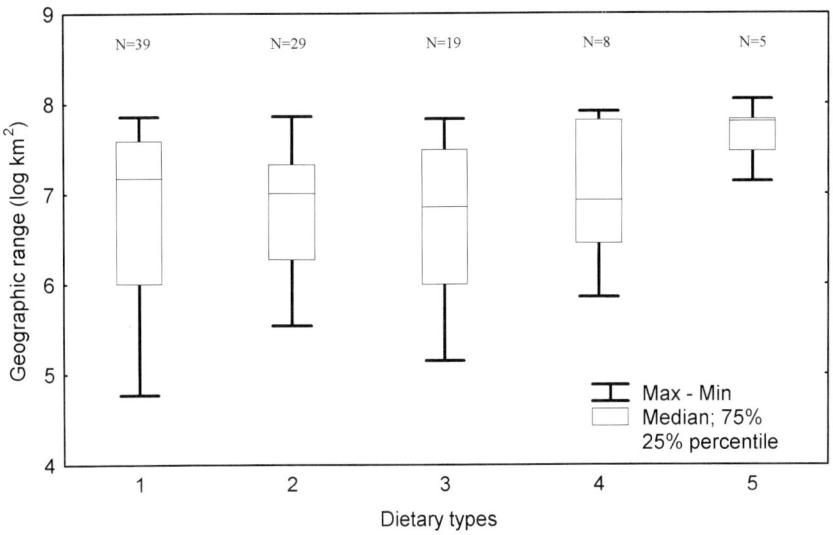

FIGURE 8 Geographic range of occurrence (km^2) in relation to dietary specialization for species of butterfly fishes. Geographic range and diet categories follow Allen *et al.* (1998). Dietary specialization is based on the number of dietary categories consumed.

about the tendencies of individual species to be common or rare throughout their geographic range. The degree to which the abundances of particular species vary in space relative to other species in the community is often unknown (but see Schoener, 1987; B. R. Murray *et al.*, 1999).

A. Is Abundance Related to Distribution?

One of the most commonly reported, although not universal (Gaston and Lawton, 1988a, 1990b), macroecological patterns is a positive correlation between local abundance of a species and its geographic range (Lawton, 1999). Consequently, species that are rare with respect to abundance are also likely to be rare with respect to geographic range (Gaston, 1994a). For most families of coral reef fishes there are presently insufficient data to estimate mean abundance–geographic range size relationships. The Chaetodontidae, however, is one family of reef fishes for which estimates of geographic range (Allen *et al.*, 1998) and reliable estimates of abundance from multiple geographic locations are available for a number of species. For this family of reef fishes there is no apparent relationship between mean abundance and geographic range size (Fig. 10). In other words, butterfly fishes that are rare in terms of abundance are no more likely than other species to be rare in terms of geographic range.

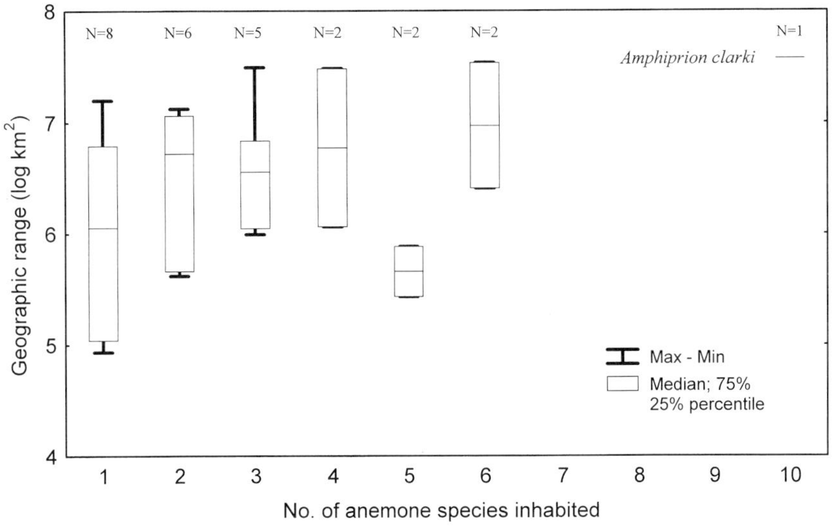

FIGURE 9 Geographic range of occurrence (km^2) in relation to habitat specialization for species of anemone fishes. Geographic range and number of anemones inhabited by each species of anemone fish follow Fautin and Allen (1992).

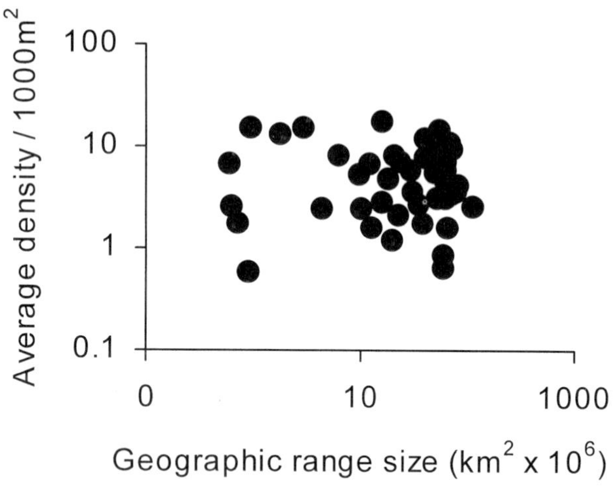

FIGURE 10 Average density versus geographic range of occurrence (km^2) for species of butterfly fishes. Mean densities were derived from estimates of abundances from Bouchon-Navaro (1980) (sites <10 m in depth only), Anderson *et al.* (1981), Bell *et al.* (1985), Bouchon-Navaro *et al.* (1985) (1979 data only), Findley and Findley (1985), Fowler (1990b), Ayling (1992) (Davies Reef only), Cox (1994), Pratchett (1995), unpublished data of G. Jones (Great Barrier Reef, Papua New Guinea, Tuvalu). Geographic range of occurrence follows Allen *et al.* (1998).

If a positive relationship between local abundance and range size were to exist for reef fishes, then endemics and other species with restricted ranges should have lower local abundances than do more widespread species. This does not appear to be the case. In Hawaii, for example, endemic fish species in a number of different families consistently rank among the most abundant species on coral reefs (Hourigan and Reese, 1987). Similarly, Hawkins *et al.* (2000) found that spatially restricted and nonrestricted species could be equally abundant on a local scale. In the literature, there are many anecdotal references to the unusual abundance of endemic species on isolated islands or in enclosed seas (e.g., Fishelson, 1977; Allen and Steene, 1988), suggesting that some species that are rare in terms of geographic range tend to be more abundant as compared to widespread species. For example, *Chaetodon miliaris*, which is restricted to the Hawaiian Islands and Johnston Atoll, is very abundant in Hawaii (Allen *et al.*, 1998). Similarly, *Chaetodon multicinctus* is restricted to the Hawaiian Islands and Johnston Atoll but is the most abundant species of butterfly fish at some locations in Hawaii (Cox, 1994). Some endemics are also very abundant at Lord Howe and adjacent islands. *Chaetodon tricinctus*, which is confined to Lord Howe, Norfolk Island, and Middleton Reef, is the most common butterfly fish at Lord Howe Island. Similarly, *Amphiprion mccullochi*, which is known only from Lord Howe and Norfolk Islands, is common in the lagoon of Lord Howe and is regularly found on other areas of reef around the island (Allen *et al.*, 1976). At Christmas Island in the Indian Ocean, *Centropyge joculator* is highly abundant (Allen and Steene, 1988), despite being restricted to Christmas Island and the Cocos–Keeling group (Allen *et al.*, 1998). Abundance–range size relationships need to be explored in other families to determine if the relatively high abundance of some endemics is a general phenomenon in coral reef fish communities. If so, this would be in contrast to terrestrial endemics, which tend to be rare in terms of local abundance (Lawton, 1999).

B. Body Size

Within local assemblages of coral reef fishes, the relationship between body size and abundance of each species is usually polygonal (Munday and Jones, 1998). In general, the upper bound of the abundance–size relationship declines with increasing body size, except among the smallest species, for which abundance declines with decreasing body size. The lower boundary of the abundance–size relationship is horizontal. Similar abundance–size relationships have been observed for most other local assemblages of animals (e.g., Lawton, 1990; Cotgreave, 1993; Blackburn *et al.*, 1993; Blackburn and Lawton, 1994; Blackburn and Gaston, 1997). Because of the shape of the abundance–size relationship, body size tends to be a poor predictor of abundance in reef fish assemblages. Despite this, large species of coral reef fishes are on average rarer than are small- to medium-sized species (Munday and Jones, 1998). Limits to the abundance of these larger species are likely to be linked with space and energetic requirements. Home range size is positively correlated with body size in many animals (Gaston and Blackburn, 1996), including coral reef fishes (Sale, 1978b). This body size–home range size relationship may occur simply because large animals need larger home ranges to satisfy their energetic requirements. Indeed, to the right of the modal size class, the upper bound of the size–abundance relationship for assemblages of reef fishes usually has a slope near −0.75 (Munday and Jones, 1998). This slope coefficient is consistent with allometric scaling patterns observed in a wide range of other taxa and can be explained by metabolic requirements (Peters, 1983).

In many local assemblages of animals, including coral reef fishes, the lower boundary of the abundance–size relationship is approximately horizontal

(Blackburn and Gaston, 1997; Munday and Jones, 1998). Populations of coral reef fishes are divided into many subpopulations occupying disjunct habitat patches, and these subpopulations are linked by larval dispersal (Doherty and Williams, 1988; Sale, 1991b; Caley *et al.,* 1996). Therefore, at the scale of local populations and because of the open demographic structure of these populations, it is likely that some species of all sizes can occur in low abundance on particular reefs because local populations do not need to be self-sustaining. Depending on the magnitude and variability of larval supply and the action of postsettlement processes, three outcomes are possible: (1) these species may be sporadically present in time and space, (2) they might always be present in low numbers, or (3) their abundance may be highly variable (Munday and Jones, 1998). All three possibilities have been observed on small patch reefs at One Tree Island monitored for over 10 years (Sale *et al.,* 1994).

The horizontal lower boundary to the abundance–size relationship in local assemblages of animals contrasts with a negative slope to the lower boundary that is usually found in analyses of data collected over larger spatial scales (Cotgreave, 1993; Currie, 1993; Blackburn and Gaston, 1997). These larger scale analyses are usually compilations of data on the abundances of a wide range of species derived from the literature. The larger minimum population sizes estimated for small species in these large-scale studies have been attributed to a number of sampling and methodological problems (see Blackburn and Gaston, 1997). However, if a negative slope to the lower boundary of the abundance–size relationship persists when sampling biases are accounted for, this suggests that small species have larger minimum viable population sizes as compared to larger species. Indeed, Lawton (1989) suggests that minimum population densities might be higher for small species because (1) small taxa tend to be shorter lived and less mobile than large taxa and therefore are less likely to find a mate at low densities, and (2) the population sizes of small species tend to fluctuate more than do those of large species and, therefore, they are more likely to go extinct at low densities. It seems unlikely that all these explanations will apply to coral reef fishes. For example, many small-bodied species of coral reef fishes are as long-lived as large species (Munday and Jones, 1998; see Chapter 3, this volume). Unfortunately, because of the metapopulation structure of reef fishes, abundance–size relationships on a very local scale (e.g., Munday and Jones, 1998) tell us little about the minimum viable population sizes of coral reef fishes. We will need to work on larger spatial scales in order to determine if small species of reef fishes do have greater minimum viable population sizes as compared to large species.

C. Specialization

Because specialists use fewer resource types than do generalists, the local abundances of the former are more likely to be limited by the availability of resources (Brown, 1984). Specialists are expected to dominate the best resource types (Fox and Morrow, 1981; Rosenzweig and Lomolino, 1997), but if the best resources tend to be rare then specialist species will almost certainly be less abundant than generalist species. Some support for the notion that specialization might limit local abundances comes from patterns of habitat use and abundance of coral-dwelling gobies (Munday *et al.,* 1997; Munday, 2000). Coral gobies that inhabit only one or two species of *Acropora* tend to be less abundant than species that inhabit a larger number of coral species (Fig. 11). There is a greater probability of finding abundant species of fishes in rarely used habitats and, therefore, the negative relationship between increasing habitat specialization and total abundance could be a statistical artifact if all coral species were included. This problem is avoided by defining habitat specialization as the number of coral species used more frequently than expected on the basis of availability.

If species specialize on the most abundant resources, then low abundance need not be associated with specialization. This appears to be the case for anemone fishes, for which a negative relationship between resource specialization and local abundance is not evident. *Premnas biaculeatus* is the most habitat-specialized species of anemone fish (Fautin and Allen, 1992) and is also the most abundant anemone fish at Lizard Island on the Great Barrier Reef and in Papua New Guinea (Fautin, 1986, 1992; Srinivasan *et al.,* 1999). *Premnas biaculeatus* lives almost exclusively in a single morph of the anemone, *Entacmaea quadricolor*, which is the most abundant anemone at both these locations. Similarly, butterfly fishes, with only one or two major dietary types, are the most abundant species at many locations (Table 1). In general, there is no relationship between dietary specialization and rank abundance. The use of such broad dietary classifications, however, must be treated with caution. Species using only one food type, such as obligate corallivores, may be further classified into specialists and generalists depending on the number of coral species they eat. Furthermore, dietary specialization of butterfly fishes may change among locations. For example, *Chaetodon unimaculatus* is classified as a generalist feeder by Allen *et al.* (1998) but appears to be a specialist corallivore

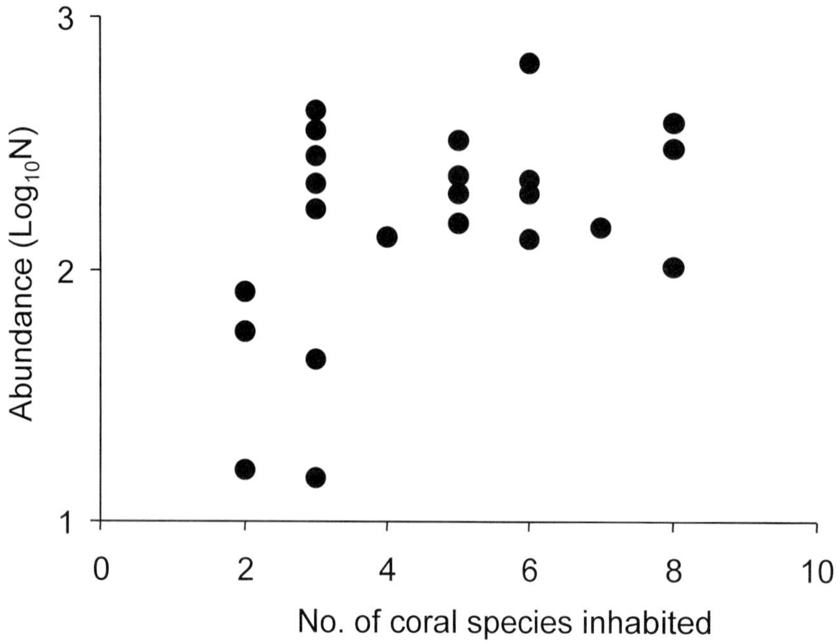

FIGURE 11 Local abundance compared to the number of coral species inhabited by species of *Gobiodon* at One Tree Island (southern GBR), Lizard Island (northern GBR), Motupore Island (southern Papua-New Guinea), and Kimbe Bay (northern Papua-New Guinea). Data from Munday *et al.* (1997), Munday (in press). Only coral species used in accordance with availability or used more frequently than expected are included. Using this method the relationship between habitat specialization and abundance is not confounded by the increased likelihood of finding abundant species in habitats that are rarely used.

in Hawaii (Cox, 1994). More work is needed on the details of dietary specialization and patterns of abundance in butterfly fishes.

Resource-specialized species may be more prone to rapid declines in abundance as compared to generalists, if they are unable to switch resources when their preferred resources become scarce. For example, habitat-specialized coral-dwelling gobies declined in abundance when the corals they usually inhabited declined in abundance at Lizard Island (Munday *et al.*, 1997). Similarly, the abundance of the hole-dwelling blenny, *Acanthemblemaria spinosa*, declined dramatically when the abundance of its preferred habitat declined (R. D. Clarke, 1996). In contrast, *Acanthemblemaria aspera*, which is able to tolerate a wider range of conditions (Clarke, 1992), became more abundant following the decline of *A. spinosa*. Experiments have confirmed that habitat space can be limiting for *A. spinosa* (Buchheim and Hixon, 1992), which is competitively superior to *A. aspera* (Clarke, 1992). Rapid declines in the abundances of specialist species rather than generalist species have also been reported among butterfly fishes. For example, *Chaetodon trifascialis*, which is a specialist feeder on plating corals, went locally extinct following a decline in the abundance of plating corals at Lizard Island due to a crown-of-thorns outbreak. Another apparent specialist, *Chaetodon baronessa*, changed its dietary preferences and became a more generalist feeder (M. S. Pratchett, unpublished data). *Chaetodon trifascialis* also suffered a major decline in abundance following the loss of live coral in French Polynesia (Bouchon-Navaro *et al.*, 1985; Findley and Findley, 1989). *Chaetodon trifasciatus*, which appears to have a more generalist coral diet, declined in abundance at both these sites, but to a lesser extent than its highly specialized congener. Williams (1986) also observed a decline in the abundance of some obligate coral-feeding chaetodontids following a reduction in live coral cover on the central Great Barrier Reef. However, in contrast to the previous examples, neither *C. trifascialis* nor *C. trifasciatus* declined in abundance. Clearly, the observed degree of specialization is not always a good indicator of a species' ability to alter resource use and maintain abundance in response to changes in habitat structure.

D. Spatial and Temporal Patterns in Relative Abundance

In studying rarity, the absolute abundance of species within and among localities is often of interest for a variety of reasons, including the estimation of minimum viable population sizes and extinction risks (Simberloff, 1988; Nunney and Campbell, 1993) and explorations of correlates of small population size (see above). However, in terms of relative abundance, most communities contain many more rare species than common ones. Reef fishes are no exception. For example, in Madagascar and French Polynesia, over 85% of the fish species each constituted less than 1% of the total number of individuals sampled (Harmelin-Vivien, 1989).

TABLE 1 Rank Abundance and Range of Dietary Types for Species of Chaetodontidae at Various Locations in the Indo-Pacific[a]

Species	Number of dietary types	Rank abundance, 10 geographic locations[b]									
		1	2	3	4	5	6	7	8	9	10
Chaetodon trifasciatus	1	—	—	—	—	—	—	—	—	—	1
Chaetodon lunulatus	1	2–3	5	4	1	5	7	3	—	—	—
Chaetodon reticulatus	1	1	—	—	—	—	—	7	—	—	—
Chaetodon baronessa	1	—	15	3	2	1	—	1	—	—	—
Chaetodon octofasciatus	1	—	—	—	—	10	—	2	—	—	—
Heniochus chrysostomus	1	9	—	11	—	2	—	—	2	—	—
Chelmon rostratus	1	—	3	5	—	—	—	—	14	—	—
Chaetodon bennetti	1	16	19	—	—	4	—	—	1	—	—
Chaetodon ornatissimus	1	5	20	—	—	13	—	8	8	—	—
Chaetodon quadrimaculatus	1	13	—	—	—	—	—	—	—	—	—
Chaetodon austriacus	2	—	—	—	—	—	—	—	—	2	—
Chaetodon assarius	2	—	—	—	—	—	3	—	—	—	—
Chaetodon decussatus	2	—	—	—	—	—	—	—	—	—	4
Chaetodon rainfordi	2	—	1	2	9	—	—	—	—	—	—
Chaetodon plebius	2	—	2	9	8	—	1	—	—	—	—
Heniochus intermedius	2	—	—	—	—	—	—	—	—	6	—
Chaetodon trifascialis	2	7	9	12	7	3	2	14	—	3	3
Chaetodon aureofasciatus	2	—	10	1	13	6	—	—	—	—	—
Chaetodon ocellatus	2	—	—	—	—	—	8	—	—	—	—
Chaetodon melannotus	2	—	6	10	12	—	—	—	—	5	—
Heniochus varius	2	—	—	—	—	—	—	4	16	—	—
Coriodon chrysozonus	2	—	—	10	—	—	—	—	10	—	—
Chaetodon ocellicaudatus	2	—	—	—	—	—	—	11	—	—	—
Chaetodon pelewensis	2	2–3	17	—	—	—	—	13	13	—	—
Coriodon melanopus	2	—	—	—	—	—	—	—	12	—	—
Chaetodon speculum	2	—	12	14	16	—	4	—	15	—	—
Chaetodon ulietensis	2	14	11	17	17	—	—	—	—	—	—
Heniochus monoceros	2	17	—	—	—	15	—	—	—	—	—
Chaetodon collare	3	—	—	—	—	—	—	—	—	—	2
Chaetodon fasciatus	3	—	—	—	—	—	—	—	—	4	—
Chaetodon citrinellus	3	4	14	18	4	8	—	9	—	—	5
Chaetodon curiosus	3	—	—	—	—	—	9	—	—	—	—
Chaetodon kleinii	3	—	23	6	5	—	—	6	6	—	—
Heniochus singularis	3		—	—	—	—	—	—	11	—	—
Chaetodon flavirostris	3	—	4	19	—	—	—	—	—	—	—
Chaetodon rafflesi	3	—	—	—	15	9	—	—	—	—	—
Chaetodon paucifasciatus	4	—	—	—	—	—	—	—	—	1	—
Chaetodon vagabundus	4	10	16	7	3	11	—	5	4	—	—
Chaetodon lineolatus	4	—	8	—	14	—	6	10	—	—	—
Chaetodon lunula	4	11	18	11	18	7	5	—	5	—	—
Chaetodon auriga	5	12	7	8	6	12	3	—	—	7	—
Forcipiger flavissimus	5	6	21	—	—	—	—	12	7	—	—
Chaetodon ephippium	5	15	13	15	10	16	—	—	11	—	—
Chaetodon unimaculatus	5	8	22	13	11	—	—	—	—	—	—

[a]Species are ranked within locations, first by the number of dietary types used (following Allen *et al.*, 1998) and then by mean rank abundance.

[b]Locations and data: 1, French Polynesia (Bouchon-Navaro *et al.*, 1985); 2, southern GBR (Fowler, 1990b); 3, central GBR, Davies Reef only (Ayling 1992); 4, Lizard Island GBR (M. S. Pratchett, unpublished data); 5, northern GBR (G. Jones, unpublished data); 6, Abrohlos Islands (G. Jones, unpublished data); 7, Kimbe Bay, Papua New Guinea (G. Jones, unpublished data); 8, Tuvalu (G. Jones, unpublished data); 9, Red Sea (Bouchon-Navaro and Bouchon, 1989); 10, Sri Lanka (Öhman *et al.*, 1998b).

Various models predict different species abundance relationships (e.g., geometric series, log series, log-normal, broken stick), depending on the underlying community-level process (see Magurran, 1988, for review). Although it is often difficult to distinguish between the different species abundance relationships, and a good fit to a particular model provides only correlative evidence of the underlying organizational processes, species abundance relationships are one of the most complete descriptions of community structure available to us. Therefore, the exploration of these relationships should be useful for understanding variations in relative abundances within and among locations. In spite of this potential of species abundance relationships, such relationships are virtually absent from the coral reef fish literature (but see McManus *et al.*, 1992).

In communities where one or few limiting resources control community structure, typically a species-poor community, species rank–abundance relationships should follow either a geometric or log series (May, 1975). In species-rich communities, where large numbers of interacting factors control the ecology of participating species, central tendency should generate a log-normal distribution of species abundances (May, 1975; Sugihara, 1980). Finally, if a single resource that controls community structure is shared equitably among species, a "broken stick" distribution is expected (May, 1974). The majority of species abundance relationships reported have been log-normal (Sugihara, 1980).

To examine which model might apply to coral reef fishes, we assembled a series of published and unpublished data sets of butterfly fish communities for which species rank–abundance relationships could be estimated. The species rank–abundance curves for butterfly fishes appear to be no exception to the general tendency toward log-normal species abundance relationships (Fig. 12). Even where diversity of these fishes is low (e.g., the Red Sea and Hawaii) there is no strong evidence of a geometric or log series. These patterns, therefore, argue that the structure of these butterfly fish communities may be influenced by a large number of interacting factors. If so, where diversity is low, these interacting factors are likely to include interfamilial interactions. However, definitive analyses of these patterns await further study. Formal testing of the fit of species abundance patterns to specific models was precluded; in most cases actual species counts were unavailable.

There is a tendency for species to have lower abundances toward the edges of their geographic ranges (reviewed by Gaston *et al.*, 1997). Ormond and Roberts (1997) noted this pattern for the Red Sea endemic *Chaetodon paucifasciatus*, which declines toward the southern end of its distribution. If generally true, we could predict, as the edge of the distribution of coral reefs is approached, that the tail of species rank–abundance relationships of communities inhabiting coral reefs may lengthen as a greater proportion of species are closer to their geographic limit. Keeping in mind the limitations of the data available, there is little evidence for thinning tails of these distributions for butterfly fishes (Fig. 12). Three of the longest tailed distributions (as a proportion of total species richness) are from communities close to the distributional limits of coral reefs (i.e., Southern Great Barrier Reef, French Polynesia, Red Sea). However, other communities of butterfly fishes located well within the geographic limits of coral reefs also show relatively long, thin tails (i.e., Midshelf, Northern Great Barrier Reef). In one case (i.e., Abrolhos Islands), a community close to the geographical limits of coral reefs exhibits a rather fat-tailed distribution, harboring several broadly distributed species that are found in abundance.

Species abundance relationships, by their nature, ignore species identity, and rare species at one location may be abundant at others. This aspect of rarity and the correlates of rarity for those species that are either "everywhere rare" or "everywhere common" have not been widely considered. It does appear, however, that some species of Australian temperate rain forest and dry sclerophyll woodland plants are "everywhere rare," but no single attribute of these species can predict their pattern of rarity (S. N. Murray *et al.*, 1999). Using data available for butterfly fishes, we examined whether any species are rare or common throughout the range of locations in which they have been sampled. In this case, we adapted Gaston's quartile definition of rarity and asked the question, "How many species are always in the bottom (rare) or top (very common) 25% of species abundances within the communities in which they have been sampled?" Excluding species found at only one location, we found only one species that was "everywhere rare" and none that was "everywhere common." The vast majority of species were very abundant somewhere and rare somewhere else. One species, *Chaetodon kleinii*, varied from being the rarest species in the most species-rich community (Fowler, 1990b) to the most abundant by nearly an order of magnitude in another (G. R. Russ, unpublished data). Hence, it is unlikely that there are many butterfly fishes that are "everywhere rare," but a broader taxonomic perspective is needed to evaluate this pattern more fully.

IV. The Conservation of Rare Species

The patterns of rarity we have described for coral reef fishes have a number of important implications that are relevant to their conservation. There is no evidence

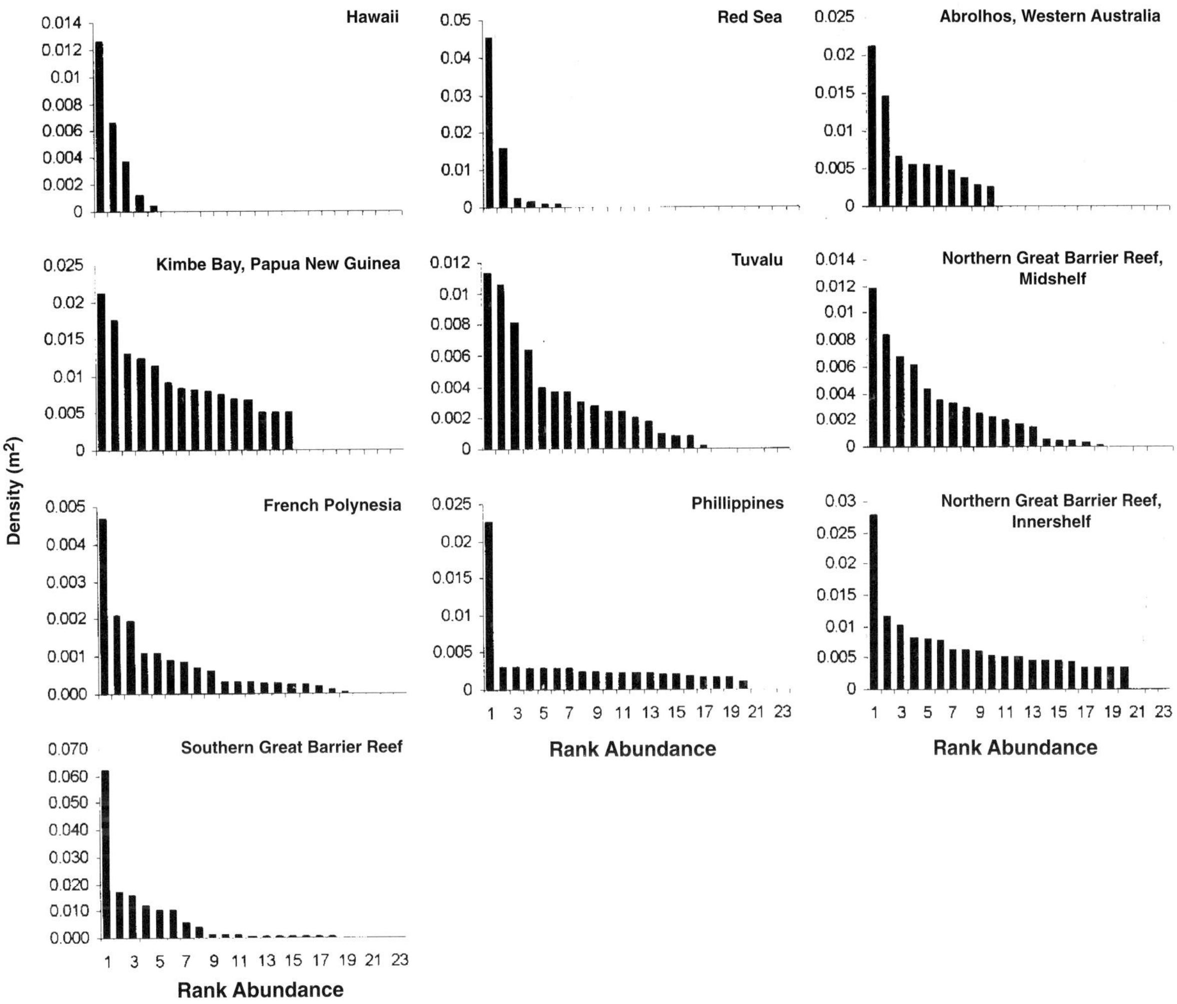

FIGURE 12 Species rank–abundance relationships for 10 communities of butterflyfishes. Two additional data sets were available for French Polynesia. The one presented here was chosen because it included the greatest number of species and is broadly similar in shape to the other data sets available. In most cases these relationships were estimated from data reported as mean abundances per transect standardized to individuals per meter squared. Note that the abundance axis has not been standardized among studies because the absolute abundances among studies are unlikely to be comparable due to differences in sampling methods. Note also that the data for Sumilon and Apo islands do not include *Chaetodon octofasciatus* nor any *Heniochus* or *Coradion* species. These exclusions should not materially affect the shape of the relationship illustrated here (G. R. Russ, personal communication). Sources: Bouchon-Navaro *et al.* (1985), Bouchon-Navaro and Bouchon (1989), Findley and Findley (1989), Fowler (1990b), G. Jones, unpublished data, M. S. Pratchett, unpublished data, and G. R. Russ, unpublished data.

for the "double jeopardy" faced by many terrestrial species that are rare both in terms of abundance and geographic range (Gaston, 1998), although there may be isolated examples. This may contribute to a reduced susceptibility to extinction. The fact that distribution and abundance do not appear to be closely correlated suggests that the factors constraining distributions differ from those that limit local abundance. Therefore it is logical to develop dual management strategies: ones for locally scarce species distributed over broad areas and others for species with restricted geographic ranges.

The widely distributed species that are often rare in terms of abundance are probably best managed by protecting multiple subpopulations in networks of marine conservation areas (Agardy, 1994; Allison *et al.*, 1998). Marine conservation areas are proliferating around the globe and there is increasing evidence that populations of exploited species can recover, albeit slowly, when areas are closed to fishing (Roberts and Polunin, 1991; Jones *et al.,* 1993, Russ, Chapter 19). Other unexploited species may also benefit. For example, due to changes in habitat structure associated with fishing, the damselfish *Pomacentrus baenschi* can be found only inside marine parks off the coast of Kenya (McClanahan, 1994). In general, the selection of appropriate reserve sites for locally rare species is probably best done on the basis of protocols that maximize, for a given area, the number of different species coming under protection. This could include choice of representative biogeographic groupings and habitats (McNeill, 1994), diversity "hotspots" (Reid, 1998) and complementary areas (Pressey *et al.,* 1993; Williams *et al.,* 1996a).

On a global scale, the greatest coral reef fish diversity occurs in Southeast Asia, which coincides with the region of greatest threat to coral reef fishes and their habitats (Hawkins *et al.,* 2000). It is unfortunate that all-purpose conservation areas are most needed in areas where they will be most difficult to establish. Networks of reserves in coral reef areas with lower human population densities may ultimately serve as a refuge for widely distributed species that are threatened over part of their geographic range.

Because large-bodied species tend to be more heavily exploited than most other fishes and also tend to be rare in terms of abundance, they will be of special conservation interest (Jones and Kaly, 1995; Morris *et al.,* 2000). The fact that many of these large reef fishes gather into spatially restricted spawning aggregations that are targeted by fishing activities compounds the problem of human impact on these species (Sadovy, 1993). Such species will benefit from the application of comprehensive management strategies that are applied in addition to marine reserve networks, which may include breeding area closures, catch quotas, and size limits to protect the breeding stock. Many large reef fishes have already been given special status under conservation and/or fisheries legislation in different parts of the world, where their status as potentially threatened species has been recognized. Maximum body size would be a useful proxy for categorizing such species on a worldwide basis.

It is worth noting that all-purpose conservation areas do not necessarily guarantee the protection of certain individual species. For example, the anemone fish species *Amphiprion allardi,* found on the coast of Kenya, is now restricted to heavily fished areas dominated by sea urchins and is locally extinct in areas that are actually protected from fishing (McClanahan, 1994). As for terrestrial species, a safety net of species-specific management strategies may be necessary, in addition to natural habitat protection.

Species with restricted geographic ranges may also be best protected by area closures, but selection of sites will be based on species-specific distributions. On a global scale, there are no endemicity hotspots or areas where species with small geographic ranges tend to be concentrated. Species restricted to small, isolated islands may be particularly vulnerable because the absolute population size is likely to be small and the isolation may limit the effectiveness of any management strategies (e.g., endemic species of Clipperton Atoll, West Pacific). The concentration of species with small geographic ranges at higher latitudes (inverse Rapoport's rule) suggests that reef areas near the northern and southern limits of reef development will be a particular conservation priority. However, perhaps the most threatened of all reef fish species are those with small geographic ranges centered on the Philippines or other areas in Southeast Asia, where reefs are being subjected to extreme habitat degradation (Hawkins *et al.,* 2000, Roberts *et al.,* 2001).

V. Conclusions and Future Directions

This analysis has exposed a number of patterns of rarity in coral reef fishes, some confirming and some contrasting with those previously identified for terrestrial animals. It appears likely that in coral reef fishes, distribution and abundance are not closely linked, casting doubt on whether this is a universal law in ecology (see Lawton, 1999). It is likely for coral reef fishes that different suites of processes limit distribution and abundance. The possibility that endemic species often reach greater abundance than do widely distributed species present at the same location requires further investigation.

Although many coral reef fishes have geographic ranges far larger than those of terrestrial animals, there are many that are more spatially restricted than their presumed dispersal abilities would imply. Clearly, many species on isolated islands have a limited range due to the absence of suitable habitat in close proximity. The pattern of geographic range in relation to latitude is the reverse of that proposed for many other kinds of organisms. The small latitudinal ranges of high-latitude species suggest that their temperate limits are being constrained by the availability of suitable reef habitat or that they are at the limits of their physiological tolerance. Body size is a poor predictor of geographic range, except that larger species tend not to have small geographic ranges. Other factors such as dispersal ability and habitat specialization do not appear to be important by themselves in determining geographic ranges.

Hence, the causes of small geographic ranges are still largely unknown.

Few factors constraining the abundance of locally rare species have been identified. Although body size appears to put constraints on the abundance of large species, it is generally a poor predictor of abundance. Although habitat specialization may be related to low abundance for some species, the interaction between the degree of resource specialization and resource availability is complex and may produce a variety of outcomes.

At this stage, coming to general conclusions about the patterns and causes of rarity in coral reef fishes would be premature. The comparative approach we have taken here is necessarily restricted to a few taxa. Although the relationships identified are preliminary at best, they have provided a useful starting point in the search for general patterns of rarity in marine species and in generating hypotheses that need to be tested using other groups. Without further information on spatial variation in abundance, and temporal variability in patterns of commonness and rarity of these species, there will be constraints on the kinds of questions that can be addressed in the future. Below we identify a number of important issues about rarity, both on ecological and evolutionary time scales, that we see as priorities for further research.

A. Ecological Patterns

Although it is our perception that at any one place, common species tend to stay common in terms of local abundance and rare species tend to stay rare, there are few data sets that span appropriate time scales to verify this pattern. How often do rare coral reef fish species become common or vice versa? This needs to be measured over time scales in which we see a turnover of all individuals in the community. In terms of range size, how often do range extensions or reductions occur? There are still too few community-level monitoring studies on coral reef fishes to draw any conclusions about the stability of patterns of commonness and rarity. In addition, there are still insufficient data on the abundance of most species, throughout their geographic range, to identify key factors limiting the abundance of rare species or to identify assembly rules that control the number of rare species in a community.

One question that we have not examined here is why some rare species appear consistently to recruit poorly. Long-term recruitment failure, a potentially important explanation of rarity in reef fish communities, warrants attention. Likewise, constraints on dispersal and the ability of juveniles to colonize and survive in areas beyond their normal geographic range need to be investigated. The degree to which patterns of commonness and rarity can be explained by the complex interaction between resource preferences, requirements, and availability needs to be more fully explored.

B. Evolutionary Patterns

Given that patterns of diversity are often the result of interactions between local and larger scale processes (e.g., Ricklefs, 1987; Ricklefs and Schluter, 1993b; Caley, 1995a, 1997; Caley and Schluter, 1997), we should recognize that restricted geographic ranges and low local abundances of reef fish species are also likely to be affected by processes operating on both ecological and evolutionary time scales (Wellington and Victor, 1989a). Patterns of speciation and local extinction will influence a species' geographical range in addition to its ability to disperse to and colonize new areas (Gaston, 1998). Answers to a number of evolutionary questions may explain the present-day distribution of species with different geographic range sizes. Do species with limited distribution derive from widespread species, as is the case of some terrestrial groups (e.g., Glazier, 1980)? This may explain the wide geographic distribution of species with small geographic ranges that we see today. If so, species with restricted ranges should often be younger. However, the average age of species with limited distributions would also be younger if they suffered greater rates of extinction than did widely distributed species. Clearly, better information about reef fish phylogeny will aid in explaining the present-day variation in geographic range size.

Acknowledgments

Special thanks to J. Eagle, who assisted in the compilation of the data base used in this review. We thank M. Pratchett for assembling the information on the abundance of butterflyfishes. Thanks also to T. Ayling, M. Pratchett, and G. Russ for access to unpublished information. Comments by N. Dulvy, J. Eagle, K. Gaston, C. Johnson, P. Nangle, G. Russ, P. Sale, and B. McArdle have contributed to the final version of this manuscript.

CHAPTER 5

The Ecological Context of Reproductive Behavior

Christopher W. Petersen
College of the Atlantic
Bar Harbor, Maine 04609

Robert R. Warner
Department of Ecology, Evolution, and Marine Biology
University of California, Santa Barbara
Santa Barbara, California 93106

I. Introduction

Over the past 20 years, population-level studies of coral reef fishes have shifted from simple comparisons between coral reefs and terrestrial systems, to focus instead on the patterns unique to marine and particularly coral reef communities. This shift in attention of coral reef ecologists is reflected in the development of truly marine ecological theory that currently emphasizes the dynamics of open populations and the sources of recruitment fluctuations, including the possibility of recruitment limitation on population size. Inevitably, as our understanding of ecology has grown, many scientists have begun to investigate the applied aspects of their work, examining the role of human impact, and especially the role of fishing, on coral reef communities.

Interestingly, during this same period, there appears to have been a decrease in studies on the behavioral ecology of coral reef fishes. During the 1970s and 1980s, much of the work in behavioral ecology was directed toward understanding the social and specifically the mating systems of coral reef fishes. It is this work, in particular, that seems to have diminished. This change may simply reflect the waning interest in studies relating ecology to social and mating systems. This topic carried behavioral ecology strongly forward ever since publication of several early seminal papers relating species ecology and mating systems (Orians, 1969; Trivers, 1972; Maynard Smith, 1977; Bradbury and Vehrencamp, 1977) began to change the analysis of mating systems, culminating in the synthesis by Emlen and Oring (1977). However, the decline in studies dealing with coral reef fish reproductive biology and behavior seems particularly severe. A quick survey of the International Coral Reef Symposium volumes of 1985 and 1996 reveals a change from 29% (12 of 41) to 4.5% (2 of 44) of coral reef fish papers that could be classified as dealing with behavioral ecology. Among those, the number of papers addressing social and mating systems dropped from five to one. In contrast, the number of papers assessing fisheries and fish conservation doubled over the same time period, from 10 in 1985 to 22 in 1996.

Behavioral ecology in general is in an important transition. Previously, the focus was almost exclusively on delineating the evolutionary mechanisms that have led to the diversity of behaviors that we observe among animals today. Increasingly, however, behavioral ecologists are integrating their studies and findings into population biology, often with an applied flavor. The merger of behavioral and population biology has been spurred by the development of individual-based models and dynamic programming methods, whereby behavior of individuals can be explicitly included.

In this chapter, we review what we believe are several interesting aspects of the behavior of coral reef fishes, seen within the larger framework of changes occurring in behavioral ecology. We start with a summary of social and mating system studies, and then focus on several areas of reproductive and behavioral ecology of coral reef fishes that can have important connections to ecology, conservation, and management. These include patterns of mate choice, the timing and location of spawning, fertilization ecology, and sex change. To reinforce this point, we then look specifically at areas where there is an obvious application to management and conservation problems, including exploitation of spawning aggregations and potential sperm limitation, the potential for Allee effects in exploited populations, and the role of individual movement in marine reserve design. Finally, we finish by asking what role behavioral ecology can play in the coming years to increase our understanding of coral reef systems, and what kinds of data we believe are important to help us gain that understanding.

II. The Evolution of Social and Mating Systems in Coral Reef Fishes: A Short Review

A major emphasis for much of the work reviewed here was intraspecific variability in social and mating systems (Shapiro, 1991; Warner, 1991), and how intraspecific and interspecific variability could be used to test hypotheses in evolutionary and behavioral ecology (Warner, 1980, 1991). Although the generalizations from this work certainly can be used in this way, they cannot necessarily be directly applied to questions in population and community ecology. Instead, the social and mating systems of species are often thought to respond to resource use, rather than vice versa. This is not to say that a consideration of social and mating systems is unnecessary in population biology. Interactions between individuals can profoundly affect movement patterns, distributions, and population density, all of which affect dynamics and management strategies (e.g., Kramer and Chapman, 1999). Also, social systems affect life history allocations (e.g., age-specific decisions about energy devoted to growth vs. reproduction), and these are reflected in growth and mortality rates (e.g., Warner, 1998). These and other links to population biology will be discussed in subsequent sections of this chapter.

Between 1975 and 1995, many studies of behavioral ecology in marine fishes were focused on the explanation of social and mating systems in tropical reef species. This culminated in a series of papers that explored the variation in mating systems in several of the genera and families of common tropical reef fishes: butterfly fishes (Hourigan, 1989), damselfishes (Godwin, 1995; Petersen, 1995), gobies (Robertson and Justines, 1982), jawfishes (Hess, 1993), labroids (wrasses and parrot fishes) (Robertson and Hoffman, 1977; Warner and Robertson, 1978; Robertson and Warner, 1978; Moyer, 1991), and surgeonfishes (Robertson, 1983). In addition, several reviews examined general patterns in particular subgroups such as hermaphroditic species (Warner, 1984b; Fischer and Petersen, 1987; Ross, 1990). These studies have shown repeatedly that the basic reproductive ecology of coral reef fishes appears to be governed by the same rules that affect terrestrial species: the distribution of resources critical to females, along with the population density of males, combine to determine the economic defensibility of females or mating sites, and both interact with patterns of parental care to determine the mating system. This essentially reaffirms the general predictions of Emlen and Oring (1977). Coral reef fishes add some interesting variations (and confirmations) to the basic rules because there can be dramatic differences *within* species from place to place due to changes in resource distribution and population density (Warner, 1991), and individuals often have the option of changing sex or sexual allocation over the course of their lives.

A. Population Density and Social and Mating System Evolution

One of the first generalizations to come out of studies in behavioral ecology of reef fishes was the importance of local density in determining mating systems (reviewed in Warner, 1991). This work included the importance of density in spawning site defense by individual males (Warner and Hoffman, 1980), and the switch from monogamous to more polygynous mating systems with increasing density (Fischer and Petersen, 1987; Hourigan, 1989). In a mating system in which either females or the resources used by females in reproduction were predictable in time and space, the ability of a male to defend those individuals or resources from other males played a critical role in the mating system that emerged. For the bluehead wrasse, *Thalassoma bifasciatum*, Warner and Hoffman (1980) showed that as density increased the mating success of individual males at first increased, but then decreased at high densities when those males were unable to defend successfully those sites from other males. In other studies, Fischer and Petersen (1987) showed an interspecific trend from monogamy to harem polygyny with increasing density in site-attached simultaneously hermaphroditic sea basses, and Petersen (1990a) showed that the same pattern existed intraspecifically in one of the species, *Serranus psittacinus* (= *Serranus fasciatus*).

Population density also appears to interact with female spawning locations, especially whether females migrate to spawn or spawn within feeding territories, in its effects on the mating system. For small reef fishes that free-spawn (release pelagic eggs), two general patterns emerge. For many species, individuals migrate daily to the downcurrent edge of the reef to spawn for a limited time, determined by tide, time of day, or both. For some species, especially on patch reefs, if the number of individuals is relatively small, this group operates as a harem, with one dominant male and a group of females. Furthermore, in families capable of hermaphroditism, protogyny (sex change from female to male) is often observed [e.g., *Bodianus rufus* (Hoffman, 1983, 1985)]. For species at higher densities, multiple males often establish spawning territories at downcurrent reef edges, with females mating with individual males. This system more closely resembles the resource defense model of Emlen and Oring (1977), and sperm competition is virtually absent. At very high densities, small males form large groups at preferred spawning sites, and group spawning appears to be the most common form of reproductive behavior. Here, the mating system is promiscuous, with a high degree of sperm competition.

Some social systems appear to be more common in certain habitats. Where moderate densities of individuals are site attached and can be defended by an individual male, harems are commonly observed [*Labroides dimidiatus* (Robertson, 1972; Kuwamura, 1984), *Bodianus rufus* (Hoffman, 1985), *Centropyge bicolor* (Aldenhoven, 1986), *Serranus psittacinus* (Petersen, 1987, 1990a), *Serranus baldwini* (Petersen and Fischer, 1986), *Canthigaster valentini* (Gladstone, 1987a,b), *Xyrichtys martinicensis* (Victor, 1987), and *Malacanthus plumeri* (Baird, 1988)]. Haremic mating systems appear to be more common in species that live in the edge habitats between reef and sand, where many reef species migrate to spawn (Baird, 1988). This is probably because the preferred spawning habitat can cooccur within a series of small feeding sites for females, and they in turn can be successfully defended by a male. Female site attachment is hypothesized to be due to the predictable availability of a resource, such as food or a shelter site (Robertson and Hoffman, 1977). Monogamy often appears to be a special case of harem polygyny, whereby individual males are unable to sequester more than one female successfully (Barlow, 1984; Hourigan, 1989; Hess, 1993).

In some species, food resources, such as plankton, are not associated with sites, but are mobile, and do not restrict female movement or allow defense of the resource. Planktivorous species with pelagic eggs tend to have either pair spawning with variable levels of mate fidelity [butterfly fishes (Hourigan, 1989), *Clepticus parrae* (Warner and Robertson, 1978), *Serranus tortugarum* (Fischer, 1984)] or large group spawning aggregations (Robertson, 1983; Domeier and Colin, 1997). Site-attached planktivores, which may have specific shelter sites on the reef, often occur at intermediate or high densities and typically have mating systems more closely resembling harems [*Anthias squamipinnis* (Shapiro, 1979)].

B. Mate Choice and Offspring Fitness

An expectation of natural and sexual selection is that females will prefer spawning circumstances (males or sites) that maximize offspring fitness, with several qualifying statements. These qualifiers include that there are no differential costs for the various choices (e.g., predation risk, energetic costs, parasite acquisition), that the benefits may not be immediately measurable in terms of offspring survivorship (indirect benefits: good genes, or sexy sons), and that females can detect cues that indicate offspring survivorship. There are several points from gamete release to the production of subsequent generations at which these fitness differences could occur. Fitness components include fertilization rate of a clutch, immediate survivorship of the clutch (egg predation immediately postspawning), survivorship during larval development, and successful recruitment. Additional fitness differences might occur from genetically based differential vigor of offspring leading to differences in growth, survivorship, or reproduction during its lifetime. Evidence for "adaptive" female choice is highly variable in quality, and support for the adaptive consequences of female choice appears to be quite different between species with demersal and pelagic eggs (treated separately below).

1. Species with Demersal Eggs

In a review of the territorial damselfishes, Petersen (1995) found that within a species the traits used in mate choice were correlated with offspring survivorship, whereas traits not used lacked this correlation. Territorial damselfishes have male parental care of eggs laid at an oviposition site within a male's territory that is used for both feeding and reproduction. Eggs typically take 4–5 days to hatch, and males often care for several clutches at once. Females may spawn several times over an extended period every month (Gronell, 1989; Petersen and Hess, 1991; Knapp, 1993) that typically follows a species-specific lunar phase (Robertson *et al.*, 1990). Females have territories near males, and visits by females to male territories have been recorded in several species (Keenleyside, 1972; Thresher, 1980; Bartels, 1984; Gronell, 1989), so the potential for direct assessment of nest status and egg survivorship by females is possible. Out of nine cases in which a trait was examined for both its correlation with male mating success and egg survivorship, there was a one-to-one correspondence between the relationship of the trait with mating success and egg survivorship (Petersen, 1995). In five cases the trait was positively correlated with egg survivorship and appeared to be used by females in mate choice, and in four cases the trait was not correlated with egg survivorship and did not appear to be used by females.

Other work on damselfishes suggests that the relationship between mate choice and offspring survivorship is not as consistent as in the earlier papers reviewed by Petersen (1995). In a study of bicolor damselfish, *Stegastes partitus,* Cole and Sadovy (1995) reported a correlation for increased male mating success with male size with no coincidental increase in egg survivorship with male size. This differed from an earlier study (Knapp and Kovach, 1991) of a different population in which no correlation was found between male size and mating success, and no correlation between male size and egg survivorship. In an even earlier third study of this species, Schmale (1981) found a positive correlation between male size and mating success, but the relationship disappeared when he discarded the data from small, possibly immature males that received no matings. Cole and Sadovy (1995) suggested that some of the differences among these studies might have been due to the use of artificial substrates in the Schmale and Knapp and Kovach studies. These might reduce variance in oviposition site parameters and remove the mating advantage for larger males. The result in this species is suggestive, but in his review Petersen (1995) found no differences in the variance in male mating success or in the characteristics used in female choice in species in which both natural and artificial oviposition sites had been used.

Support for the mate choice–offspring fitness correlation occurs in several other species with demersal eggs. In the redlip blenny, *Ophioblennius atlanticus,* females prefer larger males, and larger males have higher clutch survivorship (Côté and Hunte, 1989). In addition, this correlation appears to occur in several fish species not associated with coral reefs (Marconato, 1985; DeMartini, 1987; Sargent, 1988; Forsgren *et al.*, 1996; Kraak, 1996).

One of the most common forms of female choice is to spawn where other females are spawning or where there are newly laid eggs (Ridley and Rechten, 1981; Constantz, 1985; Bisazza and Marconato, 1988; Unger and Sargent, 1988; Gronell, 1989; Petersen, 1989; Petersen and Marchetti, 1989; Sikkel, 1989; Kraak and Videler, 1991; Petersen and Hess, 1991). In species with demersal eggs, female copying will result in the few males that are successful guarding relatively large clutches. There is evidence from a wide variety of species that larger clutches have higher survivorship (Petersen, 1995; Jamieson, 1996). Jamieson draws an important distinction between this and other forms of mate copying, which occur in a wide variety of animals for which male care is absent (Gibson and Höglund, 1992). In those systems mate copying is sometimes seen as driven by sexual selection, and is often done by young females, with the only potential benefits being genetic quality of offspring. In fishes with demersal eggs, the benefit for female copying appears to be increased survivorship of offspring (either from increased parental commitment from the male, or from a simple dilution of predation). In both cases, copying can intensify sexual selection by magnifying the rewards to a male of initially being the object of female choice.

It is unclear if the cues used in a local population represent a fixed cue used by the species in general or a cue that is used because females, by assessing egg survivorship, have found that the cue is reliable. It might be possible to test this idea by changing patterns of egg survivorship in a species, and then following patterns of male choice to see if they change accordingly. For example, in a species in which male size is correlated with egg survivorship, will females stop using the trait if the correlation between male size and clutch survivorship is removed? Alternatively, and perhaps more difficult to test, can female choice for a trait be induced by manipulating egg survivorship to produce a correlation with a trait that is used in other species or populations but not by the focal population? These questions might help us to understand the role of local cues in determining individual decisions in species with extremely open populations (see Warner, 1997a).

Several potential costs of mate choice have also been realized in species with paternal care of demersal eggs. The potential costs of movement of females to oviposition sites have been studied in a damselfish by Khoda (1988) and in a blenny by Reynolds and Côté (1995). Khoda (1988) found that while females were away spawning, fishes entered their territories and took bites of algae in typically defended areas. Females that spawned at a greater distance made fewer trips back to their territories, and their territories suffered more bites from intruding fishes. In *Ophioblennius atlanticus,* females prefer larger, older males that have higher probabilities of clutch survivorship (Côté and Hunte, 1989, 1993). The distances females travel to spawn vary from 1 to 12 m, and females that travel farther spawn with larger males. These females are sometimes attacked as they move through damselfish territories on their way to spawn, and females that spawned with larger, more distant males were attacked at a higher frequency by damselfishes. Although not fatal, these injuries could constrain the movement of females and restrict female choice in this species.

2. Species with Pelagic Eggs

Much less is known about mate choice in species with pelagic eggs, compared to species with demersal eggs. Female copying in choice of spawning sites can be inferred from the study of traditionality in spawning-site selection in the bluehead wrasse, *Thalassoma bifasciatum* (Warner, 1988a, 1990a,b). In this species, young females copy the choice of the current females on the reef, leading to long-term stability of spawning sites through cultural transmission. When tradition was broken by removing all of the females from a reef, new females chose a set of sites that were a different mix from the previous subset among all available sites (Warner,1988a). Although site choice tended to be consistent after the first manipulation, Warner (1990a) found no direct evidence for any immediate fitness benefits for this mechanism for mating-site selection. The new set of sites tended to be the shallowest projections on the downcurrent end of the reef, which may offer better protection (Robertson and Hoffman, 1977). Traditionality may convey benefits in safety for the female (Warner, 1988a). These benefits could include the time saved and safety gained through using other females to find spawning sites instead of individuals assessing different sites. Traditionality may also be more likely in cases in which there are few or no fitness differences among potential sites (Warner, 1988a; Hensley *et al.*, 1994; Appeldoorn *et al.*, 1994). This example is different from the mate choice copying in demersal egg species in that there are no obvious fitness benefits in terms of increased egg survivorship. To our knowledge, mate copying has not been investigated in any other species that spawns pelagic eggs.

In addition to mate choice copying, bluehead wrasses exhibit female choice of male phenotypic traits. By removing males at sites and monitoring the change in mating success with a new male present, Warner and Schultz (1992) were able to show that females selected males with a larger white bar on the flank, a trait that tends to make terminal-phase males more conspicuous. Similar choice of male coloration occurs in the damselfish *Chrysiptera cyanea* (Gronell, 1989). Warner and Dill (2000) suggest that bright male coloration and pelagic courtship displays function to indicate to females the current safety of a mating site relative to transient predators. This could be a form of direct benefit. Although their observations and experimental results are consistent with this interpretation, they did not attempt to measure any actual survival benefits to females.

To date, no published papers demonstrate a positive fitness effect for mate choice in any species spawning pelagic eggs. It is unclear whether this represents a real difference between pelagic and demersal spawners, or just the difficulty of quantifying fitness effects for pelagic eggs. Although differences in fertilization rate within a species have been documented (see below), we have no way to quantify possible differences in offspring survivorship directly, and indirect methods such as differences in egg predation rates among spawning sites are unlikely to yield much information, given the low rate of observed egg predation in most studies. Nevertheless, the lack of subsequent contact with eggs by pelagically spawning species suggests that these species might be less able to affect egg survivorship or change behavior in response to changes in offspring fitness.

C. Fertilization Success

Over the past decade, several studies have investigated aspects of fertilization success in tropical reef fishes (Petersen, 1991a; Petersen *et al.*, 1992; Warner *et al.*, 1995; Marconato *et al.*, 1995; 1997; Marconato and Shapiro, 1996; Kiflawi *et al.*, 1998). This work grew out of theoretical predictions that it would not always be optimal for males to produce as much sperm as needed to fertilize an entire female's clutch (Petersen, 1991b; Shapiro and Giraldeau, 1996). Although this theoretical work predicted that fertilization success might be less than 100%, without an actual fertilization curve there was no way to predict the sperm allocation of an individual male and the resulting fertilization success of a spawn. Generally, previous work on sperm competition predicted that sperm would be less limiting in species for which sperm competition was more prevalent (reviewed in Petersen and Warner, 1998).

Early measurements of fertilization success produced relatively high estimates of success [means in the range of 75–88% (Petersen, 1991a; Petersen *et al.*, 1992)], and subsequent work using more refined techniques revealed that fertilization success in pelagic spawners is typically above 90% (Warner *et al.*, 1995; Marconato *et al.*, 1995, 1997; Marconato and Shapiro, 1996; Kiflawi *et al.*, 1998; Petersen *et al.*, 2001).

For fertilization success to influence mate choice, females must have ways to assess directly or indirectly the expected fertilization success of a male. In bluehead wrasse, Warner *et al.* (1995) found that males with higher mating success produced less sperm per spawn and had decreased average fertilization success compared to males with lower mating success. In subsequent analyses, average water flow at a spawning site has also been shown to be negatively correlated with fertilization success (Petersen *et al.*, 2001). Although females cannot directly assess fertilization success in pelagic spawners, the two factors above seem measurable by females and could be used to predict sites or males that would provide higher fertilization success. Despite this potential predictability, there is no evidence that females use these cues when choosing a spawning site (Warner *et al.*, 1995; Petersen *et al.*, 2001).

Variance in number of males releasing sperm in a spawn should lead to high variance in sperm numbers in the vicinity of eggs, and could lead to differences in fertilization rate among spawns. This possibility has been investigated in three species. Petersen (1991a) found higher fertilization rates for spawns when a second male individual joined a pair spawn (streaking) in the slippery dick, *Halichoeres bivattatus* (Fig. 1A). In a second study, group-spawning bluehead wrasses were shown to have equal or higher fertilization success compared to pair-spawning males on the same patch reef (Fig. 1B) (Marconato *et al.*, 1997). The difference in results between reefs appeared to be due to differences in the success of pair-spawning males; on reefs where males had high mating success, group spawns were significantly better, whereas on reefs where territorial pair-spawning males had lower mating success, pair and group spawns had similar fertilization success. Finally, Kiflawi *et al.* (1998) compared what they called "group" vs. "mass" spawning in the brown surgeonfish, *Acanthurus nigrofuscus*. This species forms large spawning aggregations, and the authors compared eggs collected from single group spawns with an estimated 3–14 males to eggs from spawns during mass spawning, when several group spawns occur in a quick flurry, leaving a highly visible gamete cloud that can last for several minutes. The authors tested the hypothesis that during mass spawning, sperm from other group spawns would increase fertilization success, but found no difference in fertilization success between the two types of spawns (Fig. 1C) (Kiflawi *et al.*, 1998).

There are no studies of paternity for any tropical reef fishes when sperm competition has occurred. This is in sharp contrast to avian and mammalian studies, for which hypervariable microsatellite DNA is regularly used in studies of parental care and mating systems. However, work is beginning to be done on coral reef fishes to identify and use these or other highly variable components of DNA to gather information on how fertilization is partitioned among males in multimale spawns (Wooninck *et al.*, 1998).

Although there are still only data from a few species, the results of these studies suggest that fertilization success is not used in female mate choice in coral reef fishes. Fertilization rates for species with demersal eggs are very high (e.g., van den Berghe *et al.*, 1989), and observations showed nearly complete development of eggs in all clutches (Cole and Sadovy, 1995; C. W. Petersen, unpublished data), implying that there is little variance in these spawners. These results also suggest that, in tropical reef fishes, variance in fertilization success among or within populations may not be an important contributor to variance in zygote production.

D. Timing and Location of Spawning

Interspecific variation in both the spatial and temporal patterns of spawning of species within a local area has begged explanation for over a decade (Robertson, 1991a). Within a reef, several species often use the same general or specific area for spawning (Johannes, 1978; Moyer, 1989; Colin and Bell, 1991; Sancho *et al.*, 2000a), although different species may have highly different times for spawning. The factors that affect this choice should be important in understanding the dispersal of eggs and larvae, how different environments contribute to sources and sinks in populations, and speciation in coral reef fishes.

1. Diel Timing

The timing of spawning for species appears to change between locations, and areas with stronger tidal currents tend to have stronger between-day variation in spawning times (Robertson, 1991a). This suggests that spawning times might be influenced by two factors, a fixed time that may be species specific, and a current component that corresponds to some tidal signal. However, the variation in spawning times between species with similar eggs and spawning behavior has never been systematically investigated, perhaps because no clear trends emerge from an initial examination of the data.

In a variety of pelagic spawners, egg biology is similar: positively buoyant eggs up to 1 mm in diameter that hatch in slightly less than 24 hours to begin a larval life with a minimum duration of 2 weeks. Given this similarity, it is difficult to understand what factors might have led to the diversity of spawning times used by various species of tropical reef fishes. Several hypotheses deserve testing and have implications for how we view reproduction and speciation in coral reef fishes. These hypotheses include species displacement, differential egg or adult predation, and a null hypothesis that spawning times are not significantly dispersed.

Species displacement can be defined as the presence of one species spawning at a site, reducing the

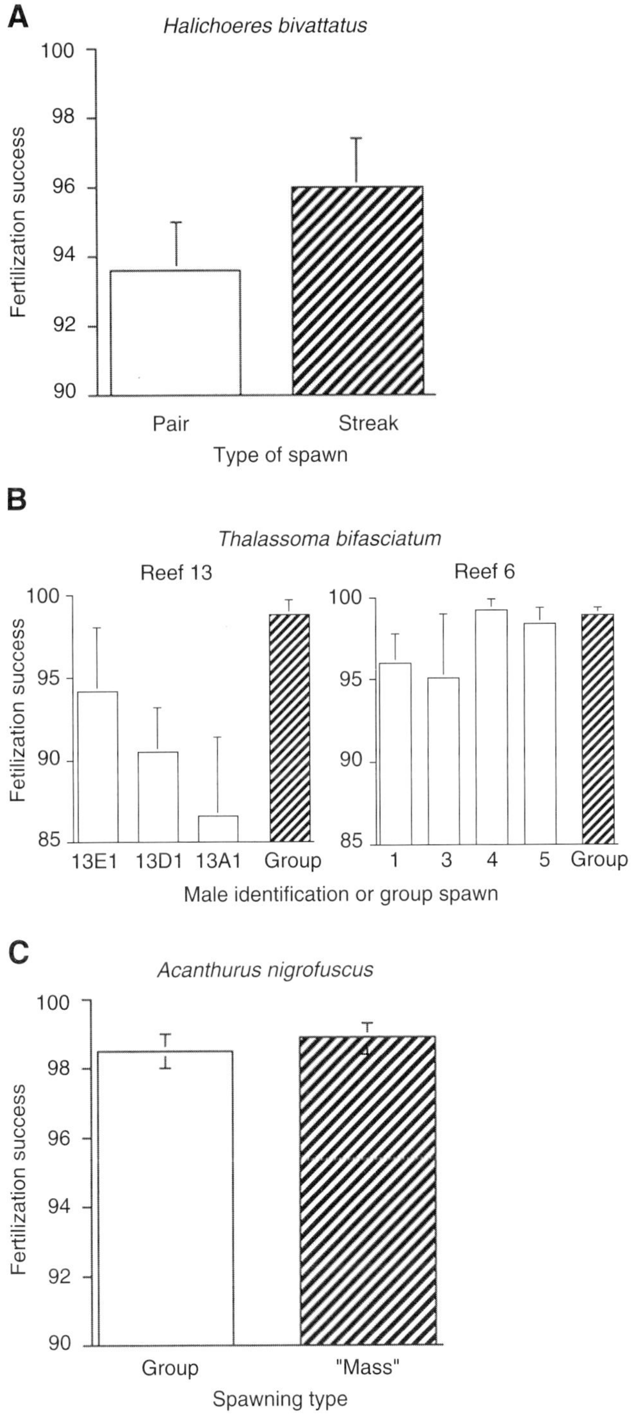

FIGURE 1 Effect of number of males involved in spawning on fertilization success (percentage of eggs fertilized) for three species of coral reef fishes. (A) Fertilization success for slippery dicks, *Halichoeres bivattatus*, in pair spawns without (open bar) and with (hatched bar) additional males (streakers) joining the spawn. Spawns with streakers averaged significantly higher fertilization success [mixed-model ANOVA, $P < 0.05$; data from Petersen (1991a)]. (B) Comparison of fertilization success for pair (open bar) and group-spawning (hatched bar) bluehead wrasses, *Thalassoma bifasciatum*, from two patch reefs in St. Croix. For reef 13, group spawns have significantly higher fertilization success; for reef 6, the difference is in the same direction but not significant [data from Marconato *et al.* (1997)]. (C) Fertilization success for group spawning (open bar) and "mass spawning" (multiple group spawns in an aggregation; hatched bar) for brown surgeonfishes, *Acanthurus nigrofuscus*. There was no difference in the two groups [data from Kiflawi *et al.* (1998)]. Bars in all figures represent 95% confidence limits.

probability of a second species spawning at the same site and time. This could occur because one species aggressively excludes the first, or because the presence of one species decreases the success of a second species spawning at the same time and location. The presence of group spawning, or extensive contributions of sperm by peripheral males in one species, might reduce the success of spawns by a second species, particularly if the presence of heterospecific sperm in spawns decreases successful fertilization. Individuals that streak or group spawn are extremely opportunistic and may move toward any upward movement on the reef that might be a conspecific pair spawn. Such heterospecific matings might incur little costs to the heterospecific males, but might decrease fertilization success by sperm interfering at the egg micropyle and reducing fertilization. Heterospecific streaking has been observed in the field (C. W. Petersen, personal observation), but the consequence of these spawns is not well known. Apparent hybrids are uncommonly observed in nature, suggesting that the costs of heterospecific fertilization may be immediate.

If there are differences between species in vulnerability of either spawning adults or their eggs to predation at particular times, then this could lead to selection for divergence in spawning times. Deeper bodied species and species with chemical defenses or morphological adaptations that reduce susceptibility to predation might spawn at times with more active predators if those times produce increased survivorship of young.

The overall low rate of predation observed by most reef fish behavioral ecologists, and the variation in predation intensity observed between locations, make it difficult to gauge predation's effect on the evolution of specific spawning behaviors or choices. Interpretations may be further confounded by human activities that have reduced the historical abundance of predators (Jackson, 1997). If fishes have evolved fixed behaviors to avoid predators and current predator numbers are reduced compared with historical levels, the population might maintain risk-averse behavior for a long period.

However, predation clearly has influenced reproductive behavior of reef fishes. We are always impressed with the simultaneous, rapid darting to cover of what seems to be an entire reef of active fishes, and then turning and seeing a jack or other piscivore slowly become visible in the distance. Courting or spawning individuals have also been shown to be more vulnerable to predation (Thresher, 1984; Moyer, 1987; Clifton and Robertson, 1993; Sancho *et al.*, 2000b). At Johnston Atoll, Sancho *et al.* (2000a) found the highest activity of piscivores during the daytime, when abundant group-spawning species were reproductively active. The abundance of a jack, *Caranx melampygus,* was positively correlated with spawning prey numbers at this location. Interestingly, piscivore activity at this location was lowest at dusk, when the most species were observed spawning. However, these species represented a low number of total individuals, so fish density at dusk at the spawning site was substantially lower than during the daytime.

A final explanation for variation in spawning times is that they are largely phylogenetically constrained, and that most species show little variation from the trends of related species. The divergence of species could be tested once the phylogeny of species in a data set was known. Unfortunately, we have little information on the phylogeny of species, especially within families, to develop powerful comparative tests to determine how factors influence the evolution and constraints on diel patterns of spawning.

It is important to remember that the open nature of many reef fish populations may place a constraint on local adaptations (Warner, 1997a). Thus many behaviors, including the timing of spawning, may represent the average best response but may often be locally maladaptive. As a hypothetical example, imagine two species, one predominantly a fore-reef species, one predominantly a lagoonal species, but with both species occurring sympatrically over a large area of habitat. The lagoonal species may have been selected to spawn during outgoing tides, whereas the fore-reef species evolved to spawn at dusk. In the lagoon, the fore-reef species continues to spawn at dusk, and selection does not act strongly to change its spawning time, because its offspring are most likely to recruit to fore-reef environments, where the dusk spawning time is most adaptive. In general, behavioral ecologists working with marine species must be very cautious in assuming local adaptation (Warner, 1997a).

2. Periodicity of Spawning

Relative fecundity, measured either as egg output per unit weight or as gonosomatic index, decreases interspecifically with female size in fishes (Sadovy, 1996). The frequency of spawning also decreases with species size, with smaller species much more likely to be daily spawners. One way to consider more periodic spawning, whether it is restricted to a lunar period or a yearly period, is that the benefits for spawning at a particular time of year outweigh two potential costs: not surviving to the next reproductive period, and morphological limitations on how many eggs can be developed at a given time. For larger species, lower instantaneous mortality rates could lessen any costs of delaying reproduction, tipping the balance in favor of seasonal reproduction. This topic has not been much pursued and represents a profitable area of collaboration between behavioral biology and physiology.

3. Location of Spawning

Randall and Randall (1963) were the first to hypothesize that reef fishes picked specific spawning locations that maximized the probability of their eggs successfully being carried away from the reef. There are two possible advantages to spawning at specific locations in terms of dispersal of eggs and larvae from a reef. There could be short-term benefits from getting eggs immediately away from reef-based egg predators, and there could be long-term benefits of getting offspring into water masses that have higher probabilities of producing successful recruits to the reef. There is substantial indirect evidence that the short-term benefits exist; fishes tend to spawn at downcurrent ends of reefs, and dye released from spawning locations on one reef was shown to spend less time over the reef than dye released from randomly placed locations (Hensley *et al.*, 1994). The only available evidence suggests that there is no difference among local spawning sites in the long-term location of propagules (Appeldoorn *et al.*, 1994). Again, without intergenerational feedback on any particular reef, the evolution of such local adaptations may be impossible.

Predators have also been thought to influence the location of reef fish reproduction (Robertson and Hoffman, 1977). For example, Warner (1988a) quantified physical characteristics of spawning sites for bluehead wrasse and found that females preferred locations where there were shallow upward projections at the downcurrent edge of the reef. These characteristics presumably allowed individuals to spawn higher in the water column for a given predation threat, reducing reef-based predation on them and their eggs. Sancho *et al.* (2000a) found that species were more likely to spawn over a complex substrate that offered potential refuges for spawning individuals.

4. Spawning Migrations

The behavior and reproductive biology of fishes that migrate to spawn have been of intense interest to both behavioral ecologists and conservation biologists over the past decade. Many individual species migrate to specific spawning sites, with the migration distance varying from tens of meters to kilometers. These migrations can be to the edge of or extend well beyond the normal home range of individuals.

The possible benefits from migrating to spawning sites are probably consistent among all species with pelagic eggs. The most likely benefit is enhanced off-reef transport of eggs, because large spawning aggregations often form at the edge of large drop-offs at reefs. The use by multiple species of the same site supports the idea that these sites have intrinsic value for spawning (Domeier and Colin, 1997). Additional benefits to these large spawning aggregations might be predator dilution or satiation, whereby large numbers of adults or zygotes might decrease the per capita predation risk.

Some authors have suggested other potential benefits for spawning aggregations, including enhancement of fertilization rate, although this hypothesis was first put forward for free-spawning marine invertebrates. Given the lack of a great difference between pair and group spawning in the one species studied (Fig. 1B) (Marconato *et al.*, 1997) and the lack of a difference between single group spawns and multiple group spawns with large aggregations (Fig. 1C) (Kiflawi *et al.*, 1998), this difference is unlikely. Domeier and Colin (1997) reported unpublished fertilization rates averaging 85% for a small pair-spawning grouper, *Epinephelus fulva*, and suggested that this lower fertilization rate might suggest benefits for mass spawning. However, given the damage caused by some egg collection methods (Marconato *et al.*, 1997) and the differences in egg damage among species caused by techniques (C. W. Petersen and D. R. Levitan, unpublished data), more complete data on fertilization rate in groupers in particular and aggregating species in general will be needed before evidence for fertilization benefits can be inferred from spawning in large groups.

The costs of migrating to spawning aggregations are several, and include loss of feeding opportunities while migrating, increased predation risk while migrating or spawning, and energetic costs of traveling. Given these costs and benefits, we would expect species that are less vulnerable to predation and spawn less frequently to be more likely to migrate the tens of kilometers that are typical of large spawning aggregations. These traits are consistent with large, reef-based piscivores such as groupers and snappers, and these are the major groups of species that do form spawning aggregations (Domeier and Colin, 1997). Individual size also appears to be important in interspecific trends within the groupers; smaller groupers in the Caribbean are less likely to migrate and aggregate than are larger species (Sadovy *et al.*, 1994). The differences between species in spawning migrations can best be shown by looking at the differences in two species that represent the most thoroughly studied species for their type of spawning mode: Nassau grouper, *Epinephalus striatus*, as an example of a long-distance migrator, and bluehead wrasse, *Thalassoma bifasciatum*, as an example of a species that typically migrates short distances within a reef to spawn.

Nassau groupers are large piscivorous fishes that are common in lightly fished areas of the Caribbean, although they are often absent or at very low density in moderately or heavily fished areas. They can reach 25 kg, and migrate up to 240 km to reach spawning aggregations, typically at or near the continental shelf break (reviewed in Domeier and Colin, 1997). Within any population, the spawning aggregation typically occurs for a few days during the full moon for 2 months every year.

In contrast, bluehead wrasses are small carnivorous fishes that are common throughout the Caribbean. They reach about 20 g, and migrate varying distances to spawn. On a continuous fore-reef, an individual fish might migrate as much as 1.5 km to spawn daily, but on patch reefs individuals migrate tens to hundreds of meters to spawn, and individuals on back-reef environments may spawn within their normal feeding range (Fitch and Shapiro, 1990; Warner, 1986, 1995). Spawning occurs daily, with the average female spawning on about 2 of every 3 days (Schultz and Warner, 1989).

Warner (1995) attempted to study the potential costs of migration in a population of bluehead wrasses in a fore-reef environment, where some fishes spent nearly 2 hours a day in migration. However, he was unable to find differences in spawning frequency or fecundity between groups of individuals with very different migration times. The only difference found was surprising: long-distance migrators grew *faster* than those with more foraging time. The result appeared to be caused by differences in foraging between the populations, with the upcurrent (long-migration) fishes having significantly more food available. This is the only study we are aware of that attempted to quantify costs of migration in a coral reef fish. The bluehead wrasse, with its more frequent spawning and shorter migrations to spawning sites, is a good example of a small species that produces pelagic eggs, with relatively high reproductive effort and a relatively short lifespan. At the other extreme is the Nassau grouper, with its much less frequent spawning and extensive spawning migrations. It epitomizes a large species with a life history that matches what we would predict from a species with lower reproductive effort and high adult survivorship.

Many species spawn in a pattern related to the lunar cycle, and these lunar spawners can be migrators or not, including species with demersal and pelagic eggs (Robertson *et al.*, 1990; Robertson, 1991a). Certainly, most species that migrate long distances spawn within a fixed period of the lunar cycle. Despite considerable work in this area (reviewed in Robertson, 1991a), it is unclear whether lunar periodicity reflects responses to tidally related current regimes (e.g., facilitating off-reef transport), changes in nocturnal lighting (e.g., facilitating nocturnal spawning migrations of adults, or settlement in low-light periods), or merely a cue used to synchronize reproduction.

E. Hermaphroditism in Coral Reef Fishes

Until the 1990s, most work on hermaphroditism in coral reef fishes focused on how the mating system affected fitness associated with different sexual life histories (Warner, 1988b,c; Shapiro, 1989). Over the past decade, several advances in our understanding have been made concerning both the type of hermaphroditism that exists in fishes, and about the ways that sex change is induced. Perhaps the biggest surprise in studies of the evolution of hermaphroditism in fishes is the observation that multiple sex reversals are possible in several species of gobies and probably in some other families in which protogynous hermaphroditism is common (reviewed by Nakashima *et al.*, 1995; Kuwamura and Nakashima, 1998).

In most protogynous hermaphrodites, sex reversal back to a female that was once male was not thought likely due to two reasons. Morphologically, when most fishes change sex from female to male, the ovarian tissue in the gonad appears to degenerate completely, whereas male tissue exists as observable small pockets of tissue in females in most protogynous species. Evolutionarily, the explanation for sex change has traditionally involved changes in each sex's reproductive value with individual size, and in most cases once an individual becomes more successful as a male due to large size, it was not thought likely that this pattern would reverse as the individual aged (Warner, 1988b).

In bidirectional sex changers, both of these reasons do not seem to hold. After protogynous sex change, there are still observable pockets of female tissue in some hermaphrodites, to the point that some investigators refer to them as simultaneous hermaphrodites (cf. St. Mary, 1993). In addition, growth rates in these species appear to be sex specific, with the males growing at lower rates than females, so the possibility exists for individuals to leapfrog each other in size or dominance, and this may favor multiple sex change. Finally, mortality rates of individuals in the field may lead to frequent reforming of social groups, and a large individual in one group may be a smaller individual in the next, selecting for sex reversal back to a female.

Similar degrees of sexual lability occur in simultaneous hermaphrodites in response to changes in levels of sperm competition and levels of male and female reproductive opportunities. There are both interspecific (Petersen, 1991a) and intraspecific (Petersen, 1990b; Petersen and Fischer, 1996) changes in the percentage of gonadal tissue allocated to male versus female gamete production. As the amount of sperm competition increases, individuals allocate a higher proportion of gonadal tissue to sperm production (Petersen, 1990b; Petersen and Fischer, 1996). In addition, individuals appear to modify allocation depending on the fecundity of their partners, increasing testicular allocation when they have opportunities to fertilize larger batches of eggs (Petersen and Fischer, 1996).

All of these examples point out the extreme sexual lability of fishes, and their ability to respond ontogenetically and evolutionarily to changes in their social status and mating opportunities. In addition, coral reef fishes have been shown to respond to a wide variety of environmental cues to induce sex change; Cole and Shapiro (1995) demonstrated that social cues as well as water-borne cues can affect the probability of sex change in the goby *Coryphopterus glaucofraenum*.

III. Applied Behavioral Ecology: Some Examples

A. Exploitation of Sex-Changing Species

Bannerot *et al.* (1987) modeled life history characteristics of protogynously hermaphroditic groupers and found that under the most realistic assumptions, these species should be less vulnerable to exploitation, as compared to a similar gonochoristic (separate-sexed) species. In some ways, protogynous hermaphroditism has the potential to be especially resilient to fisheries pressure. The largest individuals are males, so taking large individuals should not greatly influence the immediate fecundity of the population. Also, if individuals in populations are able to assess changes in sex ratio, then successful sex change would provide for increased numbers of mature males to replace removed males, although at some cost to population fecundity.

Because exploitation of larger fishes on coral reefs is expanding rapidly worldwide, and because many of these species are sex changers, it is worthwhile to pursue this topic in detail, using groupers as an example. Even in the earliest published observations of spawning aggregations, there was concern for a possible shortage of males at spawning aggregations leading to reduced fertilization rate and a resulting reduced recruitment into the fishery (Smith, 1972; Shapiro, 1984a; Bannerot *et al.*, 1987; Huntsman and Schaaf, 1994). The possibility of this failure depends on critical aspects of the reproductive biology of groupers: the proximal mechanisms of sex change and their timing, the ability of males to fertilize multiple females at a rate perhaps more than is historically typical, and the degree to which males are fished relative to females.

Little is known about the mating and fertilization success of groupers at aggregations. Several, such as the Nassau grouper, are known to group spawn (Carter, 1989; Domeier and Colin, 1997), whereas species such as coral trout, *Plectropomus leopardus,* pair spawn (Samoilys, 1997a). Given that protogynous sex change is most adaptive under pair-spawn situations, whereby large males can monopolize matings (Warner, 1975, 1988b,c), it may be that species characterized as group spawners have simply suffered the loss of relatively large males. There is substantial evidence that there is a decreased proportion of reproductively active males at spawning aggregations in some species, leading at least to the possibility of sperm limitation. Vincent and Sadovy (1998) highlight two examples, the porgy *Chrysolephus puniceus* and the gag grouper *Mycteroperca microlepis*. In both species there has been heavy fishing pressure, especially on large individuals, and in both species the sex ratio has become more female biased. For *M. microlepis,* the proportion of males and sex-changing individuals (fishes with transitional gonads) in the adult population has dropped significantly in fisheries samples, from 17.5 to 2.5% in the Gulf of Mexico (Fig. 2) (Koenig *et al.,* 1996) and from 21 to 5.5% in the southeast Atlantic (McGovern *et al.,* 1998); this latter paper suggested that the more recent sex ratio might even be an overestimate of the proportion of males. Using the current estimates of each sex in the adult population, the sex ratio appears to have changed in approximately 20 years from about 4:1 in favor of females to 17:1 in the southeast Atlantic and 39:1 in the northeast Gulf of Mexico. It seems reasonable to assume that sperm limitation is potentially important when the sex ratio increases 4- to 10-fold, but how would we expect this to be manifested in changes in fertilization rate? Data from two smaller species of reef

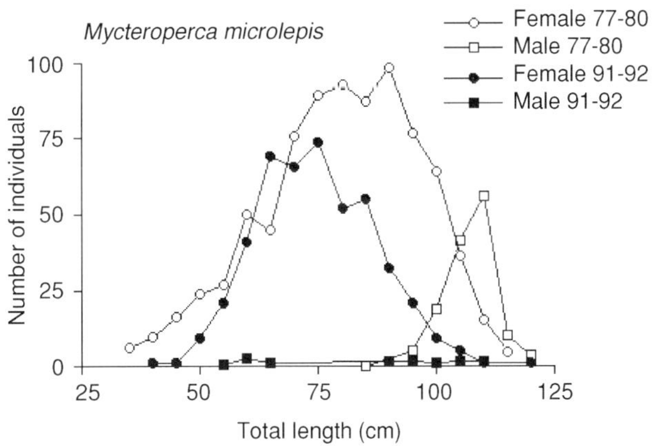

FIGURE 2 Size frequency distributions by sex for gag grouper (*Mycteroperca microlepis*) from the Gulf of Mexico. Historical data are from Hood and Schleider (1992, as cited in Koenig *et al.*, 1996) in 1977–1980; additional data from 1992 come from Florida State University. Figure redrawn with permission from Koenig *et al.* (1996).

fish may be able to shed some light on how fertilization might be influenced.

Marconato *et al.* (1995) studied the fertilization rate in *Xyrichthys novacula,* a Mediterranean wrasse with a haremic mating system. The number of females spawning per day is highly predictable for a male, because harem size is relatively constant and all females spawn daily during the reproductive period. During the study, one male disappeared and a male enlarged his harem to include the second harem, obtaining a mating success two to three times higher than the other two spawning males in the study. The fertilization success of the male was approximately 10% less than two other males that had not experienced changes in mating success (86.6 vs. 96 and 97%), and he appeared to release about one-quarter the number of sperm per spawn relative to these other males.

The relationship between the amount of sperm released in a spawn and fertilization rate of eggs should be asymptotic, with the shoulder of the curve near 100% if all eggs are viable (Petersen, 1991b; Warner *et al.*, 1995; Shapiro and Giraldeau, 1996). If a species' average spawn has a sperm number well beyond this shoulder of the fertilization curve, then substantial decreases in sperm per spawn will have little effect on fertilization rate. In contrast, the non-linear relationship between sperm release and fertilization success can lead to substantial decreases in fertilization success with small declines in total sperm released when sperm numbers are initially low (Fig. 3).

In the bluehead wrasse we have altered male mating success in field populations, and have been able to induce sperm depletion and declines in fertilization by increasing a male's mating success to unexpectedly high levels (C. W. Petersen, R. R. Warner, and D.Y. Shapiro, unpublished data). In these experiments males quickly changed their sperm allocation per spawn, decreasing it at the beginning of the spawning period, and in a period of a few days were allocating sperm independent of order during the spawning period. The mating success of individuals in this species is normally quite predictable because females tend to show high site fidelity (Warner, 1987). However, the success of a site can vary from day to day due to changes in current direction (Warner, 1986) and to the proportion of females spawning on a given day (Schultz and Warner, 1989).

Although fertilization rates declined in these two examples, it is important to note that both were short-term responses of males that normally have very predictable spawning prospects. The spawning rate for a male grouper joining a spawning aggregation should be less predictable than that for a resident reef fish. Because there is undoubtedly a larger cost for running out

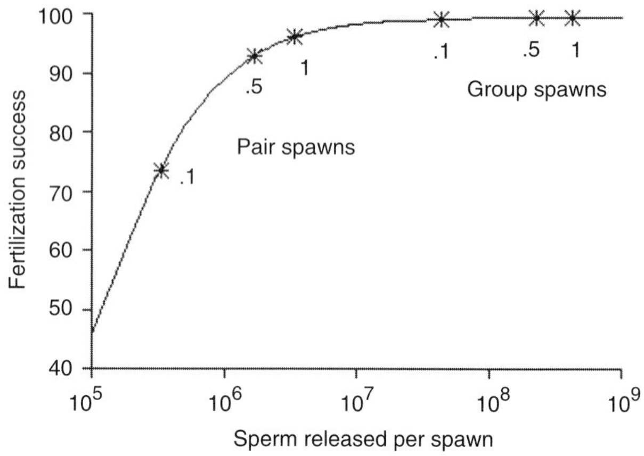

FIGURE 3 Fertilization curve for bluehead wrasses, showing the predicted effect of reduced sperm output on fertilization success in spawns. For group spawns, reducing the mean number of sperm released per spawn (427 million) to 50% (0.5) or 10% (0.1) has virtually no effect on predicted fertilization success, but the same proportional decreases in pair spawners (mean of 3.3 million sperm released per spawn) substantially reduce fertilization success. Data from Shapiro *et al.* (1994b) and Warner *et al.* (1995).

of sperm when reproductive opportunities exist than there is for producing excess sperm, life history theory would predict some bet hedging in sperm production by males to reduce the probability of sperm depletion in an unpredictable mating situation.

Also, if group spawning is the common spawning mode in some species, then sperm limitation may be less likely. Because of sperm competition, the number of sperm released in a group spawn is substantially higher than it is in pair spawns (Shapiro *et al.*, 1994b; Warner, 1997b; Petersen and Warner, 1998), suggesting sperm limitation of fertilization rate will not occur even if sperm release is reduced severalfold. In the bluehead wrasse, group spawns average at least 10 times the sperm output compared with pair spawns in the same species (Shapiro *et al.*, 1994b). This difference in sperm output has only minimal effects on fertilization success (Fig. 1B) (Marconato *et al.*, 1997). If reducing the proportion of males merely reduces the average number of sperm in a group spawning species 5- to 10-fold, fertilization rates may be only moderately affected (Fig. 3). If true, then at least for aggregating species that group spawn, substantial increases in the female:male sex ratio may have little effect on the percentage of fertilized zygotes produced by females.

There may be several reasons to doubt this optimistic scenario. First, as sex ratios increase to levels near 50:1, it is increasingly probable that in some smaller spawning aggregations no males will arrive, leading to at least a temporary disruption in spawning. Second, the studies showing that males that can control the amount of sperm released (relative to factors such as female size) are from pair spawnings (Shapiro *et al.*, 1994b, Marconato and Shapiro, 1996). It is not known if group-spawning males, with their larger sperm production and output, have the ability to regulate sperm release. If they do not, then these individuals may not be capable of lowering their sperm output per spawn and may experience sperm depletion during their tenure in a spawning aggregation.

Finally, in species that pair spawn in spawning aggregations, sperm limitation of zygote production is much more likely than in group-spawning species. In pair-spawning species, we expect the relationship between male mating success and size to be stronger, and overall there should be a lower proportion of males in the total population. These factors, combined with the smaller amounts of sperm released in pair spawns, will make pair-spawning species more vulnerable to intense fishing, especially of large size classes. The lower sperm output per spawn will tend to be associated with smaller testis size, and these males would potentially be unable to compensate for large increases in spawning rate. Unfortunately, as noted above, there is reason to believe that most sex-changing species evolved under conditions in which pair spawning controlled by large males was the predominant mating mode, regardless of whether group spawning also existed.

One problem with predicting the long-term response of individuals to intense fishing is that we do not understand the proximal basis of sex change in any large hermaphroditic species. Our best examples, owing to the excellent field experiments by Robertson (1972) and Shapiro (1980), showed a 1:1 immediate replacement of males by the largest females within a social unit, but these were in permanent social groups. In species that form temporary spawning aggregations, there appears to be little or no correspondence between the social structure at the aggregation and the interaction among individuals throughout the rest of the year (Shapiro *et al.*, 1994a). The two sexes are often found in different habitats during the nonreproductive season (McGovern *et al.*, 1998). Thus, if sex is influenced by social factors, such as the relative size of conspecifics and sex ratio, most of that information is available only at the spawning aggregation. In one large species, transitional individuals tend to be found immediately after the spawning season, suggesting that sex change is induced after spawning (McGovern *et al.*, 1998). Despite the presence of protogynous hermaphroditism, sex change appears to vary among species. In at least one species that now tends to group spawn, the Nassau grouper, sex change occurs over a wide range of sizes, with some potential prematurational sex change, whereas in species in which pair spawning predominates sex change appears to be more closely related to individual size.

The data on depletion of grouper stocks clearly show that they have been or are being overfished in many locations (Sadovy, 1994). Spawning aggregations at traditional sites and times make these species particularly susceptible to overfishing. Although protogynous hermaphroditism can increase the resiliency of a stock to depletion, the extreme levels of male mortality and the low level of compensation through sex change in the species studied suggest that a shortage of males and potential sperm depletion at spawning aggregations is a real possibility. What is missing from the current analysis is the pattern of spawning rate, sperm output, and fertilization success for either group or pair spawns in these species. Although the data on spawning rate might be quite difficult to obtain, data on fertilization success, and the occurrence of decreased fertilization success over the duration of a day or of the spawning aggregation, might be obtainable with divers or stationary nets collecting eggs after spawning.

B. Exploitation and the Allee Effect

Exploitation of tropical reef fishes tends to have immediate consequences for reef fish communities. The most noticeable effect of this exploitation is that the number of large, mobile, often piscivorous fishes is greatly reduced. There are several reasons that fishes are exploited, including their use as food and for the aquarium trade and as souvenirs or medicines (Vincent and Sadovy, 1998). Our discussion here is independent of the type of exploitation, and solely relies on the direct effect of reduced densities on the population biology of exploited species.

If an individual fish gains advantages by aggregating or through other social behaviors, then decreases in population density have the potential to decrease individual fitness, the opposite effect of what we typically expect in a population, in which competition causes decreases in individual fitness with *increasing* population density. This positive density dependence, the Allee effect, has the potential to decrease per capita growth rates of populations at low densities, potentially leading to their extinction (at least in closed populations) (Courchamp *et al.*, 1999; Stephens and Sutherland, 1999; Petersen and Levitan, 2001).

In tropical reef fishes, there are three possible manifestations of the Allee effect that we will consider here. By reducing the number of fishes in a local area, the effectiveness of schooling as a foraging strategy to obtain resources in territories defended by individual fish may be greatly reduced. Two major examples of this foraging strategy have been documented in coral reef fishes: schools of herbivorous fishes gaining access to algae in territorial damselfish territories (Robertson *et al.*, 1976; Foster, 1985) and schools of omnivorous fishes gaining access to eggs defended by territorial males (Foster, 1987). By reducing numbers of the schooling species, the access to defended food for individuals from the school could be greatly reduced. This positive frequency dependence in foraging ability with local population size is a form of an Allee effect. Overfishing is most likely to reduce intermediate- to large-sized fishes such as the surgeonfishes and parrot fishes found to enter damselfish territories in the Caribbean, rather than small omnivorous egg predators.

A second form of the Allee effect has been proposed in marine organisms: the potential lowered fertilization success of eggs due to low densities of spawning males in externally fertilizing species that have been exploited. This possibility has been most often associated with shallow-water temperate sea urchins and abalones (Quinn *et al.*, 1993; Levitan and Sewell, 1998; S. N. Murray *et al.*, 1999), examples of highly exploited species. Theoretical models and experiments *in situ* have suggested that fertilization can be limited by the population density of males in these species (Levitan and Sewell, 1998).

In contrast, the high fertilization success recorded from pair spawns in several species of coral reef fishes suggests that local density is not an important factor controlling fertilization success in these species. Although the possibility exists that individuals in some species would not spawn without a minimal number of males in an area, observations from nonaggregating species suggest that females will spawn with any courting male. For example, when we have removed large territorial males from spawning sites of the bluehead wrasse, small males (that are not normally territorial and typically do not actively court females) will enter the area, court, and sometimes spawn with females until excluded by larger males (R. R. Warner and C. W. Petersen, unpublished observation). Thus, the Allee effect is unlikely to be important through its effects on fertilization failure in coral reef fishes, although it may be important in other marine organisms (Levitan and Sewell, 1998; Petersen and Levitan, 2001).

A third consequence of exploitation in coral reef fishes is reduced effective population size, which has the potential to increase the effects of genetic drift and lead to reduced genetic diversity in populations of reef fishes. Effective population size in exploited coral reef fishes is reduced for two reasons. Most obviously, population size decreases. Less obviously, populations at lower population size are more likely to have mating systems with highly skewed male mating success, such as the haremic or resource defense mating systems seen in many species at intermediate densities (Warner, 1991). This increased mating success of a few males results in the offspring having the equivalent genetic diversity of a randomly mating population of much smaller size. These decreases in local effective population size will be less of a problem if local populations are embedded in much larger metapopulations that are connected via larval dispersal. Interestingly, the first two direct measurements of local retention vs. dispersal of young in coral reef fishes indicate much higher retention rates than previously expected (Jones *et al.*, 1999; Swearer *et al.*, 1999).

C. Marine Reserves

It is clear that conservation efforts in both temperate and tropical marine ecosystems are turning toward the establishment of marine reserves (Carr and Reed, 1993). Kramer and Chapman (1999) point out that much of the function of reserves depends on the behavior of the species contained within such protected areas. Even simple considerations of whether adults are likely to move out of reserves (and thus be available for capture) are often ignored in models (e.g., Sladek Nowlis and Roberts, 1997). It is clear that some species have behaviors that enhance take of adults on the reserve borders (Russ and Alcala, 1996a; Kramer and Chapman, 1999). We remain ignorant of the extent to which larval production is successfully exported from reserves. Equally, we know little about habitat selection behavior of settling young, and this can be a critical bottleneck in population dynamics of any species for which local production is decoupled from local recruitment (Warner *et al.*, 2000).

It is imperative that reserves be designed with the proper criteria appropriate to the goal of the reserve. If they are to be biodiversity reserves, then we need knowledge of the home-range characteristics and settlement preferences of the charismatic fauna and those species that are key components of local food webs. If they are to be fishery management tools, we need information about the commercially important species. Although local settlement aids both goals of reserve design, the nature of the ultimate consumer begins to play a role at this point. Sportfishing interests may be content with the occasional export of very large adults, so an ideal scale of a reserve might represent where an increase in home-range size with length brings large adults outside a reserve. For commercial clients, the complete protection of sedentary, large adults might be ideal for the massive export of seed stock to aid recruitment over a larger geographic range.

IV. Future Directions

A. Scaling up Observations

Due to their higher density, lower mobility, and more frequent and predictable spawning behavior, we know much more about the reproductive ecology of small reef fishes than about the larger reef fishes that tend to be the focus of most fisheries. Although our references are certainly a biased sample, we cite more papers for the bluehead wrasse (22), a small reef fish with no value as a fisheries species, than we cite for major groups of coral reef fishes that are commercially important (groupers and snappers combined, 14). This creates a potential problem if scientists or managers attempt to generalize knowledge from coral reef fishes to larger species that are most often exploited for food in coral reef fisheries. Some interspecific patterns appear to be correlated with species size (e.g., Sadovy *et al.*, 1994, Sadovy, 1996), so the generality of observations on small species may not be completely transferable to larger species. Some of these patterns include a decrease in fecundity per weight with increasing size, a decrease in instantaneous mortality with size, and changes in spawning migration and aggregation behavior with individual size.

An additional consideration of scale is the spatial and temporal patterns of selection and gene flow for any behavioral adaptation. Behaviors may change dramatically from one population to the next in coral reef species, suggesting local adaptation or phenotypic plasticity. Other behaviors may be stable over wide geographic areas, suggesting selection for an average response that works well over the dispersed range of the species. We do not know the extent of geographic variation in behavior in any coral reef species, and this knowledge relates directly to our ability to manage them effectively. In other words, comparative behavioral observations and experiments can help to identify the extent to which local populations can achieve local adaptation. This relates directly to the next point, discussed in Section IV,B.

B. The Role of Behavior in Linking Production and Recruitment

To what extent does local production contribute to local recruitment? This is a central question in marine ecology, and forms much of the current focus in marine conservation and management. Research now suggests that locally produced larvae may sometimes recruit to their parental reefs, the result of accumulation over many spawning events (Jones *et al.*, 1999; Swearer *et al.*, 1999; reviewed in Warner *et al.*, 2000). However, we know very little about when, where, or how the larvae accomplish this task. Any behavioral rules would have to be broadly applicable because of the diversity of physical environments that marine organisms face. Understanding how larvae influence their own dispersal or retention will let us know how much individual biology impacts future population size. A rich area for future research will combine larval behavior studies, intensive near-shore surveys, and coastal oceanography to reveal the mechanisms of retention, dispersal, accumulation, and eventual recruitment.

V. Conclusions

The decline in the study of the behavioral ecology of coral reef fishes appears to be due, at least in part, to an increase in applied work, including conservation and management of coral reef fishes. This pattern is particularly noticeable in reproductive ecology and behavior of coral reef fishes, in part because there may be few connections between reproduction and changes in population size at the level of individual coral reefs. Despite this decline in effort, we believe that knowledge of the advances in the reproductive biology of coral reef fishes is centrally relevant for conservation biology and management.

For much of behavioral ecology, the link with conservation biology has come through a greater understanding of how patterns of individual behavior can affect population growth through patterns of individual growth, survivorship, and reproduction. However, coral reef fishes and many other marine populations may be relatively open, with some larval transport between populations. In this case, the link between local reproduction and population growth may either not exist or be very weak. Our ignorance here points to the importance of answering this question. If coral reef fish populations are open, then it is difficult to make the connection between production and population size. However, even if production has little to do with local population size, reproductive ecology still has strong connections to population biology. This is because the effects that we have identified as important (particularly population density and size structure changes) are occurring universally. Isolated changes are interesting in the context of local experiments, but wide-scale alterations play a major role in management and conservation. We especially need to turn our attention to the behavioral ecology of larvae and juveniles, because recruitment is the key to much of marine population biology.

Interestingly, coral reef fishes also have promise as the best underexploited source of information in pure behavioral ecology. They are often fearless, colorful, conspicuous, and behavior rich, mating and feeding much of the day. They are much easier to watch and manipulate than are birds or mammals, and, given the thousands of species, they show unimaginable variety and plasticity. All it takes is a good knowledge of the outstanding problems in behavioral ecology, and a willingness to get wet.

Acknowledgments

We thank Ken Clifton, Isabelle Côté, John Reynolds, Yvonne Sadovy, and Paul Sikkel for sharing ideas concerning many of the topics discussed in this chapter. Helen Hess, John Anderson, John Reynolds, and Peter Sale commented on the manuscript. This work was supported in part by the Partnership for Interdisciplinary Studies of Coastal Oceans (funded by the Packard Foundation), by the National Center for Ecological Analysis and Synthesis (University of California, Santa Barbara), and by the Davis Center for International and Regional Studies (College of the Atlantic).

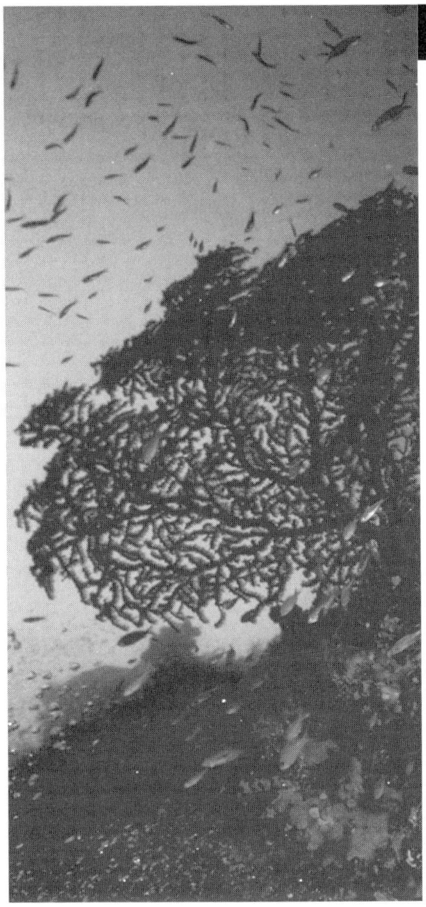

SECTION II

Replenishment of Reef Fish Populations and Communities

Introduction

Few organisms are eternal, and populations composed of organisms cease to exist if they are not replenished by the addition of new juveniles to take the place of individuals that die. The five chapters in this part, and the five in Section III, all have some relationship to the dynamics of reef fish populations and to this fundamental process of replenishment. The division of Chapters 6–15 into two parts is a convenience for the reader; although my intention originally was to group in Part II all chapters that related directly to larval ecology, to settlement, and to early, postsettlement survivorship. The subdivision enhances focus on intricate processes.

In 1991, Jeff Leis argued that reef fishes were pelagic animals with a demersal adult phase, in order to make the point that we could not afford to neglect studies of larval ecology. There have been significant advances in knowledge of larval ecology and in behavior of reef fishes in the past decade, and these advances are well reviewed in the first three chapters of this part. Behavioral studies from several labs have demonstrated that larval reef fishes may be quite capable swimmers, able to get themselves to reefs at the end of larval life, and have identified possible cues to the presence of reefs that are used to facilitate this active movement. Further advance in understanding how reef fishes may actively find their way to reefs demands an understanding of the sensory capabilities these fishes possess. Art Myrberg and Lee Fuiman have provided, in Chapter 6, a masterful review of the sensory capabilities of reef fishes, with special attention to what is known about the abilities of larval stages. Their chapter provides an up-to-date statement of what is known to be possible, and identifies the gaps in knowledge. It should be essential reading for anyone contemplating behavioral studies with larval reef fishes.

In Chapter 7, Bob Cowen reviews the range of physical processes that may influence the distribution and dispersion of reef fish larvae in the pelagic environment. He considers each oceanographic process in turn, providing examples that illustrate its interaction with larval distribution. He also considers what will be needed to advance our understanding of the biophysical interactions during larval life. That his chapter focuses on ways in which physical mechanisms may act to slow advection, or to retain larvae near natal sites, rather than to disperse them, is a clear demonstration of the changes in thinking that have occurred in the past decade.

Jeff Leis and Mark McCormick team in Chapter 8 to review the biology, behavior, and ecology of reef fish larvae. After a polite admonition to their colleagues to read the temperate fish literature (with an acknowledgement that reef fish larvae differ in a number of ways from temperate ones), and a survey of technological advances that have helped explore aspects of larval biology, they follow a logical path from production of larvae, to the lives of larvae in the pelagic realm, to the return to the reef, settlement, and metamorphosis. Recent years have amplified our understanding of these critical ontogenetic stages, but there remain impressive gaps and this chapter highlights where they exist.

Our need to better understand larval biology arises from the role of larvae as the link between neighboring local populations. Though we have learned a lot, we still do not know, for any reef species, the extent of "effective" larval dispersal, i.e., the spatial extent over which sufficient numbers of larvae travel to maintain connectivity among local demersal populations. Knowing the spatial scale of connectivity is important both for building a better understanding of reef fish demography and for more effective management of what are increasingly important resources. Solving the question of just how far apart local populations can be, and still be connected, will be achieved partly by continued, intensive study of larval biology. However, the question is sufficiently difficult and important to warrant application of additional approaches. One of these is the use of population genetics, because populations that are not connected do not exchange genes. Modern molecular genetics provides a broad suite of techniques for resolving genetic relatedness among populations. Serge Planes uses Chapter 9 to review studies of population genetics, with particular reference to what these have told us about the geographic scales of connection among reef fish populations. As well as surveying studies completed to date on reef fish species, Planes uses the genetic information to generate biogeographic and evolutionary hypotheses concerning reef fishes of the Indo-West Pacific region.

Chapter 10, by Geoff Jones and Mark McCormick, draws attention to the importance of replenishment of the young organism's ability to fill needed energetic requirements. The processes of metamorphosis, growth, and ontogenetic development require energy, and the gaining of that energy brings the young fishes into contact with competitors and predators. Fishes that settle after a well-fed larval life may be better equipped for the rigors of juvenile life, and fishes that are doing well energetically will better withstand the challenges of their demersal life. Jones and McCormick argue that we need to understand the energetic requirements and processes if we are to make sense of demography, and that this interplay between energetics and demography extends throughout life to include success in reproduction. Their chapter amplifies this argument with a review of a broad range of chiefly experimental studies, and imbeds it in a thoughtful discussion of the regulation of population size. In this way, their chapter leads logically into the chapters to follow in Section III.

CHAPTER 6

The Sensory World of Coral Reef Fishes

Arthur A. Myrberg, Jr.
Rosenstiel School of Marine and Atmospheric Science
University of Miami
Miami, Florida 33149

Lee A. Fuiman
Department of Marine Science
University of Texas at Austin
Marine Science Institute
Port Aransas, Texas 78373

I. Introduction

The fundamental goals of survival and reproduction require, throughout the life of any animal, countless decisions relative to given courses of action. These decisions depend, in turn, on accurate and relevant information from the sensory world to which each animal is attuned. The interface of that world with behavior and ecology is essential for explaining or predicting the mechanisms that regulate responses of animals to aspects of their environment and their distribution in space and time. Where adaptive radiation has been extensive, such as that evidenced by fishes, it is reasonable to expect that different species will vary as to the degree that each relies on specific sensory modalities to provide the information needed to answer specific sets of problems.

Fortunately, the major sensory systems use different channels of physical/chemical energy to obtain the information needed for appropriate response; through such channels, each system can be individualized, so to speak, by appropriate experimental manipulation, to reveal its respective capabilities relative to possible information carried by a given channel. A wide variety of behavioral and neurobiological techniques is used to examine such capabilities, but these techniques possess important caveats. The worker, using behavioral techniques, must be concerned that, during examination, animals confront conditions appropriate to their natural environment. Neurobiological results usually center on "avenues" lying between sensory receptors and the central nervous system, and the information passing along such avenues, although coherent, may not necessarily be "meaningful" to the animal (Blaxter, 1988a). Such techniques, when cautiously applied, in combination with appropriate ecological information, can provide enormous insight into the likely ways that a subject may well answer specific problems faced during its lifetime.

Although there exists much knowledge about many sensory systems of fishes, the information available clearly centers on those species that either are relatively easy to obtain and/or maintain in the laboratory or have high economic value. The former are dominated by freshwater species and the latter are dominated by marine species from the Baltic or the English seas. With few exceptions, relatively little is known about the sensory biology of coral reef fishes. This is likely due to difficulties in obtaining reasonable sample sizes, maintaining species in laboratory settings for reasonable periods, and traveling to appropriate facilities near coral reefs.

With this in mind, we have decided that along with information presently available on the sensory biology of fishes from the coral reef, we shall provide additional information about the sensory biology of fishes from other habitats that may well be relevant to future studies of the sensory world of coral reef species. Fortunately, the physical principles controlling the flow of energy used by sensory systems are not constrained by the staggering array of habitats populated by fishes. And though variation clearly occurs in energy dynamics, it follows predictable rules arising from the physical principles. We hope that our summaries will instill interest in extending findings from fishes of other habitats to species of the coral reef, because only through such efforts can we truly understand the ecology of coral reef fishes. Based on recent interest in the abilities of larval and early juvenile stages of many fishes of the coral reef, we have also summarized information on the major senses of the early life stages of fishes from the coral reef as well as those found in fishes from other habitats.

We review each major sensory system separately. As more knowledge accrues, it appears that often a course of action is initiated by information obtained from one sensory modality; however, cross-modality compliance often dictates further steps along the course, and when complex activities occur, mediation via many, if not all, senses is needed to achieve a given goal.

II. The Visual System

The clear waters around coral reefs provide for a well-lit daytime environment and the potential for useful illumination during dawn, dusk, and moonlit nights. It is obvious that a great many of the interactions of coral reef fishes with their predators, prey, competitors, mates, and physical environment rely on light and especially on vision. McFarland's (1991) contribution to the previous edition of this book is a comprehensive description of the visual world of a coral reef, from the characteristics of light in the tropical reef environment to the visual capabilities and adaptations of its fishes. It remains a valuable resource. Here, we summarize parts of this earlier contribution and supplement it with recent findings and information on related topics.

A. Light in the Aquatic Medium

Light can be characterized by its intensity and spectral composition, both of which change as light passes through water. These changes occur because water molecules, suspended matter, and dissolved substances scatter and absorb light. In water containing little suspended and dissolved matter, the character of light changes rather predictably with depth. The intensity of downwelling light decreases by about 90% for every 75 m of depth in clear oceanic waters (Denton, 1990). However, the longer wavelengths (yellow, orange, and red hues) are selectively absorbed and so do not penetrate as far as shorter wavelengths, so that the blue and green regions of the spectrum dominate below a few tens of meters (see McFarland, 1991). Shallow coastal areas present a much more complex light environment. There, the higher load of suspended material in the form of sediments, plankton, and dissolved organics increases the scattering and absorption of light, resulting in reduced clarity, a broader spectrum, and a shift toward longer wavelengths (green–yellow) (McFarland and Munz, 1975a). In coral reef environments, different habitats can have distinct irradiance spectra, grading from the blue waters of the outer reef, through the progressively greener waters of the middle and inner reef and estuary (Lythgoe *et al.*, 1994). Further, photic conditions on a reef change during the day due to the angle of the sun and as a result of the activities of the fish community. McFarland (1991) attributed observed changes in water clarity and hue (from blue to greenish) on coral reefs between morning and midafternoon to the accumulation of excretions by herbivores such as parrot fishes. These differences in the ambient light field directly influence a fish's visual perception.

B. Morphological Features

The eyes of most teleosts are oriented and designed so that light can enter from almost any direction—above, below, forward, or behind. In front, a fish can view an object with both eyes to create binocular vision and improved depth perception. Frontally positioned eyes, such as those depicted for *Halophryne diemensis* (Batrachoididae) by Collin and Pettigrew (1988a), greatly increase the size of the binocular field. Many species can use ocular movements to adjust the size of the binocular field. Sea horses, pipefishes, and at least one creediid sand lance (*Limnichthys fasciatus*) have an extraordinary degree of ocular mobility, being able to rotate each eye independently through large arcs. Most fishes can see objects in the lateral fields with only one eye, where depth perception is presumably compromised. However, *L. fasciatus*, a predator that lies in wait buried in the sand, appears to be able to judge the distance of its prey monocularly (Pettigrew *et al.*, 1999).

Light entering the eye is focused on the retina by a spherical lens that moves away from or toward the retina to accommodate objects near or far. The neural circuitry from the brain extends through the optic nerve to the retina, which is composed of layers of distinctive neural cells (ganglion, bipolar, photoreceptors). The major classes of photoreceptors are cones and rods. Cones are generally larger but less numerous than rods, although labrids have very small cones and young larvae of most species lack rods altogether (see below). Vision under bright conditions (photopic vision) is mediated by cones, and rods take over the visual function when light levels are low (scotopic vision). Both types of photoreceptors operate by photoactivation of visual pigments, principally rhodopsin in marine fishes, which are contained in the outer segment of each photoreceptor. Visual pigments generally respond to a broad range of wavelengths but have a peak sensitivity, identified as the wavelength of maximum absorption, λ_{max}. Up to four classes of visual pigments, each with a different λ_{max}, can be found in the different types of cones. This is the basis for color vision. Species with a single pigment discriminate objects from the background by differences in contrast; wavelengths closer to their λ_{max} appear brighter than other wavelengths.

Variations to this basic design allow some species to exploit different habitats and niches more effectively (Muntz, 1990; Bowmaker, 1990, 1995; Collin, 1997). We discuss the major features that improve the spatial resolution (acuity), scotopic sensitivity, and spectral sensitivity of fishes in reef habitats. Space limitations do not permit us to examine other aspects of visual performance, such as the perception of movement, size and distance of objects, shapes, and polarized light (see Douglas and Hawryshyn, 1990).

C. Spatial Resolution

Visual acuity, the degree to which a fish can resolve detail within its visual field, is largely dependent on eye size and the density of retinal photoreceptors and ganglion cells. In larger eyes, light from a given visual angle is projected onto a larger area of the retina and potentially more photoreceptors to produce greater acuity. Carnivorous fishes generally have larger eyes as compared to herbivores, probably because foraging on moving and elusive prey is more demanding of the visual system (Pankhurst, 1989). Acuity is also directly related to the density of photoreceptors, which varies among species. Diurnally active species, especially carnivores, generally have better acuity as compared to nocturnal fishes (Pankhurst, 1989).

The distribution of photoreceptors and ganglion cells across the retinal hemisphere is often uneven, and retinal topography in coral reef fishes can be related to ecological niche (Collin, 1999). Often, there are one or more regions (areae) in which photoreceptors or ganglion cells are densely packed to produce acute vision. Such regions are most common in the temporal (posterior) retina (area temporalis), so that forward vision benefits, especially the binocular field. The implications for feeding are obvious. Species living in complex habitats, such as among the coral rubble or branches and the overhangs or pockets in reefs, often have multiple areae. For example, the coral cod (*Cephalopholis minimatus:* Serranidae) has an area nasalis in addition to an area temporalis. Staghorn damselfishes (*Amblyglyphidodon curacao*) and blue angelfishes (*Pomacanthus semicirculatus:* Pomacanthidae) have both of these plus a third region of concentrated ganglion cells (Collin and Pettigrew, 1988a). These accessory areae help fishes maneuver in complex environments or fix on prey outside the binocular field (Collin, 1999). In addition, active predators with well-defined areae often exhibit increased eye mobility (Collin, 1999), which allows them to align their target for better vision.

Fishes that frequent open habitats have an uninterrupted view of the sand–water horizon. They use an area temporalis for viewing the binocular field, but they also have a band of concentrated ganglion cells across the meridian of the retina (see Collin and Pettigrew, 1988b). Examples include blue tusk fishes (*Choerodon albigena:* Labridae), red-throated emperors (*Lethrinus chryostomas:* Lethrinidae), and collared sea breams (*Gymnocranius bitorquatus:* Sparidae) The horizontal streak provides a low threshold for perceiving movement and a detailed panoramic view of the environment without the need for eye movements (Collin and Pettigrew, 1989). It is an efficient way to maintain surveillance for predators.

The quality of the visual image on the retina can be improved by eliminating "visual noise," which may result from scattered light entering the eye or from chromatic aberration. These types of noise are generally produced by short wavelengths of light, so that a yellow filter can improve visual contrast. Some diurnal reef fishes, such as wrasses and parrot fishes, have in the cornea or lens yellow pigments that would act as a filter. However, such a filter has its drawbacks. It would reduce sensitivity to those wavelengths it filters, thereby modifying color vision, and it may reduce scotopic sensitivity. Muntz (1990) pointed out that many species with yellow filters are immobile at night and may not be competitive under low-light conditions. Corneal iridescence may have the same result on vision by reflecting downwelling light out of the eye by constructive interference (Shand and Lythgoe, 1987).

D. Scotopic Sensitivity

Scotopic sensitivity refers to a fish's ability to use vision when light levels are low, such as at night, in caves, or at depth. Specializations that allow more light to enter the eye or increase the likelihood that a photon will trigger an impulse to the central nervous system

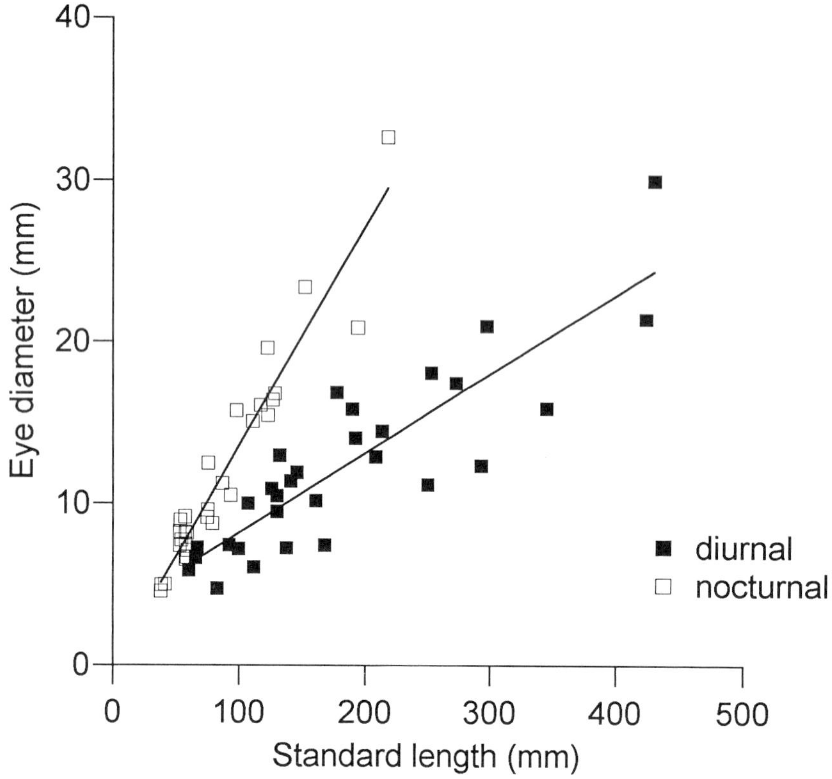

FIGURE 1 Eyes of nocturnal coral reef fishes are generally larger than those of diurnal species. The difference is greatest in large species, but converges in smaller species. Data are from illustrations in Böhlke and Chaplin (1968) for 30 nocturnal species (Apogonidae, Priacanthidae, Pempheridae, and Holocentridae) and 30 diurnal species (Chaetodontidae, Pomacentridae, Labridae, Scaridae, Acanthuridae, Balistidae, Ostraciidae, and Tetraodontidae). Each point represents a different species.

will enhance scotopic vision. Many of the adaptations found in deep-living species are also present in nocturnal reef fishes. For example, nocturnal fishes have larger eyes than do diurnal reef fishes (Pankhurst, 1989), with diameters as much as 30–110% larger in apogonids, holocentrids, pempherids, and priacanthids (Fig. 1). This suggests that the pupils are also larger and that more light enters their eyes. A larger eye improves the ability to detect point sources of light in a dark environment. However, by itself, a larger eye does not provide a brighter retinal image of the extended light source from an underwater scene because the image is projected over a larger, more distant retinal surface (Fernald, 1988, 1990). It may be that the larger eye provides more space for larger photoreceptors, which would improve sensitivity. We are not aware of a comparison of diameters of photoreceptor (rod) outer segments, but McFarland (1991) reported that a holocentrid and an apogonid (nocturnal species) had longer outer segments as compared to diurnal fishes.

The trends in eye sizes of reef fishes, especially their divergence with increasing body size (Fig. 1), raise other interesting questions. Do larger nocturnal fishes require disproportionately larger eyes, as compared to diurnal fishes, to achieve a certain level of sensitivity? Or, is it that as head size increases, the constraints on eye size are eased? The trends converge at a standard length of about 40 mm. Are there very small species that are nocturnally active? Of 13 species from five families active at night on Hawaiian reefs, all but one (*Scorpaena coniorta,* length 26 mm: Scorpaenidae) were longer than 52 mm (Hobson, 1972). Does eye size constrain the age at which a species may assume a nocturnal habit? Experimental results show that apogonid larvae can feed at light levels that are one-tenth of the minimum required by pomacentrids of a similar developmental stage (Job and Bellwood, 2000), although these levels may be considerably higher than those required by adults. Gut contents of recently settled *Apogon doederleini* (Apongonidae) suggest that they feed at night even when they are less than 20 mm in length (Finn and Kingsford, 1996). The value of a larger eye for nocturnal reef fishes remains unclear, unless, of course, some of the visual behaviors of nocturnal reef fishes are based on point sources of light after all. Nevertheless, one benefit of the larger eye and its larger projection area is magnification—the larger number of photoreceptors subtended by the image increases spatial resolution, which is probably important for nocturnal planktivores. But sensitivity must be more limiting to nocturnal activity than resolution, so how else do nocturnal reef fishes improve sensitivity? Retinal illuminance for an extended source is inversely proportional to the square of Matthiessen's ratio, the ratio of the distance from the center of the lens to the retina and the lens radius. Matthiessen's ratio was traditionally thought to be constant (2.55 in fishes), but has been found to vary by about 30% among species (Fernald, 1988). If nocturnal fishes have consistently lower Matthiessen's ratios (relatively larger lenses), they could make better use of available light at night. At least two holocentrids have a relatively low Matthiessen's ratio of about 2.1, resulting in a 10–20% increase in retinal illumination (McFarland, 1991). Matthiessen's ratios for five species of nocturnal tropical fishes (in four families, the Apogonidae, Holocentridae, Ambassidae, and Teraponidae) range from 2.3 to 2.4, whereas nine diurnal species have values of 2.4 to 2.6 (Shand, 1994b).

Specializations of nocturnal fishes (not necessarily reef fishes) that increase the chance of stimulating the central nervous system include rods spread over more of the retina (Munz and McFarland, 1973), higher densities of rods (Pankhurst, 1989), multiple layers of rods (McFarland, 1991), and, in or behind the retina, a reflective layer (tapetum) that sends light back through the photoreceptors (Nicol, 1989). In both nocturnal and diurnal fishes, many rods are usually connected to each ganglion cell, so that multiple photoreceptor signals can be summed to exceed the threshold. McFarland's (1991) calculations suggest that even with summation ratios (rods:ganglion cells) of 100, diurnal reef fish vision is probably light limited at twilight. Pankhurst (1989) showed that nocturnal and crepuscular fishes are better equipped for scotopic vision than are diurnal fishes because of differences in the length of the light path through visual pigments and the density and summation of rods. Exceptions to this were nocturnal species that used other sensory systems in darkness. The " twilight changeover" (Hobson, 1991), the dramatic daily change in community structure on coral reefs, attests to the ecological importance of visual sensitivity.

E. Spectral Sensitivity

Discrimination of wavelengths (color vision) requires the presence of at least two visual pigments with overlapping absorbance spectra (Bowmaker, 1995). Additional pigments provide even more color information, especially in typical shallow coral reef habitats, where the daytime spectrum can be quite broad. Several types of cone photoreceptors, commonly single cones and a variety of double cones, differ in their morphology and pigment content. Typically, double cones are sensitive to longer wavelengths than are the single cones, and their pigments vary with habitat to maximize sensitivity to the daytime ambient light. Thus, on the Great Barrier Reef, the double cones of lutjanids from the bluer outer reef habitats contain rhodopsins with λ_{max} at 487–518 nm, whereas congeners from an estuary exhibit cones with λ_{max} at 527–575 nm, and other species from middle and inner reef areas have intermediate values. The single cones in these snappers are maximally sensitive in the range of 420–450 nm (Lythgoe *et al.,* 1994). Likewise, *Pomacentrus coelestis* has double cones containing pigments with λ_{max} at 467 and 532 nm and two more pigments in single cones with λ_{max} of 493 and 354 nm (McFarland and Loew, 1994), demonstrating great potential for color discrimination.

The lower λ_{max} in *P. coelestis* (354 nm) is outside the visible spectrum for humans and in the near ultraviolet (UV). The discovery of UV-sensitive cones (λ_{max} at 350–390 nm) in fishes is relatively recent and considerable research attention has been brought to bear on this matter. Most reports of UV-sensitive cones have been for the younger stages of freshwater fishes. However, at least three adult pomacentrids, *P. coelestis, Dascyllus trimaculatus,* and *Chromis punctipinnis* (McFarland and Loew, 1994), and two larval atherinids (Loew *et al.,* 1996) have joined the list. Although UV light is readily absorbed and scattered, it does penetrate significantly in clear waters and may be adequate for vision to a depth of 100 m (McFarland, 1986; Losey *et al.,* 1999). It might be used to detect zooplanktonic prey by creating contrast with the background (i.e., dark, UV-opaque prey against a bright background), but it may also impart novel information about the species or condition of zooplankton (Loew and McFarland, 1990; Losey *et al.,* 1999). Behavioral experiments on young freshwater fishes that have UV-sensitive cones support the notion that UV light may be used in planktivory (Loew *et al.,* 1993; Browman *et al.,* 1994). Aside from foraging, these cones may be involved in communication (species recognition, signaling) via the iridescence and bright colors of reef fishes (Loew and McFarland, 1990). In fact, 100 species from 22 families of reef fishes, including labrids, chaetodontids, acanthurids, and pomacentrids, reflect large amounts of UV light from their skin (Losey *et al.,* 1999).

Visual pigments in the rods of coral reef fishes exhibit far less variation than do the pigments in cones. A large survey of coral reef species showed that rod pigments are maximally sensitive to light between 480 and 502 nm (McFarland and Munz, 1975a,b). This is supported by additional data for 12 lutjanids from the Great Barrier Reef in which λ_{max} ranges from 489 to 502 nm, despite differences in spectral irradiance of the daytime habitat (Lythgoe *et al.*, 1994). Peak sensitivity is highest for species that reside in the upper 5 m of the water column, shifting to the low end of the range for species living below 20 m (McFarland, 1991). The reason for the narrow overall range of rod sensitivity (480–502 nm) among coral reef fishes appears to be that it matches the optimal range (480–495 nm) for photoabsorption during twilight and night along most lines of sight (McFarland, 1991). This match of rod sensitivity to the underwater twilight spectrum produces a relatively bright background against which objects, such as predators or prey, appear in silhouette.

F. The System during Early Development

Development usually produces substantial modifications to the functionality of organ systems as a result of structural changes and growth, and the resulting changes in performance can be critical to survival and recruitment (Fuiman and Higgs, 1997). Much of the work on the development of the visual system in fishes has been based on freshwater species, which are generally larger and more developed at important ecological transitions (e.g., hatching, first feeding), as compared to larvae of coral reef fishes. There are few published studies on the development of the visual system in reef fish larvae, but the somewhat larger body of literature on temperate marine larvae provides useful insight. A more complete review of the lifelong changes in eye morphology was prepared by Powers and Raymond (1990).

At hatching, the visual system of most marine fishes with pelagic larvae is poorly developed and probably not functional. A distinct cornea and lens are apparent but the almost transparent eye cup lacks melanin and the retina is not differentiated into distinct layers. However, development is rapid, so that at least a minimal level of visual performance is achieved to meet the demands of foraging. By the time of first feeding, the eye is opaque (appearing black), the retina has distinctive layers, and the only photoreceptors present are cones. Rods begin to differentiate late in the larval period. The lentis retractor muscle, which moves the lens to focus the image on the retina, develops after first feeding. Prior to that, larvae are probably myopic in the forward field of view (Pankhurst *et al.*, 1993), but may be able to focus on more distant objects along other lines of sight (Shand *et al.*, 1999). In addition to these structural changes, the lens and the entire eye increase in size throughout the larval period and beyond, which have important consequences for visual capabilities.

1. Spatial Resolution

The continuous addition of cones to the larval retina occurs as the eye is growing. Ocular growth stretches the retina so that the density of cones in a given area of retina decreases during development (Shand, 1994a, 1997; Margulies, 1997; Higgs and Fuiman, 1998a; Poling and Fuiman, 1998). This suggests that visual acuity decreases during development, but the increasing focal length of the growing lens counters the effect of decreasing cone density. Thus, the density of cones within a given visual angle actually increases during development (Pankhurst *et al.*, 1993; Pankhurst and Butler, 1996; Poling and Fuiman, 1998). Acuity, typically expressed as minimum separable angle calculated from histological measurements, improves through the larval period in a wide range of species (see preceding references), including coral reef fishes (Job and Bellwood, 1996; Shand, 1997; Lara, 1999, 2001). It often reaches an asymptote near metamorphosis, which sometimes coincides with a habitat shift, such as settlement (Shand, 1994a, 1997).

Visual acuity can also be determined from behavioral experiments, a method that combines optics with higher order neural processing and one that may be more ecologically relevant. Several studies during the past decade have compared behavioral and morphological assessments of visual acuity during the larval period and found that the morphological approach overestimates behavioral acuity (Pankhurst *et al.*, 1993; Pankhurst, 1994; Higgs and Fuiman, 1996; Job and Bellwood, 1996). This difference has been attributed to the poor development in larvae of the brain's optic tectum (Schmitt and Kunz, 1989; Shand *et al.*, 1999) and the lentis rectractor muscle (Pankhurst *et al.*, 1993).

2. Scotopic Sensitivity

Rod photoreceptors first appear in the retina late in the larval period and proliferate so rapidly that their density increases on both an areal (Shand, 1994a, 1997; Higgs and Fuiman, 1998a; Poling and Fuiman, 1998) and visual angle (Pankhurst *et al.*, 1993; Poling and Fuiman, 1998) basis. Summation of rods onto ganglion cells increases rapidly once the rods begin to form, but by the end of the larval period there may be only 10 rods per ganglion cell (Higgs and Fuiman, 1998a; Poling and Fuiman, 1998), leaving considerably more neural connections to be made during the juvenile period. There is mounting evidence that Matthiessen's ratio decreases during ontogeny in at least some species. During the first 10–14 days after hatching the ratio decreases from 2.8 to 2.2 in *Forsterygion varium* (Tripterygiidae) (Pankhurst *et al.,* 1993) and from 3.6 to 2.3 in *Acanthopagrus butcheri* (Sparidae) (Shand *et al.,* 1999). In *Sciaenops ocellatus* (Sciaenidae), it decreases from 2.5 to 1.9 as larvae grow from 3 to 21 mm total length (based on data in Fuiman and Delbos, 1998). However, the pattern may be more complex than a simple monotonic decline, because adults of these species have higher values. Additional data for pomacentrid and apogonid larvae show increases and decreases (or broad variability) in Matthiessen's ratio during the larval period (Job and Bellwood, 2000). As discussed above, the retina is better illuminated when Matthiessen's ratio is low, so this may be important for visual sensitivity when larvae have few rods.

Comparative studies of visual system development are just beginning to appear. Shand (1997) examined four reef fish species around the time of settlement and found patterns in development that corresponded to their ecological niche. Rod density in *Apogon doederleini* (Apogonidae), a nocturnal planktivore, increased much faster and reached a much higher plateau immediately after settlement, as compared to two diurnal planktivores, *Stethojulis trigiventer* (Labridae) and *Pomacentrus moluccensis*. *Apogon doederleini* also achieved a much higher summation ratio (10:1) than did the others (3–4:1). The diurnal species maintained much higher densities of cones after settlement compared to *A. doederleini*. Despite these differences, visual acuity (minimum separable angle) for a given size of lens was the same in all species.

How do these developmental changes influence scotopic vision? The general pattern of retinal development suggests poor scotopic sensitivity for most of the larval period, followed by rapid improvement when rods begin to differentiate and summation begins. Several studies have demonstrated that scotopic sensitivity increases during the larval period (Champalbert *et al.,* 1991; Stearns *et al.,* 1994; Pankhurst and Hilder, 1998). The light intensity threshold for feeding in pomacentrid and apogonid larvae increases by 3–4 orders of magnitude during the larval period (Job and Bellwood, 2000). Likewise, scotopic sensitivity of *Sciaenops ocellatus* increases by a factor of 5000 during the larval period, but in two distinct phases (Fuiman and Delbos, 1998). The early phase is one of gradual improvement, representing 20% of the total change in sensitivity. The second phase begins at 8 mm total length, which coincides with the appearance of rods in the retina. Sensitivity during the latter phase appears to be limited by rod density, an observation that parallels previous work on the goldfish (Powers *et al.,* 1988). However, the more gradual improvement in sensitivity of *S. ocellatus* during the earlier phase seems to mediated by the cones. Pankhurst and Hilder (1998) attributed an increase in sensitivity of striped trumpeter (*Latris lineata:* Latridae) larvae to the appearance of double cones.

3. Spectral Sensitivity

Developmental changes in visual pigments and spectral sensitivity have scarcely been examined in marine fishes. Shand *et al.,* (1988) and Shand (1993) used microspectrophotometry to assay the visual pigments in the photoreceptors of the pollack (*Pollachius pollachius:* Gadidae) and the goatfish (*Upeneus tragula:* Mullidae) during the transition from a pelagic larva to a benthic juvenile. In the pollack, she found no change in the spectral sensitivity of rods (λ_{max} at 498 nm) or double cones (521 nm). Single cones shifted very quickly from a λ_{max} at 420 nm for pelagic juveniles of 50 mm standard length to 450 nm for benthic juveniles of 80 mm. In the goatfish, there was no change in rods ($\lambda_{max} = 498$ nm) or single cones (400 nm). The double cones were composed of a longwave-absorbing member and a shortwave member. In pelagic larvae, the members of the double cones had λ_{max} at 580 and 487 nm, respectively. The longwave member shifted to 530 nm after settlement, and the shortwave member shifted toward one of two classes (487 or 515 nm). Thus in goatfishes there was a rapid loss in sensitivity to longer wavelengths (>530 nm) at the time when larvae leave the surface waters for a benthic existence. In flatfishes, the transformation from a pelagic to benthic lifestyle is accompanied by major changes in the visual system. Pelagic larvae of the winter flounder (*Pleuronectes americanus*) have only single-cone photoreceptors with a pigment having a λ_{max} at 519 nm. After metamorphosis, the single cones shift to 457 nm, but double cones develop, having maximal sensitivity at 531 and 547 nm. Rods are maximally sensitive at 506 nm (Evans *et al.,* 1993). The differences in the direction of the λ_{max} shift and the photoreceptors undergoing the change during a vertical habitat shift invite further investigation of other species.

Early behavioral studies on the spectral sensitivity of marine fish larvae have not been superseded. Blaxter's (1968) experiments, based on the threshold for a photonegative response, suggest two peaks in sensitivity (at approximately 460 and 520 nm) in *Clupea harengus:* Clupeidae), with little change during the first couple of weeks after hatching. Results from his parallel experiments on flatfish larvae (Blaxter, 1969) were less clear, but *Pleuronectes platessa* (Pleuronectidae) appeared to exhibit a peak at 520 nm whereas *Solea solea* (Soleidae) had no obvious peaks in sensitivity.

The action spectrum for feeding by 10- to 15-mm long *Engraulis mordax* (Engraulidae) larvae also peaks between 510 and 550 nm (Bagarinao and Hunter, 1983). Unpublished experiments on the feeding of *Apogon compressus,* and the pomacentrids *Pomacentrus amboinensis* and *Premnas biaculeatus* (Pomacenrtridae) (S. D. Job and J. Shand, personal communication), show maximum sensitivity (λ_{max}) and median wavelength (λ_{P50}) near 500 nm. The λ_{P50} of the pomacentrids shifted toward longer wavelengths during development. The longer wave sensitivity of temperate species would be appropriate to the greener coastal waters.

III. The Acoustical System

A. Underwater Sound

Sound obeys the same physical principles whether traveling through air or water, but quantitative differences in its structure and behavior occur due to differences in the density and compressibility of the two media (see Urick, 1975). These differences include speed (4.8 times faster in water than in air) and wavelength for any given frequency (4.8 times longer in water than in air). The physical phenomena associated with sound propagation in water, as it applies to hearing in fishes, have been detailed since the mid-1960s, but with varying degrees of clarity to biologists (Harris, 1964; van Bergeijk, 1964; Parvulescu, 1964; Hawkins, 1973, 1993; Hawkins and Myrberg, 1983; Rogers and Cox, 1988; Schellart and Popper, 1992; Kalmijn, 1988a, 1997). This precludes the necessity to cover many of the phenomena here, except to clarify several aspects, because the sensory biology of this system has been reasonably explored in coral reef fishes.

As sound passes through water, its pressure is accompanied by oscillations (successive compressions and rarefactions) of water particles along the axis of propagation. Pressure, being a scalar quantity, provides information about sound intensity, but no information about the direction of propagation. Particle motion, on the other hand, has three interrelated vector quantities, displacement (distance each water particle moves), velocity (speed of such movement), and acceleration (rate of change in velocity), each of which provides information about the intensity of a sound as well as its direction. Sound pressure and the resulting particle motion are related to each other and to the distance of the sound source. Under idealized conditions, when the sound is near the source, the sound pressure falls off as $1/r$, where r = distance from the source, whereas particle motion falls off as $1/r^2$ for a pulsating source (volumetric change) or $1/r^3$ for a vibrating source (no volumetric change). Sound, as it travels away from the source, conforms, initially, to a spherical wave. When the sound is far from the source, its pressure and particle motion both fall off as $1/r$ and the sound is considered a plane wave. As the source of a spherical wave is approached, particle motion increases at a much faster rate than does its corresponding pressure, because additional particle motion is caused by the motion of the source (the acoustical near-field). At a distance where the ratio of pressure and particle motion remains constant, i.e., in a plane wave, the acoustical far-field exists. The distance where the spherical wave becomes a plane wave is termed the near-field/far-field boundary. The boundary is frequency dependent: the lower the frequency, the larger the near-field. Although the near-field has little importance in air, its importance to fishes is due to the relative noncompressibilty of water. For example, the near-field for a sound of 500 Hz extends about 3 m from the source in water, whereas in air that distance is only about 0.6 m.

Although, experimental evidence has demonstrated that fishes can indeed orient to sound in the far-field (Schuijf *et al.*, 1972; Schuijf and Siemelink, 1974), the near-field is of primary importance to fishes because low-frequency sounds, which most species attend to, have near-fields that can extend many meters from their sources. Also, when considering vibrating sources (dipoles), such as a fish tail, sounds are likely restricted to the near-field, because they do not have much of a compressional wave and therefore little sound propagation. Even pulsating sources (monoples), e.g., swim bladders, require considerable energy to produce compressional waves capable of reaching the far-field. Finally, although little incident energy passes across the water–air interface, the ratio between pressure and particle motion will change at that interface as well as at the water–bottom interface, and that includes complex structures such as the coral reef (Hawkins and Myrberg, 1983).

B. The Inner Ear

Hearing in teleost and chondrichthyan fishes involves the inner ear, along with accessory organs (when present) such as the swim bladder or analogous structures, and neural components of the inner ear (Popper, 1977, 1983; Fay and Popper, 1980; Popper and Coombs, 1980a,b; Corwin, 1981a,b, 1989; Kozloski and Crawford, 1998). The inner ear consists of dorsally positioned semicircular canals that use fluid inertia to detect angular head motions, and three sac-like organs termed the sacculus, the lagena, and the utriculus, which respond to sound and head position (Platt, 1988). Each of these latter three organs contains masses of calcium carbonate, either as a single, dense otolith in teleosts or as a gelatinous mixture containing numerous otoconia in chondrichthyan fishes. Each mass rests on a sensory epithelium, termed a macula, containing a large number of ciliary bundles, each of which possesses sensory hair cells and surrounding supporting cells. Associated membranes allow the sensory cells and the respective masses to move independently of one another.

The base of each ciliary bundle synapses with neurons of the eighth cranial nerve, and the apex of each bundle holds a cluster of cilia. A single, true cilium, the longest structure in the cluster, is termed the kinocilium. It is positioned at the periphery of the remaining cilia, collectively termed stereocila, numbering 40–70. The latter are graded in size, with the longest closest to the kinocilium. This arrangement provides the physiological polarization of each cell. Bending of the stereocilia toward the kinocilium causes a depolarization of the associated afferent neuron and, accordingly, the pulse rate increases above its spontaneous activity; bending of the kinocilium toward the stereocilia causes a hyperpolarization of the associated afferent neuron, with a corresponding brief inhibition of its spontaneous pulses directed at the brain centers (Flock, 1967, 1971; Roberts *et al.*, 1988). Response varies as a cosine function of the direction of stimulation and thus bending in an orthogonal direction has no influence on pulse rate.

Distinctly different ciliary bundle types are found in different regions of the maculae (Popper and Hoxter, 1981; Platt and Popper, 1984), suggesting differences in detection. Much evidence also exists of a species-specific pattern of ciliary bundle orientations in different parts of the maculae [see Popper (1983), for examples from coral reef fishes] and, thus, direction of a sound source can be ascertained, depending on which part of a macula is most stimulated (Enger, 1976; Popper, 1977, 1983; Lu and Popper, 1998). One theoretical problem exists, however, with such a possibility: hair cells oscillate and this causes a 180° ambiguity in determining direction. Evidence has demonstrated, however, that no such ambiguity actually exists for experimental subjects, which has resulted in several hypotheses that attempt to explain how fishes accomplish what they should not be able to do (Schuijf and Buwalda, 1980; Schuijf, 1981; Corwin, 1981b; Rogers *et al.*, 1988; Kalmijn, 1988a, 1997). Unfortunately, none of these hypotheses has yet been empirically examined. Precise localization of a nearby source appears to require free and canal neuromasts of the lateral line system.

C. The Adequate Stimulus

The response of the receptor cells points to the ultimate auditory stimulus in a sound field, i.e., particle motion. Because the density of a fish is similar to that of water, and fish length is normally small relative to the wavelength of low-frequency sound, the fish actually moves in concert with the particle motion being imparted by the passing sound wave. The otoliths and otoconia, being discontinuities due to their greater density as compared to the rest of the fish, possess inertia and thereby, because they are not attached to the body, will lag behind the fish's movement while in the sound field. This lag results in bending of the cilia resting below the calcareous masses, with appropriate information being imparted to the auditory nerve. Without some other discontinuity, such as gas in a swim bladder, particle motion, arising directly from the incident sound wave, is the sole stimulus that can be sensed by the auditory system of fishes. Many teleosts, however, possess swim bladders or analogous structures, and though fishes possess no apparent sensory cell that responds to pressure changes relative to such structures, any such "bubble" will pulsate in sympathy with pressure changes in the sound field. This will result in a secondary near-field, which can also stimulate the ciliary bundles of the inner ear by its associated particle motion. Fishes possessing such bubbles have better hearing, in both frequency range and sensitivity, than do fishes without such structures. And the closer the discontinuity is to the inner ear (e.g., anterior projections or directly connected), the better the hearing; however, if the bubble is deflated, hearing is impaired.

Reserchers must be aware of whether fishes possess or do not possess pressure-to-particle-motion transducers, such as swim bladders, before attempting to determine hearing abilities. Early audiograms (graphics showing auditory sensitivity and frequency range) were established by using a pressure-sensitive device such as a hydrophone; this resulted in sensitivity curves for fishes that were actually incapable of responding to pressure changes. Hearing should have been examined by using a particle-motion sensor such as an accelerometer. Because hydrophones are simpler to use and were available to aquatic physiologists before appropriate accelerometers were available, it is fortunate that most fishes that were studied possessed swim bladders and thus were capable of pressure-sensitive responses. Also, when measuring both acoustical variables, such fishes have demonstrated that sensitivity is indeed based on

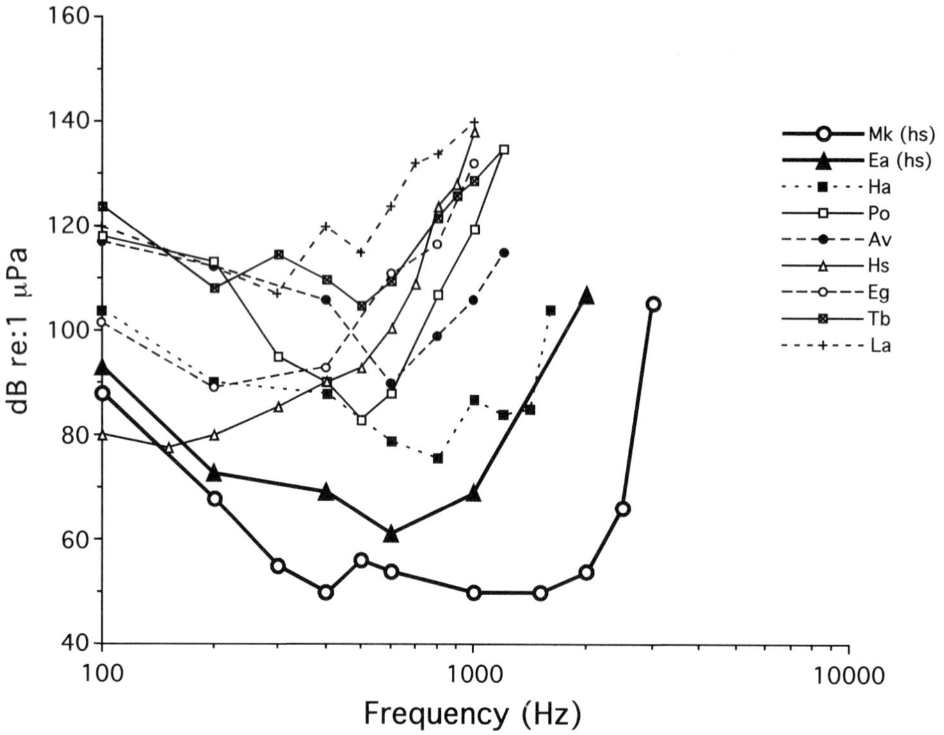

FIGURE 2 Sound pressure audiograms of teleost species from the coral reef. Mk, *Myripristis kuntee;* Ea, *Equetus acuminatus;* Ha, *Holocentrus acensionis;* Po, average for six species of *Pomacentrus,* Av, *Adioryx vexillarius;* Hs, *Haemulon sciurus;* Eg, *Epinephalus guttatus;* Tb, *Thalassoma bifasciatum;* La, *Lutjanus apodus;* hs, hearing specialist. Sources of data: Mk, Coombs and Popper (1979); Ea, Ha, Av, Eg, Tb, and La, Tavolga and Wodinsky (1963); Hs, Tavolga and Wodinsky (1965); Po, Myrberg and Spires (1980).

pressure at frequencies above about 100 Hz, but at frequencies below 100 Hz the same subjects showed that sensitivity is due to particle motion (see Myrberg and Spires, 1980). This fact, now established, has been even more clearly defined. Of the three quantities comprising the particle-motion variable (displacement, velocity, acceleration), compelling evidence has shown that the otolith organs are acceleration detectors (Sand, 1974; Hawkins and Horner, 1981; Fay, 1984; Kalmijn, 1988a; Enger *et al.*, 1989).

D. Hearing Capabilities

Although hearing thresholds of greatest sensitivity and/or upper frequency limits of hearing for several species of fishes (none from the coral reef) were published prior to 1960, most studies did not provide audiograms, the basis of any hearing study, and many provided few data and little or nothing about methods (reviewed by Tavolga and Wodinsky, 1963). Since then, audiograms have been established for more than 50 species of fishes, with inhabitants of the coral reef accounting for about 20 (Fig. 2). Most of the latter are congeneric damselfishes (seven *Pomacentrus* species: Pomacentridae) (Ha, 1973; Myrberg and Spires, 1980) and squirrelfishes (four species of *Adioryx, Holocentrus,* and *Myripristis:* Holocentridae) (Coombs and Popper, 1979; Tavolga and Wodinsky, 1963).

The damselfishes showed moderate hearing, similar to most other species that possess swim bladders that are not in close association to the inner ear. They did not demonstrate the sensitivity or the frequency range shown by species possessing swim bladders with such associations, the so-called hearing specialists. The squirrelfishes showed the importance of such associations. *Adioryx* has no such association and its hearing is similar to that of damselfishes; *Myripristis* has a direct connection between the swim bladder and the inner ear and accordingly its abilities are consistent with those of the hearing specialists. *Holocentrus* has anterior projections of the swim bladder, but no direct connections to the inner ear (see Hawkins, 1993). Its sensitivity and bandwidth are, accordingly, intermediate to those of the other two genera. Unfortunately, few other coral reef fishes have been examined and data on their hearing must be considered preliminary, at best.

Tavolga and Wodinsky (1963, 1965) examined seven species of coral reef fishes. The audiogram of *Equetus acuminatus* (Sciaenidae) showed it to be a hearing specialist; the audiograms of the remaining species; *Holocentrus ascensionis, Adioryx vexillarius,* and *Haemulon sciurus* (Haemulidae), *Lutjanus apodus* (Lutjanidae), *Thalassoma bifasciatum* (Labridae), and *Epinephalus guttatus* (Serranidae), showed wide dissimilarity and, depending on the species, good to relatively poor hearing.

Audiograms for fishes known to inhabit coral reefs but lacking swim bladders or analogous structures exist only for sharks. Unfortunately, the audiograms for *Carcharhinus leucas* (Kritzler and Wood, 1961) and *Negaprion brevirostris* (Carcharhinidae) (Nelson, 1967) were established in the pressure mode. Later evidence showed that sharks are responsive solely to particle motion (Banner, 1967; Kelly and Nelson, 1975).

Although auditory sensitivities of fishes have long been shown to be limited to frequencies between 50 and about 3000 Hz, evidence has since shown that some fishes are capable of hearing infrasounds, i.e., below 20 Hz, e.g., the cod, *Gadus morhua,* the perch, *Perca fluviatilis,* and the plaice, *Pleuronectes platessa* (Sand and Karlsen, 1986; Karlsen, 1992a,b; Enger *et al.,* 1993), although some clupeids (*Alosa* spp.) can sense sounds of ultrahigh frequency (Dunning *et al.,* 1992; Nestler *et al.,* 1992; Ross *et al.,* 1993; Mann *et al.,* 1997). Evidence has shown that infrasound detection is mediated by the inner ear, but mediation of ultrahigh frequencies is unknown. Coral reef fishes have not yet been examined regarding such capabilities, and additional data require new studies.

Space limitations preclude discussing further points about the hearing capabilities of fishes, e.g., frequency and intensity discrimination, sound masking, and critical bands. Interested readers are encouraged to peruse these subjects and others in the impressive summaries by Fay (1988) and Hawkins (1993).

E. Sound Production

Fishes produce epiphenomenal sounds during sudden movement and when feeding (Colson *et al.,* 1998), and many species also produce a variety of voluntary sounds, often correlated with specific behavioral actions associated with aggression and reproductive activities (Schneider, 1961; Fine *et al.,* 1977; Myrberg, 1981; Ladich, 1997).

Sounds produced by biological entities have rarely been studied directly on or near coral reefs. The most comprehensive studies were accomplished at a video–acoustic installation located on coral reefs west of Bimini, Bahamas, during the early 1960s (Cummings *et al.,* 1964; Kumpf, 1964; Steinberg *et al.,* 1965). Findings included descriptions of numerous sounds, their frequency of occurrence throughout the day and night, evidence of lunar (several sounds) and seasonal cycling (one sound), and changes in ambient noise levels apparently due to biological sources. Although relatively few sound sources were identified because of the stationary video apparatus, eleven sources were identified and four more were tentatively identified. The former included a damselfish (chirp), a grunt (burst), a surgeonfish (pop), a squirrelfish (pops), a triggerfish (crunch and rattle), and a wrasse (rasp).

Large invertebrates, e.g., gastropods (*Strombus* sp. and *Cassis* sp.), were also found to be the sources of a common sound (roar). One highly distinctive sound (click), commonly heard as a chorus during the evening and which possessed a clear lunar periodicity, was identified several years later as originating from the cowfish, *Lactophrys quadricornis* (Ostraciidae) (A. A. Myrberg, Jr., personal observation). Because cowfish are almost invariably seen moving, individually, over the reef during the day, their chorusing and lunar-based activities during the nocturnal period, although not yet examined, suggest caution when considering sound sources based on limited knowledge. Other fishes known to be sound producers are either from the coral reef, e.g., tiger fishes, clown fishes, and angelfishes, or have family members as reef inhabitants, e.g., batrachoidids (toadfishes and the midshipman), blennies, and gobies (Tavolga, 1958a,b; Winn, 1964; Mok, 1981; Zhang and Takemura, 1989; Bass, 1990; Brantley and Bass, 1994).

F. Functions

After source identification, studies have often attempted to identify or suggest functional associations of sound production by fishes. Such associations were generally lumped together under the rubric of communication, a popular function that has rapidly become synonymous with social behavior. Also, although voluntary sound production is widespread, it has long been known that many fishes produce no sounds, even though they often possess good to excellent hearing. These troubling points caused one of us (A. Myrberg) to reconsider the concept of communication, which resulted in a more precise definition. In essence, communication must provide adaptive advantage to the sender in an interaction (the receiver may also benefit, but such has no relevance to the concept). This important restriction allowed a new process to be introduced, i.e., interception. This process provides the adaptive advantage solely to the receiver (in the case of sound, an attentive listener) if it enters into an interaction; and often the sender is disadvantaged (Myrberg, 1981, 1997a), We shall use these terms below, where we hope their differences will become clear.

Acoustically based species recognition has been demonstrated by damselfishes on coral reefs of the Bahamas and southern Florida and is accomplished by discerning the species-typical, pulse-repetition rates of sounds produced by sympatric congenerics (Myrberg and Spires, 1972; Myrberg *et al.*, 1978; Spanier, 1979). The consistent energy distribution among the spectral frequencies of the sounds of differently sized conspecifics (Myrberg *et al.*, 1993) also likely holds the key to the acoustically based, individual recognition among colony members, e.g., *Pomacentrus partitus* (Myrberg and Riggio, 1985). Although such a relationship has also been found in other teleosts, exceptions exist. The latter include species that do not possess swim bladders, e.g., gobies (Mok, 1981; Torricelli *et al.*, 1990), or those that produce sounds with high pulse rates, e.g., toadfishes (*Opsanus* spp.), the related plainfin midshipman (*Porichthys notatus*) (Ibara *et al.*, 1983; Bass, 1990; Brantley and Bass, 1994), and the marine catfish (*Bagre marinus*) (Tavolga, 1960). It is difficult to consider that site-attached species such as gobies and batrachoidids are less capable than damselfishes of acoustically recognizing their neighbors. It would be fascinating to discover how acoustically based recognition is accomplished.

Members of many families of both marine and freshwater species produce intense sounds when held or prodded mechanically (Myrberg, 1981; Smith, 1992). Studies have suggested that such sounds are used to startle predators such that they release a captured prey or, if capture has not occurred, are driven away. Unfortunately, no evidence has ever substantiated such reasoning. Although an altruistic goal, informing conspecifics about the presence of a nearby predator is a possibility; a more reasonable suggestion has arisen from experimental studies of the chemical alarm pheromone, Schreckstoff, of ostariophysan fishes, which has long been known to cause flight responses in conspecifics and other taxonomically related species. Studies have now demonstrated that Schreckstoff also attracts additional predators to the capture site, and instances of prey escape have been witnessed while the greater numbers of predators competed among themselves for the fewer numbers of prey (Smith and Lemly, 1986; Mathis *et al.*, 1995). Research suggests that, on capture, prey have little to lose regarding immediate survival and, thus, although they inform others of nearby danger, their primary goal is to attract predator-competitors whose actions increase chances of escape in the manner first proposed by Högstedt (1983; see Smith, 1992). The compelling evidence of survival in the face of imminent death suggests that the "startle" sounds of fishes may indeed hold a similar communicative function; further studies are required.

Sounds have been monitored during aggressive interactions among numerous fishes, and often both males and females produce such sounds (Schneider, 1964a,b; Luh and Mok, 1986; Myrberg, 1981, 1997a; Lobel, 1996; Ladich, 1997). Inhabitants of the coral reef include triggerfishes, squirrelfishes, trunkfishes, wrasses, damselfishes, and tiger fishes. Other marine species include bactrachoidids, gobies, and blennies (Tavolga, 1956, 1958a,b; J. F. Fish, 1972; Mok, 1981; Bass, 1990, 1993). Because sudden movements in water, if sufficiently intense, will produce sonic epiphenomena, the function of such sounds must be demonstrated by controlled experiments employing sound playback. But such a technique is not infallible, because response may require complex signals, involving various senses (Myrberg, 1997a). Such reasoning, when response is absent during playback, suggests that sounds produced during aggression probably enhance visual displays, which they often accompany. One sound, the chirp, produced by territorial males of the bicolor damselfish *Pomacentrus partitus,* has been shown by the playback technique to possess a clear communicative function. It is a territorial "keep-out" signal (Myrberg, 1997b), not unlike the territorial song of many birds.

Sounds also accompany reproductive actions of fishes, particularly during courtship, and in such cases males are almost invariably the sound producers (Hawkins, 1993). On the coral reef, damselfishes produce sounds during courtship (Myrberg, 1972) and damselfishes (Lobel and Mann, 1995), trunkfishes (Moyer, 1979; Lobel, 1996), a serranid (the hamlet, *Hypoplectrus unicolor*), and a parrot fish (*Scarus iserti*) (Lobel, 1992) produce sounds during spawning. In the latter case, the sounds were apparently those expected to accompany intense spawning movements by parrot fish groups in the water column. Other marine fishes known to produce courtship sounds are bactrachoidids, gobies, and blennies. Sound playback has demonstrated clear communicative function for the courtship sounds of toadfishes (J. F. Fish, 1972; Winn, 1972) and damselfishes (Mohler, 1984; Myrberg *et al.*, 1986). Interestingly, female damselfishes are responsive to such sounds only when their eggs are ripe and during early morning hours. In that case, response to sound playback required not only a physiological basis, but a temporal one as well.

Sound production by fishes, though clearly capable of communicative function in specific situations, has been found at times to carry significant costs for signal producers. And often, the signal receiver could not have been the intended receiver because the sender was disadvantaged, while the receiver clearly benefited. Examples of the process of interception include interspecific as well as intraspecific interactions. Nearby colonial males of the damselfish *Pomacentrus partitus* are attracted to the nests of other males when the latter produce specific courtship sounds (grunts) just prior to spawning. The resulting territorial invasions and nest intrusions by such reproductive competitors cause females to leave the nests, with clear costs accruing to the sound producers (Kenyon, 1994). A case of interspecific interception involves chirps, a sound produced during early courtship by congeneric males of *Pomacentrus*, and predatory trumpet fishes, *Aulostomus maculatus*. The latter are rapidly attracted to specific locations where chirps are transmitted (Myrberg *et al.*, 1986) and, again, the sound producers lose when predation is successful. These same sounds also attract spawn-ready females to the sound producers and thus it is clear that communicative and interceptive processes can occur simultaneously.

Other instances of sonic interception involve sharks, numerous species of which inhabit coral reefs throughout the world. Observers have seen such predators attracted to sounds either directly produced by prey or closely mimicking sounds produced by prey (Myrberg *et al.*, 1969, 1976; Nelson *et al.*, 1969; Nelson and Johnson, 1972; Banner, 1972; Myrberg, 1978). Understanding sound production and sound reception in the context of interception clarifies, in that example, how some fishes, although known not to be sound producers, hear quite well. Another such case involves a small wrasse, *Halichores biviattatus*, common on coral reefs throughout the western Atlantic. Although it is also not known as a sound producer, it can certainly hear the intense sounds produced by the rapid turning movements of the blue runner *Caranx crysos* (Carangidae) as these predators approach the reef. Accordingly, these small prey rapidly dive into the sand shortly before the runners arrive on the scene and only reappear several minutes after the runners depart (Steinberg *et al.*, 1965). In this instance, the attentive listeners are again benefited and the sound producers are again disadvantaged. The above cases demonstrate that auditory systems can be used to detect signals of opportunity or danger by interceptors and their owners alone benefit when they act accordingly. Interception might well have been the initial function of the earliest auditory systems and from which acoustical communication later evolved in groups capable of producing voluntary acoustical signals.

G. The System during Early Development

Little information is available about the acoustic modality of young fishes, and what little is known concerns animals from habitats other than the coral reef. Regardless of source, such information can provide insight as to future experiments and observations regarding the probable importance of the acoustical modality to recruitment processes, be it orienting to the reef or survival after settlement.

The labyrinth and otoliths are present at the time of hatching, the latter being used to age larvae of many fishes (Blaxter, 1986). The hair cells of the saccular macula of embryos of *Opsanus tau*, several days before hatching, are identical to those of the adult (Sokolowski and Popper, 1988). Proliferation of such cells and increase in macular size continue to adulthood in the sacculi of the freshwater cichlid *Astronotus ocellatus* (macula size, 2–19 cm) (Popper and Hoxter, 1984) and the coral reef fish *Pomacentrus partitus* (12–48 mm) (Kenyon, 1996) (Fig. 3). Such proliferation has been extended to all the acoustical end organs and throughout virtually the whole life of the European hake *Merluccius merluccius* (Gadidae) (macula size, 7–75 cm) (Lombarte and Popper, 1994). Ganglion cell proliferation in the eighth nerve has also been found in *Astronotus ocellatus* (macula size, 2–19 cm) (Popper and Hoxter, 1984). Such postembryonic growth of hair cells has been found also in the maculae neglectae of condrichthyan fishes, e.g., the blacktip reef shark *Carcharhinus melanopterus* (Corwin, 1981c, 1983). Proliferation of hair cells in the macula neglecta of the skate *Raja clavata* (Corwin, 1983) and in the sacculus of the damselfish *Pomacentrus partitus* (Kenyon, 1995) has been correlated with increasing hearing sensitivity, but it should be mentioned that the minute maculae neglectae of teleost fishes have never been implicated as acoustical end organs. Finally, larval herring *Clupea harengus* (Clupeidae) responded to acoustical stimuli only after their auditory bullae (located near the utriculi) filled with gas (22–30 mm); if the bullae contained liquid, larvae responded solely to tactile stimuli (Blaxter and Batty, 1985).

Interest in the acoustical sense as a possible mechanism for presettlement orientation to the coral reef by fishes is due to the improbable use of other sensory

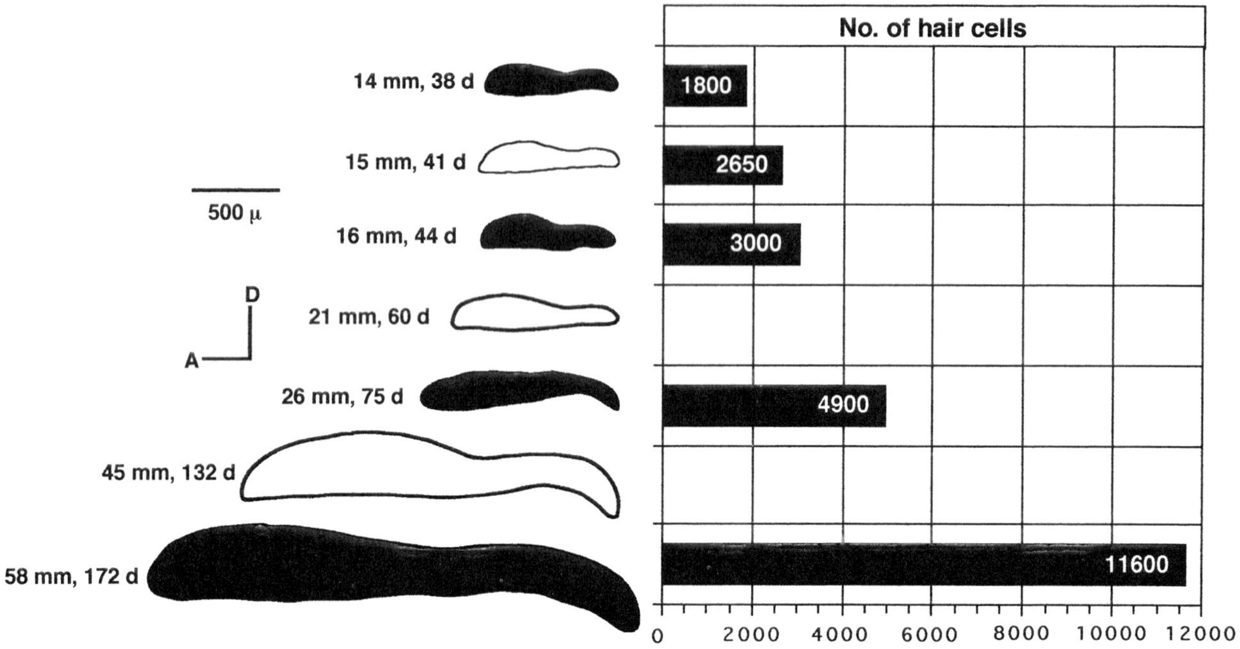

FIGURE 3 The relative sizes and shapes of the saccular maculae of differently aged members of *Pomacentrus partitus*. Total hair cell counts are plotted adjacent to associated maculae. Standard length (mm) and estimated age are listed next to each macula (with permission from Kenyon, 1995).

systems at the distances involved (Leis *et al.*, 1996; Stobutzki and Bellwood, 1998). No doubt exists that levels of ambient noise on coral reefs are often characterized by the sonic activity of its numerous inhabitants (e.g., snapping shrimp) (Fish, 1964; A. A. Myrberg, Jr., personal observation) and such activity has been shown to produce much higher sound levels at certain times of the diurnal period, particularly at night, due to distinctive choruses [10–15 dB above broadband ambient level (Cummings *et al.*, 1964) and 30 dB above a bandwidth of 1600 Hz, centering on 1600 Hz (DeFerrari and Green, 1968)]. The importance of such choruses to levels of ambient noise was initially demonstrated during the second World War (Love and Proudfoot, 1946) and since then, in several publications (e.g., Cato, 1978, 1980; Myrberg, 1980) (Fig. 4). McCauley and Cato (1998) reported nocturnal choruses of a high-level, poplike sound around various coral reef systems in northern Australia. The source of the choruses was believed to be holocentrids or priacanthids and sounds were produced year round in midwater over depths less than 50 m. Additional studies are needed to compare ambient noise levels on coral reefs with those in nearby waters, but away from the reefs. The importance of noise being a "beacon" for homing depends on at least two important aspects: (1) whether subjects are capable of hearing such sounds, and if so, (2) whether the noises possess, within the frequency spectrum of audibility, sufficient energy to reach above the threshold of hearing (usually 15–20 dB above the spectrum level noise) (Myrberg, 1980).

Using the auditory sense for such a purpose is unlikely for at least two species found on the coral reefs of the Caribbean and nearby waters. The hearing sensitivity of two species of damselfishes, *Pomacentrus partitus* and *P. variabilis,* which were 12 mm long and likely within a day or two of settling, was examined and all subjects showed similar findings. They exhibited extremely poor hearing, with detection limens ranging from 54 dB (at 300 Hz) to 15 dB (at 1500 Hz), higher than known adult thresholds. Thresholds decreased, however, with increasing age, and approached adult levels after several weeks. The flat, untuned audiograms exhibited by the youngest juveniles and the appearance of a best frequency and a pattern of increasing acuity progressed in the manner observed in mammals and birds (Kenyon, 1996). On the other hand, a study off the coast of New Zealand, using extremely high-level, ambient sounds (180 dB re 1 μPa) from the reef, has shown that significantly more triplefin (Tripterygiidae) larvae were caught in light traps adjacent to the underwater loudspeaker broadcasting the reef sounds than in "silent" light traps (Tolimieri *et al.*, 2000). The results are suggestive of a role for audition in the settlement process. There clearly exists great need for examining the hearing capabilities of many other coral reef fishes during early ontogeny so that the role of the auditory sense can be appropriately assessed as to its importance

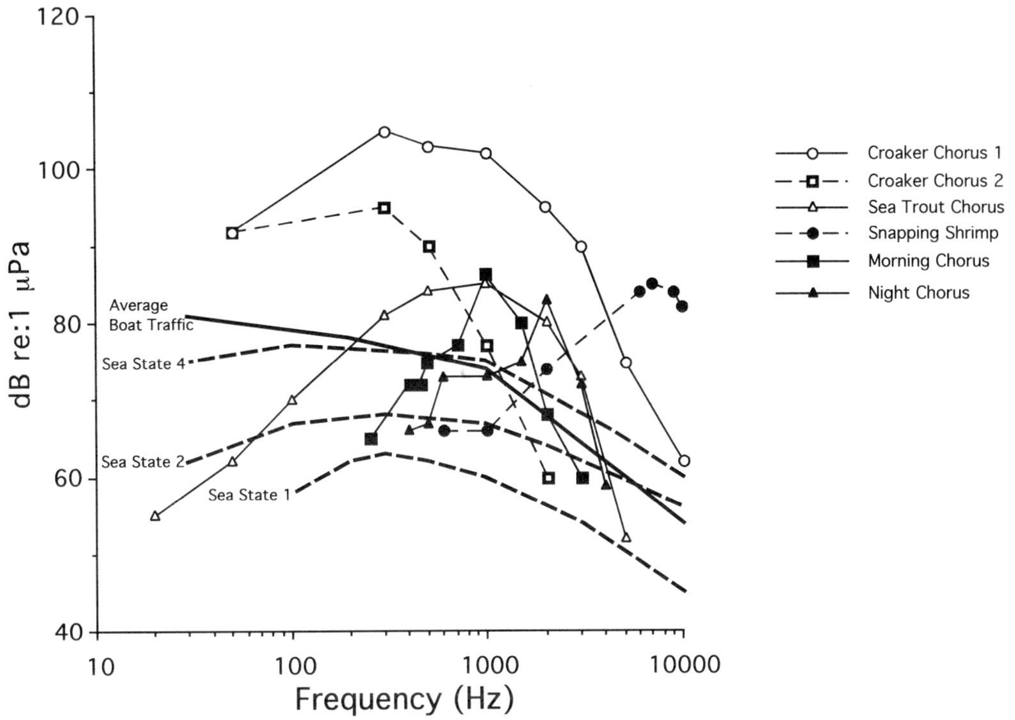

FIGURE 4 Shallow-water (<70 m) ambient noise (spectrum level). Data have been extracted from various sources and redrawn: boat traffic (Urick, 1975), sea state (Piggott, 1964), croaker choruses (Naval Ordinance Laboratory, 1942), sea trout chorus (Fish and Cummings, 1972), snapping shrimp (Tavolga, 1974), morning and night choruses (Cato, 1980) (modified from Myrberg, 1980).

to settlers in locating reefs, as well as surviving after reaching them.

IV. Mechanoreception via the Lateral Line System

A. Morphological Features

Fishes sense the innermost part of the near-field, the so-called local-flow field (see below), through the mechanosensory lateral line system. The term *mechanoreception* is used because components of the lateral line system, including the free neuromasts of the body surface, can be associated with other sensory systems (e.g., electrical, chemical) in some fishes. The latter are not considered here. The most visible parts of this system are the canals or grooves on the head and trunk of most species, which are punctuated with pores. The principal lateral line canals of the head include the supraorbital, infraorbital, and mandibular (the latter includes canals on the cheek and lower jaw). These canals may be interconnected and elaborated to varying degrees among species. The main trunk canal, when present, begins just posterior to the eye and extends along the body, usually to the base of the caudal fin. The trunk canal is sometimes strongly arched near the pectoral fin. It may be broken into segments, or there may be multiple, parallel canals along the body. Within the canals and between each pair of pores are canal neuromasts, the end organs of the lateral line system. Neuromasts are also present on the body surface separate from canals or grooves. These so called free, or superficial, neuromasts can be found on almost any part of the body and can be highly concentrated in certain areas. Free neuromasts may project well beyond the surface of the body or they may lie in pits.

An individual neuromast consists of several hair cells surrounded by supporting cells. The structure of the hair cells, the arrangement of their cilia, and the mechanism of neural excitation are the same as was described for hair cells of the inner ear (see p. 131). Neuromasts are innervated by lateral line nerves, which are a distinct series, separate from dorsal cranial nerves (Northcutt, 1989). The anterior lateral line nerve innervates the free and canal neuromasts of the head, and the posterior lateral line nerve innervates neuromasts on the trunk. Unlike hair cells of the sac-like organs of inner ear, the cilia of neuromast hair cells project into an acellular, gelatinous cupula. Cupulae of free neuromasts are typically cylindrical; those of canal neuromasts are often dome shaped or bladelike and oriented parallel to the axis of the canal. Cupulae are flexible distally so that water movements bend the cupula, which in turn bends the cilia and changes the discharge rate of the hair cells.

Clupeoid fishes, principally herrings and anchovies, have a unique adaptation that merges the functions of the inner ear and lateral line canals to increase their dynamic range and sensitivity to underwater sounds. Although clupeoids lack a lateral line canal on the trunk, they have an elaborate network of cephalic lateral line canals that converge at a point posterior to the eye, the lateral recess. A compliant membrane on the inner surface of the lateral recess separates the canals from the endolymph of the inner ear. Also within the otic capsule is a gas-filled extension of the swim bladder, the prootic bulla. This functional link between the swim bladder, inner ear, and lateral line allows the pressure component of sounds to be transduced by the bulla into displacements of endolymph, which then displace the otoliths and the lateral recess membrane, in turn displacing the water and neuromasts inside the cephalic lateral line canals (Denton and Blaxter, 1976). All of this imparts extraordinary acoustical sensitivity and dynamic range to clupeoids.

B. The Adequate Stimulus

Like the otolith organs of the inner ear, neuromasts are essentially acceleration detectors, but free neuromasts are also able to sense water velocity (Münz, 1989; Kroese and Schellart, 1992; Montgomery *et al.*, 1997). A significant difference between the lateral line system and the inner ear is the spatial array of lateral line end organs (neuromasts) distributed over the body, as opposed to being confined to a discrete organ. This allows a fish to sense spatial differences in the local-flow field of the larger acoustical near-field (Kalmijn, 1989). However, because accelerations in a flow field diminish rapidly with distance from the source, the lateral line system is most useful when a fish is near the source of the stimulus, e.g., within a few body lengths of moving prey (Denton and Gray, 1988). The frequency response and sensitivity of neuromasts are determined by their placement, the shape of their cupulae, whether they are in a canal, and the dimensions of the canal (Bleckmann, 1993). Generally, neuromasts are sensitive to low frequencies, below a few hundred hertz. Canal neuromasts are most sensitive in the range of 50–200 Hz, whereas free neuromasts are most sensitive to frequencies below about 70 Hz (Münz, 1989). Both components of the lateral line system are very sensitive. Canal neuromasts can detect accelerations of 0.3–20 mm sec^{-2} and free neuromasts are sensitive to velocities as small as 0.03 mm sec^{-1} (Bleckmann, 1993).

C. Functions

The characteristics of the stimuli for the lateral line system—low frequency, close proximity—and the concentration of end organs on the head suggest an ecological function associated with feeding. Many prey organisms generate low-frequency vibrations. For example, the swimming and feeding movements of various types of zooplankton produce frequencies in the range of 10–50 Hz (Montgomery, 1989). Likewise, routine swimming of fishes (e.g., forage species) results from low-frequency movements of their fins. Behavioral experiments on a variety of species have shown that fishes can use the lateral line to sense the location or trajectory of prey over short distances, whether the prey is planktonic [e.g., clupeid feeding on crustaceans (Janssen *et al.*, 1995)], benthic infauna [e.g., cottid feeding on buried annelids (Janssen, 1990)], or piscine [e.g., stargazer feeding on a fish (Montgomery *et al.*, 1995)]. Species that appear to specialize in the use of the lateral line for prey detection, such as the piper (*Hyporhamphus ihi:* Hemirhamphidae), a reef-dwelling nocturnal planktivore, have an elaborate system of canals on the head, with wide canals and neuromasts placed near constrictions in the canals (Montgomery and Saunders, 1985). These modifications may amplify flow in the canals and expand the frequency response of the lateral line. The ability to locate and strike prey using only the lateral line means that these fishes can feed at night and under other conditions of low light.

In addition to fin movements, the vortex trails behind a swimming fish can betray its recent location. The vortices contain frequency components up to at least 100 Hz (Bleckmann, 1993). This type of disturbance in the water column can persist for several seconds and at a distance of several meters from the organism that generated it. This function of the lateral line is, at present, speculative because definitive behavioral experiments have not been performed. However, such information about the position of a nearby fish could be useful for interspecific and intraspecific interactions.

Maintenance of position within a school is another function of the lateral line system. The close proximity of neighbors and the relatively low frequency of their tail beats are conditions ideally suited to the lateral line sense. Gadids have been shown to rely, at least in part, on the posterior lateral line to maintain position within schools (Partridge and Pitcher, 1980). Clupeoids, which lack a posterior lateral line, have an especially elaborate network of cephalic lateral line canals, and the unique connection between the swim bladder, inner ear, and lateral line is probably critical for rapid adjustments of swimming speed and position required of schooling individuals (Denton and Gray, 1988).

The lateral line system may play a role in escape behavior. The Mauthner cell is a large neuron that mediates the startle response in fishes and other vertebrates. It receives inputs from several sensory systems, including the posterior lateral line. It has been demonstrated that auditory and visual inputs can stimulate the Mauthner cell beyond threshold to produce a startle response. It has been suggested that lateral line inputs contribute to the general excitatory state of that neuron but without exceeding the threshold (Zottoli and Danielson, 1989).

In addition to low-frequency mechanical stimuli originating from other organisms, the lateral line system is capable of detecting stationary objects. When gliding in open water, a symmetrical flow field surrounds a fish. The flow will be distorted when the fish nears any object. The nature of the distortion depends on the size and geometry of the object. Blind fishes, such as the blind cave fish (*Astyanax mexicanus:* Characidae), detect and discriminate stationary objects using the lateral line system, a process sometimes called hydrodynamic imaging (Hassan, 1989; Teyke, 1989). Other species, even those that are not naturally blind, are also able to avoid stationary objects without the use of vision.

The notion that free neuromasts might sense velocity in addition to accelerations has been confirmed by behavioral experiments. In three unrelated and ecologically distinct species, *Cheimarrichthys forsteri* (Pinguipedidae), *Pagothenia borchgrevinki* (Nototheniidae), and *Astyanax fasciatus,* the threshold water velocity for rheotaxis increased significantly when free and canal neuromast function was pharmacologically blocked. The same result occurred when free neuromasts were selectively blocked but canal neuromasts were left intact (Montgomery *et al.*, 1997). This role of free neuromasts in sensing water velocity suggests that fishes not only assess and orient to currents, but also sense their own swimming speed relative to the surrounding water.

D. The System during Early Development

The anatomy of the lateral line system undergoes a substantial amount of development after hatching, resulting in a corresponding change in functionality. Thus, the mechanosensory abilities of larvae not only differ from those of adults, but they change rapidly over a brief period of time. At hatching, all species studied to date have at least a few free neuromasts, usually located on the head. The number and distribution of free neuromasts increase during the larval period. Lateral line canals are not present in young larvae. Canal neuromasts first appear as series of free neuromasts that descend into a grooves during the latter part of the larval period. The grooves eventually close over to form the definitive canals. Complete canal fomation takes place around the time of settlement in labrids and somewhat later in scarids (Lara, 1999). It has been suggested that free neuromasts operate within the relatively thick hydrodynamic boundary layer of young larvae and that the late development of canals occurs when the boundary layer becomes much thinner due to the larger size and faster swimming speeds of older larvae (Blaxter and Fuiman, 1989). Another possibility is that a canal system could not function in small larvae because water viscosity would restrict flow through such tiny canals (Lara, 1999). Several researchers have used the late development of lateral line canals as a "natural experiment" from which to infer the separate functions of free and canal neuromasts.

Experiments on clupeoid, gadid, sciaenid, and flatfish larvae show that free neuromasts play a role in evasive behavior (Yin and Blaxter, 1987; Blaxter and Fuiman, 1989; Higgs and Fuiman, 1998b; Poling and Fuiman, 1999). Even prior to canal formation, larvae startle when confronted with a rapidly approaching probe in the dark. This responsiveness decreases significantly when the neuromasts are pharmacologically ablated (Fig. 5). Reactive distance to the probe, a measure of sensitivity, increases ontogenetically as the number of free neuromasts increases (Fig. 5). There is no obvious change in the ontogenetic trend for responsiveness or reactive distance when lateral line canals form. When the stimulus is a larger fish, Atlantic herring (*Clupea harengus*) larvae become much more responsive after the prootic bulla becomes functional and the cephalic lateral line canals close (Blaxter and Fuiman, 1990). Apparently, the gas-filled bulla amplifies the stimulus beyond the threshold of the canal neuromasts.

Orientation behaviors, such as rheotaxis and schooling, rely on mechanical stimuli, but observations of, and experiments on, larvae implicate different components of the lateral line system. Striped jack (*Pseudocaranx dentex:* Carangidae) larvae orient into a current from as early as 4.5 mm total length (TL), when free neuromasts are present on the head but no canals are developed (Masuda and Tsukamoto, 1996). The first canal, the supraorbital, begins to form at about 6 mm TL and begins to close at about 8 mm TL. On the other hand, schooling behavior may depend on functional canals. In an atherinid and a clupeid, schooling does not appear until after lateral line canals form (Cahn *et al.*,

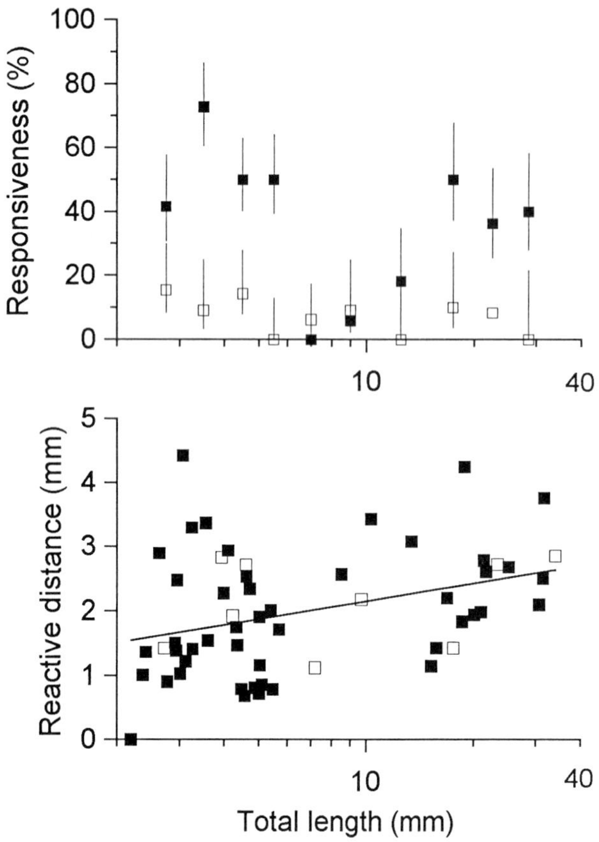

FIGURE 5 Responses of Atlantic croaker (*Micropogonias undulatus*) larvae at different stages of development to an approaching probe. Experiments were performed in darkness (using infrared video techniques) to eliminate visual responses. Treatment with streptomycin sulfate (open symbols) ablates neuromasts and results in lower responsiveness (top: symbols show means for several individuals) but has little effect on the distance at which the larvae respond to the probe (bottom: symbols represent individual responses). Larger larvae have progressively more neuromasts, explaining the increasing reactive distance with size. Vertical lines represent 95% binomial confidence intervals (modified from Poling and Fuiman, 1999).

1968; Gallego and Heath, 1994). These results from larval stages corroborate the conclusions about the functions of free and canal neuromasts derived from theory and experimental work on adult fishes.

V. The Electrosensory System

A. Systematic Distribution

Electroreception is known in teleost fishes from only three primarily freshwater orders [Gymnotiformes, Siluriformes, and two osteoglossiform families, Mormyridae and Notopteridae, following Nelson (1994)]. Thus, little can be stated directly about the electrosensory system of coral reef fishes, because it has not been described in any marine fishes other than elasmobranchs and one marine species in the Plotosidae, a small family of catfishes with several marine representatives. Nevertheless, because numerous elasmobranchs inhabit coral reefs throughout the world and many findings, though centering on freshwater fishes, should not be considered as irrelevant to marine fishes, we provide a brief summary about the electrosensory system.

B. Major Sources of Information on the System

Much of our present knowledge about the field of electroreception in fishes has been admirably summarized in the massive volume edited by Bullock and Heiligenberg (1986), which covers all levels of the subject from peripheral to central, including a chapter on electric organs, stressing the electrical organ discharge (EOD) (Bass, 1986). The previous major review of the field, including electric organs, was by Bennett (1971a,b). Bleckmann (1993) reviews the system briefly as a specialized "offshoot" of the lateral line organs. One subject not covered by these reviews is the physical nature of weak electric fields in the natural environment, be they electrogenic, i.e., produced by animals (bioelectric fields), or environmental, i.e., electrochemical (mainly freshwater) or electromagnetic (mainly oceanic) in origin. Fortunately, Kalmijn (1988b) provides such information, emphasizing bioelectric fields, along with a brief history of the subject primarily as it relates to elasmobranch fishes. Further insight into the physics of underwater electrical fields, including mechanisms and theories of orientation, using such fields, are provided by Kalmijn (1981, 1984, 1997) and by Carl Hopkins and his students (Schluger and Hopkins, 1987; Hopkins, 1988; Kwang-Tse *et al.*, 1996; Hopkins *et al.*, 1997).

C. The Receptors and the Adequate Stimulus

The two major types of receptors, along with numerous subtypes, found in the electrosensory systems of fishes are described by Zakon (1988). One major type, the tuberous receptor, is found only in gymnotid and mormyrid fishes and little more will be mentioned about it. The second major type, the ampullary receptor, is found in all electrosensitive fishes. Found in elasmobranch fishes, the ampullary system, termed the ampullae of Lorenzini, has its receptor cells, numbering in the hundreds, within the epithelium of a single or multilobed pouch located at the terminals of canals of varying length, which open via pores on the skin surface of the head (sharks) or over the head and body (skates and rays). Often, these canals lead to clusters of ampullae. The ampullary system found in the head of catfishes is similar to that seen in elasmobranchs and in the case of the marine catfish, *Plotosus anguillaris* (Plotosidae), the system has long canals like those of marine elasmobranchs. This catfish has been trained to orient to fields as weak as 0.08 μV cm^{-1} and, as in marine elasmobranchs, possesses receptors 10 to 100 times more sensitive as compared to freshwater fishes, i.e., about 1 μV cm^{-1} for the gymnotid, *Apteronotus albifrons* (Knudsen, 1974). The lower receptor sensitivity in fresh water compared to sea water is further seen in the freshwater stingray, *Potamotrygon laticeps*, which is about as sensitive as the freshwater teleosts and clearly less sensitive than its marine relatives (Kalmijn, 1988b; Hopkins, 1988). Ampullary organs are specialized to detect low-frequency (<50 Hz to direct current) electric fields primarily of biological origin (produced by prey); in elasmobranchs, the frequencies are much lower (8 Hz to direct current) and these marine species (e.g., *Mustelus canis:* Carcharhinidae) demonstrate the greatest electrical sensitivity known in the animal kingdom, i.e., 0.005 μV cm^{-1} (Kalmijn, 1982).

Ampullary organs face a difficult problem, however, in that the respiratory and swimming movements of a fish possessing ampullary receptors will stimulate the fish's own receptors, interfering with detection of extrinsic signals. One way that the problem has been solved is by "common mode rejection" as predicted by Kalmijn (1974). Montgomery (1984) showed in the thornback ray *Platyrhinoidis triseriata* (Rhinobatidae) that afferent fibers from canals, oppositely oriented, have ampullae in the same cluster and will see the same internal potential. Thus, internal potentials, such as respiratory potentials, will be expressed identically across both ampullae, driving afferent fibers from both. One afferent makes an excitatory contact with the postsynaptic cell; however, the other is inhibitory. Thus, the effect of any large internally generated potentials is effectively canceled, whereas smaller local external fields that uniquely or preferentially stimulate one ampulla affect central neurons (Zakon, 1988).

D. Functions

One of the primary functions of electroreception is to sense weak electrical fields in the near vicinity and to locate the source. Electrolocation can be of two types: active and passive. Active electrolocation (Lissman, 1958; Lissman and Machin, 1958) is the ability to detect the presence of objects of varied conductivity that distort the reafferent electrosensory input from the animal's own electric organ discharge (Yager and Hopkins, 1993), which forms, in essence, a three-dimensional field around its body, analogous to hydrodynamic imaging (see p. 139). To accomplish such a feat requires receptors widely distributed over the body to measure the instantaneous shape of the electric field. A variety of such receptors has been characterized (Bullock, 1973). Another function of electroreception in animals possessing electric organs is social signaling, which is accomplished through species-specific electric organ discharges (EODs) that provide information on the location, species type, sexual identity, and individual identity for a receiver capable of handling the information (Hopkins, 1972, 1988; Davis and Hopkins, 1988).

Passive electrolocation occurs when an electroreceptive animal detects and localizes objects or organisms based on information originating from their respective minute electric fields. In the case of animals, the latter are basically direct current fields of millivolt potentials presumably generated between body fluids and the water and between different parts of the body (Kalmijn, 1974). This mode, requiring no electric organ, is used by elasmobranchs and catfishes in locating prey (Bullock, 1973; Kalmijn, 1971, 1974; A.J. Kalmijn *et al.*, 1976; V. Kalmijn *et al.*, 1976). Electrolocation of prey, first demonstrated by captive skates and sharks (Kalmijn, 1971) and then by free-ranging sharks in the field (Kalmijn, 1982), using actual prey as well as electrodes simulating prey fields, clearly demonstrated the use of electroreception during the final phases of feeding, i.e., precisely locating the prey followed by attack. Additional laboratory experiments with the round stingray, *Urolophus halleri*, provided further evidence of the electrical sensitivity of these animals. Subjects clearly responded to changes in the Earth's magnetic field, using it to locate specific regions in the test enclosure (Kalmijn, 1977, 1978). Based on the physics of electromagnetic fields, Kalmijn (1977, 1984) proposed a theory of geomagnetic orientation, leading to navigation. Thus, the long-distance movements of numerous elasmobranchs may well not be random swimming events, but well-oriented activities with specific goals.

Additional evidence has shown that the electrosensory system serves another biological function in elasmobranchs, i.e., social behavior. Tricas *et al.* (1995) have experimentally demonstrated in the field that the weak electric fields of females of the ray, *U. halleri*, can be used by reproductively active males to locate buried mates and by other females to locate consexuals. Tricas and New (1998) have also shown that the electrosensory primary afferents of *U. halleri* have a relatively high discharge rate and entrain to 1 Hz sinusoidal peak field gradients of 0.02 μV cm^{-1}, and argue that such properties reflect evolutionary adaptations to detect prey and for social interactions and electric-mediated geomagnetic orientation.

E. Orientation to Electric Fields

Weakly electric fishes also use passive electrolocation. For example, territorial gymnotid fishes locate dipole electrodes not by swimming directly to them, but rather by bending their bodies to follow the local current vectors until they reach the source. Interestingly, when the electrodes are switched off in midtrack, approach to the target slows and subjects become disoriented and drift away from the target in a few seconds. Such results suggest that subjects do not form cognitive maps so as to locate the position of the stimulus in space (Shieh *et al.*, 1996). Following local current lines to an electrical dipole source is also the way Kalmijn (1997) visualizes approach to prey by elasmobranchs and, interestingly, he postulates the same basic maneuver when sharks are confronted by acoustical dipole signals in the acoustical near-field.

F. The System during Early Development

Little is known about the electrosensory system of fishes during early development. Sisneros *et al.* (1998) have examined the electrosensory system of the clearnose skate, *Raja eglanteria,* during ontogeny. Afferent neurons of prehatch embryos (8–11 weeks) can detect weak electric fields outside their egg case, showing greatest sensitivity at 1–2 Hz. Immediately after detecting weak uniform fields of 0.5 and 1 Hz, the embryos "freeze" all movement. This phasic electrical stimulus contains spectral information found in the potentials produced by natural fish predators, and the authors interpret such behavior as decreasing the likelihood of mechanoreceptive detection by potential predators. Vischer *et al.* (1989) have shown that the nerve fibers of the anterior lateral line system, which will innervate the electroreceptors in the gymnotid (presumably *Eigenmannia lineata*), are present in the embryonic and larval stages before electroreceptors have developed. Their results suggested that tuberous organs, innervated by the anterior lateral line ganglion, are not functional prior to 18 days following fertilization. Different weakly electrical organs of larvae and adults have been identified in the gymnotid, *Apteronotus albfrons;* the larval organ is myogenic, whereas the adult organ is neurogenic, yet both apparently have the same pacemaker nucleus (Bass, 1986). Distinctive larval and adult organs have also been described in several species of mormyrids. The organ, identified in larval *Pollimyrus isidori* (Mormyridae), is replaced by the adult organ between 60 and 80 days following fertilization (Postner and Kramer, 1995). The EODs of the larval organ were found to be stabile in 10- to 15-day-old larvae and their possessors were exceedingly sensitive to weak electrical stimuli, down to 10 μV cm^{-1}. Response to various stimulus types showed that they were most responsive to those reflecting larval EODs and appeared perfectly adapted for sensing larval rather than adult EODs. Older larvae (54 to 60 days) showed even greater sensitivity (2.4 μV cm^{-1}) and increased sensitivity to characteristics noted in adult EODs, while remaining sensitive to their own EODs. Apparently, development anticipates the tuning necessary for the reception of their own future adult EODs. Development of the medullary electrosensory lateral line lobe (ELL) in the channel catfish *Ictalurus punctatus* (Ictaluridae) was examined by Lannoo and Lannoo (1996) in reference to the onset of three behaviors: undulatory swimming, schooling, and exogenous feeding. All components of the lobe were morphologically mature 15 days postfertilization (hatching occurred 6–8 days postfertilization) and this meant that the ELL does not mature in time to support the first two behaviors, but does mature in time to support exogenous feeding, assuming that electroreception is required for feeding by larvae after yolk absorption.

Although only indirectly related to the present subject, findings from different levels of organization have suggested a clear magnetic sense in juvenile rainbow trout, *Oncorhynchus mykiss* (Salmonidae) (Walker *et al.*, 1997), which agrees with limited behavioral evidence obtained on smolt of sockeye salmon, *Oncorhynchus nerka* (Quinn *et al.*, 1981; Quinn and Brannon, 1982). The young trout first performed a learned magnetic discrimination. This was followed by demonstrating response to changes in magnetic field intensity by the ros V nerve, which is a branch of the trigeminal nerve that enters the anterior dorsal area of the medulla oblongata. Subsequent findings showed that the ros V ended in the laminar propria layer within olfactory lamellae, which contained cells holding iron-rich crystals of a size and shape almost identical to single-domain magnetite, previously extracted from sockeye salmon (Mann *et al.*, 1988). Walker *et al.* (1997) considered these cells as candidate magnetoreceptors, the first cells ever so designated based on the preponderance of evidence. The crystals have been conclusively identified, using confocal and atomic force macroscopy imaging, as single-domain magnetite (Diebel *et al.*, 2000; Walker *et al.*, 2000). It appears that the long-sought magnetoreceptor cell in vertebrates has been found in a fish.

VI. The Chemosensory Systems

A. Chemicals in the Aquatic Medium

The physical processes controlling chemical dispersal in the aquatic medium, ranging from ocean basins to molecular diffusion, have been clearly described by Atema (1988). Chemoreception, the ability to sense and respond to a local concentration change of a particular chemical compound or mixture, can be critical for survival in fishes, which live in essentially turbulence-driven and, thus, patchy environments (Atema, 1988). Although the same fluid dynamic processes occur in water and in air, studies of stimulus dispersal related to chemoreceptive responses are more easily accomplished in water because of smaller spatial scales and because tracers of dispersal, such as dyes, are easier to measure than aerial tracers, such as ions. Water-soluble, organic stimulus compounds are also relatively easy to measure and quantify at biologically relevant concentrations. Chemical compounds or mixtures, capable of interacting with a receptor cell and at a delivery rate to which the cell responds, originate from a source and are removed by various physical, chemical, and biological processes. They are distributed by diffusion, laminar flow, and turbulence at various scales. For each organism, the spatiotemporal scale of interest is bounded on the lower end by the sampling volume and the time constants of its most phasic receptors, and on the upper end by the adaptation rate of tonic receptors as well as the integration characteristics and search strategies of the organism. Although turbulence characterizes the aquatic medium of fishes, chemoreceptor organs demonstrate adaptations to reduce such chaos by being capable of utilizing viscous and laminar flow fields (Døving *et al.*, 1977).

Atema (1988) emphasized the vertically stratified density layers often found in the ocean and in freshwater lakes and whose horizontal dimensions are some 100 times their vertical dimensions (1–10 m). Such layers likely contain different populations of biota and dissolved organic material and can exist for days to weeks before losing their identity in mixing processes. Atema considered that such patterns of vertical structure could permit a discriminating nose to sample features in an extremely large horizontal scale (see Atema, 1988). In theory, a fish can sample several layers in one vertical dive, choose the one that smells right, and swim in that layer until it ends and then sample once again. This may actually be occurring in waters near coral reefs, where presettlement fishes have shown such behavior on their way to settlement sites (Leis *et al.*, 1996).

B. The Chemical Senses

The so-called common chemical sense has been attributed to spinal nerves (Hara, 1971), but little is known thereof. One exception has dealt with the spinal innervation of the ventral rays of the pectoral fins of sea robins, genus *Prionotus* (Triglidae), which lack taste buds. These ventral rays, used to locate food, have been classified as elements of the common chemical sense based on their innervation (Finger, 1988). This contrasts sharply with the situation of cods (*Gadus:* Gadidae) and freshwater gouramis (*Trichogaster:* Anabantidae), in which specialized fin appendages contain taste buds and are innervated by branches of the facial nerve. Finger (1997) provided new morphological evidence that the chemical sensors on the pectoral fin rays of *Prionotus* and on the anterior dorsal fin of rocklings (*Ciliata:* Gadidae), and the oral papillae of the lamprey, *Lampetra planeri,* form a separate chemical system termed solitary chemoreceptor cells (SCCs). The latter system has similarities with one type of taste receptor cell (the light cell), and Finger suggested that taste buds may be compound organs that include some cells related to SCCs. SCCs are secondary receptor cells that are innervated by either cranial or spinal nerves appropriate to the epithelium in which they reside. Additional chemosensory structures include pit organs of the nurse shark *Ginglymostoma cirratum* (Katsuki *et al.*, 1969) and free neuromasts associated with the lateral line system in some teleosts (Katsuki and Yanagisawa, 1982). These small organs are sensitive solely to cations in the external environment. However, this sensitivity may simply be an incidental by-product of receptor physiology (Bleckmann, 1993).

Fish taste buds, located in the epidermis, vary between 30 and 80 μm in length and 20 and 50 μm in width. Generally, two types of microvillous cells, along with a basal cell, are routinely described in the taste buds of teleosts. One of the former, the so-called light cell, possesses a large club-shaped microvillus at its apical surface and is generally accepted as a gustatory receptor cell because of its synaptic contacts with nerve processes within the taste bud (Caprio, 1988). The other type of cell, the so-called dark cell, possesses numerous small microvilli and has been considered a supporting cell. Reutter (1978) found, however, that dark cells in ameiurid catfishes also form synapses with nerve terminals within taste buds. Basal cells, two to five per taste bud, are located at the base of the bud, resting against the basal lamina. Their function is unknown, but they may modify gustatory receptor activity through synapses with both light and dark cells, at least in ameiurid catfishes (Reutter, 1982). Although the gross morphology of the taste buds of fishes is similar, caution about generalizations regarding vertebrate taste buds is suggested based on differences in their ultrastructure (Reutter and Witt, 1993). Taste buds are generally distributed within the oral cavity, pharynx, and gill rakers. In some fishes, taste buds extend to the external surface. Those on the surface, lips, and rostral palate are innervated by the facial (VII) nerve, whereas the glossopharyngeal (IX) nerve innervates a relatively minor intraoral field in most fishes and the vagal (X) nerve innervates the bulk of the oropharyngeal region, gill arches, and palate (Atema, 1971; Finger, 1988). These nerves terminate centrally in the visceral sensory column of the rostral medulla oblongata and, from there, further projections extend to other regions of the brain (Finger, 1988).

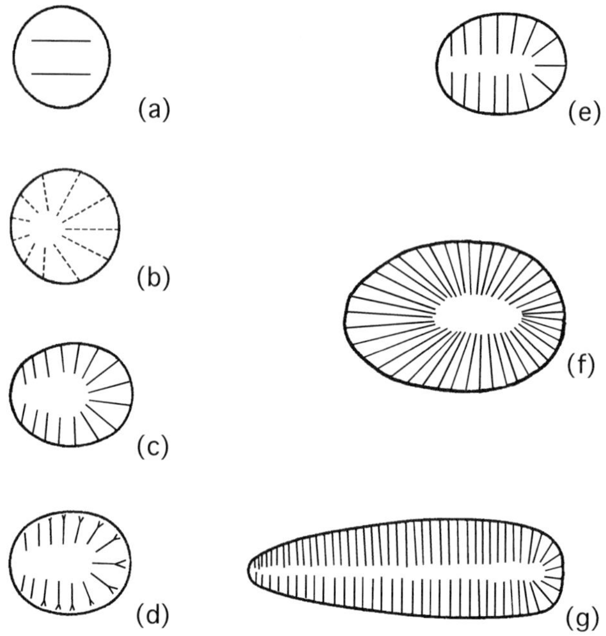

FIGURE 6 Schematic representations of numbers and arrangements of lamellae in the olfactory rosette: (a) stickleback (*Gasterosteus*), (b) pike (*Esox*), (c) rainbow trout (*Oncorhynchus*), (d) perch (*Perca*), (e) minnow (*Phoxinus*), (f) bullhead (*Ameiurus*), (g) eel (*Anguilla*) (modified from Bardach and Villars, 1974).

The olfactory organ in fishes, as in other vertebrates, is located in the snout, where water, often aided by a groove or skin flap, passes over receptors in the epithelium, which is arranged in folds or lamellae, often resembling a rosette (Kleerekoper, 1969). Water passes through the organ either by movements of cilia located on supporting cells in the epithelium or by muscular pumps associated with accessory sacs (Bardach and Villars, 1974; Døving *et al.*, 1977). The ultrastructure of the epithelium, bulb, and tract of the olfactory system is reviewed in detail by Hara (1993). The sensory regions of the epithelium, in the majority of teleosts studied, contain at least two types of bipolar neurons, ciliated and microvillous, which are classified on the basis of their dendrites. The former have knob-like dendritical endings containing 4 to 12 cilia that are 3–8 μm long, whereas the latter possess 30 to 50 microvilli, 2–5 μm long, on a protruding apical ending ("tufted" cell) or 15 to 25 microvilli, 3–6 μm long, radiating laterally from a flat, disklike apical surface ("stellate" cell) (Laverack, 1988). Elasmobranchs, long considered to possess only microvillous receptors in the olfactory epithelium, have been shown in the Oman shark, *Iago omanensis* (Carcharhinidae), to possess also ciliated neurons (Fishelson and Baranes, 1997). Differences in physiological function among the cell types are unclear. The axons of these bipolars, forming the olfactory nerve fibers, lead to the rostral end of the olfactory bulb, where they spread across its surface. The bulb, thought to process the bulk of olfactory signals (Bruckmoser, 1973), projects via the olfactory tract to the telencephalon and minor components to the ventral diencephalon (Finger, 1988). Olfactory acuity appears to be correlated with the number and arrangement of olfactory lamellae in the nasal pit (Fig. 6) as well as the area covered by sensory versus supporting units in the epithelium (Bardach and Villars, 1974).

Although a reasonable amount of information exists regarding the chemical senses of fishes, little information exists regarding chemoreception in marine teleosts and practically nothing is known about the roles that the chemical senses play in the lives of coral reef fishes.

C. The Adequate Stimulus

Responses by the common chemical and gustatory senses and presumably SCCs are apparently elicited by a simultaneous duality of tactile and chemical stimulation (Laverack, 1988), except in cases, exemplified by catfishes, where, in the absence of tactile stimulation, taste receptors are used in locating distant food sources, i.e., distant taste. Although the common chemical sense is often highly associated with detecting irritants, SCCs appear to possess diverse chemical responsiveness even among the few species that have been examined. Adequate stimuli for these systems are not unlike those of the olfactory system, but generally possess higher thresholds. Hara (1993) points out, however, that this distinction is no longer strictly tenable.

Far more is known about the stimuli that mediate responsiveness by the olfactory system in fishes. Important chemical stimuli include bile salts, nucleotides, and fatty acids. However, to date, no class of compound stimulates olfactory and gustatory systems as strongly as do amino acids, the breakdown products of peptides and proteins (Jónsson, 1980; Caprio, 1988). A sensitive amino acid detector can certainly help in gaining vital information about biological activity underwater (Holland, 1978). The highly specific findings often obtained in studies, particularly involving olfaction, attest to the specificity of these and other detectors. For example, L-serine is the salmon repellent found in mammalian skin (Idler *et al.*, 1956). There are other cases of specificity: (1) Atema *et al.* (1980) showed discrimination between Hawaiian and California anchovies of the same species during feeding trials by yellowfin tuna, *Thunnus albacares;* (2) a synthetic mixture of 20 amino acids and betaine was as effective a stimulus as shrimp extract for the pigfish, *Orthopristis chrysoptera* (Haemulidae), whereas in the pinfish, *Lagodon rhomboides* (Sparidae), it had only about 25% of the potency of shrimp extract (Carr, 1976; Carr and Chaney, 1976; Carr *et al.*, 1977); (3) catfishes reacted differently to the same conspecific before and after social stress (Todd, 1971); and (4) Schreckstoff (fright substance) among ostariophysan fishes has shown that levels of response depend on taxonomic position (Schutz, 1956; Pfeiffer, 1960, 1977). The above findings lead one to suspect that biologically meaningful scents, triggering feeding as well social behavior, are often mixtures of several compounds, each at its own level.

D. Functions

The behavioral significance of the solitary chemoreceptor cell system is, in most instances, still enigmatic. It clearly functions in feeding in sea robins and suggestion has been made that it functions in predator avoidance in rocklings (Finger, 1997). The gustatory system clearly mediates feeding responses (but see below), but the common chemical sense appears to mediate aversive responses. Hasler (1957) reviews many of the early experiments dealing with the functions of gustation and the common chemical sense in fishes and precious little more is known about them. Freshwater catfishes (*Ameiurus nebulosus* and *A. natalis:* Ictaluridae) find food by taste only (distant taste), based on findings that destruction of peripheral smell does not interfere with food finding (Bardach *et al.*, 1967; Atema, 1971). Atema (1971) also showed in *Ameiurus* catfishes that the facial lobes control the "pick-up" component of feeding, whereas the vagal lobes control the "swallowing" component. Distant taste has also been found in the Hawaiian goatfish, *Parupeneus porphyreus* (Holland, 1978). When barbels (part of the gustatory system) were removed from that species, arousal was greatly reduced and food search was eliminated, but normal food search occurred after cauterizing the nasal rosettes of the olfactory organ (Holland, 1975). In marine species, toxins, e.g., tetrodotoxin and saxitoxin, are detected through the gustatory system and seem to play a significant role in avoiding noxious substances in the environment (Yamamori *et al.*, 1988; Hara, 1993).

Far more is known about the varied functions of the olfactory system, centering for the most part on feeding, but extending to social functions as well in specific circumstances. The olfactory organ is usually the system that mediates food location by sharks (Hobson, 1963; Tester, 1963, Gilbert *et al.*, 1964) and by many teleosts (Bardach and Villars, 1974). Studies, including field experiments, have shown, however, clear differences in the importance of olfaction in food search by similar taxa of moray eels. Species within the muraenid genera *Siderea* and *Echidna* were macrosmatic (smell acutely developed) and almost totally dependent on the sense of smell for food detection, whereas species within the muraenid genera *Gymnothorax* and *Lycodontis* were microsmatic (smell feebly developed), using vision as the primary sense for food detection (Tannenbaum *et al.*, 1992; Fishelson, 1995, 1997). The latter findings, using a species of *Gymnothorax* from the Gulf of Aqaba, contrast with those of Bardach *et al.* (1959), who found two species of *Gymnothorax* from the waters off Bermuda to be extremely sensitive to olfactory stimuli.

The location of chemical stimulus sources may well be difficult to determine, because the source may move away but the stimulus it produced may linger for a long time. Also, currents may carry the stimulus along complex paths of dilution, obscuring any kind of linear concentration gradient that would provide the direction of the source. Nevertheless, chemical trails of food stimuli have been followed by the eel, *Anguilla anguilla* (Anguillidae), and by the nurse shark, *Ginglymostoma cirratum,* by successive comparison of concentrations along the trail (Teichmann, 1959; Matthewson and Hodgson, 1972) or by positive rheotaxis as long as the chemical stimulation persists and sensory adaptation does not occur [as in the lemon shark, *Negaprion brevirostris* (Matthewson and Hodgson, 1972)].

Atema (1980) has likened fishes to leaky bags that slowly, through the skin and gills, or in large pulses, through urine and feces, release various chemicals, including the above-mentioned amino acids. These chemicals form the equivalent of body odor, which can be used to identify species, sex, stress level, and perhaps size and individuality. One can readily envision how predators detect prey by such means, but the opposite probably can also occur, i.e., prey can become aware of a predator's "leaky bag." A pheromone, a chemical substance secreted externally by an individual and received by a conspecific, elicits a specific reaction, such as a definitive behavior or a developmental process. Such communication has been shown during various stages in the life history of fishes (Jónsson, 1980). Some chemical stimuli, mentioned above, are clear examples, e.g., Schreckstoff. One of the most remarkable examples of the use of pheromones in fishes concerns their intimate relationship with numerous aspects of social behavior in ameiurid catfishes (Todd, 1971). Other examples of their use in marine species include species and gender recognition in gobies [*Typhlogobius californiensis* (MacGinitie, 1939)] and blennies [*Hypsoblennius* spp. (Losey, 1969)], as well as sex pheromones in gobies [*Bathygobius* spp. (Tavolga, 1956; Stadler, 1999)] and damselfish [*Pomacentrus partitus* (Kenyon, 1994)].

E. The Systems during Early Development

The early ontogeny of the olfactory organ in elasmobranchs has been examined solely in the Oman shark, *Iago omanensis* (Fishelson and Baranes, 1997). In general, it was found that the macrostructure of the nasal pits and the sensory rosettes during development resembled that found in teleosts (Zeiske *et al.*, 1992). It was also found that growth of the olfactory organs in these sharks is allometric to growth of the central nervous system and, in adult animals, that the mass of the brain is less than that of the olfactory organs, indicating the importance of smell in foraging by these deep-water sharks. For teleost larvae, several studies have demonstrated that the receptor cells in the olfactory epithelium develop quickly (summarized by Døving and Knutsen, 1993). Around the time of settlement, labrid larvae have high densities of both microvillous and ciliated receptor cells, densities as high as those found in adults of some macrosmatic species. Both incurrent and excurrent nares are formed at this stage in labrids. Scarid larvae at settlement are somewhat less developed, lacking the tissue that separates the two nares (Lara, 1999).

The importance of chemoreception during development in fishes has not received enough attention. Kühme (1963) and Myrberg (1966) found that chemical stimuli from larval and young juvenile cichlids influence parental behavior, and that young juveniles of *Cichlasoma* spp. are also attracted by chemical stimuli from adults (Myrberg, 1975; Barnett, 1977). Experiments have demonstrated that larvae of several species such as *Balanus* nauplii, change their swimming behavior in the presence of washings and extracts of prey organisms, as well as in the presence of predominant amino acids found in those solutions (Dempsey, 1978; Knutsen, 1992; Døving *et al.*, 1994). Patches of phytoplankton and zooplankton probably contain elevated concentrations of metabolites that larvae can detect. The behavioral changes observed in the laboratory are thought to increase the likelihood of finding a patch of food and staying within it once encountered (Døving and Knutsen, 1993). Sweatman (1988) provided field evidence that settlement onto coral reefs by young of the damselfish, *Dascyllus aruanus,* was due to chemical cues from conspecific residents and additional evidence suggested that two species of damselfish of the genus *Pomacentrus* avoided reefs possessing chemical cues of *Dascyllus*. Elliott *et al.* (1995) found that settling anemone fishes of the genera *Premnas* and *Amphiprion* (Pomacentridae) use chemical cues to identify and locate appropriate species of host anemones in the field from a maximum distance of 8 m downstream. Trials suggested that the young fishes are not visually attracted to anemones from distances of 0.5 m or more and that the fishes must be downstream from a host in order to locate it. Further experiments by Arvedlund and Nielson (1996) showed that laboratory-raised *Amphiprion ocellaris* do not innately recognize their species-specific host anemone, *Heteractis magnifica*. Instead, they apparently imprint olfactorily to such hosts during a brief period around the time of hatching. After 60 days in water free of anemones, they are rapidly attracted to the hosts's odor trail and even after 1.5 years in such water, the individuals begin the acclimation process with host anemones within 10 minutes. On the other hand, individuals having never been in water holding dissolved chemicals from such anemones showed no interest in water holding hosts after 60 days posthatch, and after 1.5 years, acclimation to host anemones took 48 hours.

Studies of the morphological changes in the barbel system (an outgrowth of the gustatory system) of the goatfish, *Upeneus tragula* (20–34 mm standard length), around the time of settlement showed that sudden changes in structures occurred precisely during the 6- to 12-hour period of settlement (McCormick, 1993; McCormick and Shand, 1992). Although considerable variability was found, various changes included rapid growth of barbels (increasing up to 52% of length vs.

slightly exponential growth during the earlier pelagic stage), rapid increase in cross-sectional area of the barbel with a simultaneous asymptotic increase of its epidermal region and an asymptotic decrease of its dermal region, and also an increase in size of taste buds by up to 100%. Such rapid changes indicated the functional importance of the system at the time when demersal existence replaced a pelagic way of life. Associated laboratory experiments during the settlement period also showed that the size and internal structures of the barbels were influenced by food availability, i.e., barbels and taste buds were both larger among individuals receiving less food, but were not influnced by slight temperature change.

VII. Sensory Systems and Settlement

One of the great puzzles of coral reef fish ecology is how the young, pelagic stages locate the relatively rare patches of coral reef habitat on which they settle and ultimately reside as adults. The answer must lie in the sensory world of these fishes, because it seems unlikely that successful settlement is solely a matter of chance. Presettlement stages of reef fishes have excellent locomotor capabilities that may even exceed those of recently settled individuals (Stobutzki and Bellwood, 1994; Leis and Carson-Ewart, 1997; Stobutzki, 1998). Further, their swimming direction in the sea appears to be oriented, although the specific cue or cues remain unknown (Leis *et al.,* 1996; Stobutzki and Bellwood, 1998; Leis and Carson-Ewart, 1999). Given the sensory capabilities of coral reef fishes, which cues are suitable candidates for orientation toward reefs? A primary consideration is the large distance over which the signal must be detected, on the order of kilometers or perhaps tens of kilometers. This essentially eliminates lateral line and visual cues. The idea that vision is probably unimportant for orientation is reinforced by observations that settlement-stage fishes move toward the reef at night (Stobutzki and Bellwood, 1998) and away from it during the day (Leis *et al.,* 1996). Transplantation experiments coupled with behavioral observations suggest that these fishes are not using tides, currents, sun compass, or geomagnetic fields (Leis *et al.,* 1996).

Layers of the water column with specific chemical cues that can be followed for long distances (Atema, 1988) may guide presettlement fishes toward a reef. In theory, a fish can sample several layers in one vertical dive, choose the one that smells right, and swim within that layer until it ends and, then sample once again. Because chemical cues may be moving with currents, larvae may also use their superficial or canal neuromasts to orient to local currents (Lara, 1999). In most situations in which the cue is chemical, larvae would be expected to arrive at the reef from downstream locations. It is important to note that although the functionality of the olfactory system of coral reef fish larvae has not been adequately studied, the system may be sufficiently developed to serve orientation behavior. The density of olfactory receptor cells in many labrids at settlement is approximately equal to that of adults fishes that have acute olfactory capabilities (Lara, 1999).

The sounds of a coral reef constitute another potential cue that could steer settlers. Acoustical signals are even less limited than chemical cues. They travel greater distances and could attract settlers from all directions. Unfortunately, experimental data are available for only two species (*Pomacentrus*), and these indicate that prior to settlement the fishes have poor hearing (Kenyon, 1996). Even if other species are shown to be rather sensitive to sound, can they hear the reef sound over the ambient levels they experience at the same time? All evidence from fishes indicates that the reef sounds need to be 15–20 dB above the ambient level for the sound to be heard! So the reef sounds cannot be only a bit above ambient, but must be considerably more intense. A field study has shown that significantly more triplefin (Tripterygiidae) larvae were caught in light traps adjacent to an underwater loudspeaker broadcasting recorded reef sounds than were caught in "silent" light traps (Tolimieri *et al.,* 2000). Although the sound source levels were extraordinarily high (180 dB re 1 μPa) in this experiment, the results are suggestive of a role for audition in the settlement process. Further experimentation will be necessary to unravel this mystery. It is likely that presettlement fishes will use more than one sensory system to orient to a reef and that the role of different senses will change as a fish draws nearer to its destination.

After settlement, a fish resides in a habitat that differs greatly from the water column and likely requires different sensory skills for survival. Because settlement takes place relatively early in life, it is reasonable to expect that a larva that settles too soon may not have developed the appropriate sensory (and other) abilities for survival on the reef. The fact that settlement takes place over a much narrower range of sizes than ages, within a species, suggests that developmental stage (which is closely linked to size, not age) helps determine competency to settle. Even our cursory review of development demonstrates that reef fishes settle only after most sensory systems (visual, olfactory, lateral line) are well developed. However, it is possible that natural selection may have countered this ontogenetic constraint on settlement by tailoring the ontogenies of reef species differently from other species, accelerating development of critical systems so that settlement can take place sooner. A careful comparison with species that never become closely associated with a substrate may provide useful insight into reef fish ecology and, in particular, the special challenges for fishes living on reefs. A similar analysis, performed on flatfishes, indicated that lateral line canals seem to form at about the time of settlement and at an earlier stage of development than in pelagic species (Fuiman, 1997). Correspondingly, most cephalic lateral line canals are present at settlement in 12 labrid species (Lara, 1999). Whether the appearance of these structures is accelerated, compared to pelagic species, has yet to be determined.

VIII. Conclusion

Fishes perceive their world through many and diverse senses. We have come to understand the anatomy and functional mechanisms of most of the known sensory systems quite well. The fact that this understanding is based largely on species from freshwater or temperate marine habitats is probably unimportant. However, the coral reef environment differs from these other habitats in important ways, and so the sensory capabilities of reef fishes and the ways they use their senses cannot be presumed equivalent. We need direct knowledge of the limits of sensory function in coral reef fishes because inferences from morphology and phylogeny cannot lead to definitive answers. Further, the time is now ripe to begin a serious quest for behavioral evidence of sensory involvement and how various systems, alone and in aggregate, aid fishes, regardless of age, stage, or way of life, in answering the ceaseless problems dealing with avoiding predation, obtaining food, locating appropriate conditions and shelter, and last, but certainly not least, locating mates. It is highly likely that such information, when obtained from a goodly number of coral reef fishes, will be essential to more realistic management practices.

CHAPTER 7

Larval Dispersal and Retention and Consequences for Population Connectivity

Robert K. Cowen
Rosenstiel School of Marine and Atmospheric Science
University of Miami
Miami, Florida 33149

I. Introduction

The larval stage of most coral reef fishes is spent in the pelagic environment, potentially well away from the reef proper. Survival of this stage is tenuous, being mediated by factors such as food availability, predator abundance, and physical conditions. The complex biological and physical interactions of these factors can result in a seemingly stochastic larval supply that drives temporal and spatial variation in recruitment intensity (Cowen, 1985; Roughgarden *et al.*, 1988; Choat *et al.*, 1988). This variation can be a major determinant of adult population sizes (Cushing, 1975; Rothschild, 1986; Sinclair, 1988). A growing awareness of this perspective has placed a premium on the need to study the early life history stages of marine organisms, forcing us to peer into the black box of larval biology and ecology. The following discussion will outline where advances have occurred in our understanding of how, and to what effect, coral reef fish larvae interact with the pelagic environment.

Without denying the importance of food and predation to larval survival (Houde, 1987; Bailey and Houde, 1989), it is clear that many species of coastal marine fishes are particularly dependent on directed transport to their juvenile or adult habitat. For example, coral reef fish larvae will not survive the dispersive stage of their life history without appropriate transport back to a reef (Leis, 1991a). Similar arguments can be made for many temperate species that are estuarine dependent or benthic oriented (Boehlert and Mundy, 1988). In contrast, many pelagic species do not require directed transport to complete their larval life, although some species may recruit only if their larvae are retained within a specific area [e.g., herring populations (Sinclair, 1988)].

Although fish larvae are often (erroneously) considered strict constituents of the zooplankton community, evidence suggests that many species exhibit some form of active behavior during their pelagic stage (Neilson and Perry, 1990; Leis, 1991b). Although some larval distributions can be explained purely through a combination of spawning sites and passive larval transport [i.e., constant buoyancy (Boehlert and Mundy, 1994)], other larval distributions are influenced by larval behavior interacting with the local current regime. The importance of vertical larval movements in affecting horizontal movements in estuarine systems (Boehlert and Mundy, 1988; Epifanio, 1988) or retention in bank systems (Werner *et al.*, 1993; Tremblay *et al.*, 1994) is well documented. There is also evidence of impressive horizontal swimming abilities in late-stage larvae (see Chapter 8). Thus there is a continuum of possible behaviors (e.g., passive to vertical to horizontal movement), as well as the timing of the onset or ontogeny of such behaviors. Such behavioral capabilities may strongly impact the survival and recruitment of coral reef fish larvae to benthic habitats.

The complexity introduced into biological systems by the physics of water movement contributes significantly to the variability associated with the recruitment of marine species. Variability in larval transport will be determined by the interaction of water masses and the effects of external forcing such as winds and tides. Active behavior by larvae may modulate some of this variability, yet a strong change in the direction or intensity of flow of a particular water mass may result in a substantial change in larval supply to a given reef or island, or may carry larvae away from a suitable food environment. Because recruitment integrates the effect of physics across all life history stages, this variability is large and complex. Though it is unlikely that we will ever measure the importance of *all* processes contributing to variable recruitment patterns, we potentially can isolate the most important processes that are operating. By doing so, we stand a better chance of explaining observed recruitment patterns, predicting future recruitment, and ultimately forging a general theory of recruitment dynamics.

In such an effort, we also need to consider the evolutionary implications of how larvae have adapted to the physical environment to ensure their survival and successful transport or retention. How well can larvae match their behavior to the particular system in which they live? At face value this seems like an elementary question for any adaptationist; presumably a fish is adapted to the set of environmental variables it typically encounters. However, the larvae of a single species may encounter a broad range of coastal environments, ranging from continental shelves to island chains to isolated oceanic islands (e.g., most Caribbean or Indo-Pacific species), and therefore the question becomes one of how specific are the features to which larvae are adapted. If larval behaviors are adaptations to promote successful return to adult habitat, are these adaptations more likely to be keyed to the specifics of a particular location, or to features common to many systems? To address this, we need to determine the generalities of larval transport in terms of both the specific circulation features important to larval transport and the larval responses to these features.

In addressing the above topics, the emphasis here is on processes contributing to larval transport, although some discussion is devoted to processes contributing to survival as well. However, it is impossible to do full justice to this multifactor issue of larval survival in this chapter. To constrain the topic, this chapter is not a detailed review of physical oceanography, but there are sufficient citations to physical oceanographic studies that an interested reader can pursue the trail in greater detail. In most cases, the focus is on recent literature, particularly on those topics covered previously in Leis (1991a). Finally, although the primary focus is on coral reef fishes, the literature on temperate species is examined where necessary (see also Cowen and Sponaugle, 1997).

II. Scale

Embedded in any discussion of larval transport and its role in marine population dynamics is the issue of scale. Marine systems are highly dynamic largely because of the interaction of various physical processes operating on a variety of spatial and temporal scales (Haury *et al.*, 1978; Denman and Gargett, 1983; Mullin, 1993). Determining the relevant scales of physical processes will depend on the biological questions being asked as well as on the environments being studied.

The scale over which larval transport (or retention) occurs varies substantially among species and even among locations within a single species. For some species, larval dispersal is minimal and possibly nonexistent [e.g., the damselfish, *Acanthochromis polyacanthus* (Robertson, 1973; Thresher, 1985)], whereas at the other extreme dispersal can occur over thousands of kilometers [e.g., *Anguilla rostrata* and *A. anguilla* (McCleave *et al.*, 1987)]. However, for most coral reef fish species, with larval durations of weeks to months, transport will usually be on the scale of tens to hundreds of kilometers. It may ultimately be shown (see Section VII) that local physical effects aid retention, thereby placing a greater emphasis on smaller scale processes as influencing local population dynamics. If so, the sampling constraints that have led to focusing on smaller scales will have been fortuitously important in providing evidence of greater retention of young than was previously expected.

In addition to spatial scales, temporal scales are crucially important in the issues we study. For example, population exchange (or connectivity) is often indicated via population genetics studies (e.g., Lacson, 1992; Doherty *et al.*, 1995; Shulman and Bermingham, 1995; Planes, 1993, 1998). In general, these studies have suggested that a fair amount of genetic exchange occurs over rather extensive (thousands of kilometers) distances. Yet genetic exchange operates at a very different temporal scale, as compared to biological and physical processes contributing to larval transport. The level of exchange required to sustain genetic homogeneity is many orders of magnitude lower than that required to sustain a population ecologically (Avise, 1992; Lacson, 1992) because measurable genetic exchange occurs over many generations and often does so in a stepping-stone fashion. Knowledge of the extent of ecological time scale larval exchange is of fundamental importance to the study of population dynamics as well as to the design of marine reserves. Decisions predicated on the assumption that larvae are widely dispersed may lead to erroneous actions if the assumption is wrong. However, lack of evidence of retention has generally resulted in the acceptance of the concept of well-mixed populations. This, in turn, has been interpreted to infer that mean current trajectories and larval duration may predict exchange between populations (e.g., McManus and Menez, 1997; Roberts, 1997b; but see Sale and Cowen, 1998; Bellwood *et al.*, 1998; Cowen *et al.*, 2000).

Thus, when examining processes affecting larval exchange, we need to examine time scales relevant to the ecological sustainability of coral reef fish populations: Larval stages last weeks; reproduction may occur once per year to daily; life-spans range from 1+ to 50+ years. A suite of physical processes fall within this temporal context, leading both to physical transport of larvae and to conditions directly affecting their survival. Recruitment of fishes with very protracted spawning seasons (e.g., year round) can be influenced by a broader spectrum of processes than is likely with fishes with short spawning seasons. A species that spawns over a very short period will exhibit interannual trends, whereas a species with more protracted spawning seasons will likely exhibit intraannual and even intraseasonal patterns as well. In the latter case, there is greater opportunity for small to mesoscale processes to introduce considerable variability.

On a larger spatiotemporal scale, maintenance of populations is affected by the longevity of individuals. Short-lived species will require more frequent input and survival of new recruits to sustain their populations than will long-lived species (Cowen, 1985; Warner and Chesson, 1985). The scale of important physical processes also depends on the type of environment being studied. Particular environments will determine which physical processes likely operate and, importantly, which do not operate. This effect of the physical characteristics of the system becomes particularly important when different types of environments are compared (e.g., shelf vs. oceanic environments). This latter point is often overlooked when attempting to compare among studies.

Another aspect of the temporal linkage between biological and physical processes relates to the capacity of organisms to respond to a given physical process. For example, availability of food obviously impacts all life-cycle stages, though it may do so in different ways. Increased nutrients may be rapidly transferred into potential prey (1° and 2° plankton production) and therefore a relatively short-duration event associated with coastal upwelling or internal waves (e.g., Leichter *et al.*, 1996) may be an important contributor to the success of a cohort of larvae. Such short-duration, local nutrient pulses may not be immediately transferred into available food for adults, but longer (i.e., seasonal), more spatially extensive pulses may be, ultimately influencing reproductive output (e.g., Robertson, 1992; Clifton, 1995). However, if short-duration, localized nutrient pulses serve as an environmental cue for reproduction (i.e., ensuring food availability for newly hatched larvae, or recently settled juveniles), then similarly scaled processes may be important to both life history stages. In this case, the behavior of adults has increased the potential number of relevant scales over which physical processes may operate.

III. Biological Pattern

Much of the initial work on larval fishes focused explicitly on describing distribution patterns. More recently, the effort has shifted to identifying the processes contributing to the observed patterns of distributions, and, more importantly, trying to link these processes to observed patterns of recruitment variability. Below is a brief review of work on the horizontal, vertical, and temporal distribution patterns of coral reef fish larvae, with a strong emphasis on the factors contributing to the observed pattern.

A. Horizontal Patterns

Early work on the larval distributions of coral reef fishes described patterns with regard to the distances from shore certain taxa were distributed (see review by Leis, 1991a). To this end, Leis (1978) described three main assemblages of coral reef fish larvae with respect to proximity to shore: nearshore, neritic, and oceanic. Subsequently, a distinct fourth, and occasionally completely isolated, embayment (or lagoonal) assemblage was identified (Leis, 1986a, 1991b, 1994; Boehlert and Mundy, 1993; Leis *et al.* 1998). Leis (1991a) reviewed work on the horizontal distributions of coral reef fish larvae, demonstrating that taxon-specific spatial patterns exist for many taxa with respect to distance from shore. He concluded that such taxon-specific patterns clearly indicate that larval distributions are not merely the result of passive dispersion by currents; larval behavior must play an integral role.

Beyond determining that larvae of certain taxa typically are found at distinct distances from shore, several studies have been of sufficient spatial and/or temporal scope and resolution to identify larval assemblage dynamics. These broad-scale studies have been coupled with detailed physical oceanographic data that indicate that far more complex mechanisms are operating in determining assemblage makeup and location than simply distance from shore. For example, Boehlert *et al.* (1992) identified high concentrations of pomacentrid/gobiid larvae nearshore of Johnston Atoll in the Pacific Ocean in the vicinity of an onshore flow. Though generated by different forcing mechanisms, similar nearshore distributions of pomacentrid/gobiid larvae were identified along the coast of Barbados, West Indies, where flow was directed onshore [at the depth of maximum larval concentration (Cowen and Castro, 1994)]. These same larval taxa were absent nearshore where such flow conditions were lacking.

It is now clear that larval assemblages are dynamic with respect to both the spatiotemporal distribution of and the species membership within each assemblage (Cowen *et al.,* 1993). Boehlert and Mundy (1993) review factors that may contribute to ichthyoplankton assemblages, noting that such factors may be separated based on their influence on the formation, maintenance, and/or disruption of assemblages (Table 1). Not only does this breakdown enable one to look at the causes of patterns and dynamics of the assemblages, it also identifies the physical processes to which species may adapt.

First, the spatial (and temporal) formation of larval assemblages is strongly affected by the interaction of adult reproductive behavior with coastal currents. Regardless of whether optimal spawning strategies have evolved to maximize larval survival or minimize adult loss (Robertson *et al.,* 1990), the physical conditions present at the time and location of spawning will affect where the larvae end up. The location and time larvae enter the pelagic realm is further affected by egg development and any early swimming behaviors and/or buoyancy characteristics of the larvae. Second, the interaction of larval behavior with the physical environment enables assemblages to persist in time (Table 1). This may occur either through larvae seeking a particular environmental cue and/or prey, or via physical transport mediated by vertical larval movement. In either case, the larvae may be grouped by a variety of vertically structured features. Finally, assemblage membership may change over time (Cowen *et al.,* 1993). Quantification of such changes may provide insight as to how different larval behaviors may contribute to

TABLE 1 Factors of Potential Importance in the Formation, Maintenance, and Disruption of Ichthyoplankton Assemblages around Islands and Seamounts[a]

Type of factor	Influencing formation[b]	Influencing maintenance[c]	Influencing disruption[d]
Adult behavior	Spawning location Spawning time (tidal, diel, lunar, seasonal)	None	Predation (cannibalism) Influences on settlement
Egg development	Buoyancy (demersal vs. pelagic eggs, egg density) Incubation time Hatching location Hatching time	None	None
Larval behavior	Buoyancy at hatching Swimming ability at hatching	Habitat preference (temperature, salinity, light, depth) Vertical migration	Ontogenetic changes in vertical distribution Metamorphosis (settlement, schooling)
Biotic environment	Predation (absence will create "ecological safe sites")	Prey presence	Predation Prey absence (starvation)
Abiotic environment	Convergences Fronts	Eddies (Taylor columns) Upwelling Downwelling Internal waves Convergences Fronts	Diffusion Mixing Advection

[a]Reprinted with permission from Boehlert, G. W., and Mundy, B. C. (1993). Bull. Mar. Sci. 53, pg. 337.
[b]Factors promoting spatial and temporal co-occurrence.
[c]Factors promoting integrity of assemblages over time.
[d]Factors causing eventual breakdown of assemblage structure.

their horizontal distribution within the same physical environment, and/or how larval behavior may facilitate the movement of larvae from one water mass (or flow feature) to another. Assemblages also may be disrupted by strong predation, the need to return to coastal, benthic habitat (i.e., settlement), or via strong diffusive and/or advective events [e.g., effect of transient, mesoscale eddies sweeping larvae away from the coastal environment (Cowen and Castro, 1994)]. Other recent studies of larval distributions (several of which include larvae of coral reef fishes) continue to shed light on the dynamic nature of larval assemblages (Sabatés and Olivar, 1996; Dempster *et al.*, 1997; Sanvicente-Añorve *et al.*, 1998).

Finally, the existence of predictable larval assemblages suggests that it should not be surprising to find multispecific recruitment events. Spatial coherence in recruitment events (especially over scales of tens of kilometers) has led to the suggestion of a single, mesoscale larval patch (Victor, 1986a; Williams and English, 1992; Thorrold, 1993). However, the behavioral and physical dynamics leading to larval assemblages, coupled with synchrony in spawning (and settlement) of many species (Robertson *et al.*, 1988, 1993; Milicich and Doherty, 1994; Sponaugle and Cowen, 1996a), may preclude the requirement that a single, mesoscale patch exist. At small scales, increased concentrations of larvae are often associated with frontal features (see reviews by Kingsford, 1990; Govoni and Grimes, 1992; Govoni, 1993; Olson *et al.*, 1994; Grimes and Kingsford, 1996). At larger scales, coral reef fish larvae have been found to be entrained in spin-off eddies (Limouzy-Paris *et al.*, 1997) and around offshore currents [e.g., Loop Current (Richards *et al.*, 1993)], or in jets of water steered by topographic features (Cowen and Castro, 1994). Such physically driven distributions can interject significant variability into general patterns of onshore/offshore distribution.

B. Vertical Patterns

Attempts to explain the above observed horizontal distributions of larvae have often required invoking an interplay between the vertical distribution of larvae and the physical environment. Although we have some evidence for vertically mediated larval transport for a few species of coral reef fish (Hare and Cowen, 1991; Cowen *et al.*, 1993; Cowen and Castro, 1994), there is a surprising dearth of information on the detailed vertical distribution of larval coral reef fish, particularly with respect to both diel and ontogenetic vertical migrations or movements. This is largely due to sampling capabilities that have only recently enabled more than a coarse resolution of vertical distributions (i.e., surface/nonsurface) and the problem of labor-intensive sorting and identifying of samples taken from multiple depths at each sampling station. In contrast, there is a substantial literature on larval vertical distribution of temperate species (see review by Neilson and Perry, 1990).

As with horizontal larval distributions, early work reviewed by Leis (1991a) described certain taxon-specific vertical distributions, though the relatively few available studies made it difficult to generalize. That same year, Leis (1991b) published the first detailed study of the vertical distribution of coral reef fish larvae from a vertically stratified sampling program of the upper 20 m over the Great Barrier Reef (GBR) lagoon shelf. His general findings indicated a stronger vertical pattern in larval distribution during the day than at night [a finding similar to that of Gray (1998) for more temperate larval fishes, but see also Olivar and Sabatés (1997)], with most taxa avoiding the upper 6 m during the day. Several species, however, were daytime inhabitants of the near-surface (i.e., neuston) layer, a finding that has been observed in other areas [e.g., in Barbados, late-stage dactylopterids (R. K. Cowen and C. Paris, unpublished data)].

Subsequent studies have incorporated various vertically stratified sampling protocols to examine the distribution of tropical fish larvae in coastal and insular environments with relatively narrow shelves. In these areas, sampling occurred in water that usually exceeded depths of 100 m. Working in the waters around Johnston Atoll in the Pacific, Boehlert *et al.* (1992) found coral reef fish larvae to be most abundant between 50 and 100 m, with high concentrations also in the upper 50 m. Cha *et al.* (1994) reported on the nocturnal larval vertical distribution patterns for a variety of fish families, including mesopelagic and coral reef fish families collected off of the Florida Keys reef tract (water depth >150 m). Generally, the highest diversity of taxa were in the upper 50 m, with a drop in diversity with depth to very few taxa collected below 175 m (though it is difficult to separate coral reef fish families from the others in this particular analysis). Of the more common taxa in their samples, they found labrids, carangids, and bothids to be most abundant in the upper 25 m, gobiids, serranids, and scorpaenids to be most common in the upper 50 m, and synodontids to be most abundant at depths greater than 50 m. Cowen and Castro (1994) and Castro (1995) found similar patterns. Of particular note was the linkage between the shallow larval distribution of carangids (especially a preflexion *Caranx* sp.) and their horizontal distribution associated with surface convergences to the southwest and northwest of the island (Fig. 1A);

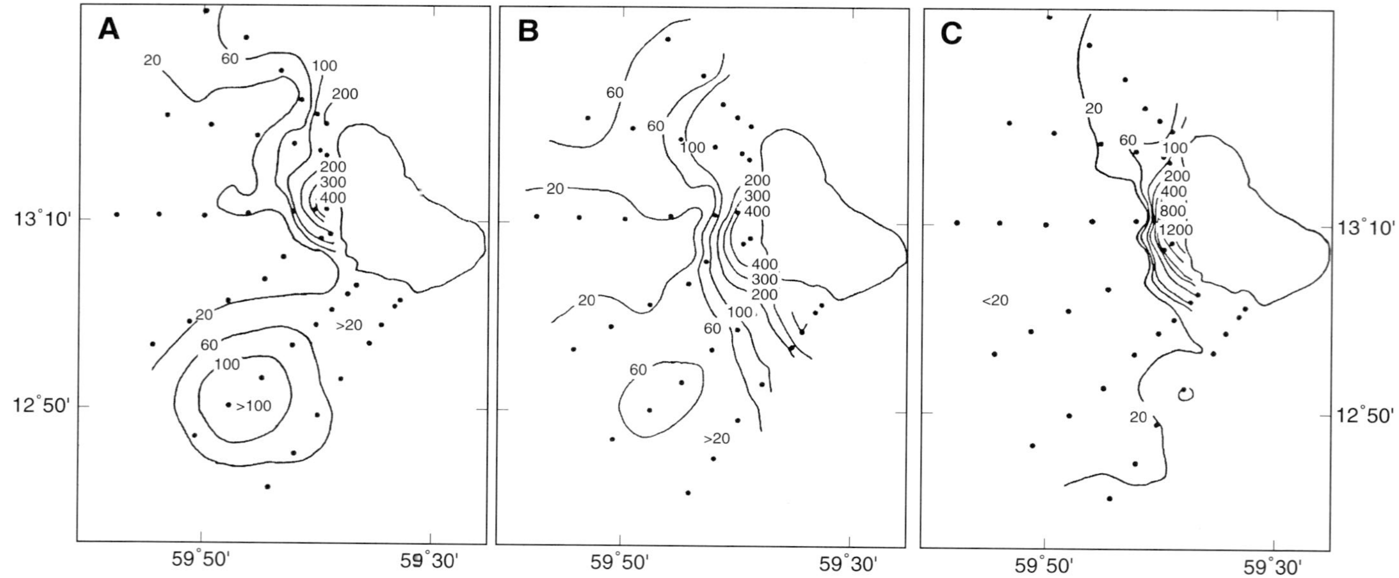

FIGURE 1 Distribution by depth intervals of larval coral reef fish (N per 1000 m^3) around the western side of Barbados during May, 1990. (A) 0–10 m; (B) 10–30 m; (C) 30–60 m. Modified with permission from Cowen, R. K. and Castro, L. R. (1994). Bull. Mar. Sci. **54**, pg. 238.

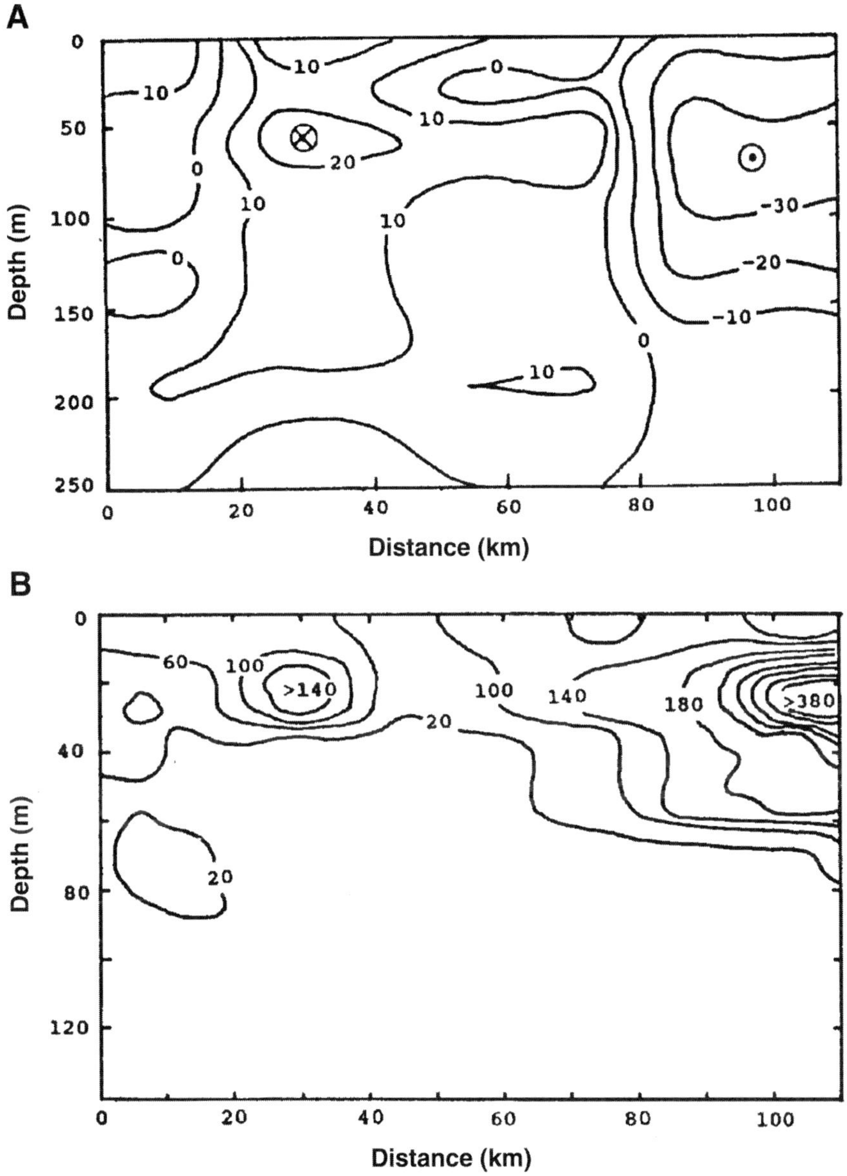

FIGURE 2 Vertical section of current (A) and coral reef fish larval density (B) of a section crossing the Barbados ridge, about 50 km northwest of Barbados. Currents are derived from a shipborne (bottom-tracked) acoustic Doppler current profiler; positive contours represent flow into the page, negative contours represent flow out of the page: Larval concentrations are N per 1000 m^3. The Barbados ridge is located centrally along the section (i.e., near 50 km) and has a minimum depth of 275 m. Modified with permission from Cowen, R. K., and Castro, L. R. (1994). Bull. Mar. Sci. **54**, pgs. 236 and 238.

pomacentrid and gobiid larvae were concentrated in the middepths (20–60 m) where onshore flow was evident (Fig. 1B,C). In general, few coral reef fish larvae were found deeper than 90 m, and a complex of coral reef fish larvae was associated as a vertically discrete (15–50 m) and horizontally stretched patch immediately above a jet of water that was streaming around the Barbados ridge (Fig. 2).

From a recent study of the vertical and horizontal distribution of larvae around Barbados, R. K. Cowen and C. Paris (unpublished data) have discerned a variety of taxon and ontogenetic (stage-specific) vertical patterns (Fig. 3). These preliminary data reveal that few coral reef fish larvae, even as early preflexion stage, are found near the surface (i.e., less than 10–15 m), nor below 80 m. Of those species examined thus far, most occur between 15 and 60 m, the main exception being postflexion larval *Synodus* in slightly deeper water (50–80 m), a finding that is consistent with Cha *et al.* (1994). There were also several groupings of larvae, based on their ontogenetic vertical distribution patterns, for which (1) preflexion stage larvae are shallower than postflexion stage larvae (the most common pattern), with the difference in mean (center of mass) depth ~10–20 m; (2) preflexion stage larvae are at the same depth as postflexion stage larvae, i.e., no ontogenetic change in vertical distribution; and (3) preflexion stage larvae are deeper than postflexion stage larvae

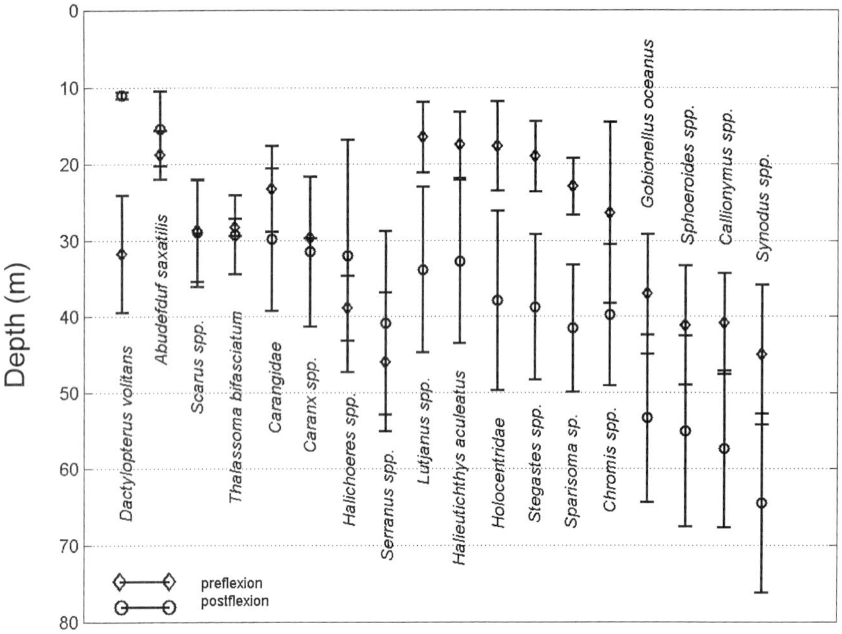

FIGURE 3 Stage-specific vertical distribution of a variety of a coral reef fish larvae from the eastern Caribbean island of Barbados (May, 1996). For each species, the mean (center of mass ± SE) is shown separately for the preflexion and postflexion stages. Larvae were collected with a 1-m² frame Multiple Opening/Closing Net and Environmental Sensing System, with nets fished at 0–20, 20–40, 40–60, 60–80, and 80–100 m. Data source: R. K. Cowen and C. Paris, unpublished data.

(obvious only in *Dactylopterus volitans*; later stage larvae of *D. volitans* move upward to the neuston).

We are still lacking examination of alternative explanations of the vertical distribution of larvae. For example, the 15- to 60-m depth range of many larvae around Barbados corresponds with the depth of mean onshore flow that may maximize local retention (Cowen *et al.*, 2000). Alternatively, this same depth range falls within the depth range of the chlorophyll maximum layer (R. K. Cowen and K. Lwiza, unpublished data), which could serve to concentrate potential food sources for the larvae. Similarly, for preflexion larvae to be concentrated between 20 and 40+ m suggests that preflexion stage larvae may either be swimming vertically and/or they are neutrally buoyant below the surface layer. If the latter is true, then periodic, low-salinity intrusions (Cowen and Castro, 1994; Kelly *et al.*, 2000) might alter the depth of preflexion larvae (i.e., they will move deeper) when low-density water is present in the upper 10–40 m. Obvious alternatives include that of seeking strata of high prey concentrations and/or avoiding predators. Also, and not necessarily mutually exclusively, is the potential need to avoid high ultraviolet (UV) levels in the upper layers (though this would probably require being only a few meters deep at most). There still are insufficient published data to assess these various hypotheses.

In summary, based on relatively few studies, several patterns of larval vertical distributions are emerging or have been confirmed. First, vertical stratification of larvae seems to be greater during daylight hours than at night (though additional studies directly assessing diel vertical migrations are needed). Second, there appear to be changes in vertical distribution with ontogeny for many taxa, the norm being that preflexion stage larvae are shallower than postflexion larvae, though notable exceptions exist. Third, most larvae appear to avoid the near-surface layer, whether over relatively shallow shelves (e.g., GBR) or in deep water. Fourth, most reef fish larvae are restricted to the upper 80 m.

C. Temporal Patterns

Few existing studies have examined the temporal pattern of coral reef fish larval abundance. However, there have been a number of studies examining temporal patterns of reproduction, which clearly translate into temporal patterns in larval abundance (at least for early-stage larvae). There have also been a variety of studies examining the temporal patterns in the abundance of late-stage larvae (i.e., presettlement) or recently settled coral reef fishes. Without pursuing an exhaustive review of reproductive patterns, it can be said that there exist several obvious patterns at seasonal, monthly (lunar/tidal), and daily time frames (e.g., Munro *et al.*, 1973; Lobel, 1989; Robertson, 1990, 1992; Robertson *et al.*, 1990; Colin and Bell, 1991; Hunte von Herbing and Hunte, 1991; Colin, 1992; Meekan *et al.*, 1993; Danilowicz, 1995; Crabtree, 1995; see review by Sadovy, 1996). An example of lunar periodicity in spawning is provided in Fig. 4 for the Caribbean pomacentrid, *Stegastes partitus*. This

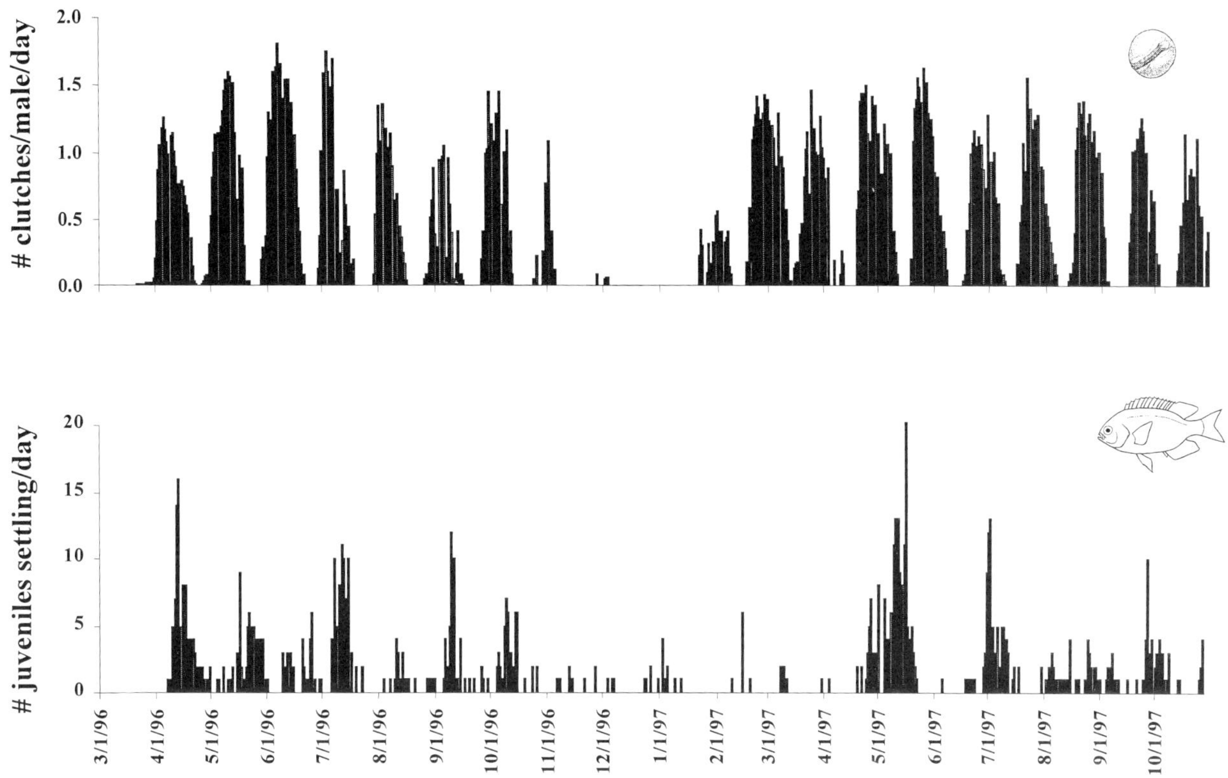

FIGURE 4 The 18-month time series of the daily pattern of spawning and settlement for the Caribbean pomacentrid, *Stegastes partitus*, at Barbados, West Indies. Daily spawning was estimated from 75 male nests and measured as number of egg clutches/male/day. Daily settlement pattern was back-calculated from otoliths of newly settled fish collected on biweekly surveys ($N = 15$ transects per survey). The time series was continual except during brief periods of extreme wave conditions. Data source: S. Dorsey and R. K. Cowen, unpublished data.

species exhibits strong synchrony in spawning with the new moon and a seasonal signal is also evident. Cues to spawning include water temperature, lunar phase, tidal phase, and even wind (see Sadovy, 1996). These cues are prevalent throughout the range of all species, though the relative importance of each cue may vary, as will be discussed later.

Similarly, many species exhibit periodic settlement patterns synchronized with particular tidal or lunar phases (e.g., Dufour, 1991; Milicich *et al.*, 1992, Robertson, 1992; Shenker *et al.*, 1993; Meekan *et al.*, 1993; Sponaugle and Cowen, 1994, 1996a; Thorrold *et al.*, 1994b,c; Robertson *et al.*, 1999). Relationships to specific phases of the moon have been found for many species, with new moon peak settlement dominating in many regions (Dufour, 1991; Robertson, 1992; Meekan *et al.*, 1993; Thorrold *et al.*, 1994b,c; Sponaugle and Cowen, 1996a, 1997), but exceptions do occur. Peak settlement for certain labrids, scarids, pomacentrids, acanthurids, blenniids, and gobiids to Barbados occurs on the third-quarter moon and with minimum-amplitude tides Sponaugle and Cowen (1994, 1996a,b, 1997). This third-quarter lunar and minimum-amplitude tidal periodicity also is shared by a variety of decapod crab larvae at Barbados (Reyns and Sponaugle, 1999). More unusual, on the full moons, two pomacentrids recruit to St. Croix (Booth and Beretta, 1994). A variety of workers also have identified semilunar periodicity (Hunte von Herbing and Hunte, 1991; Robertson, 1992; Thorrold *et al* 1994b,c). In general, the temporal patterns of settlement are more difficult to discern than are those in spawning due to substantial variability inserted during the pelagic period (see Fig. 4). Long time series are often needed to isolate patterns.

The tidal amplitude phase is closely linked to the lunar phase, thereby complicating the separation of lunar from tidal synchrony. Yet, several workers have discussed larval supply and settlement patterns of reef fishes in terms of tidal amplitude phase (Sponaugle and Cowen, 1997; Robertson *et al.*, 1999), and some semilunar periodicity observed in larval supply (e.g., Shenker *et al.*, 1993) may be tied to the tidal amplitude cycle. Morgan and Christy (1994), Morgan (1995), and Reyns and Sponaugle (1999) discuss the linkage between lunar and tidal amplitude cues for marine invertebrates, particularly crab postlarvae (see below).

Finally, although a level of predictability occurs in the *timing* (and even spatial pattern) of the early life history stages of coral reef fishes, the *magnitude* of temporal patterns is not so predictable. This is due in large part to the overall influence of the processes operating during the pelagic phase. Many of the processes operating in the pelagic environment appear chaotic in nature and can decouple reproductive output from subsequent recruitment (Robertson *et al.*, 1988, 1993; Meekan *et al.*, 1993; Danilowicz, 1997) (see Fig. 4). However, when viewed in the appropriate context, these apparently chaotic processes may possess predictable characteristics that enable us to explain both variability in larval distribution and abundance, and adaptation by larvae to utilize specific physical processes.

IV. Physical Processes

A. Wind

Wind is the major driving force in the circulation of the ocean. As the wind blows across the surface, frictional forces drag the surface water along, albeit at approximately 3% of the wind speed. Similarly, this surface layer of water drags the next layer of water along, and so forth, with speed exponentially decreasing with depth until no additional wind-generated flow occurs. This wind forcing, coupled with Coriolis deflection (due to the rotation of the Earth), results in the actual direction of surface water movement to be at a 10–45° angle to the wind (clockwise of the wind in the Northern Hemisphere, counterclockwise in the Southern Hemisphere). Similarly, lower water layers move at an angle to the forcing layer above. This results in an "Ekman spiral," with the maximum depth of wind penetration effect called the Ekman depth (the mixed-layer depth). The depth of the Ekman layer is dependent on both the strength of the wind and the thermohaline structure of the water column. The stronger the stratification of the water column (i.e., the greater the change in density with depth due either to temperature and/or to salinity), the stronger the wind must be to affect flow to any given depth. The wind speed required to mix the water column to a particular depth is also dependent on the duration of the wind event; a long-duration wind event of a given speed will penetrate deeper, compared to a short-duration event.

In addition to the Ekman spiral, persistent winds blowing across the surface of the ocean can create small-scale circulation features called Langmuir circulation cells. These cells are parallel, typically penetrate 3–10 m depth, and are up to 50 m across and hundreds to thousands of meters long. Each cell has a spirallike circulation that alternates in direction from cell to cell, resulting in alternating zones of convergence (downwelling) and divergence (upwelling). The role of such features in generating plankton and larval fish distributions is discussed by Kingsford *et al.* (1991), Werner *et al.* (1997), and MacKenzie (2000).

Another wind effect of relevance to coastal areas is that the surface layer of water (typically the mixed layer down to the thermocline or halocline) can respond rapidly to changes in wind direction. Wind shifts are common along coastlines, where they may be affected by diurnal warming of the landmass (daily cycling), passage of meteorological fronts (weekly to monthly cycling), and seasonality. Where winds are offshore or poleward on an eastern coast (opposite on the western coast), the net effect is offshore flow of the surface layer, which may result in upwelling of deeper, cooler water near the coast. Conversely, where the wind is blowing onshore or equatorward on an eastern coast, the surface flow will be onshore. Where inlets to coastal embayments occur, such onshore flow may accentuate flow into the embayment. Where such inlets do not occur, the onshore flowing water may pile up along the coast, setting up a pressure gradient wherein the water tends to flow downgradient. This flow, in concert with Coriolis deflection, then turns right or left (hemisphere dependent) to flow along the coast, generating a coastal current. For reefs, winds may pile water up until it flows over the reef crest into the back reef (or moat), returning back out via reef channels (Yamano *et al.*, 1998). Such winds also force the buildup of waves, which may further induce inflow by overtopping of the reef crest (Roberts *et al.*, 1992; Symonds *et al.*, 1995). Response to a buildup or decrease in the winds can be immediate, whereas if sufficient wind-generated waves exist, the reduction in inflowing water over the reef crest may persist for some time after winds decrease or even reverse (Yamano *et al.*, 1998). Orientation of a particular reef or coastline to prevailing winds will therefore have important impact on flow conditions.

Larger scale circulations are also strongly affected by the winds, but these are predominantly on a basin-wide scale largely associated with the major global wind bands (e.g., trade winds, westerlies, easterlies). Strong seasonal, interannual, decadal, or longer time scale variations in the wind can affect the intensity of current flows or even alter the direction of major ocean currents. The best example of such major changes is the effect of the El Niño/Southern Oscillation (ENSO) wind pattern on the equatorial currents of the Pacific Ocean (Philander, 1992). Although seasonal shifts in the wind patterns are relatively easy to capture in a typical study, longer time scale variations are not. Yet the importance of such variation cannot be overemphasized. Even if a single investigator cannot conduct long time series work [time series analysis typically requires a time series to be at least four times longer than the longest cycle being investigated (Chatfield, 1979)], when comparing among studies of 10+ years apart, potential differences in the meteorological and resultant oceanographic conditions must be considered.

Wind can have a positive effect when associated with delivery of late-stage larvae to shore. Shenker *et al.* (1993) found a strong correlation between the occurrence of two major recruitment peaks of the Nassau grouper (*Epinephelus striatus*) and two storm-related wind events that increased onshore flow during winter months. A similar positive relationship with wind was identified for other late-stage larval coral reef fishes passing through reef channels to benthic habitats on the Great Bahamas Bank (Thorrold *et al.*, 1994a). In this latter study (during summer months), a large, multispecific pulse of larvae was associated with a wind-relaxation event that resulted in a reversal of the coastal current and onshore flow. Studies on the GBR also have indicated that favorable winds generate strong recruitment pulses (Milicich, 1994; Kingsford and Finn, 1997).

Despite these positive results, the effects of wind on delivery of late-stage larvae to settlement habitat are not always temporally consistent. Whereas Shenker *et al.* (1993) found a positive relationship in supply with onshore (storm) winds, the reverse was true several years later (Grover *et al.*, 1998). Elsewhere, either no relationship between wind and recruitment has been found, or negative effects have been identified (e.g., Sponaugle and Cowen, 1997; Robertson *et al.*, 1999). Thus caution must be exercised when generalizing from one study to the next with respect to specific relationships between the wind field and delivery of young (late-stage) larvae to settlement habitat. These results suggest it is likely that onshore favorable winds may enhance the delivery of late-stage larvae to coastal habitats; however, such winds may serve to complement other behaviorally mediated onshore movements (e.g., swimming, tidal flows), and are not necessarily a prerequisite for recruitment.

Wind can also influence initial dispersal of larvae. Appeldoorn *et al.* (1994) found that spawning of the bluehead wrasse (*Thalassoma bifasciatum*) co-occurred with periods of maximum wind-generated flow (on both a diel and seasonal basis), which rapidly carried the eggs off the reef patch. However, contrary to standard belief, the eggs were carried along and slightly toward shore, an area hypothesized (Johannes, 1978; Shapiro *et al.*, 1988) to result in high mortality of eggs by planktivores concentrated in coastal habitats. Similar evidence of primarily alongshore dispersal (versus offshore) of spawned eggs was obtained by Colin (1992) for the Nassau grouper (*E. striatus*) in the Bahamas. In the extreme, strong wind events associated with large storms may preclude successful spawning altogether due to extreme turbulence from large waves in shallow water (Petersen *et al.*, 1992; Robertson *et al.*, 1999).

Beyond transport, wind can affect food availability via upwelling, vertical mixing and even microturbulence. Several of these features may require time to manifest their impact on larval feeding conditions. For example, upwelling may inject surface waters with a pulse of nutrients, which is translated over several days into primary and secondary production. Other features, such as the effect of microturbulence enhancing prey encounter rates by larval fish (Rothschild and Osborn, 1988; MacKenzie *et al.*, 1994; Dower *et al.*, 1997; but see Dower *et al.*, 1998), may be almost immediately responsive to changes in wind conditions. Therefore, the relative timing of a wind event is likely important and can result in effects at a variety of time lags from settlement. Such interactions add to the complexity of the patterns of settlement, though utilization of nonlinear models with appropriate time lags may reveal linkages between recruitment and events occurring early in the pelagic phase (Dixon *et al.*, 1999). Spatial variability in such processes can be similarly visualized to lead to spatial patchiness in larval condition (Suthers, 1996; McCormick, 1998a) and, potentially, survival (Searcy and Sponaugle, 2000). Searcy and Sponaugle (2000) detected among- and within-cohort variability in the condition (measured as otolith growth) of newly settled labrids (*Thalassoma bifasciatum* and *Halichoeres bivittatus*) that appeared to be related to their survival potential through the early postsettlement days on the reef.

In summary, where winds are positively correlated with settlement, their effect may be important only in the final delivery of the young, not necessarily with respect to early retention or transport processes. For the later stages, wind might enhance settlement but not be a prerequisite for successful settlement (e.g., cross-shelf transport may occur when larvae are rapidly advected across the shelf with favorable winds, but such transport may also occur in the absence of these winds). The same may be true for wind and for wave-generated flow across reefs (Roberts *et al.*, 1992; Symonds *et al.*, 1995; Yamano *et al.*, 1998) that deliver larvae into atoll lagoons (Dufour, 1991; Dufour and Galzin, 1993). Over longer time scales, seasonal winds may directly influence the settlement success, either by reducing spawning success during periods of high winds and turbulent conditions on the reef or by enhancing the likelihood of coastal retention during periods of minimal wind (Johannes, 1978; Lobel, 1989; Colin, 1992).

B. Tides

From shore, the changing tide is observed to bring water either toward or away from the observer. Yet the flow generated by tidal movement turns out to be much more complex (Barnwell, 1976). In its simplest form, a tidal current along a constant shelf bottom may result in elliptical movement, moving on scales of kilometers with each tidal cycle. If the water column is strongly stratified (such as near a river/estuarine outlet), flow along the bottom may move quite independently of the lighter, fresher surface water. However, coastal and island reef environments are not topographically simple, and therefore they impose a complex frictional (or topographic) effect on flow (Hogg *et al.*, 1978; Hogg, 1980). This effect is scale dependent, and results from water being forced from one area to the next, while being squeezed past obstacles (e.g., reefs, headlands). Along the bottom, the flow is reduced due to friction, which ultimately steers the water (typically along contours); closer to the surface, the water may accelerate because the same amount of water is passing through a reduced cross-sectional area (e.g., water flows out of a hose more rapidly when you place your finger over part of the opening). Once on the other side of the obstacle, the flow quickly slows (depending on the topography), resulting in adjacent areas of decelerated and accelerated flows. Water inshore of regions of flow deceleration will be pulled along by frictional forces. These frictional forces will also rapidly decrease the flow closer to shore, potentially setting up a zone of convergence between the fast- and slow-flowing water and even an eddy in the lee of the obstacle. What becomes particularly complex about tidal flows, however, is that the flow then reverses. Yet rather than just simply reversing this effect, the picture is further complicated by both the momentum of the water and (at nonequator latitudes) Coriolis forcing. Numerical models demonstrate the overall effect of topography on tidal flow with these two added factors (e.g., Pingree and Maddock, 1980).

Tidal circulation around islands and headlands has been shown to generate occasionally strong, eddying motions that may serve to slow the offshore advection of larvae (e.g., Hamner and Hauri, 1981; Black and Gay, 1987; Wolanski *et al.*, 1984a, 1989; Black, 1994; Furukawa and Wolanski, 1998). Scaling effects, however, tend to limit tidally generated eddies [or phase eddies; sensu Black and Gay (1987)] to smaller islands and reefs in shallow water (versus larger, oceanic islands in deep water). Over time, these phase eddies grow and decay with changing tidal phases and, coupled with Coriolis effects, result in a nonuniform residual (net) flow around the island [anticlockwise in the GBR; reviewed by Black (1994)]. This asymmetry in flow results in potentially higher retention in certain regions around a reef, as opposed to others. Efforts to verify the retention of larvae in these eddies have been based on both field and model estimates, but largely for invertebrate larvae of relatively short duration, [e.g., coral and crown of thorn starfish (reviewed in Sammarco, 1994; Gay and Andrews, 1994; Black, 1994)].

Additional work suggests that retention of reef fish larvae around a reef (or small island) does indeed occur. Jones *et al.* (1999) marked the otoliths of prehatching pomacentrids at Lizard Island, GBR, then, after a lag phase for larval duration, collected presettlement larvae with light traps around the same reef. They successfully "recaptured" tagged larvae and estimated that between 15 and 60% of the hatched fish were retained, a rather remarkable finding! Missing from this study, however, is a mechanistic explanation of what retention mechanisms were responsible. Physical retention mechanisms associated with tidal and current flow past and around a reef may well have played a strong role by slowing or countering the outward advective and diffusive flow of the shelf currents. Also, late-stage larval behavior (Stobutzki and Bellwood, 1994; Leis *et al.*, 1996; Leis and Carson-Ewart, 1999) and sensory capabilities interacting with tidally generated frontal features (Kingsford *et al.*, 1991) may have served to augment any physical retention mechanisms.

Overflow of the reef crest also can occur with flooding tides (e.g., Parnell, 1988). Such overflow may serve as an important delivery mechanism of settlement-stage fish, especially in reef systems with few other avenues of entry to backreef/lagoon habitats [Dufour and Galzin, 1993; Dufour *et al.*, 1996; but see Sancho *et al.* (1997) for an example of larval acanthurids swimming upcurrent through a reef channel]. The relative importance of this sort of tidal flow, as compared to wind- and wave-generated overtopping flow (*sensu* Roberts *et al.*, 1992; Kamano *et al.*, 1998), will clearly depend on local tidal ranges and reef orientation.

Tidal flows also may have strong impacts on reefal systems located near shelf edges via tidally mediated upwelling, which can occur in two different ways. First, tidal flows near the shelf edge can generate large internal waves that propagate toward shore. When formed where the thermocline is relatively shallow, larger amplitude internal waves (e.g., 5- to 10-m excursions) can move cooler, nutrient-rich subthermocline waters into shallow depths (Wolanski and Delesalle, 1995). When these internal waves break, nutrient-rich water may be mixed upward and carried onto the reef environment, as has been shown in the Florida Keys by Leichter *et al.* (1996, 1998). Second, tidal movement through narrow passages between reefs near the shelf edge of the GBR has also been implicated as a mechanism creating the upwelling of nutrient-rich deeper waters (and associated organisms). Such sources of increased nutrients, and associated zooplankton (see below), may serve as an important food source for planktivorous coral reef fishes, perhaps even influencing their reproductive success (Robertson *et al.*, 1990; Schultz and Warner, 1991; Tyler and Stanton, 1995).

In temperate systems, internal waves and tidal bores can transport larvae across shelf environments (Shanks, 1983, 1986; Kingsford and Choat, 1986; Pineda, 1991, 1994, 1999), but their importance as a cross-shelf transport mechanism has not been examined for coral reef fish larvae [but see Leichter *et al.* (1998) with respect to tropical zooplankters]. While early interpretations of internal wave-mediated surface transport invoked onshore movement of neustonic larvae in surface convergences, or “slicks” (Shanks, 1983, 1986; Kingsford and Choat, 1986), Pineda (1991, 1994, 1999) and Leichter *et al.* (1996, 1998) have proposed a more dynamic explanation with respect to how internal tidal bores function.

Of special relevance to some of the temporal settlement patterns observed for coral reef fishes, internal tides (and their associated tidal bores) may be lunar cyclic, but not necessarily in direct association with the spring-neap tidal cycle. Specifically, Pineda (1991, 1995) found that when the water column is stratified, peak internal tide activity occurred approximately 7–12 and 19–24 days after the new moon (i.e., on the quarter moons). Such timing corresponds to the third-quarter peak of settlement observed for a variety of coral reef fish species (Sponaugle and Cowen, 1994, 1996a, 1997) and decapod crab larvae (Reyns and Sponaugle, 1999) around Barbados. Reyns and Sponaugle (1999) suggest that this temporal relationship at Barbados, but not, for example at Panama (e.g., Victor, 1986a; Robertson *et al.*, 1999), may be due to the relative importance of internal tides in the two locations.

In addition to direct transport and nutrient-related effects, tidal flows may also serve as important environmental cues for reproduction and timing of settlement. As discussed above, a variety of coral reef fishes exhibit lunar cyclic recruitment patterns, but there is evidence of geographic variation in terms of in which specific lunar phase the recruitment is occurring. Timing to lunar phase is therefore not likely a simple function of light/dark as has often been suggested (see Reyns and Sponaugle, 1999). Rather, flows associated with the tidal amplitude cycle may also be important, leading to complex interactions between recruitment patterns and lunar/tidal cycles (Morgan, 1995; Sponaugle and Cowen, 1997; Robertson *et al.*, 1999; Reyns and Sponaugle, 1999) (see Fig. 5). These patterns lead to an important final point: given the potential for geographic variation in the temporal pattern of reproduction, and especially settlement, sampling should encompass the *entire* lunar/tidal cycle rather than focus on specific phases, at least until the temporal patterns are definitively determined.

C. Freshwater Input

In the immediate vicinity of a river, both the low salinity and typically high turbidity of the outflowing water will have severe negative impacts on the survival of coral reef fish (and on the coral reefs). Yet, freshwater input is pervasive throughout the wet tropics, at least in a seasonal context, both from direct rainfall and as river runoff. Direct rainfall may lower surface salinity, but generally not by much because the rainfall is spread over a wide surface and mixed with the seawater (e.g., 100 mm of rain can mix with several meters of seawater without changing the salinity more than a few practical salinity units) (Andrews and Pickard, 1990). However, over a season, the input of freshwater may create a strong seasonal signal in lowered surface salinity.

The extent of impact of freshwater river input, outside of the extreme case mentioned above, is related to the outflow, local currents, and topography. Some rivers of small, local watersheds will input sufficiently small amounts of freshwater such that mixing with seawater will eliminate the signal even very close to the source. Other rivers, such as the Orinoco River of Venezuela, can be observed to spread their effect over very large areas ($>10^6$ km^2) (see Müller-Karger *et al.*, 1989). At varying distances downstream from such

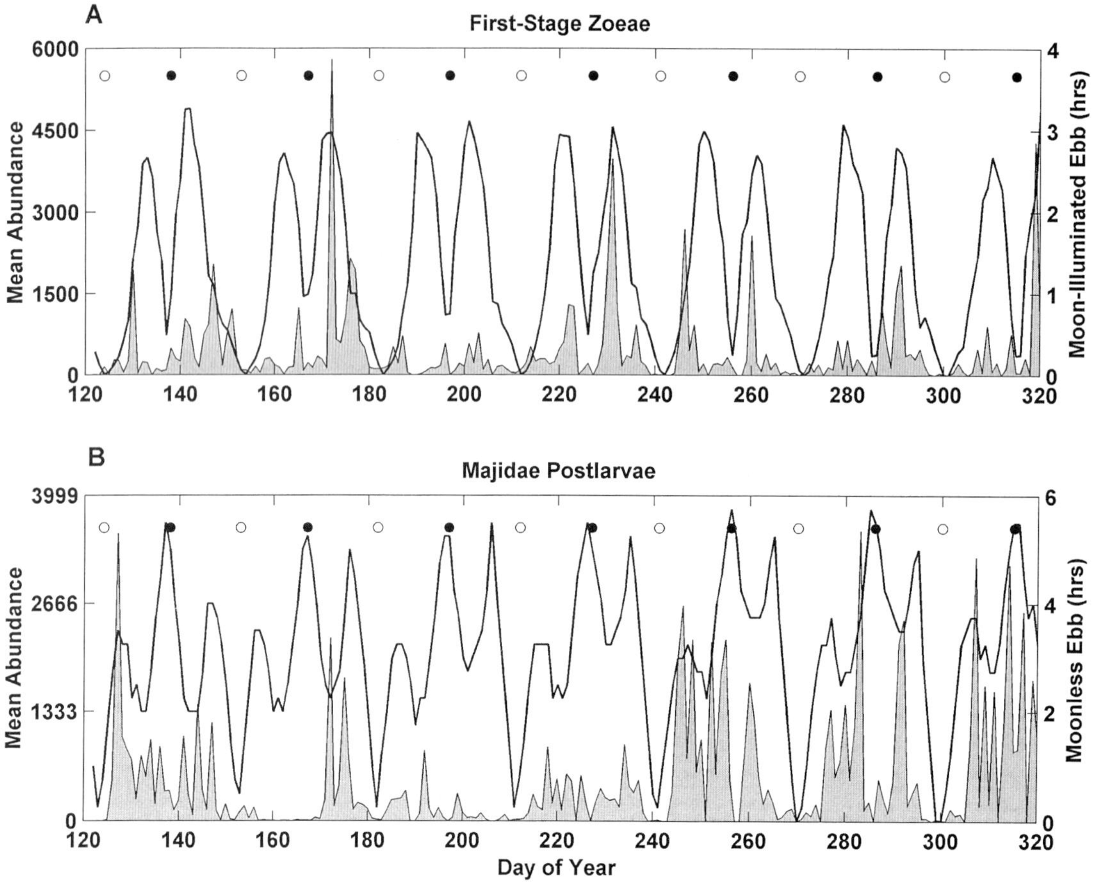

FIGURE 5 Time series (200 days) of the relationship between larval crab abundance (shaded) and proximate environmental cues (solid line) along the west coast of Barbados during 1996. Shown relationships are significant cross-correlations at $p < 0.05$. (A) Crab zoeae abundance vs. moon-illuminated ebb tides (i.e., the number of hours ebbing tides occur in relative darkness due to nightly change in moon phase). (B) Majid crab postlarve vs. moonless ebb (i.e., total darkness). Modified from Reyns and Sponaugle (1999).

freshwater sources, the river can be a positive nutrient source, possibly a long-distance transport process, or an episodic detriment to recruitment or coral reef fish survival. A recent example of the latter was the transport of a freshwater-borne pathogen to the southern Lesser Antilles Islands, apparently via an unusual turn of the Orinoco River toward the east (H. Oxenford, University of the West Indies, personal communication). This pathogen (*Streptococcus iniae*) has been implicated in extensive fish kills of coral reef fish around the islands of Tobago, Grenadines, Grenada, Dominica, and Barbados.

An important role of freshwater input into the coastal ocean is its effect on the baroclinic (thermohaline) circulation. Because freshwater is less dense than seawater, it floats on the surface, adding to the stratification of coastal waters. This can result in a plume of relatively fresh water flowing away from the source, in some cases hundreds of kilometers (or more) distant (Müller-Karger *et al.*, 1989; Gilbert *et al.*, 1996). Depending on the spread of such plumes, their impact on larval fish distributions (and corals) can range from minimal to great (Murray *et al.*, 1982; Roberts and Murray, 1983; Andrews and Gentien, 1982; Andrews, 1983; Wolanski and Van Senden, 1983; Wolanski *et al.*, 1984b; Sabatés and Maso, 1992).

Associated with such plumes are fronts, where the salinity changes rapidly and flow tends to converge. Fronts may serve to concentrate organisms either through convergent mechanisms or by the attraction of organisms to areas of high prey concentrations (e.g., Alldredge and Hamner, 1980; Denman and Powell, 1984; Mackas *et al.*, 1985; LeFerve, 1986; Wolanski and Hamner, 1988; Govoni *et al.*, 1989; Kingsford, 1990; Kingsford *et al.*, 1991; Govoni and Grimes, 1992; Olson *et al.*, 1994; see also review of Grimes

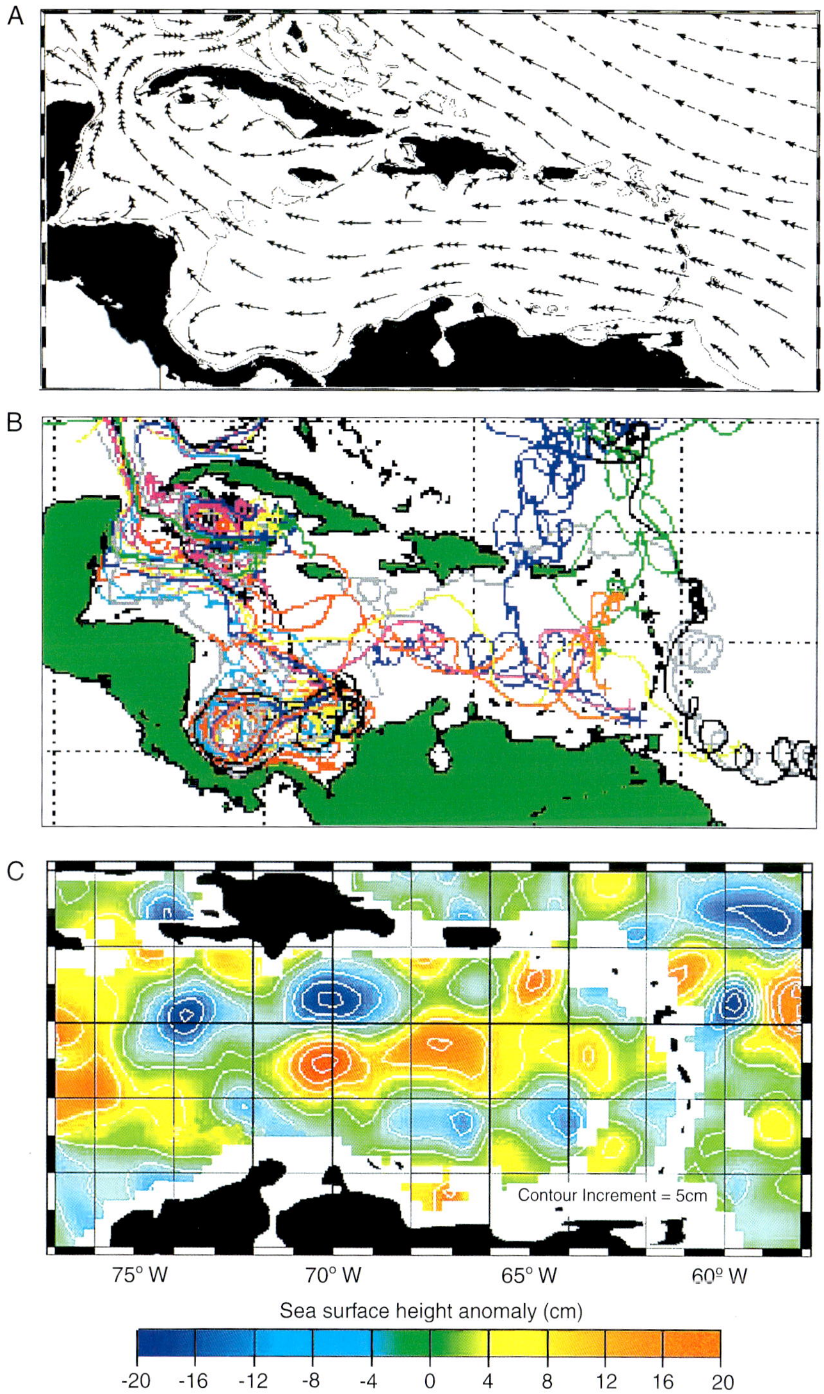

FIGURE 6 Three perspectives of the currents of the Caribbean Sea. A) Mean current estimated from reports of ships' drift (after Wust 1964); B) Lagrangian flow exemplified from surface drifters released as part of the 1998 NOPP-YOTO Drifter Program. Each drifter track is plotted as a separate color with the starting point marked with a '+'. Drifters reported GPS position via satellite every six hours, duration varied from several days to over 6 months (data source: **http://www.iaslinks.org**); C) Eulerian current field based on sea surface altimeter data (TOPEX-POSEIDEN satellite) during July 1997. Flow is clockwise around highs (red) and counterclockwise around lows (blue). (Data source: University of Colorado, Colorado Center for Astrodynamics Research).

and Kingsford, 1996). These processes have important implications for understanding the roles riverine and estuarine plumes, and their associated fronts, may have in determining the distribution and abundance of larval fishes. Despite the potential importance of riverine plumes, however, few studies exist that document the effect of riverine plumes on coral reef fish larvae [but see Govoni (1993), Grimes and Kingsford (1996), and Allman and Grimes (1998) for examples of plume effects on other fish species]. One exception is the study by McKinnon and Thorrold (1993) and Thorrold and McKinnon (1995) on a plume (resulting from heavy rainfall from Cyclone Joy) associated with river outflow into the GBR lagoon north of Townsville, Australia. They found very high zooplankton biomass associated with the freshwater plume that spread up to 20 km offshore (roughly into the middle section of the GBR lagoon). Coastal (offshore of the plume) species of larval fishes (e.g., some holocentrids and pomacentrids) were apparently aggregated at the front as the plume propagated offshore, though their numbers did not increase within the plume, per se. Whether these offshore taxa actually benefited from the increased food environment associated with the increase of zooplankton biomass and production is not clear, but is suspected. Though not explicitly examined, the vertical distribution of larvae is important, because these plumes do not generally penetrate to particularly great depths (e.g., less than 5 m). Except for the neustonic-oriented mullids (which were concentrated in the plume front), other typically deeper larvae may benefit from the vertical mixing of zooplankton-rich plume water as it passes over coastal waters.

D. Ocean Currents

There have been many efforts to use long-term mean currents as indicators of larval transport potential and/or pathways of population connectivity. However, within ecologically meaningful time scales, there is really no such thing as a "mean" current field. Neither are there purely random motions of water. Rather, oceanic flows tend to be coherent and exhibit a variety of mesoscale features (Robinson, 1983). Organized flow results in a multitude of mesoscale features, including eddies and fronts that can act independently or may interact with topographic features or other currents potentially to enhance the retention or dispersion of larvae (e.g., Owen, 1980; Hamner and Hauri, 1981; Olson and Backus, 1985; Wolanski and Hamner, 1988; Okubo, 1994; Olson *et al.*, 1994). Kingsford (1990) has summarized manifestations of these interactions (as well as smaller scale features).

Our perception of the true nature of oceanic flow has changed over the past few decades (Fig. 6. See color insert). For example, the prevailing view of circulation within and through the Caribbean has been characterized by Wust (1964) based on mean estimates of ships' drift data (Fig. 6A). Such views continue to be expressed in terms of estimating larval drift pathways (e.g., Roberts, 1997b). However, over shorter time frames, flow is much more dynamic than that indicated by climatological (i.e., mean) data. From a Lagrangian perspective, based on drifter tracks released and followed over an 18-month period, it is clear that there is considerable flow variability within the Caribbean, often occurring either orthogonal or opposite to the mean flow (Fig. 6B). Particular pathways of individual drifters released from a similar location can end up thousands of kilometers apart after just a few months, whereas others may remain remarkably coherent where strong recirculations occur (e.g., in the Panama/Colombia gyre). Even at a single point of time, high spatial variability in flow is apparent (Fig. 6C). The circulation within the core region of the Caribbean is characterized by the presence of a variety of mesoscale features, including eddies and meanders ranging in size from 50 to 500 km. Further, closer to shore, it is apparent that the oceanic circulation features are interacting with coastal flow to create a variety of regions of nearshore recirculation.

How the coastal flow field is affected by the larger scale oceanic circulation is clearly illustrated in the work of Lee and others within the vicinity of south Florida and the Florida Keys. Their work over the past two decades demonstrates how interactions of physical flows over several scales result in a series of potential larval retention mechanisms (Fig. 7) (summarized in Lee and Williams, 1999). Briefly, warm, highly saline flow exits the Caribbean Sea through the Yucatan Straits into the Gulf of Mexico and transits northward. The flow then anticyclonically reverses direction to form the Loop Current, flowing south and exiting the Gulf through the Straits of Florida (SSF) to become the primary source of the Gulf Stream (Maul, 1977). Periodically (every 7–19 months), the entire Loop Current pinches off and is translocated westward into the Gulf of Mexico (Maul, 1977; Sturges, 1992; Maul and Vukovich, 1993). When this happens, water exiting the Yucatan Straits flows directly into the SSF. In either situation, the Florida Current (FC) follows the steep topography associated with the shelf break/slope occurring between 7 and 10 km offshore of the Florida Keys at a speed of 150–250 cm sec^{-1}.

When the Loop Current is present, its outer edge spawns smaller scale meanders and cyclonic eddies that

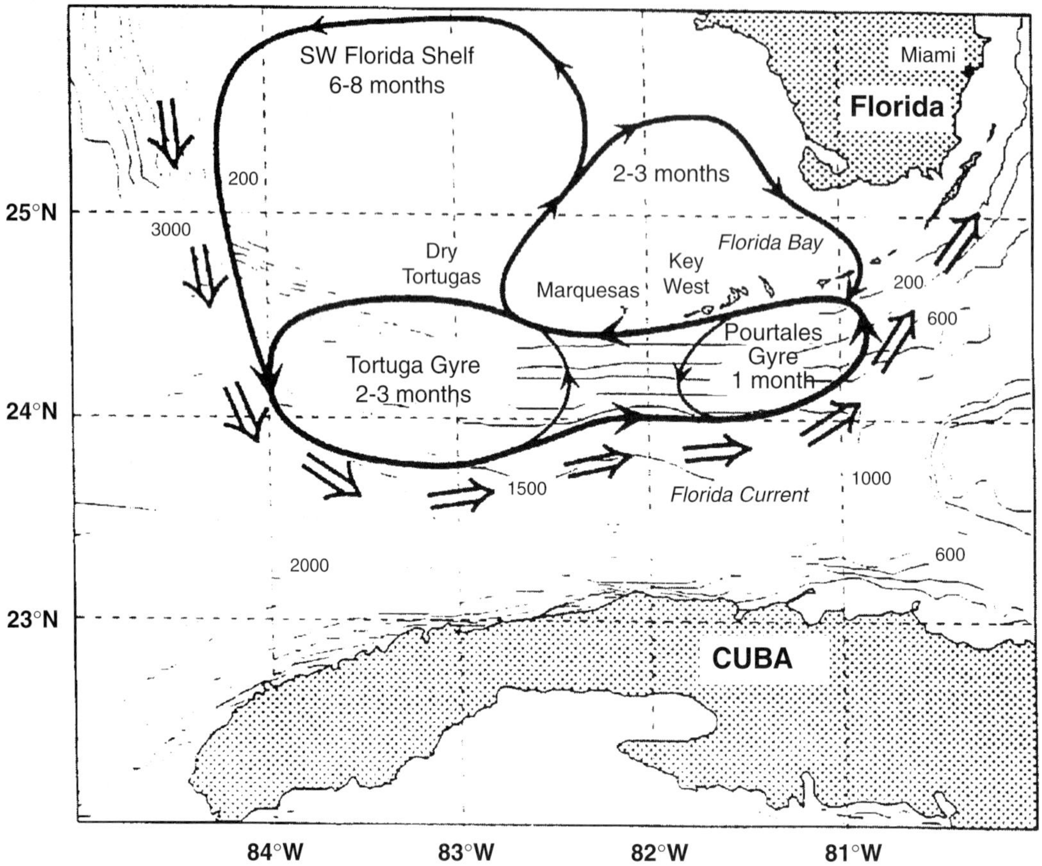

FIGURE 7 Hypothesized larval transport pathways of varying duration based on observations of circulation and hydrographic structure in waters south and west of the Florida Keys. Reproduced with permission from Lee, T. N., and Williams, E. (1999). Bull. Mar. Sci. 64, pg. 51.

are carried anticyclonically around the Loop Current (Fratantoni *et al.*, 1998). Once one of these smaller eddies makes the full transit around the Loop, it may get trapped between the northern edge of the FC and southwest of the Dry Tortugas (westernmost key of the Florida Keys). These quasistationary cyclonic "Tortugas gyres" are of the scale of 50–100 km in size, and may reside for between 50 and 140 days, maintaining their position until being pushed out by the arrival of the next gyre transiting around the Loop Current. Once displaced into the Straits of Florida, the Tortugas eddies are deformed (and shrink) by the topography and the curvature of the Middle Keys as they propagate downstream between the shelf/slope topography and the FC at a rate ranging from 5 to 16 km/day. Circulation velocities of the gyres range from 20 to 50 cm sec^{-1} (Lee *et al.*, 1994) prior to being absorbed beyond the Pourtales terrace, where the shelf narrows and the FC moves close to shore.

The passage of a Tortugas gyre into the Straits of Florida contributes to cross-shelf flow (between the Florida Current and the shallow shelf) and is a potential source of enhanced nutrients in the Florida Keys system (Lee *et al.*, 1994). Further, the inshore flow associated with these eddies provides a countercurrent circulation to the strong FC. Their infrequent (up to two to three per year) (Fratantoni *et al.*, 1998) but relatively slow (~months) passage (Lee and Williams, 1999) interjects considerable variability in coastal flow, which is counter to the flow conditions of the FC and may serve as a retention mechanism for coastal-derived larvae that would otherwise be carried away by the strong, unidirectional FC (Lee *et al.*, 1992, 1994; Lee and Williams, 1999).

At a smaller scale, submesoscale eddies (SMEs) are periodically trapped and advected along the inshore edge of the FC due to strong horizontal shears caused by the strong flowing FC encountering the steep topography of the Florida Shelf (Lee and Mayer, 1977; Lee *et al.*, 1991). These small, generally cyclonic eddies (~5–10 km in diameter) are rapidly translated along the shelf break, bringing cool, upwelled water onto the shelf (Lee *et al.*, 1991) and potentially aggregating larval fish and their prey (Limouzy-Paris *et al.*, 1997). The circulation associated with these SMEs and the larger Tortugas gyre has been implicated with the onshore transport (Limouzy-Paris *et al.*, 1997) and retention of locally spawned organisms (Yeung, 1996; Lee *et al.*, 1992, 1994; Criales and Lee, 1995; Porch, 1998), though direct linkages between these features and recruitment patterns are lacking.

Thus, the various mesoscale circulations associated with larger scale flows, and direct interactions of larger scale flows with coastal topography, can interject a strong influence on the coastal circulation. Such interactions create not only strong variability in the coastal flow, but also multiple mechanisms for retention (and nutrition) of coastal larvae that would not be apparent through consideration of mean (large-scale) currents alone. The variously scaled mechanisms exemplified in the Florida Keys system provide a variety of mechanisms that may promote localized retention of coral reef fish larvae (as well as larvae of other coral reef organisms such as lobsters, shrimp, and conch) (Munro *et al.*, 1968; Yeung, 1996; Criales and Lee, 1995; Stoner *et al.*, 1997; Limouzy-Paris *et al.*, 1997). Similar interactions of oceanic and coastal currents are indicated in other regions of the Caribbean [e.g., Barbados (Cowen and Castro, 1994) and Cuba (Lindeman *et al.*, 2001)], as well as other coastal regions [e.g., United States east coast (Hare and Cowen, 1991, 1996) and the western Australia coast (Hutchins and Pearce, 1994)], though none perhaps so well described in terms of the physics. Despite this detailed knowledge of the currents and supportive larval distribution data (e.g., Limouzy-Paris *et al.*, 1997; Criales and Lee, 1995), hypothesis-driven studies that link larval distribution and behavior to such retention mechanisms are still lacking.

E. Insular versus Continental Environments

Along the leeward side of many oceanic islands, a strong, nutrient-enriched patch of water may exist, termed the "island-mass effect" (Doty and Oguri, 1956; Gilmartin and Relevante, 1974). This effect has been attributed to local rainwater as either runoff or groundwater; local benthic production enhancing overlying, nearshore waters; wind-induced upwelling bringing up deeper, nutrient-rich water; and/or vertical oscillations from internal waves (Hamner and Haury, 1981; Sander, 1981). Theoretical studies, laboratory experiments, and field studies of flow around islands and seamounts have also indicated the effect of flow disturbances producing eddying motions capable of upwelling nutrients above the thermocline (e.g., Hogg, 1972; Emery, 1972; Hogg *et al.*, 1978; Pingree and Maddock, 1980; Gorden and Hughes, 1981; Simpson *et al.*, 1982; Genin and Boehlert, 1985; Roden, 1987). Heywood *et al.* (1990, 1991) found field evidence of enhanced nutrient levels in the lee of a small island, Aldrabra, in the Indian Ocean. During periods of moderately high flow past the island, isopycnals were elevated in the lee of the island, suggesting the existence of a trapped eddy. However, during periods of slower ambient flows, no such conditions existed. Other studies (in addition to several of those cited above) have identified increased zooplankton biomass around islands (LeBorgne *et al.*, 1985; Hernandez-Leon, 1988, 1991; Rissik *et al.*,1997). Such effects on the primary and secondary production around islands may be important in providing adequate food sources for the larvae of island populations of coral reef fishes (Suthers, 1996), particularly in the highly oligotrophic oceanic waters surrounding many of these islands. Further, the doming effect of the nutricline, and related production, bring these potential prey sources into depths [e.g., 50–70 m (Rissik *et al.*, 1997)] that correspond to depths at which larvae are typically found (see Section III,B).

In addition to increased nutrients, these island effects can act to magnify the vertical stratification of the water column. For example, where winds generate upwelling, surface flows are typically directed in the opposite (offshore vs. onshore) direction from the deeper layers. Such stratified flows may be only a few tens of meters apart. Clearly, a larva could benefit by moving into deeper water to enhance local retention (e.g., Cowen and Castro, 1994). Even without enhancement of nutrients via upwelling, winds passing around and over an island can create a wind shadow close to the leeward shore (the offshore extent of this shadow is in part dependent on the size and altitude of the island). Such shadows in the lee of an island may serve to enhance local retention of larvae because of limited wind-generated flow (Barton *et al.*, 1998).

Wind-generated upwelling also occurs in noninsular areas. For example, where coastlines are irregular and/or headlands and hills occur, the windfield can be significantly interrupted, resulting in considerable variability in alongshore wind conditions. Variation in the coastal bathymetry will also have strong impact on how the wind field influences the flow of water. Additionally, orientation of coastlines relative to prevailing wind directions can create opportunities for seasonal upwelling (e.g., along the Caribbean coastline of Colombia and Venezuela). Where such upwelling occurs, surface flow is offshore within the field of the wind, yet "upwelling shadows" may form (sensu Graham *et al.*, 1992) in the coastal recess equatorward of the headland. The water within the shadow remains nearshore and may form a sharp front, where it grades into the upwelled water. Depending on the persistence of the upwelling, the surface-oriented species associated with the upwelled water may be prevented from reaching the coast except during periods of relaxation (e.g., Wing *et al.*, 1995a,b). Conversely, larvae within upwelling shadows may experience reduced offshore advection and thus increased retention.

Island wake eddies, eddying motions in the lee of an island, have been implicated as important retention mechanisms for larvae (e.g., Boden, 1952; Sette, 1955; Sale, 1970; Emery, 1972; Hamner and Hauri, 1977, 1981; Pingree and Maddock, 1980; Lobel and Robinson, 1986; Black and Gay, 1987; Wolanski and Hamner, 1988). Generally forced by topographic influence on flow past the island, eddies have been identified in both deep-water (steep topography) and shallow-water (shelf) environments (e.g., Simpson *et al.*, 1982; Wolanski *et al.*, 1984b; Pattiaratchi *et al.*, 1986; Heywood *et al.*, 1990, 1991; Wolanski, 1994). Formation of such eddies is dependent on the Reynolds number, a combination of flow speed, island diameter, island shape, depth of surrounding seas, and water column stratification, which may lead to temporal and spatial variability in the presence and duration of island wake eddies (Batchelor, 1967).

Perhaps as a result of such variability, at least for oceanic islands, the evidence supporting the eddy-retention hypothesis is equivocal. For example, although a trapped, von Karman-type eddy was hypothesized for the island of Barbados (Emery, 1972; Powles, 1975), no such eddies were found in the lee of the island (Cowen and Castro, 1994; Bowman *et al.*, 1994; Stansfield *et al.*, 1995). Instead, the flow passing the island was clearly steered by topography, resulting in flow running predominantly parallel to shore, with nearshore, vertically stratified, onshore (at depth), and offshore (at surface) components (Cowen and Castro, 1994). Elsewhere, as reviewed by Leis (1991a), the importance of semipermanent eddies in the lee of several of the Hawaiian Islands (e.g., Oahu and Hawaii) (sensu Patzert, 1970) as retention mechanisms for coral reef fish is equivocal. Despite findings of a drifter being ejected from the eddy and back to the island (Lobel and Robinson, 1986), the waters around and within the proposed retention eddy contained very few coral reef fish larvae (see Lobel and Robinson, 1988). Finally, Farmer and Berg (1990) review a variety of mechanisms, including eddies, that have been hypothesized as facilitating retention of Bermuda's fish and invertebrate larvae. Although no direct evidence of trapped eddies exists for Bermuda, inferential evidence supports the contention that local retention of coral reef fish larvae is the norm (Schultz and Cowen, 1994). Instead of topographically generated eddies, a variety of circulation features may contribute to larval retention around Bermuda. Thus, island-wake eddies may add, perhaps at the "event" level, some retention potential to locally derived larvae, but they are typically quite transient and likely do not serve as a major or permanent "retention" mechanism.

We may wonder whether the various physical processes affecting larval reef fishes differ in importance across different physical environments, in particular, insular versus continental coastlines. These are clearly different environments, yet, for example, most Caribbean coral reef fishes successfully exist (i.e., they recruit and survive to reproduce) in both classes of environments. Do the larvae of a given species rely on very different physical processes and/or respond in very different ways depending on their environment? To do so would require considerable local adaptation (with a bewildering array of both sensory capabilities and behavioral responses) to ensure successful onshore transport of larvae in many different environments. A more parsimonious alternative is that larvae are utilizing physical features common to a broad diversity of coastal environments. Despite the various forcing mechanisms operating in these different environments, the resultant flows (particularly the vertically stratified flows) may be quite similar. This would enable relatively simple sets of (or even a single) behavioral responses to result in successful cross-shelf movement of the larvae.

One example of how a species might utilize a simple vertical behavior to move onshore (i.e., across the shelf) in a variety of physical environments is that of the labrid, *Xyrichtys novacula*. The adult range of this species spans from Cape Hatteras, North Carolina, throughout the Caribbean to northern Brazil. Larval *X. novacula* are commonly found in a variety of different physical environments: around islands (e.g., Barbados) and in continental shelf environments such as the New York Bight (NYB) and South Atlantic Bight (SAB) (Hare and Cowen, 1991; Cowen *et al.*, 1993). Around Barbados there is an upwelling-generated subsurface onshore flow occurring at a depth of between 20 and 40 m. The NYB is often a three-layered system in the summer when larvae are present. During wind conditions that force the surface waters offshore and to the north, the offshore, warm salty Slope Sea water can flow onshore along the thermocline at middepth (20–40 m) over the very cold and dense remnant winter water, and under the relatively fresh surface layer.

The SAB is essentially a two-layered system in which the water below the thermocline flows gently onshore. In all three cases, an ontogenetic vertical migration from near-surface waters to middepths (e.g., 20–40 m) results in the onshore movement of larvae.

Some circulation features are ubiquitous across a broad spectrum of otherwise different physical environments. One such important and common feature are the relatively short vertical distances separating water masses flowing in different directions. As a result, a larva that is adapted to swim vertically (or alter its density) to utilize a specific flow feature at a particular stage of its development may be able to keep itself within a particular prey environment or be transported nearshore, whether it is located along a coastal headland of continental South America or near an isolated island such as Barbados. Variability in a specific physical feature may drive year-to-year variability in recruitment for a given species at various locations, but it is the general prevalence of that feature that results in species-adaptive behavior. Viewed from a different perspective, a species may persist best in those locations where there are similar physical features that either successfully retain larvae or aid in their transport to coastal habitats.

Studying the commonalties among transport processes, including the physical features involved, how external forcing influences these features, and, finally, how organisms (larvae and adult) behave in response to these features, may facilitate our understanding of how these processes can contribute to fluctuations in larval supply. The variable nature of recruitment to some populations may be a result of more frequent interruptions of specific flow regimes. This will be particularly apparent in areas that are near the upstream portion of the larval supply, where sporadic recruitment occurs only during significant alterations in typical flow patterns (Cowen, 1985) and populations persist on the basis of storage of strong year classes (Warner and Chesson, 1985). An improvement in our understanding of how larvae are transported can help us better predict when the conditions are favorable (or not) for recruitment, as well as improve our understanding of biogeographic relationships and stock maintenance (Sinclair, 1988). Although we risk overemphasizing physics in lieu of studying larval behaviors, it is important to search for general patterns in the interaction of larvae with their physical environment.

V. Biophysical Interactions

The preceding review of transport mechanisms demonstrates that our understanding of the physical regime and how it may influence the movement of larvae is considerable. The various mechanisms and processes discussed so far may all contribute to the successful survival, retention, and transport of larvae to appropriate settlement habitats. Many of the physical mechanisms can affect multiple biological factors such as reproductive timing, location and output, larval feeding and transport, and onshore movement leading to settlement. Currently, recruitment studies have been able to identify both temporal and spatial patterns in recruitment, and to a lesser extent, recruitment events have been correlated with specific physical features (e.g., wind, upwelling flow, eddies). Further, larval distributions have been identified that are best explained by the interaction of larval behaviors and physical mechanisms. Such interactions suggest that temporal and spatial patterns of larval distribution may be predictable.

Where we still have a large gap in our knowledge, however, is in explaining the magnitude of recruitment events with any predictability. Perhaps one reason we are still limited in our ability to predict recruitment strength is that particularly large (successful) events occur as a result of the co-occurrence in time and/or location of the full suite of (or some significant subset of) mechanisms favorable to recruitment. For example, onshore winds may enhance recruitment only when other factors (e.g., good food environment for larval survival, low predator field, favorable spawning conditions) all are conducive to forming a relatively large larval pool. In the absence of the winds, some low-level recruitment may occur, or vice versa; but in the absence of favorable spawning conditions, even with strongly favorable onshore winds, no measurable recruitment event will occur. The end result is only a partial explanation associated with the single variable, wind. The requirement of cooccurrence of multiple factors to elicit a strong recruitment event is akin to the "rogue wave" theory of Neill *et al.* (1994), whereby the factors affecting recruitment may vary with differing periods, but occasionally the cycles (and aperiodic factors) come together in favorable harmony to produce particularly large cohorts. Thus, our predictive abilities will be limited if we examine a reduced set of causative variables. This is clearly a multivariate problem spanning a broad range of physical and biological variables and scales.

VI. Modeling

A well-designed field study focusing on larval transport may define how biology and physics interact to explain a given distribution of larvae, but such a study is static. It is not realistic to conduct a field study under all possible field conditions at even a single site, let alone all possible sites and scales. For this, modeling is required. The system we are attempting to understand (the pelagic world of larval fishes) is very complex, but not necessarily chaotic. This complexity, at least the dominant components of it, potentially can and needs to be modeled. Simple models can serve as conceptual frameworks or can demonstrate the sensitivity of a given process to particular conditions. However, more sophisticated models should be able to help us predict where, when, and how many larvae are going to recruit to a given shoreline or coral reef. It is important to stress, though, that modeling efforts require empirical field studies to provide parameterization and identify the important biophysical processes. Empirical studies should also serve as important field tests of model predictions.

Physical models are becoming more and more sophisticated. For example, general ocean circulation models that operate at the basin scale or larger (global) are remarkably high resolution (12–20 km); examples include the S-coordinate Primitive Equation Model (SPEM), the Princeton Ocean Model (POM), and the Miami Isopycnic Coordinate Ocean Model (MICOM) (Song and Haidvogel, 1994; Bleck and Chassignet, 1994). When applied to smaller spatial scale coverage, the resolution may become finer (e.g., 6 km), enabling the resolution of submesoscale features. For small-scale, shallow-water circulation studies, hydrodynamic models are available for resolution on the scales of tens to hundreds of meters (e.g., Black *et al.*, 1991; Wang, 1990). Just as for the descriptive oceanographer working at the interface between open ocean and coastal environments, the challenge for physical modelers is linking basin-scale ocean circulation models with coastal circulation models where complex topographic effects occur.

Biological models of various levels of sophistication exist that can be linked to physical models for a more complete biophysical understanding of larval dynamics. Such models can focus on larval behaviors (e.g., vertical distribution, directional swimming, or simply random walks) (Dutkiewicz *et al.*, 1993; Werner *et al.*, 1993; Wolanski *et al.*, 1997) or vital rates (e.g., growth and mortality rates) (Werner *et al.*, 1996; Hofmann and Lascara, 1998). These models can be relatively simple linear models (e.g., fixed, size-specific rates) or more complex (and realistic) models that incorporate a range of biological responses to environmental conditions [e.g., individual-based models (see also Eckman, 1996)]. For example, models that link larval growth and mortality rates to plankton productivity, which is responsive to the physical environment, can provide important means of coupling biological and physical processes toward an understanding of larval survival. Linking such processes to a larval transport model would provide information on both the distribution and the abundance of a cohort of larvae.

However, these models are not perfect and there are, at times, considerable differences among models (e.g., see Wolanski, 1993; Black, 1995) that stem from either the model assumptions and/or how they are parameterized. Nonetheless, circulation models allow us to examine how larvae may be transported under a suite of physical conditions. Further, when coupled with biological models that are responsive to the physical conditions set forth by the physical model, these models allow us to examine the range of potential outcomes. We also can use models to make field-testable predictions (e.g., how rapidly larvae are dispersed away from a coastline or along a frontal feature, or how well larvae grow and survive under certain physical conditions), as well as to examine probabilities of the occurrence of certain events (e.g., the "rogue wave" situation—high food level, low predator level, and high larval retention level all at once). The overriding caution, however, is that all models must be field parameterized and field tested. All too often, models are utilized, even with caveats that certain "oversimplifications" were made for computational simplicity, without appropriate validation. The results of such exercises often are published and then all too readily accepted. The potential error in this approach can be exemplified in the model results whereby, for example, a slight change in (or inclusion of) larval mortality rates or diffusivity constants can have order-of-magnitude effects on the model outcome.

VII. Implications for Population Connectivity

Work in a variety of systems has indicated that larval export may not be as widespread as previously thought (e.g., Schultz and Cowen, 1994; Polovina *et al.*, 1999; Jones *et al.*, 1999; Swearer *et al.*, 1999; Cowen *et al.*, 2000). This issue is particularly relevant on the ecological time scales over which most recruitment studies are conducted. Further, when considering the management of fished or otherwise impacted stocks, knowledge of the source of new recruits is of paramount importance. Yet, until recently, the general paradigm was that coral reef fish populations (like most marine populations) are broadly open due to the potential for long-distance dispersal by larvae (Thorson, 1950; Scheltema, 1986). Such perceptions have been reinforced by simple models of larval dispersal by mean currents and mean pelagic larval durations (e.g., McManus and Menez, 1997; Roberts, 1997b). Further, many genetic studies have found limited genetic heterogeneity among local populations (Shaklee, 1984; Lacson, 1992; Planes, 1993; Shulman and Bermingham, 1995; Doherty *et al.*, 1995) except in highly isolated oceanic islands (Johnson *et al.*, 1994; Planes *et al.*, 1998a,b).

One cause of this dichotomy is the mixture of relevant scales. Population genetic studies measure gene flow, which requires only very limited exchange to reduce greatly the genetic differential between local populations, whereas the levels of exchange required to sustain *ecologically* a "downstream" local population from some given source population is many times greater. The other main cause is that the oversimplified models (mean flow and primary layer depth) significantly overemphasize the true level of exchange [by about nine orders of magnitude (Cowen *et al.*, 2000)]. On an even more basic level, it does not make biological sense to send all of one's larvae downstream, because there is no means of evolutionary feedback of selective pressures to the source population required to generate specific life history traits (Strathmann, 1982).

This review describes how a variety of biophysical processes may operate to slow significantly the loss of larvae away from the local population. The interaction of deep-water mesoscale and coastal flows creates high variability in local flow conditions. Within this variation are common features, e.g., vertically stratified flows, to which larvae may respond behaviorally, resulting in enhanced retention relatively close to shore and to the source population. Nearshore environments may also convey increased probability of survival via increased productivity. Thus the biophysical processes are in place to favor local retention and survival over long-distance dispersal. This is not to say that larvae are not dispersed long distances, rather that such dispersal may be relatively minor in comparison to the number of larvae that are retained, or the frequency with which they are retained. Resolving this issue will require a multipronged approach, including larval transport dynamics, population genetics, and larval tagging (e.g., elemental fingerprinting; see Chapter 11, this volume).

VIII. Future Directions

Though there have been some significant excursions into the larval black box over the last decade, it is still dusk inside the box. By exploring the details of larval behavior, in conjunction with the physical world, we have made significant strides in understanding the dynamics of the pelagic system and how larvae have adapted to survive this pelagic stage. We also continue to be rewarded with examples of behavioral and sensory capabilities that were beyond belief only a short time ago (e.g., see Chapters 6 and 8, this volume). Similarly, that there are more and more examples of rather restricted larval exchange should not be surprising, given what we know of the pelagic stage in the life history of coral reef fishes. Linking the dynamics of this stage to recruitment and population dynamics overall, however, continues to be a challenge.

There are directions of research, as well as applications of new technologies, that will continue to shed light into the box. First, there is a need to continue to explore all aspects of the biology and ecology of larvae in the pelagic environment. One immediate approach should be to expand taxonomic coverage. For tropical species, most effort, particularly that tied to late-stage settlement/recruitment dynamics, has been focused on labroids, specifically, labrids and pomacentrids. Several recent efforts have expanded on the number of taxa studied to include mullids, acanthurids, and gobiids. Further, given their economic importance, more work is being focused on haemulids, lutjanids and serranids. This trend must continue. With greater taxonomic coverage, more general theories of larval dynamics and ecology can be forged. There is also a need to study all stages of development. Most plankton studies collect preflexion and early postflexion larvae, but relatively few (or no) later stage postflexion larvae. Light traps have added significantly to our knowledge of late-stage larvae at or near settlement, and recent behavioral studies of these late-stage larvae have revealed some remarkable swimming capabilities. Similar insights will likely be gained when statistically meaningful quantities of all stages of larvae are successfully collected.

There is also a clear need to continue to study the full range of potential larval behaviors. We need to know under what conditions they are expressed and how they interact with the physical conditions in a given system. The potential range of larval vertical behaviors of different taxa is just beginning to be understood. There is still much work to be done to understand how such behaviors change under different physical conditions (e.g., changes in hydrographic conditions), and to what cues larvae are responding and how sensitive larvae are at detecting these cues. Finally, within the context of larval biology and ecology, a more thorough examination of all factors controlling larval and juvenile survival is needed. For example, how does survival differ under different prey conditions? This information is required to comprehend fully biophysical interactions that determine recruitment success. Merely demonstrating that a particular larval advective pathway is feasible does not mean that larvae might survive the trip. We do not know whether oceanic (versus coastal) conditions can provide sufficient food for larvae to grow and survive to metamorphosis (see Searcy and Sponaugle, 2000).

Such studies (as described above) will require more extensive sampling, and as sampling frequency or duration increases, so does the effort required for processing the samples. A separate problem is that it is difficult to catch older and larger larvae with conventional nets because their density decreases and their swimming capability increases. Development of real-time or near real-time sampling technologies would be incredibly beneficial in efforts to couple synoptic physical surveys with biological studies. The Video Plankton Recorder (VPR) of Davis *et al.* (1996) shows some promise in this direction, though its field of view is still too small to sample adequately the volume of water required to quantify larval fishes. Tied to such rapid sampling techniques is the requirement for rapid identification of larvae to the species level. The state of knowledge in this regard is still limited, even with the significant progress being made (e.g., Leis and Carson-Ewart, 1999; Richards, 2001, and the many contributors to these volumes). With the advent of rapidly evolving molecular techniques (mitochondrial DNA, molecular probes, microarray techniques), rapid and potentially semiautomated, identification and eventually quantification may be possible.

Other technologies are worth exploring or developing to further enhance our ability to examine larvae *in situ*. For example, S. Sponaugle and J. S. Jaffe (personal communication) are exploring the use of new three-dimensional sonar technology (Jaffe *et al.*, 1995) to "visualize" the interaction of larvae with their environment. This technology allows individual larvae and zooplankton to be tracked in a three-dimensional volume, enabling the observation of *in situ* swimming behaviors and predator–prey interactions during the transition from pelagic to benthic environments. Technologies also exist that will enable the identification of larval sources and transport pathways. These include tracking of larvae via dye injection to track water mass movement (e.g., Houghton and Visbeck, 1998), release of larvae artificially tagged with either otoliths marking (e.g., Secor *et al.*, 1995b; Jones *et al.*, 1999) or molecular marking (Hitchcock *et al.*, in review), using natural tags (microchemistry of constituents of otoliths; see Chapter 11, this volume), and using new molecular genetic techniques such as microsatellite DNA (O'Reilly and Wright, 1995; O'Connell and Wright, 1997; Carvalho and Hauser, 1998).

Once we are able to address these questions, it will be possible to make significant advances into the larger issues of recruitment and population dynamics. Our current advances have greatly enhanced our understanding of the mechanisms creating temporal and spatial patterns in larval distributions and settlement. Moreover, details of biophysical processes have contributed to our ability to explain larval retention and the realization that marine populations may not be as "open" as previously thought. Linking new techniques with knowledge of these biophysical processes should lead to a better explanation and prediction of settlement amplitude, especially when coupled with studies of postrecruitment processes, recruitment variability, and ultimately population dynamics.

Acknowledgments

The ideas presented in this chapter have developed over time through reading the literature of, and through discussion with, many colleagues, including several of the authors of other chapters in this volume. I acknowledge several individuals who have been particularly influential in my thinking on this topic: Su Sponaugle, Jon Hare, Jeff Leis, Kamazima Lwiza, Mike Fahay, Bill Richards, Peter Sale, Eric Schultz, and Don Olson. I also acknowledge the help and insight of Claire Paris, Suzanne Dorsey, and Leonardo Castro. Cara Dickman helped organize the bibliography, C. Paris and Mark Sullivan helped prepare several of the figures, and Jack Harlan provided Fig. 6C; S. Sponaugle, C. Paris, M. Sullivan, John Purcell, and Mark Steele all provided valuable comments on earlier drafts. Finally, I appreciate Peter Sale's encouragement and patience during the writing of this chapter.

CHAPTER 8

The Biology, Behavior, and Ecology of the Pelagic, Larval Stage of Coral Reef Fishes

Jeffrey M. Leis
Ichthyology, and Center for Biodiversity and Conservation Research
Australian Museum
Sydney, New South Wales 2010, Australia

Mark I. McCormick
School of Marine Biology and Aquaculture
James Cook University
Townsville, Queensland 4811, Australia

I. Introduction

Reef fish biologists are keenly aware that nearly all bony fishes on coral reefs have a pelagic larval phase that is potentially dispersive, and that this has major implications for reef fish populations not only at evolutionary (or biogeographic) scales, but also at ecological (or demographic, including management) scales. The literature is full of statements of how important this type of life history is for reef fishes, and for study and management of them. However, this realization has not been accompanied by a major shift in research effort to studying this pelagic phase, what one might refer to as "prerecruitment" studies. Neither has it led to a widespread view of the pelagic phase as much more than a "black box" that results in open populations and large fluctuations in recruitment. Even attempts to assess the population connectivity that presumably results from larval dispersal typically make simplifying assumptions, either explicitly or implicitly, that portray the larvae as little more than passive tracers of water movement that "go with the flow," doing nothing much until they bump into a reef by chance and settle at once.

Are larvae really as simple and as uninteresting as the assumptions made by this "black box" view of larval biology? We think not. The work reviewed here reveals larvae of coral reef fishes to have remarkably good swimming abilities, good sensory systems that develop early in ontogeny, and sophisticated behavior that is very flexible. Little of this would have been predicted from the much better known larval biology of temperate, non-reef species such as herring, cod and plaice. We explore some of the reasons for this. The interaction of larval distributions with oceanography is the subject of Cowen's chapter in the present volume, and we do not address that subject area. This chapter is not a revision of either Leis' (1991a) review chapter in the first "Sale Book", nor do we cover ground already dealt with by reviews of coral reef fish larval biology Boehlert (1996) or Cowen and Sponaugle (1997). Instead, we focus on recent research that examines reef fish larvae as animals interacting with their environment. We emphasize a perspective from the pelagic environment toward the demersal reef environment. The larvae have a similar perspective. Other studies take the opposite view, and indirectly examine the pelagic stage from the reef. These utilize information gleaned from otoliths of recruits or from abundance patterns either of settlement-stage larvae captured by reef-edge light traps and reef-based nets, or of recruits on the reef (e.g., Dufour and Galzin, 1993; Milicich, 1994; Sponaugle and Cowen, 1994; Thorrold *et al.*, 1994b,c; Robertson *et al.*, 1999). Studies of this sort provide valuable insight, but they are largely beyond the scope of the present review. We review new information on the pelagic stage, from spawning to settlement, including metamorphosis, but not post-settlement issues.

There are large differences in biology between the larvae of coral reef fishes and those of the temperate fishes that dominate the literature. Predictions based on these temperate species have often badly misled biologists working on reef fishes. Below, we examine some of the reasons why such large differences exist. We conclude that such predictions should be made very cautiously indeed, and have tried to avoid over-reliance on the temperate literature. We can learn much from the excellent and extensive research done on larvae of temperate fishes, but recent experience shows that everything must be ground-truthed with larvae of reef fishes. Coral reef fishes are overwhelmingly of the order Perciformes, or perchlike fishes (here considered to include the Scorpaeniformes—see Chapter 1, this volume). In contrast, most research on temperate marine fish larvae has been on other orders, in particular the Clupeiformes (herrings, sardines, and anchovies), Gadiformes (cods), and Pleuronectiformes (flatfishes). These four orders have been distinct for at least 50–60 million years, and the Clupeiformes for much longer (Carroll, 1988). Further, the nonperciform fishes include few species that live on either temperate or tropical reefs as adults. Few would expect the biologies of the adults of these distantly related orders from different habitats to be similar, but similarities are somehow expected among the larvae.

There is a flip side to this coin, however. Tropical researchers are guilty of emphasizing the dispersal capabilities of reef fishes and not paying enough attention to what influences larval survival—a theme that has been the major thrust of temperate fisheries research for the last 100 years (Cowen and Sponaugle, 1997; Kingsford, 1998). Despite the taxonomic differences in study species, coral reef fish research can gain much from temperate research regarding the processes influencing larval mortality schedules, although the specifics, as emphasized above, will undoubtedly differ. Many of the recent advances in larval research have been through the application of techniques and philosophies gleaned from temperate fisheries research, such as the use of otolith analyses to examine larval mortality schedules and dispersal.

Some definitions and abbreviations are required. Nomenclatures applied to the early life history stages of marine fishes may be based on morphological criteria, ecological criteria, or some combination. Some terms were developed for use with pelagic fishes, such as clupeids, others for benthic fishes with pelagic larvae, such as reef fishes, and they reflect the different priorities and perspectives of the researchers, or often, tradition. Standardization is unlikely, if for no other reason than that the diversity of fish ontogenies makes no one system entirely satisfactory. To avoid these confusing nomenclatural issues, we use a broad definition of "larva" that includes everything from hatching to acquisition of full fin rays and scales, and loss of "specializations for pelagic life" (such things as transparency, elongate or ornamented fin spines, or head spination) not found in the adult. In some species, the ecological transition of settlement coincides with the morphological transition of metamorphosis. In others, it does not. The morphological transition may be gradual or abrupt, whereas the ecological transition is usually abrupt.

Lengths referred to here are "standard length" (SL; snout to base of the caudal fin rays, or to the tip of the notochord before the caudal fin forms), but total length (TL) is used in some of the cited literature. Swimming speeds are often standardized as "body lengths per second" (BL sec^{-1}). "Effective swimming" is swimming faster than ambient currents (Leis and Stobutzki, 1999). Pelagic larval duration (PLD) is the length of time the propagule is pelagic, i.e., the time between the leaving the reef and settlement. Lizard Island is referred to frequently—it is a midshelf reef complex on the northern Great Barrier Reef.

II. Near Ubiquity of a Pelagic Stage

A pelagic early life history stage is found in nearly all bony reef fish species (Leis, 1991a; Leis and Carson-Ewart, 2000). This is true regardless of whether eggs are pelagic, demersal, or brooded, or if the adults are viviparous. The pelagic stage may last as little as a week in some damselfishes (Pomacentridae) (Thresher *et al.*, 1989), or more than 64 weeks in some porcupine fishes (Diodontidae) (Ogden and Quinn, 1984).

Leis (1991a) was able to list only a few species of bony reef fishes that lack a pelagic stage, but in the past few years, three more have been discovered in the Indo-Pacific. The Banggai cardinal fish, *Pterapogon kauderni* (Apogonide), orally broods both eggs and hatched young until the latter reach about 10 mm (Allen and Steene, 1995). The young remain on the reef and bypass the pelagic phase. In two species of the recently described chromine pomacentrid genus *Altrichthys*, larvae hatch from demersal eggs, and are tended by both parents for "an extended period" (Allen, 1999), thereby avoiding the pelagic phase.

Thus, we should perhaps expect the discovery of a few more reef fish species lacking a pelagic stage. These will most likely be apogonids and chromine pomacentrids, and will prove useful models to compare to species that retain a pelagic stage, as has *Acanthochromis polyacanthus*, previously the only pomacentrid known to lack a pelagic stage (Kavanagh, 2000). However, compared to other groups of marine animals (Strathmann, 1986; Strathmann and Eernisse, 1994), very few coral reef fishes have lost or greatly abbreviated the pelagic stage.

Why Are Temperate Fish Larvae So Different from Reef Fish Larvae?

Most of the literature on fish larvae from temperate waters concerns species that are pelagic or live on soft (sand or mud) bottoms as adults rather than species from (rocky) reefs. This alone may confound temperate/coral reef comparisons. Adult pelagic fishes usually have habitat requirements that differ from those of their larvae, but they never make the abrupt changes entailed by settlement out of the pelagic environment and into the benthos. Except on oceanic islands, soft-bottom habitats are usually far more extensive and less discrete than reef habitats, so larvae of reef fishes have a much smaller target to find at the end of the pelagic phase than do fishes of soft bottoms. Further, it is thought that most reef fishes are relatively sedentary as adults, whereas adults of many pelagic and soft-bottom species undertake extensive migrations: thus, in contrast to most pelagic and soft-bottom species, dispersal in reef fishes is thought to be limited largely to the pelagic larval phase.

A major difference between coral reef fishes and temperate marine fishes is the incubation period of their eggs. Pelagic eggs of most coral reef fish species hatch within 1 day (Watson and Leis, 1974; Tucker, 1998), far more rapidly than pelagic eggs of temperate fishes, which commonly have incubation periods of 3–20 days (Russell, 1976).

A second major difference between coral reef fish and most temperate fish larvae is that at any given size, the reef fish larvae are more developed. This is particularly apparent if one compares the state of development of well-studied temperate larvae such as herring and cod at the sizes at which reef fish larvae commonly settle (1–2 cm) (Fig. 1). At all sizes, the reef fish larvae have more complete fins. They develop scales at a smaller size, seemingly have better developed sensory apparatus at any size, and are morphologically equipped for effective feeding within a few days of hatching, and at a smaller size than the herring and cod (Table 1). Temperate perciform fishes (including many reef fishes), in contrast, reach developmental milestones at sizes similar to those of coral reef fishes (Moser, 1996; Fuiman *et al.,* 1999).

Reef fish larvae are much better swimmers at these sizes (1–2 cm) than are larvae of cod or herring (see Section IV,B,2). The few data on larvae of temperate perciform fishes indicate that although they are better swimmers than temperate gadiform or clupeiform larvae, they are apparently much slower than coral reef fishes of similar size.

So, why are reef fish larvae better swimmers? As implied above, there is undoubtedly a taxonomic component, with perciform larvae apparently superior performers than gadiform or clupeiform larvae. Second, swimming may be more efficient in tropical waters (25–29°C) than in temperate waters (10–20°C) for both physiological and hydrodynamic reasons (Fuiman and Batty, 1997; Wieser and Kaufmann, 1998). Third, most of the temperate species were lab-reared larvae, and were usually evaluated in the laboratory, whereas the reef species were wild larvae evaluated in the field. Lab conditions may not be conducive to peak swimming performance, and lab-reared larvae and juveniles often swim much more slowly than do wild larvae or juveniles (citations in Leis and Carson-Ewart 1997; Leis and Stobutzki, 1999).

Other factors may favor better swimming by tropical reef fish larvae. As one approaches the equator, the magnitude of the Coriolis force decreases (Bakun, 1986). Some of the mechanisms that have been proposed to help structure "retention areas" in temperate systems involve Coriolis-force-mediated phenomena such as coastal upwelling (Bakun, 1986). To the extent that these phenomena might be less active in the tropics, there may be reduced scope for passive larval retention in the tropics, and therefore more selective pressure to

TABLE 1 Comparison of Size at Various Developmental Milestones for Temperate and Tropical Fish Larvae[a]

Taxon	Approximate size at first feeding (mm)	Size at caudal fin formation (mm)	Size at complete fin formation (mm)	Size at settlement (mm)
Herring (*Clupea*)	5–11	16–20	25–27	Not applicable
Cod (*Gadus*)	4–6	9–11	26–30	25–50
Benthic, temperate perciform fishes	2–3	3–5	4–8	8–20
Coral reef perciform fishes	2–3	3–5	4–10	7–75

[a]Information on cod, herring, and other temperate species from Moser (1996) and Fahay (1983) and on coral reef species from Leis and Carson-Ewart (2000).

A

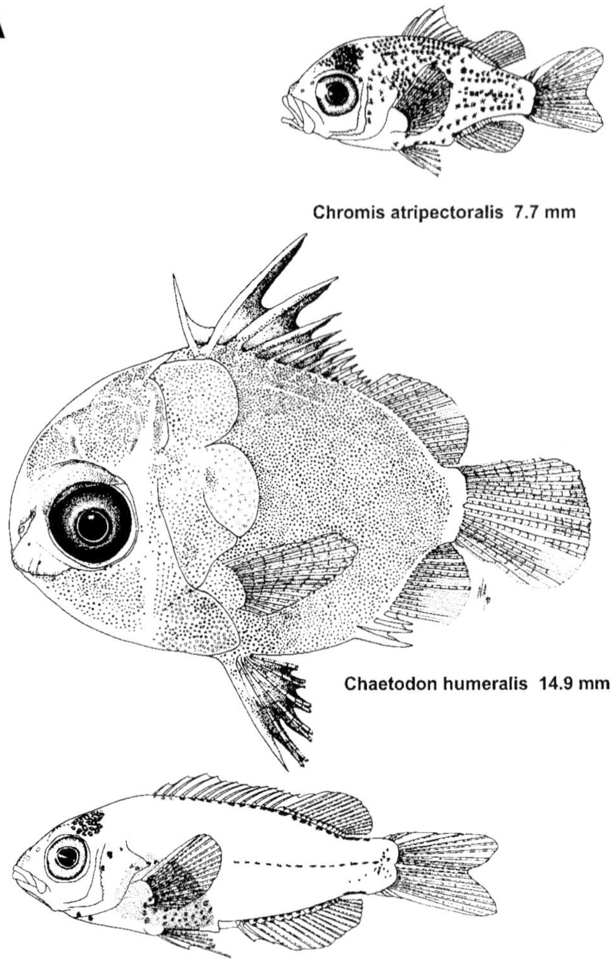

FIGURE 1 Comparison of developmental state in larvae of three species of reef fishes at settlement, with that of herring and cod at similar sizes. (A) Settlement stage larvae of *Chromis atripectoralis*, ~8 mm; *Chaetodon humeralis*, ~15 mm; and *Pomacentrus* sp., ~12 mm. *Chromis* and *Pomacentrus* from Leis and Carson-Ewart (2000); *Chaetodon* from Watson (1996). (B) Developmental series of larvae: Atlantic cod, *Gadus morhua*, 9, 11, and 16 mm (TL) (with permission from Fahay, 1983) and Pacific herring, *Clupea pallasii*, ~8, 10, and 19 mm (with permission from Moser, 1996).

attain active means of larval retention, including swimming. Finally, the selective regime in which reef fishes live—requiring, as it does, the location of small, scattered bits of habitat at the end of a pelagic stage—is far different from that faced by a cod or herring, and superior swimming performance by the larvae may be adaptive.

III. Recently Developed Research Tools

A. The Light Trap

The semiautomated light trap (Doherty, 1987b) is now widely used for sampling late-stage larvae for distributional studies. Little is known of how light traps actually catch larvae, why they usually capture almost exclusively late-stage (settlement stage) larvae, or what their catches actually represent in terms of larvae in the water column (Choat *et al.*, 1993). No one overtly claims that light-trap catches are quantitative, but, light traps are widely used for relative comparisons among locations or times. How catches might be affected by variations in ambient current speed, ambient light (either the moon, or cloud cover), or water turbidity is unknown. The radius from which a light trap draws its catch was estimated based on water clarity arguments to be up to 90 m (Milicich *et al.*, 1992), but this is untested, and seems excessive. Guessing a much smaller radius of 7–50 m, and depending on current speed, Choat *et al.* (1993) estimated that a light trap could sample 40,000 m^3 hr^{-1}, and argued that the ability to sample such huge volumes was a major advantage. Meekan *et al.* (2000) showed, primarily for pomacentrids, that capture efficiency of two light-trap designs was less than 10%, but that escape rates were relatively low and dependent on taxon: 5% of reef fish larvae, but 27% of clupeid larvae, escaped in 1 hour. They argued that low efficiency, provided it is relatively constant, is not an obstacle to use of light traps to study spatial and temporal patterns.

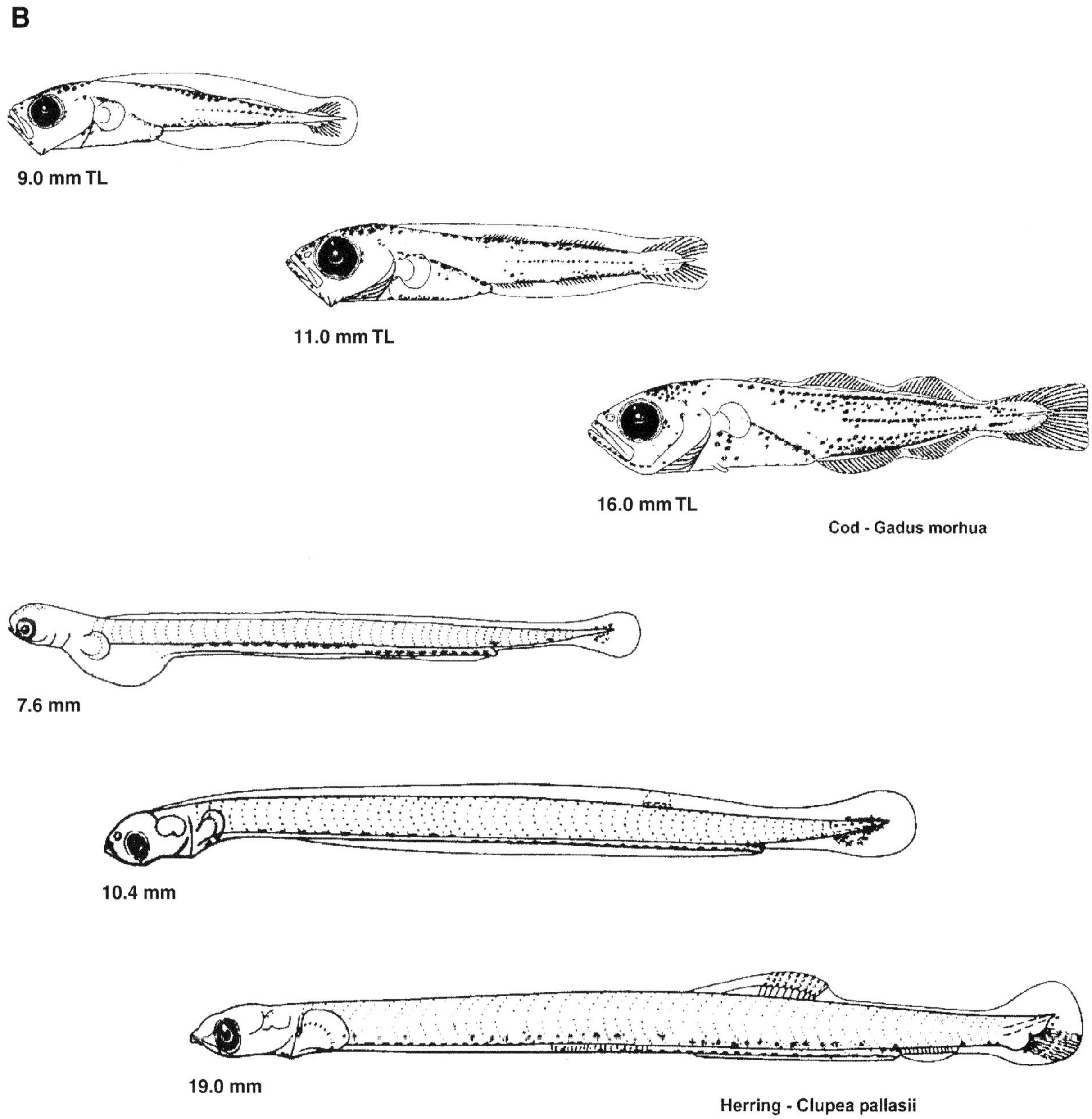

FIGURE 1 (*continued*)

Whether the range of efficiency values reported (±50% of the mean) amounts to "relatively constant" depends on the research question. Larger fishes (clupeoids, scombrids, terapontids, etc), decapods, and cephalopods commonly enter traps, sometimes in large numbers, and may eat the captured larvae (personal observations), but nothing is known about the level of this "in-trap" predation and its effects on larval catches.

Light traps are highly selective on a taxonomic basis, working best with clupeids, lethrinids, pomacentrids, and some blennioids, serranids, and chaetodontids (Milicich and Doherty, 1994; Choat *et al.*, 1993; Brogan, 1994a; Hernandez and Lindquist, 1999; J. M. Leis and M. I. McCormick, unpublished results), but even within these families some species are virtually absent from catches. Light traps located directly over reefs or seagrass beds in shallow water sometimes capture large numbers of small, often just hatched, larvae from demersal eggs, primarily gobioids and blennioids (Brogan, 1994a,b; Riley and Holt, 1993; A. Cabanban, personal communication). Traps in slightly deeper water capture almost exclusively late-stage larvae ready to settle (Choat *et al.*, 1993; Milicich and Doherty, 1994; Hernandez and Lindquist, 1999), although at times light traps capture large numbers of pomacentrid larvae midway through larval development (Brogan, 1994a,b; J. M. Leis, unpublished data). Light-trap selectivity has led to a focus on the study of distributions of larvae near the end of their pelagic phase (Kingsford, 1998).

Light traps also provide a convenient means to capture larvae without the damage often inflicted by towed nets. Light-trap-captured late-stage larvae can be used for a variety of both field and laboratory investigations (Stobutzki and Bellwood, 1997) on swimming abilities, settlement behavior, sensory abilities, orientation, physiology, and metamorphosis (see later). Light-trap larvae that have been allowed to metamorphose can be used in other field and laboratory studies of postsettlement issues that are beyond the scope of this chapter. All these studies have led to major advances in our understanding of larvae as animals, rather than as biologically inert drift cards. It is arguable that light traps have made their most significant contribution to reef fish biology in this way.

A critical unknown is the possible alteration to behavior or physiology that may result from attraction to, and retention within, a light trap. Usually, light traps are fished all night, so larvae may have been in the trap for several hours on recovery. Larvae of some species appear to be irreversibly "benthic" when removed from the trap in the morning, and refuse to swim in the pelagic environment (Leis *et al.*, 1996). What other, more subtle, alterations may have taken place in other species while in the trap, or while being held in the lab, remain to be determined.

Modified light traps that direct their catch to a nearby reef can enhance recruitment by two- to threefold (Munday *et al.*, 1998). This is potentially useful not only in experimental work, but also in management of reefs. Interestingly, the taxa that recruited on the reefs differed in proportion from the catch in nearby conventional light traps. This might be the result of habitat selection at settlement by the larvae, as indicated by Munday *et al.* (1998), or an indication that among the taxa attracted to the light, entry into the trap is more selective than is settlement (Choat *et al.*, 1993). Light traps may have still other innovative uses not anticipated by their designers.

B. Larval Rearing

In the past 10 years aquarists and researchers have managed to rear a wide range of tropical reef fishes, over 100 species from at least 24 families. In some instances this may have reduced some of the pressure from aquarium collectors on the reef communities. Excitingly, a handful of dedicated researchers have reared reef fishes in sufficient quantities to allow the description of larval development and exploration of larval processes under controlled conditions. Some success has been obtained by rearing larvae collected from the wild midway through their larval phase (e.g., McCormick and Molony, 1992, 1995), but most workers have had to rear larvae from eggs of captive brood stock. Initial success was with species, such as clown fishes, that spawn benthic egg clutches. Additional success has been achieved in Japan with the commercially important broadcast spawner, the coral trout, *Plectropomus leopardus* (Masuma *et al.*, 1993), and a number of wrasse species (Kimura and Kiriyama, 1993; Kimura *et al.*, 1998). Although some of the damselfishes, particularly the clown fishes, have been reared for decades, it is only recently that rearing has been used to elucidate important processes in the larval stage. Much of the work that we review was based on these studies. The ability to rear larvae from known parents, under manipulated environmental conditions, will undoubtedly yield our most detailed insight into the black box in the decade to follow.

C. *In Situ* Studies of Behavior

An approach that has proved fruitful in casting light into the black box of reef fish larval biology is the *in situ* observation of behavior. Divers can simply look for larvae and observe them. This has been successful in temperate waters (Breitburg, 1991, and references therein), but not generally in coral reef waters (but, see Kaufman *et al.*, 1992). However, wild larvae can be captured either by light trap or by fixed net, and then released and observed individually by divers in the pelagic environment, either in lagoons, over the continental shelf, in open ocean conditions, or in more shallow waters adjacent to reefs (Leis *et al.*, 1996). One diver concentrates on and follows the larvae, and does nothing else, while the other diver follows and notes depth and direction every 30 seconds and measures distance traveled with a flowmeter. Thus, three-dimensional trajectories of individual larvae can be constructed. When released near reefs, the reaction of the larva to the reef and its residents, and its settlement behavior, can be observed. In addition, "unplanned" observations on predation on or feeding by the larvae can be made. Many, but not all, species are amenable to this approach. Most species are seemingly not bothered by the nearby observer divers: they swim at a speed far below their maximum, and alter it to suit different situations, undertake no obvious avoidance behavior, frequently feed, may retreat to the shelter of the observer diver when threatened, occasionally try to "settle out" on the diver or his/her gear, and deliberately and carefully examine and often reject potential settlement sites over the reef before swimming on (Leis *et al.,* 1996; Leis and Carson-Ewart, 1997, 1998, 1999).

This methodology has limitations: not all species are amenable to its use, it is very labor intensive, any bias in behavior of the larvae caused by the presence of the divers is unknown, it cannot be applied at night, and safety considerations limit *in situ* observations to depths less than 20 m. To date, the number of taxa examined *in situ* in this manner is relatively small, being largely a subset of the relatively small number of species captured by light traps. Capture of wild larvae by other methods that do not share the biases of light traps, or use of reared larvae *in situ,* can help overcome the first limitation. The method is inherently labor intensive, so only alternative methodologies can solve this problem. The use of high-tech equipment such as the "fish TV" sonar (Jaffe *et al.,* 1995) could help to assess diver induced bias. Because such equipment can be used at night and in unlimited depths, this should help to resolve some of the other limitations.

IV. Off the Reef and into the Blue

A. Egg Loss: Benthic versus Pelagic Spawners

There are two broad modes of spawning among coral reef fishes: benthic spawning of eggs followed by varying levels of parental care, and broadcast (or pelagic) spawning whereby eggs are released directly into the water column, with no subsequent parental care. Spawning mode does not appear to influence female fecundity or traits that may affect larval survival. Thresher (1984) showed that pelagic and benthic spawners do not differ in fecundity when adjusted for size. Cowen and Sponaugle (1997) reviewed the available evidence supporting the idea that egg and larval traits are related to which spawning mode is adopted. They found no support for the suggestion that benthic spawners should have larger eggs than pelagic spawners, or shorter larval durations.

Spawning mode does appear to influence the initial mortality of the eggs prior to the embryos dispersing from their natal reef. The mortality levels of eggs from broadcast spawners is typically low immediately after egg release. Detailed observations over 2 years at a spawning site on Johnston Atoll showed that only 0.6% of broadcast spawnings were preyed on by planktivores (out of a total of 7448) (Sancho *et al.,* 2000a). Observations suggest that planktivores are attracted to the gamete cloud for only a few seconds immediately after a spawning rush and interest is lost soon after as the eggs are quickly dispersed by the prevailing water currents (e.g., Colin and Bell, 1991). Patterns of predation vary among localities on the same reef, from none (out of greater than 1000 observed spawning ascents) to low levels at other sites where pomacentrids were a more common component of the fish assemblage (Colin and Bell, 1991). Although generally low, the impact of planktivores on the gamete clouds was also found to vary for different species of spawners at the same site (Colin and Clavijo, 1988). Whether spawning is conducted in pairs or groups also influences the rates of gamete predation, with lower levels of predation after group spawning events (Sancho *et al.,* 2000a). Cannibalism can also contribute greatly to the loss of gametes immediately after release in some species of broadcast spawners. One field study found that males of two of three species of razor wrasses (*Xyrichthys*) eat ova from 40% of spawnings (Nemtzov and Clark, 1994).

The only predation study that has come to our attention that measured predation rates on the eggs of a benthic-spawning reef fish was conducted on a damselfish (*Pomacentrus amboinensis*), at Lizard Island (Emslie and Jones, 2001). As in many benthic spawners, the male guards the egg clutches until they hatch (4.5 days after spawning). Daily censuses of mapped clutches indicated that predation by fishes on eggs was high and varied among sites around the island. The mean egg loss from 20 monitored nests in each of four sites ranged from 14 to 40% per clutch, suggesting that for at least some benthic spawners, egg predation can be an important source of offspring mortality.

B. Out in the Blue: The Pelagic Biology of Reef Fish Larvae

1. What Influences Larval Survival?

Studies of temperate fisheries species have placed emphasis on the early life history traits that affect the ability of larvae to acquire food and avoid predation, whilst most research on recruitment variability in coral reef fishes has focused on abundance patterns of larvae and the processes responsible for delivering them back to the reef (reviewed by Leis, 1991a). Recent research suggests that early life history traits, such as growth and length of larval life, influence recruitment variability in coral reef fish species as they do in temperate fish species.

Most mortality occurs during a pelagic phase that may range in duration from 8 days in some anemone fishes to greater than 120 days in some wrasses (Victor, 1991). It is well known that small changes in this initial mortality rate can lead to large changes in the numbers of larvae surviving to settle to the reef population, even if reproductive output remains constant (Houde, 1987; Underwood and Fairweather, 1989). Any factors that influence these mortality rates will have an influence on the magnitude of recruitment of that cohort. It is becoming apparent that a complex of interacting factors determines the mortality schedules of a larval cohort. The quality of the eggs will influence the viability of the larvae. Larval growth rates have been shown to have a direct bearing on mortality schedules, with growth being influenced by features such as egg quality, larval density, food availability, water temperature, and the number of predators for the ontogenetic stage in question. Research on temperate species has stressed the importance of some of these factors for decades, but it is only recently that a handful of researchers started to explore their importance to the dynamics of tropical reef fishes.

a. Maternal Characteristics Influence Larval Quality Egg quality may have a profound influence on subsequent larval growth rates and mortality schedules. Very little research has been done on the importance of egg quality in coral reef fishes. Many of these fishes are protogynous hermaphrodites that are serial spawners, meaning individual females spawn numerous times during a spawning season. Many of these fishes are remarkably fecund, releasing hundreds of thousands to millions of eggs in a single spawning season (Sale, 1980). Breeding ambon damselfish females, *Pomacentrus amboinensis*, spawn every second day of an approximately 3-month spawning season, laying a monolayer of ~6600 eggs (Kerrigan 1994; Kerrigan and McCormick, unpublished).

It is unknown how the nutritive quality of the eggs changes through the spawning season. However, information from following the egg output of individual *P. amboinensis* throughout the main portion of the 3-month spawning period suggests that there may be a general trend for a reduction in the size of eggs over the spawning season (Fig. 2). Interestingly, there was no reduction in clutch size over the same period, suggesting that clutch size was being maintained at the expense of egg quality in this species (M. I. McCormick, unpublished data). The trend for reduced egg sizes later in spawning season has been found for serially spawning temperate fish species such as the Atlantic cod, *Gadus morhua* (Chambers and Leggett, 1996).

The size of larvae that hatch from benthic eggs can vary considerably among local populations of a coral reef fish. McCormick (1999b) and Kerrigan and McCormick (unpublished) showed that the size of

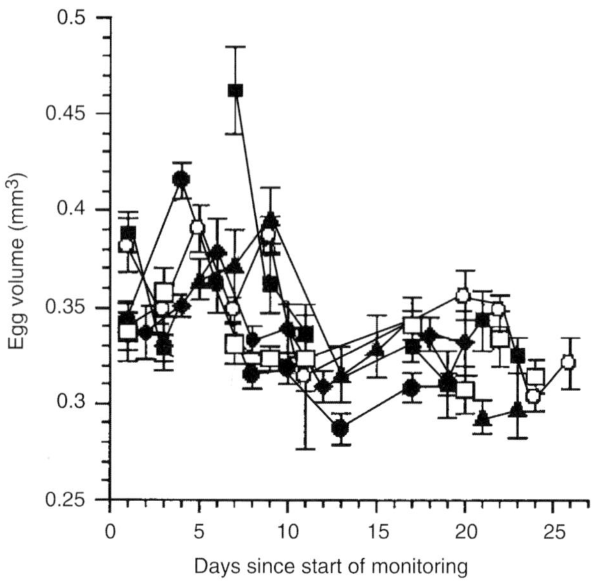

FIGURE 2 Trends in mean egg volume (± SE) among consecutive clutches for five females of the damselfish *Pomacentrus amboinensis* over a 4-week period in the middle of the breeding season.

newly hatched larval *P. amboinensis* ranged between 2.56 and 3.39 mm SL (mode 3.1 mm), with more variability attributable to differences among clutches within a location (47%) than within individual clutches (29%) or differences among localities (24%). Similarly, yolk sac size at hatching varied over a threefold range (0.04–0.13 mm^2), with most variability found at the local scale and attributable to differences within clutches (64%) and among clutches (35%) rather than to those at the locality level (2%). Much of the variability in these qualitative attributes of early stage larvae can be attributed to differences (among individuals) and changes (within an individual) in the maternal energy investment in offspring.

The nutritional condition of the mother at the time of gametogenesis can influence the quality of the offspring she produces, and this will directly influence larval mortality rates. Kerrigan (1997) experimentally manipulated food availability for a 4-week period for two size classes of breeding female *P. amboinensis* at Lizard Island. To enable the egg output to be quantified, these polygamous damselfishes were placed in breeding pairs on isolated patch reefs. Kerrigan found that, regardless of female size, supplementary fed females produced larvae that were larger at hatching and possessed larger nutritive yolk sacs than did non-supplementary-fed females. Although supplementary-fed females produced larger and more nutritively rich offspring, clutches of eggs were the same size as non-supplementary-fed females of the same size. Maternal liver size, the metabolic powerhouse, was found to be the best predictor of size of the larvae that the females produced (Kerrigan and McCormick, unpublished).

The behavioral interaction the mother experiences during egg production profoundly influences the characteristics of her offspring. Kerrigan (1997) found that when an additional female, restrained in a glass container, was introduced into the patch reefs occupied by a pair of breeding *P. amboinensis,* then the unrestrained females produced larger offspring, regardless of whether the females had been supplementary fed. Maternal behavioral interactions appear to influence progeny characteristics through alterations in the maternal hormonal system. Maternal hormones play an important role in development of fish offspring, as they do in other vertebrates (Clark and Galef, 1995, 1998). Developmental hormones aliquoted to the egg during gametogenesis govern developmental rates prior to the production of these hormones by the embryo. Current evidence suggests a direct link between hormones in the maternal plasma and the eggs the female produces, with developmental (e.g., thyroxine, triiodothyronine), reproductive (e.g., testosterone, estrodiol), and metabolic (e.g., cortisol) hormones transferred passively (Schreck *et al.,* 1991; Hwang *et al.,* 1992; Mylonas *et al.,* 1994).

All the behavioral interactions and environmental alterations are interpreted through the fishes' sensory systems and then transformed into hormonal signals by the action of the hypothalamus and pituitary gland. Hormones associated with sensory responses and body maintenance form a potent cocktail that is transported through the plasma of the circulatory system and is in direct contact with the developing oocytes in the ovary during egg formation. During gametogenesis these hormones are incorporated into the yolk sac, which forms the nutritive store prior to the start of larval feeding. Two steroid hormones that are of particular ecological interest are the corticosteroids cortisol, which has been used as an endocrinological indicator of stress (Pankhurst and Van der Kraak, 1997), and testosterone.

Cortisol can directly influence larval morphology. McCormick (1998b) found a correlation between levels of cortisol in the ovaries of breeding females and the size of their larvae in a natural population of tropical damselfishes. In this instance, female cortisol levels appeared to be elevated by high densities of egg predators around the nest site. Elevation of maternal cortisol to levels within the natural range for a species, using cortisol implants, resulted in the production of smaller larvae. Detailed laboratory experiments confirmed this and found that cortisol directly affects larval growth and developmental rates (McCormick, 1999a, also unpublished data). Further, a small increase in the amount of testosterone in an egg increases the size of the yolk sac at hatching (McCormick, 1999a). It is thought that testosterone may improve the efficiency with which nutrients within the yolk are used. The importance of these hormonal mechanisms in influencing variability in larval characteristics is highlighted by the finding that larval morphology can be made to span the whole range of natural variability in body attributes by simply altering the levels of cortisol and testosterone that embryos are exposed to, within naturally occurring limits (McCormick, 1999a).

Maternal condition appears to have an important and as yet poorly studied influence on the initial morphology of coral reef fish larvae. Many questions remain unanswered. Is this variability accentuated over the rest of the larval phase by the selective forces, such as size- or growth-selective processes (see Chapter 10, this volume)? Are there long-term postsettlement effects on individuals? Studies of terrestrial vertebrates, such as rodents and birds, suggest that the embryonic hormonal regime influences important milestones, such as size at maturity, success in competing for mates, and overall reproductive fitness and survival (e.g., Clark and Galef, 1998; Price, 1998). Studies are currently being conducted by Mark McCormick to determine the extent to which maternal characteristics influence postsettlement events in coral reef fishes.

b. Importance of Larval Growth The theory that larvae that grow and develop faster have higher survival and enhanced recruitment (i.e., "stage-duration" hypothesis, Chambers and Leggett, 1987; Houde, 1987; Hare and Cowen, 1997) is now supported by several field studies of temperate species (Campana, 1996; Meekan and Fortier, 1996; Fortier and Quiñonez-Velazquez, 1998; Meekan *et al.,* 1998) and by recent work on tropical species. Bergenius *et al.* (2002) back-calculated larval growth history from daily collections of newly settled Caribbean surgeonfish, *Acanthurus chirurgus*. They found a strong correlation between the growth rate immediately after the start of the first feeding and the magnitude of the settlement pulse ($r = 0.76$; see Fig. 3). This period, between 9 and 12 days after hatching, is the period of exponential growth for this species and appears to represent a "critical period" in the larval life history. Average larval growth rates in the Caribbean damselfish, *Stegastes partitus,* accounted for 83% of the variability in the magnitude of light-trap catches of this species on a monthly basis (Wilson, 2000).

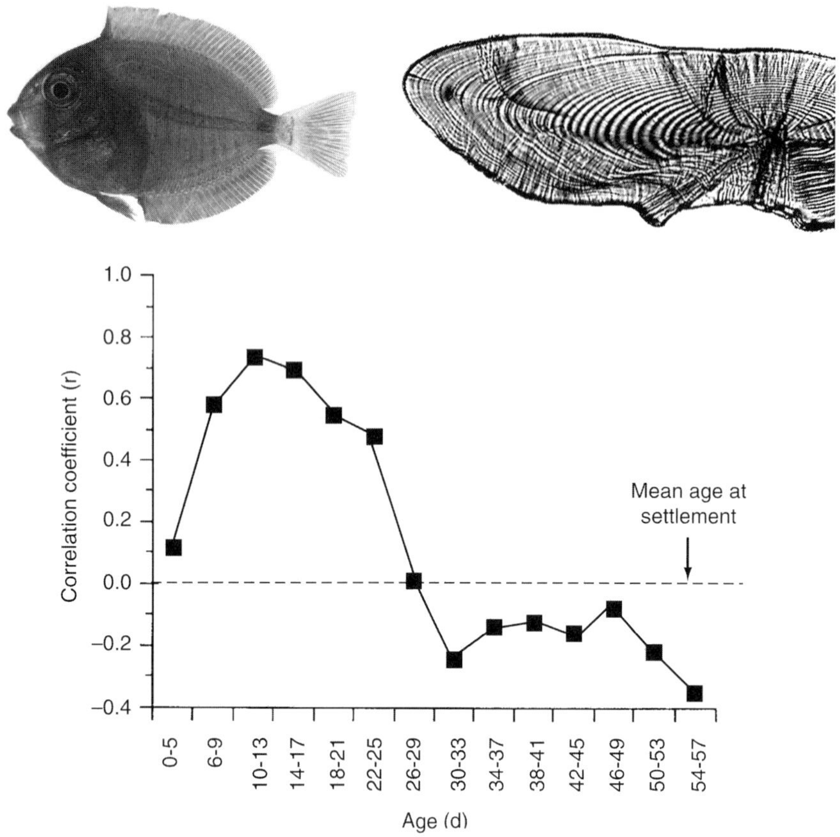

FIGURE 3 Importance of larval growth to survival. Coefficients (*r*) from correlation analyses between average growth rate during 4-day intervals of pelagic life and monthly settlement of a Caribbean surgeonfish, *Acanthurus chirugus,* to a San Blas (Caribbean) reef. Newly settled fish were collected over 44 consecutive lunar months (1984–1987) and their growth history was determined using the microstructure of their sagittal otoliths (adapted from Bergenius *et al.*, 2002).

At the time of writing, these are the only attempts to examine the influence of growth history on larval mortality rates. The use of otoliths to delve into the larval growth history of tropical fishes is a useful and underutilized approach (see Chapter 11, this volume). Decades of larval research on temperate fisheries species have led to identification of a number of key or critical periods in the life phase that may prove to be useful foci for tropical larval research. Critical periods have been identified at first feeding, during swimbladder inflation, and during metamorphosis at the end of the larval phase (Blaxter, 1988b).

Starvation has been suggested as a potentially important source of larval mortality in nutrient-poor tropical waters. To date, however, very little information is available on the influence of food availability on larval performance and mortality schedules. This is in part because of the difficulty of identifying tropical larvae to species. The only study to illustrate that starvation can be an important source of mortality in tropical larvae was conducted on three scombrid species in the Panama Bight (Margulies, 1993). Using a combination of field and laboratory studies, it was shown that susceptibility to starvation was stage specific and that preflexion larvae could suffer significant daily losses (62–63% day^{-1}).

Food availability influences growth rates and larval duration in reared larvae. Laboratory-based larval rearing experiments that have manipulated food availability into three or four levels have been conducted using a tropical goatfish [*Upeneus tragula* (McCormick and Molony, (1992)] and two damselfishes [*Amphiprion melanopus* (Green and McCormick, 1999) and *Pomacentrus amboinensis* (M. I. McCormick and B. Green, unpublished)]. For all species, greater access to food increased developmental rate (reduced pelagic larval duration) and increased the size of fishes at metamorphosis. Furthermore, when exposed to the same densities of prey, larvae from the high-food-density treatments fed at a higher rate than did larvae from the low-food-density treatments. This suggests that feeding history influences both how a larva will react when it encounters food and the larva's subsequent growth potential. It is presently unclear how feeding history influences metabolic efficiency, but this will undoubtedly be an important feature linking food intake to somatic growth and development.

Interestingly, McCormick and Molony (1992) found that if given the opportunity, midway through their pelagic stage, larvae that had been starved almost to the point of no return were able to compensate for poor growth. Larvae of the goatfish,

Upeneus tragula, starved for 3 days and then fed *ad libitum,* settled at the same size, weight, and lipid content as fish that had been continuously fed. Evidence suggests that these fishes are physiologically well suited to coping with a patchy food resource. Starved fish did, however, have a lower developmental rate and overall growth rate as compared to the fully fed fish, suggesting that the former may have sustained lower survival if mortality was growth or size-selective in the field.

Water temperature has a direct influence on larval growth and developmental rates in both temperate and tropical fishes (Benoît *et al.,* 2000). The only study of the effects of temperature on larval developmental rates in a tropical reef fish (McCormick and Molony, 1995) supported the large body of information on the influence of temperature on vital rates in temperate fishes (e.g., Chambers and Leggett, 1987; Francis, 1994; Brander, 1995; Rutherford and Houde, 1995; Hunt von Herbing *et al.,* 1996). Higher temperatures sped up development, leading to shorter PLDs. Moreover, for the tropical goatfish, water temperature did not affect the nutritional condition or muscle development at metamorphosis when food was unlimited.

Wind has been suggested to affect the feeding ability of larvae and therefore to have secondary effects on larval growth (MacKenzie *et al.,* 1994; Gallego *et al.,* 1996; Dower *et al.,* 1997). Dixon *et al.* (1999) found a dome-shaped relationship between wind speed (4 to 8 m sec^{-1}) and the numbers of competent larvae of damselfishes (*Pomacentrus amboinensis*) caught in light traps around Lizard Island. There was a 16-day lag between wind stress and impact on larvae. This relationship is believed to be a product of wind-generated turbulence [a cubic function of wind speed (Oakley and Elliott, 1982)]. Turbulence above or below a certain optimum is thought to reduce the effectiveness of prey capture and consequently results in reduced growth (Dower *et al.,* 1997).

2. Swimming Abilities

The late larvae of coral reef fishes are strong, "effective" swimmers, capable of sustained speeds higher than ambient currents and of swimming nonstop for tens of kilometers over tens of hours. Further, they are much faster than larvae of temperate species of similar size, particularly the well-studied Clupeiformes and Gadiformes. This surprising information on settlement-stage larvae is the result of laboratory swimming chamber (or flume) measurements of swimming endurance [51 species of 9 families (Stobutzki and Bellwood, 1994, 1997; Stobutzki, 1998)] and *in situ* speed measurements of free-swimming larvae [over 50 species of 15 families (Leis and Carson-Ewart, 1997)]. A comparison of the two approaches is found in Leis and Stobutzki (1999). Speeds mentioned herein are sustained speeds, unless otherwise stated.

Temperate fish larvae of 1–2 cm apparently swim at 1–5 BL sec^{-1} (Blaxter, 1986; Miller *et al.,* 1988; Meng, 1993; Williams *et al.,* 1996b), whereas similar-sized larvae of coral reef fishes swim at a mean speed of 13.7 BL sec^{-1} with some as fast as 34 BL sec^{-1} (Leis and Carson-Ewart, 1997, 1999, 2001). Put into perspective, a freestyle swimmer capable of 13.7 BL sec^{-1} would swim the 100-m race in 3.6 seconds; the Olympic record is 48 seconds. The actual speeds of reef fish larvae average 20.6 cm sec^{-1} with some as fast as 65 cm sec^{-1}. This is faster than the average ambient current speed in the study area, so the average reef fish larva near the end of its pelagic stage is indeed an "effective" swimmer. As an aside, we note a tendency for biologists to assume that average currents are stronger than they really are. For example, Barlow (1981) considered 5 km hr^{-1} (1.38 m sec^{-1}) to be a "typical" current speed. This is a typical speed for the Gulf Stream, and perhaps tidal jets, but average speeds in the vicinity of coral reefs where propagules are actually put into the water are rarely more than 10–20% of that value (e.g., Andrews and Pickard, 1990; Burrage *et al.,* 1994; Young *et al.,* 1994). This means that effective speeds are often lower than is widely assumed.

Just as remarkable as their swimming speeds, these larvae have great endurance, being able to swim an average of 40.7 km (some up to 140 km) in the lab unfed and without rest before exhaustion (Stobutzki and Bellwood, 1997). The mean time to exhaustion was 83.7 hours (maximum 288.5 hours). On a per-size basis, this is equivalent to a human swimming roughly 4000 km. The *in situ* swimming speeds of these larvae are strongly correlated with their lab endurance ($R^2 = 0.60$ to 0.77, depending on whether the comparison is at the family or species level) (Leis and Stobutzki, 1999), indicating that good swimming performance is independent of the means used to characterize it.

Swimming far and fast may simply increase the possibility of encountering a reef by chance alone, but pelagic reef fish larvae in blue water may be able to detect and orient to reefs (see Section IV,B,5). Orientation combined with effective swimming abilities would greatly increase the capacity of larvae to find a reef. In either case, this capacity would vary among species because swimming abilities vary among species.

In settlement-stage larvae both swimming speed and endurance are positively correlated with the size among taxa (Stobutzki and Bellwood, 1997; Leis and Carson-Ewart, 1997), although the relationships are not strong—explaining less than half the variation in swimming performance among taxa. However, it must be emphasized that swimming performance is known for only a small fraction of reef fish species or families (Leis and Stobutzki, 1999). For example, we have no information on larvae of eels, scarids, labrids, most blennioids, and most gobioids. These groups are among the most speciose on coral reefs, so this represents a large hole in our understanding of swimming performance (and other behaviors) of larvae, and in our attempts to generalize. Even within families for which some data are available, the numbers of species studied are usually small. Approximately 20 species of the lutjanid subfamily Lutjaninae (shallow-water tropical snappers) occur in the Great Barrier Reef region (Randall *et al.*, 1997), but only three species are frequently caught in light traps, and one of these is unsuitable for study (J. M. Leis, unpublished). Equivalent numbers for serranid groupers are 50 and 3. Aside from the Pomacentridae and Apogonidae, the situation is similar for other studied families in the Great Barrier Reef region. Finding a means to obtain and study underrepresented taxa would be very valuable.

Larvae begin their pelagic phase as plankton (weak, ineffective swimmers, with little or no control over trajectory, except, perhaps, vertically) but end it as nekton (strong, effective swimmers, with great control over trajectory). Here we have a key question: When during the pelagic phase do the larvae change from being plankton to being nekton? Nearly all information on swimming abilities currently available is on late- or settlement-stage larvae. Leis and Carson-Ewart (1997) reported that small (6–8 mm SL) pomacentrid (*Pomacentrus*?) larvae about midway through development, but with caudal fins formed, had *in situ* swimming speeds of 3–7 cm sec^{-1} (about 6 BL sec^{-1}). This is higher than speeds of temperate larvae of similar size, but not "effective" near Lizard Island, where average current speeds are 10–15 cm sec^{-1} (Frith *et al.*, 1986). Unfortunately, obtaining such small, wild larvae in good condition for *in situ* studies is very difficult, and it is likely that the answer to the key question about change in swimming abilities will be provided with laboratory-reared larvae. However, swimming abilities of lab-reared larvae and wild larvae must be compared, because it is by no means assured that they are equivalent (Duthie, 1987; Cobb *et al.*, 1989). Based on laboratory measurements of reared larvae, Fisher *et al.* (2000) provided the first report of swimming ability development in reef fish larvae. They found that swimming speed increased steadily with age, but that swimming endurance increased abruptly, and that both swimming performance and the timing of improvements in it varied among the three species they studied. The two pomacentrids had attained swimming ability sufficient to "significantly influence dispersal" only 1–2 weeks following hatching (about 70% of the way through the pelagic phase). The apogonid was a much poorer swimmer.

Swimming ability is more dependent on state of development than it is on size per se. Fisher *et al.* (2000) found this to be the case. This is also suggested from the fact that the mean *in situ* speed of settlement-stage *Chromis atripectoralis* and *Chromis viridis* (7–9 mm) was 19–24 cm sec^{-1}, whereas the swimming speed of similar-sized but less developed (not settlement stage) *Pomacentrus* (?) larvae was 3–7 cm sec^{-1} (Leis and Carson-Ewart, 1997). Many larvae are much larger than this at settlement, and it is tempting to predict that they will be "effective" swimmers for most of their pelagic stage, i.e., that portion of development following full development of their fins. Fins are essentially fully developed in most reef fish larvae at 5–10 mm (Leis and Carson-Ewart, 2000). Although growth rates during different portions of the pelagic phase are not well known, it appears the majority of the pelagic phase of most reef fish species occurs after the fins form. However, the data required to test this prediction are essentially lacking and the critical factor might be muscle development rather than fin development, per se. According to Matsuoka (1998) "true sustained swimming can occur only after the stratification of the red [muscle] fibres." Development of red muscle can vary among species. There are no studies of muscle development in reef fish larvae, but in temperate sparid and serranid larvae, pink muscle formed "towards the end of the larval life" (López-Albors *et al.*, 1998) and red muscle "at the end of larval development" (Ramírez-Zarzosa *et al.*, 1998). In a subtropical sparid, red muscle appeared at 7 mm (TL), and only three to four layers were present at 11.2 mm (Masuda and Tsukamoto, 1996). In contrast, in a subtropical carangid, red muscle appeared at 4.3 mm (TL), there were five to six layers at 8 mm, and 10 layers at 12.4 mm (Masuda and Tsukamoto, 1996, 1998). Because of accelerated muscle development, the carangid should be a better sustained swimmer than the sparid is at all sizes. Histological studies of muscle development in larval reef fishes might shed considerable light on the question of swimming ability. Because of their superior swimming performance, we expect that reef fish larvae would develop red muscle at a small size, and would rapidly add muscle layers with growth.

At least some species rapidly lose some of their swimming abilities on settlement (Stobutzki and Bellwood, 1994). Juveniles of three pomacentrid species a few days following settlement were able to maintain top speeds [critical speeds; see Leis and

Stobutzki (1999)] that were on average only 69% of those of settlement-stage larvae. This implies large changes in muscle physiology and probably morphology over a very short time.

Settlement stage reef fish larvae are fast enough that they should cruise swim in an inertial hydrodynamic environment (i.e., high Reynolds number, as do adults), whereas smaller larvae cruise swim in a viscous environment (low Reynolds number) (Webb and Weihs, 1986). However, the somewhat lower viscosity of seawater at tropical temperatures might lower the size at which larvae move from a low to high Reynolds number (Fuiman and Batty, 1997). This is supported by Fisher *et al.* (2000), who conclude that only the smallest reef fish larvae operate in a viscous environment. This means that reef fish larvae, because of their rapid development and high swimming speeds, pass into the more energy-efficient realm of inertial swimming when younger and at smaller sizes than do temperate larvae. This might result in greater growth efficiencies that would further accelerate development rates. It also might mean that the high-drag "chaetodontiform" shape [truncated, gibbose body; see Webb and Weihs (1986)] of many late-stage reef fish larvae does not carry a high cost in swimming efficiency.

Not only does swimming speed differ among taxa, and within taxa ontogenetically, but it can also can differ among different environments or with swimming directions. Reef fish larvae are flexible and adopt different swimming speeds in different situations. Larvae of the coral trout (Serranidae: *Plectropomus leopardus*) swim two to three times faster when leaving the vicinity of a reef than when approaching the reef (Leis and Carson-Ewart, 1999). They also swim at the higher speed in open water >1 km from a reef. Similar differences were evident in the swimming speeds of butterfly fishes and damselfishes (J. M. Leis and B. M. Carson-Ewart, unpublished data). The damselfish, *Chromis viridis*, swims 26% faster in an atoll lagoon than in the open ocean 1 km outside the atoll (Leis and Carson-Ewart, 2001). However, two other species show no difference in swimming speed in either of the two environments. This flexibility shows how much control these late-stage larvae have, and indicates that modeling of swimming will be complex, and dependent on where the larvae are and even in which direction they swim.

3. Schooling

Schooling in clupeoids, carangids, and some other temperate species begins at sizes smaller than those at which many reef fishes settle (Leis, 1986b; Blaxter, 1991; Kingsford and Tricklebank, 1991; Masuda and Tsukamoto, 1996, 1998), and the clumped distribution of newly settled reef fishes has led to speculation that schooling by reef fish larvae prior to settlement is likely (see Leis, 1991a). The observations of mass movement of acanthurid larvae along the reef bottom by Sancho *et al.* (1997) also imply schooling prior to settlement. Presettlement, pelagic schooling occurs in mullids, and in at least two species of pomacentrids, a microdesmid, and a caesionine lutjanid (McCormick and Milicich, 1993; Leis and Carson Ewart, 1998). One of the settlement-stage pomacentrids was schooling at 7–10 mm. All these species school to some extent as reef-associated juveniles, so schooling prior to settlement should not be surprising. In contrast, limited observations on chaetodontid larvae showed no tendency to school (J. M. Leis, unpublished). In addition, larvae of several temperate reef and coral reef fishes school pelagically (Kingsford and Tricklebank, 1991) or epibenthically prior to settlement (Breitburg, 1991; Lindeman *et al.*, 2001), so this could be a widespread behavior. In principle, because fins, eyes, and other sensory systems are present, schooling seems morphologically feasible for much of the pelagic period of reef fishes. An unanswered question is how the larvae in the pelagic environment find each other, to begin to school. Average densities of larvae are low in tropical waters (Leis, 1991a), so on average, huge volumes of water must be searched before a larva will encounter a conspecific. Perhaps the concentration of larvae at physical interfaces (e.g., the surface, thermoclines, the shoreline) or in hydrographic structures such as fronts or conversion zones plays a role in initiation of schooling.

Presettlement schooling by larvae, if widespread, has obvious implications for patterns of settlement and recruitment, for our interpretation of these, and for sampling design. Not the least of these is that settlement of a schooling larva would not be independent of the other individuals in the school (Breitburg *et al.*, 1995). Schooling in larvae might also have the antipredator advantages that are normally attributed to schooling in adults, and could act to enhance the orientation of larvae (many sets of sense organs may be better than one) and, therefore, the ability of larvae to locate reefs and settle onto them.

4. Vertical Distribution

The vertical distribution of fish larvae largely determines the physical and biological conditions in which they actually live, because gradients in these factors are frequently stronger in the vertical plane than in the horizontal. Currents differ in both direction and speed with depth, and because currents influence distributions of larvae, it is important to understand by which currents the larvae are actually affected.

Similar arguments apply to food concentrations, predator abundance, light levels, and a host of other factors. Leis (1991a) noted a paucity of information on the vertical distribution of reef fish larvae, particularly in the open ocean, but progress has been made in this area, especially in studies using plankton nets and midwater trawls (see Clarke, 1991; Boehlert *et al.*, 1992; Cha *et al.*, 1994; Cowen and Castro, 1994; Boehlert, 1996; Boehlert and Mundy, 1996; Cowen and Sponaugle, 1997). Such towed nets capture primarily younger larval stages (Choat *et al.* 1993).

Light traps have the potential to measure vertical distribution of older, larger larvae that are poorly sampled by towed nets, but only at night. A single study using this methodology has been published (Doherty and Carleton, 1997). Deep traps captured an assemblage of larvae distinct from that found in surface traps, and this was primarily due to differences in the relative abundance of clupeoids, monacanthids, and a pomacentrid (*Dascyllus*). Larvae of most species were more abundant in surface traps, although the picture was complicated by differences in ontogenetic stage between surface and deep catches. Large settlement-stage individuals of several species in at least five families were captured only in deep traps.

Studies of vertical distribution of larvae using *in situ* behavioral methods reveal both species-specific behaviors and behavioral flexibility (Leis *et al.*, 1996; Leis and Carson-Ewart, 1999, 2001). This approach complements towed-net studies by providing data on late-stage larvae, and light-trap studies by providing daytime data.

The most striking thing evident to date from the *in situ* work is the flexibility in vertical distribution behavior. In one study, larvae of four species of four different families selected shallower depths in an atoll lagoon than they did in the ocean 1 km offshore (Fig. 4) (Leis and Carson-Ewart, 2001). Similarly, larvae of a serranid and a chaetodontid swam much deeper 1 km off the windward side of an island than they did 1 km off the leeward side (Leis and Carson-Ewart, 1999, also unpublished). The serranid swam at similar depths in the immediate vicinity of both windward and leeward reefs, a depth that was similar to that chosen 1 km off the leeward side of the island. It is not obvious why swimming depth differs among locations, although one influence may be the light regime. In the areas where swimming depth was great, there was little upwelling light, either because of depth (the ocean) or because of dark bottom color (algal beds on the bottom off the island). In the areas where swimming depth was less, sandy bottoms resulted in upwelled, reflected light, even when the bottom could not be seen. In contrast, in the

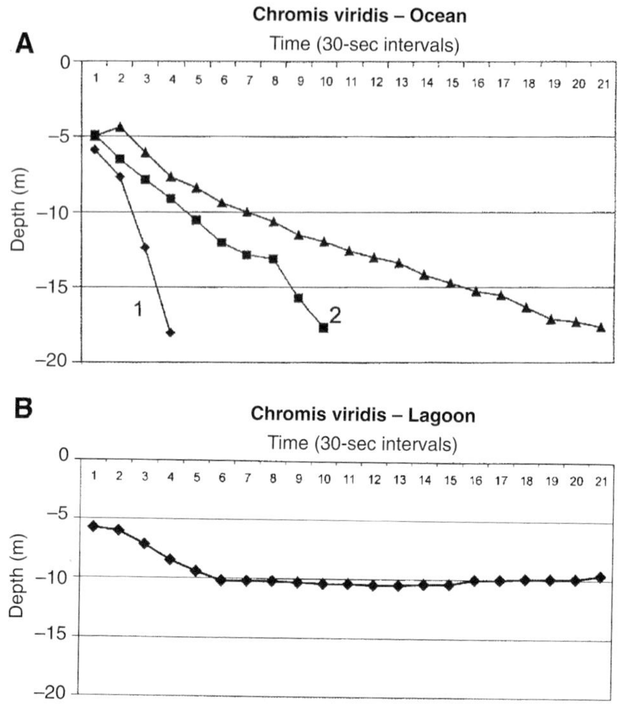

FIGURE 4 Comparison of *in situ* vertical swimming behavior of the pomacentrid, *Chromis viridis* at Rangiroa Atoll, Tuamotu Islands (from Leis and Carson-Ewart, 2001, copyright Springer-Verlag). All fish were released at a depth of 5 m and were followed for 10 minutes. (A) In the ocean, 1 km from the reef crest on the north side of the atoll, at a water depth of ~300 m. Lines are trajectories of individual larvae, and larvae were not followed below 18–20 m. Between lines marked 1 and 2 are 13 other lines not shown. (B) In the atoll lagoon, 1 km offshore, at a water depth of ~25 m. The line is the mean trajectory of 16 larvae. The standard error of each point on the line varies from 0.5 to 0.8 m.

lab, larvae of a temperate clupeid have a different vertical distribution behavior when placed in estuarine water as compared to ocean water (Forward *et al.*, 1996), implying a change in behavior due to a chemical sense. Further investigation is clearly warranted, and other factors such as directional stimuli emitted by the reefs (see below), or variations in turbulence due to wave action, should not be ignored. Turbulence affects the vertical swimming behavior of temperate crab larvae (for example, Welch *et al.*, 1999).

Variations in vertical distribution behavior among species are evident (Leis *et al.*, 1996), as studies using towed nets would lead one to expect. Some species at least in some situations had highly predictable vertical distributions (Fig. 4), whereas other species had large variations among individuals. What is perhaps less expected was the variation in depth range (amplitude) that individual larvae displayed. Amplitudes of >9 m for individuals were not uncommon, as were numerous descents and ascents of several meters during the 10-minute observation period (Leis *et al.,* 1996; J. M. Leis, unpublished). This highlights one of the shortcomings of traditional vertical distribution studies using nets or even light traps: they measure population vertical distributions, without any means of identifying what individual larvae are doing (Pearre, 1979). For example, if 60% of larvae are captured at a particular depth, traditional studies provide no information as to whether the same 60% of individuals remain at that depth, or if there is movement of individuals among depths. The *in situ* studies suggest the latter may be the case, and that vertical movement of individuals may be large and frequent. However, it is also possible that such vertical excursions are limited to older larvae, and not found in the smaller, younger larvae that dominate net samples (but not light-trap samples). Such excursions could be a means of determining if the bottom is suitable for settlement, a means of locating vertically stratified concentrations of food, or even a means of obtaining orientation cues. The last could involve ascent to assess sun angle, descent to evaluate movement over the bottom, or both to locate vertically stratified water masses that could differ in a variety of physical, chemical, or biological cues or in flow characteristics.

Larvae may also occur near the bottom, in or near the benthic boundary layer, and if so, this would greatly limit passive dispersal. Epibenthic larval distributions have been identified by *in situ* observations in temperate systems (Breitburg, 1991), but there is little information on them in coral reef systems. Few larvae actually swam near the interreefal bottom for anything but very short periods (<1 minute) before attempting to settle near Lizard Island (Leis and Carson-Ewart, 1998). No larvae of coral reef fishes were identified as being in high abundance near the bottom in net tows taken in the same area (Leis *et al.,* 1989). Predation rates on larvae are high in the epibenthos near Lizard Island reefs, with lizard fishes particularly active (Leis and Carson-Ewart, 1998, also unpublished). This could explain why the scant available information indicates that few coral reef fish larvae have epibenthic distributions.

The visual capabilities of larvae can influence vertical distributions. In the laboratory, visual sensitivity in larvae of three apogonid and four pomacentrid species increased with age, and the apogonid larvae had greater sensitivity than did the pomacentrids (Job and Bellwood, 2000). This means that younger larvae should be restricted to shallower depths than older larvae, and that pomacentrid larvae should be restricted to shallower depths than apogonid larvae. This is broadly in accord with the vertical distributions of these taxa found in the field. Pomacentrid larvae are typically most abundant at very shallow depths (references above) that during the day have considerably higher light levels than the minimum required for feeding as identified by Job and Bellwood (2000). The results of such studies should not be expected to predict the details of vertical distributions, but they should specify depth floors below which larvae would not be anticipated because they do not have enough light to feed.

5. Orientation in the Pelagic Environment

Orientation is required if a pelagic reef fish larva is to find a reef by other than chance, and orientation requires not only cues and the sensory means to detect reefs, but also the ability to determine the direction from which the cues emanate. The existing information on the swimming behavior of reef fish larvae in the pelagic environment indicates that larvae do orientate rather than merely cruise about randomly. What cues they can detect and use to achieve this is less clear.

Settlement-stage larvae in open water tend to have roughly linear horizontal trajectories (Fig. 5), or at least trajectories that are significantly different from random (Leis *et al.,* 1996; Leis and Carson-Ewart, 1999, 2001). For example, 89% of over 100 *Chromis atripectoralis* larvae released in open water near Lizard Island had nonrandom trajectories (J. M. Leis, unpublished). In the vicinity of reefs, trajectories are more complex. Larvae that swim away from reefs when released adjacent to them most commonly had linear trajectories. More complex trajectories that are clearly influenced by the structure of the reef and the distribution of reef habitats and resident fishes are noted in larvae swimming toward or over reefs. Some individuals swim away from the reef, out of sight of it, and then return (Fig. 5) (J. M. Leis, unpublished). This behavior implies either a good memory for reef location, or the ability to detect a reef remotely and return to it.

Thus far, studies of trajectories of larvae in open water have not clearly shown that larvae can detect remote reefs. At Lizard Island, Leis *et al.* (1996) analyzed the open water trajectories of a group of larvae of several species, each released individually, and showed that individual trajectories of most were nonrandom, and that on average the trajectories differed among three locations on different sides of the island, and were offshore at each location. This implied that the larvae could sense the island from >1 km offshore. At an oceanic atoll, nearly all trajectories of four species were nonrandom and usually linear regardless of location (Leis and Carson-Ewart, 2001). However, a difference

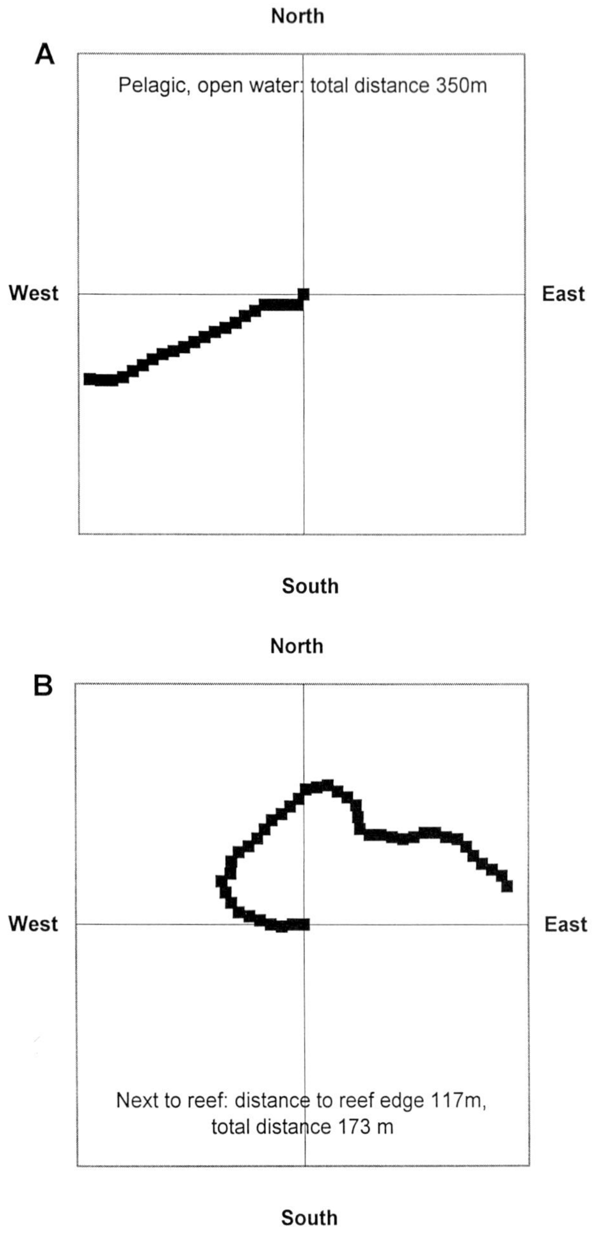

FIGURE 5 *In situ* horizontal swimming trajectories of reef fish larvae. Constructed from bearings taken every 30 seconds and plotted assuming constant speed throughout. Trajectories are relative to a possibly moving water column, not relative to the bottom. A and B not to same scale. Fish released at the origin. (A) A linear trajectory of a larval acanthurid, *Acanthurus triostegus* (~25 mm), in the open, pelagic environment in the lagoon of Rangiroa Atoll, Tuamotu Islands. Total distance swum was 350 m (average speed, 58 cm sec^{-1}). (B) A complex trajectory of a larval pomacentrid, *Chromis atripectoralis* (~9 mm), adjacent to and over a reef at Lizard Island. Slow current was running to the north. Larva released over sand bottom about 5 m to the west of the reef edge, which ran approximately along the north–south axis. Larva initially swam offshore, possibly in response to a school of fusiliers, out of sight of the reef. It then swam clockwise back to the reef, crossing the reef edge after 9 minutes (117 m swum; average speed, 22 cm sec^{-1}). From there, it swam over the reef for another 12 minutes, closely inspecting three coral heads, and was chased away from settling into a school of conspecifics at the first by a resident *Dascyllus auranus* (56 m swum; average speed, 8 cm sec^{-1}). Observations ceased at 21 minutes, with the fish still swimming over the reef. Note that due to the difference in speed before and after crossing the reef, these two portions of the trajectory are not to the same scale.

in swimming direction between the atoll lagoon and the ocean north of the atoll was not detected, nor was the distribution of average swimming directions significantly different from random at either location for any species. Other investigations specifically aimed at the orientation of larvae in open, pelagic waters are underway at Lizard Island (J. M. Leis, unpublished).

Somewhat more equivocal support for long-distance orientation by larvae comes from studies of distribution utilizing light traps. The distribution patterns of late larvae of coral reef fishes detected by light traps in the vicinity of a coral reef led Doherty and Carleton (1997) to suggest that "habitat discrimination begins while reef fish are still in the plankton." They speculated that sound was the most likely orientation cue. In contrast, Doherty *et al.* (1996) speculated that warm surface plumes emanating from a lagoon of another reef provided the cues that enabled larvae of a pomacentrid to avoid that lagoon (as shown by their light-trap catches).

Closer to shore, Stobutzki and Bellwood (1998) moored "behavioral cages" >50 m off the Lizard Island reefs at night, and showed, in both pomacentrids and apogonids, that most (64–67%) larvae swam toward the reef, regardless of the location of the cage. Based on the position of the cages, Stobutzki and Bellwood (1998) suggested that the larvae were attracted by reef sounds. Other attempts to combine larval swimming and orientation behavior with hydrography are promising, but show how far there is to go. Porch's (1998) numerical study of the southeast Florida coast showed that orientated swimming by larvae at even low speeds (1–10 cm sec^{-1}) resulted in 3- to 10-fold increases in reef fish settlement. Armsworth (2000) modeled the swimming responses of weak- and strong-swimming larvae (ineffective and effective swimmers, respectively) responding to current-dependent (e.g., odor) and current-independent (e.g., sound) cues emanating from a reef. He concluded that larval swimming and sensory abilities are far more important than the fine-scale detail of circulation in determining enhancement of settlement. Strong swimmers are able to utilize both current-dependent and current-independent cues for orientation, although the latter result in higher levels of settlement enhancement. Finally, Armsworth (2000) concluded that weak swimmers, in contrast, are able to employ only current-independent cues to navigate to reefs, and the enhancement of settlement due to these cues is less than that in strong swimmers. These results will be useful in directing future research effort.

Wolanski *et al.* (1997, 1999) and Wolanski and Sarenski (1997) combined numerical models of circulation with various scenarios of larval behavior, and concluded that only directed swimming by settlement-stage larvae beginning 2–3 km upstream of a reef could duplicate the distribution of larvae they observed with light traps (they assumed response to the sound of waves breaking on the reef). If the larvae were assumed to be passive, the model-predicted distributions were very different from observed distributions. When Wolanski and co-workers assumed larger larvae swam at 20 cm sec^{-1} whereas smaller ones swam at half that speed, the model predicted different distributions around the reef for faster larvae than for slower larvae. They then attributed the observed differences found between the distribution of two large (~14 mm) *Pomacentrus* species and two smaller (~10 mm) species (*Chromis atripectoralis* and *Chrysiptera rollandi*) to differences in swimming abilities. This is an interesting result and an admirable attempt to incorporate behavior into physical models. However, a difference in body size of about 5 mm would be expected to result in an increase in swimming speed of only 4 cm sec^{-1}, not the 10 cm sec^{-1} assumed, and, in fact, *Pomacentrus* spp. swim at 11–22 cm sec^{-1}, no faster than the other two, smaller, species (12 and 24 cm sec^{-1} for *C. rollandi* and *C. atripectoralis,* respectively) (Leis and Carson-Ewart, 1997). Clearly, more than just numerical models of currents combined with predictions of swimming speed based on body size are required to explain the differences observed between the distributions. However, more significant for such an interdisciplinary approach, other numerical models of circulation around the same reef led to different results. The other models predicted that passive larvae would have distributions similar to those observed by Wolanski and co-workers in the field (Black and Moran, 1991; Black, 1988, 1998), therefore supplying no support for long-distance orientation. Biologists cannot be expected to choose among such extremely complex numerical models of circulation, but, obviously, the veracity of the numerical model predictions is critical to the outcome. Where the modelers disagree (see also Wolanski, 1993; Black, 1995), the biologists are left in the lurch. This makes all the more significant Armsworth's (2000) conclusion that details of circulation are less important than larval swimming and sensory abilities.

Many possible cues associated with reefs could provide clues for orientation. These include smells and sounds emanating from reefs; differences in wind- or wave-induced turbulence; gradients in abundance of fish, plankton, or reef detritus; and differences in temperatures of lagoonal or reef flat water flowing from a reef. In some cases, a magnetic compass or sun compass could aid in increasing chances of larvae encountering a reef (e.g., a larva in the Coral Sea would increase its chances of encountering one of the reefs on the Great Barrier Reef by swimming to the west), but it seems unlikely either could assist orientation toward a particular reef. One possible exception is that a magnetic sense could allow a larva to detect an oceanic basalt island (or, some volcanic islands on continental plates) on which reefs were growing, because basalt islands have a magnetic anomaly (Smith, 1981; F. L. Sutherland, personal communication). Although fish can sense via the lateral line that they are moving through water when swimming, unless they have an external reference, such as a view of the bottom, they will be unable to determine that they are being moved by and with the water, as when being carried along with a current (Leis *et al.*, 1996; Montgomery *et al.*, 1997). Therefore, currents are potentially detectable using vision near the bottom or near a reef, but it is unlikely that currents or movement by them will be detectable in "blue water," and thus they are unlikely to be an aid to orientation. Some of these possibilities seem inherently more general, and therefore more likely in a evolutionary sense to have been utilized (see Chapter 6, this volume). For example, sound is virtually current independent, travels in all directions from the source, and propagates over long distances, so it could be a very general cue (Popper and Carlson, 1998). In contrast, smells are current dependent, must travel with water movement, and would be of little use "upcurrent" of any reef (Armsworth, 2000). However, where currents are weak, each reef might be surrounded by a diffusion-maintained "halo" of smell that could provide cues that a reef was near, and a similar halo could be established by current reversals such as those caused by tides. Magnetic anomalies are current independent, and more likely to be associated with reefs on oceanic islands than with continental-shelf reefs. Most reef fish species have wide distributions and live on a variety of island and shelf habitats, and in a variety of current regimes that differ in their predicability over many scales. In addition, changes in sea level over time can result in radical changes in reef systems and associated currents. Therefore, it seems likely that any cues to which reef fish larvae have become adapted to use in finding reefs would be general ones, useful over much or all of the range of the species. Such predictions based on theoretical arguments should be treated with caution until they can be tested with reef fish larvae. Researchers have been misled by similar theory-based predictions in the past.

Another important consideration is the scale of the orientation involved. Vision is critical to settlement and close approach to the reef (see Section IV) (Leis and Carson-Ewart, 1998, 1999, also unpublished), but this is at scales of meters to perhaps tens of meters. Even at night, some light is normally available, although the minimum levels necessary for larvae to see, and to what distance, are generally unknown, except for feeding (Job *et al.*, 2000). It is unlikely vision plays a role in orientation toward reefs at a scale of hundreds to thousands of meters unless orientation using sun angle or polarized light is involved (see below). Similarly, olfactory cues are important for some species over the same scales as vision. Unlike vision, olfaction has the potential to operate over larger scales, as with salmon (Helfman *et al.*, 1997). If odors are transported by currents and structured by fronts between water masses, olfaction operating at a small scale could result in orientation over larger scales (M.J. Kingsford, personal communication). Similar arguments apply to temperature differences (Doherty *et al.*, 1996). Reefs are noisy places (Cato, 1978, 1992) and sound has the potential to provide orientation cues over a wide range of scales (Popper and Carlson, 1998). The lateral line is sensitive to water movement, but is capable of detecting this over only small distances, on the order of 1–3 body lengths (Popper and Carlson, 1998), unless augmented by connection to other systems (e.g., the "laterophysic connection" in some butterfly fishes, Webb, 1998). A magnetic sense could potentially operate over a variety of scales, from very large (oceanic), as shown in tunas and salmon (Walker, 1984; Walker *et al.*, 1997), to relatively small (local), as shown in hammerhead sharks (Klimley, 1993). It is likely that different cues are used at different scales even by a single individual: a possible scenario is use of sound to locate the reef, vision and the lateral line to avoid predators near the reef, smell to locate the settlement habitat, and vision to locate the settlement site in the habitat (see Chapter 6, this volume, for an examination of the available evidence on the capability of larvae to detect and utilize such sensory input). It is fair to say that aside from olfaction, hearing, and vision, none of these has been shown to be used by reef fish larvae for orientation, and even with these, the use has been at either relatively small or unknown scales. However, the auditory, visual, and olfactory systems of fish larvae are apparently functional by the time the fins form, so they should be available early in the pelagic phase of most reef fishes. Here, we explore some of the evidence on smell and hearing and the implications for orientation.

Field and laboratory experiments show that some pomacentrids utilize smell to locate settlement sites over relatively small distances. The only species so far shown to utilize such cues are three genera of pomacentrids (*Amphiprion*, *Dascyllus*, and *Pomacentrus*), two of which are either exclusively symbiotic, or have species that are symbiotic, with anemones (Sweatman, 1985b, 1988; Elliot *et al.*, 1995; Arvedlund and Nielson, 1996; Danilowicz, 1996; Holbrook and Schmitt, 1997, 1999). Are larvae of taxa that are symbiotic with anemones particularly good at using olfactory cues to select settlement habitat? Two points about these experiments are important: first, they show that olfaction can operate over scales of up to a few tens of meters; second, they were done over the reef habitat, and did not show that olfaction can be used in the pelagic environment in the find reefs. Olfaction is clearly important in the location of specialized habitats such as anemones or corals, and in the location of conspecifics or in the avoidance of heterospecific congeners, all over small scales within reef habitats. Use of olfaction for orientation by other than a few species of pomacentrids, or over larger scales, or in the pelagic environment, is a real possibility given the results to date, but this has not yet been demonstrated. The fact that anemone fish still in the egg stage (or just hatched) can be imprinted with the smell of the host anemone (Arvedlund *et al.*, 1999) opens a number of possibilities. These include the imprinting of larvae that hatch from nonpelagic eggs with the olfactory cues of either the natal reef, or a particular type of reef (outer shelf, for example). This could aid in retention near, or return to, a natal reef, or in location of a particular reef habitat for settlement, respectively.

The case for sound being used as a cue is, except for one study, mostly inferential, theoretical, or inconclusive. Several investigators have interpreted their field results to be best explained by sound orientation, but tests of this hypothesis are few. A field experiment (Leis *et al.*, in press) to test the hypothesis that reef sounds enhance recruitment to artificial reefs >1 km from natural reefs was compromised by general recruitment failure of all but apogonids (mostly *Rhabdamia* species), and these were recruited in an extremely patchy manner. More reef fishes did recruit to the "sound-enhanced" reefs as compared to control reefs, but the difference, at $p = 0.11$, was inconclusive due to low power, so the null hypothesis of no enhancement could not be rejected. Little is known of the abilities of reef fish larvae to detect or localize sound (see Chapter 6, this volume), but the larvae of two subtropical perciform fishes have a "startle response" to sound at a size of 5–8 mm, before all fins are formed (Ishioka *et al.*, 1987; Fuiman *et al.*, 1999). So, in at least some fish larvae, hearing is operational at a small size, but that is not equivalent to being able to use hearing for orientation. In an important study, Tolimieri *et al.* (2000) showed that larvae of tripterygiids (a family of blennioid fishes) of 20 mm mean length used reef sounds for orientation. Far more tripterygiid larvae were caught in light traps adjacent to areas from which recordings of natural sounds from rocky reefs were broadcast than were caught in light traps where no sound had been broadcast. Clupeid larvae, however, showed no such difference. This constitutes the first clear demonstration that reef fish larvae, albeit temperate ones, can use sound for orientation; however, there was no indication over what scale the orientation was taking place. Innovative means of studying sound as a cue are required, but working with sound as a cue for orientation in the laboratory is extremely difficult (Popper and Carlson, 1998).

6. Predation on Larvae

Due to the enormous disparity between numbers of eggs spawned and newly recruited juveniles, predation on the pelagic stages of reef fishes is assumed to be intense. Yet, there have been few data to support this quite reasonable assumption. Johannes (1978) speculated that predation was more intense over the reef than in open water, and suggested this was a major reason why reef fish larvae were found away from the reef in open water. Leis and Carson-Ewart (1998) observed a higher rate of predation on settlement-stage larvae in the vicinity of and over coral reefs than in open water. The absolute rates of predation they observed were not meaningful due to the bias introduced by the presence of divers, but the data were useful in a relative sense to make comparisons between regions. This relative difference in predation was the first empirical support for Johannes' hypothesis.

In open water, settlement-stage larvae of several families react to schools of potential predatory fishes at distances of 3–6 m, and similar reaction distances were seen in near-reef waters, generally with more solitary predators (Leis and Carson-Ewart, 1998, 2001). The reactions included changes in swimming direction or speed, changes in swimming depth, and retreat to the shelter of the observer diver. Leis and Carson-Ewart (1998, 2001) concluded that larvae were alert to the threat of predation and could detect larger fishes at a distance of several meters. This distance is great enough that it was unlikely the larvae would have been seen, and this allowed them to react in ways that reduced their chances of coming close enough to the predators to be seen or attacked. In short, settlement-stage larvae in open waters behaved in a way that was effective in avoiding predators. The larvae reacted to the predatory fishes several meters away, so it was assumed the larvae detected these fishes visually, although detection by sound or mechanosensory means could not be ruled out. Whether the larvae reacted to visual or other stimuli, they did so at distances much greater than those reported in laboratory studies of subtropical fish larvae of similar size (e.g., Higgs and Fuiman, 1998b; Poling and Fuiman, 1999). In fact, the reaction distances observed *in situ* were much greater than the dimensions of the containers used in those laboratory studies. This type of close reaction reported in the lab is thought to indicate that larvae avoid predators by dodging at the last moment, when the predator is too close to react effectively (Blaxter, 1991). This may well be an effective last-ditch defense, but the *in situ* observations combined with the lab results imply that larvae may have a layered defence. One set of behaviors operates at distance to take advantage of the small size and low visibility of the larva to decrease the likelihood of being detected by the predator, and the other set of behaviors operates at small separation distances if the first set fails and the larva is detected. This implies considerable behavioral sophistication by the larvae, but no more than many adult fishes possess. Blaxter (1991) suggested a similar layered defense in herring larvae, with the key difference in herring being that passive, rather than active, behavior in the outer layer reduced detection. Further, Blaxter suggested that the outer, passive layer of defense was predominant in "small" larvae, whereas the active, inner defense was predominant in older larvae. In fact, in the field, a temperate mugilid, when small (< 9 mm), reacted to disturbance by aggregation, and, when larger, by scattering at high speed (Kingsford and Trickelbank, 1991). Neither fits comfortably in the layered defense model, which indicates that species-specific behavior may play a strong role.

Over the reef, Sancho *et al.* (1997) reported predation by eight fish species on settlement-stage acanthurids swimming along the bottom of the reef or on just-settled individuals in caves and crevices. Predation on pomacentrid larvae approaching the reef was primarily by synodontids and labrids and predation rate differed among species (Leis and Carson-Ewart, 1998). This difference was attributed to the swimming behavior of the larvae: species that swam more or less directly toward a settlement site on the seaward edge of the reef suffered more predation than did species that swam high above the reef before descending to examine settlement sites, or than species that turned and swam offshore. Predation on settlement-stage serranid larvae as they approached the reef was also primarily by synodontids and labrids and differed in intensity between locations (Leis and Carson-Ewart, 1999).

Predation on and near the reef is generally thought to be highest during crepuscular periods, intermediate during the day, and lowest at night (Hobson, 1972, 1991; Johannes, 1978). It is argued that this diel predation pattern shapes the diel patterns of many other activities, including settlement of larvae (e.g., Hobson, 1991; Victor, 1991). However, this assumption has been tested only once, and the results were not in accord with the predicted diel structuring of predation. Predation rates on juvenile grunts (Haemulidae) that were 3–4 cm TL, the settlement size of some other species, were highest at night and dusk and lowest during the day (Danilowicz and Sale, 1999). As pointed out by Danilowicz and Sale, this brings into question the argument that temporal patterns in larval settlement on the reef are driven by temporal patterns in predation, but the reality is that diel patterns of settlement are not well known (see Section V,C).

V. Back to the Reef: The End of the Pelagic Stage

A. Connectivity: Dispersal or Retention?

The dominant view among biologists is that reef fish populations are demographically open, meaning that settlement and recruitment of propagules are overwhelmingly not back into the natal population (Sale, 1991b; Caley *et al.*, 1996), but this is rarely tested. Some have tried to test this by indirect methods, such as attempting to match spawning cycles to recruitment cycles (e.g., Danilowicz, 1997), thereby introducing assumptions that are difficult to evaluate. Indirect biochemical genetic methods have also been used. Where these show differentiation, and they often do (e.g., Planes, 1993; Planes *et al.*, 1998b), important demographic boundaries among populations are identified. Where these genetic methods show no population differentiation (panmixia), there is great relevance at evolutionary or biogeographic scales, but little relevance at the scale of most ecological or demographic questions (Shulman, 1998). Genetic panmixia does not indicate that demographically significant connectivity exists. This is because the amount of larval connection between populations required to eliminate genetic differences detectable by most methods is so low [an average of a few individuals per generation (Shulman, 1998)] as to be demographically meaningless. Neither a fishery nor a marine reserve is likely to be maintained by such a low level of recruitment input. In other words, management must be done at ecological and demographic scales.

Models of larval connectivity typically make the "simplifying assumption" that larvae are passive (Roberts, 1997b), and utilize hydrographic models, usually of far-field currents, to estimate where propagules will end up at the end of "average" pelagic larval durations. At ecological and demographic scales, it is appropriate to use average PLDs, whereas at evolutionary or biogeographic scales, the appropriate measure is maximum PLD (Leis, 1991a). Unfortunately, such models ignore the reality that reef fishes do not normally spawn into far-field currents, and that reefs are more or less "buffered" from contact with far-field currents by hydrographic complexity at micro- to mesoscales. For example, Black (1988, 1995, 1998; Black and Moran, 1991) showed that retention of even passive larvae in the immediate vicinity of reefs can be remarkably high for periods of several days, and that this varies with depth. At a larger scale, the reefs of the Florida Keys are "buffered" from direct contact with the Florida Current/Gulf Stream system by a set of gyres with periods of several days to several months (Lee and Williams, 1999). Because of this, even passive larvae may take several days to weeks to come into contact with far-field currents. It is in the first few days of the pelagic period that the "simplifying assumption" of passive drift is most likely to apply due to the ontogeny of swimming abilities, and least likely to involve far-field currents. As shown above, the assumption of passive behavior by larvae is untenable. Therefore, the simplifying assumption of passive drift in far-field currents looks less defendable with time (see above).

Cowen *et al.* (2000) modeled dispersal of reef fish larvae from Barbados and concluded that long-distance dispersal is unlikely to be demographically relevant. They showed that at the scale considered (hundreds of kilometers), the twin effects of mortality and diffusion act to ensure only very small numbers of passively dispersed larvae ever reach downstream settlement sites. Further, the species-specific PLD limits the opportunity for a given species to settle at any reef intersected by a larval trajectory. Larvae must first be "competent" before they can settle on any reef they encounter. This study shows that the numbers of larvae reaching a distant reef are likely to be too small to be able to support a fishery or a reserve.

In some special cases, particularly atoll lagoons, it can be shown directly that the entire life cycle is completed in the vicinity of the natal reef, thus indicating that the population is closed, at least at the scale of the lagoon (Leis, 1994; Leis *et al.*, 1998; J. M. Leis, unpublished). However, in such cases, although the numbers of larvae are large, the number of "completer" species is small and is confined to a few families (often only one or two species in each), particularly Apogonidae, Blenniidae, Callionymidae, Carangidae, Clupeidae, Gobiidae, Hemiramphidae, Microdesmidae, Mullidae, Pomacentridae, and Schindleriidae. Similarly, in more open waters, larvae of some taxa may be retained over or near natal reefs, and although longshore dispersal cannot be entirely ruled out in such cases, the populations are probably relatively closed (Leis, 1986a, 1993; Brogan, 1994b). On the Great Barrier Reef in such conditions, retained taxa (Leis, 1986a) were a subset of the "lagoon completers," and in the Gulf of California they were bythitids, blennioids (four families), and gobiesocids (Brogan, 1994b). Most of these "completer" taxa spawn nonpelagic eggs (only the carangids, callionymids, and mullids do not). Other than these few completer taxa, lagoon species either spawn elsewhere or seem incapable of completing their pelagic larval phase in atoll lagoons if they do spawn there, and this is particularly true of species that spawn pelagic eggs. Although they hatch from pelagic eggs, there are indications that larvae of some haemulids and reef-associated sciaenids may never leave the vicinity of the natal reef, possibly by staying near the bottom (Lindeman *et al.*, 2001). If so, this would enhance the likelihood of self-recruitment and more closed populations for these species, although self-recruitment remains to be shown.

Some flexibility in life history pattern and a resultant geographical variation in population openness is evident in the extreme case of an atoll lagoon cut off from the ocean (Taiaro Atoll), where some species that do not normally either spawn or complete their pelagic phase in lagoons did so (Galzin *et al.*, 1998). At Taiaro, genetic evidence indicated that an acanthurid and a chaetodontid were completing their life cycles in the lagoon (Planes *et al.*, 1998c). The concentration of pelagic fish eggs in the lagoon was at least as high as that in the nearby ocean, indicating substantial spawning in the lagoon (Leis *et al.*, 1998), in contrast to other atolls (J. M. Leis, unpublished). Finally, the presence of a range of sizes (although not the full range) of scarid and labrid larvae in the lagoon indicated that some species of these two families may complete their pelagic phase in Taiaro Lagoon, again, in contrast to other lagoons. In contrast, some taxa that complete their life cycles in other lagoons were absent as both adults and larvae at Taiaro: most notable were a blenniid, a carangid, a microdesmid, and a schindleriid. Such flexibility indicates that an oceanic pelagic stage is facultative in a number of reef fishes, and that these species can achieve life cycle closure in lagoons, at least in extraordinary situations. If this is possible, then it might occur in other, less extreme, situations. Perhaps the pelagic stages of these species survive better in the ocean than in the lagoon, and this is a factor driving behavioral adaptations that normally place propagules outside the lagoon. However, survival, albeit at a lower level, may still be possible in the lagoon.

A more general direct test is to measure the proportion of self-recruitment that takes place at given reefs. Open populations, by definition, have low levels of self-recruitment, so significant levels of self recruitment would indicate a population that is not totally open demographically. One such test is simply the presence of endemic species at small, isolated islands. That there are a number of these, and that their life history characteristics do not appear to be out of the ordinary compared to other reef fishes, are proof of absolute self-recruitment of these species on the scale of the island, implying that the populations of other reef fish species at these islands may be similar (Robertson, 2001).

Two research programs using different otolith-based methods have directly demonstrated remarkably high levels of self-recruitment in two reef fish species. At Lizard Island, 15–65% of the settlement-stage larvae of *Pomacentrus amboinensis* approaching the island at the end of their pelagic sojourn originated at that same island (Jones and Milicich, 1997; Jones *et al.*, 1999). At St. Croix in the Virgin Islands, more that 50% of the recruits of *Thalassoma bifasciatum* during time of peak recruitment had not spent time outside of coastal waters, and therefore originated at St. Croix (Swearer *et al.*, 1999). The scales here are very different: the Lizard Island reef complex is about 7 km long, but many other reefs are present in a 20-km radius, some as close as 2 km away; St. Croix is about 35 km long, and over 50 km from the nearest island reefs. Given the surprisingly high levels of self-recruitment in two such different reef systems at two different spatial scales and involving species with different reproductive characteristics (demersal eggs in the pomacentrid and pelagic eggs in the labrid) and with mean PLDs that differ by 25–30 days, we may anticipate that these results are fairly general.

Indications from several different approaches are that reef fish populations are probably more toward the closed than the open end of the demographic spectrum, although this almost certainly varies among species and locations, and probably temporally. This has important implications not only for our view of reef fish populations, but also for our attempts to manage them. The design of marine protected areas (MPAs) and the push for their use a conservation and management tools are both predicated on the open population paradigm. Although, ironically, if populations are more closed than open, MPAs might be more favorably regarded by local stakeholders (K. C. Lindeman, personal communication). If populations on ecological time scales are more closed than open, then populations are probably more spatially subdivided than previously suspected. This means that different designs of MPAs are required (i.e., for a given area of MPAs, more and smaller reserves may be optimal, although the optimal size remains to be determined) and that the "dual function" of MPAs—both to maintain biodiversity and breeding stocks within MPAs and to replenish exploited populations outside them—may be difficult to attain over anything other than small to medium scales. Sladek Nowlis and Roberts (1999) confront this dilemma when they conclude on one hand that "without larval transport [outside MPA borders], the potential for fisheries benefits from reserves is more limited" (the open population view), while on the other, that without self-recruitment, MPA "benefits to the management area are likely to be much more limited" (the closed population view). The success of MPAs in achieving either of the proposed benefits, let alone both, is critically dependent on the location of reef-fish populations along the open-closed continuum, the scales over which ecologically-relevant dispersal or retention operate, and the spatial design of the MPA. We agree with Sladek Nowlis and Roberts (1999) that "we need to understand the movement dynamics of larvae" for effective MPA design.

B. Timing of Settlement

Near the end of the larval phase fish reach a developmental stage when they are ready to leave the water column and join the demersal reef population, if environmental and biotic cues indicate that appropriate habitat is available. This developmental stage is known as "competence." Thus reef fish have a larval phase made up of two parts: (1) a precompetent phase, during which rapid development and growth occurs, and (2) a competent phase, characterized by reduced growth and body maintenance. The relative contribution of each phase to the overall PLD is presently unknown because the identification of morphological and physiological characteristics that signify competence has proved elusive (McCormick, 1999b).

Some widely distributed families, such as the wrasse, surgeonfish, and flatfish, can apparently extend the period of competence, thereby delaying metamorphosis and settlement (Randall, 1961; Victor, 1986c; Evans and Fernald, 1990; Cowen, 1991; Jenkins and May, 1994; Cowen and Sponaugle, 1997; Sponaugle and Cowen, 1997; McCormick, 1999b). In doing so, they increase their chances of finding suitable settlement habitat. Evidence for the delay of settlement comes from two sources, one circumstantial, the other experimental. A number of studies have found reduced otolith growth in the late larval stages of individual fishes with greatly extended larval durations (e.g., Victor, 1986c; Cowen, 1991; Jenkins and May, 1994; Fowler and Short, 1996; Masterson *et al.*, 1997; Sponaugle and Cowen, 1997). McCormick (1999b) provided the first experimental evidence that the convict surgeonfish *Acanthurus triostegus* could delay settlement once a period of competence was reached. He caught late-larval-stage fishes at night as they colonized a reef and placed them in one of two treatments: benthic cages in the shallow backreef, or fine monofilament cages suspended between 3 and 6 m in a 50-m water column on the outer reef slope. Fishes in benthic cages attained a juvenile pigmentation and shape within 5 days. In contrast, 24% of fishes in pelagic cages (8 out of 34) retained their pelagic body shape and remained transparent.

Of the studies that have shown some evidence of delayed settlement, most have found that larval durations well beyond the average are rare. Victor (1986c) and Masterson *et al.* (1997) found that less than 1% of the bluehead wrasse *Thalassoma bifasciatum* examined had extended larval durations. The majority of coral reef fish species examined to date display low variability in larval duration (Cowen and Sponaugle, 1997), and this may be a function of either the small sample sizes on which larval durations are based for many species, or indicative that their period of competence is short and measured in hours rather than days. Another alternative is that by the time fishes reach competence most larvae have positioned themselves, through a combination of active and passive means, close to a suitable settlement habitat, so extensive delays are not required. In studies in which extended larval durations are more common, fishes have usually been collected far from their spawning grounds (Cowen, 1991; Jenkins and May, 1994; Fowler and Short, 1996) or have very specific temporal settlement cues [e.g., specific lunar phase (Sponaugle and Cowen, 1994)]. These studies indicate that there is a window of opportunity for settlement, with the lower bound governed by the rate of development and the upper bound determined by the extent to which a delay of settlement is possible.

It is presently unknown for fishes whether there is a cost to delaying metamorphosis and settlement. Delays in metamorphosis of marine invertebrates are commonplace and at times extensive (Pechenik, 1990), but often there is a cost associated with them. This cost can be manifest as reduced growth once settled, delayed maturation, reduced reproductive output, or lower survival (Pechenik *et al.*, 1998). McCormick (1999b) found that the surgeonfish he induced to delay metamorphosis for 5 days had reduced survival in tanks compared to control fish, but this may have been a feature of the low sample size or an artifact of the experimental procedure. The low incidence of delays in teleost fishes found to date may be a product of the potentially high costs associated with delay, and consequently their low survival.

The extent to which settlement to a benthic environment and metamorphosis are linked in time is largely unknown for coral reef fishes, but appears to be species specific. Some reef fishes, such as the dartfish, *Ptereleotris evides,* settle in schools into shallow caves below the reef crest and have poor mobility and are much less developed than other reef fishes, even other *Ptereleotris* species, at settlement (McCormick and Makey, 1997). Metamorphosis occurs over an approximately 12-day period involving a major change in body dimensions and pigmentation (McCormick and Makey, 1997). At the other end of the spectrum, the sergeant major, *Abudefduf saxatilis,* undergoes a gradual metamorphosis while associated with floating debris in the neuston and joins the reef community at a variable age and size as a fully metamorphosed juvenile (Druce and Kingsford, 1995; M. I. McCormick, unpublished data).

C. Settlement: A Pelagic Fish Becomes a Benthic Fish

When a ready-to-settle larva encounters a reef, it may either attempt to settle or reject the reef and move away in search of a more suitable settlement site. Although it is widely considered that larvae show strong habitat selection at settlement (Victor, 1991; Roberts, 1996; Booth and Wellington, 1998), connectivity models usually assume that larvae settle onto the first reef they encounter once they are competent to settle. This assumption is probably made because of the difficulties of inserting any other settlement behavior into models, but recent work shows it is far from realistic. A substantial proportion of competent larvae released adjacent to reefs during the day reject the nearest reef, and swim away into open water [no nighttime data are available, and, of course, results at night may differ (Leis and Carson-Ewart, 1998, 1999, unpublished)]. For 14 species of pomacentrids, about 30% (range 0–50%) of larvae swam away offshore without reaching the adjacent reef (although many initially swam toward the reef before turning offshore), a further 12% (range 0–50%) swam over the reef edge, and some closely examined potential settlement sites, before swimming away offshore. About 2% (range 0–11%) of the pomacentrid larvae swam offshore and then turned back to the reef after offshore excursions of a few to several minutes, usually out of sight of the reef. This may be a means of testing another portion of the same reef for suitability. Similar values were found for a serranid (Leis and Carson-Ewart, 1999). The percentage of larvae of a particular species that rejects any given reef varies from zero to 100%, depending on species and reef. For example, 100% of *Caesio cuning* larvae reject leeward and windward reefs at Lizard Island, whereas about 50% settle on the first lagoonal reef they encounter (J. M. Leis, unpublished). At present, the reasons for such rejection are only speculative, but it appears that the presence of potential predators accounts for only a small percentage of this, although this may vary with species. Other apparent factors are aggressive resident fishes, which often "rush" settlers, or the reef type. In fact, at least during the day, aggressive nonpredatory resident fishes appear to discourage settlement at least as much as does the presence of predators. Whatever the reason for swimming back into the pelagic environment, one would not expect such a high percentage of larvae to do so if reefs were difficult to locate from open water.

Settlement occurs when a larva leaves the pelagic environment to become closely bottom associated. Although postsettlement issues other than metamorphosis are beyond the scope of our review (see Victor, 1991; Roberts, 1996; Booth and Wellington, 1998), it is worth reiterating that most reef-based studies of "settlement" or patterns resulting from settlement are, in fact, studies of some unknown combination of settlement and postsettlement processes (Richards and Lindeman, 1987) [for notable exceptions, see Holbrook and Schmitt (1997, 1999) and Schmitt and Holbrook (1999a)]. Studies of "settlement patterns" are often of juveniles weeks to months older than settlement-stage larvae and are made with no attempt to examine what really happened during those weeks or months. Such studies assume that larvae settle into the places where the "recruits" were observed, that there has been no postsettlement redistribution, and that any postsettlement mortality has no effect on whatever is being studied. In fact, the assumption of no redistribution is rarely tested (but see Finn and Kingsford, 1996), and very little is known about this, or the related phenomenon of "relaunch," or reentry into the pelagic environment following settlement. Further, mortality can be very high in the first few days after settlement (Holbrook and Schmitt, 1997, 1999). More work on the early postsettlement interval is required (e.g., Kaufman *et al.*, 1992; Frederick, 1997; McCormick and Makey, 1997), but here we review studies that actually investigate settlement.

Most reef fish biologists apparently believe that the vast majority of, or even all, settlement occurs at night. So pervasive is this idea that evidence of daytime settlement is sometimes dismissed as methodological error (e.g., Robertson *et al.*, 1988), and vision is often dismissed as unimportant for orientation and habitat selection at settlement (see Chapter 6, this volume). However, the evidence supporting the notion of nocturnal settlement is actually based on very few studies involving few species—principally pomacentrids (reviews in Sweatman, 1985a, and Leis and Carson-Ewart, 1999). There is simply no information on diel settlement timing for the vast majority of species, genera, or families. When the timing of settlement has been carefully examined, the majority does seem to take place at night, but substantial portions of total settlement (up to 66%) frequently take place during the day. Nearly all studies report some level of daytime or crepuscular settlement, and many studies do not eliminate the possibility of crepuscular settlement, when substantial light is present. In fact, settlement-stage larvae of most of the families that are released within view of reefs during the day are likely to settle (Leis and Carson-Ewart, 1998, 1999, also unpublished). Larvae are relatively easily put off from their attempts to become reef animals during the day by potential predators or by aggressive reef residents and may swim away offshore (Sweatman and St. John, 1990; Leis and Carson-Ewart, 1998, 1999, also unpublished). This behavior is variable among species: for example, when a school of fusiliers is encountered off a reef edge, *Chromis atripectoralis* larvae are likely to reach the reef by ascending over the school, whereas *Chaetodon aureofasciatus* larvae usually swim offshore (J. M. Leis, unpublished). Most of the potential predators and aggressive residents are not active at night. So, it is possible that attempts by larvae to approach and settle on a reef are of equal frequency day and night, and that any difference in the amount of settlement between day and night is simply due to attempts that are foiled by interaction with residents (Leis and Carson-Ewart, 1998, 1999). An implication of this is that patterns of settlement and subsequent recruitment could be strongly influenced by the distribution of predators, especially those in the "wall of mouths" that fronts many coral reefs, and of aggressive residents.

Observations of settlement behavior in several species show that settlement is a complex process, that larvae are frequently highly selective about where they will settle at a variety of scales, and that the larvae are behaviorally sophisticated. One approach is to take settlement-stage larvae (usually captured in light traps, but sometimes reared in the lab) into the laboratory and provide choices of settlement substrates (e.g., Öhman *et al.*, 1998a; Danilowicz, 1996). This enables good control of many possible confounding factors, but investigators are frequently forced to define settlement as the position of the larvae in the morning or some other convenient time, rather than where settlement actually first took place. This allows postsettlement redistribution over up to 10 hours to affect the patterns observed. Lab situations can duplicate only a limited range of field conditions and the bias this might introduce is unknown. For instance, *in situ*, larvae of the serranid *Plectropomus leopardus* settled only in depths greater than can be duplicated in most laboratories (Leis and Carson-Ewart, 1999), so the value of laboratory settlement selection experiments in this species might be questionable.

In the field, examination of distribution patterns in the morning has many of the same drawbacks regarding postsettlement processes as the lab situation, but without the same level of control of other factors. Remote video recording (Holbrook and Schmitt, 1997, 1999) enables studies of the actual settlement process at any time, at least on the scale of a few meters, without other confounding lab or temporal effects. A final approach is to release larvae adjacent to or over reefs and observe *in situ* the behavior of the larvae as they attempt to settle and to deal with predators or residents (Leis and Carson-Ewart, 1998, 1999). This can only be done during the day, so can provide only half the picture, and the presence of the divers adds unknown biases. Investigators may hold light-trapped larvae in aquaria for several to 36 hours before initiating an experiment. Given the rapidity with which the morphological and behavioral changes associated with settlement take place (Randall, 1961; Stobutzki and Bellwood, 1994; McCormick, 1999a), this holding delay could affect the results, but this has not been tested.

In the lab, settling pomacentrids of four species made distinct habitat choices that varied widely among species, and postsettlement redistribution was detected in some species (Öhman *et al.*, 1998a). *In situ* observations of settlement behavior of three of the species reveal some differences from and some similarities to the lab study (J. M. Leis, unpublished). This indicates that use of multiple approaches to study settlement could help identify which results are affected by method-induced biases.

In situ approaches show remarkable selectivity by larvae of some species at settlement (J. M. Leis, unpublished). Larvae of the fusilier *Caesio cuning*, in contrast to congeners, rejected windward and leeward exposed fringing reefs, but settled with high frequency on lagoonal platform reefs (total $n = 38$). Further, they settled only into mixed-species schools of similarly sized (20 mm) juvenile planktivores on the reef edges. Interestingly, a light trap placed near, but outside and downstream of, the lagoon entrance captured many more *C. cuning* larvae than did a trap placed 2 km away, implying that some of the selectivity can begin at some distance from the selected habitat. Patterns of settlement abundance can show similar distributions: i.e., high settlement density of apogonids in lagoons and low density on outside reef slopes (Finn and Kingsford, 1996). In contrast, larvae of two species of butterfly fish (*Chaetodon*) settled on both windward and lagoonal reefs, but only into live coral, regardless of presence or absence of other species, and selected corals were located on all parts of the reefs (total $n > 70$). The damselfish, *Chromis atripectoralis*, settled on all tested reef types, but only into schools of similarly sized (10 mm) juveniles. The location of the school on the reef seemed unimportant (total $n > 50$).

There are indications that some acanthurids may initially leave the pelagic environment in deeper water and migrate upslope or through reef channels into shallow reefal waters, at times *en masse*, before settling into caves and crevices on the reef (Sancho *et al.*, 1997). Larvae of an acanthurid in oceanic conditions about 1 km off the reef crest usually swam rapidly downward, and this led Leis and Carson-Ewart (2001) to speculate that these larvae might have been attempting to find the reef slope bottom far below and to then follow this up toward the shallow reef, where they settle. More work on such behavior is needed, but if the indications are borne out, at least two possible advantages are obvious. This could be an effective means both of finding a relatively nearby reef and of avoiding predation by the "wall of mouths" (Hamner *et al.*, 1988) that exists off most shallow reefs, but predation by other fish on the larvae as they move over the reef surface can be intense (Sancho *et al.*, 1997). Such behavior also indicates that the location of initial "touchdown" on the bottom may be some distance away from where the larva finally shelters. Whether the time and site of initial touchdown or the time and site of sheltering should be considered "settlement" is debatable, but it does illustrate the potential complexity of the transition from the pelagic to benthic environment. Not every species simply slips into juvenile habitat directly from the pelagic environment as does *Chromis atripectoralis*, for example.

Given the settlement selectivity shown by larvae and their readiness to leave the immediate vicinity of reefs and swim back out to open water, combined with their ability to swim great distances at relatively high speeds, it is entirely possible that these behavioral inputs alone can determine the mesoscale distributional patterns of adults documented in places such as the Great Barrier Reef (e.g., Williams and Hatcher, 1983). For example, an average larva that rejected the midshelf reefs at Lizard Island would need to swim about 20 km [or about 28 hours at the mean speed found by Leis and Carson-Ewart, (1997)] to reach the outer-shelf barrier reefs. This is well within the endurance values for larval swimming found by Stobutzki and Bellwood (1997). Thus, settlement of this "average" species into the outer-shelf fish assemblages (Williams and Hatcher, 1983) rather than the midshelf assemblage at Lizard Island is feasible (assuming the appropriate orientation) based on proved behavioral capabilities alone.

D. Metamorphosis

A species-specific change in body structure and physiology, called metamorphosis, occurs at or around the time of settlement, during which fish lose many of the characteristics that enhance survival in the plankton while developing other features suited to their new benthic environment (e.g., Randall, 1961; Norris, 1963; Markle *et al.*, 1992; Keefe and Able, 1993; McCormick, 1993, 1999b; McCormick and Shand, 1992; Shand, 1993, 1994a, 1997). Information on what metamorphosis represents is sparse for fishes generally, and for tropical reef fishes in particular. There has been surprisingly little advance in this field since Youson's (1988) review of the subject.

The classical example of metamorphosis in fishes is that of the flatfishes (Pleuronectiformes). Temperate species that have been studied undergo a dramatic metamorphosis involving a reduction in body depth, a change from being bilaterally symmetrical to asymmetrical (one eye migrates to the opposite side), changes in dentition and jaw structure, reductions in the lengths of spines, development of scales, and changes in body pigmentation (e.g., Markle *et al.*, 1992; Keefe and Able, 1993). The speed of this metamorphosis is temperature dependent (Chambers and Leggett, 1996; Keefe and Able, 1993) and can encompass a substantial proportion of the larval duration of the species. For instance, off the Oregon coast, the Dover sole, *Microstomus pacificus,* may reach the start of metamorphosis with eye migration only 1 month into a larval phase that may last 2 years (Markle *et al.,* 1992). This is somewhat of an extreme case and more typically metamorphosis may encompass the last 50% of the larval phase for flatfishes (e.g., Kvenseth *et al.,* 1996). Tropical members of this order undergo the same magnitude of change, although metamorphosis may occur over a less protracted size range (Leis and Carson-Ewart, 2000).

Members of the Elopomorpha (which includes tarpon, bonefish, and eels) have a long, ribbonlike leptocephalus larval stage that is followed by an elaborate remodeling stage. The bonefish *Albula* sp. loses half its body length during an 8- to 12-day period (Pfeiler and Luna, 1984). The gelatinous matrix that makes up much of the leptocephali is resorbed and development of the bone and muscle framework characteristic of the juveniles occurs. The Pacific tarpon, *Megalops cyprinoides,* has a 25-day leptocephalus stage and then undergoes similar shrinkage for 12 days. This is followed by a month of slow growth, including remodeling into the true juvenile form (Tsukamoto and Okiyama, 1997).

In general, the changes that occur in tropical reef fishes (mostly Perciformes, as noted above) at the end of the larval phase are not as dramatic and occur rapidly. The convict surgeonfish *Acanthurus triostegus* comes onto the reef as a transparent discoid-shaped larva (Randall, 1961; McCormick, 1999b). Over a 5-day period the mouth, which was terminal at settlement, orients ventrally to facilitate reef-associated herbivory. Concomitant with mouth migration is a 12% reduction in body depth (McCormick, 1999b). During this period alimentary-tract length increases threefold (Randall, 1961). The Caribbean blenny, *Ophioblennius atlanticus,* undergoes a substantial metamorphosis from being a pelagic predator with long fangs and a short gut to being a grazing herbivore with fine comb-shaped teeth for nipping and scrapping algal turf and bacteria, and a long, coiled gut (Labelle and Nursall, 1985). During this metamorphosis the head profile becomes steep, the eyes migrate dorsally, and the adult coloration is assumed. This change involves at least a 7% decrease in standard length (Nursall and Turner, 1985), but over what time is unknown.

McCormick and Makey (2002) used sequential video images to examine trends in the morphological change associated with settlement in 34 species from 13 families of tropical reef fishes. They found that within a family, the extent of morphological change varied greatly among species, and this prevented generalization. Typically, changes over the first 4 days involved changes in pigmentation, snout angle, fin spination, and body depth. This emphasises the species-specific nature of the magnitude of the changes that occur at the end of the larval phase and of the rates of those changes.

Limited evidence suggests that those species that undergo a substantial metamorphosis stop feeding during this period and survive on endogenous reserves. In bonefish, metamorphosis is fueled by the breakdown of triglycerides and nitrogen-containing compounds (Pfeiler *et al.,* 1998). Similarly, reared temperate bass larvae, *Morone saxatilis,* used stored lipids during a nonfeeding metamorphosis (Chu and Ozkizilcik, 1995). The Caribbean redlip blenny also stops feeding as it comes into the vicinity of the reef and its lipid stores fuel metamorphosis (Labelle and Nursall, 1985; Nursall and Turner, 1985). Randall (1961) collected the convict surgeonfish, *Acanthurus triostegus,* from tide pools on the day of settlement and found no food in their stomachs. How common this nontrophic period is and its duration are unknown for tropical reef fishes. It does suggest that the nutritional stores available to a fish at the start of metamorphosis may help determine its survival potential because as it enters the demersal phase of its life, nutrition influences growth and performance.

Little is known of the sensory changes that reef fishes undergo as part of the metamorphosis associated with settlement. The visual and gustatory systems of one coral reef fish species have received the most attention. The goatfish, *Upeneus tragula,* is common component of ichthyoplankton samples from the Great Barrier Reef. From early in the larval phase of this species, it forms schools with a broad size range of its own and compared to other goatfish species, and is diurnally active in the top few meters of the water column (McCormick and Milicich, 1993). During the late larval phase these fish have a double layer of cone photoreceptors in their dorsal retina and high bipolar (processing) cell densities. This may greatly improve vision in shallow water. Over a 6- to 12-hour period around settlement the retina undergoes extensive reorganization; one layer of cone cells slots into the other to form a monolayer, and cone and bipolar cell densities greatly reduce (McCormick and Shand, 1992; Shand, 1994a). Interestingly, the absorbance spectra of these cones showed an abrupt shift toward shorter wavelengths, losing red sensitivity, a characteristic typical of deeper bottom-dwelling fishes (Shand, 1993). Over the same period of 6 to 12 hours, the barbels, which are used to find and excavate benthic prey, rapidly migrated toward the mouth, increasing in length by 50%, and the taste buds that cover the barbel surface doubled in size (McCormick, 1993). The speed and extent of these structural changes are remarkable. However, it is unlikely that sensory changes of this magnitude are typical at settlement. In three other species of reef fish (a damselfish, wrasse, and apogonid) changes in the densities of various cell types in the eye occurred earlier in development, and were not coincident with settlement (Shand, 1997). This emphasizes the species-specific nature of the timing and magnitude of changes that occur in the visual system during larval development.

In most reef fishes, settlement also coincides with a change in both the rate of growth of the earstones or otoliths and their axes of growth (Wilson and McCormick, 1997, 1999). Otoliths are an integral part of the membranous labyrinth system that is responsible for balance, orientation, and hearing, so these changes imply that the function of these senses may change slightly with the necessity for precise orientation in a complex three-dimensional reef environment.

VI. Future Research Directions

It is now well established that at least the later pelagic stages of many reef fishes are far from passive, and that they are real animals—not simply biological driftcards—that have complex behaviors and considerable ability to control where they are and where they will settle. The next step is to extend behavioral studies to a wider range of species, in particular such important groups as labrids, scarids, and eel leptocephali. The first two families have much smaller, apparently less-well developed larvae than do most of the groups studied thus far (Leis and Carson-Ewart, 2000), and it will be of interest to see how their behavioral capabilities compare. The marine larvae of freshwater eels have been intensively studied (e.g., Tsukamoto, 1992; McCleave *et al.*, 1998). There is reason to believe they have good sensory abilities and these could be present in the larvae of reef eels. However, the morphology, mode of swimming, and, apparently, physiology of these eel leptocephalus larvae differ considerably from those of other reef fish larvae, so some surprises may be in store. Otherwise, the most urgent need is to determine when during the pelagic phase larvae are sufficiently developed to begin to control their trajectory. It is difficult to see how this can be done with other than reared larvae, but measurements should not be confined to the lab, and *in situ* work is essential. In addition, means must be found of ensuring that results on lab-reared larvae are representative of capabilities and performance of wild larvae.

The great gap in our understanding of the sensory capabilities of reef fish larvae, particularly in the areas of hearing and olfaction, must be filled (see Chapter 6, this volume). This should probably proceed on both a morphological and a behavioral level. Because "the strength of the sensory facility" was *the* critical parameter in modeling of larval swimming responses to different cues, Armsworth (2000) concluded that determining the sensory abilities of larvae "must become a fundamental research objective of those involved in management of reef fish populations." Innovative means must be found to study the behavior of larvae at night, both in open waters and over the reef. Attempts to integrate larval behavior with hydrological models should provide much insight into the realities of dispersal and retention. This is critically important to our understanding of population connectivity and to management decisions, including design of MPAs.

Hand in hand with the need to learn about sensory capabilities are requirements to understand the performance characteristics of larvae, their ability to detect and catch prey, and how these develop. More research should be devoted to the role that starvation plays in tropical waters, and this will require some researchers to revisit their oceanographic roots to explore the link between the patchiness of prey and fish larvae of all developmental stages. To understand the link between reef populations and larval settlement, information on mortality schedules is desperately needed. This will likely lead to the detection of critical intervals in the life phase that are marked by enhanced sensitivity to environmental or biological conditions. These will serve to focus future research.

From an evolutionary perspective, it will be of interest to determine whether larval mortality is size or growth selective and what processes influence which individuals survive the larval period. Conditions of the breeding population may influence larval mortality levels on a range of temporal and spatial scales through the influence of nongenetic contributions from the mother (i.e., maternal effects). Recent modeling has shown that maternally affected traits may have enhanced rates of mutation, may evolve in a direction opposite to that favored by selection, and may have faster rates of speciation (Mousseau and Fox, 1998, for review). This may be particularly pertinent given the high levels of self-recruitment found in some studies (Jones *et al.*, 1999; Swearer *et al.*, 1999). With advances in larval rearing, we are now reaching a stage at which some of these influences on individual survival and success can be assessed in lab conditions, at least for a restricted group of species.

Cowen and Sponaugle (1997) and Kingsford (1998) point out that research on coral reef fishes has emphasized the retention/dispersal question, whereas studies of temperate fish larvae have focused on feeding and predation (mortality). Without decreasing work on the retention/dispersal question, increased work on feeding and predation would be welcome. Taxonomic work on the pelagic stages of reef fishes is still needed for most groups. Inability to identify larvae to the species level, particularly early stages, is a major impediment to understanding of distributions, feeding, condition, and many other subjects.

Leis (1991a) listed eight areas for future research on reef fish larvae, and in four areas there have been significant advances. Wolanski, Black, and others have made great strides in examining the fine scale of water circulation, particularly through the use of numerical models, and coupling this with larval distributions. However, there is an urgent need for better ground-truthing of numerical models, as pointed out by the different results of two of these for the same reef. Integration of larval behavior with this physical oceanographic input to understand dispersal will be challenging. Vertical distribution is now better known, but more knowledge is needed about what individual larvae are doing and how vertical distributions vary in different conditions. A better understanding of how smaller larvae interact with vertical structure in currents, other physical factors, and food distribution is needed. Much more is now known of the swimming capabilities of reef fish larvae, but only in the later portions of the pelagic stage, and some important taxonomic groups remain unstudied. Research on the ontogeny of swimming and sensory abilities is a top priority. Behavior of reef fish larvae is now understood as complex and flexible, but we have only just scratched the surface here, and, again, only with the later portions of the pelagic stage.

In contrast, the other four research areas identified by Leis (1991a) have received far less attention. The first question (Where and when are larvae put into the pelagic system?), aside from the intensive studies of Hensley *et al.* (1994) and Appeldoorn *et al.* (1994), remains largely unanswered. Future studies in this area need to be mindful of Warner's (1997a,c) cautions about studying adaptation. It may be profitable to focus on spawning aggregation sites, and to examine their characteristics. However, spawning sites are not necessarily chosen for their adaptive value (Shapiro *et al.*, 1988; Warner, 1988a), and many species do not participate in spawning aggregations. As to what generality there is among different regions and topographic situations, this has been largely ignored (again, see Warner, 1997a,c). The nature of the physical and biological requirements of reef fish larvae might be usefully approached by aquacultural methods (Tucker, 1998), by correlative studies between distributions of larvae and physical or biological variables, or by "hindcasting" studies based on potential information stored in the otolith. More work is required here, and cautions about extrapolation from aquacultural studies must be kept in mind (e.g., Olla *et al.*, 1998). The final question (Are distributions of larvae adaptive?) will remain unanswered until more is known about where the larvae that successfully settle on reefs have been, and this knowledge is compared with the distribution of "unsuccessful larvae."

Finally, we must point out once again that the majority of the available research is on pomacentrids. Although pomacentrids are common and are in many ways ideal subjects for experiments, it is highly desirable that future work includes a wider representation of the extreme taxonomic diversity found on coral reefs. Pomacentrids are certainly not entirely representative of the biology of the rest of that diversity. However, we do not share the view that research on "toy fishes" [i.e., small species, such as pomacentrids, not of direct commercial interest (Roberts, 1996; Polunin *et al.*, 1996)] is not of importance to fishery biologists and managers. Especially during the pelagic stage, all species are "toy fishes" and face the same problems of a small animal in a large pelagic environment: first, they must survive and grow, and second, they must find and make a radical transition to another very different environment.

VII. Conclusions

Recent research on the pelagic stage of reef fishes has given us a good look into the black box of larval biology. This look reveals that these pelagic stages are real fishes with capabilities in excess of the larvae of well-studied temperate fishes. This has changed thinking about the way larvae fit into the dispersal/retention concept that has dominated work on reef fish larvae. We now know that larvae and their behavior have a major influence on the positioning of reef fish populations along the open–closed continuum of population demography. In the foreseeable future, we should have a firm and defendable basis for design of marine reserves and of the geographical size of reef fish population units for management purposes, based on an increased understanding of larval biology rather than the simplifying assumptions of the recent past.

Acknowledgments

We thank our colleagues for sharing their unpublished work with us, and the numerous people who helped us gain the information reported here. We acknowledge ARC funding through Grants A19530997 (JML), A19804335 (JML), F19600180 (MIM), and A19701068 (MIM), which supported both the preparation of this manuscript during 1999 and most of the unpublished work reported here. Thanks to S. Bullock and B. M. Carson-Ewart for help with editorial matters, and to M. J. Kingsford and K. C. Lindeman for constructive reviews. Thanks to authors and publishers of illustrations in Figure 1 for allowing us to reproduce them.

CHAPTER 9

Biogeography and Larval Dispersal Inferred from Population Genetic Analysis

Serge Planes

Centre de Biologie et d' Écologie Tropicale et Méditerranée
EPHE ESA 8046 CNRS, Université de Perpignan, Perpignan 66860, France

I. Indo-Pacific Coral Reef Fishes: Model

Coral reef fishes, like most shallow-water marine organisms, are restricted as adults to shallow, inshore water. This favors the isolation of populations by expanses of deep water or areas of otherwise unsuitable habitat. On the other hand, most species have a pelagic larval stage (Leis, 1991a), lasting several weeks or months (Brothers *et al.*, 1983; Wellington and Victor, 1989a), and thus have the potential for long-distance transport by ocean currents. The consensus is that planktonic larvae are dispersed from their spawning sites (e.g., Hedgecock, 1986; Kay and Palumbi, 1987). This transport is fundamental in maintaining the integrity of species and avoiding speciation that could result from isolation in local patches of the adult habitat.

Pelagic and benthic stages differ in almost all characteristics, from morphology to size, habitat, food, and behavior. The pelagic stage of coral reef fishes is certainly the least understood phase of their life cycle. Because of this lack of knowledge, many conflicting ideas have been proposed concerning their potential to disperse. Questions related to the effective dispersal of larval pools in the open water are relevant not only to evolutionary biologists seeking to understand the processes of differentiation and speciation (Shaklee *et al.*, 1982; Burton, 1983), but also to managers who must determine the degree to which geographic stocks correspond to independent reproductive units for fisheries purposes (Roberts and Polunin, 1991; Russ, 1991), and to conservation biology (Smith *et al.*, 1991).

II. Estimating Larval Dispersal

Several approaches have been used to estimate larval dispersal range. Monitoring currents has been considered as the preferred way to predict dispersal and migration, based on the assumption that larvae drift passively with currents (e.g., Williams *et al.*, 1984; Hourigan and Reese, 1987; Leis, 1991a; Roberts, 1997b). Many additional assumptions are made when deducing effective larval drift from current patterns, but these may limit the value of this approach. First, this indirect estimate of larval drift in oceanic waters is based on long-term mean current patterns; data may be misleading if successful dispersal is actually due to anomalous conditions that occur infrequently. The variability in recruitment success that often appears episodic suggests this is important (Dixon *et al.*, 1999). Second, the assumption of passive drift is probably not valid. Work on the diurnally and developmentally varying vertical distribution of larvae (Leis, 1991a), and on the swimming activity and the sensorial capacity of larvae (Leis *et al.*, 1996; Stobutzki and Bellwood, 1994), indicates much greater larval motor capabilities than previously believed. Clarke (1995) suggested that the extremely low density of shorefish larvae in the open ocean is due to undersampling by standard techniques, which is compounded by a tendency of pelagic juveniles to avoid towed nets (Choat *et al.*, 1993).

Progress in using otoliths to estimate larval stage duration, by counting daily rings up to the "settlement mark," gives more reliable data to address the relationship between planktonic duration and population structure (Victor, 1986b). Up to now, studies developed on coral reef fish fauna of the Indo-Pacific (Thresher and Brothers, 1985; Brothers and Thresher, 1985; Thresher *et al.*, 1989) did not show any correlation between the mean planktonic duration of a genus and either species richness or degree of endemism in that genus. The only significant result indicated that species with mean planktonic stages longer than 45 days were almost invariably broadly distributed across the Pacific, whereas those with shorter planktonic stages varied in the extent of their distributions (Brothers and Thresher, 1985). However, Thresher *et al.* (1989) did not find any correlation between planktonic larval duration and the distribution of the Pomacentridae.

A tagging survey, although not especially well suited to the study of larval drift, was attempted by tagging otoliths of embryos for species with benthic eggs (Jones and Milicich, 1995; Jones *et al.*, 1999). Such an approach is limited because of the high mortality suffered by larvae in the plankton and their large dilution in the ocean. An estimated survivorship of 10^{-6} requires tagging thousands of eggs with very little chance of recovery. In their study, Jones *et al.* (1999) did tag several hundred spawns of about 250 eggs each and recovered only 7 (0.005%) tagged new recruits. They were able to conclude that it is possible for larvae to disperse between closely located reefs on the Great Barrier Reef, but their approach gives only a minimal dispersal range.

III. Population Genetics Approaches to Estimate Larval Dispersal

A. Interest in Population Genetics

As early as 1975, Ehrlich suggested that assessment of allozymic variation among geographically remote populations of reef fish is the optimal approach for testing dispersal and migration capabilities of larvae. Population genetics can contribute to this debate, in the sense that using intraspecific markers, such as allozymes of DNA haplotypes, provides a new opportunity to distinguish populations of different islands and/or archipelagos. Detectable allele frequencies can be used to estimate levels of migration if it can be assumed that these frequencies reflect a balance between the opposing forces of migration (gene flow) and random divergence of allele frequencies (genetic drift). Although this theory is silent about allopatric populations with identical genetic profiles (Shaklee *et al.*, 1982), differentiation provides unequivocal evidence of reproductive isolation. The main difficulty with this approach is that other forces, notably natural selection and historical contact, can influence allele frequencies, and the relative importance of these forces in natural populations has proved extremely difficult to evaluate directly. Because selective force is often evoked to explain some pattern of differentiation, a special discussion on this topic is developed later in Section V with respect to regional-scale population genetics.

The population genetics approach also gives a new perspective on biogeographic research, because it reveals the structure of genetic variability of marine species. Allelic variation within a species can be used to track patterns of migration. Concordance of species-level and allele-level biogeography has already been demonstrated in some places in the sense that processes determining species distribution are similar to those controlling gene flow (Avise *et al.*, 1987; Kirkpatrick and Slatkin, 1993). Consequently, patterns of gene flow, computed from allelic variations within a species, can contribute to explaining the origin of species in an evolutionary perspective. In addition, the analysis of genetic markers can help (1) to distinguish sibling species (Knowlton, 1993) and to provide new data on species diversity, (2) to determine divergence time between species or populations once appropriate calibration is made (Martin *et al.*, 1992), and finally (3) to calibrate relationships between species (phylogeny) and to identify derived species.

However, one effective migrant exchange between subpopulations per-generation is sufficient to stop fixation of alleles, and 1 to 10 migrants will give a 95% chance of detecting no significant differentiation between subpopulations (Allendorf and Phelps, 1981). Such a model indicates that gene flow is a powerful means of homogenizing subpopulations, even with a small number of migrants. Any significant divergence between allelic and/or genotypic frequencies of different samples indicates at least some restriction of gene flow, provided that genes are not subject to local selection. The fact that a small amount of gene flow is sufficient to homogenize populations limits the power of genetic markers. This makes it difficult to differentiate populations that have essentially separate dynamics but are linked by trivial amounts of gene flow, from populations that exchange numerous migrants and are effectively a single stock.

In this synthesis, the discussion moves from a broad- to a local-scale survey, including a comparison of the genetic diversity of coral reef fishes and other marine fish species. This section is followed by a review of reef fish biogeography, mostly concentrating on results gained from genetic surveys. The biogeographic review is centered on the Pacific Plate, for which most data are available. The Pacific Plate is defined as the area including the Pacific Ocean and the western seas in the regions of Malaysia, Indonesia, and the Philippines. A special discussion (Section IV,B) focuses on results of the founder effect on genetic markers and the consequences for interpreting biogeographic genetic patterns. Following the discussion of biogeographic scale, there is a summary of works on population genetics of coral reef fishes conducted over regional scales. The last section is devoted to intrapopulation genetic analyses. This offers a challenge for future understanding of the maintenance of diversity and the structure of populations.

B. What Molecules? Choosing a Molecular Marker

The development of molecular techniques offers a palette of technical approaches for population biologists interested in a wide range of questions. For example, the tools of molecular biology can be used to determine individual reproductive success or to measure rates of genetic divergence among populations. Which technique is the most appropriate for a particular question depends on the extent of genetic polymorphism required to best answer the question, the analytical or statistical approaches available for the technique, and the pragmatics of time and cost of materials. Empirical studies that employ allozyme polymorphisms as genetic markers have contributed a great deal to our understanding of population processes (Smith *et al.*, 1990). The hope of obtaining highly informative genetic markers for tracking individuals and/or their genes under field conditions had led many population biologists to consider switching to DNA-based techniques. The actual rate at which new molecular techniques are being developed far outstrips their efficient and reasoned incorporation into studies of population ecology (Parker *et al.*, 1998). This has contributed to confusion among population biologists about various new techniques, often causing people to abandon prematurely standard techniques (such as allozyme electrophoresis) that provide readily interpretable data. The expectation of exploring variation in DNA sequences needs to be tempered while considering the drawbacks associated with different techniques and the types of data obtained from them.

We should remember here that the choice of marker also has implications on evolutionary processes that will be investigated. Choosing markers under genetic selection, such as allozymes or the cytochrome *b* gene of mitochondrial DNA (mtDNA), may provide different results than choosing neutral markers, such as microsatellites. Differences in results may not be only a consequence of the higher variability investigated, but more likely relate to the fact we are looking at different evolutionary processes (selection vs. genetic drift). This fact has been often omitted in previous studies looking at allozymes, and assuming that they are neutral. It will be necessary to demonstrate the neutrality of markers by looking at relationships such as those with distance.

When it comes to choosing a technique, there are many currently available: protein electrophoresis, nuclear and mitochondrial restriction fragment-length polymorphism (RFLPs), minisatellite and microsatellite variable-number tandem repeats (VNTRs), random amplified polymorphic DNA (RAPD), and DNA sequencing [techniques are described in Parker *et al.* (1998)]. This section presents details and advantages of techniques already applied to describe biogeography and larval dispersal of coral reef fishes.

Gel electrophoresis is a powerful yet relatively simple technique that separates proteins in an electrical field according to their net charge and size. The key to the electrophoretic method is that each specific protein is encoded by a single gene, and differences in proteins give an indirect estimate of differences in the sequence of the DNA. Because some changes in the DNA sequence do not result in change of charge and/or size of protein, it has been estimated that electrophoretic techniques detect only about one-third of base substitutions in DNA (Lewontin, 1974). Moreover, only about 10% of the total genome will encode proteins. Consequently, the variations detected at protein gene loci may not be representative of the genome as a whole, and some species may lack allozyme polymorphism. A further limitation of using allozymes (rather than noncoding DNA) as genetic markers to separate populations is that allozymes may differ in metabolic function. Many statistical models in population genetics assume that phenotypic differences among allozymes are minimal and selectively neutral, but exceptions are known even in marine fishes wherein selection can maintain genetic variation (DiMichele and Power, 1982). Selection must be considered a possibility when differentiation originates from a single locus. However, most multilocus differentiation patterns are probably robust with regard to concerns about ongoing selection. There is substantial evidence that the majority of observed variation in enzymes is neutral (Kimura, 1983). In spite of these limitations, protein electrophoresis remains a powerful technique to exhibit genetic variation (Smith *et al.*, 1990). It has been widely used in coral reef fish surveys (>15) and provides most of the results that will be discussed in this chapter.

RFLP techniques use restriction enzymes to cut the DNA at a specific nucleotide sequence recognition site, thus generating sets of DNA fragments that differ in size when mutations have created or destroyed restriction sites. The most extensive use of restriction enzymes in population studies has been for the survey of allelic diversity and population differentiation in mitochondrial DNA. Enzymatic digestion of a small mtDNA molecule results in a number of DNA fragments that can be resolved and identified on agarose gel. Allele frequencies are estimated from the absence/presence of the restriction site among individuals, and several restriction sites (>100 restriction enzymes are commercially available) can be tested. Larger molecules such as nuclear DNA cannot be analyzed following this protocol because they produce too many segments to be separated on gels. The small size of the mitochondrial genome, its predominately maternal inheritance, and its relatively high rate of base-pair substitutions (especially in the control region) make it a valuable tool for studying relationships among populations (Avise *et al.*, 1987). This technique has been used in coral reef fishes (Shulman and Bermingham, 1995).

The most detailed analysis of DNA differentiation can be obtained by sequencing the region of interest from different individuals. The most difficult problem for analyzing genetic markers lies in the selection of the region of the genome that both reveals allelic variation and can be sequenced efficiently. Some genes, such as cytochrome *b* in the mtDNA, have been time calibrated and allow accurate estimates of time divergence between populations. Analysis of sequence differences is based on mutation of nucleotides within a position. Thus it looks only at mutation rate and does not take into account genetic drift, mostly affected by effective population size. Consequently, sequence analysis often shows lower levels of differentiation between populations, as compared to other methods. This approach has been applied by sequencing the cytochrome *b* and the control region (D-loop) of the mtDNA of chaetodontids in a phylogenetic and biogeographic perspective (McMillan and Palumbi, 1995).

Among other available techniques not used up to now in coral reef fish, the microsatellite VNTRs seem certainly the most promising approach (O'Connell and Wright, 1997), for several reasons: (1) they provide information for individual loci and therefore are amenable to calculations of allele frequencies necessary for population studies, (2) they typically exhibit high levels of gene diversity and show high levels of polymorphism, and (3) once primers have been developed, the protocol consists only of polymerase chain reaction (PCR) and electrophoresis. Some new technical advances are even starting to eliminate the electrophoresis step by looking at fluorescence or weight analysis of microsatellites. Such approaches during the PCR process may get around the problem of slippage, which can make individual alleles difficult to identify among false bands. In terms of analysis, interpretation of variation of allelic frequencies remains problematic. Measures of population subdivision based on microsatellite allele frequencies have been highly conflicting because *F* statistic (F_{st}) estimates are based on a model with low mutation rates, a postulate that is reasonable for enzymes but not for microsatellites. New models have now been proposed to estimate differences (Slatkin, 1995) and distances (Goldstein *et al.*, 1995) using microsatellites. However, the fundamental question remains open because it still is not clear whether all mutations at microsatellite loci involve changes of only one or two repeats, as assumed in many models (Valdes *et al.*, 1993), or whether mutations of larger effect occur occasionally (Di Rienzo *et al.*, 1994). Despite some estimates using genetic divergence from microsatellites to estimate coalescence times (Slatkin, 1995), this statistic must be cautiously interpreted. Technically, microsatellite analysis requires setting up primers for each locus. This is a highly time-consuming but necessary step. Because this technical step is unique for each species, microsatellite VNTRs can only be viewed as markers to be used in projects involving few species (1 to 3) and numerous samples. Presently, microsatellites are used only in commercial species studies that require large sample-size analyses and in long-term projects focused on one species. This is the case for salmon (Tessier *et al.*, 1995), cod (Bentzen *et al.*, 1996), and herring (O'Connell *et al.*, 1998). In these cases, it will save time in processing all individuals compared to other DNA approaches, and it is as easy to use as allozymes despite the time to set up primers for each locus. Among other possible developments, the microsatellite approach allows survey of oceanic larvae (Ruzzante *et al.*, 1996), which may identify the origin of larvae caught directly in the ocean. However, this technique has not yet been developed for any coral reef fishes.

IV. Biogeographic and Phylogeographic Perspectives

A. Genetic Diversity of Coral Reef Fishes

Diversity among organisms stems from a combination of DNA sequence variation and adaptive responses to the environment. Direct screening of DNA sequences is a new field of research and most of our knowledge of genetic variation comes from indirect examination of DNA through analysis of enzymatic proteins. During the past 25 years, this has been a popular technique for surveying genetic variability, with general agreement that the use of allozyme variation is a reliable tool to estimate levels of genetic variation with random samples of neutral genes. The polymorphism value (percentage of polymorphic loci) and the heterozygosity value (percentage of polymorphic genotypes) are usually used to estimate the genetic diversity of species.

Several hypotheses, mostly based on selectionist models, have been proposed to predict differences in genetic variation among species (Table 1). Various models have been described based on stability in environmental heterogeneity (Levins, 1968), time of divergence between species (Soulé, 1972), size and mobility variation (Selander and Kaufman, 1973), predictability of trophic resources (Valentine and Ayala, 1975), and ecological heterogeneity between habitat specialists and generalists (Nevo, 1978; Smith and Fujio, 1982). Most of the previous models gave a single causative explanation for the variation of genetic diversity. Such models appear unrealistic and there has been a tendency toward use of hybrid models, which consider a mosaic of relationships between species and their environment (Nelson and Hedgecock, 1980; Mitton and Lewis, 1989).

Contrary to selectionist models, Nei and Graur (1984) showed that genetic variation increases with the product of effective population size and generation time, in accordance with the neutral theory of evolution (Kimura, 1983). However, many problems remain for neutral models in explaining genetic variation, such as knowing whether species are in drift/migration/mutation equilibrium at the time of the experiment. Such a problem affects all analyses undertaken in general reviews of genetic diversity and will be discussed later.

Most studies show a decrease in genetic diversity with an increase in latitude (Hamrick *et al.*, 1979; Nevo, 1978). Nevo (1978) found a polymorphism level of 0.25 for plants, 0.36 for invertebrates, and 0.18 for vertebrates in temperate environments, and 0.57, 0.42, and 0.19, respectively, for similar groups in a tropical environment (heterozygosity was similar). In marine fishes, Smith and Fujio (1982) failed to find significant differentiation in polymorphism

TABLE 1 Major Models Proposed to Explain Variations of High and Low Genetic Diversity in Species

Model	High genetic diversity	Low genetic diversity	Parameter used to test
Selectionist models			
Environmental heterogeneity (Levins, 1968)	Heterogeneous	Homogeneous	Home range, depth range
Size and mobility (Selander and Kaufman, 1973)	Small, sessile	Large, mobile	Maximum size and ecology
Trophic resource stability (Valentine and Ayala, 1975)	Low seasonality	High seasonality	Home range
Ecological heterogeneity (Nevo, 1978)	Habitat generalist: large, continuous outbreeding species	Habitat specialist: rare, isolated, geographically restricted species	Distribution area, ecology, and egg type
Habitat heterogeneity (Smith and Fujio, 1982)	Habitat specialist: rare, isolated, geographically restricted species	Habitat generalist: large, continuous outbreeding species	Distribution area, ecology, and egg type
Rate of increase in demography (Mitton and Lewis, 1989)	Small length at maturity and small eggs	Large individuals at maturity and large eggs	Maximum fecundity, egg size, and life-span
Neutralist models			
Time divergence (Soulé, 1972)	Old	Young	Fish systematics and species diversity within orders and families
Population size (Soulé, 1976)	Large populations	Small populations	Distribution area, dispersal capabilities (egg type, larval duration)
Population size and generation time (Nei and Graur, 1984)	Large and fast	Small and long	Life-span, distribution area

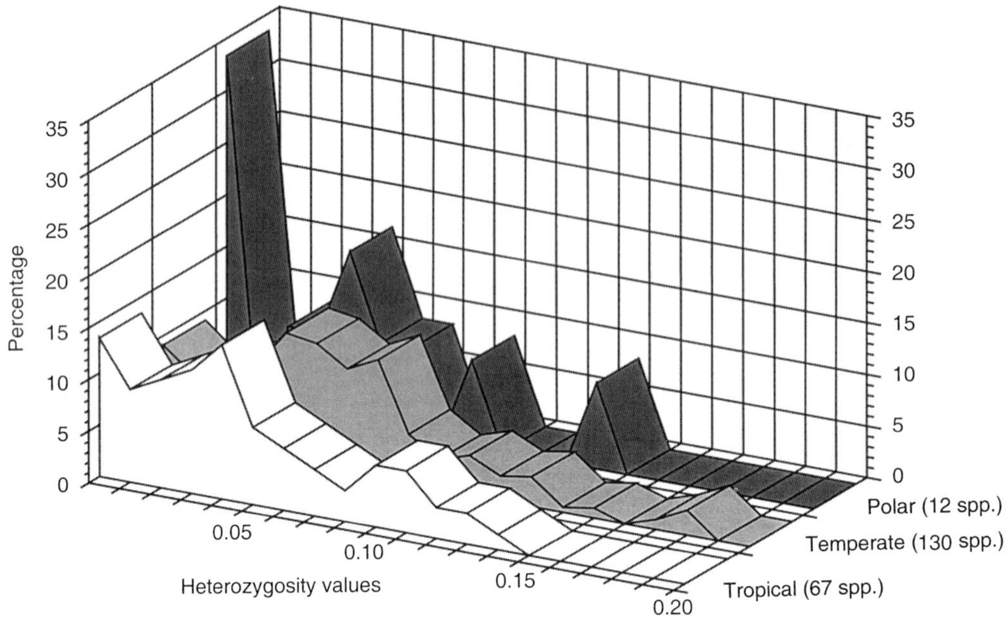

FIGURE 1 Distribution (%) of heterozygosity values for 209 species of marine fishes separated into temperate, tropical, and polar regions.

and heterozygosity between tropical species and others. Subsequently, however, comparison of heterozygosity values in 209 species showed significant variation, with higher heterozygosity in temperate fish (0.063) than in tropical (0.054) and polar species (0.039) (Planes, 1998) (Fig. 1). However, such a pattern should be treated with caution, because there is no significant difference between tropical and temperate species based on Fisher's least significant difference (LSD) test ($P = 0.250$), despite a divergence of about 10% in their heterozygosity. It has been suggested that, in long-term stable environments, speciation occurs through species becoming specialized and genetically depauperate. This lack of genetic variability may remove the ability to adapt to changing conditions and consequently may result in mass extinction when conditions change (Bretsky and Lorenz, 1970). Nevertheless, a correlation of low genetic diversity with a stable environment has often been postulated (Levins, 1968; Van Valen, 1965). The results for marine fishes conform to this pattern, with lower genetic diversity in tropical (0.054) and polar (0.039) marine fishes, compared to temperate (0.063) species.

This result came out of a multifactorial analysis, that demonstrated that a species' heterozygosity was related to the geographical zone (5.5% of the total variance, $p = 0.004$), the distribution area range (12.6% of the total variance, $p = 0.0002$), the egg type released (12.3% of the total variance, $p = 0.0001$), and the fecundity range (9.2% of the total variance, $p = 0.002$). Overall, high heterozygosity appears to be associated primarily with large species occurring over large areas and laying pelagic eggs. Large adult size and high fecundity of pelagic spawners suggest extensive dispersal capabilities, whereas small adult size and low fecundity of demersal spawners imply relatively low average dispersal capabilities. Thus, the heterozygosity relationships found with egg type are likely to be related to variability in dispersal capabilities, and suggest that high genetic diversity is associated with higher dispersal capability.

Similar surveys regarding variation of genetic diversity have been conducted on plants and have demonstrated that predominantly outbreeding species maintain a higher level of intrapopulation genetic diversity (Brown, 1979). In fact, plant surveys show that species with a large range, high fecundity, an outcrossing mode of reproduction, wind pollination, long generation time, and from habitats representing later stages of succession have higher genetic diversity than do species with other combinations of characteristics (Hamrick *et al.*, 1979). As in marine fishes, it appears that high genetic diversity in plants is correlated with dispersal capabilities. Dispersal capability appears to be a major parameter acting on genetic diversity, because high genetic diversity correlates with high dispersal capabilities.

Such a mechanism differs from previous models (Table 1), because biological processes such as dispersal capabilities, more than environmental conditions, are suggested to affect genetic diversity.

B. Founder Effect Consequence for Biogeography

Biogeography starts at the colonization event. The term "founder effect" was first introduced by Mayr (1954) to describe the genetic accidents inherent in small population size that may be important during colonization. To explain the divergence of peripheral populations, Mayr (1954) proposed that genetic change throughout the genome could be extremely rapid in small localized populations that are founded by a few individuals (genetic drift) and are cut off from gene exchange with the main body of the species. Moreover, the selective pressures acting on the population are likely to be different, because the environment of a small peripheral region might present different characteristics. Both neutral and selective processes can induce changes in the genetic structure of isolated populations, and the use of both markers under genetic selection (allozymes) and neutral ones (microsatellites) would allow us to characterize evolutionary processes driving genetic differentiation.

In the late 1950s and early 1960s the Bureau of Commercial Fisheries, at the request of the Division of Fish and Game of the State of Hawaii, undertook an introduction program, which was supposed to import new species of groupers and snappers for fisheries purposes. Such official programs of introduction of new species provide unique opportunities to follow *in situ* founder effects in terms of genetic evolution. Between 1955 and 1961, 11 species of Serranidae and Lutjanidae were introduced from several geographic locations. Among these species only 3 are known to be established (Oda and Parrish, 1981; Randall, 1987a):

1. *Cephalopholis argus* is a widespread grouper that occurs from the east coast of Africa to French Polynesia. In 1956, 571 small individuals were transported from Moorea (Society Archipelago) to the Hawaiian Islands. One lot of 171 was released in Oahu and the remaining 400 were released in Hawaii (Big Island). This fish has not become abundant and it is encountered mainly in Hawaii.
2. *Lutjanus fulvus* occurs throughout the Indo-Pacific region. A first introduction in 1956 brought 239 individuals from Moorea to Kaneohe Bay in Oahu. About 3000 additional individuals from the Marquesas Islands were subsequently introduced in the bay. This species became established in Oahu and spread to all other islands of the Hawaiian Archipelago.
3. In 1958, 2435 *Lutjanus kasmira* were transported from the Marquesas Islands and released in Kaneohe Bay. The blue-lined snapper has spread throughout the Hawaiian Islands and reaches high abundance in some places. Because of its high abundance and its consequent competition with native species, Randall (1987a) considered this introduction as an unfortunate one.

Such introduction programs can be viewed as founder effects from a population genetics perspective because, for the species that became established, the population started from a small number of individuals.

Despite the fact that only 571 *C. argus* and 2435 *L. kasmira* were released, no major change in polymorphism and heterozygosity was observed between ancestral and introduced populations (Planes and Lecaillon, 1998). However, the change in allelic frequencies allows an estimate of the effective population size for the Hawaiian Islands populations of between 1 and 5% of the total population size. Such reduced effective population size explains why most of the species introduced in Hawaii (8 out of 11) failed to establish. These planned introductions also permit a test for the selective neutrality of allozymes under extreme conditions—the transport of fish from French Polynesia to Hawaiian Islands must involve a significant change in environment. Comparisons of raw data to models including selective change suggest that the difference observed over the short period since introduction should lead to fixation of alleles within 60 to 200 generations. This seems unlikely when looking at polymorphism found in other Hawaiian species (Shaklee, 1984; Shaklee and Samollow, 1984).

The establishment of introduced species appears to be determined by some sort of lucky event that will favor the reproduction of adults and the return of oceanic larvae. This result suggests that colonization events may occur more often than we expect, but that the establishment of the new species into a new biogeographic area is less frequent. Success of the first reproductive events in a small population appears to be hazardous. Certainly, many species have managed to colonize new areas and modern species distributions are a result of those species that managed to establish at a site because they colonized, found a suitable adult habitat, and were able to use the surrounding ocean to complete the life cycle.

C. Biogeographic Patterns Inferred from Genetic Data

The Indo-Pacific region, which stretches from the Red Sea to the Polynesian Islands, contains the world's largest shorefish fauna, estimated at over 4000 species (Springer, 1982; Myers, 1989). The distribution of species richness shows higher diversity in the Indonesian–Philippines area, and decreases further east, toward the Pacific Islands (Fig. 2). According to species lists, we can observe four major steps in terms of diversity: (1) about 2500 species exist in the Indonesian–Philippines waters, (2) around 1000 to 1300 species are found around Micronesia and Australia, (3) about 600 species exist further east in the Hawaiian and French Polynesian islands, and (4) only 200 species remain around isolated islands such as Pitcairn, Easter, or Johnston islands.

Among the several hypotheses that have been proposed to explain biogeographic patterns in the Pacific Ocean (see review in Rosen, 1988), three major theories can be distinguished (Palumbi, 1996). These are the center of origin, center of accumulation, and center of overlap theories. Basically, they each propose mechanisms that lead to a high diversity in the Indonesian–Philippines area (Fig. 2). The center-of-origins theory suggests that the Indonesian–Philippines area is the place where speciation takes place and that new species diffuse to peripheral habitats such as the Pacific archipelagos (Ekman, 1953). As an alternative view, it has been proposed that speciation has more chance to occur in isolated peripheral archipelagos and that new species would be transported to the Indonesian–Philippines area via prevailing currents (Ladd, 1960). In this case the Indonesian–Philippines area can be viewed as a center of accumulation, as a consequence of migrations of new species from peripheral archipelagos. Finally, the third theory describes a center of overlap whereby the high diversity of the Indonesian–Philippines area corresponds to a crossroad of several biogeographic provinces. These biogeographic provinces could result from past geological events or present environmental conditions and overlap in the Indonesian–Philippines area, where they create high species diversity. Many studies on corals, seagrasses, molluscs, and fishes (Salvat, 1971; Springer, 1982; Vermeij, 1987; Rosen, 1988; Wallace *et al.*, 1989; Kay, 1990; Jokiel and Martinelli, 1992; Mukai, 1992; Pandolfi, 1992; Stoddart, 1992; Veron, 1995; Briggs, 1999) have investigated species distribution and support one or the other of these theories. However, it appears that there is no consensus on which model may be more generally suitable.

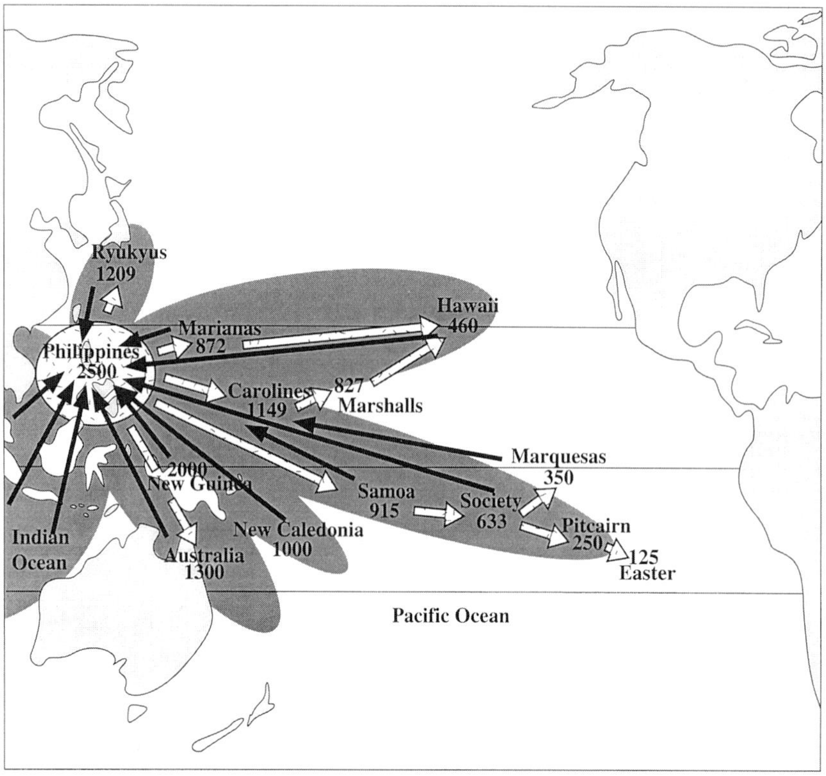

FIGURE 2 Map of the Pacific Ocean area indicating species richness of shorefish. The three major theories proposed to explain the species distribution and, more especially, the high diversity in the Indonesian–Philippines area are also represented: center-of-origins theory (white arrows), center-of-accumulation theory (black arrows), and center-of-overlap theory (gray ellipses).

Some genetic data covering the entire Pacific Plate are available for some coral reef invertebrates. All surveys show differentiation between populations despite the dispersal capacities inferred from the duration of the pelagic larval phase. An allozyme study of the crown-of-thorns starfish indicates a decreasing gradient of genetic diversity linked to the decreasing gradient of species in the Pacific (Nichida and Lucas, 1988). A giant clam (*Tridacna maxima*) study showed that gene flows were favored along an east–west gradient in the Indo-West Pacific but were limited along a north–south gradient (Benzie and Williams, 1997). Populations appeared homogeneous within similar latitudes, yet major differences occurred between samples collected at different latitudes. This pattern appears to contradict the diversity clines in the entire fauna, which show a gradient of declining species richness going east along a single latitude in the Pacific.

The distribution of number of alleles and heterozygosity within populations appears to be higher in the Great Barrier Reef and in Fiji than elsewhere in the Pacific. This does not favor the hypothesis of the Indonesian–Philippines area being a center of dispersal (Benzie, 1994; Benzie and Williams, 1997). Research on sea urchins (*Echinometra* sp.) revealed a pattern opposite to that for the giant clam. The cline of mtDNA variation within populations, taken as an indicator of genetic diversity, is similar to the cline of total species diversity, supporting the center-of-dispersal hypothesis (Palumbi and Metz, 1991; Palumbi, 1996, 1997; Palumbi *et al.*, 1997). Distribution of haplotypes from restriction enzyme analysis of mitochondrial DNA of the coconut crab (*Birgus latro*) shows significant differentiation among Pacific populations. The stem phylogeny of haplotypes is located in the Pacific lineage, suggesting that coconut crabs dispersed through the Pacific Ocean and the Indian Ocean from the Indonesian–Philippines area (Lavery *et al.*, 1995, 1996).

Prior to analyzing single-species phylogeographic patterns, data on their origin should be considered. Speciation in Pacific coral reef fishes seems recent according to analysis of cytochrome *b* in the mitochondrial DNA of several chaetodontid species through the Indian and Pacific oceans (McMillan and Palumbi, 1995, 1997; McMillan *et al.*, 1999). The similar levels of sequence differences in the two species complexes in this family suggest that they arose contemporaneously and that similar historical events have shaped their evolution. Finally, the low intraocean levels of mtDNA variation suggest that the divergence within the Pacific must have been recent in the Chaetodontidae.

When looking at single-species phylogeographic patterns, no large data set covering the Pacific is available in the literature for coral reef fishes, and most studies are restricted to smaller regional areas. A recent survey on *Acanthurus triostegus*, including 18 sample sites widespread on the Pacific Plate, also shows a different pattern than the ones previously described for giant clam and sea urchin (S. Planes, unpublished data). Enzyme analysis based on 17 polymorphic loci demonstrated significant differences among all sites. Four major groups came out of the analysis: (1) Polynesia and Clipperton Island, (2) central Pacific (Palau, Fiji, and the Solomon Islands), (3) Philippines, Great Barrier Reef, New Caledonia, and Guam, and (4) Hawaii and Marquesas (Fig. 3). The distribution of heterozygosity values was different from the expectation of the center-of-dispersal hypothesis because the values were higher in Polynesia and Clipperton (14.3–18.1%) compared to all other sites from the other groups (4.5% in the Marquesas to 12% in the Solomon Islands, with average values of 10.2%) (Fig. 3). Distribution of the number of alleles is directly correlated to the heterozygosity values, with higher values of heterozygosity associated with higher total numbers of alleles over all polymorphic loci. Assuming that we can discount colonization events when considering allele biogeography (Avise, 1994), the distribution of the genetic diversity and the differentiation between populations suggest that Polynesia is the ancestral region from which *Acanthurus triostegus* spread over the entire Pacific and Indian oceans. The Clipperton population shows unexpectedly high heterozygosity for a geographically isolated population. Any explanation of this result will be highly speculative without data on the genetic structure of the *Acanthurus* population along the American coast. The Marquesas population appears to be related to Hawaiian populations, but also very close to populations from southern Polynesia. The low diversity values in the Marquesas population compared to the Hawaiian one suggest colonization from Hawaii. This colonization arose from rare events leading to very limited numbers of migrants and consequently carrying only a small amount of the diversity. Finally, the central Pacific populations (Palau, Solomon Islands, and Fiji) seem to diverge from the western ones but are geographically intermediate between Polynesian and western populations. Similarity in allelic frequencies between the Caledonian and western populations suggests that the colonization of the western areas from Polynesia occurred through the southern islands such as New Caledonia. Such east to west colonization is concordant with the phylogeography of the Ridley sea turtle (*Lepidochelys* spp.) inferred

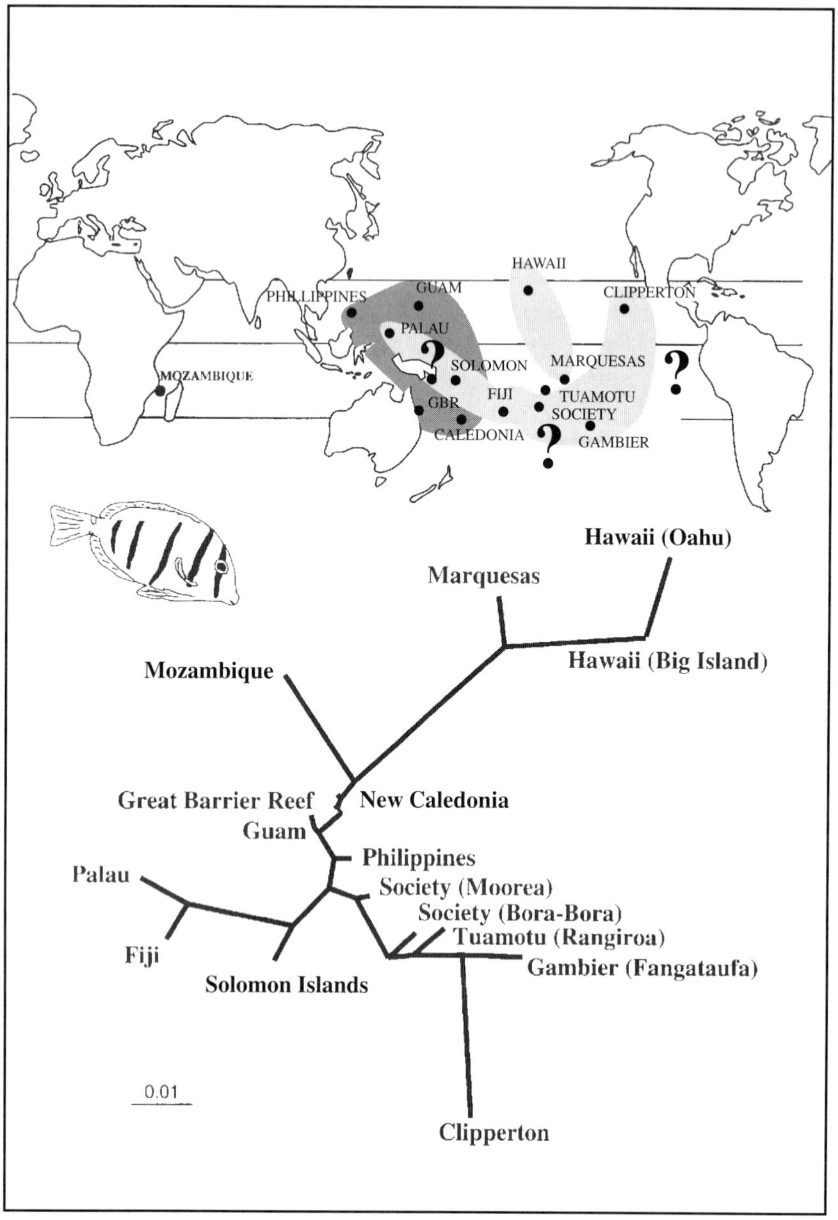

FIGURE 3 Phylogeography for *Acanthurus triostegus* in the Pacific Ocean. Populations are clustered using the neighbor-joining algorithm on nonbiased Nei's distances. Nei's genetic distances are computed from allelic frequencies of each population. The map at the top indicates closely related populations within units of similar genetic structure.

from mitochondrial DNA sequences (Bowen *et al.*, 1998).

Although not yielding a clear picture, genetic data exhibit the complexity of a system that has been underestimated up to now in studies based on species distribution records. Biogeographic genetic patterns deduced from the only three studies available on coral reef species (giant clam, sea urchin, and fish) all differed, suggesting that there is no general model to explain the distribution of species in the Pacific. There is now a need for more large-scale genetic surveys to determine how variable such patterns can be among coral reef species. Such future works should also take into account the phylogeny when comparing species, to correct correlation according to the relationship between species compared, but also to take into account the age of each species and its consequence on the related structure.

V. Population Genetics Perspectives

A. Spatial Structure Deduced from Genetic Markers

Fourteen studies have attempted to describe intraspecific patterns of genetic differentiation in coral reef fishes at various spatial scales (Table 2). Table 2 does not include studies that analyzed temporal patterns (e.g., Avise and Shapiro, 1986; Lacson and Morizot, 1991; Planes and Lecaillon, 1998), because such studies were designed with different goals. Analyses of the genetic structure of coral reef fish populations were initiated in 1982 in Japan (Bell *et al.*, 1982) and in 1984 in Hawaii (Shaklee, 1984; Shaklee and Samollow, 1984), both studies looking at allozyme variations. With the exception of a remote oceanic population of *Amphiprion clarkii* in the Bonin Islands (Bell *et al.*, 1982), these first studies undertaken in the Pacific did not reveal significant genetic differentiation among populations. The result was that genetic approaches remained unexplored over the next 10 years. The straightforward conclusion and general belief during that period was that coral reef fish populations exchanged genes through the pelagic stage at rates sufficient to homogenize the genetic structure over broad geographic ranges. Many other studies of genetic differentiation have emphasized the fact that marine organisms maintain a certain homogeneity over a much larger scale as compared to freshwater organisms (Gyllensten, 1985; Ward *et al.*, 1994). In fishes, Ward *et al.* (1994) have used an extensive survey of the literature concerning 113 species to show that the average values of interpopulation divergence as measured by Nei's (1987) statistic (G_{st}) were much smaller for marine fishes (0.062) than for freshwater fishes (0.202), with anadromous fishes such as salmonids showing intermediate values (0.108). The apparent openness of the marine environment and the large dispersal capabilities of its inhabitants have been the features proposed to explain this now widely accepted trend.

In the early and mid-1990s, several genetic surveys were conducted both in the Pacific and in the Caribbean, again using allozyme variation but developed using comparative approaches (Lacson, 1992; 1994; Johnson *et al.*, 1993; Planes, 1993; Planes *et al.*, 1993, 1998a; Doherty *et al.*, 1995; Lacson and Clark, 1995). In the following years, several studies undertook similar surveys using DNA techniques involving mtDNA RFLPs (Shulman and Bermingham, 1995; Elliot, 1996) and sequences of mtDNA (McMillan and Palumbi, 1995).

All together, there are now 45 separate sets of genetic data for 38 different species (because some species were studied in more than one location or by more than one technique). They represent 12 families of coral reef fishes but 50% are Pomacentridae and only one or two species have been investigated in any other family. It is surprising to note that some of the major coral reef fish families, such Scaridae or Balistidae, were never surveyed. Among the 45 cases, 33 were investigated with allozymes, 9 using mtDNA RFLPs, and 3 looking at sequence variations of the control region and the cytochrome *b* gene of mtDNA. Techniques, sample size, number of polymorphic loci and/or number of restriction enzymes varied enormously between studies. For example, the number of polymorphic allozyme loci ranged from 1 to 17. The average number of individuals processed per population varied from 2 to 10 in mtDNA sequence analysis, from 8 to 81 in mtDNA RFLP analysis, and from 14 to more than 100 in allozyme surveys. As discussed earlier (Section III,B), the sample size and the number of markers significantly affect the statistical power of any subsequent analyses.

Because these studies were developed separately, concentrating on questions from fisheries to evolutionary perspectives, the 45 cases investigated different spatial scales, in various regions. One-third (15) of the 45 species cases were devoted to western Atlantic tropical coral reef fishes. Among the studies on Indo-Pacific fishes (30), 4 examined Indian Ocean species (Western Australia), 6 looked at Indian vs. Pacific Plate populations, and most others looked at Pacific species within a single archipelago.

Over all 45 cases, 27 exhibited genetic differentiation in either allozyme or haplotype frequencies. For the 33 studies processed with allozymes, we have homogenized data on F_{st} departure from 0 for the present review using the Workman and Niswander (WN) χ^2 algorithm (Workman and Niswander, 1970). F_{st} values were computed either from Wright's algorithm [F_{st} (W)] (Wright, 1978) or from Weir and Cockerham estimates [F_{st} (WC)] (Weir and Cockerham, 1984). Once data were standardized, 24 out of 33 cases found significant genetic differentiation in allozyme frequencies. Over the 9 cases showing homogeneous allozyme frequencies, 6 were conducted with small sample sizes [*Epinephelus merra, Stegastes dorsopunicans, Stegastes partitus* (1), *Stegastes planifrons,* and *Thalassoma bifasciatum* (2)] and/or with a very low number of polymorphic loci [*Lutjanus sebae, Stegastes dorsopunicans,* and *Thalassoma bifasciatum* (2)]. Because reduced numbers of polymorphic loci and/or low sample sizes limit the power of any population genetic survey, results of these surveys should be considered suggestive. At minimum, they should be repeated with larger samples prior to concluding on panmixus. Finally, three species showed unambiguous homogeneity of their allozyme frequencies (*Lutjanus malabaricus, Caesio chrysozonus,* and *Ctenochaetus striatus*). Because *L. malabaricus* showed significant genetic variation when analyzed using mtDNA RFLPs, there are only two species that have been adequately studied that exhibit panmixus over broad regional scales. The problem of sample size also rises in studies using mtDNA

TABLE 2 Summary of Genetic Studies of Coral Reef Fishes[a]

Species	Genetic markers	Poly·loci	Geographical range	No. of fish (no. of samples)	Differentiation indexes and significance tests	References
Holocentridae						
Holocentrus ascensionis	mtDNA RFLPs	11	Belize, Panama, Curacao, Barbados, St. Croix, San Salvador (Caribbean Plate)	61 (6)	0.0^{ns}; % haplotype var, AMOVA	Shulman and Bermingham (1995)
Serranidae						
Epinephelus multinotatus	Allozymes	4	Western Australia (Indian Plate)	454 (5)	0.0122^{***}; F_{st} (WC), WN	Johnson *et al.* (1993)
Epinephelus merra	Allozymes	13	New Caledonia (Pacific Plate)	88 (3)	-0.005^{ns}; F_{st} (WC), WC	Planes *et al.* (1998a)
Lutjanidae						
Lutjanus malabaricus (1)	Allozymes	10	Great Barrier Reef (Pacific Plate) to Western Australia (Indian Plate)	182 (4)	0.0101^{ns}; G_{st}, Monte Carlo	Elliot (1996)
Lutjanus malabaricus (2)	mtDNA RFLPs	5	Great Barrier Reef (Pacific Plate) to Western Australia (Indian Plate)	271 (4)	0.0444^{***}; G_{st}, Monte Carlo	Elliot (1996)
Lutjanus sebae	Allozymes	2	Western Australia (Indian Plate)	461 (5)	0.0028^{ns}; F_{st} (WC), WN	Johnson *et al.* (1993)
Pristipomoides filamentosus	Allozymes	5	Hawaiian Archipelago (Pacific Plate)	1118 (8)	0.0050^{***}; F_{st} (W), WN	Shaklee and Samollow (1984)
Caesionidae						
Caesio chrysozomus	Allozymes	12	Great Barrier Reef (Pacific Plate) (Cairns vs. Swain areas)	310 (6)	0.0023^{ns}; F_{st} (WC), WN	Doherty *et al.* (1995)
Lethrinidae						
Lethrinus choerorynchus	Allozymes	8	Western Australia (Indian Plate)	569 (6)	0.006^{***}; F_{st} (WC), WN	Johnson *et al.* (1993)
Lethrinus nebulosus	Allozymes	7	Western Australia (Indian Plate)	603 (6)	0.0068^{**}; F_{st} (WC), WN	Johnson *et al.* (1993)
Haemulidae						
Haemulon flavolineatum	mtDNA RFLPs	11	Belize, Panama, Curacao, Barbados, St. Croix, San Salvador (Caribbean Plate)	65 (6)	0.1^{ns}; % haplotype var, AMOVA	Shulman and Bermingham (1995)
Chaetodontidae						
Chaetodon multicinctus	mtDNA sequences	—	Hawaii and Johnston Atoll (Pacific Plate)	15 (2)	0.6 (Cyt *b*), 13.6 (D-loop), % seq difference	McMillan and Palumbi (1997)
Chaetodon punctatofasciatus	mtDNA sequences	—	Indonesia, Philippines, Palau, Guam (Pacific Plate)	12 (2)	0.9 (Cyt *b*), 15.5 (D-loop), % seq difference	McMillan and Palumbi (1997)
Chaetodon pelewensis	mtDNA sequences	—	Fiji, Cook, Tahiti (Pacific Plate)	11 (3)	0.6 (Cyt *b*), 17.2 (D-loop), % seq difference	McMillan and Palumbi (1997)
Pomacentridae						
Abudefduf saxatilis	mtDNA RFLPs	11	Belize, Panama, Curacao, Barbados, St. Croix, San Salvador (Caribbean Plate)	67 (6)	0.0^{ns}; % haplotype var, AMOVA	Shulman and Bermingham (1995)
Acanthochromis polyacanthus	Allozymes	7	Great Barrier Reef (Pacific Plate) (Cairns vs. Swain areas)	354 (8)	0.7919^{***}; F_{st} (WC), WN	Doherty *et al.* (1995)
Amphiprion clarkii	Allozymes	7	South Japan and Bonin Islands (Pacific Plate)	171 (6)	0.0990^{***}; F_{st} (W), WN	Bell *et al.* (1982)
Amphiprion melanopus	Allozymes	8	Great Barrier Reef (Pacific Plate) (Cairns vs. Swain areas)	394 (7)	0.0497^{*}; F_{st} (WC), WN	Doherty *et al.* (1995)
Chromis atripectoralis	Allozymes	13	Great Barrier Reef (Cairns vs. Swain areas)	169 (4)	0.0377^{*}; F_{st} (WC), WN	Doherty *et al.* (1995)
Chromis cyanea	Allozymes	5	Puerto Rico vs. Jamaica (Caribbean Plate)	42 (2)	0.0290^{*}; F_{st} (W), WN	Lacson (1992)
Chrysiptera biocellata	Allozymes	9	Maldives (Indian Plate) vs. Guam (Pacific Plate)	45 (2)	0.7400^{***}; F_{st} (WC), WN	Lacson and Clark (1995)

Chrysiptera cyanea	Allozymes	15	Palau vs. Okinawa (Pacific Plate)	46 (2)	0.708***; F_{st} (W), WN	Lacson (1994)
Chrysiptera glauca	Allozymes	8	Maldives (Indian Plate) vs. Guam (Pacific Plate)	60 (2)	0.3500***; F_{st} (WC), WN	Lacson and Clark (1995)
Chrysiptera leucopoma	Allozymes	9	Maldives (Indian Plate) vs. Guam and Palau (Pacific Plate)	75 (3)	0.4500***; F_{st} (WC), WN	Lacson and Clark (1995)
Dascyllus aruanus	Allozymes	12	French Polynesia (Pacific Plate)	504 (13)	0.0076*; F_{st} (WC), WN	Planes *et al.* (1993)
Pomacentrus coelestis	Allozymes	10	Palau vs. Okinawa (Pacific Plate)	34 (2)	0.605***; F_{st} (W), WN	Lacson (1994)
Pomacentrus molucensis	Allozymes	9	Great Barrier Reef (Pacific Plate) (Cairns vs. Swain areas)	96 (2)	0.1124*; F_{st} (WC), WN	Doherty *et al.* (1995)
Stegastes dorsopunicans	Allozymes	1	Puerto Rico vs. Jamaica (Caribbean Plate)	29 (2)	0.0000^{ns}; F_{st} (W), WN	Lacson (1992)
Stegastes faciolatus	Allozymes	5	Hawaiian Archipelago (Pacific Plate)	325 (6)	0.0030***; F_{st} (W), WN	Shaklee, 1984
Stegastes leucostictus (1)	Allozymes	2	Puerto Rico vs. Jamaica (Caribbean Plate)	44 (2)	0.056**; F_{st} (W), WN	Lacson (1992)
Stegastes leucostictus (2)	mtDNA RFLPs	11	Belize, Panama, Curacao, Barbados, St. Croix, San Salvador (Caribbean Plate)	61 (6)	17.2 ***; % haplotype var, AMOVA	Shulman and Bermingham (1995)
Stegastes nigricans (1)	Allozymes	9	Maldives (Indian Plate) vs. Guam (Pacific Plate)	52 (2)	0.7300***; F_{st} (W)	Lacson and Clark (1995)
Stegastes nigricans (2)	Allozymes	10	Great Barrier Reef (Pacific Plate) (Cairns vs. Swain areas)	300 (6)	0.0516*; F_{st} (WC), WN	Doherty *et al.* (1995)
Stegastes nigricans (3)	Allozymes	14	New Caledonia (Pacific Plate)	120 (3)	0.0489*; F_{st} (WC), WC	Planes *et al.* (1998a)
Stegastes partitus (1)	Allozymes	7	Puerto Rico vs. Jamaica (Caribbean Plate)	44 (2)	0.0120^{ns}; F_{st} (W), WN	Lacson (1992)
Stegastes partitus (2)	Allozymes	8	Florida Keys (Caribbean Plate)	194 (4)	0.1000***; F_{st} (W), WN	Lacson *et al.* (1989)
Stegastes planifrons	Allozymes	4	Puerto Rico vs. Jamaica (Caribbean Plate)	43 (2)	0.0060^{ns}; F_{st} (W), WN	Lacson (1992)
Labridae						
Halichoeres bivittatus	mtDNA RFLPs	11	Belize, Panama, Curacao, Barbados, St. Croix, San Salvador (Caribbean Plate)	57 (6)	7.9 **; % haplotype var, AMOVA	Shulman and Bermingham (1995)
Thalassoma bifasciatum (1)	mtDNA RFLPs	11	Belize, Panama, Curacao, Barbados, St. Croix, San Salvador (Caribbean Plate)	89 (6)	0.0^{ns}; % haplotype var, AMOVA	Shulman and Bermingham (1995)
Thalassoma bifasciatum (2)	Allozymes	2	Puerto Rico vs. Jamaica (Caribbean Plate)	45 (2)	0.0140^{ns}; F_{st} (W), WN	Lacson (1992)
Blenniidae						
Ophioblenius atlanticus	mtDNA RFLPs	11	Belize, Panama, Curacao, Barbados, St. Croix, San Salvador (Caribbean Plate)	64 (6)	0.3^{ns}; % haplotype var, AMOVA	Shulman and Bermingham (1995)
Gogiidae						
Gnatholepis thompsoni	mtDNA RFLPs	11	Belize, Panama, Curacao, Barbados, St. Croix, San Salvador (Caribbean Plate)	61 (6)	8.2 **; % haplotype var, AMOVA	Shulman and Bermingham (1995)
Acanthuridae						
Acanthurus triostegus (1)	Allozymes	13	French Polynesia (Pacific Plate)	409 (11)	0.0886***; F_{st} (WC), WN	Planes (1993)
Acanthurus triostegus (2)	Allozymes	17	New Caledonia (Pacific Plate)	120 (3)	0.0380*; F_{st} (WC), WC	Planes *et al.* (1998a)
Ctenochaetus striatus	Allozymes	11	Great Barrier Reef (Pacific Plate) (Cairns vs. Swain areas)	100 (2)	0.0055^{ns}; F_{st} (WC), WN	Doherty *et al.* (1995)

[a]The table details the markers used [allozymes, restriction fragment-length polymorphisms (RFLPs) of mitochondrial DNA (mtDNA), sequences of microsatellite DNA], the number of polymorphic loci assessed (Poly-loci), the geographical range of each study, and the number of fish and samples analyzed. Detailed information can be found in references. Results are synthesized through a divergence index varying according to the marker used: F_{st} for allozyme, AMOVA for RFLPs of mtDNA, and divergence (%) among sequences for sequences of msDNA. The F_{st} values were computed either from Wright algorithm [F_{st} (W)] (Wright, 1978) or from Weir and Cockerham estimates [F_{st} (WC)] (Weir and Cockerham, 1984). Signficance of F_{st} values (e.g., departure from 0) were computed using the Workman and Niswander (WN) χ^2 algorithm (Workman and Niswander, 1970) or Weir and Cockerham (WC) method. Significance is denoted as follows: ns, not significant;*, 0.05; **, 0.01; ***, 0.001.

RFLPs, and sequences of mtDNA (control region and cytochrome *b*). Both existing studies used 10 or fewer individuals per location. In the analysis of Shulman and Bermingham (1995), there was only a rare case of evidence for fixation of haplotypes in any population, and a larger sample size would have been necessary to obtain statistical analysis with higher levels of confidence. For example, an analysis of mtDNA RFLPs of *L. malabaricus* (Elliot, 1996) in the eastern, northern, and western areas of Australia was processed using average sample size of 68 individuals per population. Such a high sample size makes it possible to show significant differences in frequencies of haplotypes between populations, despite the fact that they were sharing most haplotypes. Finally, an analysis of chaetodontids (McMillan and Palumbi, 1995) was oriented to phylogenetic relationship between closely related species rather than population structures. Because of this perspective, sample sizes were very small and analysis did not specifically look at population structure within each species.

Overall, recent detailed studies, mostly on allozymes, have revealed that spatial differentiation is very often more important for genetic differentiation than what might be expected from the duration of the pelagic phase and subsequent dispersal capacities of larvae through oceanic currents. Local differentiation may sometimes be due to temporal changes, such as in the *Stegastes partitus* populations of the Florida Keys that were significantly divergent in a preliminary sampling, and similar 2 years later (Lacson and Morizot, 1991). However, in most cases, this local differentiation seems most easily explained by a lack of gene flow, such as in populations of *Acanthurus triostegus* between Moorea and Tahiti (18 km apart) (Planes, 1993; Planes *et al.*, 1996), or in those of *A. triostegus* and *Stegastes nigricans* around New Caledonia (Planes *et al.*, 1998a). This suggests that the immense majority of larvae, if they are to recruit, will recruit on their natal island, even if they are initially transported several hundred kilometers away. Among the few studies devoted to this topic in coral reef fishes, most have found that effective dispersal appears much smaller than potential dispersal extrapolated from current speed and duration of the larval stage. This leads to the hypothesis that dispersal-limiting mechanisms are at work, at least in fishes (cf. Chapter 8, this volume). This fresh view of the population structure of coral reef fishes must be considered in future management strategies. It also opens new perspectives for future research because it is not possible to assume that the presence of a pelagic larval stage always ensures large panmictic populations.

B. Examples of Population Genetics Results

Very few surveys have attempted multispecies large-scale analyses on numerous sample sites and/or high numbers of individuals processed per site. This section summarizes three major examples and emphasizes the difficulty in separating biological and historical parameters that shape the observed population structure.

One of the largest studies in terms of number of sample sites was a survey of the genetic structure of *A. triostegus* (Planes, 1993) and *Dascyllus aruanus* (Planes *et al.*, 1993) over all areas of French Polynesia. These species showed contrasting genetic structures. *Dascyllus aruanus* populations appeared almost homogeneous across Polynesia (minor genetic differences could be detected when comparing remote archipelagoes), whereas *A. triostegus* populations were highly structured, with large genetic differences between islands. The genetic structure of populations of *A. triostegus* was directly linked with the major oceanic currents within the area. Comparison of results obtained on *D. aruanus* and *A. triostegus* did not fit with the idea that species with long pelagic larval stages exhibit greater dispersal, and consequently show higher gene flows between populations. In fact, *D. aruanus* larvae spend about 25 days in the ocean whereas *A. triostegus* larvae remain in the plankton for 60 days. We proposed that this contradiction (species with longer larval duration showing restricted gene flow) is due to historical factors. During the last glaciation, 8000–10,000 years before present (BP), sea level was 120 m below present sea level and lagoons of classic darwinian atolls that are 50–70 m deep at the maximum were completely dry. Because *D. aruanus* only inhabits lagoons, and such habitats were not available in French Polynesia during the last glaciation, we proposed that this species became extinct in this region at that time. The present population is a result of a unique and recent recolonization. Consequently, there has not been enough time for the appearance of genetic structure as a result of genetic drift. By contrast, *A. triostegus* is more ubiquitous and inhabits lagoons as well as oceanic slopes. Because this fish could be maintained on oceanic slopes when the sea level was lower, the structure of *A. triostegus* populations is older and expresses longer evolution, showing genetic differentiation due to long-term genetic drift and limited gene flow. This study suggests that oceanic currents as well as historical events affect the population structures between oceanic islands. In these interpretations, historical factors played a more crucial role than did the pelagic larval duration.

In another context, Doherty *et al.* (1995) selected seven species of coral reef fishes, to provide a range of pelagic larval duration from 0 to 55 days, for genetic comparison between two regions of the Great Barrier Reef separated by 1000 km. Only one species, an anemone fish (*Amphiprion melanopus*), did not show significant heterogeneity among reefs within regions, indicating chaotic genetic patchiness at local scales, probably arising from patchy recruitment limited to sea anemones. After pooling within-region variation, all but two species showed significant regional differences (Table 2). The two exceptions were a caesionid (*Caesio chrysozonus*) and a surgeonfish (*Ctenochaetus striatus*), which had the longest pelagic larval durations. The logarithm of genetic differentiation between regions, expressed as F_{st} values, was correlated with pelagic larval duration by an inverse linear relationship that explained 85% of the variance among species (Fig. 4). On this basis, it was estimated that larval periods longer than 1 month should result in enough gene flow to homogenize fish populations over 1000 km along the Great Barrier Reef. Comparisons with invertebrate taxa sampled over the same sites show that fish populations are more differentiated for a given pelagic larval duration. Such a difference between invertebrates and fishes suggests that fish larvae may use their greater mobility to resist passive dispersal, using either their swimming abilities or favorable currents to limit dispersal (cf. Chapter 8, this volume). This study clearly demonstrates that within the fishes, in some cases, pelagic larval duration may explain larval dispersal and subsequent gene flow. At the same time, Doherty *et al.* (1995) compared populations of *Amphiprion melanopus* of the Great Barrier Reef with an out-group (Chesterfield Reefs). They confirmed the genetic homogeneity of mainland populations but demonstrated that the relationship between pelagic larval duration and genetic differentiation does not apply to island systems. In the Great Barrier Reef, which is a nearly continuous reef matrix along a continental shelf, pelagic larval duration

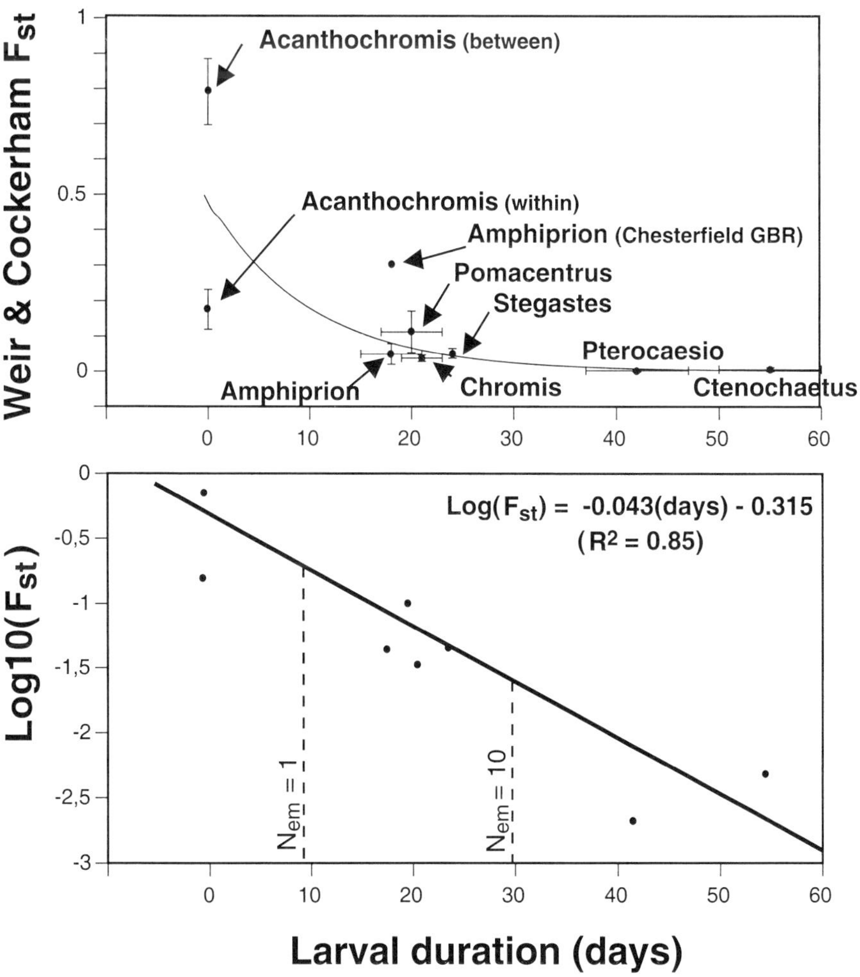

FIGURE 4 Relationship of genetic variance (F_{st}) vs. pelagic larval duration for the seven Australian coral reef fish species, plotted on arithmetic scales and after logarithmic transformation of F_{st} [F_{st} is a measure of a standardized variance among allelic frequencies that is calculated according to the method of Weir and Cockerham (1984) in the present work]. The horizontal error bars are 95% confidence intervals. The values for *Acanthochromis* (within) and *Amphiprion* from the Chesterfields are for comparison only and were not included in the least-squares calculations [see detail of analysis in Doherty *et al.* (1995)]. The N_{em} values (number of migrants exchanged per generation at equilibrium) were calculated from F_{st} values. Populations to the left of the first dashed line have the potential to speciate ($N_{em} < 1$). Populations to the right of the second dashed line should be genetically homogeneous ($N_{em} > 10$).

is the major factor explaining population structure. The Great Barrier Reef also was affected by sea level variations during the Pleistocene, but because it is a geomorphologic unit, the evolution of different species was parallel in that region. Within a geomorphologic unit, the biology of each species becomes the key factor to determine population structure. Evidence that historical events are predominant over biological ones can be found in the population structures observed in the Capricorn and Bunker reefs (located at the extreme south of the Great Barrier Reef, where reefs are at their current limit of extension). They show large genetic differences when compared to populations of the central Great Barrier Reef for *Acanthochromis polyacanthus* (Doherty *et al.*, 1994a). Because it lacks a pelagic larval stage, *A. polyacanthus* can be used to draw historical patterns (Planes and Doherty, 1997a,b).

Finally, population genetic structure of *Stegastes nigricans, Epinephelus merra,* and *Acanthurus triostegus* was investigated throughout New Caledonia (Planes *et al.*, 1998a). Given the size and the geomorphology of the New Caledonia lagoon, we expected to find substantial gene flow between the sampling sites, resulting in genetic homogeneity throughout the lagoon. The survey examined the three species at three sites, Noumea, Thio, and Point Game, roughly separated by 400, 200, and 350 km, respectively, around New Caledonia. Genetic analyses were conducted in order to test Crisp's (1978) hypothesis that the magnitude of dispersal of a species may be predictable from the duration of its pelagic larval stage [an approach similar to that of Doherty *et al.* (1995)]. The high degree of genetic differentiation in two species (*S. nigricans* and *A. triostegus*) is not consistent with what we expected. Our results contradict Crisp's (1978) hypothesis because the duration of the pelagic larval stage was not correlated with genetic differentiation. The duration of the larval stage differs greatly between *S. nigricans* and *A. triostegus* (24 and 60 days, respectively) but the genetic differentiation was similar. On the other hand, *E. merra,* whose larval stage is of intermediate length (39 days), displays no genetic differentiation among these three sites. The F_{st} values for *A. triostegus* (0.028 to 0.044) in New Caledonia were much higher than those found in the Great Barrier Reef for *Ctenochaetus striatus* (0.006), a member of the same family with a larval stage of similar duration. Evidence of differentiation between populations of *S. nigricans* and *A. triostegus* on a small spatial scale suggests that larvae, juveniles, and adults do not disperse. Whereas dispersal capacities of juveniles and adults depend mainly on behavioral factors, the dispersal capacity of larvae is affected both by their behavior and by current flows. There are no data on the larval behavior of these two species but similarity of genetic differentiation (uniform F_{st} values) suggests that a similar barrier to gene flow applies to both species, whatever their larval behavior. Both Jarrige *et al.* (1975) and Dandonneau *et al.* (1983) suggested that the climate of New Caledonia, which is largely determined by the south–east trade wind inducing currents and waves flowing from the east, has led to hydrologically separated reef systems on the east and west coasts of New Caledonia, with little interconnection in terms of water flow. This strong difference between the eastern and western reefs combined with the limited degree of exchange between the ocean and the lagoon may explain the observed isolation and genetic differentiation in *S. nigricans* and *A. triostegus,* regardless of the length of the larval stage. In the context of these hydrodynamic parameters, the *E. merra* data are contradictory because they indicate genetic homogeneity across the New Caledonia lagoon. However, the sampling of *E. merra* was limited in one site, which did not provide enough replicates for accurate estimates of allele frequency (increased variance of frequencies) and restricted comparisons with other sites. The absence of genetic structure may be an artifact of the difference in the sampling between this species and the other two. On the other hand, if the absence of genetic differentiation in *E. merra* does not come from sampling artifacts, then we must consider the possibility that the observed genetic homogeneity resulted from an ecological feature. In particular, the observed lack of genetic differentiation could have resulted from a more recent colonization of the New Caledonia lagoon by *E. merra,* in contrast to the other two species, and in consequence an insufficient number of generations for genetic drift to lead to significant differentiation. In addition, groupers are well documented as forming large aggregations at spawning grounds, with individuals migrating up to several miles (Thresher, 1984). Such behavior may favor gene flow and would homogenize populations to a single panmictic stock.

Our three examples allow us to divide factors affecting genetic structure of coral reef populations into three majors groups: (1) functional factors related to the ecology and the biology of each species, including parameters such as reproduction, behavior, or pelagic larval duration; (2) historical factors such as colonization or extinction consequent to sea level variation; and (3) physical factors, including the effect of oceanic currents and/or the geomorphologic structure of the reef, limiting the potential to disperse. We have shown that the patterns observed along the Great Barrier Reef are not similar to those found around isolated island systems. The three major factors acting on the genetic structures are not equivalent: historical factors are certainly predominant over the other ones. This emphasizes the necessity to determine the historic origin of a species prior to attempts to understand actual mechanisms driving genetic differentiation.

C. Metapopulation Perspectives

The metapopulation concept describes assemblages of populations that breed and evolve sufficiently independently of one another. Independent evolution of local populations may occur if migration rates and concomitant gene flow are not very high (Olivieri *et al.*, 1990). Although the metapopulation concept has been used in several areas of ecological and evolutionary theory, many empirical questions remain about how metapopulations function. The extinction of local populations is central to most metapopulation models. In fact, initial models described populations subject to extinction with equal and temporally independent probabilities (Levins, 1969, 1970). However, some authors have proposed a model of "patchy populations" in which extinction is absent or unimportant (Harrison, 1991). This model could be more common in marine ecosystems, because the life cycle of marine organisms gives many opportunities for migration, providing a mechanism for resistance against extinction events. For example, many sessile marine organisms appear to spread their offspring widely in comparison to the scale of patches in which recruitment and growth occur (Strathmann, 1974). Random dispersal may be a very general means of population persistence in spatiotemporally varying environments (Goodman, 1987), but is also an extremely powerful force likely to produce large-scale genetic homogeneity. Moreover, field evidence that such a metapopulation model (patchy populations) can produce local genetic differentiation remains infrequent. Two kinds of genetic models for subdivided populations under restricted dispersal have been developed: the island model by Wright (1943, 1946), and the stepping-stone model by Kimura and Weiss (1964) and Malècot (1968).

Studies completed on *Acanthurus triostegus* in French Polynesia did show significant amounts of genetic differentiation (Planes, 1993; Planes *et al.*, 1994). These data were used to determine which model of dispersal best describes the relationships among these populations—that is, whether the island model or a model of isolation-by-distance is more appropriate, depending on the scale surveyed. The significant correlation between logarithmic transformation of genetic divergence and geographic distances, at the within-archipelago level ($r = 0.709$, $p = 0.024$) indicates exchange of individuals mainly between neighboring populations (Fig. 5). The correlation was, however, not significant at the among-archipelagos level ($r = 0.325$, $p = 0.330$), suggesting that long-distance migrations are more sporadic. At the archipelago scale, exchange among neighboring islands seems to prevail; however, no clear link with geographic distances appears on a larger scale. This results in an overall relation between geographic distances and genetic differentiation in accordance with a stepping-stone model at the within-archipelago scale. At the scale of French Polynesia, three main seasonally varying currents occur. Fronts set up by these currents could act as barriers, decreasing the possibility of communication between different water masses (Rougerie and Wauthy, 1986). This could explain the lack of a clear relationship between geographic distance and genetic differentiation among archipelagos. This hypothesis is based on present oceanographic and biological data, but the genetic structure represents several thousand years of evolution during which the surrounding conditions may have been different. Knowledge of these different past conditions might point to other theories. Biological patterns and processes can be quite different when viewed at different spatial scales. Full understanding requires an appreciation of how patterns and processes vary with scale. These genetic studies show that according to the spatial scale of analysis, results can

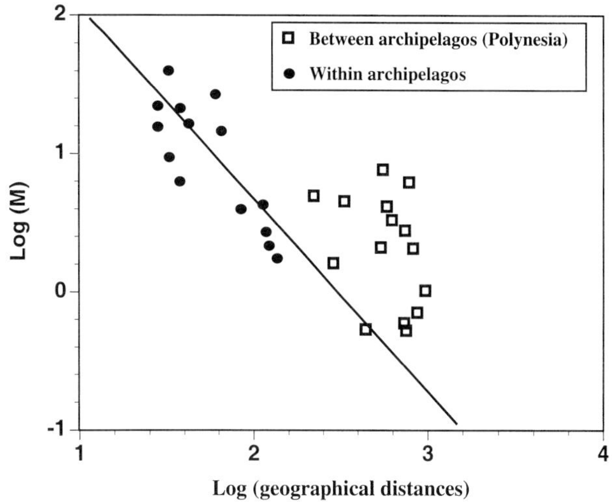

FIGURE 5 Relationship in *Acanthurus triostegus* of French Polynesia between F_{st}-transformed values, $\log[(1 - F_{st})/4F_{st}]$, and geographic distances for two spatial scales. Dark circles represent 15 intraarchipelago pairwise population comparisons, and the linear relationship drawn corresponds only to this spatial scale ($r = 0.810$). Open squares are interarchipelago pairwise population comparisons.

conform to an island model, with no relation between genetic differentiation and geographical distances between archipelagos, or can be better described by an isolation-by-distance model within archipelagos.

The *Acanthurus triostegus* results led us to propose a genetic definition of the "patchy population" model, in which all patches are occupied and reproductively active, with few successful migrations between neighboring populations. According to this model, the island populations represent smaller parts of a larger metapopulation. Each island contains an established population that contributes migrants to, and recruits migrants from, other islands on a similar time scale, with migrants corresponding to a very small fraction of the total recruitment. Although this model is one particular case of a subdivided population, we call it "metapopulation" because it resembles the "patchy population" model proposed by Harrison (1991) on a demographic basis. In the present case, however, migrants do not ensure the replenishment of local populations, but they are abundant enough to prevent complete genetic isolation. This creates the interesting properties of our model: it differs from the usual metapopulation models in that it does not require extinction events at the time scale proposed. Exchanges are not based on reproductive success, as in Gill's model (1978), or on the equilibrium between extinction and immigration (Levins, 1970; Boorman and Levitt, 1973). They depend only on success in migration during the period of dispersal. Migration is not important in terms of population dynamics because so few recruits are migrant, but it maintains the genetic cohesion of the species. This allows evolutionary novelty to circulate within the metapopulation despite a certain amount of local differentiation.

This model describes a subdivided population that is stable through time, with an amount of gene flow small enough to allow significant local differentiation in neutral gene frequency, but high enough to prevent differential fixation over the long term, and therefore preserving the genetic cohesion of the species.

D. A New Challenge Regarding Population Dynamics and Population Genetics

Variability in the supply of new individuals from the plankton may have a major influence on the dynamics of populations of reef fishes. Larval supply is usually episodic, with the arrival of young fishes restricted to a particular season or part of the year. Within this time, year classes may be formed from discrete pulses of settlement that are generally unpredictable in their timing and magnitude. It is often assumed that individuals that comprise a settlement pulse are unlikely to be closely related. Local populations of many species spawn *en masse,* at places and times that are argued to maximize the dispersal of their offspring away from natal reefs (Johannes, 1978; Domeier and Colin, 1997). The small size of the eggs and larvae of reef fishes, and the large volumes of water in which they are quickly diluted, are thought to promote the mixing of progeny, so that a settlement pulse is likely to represent a random sample of the reproductive output of a wide variety of adults.

Despite this view, some studies suggest that the possibility of the progeny of an individual fish remaining together during the larval phase cannot simply be dismissed (Shapiro, 1983). Experimental studies show that the late larval stages of reef fishes are very capable swimmers (Stobutzki and Bellwood, 1994) that may be able to detect and respond to the presence of reefs (Leis *et al.,* 1996). Wolanski *et al.* (1997) demonstrated that late-stage larvae direct their swimming toward reefs, and thus may be able to restrict their dispersal. Furthermore, we have seen that reef fishes can exhibit distinct regional and local genetic patterns. These studies imply that the assumption of widespread dispersal and mixing of the progeny of reef fishes within the plankton should be treated with caution.

To date, only one study has examined the possibility that progeny of the same parents may remain together during the larval phase. Avise and Shapiro (1986) analyzed the genetics of new recruits of *Anthias squamipinnis* from the Red Sea (Eilat, Gulf of Aqaba). As for many coral reef fishes (Jones, 1991), this species settles in small groups (10–20 individuals) into a discrete microhabitat on the reef. Groups of young *A. squamipinnis* (about 1 month post-settlement) were collected from benthic habitats and the genetic structure was described from allozymes. Avise and Shapiro (1986) concluded that there was no evidence that groups of newly settled fishes of this species were the progeny of the same pair of adults. However, their result may have been strongly influenced by the sampling of settled fishes. The young fishes used in their analysis had been present in adult habitats for up to a month prior to their collection. Consequently, it was impossible to determine if the individuals in a group had arrived in a single pulse or if they were in fact the accumulation of several discrete settlement events. The latter possibility seems highly likely, given that schooling species may migrate tens to hundreds of meters after settlement to form groups. Additionally, the presence of resident fish is known to attract newly settling individuals in some species (Sweatman, 1988; Booth, 1991).

A second important assumption made by Avise and Shapiro (1986) was that any mortality that affected cohorts prior to collection occurred in a random fashion. In the few weeks immediately following settlement, newly settled reef fishes typically suffer very high rates of mortality (up to 90%), most of which is attributed to predation (Carr and Hixon, 1995). If this occurs selectively, so that larger and/or faster growing recruits (Anderson, 1988; Meekan and Fortier, 1996) or those in better condition (McCormick, 1998c) are favored, then the genetic structure of a cohort may be altered from the pattern exhibited at the time of settlement. Such selective mortality of newly settled fishes has been recorded in other marine fishes (e.g., Tsukamoto *et al.*, 1989; Ellis and Gibson, 1997; see review by Sogard, 1997).

In a recent study, the opportunity to collect settling larvae and juveniles from the same cohort allowed us to compare the intracohort genetic structure of adults, larvae, and new recruits of a coral reef fish, *Naso unicornis* (Planes and Lecaillon, 2002). These comparisons aim to examine the relative importance of random, selective, and kin processes during the pelagic and the settlement phases. The most significant result comes from the difference in the genetic structure we observed between larvae and juveniles. A parallel survey of otoliths ensured that the juvenile sample caught in the lagoon came from the same cohort as the larvae that colonized the reef 10 days earlier. During these 10 days we observed (1) a significant multilocus differentiation ($F_{st} = 0.0164$), (2) a significant increase in the number of linkage disequilibria, and (3) changes in the loci exhibiting disequilibrium with an overall increase of F_{is}. Such important genetic changes affecting the same cohort over 10 days, combined with a mortality of up to 95% over the same period (Doherty and Sale, 1985; Victor, 1986b; Carr and Hixon, 1995; S. Planes, unpublished data), indicate that selective processes are likely to be responsible for the genetic change of the population. The high mortality of the larvae recruiting to the reef provides a mechanism for selection (Smith, 1979; Smith *et al.*, 1991), and the genetic markers (F_{is} values and linkage disequilibrium) indicate that the genetic structure of the population of larvae entering the lagoon is likely to be modified by selective processes. Selective phenotypic mortality in larvae and new recruits has also been indicated in other fish, with survivorship related to the size and the condition of the fish (Anderson, 1988; Rice *et al.*, 1993; Bertram and Leggett, 1994; McCormick and Kerrigan, 1996; Meekan and Fortier, 1996).

When looking at the age distribution of larvae in our samples, we found that larvae recruiting the same night were between 67 and 94 days old (27-day range), and could not have derived from a single spawning event. Despite the fact that 70% of the larvae were closely similar in age—between 71 and 77 days—there were clearly a number of different days of spawning included. The hypothesis that siblings remain together during the oceanic phase remains a possibility, but our sampling was not extensive enough to be able to test whether fishes of the same age were siblings.

Overall, this is the first time that a variance in the genetic structure between different stages of the same cohort has been demonstrated. Predominance of selective processes in the mortality of newly recruiting larvae has been proposed for sessile invertebrates that would produce a maximum variety of brood genotypes (high fecundity) to ensure that some allelic combinations would be adapted to the environment where they will settle (Blanc and Bonhomme, 1987). Because adults cannot predict the environment (either pelagic or benthic), they produce high numbers of genotypes (e.g., due to the large fecundity of most marine species), with the strategy that some combinations will be selectively adapted to the oceanographic conditions they will encounter during the larval stage and to the environment in which they will settle. Results on *Naso unicornis* open a new research area that allows us to link demographic studies and studies of the genetic evolution of cohorts. Understanding genetic processes occurring during the settlement stage will certainly contribute to understanding parameters affecting the dynamics of coral reef fish cohorts.

VI. The Future Role of Genetics for Coral Reef Fish Studies

The genetic approach has been adopted by reef fish biologists as a way of assessing variation among geographically remote populations of coral reef fishes, and therefore for testing effectiveness of larval dispersal. Most results to date suggest that spatial genetic differentiation in coral reef fishes is very often more pronounced than what would be expected from the duration of the pelagic phase, and the dispersal capacities of larvae through oceanic currents. Selective processes could also contribute to such a pattern by generating genetic variations resulting from adaptation and not from the equilibrium genetic drift migration.

Nevertheless, the number of genetic studies completed remains very small [12 publications since Bell *et al.* (1982)], making it impossible to draw any general conclusions. Available data give some trends, but genetic surveys have to contend with interactions between biological parameters, historical features, hydrodynamics, and geomorphological characteristics of coral reefs that vary among places and species. These make any generalization impossible without many more studies. Dealing with such numbers of variables requires multispecies and multiscale intensive studies on numerous individuals if a synthetic general pattern is desired. Only a few studies have been done from that perspective, and the only point of satisfaction is that they all give significant results that provide key information for understanding persistence and maintenance of coral reef fish populations. Clearly, more data are needed, and not just data for more species, but data obtained from well-designed studies.

A second research perspective, in which genetics will certainly have a significant input, concerns questions relevant to biogeography and phylogeography. A comparison of reef fish assemblages over large scales requires knowledge of the establishment of each assemblage. Genetic markers can surely provide good data on the time of origin of species and of colonizations of particular regions if some suitable calibration from the fossil record is available. Again such approaches require large-scale surveys. A few are now available for some coral reef invertebrates, but this field remains virtually unknown for coral reef fishes. For the few species that have been studied, several models seem to be appropriate, depending on the species analyzed. No general patterns are yet evident, but general trends may emerge as more data become available.

Finally, the genetic/demographic link is starting to be investigated. Species with high fecundity, such as most coral reef fishes, often display two paradoxes: (1) much lower genetic variation than expected under neutrality theory based on their abundance (Avise, 1994; Nei and Graur, 1984), and (2) "chaotic patchiness" involving seemingly stochastic genetic heterogeneity over small spatial and temporal scales. Variance in the reproductive success among spawning groups may account in part for both observations (Hedgecock *et al.*, 1992). This hypothesis was tested in coral reef fishes (Avise and Shapiro, 1986; Planes and Lecaillon, 2002) and did not reveal a significant pattern. Further work needs to be developed in this direction in order to couple genetics and demography during the larval stage and the recruitment transition.

The techniques used for genetic analysis are changing rapidly. Allozyme techniques have been the most frequently used approach up to now, and have provided significant results. In general, if it is feasible to use variation in allozymes rather than nucleic acids for a particular question, one should do so (Parker *et al.*, 1998). This is especially the case when looking at multispecies surveys, because setting up an allozyme protocol can be rapid compared to DNA approaches. However, we should keep in mind that the choice of marker also has implications on evolutionary processes that will be investigated. Choosing markers under genetic selection (such as allozymes or the cytochrome *b* gene of mtDNA) may provide a different result than choosing neutral markers (such as microsatellites). Differences in results may not only be a consequence of the higher variability investigated but more likely are related to the fact we are looking at different evolutionary processes (selection vs. genetic drift).

In addition, the use of hypervariable codominant microsatellite systems can be viewed as the future tool for population genetics. Because they express many alleles for each locus (up to 50), these systems allow more precise genetic discrimination of populations and even of individuals. Five microsatellite loci with an average of 20 alleles per locus provide 95 independent variables, yet it requires at least 25 polymorphic allozyme loci with an average of 5 alleles per locus to get almost the same number of independent variables. No allozyme study done on fish has provided such a level of polymorphism. The disadvantage of the use of microsatellites for population genetics is that it requires significant effort and time to set up the protocol. This has heretofore limited any multispecies approach. In the present context, microsatellite systems seem an ideal tool when effort is centered on one or two species with numerous individuals and samples to process and compare. Finally, microsatellite analysis offers the possibility of genetic surveys of small larvae (Ruzzante *et al.*, 1996). This surely represents a new area of interest that can link genetics and demography during the larval stage and recruitment.

Acknowledgments

Many thanks to those with whom I have discussed these ideas and to those who have commented on this manuscript, including E. Bermingham, G. Bernardi, M. Carr, P. Doherty, R. Galzin, and B. Salvat. Special thanks to P. F. Sale, who challenged me to write this review, for encouragement and review of the manuscript. Most of my work cited here was supported by the Programme National Biodiversité, the Programme National Récif Corallien, and funds of the EPHE and CNRS institutions.

CHAPTER 10

Numerical and Energetic Processes in the Ecology of Coral Reef Fishes

Geoffrey P. Jones
Centre for Coral Reef Biodiversity
School of Marine Biology and Aquaculture
James Cook University
Townsville, Queensland 4811, Australia

Mark I. McCormick
School of Marine Biology and Aquaculture
James Cook University
Townsville, Queensland 4811, Australia

I. Introduction

One of the most fundamental tasks in ecology is to explain why populations behave the way they do (Royama, 1992). Despite many years of research on population dynamics there is little consensus, not only over the answer to this question, but also as to exactly what the question is (Turchin, 1995, 1999; Wolda, 1995; den Boer and Reddingius, 1996; Murray, 1999). When populations are highly variable, researchers will naturally ask why they vary (Sissenwine, 1984). If they remain relatively constant, it is more likely that they will ask why they do not vary (Lack, 1954). In the rarer instances when populations exhibit cycles of abundance, then the challenge is to determine what generates regular fluctuations (Boonstar *et al.*, 1998). The exact question we ask and our initial assumptions can have profound effects on the research programs that we establish, the variables we measure, and ultimately the processes we identify as important. Although population ecologists broadly seek an explanation of numerical patterns, the underlying processes may at first sight have very little to do with numbers at all. A comprehensive explanation of numerical patterns will require an assessment of all the processes that act directly or indirectly to determine the number of individuals in the population.

The lack of consensus among population ecologists is illustrated by the emergence of opposing schools of thought over the processes that regulate population numbers (Krebs, 1985). In historical context, dichotomous views highlight the search for generalizations based on single factors and the ultimate failure to find them. In hindsight, they serve to illustrate the variety of antagonistic processes that may be acting on populations at any one time. The first dichotomy in population ecology involved the existence and importance of density-dependent population regulation, a debate that continues to this day (see Murray, 1999; Sale and Tolimieri, 2000). It initially involved the dichotomy between researchers arguing that insect populations are limited by climatic or other density-independent factors (Andrewartha and Birch, 1954) and those arguing that they are regulated by density-dependent biotic interactions such as competition (Nicholson, 1954). Later, another dichotomy developed over whether intrinsic or extrinsic mechanisms are involved in the regulation of animal populations (Lidicker, 1978; Tamarin, 1983; Wolff, 1997). Intrinsic mechanisms that may regulate populations include territoriality, dispersal, and reproductive suppression, which reduce population growth before extrinsic factors start operating. Alternatively, populations may be regulated by extrinsic factors that are external to the population, e.g., resource limitation, predators, disease, and weather. A more recent dichotomy has concerned the processes limiting demographically "open" populations that are linked by long-distance dispersal. The traditional view is that regional population size is limited by biological interactions occurring "within" populations. The more modern counterargument holds that they are limited by the supply of juveniles that have dispersed from other source populations (Roughgarden *et al.*, 1987). Although new alternative theories can expand our views and result in a flourish of research from both supporters and detractors, the resulting dichotomies unnecessarily polarize opinion over what are not essentially mutually exclusive processes.

None of the variations on the search for a single, all-encompassing explanation of what determines population numbers has been successful. Population ecologists now speak in terms of "pluralism" and the relative importance of many different processes (e.g., McIntosh, 1987; Olaffson *et al.,* 1994; Wolda, 1995; Caley *et al.*, 1996; Osenberg and Mittelbach, 1996; Schmitt *et al.*, 1999; Turchin, 1999). There are many processes that cause numbers to increase and many others that cause numbers to decline. A combination of processes may act to bring about stasis, while others will act to bring about change. Different factors may be important at different life history stages and different factors may operate over different spatial and temporal scales. Some processes may act directly and have immediate effects on the "quantity" of individuals in the population. Other processes act on population "quality," but in doing so, indirectly impact on population numbers in the future. That is, current population size and trends may not just be a product of current ecological processes, but may represent the lagged effects of historic conditions (Turchin, 1999). Comprehensive models to explain the behavior of populations must account for all the major direct and indirect effects that determine long-term trajectories in population size. In dispensing with single-factor theories, emphasis has shifted toward identifying predictable ecological circumstances or kinds of organisms whereby particular factors are relatively important (Wolff, 1997).

The core of the debate in population ecology is still centered on the issue of "population regulation" and the relative importance of density-dependent and density-independent processes (Sinclair, 1989; Murdoch, 1994; Krebs, 1995; Wolda, 1995; Murray, 1999). Turchin (1999) distinguishes between exogenous factors (those that affect population change, but are not themselves affected by population numbers) and endogenous factors (those that involve "dynamic feedbacks" affecting population numbers). A similar distinction has been drawn between "limitation" (the extent to which a process depresses population growth rate) and "regulation" (whereby a process leads to a local equilibirum through density-dependent effects) (see Osenberg and Mittelbach, 1996). Evidence for and against population regulation and the central role of density-dependent factors is still being compiled and continues to divide ecologists. However, Turchin (1999) emphasizes that both exogenous and endogenous factors affect population change, and their relative strength varies for different population systems. An additional problem is that the same factor may have both exogenous and endogenous effects, and therefore defy such classifications. Hence, rather than categorizing different processes as either density dependent or independent, it is more productive to identify processes that cause stasis or change in population numbers and determine the direct and indirect mechanisms by which these effects are manifest.

Dichotomous views over the processes limiting populations of coral reef fishes have been particularly extreme (see reviews by Doherty, 1991; Hixon, 1991; Jones, 1991). At first, all researchers assumed that space on coral reefs was a limiting resource. The ensuing debate centered on whether individual species were at the carrying capacity of the environment (Smith and Tyler, 1975) or whether chance recruitment patterns determined the population size of individual species (Sale, 1977). Subsequent research showed that there was not a lot of evidence for space limitation and the argument shifted to a debate over whether populations do not reach the carrying capacity because of recruitment limitation (Doherty, 1983a) or predation (Talbot *et al.,* 1978). Consistent with studies of other groups of organisms, modern theory and empirical research suggest that all of these processes, including recruitment, competition, and predation, can contribute to determine population size (Jones, 1987b, 1990, 1991, 1997; Forrester, 1990; Hixon, 1991; Caley *et al.,* 1996; Schmitt and Holbrook, 1999a,b; Schmitt *et al.,* 1999). These "multiple causes" may act independently, or may strongly interact to determine the patterns we see (Steele, 1997a). However, despite considerable research, the full extent of the interactions between recruitment, competition, and predation is only beginning to emerge.

Most local populations of marine organisms are demographically "open" and coral reef fishes are no exception (Caley *et al.,* 1996). By definition, local population changes will be to some degree driven by variation in the supply of juveniles from the plankton, although this may be modified by postrecruitment interactions. It has been well recognized that factors influencing recruitment and adult reproductive output are "decoupled" in open marine populations such as coral reef fishes (Mapstone and Fowler, 1988; Wellington and Victor, 1988). Because factors affecting reproductive output could have little or no effect on population size, it has been argued that they should be disregarded in population studies (Doherty, 1991; Doherty and Fowler, 1994a). However, this assumes that (1) the survival of individuals in a population is totally unrelated to their historic or current condition, and (2) that the recruitment of individuals is at no scale in space or time related to the size and combined reproductive output of the parent stock. In this review we provide evidence to challenge these assumptions.

II. Integrating Numerical and Energetic Processes

A. The Numerical–Energetic Dichotomy

Although we have been critical of dichotomies, in the ecology of open populations it is constructive to distinguish between two potentially distinct patterns and processes that may be important in different ways. The first category concerns "numerical" patterns, or the "quantity" of individuals in the population. Numerical processes will be those that directly add individuals to the population (factors affecting recruitment) or those that subtract individuals from the population (factors affecting mortality). Recruitment variation and lethal factors will have immediate effects on population numbers. On the other hand, "energetic" processes concern the "quality" of individuals, either in terms of their size, condition, or reproductive success. These processes concern the acquisition, allocation, and expenditure of energy to different morphological, behavioral, or physiological functions (feeding rate, growth, condition, maturation, fecundity). At an individual level, energetic features represent important fitness parameters because they can be an important component of reproductive success. At the population level they do not directly cause numerical changes because they have only sublethal effects on population members . However, if energetic parameters affect the long-term probability of survival, they may contribute lagged effects on population changes as a result of indirect interactions.

In this review we argue that both numerical and energetic processes, and their interactions, are critical in explaining long-term patterns in population size and structure in coral reef fishes. Our main thesis is that fundamentally different processes affect numerical and energetic parameters, but each one can have important indirect effects on the other. We do not deal specifically with evidence for and against density dependence and population regulation in coral reef fishes (see Caley *et al.*, 1996; Schmitt *et al.*, 1999; see Chapter 14, this volume, for discussions). Here we are concerned with identifying particular processes and determining if they primarily limit numerical or energetic parameters, and whether there are indirect effects that have secondary consequences for other demographic patterns. Discussions of the relative importance of different processes such as recruitment, competition, and predation are meaningless, unless the parameters they affect are clearly measured over appropriate time scales. An integrated model of reef fish populations will require a separate understanding of the two suites of processes affecting numerical and energetic patterns and their interactions.

Although counterintuitive, it is particularly important to consider energetic processes, such as growth and reproductive condition in open populations. Open populations are expected to exhibit greater phenotypic plasticity in these life history traits than are closed populations, because the environment experienced by offspring will be less predictable (Warner, 1997a,c). Indeed, the growth, maturation, and fecundity of marine fishes are extremely flexible, resulting in considerable spatial and temporal variation in these parameters. Thus open populations tend to be "size structured" (*sensu* Werner and Gilliam, 1984; Ebenman and Persson, 1988; Olson, 1996), meaning that shifts in ecological relationships and between life history stages appear to be more closely related to size than to age. Even though recruitment may not be related to adult stock, actual numbers of individuals reaching particular life history stages (e.g., breeding population) can be directly linked to "bottlenecks" affecting growth and condition during previous life history stages (Jones, 1984; Bystrom *et al.*, 1998). The potential roles of both numerical and energetic processes in reef fish populations will not be understood unless we examine the dynamics relating to both age and size.

B. A Model Linking Numerical and Energetic Processes

In this review we put forward and assess a model that links numerical and energetic processes in the ecology of reef fish populations (Fig. 1). Initially we examine the potentially independent effects of processes

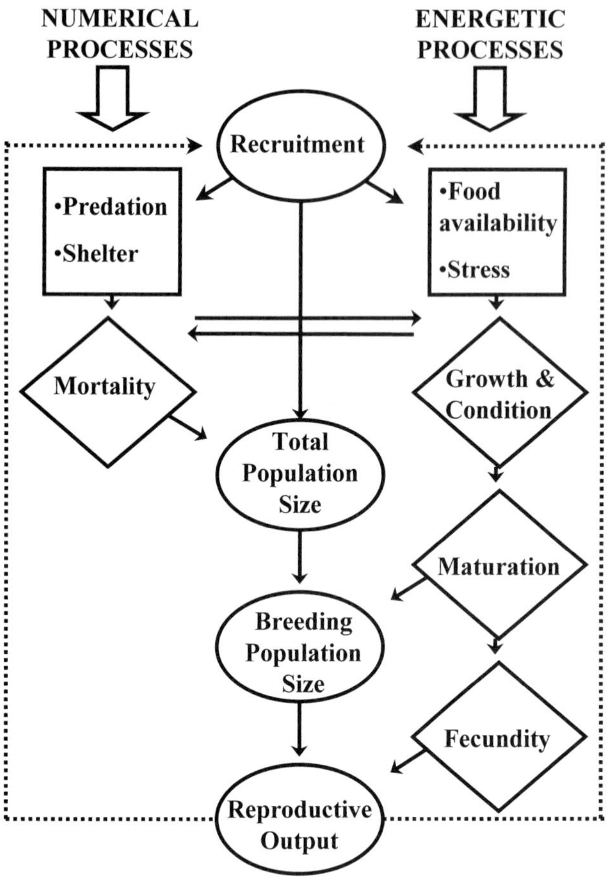

FIGURE 1 Model illustrating linkages between numerical and energetic processes and their consequences for the ecology of reef fish populations. Numerical processes (e.g., predation) directly determine the "quantity" of individuals in the population (Section III); energetic processes (e.g., competition for food) determine their "quality" (Section IV). Predator pressure can indirectly determine many energetic parameters (Section V); conversely, competition and behavioral interactions can have an indirect effect on mortality rates (Section VI). Habitat selection may be a trade-off between the numerical and energetic costs of living in different habitats (Section VII). Population structure and breeding population size will be determined by a combination of numerical and energetic processes (Section VIII). The size of the breeding population and the interaction between numerical and energetic processes in the larval phase will determine the recruitment into the reef-based population every generation.

affecting numerical and energetic parameters in reef fishes. We assess the predictions that recruitment, predation, and competition for shelter play can play important roles in determining absolute population size and the short-term dynamics of reef fish populations, and that competition for food and behavioral interactions can play a major role in determining parameters. We also review the evidence that the reciprocal interactions between numerical and energetic processes (indirect effects) are vital to an understanding of reef fish population dynamics (Fig. 1). We examine the hypotheses that recruitment and predation indirectly affect the acquisition and allocation of energy, as a result of predator avoidance, restrictions on foraging, and predator-induced stress. We also examine how food resources and competition may indirectly affect recruitment and mortality, and thus produce lagged effects on long-term population changes. The evidence that energetic effects in the early life history can subsequently determine the numbers of individuals reaching adult life history stages will be evaluated. That is, we determine whether numerical and energetic processes can interact to produce the secondary characteristics of reef fish populations, such as population size and age structure, biomass, adult population size, and the reproductive output of local populations. Finally, we examine the potential for numerical and energetic links between the adult population and the subsequent recruitment of reef fishes.

C. Where Does Movement Fit in?

There is increasing evidence that movement after settlement can have a major influence on local population size (Robertson, 1988a; Lewis, 1997; Ault and Johnson, 1998a). Whether movement is considered an energetic process or a numerical one depends on the scale at which a local population is defined. Immigration and emigration will have direct numerical effects, but these may become less important at larger spatial scales. Also, movement is associated with a risk of predation and so may result in a reduced life expectancy. However, movement is fundamentally an energetic process, because it exacts a cost in terms of energy expenditure. The energy gains associated with movement depend on the quality of the habitat that is being selected. Clearly, both numerical and energetic processes will be of importance in explaining the patterns of movement of fish populations to and from reefs.

D. Recruitment versus Postrecruitment Processes

Our model of reef fish populations begins with recruitment, the parameter that determines the initial size of each cohort entering the reef-based population

(Fig. 1). Recruitment, defined here as the number of juveniles settling out of the plankton into the reef-associated population, is the major factor replenishing populations. Its theoretical role in determining local population size in open systems, as a result of variations in larval supply, has been well documented (Doherty, 1991; Caley *et al.*, 1996). In support, there is ample empirical evidence that recruitment can be a major factor affecting population size (e.g., Jones, 1990; Doherty and Fowler, 1994a,b). Unless density-dependent mortality is so strong that there is complete compensation (i.e., one death for every new recruit), variation in recruitment will lead to a change in population size (Caley *et al.*, 1996). However, there is also considerable evidence that patterns established at the time of recruitment can be substantially modified by postrecruitment processes (Jones, 1990, 1991, 1997; Forrester, 1990, 1995; Hixon and Carr, 1997; Steele, 1997a,b, 1998; Holbrook and Schmitt 1999; Schmitt *et al.*, 1999). Some of these act almost immediately after recruitment (e.g., Holbrook and Schmitt, 1999), whereas others are noted only after several years of observation (Jones, 1991; Robertson, 1996). Both predation and competition have been implicated, but a general understanding of where and when each process will be important has not been reached. In our model a general understanding of how recruitment is modified requires two clear distinctions to be drawn. First, direct and indirect effects of the actions of predation, competition, and other processes need to be clearly distinguished. Second, competition cannot be treated as a single cohesive process. Competition for shelter or refuges from predators has a fundamentally different effect compared to competition for food. Hence, to achieve an understanding of the effects of competition on predation rates, it is critical to identify the resources that are the object of competition.

III. Direct Effects on Numerical Parameters: Predation and Competition for Shelter

A. Predation

Since Hixon (1991) reviewed the role of predation in coral reef fish communities, evidence has continued to accumulate showing predators can have a major influence on the abundance of reef fish (Caley, 1993; Carr and Hixon, 1995; Beets, 1997; Beukers and Jones, 1997; Steele, 1998, 1999; Steele *et al.*, 1998; Connell, 1998a, 2000). For example, removal of the predatory squirrelfish *Holocentrus adscensionis* in the United States Virgin Islands has had a significant effect on the juvenile densities and adult abundance of other species (Beets, 1997). The effects of predators on juvenile densities or adult numbers can be directly attributed to an increase in mortality rates (Carr and Hixon, 1995; Hixon and Carr, 1997; Beukers and Jones, 1997; Connell, 1998a; Steele, 1999; Forrester and Steele, 2000; Emslie and Jones, 2001). Connell (1998a), for example, showed that juvenile *Acanthochromis polyacanthus* survived substantially better when predatory fishes were excluded by cages (Fig. 2). Small reef fish species may be particularly susceptible to predation, exhibiting generally higher mortality rates as compared to large species (Munday and Jones, 1998).

It is almost considered a general rule that juvenile mortality rates in fishes increase in relation to density, whereas adult mortality is density independent (Myers and Cadigan, 1993a; Valiela, 1995; Bjornstad *et al.*, 1999). Mechanisms put forward to account for this include increased predation (including cannibalism) at high densities and competition for food and habitat. However, there is little evidence that competition for food causes mortality directly. In coral reef fishes, juvenile mortality rates are density dependent, and this can largely be attributed to the effects of predation (Forrester, 1995; Beukers and Jones, 1997; Hixon and Carr, 1997; Steele 1997a,b; 1998; Connell, 1998a, 2000). However, the role of predation in population regulation has yet to be tested via long-term predator removals. That is, we do not know if populations in the long term reach a carrying capacity in the absence of predators.

B. Availability of and Competition for Shelter

Coral reef fishes frequently take refuge from predators in the branches of corals or in holes in the reef matrix. Under conditions of high predation pressure, the availability of shelter or predator-free space will often influence the abundance of reef fishes through effects on mortality rates (Fig. 1). Recent manipulations of shelter availability have been shown to affect reef fish abundance (Hixon and Beets, 1989, 1993; Connell and Jones, 1991; Buchheim and Hixon, 1992; Caley and St. John, 1996; Holbrook *et al.*, 2000; Syms and Jones, 2000; Schmitt and Holbrook, 2001). In addition, orthogonal manipulations of both shelter and predators (e.g., Beukers and Jones, 1997; Eggleston *et al.*, 1997;

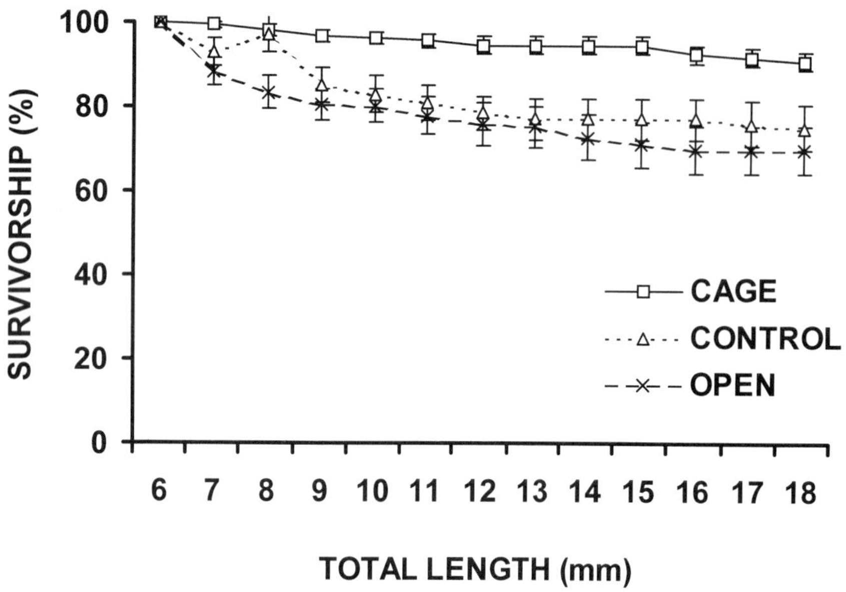

FIGURE 2 Effects of piscivore exclusion on the survival of juvenile *Acanthochromis polyacanthus* at One Tree Reef, Great Barrier Reef (with permission from Connell, 1998a).

Steele, 1999) have shown that survival varies with shelter treatments when fishes are exposed to predators, but less so when predators are reduced or absent (e.g., *Pomacentrus moluccensis*; Fig. 3).

There is a close relationship between predation and competition for shelter. When shelter from predators is in short supply, individuals are expected to compete for shelter, provided that predation pressure is sufficient to lead to greater mortality when shelter is poor or absent. Conversely, in the absence of predation pressure, competition for shelter may not occur at all. Thus we might expect competition for shelter to occur at a threshold below which there is ample shelter for all individuals in the population.

There is increasing evidence that competition for shelter, both within and among species, directly affects mortality rates and population size, particularly in species that are highly dependent on specialized shelter sites. Anemone fishes exhibit strong competition for anemones, both within and among species, although competitive hierarchies do not appear to explain differences in anemone use among species (Srinivasan *et al.*, 1999).

Schmitt and Holbrook (1999b) measured strong density-dependent mortality in *Dascyllus* species associated with both anemones and live coral heads in Tahiti. The effect of predation appeared to be mediated through competition for shelter, with behavioral

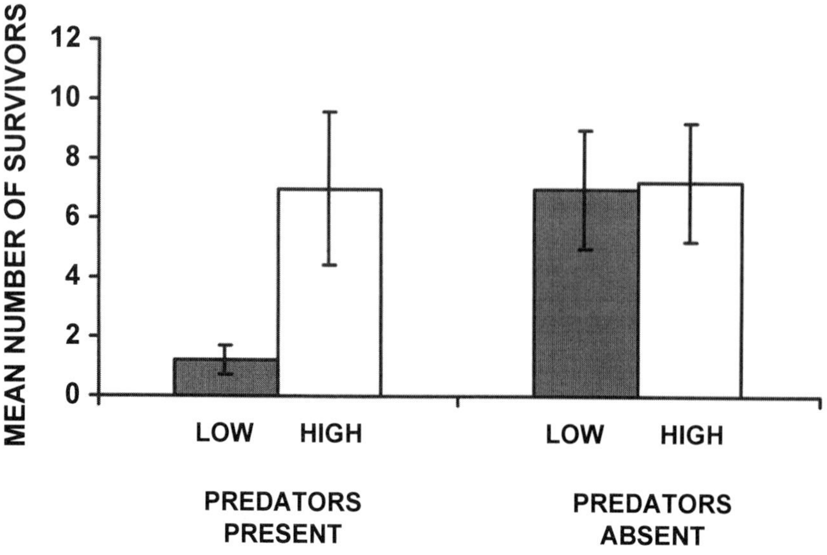

FIGURE 3 Interaction between predator pressure and habitat quality in determining the survival of the damselfish *Pomacentrus moluccensis* at Lizard Island, Great Barrier Reef. The graph shows the mean number surviving 2 months on patch reefs made from low shelter (*Acropora nobilis*) and high shelter (*Pocillopora damicornis*), both in the presence of two predatory fishes (*Cephalopholis boenak* and *Pseudochromis fuscus*) and where they were absent (from Beukers and Jones, 1997).

interactions reducing the access of smaller fish to shelter, making them more susceptible to predation. Competitive interactions between two species inhabiting the same coral species, *Dascyllus aruanus* and *Dascyllus flavicauda,* were asymmetrical. *Dascyllus flavicauda* appeared to be a superior competitor where the two species cooccurred, having a greater effect on the mortality of *D. aruanus* than vice versa. Munday *et al.,* (2001) have shown that interspecific competition is a major determinant of the population densities of *Gobiodon* species, which are the specialized inhabitants of certain acroporid corals. The larger, dominant species *Gobiodon histrio* has a major influence on the recruitment and survival of other species with similar habitat preferences, and excludes these species from preferred shelter sites. However, effects of competition for shelter are not universal. Steele (1997a,b), for example, detected no effect of interspecific competition for shelter in two temperate reef gobies. Species that evolve either specialist or generalist shelter requirements may be trading off the risks of predation and competition. That is, adaptation to specific shelters may reduce the risk of predation, but increase the likelihood of competitive interactions over shelter or other resources.

IV. Direct Effects on Energetic Parameters: Competition for Food, Behavioral Interactions, and Stress

A. Food Availability and Competition for Food

There is ample evidence that food availability and competition for food can have major effects on growth rates and the onset of key life history events in fishes from a variety of habitats (Abbott and Dill, 1985; Metcalfe, 1986, 1993; Thorpe, 1989; Simpson, 1992; Levin *et al.,* 1997b). The actual amount of food energy available for maintenance, growth, and storage will be determined by the absolute amount of food available and how many individuals are sharing that resource. When food declines and/or densities of fishes increase to a point at which the supply of food does not meet demand, competition will affect the growth of individuals. As a consequence, growth is typically density dependent in fishes, especially prior to maturation, when individuals are growing most rapidly (Valiela, 1995). Evidence also suggests that, after settlement, food availability and competition for food seldom cause mortality directly (Brown *et al.,* 1989; Jenkins *et al.,* 1999). It appears that fishes are resilient to variation in the acquisition of energy that might arise from fluctuations in food abundance or population density.

Studies on coral reef fishes appear largely to support these generalizations. When juvenile densities have been manipulated, growth is almost always negatively related to density, which may be explained by competition for food (Doherty, 1982; Jones, 1987b, 1988b; Forrester, 1990; Booth, 1995; Webster and Hixon, 2000). However, density-dependent growth may also result from behavioral interactions or may be an indirect response to predation pressure, if predators aggregate at sites of high density (see Section VI). One obvious way density influences growth is through food limitation to less competitive parts of a group or population. For example, Jones (1987b) manipulated the densities of recruited *Pomacentrus amboinensis* in small patch reefs from natural to three times natural densities and found that growth was slower at high densities. In addition, juvenile growth was considerably slower in the presence of adult individuals. In these experiments, mortality did not decline in relation to density, suggesting that individuals can survive a food reduction associated with high fish density. Among coral reef fishes there are no documented examples of competitors directly killing one another in the process of acquiring food resources. Observed declines in the abundance of food seldom lead directly to increased mortality rates, except in rare instances when starvation results from dramatic environmental perturbations (e.g., Tsuda and Bryan, 1973).

Food is obviously going to be important for growth, but is it in short supply and therefore a resource that will be competed for? One source of circumstantial evidence suggesting that food is at a level that limits growth comes from studies of planktivorous fishes. Several studies have shown that the abundance and growth of planktivores are correlated with current speed, and hence the supply of food to reefs (Thresher, 1983a,b; Kingsford and MacDiarmid, 1988; Anderson and Sabado, 1995). Food availability may also be important to the growth of reef fishes in other trophic groups. A unique study by Clifton (1995) suggested the overwhelming importance of food to the Caribbean parrot fish, *Scarus iserti.* He examined two populations, separated by only 3 km, that were exposed to very different patterns of seasonal food availability. Both growth and reproduction tracked the availability of food and were very different for the two populations. Similarly, Stewart (1998) showed that the growth of small piscivorous fishes on the northern Great Barrier Reef was correlated with the abundance of prey.

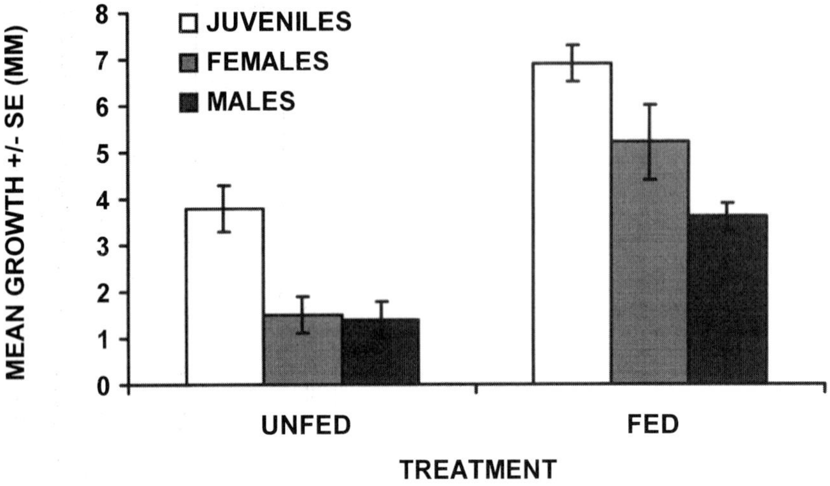

FIGURE 4 Influence of supplemental feeding on the growth of subadult juveniles, adult females, and adult males of *Pomacentrus amboinensis* in field populations at One Tree Reef. Fish were fed on freeze-dried euphausids, twice a day over a 6-week period. Growth is compared to fish exposed to natural prey levels.

Studies that have directly manipulated food availability on coral reefs provide the strongest evidence that it influences juvenile growth rates (G. P. Jones, 1986; Forrester, 1990; Kerrigan, 1994). G. P. Jones (1986) manipulated the diets of recently settled *Pomacentrus amboinensis* transplanted to patch reefs at two depths in One Tree Lagoon, using daily zooplankton supplements. Food supplementation was found to increase growth rates, regardless of depth, but did not influence mortality rates. Kerrigan (1994) manipulated food availability to social groups of the same species in mesocosms and found that growth tracked food availability. Similarly, Forrester (1990) found that supplemental feeding enhanced the growth of *Dascyllus aruanus*. G. P. Jones (unpublished data) has provided supplemental food to older individuals of *P. amboinensis* and found that it promotes growth throughout the life cycle, impacting large juveniles, mature females, and mature males (Fig. 4).

Food availability and competition for food may affect other life history parameters that are closely related to growth, such as time to maturity and fecundity (Jones, 1984, 1987b; Ma *et al.*, 1998; Wootton, 1998; Ali and Wootton, 1999). In size-structured populations this has major implications for determining the size of the breeding population and the reproductive output of the population (Fig. 1). Because the attainment of sexual maturity in reef fishes is often size based, time to maturity is often found to be density dependent, as a consequence of growth rates and ultimately food consumption. Booth (1995) found that juvenile *Dascyllus albisella* took longer to reach maturity in larger groups. Alternatively, Jones (1987b) found that numbers of *Pomacentrus amboinensis* reaching the size of maturity were not related to the initial juvenile densities, but rather were strongly influenced by the presence of adult conspecifics. Both results could relate to competition for food. Both Forrester (1990) and Jones (Fig. 5) have found that supplemental feeding can advance the onset of maturation in reef fishes. In *P. amboinensis*, adult females given supplemental food produced a significantly greater number of clutches during the breeding season (Fig. 6).

Competitive interactions can have a negative influence on the growth, maturation, and fecundity of inferior competitors, particularly when they are displaced

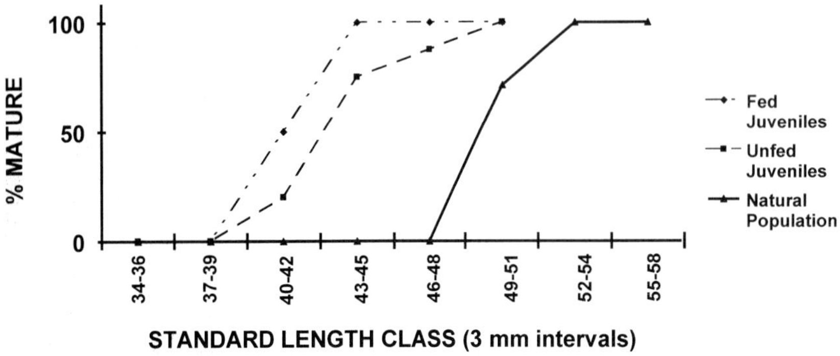

FIGURE 5 Influence of supplemental feeding on the onset of maturation in juveniles of *Pomacentrus amboinensis* in the absence of adults. Size at maturation for the fed group is compared with unfed juveniles on control experimental patch reefs and with juveniles collected from undisturbed populations.

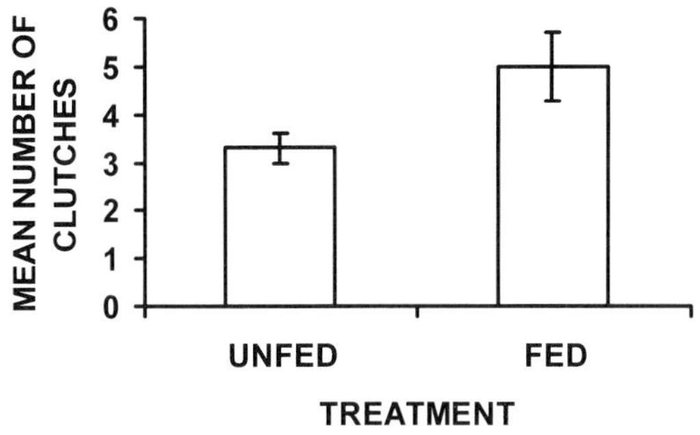

FIGURE 6 Influence of supplemental feeding on the frequency of spawning by marked female *Pomacentrus amboinensis* on experimental patch reefs. Frequency of spawning is compared to that for females exposed to natural food levels.

into lower quality habitats. For example, Munday *et al.* (2001) found that among coral-dwelling gobies at Lizard Island, a superior competitor (*Gobiodon histrio*) limited access of an inferior competitor (*Gobiodon brochus*) to a high-quality coral habitat (*Acropora nasuta*). Both goby species prefer *A. nasuta,* where they exhibit their highest growth rates (Munday, 2001) (Fig. 7). *Gobiodon brochus* grew at approximately half the rate in the low-quality coral habitat (*Acropora loripes*) it is normally found in compared to the high-quality habitat it is competitively excluded from (Munday, 2001). Reproductive success is closely linked to body size in coral-dwelling gobies (Kuwamura *et al.*, 1993). Therefore, habitat-related differences in growth rate are predicted to have substantial effects on lifetime reproductive success. Similarly, Clarke (1989, 1992)

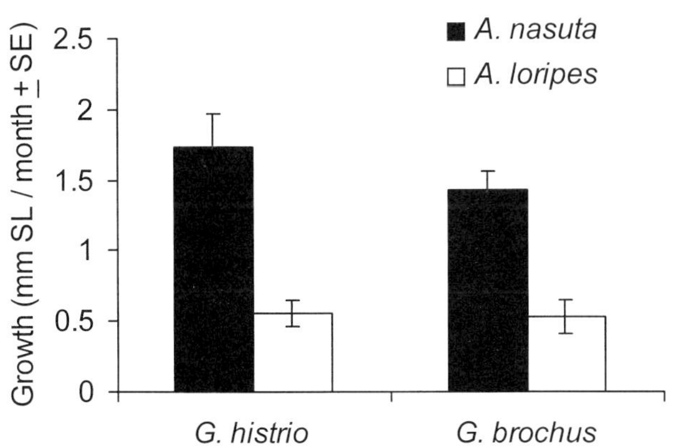

FIGURE 7 The coral-dwelling gobies, *Gobiodon histrio* and *Gobiodon brochus,* both grow faster in colonies of *Acropora nasuta* than in colonies of *Acropora loripes* at Lizard Island (P. L. Munday, unpublished data). However, *G. brochus* is excluded from colonies of *A. nasuta* and is forced to use the inferior habitat as a result of competition with *G. histrio* (see Munday, 2001).

found that interspecific competition resulted in habitat partitioning in two planktivorous blennies of the genus *Acanthemblemaria* living on dead corals. The dominant species (*Acanthemblemaria spinosa*) occupies the tips of the coral and displaces *Acanthemblemaria aspera* downward. Both prefer to be higher, where they exhibit higher feeding rates, faster growth, and higher fecundity. This seems to be a case of competition for food, with the dominant species having a major effect on the energetic parameters of a subordinate species, but having no direct influence on survival. Clarke argued that the subordinate species could tolerate less food because it has a lower metabolic rate.

Thresher (1983b) produced a conceptual model of the factors that influence reproductive success in the only damselfish that broods its larvae, *Acanthochromis polyacanthus* (Fig. 8). He found reduced reproductive output correlated with increased numbers of other planktivorous species who were apparent food competitors. Selective removal of these competitors supported this idea. Moreover, he showed that juvenile growth rate was related to the strength of the current in the vicinity of the brood of juveniles, the date that the juveniles were spawned, and the size of the initial brood. Together these three variables explained 73% of the variability in juvenile growth. Juvenile mortality was positively related to the initial size of the brood and the mean number of fishes of all species present within the area.

Competition influences not only the growth, but also other aspects of performance such as body condition, which in turn can have direct effects on maturation and reproductive output. Condition has been shown to influence maturation and fecundity of females in a variety of fish species (Hislop *et al.*, 1978; Kjesbu *et al.*, 1991,1998; DeMartini, 1991; Koslow *et al.*, 1995). Lipid storage appears to be particularly important for gonad development (Ballantyne *et al.*, 1996). Duston and Saunders (1999) showed that seasonal food deprivation led to reduced growth and condition, which reduced maturation in Atlantic salmon. They showed that, regardless of body size, a high condition factor was necessary in late winter for maturation to occur. The physiological decision to mature in these fishes may be dependent on some measure of either the rate of storage or the turnover of surplus energy (Thorpe, 1986; Thorpe *et al.*, 1990). A reduction in food intake prior to the spawning season or during oocyte development can reduce fecundity and egg size (Kjesbu *et al.*, 1991). These potentially important processes have received little attention in the coral reef fish literature.

Factors that influence growth trajectories during the juvenile phase are likely to have major lifetime

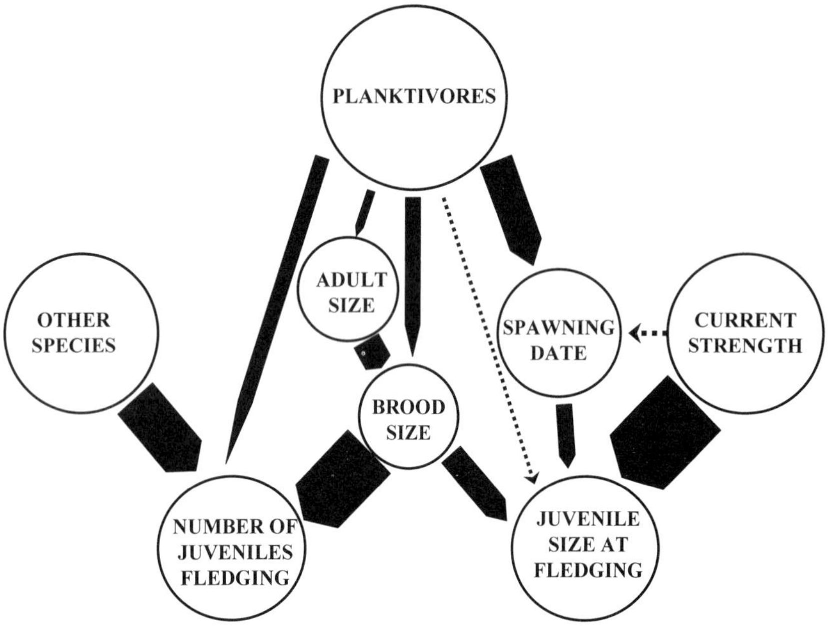

FIGURE 8 Summary of the determinants of reproductive success in a brooding damselfish *Acanthochromis polyacanthus* at One Tree Reef, Great Barrier Reef. Width of the arrows is proportional to the amount of variance accounted for by the variable; direction of the arrow indicates the direction of the causal relationship. Solid arrows indicate statistically significant relationships; dashed arrows indicate trends in data (with permission from Thresher, 1983b).

fitness implications for individuals in size-structured populations. There is some evidence to suggest that size and growth advantages that are present at or shortly after settlement are maintained and reinforced in later life. Tupper and Boutilier (1995a) monitored the growth of recently settled Atlantic cod *Gadus morhua* and found that size differences that were present at settlement were accentuated through time. Likewise, for a reef fish, Forrester (1990) monitored the sizes of groups of juvenile humbug damselfishes (*Dascyllus aruanus*) and found that initial differences in size were maintained and the variance increased over an 8-month period. The formation of social hierarchies within species magnifies variability in growth-related traits, as dominant individuals garner a disproportionate amount of the food resource (Forrester, 1991). Also, when there are size-related changes in diet, larger individuals may be able to access higher energy food sources, which will contribute to faster growth. For example, large piscivorous juveniles rapidly outgrow nonpiscivorous members of the same cohort (e.g., Buijse and Houthuijzen, 1992; Juanes and Conover, 1994).

So does this mean that a fish's fate is predestined by its growth history? Well, possibly not. Fishes do appear to undergo rapid physiological responses when conditions improve (e.g., Cowen, 1991; McCormick and Molony, 1992), which suggests that they have the ability to compensate for poor growth history (McCormick, 1998a). Growth compensation after settlement has been shown to occur in the laboratory for the winter flounder *Pleuronectes americanus* (Bertram *et al.*, 1993), although there are presently no examples of it occurring in tropical reef fishes. Given the differences in the ecological and physiological environment between the larval and benthic life phases, it is unlikely that traits promoting high growth in the larval phase will be the same as those promoting growth after settlement.

B. Behavioral Interactions and Stress

There are three ways that behavioral interactions can exert a negative influence on energetic parameters. First, dominant individuals may limit access to food by subordinates, especially if food is in short supply. However, behavioral interactions may limit growth though limiting access to an important resource even if there is a plentiful supply. For example, Adams and Huntingford (1996) found that at high stocking densities dominant Atlantic salmon, *Salmo salar,* monopolize food and cause social growth suppression even when food is abundant. Clifton (1990) found that dominant members of territorial groups of the Caribbean parrot fish, *Scarus iserti,* reduced the amount of time subordinates spent feeding and limited a subordinate's access to high-quality resources.

The second way that behavioral interactions can influence fitness is by causing or elevating physiological stress (Billard *et al.,* 1981). The aquaculture literature is replete with studies demonstrating that high densities cause stress and lead to greater variance in life history characteristics. Elevation of stress, whether from behavioral interactions or environmental stress, inhibits growth by exerting effects on metabolic and endocrine pathways that regulate growth (Pankhurst and Van der Kraak, 1997). In fishes, stress leads to elevations of the hormone cortisol, which plays an important role in glucose production (Vijayan *et al.*, 1996). It also interacts with other hormones that are responsible for growth and development. Elevated levels of cortisol have been shown to have diverse and consistent effects, including reduced somatic growth, reduced reproductive output, and, in extreme cases, reproductive senescence (Pankhurst and Van der Kraak, 1997). Stress has inhibitory effects on all stages in the reproductive process, including gamete development and quality (Carragher and Pankhurst, 1991), egg and larval quality (McCormick, 1998b, 1999a; Morgan *et al.*, 1999), and survival (Campbell *et al.*, 1992, 1994).

Physiological stress can also influence food intake directly. An individual's foraging success can be influenced by its appetite, visual and chemosensory abilities and ability to capture, handle, and ingest prey. Any stress that impacts on this sequence will adversely affect an individual's energy intake and growth. For instance, handling stress has been found to result in the loss of feeding behavior for a length of time that correlates with the reestablishment of normal physiological status (Schreck *et al.*, 1997). The interactions between behavioral and physiological processes that lead to stress-related effects on foraging efficiency are poorly understood.

A third way in which density-related behavioral interactions can influence fitness is by changing the energy budget of individuals. Elevated population densities force individuals of all levels of a social hierarchy to put more energy into aggression and social interactions, and less into growth, storage, and reproduction. For example, Marchand and Boisclair (1998) found that increasing the stocking density of juvenile brook trout (*Salvelinus fontinalis*) caused a reduction in growth rate, but did not reduce food consumption. At high densities fish were more aggressive and mobile, leaving less energy available for growth.

Few studies have attempted to weigh the relative importance of these three mechanisms (differential access to food, physiological stress, and changes in activity) by which behavioral interactions can influence energetic of individuals. In the freshwater fish *Tilapia zillii*, Koebele (1985) ran a series of five experiments to determine their relative importance in influencing the growth of juveniles. He found that disproportional food acquisition, driven by established dominance hierarchies, was the primary mechanism responsible for the high variance in growth within social groups. Dominant fish ingested more food, either by acquiring a limited ration first and preventing access to food by subordinates, or by inhibiting a subordinate's feeding behavior. These mechanisms have not been investigated in coral reef fishes.

Whatever the mechanism, social interactions appear to regulate reproductive function in many vertebrates. Dominant individuals can regulate access to mates, reproductive condition, timing of maturation, and the determination of gender in subordinates. Such effects are widespread in site-attached reef fishes living in small social groups. Interactions appear to occur most frequently between individuals of near similar size or social status. Jones (1987b) found that adults did not affect the growth of 0+ juveniles up until just prior to maturity, at which time adults began to suppress juvenile growth. This effect on growth was particularly marked in the larger, dominant juveniles. In species with polygynous mating systems, dominant males often suppress the growth of females, which also inhibits sex change from female to male (Warner, 1988c). The reverse occurs in clown fishes; the breeding pair of this species consists of a large female that represses growth of and sex change by the smaller, subordinate male (Fricke, 1983).

Lifetime fitness can be highly dependent on a juvenile's initial position in a social hierarchy. For many fishes that have seasonal cycles of recruitment and growth, the exact timing of settlement can affect social status, which will in turn influence rate of growth and maturation. For example, in salmonids Metcalfe *et al.* (1990) found that juveniles high in social status are more likely to smolt (i.e., metamorphose into the migratory marine phase) in their second year, whereas those of low status are subject to growth suppression by dominant fish and smolted in their third year. Similarly, juveniles of the reef fish *P. amboinensis* settling early in the recruitment season exhibit a growth advantage over late settlers, which is accentuated over time (Jones, 1987a). Early settlers may mature in their first year, but slow-growing late settlers must wait another year before reaching the critical size (Jones, 1987a).

V. Predation: Indirect Effects on Competition and Energetic Processes

The intensity of stress or competitive interactions over food will initially be set by recruitment levels, and will be reduced in proportion to subsequent losses. However, predation can indirectly result in a number of nonlethal effects on prey fish populations. This can include an increase in competition for food when fishes are concentrated in areas serving as refuges from predators (Mittelbach and Chesson, 1987). Predation pressure and crowding effects may be strongly correlated, if increasing predation pressure forces individuals into smaller areas of safe habitat. There is evidence for strong density-dependent growth of juveniles in species occupying specialized shelter sites, which results in delayed maturation (e.g., Jones, 1987b; Forrester, 1990; Booth, 1995; Steele, 1998). Hence, predator-mediated habitat use can exact a cost in terms of growth and reproduction, but this appears to be a trade-off against improved survivorship in shelter sites and in schools. Habitat structure appears to influence the growth of some species that use the habitat primarily for shelter (e.g., Jones, 1988b). Because this cannot be explained by food availability, the differences are likely to be related to differences in the threat of predation on different substrata. For example, gobies of the genus *Gobiodon* appear to grow much more slowly in less preferred coral shelter sites, in which coral branches are closely spaced (Munday, 2001). This may relate to an increase in the time spent on predator avoidance on some corals or to a mechanical restriction on their ability to forage successfully.

Few studies have directly investigated the effects of predation risk and the costs of predator avoidance on energetic processes in coral reef fishes. However, Connell (1998a) found that the growth of juvenile *Acanthochromis polyacanthus* was higher when predators were experimentally excluded (Fig. 9). A similar result was obtained for the temperate reef fish *Lythrypnus dalli,* which reduced its foraging rate and grew more slowly in the presence of predators (Steele, 1998). *Pomacentrus amboinensis* appears to grow more slowly in shallow water, where this species appears to spend more time sheltering from predators (Jones, 1997). If predators aggregate at sites of greater abundance, a density-dependent reduction in growth rates may occur because individuals must devote more time to predator avoidance.

Predation risk is known to affect energetic processes in a variety of animals (Prejs, 1987; Magnhagen, 1993; Lima 1998). For example, the presence of the largemouth bass, *Micropterus salmoides,* causes small bluegill sunfish, *Lepomis macrochirus,* to spend longer periods in dense cover where there were fewer planktonic prey, leading to slower growth rates (Werner *et al.,* 1983). In the freshwater guppies, intimidation by predators reduces the growth of adults and causes a 50% reduction in egg production (Fraser and Gilliam, 1992). A decline in growth and reproductive output associated with an increase in predation pressure is likely to be due to the higher level of harassment, which may interrupt feeding activity (Prejs, 1987). Holbrook and Schmitt (1988) experimentally varied predation pressure on juvenile black surfperch *Embiotoca jacksoni* and showed that they are more closely associated with shelter and feed on less preferred algae at higher predation pressure. However, the opposite effect has been reported for juvenile coho salmon (*Oncorhynchus kisutch*). In the presence of predators, the social hierarchy in groups of juveniles breaks down, allowing smaller fishes to grow faster than they do in the absence of predators (Reinhardt, 1999).

Predation can also influence the use of space by fishes. Individuals of a prey species may avoid areas without actually encountering the predator by the detection of chemicals released by predators or injured

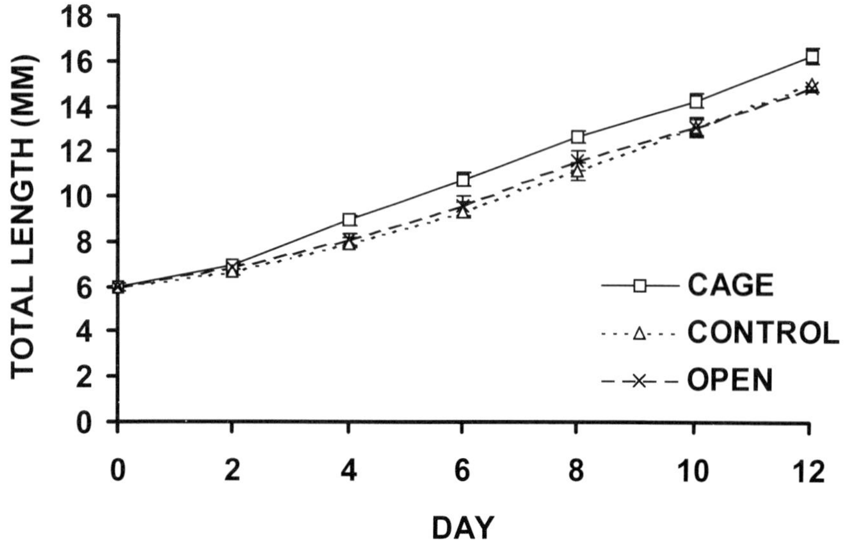

FIGURE 9 Effects of piscivore exclusion on the growth of juvenile *Acanthochromis polyacanthus* at One Tree Reef, Great Barrier Reef (with permission from Connell, 1998a).

prey. The fathead minnow, *Pimephales promelas* (Mathis and Smith, 1992), and the European minnow, *Phoxinus phoxinus* (von Frisch, 1941, cited in R. J. F. Smith, 1997), have been shown to avoid regions where alarm pheromones are present. These alarm pheromones are released only by mechanical injury (Smith, 1992) and are therefore a reliable indicator of the action of a predator for other potential prey. Alarm pheromones in the feces of predators may also indicate high-risk habitats (G. E. Brown *et al.*, 1995a) and the predatory pike counter this by defecating away from their foraging areas when eating minnows (G. E. Brown *et al.*, 1995b). It is currently unknown whether chemicals released by predators influences the distribution of small fishes on coral reefs. It would be of particular interest to know whether the presence of pheromones from predators influences the choices made at the time of settlement, when prey species are naïve and most vulnerable.

VI. Competition and Stress: Indirect Effects on Predation and Numerical Processes

Competition for food directly affects energetic parameters via the acquisition of energy, but in itself is not an agent that directly causes mortality (Section IV). Food availability and crowding can reduce the average (compensation) and increase the variance (depensation) in important energetic parameters. At the population level, reductions in growth or condition can lead indirectly to greater mortality rates under selective predation regimes. In addition, enhanced variability in growth rates, body size, and nutritional condition means that some individuals are more susceptible than others to the actions of selective predators.

A. Body Size and Growth: Susceptibility to Predators

If either competition for food or behavioral interactions lead to a reduction in body size or growth, a greater mortality rate can result from the greater susceptibility to size-selective predation. Being larger than average seems to be a distinct advantage if you are a fish trying to avoid being eaten. Sogard (1997), in a review of the importance of body size to the survival of juvenile fishes, found general support for the "bigger is better" hypothesis. Among reef fishes, Jones (1997) found that small *Pomacentrus amboinensis* exhibited a higher mortality as compared to large juveniles over the first year of life. Likewise, Carr and Hixon (1995) compared the sizes of recruits exposed to predators for a month to those of recruits living on isolated patch reefs without benthic predators. For one species of wrasse, *Halichoeres pictus*, there was no selective mortality. However, for the damselfish, *Chromis cyanea*, mean size was larger where predators were absent. This was either due to density-dependent growth or to size-selective mortality removing the smallest individuals. Manipulations of adult density of the temperate wrasse *Tautogolabrus adsperus* were found to affect growth and survival of newly settled conspecifics (Tupper and Boutilier, 1995a). An elevation in adult densities reduced juvenile growth and survival, whereas a reduction in densities led to an increase in growth and enhanced survival. A relationship between body size and survival suggests that the mechanism causing variation in survival is related to growth. However, mortality in fishes is not always size selective (McCormick and Kerrigan, 1996; Sogard, 1997). Whether faster growth in prey fishes is an advantage may depend on the characteristics of the predator(s).

The situation in which juveniles suffer intense competition that affects growth, and as a consequence extends the period that they are susceptible to predators or starvation, has been referred to as "competitive bottleneck" (Bystrom *et al.*, 1998). This situation is most well known in freshwater fish, particularly those that have a limited growing season (Post and Evans, 1989; Olson *et al.*, 1995; Bystrom *et al.*, 1998). For example, interspecific competition between the perch (*Perca fluviatilis*) and the roach (*Rutilus rutilus*) increases the mortality of the roach by increasing the time during which it is subject to gape-limited predation and reduces the numbers reaching a size and condition capable of overwintering (Bystrom *et al.*, 1998). In the same way, competition among juvenile reef fishes represents a bottleneck that limits entry into the adult population (Jones, 1987b, 1990). As competition intensifies, this increases both the period over which competition occurs and the period over which juveniles are exposed to size-selective predators.

The size distribution of the main predators will also have a marked influence on the nature of any selective removal of prey that occurs. Predators are usually gape limited, and the size distribution of prey they target is a function of their size. Where recruitment of prey and predators is seasonal, such as along the Great Barrier Reef and in temperate waters, the selectivity of loss may be a function of the timing of recruitment in relation to their key predators. The optimal growth rate of prey will depend on their position relative to the prey selection curves of the predators (Rice *et al.*, 1997).

For instance, newly recruited snappers (*Lutjanus quinquelineatus*) are known to be voracious predators of newly settled fishes around patch reefs at Lizard Island on the Great Barrier Reef (Sweatman, 1993). Fishes that recruit early in the season would increase their survival probabilities by growing rapidly, thereby remaining ahead of the optimal size targeted by the predators.

B. Condition and Susceptibility to Predation

Fish do not enter the demersal phase of their life cycle at settlement in the same condition, with an equal probability of survival. At settlement individuals exhibit subtle differences that may affect their ability to survive (McCormick, 1998a). These differences are in part due to genotype, but in the main are due to the myriad of other biological and environmental factors that impact on the larval phase (see Chapter 8 for details). Kerrigan (1994) found that temporal differences among settlement pulses (between and within years) explained 38% of the variability in size and only 23% of the variability in larval duration for the common damselfish, *Pomacentrus amboinensis,* from the Great Barrier Reef. The rest of the variability in these attributes was explained by differences among individuals within a pulse. In virtually all physical and biochemical facets measured there have been ecologically relevant levels of variability. This includes growth rates (McCormick, 1994), proximate body composition [total lipids, proteins, carbohydrates (e.g., McCormick and Molony, 1992, 1993; Kerrigan, 1996)], sensory development (McCormick, 1993), burst swimming speed (McCormick and Molony, 1993), and sustained swimming capacity (Stobutzki, 1998). It has been suggested that much of the variation in life history traits, both within and between cohorts, is driven by density-dependent processes acting during the larval phase (McCormick and Molony, 1992). Interestingly, there is little relationship between different measures of condition, with the exception of morphological measures that usually correlate well with fish length (McCormick and Molony, 1993; Kerrigan, 1996). This lack of correlation between measures suggests that the action of selective predation directed toward one trait, such as fish length will have little influence on the levels of variability in other traits (McCormick, 1998a).

These energetic or qualitative differences among individuals can have a marked influence on the fate of individuals. Mesa *et al.* (1994) reviewed 37 experimental papers since 1960 on the effects of "substandard" prey condition in predator–prey interactions. Categories of "substandard" prey included fishes subjected to disease, physical stressors, thermal shock, or chemical pollutants. Overall, 73% of the experiments showed that prey of lower condition were captured in higher than expected proportions. Mesa and colleagues noted, however, that "most were simplistic, empirical laboratory studies."

Two studies have examined the influence of variable body condition at settlement on the survival of two coral reef fishes. Booth and Hixon (1999) collected juveniles of a Caribbean damselfish *Stegastes partitus,* brought them back to the laboratory, and fed them either high or low rations. After 7 days fishes were placed in pairs on coral heads, one fish from each feeding regime. Survivorship of the high-ration fishes was double that of low-ration fishes over the next 10 days. Laboratory experiments suggested that selective predation by fishes on low-ration prey was the mechanism for the survival differences. Similar results were found in a field study of the damselfish *Pomacentrus amboinensis* from the Great Barrier Reef. Hoey and McCormick (unpublished) collected naive newly metamorphosed fishes from light traps, tagged them using a fluorescent elastomer, and then assigned them randomly to groups on different patch reefs on a shallow sand flat, which is part of their natural habitat (Jones, 1990). Fishes were left for 2 days and then collected. Whilst migration was negligible, fishes on some patch reefs sustained very high mortality but others had minimal losses. Physical and biochemical measures of condition of the survivors of four patch reefs that sustained high mortality (50%) were compared to those of fishes from four patch reefs that sustained low mortality (<5%). Results clearly showed that it was the juveniles with low growth rates and lipid levels that were being selectively removed by predators on the high-mortality reefs (Fig. 10).

C. Social Hierarchy and Predation

The social hierarchies that form in small social groups can lead to differential probabilities of mortality for group members. Often mortality rates of subordinates are higher than they are for dominant individuals. Post *et al.* (1997) monitored the fate of high- and low-density cohorts of yellow perch (*Perca flavescens*) in a Canadian lake. They found that at high density, recruit cohorts split into a fast-growing group that inhabits the normal littoral area of the lake, and a slow-growing component that remains pelagic. It appears that interference competition by dominant (littoral) individuals at high densities forces the subordinate group into suboptimal habitats, where they sustain twice the mortality of the faster growing group. There are examples

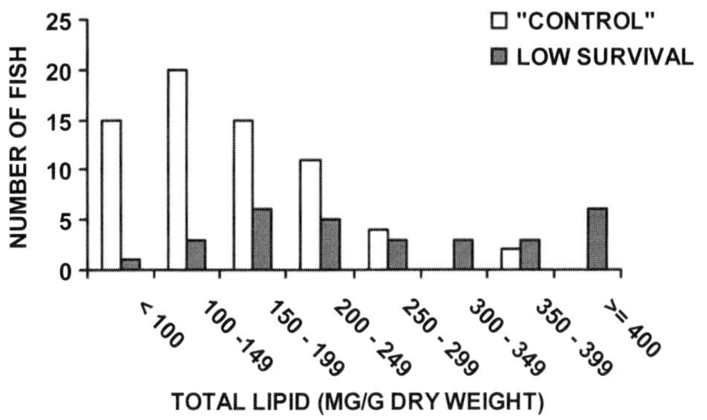

FIGURE 10 Selective predation on low-fat fish. Results of an experiment illustrating the selective nature of predation on newly settled *Pomacentrus amboinensis* at the reef edge. Fishes that were placed on patch reefs and that sustained low mortality levels over a 2-day period (control) have lower levels of total lipids, compared to those fishes from patch reefs that sustained high mortality levels (from Hoey and McCormick, unpublished).

of juvenile coral reef fishes occupying marginal habitats during years of extremely high recruitment [e.g., siganids (Tsuda and Bryan, 1973), surgeonfish (Pillai *et al.*, 1983), and cod (Dufour *et al.*, 1996)], but it is not known whether subordinate individuals are forced into marginal habitats.

The impact of behavioral interactions tends to be most accentuated when there are large size differences, at which time mortality can increase. A number of studies have shown that the growth and subsequent survival of juveniles are reduced in the presence of adults (e.g., Jones, 1987b; Tupper and Boutilier, 1995a). For example, Tupper and Boutilier found that growth and mortality of juvenile cunner *Tautogolabrus adspersus* were influenced by elevated densities of adults; removal of adults resulted in enhanced growth and recruitment success. Heightened levels of mortality may occur because adults force juveniles into lower quality habitats, where the predators are more effective.

In some circumstances it may be the dominant individuals that suffer higher mortality rates. For example, dominant males can be exposed to risk of higher mortality during spawning, by the nature of their higher activity level and gaudy color patterns. This appears to be the case in the Caribbean scarid, *Sparisoma radians*, in which the terminal males, who are much more active during courtship than are other members of the social group, are exposed to higher levels of predation by yellow jacks, *Caranx bartholomaei* (Clifton and Robertson, 1993).

Interestingly, changes in foraging behavior associated with a change in the availability of a food resource can lead to density-dependent loss. Experiments with anuran tadpoles have shown that activity patterns associated with foraging increase with the depletion of an important food resource (reviewed by Werner and Anholt, 1993). Food depletion may be due to density or environmental changes. As searching activity increases in response to reduced food availability, encounter rates with predators and therefore per-capita mortality rates increase (Anholt and Werner, 1995). This model suggests that food availability, behavior, and predation risk are inextricably linked and the effect of one on survival cannot be understood without the others. Given the generality of the trade-off between growth and mortality rate, this mechanism may also be of importance for reef fishes.

In some instances, behavioral interactions lead to positive density effects on energetic and population processes [so-called Allee effects (Stephens *et al.*, 1999)]. Many of these effects appear to relate to the advantages of living in groups. Individuals in large broods of *Acanthochromis polyacanthus* grow faster compared to those in small broods (Connell, 1998a). Either this is because larger broods are more effective at finding patches of prey or because individuals can afford to spend less time on vigilance or more time on foraging in larger broods. Similar explanations may apply to species in which juveniles preferentially settle into coral heads containing conspecifics (Sweatman, 1985b; Jones, 1987a), and survival is better in larger social groups (Jones, 1987a; Booth, 1995).

VII. Role of Habitat Structure and Habitat Selection

There is increasing evidence that the structure of coral reef habitats can have a major influence on the abundance of reef fish populations and the structure of reef fish communities (e.g., Ormond *et al.*, 1996; Munday *et al.*, 1997; Munday, 2000; Jones and Syms, 1998; Syms and Jones, 2000). Correlations between fish abundance and habitat availability suggest that some resource provided by the habitat must play a role in limiting populations. Habitat structure can affect the magnitude of recruitment in space and time, particularly in species that exhibit strong habitat selection at the time of settlement onto the reef (Tolimieri, 1998a; Srinivasan *et al.*, 1999; Schmitt and Holbrook, 1999b; Syms and Jones, 2000).

Habitat structure can also modify patterns of juvenile growth (Jones, 1988b, 1997) or mortality (Jones, 1988b; Beukers and Jones, 1997). However, at the population level there appear to be no generalizations emerging as to which parameters—numerical or energetic—will be affected. To extend the work on the significance of habitat structure, we need to be able to identify the actual ecological processes that are responsible for interactions between habitat structure and demography. This requires that we isolate the critical resources that are provided by the habitat and whether these include shelter from predators, food resources, living space, or a combination of these factors.

There is increasing evidence that habitat selection at settlement may be a trade-off between the numerical and energetic consequences of particular choices. Growth and life expectancy will both contribute to differences in fitness associated with different habitat choices. Shelter-site selection may be critical, because an individual may immediately pay the ultimate price for a bad choice. Not surprisingly, many species are exceedingly selective when it comes to recruiting into shelter sites (e.g., Ohman *et al.*, 1998a), particularly where predation pressure is high (Steele, 1999). Species such as *Pomacentrus amboinensis* appear to settle preferentially into habitats that improve chances for both growth and survival (Jones, 1997). In other instances there will be a trade-off between growth and mortality. For example, in *Dascyllus albisella,* juveniles appear to settle preferentially into larger groups because the improved survivorship rate offsets the cost of slower growth and delayed maturation (Booth, 1995). Conversely, pinfish (*Lagodon rhomboides*) seem to choose seagrass beds over open sandy areas because higher growth rates allow them to achieve more quickly a size that protects them from predation (Levin *et al.*, 1997b). The relative contributions of population and energetic parameters to the fitness and survival require further investigation.

VIII. Secondary Population Characteristics: Interaction between Numerical and Energetic Processes

Generally, the factors affecting spatial and temporal patterns in the structure of reef fish populations are poorly understood. Patterns of recruitment and mortality will be the proximate determinants of the age structure of populations (e.g., Doherty and Fowler, 1994a), whereas patterns of growth and size-specific mortality will determine the size structure (e.g., Jones, 1984). To explain the dynamics of fish populations it is necessary to understand the processes that explain both age and size structures, and the consequences of having different age or size structures (Jones, 1991; Trippel *et al.*, 1997; Robertson, 1998a). In animals for which key life history transitions are more closely related to size than age, recruitment to these life history stages must be a function of both mortality, and growth. Hence, recruitment, mortality, and growth will determine the size of the breeding population (Jones, 1984). The overall biomass of the population and its reproductive potential will also be determined by both numerical and energetic processes (Fig. 1), but their relative importance has not been evaluated. The longer an animal lives, the greater the potential stability in the face of fluctuations in recruitment or intercohort variation in survival [i.e., the storage effect (Warner and Chesson, 1985)]. In addition, the larger an animal grows and the more flexible the growth rate, that greater the stability in breeding population size.

IX. Numerical and Energetic Links between Stock and Recruitment

A. Interactions between Growth and Mortality in Larvae

Decades of research on temperate fish larvae have emphasized that recruitment is the product of both numerical and energetic processes occurring during the larval phase. Both the importance of food availability and the way this interacts with density have been shown to be key factors influencing the growth and survival of larvae through to recruitment. The availability of planktonic prey of the correct size, at key life stages such as first feeding, appears to drive cohort success (Hjort, 1914; Cushing, 1972; Leggett and Deblois, 1994). At this stage, food availability acts not only directly on mortality levels through starvation, but also indirectly through growth rate effects. Food abundance has been shown to elevate larval growth and development rates, thereby reducing larval-stage duration. In this way, fast-growing larvae, in contrast to slow-growing larvae, are exposed for a shorter period of time to the high levels of predation characteristic of the early larval phase (the "stage-duration" hypothesis). Bigger fish have been shown to be more successful at capturing prey (creating a positive feedback, elevating growth further), and better at evading predators (the "bigger is better" hypothesis). It is generally regarded that larvae that grow slowly or are in poor physical condition will be more susceptible to predation. Predation therefore works in concert with processes that influence growth and body condition. Predation may be the proximate factor that removes juveniles from the population, but the ultimate cause may be related to energetic processes (Hunter, 1981; Bailey and Houde, 1989). Although most of this scenario is based on temperate nonperciform species, it has been assumed that larval mortality in tropical species is governed by a similar suite of numerical and energetic processes (Houde and Zastrow, 1993).

Because of the almost absolute levels of mortality during the larval phase, small changes in larval mortality rates can lead to order-of-magnitude differences in recruitment (Houde, 1987, 1989; Fogarty, 1993). Any biological or physical factor that reduces larval growth will influence mortality by increasing the period that a larva is exposed to the high predation pressure characteristic of the larval environment. This relatively simple view is supported by a large volume of empirical evidence and simulations (see reviews of Cowen and Sponaugle, 1997; Houde, 1997). The little evidence that is available for tropical reef species is reviewed in Chapter 8 (this volume) and generally supports the importance of factors that influence larval growth and the interactive nature of energetic and numerical processes in determining recruitment success.

B. Is Recruitment Related to Parent Stock?

The relationship between the abundance of a spawning population and recruitment of juveniles is one of the fundamental questions of fisheries biology. Despite this, there is still wide opinion as to the importance of the size of the breeding population in determining recruitment (Myers, 1997). Myers and colleagues have examined this question with a compilation of data encompassing over 300 temperate fisheries stocks (Myers and Barrowman, 1996). Using a meta-analysis they found that when there was a large range in spawning biomass in the data set, the largest recruitment tended to occur when spawner abundance was large, and the lowest recruitment tended to occur when the spawner abundance was low (Myers, 1997). If the range of the spawner abundance was low in the data set then there was no clear relationship between spawner abundance and recruitment. Both spawning stock size and condition have been implicated. For example, Marshall and Frank (1999a,b) found that recruitment in haddock (*Melanogrammus aeglefinus*) correlated with indices of growth, condition, and adult body size (a proxy for egg production). Stock-recruitment relationships may also be influenced by direct links between female condition and offspring quality/survival (Kerrigan, 1997; Laine and Rajasilta, 1999).

It is generally considered that coral reef fishes have a highly dispersive larval stage. Consequently, most studies on reef fishes have been carried out at too small a scale to examine the potential for the size of the breeding stock to determine subsequent recruitment onto the reef. However, there is increasing evidence that not all larvae disperse long distances and some may not disperse at all. Ecologically significant amounts of self-recruitment may occur on individual reefs at scales of kilometers to tens of kilometers (Jones *et al.*, 1999; Swearer *et al.*, 1999; Cowen *et al.*, 2000; Palumbi, 2001). This raises the possibility that local processes affecting the size of the breeding population may exert some influence on subsequent recruitment to the reef. However, only one attempt to link spawning output to recruitment has been undertaken for a tropical reef fish. Meekan *et al.* (1993) measured the spawning output and recruitment of the common damselfish, *Pomacentrus amboinensis*, at Lizard Island on the northern Great Barrier Reef. Using time-series analysis, they found that although the temporal sequence of spawning output matched that of recruitment, the fluctuations in the magnitude of spawning did not account for fluctuations in the abundance of recruits. This is despite the fact that a significant number of larvae spawned at Lizard Island do not disperse away from this reef (Jones *et al.*, 1999). This suggests that lack of a relationship between spawning output and the magnitude of recruitment may have been due to variation in larval mortality.

X. Conclusions

A pitfall of the multifactorial view of population limitation in populations of coral reef fishes has been its tendency to reduce the field to a series of case studies, each with their own interesting peculiarities. If the only generalization is that there are no generalizations, how far have we progressed? If it is really true that nothing is important all the time and everything is important some of the time, will we ever have the foundation for models that explain population limitation for more than one or two well-studied species? You cannot escape the conclusion that any attempt to formulate general models, even today, will be premature. Such a small proportion of known reef fish species have been studied and only a small proportion of those have been studied to the necessary depth. Model species such as *Pomacentrus amboinensis* have the characteristics necessary to test hypotheses using manipulative experiments, but we are only too aware that model species may not be representative of reef fishes as a whole. The processes limiting the vast majority of species that are relatively rare have not been examined at all (see Chapter 4, this volume).

On the other hand, the realization that many factors may be important has opened up what was once a very narrow field. Studies on nearly every life history stage and every process are now potentially worthwhile. At the very least, we now have some information on nearly everything that "might" be important. Up until 10 years ago reviews of the subject were dominated by studies on damselfishes (see Doherty, 1991; Hixon, 1991; Jones, 1991). Although damselfishes are still the most popular choice for testing new hypotheses, in recent years there has been a dramatic increase in the taxonomic breadth of the field. We have a greater understanding, not only of some of the larger, more mobile reef fish families, but also the even more numerous smaller, specialized reef fish families that were once considered an anomaly (Munday and Jones, 1998). So even though generalizations may be premature, there has never been a better time to formulate them.

We suggest that the new breadth of information supports the view that different ecological processes (competition, predation, etc.) act in fundamentally different ways, and each one has an indirect influence on the other. Different processes cannot be directly compared in terms of relative importance in the short term, because to detect them in the short term we have to measure different variables. Thus, at this stage we cannot test models that predict the circumstances under which resource limitation might be more important than predator limitation or vice versa (cf. Osenberg and Mittelbach, 1996). Our prediction is that in the long term, energetic processes associated with resource limitation will play an important role in explaining population patterns. However, almost no studies have been carried out long enough for the lagged effects of density-dependent energetic processes to be evaluated.

The different processes that ultimately govern population trends operate on and can only be detected on different time scales. Factors affecting the survival of new settlers may be detected in a matter of days. The effects of processes directly affecting juvenile growth and mortality can often be detected in a matter of weeks to months. Patterns of recruitment and predation can set both the short-term and the long-term dynamics of populations. The indirect effects of competition and stress on growth-mediated predation rates could take years to observe. Robertson's (1996) removal experiment, which detected the competitive dominance of *Stegates planifrons*, was carried out over 10 years, with the full effect of competition taking 4 years to observe. This experiment is one of the few adequate demonstrations of interspecific competition among reef fishes. The effects of competition on reproductive output and its role in determining recruitment patterns could take decades to observe (Gurney and Nisbet, 1985; Nisbet and Onyiah, 1994). To measure the relative effects of any two processes requires that both are measured over time scales commensurate with the slowest acting process. Otherwise the importance of slow-acting processes will be underestimated.

At this stage, our greatest understanding of the dynamics of reef fish populations concerns variation in recruitment, the influence of predation on mortality, and the influence of competition on growth. Our understanding of the interdependence of numerical and energetic processes has lagged behind other disciplines. To define the kinds of indirect processes that may be potentially important we frequently have had to draw from the literature on freshwater and other marine fishes. Much greater attention must be given to additional factors, such as the energetic consequences of the risk of predation and the long-term numerical consequences of competition, behavioral interactions, and stress. Population structure and the size of the breeding population will reflect ontogenetic changes in the suite of processes affecting the quantity and quality of individuals in successive cohorts.

Whether generalizations ultimately turn out to be right or wrong they all have one thing in common. They encourage research by both supporters and detractors, which inevitably directs attention to where it is needed most. In our model we have highlighted the lack of information on the lagged effects of energetic processes on population dynamics, and we hope this stimulates research in that area. However, unless the average time span of population studies on reef fish increases substantially, the importance of these processes will not be recognized. Our lack of information on the quantitative and qualitative link between stock and recruitment is a major impediment to a complete understanding of population regulation in reef fishes.

Acknowledgments

Special thanks to Julian Caley, Janelle Eagle, Mark Hixon, Philip Munday, and Peter Sale for their many and various contributions to this manuscript.

SECTION III

Dynamics of Reef Fish Populations and Communities

Introduction

The study of reef fish ecology turned very early to consideration of questions about community structure and dynamics. Hiatt and Strasburg (1960) described the orderly partitioning of reef zones with quite different assemblages of species occupying each, and many subsequent researchers through the 1960s and 1970s explored the processes that determined these community patterns. During the 1970s, and more particularly during the 1980s, attention shifted toward determinants of population dynamics, perhaps because a more Gleasonian view of the ecological community, as simply the sum of its species, began to take hold (Sale, 1980, 1991b). Questions about the determinants of population dynamics have driven much of the best ecological research on reef fishes through the 1990s, and may continue to do so through the next few years.

The five chapters in Section III include one methodological review for a suite of techniques that have been, and will continue to be, very important for answering questions about dynamics, one chapter focused at the community level proposing a new hypothesis to account for community composition and richness on tropical reefs, and three chapters dealing with the dynamics of reef fish populations. In Chapter 11, Simon Thorrold and Jon Hare ably review otolith techniques used in reef fish ecology. Reef fish biologists have been continually rewarded by their perusal of these tiny calcareous structures. Beginning with the use of otolith microstructure to provide age estimates of newly settled juveniles, it was quickly realized that careful study of otolith structure could also provide information on growth rate, especially during larval and juvenile life, could sometimes signal the precise date of settlement or other demographic transitions, and, with care, could reliably age adult reef fishes. More recently, it has been learned that chemical analysis of otoliths provides additional information, because the trace elements trapped during increment formation provide a trace of the environment in which the fish was living at each stage in its life. (Along the way, ecologists have almost forgotten that the fish uses its otoliths for purposes that have nothing to do with providing a detailed diary!) Otolith analysis, like molecular genetics, is a rapidly advancing cognate field of considerable importance for reef fish ecologists. Thorrold and Hare are well equipped to review it, and have produced a thorough review that will help us all interpret the exciting results that now are appearing.

In Chapter 12, Mireille Harmelin-Vivien reexamines the question of the diversity of reef fish communities, and presents a new evolutionary hypothesis based on energetics to account for it. Her thesis is that the warm and relatively stable climate of the tropics has permitted the evolution of fish species that can make use of energetically less rich sources of food (such as algae) and sessile invertebrates (such as sponges and corals). Her chapter demonstrates that the old question of why reefs are so rich in species is still capable of provoking good scientific inquiry, and it is particularly interesting to read this chapter in conjunction with Chapters 1 and 2 by Bellwood and Wainwright.

Chapters 13, 14, and 15 each address the regulation of population dynamics in reef fishes. They take three quite different approaches, and draw some different conclusions. They were written without the benefit of formal cross-comparison, partly because I wanted separate perspectives rather than the thrust and parry of manuscripts written to defend one view while mortally wounding another. Taken together they provide ample ammunition for an in-depth graduate seminar, and there still may not emerge a final answer.

That reef fish ecologists have been spurred by this question to do good science is well illustrated by these chapters. That reef fish ecologists cannot agree on the nature of population dynamics will be no surprise to anyone who has followed the wider literature on the subject over the past half century. Fundamental to the difficulty surrounding the regulation of populations are (1) problems with definitions of terms, and in collecting definitive data that will discriminate hypotheses (Sale and Tolimieri, 2000), and (2) the varying philosophical positions ecologists hold with respect to the organization of nature. Reef fish ecologists have been reasonably careful about terms, and reef fishes are ideal for demographic experiments, but philosophical differences remain. Some of us see abundant evidence of an orderly world structured by strong ecological processes. Others are comfortable with a less orderly world structured by more feeble processes. (There is, perhaps, a parallel, between those evolutionary biologists who see a world tightly structured by selection and a resulting, continual, gradual evolution, and those who think in terms of punctuated equilibria in a world in which selection occurs from time to time, but is neither uniform nor continuous.) For ecologists, philosophically disposed to expect orderliness, population regulation (in the sense of density-dependent control of population growth) is expected to be both important and continuous. The suggestion that population dynamics may be unregulated at some times and places seems to threaten the philosophical place on which these ecologists stand. Conversely, those ecologists who are philosophically more flexible may go so far as to believe population regulation to be unimportant, as well as noncontinuous. Their relaxed (some will say superficial) attitudes serve to further energize the debate, and the argument continues. (That the hypothesis that population regulation is not universal is, itself, not falsifiable ensures that this argument can continue, and may deteriorate into triviality.)

In 1991, I sensed that reef fish ecologists were reaching for a middle ground by acknowledging that a suite of processes, some density dependent and some not, acted together to determine population dynamics (Jones 1991), with certain processes important at one and others at another time and place. However, in the past decade, there has not been a lot of progress to develop this "multifactorial" view, perhaps because advocating such a viewpoint may appear to some to be a cop-out. In their conclusion to Chapter 10, Jones and McCormick allude to its "pitfall" of tending to reduce the study of population dynamics to a series of case studies each with their own peculiarities. They ask, "If the only generalization is that there are no generalizations, how far have we progressed?" In my view, to recognize that there are different processes operating at different times and places is a real advance in understanding, even if it leaves you grasping for "general principles." Ecological principles are unlikely to be simple ones, and we must stop expecting to find simple and elegant general rules.

Chapters 13, 14, and 15, then, should be read as three current views on this topic. In the first, Graham Forrester, Rick Vance, and Mark Steele use their extensive empirical data on small gobies to develop a simulation model to explore the dynamics of groups of local populations that form "mesopopulations," which in turn form a metapopulation. Their particular concern is whether the density-dependent processes that can be shown at the scale of manipulative field experiments will "scale up" and influence dynamics at larger scales. Their chapter is particularly important for the way it grapples with the problem of spatial scale, and may provide a guide for new modeling efforts with other reef fish species organized into metapopulations.

In Chapter 14, Mark Hixon and Michael Webster carefully survey the reef fish literature to demonstrate that density-dependent processes have been documented in many instances and in many taxa of reef fishes. Their thorough review points to the inadequacy of some studies, and the effectiveness of others, and is an excellent entry to the many field experimental studies of reef fishes. By contrast, in Chapter 15, Peter Doherty uses a similarly careful review of literature to demonstrate the frequent cases in which patterns of local population growth are directly attributable to patterns of local replenishment, often in the absence of any density dependence. His data are also convincing, at least to this philosophically flexible ecologist. Both chapters end with comments about a middle ground, and surely it is there. Sometimes populations grow and diminish due to non-density-dependent processes such as variable replenishment, and sometimes density-dependent processes act to prevent a population from growing ever larger. The interesting question for the future must be, "How often, and under what circumstances, is sometimes?"

CHAPTER 11

Otolith Applications in Reef Fish Ecology

Simon R. Thorrold
Biology Department
Woods Hole Oceanographic Institution
Woods Hole, Massachusetts 02543

Jonathan A. Hare
National Oceanic and Atmospheric Administration
National Ocean Service, National Centers for Coastal Ocean Science
Center for Coastal Fisheries and Habitat Research
Beaufort, North Carolina 28516

I. Introduction

Age estimation of fish from periodic rings in otoliths has a long history in fisheries ecology, dating back a full century to Reibisch's observations of otolith annuli in 1899 (Campana, 1999). However, this long tradition could easily be missed after perusal of the literature on age studies of reef fishes. An understandable preoccupation with the development and verification of ecological theory (reef fishes are excellent models for such work) has been at the expense of basic demographic information. Part of this apparent lack of interest may be explained by a prevailing dogma that tropical fishes are difficult to age (Longhurst and Pauly, 1987; Polunin and Roberts, 1996). However, the oversight is puzzling given the importance of age and growth data in the management and conservation of reef fish resources. For instance, estimates of growth rates and age at first maturity are both minimum requirements for effective single-species fisheries management. Age-related differences in factors such as fecundity and migration patterns similarly indicate that age is a critical parameter in both demographic and spatially explicit population models.

Although slow to develop techniques for age determination of adult fishes, reef fish researchers were much quicker to see the potential of Panella's (1971) discovery of daily increments in larval and juvenile fish otoliths (Victor, 1991). Daily increments in otoliths of temperate fish species have been routinely used to determine size-at-age of both individuals and populations. In tropical reef environments, workers have made much more use of optical transitions in otolith microstructure that are coincident with the timing of larval settlement from the pelagic realm into benthic environments ("settlement marks"). These microstructural features can obviously be used to determine pelagic larval durations (PLDs), but also confer the ability to reconstruct daily spawning and settlement patterns of those offspring that survive to recruit into suitable reef habitats. It is also possible, at least in theory, to reconstruct instantaneous daily growth rates of larval fishes from increment width trajectories, based on an empirical relationship between otolith size and fish size. Taken together, daily increments in otoliths can provide a wealth of information on the early life history stages of coral reef fishes.

The amount and utility of information recorded in the microstructure of otoliths may be matched by the data recorded in otolith chemistry. Otoliths are unique among calcified structures in fish skeletons because they are metabolically inert. Material laid down in the otolith is not reworked or absorbed after deposition. Further, otolith chemistry of at least some stable isotopes and trace elements is determined by the physicochemical properties of ambient waters. It is therefore possible to determine retrospectively the physical and chemical characteristics of the water mass in which a fish has resided, by combining the chronological records of either annual or daily growth increments with chemical analyses at the appropriate spatial scale (Thorrold *et al.*, 1997a). Studies of otolith chemistry have undoubtedly been technology limited, and recent developments in analytical instrumentation may explain the exponential increase in published papers on the topic over the past 10 years (Campana and Thorrold, 2001). The recent heightened interest in the field, and the observation that the first paper examining otolith microchemistry of a coral reef fish was published 17 years ago (Radtke, 1985), suggest a critical assessment of the technique as applied to reef fishes is timely.

In this chapter, we review the development and application of otolith methodologies in reef fish ecology over the past 10 years, and highlight new directions that show promise for the future. We start with a brief description of otolith mineralogy and structure. The need for age data, and the demographic implications of age-structured models, are covered elsewhere in this volume (Chapter 3, this volume), and therefore will not be discussed in detail here. Other topics, such as the use of otolith morphology in stock identification (Campana and Casselman, 1993), are not dealt with owing to their limited application in the field of reef fish ecology. Rather, we will concentrate on the ability of otoliths to provide retrospective information on size-at-age, timing of life history transitions, and environmental conditions experienced by individual fishes.

II. Otolith Mineralogy and Structure

Otoliths are calcareous accretions found within paired otolithic organs (the sacule, lagena, and utricle) that, together with the semicircular canals, make up the inner ear of teleost fishes. All three otolithic organs are believed to be responsive to sound and vestibular movement of the head, whereas the semicircular canals detect angular acceleration (Popper and Fay, 1993). Sagittal and lapillal otoliths (located within the sacule and utricle, respectively) are normally composed of calcium carbonate in the form of aragonite, and asterisci (located within the lagena) appear to be predominantly vaterite (Campana, 1999).

Although it is likely that water-soluble proteins play an important role in determining otolith formation (Belcher *et al.*, 1996; Falini *et al.*, 1996), hydrophobic, high-molecular-weight proteins are more obvious components of sectioned otoliths. Similar to bivalve shells, otoliths are microlaminates composed of alternating layers dominated by these insoluble proteins [termed otolin by Degens *et al.* (1969)] or inorganic $CaCO_3$. Daily formation of increments has now been widely documented for numerous species spanning most aquatic environments (Campana and Neilson, 1985). The formation of daily increments occurs under most conditions, including food deprivation and in the absence of any somatic growth (e.g., Marshall and Parker, 1982; Campana, 1983), due at least in part to the ability of fishes to buffer endolymph chemistry from the blood plasma (Payan *et al.*, 1998). These observations, along with related evidence that otolith material is neither resorbed or reworked after deposition (Campana and Neilson, 1985), form the empirical basis for retrospective analyses combining otolith microstructure and microchemistry.

A second category of increments is typically found in otoliths of fishes after their first year of life. These increments are termed *annuli*, and although no temporal periodicity is implied [the term is derived from the Latin *anus*, meaning ring, (Kalish *et al.*, 1995)], they are frequently laid down on an annual basis. Annuli typically consist of a translucent zone and an opaque zone, although the optical appearance of the zones is variable among species and even within species from different geographic areas (Fowler, 1995). The mechanisms generating seasonal variation in otolith microstructure leading to annulus formation are not well understood. Annuli may be formed by variations in protein to Ca ratios and/or by differences in the dimensions of aragonite crystals. Both mechanisms are probably linked to changes in somatic growth, and this helped to perpetuate the dogma that annuli are not present in fish otoliths from low latitudes, where seasonal variations in growth were believed to be minimal (Polunin and Roberts, 1996). However, it is now well established that the otoliths of coral reef fishes contain identifiable annuli (Fowler, 1995), even from fishes close to the equator (Meekan *et al.*, 1999), and that the periodicity of annulus formation is approximately annual in a number of species (Chapter 3, this volume).

The chemistry of otoliths is determined, in large part, by thermodynamic properties of $CaCO_3$ crystal formation. Differences in the lattice structure of each of the three $CaCO_3$ polymorphs influence incorporation of trace elements and stable isotopes in otoliths. For instance, differences in Sr levels in aragonitic and vateritic otoliths can be explained by the relative ease with which Sr ions substitute for lattice-bound Ca ions in the two polymorphs (Gauldie, 1996a; Gauldie *et al.*, 1997). Determining the significance of more subtle differences in otolith chemistry requires an understanding of the source of inorganic ions that are found in otoliths and the transport of these ions to depositional sites on the otolith within the endolymph. Using radioactive tracers, Farrell and Campana (1996) found that more than 80% of the Ca and Sr ions in otoliths is sourced from the ambient water rather than from the diet. It is not known, however, if these findings can be generalized to other minor and trace elements found in otoliths.

Assuming most of the inorganic constituents in otoliths are derived from the ambient water, otolith composition must be some function of water chemistry. However, this relationship is in all probability a complex one. Inorganic ions must pass from the water into the blood plasma via the gills or the intestine, and then cross another membrane into the endolymph, before precipitating out of solution on the otolith surface. There is obviously potential for decoupling of ion concentrations across the branchial and intestinal membranes, because ion barriers are essential for any organism with high osmoregulatory requirements. As expected, most minor and trace elements are found in significantly lower concentrations in the otolith than in the ambient water (Campana, 1999). However, debate remains as to the exact mechanism by which ions cross the membranes. Based on the observation that Sr:Ca ratios were similar in both blood plasma and the endolymph, Kalish (1991a) suggested that both Ca and Sr were transported across the macula via a paracellular pathway. Subsequent experimental data indicated that movement of at least Ca ions from the blood plasma to the endolymph was achieved through a transcellular route (Mugiya and Yoshida, 1995). Such a pathway provides an individual fish with a mechanism for more direct regulation of endolymph chemistry, but it is necessary to argue that this transcellular route does not discriminate between Ca and Sr to explain Kalish's (1991a) data. Finally, Kalish (1991a) has pointed out that the presence of specific metal-binding proteins may alter free ion concentrations in the blood plasma, and presumably in the endolymph, without any changes in dissolved metal levels in the environment. The generality of these findings remains to be tested, but serves to underline the difficulties interpreting chemical records in biogenic carbonates.

III. Age and Growth

A. Daily Increments

Otolith microstructure holds a wealth of information on daily age, size, growth, and ontogeny that has broad application to the study of coral reef fish ecology. But before this information can be extracted a researcher is faced with several seemingly straightforward methodological issues.

1. Preliminary Issues

a. Which Otolith to Use One must first decide which of the three pairs of otoliths to use, and then determine the appropriate preparation method. Most studies of coral reef fishes have used either sagittae or lapilli. Sagittae have been used primarily for gobiids, labrids, and monocanthids (e.g., Brothers *et al.*, 1983; Victor, 1986a; Kingsford and Milicich, 1987; Kishiro and Nakazono, 1991; Sponaugle and Cowen, 1994, 1997), whereas lapilli have been used primarily for chaetodontids, holocentrids, pomacentrids, scarids, and serranids (e.g., Brothers *et al.*, 1983; Keener *et al.*, 1988; Wellington and Victor, 1989a; Fowler, 1989; Thorrold and Milicich, 1990; Lou, 1993; Tyler *et al.*, 1993; Danilowicz, 1997). Whole otoliths mounted in the sagittal plane with little or no polishing are most commonly used (e.g., Victor, 1986a; Wellington and Victor, 1989a; Sponaugle and Cowen, 1994, 1997), but some studies have used polished transverse sections (e.g., Fowler, 1989; Wilson and McCormick, 1997, 1999). Secor *et al.* (1992) provide a good description of sectional planes and preparation methods.

The decision between sagittae or lapilli, and which preparation technique to use, is often based on a subjective determination of the ease of preparing otolith sections and discriminating increments. However, the relevant criteria are the precision and accuracy of otolith increment counts (Campana and Moksness, 1991). Precision refers to the repeatability of increment counts on a given otolith. Precision is easily measured and many researchers have set criteria whereby the otolith is excluded from further analysis if the coefficient of variation of repeated counts exceeds some level (typically 10–15%). However, precision cannot be used as a proxy for age estimation accuracy—the relation between increment count and true age (Campana and Moksness, 1991). Accuracy is most likely influenced by adequate resolution and preparation for resolving increments near the otolith core, and correct identification of daily and subdaily increments (see Victor, 1982). The otolith used for age estimation is also important, because several studies have found that lapillar and sagittal increment counts differ (Fowler, 1989; Schultz and Cowen, 1994). Examination of the hypothesis of daily increment formation is one part of accuracy (see below), but training of readers to count daily increments is also important (Campana and Moksness, 1991). Fowler (1989) provides a good framework for deciding which otolith and preparation technique to use.

b. Deposition Rate Determination All analyses of daily otolith increments start with the null hypothesis that increments are deposited daily. In the coral reef fish literature, we found no examples in which the null hypothesis of daily ring formation was rejected. In several situations in which this null hypothesis was rejected in other fish species, poor sample preparation and inadequate resolution of light microscopy were invoked as possible causes (Campana *et al.*, 1987; Neilson, 1992). However, well-documented examples of nondaily increment formation (e.g., Geffen, 1982; Szedlmayer, 1998) argue that the null hypothesis of daily increment formation should be rigorously tested or the assumption of daily increment formation should be clearly stated.

Examinations of daily increment formation in reef fishes are rife with poor experimental design, i.e., it is unclear if a hypothesis is being tested or a reader is being trained to count daily increments. Owing to the presence of subdaily rings and discontinuities in increment patterns (Victor, 1986a; Fowler, 1989; Wilson and McCormick, 1999), training of readers is not trivial, but should not be confused with testing the null hypothesis of daily increment formation. There is no ideal experimental design, because each species often presents unique problems, and laboratory facilities and field realities vary widely (see Geffen, 1992). However, the studies of Szedlmayer (1998) and Ahrenholz (2000) provide good examples of thorough tests of the hypothesis of daily increment formation. If the null hypothesis is not rejected, the same otoliths can then serve as a tool for training readers to discern daily rings.

c. First Increment Validation Another basic aspect of daily increment analysis is determining when the first increment is deposited. All reef fish studies of which we are aware found first increment formation to occur at or around hatching [Monacanthidae (Kingsford and Milicich, 1987), Labridae (Kishiro and Nakazono, 1991), Pomacentridae (Wellington and Victor, 1989a; Thresher, 1984; Kawase *et al.*, 1993)]. Formation of the first increment generally occurs at hatching, yolk-sac adsorption, or first feeding in most teleost fishes (Jones, 1986). Because hatching, yolk-sac adsorption, and first feeding occur in rapid succession in many reef fishes (Thresher, 1984), the error introduced by not knowing exactly when first increment formation occurs is probably small (±2 days).

d. Settlement Marks Victor (1982) first indicated a settlement mark in coral reef fishes: "At this transition, the previously prominent dark lines delineating each increment abruptly disappear. Regular increments only reappear after a band without discrete increments is formed." He found larval *Thalassoma bifasciatum* (Labridae) without such a settlement mark, newly settled fish with a mark at the edge of the otolith, and postsettlement fish with a mark 38 to 70 increments from the primordium (Fig. 1). The initial description of settlement marks led to a flurry of studies using settlement marks to estimate age-at-settlement, to back-calculate settlement patterns, to back-calculate size-at-settlement, and to estimate growth before and after settlement.

Despite widespread use of settlement marks by reef fish ecologists, very few studies have verified that the marks are indeed temporally or functionally linked to settlement. In a most thorough study, Wilson and McCormick (1997) examined the otoliths of new recruits, fishes with known settlement history from caged patch reefs, and fishes with known settlement history from visually censused, noncaged patch reefs. The

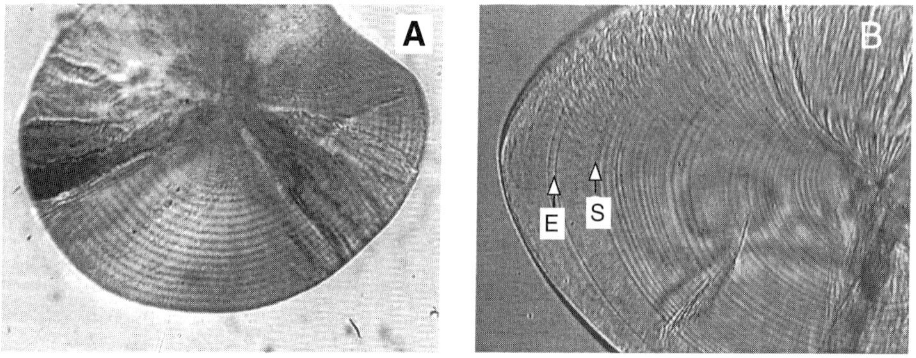

FIGURE 1 Sagittal otoliths of (A) larval and (B) juvenile *Thalassoma bifasciatum* (Labridae) (photomicrograph provided by Su Sponaugle). There is no evidence of an opaque transition zone in the larval otolith. In the juvenile otolith, the beginning of the opaque zone indicates settlement (S) and the end of the zone indicates exit (E) from the sand (Victor, 1983a).

coincidence of settlement with the settlement mark was unequivocal in *Pomacentrus amboinensis* and *Pomacentrus nagasakiensis* (Pomacentridae). Wilson and McCormick (1997) also presented evidence for a temporal link between settlement and the formation of a settlement mark in 14 other species representing a total of seven families. Wilson and McCormick (1999) went further and classified and carefully described four types of settlement marks, thereby providing an excellent frame of reference for future studies.

Although many species exhibit settlement marks, several studies have not found settlement marks in all postsettlement individuals examined. Fowler (1989) examined the otoliths of 14 *Chaetodon rainfordi* (Chaetodontidae) of known settlement history and found a settlement mark at the time of settlement in only 6 fish. Moreover, of the 92 field-collected *C. rainfordi* juveniles, 76 possessed a settlement mark. Of a total of 52 *Chaetodon plebius* (Chaetodontidae), 33 possessed a settlement mark. All 14 *Chelmon rostratus* (Chaetodontidae) juveniles possessed a settlement mark. Rogers *et al.* (2001) identified a population-level change in increment widths coincident with the transition from larval to juvenile life stages in *Stephanolepis hispidus* (Monacanthidae), but were able to identify this increment width change objectively in less than half of the individual records.

Variable patterns in increment widths, and the enigmatic nature of settlement marks in some species, raise the question: What is the exact nature of this transition zone if all individuals of a given species undergo similar life history stage events, yet a settlement mark or a specific pattern in otolith increment widths cannot always be discerned? Even when a settlement mark can be identified in all individuals, the physiological cause of the mark remains unknown. Although studies that have elucidated and classified settlement marks strengthen the validity of approach (Wilson and McCormick, 1997, 1999), we clearly have much to learn about the factors generating settlement marks in the otoliths of reef fish.

2. Estimation of Age and Timing of Settlement

Estimates of larval durations provided one of the first views into Sale's (1980) "black box" of the larval stage of coral reef fishes. Larval durations of > 200 coral reef fish species have been determined (e.g., Thresher and Brothers, 1985; Victor, 1986a; Wellington and Victor, 1989a), and, subsequently, temporal and spatial differences in larval durations have been found (McCormick, 1994; Wellington and Victor, 1989b). The processes responsible for variable larval duration remain unclear, as do the consequences to postsettlement ecology. Researchers have also used settlement ages to back-calculate settlement dates, resulting in daily time series of settlement. Back-calculated settlement dates in addition to daily measurements of settlement have provided insights into the relative importance of pre- and postsettlement processes in determining recruitment (e.g., Victor, 1986b; Meekan *et al.*, 1993; Sponaugle and Cowen, 1996a,b). McCormick (1994) outlined the advantages and disadvantages of the various methods that can be used to estimate settlement age.

a. Back-Calculated Settlement Dates The most widely used method for estimating pelagic larval duration and patterns of settlement is based on back-calculation from settlement marks in otoliths. Collecting fish on the day of settlement is more direct, but requires a large effort, especially when sampling over space and time. Back-calculation of temporal settlement patterns obviates the need for daily sampling, but assumes that back-calculated settlement patterns accurately reflect daily settlement histories. Pitcher (1988) compared settlement records of *Pomacentrus wardi* (Pomacentridae) based on daily censuses, collections made every 2 weeks, and an end-of-the-season collection (Fig. 2). He found general congruence among the three methods and from three different locations, with cross-correlations of 0.75–0.86 between the back-calculated time series and the daily and biweekly censuses. In a similar study, Meekan (1992) compared settlement records based on back-calculation from weekly, monthly, and end-of-the-season collections. In one year Meekan (1992) found cross-correlations of 0.74–0.77 between back-calculated settlement records based on the three collection types. In another year, however, settlement records from weekly and monthly collections were similarly cross-correlated ($r = 0.72$), but cross-correlations with end-of-the-season settlement records were lower ($r = 0.42$). Pitcher (1988), Doherty (1991), and Meekan (1992) discuss several potential problems with settlement date back-calculation, which also apply to some degree to back-calculation of settlement age.

1. *Age estimation errors.* In the age estimates, there can be measurement errors, which often increase with age (Pitcher, 1988; Doherty, 1991; Meekan, 1992). This bias decreases the ability to back-calculate settlement dates from older individuals and therefore limits the time between sampling events.

2. *Sample sizes.* There are logistic constraints of time and effort to estimate the age of a large number of postsettlement individuals for back-calculation.

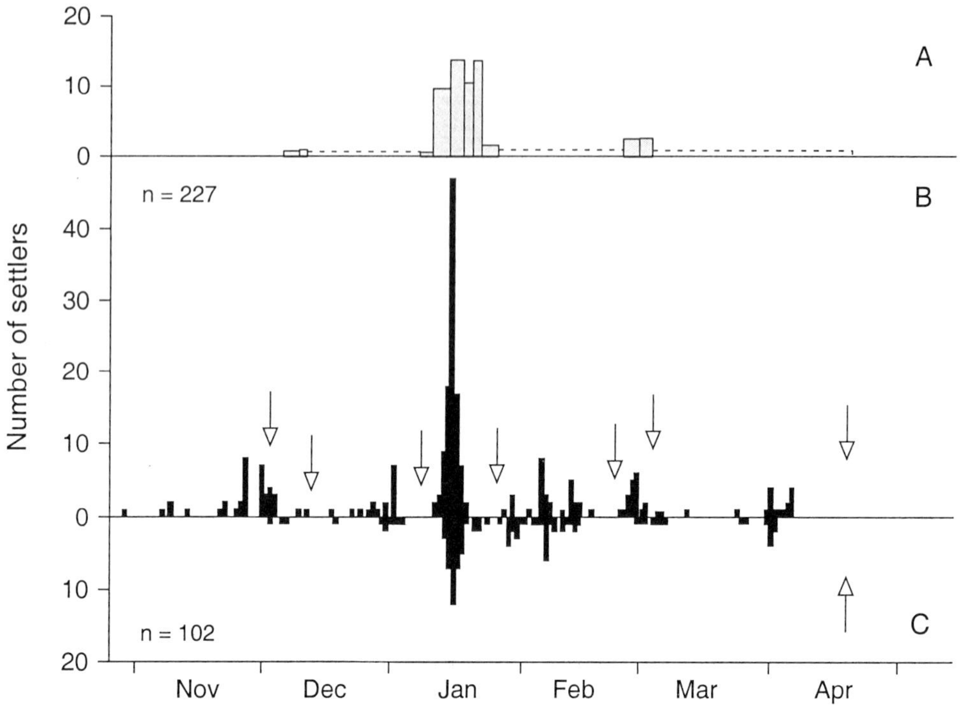

FIGURE 2 Settlement patterns of *Pomacentrus wardi* (Pomacentridae) to Heron Reef in 1983/1984 as determined by (A) visual censuses, (B) back-calculated settlement dates from collections made seven times during the summer (marked by arrows), and (C) back-calculated settlement dates from a collection made at the end of the summer (marked by arrow). Data from Pitcher (1988).

Pitcher (1988) suggested that 100 individuals would be adequate to resolve settlement patterns over 3–4 months from one location, but both Doherty (1991) and Meekan (1992) indicated that this number is probably too small.

3. *Mortality.* Back-calculated settlement dates will be affected by postsettlement mortality. For a given collection, fishes settling more recently will have experienced a lower cumulative probability of mortality than fishes settling further from the time of collection. Because mortality rates of postsettlement fishes vary across spatial and temporal scales, with time from settlement and among species (Victor, 1982; Meekan, 1992; Sponaugle and Cowen, 1997; Schmitt and Holbrook 1999a), correcting a back-calculated settlement record requires an accurate mortality schedule. Using inappropriate mortality rates in any correction risks creating error in the settlement record (Doherty, 1991). If, however, the mortality bias is left uncorrected, back-calculation of settlement dates will underestimate the abundance of earlier settling (i.e, older) individuals relative to later settling (i.e., younger) individuals (Doherty, 1991). Gear avoidance and movement by older or larger fish will exacerbate the mortality bias. Underestimation of earlier settlers is readily apparent in Pitcher's (1988) data when comparing settlement patterns from the multisampling and end-of-season sampling schemes (Fig. 2B and C).

4. *Selective processes.* Back-calculated settlement dates encompass selective processes that occur between settlement and capture. Mortality-related events may act differentially on individuals that settled on the same day owing to some individual trait (e.g., Searcy and Sponuagle, 2001). The effects of selective mortality on back-calculated settlement dates have not been considered, but further complicate the effects of mortality rates that may vary dramatically in both time and space.

5. *Catch per unit effort.* If multiple collections are made over time, earlier settling cohorts may be sampled more frequently than late-settling cohorts, thereby overemphasizing their abundance in any back-calculated settlement record. This problem may be corrected in two ways. (1) For a given collection date, include only fishes that settled after the previous collection date in the back-calculation. (2) Correct the number of fishes settling on a date by dividing by the number of sampling dates that contributed those fishes. To our knowledge, neither of these approaches has been applied and therefore effects on back-calculated settlement records will need to be investigated.

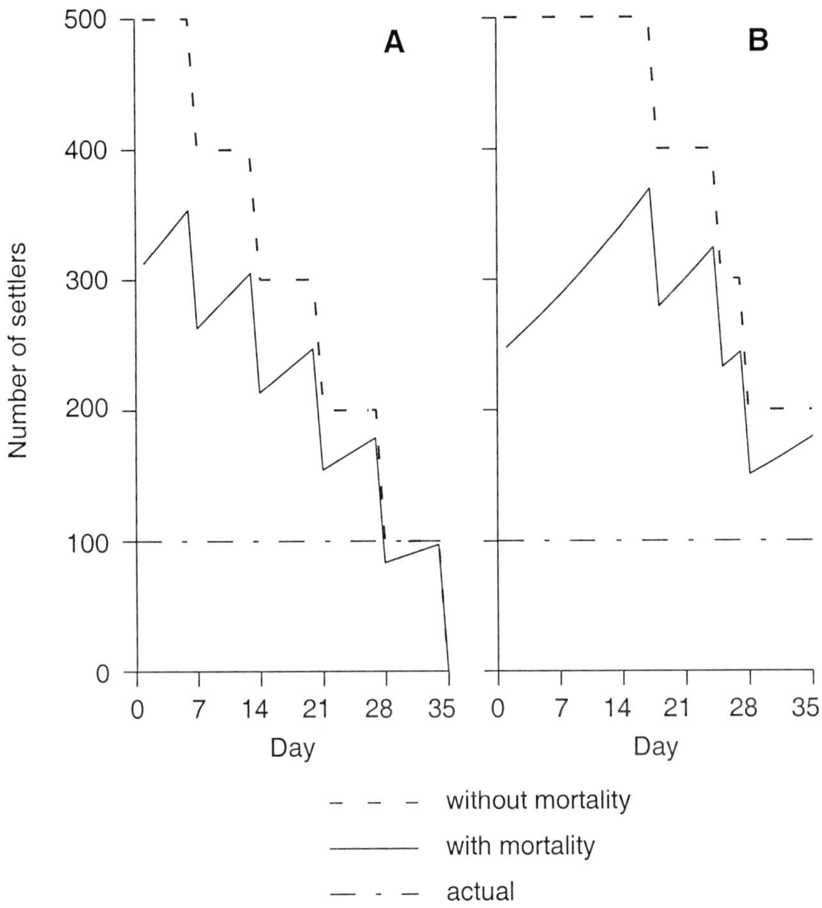

FIGURE 3 Model simulations of back-calculated settlement dates. (A) Samples were collected on five evenly distributed dates (days 7, 14, 21, 28, and 35) and settlement dates were calculated for all sampled fishes. (B) Samples were collected on five unevenly distributed dates (days 3, 9, 14, 28, and 35) and settlement dates were calculated for all sampled fishes. We used a mortality rate of 0.025, based on empirical estimates of postsettlement mortality in juvenile reef fish (Victor, 1986b; Sponaugle and Cowen, 1997; Schmitt and Holbrook, 1999a).

6. *Subsampling.* Often age and settlement date are estimated for a subsample of fishes collected on a given day. If collection of fishes is quantitative, then age and settlement date records need to be estimated for the entire sample before further analyses can be conducted. Issues of subsampling are well beyond the scope of this chapter, but in general have not been thoroughly dealt with regarding the estimation of settlement date or hatch date data. Sponaugle and Cowen (1994) measured all fishes and estimated postsettlement ages for a subsample. They then used an age–length regression to predict postsettlement ages for fishes that were only measured to derive back-calculated settlement dates. A second option is to use an age–length key, where for each length class of a subsample of fishes, the percentage age distribution is enumerated. Lengths of fishes for which age was not estimated can then be distributed over age groups according to the key (Campana and Jones, 1992).

Results from a simulation model illustrated the problems associated with mortality and resampling of cohorts when back-calculating settlement. Weekly sampling produced a weekly pattern in abundance, owing to lower cumulative mortality of fishes settling closer to the day of sampling (Fig. 3A). The overemphasis of early settlers was also clear; the cohort settling on day 5 was sampled six times, whereas the cohort settling on day 34 was sampled only once. Sampling interval also affected the appearance of the back-calculated settlement record (Fig. 3B). Variable mortality, increasing avoidance of capture with age, and postsettlement movement would have further complicated the back-calculated settlement record. The modeling results, and the studies of Pitcher (1988), Doherty (1991), and Meekan (1992), suggest that more caution should be used in the back-calculation of settlement dates and that researchers should assess the magnitude of any biases in their particular situation.

b. Back-Calculated Hatch Dates Back-calculated hatch dates are used to determine the timing of spawning and to compare the hatch dates of individuals surviving to settle on the reef to records of reproductive output (McFarland *et al.*, 1985; Robertson *et al.*, 1988;

Kishiro and Nakazono, 1991; von Herbing and Hunte, 1991; Colin *et al.*, 1997; Danilowicz, 1997; Masterson *et al.*, 1997). Many of the same caveats regarding back-calculation of settlement date apply to back-calculation of hatch date. An excellent discussion of hatch date back-calculation is provided by Campana and Jones (1992). They indicate that hatch date distributions derived from settled juveniles should be relatively immune to the mortality problems discussed above, because most of the mortality between hatching and the juvenile stage occurs in the early larval stage. However, if multiple collections are made it will still be necessary to correct for uneven sampling of cohorts.

3. Estimates of Size and Growth

The ability to estimate size and growth provides a powerful application of otolith microstructure to the study of coral reef fishes. Temporal and spatial patterns in postsettlement growth have been documented from models of length-on-age (e.g., Kingsford and Milicich, 1987; Thorrold and Milicich, 1990; Sponaugle and Cowen, 1994, 1997). Masterson *et al.* (1997) used otolith size of newly settled *Thalassoma bifasciatum* (Labridae) to back-calculate body size at settlement and found little difference among sites and years. In a similar study, Searcy and Sponaugle (2000) found intercohort differences in larval growth as inferred from otolith increment width patterns. The cohort settling at smaller sizes also grew faster during the larval stage and spent less time in the plankton. Further, Searcy and Sponuagle (2000) found fishes with a wider metamorphic band (see Fig. 1) had a higher probability of survival during the juvenile stage, suggesting general condition at settlement may be important to postsettlement survival. As more studies make use of the otolith as a recorder of age, size, growth, and development, we will undoubtedly develop a better understanding of the role that processes occurring within the pelagic environment and at settlement play in the regulation of reef fish populations.

a. Back-Calculation of Individual Size and Growth Two approaches have been used to back-calculate size-at-age from otolith size of individual larval and juvenile fishes (Francis, 1990). Regression methods use parameters from a fish size-at-capture/otolith size-at-capture regression to estimate fish size-at-settlement from otolith size-at-settlement. Proportional methods use regression parameters but assume that if at capture an otolith was 10% larger than the average otolith for a fish that size, the otolith would be 10% larger at every size (this is a scale proportional back-calculation; there is an analogous body proportional back-calculation).

Campana (1990) demonstrated that many regressions of fish size-at-capture and otolith size-at-capture misestimated true fish size at otolith formation, which led to significant errors in back-calculated lengths. To correct these errors he presented a proportional approach (the "biological intercept method") that used independently derived estimates of fish size and otolith size at the biological intercept, rather than regression parameters. Francis *et al.* (1993) concluded the biological intercept method was the preferred form of back-calculation.

Secor and Dean (1989) and Hare and Cowen (1995) argued against proportional methods because they found that that fish growth and otolith growth were not in constant proportion due to growth rate effects and ontogenetic effects on the fish size/otolith size relation. Rogers *et al.* (2001) provided a useful illustration of potential complexity of relations between otolith size and fish size in *Stephanolepis hispidus*. The population-level relation between otolith growth and somatic growth from the early larval into the juvenile stages exhibited significant nonlinear trends (Fig. 4). These data demonstrate that otolith growth cannot be used to compare individual somatic growth rates across ages or life history stages (see also Thorrold and Milicich, 1990; McCormick, 1994).

Many studies erroneously interpret a positive correlation between fish size and otolith size as evidence

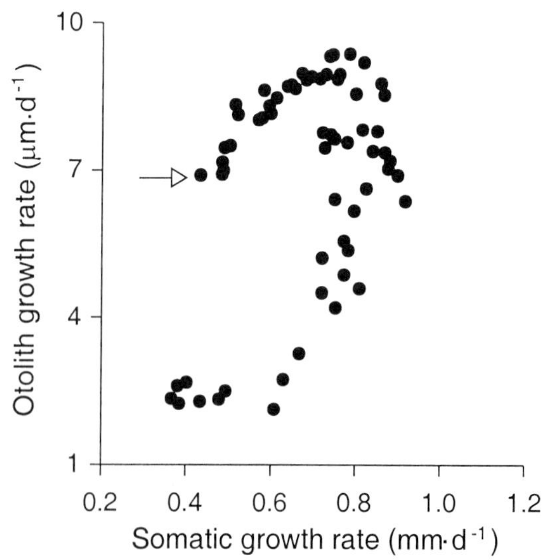

FIGURE 4 Relation between otolith growth and fish growth at different ages in juvenile *Stephanolepis hispidus* (Monacanthidae). The youngest age in the trajectory is labeled with an arrow, and age increases away from this point. Data from Rogers *et al.* (2001). Growth rates determined from moving regressions of otolith radius and standard length on age.

of a positive relation between otolith growth and fish growth. However, fish size and otolith size are positively correlated simply because both are growing, with or without a positive relationship between otolith growth and fish growth. To illustrate the problem, we simulated an extreme example in which otolith growth and fish growth were negatively correlated (Fig. 5A). However, both fish size (Fig. 5B) and otolith size (Fig. 5C) increased with age, simply because both continued to grow. The important point was that fish size and otolith size were positively correlated (Fig. 5D), even though their growth was negatively correlated (Fig. 5A). When the residuals of an otolith size-on-age model and a fish size-on-age model were compared, the true relation between fish growth and otolith growth was apparent (Fig. 5E) (see Hare and Cowen, 1995).

Ontogenetic and interpretative problems aside, several studies found a positive relation between otolith growth and somatic growth of individuals within a life history stage and of similar ages. Szedlmayer (1998) provided the best example from a coral reef fish. Otolith growth was positively related to somatic growth in postsettlement *Lutjanus campechanus* (Lutjanidae). Victor (1982) found wider otolith increments in postsettlement *Halichoeres bivittatus* (Labridae) corresponding to increased ration. Studies on temperate fishes also support a positive relation between otolith growth and somatic growth (e.g., Barkman and Bengston, 1987; Paperno *et al.*, 1997).

Our intent is not to suggest that the promise of reconstructing size and growth throughout the life of an individual fish from daily increments in otoliths is empty. Otolith growth and fish growth are strongly correlated in at least some reef fish species (Szedlmayer, 1998), and clearly back-calculation of fish size and growth from otolith size and growth has merit. Rather, our purpose is to acknowledge some of the pitfalls of size back-calculation to lead to better science and to generate interest in solving some of these problems so that the potential of size back-calculation can be achieved in the study of coral reef fishes.

b. Cohort/Population Level Size–Age Models The growth rate of a fish population, as opposed to that of an individual, can also be estimated from otolith-derived ages and the size of fishes at capture. Comparisons can then be made between groups of fishes (collected at different times, different locations) to determine if growth rates differ. Several studies have used size-at-age data to estimate a population-level postsettlement growth and size-at-settlement (Kingsford and Milicich, 1987; Thorrold and Milicich, 1990; Sponaugle and Cowen, 1994, 1997). Here we review the application of linear models and then discuss the use of nonlinear models.

Most studies of reef fish growth have used a linear least-squares regression (Kingsford and Milicich, 1987; Fowler, 1989; Thorrold and Milicich, 1990). However, some workers have used a geometric-mean regression approach (Sponaugle and Cowen, 1994, 1997). There remains active debate over which of these approaches is most appropriate. One assumption of Model I regression is violated by the nature of length and age data derived from the field—the independent variable is assumed to be under the control of the investigator. Ricker (1973, 1975) suggested the use of a Model II regression, specifically the geometric-mean regression (also termed the standard major axis). Laws and Archie (1981) provided compelling examples of how conclusions based on geometric-mean regression and least-squares regressions can differ. Jolicoeur (1975) argued that an ordinary major-axis regression (also called principal-axis or major-axis regression) was a more appropriate Model II regression (see Campana and Jones, 1992).

Snedecor and Cochran (1980) discuss the effect of measurement error in an independent variable on a least-squares regression slope. They provide a correction, if measurement error can be quantified,

$$b_{\text{corr}} = b \cdot (1 + s_E^2/s_A^2), \tag{5}$$

where b is the slope estimated with least-squares regression, s_E^2 is the variance of measurement error, and s_A^2 is the variance in estimated ages used in the regression. The levels of measurement error required to create a visible downward bias in b are large ($s_E^2 > 10\%$ of s_A^2). Many researchers quantify measurement error in their age estimation (e.g., CV $< 10\%$), making the correction of Snedecor and Cochran (1980) possible. Such a correction still leaves open the question of which regression model to use, but to provide some solace, as the r^2 of a bivariate relation approaches 1, the parameter estimates from the different regressions converge. We offer no solution as to which regression model to use, but encourage future research to provide a definitive answer.

A number of researchers have used nonlinear models to describe fish size-at-age data and the different forms are well covered in Campana and Jones (1992). Standard growth models such as the Gompertz and the von Bertalanffy models are often used. Other studies have used an asymptotic regression (Fowler, 1989), exponential models (Kishiro and Nakazono, 1991; Lou, 1993), and polynomials (Lou, 1993; Rogers *et al.*, 2001). One benefit of commonly used models is that the parameter estimates are readily interpretable by other workers (Campana, 1992). However, the use of a model

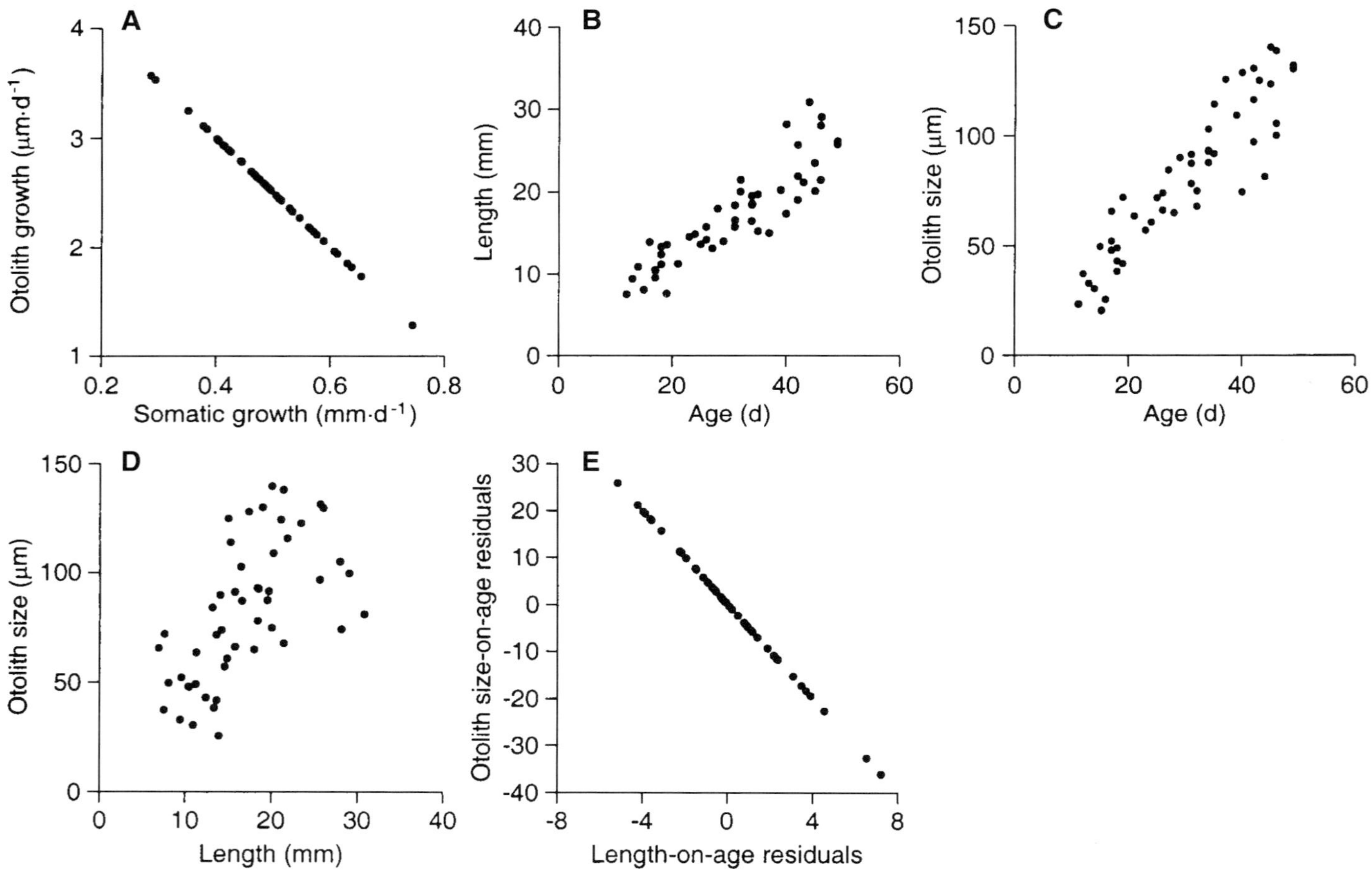

FIGURE 5 (A) Otolith growth and fish growth relation used in model simulations. Individuals were randomly assigned a somatic growth rate using a normal distribution, with $\mu = 0.5$ mm day^{-1} and $\sigma = 0.1$ SD. Otolith growth was the inverse of somatic growth $\times$ 10 to represent μm day^{-1}. (B) Relation between somatic size and fish age. Individuals were randomly assigned an age based on a uniform distribution and age $\times$ somatic growth rate was used to calculate fish size. (C) Relation between otolith size and fish age. Otolith size was calculated from age $\times$ otolith growth rate. (D) Relation between otolith size and fish size was positive ($r = 0.70$), simply because both otolith size and fish length are increasing with age (C, D) and despite the negative correlation between otolith growth and somatic growth. (E) Residuals of the otolith size–age and fish size–age relations revealed the true otolith growth and fish growth relation (*sensu* Hare and Cowen, 1995).

with easily interpretable and comparable parameters but that misrepresents the fish size–age relation (nonrandomly distributed residuals or nonconstant variance across the range) inevitably creates more problems than the use of a more complex model that better fits the relation. Age-dependent growth rate can be estimated from the first derivative of most nonlinear size-on-age models (Fowler, 1989; Rogers *et al.,* 2001). Similarly, moving regressions can be used to estimate growth rates over successive, short intervals of the age range (Rogers *et al.,* 2001), allowing for statistical comparisons of growth rates among different models.

4. Delayed Metamorphosis

Several studies have concluded that larvae of some coral reef fishes can delay metamorphosis until they reach appropriate settlement habitat (Victor, 1986c; Cowen, 1991; Sponaugle and Cowen, 1994). Much of the evidence supporting the ability to delay metamorphosis has been inferred from the otolith microstructure of settled juveniles. Larvae that delayed metamorphosis exhibited a right-skewed distribution of pelagic larval duration (Victor, 1986c; Masterson *et al.,* 1997; Sponaugle and Cowen, 1994, 1997), and the period between attainment of competency and settlement was witnessed in the otolith as narrow increment widths (Fig. 6) (Cowen, 1991). Taken together, the increment width patterns in the above studies suggested that the time to competency is relatively fixed, that postcompetent larval growth is very slow, and that larval development is "on hold" until settlement is possible.

McCormick (1999b) used an experimental approach and found that 24% of settling *Acanthurus triostegus* (Acanthuridae) were able to maintain their late-larval morphology and characters when returned to the pelagic environment, thus demonstrating the ability to delay metamorphosis. He found that otolith increment widths peak at about day 20 in *A. triostegus* and subsequently decreased until settlement. Wilson and McCormick (1999) found qualitatively similar patterns of otolith increment widths in many of the 44 species that they examined. McCormick (1999b) argued that either the ability to delay metamorphosis is widespread among reef fishes or researchers should be cautious in inferring the attainment of competency from otolith increment width patterns. Caution in the interpretation of otolith increment widths across ages and life history stages is also indicated by age-specific patterns in the otolith growth and somatic growth relation (Fig. 4).

The interpretation of delayed metamorphosis from a skewed distribution of larval duration is also not straightforward. McCormick (1999b) demonstrated the ability to delay metamorphosis in *Acanthurus triostegus* but found larval duration to be normally distributed. Hare and Cowen (1997) illustrated that a settlement age distribution (larval to juvenile transition in their example) can become skewed owing to differential

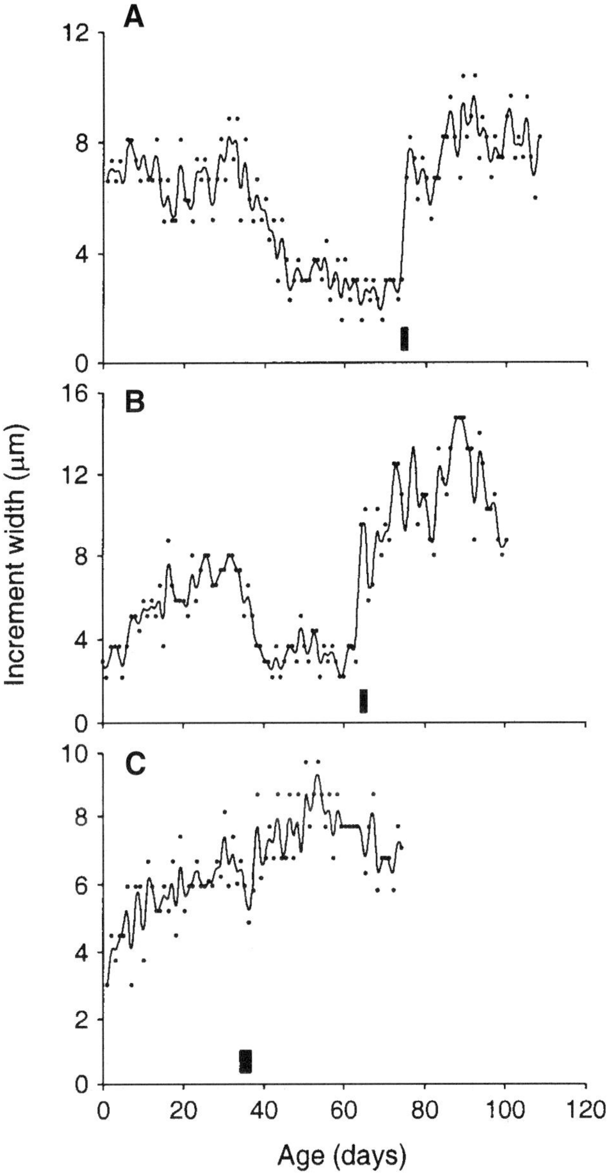

FIGURE 6 Daily increment widths in otoliths of *Semicossyphus pulcher* (Labridae) recruits. Black bars represent the location of the settlement mark. (A) Individual with a precompetency period of approximately 40 days and a competency period of 35 days. (B) Individual with a precompetency period of approximately 40 days and a competency period of 20 days. (C) Individual with a precompetency period of approximately 40 days and no apparent competency period. Figure modified with permission from Cowen (1991).

larval and juvenile survival. Individuals settling at an earlier age left the high-mortality environment of the plankton ($z = 0.1–0.3$) for a lower mortality environment ($z = 0.01–0.1$) and thus had a higher probability of survival.

5. Selective Processes Inferred from Otolith Growth

Evaluating the processes generating selective mortality of individuals is at the core of much of fish ecology. Is mortality dependent on the size at which a fish settles? Do faster growing fish have a higher probability of survival? Do certain characteristics of the larval stage convey some survival advantage to postsettlement individuals? Do larvae settling at different times experience different mortality? All these questions can be examined through analysis of otolith microstructure. For instance, techniques exist for comparing growth rates of pre- and postselection samples [cross-sectional analyses (Anderson, 1995; Hare and Cowen, 1997)] and for comparing growth rates among groups over time [longitudinal analyses (Thorrold and Williams, 1989; Chambers and Miller, 1995; Meekan and Fortier, 1996; Jones, 2000)]. Because these techniques are relatively new, the application to the study of coral reef fishes is limited. However, Searcy and Sponaugle (2000, 2001) used both cross-sectional and longitudinal analyses to examine the effect of various characters (larval otolith growth, larval otolith size-at-age, width of the metamorphic band) on postsettlement survival in *Halichoeres bivittatus* and *Thalassoma bifasciatum*. They used otolith characters rather than calculating size owing to the problems discussed above and made the assumption that otolith size (growth) was positively related to fish size (growth). Survivors collected a period of time after settlement had a wider metamorphic band than did newly settled fish. In addition, surviving individuals of *H. bivittatus* grew faster as larvae. This work highlighted the importance of larval and settlement traits to the subsequent survival of fish on the reef.

6. Larval Transport

Many studies have proposed that reef fish populations are recruitment limited (e.g., Victor, 1986b; Milicich *et al.*, 1992; Doherty and Fowler, 1994a) and thus processes controlling the supply of larvae to reefs are important to the population dynamics of reef fishes. Larval supply is controlled by four processes: spawning, survival, transport, and behavior (e.g., Robertson *et al.*, 1988; Milicich, 1994; Sponaugle and Cowen, 1996a,b; Cowen *et al.*, 2000). Otoliths can be used to infer spawning distributions and to examine processes affecting survival in both the plankton and on the reef (see above). Otoliths can also provide information on dispersal pathways of coral reef fishes during pelagic stages.

From the perspective of a physical oceanographer, otoliths allow fish larvae to be used as time-constrained physical tracers. Daily age estimates from otoliths provide a temporal scale that can be superimposed on the changing distributions of fish larvae in space. Cowen (1985) determined the pelagic larval duration of *Semicossophyus pulcher* (Labridae) for settled juveniles and then used these ages as a time scale within which to evaluate larval transport processes acting along the southern California coast. Schultz and Cowen (1994) coupled pelagic larval durations with a probabilistic model of larval transport to calculate the recurrence interval for larvae to be transported from the United States mainland and Caribbean to Bermuda. Hare and Cowen (1991) used larval ages to evaluate several transport mechanisms responsible for the northward expatriation of *Xyrichtys novacula* (Labridae) larvae along the United States east coast. In the next 10 years, further coupling of larval ages, larval distributions, empirical physical observations, and three-dimensional circulation models will allow researchers to develop adequate models of the physical processes determining larval supply and settlement to coral reefs. Moreover, the inclusion of larval growth estimates and information from otolith chemistry (see below) will provide an unprecedented view of the mechanisms of larval transport and their role in shaping coral reef fish ecology.

B. Annual Increments

Age estimation based on annuli in otoliths is a routine component of fisheries research in most temperate systems, but has seen widespread interest from reef fish ecologists only in the past 10 years (see Chapter 3, this volume). At least part of this attentiveness has been due to reports of surprising longevity in some temperate fishes. Based on validated otolith analyses, rockfishes of the genus *Sebastes* (Scopaenidae) regularly live more than 75 years (Campana *et al.*, 1990), and sciaenids such as *Pogonias chromis* in the western Atlantic and Gulf of Mexico commonly attain maximum ages of 50 years or more (Campana and Jones, 1998). Claims of such longevity have not been without controversy. For instance, rigorous debate surrounds assertions that orange roughy, *Hoplostethus atlanticus* (Trachichthyidae), at lengths of 38–40 cm may be up to 150 years old (Fenton *et al.*, 1991; West and Gauldie, 1994). Of course, the presence of such old fishes implies that natural mortality rates in unfished populations are extremely low, and hence the stocks are likely to support only minimal rates of exploitation. These debates underscore the importance of adequate validation of presumed annual increments, especially given recent estimates of longevity in some reef fish populations (e.g., Newman *et al.*, 2000a,b).

1. Validation Techniques

A number of validation techniques have been applied to determine the periodicity of annulus formation, of which three have been used routinely with coral reef fishes (Fowler, 1995). Marginal increment analysis remains the most common approach (e.g., Hood and Schlieder, 1992; Potts and Manooch, 1995; Crabtree and Bullock, 1998; Potts *et al.*, 1998; Rocha-Olivares, 1998; Hood and Johnson, 1999; Wyanski *et al.*, 2000), even though it is generally not suitable for validating estimates of extreme longevity. Tetracycline tagging has been used to validate the periodicity of annulus formation in species from a number of families in the Great Barrier Reef region (e.g., Fowler, 1990a; Choat and Axe, 1996; Choat *et al.*, 1996; Hart and Russ, 1996; Newman *et al.*, 1996, 2000b; Cappo *et al.*, 2000). As with marginal increment analysis, chemical tagging studies are rarely able to validate adequately the periodicity of annulus formation over the entire age range of most fishes. Finally, scanning electron microscopy has been used to enumerate daily increments between adjacent annuli in otoliths (reviewed by Fowler, 1995). The method has, however, been applied only sparingly over the past decade.

a. Radioisotopes Logistic difficulties associated with age validation using each of the three approaches listed above provided the impetus for validation techniques using radiometric dating. Put simplistically, this approach measures the abundance of both a parent and a daughter radioisotope in a sample, and then calculates the age of the sample based on the rate of exponential decay (half-life) of the parent isotope with respect to the daughter isotope. More accurately, age determination is based on the extent of disequilibrium between the activities of the parent:daughter pair (Fenton and Short, 1992). For instance, the activity ratios of ^{210}Pb:^{226}Ra reach secular equilibrium in about 150 years, which explains the prevalence of this parent:daughter pair in age validation studies (Bennett *et al.*, 1982; Campana *et al.*, 1990; Fenton *et al.*, 1990, 1991; Fenton and Short, 1994; Kastelle *et al.*, 1994; Milton *et al.*, 1995; Stewart *et al.*, 1995; Andrews *et al.*, 1999). A second pair (^{228}Th:^{228}Ra), which achieve secular equilibrium in approximately 8 years, was used by Campana *et al.* (1993) to validate the periodicity of annulus formation in flying fish (*Hirundichthys affinis*) otoliths.

Radiometric age estimation has been applied to coral reef fishes in only a single study to date. Milton *et al.* (1995) used ^{210}Pb:^{226}Ra activity in three *Lutjanus* (Lutjanidae) species to validate age estimates based on annuli in whole and sectioned otoliths. They concluded that sectioned otoliths significantly overestimated ages in *Lutjanus malabaricus, L. erythropterus,* and *L. sebae,* based on results of radiometric analyses of otolith cores. The finding was noteworthy because all other radiometric analyses to that point, from long-lived temperate fishes, had generally confirmed the accuracy of age estimates based on annuli in sectioned otoliths. However, a more recent study validated the annual deposition of annuli in sectioned otoliths of *L. malabaricus*, *L. erythropterus* and *L. sebae* from the central Great Barrier Reef (Cappo *et al.*, 2000). Based on these validated annuli, Newman *et al.* (2000a) found maximum ages of more than 20 years for all three species. It therefore seems likely that there may have been significant errors in the radiometric study.

b. Radiocarbon Produced by Atomic Weapons Testing Atmospheric tests of thermonuclear weapons produced large increases in the levels of a number of radioisotopes in the environment, including ^{3}H, ^{14}C, ^{36}Cl, ^{90}Sr, ^{129}I, ^{185}W, and ^{230}Pu, over a relatively short period of time (1945–1970). For instance, atmospheric levels of bomb-derived ^{14}C effectively doubled from pre-bomb levels in the early 1950's to peak values in the mid 1960's (Fig. 7). The rapid increase in $\Delta^{14}C_{DIC}$ in tropical and sub-tropical waters between the mid-1950's and early 1970's was recorded in the aragonitic skeletons of hermatypic corals at numerous locations including the Florida Keys (Druffel and Linick, 1978), Heron Island on the Great Barrier Reef (Landman *et al.*, 1988), Nauru Island (Guilderson *et al.*, 1998), and the Galapagos Islands (Guilderson and Schrag, 1998). Kalish (1993) demonstrated that similar records could also be reconstructed from fish otoliths, because most if not all of the carbon in otoliths comes directly from DIC. By comparing the time series of bomb-derived ^{14}C input to the oceans from otolith chronologies premised on the annual deposition of increments with those from nearby corals, Kalish was able to validate the periodicity of the increment formation in the otoliths of *Pagrus auratus* (Sparidae). This approach has now been used to validate the periodicity of annulus formation in both pelagic and demersal species from the Southern and Northern Hemispheres (Kalish, 1995; Kalish *et al.*, 1996, 1997; Campana, 1997; Campana and Jones, 1998).

Age validation based on bomb radiocarbon chronologies relies on few assumptions. The technique does require that the otolith core, consisting of material laid

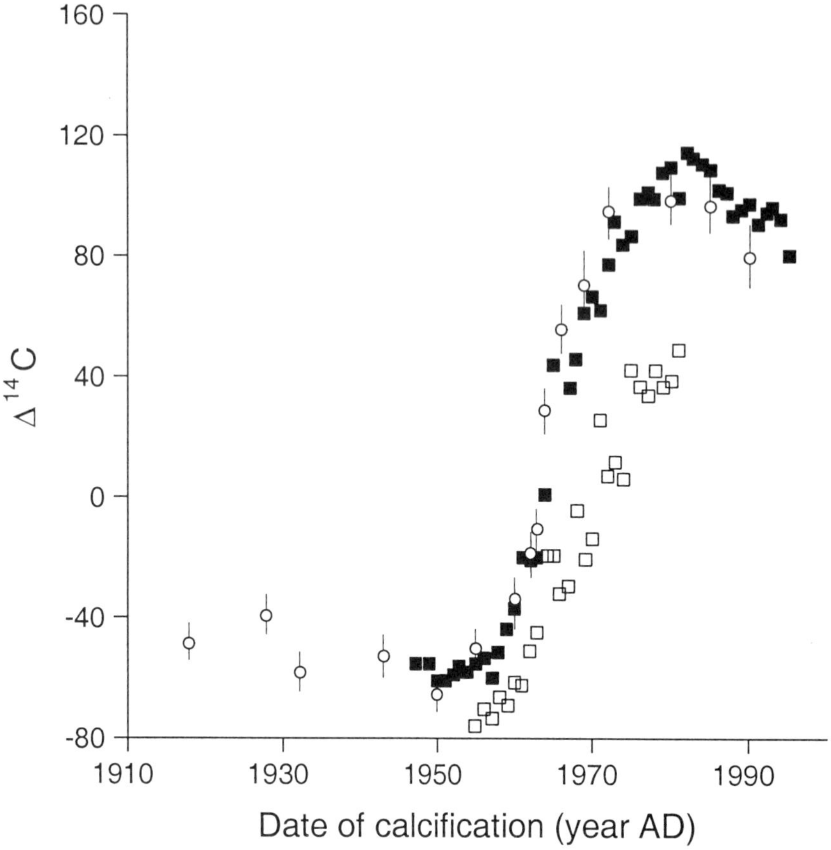

FIGURE 7 Reconstructions of mean surface ocean water $\Delta^{14}C_{DIC}$ versus time based on *Pagrus auratus* otoliths (○, ±SD), a coral of the genus *Porites* from Nauru Island in the western equatorial Pacific Ocean (□), and a *Porites lobata* core from Urvina Bay in the Galapagos Islands (■). Fish otolith data from Kalish (1993), Nauru Island coral data from Guilderson *et al.* (1998), and Galapagos Island coral data from Guilderson and Schrag (1998).

down during the first year of life, be removed for subsequent analysis. Although contamination of the core with material deposited later in life is probably common, violation of the assumption will lead to underestimates of actual age. It is also necessary to have an accurate reference $\Delta^{14}C_{DIC}$ chronology that reflects the environment inhabited by the juveniles so as to avoid post-hoc explanations for phase differences between reference and otolith-based time series (e.g., Campana and Jones, 1998). This should present few difficulties in most reef systems, given the number of coral-based records that are currently available (see references above). Logistic requirements may be more problematic when applying the validation technique to coral reef fishes. Otoliths are required from fishes that were spawned during the rapid increase in $\Delta^{14}C_{DIC}$, from the early 1950s to the early 1970s. Unless archived material is available, the approach will be applicable only for fishes that live for 50 years or more.

2. Application of Annual Increments in Otoliths

The need for age data from coral reef fishes has been proselytized in an earlier chapter (see Chapter 3, this volume) and will not be repeated here. However, several innovative applications of the data generated by age estimation from validated otolith annuli deserve mention. Ferreira and Russ (1995) used annual increments in otoliths of coral trout, *Plectropomus leopardus* (Serranidae), to investigate the effectiveness of marine protected areas on the Great Barrier Reef. No significant differences in size structure or sex ratios were found on protected and unprotected reefs. The effects of protection from fishing pressures were, however, obvious in the age structures on the two protected reefs. Age structures from protected reefs were dominated by a single year-class that was not apparent on the unprotected reefs. Ferreira and Russ (1995) concluded that age structures of *P. leopardus,* and presumably other Epinepheline serranids and similarly long-lived species, were more sensitive than size structure to the effects of fishing. These results will hopefully not be lost on fisheries managers charged with the monitoring of an increasing number of marine protected areas in coral reef ecosystems.

A second example of the application of age data to a fundamental ecological question comes from a natural experiment in which crown-of-thorns (COT) starfish outbreaks caused large increases in algal biomass on some, but not all, reefs in the central Great Barrier Reef. In an earlier study, Williams (1986) found no detectable numerical response of herbivorous fish populations, including acanthurids and scarids, to considerable increases in food supply, and concluded that larval supply, rather than food, appeared to be limiting population sizes in this region. However, Williams also noted that the effects of increased algal biomass may have been manifested in more subtle ways, such as increased growth rates of herbivorous fish after COT outbreaks. Hart and Russ (1996) tested this hypothesis by collecting age data, via annual increments in otoliths, from *Acanthurus nigrofuscus* (Acanthuridae) on both outbreak and control reefs. They found, as predicted, increased mean size, and mean size-at-age, on the COT-impacted reefs.

Finally, in an elegant test of the recruitment limitation hypothesis, Doherty and Fowler (1994a) used annual increments in otoliths to determine age structures of *Pomacentrus moluccensis* (Pomacentridae) on patch reefs in the Capricorn-Bunker section of the Great Barrier Reef. A 9-year time series of recruitment to each of seven reefs in the region allowed the authors to correlate estimates of recruitment with age structure. These data showed that year-class strength was primarily, if not singularly, determined by density-independent postsettlement mortality superimposed on highly variable patterns of larval supply. Recruitment rates to at least some reefs in the Capricorn-Bunker section are, on average, lower than in more northern sections, and hence the level of recruitment limitation found in these populations may not be more generally applicable. However, the study emphasized the power of combining long-term monitoring with age-based demographic information to test complex ecological hypotheses.

IV. Otolith Chemistry

Chemical analyses of calcified structures such as coral skeletons and mollusc shells have a long history in coral reef research. For instance, the chemistry of aragonite in coral skeletons has been used to reconstruct temperature, salinity, and pollution levels in a number of modern and paleoenvironments (e.g., Lea *et al.*, 1989; Scott, 1990; Guilderson *et al.*, 1994; Leder *et al.*, 1996). Several workers were quick to recognize that trace element and stable isotope analyses of otoliths could, when coupled with otolith microstructure, provide retrospective information on the exposure of individual reef fish larvae to different water masses (Radtke, 1985; Radtke and Schafer, 1992). Unfortunately, the promise of the technique has, at least until very recently, gone unfulfilled. At the time of publication, we know of only a handful of publications that deal explicitly with otolith chemistry in coral reef fish (Radtke, 1985; Sadovy and Severin, 1992, 1994; Dufour *et al.*, 1998; Patterson *et al.*, 1999; Jones *et al.*, 1999; Swearer *et al.*, 1999). This lack of productivity has been largely due to analytical limitations. However, recent advances in instrumentation in both trace element and stable isotope analysis should afford otolith chemistry a number of applications in reef fish ecology in the near future.

A. Trace Elements in Otoliths

If geochemical signatures in otoliths are to be useful as tracers of water chemistry, the elements of interest must be deposited in proportion to dissolved concentrations in the ambient environment. This assumption has only recently been addressed in any comprehensive manner (Campana, 1999). Any relationship between seawater composition and otolith chemistry will be a function of the mechanism by which the trace elements are incorporated into otolith aragonite. A number of divalent metal ions, including Mg, Mn, Sr, Cd, Ba, and Pb, may substitute for Ca in the aragonite matrix according to the following equation:

$$Me^{2+} + CaCO_3 \rightleftharpoons MeCO_3 + Ca^{2+} \qquad (6)$$

The trace element composition of otoliths ($[Me/Ca]_{otolith}$) can be related to that of the water ($[Me/Ca]_{H_2O}$) by way of a partition coefficient (D_{Me}), where

$$\left[\frac{Me}{Ca}\right]_{otolith} = D_{Me}\left[\frac{Me}{Ca}\right]_{H_2O} \qquad (7)$$

Reconstruction of water chemistry from trace element concentrations requires, therefore, validation of the proportionality between seawater and otolith chemistry, along with an estimate of D_{Me}. Although neither of these components have been provided for any coral reef fish, a recent example from a temperate shorefish suggests that the approach has merit. Bath *et al.* (2000) reared larval spot, *Leiostomus xanthurus* (Sciaenidae), in the laboratory at four dissolved Sr/Ca and Ba/Ca levels within the tank waters, and at two temperatures (20° and 25°C). They found high ($r > 0.9$) correlations between seawater and otolith chemistry for both elements. Estimates of D_{Sr} (0.18–0.2) and D_{Ba} (0.08) were both considerably less than 1, consistent with the idea that compared to Ca, both Sr and Ba are discriminated against during uptake from seawater. However, Kalish (1989) calculated a similar value for D_{Sr} based on Sr/Ca measurements of endolymphatic fluid rather than seawater. This suggests that most of the Sr discrimination may be occurring within the endolymph, rather than at the branchial or intestinal membranes.

Substitution of Ca ions by at least some metal cations may, then, be proportional to the concentrations of the ions in the ambient environment. However, the partition coefficient, D_{Me}, may also be a function of physical parameters such as temperature and $CaCO_3$ precipitation rate. Temperature is perhaps the most widely studied of these parameters in biogenic carbonates. Negative relationships between Sr/Ca ratios and temperature have been reported for coral skeletons in a number of species (e.g., Beck *et al.,* 1992; Shen *et al.,* 1996). In contrast, results from earlier studies on the effect of temperature on Sr/Ca ratios in fish otoliths are contradictory, both in the direction and magnitude of the temperature dependence. Negative (e.g., Radtke *et al.,* 1990; Townsend *et al.,* 1995), positive (Kalish, 1989; Arai *et al.,* 1996), and no relationship (Gallahar and Kingsford, 1996; Tzeng, 1996) between Sr/Ca and temperature have been reported. Two studies found that Sr/Ca ratios were positively related to temperature in larval and juvenile *Micropogonias undulatus* (Sciaenidae) (Fowler *et al.,* 1995) and *Leiostomus xanthurus* (Bath *et al.,* 2000). Both experiments suggested that D_{Sr} in otoliths had a temperature dependence of approximately 3% per degree. Although this dependency was greater than for corals [$\sim -0.7\%$ (Shen *et al.,* 1996)] or foraminifera [$\sim 1\%$ (Lea *et al.,* 1999)], high-precision measurements were still needed for accurate reconstruction of temperature profiles from otoliths. With proper validation and the use of appropriate instrumentation, Sr/Ca ratios may indeed record ambient temperatures experienced by larval reef fish, as suggested by Radtke (1985).

A number of elements, including most that form mono- and divalent cations (Na^+, K^+, Cr^{2+}, Cu^{2+}, Ni^{2+}) and anions other than CO_3^{2-} (e.g., S, Cl), are also found in otoliths. Some of these elements (e.g., Cr, Cu), are strongly complexed in marine waters and thus are not readily bioavailable except possibly through the food chain (Turner *et al.,* 1981). Others are involved in osmoregulatory processes that probably decouple environmental levels of the element and subsequent concentrations in the otolith (e.g., Na^+, K^+). Some of these elements may also be associated with either structural proteins or their constituent amino acids and thus are subject to physiological processes that will also alter their transport from the water to the otolith. There would seem little opportunity for concentrations of such elements in otoliths to be correlated with environmental levels (see Hanson and Zdanowicz, 1999). It is therefore unlikely that concentrations of these elements in otoliths will reveal anything meaningful about past environmental conditions, given that they do not reflect either water chemistry or temperature in any predictive fashion. Not surprisingly, these elements also appear particularly sensitive to postcollection modification during and after preservation (Proctor and Thresher, 1998).

B. Instrumentation

A thorough discussion of the instrumentation used in the analysis of otolith chemistry is beyond the scope of this chapter. Several reviews in the primary literature provide excellent discussion of the many techniques available (e.g., Jackson *et al.,* 1993; Campana *et al.,* 1997). However, a brief description of techniques that are available to reef fish researchers is warranted given recent developments in the field.

The electron microprobe has been widely applied to the routine analysis of otolith chemistry (e.g., Radtke, 1985; Sadovy and Severin, 1992, 1994). Electron microprobe microanalysis (EPMA), as applied to analyses of otolith chemistry, is described in detail by Gunn *et al.* (1992). Although spatial resolution of the instrument is excellent, with beam diameters of $<10\ \mu m$, a lack of sensitivity limits quantification to no more than six elements present in the otolith at minor percentage levels (Na, S, Cl, K,Ca, and Sr). Unfortunately, this list includes only one element, Sr, whose concentration in the otolith is likely to be reflective of environmental availability. Further, EPMA analysis is generally not considered sufficiently precise for use in Sr/Ca thermometry, despite several applications for this purpose in the literature. Otolith chemistry applications that rely on relatively large differences of water mass chemistries, such as tracing movements of fish from oceanic spawning grounds to low-salinity estuarine nursery areas through differences in Sr/Ca profiles, can be adequately addressed using EPMA (e.g., Radtke *et al.,* 1988, 1999; Radtke and Kinzie, 1996; Kawakami *et al.,* 1998; Arai *et al.,* 1999). However, given the increase in availability of other instruments, we do not envisage that EPMA will continue to be widely applied in such studies.

A number of studies of otolith chemistry have been published based on assays using inductively coupled plasma mass spectrometry (ICP-MS). The technique is very sensitive, with limits of detection in solution-based mode of 1–10 ng g^{-1} otolith weight. This, in turn, means that it is possible to assay a number of elements in otoliths that cannot be quantified using EPMA, including Li, B, Mg, Mn, Ni, Cu, Cd, and Pb. By coupling the ICP-MS with a laser ablation system, probe-based analyses of sectioned otoliths can be performed with a spatial resolution that is now, with the use of excimer lasers, approaching that of EPMA (Sinclair *et al.*, 1998).

ICP-MS instrumentation is less than two decades old and development of the technique is still proceeding rapidly. For instance, most ICP-MS systems sold to date have been single-collector instruments based on quadrupole mass analyzers. However, the newest generation of ICP-MS instruments use sector-field mass analyzers. Single-collector sector-field instruments have several advantages over commercial quadrupole-based instruments, including the ability to operate in high mass resolution. This, in turn, allows molecular interferences on isotopes of interest to be resolved and eliminated. Molecular interferences are particularly common in high-Ca matrices such as otoliths, and appear the likely reason for anomalously high Fe concentrations that have been reported in some studies (e.g., Dove *et al.*, 1996). These instruments can also assay interelement ratios, such as Sr/Ca, with precision (0.05% RSD), approaching that of thermal ionization mass spectrometry (Rosenthal *et al.*, 1999). Another variant of ICP-MS uses a multicollector array to produce extremely precise isotopic measurements. The technique is rapidly becoming the instrument of choice for isotope ratio measurements of a number of elements in geochronology and paleoceanography (Halliday *et al.*, 1998).

A number of other instruments are also suitable for quantifying otolith chemistry. For instance, the sensitivity of the ion microprobe approaches that of laser ablation ICP-MS (e.g., Friedland *et al.*, 1998), and appears particularly suited to analysis of elements lighter than Ca. Synchrotron radiation X-ray fluorescence (XRF) has spatial resolution superior to that of any of the techniques mentioned above, and extremely low detection limits, but systems are scarce and in high demand by the research community (Reed, 1990). The prohibitive cost of the XRF instrumentation suggests, however, that the lack of availability is unlikely to be remedied in the near future.

C. Stable Isotopes

The use of stable isotopes in carbonates has a long history in paleoceanography and climatology. For instance, carbon and oxygen isotope ratios in coral skeletons have been used to reconstruct temperature, salinity, and current regimes of modern and ancient oceans (e.g., Guilderson *et al.*, 1994; Leder *et al.*, 1996; Swart *et al.*, 1996). A number of studies have confirmed that stable isotopes of carbon and oxygen in otolith aragonite may also contain significant information on parameters such as temperature, salinity, diet, and growth rate experienced by an individual fish (e.g., Mulcahy *et al.*, 1979; Kalish, 1991b,c; Patterson *et al.*, 1993; Iacumin *et al.*, 1992; Edmonds and Fletcher, 1997; Thorrold *et al.*, 1997b, 1998; Schwarcz *et al.*, 1998; Edmonds *et al.*, 1999; Newman *et al.*, 2000c; Weidman and Millner, 2000). There is, therefore, little doubt that analyses of stable isotopes in otoliths have a number of applications in reef fish ecology.

1. Oxygen Isotope Fractionation

The oxygen isotope composition of biogenic aragonite, i.e., the ratio of ^{18}O to ^{16}O (or more conveniently $\delta^{18}O$), is primarily a function of the isotopic composition of the environment in which the aragonite is precipitating. The oxygen isotope composition of the endolymphatic fluid appears to be very close to that of the ambient water, despite the physical separation of this fluid from the external environment (Iacumin *et al.*, 1992; Thorrold *et al.*, 1997b). Fractionation of oxygen isotopes is also a function of ambient temperature, and this temperature dependence forms the basis for oxygen isotope thermometry. The fractionation factor, α, is given by

$$\alpha = \frac{\delta_{\text{otolith}} + 1000}{\delta_{H_2O} + 1000}, \tag{8}$$

where δ_{otolith} is the $\delta^{18}O$ value of the otolith aragonite, and δ_{H_2O} is the $\delta^{18}O$ value of the ambient water. The degree of fractionation in inorganic aragonite is on the order of 0.22‰ per degree, which, when coupled with the precision of state-of-the-art isotope ratio mass spectrometers, suggests that temperatures could be reconstructed from otolith aragonite with an accuracy of approximately $\pm 0.2°C$.

In practice, the accuracy of oxygen isotope thermometry relies on accurate calibration of the relationship between temperature and the fractionation factor. Several studies have estimated this relationship in otolith aragonite (Patterson *et al.*, 1993; Radtke *et al.*, 1996; Thorrold *et al.*, 1997b). In all cases, oxygen isotopes were deposited close to isotopic equilibrium with the ambient water. Although the slopes of all the relationships were statistically indistinguishable from that of inorganic aragonite (Kim and O'Neil, 1997), the intercepts of the lines were significantly different (Fig. 8). Reconstructed temperature histories would, therefore, be different depending on which temperature fractionation relationship was used. To illustrate that these differences are not trivial, we calculated average

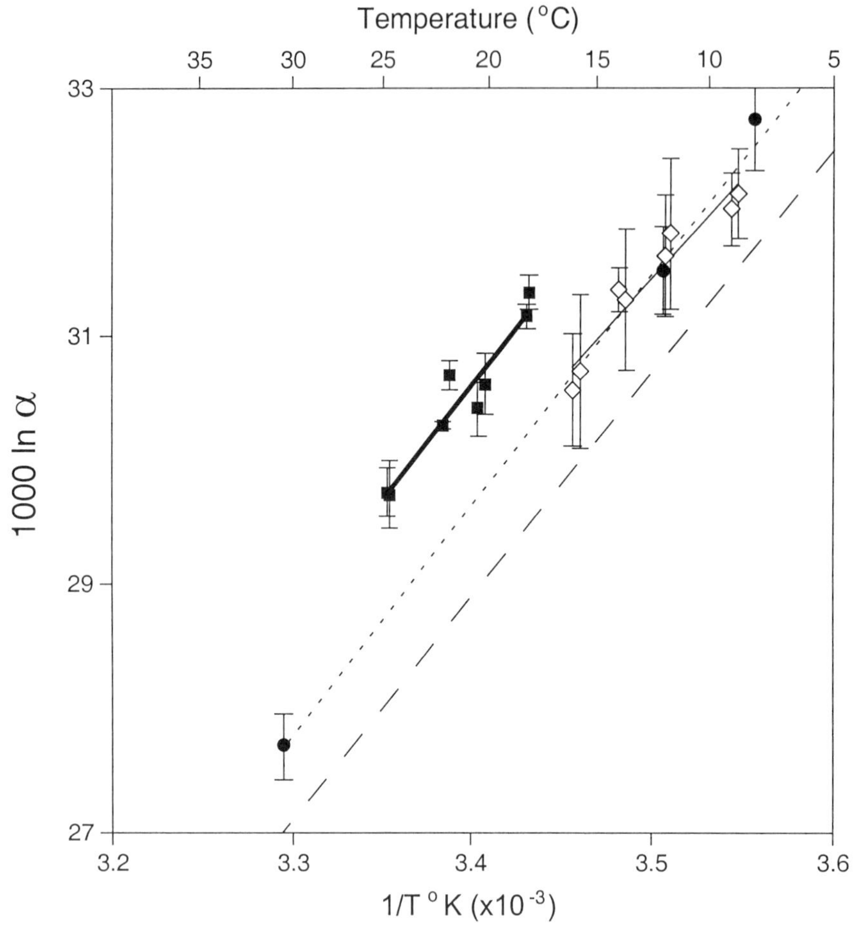

FIGURE 8 Oxygen isotope fractionation (1000 ln α) in fish otoliths as a function of temperature from laboratory studies on *Micropogonias undulatus* (■; thick solid line), *Gadus morhua* (◇; thin solid line), several freshwater fish species (●; short dashed line), and for inorganic aragonite (long dashed line). Data for *M. undulatus* from Thorrold *et al.* (1997b); data for *G. morhua* from Radtke *et al.* (1996), corrected for errors in the original paper; freshwater fish data from Patterson *et al.* (1993); and inorganic aragonite relation from Kim and O'Neil's (1997) data on inorganic calcite with a 0.6‰ enrichment.

temperatures experienced by *Chaetodon ulietensis* (Chaetodontidae) and *Acanthurus triostegus* (Acanthuridae), both inside and outside of the lagoon of Taiaro Atoll, based on $\delta^{18}O$ values of otoliths from Dufour *et al.* (1998). We initially used the inorganic relationship of Kim and O'Neil (1997) as recommended by Campana (1999),

$$1000 \ln \alpha = 18.03(\times 10^3 T \quad {}^\circ K^{-1}) - 31.82, \quad (9)$$

where T (°K) is the temperature estimated from observed $\delta^{18}O$ values of both ambient seawater and otoliths. Temperature estimates were, at 25°C, some 4° below the values of between 29° and 30°C given by Dufour and co-workers (Fig. 9). However, if the empirical relationship given in Thorrold *et al.* (1997b) is used, i.e.,

$$1000 \ln \alpha = 18.56(\times 10^3 T \quad {}^\circ K^{-1}) - 32.54, \quad (10)$$

temperature estimates increased to approximately 30°C. The exact reasons for this discrepancy are not known, and therefore we suggest that it is premature to recommend any one relationship for reconstructing temperature based on $\delta^{18}O$ values of reef fish otoliths. More empirical studies will be necessary if the application requires an accurate measure of absolute temperature. However, $\delta^{18}O$ values can provide information on relative differences in temperature exposure, even if the exact relation between temperature and oxygen isotope fractionation is unknown. This level of detail has proved sufficient in several studies (Dufour *et al.*, 1998; Ivany *et al.*, 2000).

2. CARBON ISOTOPE FRACTIONATION

Interpretation of carbon isotopes (the ratio of ^{13}C to ^{12}C, or $\delta^{13}C$) in otoliths is more problematic than that of oxygen isotopes. Carbon isotopes in otoliths are typically not deposited in isotopic equilibrium with the surrounding water (Kalish, 1991b,c; Gauldie, 1996b; Thorrold *et al.*, 1997b; Schwarcz *et al.*, 1998). However, unlike corals, in which isotopic disequilibrium is generated by kinetic effects (Swart *et al.*, 1996), the carbon deposited in otoliths apparently comes

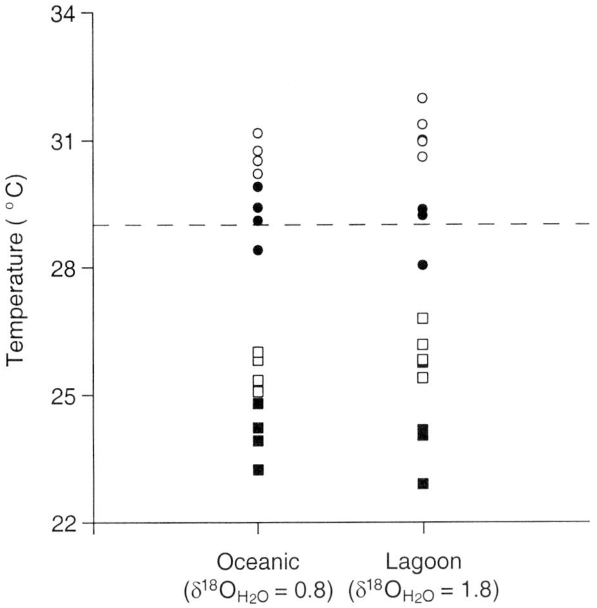

FIGURE 9 Temperature estimates from $\delta^{18}O$ values in otoliths of *Chaetodon ulietensis* (closed symbols) and *Acanthurus triostegus* (open symbols) collected inside and outside of the enclosed lagoon of Taiaro Atoll, French Polynesia. Estimates are based on the relation between temperature and oxygen isotope fractionation from inorganic aragonite (squares) or from an empirical study (circles) using lab-reared fishes (Thorrold *et al.,* 1997b). Dashed line at 29°C represents the average temperature in both the lagoon and adjacent oceanic waters at the time of fish collection. Data from Dufour *et al.* (1998).

from both dissolved inorganic carbon and metabolic carbon (Kalish, 1991b). These two sources have very different isotopic signatures. Dissolved inorganic carbon in surface ocean waters has an average $\delta^{13}C$ value of 1%, whereas the $\delta^{13}C$ of metabolic carbon ranges from −17‰ to −20‰ (Schwarcz *et al.,* 1998). Carbon isotope disequilibrium in otoliths is, then, a function of the relative amounts of carbon incorporation from DIC and metabolic sources, which varies considerably among species. For instance, Kalish (1991c) found that metabolic carbon contributed approximately 30% to otolith carbon, but Thorrold *et al.* (1998) found little indication of metabolic carbon in the otoliths of juvenile weakfish *Cynoscion regalis* (Sciaenidae). There is also evidence that the $\delta^{13}C$ values in otoliths are correlated with growth rate in juvenile croaker (Thorrold *et al.,* 1997b), suggesting that the proportion of metabolic carbon in otoliths may change with physiological factors that are not yet understood.

The carbon isotopic composition of metabolic carbon is a function of the $\delta^{13}C$ values of diet. A dietary change across one trophic level will alter $\delta^{13}C$ values of metabolic carbon by approximately 1% (Peterson and Fry, 1987), suggesting that carbon isotopes in otoliths may provide information on trophic interactions in fish communities. However, this would seem to be difficult to achieve without either assuming that the proportion of metabolic carbon incorporated in otoliths is constant, or determining this value from some independent measure of fish metabolism. An independent measure of metabolic carbon incorporation based on the width of individual increments in otoliths may be possible if $\delta^{13}C$ values are correlated with fish growth.

D. Otolith Chemistry Applications

1. Population Connectivity

The degree of larval connectivity among geographically separated reef fish populations is a critical parameter in any attempt to model or manage reef fish populations. These rates have traditionally been considered to be high, based on genetic studies that were unable to reject the null hypothesis of no genetic structure over large, basin-wide, spatial scales (Shulman and Bermingham, 1995), and simple numerical models of larval dispersal (Roberts, 1997b). The paradigm of extensive dispersal has been challenged by studies that have documented the formidable swimming abilities of reef fish larvae (Stobutzki and Bellwood, 1994, 1997; Leis and Carson-Ewart, 1997) and vertical distribution patterns that promote near-reef retention (Cowen and Castro, 1994). Coupled physical–biological models have also suggested that there may be little chance of downstream seeding of reefs by upstream populations (Cowen *et al.,* 2000). Although these studies undoubtedly highlighted the limits of physical models that neglect larval mortality and behavior, they provided no direct evidence for the return of larvae to natal reefs.

Given the difficulties of following individual larvae from hatching until settlement, tagging approaches appear to be the only direct method of quantifying larval connectivity. Although simple in concept, it is not immediately obvious how one might first tag millions of early-stage larvae at or before hatching, and then recapture enough marked individuals at settlement to ensure the statistical robustness of the result. Two studies, both based on otolith chemistry, outline approaches that provide the first direct estimates of the proportion of larvae that are locally retained around natal coral reefs. In what will surely become a classic study, Jones *et al.* (1999) reported on the development and application of a method for tagging large numbers of developing embryos of a damselfish, *Pomacentrus amboinensis* (Pomacentridae). The otoliths of over 10 million embryos were tagged *in situ* by immersion in tetracycline during the spawning season at Lizard Island, in the northern Great Barrier Reef. Light traps were then used to capture late-stage larvae, immediately before settlement, in several habitats around the island. An examination of 5000 late-stage larvae found 15 tetracycline-tagged individuals. Based on an estimate of total larval production, the total number of marked embryos, and the ratio of tagged to untagged late-stage larvae, as many as 15–60% of the recruits may have been spawned around Lizard Island. Although it is difficult to calculate a variance estimate on this proportion, the study has clearly altered our thinking on the likelihood of substantial self-recruitment to reefs in this region.

Jones *et al.* (1999) showed that tagging of embryos on the scale necessary to ensure adequate numbers or recaptures at the end of the larval phase is feasible. However, it is unlikely that the technique will be applicable to reef fish species that spawn pelagic eggs. Swearer *et al.* (1999) developed an approach for tracing the dispersal histories of bluehead wrasse (*Thalassoma bifasciatum,* Labridae) recruits using differences in larval growth rates and otolith chemistry as a natural tag of either local retention within near-coastal waters or larval development within open ocean waters, indicative of a distant source location. Swearer and co-workers hypothesized that locally retained larvae would both grow faster and encounter higher concentrations of trace elements in seawater, which would be reflected in the trace element composition of their otoliths. Larval growth rates were measured by counting daily increments in the otoliths of newly settled juveniles, and trace elements in otoliths were quantified using ICP-MS. Based on these "retention signatures," the authors concluded that more than 50% of the juveniles successfully recruiting to St. Croix during the summer were spawned locally. Juveniles recruiting in the fall, however, were more likely to be characterized by "dispersal signatures," suggesting that there may be seasonal variations in the probability of self-recruitment. These results are intriguing, but would have been more convincing if the retention and dispersal signatures were ground-truthed with larvae of known dispersal histories, or with water sampling concurrent with the recruit collections. Certainly the results from studies utilizing a natural tag such as geochemical signatures in otoliths are unlikely to be as definitive as those in which a tag is physically applied, released, and then recaptured (e.g., Jones *et al.,* 1999). However, the power of natural tags is that every fish from an area is invisibly tagged and capture of a single fish spawned in that area represents a recovery. Therefore, small uncertainties in the classification of an individual fish may be overcome by much larger sample sizes ("recaptures") than can typically be achieved in conventional tagging studies.

Finally, Dufour *et al.* (1998) tested if reef fishes completing their entire life cycle within the enclosed lagoon of Taiaro Atoll, French Polynesia could be distinguished from fishes that had been exposed to oceanic waters based on stable carbon and oxygen signatures in otoliths. The carbon isotope composition of dissolved inorganic carbon ($\delta^{13}C_{DIC}$) within the lagoon (0.19‰) was lower than that of the surrounding oceanic water (1.37‰), apparently due to the input of ^{13}C-depleted DIC sourced from organic matter. This isotopic difference was reflected in the $\delta^{13}C$ values of otoliths from adult *Chaetodon ulietensis* (Chaetodontidae) and *Acanthurus triostegus* (Acanthuridae) collected inside and outside of the lagoon. Differential evaporation of $H_2{}^{16}O$ within the lagoon also generated a distinct oxygen isotope signature, with $\delta^{18}O_{H_2O}$ values that were 1‰ higher than adjacent oceanic water. Again, the distinctive isotopic signature of the lagoon water was reflected in the $\delta^{18}O$ values of otoliths from adult *C. ulietensis* and *A. triostegus* collected inside the lagoon. Both $\delta^{13}C$ and $\delta^{18}O$ signatures in the otoliths of fish collected either in the lagoon or on the outer reef slope were sufficiently different that it will be possible to determine, with little ambiguity, if recently settled juveniles spent their entire larval lives within the lagoon.

2. Contribution of Juvenile Nursery Areas

Several examples in temperate systems have suggested that geochemical signatures in otoliths may be useful natural tags of juvenile nursery areas. For instance, it has often been assumed that estuaries, and particularly seagrass areas within these habitats, are important nursery areas for many reef fish species. Gillanders and Kingsford (1996) found that trace elements (Mn, Sr, and Ba) in otoliths distinguished between *Achoerodus viridis* (Labridae) individuals that had spent their juvenile life within nearshore seagrass nursery areas, and those that had settled directly into adult habitats on coastal reefs (Fig. 10). The authors then cored otoliths from adult fish and assayed the material laid down during the larval and juvenile phases. Using discriminant functions parameterized from the juvenile data set, they suggested that over 50% of the adults examined settled directly onto coastal reefs.

Spatiotemporal variations in water chemistry are generally more pronounced in estuarine and near-coastal areas than in oceanic water masses (Millward and Turner, 1995; Donat and Bruland, 1995). These differences, if also manifested in otolith chemistry of

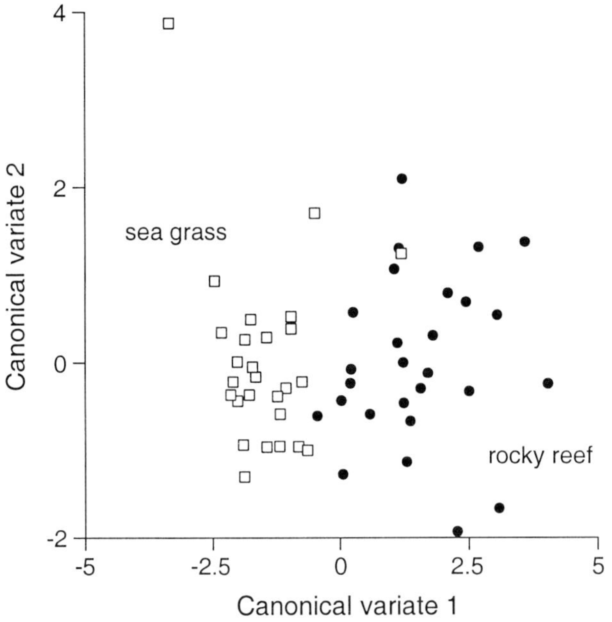

FIGURE 10 Results from a discriminant function analysis of trace element signatures (Mn, Sr, and Ba) in the otoliths of juvenile *Achoerodus viridis* sampled from nearshore seagrass habitat (□) and from coastal reef habitat (●). Data from Gillanders and Kingsford (1996).

juvenile fishes, may provide a method for determining not only if reef fishes are using estuarine nursery areas, but also for identifying the specific estuaries utilized by juveniles. Thorrold *et al.* (1998) examined spatial variations in the geochemical signatures of otoliths from juvenile weakfish *Cynoscion regalis* (Sciaenidae) collected at five locations along the United States Atlantic coast. They were able to classify individuals to natal estuaries based on $\delta^{13}C$, $\delta^{18}O$, Mg/Ca, Mn/Ca, Sr/Ca, and Ba/Ca values in otoliths, using neural network analysis, with an overall accuracy of better than 90%. Adults from the same annual cohort, and from the same locations, were collected 2 years later, and otolith cores were examined to determine the natal history of the adult spawners. Thorrold *et al.* (2001) concluded that spawning-site fidelity of the adult weakfish ranged from 60 to 81%, and was comparable to estimates of natal homing in birds and anadromous fishes. These data were in striking contrast to mitochondrial and nuclear DNA analyses, which detected no population structure throughout the species range (Graves *et al.*, 1992; Cordes, 2000). As with connectivity studies, the advantage of a natural tag such as geochemical signatures in otoliths is that every individual within the nursery area is marked. A significant disadvantage is that interannual variations in water chemistry within estuarine systems may mean that the functions used to classify older fishes will have to be recalibrated on an annual basis (Gillanders and Kingsford, 2000).

V. Conclusions

Studies of daily and annual increments in otoliths have made significant contributions to our understanding of the ecology of coral reef fishes. Annual increments in otoliths have been validated in a number of species, and have, within the past 10 years, provided significant data on age, growth, and longevity of a number of reef fish species. Daily increments in the otoliths of late-stage larval and juvenile fishes continue to be used to examine the timing of settlement and the length of larval life. Reef fish workers have been reticent about applying more sophisticated otolith analyses, particularly those based on the back-calculation of individual growth characteristics. However, studies by Searcy and Sponaugle (2000, 2001) clearly show the potential of the general approach. A lack of taxonomic resolution in larval studies may explain some of this reticence, although with the development of genetic identification methods this is no longer an excuse (Hare *et al.*, 1994). Some skepticism regarding the relationship between otolith growth and fish growth is certainly warranted. Nonetheless, examples from temperate environments suggest that we are at a point at which questions regarding the selective biological and physical processes generating year-class strength can be addressed at the individual level.

The application of otolith chemistry to ecological studies of coral reef fishes has generated considerable interest, largely due to the potential to trace the larval dispersal and hence to quantify connectivity within reef fish metapopulations (Swearer *et al.*, 1999). The prospects of tracing larval dispersal from spawning to settlement are especially intriguing given the importance of such information in the design and implementation of marine protected areas. Is such optimism justified? There seems little doubt that concentrations of at least some elements and isotopes in otoliths reflect environmental levels. However, the corollary also applies—the technique will be of little practical use in situations in which environmental differences among sites of interest are subtle or nonexistent. Many tropical reef systems lack significant runoff from adjacent continental or island landmasses that may, in turn, generate water chemistry differences over ecologically relevant spatial scales (tens to hundreds of kilometers). For instance, it seems unlikely that there will be sufficient variations in water chemistry among adjacent reefs on the Great Barrier Reef to examine the question of self-recruitment using natural geochemical signatures in otoliths. Rather, labeling of calcified structures with tetracycline or other fluorescent compounds would seem the only way to address questions of local self-recruitment at this scale (Jones *et al.,* 1999).

The application of otolith chemistry certainly expands the information that can potentially be recovered from otoliths, particularly when combined with more traditional microstructural analyses. For instance, oxygen isotope thermometry has the potential to allow temperature histories of reef fish larvae to be reconstructed. It would be particularly informative to combine temperature histories based on oxygen isotopes with increment width profiles from the same fish. This may allow workers to decouple the effects of food and temperature on daily instantaneous growth rates in the field. The potential for geochemical signatures in otoliths to provide a means for quantifying the proportion of larvae that are locally retained around natal coral reefs is now apparent (Swearer *et al.,* 1999). However, probe-based analyses of otoliths from late presettlement or early juvenile fish may also allow researchers to reconstruct larval dispersal pathways with temporal resolution approaching the daily level (e.g., Thorrold and Shuttleworth, 2000). This may, in turn, provide more relevant information on larval connectivity than is available from a binary retention/dispersal categorization.

A potential problem with the use of elemental tracers in otoliths is the likelihood of considerable physiological regulation by the fish, either at the outer membrane/water interface, or in the blood plasma. We expect, therefore, to see more attention directed to those isotopic tracers in the otolith that are unlikely to be affected by fish physiology (e.g., Kennedy *et al.,* 1997, 2000). The recent development of instruments capable of high-precision isotope ratio measurements in structures such as otoliths may prove to be a serendipitous and significant development. We hope that in a similar contribution 10 years hence, the combination of otolith microstructure and otolith chemistry will figure prominently in the examination of coral reef fish ecology.

Acknowledgments

Partial support for the preparation of this manuscript was provided by National Science Foundation Grant OCE-9876565 to SRT and JH. We thank Betsy Laban for help in preparing Fig. 1A, Su Sponaugle and Steve Searcy for providing Fig. 1B, and Bronwyn Gillanders for access to original data. The manuscript was considerably improved by the comments of Tony Fowler, Peter Sale, Jennifer Potts, and Gretchen Bath Martin. This is Woods Hole Oceanographic Institution Contribution No. 10400.

CHAPTER 12

Energetics and Fish Diversity on Coral Reefs

Mireille L. Harmelin-Vivien
Centre d'Océanologie de Marseille, CNRS UMR 6540
Université de la Méditerranée-Station Marine d'Endoume
13007 Marseille, France

I. Introduction

Coral reefs are the most complex ecosystems in the seas. Fish communities reach their highest degree of diversity in these ecosystems, and differ enormously within and between reefs in the same area (Williams, 1991; Ormond and Roberts, 1997) and between geographic regions (Briggs, 1974, 1996). The relative roles of local and regional processes in explaining community diversity in marine systems, as well as in terrestrial systems, have been hotly debated and several, most often contradictory, explanations have been proposed (Strong *et al.*, 1984; Ricklefs, 1987). The high level of diversity supported by coral reefs may be best explained as the result of various processes operating on different scales in space and time (Jackson, 1991; Kohn, 1997). At the local scale (e.g., within reef zones), the diversity observed in local fish assemblages is explained by both deterministic (interspecific competion for food and shelter; predation pressure) and stochastic (recruitment; perturbation) ecological processes (Sale, 1977, 1991a; Harmelin-Vivien, 1989). On the regional scale (e.g., Pacific vs. Atlantic, West Pacific vs. central Pacific), the diversity of extant faunas of reef fishes is explained mainly by interactions of historical hydrodynamic and geological processes with each species' life cycle characteristics, particularly larval dispersal ability (Victor, 1991). On the global scale (e.g., tropical vs. temperate), explaining why tropical regions contain so many species has been one of the central problems of community ecology since the nineteenth century (Pianka, 1966; MacArthur, 1972; Stevens, 1989; Crame and Clarke, 1997), despite intensive studies in both aquatic and terrestrial environments. Until now, no convincing explanation in terms of physiology, ecology, or evolutionary processes has been offered (A. Clarke, 1996).

In this chapter it is argued that the high diversity of fishes observed on present day coral reefs is partly related to the sustained higher temperatures in the tropics over geological time, and to the more efficient use and transfer of energy permitted by long-term temperature stability. High temperature and environmental stability have influenced evolutionary processes from the molecular level to the community level of organization. The mechanisms proposed to support this hypothesis are illustrated by comparing feeding behaviors of fishes within and between families from different regions with recently acquired data on reef fish phylogeny and biogeography.

II. Latitudinal Gradients and the Evolution of Fish Feeding

A. Trophic Structure of Tropical versus Temperate Reef Fish Communities

A characteristic pattern of fish diversity, as for many other taxonomic groups, is the general increase in number of species from high to low latitudes (Briggs, 1974; Ebeling and Hixon, 1991; Hobson, 1994). An example is given by the total numbers of fish species recorded by visual census on typical coral reef biotopes and on warm and cold temperate reefs (Table 1). A significant negative correlation exists between latitude and the total number of fish species recorded ($R = -0.610$, $p < 0.005$). Total fish species richness decreases by

TABLE 1 Total Fish Species Richness and Number and Percentage of Herbivorous Fish Species[a]

Ecosystem[b]	Geographic position (latitude–longitude)	Total number of fish species	Number of herbivorous species (%)
Coral reefs			
1. French Polynesia, Moorea	17°32′S–149°50′W	104	21 (20%)
2. Virgin Islands, St. Croix	17°46′N–64°35′W	110	13 (12%)
3. Hawaii, Kona	19°28′N–155°56′W	123	27 (22%)
4. Madagascar, Tuléar	23°21′S–43°40′E	295	39 (13%)
5. Florida Keys	24°30′N–81°24′W	89	14 (16%)
Warm temperate waters			
6. El Hierro, Canary Islands	28°00′N–13°30′W	47	4 (8%)
7. Madeira Island	32°40′N–16°45′W	31	3 (9%)
8. Santa Catalina Island, California	33°28′N–118°22′W	32	2 (6%)
9. Santa Cruz Island, California	34°10′N–119°50′W	27	2 (7%)
10. San Diego County, California	39°19′N–117°31′W	25	2 (8%)
11. Berlenga Island, Portugal	39°25′N–9°30′W	51	1 (2%)
12. Medes Islands, Mediterranean	42°16′N–3°13′E	65	1 (3%)
13. Banyuls, France, Mediterranean	42°29′N–3°08′E	54	1 (2%)
14. Port-Cros Island, France, Mediterranean	43°01′N–6°24′E	47	1 (2%)
15. Carry-Marseille, France, Mediterranean	43°20′N–5°09′E	54	1 (2%)
Cold temperate waters			
16. San Juan Islands, Washington	48°35′N–123°05′W	23	0 (0%)
17. Chile, Fuegian Island	55°04′S–67°04′W	18	0 (0%)
18. Southern Alaska, Banarof Island	56°30′N–134°50′W	14	0 (0%)
19. Gulf of Alaska, Prince William Sound	60°15′N–147°10′W	22	0 (0%)

[a]Species were recorded in visual counts during assessments of fish communities in tropical, warm temperate, and cold temperate waters, to compare fish assemblages observed with the same technique in similar habitats, i.e., hard subtrata [from Hobson (1994), enlarged and modified].

[b]*Sources:* 1, Galzin (1985); 2, Gladfelter *et al.* (1980); 3, Hobson (1974); 4, Harmelin-Vivien (1979); 5, Bohnsack and Bannerot (1986); 6, Bortone *et al.* (1991); 7, Andrade and Albuquerque (1995); 8, Hobson and Chess (1986); 9, Ebeling *et al.* (1980); 10, DeMartini *et al.* (1989); 11, Almeida (1996); 12, Garcia Rubies (1997); 13, Jouvenel (1997); 14, Harmelin (1987); 15, Harmelin *et al.* (1995); 16, Moulton (1977); 17, Moreno and Jara (1984); 18, Hobson (1994); 19, R. J. Rosenthal, in Hobson (1994).

more than one order of magnitude from coral reefs to cold rocky shores when studying similar areas. When looking at the trophic structure of these fish faunas, the major difference resides in a drastic decrease in the species number and relative importance of herbivores and sessile invertebrate browsers (fishes feeding mainly on sponges, cnidarians, and ascidians) from tropical to temperate reefs. The number of herbivorous fish species recorded is significantly negatively correlated with latitude ($R = -0.721$, $p < 0.001$), as well as their importance percentage ($R = -0.897$, $p < 0.0001$). Herbivorous species represent 12 to 22% of the total number of fish species recorded on coral reefs, 2 to 8% of species recorded on warm temperate reefs, and are totally absent from cold temperate areas. The latitudinal decrease in the relative importance of herbivores is further emphasized when fish abundance is considered. Herbivores account for more than 50% of fish numerical abundance on coral reef flats in Moorea, French Polynesia (Galzin, 1987), and only 2 to 6% on Mediterranean rocky shores (Harmelin *et al.*, 1995). On a latitudinal gradient along the eastern Australian and New Zealand coasts, Choat (1991) observed a sharp decline (over one order of magnitude) in the abundance of herbivorous fishes with increasing latitude. The abundance of herbivorous fishes recorded by visual counts in northern New Zealand is only 20 to 25% of those found at Lizard Island (Great Barrier Reef) and San Blas (Caribbean) (Meekan and Choat, 1997). In the same way, fishes feeding exclusively or mostly on sessile invertebrates are highly diversified on coral reefs, where they represent 10 to 17% of the total fish species richness (Randall, 1967; Hobson, 1974; Harmelin-Vivien, 1979), whereas exclusive sessile invertebrate browsers do not seem to exist on temperate reefs. For example, the Mediterranean sparid *Diplodus puntazzo,* which commonly feeds on sponges, hydroids, and gorgonians, does not ingest more than 50% in weight of these prey

TABLE 2 Relative Importance of Planktivorous and Herbivorous Species among Damselfishes (Pomacentridae) with Latitude

Site[a]	Geographic position (latitude–longitude)	Number of pomacentrid species	Planktivores (%)	Herbivores + omnivores (%)
Coral reefs				
Moorea (French Polynesia) (1)	17°32′S–149°50′W	38	32	68
West Indies, Caribbean (2)	18°15′N–66°30′W	16	31	69
Tuléar (Madagascar) (3)	23°21′S–43°40′E	41	34	66
Warm temperate waters				
East Atlantic (4)	33°57′N–6°50′W	15	40	60
Azores (5)	37°40′N–24°32′E	2	50	50
Mediterranean, Marseille (6)	43°18′N–5°24′E	1	100	0
Cold temperate waters				
British waters, Swansea (6)	51°38′N–3°57′W	0	0	0

[a]*Sources:* 1, Galzin (1985); 2, Randall (1968); 3, Harmelin-Vivien (1979); 4, Quéro *et al.* (1990); 5, Azevedo (1995); 6, Fischer *et al.* (1987).

(Sala and Ballesteros, 1997). Thus, in cold waters the trophic structure of fish assemblages resides in the dominance of piscivores and carnivores preying on motile invertebrates (Gibson, 1972; Hobson, 1994), whereas tropical fish assemblages display a larger trophic spectrum. In the tropics, in addition to carnivores and piscivores, numerous herbivores, omnivores, sessile invertebrate browsers, and zooplanktivores are major components of the trophic structure (Hobson, 1974; Harmelin-Vivien, 1979). Similar trends occur among freshwater fish communities on a latitudinal gradient (Lowe-McConnell, 1987).

B. Use of Low-Caloric Food Resources

The trophic structure of tropical fish communities is thus differentiated from temperate ones by the better use of low-quality food resources, such as algae, seagrasses, and sessile invertebrates (sponges, cnidarians, and ascidians). The energy content of algae and sessile invertebrates such as sponges, gorgonians, and ascidians is low compared to that of motile invertebrates and fishes (Cummins and Wuycheck, 1971; Brey *et al.*, 1988). These organisms also contain generally high concentrations of secondary metabolites that are known to deter fish predation (Fenical and Pawlik, 1991; Pawlik *et al.*, 1995; Schmitt *et al.*, 1995; Hay, 1996). The difference in trophic structure is partially due to the existence on coral reefs of fish families that feed on these food types and are mostly restricted to tropical waters (chaetodontids, scarids, or acanthurids). However, the trophic difference is also observed within families encountered both in tropical and in temperate waters. For example, the species richness of the family Pomacentridae significantly decreases with increasing latitude ($R = -0.814$, $p < 0.02$) and the relative importance of herbivores and omnivores decreases ($R = -0.893$, $p < 0.01$) (Table 2). Conversely, the relative importance of carnivorous damselfish species significantly increases with latitude ($R = 0.819$, $p < 0.05$), exemplified by the only pomacentrid found in the Mediterranean being a zooplanktivore, *Chromis chromis*. Similarly, a decrease in the mean consumption of algae is observed from tropical to temperate waters within fish families that include omnivorous species such as gobiid and blenniid families (Table 3). Algae are an important food source for most blenniids on coral reefs (mean = 96.6% ± 6.2, $n = 12$) and are used by only a few species on Mediterranean rocky shores (mean = 38.9% ± 27.4, $n = 21$), the difference in algal consumption between the tropical and temperate species being highly significant ($p < 0.0001$, *t* test). Gobiid species that rely mainly on algae for food are encountered on coral reefs, but in temperate waters most gobiids are strictly carnivorous (Gibson, 1972). The mean consumption of algae by gobiids is significantly higher ($p < 0.0005$, *t*-test) on coral reefs (29.4% ± 32.7, $n = 12$) than in temperate waters (1.1% ± 3.9, $n = 22$).

These striking differences in trophic structure and food consumption of fish faunas cannot be attributed to differences in food resources between high and low latitudes (except for scleractinian corals and associated corallivorous species), because algae, seagrasses, and sessile invertebrates (sponges, hydroids, gorgonians, alcyonarians, bryozoans, and ascidians) are abundant

TABLE 3 Percentage of Algae in the Diet of Blenniid and Gobiid Fish Species on Coral Reefs and in Warm Temperate Waters

Site[b]	Blenniidae[a]		Gobiidae	
	N[c]	% Algae mean (range)	N[c]	% Algae mean (range)
Coral reefs				
Tuléar, Madagascar (1)	8	98 (89–100)	9	25 (0–82)
Puerto Rico, Caribbean (2)	4	94 (79–99)	3	41 (0–74)
Warm temperate waters				
Azores (3)	7	45 (0–96)	3	3 (0–10)
Mediterranean (4)	14	36 (3–93)	19	1 (0–17)

[a]The carnivorous sabertooth blennies are not included because they have no counterpart in the Mediterranean.

[b]*Sources/data:* 1, Harmelin-Vivien (1979) (weight percentage); 2, Randall (1967) (volume percentage); 3, Azevedo (1995) (weight percentage); 4, Zander (1982), Bell and Harmelin-Vivien (1983), Goldschmid *et al.* (1984), and Khoury (1987) (weight percentage).

[c]*N*, Number of species studied.

and available for grazing in temperate zones. It has been argued that the abundance of herbivorous fishes on coral reefs was related to the dominance of diverse, highly productive stands of very small algae (Horn, 1989; Choat, 1991). However, reefs are rapidly invaded by large macroalgae when herbivores are removed (Hixon, 1996b), and their low abundance on healthy reefs seems to be more the result of intense grazing than on an intrinsic reefal characteristic. Conversely, small turf algae can be important components of temperate algal communities in areas subjected to high rates of herbivory (Verlaque, 1987). Plants and sessile invertebrates are of low caloric value (Brey *et al.*, 1988), can include deterrent metabolites (Hay, 1996), and are less easily digestible compared to motile invertebrates (polychaetes, molluscs, crustaceans) and fishes, which are the preferred food of fishes whatever the latitude (Harmelin-Vivien, 1981; Parrish *et al.*, 1985; Gerking, 1994). Thus, the difference in the trophic structures of temperate and tropical fish faunas resides more in a differential utilization of resources by the fish than in differences in basic food sources with latitude.

C. Feeding Behavior and Phylogeny in Coral Reef Fishes

When looking at the phylogenetic relationships of fish families living on coral reefs, it is striking to notice that herbivores and sessile invertebrate browsers are found in the most derived families, from modern perciforms to tetraodontiforms (Table 4). On the contrary, the elasmobranch and ancestral teleost families are strictly predatory carnivores (Schaeffer and Rosen, 1961). Herbivory arose in different families at different geological times with evolution, and certainly in different habitats (Choat, 1991). Although some herbivorous species are encountered in less derived perciform families, such as the Sparidae and Kyphosidae, most herbivores occur in highly derived fish families, such as the Pomacentridae, Scaridae, Blenniidae, Acanthuridae, Siganidae, and Monacanthidae. These families are recently derived; reef fishes rapidly evolved during the early Tertiary, 50 to 30 million years ago (Choat and Bellwood, 1991). Parrot fishes appear to have evolved less than 15 million years ago (Bellwood, 1994, 1996a). The ability to exploit reef algae and small colonial invertebrates was apparently restricted to post-Cretaceous perciforms (Choat and Bellwood, 1991).

Relationships between phylogenetic history and feeding behavior are also evident at the family level (between genera) with a higher consumption of algae and/or sessile invertebrates in the morphologically most derived genera. Among butterfly fishes, basal genera or subgenera are zooplanktivores or prey on noncoralline invertebrates (Fig. 1). Feeding on corals appears in more derived genera, and exclusive corallivores are found only among the most derived subgenera of the evolutionary tree proposed by Blum (1989) for the Chaetodontidae. The two most derived subgenera, *Corallochaetodon* and *Citharoedus*, each include three species that are all exclusive corallivores. Feeding on scleractinian corals is a behavior that has evolved over evolutionary time in some butterfly fishes, and is mainly related to the morphological characters of their dentary and feeding apparatus (Motta, 1989). Another observation, which is discussed in more detail later on,

TABLE 4 Main Feeding Behaviors of Reef Fish Families Recorded in Tuléar, Madagascar[a]

Family	Feeding behavior[b]	Family	Feeding behavior[b]
Carcharhinidae	(P)	Lutjanidae	(C-Z)
Torpedinidae	(C)	Sparidae	(O)
Rhinobathidae	(C)	Lethrinidae	(C)
Dasyatidae	(C)	Nemipteridae	(C)
Congridae	(C)	Kyphosidae	(H)
Muraenidae	(C)	Ephippidae	(C)
Ophichthidae	(C)	Mullidae	(C)
Xenocongridae	(C)	Malacanthidae	(C)
Moringuidae	(C)	Pomacanthidae	(B)
Clupaeidae	(Z)	Chaetodontidae	(B-C)
Plotosidae	(C)	Carangidae	(P)
Synodontidae	(P)	Cirrhitidae	(C)
Ophidiidae	(C)	Pempheridae	(Z)
Carapidae	(Z)	Pomacentridae	(H-O-Z)
Antennariidae	(P)	Labridae	(C)
Gobiesocidae	(C)	Scaridae	(H)
Atherinidae	(Z)	Sphyraenidae	(P)
Exocoetidae	(P)	Congrogadidae	(C)
Holocentridae	(C-Z)	Mugiloididae	(C)
Aulostomidae	(P)	Blenniidae	(H-O-C)
Fistulariidae	(P)	Tripterygiidae	(C)
Syngnathidae	(C)	Callionymidae	(C)
Centricidae	(C)	Gobiidae	(O-C-Z)
Scorpaenidae	(C)	Acanthuridae	(H-Z)
Caracanthidae	(C)	Zanclidae	(B)
Platycephalidae	(P)	Siganidae	(H)
Serranidae	(Z-C-P)	Scombridae	(C)
Grammistidae	(C)	Bothidae	(C)
Pseudochromidae	(C)	Pleuronectidae	(C)
Acanthoclinidae	(C)	Cynoglossidae	(C)
Pseudogrammidae	(C)	Soleidae	(C)
Plesiopidae	(C)	Balistidae	(O-B-Z-C)
Theraponidae	(C)	Monacanthidae	(H-O-B)
Priacanthidae	(Z)	Ostraciidae	(B)
Apogonidae	(Z-C)	Tetraodontidae	(O-B-C)
Haemulidae	(C)	Diodontidae	(C)

[a]Fish families are ordered from the Chondrichthyes (Carcharhinidae) to the most derived Osteichthyes (Diodontidae). From Harmelin-Vivien (1979).

[b]Key: H, Herbivores; O, omnivores; B, sessile invertebrate browsers; Z, zooplanktivores; C, carnivores preying on motile invertebrates; P, piscivores.

is that there are no exclusively corallivorous butterfly fishes in the West Indies, but only facultative corallivores (Fig. 1). Among parrot fishes, which are all herbivores, a relationship exists between the radiation of the genera and the energetic value of the food, inversely determined by the percentage of calcium carbonate ingested with plant material (Fig. 2). Bellwood (1994) has demonstrated that feeding on large fleshy algae and seagrasses was plesiomorphic, whereas feeding on turf algae was derived. Excavating and scraping scarid

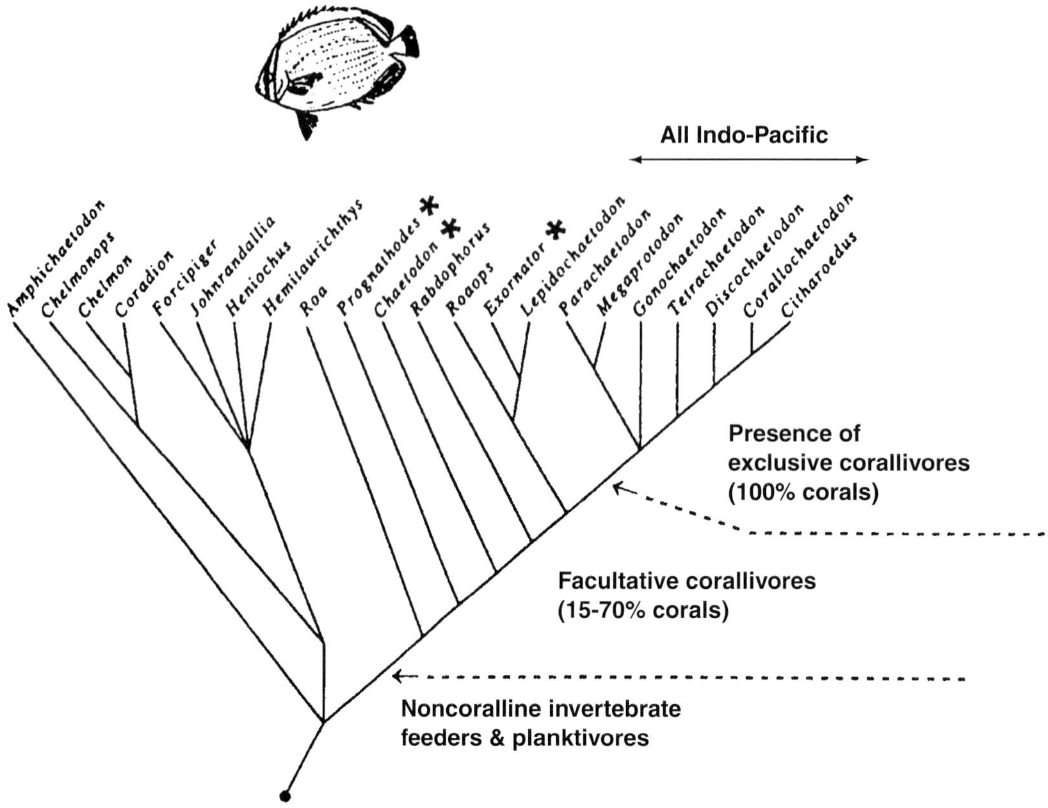

FIGURE 1 Phylogenetic relationships among the Chaetodontidae and the evolution of feeding behaviors in butterfly fish genera and subgenera. The genera present in the Caribbean are indicated by an asterisk (from Blum, 1989, with kind permission from Kluwer Academic Publishers).

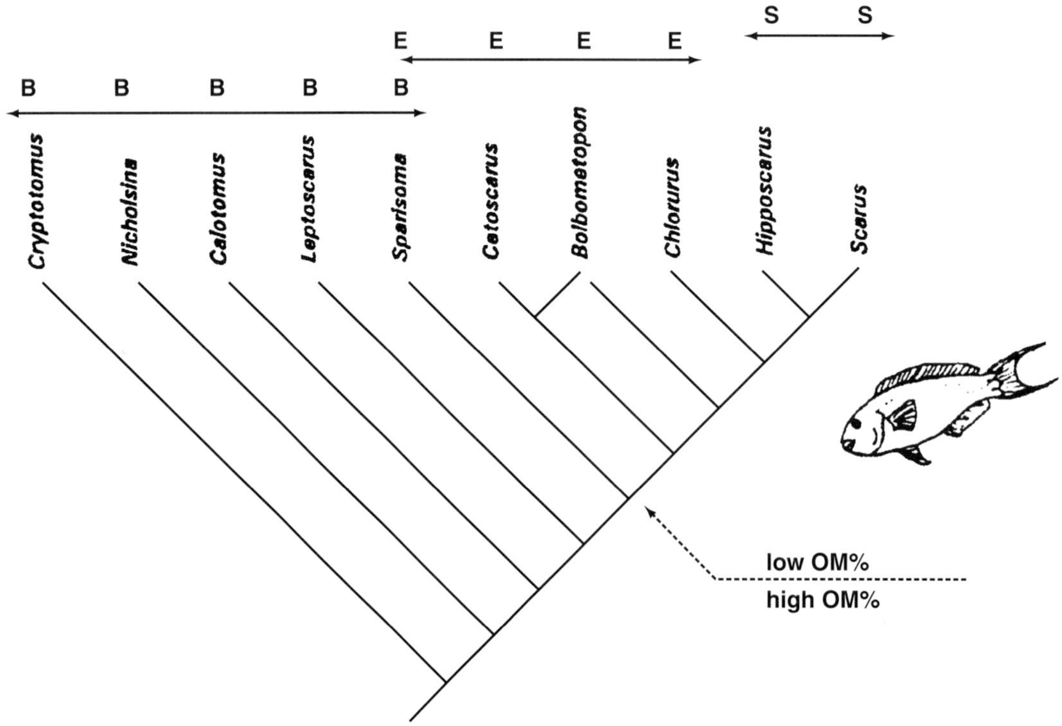

FIGURE 2 Cladogram of genera and feeding modes within the Scaridae (with permission from Bellwood, 1994). Key to modes: B, biting mode (species feeding on large fleshy algae and seagrasses), E, excavating mode (species feeding on turf algae, primary bioeroders), S, scraping mode (species feeding on turf algae, secondary bioeroders). High OM% refers to high organic matter percentage and low level of $CaCO_3$ in gut contents; low OM%, low organic matter percentage and high level (50–90%) of $CaCO_3$ in gut contents.

species ingest large amounts of calcium carbonate (up to 90%), and thus rely on the lowest quality food. Bellwood's ideas concerning the evolution of feeding patterns within the scarids have been supported within the genus *Sparisoma* by Bernardi *et al.* (2000). Using molecular data, they confirmed that browsing was the feeding mode of ancestral *Sparisoma* species, whereas the excavating mode was observed only in the more recently derived species. Similar evolutionary patterns are found among the Acanthuridae (Winterbottom and McLennan, 1993). The plesiomorphic feeding strategy for acanthurid fishes was to browse on macrophytic algae; feeding on filamentous algae arose later, and the most derived genera are grazers. Within the genus *Zebrasoma,* the basal diet consists of macroalgae, with a switch to filamentous algae in the more derived species (Guiasu and Winterbottom, 1998). In this family, zooplanktivory is a derived feeding strategy that has evolved independently in different genera—*Naso, Paracanthurus,* and *Acanthurus* (Winterbottom and McLennan, 1993; Borden, 1998). As in parrot fishes, the morphologically most derived genus of acanthurids, *Ctenochaetus,* relies on the least energetic food source (Fig. 3). *Ctenochaetus* species feed on diatoms, detritus, and filamentous algae while sucking soft

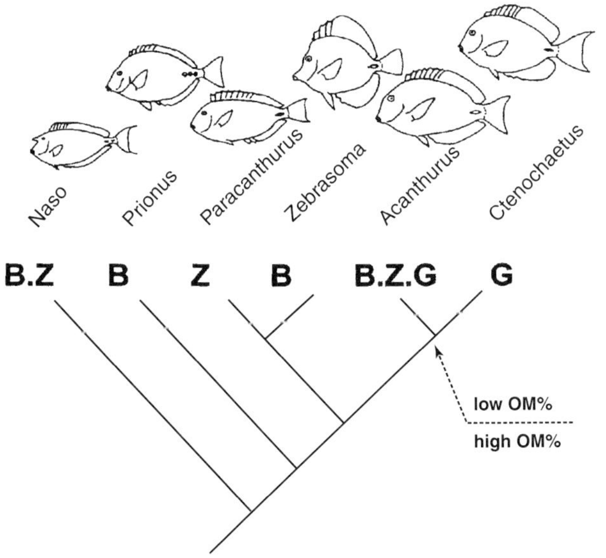

FIGURE 3 Cladogram of acanthurid genera and changes in feeding behaviors (with permission from Winterbottom and McLennan, 1993). Key to modes: B, browsers on macrophytic algae; Z, zooplanktivores; G, grazers on filamentous algae and diatoms/detritus. High OM% refers to high organic matter percentage and low level of CO_3 in gut contents; low OM%, low organic matter percentage and high level (>80%) of $CaCO_3$in gut contents.

bottom sediments or combing macroalgae, ingesting 80–90% by weight of sand particles with their food (Polunin *et al.,* 1995).

An excellent review of the problems induced by the digestive physiology and processing of plant food in marine herbivorous fishes was given by Choat and Clements (1998). The use of low-quality food sources such as algae and sessile invertebrates, which often contain diverse noxious deterrent compounds, implies specific morphological, physiological, and behavioral adaptations that would have had an energetic cost during their evolution. These adaptations are found among the most derived fish families, and within families, among the most derived genera. Feeding is an important biological force, because fishes, like most animals, usually need to eat every day, and the evolution of feeding and morphology are certainly intimately related (Motta, 1989). Among all the morphological characters studied by Bellwood (1994) to build the cladogram of genera of Scaridae, the most discriminant ones are also related to the feeding apparatus. As emphasized by Hobson (1994), "it was mostly the trend toward expanded feeding capabilities that contributed to the increasing diversity of fish species." In terms of energy, the trend tends toward a better use of lower quality food sources on coral reefs. Why are these low-caloric food sources used by fish in tropical regions and discarded in temperate ones? In the following section, I will propose a hypothesis stating that different processes have worked in synergy over evolutionary times in the tropics and especially on coral reefs, leading to the emergence in fishes of particular feeding behaviors that could not have arisen in temperate regions.

III. Evolutionary Processes

A. Synergy between Thermodynamic, Ecological, and Molecular Constraints

Several hypotheses have been proposed to explain the high degree of biodiversity observed in the tropics (Pianka, 1978; Rohde, 1978; Huston, 1979; Terborgh, 1985, among others) or the high level of diversity of herbivores on coral reefs (Horn, 1989). Stevens (1989) listed 12 hypotheses. These may be classified in two groups, depending on the two main factors invoked as ultimate causes: (1) a stable moderately high temperature, and/or (2) a high stability of the environment, both between seasons and over geological times. These two factors operate at different levels of biological organization: molecular, organism, and community levels (Fig. 4).

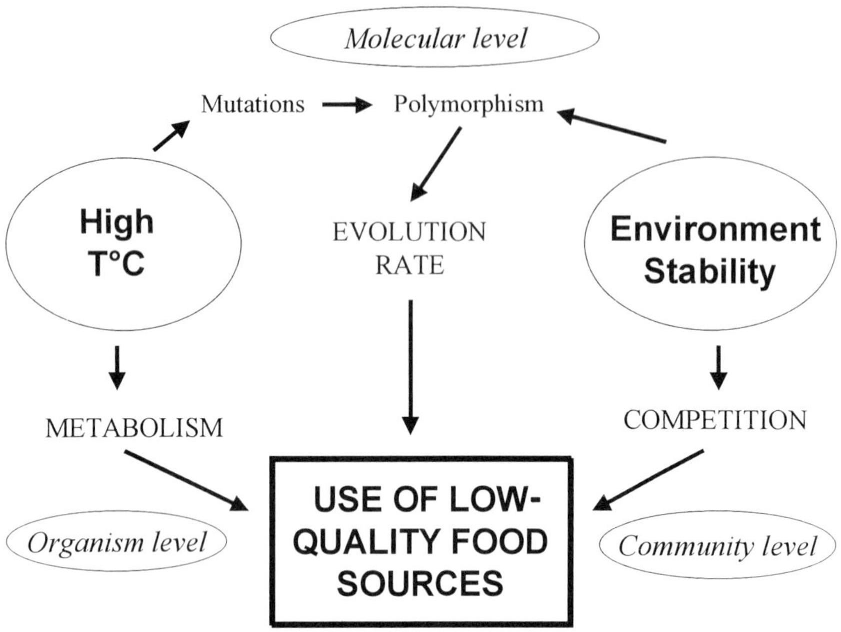

FIGURE 4 Schematic representation of the hypothetical effects of high temperature and environmental stability at different levels of biological organization to explain the higher consumption of low-quality food sources by fishes in the tropics compared to those in temperate zones.

1. Effects of a High Temperature

At the molecular level, temperature speeds up the rate of enzymatic reactions, including at the DNA level (Bernardi and Bernardi, 1986). Studies have shown that a high temperature may increase the rate of mutations. Bazin *et al.* (1997) found an increase in transcription with temperature in *Drosophila,* and the rate of mutations induced by transposon elements was much higher than that induced by nucleotide substitutions. Rate of mutation might be correlated also with metabolic rate, which is also linked with temperature. Mitochondrial respiratory processes produce free oxygen radicals, which in turn cause mutation by oxydative damage in mitochondrial DNA (Rand, 1994). Other evidence for evolutionary adaptations to temperature in poikilotherms is given by Mulvey *et al.* (1994) and Hawkins (1996). They demonstrate that fishes with a higher heterozygosity, represented by an increase in allozyme polymorphism, better resist environmental fluctuations. A positive relationship between heterozygosity and fitness was generally found, expressed as superior performances of individuals (Carvalho, 1993). Thus, temperature increases the rate of mutation, acting to increase the genetic polymorphism of organisms, which may in turn favor their adaptive ability.

At the organism level, most physiological processes are temperature dependent. The digestive physiology of poikilotherms is influenced by temperature, with a decrease in metabolic loss (Tarr, 1969) and an increase in food consumption at higher temperatures (Horn and Gibson, 1990). The maintenance metabolism, food intake, and growth are all affected by temperature, with higher rates in tropical fishes (Pandian and Vivekanandan, 1985). The time needed for feeding is much higher and digestive processes more laborious for herbivores than for carnivores. Herbivores expend more energy in extracting nutrients from their food because plant digestion is difficult and limited by temperature (Horn, 1989; Ebeling and Hixon, 1991). Thus, high temperature allows ectothermic vertebrates to remain continuously active and to be content with low-caloric diets, whereas food-processing rates are more constrained at low temperatures, which implies the need for more energetic food sources.

2. Effects of Environmental Stability

Relative stability of environmental conditions through time affects organisms at the molecular level. Long-term stability increases polymorphism, especially in noncoding genome areas (Kimura, 1991). The neutral theory of molecular evolution predicts that when functional constraints are decreasing, the evolutionary rate tends toward a maximum value determined by the total rate of mutations. Molecules submitted to weak functional constraints evolve more rapidly (in terms of allelic substitutions) than do those with strong functional constraints. Moreover, Bernardi and Bernardi (1986) suggested that noncoding sequences do play a physiological role that may concern the modulation of basic genome functions. Thus, genetic polymorphism of fish species is enhanced in relatively stable environments by random genetic drift. Environmental stability is conducive not only to increased speciation rates, but also to decreased rates of extinction (Cracraft, 1985; Ricklefs and Schluter, 1993a; A. Clarke, 1996). At the population and community levels, long-term stability increases all biotic interactions, particularly the interspecific competition for food and shelter (Elton, 1958; MacArthur, 1968). To be able to use food resources neglected by other species becomes a competitive advantage in saturated habitats, where high rates of competition reinforce the selection pressures. Temperate habitats are generally undersaturated because high fluctuations of environmental conditions have periodically reduced whole faunas (Briggs 1996), resulting in a decrease in interspecific competition.

Evolution results from dynamic interactions of several processes acting on many different levels (Dobzhansky, 1951; Brooks and McLennan, 1991). The progressive adaptation to use low-caloric food sources was possible only where allowed by high temperature and favored by long-term stability of environmental conditions, and involved mechanisms acting at different levels and on different processes, at a molecular level on genome evolution, at organismal level on digestive physiology, and at community level on competitive interactions and niche differentiation. The use of low-quality food sources by fishes in the tropics, and particularly on coral reefs, is competitively advantageous, energetically possible, and genetically achievable (Fig. 4). The darwinian principle of selection, which acts on organisms presenting reproduction, variation, and heredity, finally tends toward the emergence of new species at a higher rate in the tropics. As has been discussed by Jablonski (1993), tropical regions have been a major source of evolutionary novelty through time. This is apparent when considering the evolution of feeding behavior in fishes. Conversely, the rate of extinction is lower in tropical than in temperate and colder regions, which have suffered several historical increases in environmental harshness (Cracraft, 1985). The evolutionary nature of this process is indicated by the fact that the most derived species are those that rely on the poorest food resources. Whether this evolutionary trend is driven or passive (sensu McShea, 1994) remains a question of debate.

B. Biogeography and the Evolution of Reef Fish Feeding

The relative roles of local, regional, and historical processes, as well as unique events, in explaining community diversity are well recognized if not totally determined, and biogeographic patterns result from the evolution of organisms and their environment (Jackson and Harvey, 1989; Ricklefs and Schluter, 1993a). The role of stability in coral reef fish evolution may be indicated *a posteriori* when comparing reef fish faunas from geographically distinct tropical regions that have experienced very different geological histories. When considering the feeding adaptations of reef fish species, one notices the same trend in geographic speciation among different regions within the tropics, as the evolutionary trend in resource use observed between temperate and tropical zones (i.e., the better use of lower quality food types in historically more stable regions). The very long and chaotic history of coral reefs over geological times differed among regions (see Newell, 1972; Potts, 1985; Briggs, 1996). When considering the fluctuations of broad environmental conditions over the past 65 million years, the period during which most coral reef fish species have evolved (Choat and Bellwood, 1991), it is striking to notice that the western Pacific was more stable than the Atlantic, which suffered drastic modifications, particularly during the Pleistocene (Budd *et al.*, 1994). The Caribbean fluctuates less than the temperate regions, particularly the Mediterranean Sea. Among these geological changes, strong fluctuations of seawater temperature may have had a higher influence on fish speciation as compared to sea level changes. The ocean is a capacitor for temperature and its role was certainly less efficient in the Atlantic compared to the Pacific, due to the smaller size of the Atlantic.

The observed differences in coral reef fish diversity between the Indo-Pacific and the Atlantic are not only the result of differential extinctions (Bellwood, 1997), but also of differential rates of speciation. Among butterfly fishes, the highest number of species feeding exclusively on scleractinian corals is observed in the western Pacific, whereas no exclusive corallivore is found in the Caribbean. Exclusive corallivores belong to the most derived subgenera of Chaetodontidae, which do not exist in the Atlantic. In the same way, the most derived surgeonfish genus, *Ctenochaetus*, which ingests a high percentage of calcium carbonate with its food, does not exist in the Caribbean. The diversity of the family Scaridae is maximum in the Indo-Pacific, where the continuity of habitat availability permitted the speciation and survival of reef-associated scarids (Bellwood, 1994), and is particularly high in genera ingesting a high percentage of calcium carbonate. In the Caribbean, scarids are not only less diversified, but are dominated by basal genera that mainly browse on macroalgae and seagrasses. In the warm temperate waters of the eastern Atlantic and Mediterranean, only one scarid species is encountered, *Sparisoma cretense*, which feeds on large fleshy macroalgae and seagrasses, without scraping the substratum. Thus, the use of low-caloric food sources (high $CaCO_3$ percentages) by parrot fishes is common in the Pacific, present in a few species in the Caribbean, and absent in warm temperate waters. In these three fish families (Chaetodontidae, Acanthuridae, and Scaridae), the most derived genera are absent or less diversified in the Caribbean. As evolutionary processes have differed between regions, biogeographic comparisons of the trophic organization of fish assemblages should take into account the phylogenetic and biological relationships of species along with the geological history of the region.

IV. Conclusion

The better use of low-quality food resources is not the only process that may account for the higher diversity of fish fauna in the tropics. We have discussed here the consumption in the tropics of food resources that are largely neglected in temperate regions (algae, seagrasses, and sessile invertebrates such as sponges, cnidarians, and ascidians) or that are specific to tropical waters (scleractinian corals). But when all feeding behaviors are considered, one notices that both generalists and specialists are more diversified on coral reefs (Harmelin-Vivien, 1983). Tropical organisms have generally narrower microhabitat requirements and environmental tolerances compared to temperate organisms (Stevens, 1989), and reef fish species generally exhibit a finer distribution in space (Williams, 1991). These differences imply other evolutionary mechanisms driven by forces different than those discussed in this chapter. The initial question of why there are so many species in the tropics is far from being answered. However, the synergy between constant high temperature and a relative long-term stability may be viewed not only as a mechanism that speeds upt the rate of molecular evolution and increases the genome variability, but also allows the emergence of new species by natural selection through competitive interactions, which can afford to be energetically constrained by the environment.

To better understand the evolution of food resource use by fishes on coral reefs, more phylogenetic studies at the level of genus and species are needed. Cladograms of genera and species based on both morphological and molecular characters of Pomacentridae, Balistidae, Tetraodontidae, and other families with diversified feeding behaviors are necessary to test the hypothesis presented in this chapter of an evolutionary trend toward the use of less energetic food among fishes. A multidisciplinary approach to reef fish study, including phylogeny, biology, physiology, and ecology, will certainly give us in the next few years a better understanding of the role of fishes in coral reef functioning and how this role may have evolved through geological times.

Acknowledgments

Hypotheses developed in this chapter were presented at the OTS-ATB Annual Meeting on the Origin and Maintenance of Tropical Diversity in San Jose, Costa Rica, in June, 1997. Greatest thanks are expressed to Jorge Cortés for organizing a Symposium on Coral Reef Biodiversity during this meeting, and to John Pandolfi, Donald Potts, Marjorie Reaka Kudla, and Peter Sale for stimulating discussions. I am grateful to David Bellwood and Richard Winterbottom for providing helpful information on the ecology and evolution of coral reef fishes, and to René Galzin for field cooperation in French Polynesia. I express particular thanks to Antonio Sole-Cava, Stjepko Golubic, and an anonymous reviewer for valuable comments and corrections on the manuscript.

CHAPTER 13

Simulating Large-Scale Population Dynamics Using Small-Scale Data

Graham E. Forrester
Department of Biological Sciences
University of Rhode Island
Kingston, Rhode Island 02881

Richard R. Vance
Department of Organismic Biology, Ecology and Evolution
University of California at Los Angeles
Los Angeles, California 90095

Mark A. Steele
Department of Biological Sciences
University of Rhode Island
Kingston, Rhode Island 02881

I. Introduction

A. A Brief Overview—What Is the Focus of This Chapter?

Population ecology has been one of the main foci of research on reef fishes since the early 1980s. This research has generated considerable controversy over which processes are primarily responsible for driving population dynamics, and over the nature and strength of population regulation (for reviews, see Doherty, 1991; Jones, 1991; Booth and Brosnan, 1995; Caley *et al.*, 1996; Hixon, 1998). In this chapter we address questions about population ecology that have been at the center of the controversy: (1) What influence do density-dependent interactions among reef fishes have on fluctuations in their abundance? (2) Are the sort of density-dependent interactions observed among reef fishes likely to result in population regulation? (3) Will density-dependent interactions at small scales appreciably influence population dynamics at larger scales, i.e., does small-scale density dependence scale up? These questions are, we feel, important ones in ecology generally. In addressing them, our aim is to illustrate how some insights and approaches from other systems might help us better understand the population dynamics of reef fishes, and also to point out a few ways that the study of reef fishes can contribute to ecology in general.

The main lesson we apply from other systems is that density-dependent interactions often occur at particular scales both in space and time. Understanding the influence of density-dependent interactions, and even detecting them in the first place, depends critically on being able to identify the scale at which they occur and then being able to extrapolate their effects to other scales. We think reef fishes have a lot to offer to the study of population dynamics, in part, because they are one of the few groups of vertebrates that can be readily experimented with in the field. Field manipulations are viewed as the best way to detect density-dependent interactions, identify the stage in the life cycle at which they occur, and isolate their biological causes. Reef fishes have now been the subjects of a reasonably large collection of such experiments (reviewed in Chapter 14, this volume), and the results of these experiments may provide valuable insights into the ecology of other species that are more difficult to study, or occupy less accessible habitats. This chapter is motivated by the fact that virtually all experimental studies on reef fishes (as well as related observational studies) were done on habitat patches smaller than a coffee table and lasted for just a small segment of the fishes' generation time. There are compelling reasons to desire an understanding of population dynamics at much larger spatial scales and over multiple generations in time, and therefore the central focus of this chapter is to determine whether the results of this small-scale work on reef fishes might be expected to scale up.

B. Density Dependence and Population Regulation

A substantive understanding of population dynamics requires that we determine the causes of both fluctuations in and regulation of abundance. By regulation we mean the long-term persistence of a population within upper and lower bounds (Murdoch and Walde, 1989). Density dependence, when one or more per-capita demographic rates are related to current or past population density, is a necessary, but not sufficient, condition for regulation. Regulation occurs when density dependence has a direct effect (population growth is negatively related to density) that is strong enough to put bounds on temporal fluctuations in abundance, but not so strong or delayed in its effect as to have a destabilizing effect on abundance (Murdoch, 1994; Cappuccino and Price, 1995). Density *independence* means, of course, that demographic rates do not vary as a function of population density and the population's abundance is on an unregulated "random walk" over time (e.g., den Boer, 1968). Variation in demographic rates that is unrelated to density can, obviously, have major effects on population dynamics, but can never regulate populations. Questions about fluctuations in abundance and the regulation of population size are clearly interrelated, but the interrelation need not be a close one. For example, density-dependent interactions may have a trivial influence on dynamical patterns in abundance but still keep the population within bounds (e.g., Davidson and Andrewartha, 1948; Smith, 1961), and, conversely, intense density dependence may have an overwhelming effect on dynamics without stabilizing the population (e.g., Nicholson and Bailey, 1935).

One lesson reef fish ecologists can draw from other systems is that we have very limited ability to test directly whether reef fish populations are regulated. It is generally agreed that the best way to identify regulation is to use a time series of census data and test directly for boundedness in the observed fluctuations of abundance (Turchin, 1995). The accuracy of these tests increases dramatically with the length of the time series, and we simply do not have the sort of census data spanning multiple generations that are needed to perform these tests reliably (e.g., Hassell *et al.*, 1989; Holyoak, 1993; Wolda and Dennis, 1993). We do, in contrast, know a reasonable amount about the occurrence and nature of density dependence in reef fishes (see Chapter 14, this volume). Another lesson we should learn from the general ecological literature is that the mere existence of density dependence in some form does not tell us much about its effect on population dynamics. Our chapter is a modest attempt to improve our understanding of the role that density-dependent interactions play in population dynamics of reef fishes.

C. The Spatial Structure of Reef Fish Populations

In order to predict the outcome of density-dependent interactions, we need to recognize that there can be different sorts of density dependence that vary in the spatial and temporal scales over which they operate. We therefore begin by describing the spatial structure of reef fish populations. Reef habitats are patchy at a number of scales, but for the purposes of this chapter we will simplify things and assume that reef fish populations are organized at three basic spatial scales:

1. *A local population occupying a single patch of reef habitat.* The fishes occupying a single patch are defined as a local population. We focus on this scale first, because patch occupants interact with one another on a regular basis, and so this is the spatial scale at which fishes actually experience "density." The movement of juvenile and adult fishes among patches is often possible, but does not occur routinely. Because reef fish larvae are in the pelagic environment for weeks to months, it is likely that local populations are completely demographically open—i.e., larval input to a patch is unrelated to the reproductive output of the resident adults. The sort of habitat we have in mind here is a patch reef that is typically a few to tens of meters in extent, and is usually physically separated from other patches by inhospitable habitat (usually sand or seagrass). Such patch reefs have been the setting for virtually all field experiments on reef fishes, and they support reasonably large local populations of many smaller species that have home ranges of a few square meters or less. We note, though, that a group of regularly interacting conspecifics in any sort of habitat could be viewed as a local population.

2. *A mesopopulation occupying a large array of patch reefs.* We introduce this term to describe a collection of local populations occupying an array of neighboring habitat patches. The key feature of a mesopopulation is that, as a whole, it is sufficiently isolated from other reefs that the successful migration of juveniles and adults into, or out of, the mesopopulation does not occur. The replenishment of mesopopulations thus occurs only by the settlement of planktonic larvae, and losses

occur only through mortality (because any emigrants die). We assume that mesopopulations are demographically open to some degree, because at least some of the larvae arriving to replenish it are spawned elsewhere. For our purposes, it does not matter what fraction of larvae arrive from elsewhere. To give a realistic example, the fish occupying the reefs associated with a single island might constitute a mesopopulation, and so an actual mesopopulation might commonly be a few to tens of kilometers in extent.

3. *A metapopulation: a collection of mesopopulations.* A metapopulation is large enough so that all larvae arriving at the constituent reefs originated from within the metapopulation. In other words, a metapopulation is a demographically closed collection of mesopopulations.

D. Density Dependence at Multiple Scales

In general terms, there can be both spatial and temporal density dependence, and both of these can operate at different scales of space and time. Spatial density dependence occurs between populations and so, for reef fish populations structured in the simplified manner we have specified, spatial density dependence might occur among a set of local populations, or among a set of mesopopulations. In either case, if locations supporting greater population densities suffer higher loss rates and/or lower gain rates, then those demographic rates are spatially density dependent. Temporal density dependence refers, instead, to a single population whose loss and/or gain rates vary as a function of density such that the loss rate increases and/or the gain rate declines when population density increases (Stewart-Oaten and Murdoch, 1990). For reef fishes, temporal density dependence might occur at the scale of a single local population, a single mesopopulation, or a single metapopulation. To make matters more confusing, both spatial and temporal density dependence may occur at specific stages in the life history and therefore operate within a generation, or conversely might occur among generations (e.g., Hassell, 1986). To give a hypothetical example, larval and juvenile mortality might be negatively related to density, whereas adult mortality might be positively related to density, and the two might cancel each other out so that mortality over the entire life-span was density independent! Making these distinctions and outlining this complex set of possibilities underscore the importance of specifying carefully the spatial and temporal domain over which answers to questions about the causes of population dynamics will apply.

This hierarchical spatial structure of reef fish populations also highlights the fact that the four basic demographic rates that influence population size—birth, death, immigration, and emigration—will vary in their importance among spatial scales. For local populations, the immigration and emigration of postsettlement fishes will affect local abundance. Losses will also occur via mortality of residents, but the birth rate has no influence on future population size within the patch. Instead, the settlement of planktonic larvae effectively substitutes for the birth rate, because it results in the arrival of offspring. Biologically, though, it is more analogous to immigration, because the reproductive activity of patch residents does not influence the rate at which larvae are added to the population. At the mesopopulation scale, successful postsettlement migration does not occur, and so abundance is the balance of input by larval settlement and losses by mortality of settled individuals. The input rate via settlement may, or may not, be affected by the reproductive output of the resident females, depending on (1) the fraction of larvae spawned by mesopopulation residents that return to settle into the population and (2) the relationship between spawning output from the mesopopulation and the subsequent mortality of those larvae. Finally, at the metapopulation scale, population size is affected only by births and deaths within the population.

E. What We Know about Density Dependence in Reef Fishes

Having outlined different forms of density dependence, the spatial and temporal scales over which they operate, and which demographic rates are likely to be involved at these scales, we must point out that we have data on reef fishes for only a small subset of the possible forms and scales. First, as we have noted above, most empirical work, and all experimental work, has been done on local populations—typically using patches of reef a few or tens of meters in extent. Of course, this bias toward small plot sizes is a general one in ecology (Kareiva and Andersen, 1986). Second, most of these local populations have occupied habitat patches that were physically isolated from other reefs by inhospitable habitat (usually sand). Generally, the distance among patches was small (less than 20 m), but was in some cases hundreds of meters (e.g., Hixon and Carr, 1997). The choice of widely spaced habitat patches was usually made with the deliberate aim of minimizing two of the demographic rates that may affect local populations, immigration and emigration by settled fishes. There were good reasons for this: migration among replicate patches makes it difficult to maintain experimental treatments, and also makes it difficult to measure accurately other demographic responses (mortality and recruitment). A third, and related, bias is toward within-generation studies that focus on a limited part of the life cycle—usually the first few weeks or months after larval settlement. The paucity of among-generation studies is caused by logistical constraints that limit the duration of experimental manipulations to weeks or months, whereas many reef fishes can live for several years. Several good summaries of older literature support these generalizations (Jones, 1991; Booth and Brosnan, 1995; Caley *et al.*, 1996), as does the more recent literature (e.g., Hixon and Carr, 1997; Tolimieri, 1998b; Caselle, 1999; Schmitt and Holbrook, 1996, 1999a,b), including our own work (e.g., Forrester 1990, 1991, 1995, 1999; Steele, 1996, 1997a,b, 1998; Steele *et al.*, 1998; Forrester and Steele, 2000). In summary, then, virtually all of the tests for density dependence that we know of focus on spatial density dependence within generations, and use local populations as replicates. A few tests for spatial density dependence have used mesopopulations as replicates (e.g., Doherty and Fowler, 1994a), and a few studies have tested for temporal density dependence by tracking local populations over time (e.g., Victor, 1986b). In general, though, spatial and temporal density dependencies at both the mesopopulation and metapopulation scales remain largely unexplored.

Focusing most of our attention on local-scale, short-term studies has had some benefits and some drawbacks. Working at small scales does facilitate experimental manipulations of density, which is considered to be the best means of testing for density dependence in demographic rates (Nicholson, 1957; Murdoch, 1970; Sinclair, 1989; Harrison and Cappuccino, 1995). Experiments of this sort have revealed variable relationships between the mortality of juvenile and adult fishes and local density, but density dependence at both stages can be intense. Some experiments have gone a step further, and identified the biological interactions that cause density dependence (Forrester, 1990; Hixon and Carr, 1997; Forrester and Steele, 2000). In a few species, mortality inflicted by predators has been found to increase with prey density (Hixon and Carr, 1997; Forrester and Steele, 2000), but other agents of mortality can also be density dependent (Forrester and Steele, 2000, and unpublished). Detailed experimental evidence defining the intensity and biological cause of density-dependent interactions is not available, nor easily obtainable, for many taxa (Harrison and Cappuccino, 1995), and so it is important for us to make the best use of this understanding that we have gained by studying reef fishes.

Working at small scales has also facilitated tests of density dependence based on examining relationships between the rate of larval settlement or recruitment and the abundance of fishes remaining at the site some time later (e.g., Doherty and Fowler, 1994a; Levin, 1996; Robertson, 1988a, 1992; Caselle, 1999; Schmitt and Holbrook, 1999a,b; Shima, 1999b). Because accurately measuring larval settlement in reef fishes is very labor intensive, increases in the spatial scale of such studies have always come at a cost in terms of accuracy and precision of settlement estimates (e.g., Doherty and Fowler, 1994a; Caselle 1999). Some local-scale studies have thus been able to correlate settlement with the abundance of older fishes to test for density-dependent mortality starting at, or just after, settlement (e.g., Schmitt and Holbrook, 1999a,b), whereas studies using mesopopulations as replicates have not been able to encompass mortality acting early in reef-associated life (e.g., Doherty and Fowler, 1994a). At both local and mesopopulation scales, relationships between settlement/recruitment and the abundance of older life stages are sometimes apparently linear, implying density-independent mortality in the intervening period (e.g., Doherty and Fowler, 1994a). In other cases, there is evidence that they may be curvilinear, indicating density-dependent mortality (e.g., Caselle, 1999; Schmitt and Holbrook, 1999a,b). A synthesis of these results is difficult, however, precisely because they were done at different spatial scales and focus on different portions of the life cycle.

F. Some Reasons for Wanting to Extrapolate to Larger Scale Population Dynamics

In this chapter we will take this body of information on short-term, local-scale relationships between density and demographic rates and attempt to predict the consequences for longer term mesopopulation dynamics. Knowing the extent to which local-scale, within-generation findings can be extrapolated would be desirable for several reasons. First, and very simply, much of the field research on reef fishes has dealt with local "populations" comprising just a handful of fishes, and so are hardly large enough to constitute actual populations (colleagues working on other taxa sometimes laugh at us when we refer to them as populations). A more important practical reason for wanting to extrapolate to the mesopopulation spatial scale is that, for many reef fishes, a mesopopulation occupies a spatial region that might be subject to a local fishery, or be designated as a marine reserve. Extrapolating to multigenerational time scales would be helpful because plans for managing fisheries and conserving natural habitats are usually implemented over years rather than weeks or months (e.g., Done, 1998; Done and Reichelt, 1998). We would, therefore, be able to provide better advice to fisheries scientists and conservation biologists if we knew whether our findings might apply over appropriately large domains of space and time.

Along these lines, we also hope that our work might provide a means of forging more explicit links with research done on commercially exploited species in temperate waters (e.g., Beverton and Holt, 1957; Rothschild, 1986; Cushing, 1995; Quinn and Deriso, 1999). This extensive body of research is mostly descriptive in nature, and has been conducted at large spatial and temporal scales. In some cases these temperate populations are believed to form distinct stocks, and so are demographically equivalent to reef fish metapopulations, but in other cases the populations studied are probably demographically open to some degree, and so are directly comparable to reef fish mesopopulations. The type of detailed experimental fieldwork that we can do on reef fishes is often impossible in many other habitats where fishes are commonly harvested. These two bodies of work are therefore complementary in nature, and we would benefit tremendously from increasing the degree of overlap and exchange between them.

G. Can Small-Scale Results Be Expected to Extrapolate?

Researchers studying reef fishes have been skeptical that the results of small-scale studies will scale up (e.g., Doherty and Williams, 1988; Doherty, 1991; Doherty and Fowler, 1994a; Caley *et al.*, 1996; Caselle, 1999). A common view, well summarized by Caley and co-workers (1996), is that density-dependent interactions detected by experiments on small isolated habitat patches may be less important at large scales because "density-dependent emigration can ameliorate competition and competition may occur only at a limited number of sites." The potential for this is supported by tagging studies indicating that even small reef fishes can redistribute themselves, and their ability to do so improves in areas of continuous reef habitat, or where patches are closely spaced (e.g., Frederick, 1997). Density-dependent interactions experienced by fishes restricted to isolated reefs might, therefore, not occur if those fishes have the opportunity to disperse from high-density patches to more favorable locations. Models of predator–prey and host–parasitoid interactions in patchy habitats generally find that making patches more homogeneous has a destabilizing effect (e.g., Hassell *et al.*, 1991), but the opposite has also been found (e.g., Kareiva, 1987; Murdoch and Oaten, 1989). There is, as a result, no clear basis from which to predict the consequences of studying reef fishes mainly on isolated patches of reef. We, therefore, address the issue of whether varying the rate of migration among patches affects the outcome of local-scale density dependence.

Another important insight from the wider literature is that density-dependent interactions may be explicit to certain spatial and temporal scales, and not detectable at other scales. This has been particularly well studied in insects that are patchily distributed in space and subject to predation by parasitoids. In this system, aggregation by parasitoids to dense prey patches can be an important mechanism of density-dependent predation. The appropriate spatial scale to detect this response depends on exactly how prey are distributed in space and on the searching abilities of the parasitoids (e.g., Heads and Lawton, 1983; Freeman and Smith, 1990; Rothman and Darling, 1990; Hopper *et al.* 1991; Stiling *et al.* 1991; Ray and Hastings, 1996). Similarly, the spatially density-dependent prey mortality that results is sometimes experienced only by certain life stages of the prey population. Theoretical analyses show that this spatial density dependence within generations does not always lead to temporal density dependence over multiple generations (e.g., Hassell, 1986; Murdoch and Oaten, 1989; Stewart-Oaten and Murdoch, 1990). These theoretical analyses of insect–parasitoid interactions are not directly applicable to reef fish populations, but the general message is clear—we can not assume that relationships between density and demographic rates will remain constant when we change our scale of observation. We must, therefore, develop models for reef fishes to reconcile the varying results of tests for density dependence performed at different scales.

The last and perhaps most important reason why we wrote this chapter is that detecting density dependence says very little about its *quantitative* effect on population dynamics. It has long been recognized that, although field experiments provide a rigorous means of identifying interactions, they cannot identify the role those interactions play in controlling population size (e.g., Weinberg *et al.*, 1986). This argument applies equally to interactions that are dependent on or independent of population density. It is obvious that all of the demographic rates affecting abundance will have some influence on population dynamics, and the same is true of the environmental factors that control those demographic rates. There are now some excellent methods to evaluate the relative influence of processes occurring at different life history stages on overall population dynamics (e.g., Nisbet *et al.*, 1996; Pfister, 1996; Schmitt *et al.*, 1999). These approaches can be used with data collected at any spatial scale, but are not designed to extrapolate among different spatial scales.

Models focusing explicitly on extrapolating population dynamics across spatial scales indicate that extrapolating from small-scale results to larger spatial domains is greatly complicated by density dependence operating at the small scale (e.g., Chesson, 1996, 1998a; Anneville *et al.*, 1998; Pascual and Levin, 1999). For example, Chesson (1996, 1998a) analyzed the dynamics of a set of populations (each equivalent to a mesopopulation by our definition) that, as a group, formed a single, closed metapopulation. These analyses show that density-dependent interactions within mesopopulations can have dramatically altered consequences when viewed across the entire metapopulation (Chesson, 1996, 1998a). Chesson argued that these "scale transitions" in population dynamics result from the interaction of density dependence with small-scale heterogeneity in density. Our work tackles similar questions, but we focus instead on a set of local populations that are connected to form a single mesopopulation. We made this different choice of scales for reasons already outlined: (1) virtually all empirical data on reef fish concern local populations and (2) extrapolating to the mesopopulation scale will be helpful because marine reserves and fisheries are often managed at this scale. A third important reason was that we wanted to make quantitative predictions about population dynamics that would be empirically testable. The logistics of testing predictions about the dynamics of mesopopulations are more manageable than testing similar predictions about metapopulation dynamics!

II. Questions

We used computer simulations to examine local populations of fishes, each of which occupies a small patch of habitat (a patch reef) isolated from similar neighboring patches by expanses of inhospitable space (sand). The model simulates a large collection of neighboring local populations that, as a whole, form a mesopopulation. We are thus able to address questions about the relationships between the demography and population dynamics of a mesopopulation and its component local populations.

We employed systematic variation of selected parameter values to address four questions.

Question 1: How does local-scale density dependence in different demographic rates affect the meso-scale relationship between larval supply and abundance?

Question 2: How does local-scale density dependence in different demographic rates affect temporal fluctuations in mesopopulation abundance?

Question 3: Are relationships between density and demographic rates the same for the mesopopulation and its component local populations?

Question 4: How does the amount of adult migration between local populations influence the answers to Questions 1–3?

The qualitative features of our model fish species closely resemble attributes of the bridled goby *Coryphopterus glaucofraenum,* a fish we have studied extensively and so can provide reasonable estimates of the model parameters (e.g., Forrester 1995, 1999; Steele *et al.*, 1998; Forrester and Steele, 2000). The reason for developing a detailed, empirically parameterized, simulation model is that the questions we address do not have qualitative "either/or" answers. Instead they have *quantitative* answers based on the relative effect of different processes and interactions. The model makes detailed quantitative predictions about the answers to these questions for *C. glaucofraenum,* which can be tested in the field. It is important to stress here that the answers to these questions are likely to vary among species and circumstances. With appropriate parameterization, our model can be widely applied to any reef fish that has dispersing planktonic larvae and an adult stage that occupies patchy reefs surrounded by uninhabitable habitat. It is thus a tool that can be used to address general questions about when, and why, we would expect the processes driving population dynamics to differ.

III. Model Formulation

A. Basic Structure

The habitat that contains our hypothetical mesopopulation consists of small, discrete reefs, suitable for fish occupancy, that are surrounded by inhospitable habitat. These reefs lie at the 625 nodes of a 25×25 square array that is well isolated from any reefs outside the array. Each reef supports a population of adult fishes. In most simulations, adult fishes were allowed the possibility of moving from each reef to the nearest neighboring reefs. Because the location of each reef influences the dynamics of its population explicitly, this model is "spatially explicit" in the sense of Hanski (1996).

The population abundance of adult fishes on each reef is measured as an integer. Each population obeys a difference equation of the form

$$\begin{aligned} \text{adult abundance}(t+1) &= \text{adult abundance}(t) \\ &+ \text{larval recruits}(t) + \text{adult immigrants}(t) \quad (1) \\ &- \text{adult deaths}(t) - \text{adult emigrants}(t), \end{aligned}$$

with time t measured in weeks and a step size of 1 week. At each time step, the effect of each of the four demographic processes of larval recruitment, adult immigration, adult death, and adult emigration is described by an integer-valued random variable. Accordingly, our mesopopulation model consists of a system of simultaneous, stochastic, difference equations. These equations concern an idealized fish species whose adults are all functionally identical; that is, the model makes no attempt to distinguish between adults of different ages or sizes. To begin each simulation, we set the initial abundance of each reef's population equal to the estimated long-term average abundance. Analytical calculation of this estimate will be described below.

Our interest centers on the behavior over the long term of a collection of local populations surrounded by neighboring populations on all sides, a condition probably typical of mesopopulations in nature. To examine typical mesopopulation behavior, we ran each simulation for 12 years (624 time steps). As an attempt to eliminate effects of the initial condition, we discarded all population data for the first 2 years (104 time steps) and examined mesopopulation behavior only over the following 10 years (520 time steps). We performed all calculations for all populations in the full 25×25 reef array. However, to eliminate boundary effects, in calculating mesopopulation properties we discarded all population data from the outer 2 rows. Thus, our examination focused on the dynamics over what we view as a typical 10-year period of what we consider a typical $21 \times 21 = 441$-reef mesopopulation situated in the interior of a larger array. All calculations employed Mathematica 4.0 software (Wolfram Research Inc.).

Demographic parameters fall into two classes. The first contains parameters associated with "normal" population densities usually encountered in nature and for which we can supply reasonable empirical estimates for bridled gobies from published sources (Forrester, 1995, 1999; Steele *et al.*, 1998; Forrester and Steele, 2000) or from our unpublished data. The next subsection concerns these parameters. The second class includes parameters that describe how demographic rates depend on population density. Field data required for estimating these parameters are sparse at best for any fish species. The subsection after next describes a simple, intuitively interpretable, and yet versatile formulation of density dependence in the demographic rates of reef fishes, as applied to the bridled goby.

B. "Normal" Demographic Rates

Field population data for well-studied reef fish species can supply reasonable estimates, or at least inspired guesses, of demographic rates that usually apply when fish abundance lies within "normal" limits. Our estimates and guesses concern bridled gobies that occupy hypothetical 2- to 8-m^2 patch reefs like the experimental reefs we have studied in the Bahamas and Caribbean (Forrester 1995, 1999; Steele *et al.*, 1998; Forrester and Steele, 2000; G. E. Forrester and M. A. Steele, unpublished data).

1. Larval Recruitment

We envision larval recruitment onto a reef as proceeding in two steps, arrival of larvae into the immediate vicinity of the reef followed by recruitment of these larvae to the adult population there. Larval arrival is not a function of adult density, so the mesopopulation is effectively demographically open. We set the expected fraction of arriving larvae that successfully recruit at 0.2416, meaning that the fraction 0.7584 die while attempting to recruit. In combination with other parameter values given below, this expected recruitment success gives rise (as described below) to an estimated long-term average population abundance of 10 individuals, a figure consistent with our field observations.

Larval arrival on each reef is influenced by multiple processes that are both ecologically complex and poorly understood. Accordingly, we formulated the model's hypothetical larval arrival pattern to mimic the well-documented qualitative features of reef fish species in general (e.g., Doherty and Williams; 1988; Doherty, 1991), and our field monitoring of bridled gobies over 3 years (M. A. Steele and G. E. Forrester, unpublished). Figure 1 displays a typical random sample produced by our algorithm of 10 years of larval arrivals to the mesopopulation. This sample displays four important qualitative features. (1) Each year there occurs a larval arrival season during which most larvae arrive onto adult habitats. These larval arrival seasons have variable length, and they are separated by off-seasons

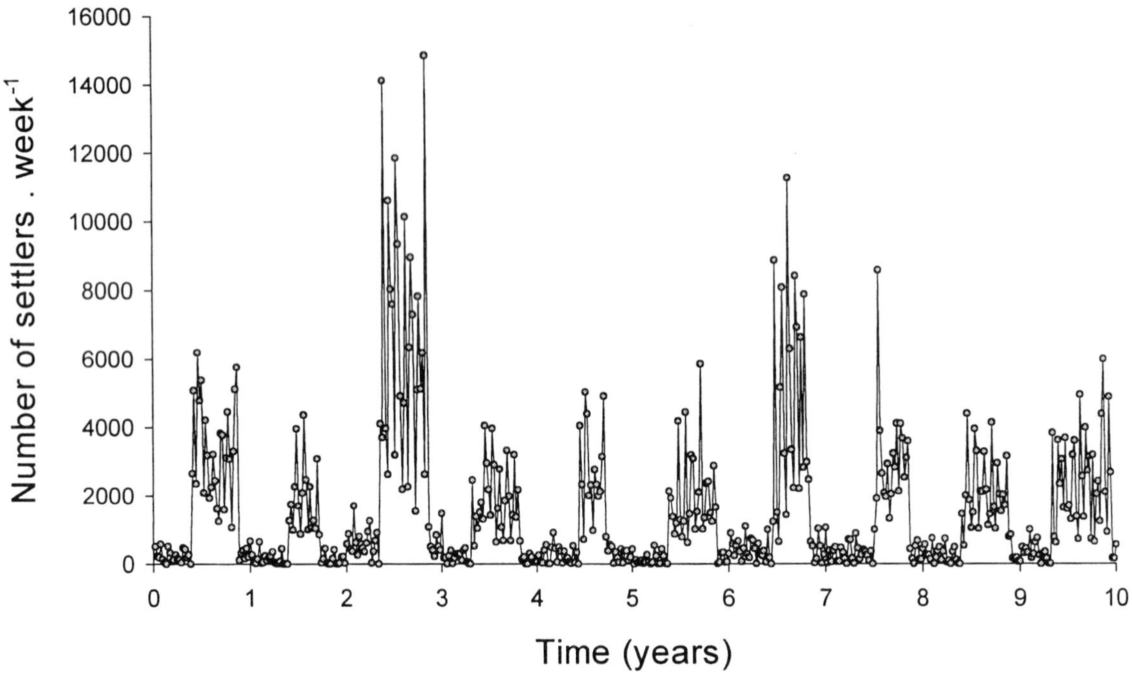

FIGURE 1 An example of the pattern of larval settlement generated by the simulation model. Note that this simulated pattern of settlement closely mirrors many empirically measured patterns (see, e.g., Doherty and Williams, 1988). Displayed is the total number of settlers to the mesopopulation each week during one 10-year run of the model. The simulation shown was selected haphazardly from those generated for this chapter.

during which only a few larvae arrive from time to time. (2) Wide and irregular variation occurs from year to year in the annual total of larval arrivals onto a reef. (3) Considerable variation in weekly larval arrival occurs through the larval arrival season but with an overall trend of fewer arrivals near its beginning and end than during its long middle period. (4) Average larval arrival rate during the 2 weeks surrounding the time of new moon exceeds the average during the 2 weeks surrounding the time of full moon. Details of the larval arrival algorithm's structure appear in the following paragraphs. Statistical distributions and parameter values were selected to produce the qualitative pattern displayed in Fig 1.

Variable length and timing of each year's larval arrival season is achieved by making the season's beginning and ending dates random variables. Specifically, the beginning week of each year is a random variate, rounded to the nearest integer, selected from a normal distribution with mean 18 (= first week in May) and standard deviation 4. The ending week is another independent random variate, rounded to the nearest integer, selected from a normal distribution with mean 44 (= last week in October) and standard deviation 4. "Pathological" years are avoided by requiring that both initial and final weeks lie between week 1 and week 52 and that the ending week not precede the beginning week.

The number of larvae that actually arrive on a reef during each week is a complicated random quantity whose overall pattern through time incorporates variation on three time scales—annual, seasonal, and lunar. Specifically, this number of larval arrivals is a gamma-distributed random variable with expectation

$$A \times S \times L,$$

where A, S, and L describe the annual, seasonal, and lunar effects. The annual contribution A is a random variate selected each year from a gamma distribution with parameters $\alpha_A = 5$ and $\beta_A = 2$. The seasonal contribution S is deterministic. It increases from $S = 0$ in the larval arrival season's first week to $S = 1$ at its midpoint, and then it decreases again to $S = 0$ in the larval arrival season's final week. This curve has a broad hump. Specifically, its shape is given by the tenth root of the positive portion of a sine wave that passes through these three points, and the value of S during each week of the season is just the height of this curve at that time. During the off-season, this variable is assigned the constant value $S = 0.1$. Consequently, the seasonal contribution to larval arrivals during the off-season is one-tenth as great as during the larval arrival season's peak. The lunar contribution L is a random variate selected each week from a gamma distribution. This gamma distribution has parameters $\alpha_N = 5$ and $\beta_N = 0.2$ during the 2 weeks of each month that surround the new moon, and it has parameters $\alpha_F = 5$ and $\beta_F = 0.1$ during the 2 weeks of each month that surround the full moon. With these parameters, the expected value of L during the two weeks surrounding the new moon is twice its expected value during the 2 weeks surrounding the full moon. Each week's values of A, S, and L apply to all reefs in the array. The actual number of larvae that arrive on each reef during any given week is specified by an independent random variate selected from a common gamma distribution with parameters $\alpha = 5$ and $\beta = A \times S \times L/\alpha$.

2. Mortality

Field observations of tagged gobies in the Bahamas and Virgin Islands have revealed that on average the fraction of adults that fail to survive from one week to the next is about 1/11 = 0.0909, which implies an average adult life length of about 10 weeks. In the model, each week the fate of each goby, anywhere in the reef array, was decided randomly and independently of the fates of all other gobies. When local density on each reef did not influence deaths there, each goby experienced probability 1/11 of dying each week.

3. Adult Emigration, Movement, and Immigration

To study effects on mesopopulation behavior of adult movement between reefs, we examined three hypothetical movement rates: zero, low, and high. The fraction of adults each week that emigrate from any reef and successfully immigrate to another is 0% in the zero-movement case, 2% in the low-movement case, and 10% in the high-movement case. The low-movement case employs the measured movement rate of marked gobies from patch reefs in the Virgin Islands (G. E. Forrester, unpublished). Because we do not know in detail what fate befell any individuals that disappeared from the reef array there, we were forced to base calculated numerical values for some model parameters on two educated guesses. The first guess is that when adults can move between reefs, half of all adult deaths are experienced by reef residents and the other half by reef emigrants. The second guess is that 90% of the deaths experienced by emigrants occur during movement across inhospitable habitat between reefs, and the remaining 10% during attempted immigration into another reef's population.

Back-calculations of primary model parameters from these guesses, which apply when population densities influence neither emigration nor immigration, give rise to different sets of parameter values for the three adult movement rates. In the zero-movement case, all adult deaths necessarily occur to reef residents, and emigration occurs with probability zero. In this case, reef residents experience death with probability 1/11 each week as explained above. In the low-movement case, each week reef residents experience death with probability 1/22 = 0.04545, each individual emigrates with probability 0.06465, each emigrant dies over inhospitable habitat with probability 0.6216, and each individual that arrives alive on a reef dies while attempting to immigrate there with probability 0.1825. In the high-movement case, resident death occurs with probability 1/22 each week, emigration with probability 0.1415, death over inhospitable habitat with probability 0.2637, and death while attempting to immigrate with probability 0.3980.

In one time step, emigrants from any reef may move only to the four immediately adjacent reefs. Each emigrant's movement is random, independent of each other emigrant's movement, and equally likely to occur in each of the four cardinal directions. All emigrants from edge reefs that move away from the reef array die, but these fatal movements constitute only 4% of all possible adult movements in the reef array.

4. Average Population Abundance

Choosing the probability of larval death during attempted recruitment, namely, the figure 0.7584 mentioned above, requires knowing the long-term average population abundance on individual reefs. This average must be determined ahead of time by some method independent of the main computer runs. Calculating its exact value by analytical means seems unlikely and perhaps even impossible, and so we employed an approximation.

The approximation takes place in two steps. The first step is estimating the average number of larval arrivals on a single reef over a full year. Our estimate involved summing the expected weekly larval arrivals over the 52 weeks of a year for which larval arrival season has the expected beginning and ending dates. This analytical estimate is 195.7 larvae per reef. To check this figure against numerical results, we used the model to generate larval arrivals on a single reef over a 1000-year period. The resulting annual average was 197.01, with a standard error of 3.05, in excellent agreement with the analytical estimate. Our field observations make this figure seem plausible.

The second step is estimating the long-term average population abundance. Our procedure consisted of calculating expected population abundance directly from the original difference equation[Eq. (1), Section III,A] with zero adult movement and with every week's larval arrivals replaced by the weekly average of the expected 195.7 larval arrivals per year. This procedure produced an algebraic expression for expected population abundance in terms of the probability of larval death during attempted recruitment into an adult population. Our field observations suggest that average population abundance on individual reefs is about 10 individuals. Inserting this figure into this algebraic equation gave rise to the calculated recruitment failure probability of 0.7584 mentioned earlier.

To check whether our computer model produces a long-term average population abundance near to the assumed value of 10 individuals, we did not perform any 1000-year computer runs. Instead, we performed many separate calculations of 10-year averages under different kinds and degrees of density dependence and adult movement as described in the next section. These 10-year averages fell close to the assumed value, and deviations were symmetrically arranged about this figure.

C. Density Dependence

Dynamical behavior of the mesopopulation arises from larval arrivals, adult movements and deaths over inhospitable habitat, and four demographic processes that take place at the level of populations on individual reefs. These reef-associated demographic processes are (1) deaths of newly arrived individuals in the week between their arrival as larvae and recruitment to the adult population 1 week later, (2) deaths of reef residents, (3) emigration from reefs, and (4) deaths of emigrants while they attempt to immigrate into a reef's population. Because of their association with individual reefs, it is possible for population density on individual reefs to influence the rates at which these four processes occur. Experiments indicate that the first three of these processes are density dependent in bridled gobies (Forrester, 1995, 1999; Steele *et al.*, 1998; Forrester and Steele, 2000; unpublished data), but in no case are the data precise enough to establish the mathematical form of density dependence. For the purpose of our exploration, we imposed density dependence on these processes of the simplest and most easily interpreted mathematical form. Though this mathematical form lacks rigorous empirical foundation, it produces an excellent fit to our field measurements of mortality rates (G. E. Forrester and M. A. Steele, unpublished data) and also arises from a mechanistic (albeit simplified) description of how spatial refuges might influence death rates (R. R. Vance, unpublished data).

This simplest possible relation between a demographic rate and population density is a straight line whose slope measures the strength of density dependence. Examining effects of varying the strength of density dependence with all other factors held constant requires that each demographic rate function retain its "normal" value whenever population size lies at its long-term mean. That is, lines of all slopes must pass through the same fixed point that specifies normal demographic rates at the long-term mean population abundance.

Explicit algebraic expression of the demographic functions clarifies this point. Let x represent population abundance on any particular reef, and let x^* denote its long-term average. Let $f(x)$represent the magnitude of a demographic rate when the population has abundance x. Let f^* denote this magnitude when the population lies at its long-term average abundance; this definition implies that f^* satisfies $f^* = f(x^*)$. The general function f can represent any one of the four reef-associated demographic rate functions, namely, the probability during one time step of (1) recruit mortality, (2) adult mortality while residing on a reef, (3) adult emigration from a reef, or (4) adult death during attempted immigration into a reef's population.

The linear character of this function $f(x)$ can be expressed algebraically in two equivalent ways that emphasize different features. The first is

$$\frac{f(x) - f^*}{f^*} = m\left(\frac{x - x^*}{x^*}\right). \qquad (2)$$

The left-hand side of this expression is the fractional deviation of the demographic rate from its long-term average value, and the parenthetical quantity on the right-hand side is the fractional deviation of population abundance from its long-term average value. The equation asserts simply that these two quantities are directly proportional to each other. The proportionality constant m cannot be negative because, by assumption, all four of the reef-associated demographic rates either increase with population abundance or remain constant at all abundances. The lower limit of m is 0 when the demographic rate does not depend on density. Its upper limit is 1 because any higher value would force $f(x)$to become a negative quantity for small values of x, a nonsensical situation because $f(x)$ represents a probability. Figure 2a displays a graph of Eq. (2).

Algebraic rearrangement of Eq. (2) produces an explicit expression for $f(x)$,

$$f(x) = (1 - m)\,f^* + \left(\frac{mf^*}{x^*}\right)x, \qquad (3)$$

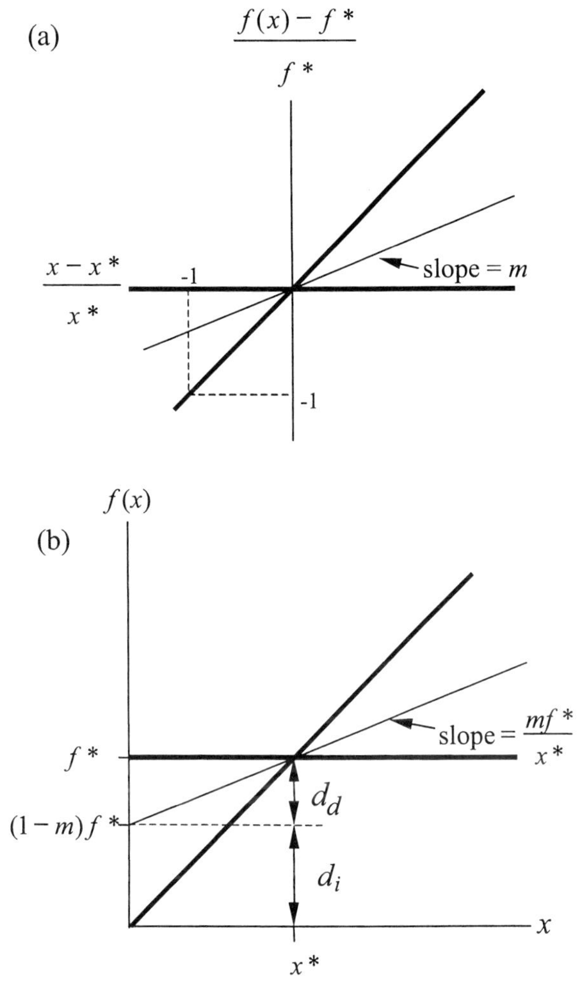

FIGURE 2 Density dependence in the generic demographic rate function f. (a) Equation (2) describes a straight line that lies somewhere between the two heavy lines and passes through the origin. In our model, we varied the strength of density dependence by varying m from 0 (no density dependence) to 1 (complete density dependence). (b) Equation (3) describes a straight line that lies somewhere between the two heavy lines and passes through the point (x^*, f^*). In a population with abundance x^*, the demographic rate f^* can be thought of as consisting of a density-independent component of magnitude d_i and a density-dependent component of magnitude d_d.

which is graphed in Fig. 2b. The graph of Eq. (3) provides a simple biological interpretation of the proportionality constant. The demographic rate f^* experienced by a population at its long-term mean size x^* can be expressed as the sum of two quantities, a portion d_i that arises from purely density-independent processes, and a second portion d_d due to density-dependent mechanisms. It follows that the fraction of demographic events that arise from density-dependent processes satisfies

$$\begin{aligned}\frac{d_d}{d_i + d_d} &= \frac{f^* - f^*(1-m)}{f^*}\\ &= 1 - (1 - m) \\ &= m.\end{aligned} \tag{4}$$

Thus, the proportionality constant m, the slope of the line in Fig. 2a, is just the fraction of demographic events in a population of abundance x^* due to density-dependent processes. This constant ranges from 0 when no demographic events result from density-dependent processes to 1 when all demographic events result from density-dependent processes.

Equation (3) has the undesirable property that whenever m is positive, $f(x)$ exceeds 1 for large enough values of x, a nonsensical feature for a quantity meant to represent a probability. Populations in our simulations reached such high abundances very rarely, but to prevent unrealistic behavior even at these few times, our model actually employs a slightly modified definition of f that lacks this undesirable property, namely,

$$f(x) = \min\left\{1, (1-m)f^* + \left[\frac{mf^*}{x^*}\right](x)\right\}.$$

Our collection of simulations examined mesopopulation consequences of density dependence in all four reef-associated demographic processes, both individually and in certain combinations. Explorations of mesopopulation variability under fixed conditions employed replicated runs with the same parameter values. For maximum contrast between sets of runs, density dependence in each demographic rate was either absent ($m = 0$) or complete ($m = 1$) throughout each set. Examinations of how mesopopulation behavior changes with alterations in the strength of density dependence employed sets of runs with graded values of m, but otherwise identical parameter values. For clarity, we describe in this chapter only a small subset of our rather large collection of numerical results that best illustrates the main features of the full collection.

It will not have escaped the reader's attention that our method necessarily rests on the tacit assumption that density dependence and adult movement do not influence average population abundance. Actually, our simulations suggest that this assumption is probably false. However, most correlations (not shown) between numerically calculated 10-year average population abundance and strength of density dependence under any adult movement regime proved to be quite weak and nonsignificant statistically. We interpret these findings as indicating that these correlations probably do not influence the model's main qualitative features to be discussed in the remainder of this chapter.

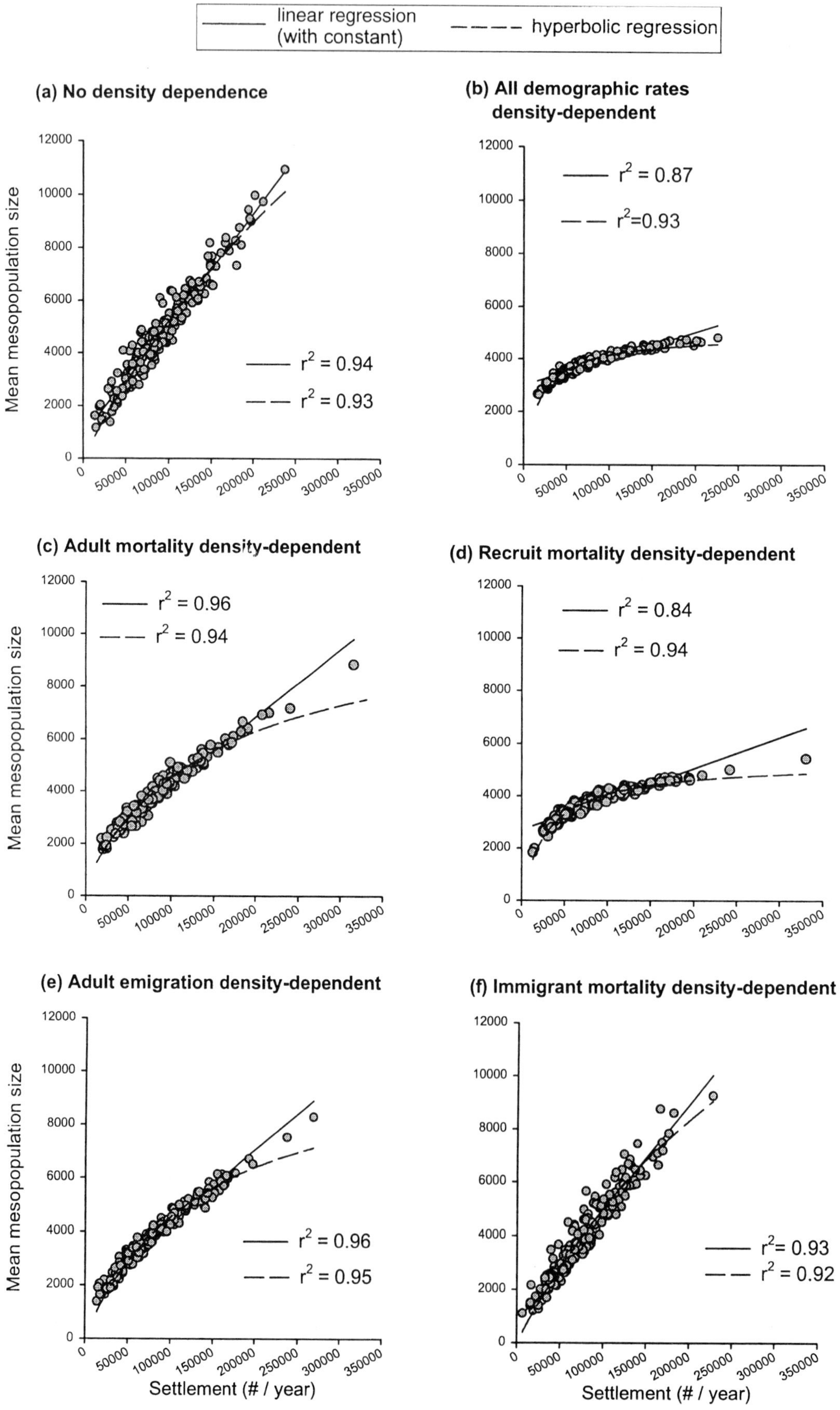

linear regression (with constant)
hyperbolic regression
(a) No density dependence
r² = 0.94
r² = 0.93
(b) All demographic rates density-dependent
r² = 0.87
r²=0.93
(c) Adult mortality density-dependent
r² = 0.96
r² = 0.94
(d) Recruit mortality density-dependent
r² = 0.84
r² = 0.94
(e) Adult emigration density-dependent
r² = 0.96
r² = 0.95
(f) Immigrant mortality density-dependent
r²= 0.93
r² = 0.92
Mean mesopopulation size
Settlement (# / year)

IV. Results from Model Simulations

A. Question 1: How Does Local-Scale Density Dependence in Different Demographic Rates Affect the Meso-Scale Relationship between Larval Supply and Abundance?

To address this question, the mean mesopopulation abundance in a given year was compared to the total settlement in that year. To evaluate the nature of the relationships between settlement and mesopopulation abundance, we fitted simple linear and nonlinear functions to the data. Our intent was to get a rough idea of how well each type of relationship fit the data, and not to provide a mechanistic description of the relationship. The nonlinear relationship we fitted to the data is the simple hyperbolic function

$$y = ax/(b + x)$$

also used to model enzyme kinetics (Real, 1977) and the type II functional response (Holling, 1959). This model fit the empirical results as well as either of two other simple nonlinear models we tried, and it can be algebraically rearranged to match the Beverton–Holt stock-recruitment function (Beverton and Holt, 1957).

For linear functions, we employed standard Model I regressions either with or without a constant. Regression without a constant is biologically appropriate because zero settlement certainly produces zero population abundance (Cushing, 1996), but it does not yield r^2 values that are comparable to those of the nonlinear functions we used. Regression with a constant yields r^2 values that are more directly comparable to the r^2 of the nonlinear model (Myers, 1986), and always fit the data better than a straight line forced through the origin. This was true even when there was no density dependence in any demographic rates. As expected under these circumstances, mesopopulation abundance appeared to be directly proportional to the settlement rate (Fig. 3). Linear relationships with and without a constant fit the data very well, and there was no improvement in r^2 shown by fitting a nonlinear function (Fig. 3a). However, the conclusion that the true relationship is actually nonlinear seems inescapable, because the best fitting straight line has a significantly positive intercept whereas the true relationship must pass through the origin. We suspect that the nonlinearity lies near the origin and has no biological significance in this case. Most likely, it arises from some combination of rounding continuous random variables to integers and lack of independence of successive years' mesopopulation averages in each run. The effects of either of these features should cause the greatest deviation from linearity at very low average settlement intensities. Understanding this issue will require further study.

In contrast, when all demographic rates were density dependent, the relationship between settlement and abundance was obviously and strongly nonlinear. Increased settlement appeared to result in progressively smaller increases in abundance as the settlement rate reached higher levels, and abundance appeared to approach an asymptote at very high settlement rates. In this case, the nonlinear regression was clearly a better fit to the data than was the linear model (Fig. 3b), and linear regression with a constant has a large positive intercept. In this case, the large positive intercept of the linear regression arises because our data sample only a small portion, located far from the origin, of an underlying function whose true shape is nonlinear.

Our results also indicate that the extent to which density dependence affects the relationship between settlement and abundance depends strongly on which demographic rates were density dependent. When the survival of immigrants was density dependent, the influence was negligible and mesopopulation size appeared to be directly proportional to the rate of larval settlement and the hyperbolic regression did not fit better than either linear regression model (Fig. 3f). When the mortality or migration of adult gobies was density dependent there was a definite but modest influence on mesopopulation size (Fig. 3c and e). These relationships were slightly nonlinear, so that any given increase in settlement resulted in a smaller increase in mesopopulation size than would be the case with no

FIGURE 3 Plots of mean mesopopulation size in a given year versus the total settlement that year. The six plots show cases in which (a) there is no density dependence or (b) all demographic rates are density dependent, and in which (c) just adult mortality, (d) just recruit mortality, (e) just adult emigration, or (f) just the mortality of adults immigrating to new reefs is density-dependent. Note that effect of density dependence (i.e., the curvature of the relationship) in cases c–f hinges on which rate is density dependent. Moreover, the effect of density dependence just in recruit mortality (d) is nearly equivalent to density dependence in all demographic rates (b). All six plots show cases in which migration among reefs occurs at normal rates. We have plotted data for each of the 10 years simulated in each of the 20 simulation runs for a given parameter set, yielding 200 points (years) per plot. Linear regression (with constant) and hyperbolic regression lines are fitted to the data.

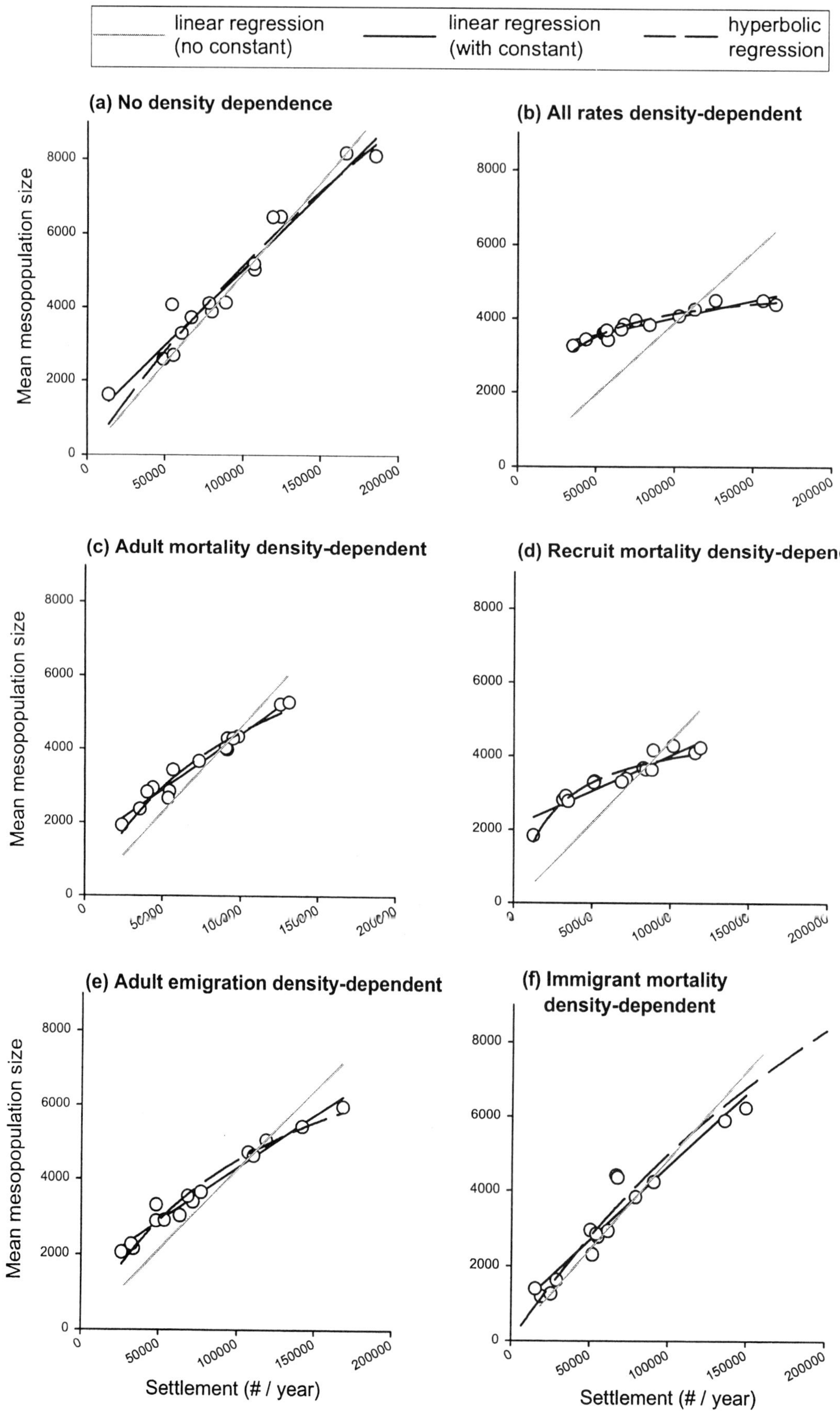

linear regression (no constant)
linear regression (with constant)
hyperbolic regression
(a) No density dependence
(b) All rates density-dependent
(c) Adult mortality density-dependent
(d) Recruit mortality density-dependent
(e) Adult emigration density-dependent
(f) Immigrant mortality density-dependent
Mean mesopopulation size
0
2000
4000
6000
8000
0
50000
100000
150000
200000
Settlement (# / year)

density dependence. In contrast, density dependence in recruit mortality strongly moderated abundance at the mesopopulation scale, and the effect was almost as great as that seen when all demographic rates are density dependent (Fig. 3d).

Using years as replicates yielded 200 data points with which to define the settlement–abundance relationship. The only published empirical study presenting such data was based on 7 data points, despite the fact that the study involved an extraordinary and impressive amount of effort on the part of the investigators (Doherty and Fowler, 1994a). We therefore also wanted to examine settlement–abundance relationships using smaller, more realistic, sample sizes. To do this, we randomly picked data for 1 year from 15 of the simulation runs for a given parameter set, so that we now had 15 data points per plot instead of 200 (Fig. 4). When we used smaller sample sizes, nonlinearity in the settlement–abundance relationship was less obvious (Fig. 4). Only when recruit mortality alone or when all demographic rates were density dependent was it still clear visually (Fig. 4, b and d). When density dependence affected only the mortality and movements of adults, curvilinearity was visually obvious with 200 data points (Fig. 3, c and e). With 15 data points, however, the departure from a straight line was minimal even when the linear regression was forced through the origin, and we believe would be difficult to demonstrate statistically (Fig. 4, c and e). In summary then, local-scale density dependence does cause curvilinearity in large-scale relationships between settlement and abundance, with the degree of curvature depending on which demographic rates are density dependent. The curvilinearity may, though, be sometimes hard to detect with the small sample sizes feasible in empirical studies.

B. Question 2: How Does Local-Scale Density Dependence in Different Demographic Rates Affect Temporal Fluctuations in Mesopopulation Abundance?

We addressed this question informally by inspecting plots showing how mesopopulation abundance fluctuates over time in the presence and absence of density dependence (Fig. 5). We conducted a more rigorous test of the stabilizing influences of density dependence by systematically varying the strength of density dependence in demographic rates. To accomplish this, simulations were run with density dependence in demographic rates set to 20 different levels, ranging from absent to complete (m was varied from 0 to 1 in increments of 0.05). One simulation was run at each level of density dependence and, as usual, each simulation yielded 10 years of data on goby abundance. Mesopopulation variability for a given strength of density dependence was measured as the coefficient of variation (CV) around the mean annual abundance (averaged across the 10 years of the simulation).

Inspection of fluctuations in abundance simulated with and without density dependence shows that goby abundance oscillated on an annual cycle under all conditions simulated because of the seasonality in larval settlement (Fig. 1) and the short life-span of the species. In the absence of density dependence, the amplitude of the oscillations was quite variable from year to year (Fig. 5a). Temporal fluctuations in mesopopulation abundance were, though, damped to varying degrees by spatial density dependence at the population level. Relationships between the strength of density dependence and mesopopulation variability were always negative (Fig. 6). The relationship was strong, and statistically significant, when all demographic rates were density-dependent (Fig. 6a) and the stabilizing effect was visually obvious in this case (Fig. 5b). When density dependence occurred in only one demographic rate, its influence depended on which rate was density dependent. Population-level spatial density dependence in recruit mortality had a visually obvious stabilizing effect on mesopopulation abundance (Fig. 5d). The stabilizing effect on mesopopulation dynamics was highly significant (Fig. 6c), and nearly as strong as when density dependence occurred in all demographic rates (compare Fig. 5, b and d and Fig. 6, a and c). A less visually obvious (Fig. 5f) but statistically significant stabilizing effect was apparent when only the survival of immigrants was related to population density (linear regression of CV on strength of density dependence; $F_{1,20} = 10.6$, $p = 0.004$, $r^2 = 0.33$). In contrast, when density-dependent mortality was experienced only by resident adults, the effect on mesopopulation dynamics

FIGURE 4 Plots of mean mesopopulation size in a given year versus the total settlement that year, as in Fig. 3, except that only 15 points selected randomly from the corresponding graph in Fig. 3 are shown in each case. Fifteen points is an optimistic estimate of the sample size possible in field studies that measure the relationship between mesopopulation size and settlement. Note in some cases in which the actual relationship is clearly curvilinear (see Fig. 3, c and e), with a sample size of 15, it is difficult to distinguish the curvilinear fit of those points from a linear fit (c and e). Linear regression (with and without a constant) and hyperbolic regression lines are fitted to the data.

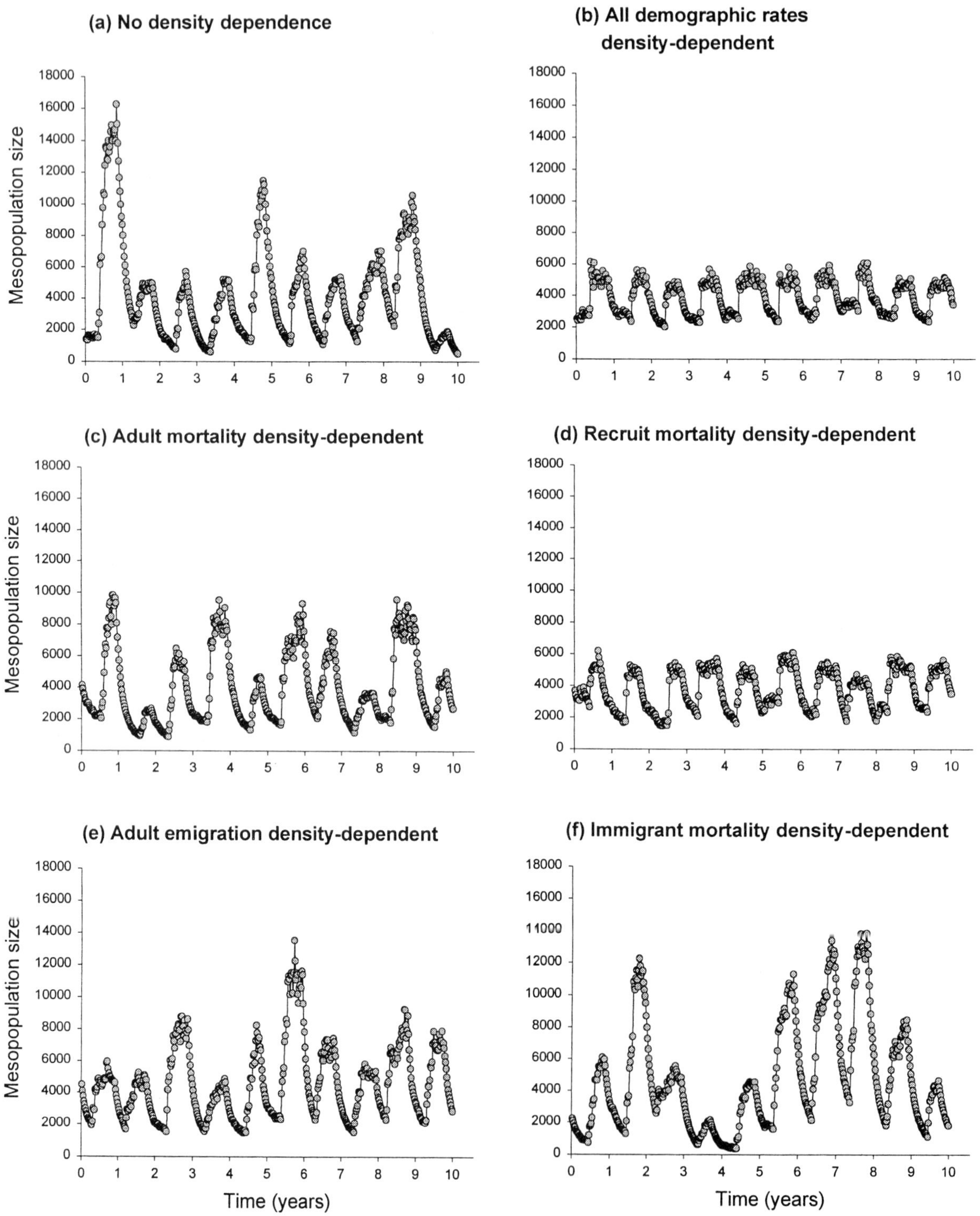

FIGURE 5 Plots of mesopopulation size over time for individual simulation runs under six different conditions. Plots differ in whether spatial density dependence at the local population level was (a) absent, (b) present in all demographic rates, (c) present in adult mortality only, (d) present in recruit mortality only, (e) present in adult emigration only, or (f) present in the mortality of immigrating adults. Note that the fluctuations in abundance were strongly dampened when recruit mortality (d) or all demographic rates were density dependent (b), and dampened to a lesser degree when other rates were related to density (c, e, and f). Each simulation run displayed was selected at random from the 20 runs generated using each parameter set. All populations experienced normal migration rates.

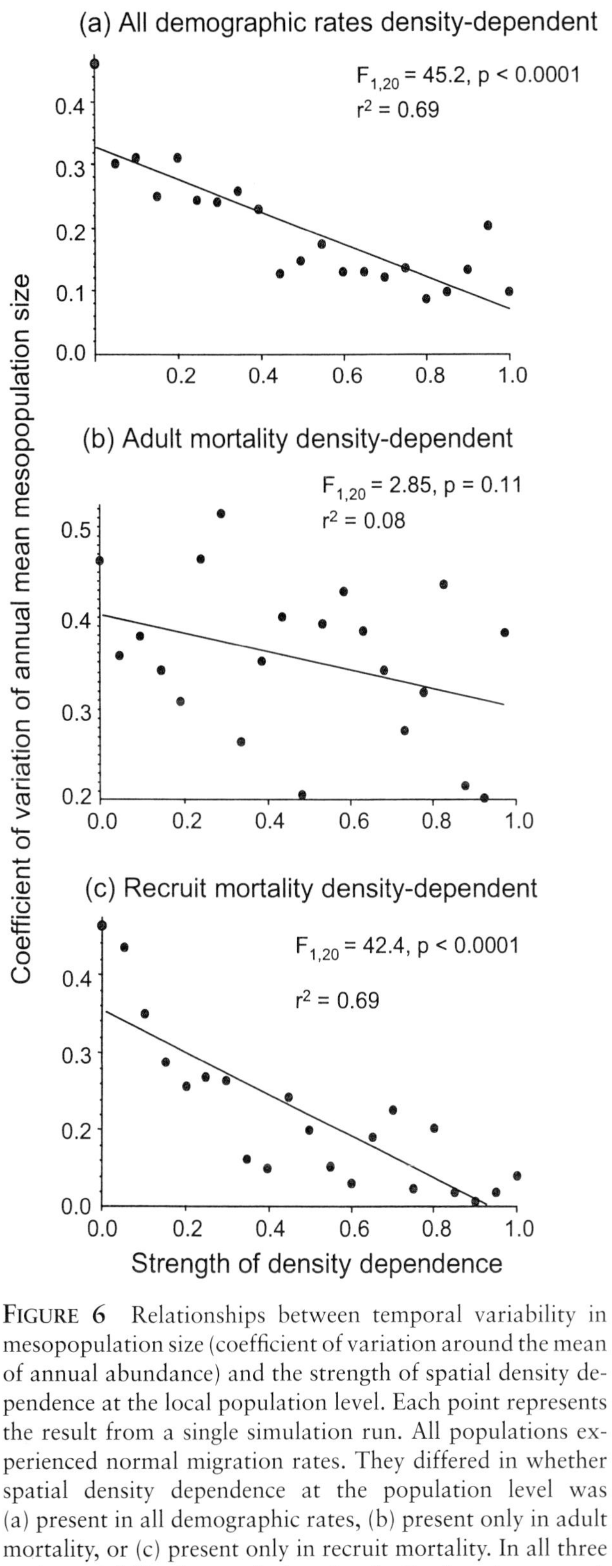

FIGURE 6 Relationships between temporal variability in mesopopulation size (coefficient of variation around the mean of annual abundance) and the strength of spatial density dependence at the local population level. Each point represents the result from a single simulation run. All populations experienced normal migration rates. They differed in whether spatial density dependence at the population level was (a) present in all demographic rates, (b) present only in adult mortality, or (c) present only in recruit mortality. In all three cases, variability in population size declined as the strength of density dependence increased; however, this decline was statistically significant only in a and c.

appeared to be slight (Fig. 5c) and, in fact, density-dependent adult mortality did not have a statistically significant dampening effect on fluctuations in abundance (Fig. 6b). Similarly, there was also no significant effect on mesopopulation dynamics of altering the degree of density dependence in adult emigration (Fig. 5e; linear regression of CV on strength of density dependence; $F_{1,20} = 2.41$, $p = 0.14$, $r^2 = 0.07$). In general, local-scale density dependence causes damping of temporal fluctuations in mesopopulation size, the strength depending on which rates were density dependent.

C. Question 3: Are Relationships between Density and Demographic Rates the Same for the Mesopopulation and Its Component Local Populations?

The demographic characteristics of the entire mesopopulation matched closely the characteristics of populations on individual reefs. This result is well illustrated by a simulation in which density dependence occurred only in adult mortality and all other demographic rates remained density independent (Fig. 7a). When density-dependent mortality occurred at the population level, it appeared also at the mesopopulation level. Per-capita mortality of adults throughout the mesopopulation increased significantly when their abundance over the entire reef array increased (Fig. 7a). The form of the function appears linear. In addition, this functional relationship between per-capita mortality rate and abundance at the mesopopulation level very closely resembles the population-level per-capita mortality rate function scaled up to the mesopopulation level (Fig. 7a). For any demographic rate in this model, this scaling up is accomplished simply by creating a new mesopopulation demographic function $F(X)$ by replacing x on the right-hand side of Eq. (3) with $X/441$, where X represents mesopopulation abundance and 441 is the number of censused reefs in the array. This extrapolation from the population to the mesopopulation level tacitly assumes that population abundances on individual reefs are always equal. Except at time 0, this condition is never satisfied, of course, and the slight difference between the fitted regression line and the scaled-up population function certainly arises from between-reef variation in population abundance.

When population-level density dependence occurred only in the adult mortality rate, all other mesopopulation-level demographic rates proved independent of mesopopulation density. For example, recruit mortality observed over the entire mesopopulation was density independent, as it was on individual

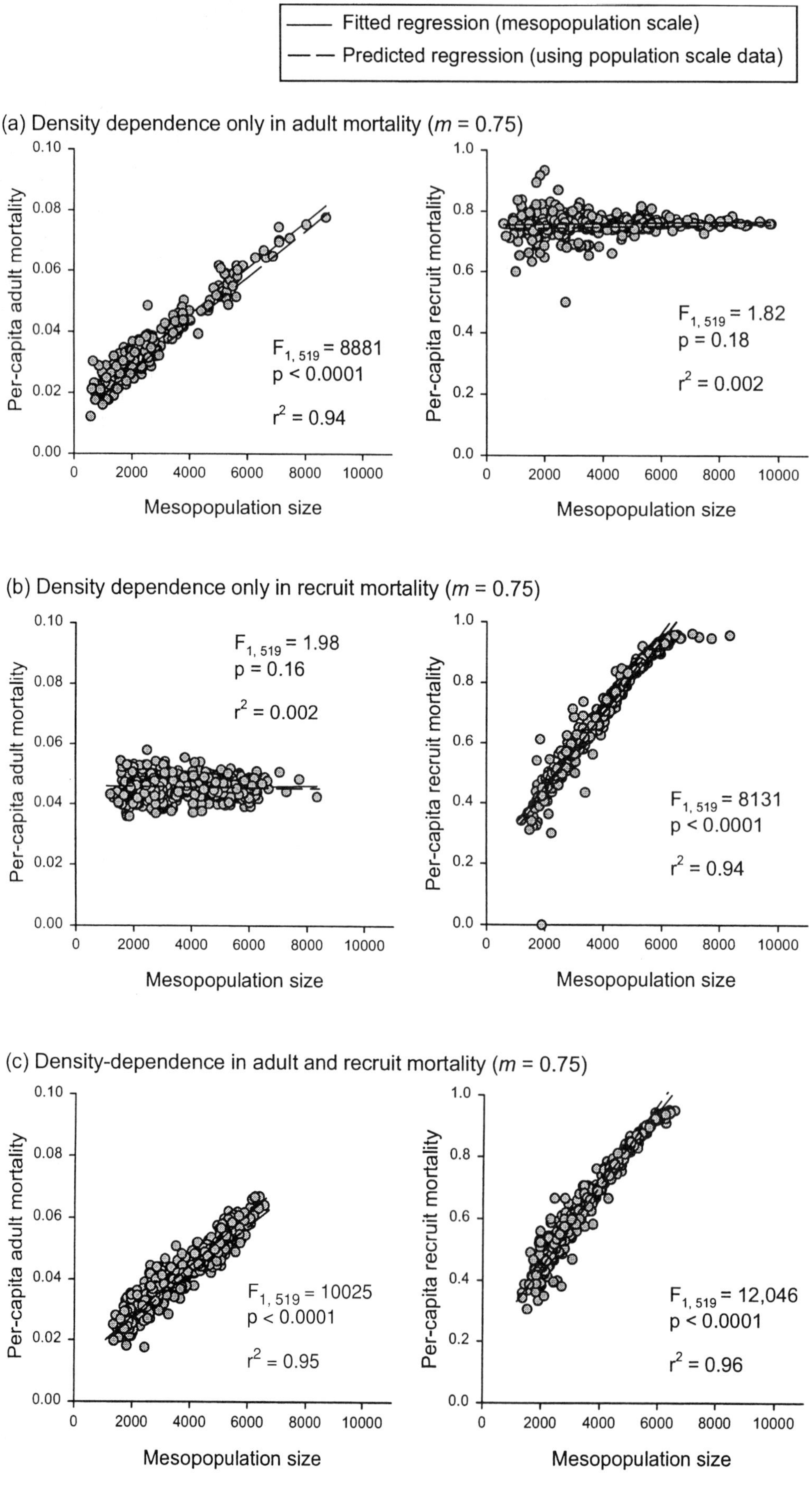
Fitted regression (mesopopulation scale)
Predicted regression (using population scale data)
(a) Density dependence only in adult mortality (m = 0.75)
Per-capita adult mortality
Mesopopulation size
$F_{1,519}$ = 8881
p < 0.0001
r^2 = 0.94
Per-capita recruit mortality
Mesopopulation size
$F_{1,519}$ = 1.82
p = 0.18
r^2 = 0.002
(b) Density dependence only in recruit mortality (m = 0.75)
Per-capita adult mortality
Mesopopulation size
$F_{1,519}$ = 1.98
p = 0.16
r^2 = 0.002
Per-capita recruit mortality
Mesopopulation size
$F_{1,519}$ = 8131
p < 0.0001
r^2 = 0.94
(c) Density-dependence in adult and recruit mortality (m = 0.75)
Per-capita adult mortality
Mesopopulation size
$F_{1,519}$ = 10025
p < 0.0001
r^2 = 0.95
Per-capita recruit mortality
Mesopopulation size
$F_{1,519}$ = 12,046
p < 0.0001
r^2 = 0.96

reefs (Fig. 7a). Again, the statistically fitted mesopopulation function closely matched the scaled-up population function. The same was true for emigration and emigrant survival (data not shown).

Similar patterns arose in simulations involving density dependence in other demographic rates. By way of illustration, we consider a simulation in which only the mortality of recruits was density dependent (Fig. 7b). When the loss during recruitment was higher on high-density patches of reef, per-capita recruit mortality across the mesopopulation was also positively related to abundance across the entire reef array. The large-scale relationship between per-capita recruit mortality and density again closely resembled the scaled-up population-level function. In this case, deviations from linearity arose at very high mesopopulation densities, at which the recruitment mortality probability in most individual populations reached its maximum value of 1, as specified by Eq. (4). As before, other demographic rates that were independent of density at the population level remained so at the mesopopulation level, and observed rates closely matched scaled-up population-level rates (Fig. 7b).

The close match between demographic rate functions at small and large scales was preserved when more than one population-level demographic rate depended on abundance. For example, the simulation involving population-level density dependence in mortality of both adults and recruits displayed similar density dependence at the mesopopulation level (Fig. 7c). This pattern prevailed with density dependence in all possible combinations of demographic rates. In all cases, observed mesopopulation per-capita rate functions closely resembled scaled-up versions of the corresponding population-level functions. The most conspicuous deviations from linearity again occurred at mesopopulation densities sufficiently high to cause complete mortality in many individual populations. Simply put, demographic functions developed using data on small habitat patches always "scaled up" accurately to populations at larger spatial scales.

TABLE 1 Linear Regression Statistics[a]

Migration rate	Regression equation	r^2	P
Zero	$y = 0.040x + 658$	0.95	<0.0001
Normal	$y = 0.042x + 877$	0.94	<0.0001
High	$y = 0.033x + 311$	0.98	<0.0001

[a]These regression statistics describe relationships between mean mesopopulation size in a given year (y) versus the total settlement that year (x) when all demographic rates are density independent ($n = 200$). Results are shown for three different rates of migration among individual reefs.

D. Question 4: How Does the Amount of Adult Migration between Local Populations Influence the Answers to Questions 1–3?

Not surprisingly, varying the rate of movement among reefs had little influence on the large-scale relationship between settlement and abundance in the absence of small-scale density dependence, (Table 1). At each of the three levels of migration we simulated, the relationship was very well described by a straight line and the regression equations best fitting the data are quite similar (Table 1).

When fishes on individual reefs experienced density-dependent mortality, an increase in migration among patches of reef always reduced the degree of curvature in the large-scale relationship between settlement and abundance, and increased the level at which the mesopopulation abundance appeared to level off at high settlement rates. The quantitative influence of migration depended on whether it was recruits or adults that suffered density-dependent mortality. To illustrate this point, we consider simulations for which small-scale spatial density dependence was occurring in either adult mortality only, or in recruit mortality only (Fig. 8). In both cases, the large-scale relationship between settlement and abundance is more curvilinear and the asymptotic mesopopulation size is lower when gobies remain on the reef to which they settle as larvae,

FIGURE 7 Relationships between demographic rates and abundance at the mesopopulation level. Per-capita demographic rates were calculated for each week during simulation runs, 10 years in duration each (yielding 520 points per graph). Gobies migrated at normal rates during all simulations, and the strength of density dependence (m) was always 0.75. Displayed are simulations run when density dependence occurred (a) only in adult mortality, (b) only in recruit mortality, and (c) in both demographic rates. Relationships between demographic rates and abundance at the mesopopulation level are defined by linear regression (solid lines), with regression statistics shown on each plot. Also shown is the mesopopulation-level relationship predicted from the underlying function at the population level (dashed line). Note that the observed relationships differ little, if at all, from the predicted relationships; in other words, density dependence at the local scale "scaled up" almost perfectly to the mesopopulation scale.

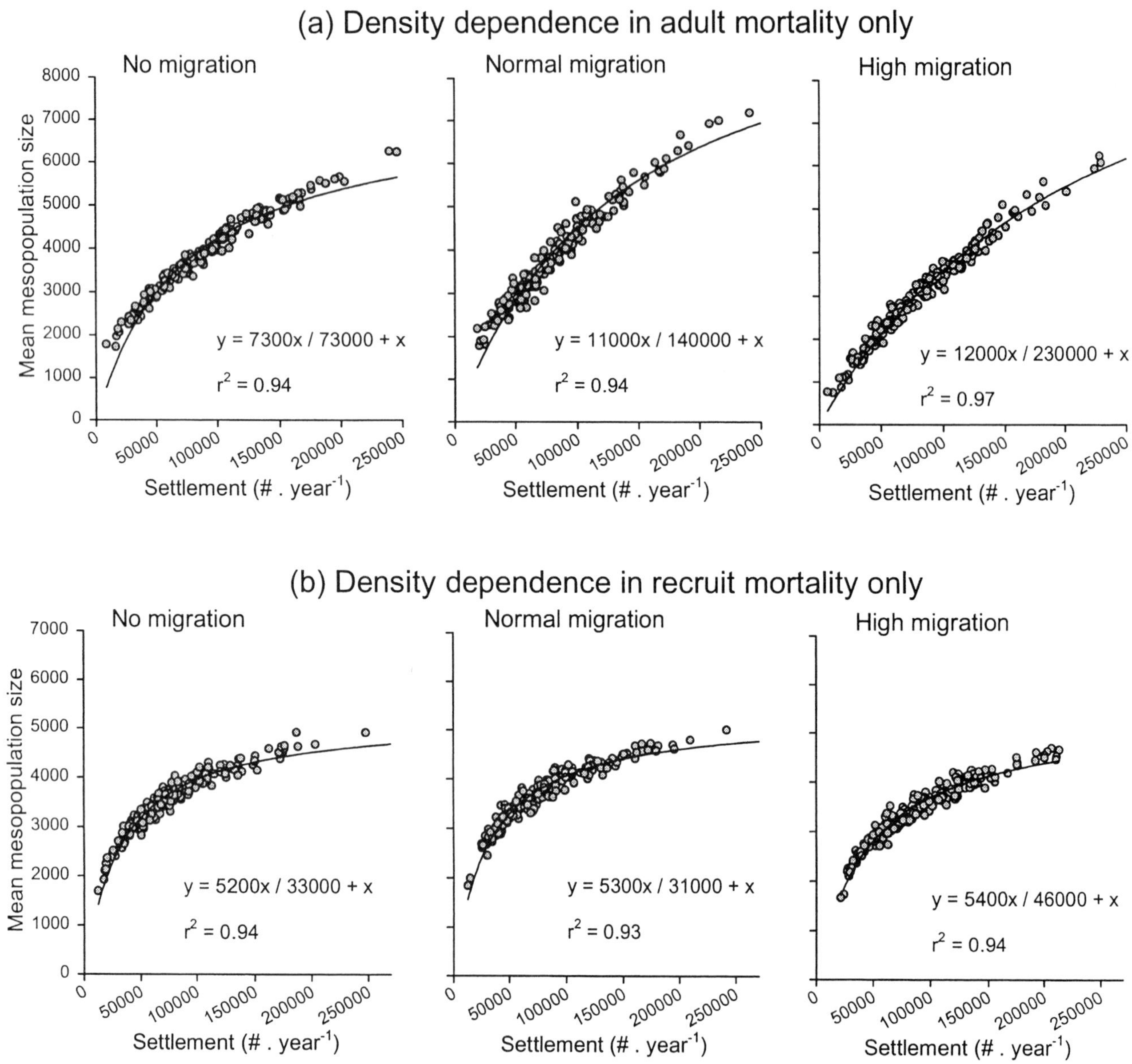

FIGURE 8 Plots of mean mesopopulation size in a given year versus the total settlement that year. The plots show cases in which there is density dependence (a) just in adult mortality or (b) just in recruit mortality. In both cases, plots are displayed for cases in which migration among reefs did not occur, occurred at normal rates measured for bridled gobies, or occurred at five times the normal rate. Note that variation in the rate of migration had no effect on the qualitative relationship between mesopopulation size and settlement rate, but it did quantitatively affect the relationships between these variables, particularly when only adult mortality was density dependent (a). Data are plotted for each of the 10 years simulated in each of the 20 simulation runs for a given parameter set, yielding 200 points (years) per graph. Hyperbolic regression lines are fitted to the data.

compared to situations when they are able subsequently to migrate among reefs. This effect of increased migration is visually obvious when density dependence was restricted to adult mortality, and is apparent from the changes in the parameters of the hyperbolic regression fitted to the settlement–abundance relationships (Fig. 8a). Increased migration among patches of reef had a much less obvious effect when density-dependent mortality occurred during recruitment. In this case, the degree of curvature in the relationship between settlement and mesopopulation size, and the abundance at which mesopopulation size appears to "level off" at high settlement rates, did not vary appreciably with the migration rate (Fig. 8b). As a result, the regression equations best fitting the relationships between settlement and abundance were similar at each of the migration rates simulated (Fig. 8b).

When movement among individual reefs was density dependent, either because emigration was related to density or because the survival of immigrants declined at high densities, increasing the overall rate of movement exaggerated the large-scale consequences of density dependence (Fig. 9). Specifically, an increase in migration among patches of reef from "normal" to "high" levels increased the curvature in the large-scale relationship between settlement and abundance, and lowered the level at which the mesopopulation abundance appeared to asymptote at high settlement rates (Fig. 9).

We evaluated the effect of migration on mesopopulation stability using simulations in which we varied both the strength of density dependence in demographic rates and the rate at which adult fishes moved among reefs. Simulations were run when density dependence occurred only in adult mortality, or only in recruit mortality. Twenty simulations were run with density dependence in the specified demographic rate varying progressively from absent to complete. One simulation was run at each level of density dependence and each simulation yielded 10 years of data on goby abundance. Mesopopulation variability for a given strength of density dependence was measured as the coefficient of variation around mean annual abundance (averaged across the 10 years of the simulation). To test the influence of migration, we ran a set of 20 simulations at each level of migration among reefs (zero, normal, and high).

The overall rate of adult migration did not markedly affect temporal mesopopulation variability, regardless of which demographic rates were density dependent. For example, when adult gobies experienced density-dependent mortality on reefs, there was a stabilizing effect on fluctuations in abundance at the mesopopulation level. This was true whether migration was simulated at the "normal" levels estimated from our empirical data, was set to zero, or was simulated at higher rate than normal (Fig. 10a). We compared the relationship between mesopopulation variability and the strength of density dependence among the different migration regimes using analysis of covariance (ANCOVA). This analysis confirmed a significant negative relationship between mesopopulation variability and the strength of density-dependent mortality (indicated by a significant effect of "density dependence" in the ANCOVA, Table 2). There was, however, no detectable influence of the movement rate on the slope or elevation of the relationship (indicated by a nonsignificant "migration" term, and a nonsignificant "interaction" term in the ANCOVA, Table 2). A similar pattern emerged from simulations in which small-scale density dependence was restricted only to recruit mortality. The stabilizing influence on mesopopulation dynamics appeared slightly stronger and more consistent under these circumstances (Fig. 10b). There was still, however, no significant effect of changing the rate of migration on mesopopulation dynamics (Table 2). Overall, then, changes in the rate at which fishes redistributed themselves among local populations had quantitative effects on mesopopulation dynamics that were sometimes very subtle and other times quite pronounced. There were, however, no qualitative changes in mesopopulation dynamics.

V. Discussion

A. Local-Scale Density Dependence Does Affect Population Dynamics at Large Scales

Overall, our model predicts that density-dependent interactions occurring among groups of fishes occupying small patches of reef, a few meters across, are not irrelevant when we expand our field of vision to areas that are kilometers in extent. As a result of local-scale density dependence, mesopopulation abundance ought to show a nonlinear relationship with settlement so that mesopopulation size will begin to level off when settlement rates are very high. Correspondingly, the mesopopulation should fluctuate within bounds that are sufficient to prevent abundance from precisely tracking variation in settlement. Our results are thus in accord with other models for open populations that incorporate variable recruitment and density dependence, but do not address multiple spatial scales (Warner and Hughes, 1988; Holm, 1990; Pfister, 1996).

B. The Effects of Density Dependence Vary Depending on Which Demographic Rates Are Related to Density

Interestingly, the model indicated that the quantitative influences of local density dependence on mesopopulation abundance were strongly conditional on which demographic rates were functions of density. These differences arose from the specifics of bridled goby demography that we used to parameterize the model. For example, density dependence in recruit mortality

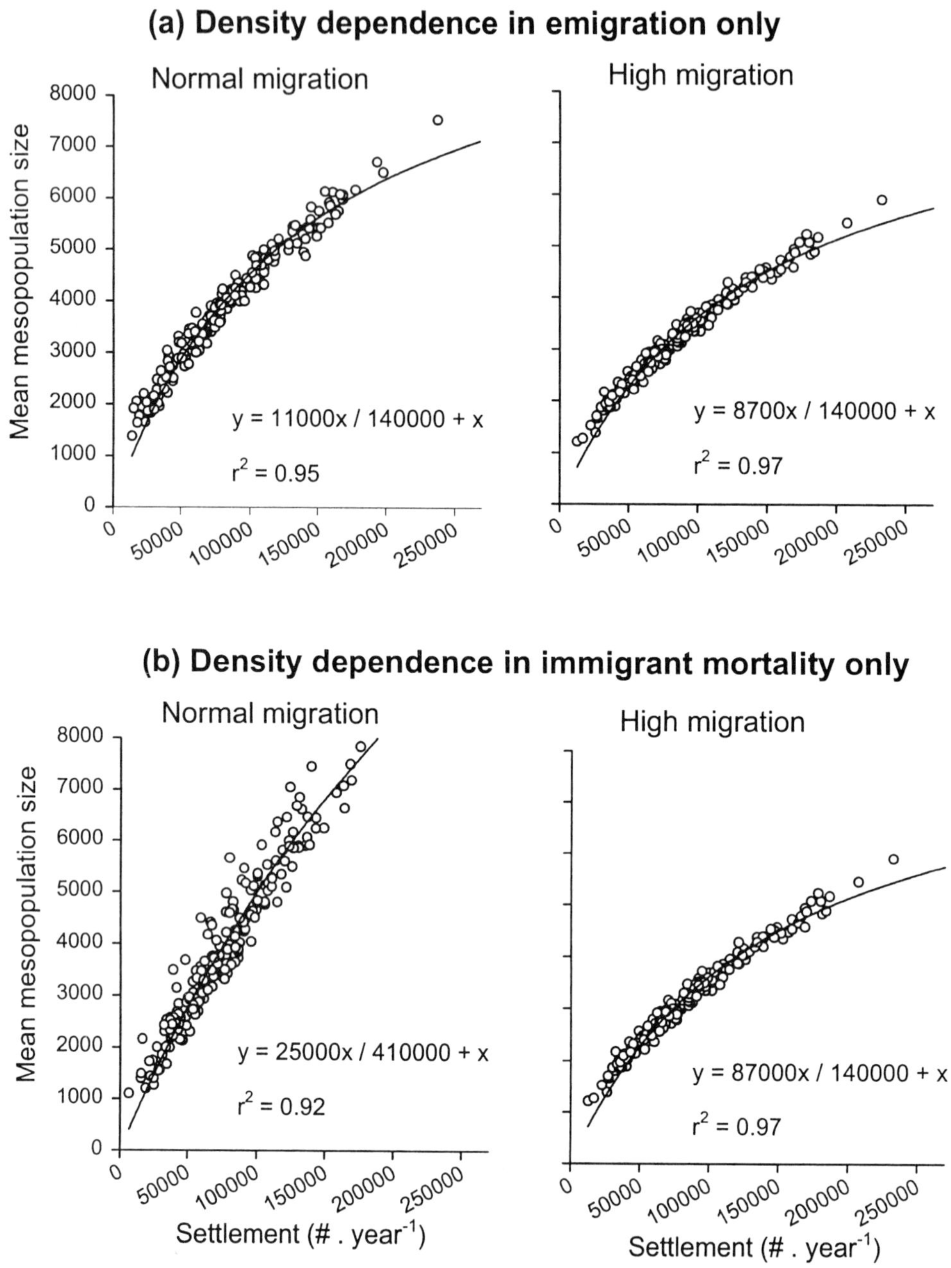

FIGURE 9 Plots of mean mesopopulation size in a given year versus the total settlement that year. The plots show cases in which there is density dependence (a) just in adult emigration or (b) just in the mortality of adults immigrating to new patch reefs. In both cases, plots are displayed for cases in which migration among reefs occurred at normal rates measured for bridled gobies, or occurred at five times the normal rate. For both cases (emigration and immigrant survival), increasing the rate of migration strengthened the effect of density dependence (i.e., lowered the asymptotic density of the mesopopulation). Data are plotted for each of the 10 years simulated in each of the 20 simulation runs for a given parameter set, yielding 200 points (years) per graph. Hyperbolic regression lines are fitted to the data.

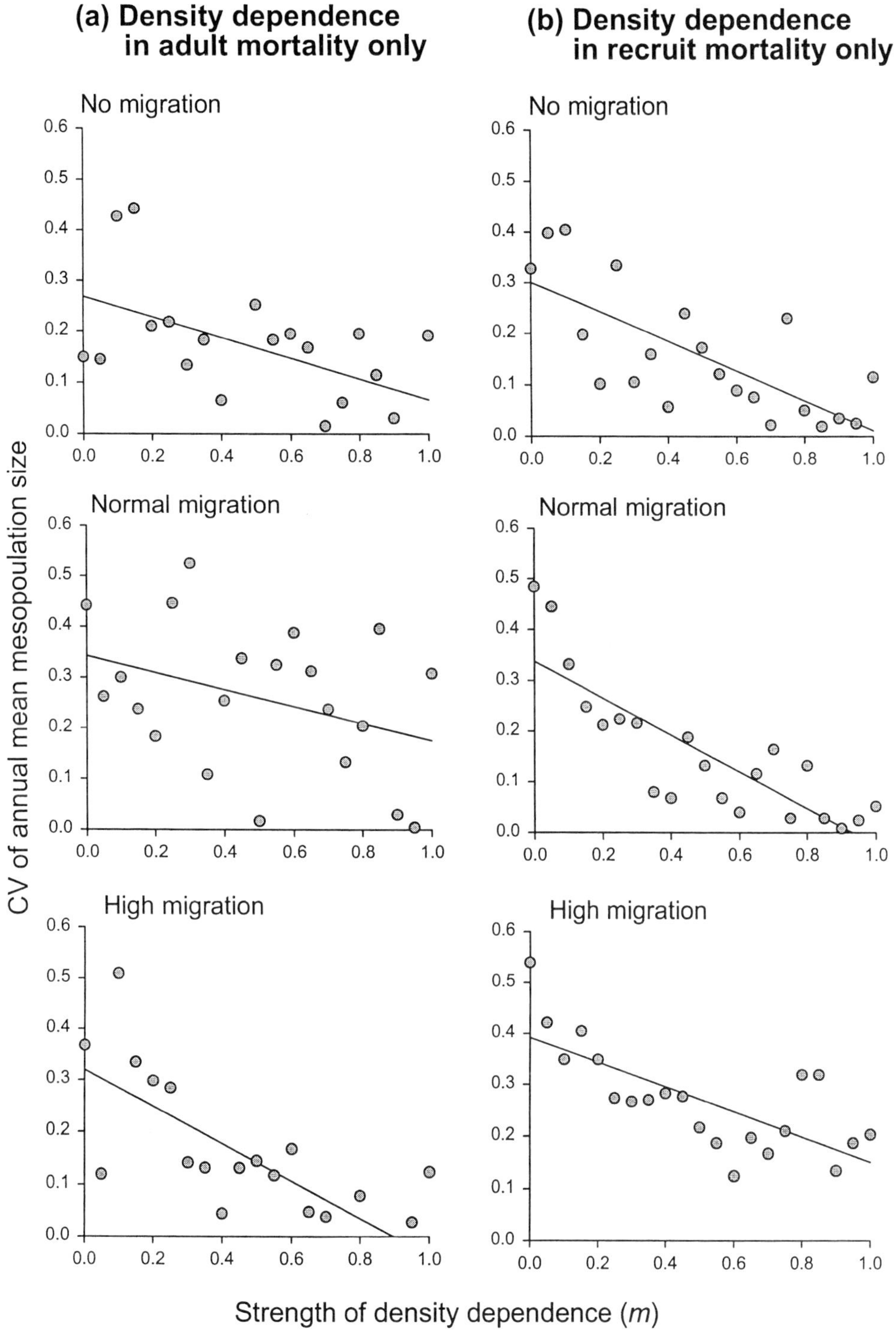

FIGURE 10 Relationships between temporal variability in mesopopulation size (coefficient of variation around the mean of annual abundance) and the strength of spatial density dependence at the population level (m). Each point represents the result from a single simulation run. Populations differed in whether spatial density dependence at the population level was (a) present in adult mortality only or (b) present in recruit mortality only. In each case, plots are displayed for simulations during which gobies migrated among reefs at three different rates: zero, normal, and high (five times normal). Variation in the rate of migration did not alter the moderating influence of density dependence on fluctuations in mesopopulation size (i.e., slopes of the relationships did not differ among the three levels of migration; see Table 2).

TABLE 2 Analyses of Covariance Testing the Effect of Migration Rate on the Relationship between Mesopopulation Variability and the Strength of Density Dependence[a]

Source	Sum-of-squares	df	Mean square	*F* ratio	*P*
	Density dependence only in adult mortality				
Migration	0.017	2	0.008	0.64	0.529
Density dependence	0.336	1	0.336	25.72	0.000
Interaction	0.039	2	0.019	1.49	0.234
Error	0.732	56	0.013	—	—
	Density dependence only in recruit mortality				
Migration	0.024	2	0.012	1.96	0.150
Density dependence	0.510	1	0.510	84.74	0.000
Interaction	0.014	2	0.007	1.17	0.319
Error	0.343	57	0.006	—	—

[a]Temporal variability in mesopopulation size was measured as the CV around the annual mean of mesopopulation size measured during a 10-year simulation. The strength of density dependence was varied systematically from absent ($m = 0$) to complete ($m = 1$). Migration of gobies among reefs was set to one of three levels: zero, normal, or high.

exerted a greater influence on mesopopulation properties than did density dependence in adult mortality, simply because the death rate of gobies is higher early in life. Similarly, density dependence in migration (when migration occurred at normal rates) exerted a relatively minor influence because adult gobies do not redistribute themselves among patches very frequently. More interestingly, density dependence in all four demographic rates together exerted a stronger moderating influence on mesopopulation abundance than did density dependence in any one rate alone. However, the combined effect was far less than additive because the number of individuals that experienced each demographic process had already been moved closer to the long-term average population abundance by density dependence in earlier demographic events. These findings are enlightening for those of us who have conducted density manipulations and measured responses over limited periods in the life cycle. They highlight the value of models in assessing the relative "importance" of such interactions for population dynamics.

C. Local Density Dependence Should Lead to Population Regulation

Most of the other open-population models for marine species are tailored to organisms for whom space is clearly limited (such as barnacles on the rocky shore or kelp plants on hard substrata), and therefore incorporate density dependence in a form that may not be applicable to most reef fishes (Hughes, 1984; Roughgarden *et al.*, 1985; Bence and Nisbet, 1989; Nisbet and Bence, 1989; Possingham *et al.*, 1994; Johnson, 2000). In these space-limited marine communities, adults inhibit settlement by occupying space necessary for settlement and this has a strong influence on population dynamics. Settlers arriving at one point in time inhibit settlement at some point in the future once they grow to become adults. The time lag between settlement and the inhibition of later settlement can introduce cyclic fluctuations in abundance into what is otherwise a stable population (Bence and Nisbet, 1989) [but see also Possingham *et al.* (1994), for an alternative explanation]. The same sort of time-lagged inhibitory interactions at the time of settlement have not been documented in reef fishes, and space limitation is unlikely to operate in the same fashion in reef fishes. We do know, however, that adult reef fishes can have other types of inhibitory effect on juveniles (e.g., Sale, 1976; Jones, 1987a,b; Tupper and Boutilier, 1995b). It would be informative to explore the consequences of these interactions through the inclusion of more stage structure in future models.

Apart from the specific form of density dependence caused by space limitation, other agents of density-dependent mortality tend to stabilize abundance in previous models of open marine populations (Gaines and Lafferty, 1995), and density-dependent mortality consistently had the same effect on the mesopopulation in our simulations. The important implication is that local density dependence in reef fishes is of the sort that ought to regulate abundance and so may contribute to the global, long-term persistence of reef fish populations.

Of course, because our simulated mesopopulation is demographically open, it shared another feature with previous models for open marine populations—that the absolute settlement rate was independent of local population density. This pattern of simulated settlement was based on our field observations of bridled gobies and on published data on other reef fishes and so is realistic. Density independence in the absolute rate of settlement also stabilizes populations because it causes per-capita rates of settlement to be density dependent (Hughes, 1984; Roughgarden *et al.*, 1985; Bence and Nisbet, 1989), but knowing this does not help us understand population regulation unless we know what controls the settlement rate (Chesson, 1996). In other words, it means we need to identify the regulatory interactions that occur at scales larger than the mesopopulation and that put bounds on the settlement rate.

D. Migration among Local Populations Does Not Obscure the Effects of Local-Scale Density Dependence

Encouragingly, the model results predict that changes in the overall rate at which adult fishes move among reefs should not cause qualitative changes in mesopopulation dynamics. For example, allowing fishes to redistribute themselves among patches did not negate the influence of local density-dependent interactions, as some workers have speculated, though the influences of density-dependent recruit mortality and adult mortality were reduced somewhat when migration rates were highest. Local interactions detected by experiments should not, therefore, be dismissed *carte blanche* as irrelevant to population dynamics at large scales. The potential influences of local movement are, in reality, more complex than to simply to homogenize patches, because the propensity to move and the ability to relocate successfully may be functions of population density. Density dependence in the rate at which fishes leave reefs, and in their chance of successfully relocating to a new reef, both had a measurable effect on mesopopulation abundance. Not surprisingly, the strength of these regulatory effects increased when movement became more frequent.

E. Local Density Dependence Scales Up Accurately

The close similarity of mesopopulation-level demographic rates in our simulations to scaled-up population-level demographic rate functions is perhaps somewhat surprising in light of Chesson's (1996, 1998a) perceptive observations about spatial averaging of nonlinear population growth rates. Chesson points out that between-population variation can potentially cause growth of an assemblage of interconnected local populations to differ qualitatively from growth of a single isolated population. Biologically, this difference arises from the fact that the average individual selected from a collection of local populations experiences a higher local density than occurs in the average local population. Mathematically, it arises because, with nonlinear functions, the operations of arithmetic averaging and function evaluation do not commute; that is, the average of a function is not in general equal to the function of the average.

Although Chesson states his observations in terms of complete population growth functions, his argument applies also to their components, namely, the rates at which various demographic processes occur. Calculations similar to his (not shown) establish that the true mesopopulation-level per-capita demographic rates in our model are actually not linear functions of mesopopulation density, even when the corresponding population-level demographic rate functions are. Rather, deviations from linearity occur, and their magnitudes increase, with the variance in local population abundance. However, our numerical results, which employ parameters we consider realistic for the bridled goby, show that deviations from linearity appear so slight as to be visually indistinguishable in appropriate graphs (e.g., Fig. 7). The sole exception to this pattern arises at very rarely achieved mesopopulation abundances that are high enough to kill most individuals present in most local populations. Except at such high densities, enlarging the perspective in our bridled goby model from the population level to the mesopopulation level preserves not just the qualitative forms of the per-capita demographic rate functions, but also their actual numerical values. That is, numerically calculated mesopopulation demographic rates fall very close to the population demographic rate functions scaled up to the mesopopulation level.

How faithfully scaling up population properties reproduces corresponding properties at the mesopopulation level (and the metapopulation level) in other fish species (and other organisms in general) is a topic that richly deserves further study. Chesson's (1996, 1998a) calculations suggest that accuracy of scaled-up demographic functions will deteriorate as local populations become more variable. Whether this subtle mathematical principle will exert a measurable effect when accompanied by other powerful natural processes that operate at larger spatial scales (such as nonrandom movement of large ocean water masses) remains to be learned.

F. Does the Form of Density Dependence in Reef Fishes Match That in Our Model?

How closely our model results resemble the properties of real mesopopulations depends entirely on whether our modeling assumptions are realistic. A key question is whether the functions relating demographic rates to population density are a good fit to relationships derived from field data. We have found that the relationship between the finite rate of mortality in bridled gobies and their population density closely approximates the function we chose for the model (G. E. Forrester, R. R. Vance, and M. A. Steele, unpublished). Migration in bridled gobies can also be density dependent (G. E. Forrester, unpublished), but the data are insufficient to define the form of the relationship. We need to better define these relationships for bridled gobies and test how well they describe these same relationships in other species. Density manipulations with this purpose in mind should treat population density as a continuous variable (e. g., Forrester, 1995; Steele, 1997b; Schmitt and Holbrook, 1999a) rather than as a categorical variable with just a few levels (e.g., Jones, 1987a,b; Forrester, 1990).

G. Do Our Findings Apply to Reef Fishes in General?

Although our efforts in this chapter were focused on simulating bridled goby populations, the model could easily be parameterized with demographic data from other species. In this way, it would be possible to make comparisons among species and ask questions about the generality of our findings. For example, how might changing factors such as the longevity of the species, or the seasonality of recruitment, influence mesopopulation dynamics? The substantial body of published research on damselfishes could easily be used for this purpose, and would provide a useful contrast to our results. Another possibility would be to utilize demographic data on larger species that are the subject of commercial and artisanal fisheries (e.g., Matheson and Huntsman, 1984; Polovina and Ralston, 1987, and references therein; Acosta and Appeldoorn, 1992; Bullock and Murphy, 1994; Kara and Derbal, 1995; Chakraborty and Vidyasagar, 1996; Rocha-Olivares, 1998). We desperately need information on the population ecology of many such species in order to more effectively manage their exploitation and conserve their populations (Russ, 1991). These species are, however, much harder to study in the field compared to the smaller species that we know the most about, and are usually studied using methods borrowed from temperate fisheries biology (e.g., Polovina and Ralston, 1987, and references therein). Along with other colleagues (e.g., Doherty and Fowler, 1994a; Hixon and Carr, 1997), we have sometimes argued that small reef fishes are good "model systems" that provide insights into the ecology of larger exploited reef fishes, and of demersal fishes in general (usually we have made this argument when trying to secure funding for our work on small species!). Frankly, we have seen no evidence that either supports or refutes this contention. If, however, gobies and damselfishes are good models for understanding groupers and snappers, then our model ought to predict qualitatively similar population dynamics when parameterized with demographic rates appropriate for both small and large species.

H. How Should We Best Test for Density Dependence at Large Scales?

Testing for density dependence at large spatial scales, and assessing its effects on dynamics, would obviously be a lot of work, and so one of our major goals was to gain some insight into how we can best perform such a test in the field. Testing for the regulation of mesopopulation abundance by examining the relationship between settlement (not recruitment) and adult abundance is feasible, though logistically demanding (M. A. Steele and G. E. Forrester, unpublished data). In our simulations, the nonlinearity in this relationship caused by local density dependence was not always visually obvious with a realistically small sample size (15 points), and the relationship would be hard to differentiate statistically from a straight line. The two key messages to field ecologists are, therefore; that (1) exploring the relationship between adult abundance and settlement may be an insensitive, and ineffective, way to test for the regulation of mesopopulations, and that (2) apparently linear relationships between juvenile and adult abundance (e.g., Doherty and Fowler, 1994a) do not rule out density dependence.

To be more optimistic, our model suggests that relationships between population size and demographic rates scale up very well. Significant relationships between mortality and density at the local scale have been detected with as few as eight replicates (e.g., Forrester, 1995), so collecting mortality data at large scales and relating death rates to population size may perhaps be a more effective way to test for density dependence in real mesopopulations. The discovery of annual growth rings in some reef fishes (e.g., Fowler and Doherty, 1992) is significant in this context, because it means that constructing an age-specific mortality schedule for a mesopopulation is feasible. It then would be possible to use the conventional life-table approach (e.g., Varley *et al.*, 1973) to test for regulation of a real mesopopulation. Ironically, one of the criticisms of this method in other systems is that it can detect temporal density dependence among generations, but often fails to detect exactly the sort of spatially localized, density-dependent interactions occurring within specific life history stages that we know to be quite common among reef fishes (Hassell, 1986, 1987). What has been a drawback in some systems is thus a boon to those of us reef fish ecologists eager to understand the long-term dynamical behavior of populations.

Of course, we really need to collect long-term census data on reef fishes to assess adequately whether our model simulations capture the essential features of population dynamics. For many species this will take a concerted and patient effort because they have lives that are not much shorter than our own. We make this point because of the rapid expansion of monitoring programs on coral reefs in most parts of the world. So far, most of the programs are being designed to collect qualitative or semiquantitative community-level data. We think there is a case for including quantitative censuses in these programs, perhaps of just a few well-chosen species, and that the information gained might improve our children's ability to manage and conserve populations of reef fishes.

Acknowledgments

Many thanks to all of those who have helped us with our empirical field work on gobies, including G. Almany, R. Buckley, S. Bull, B. Byrne, D. Canestro, M. Carr, L. Conway-Cranos, B. Evans, B. Finley, R. Frodsham, I. Greenspan, M. Hixon, L. Jarecki, A. Karpenske, E. Kintzing, A. LaBonte, L. O'Bryan, C. McKinney, J. Schinske, Y. Springer, J. Standish, C. Tinus, C. Tran, E. Tynes, D. Weisman, and C. Wormald. We are also grateful for the excellent logistical support provided by the staffs at the Caribbean Marine Research Center and Guana Island. This research was supported by grants to GEF from the National Science Foundation (OCE 96-18011), the Perry Foundation, and the Guana Island Wildlife Sanctuary and the Falconwood Corporation.

CHAPTER 14

Density Dependence in Reef Fish Populations

Mark A. Hixon and Michael S. Webster
Department of Zoology
Oregon State University
Corvallis, Oregon 97331

Counting fish is as easy as counting trees, except that fish are invisible and they move.

—J. G. Shepherd

I. Introduction

One of the greatest mysteries of the sea is the question of what processes drive and regulate the dynamics of marine fish populations. This issue has troubled researchers in the fields of fisheries biology and marine ecology since these disciplines originated over a century ago (Smith, 1994). Although numerous concepts and hypotheses related to this problem have been proposed, only in recent years have detailed demographic studies, and especially manipulative experiments, provided partial empirical answers. Many questions and major controversies remain. In fact, even whether density dependence is necessary (as opposed to sufficient) for long-term persistence of populations has been questioned by some reef fish ecologists (e.g., Sale and Tolimieri, 2000) (see also Chapter 16, this volume). Yet, in these days of fishery collapses on a global scale (National Research Council, 1999), and the heightened risk of extinction of marine fish species due to human activities (Roberts and Hawkins, 1999; Hawkins *et al.*, 2000), probing these questions has become more crucial than ever before. This is because the scientific basis for conservation of marine species, from conventional fisheries management to the implementation of marine protected areas (see Chapter 19, this volume), lies in understanding and conserving naturally regulating density-dependent mechanisms.

How (if at all) does the concept of demographic density dependence apply to marine fishes and why is this concept controversial? How can coral reef species serve as model systems for examining density dependence in marine fish populations, and how do data from reef populations provide understanding of underlying mechanisms? What is needed in the future to resolve the mystery of fish population dynamics and regulation in the sea? This chapter examines these questions, focusing in turn on essential concepts, empirical evidence, conceptual synthesis, and future research needs. We consider coral reef fishes in the context of all demersal (bottom-associated) marine fishes. Pelagic (open ocean) fish populations, although conceptually relevant, are beyond the scope of this review.

II. Essential Concepts

We believe that much of the controversy surrounding the notion of density dependence in demersal marine fishes, and especially coral reef fishes, is due to misunderstanding (or at least ambiguous definitions) of key concepts and relevant spatiotemporal scales. Indeed, although most population ecologists accept the existence of density dependence as necessary for any population to persist indefinitely (Cappuccino and Price, 1995), even regular practitioners require occasional clarification [e.g., the recent exchange between Murray (1999) and Turchin (1999)]. Thus, a review of basic concepts seems essential before the available data from reef fishes can be interpreted in a meaningful way (see also Chesson, 1998a; Hixon, 1998).

A. What Is Density Dependence?

By definition, all changes in population size are due to changes in vital or demographic rates (birth + immigration − death − emigration), the combination of which gives the population growth rate. Demographic density dependence is generally defined as an effect of present and/or past population sizes on the per-capita population growth rate, and thus at least one of the constituent demographic rates (Murdoch and Walde, 1989). Specifically, the overall input rate due to birth and immigration is (directly) density dependent, or "compensatory," when it varies negatively with population size, whereas the overall loss rate due to death and emigration is density dependent when it varies positively with population size (Fig. 1). These rates are inversely density dependent, or "depensatory," when they vary in the opposite directions, which can happen at low population sizes due to the Allee effect (Courchamp *et al.,* 1999). Note that these relationships ultimately must be caused by changes in population size to comprise true density dependence (see Section II,C,2). Royama (1977, 1992) and Turchin (2002) provide detailed discussions of the concept of density dependence.

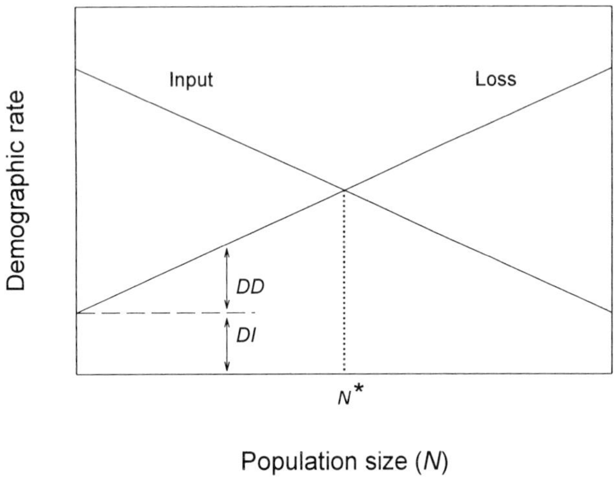

FIGURE 1 Typical textbook illustration of demographic density dependence, showing an input rate (birth, or in open local populations, settlement or recruitment) and a loss rate (typically mortality) that are both density dependent, as well as the resulting simple point equilibrium (N^*) in population size. Note that each demographic rate at each population size is actually the sum of both a density-independent component (DI, defined by the *y* intercept) and a density-dependent component (DD, defined by the slope), as illustrated for the mortality rate. This static picture belies the reality of reef fish populations, in which temporal variation in the position and slope of each curve will cause the equilibrium point to vary, and perhaps never be attained.

Demographic rates are density independent when they do not vary significantly as a function of population size. This is not to say that density-independent processes do not vary through time. In fact, interannual variation in density-independent mortality of marine fish larvae is considered to be the major determinant of year-class strength in marine fishes (Houde, 1987). It is important to note that both density-independent and density-dependent factors, combined, drive changes in population size—not simply one or the other—and both vary in time and space (Sinclair, 1989).

B. Why Study Density Dependence?

It is worthwhile to test for and study the mechanisms causing demographic density dependence for several related reasons. First, density dependence is the essence of population regulation, which is the long-term persistence of a population via a return tendency caused by negative feedbacks between the size of a population and its growth rate, such that fluctuations in abundance, however great and seemingly stochastic, are bounded above and below with a lower limit above zero (Royama, 1992; Cappuccino and Price, 1995; Turchin, 2002). Bounded fluctuations occur when the input rate exceeds the loss rate after the population size becomes sufficiently low, and vice versa when the population size becomes sufficiently high. By definition (some would say by tautology), this return tendency can occur only when either or both rates are density dependent (Fig. 1). More accurately, density dependence is a necessary but not sufficient condition for population regulation. Specifically, (1) density dependence must be direct and temporal (i.e., occur within each population rather than merely among populations; see Section IV,B,2), so that (2) if one demographic rate is inversely density dependent, another is sufficiently density dependent to counteract it; (3) density dependence must also be sufficiently strong to counteract any disruptive effects of density-independent or inversely density-dependent factors, but at the same time, (4) the strength and time lag of directly density-dependent responses must not be so great as to cause destabilizing population cycles (Turchin, 1995). Although how long a population must persist before it is considered regulated has not been defined explicitly, the general notion necessarily implies multiple generations.

Second, although long-term persistence of any population can by definition occur only via demographic density dependence (Haldane, 1953), past existence does not ensure future persistence. Indeed, all populations eventually go extinct, which occurs when their sources of density dependence break down.

Therefore, the conservation of any population involves protection of naturally regulatory density-dependent processes (or artificial imposition of density dependence). By identifying natural density-dependent processes, we learn what aspects of a population's biotic and abiotic environment are crucial candidates for protection. Such knowledge is of fundamental importance in both fisheries management and conservation, even though we can also conserve populations with minimal data by establishing sufficiently large marine protected areas (Roberts, 1997b; Bohnsack, 1998; Johannes, 1998) (Chapter 19, this volume).

Third, tests for density dependence in demographic rates, regardless of whether density dependence actually occurs, provide means of understanding what drives population dynamics (Krebs, 1995). Ultimately, population structure is the foundation on which community-level patterns emerge. Moreover, reef fishes provide some of the few cases in which mechanisms driving population dynamics in vertebrates can be examined experimentally (Sale, 1991b). Most evidence for density dependence in animals is based on observational data, especially from studies of insects and terrestrial vertebrates (reviews by Hanski, 1990; Cappuccino and Price, 1995; Lindström *et al.*, 1999), although field experiments are slowly becoming more common (review by Harrison and Cappuccino, 1995). Particularly relevant to reef fish ecology in this context is testing the recruitment limitation hypothesis, discussed in the next section.

C. Sources of Density Dependence in Reef Fishes

Whereas basic concepts regarding density dependence are straightforward, applying them to reef fishes and other demersal marine fish populations is not. Given that a population is typically defined as a group of organisms of the same species occupying a particular area, the key issues are defining the spatial boundaries and structure of the population, the nature of the demographic rates, and the spatiotemporal scales over which they can operate in a density-dependent fashion (and be detected).

1. Local Populations vs. Metapopulations

A vast majority of demersal species have a bipartite life cycle: a pelagic larval stage that is at least initially planktonic, followed by a nektonic juvenile and adult stage that is closely associated with habitat patches on the sea floor (Sale, 1980). Consequently, demersal fishes may form metapopulations, at least in a simplest sense of Levins' (1969) original definition of a "population of populations" linked by dispersal. [See Harrison and Taylor (1997) and Hixon (1998), for a discussion of whether reef fish populations fit more restrictive definitions of a metapopulation.] Therefore, there are two spatial scales for population studies of demersal fishes relevant to questions regarding demographic density dependence: the local population and the entire metapopulation.

We define a local population as a group of demersal (juvenile and adult) conspecifics meeting two criteria: (1) mature members of the group (if any) spawn only with other group members, and (2) the group is sufficiently spatially isolated from other such groups that between-group movement is demographically negligible (see Section II,C,2). For coral reef fishes with extremely limited home ranges, the local population can be as small as a single coral head or patch reef (review by Sale, 1978a). For more mobile species, the range of the local population can be several orders of magnitude greater, especially in larger fishes such as groupers, which seasonally migrate to regional spawning aggregations (Domeier and Colin, 1997). In contrast, a metapopulation comprises a group of local populations linked only by larval dispersal.

Thus, the metapopulation is the entire, reproductively closed population (i.e., only rare larval dispersal events create new metapopulations). In contrast, larval dispersal and connectivity among sites create at least partially open populations at the local scale (Fig. 2). In a completely open local population, reproductive output is entirely unrelated to subsequent input. In this case, none of the larvae produced by the local population settles back to that population, and consequently, all the larvae that do settle were spawned elsewhere. Smaller local populations at the scale of patch reefs are probably completely open, whereas larger local populations at the scale of isolated oceanic islands undergo various levels of larval retention and are thus relatively closed (Jones *et al.*, 1999; Swearer *et al.*, 1999) (see also Chapters 7–9, this volume). In other words, demographic openness is largely a function of spatial scale.

Given these concepts, when one discusses whether reef fish populations are regulated, the ultimate focus is the metapopulation. It is not necessary that every local population be regulated for a metapopulation to persist, only that density dependence of the right characteristics occurs somewhere within the metapopulation (Hanski *et al.*, 1996). If one does detect substantial density dependence within local populations, then a potentially important source of regulation has been identified.

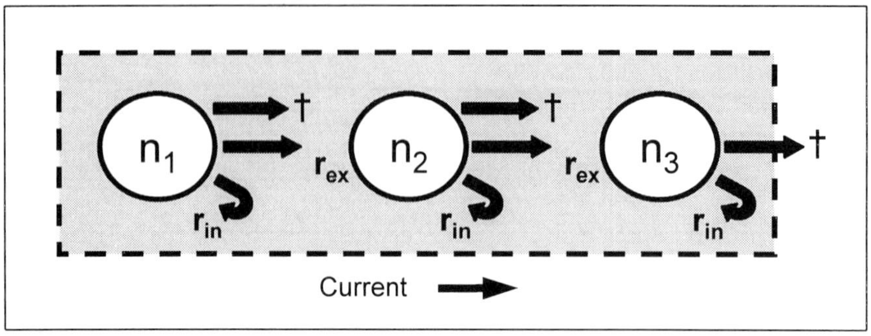

FIGURE 2 Diagram of a reef fish metapopulation (enclosed within the dashed line), illustrating possible kinds of open local populations (n_x). For each local population, dispersing larvae have three possible fates: death before settlement (†), internal retention to self-recruitment (r_{in}), or export to another local population (r_{ex}). Population n_1 depends entirely on internal retention but can export larvae, whereas population n_3 has both internal and external sources of recruitment but cannot successfully export larvae. Only population n_2 serves as both an export source and an import recipient of larvae.

2. Approaches for Detecting Density Dependence

All field studies of reef fishes reviewed here were conducted at the scale of local populations (or smaller). Because each local population is only part of a metapopulation, studies at this smaller spatial scale can provide only partial information regarding density dependence at the scale of the entire metapopulation. Thus, logistic constraints limit the spatial scale of our understanding of population dynamics in demersal marine fishes (Sale, 1998). What are the possible sources of demographic density dependence in a demersal metapopulation? There are four possibilities: (1) fecundity, (2) presettlement mortality of eggs and larvae, (3) mortality during the pelagic–demersal transition (settlement), and (4) postsettlement mortality of juveniles and adults. Given that we are presently limited almost exclusively to data from local populations, the key question is how (if at all) these possibilities can be examined with existing data.

Biological "birth" in most marine fishes occurs via spawning with external fertilization. Fecundity (egg or zygote production) is a possible source of demographic density dependence in all fishes because of two well-known relationships: (1) growth is often density dependent and (2) fecundity is a function of body size. Both these relationships have been well-documented in reef fishes, the former reviewed by Jones (1991) and the latter by Sadovy (1996). However, presently lacking and much needed are studies that integrate these patterns within a detailed demographic framework. Important contributions come from studies of the impact of maternal stress on the quality of offspring in reef fishes (McCormick, 1998b; Suthers, 1998) (see Chapter 10, this volume).

Due to larval dispersal, demographic "birth" in a local population of demersal fishes occurs by the process of settlement. Settlement is the transition from a pelagic larval stage to a demersal juvenile stage, involving an interaction between the local density of late-stage larvae competent to settle (hereafter "larval supply") and the availability of a suitable benthic environment (Kaufman *et al.*, 1992; McCormick and Makey, 1997; Booth and Wellington, 1998). Because settlement seems to occur mostly at night (reviews by Leis, 1991a; Victor, 1991; Cowen and Sponaugle, 1997), it is rarely observed directly (e.g., Holbrook and Schmitt, 1997). There is evidence that mortality at the time of settlement can be very high. Crest-net samples have shown that most nocturnally settling larvae entering the lagoon in Moorea, French Polynesia, disappear before they can be counted as new recruits the following morning (Dufour *et al.*, 1996; Doherty *et al.*, in preparation). Is mortality during settlement density dependent? To our knowledge, there are no data available to answer this important question.

Because settlement is so difficult to study directly, it is typically measured indirectly as recruitment, which is the appearance of recently settled fish (review by Doherty and Williams, 1988). Thus, "recruitment" as typically used by reef fish ecologists is an estimate—subject to variation in methodology and artifacts—of the real biological phenomenon of settlement. (This definition of "recruitment" differs from that commonly used in fisheries biology, where one estimates recruitment of larger fish to a fishery. It also differs from the general ecological definition, whereby "recruitment" is used with an explicit modifier to describe the transition between particular life history stages.) Obviously, the greater the time lag between actual settlement and measured recruitment, the greater the error in estimating settlement (due to postsettlement mortality) and the less the causal linkage between larval supply and recruitment. One of the major sources of confusion and controversy has been various authors implicitly defining recruitment based on quite different sampling frequencies, ranging from 1 day to several months (see Section III).

How relevant is measuring recruitment to the question of demographic density dependence? The answer depends on the spatial scale examined. If one wishes to understand only the dynamics of a local population, treating recruitment as a "black box" external source of settlers, then one can calculate the per-capita recruitment rate (i.e., divide the density of new recruits by the density of local residents during some time period) and examine this rate as a function of local population size. In every case we examined (see Section III,A), this procedure resulted in recruitment appearing to be density dependent, simply because each recruit represents a decreasing proportion of the total population as local population size increases (Hughes, 1984, 1990; Caley *et al.*, 1996). Such per-capita density dependence could occur in the absence of any biotic feedbacks whatsoever if per-area settlement was constant through time and new recruits simply accumulated within an area. Although an equilibrial local population size could result, this phenomenon is not true demographic density dependence (Bence and Nisbet, 1989; Sale and Tolimieri, 2000). Rather, such apparent or "pseudo-density dependence" is simply a mathematical phenomenon and is not caused mechanistically by changes in local population size.

If one wishes to understand the role of recruitment in the dynamics of the metapopulation, then the per-capita currency is inappropriate because recruits from sources outside the local population are divided by fish within that population. From the metapopulation perspective, it is better to examine the total recruitment rate (i.e., the density of new recruits appearing during some time period). If local population size affects total recruitment, due to negative interactions between settling fishes and established residents, then chronically unsuccessful settlers are doomed to die and true density dependence is possible. In such cases, the total recruitment rate will be density dependent if it declines with increasing local population size.

Direct measures of mortality are less problematic, at least conceptually. Unfortunately, loss rates of eggs and larvae during presettlement mortality are not well documented among reef fishes (review by Boehlert, 1996), but in general are believed to be highly variable and density independent (reviews by Houde, 1987; Bailey and Houde, 1989; Heath, 1992). Note that, if density dependence indeed does not occur during the pelagic egg or larval stage, then persistence of the entire metapopulation requires that density dependence must occur during the demersal juvenile or later stage, as hypothesized by Sissenwine (1984), Houde (1987), and others. Authors asserting that their models demonstrate that a metapopulation can persist indefinitely without density dependence (e.g., den Boer and Reddingius, 1996) are mistaken because these models actually incorporate density dependence (see Chesson, 1981, 1996; Hanski, 1990; Murdoch, 1994; Walde, 1995; Hanski *et al.*, 1996; Hanski and Gilpin, 1997, for details).

Postsettlement mortality of reef fishes is much better documented than egg and larval mortality. To test for density dependence, the mortality rate is calculated on a per-capita basis as a proportion of the population dying during some time period (alternatively, as an instantaneous rate), then examined as a function of the initial population size. Note that testing for density-dependent mortality by examining the number of adults (or larger juveniles) as a linear function of the initial number of recruits is not advisable because a significantly positive linear regression may mistakenly lead to the conclusion of density-independent mortality when the underlying pattern is actually density dependent (Caley *et al.*, 1996; Forrester, 1998; Hixon, 1998). For such plots, data showing a decelerating curvilinear regression (which can also produce a significant linear regression of positive slope) are indicative of density-dependent mortality, not simply linear regressions of zero slope (Caley *et al.*, 1996; Hixon, 1998). For example, the data fit by Doherty and Fowler's (1994a) density-independent linear regression is equally well fit by a density-dependent curvilinear regression (Forrester, 1998). In any case, the "initial population size" in question depends on the source of processes affecting mortality: within-cohort (a group of conspecifics that settle during the same time period), among-cohort (all conspecifics in the same local population regardless of age), or among-species (in the case of interspecific competition). The focus of this review is within-species density dependence (both within and among cohorts) because few studies have adequately examined the demographic effects of interspecific competition in reef fishes (but see Robertson, 1996). There is a clear need for more detailed studies of between-species competition. Importantly, predation—not just competition alone—can cause density-dependent mortality (see Section IV,B).

Examining whether postsettlement mortality rates are density dependent is the fundamental test of the "recruitment limitation hypothesis." Since formalized by Doherty (1981), this popular hypothesis has undergone various changes in meaning that threaten its utility (Caley *et al.*, 1996; Hixon, 1996a, 1998; Chesson, 1998a). The latest meaning depends on the definitions of key words: recruitment limitation occurs "when the relative abundance of reef fishes among replicate units of habitat reflects the spatial distribution of their recruitment accumulated over a lifetime" (Doherty, 1998, p. 129). As worded, this pattern could occur as a result of either density-dependent or density-independent processes (Warner and Hughes, 1988; Caley *et al.*, 1996; see also Holm, 1990). Fortunately, one of the explicit predictions (stated as an "assumption") of this version of the hypothesis (even though it does not follow necessarily from the above definition) is forthright: "post-settlement mortality schedules at this scale are stable (albeit may vary with age) and density-independent" (Doherty, 1998, p. 131). Therefore, one can test the recruitment limitation hypothesis by determining whether the postsettlement death rate is density dependent. However, this "either/or" approach to the issue of what drives local population dynamics is quite restrictive. A more fruitful approach is to examine the relative roles of recruitment vs. mortality in determining local population size (see Section V,D). Moreover, a less restrictive definition of recruitment limitation—occurring when increases in recruitment cause increases in population size—can be manifested regardless of whether mortality is density dependent or density independent (and thus is irrelevant to this chapter).

Given that recruitment is better examined as a total rate, whereas postsettlement mortality is better examined as a per-capita rate, how can these rates be compared? If one wishes to understand only the dynamics of the local population, including any local "equilibrium" driven externally by incoming recruits interacting with local mortality (Bence and Nisbet, 1989), then comparing per-capita rates is useful (see Section III,C). However, if the goal is to use local rates to understand metapopulation dynamics, then comparing total rates is more appropriate.

Finally, what about emigration and immigration? Our definitions clearly separate local populations from metapopulations by eliminating postsettlement movements as demographic rates (Section II,C,1). This is not to say that emigration and immigration do not occur at smaller spatial scales. Postsettlement movement of juveniles and adults between habitat patches most certainly occurs at such scales, sometimes at substantial levels (e.g., Robertson, 1988a; Schmitt and Holbrook, 1996; Frederick, 1997; Lewis, 1997; Ault and Johnson, 1998a). However, we contend that in cases in which between-patch movement is common, the linked patches comprise only parts of a single local population (rather than multiple local populations). Studying spatial scales much less than (or even much greater than) that of the local population inhibits the ability to detect biologically meaningful density dependence (Ray and Hastings, 1996; Anneville *et al.*, 1998). At the same time, between-patch emigration and immigration can be demographically relevant. If emigration leads to death, then it is a mechanism of mortality. In addition, measuring between-patch movement is essential for defining the boundaries of local populations. Therefore, tracking movements of juveniles and adults, typically via tagging, is an important part of any meaningful demographic study. Of course, at the scale of the metapopulation, immigration and emigration have no meaning, because the entire population occurs within its boundaries by definition.

Given the above considerations, this review focuses on the two possible sources of demographic density dependence that have been reasonably well studied in reef fishes: local recruitment and natural postsettlement mortality. Note that studies of reef fishes, especially in an experimental context, comprise the most detailed data available on these demographic rates in marine fishes. Therefore, we believe that reef fishes are excellent model systems for understanding mechanisms driving and regulating population dynamics of demersal marine fishes in general (see Section V,B).

III. How Common Is Density Dependence in Reef Fishes?

We searched the literature for evidence published through 1999 regarding the presence or absence of demographic density dependence in coral reef fishes. Specifically, we compiled data sets that examined recruitment and mortality rates as a function of local population size (or density). Obviously, a reasonably broad range and number of population sizes are required to test for density dependence statistically. Therefore, we did not consider studies that examined recruitment or mortality only in the presence vs. absence of previously settled fish (i.e., two points), even though such studies provide mechanistic information on the effects of residents (e.g., Sale, 1976; Williams, 1980; Doherty, 1983a; Tolimieri, 1995; Gutierrez, 1998). Also, the design of some studies that reported density dependence did not allow determination of which demographic rates were involved, so could not be categorized (e.g., Schmitt and Holbrook, 1999a).

Published evidence regarding demographic density dependence includes both observational studies, which provide pattern only, and experimental studies, which can additionally elucidate causation if the causative processes are appropriately manipulated. For both kinds of studies, the basic analytical design has been to examine, by regression (linear or curvilinear) or correlation, a demographic rate as a function of local population size (or density) at the beginning of the time period over which the rate was measured. We know of no long-term observational studies of reef fishes that have employed the time-series analyses so prevalent in terrestrial studies (reviewed in Cappuccino and Price, 1995), and regardless, detecting density dependence from time series of abundance alone is problematic (Shenk *et al.,* 1998). In the case of well-designed experimental studies, initial population sizes were manipulated, with appropriate controls and replication, over a range of natural densities. To examine causation, the initial-population-density treatment was cross-factored with one or more treatments that manipulated putative causative processes (e.g., predators present or absent).

For both observational and experimental studies, the spatial scale has thus far been constrained to local populations (Section II,C,1). The appropriate temporal scale for testing for density dependence is a sufficient period for any existing density-dependent process to operate, which may be as long as a full generation when mechanisms are immediate (Harrison and Cappuccino, 1995) or longer if density dependence is delayed by time lags (Turchin, 1990). In such tests, both observational and experimental studies can employ a spatial design (i.e., compare multiple populations over the same time period) or a temporal design (i.e., compare multiple time periods within a single population). Although both designs are valid and provide insight, spatial density dependence (among populations) may not necessarily translate into the temporal density dependence (within a population) required for regulation (see Section IV,B,2). All but three studies reviewed here used spatial designs, and therefore could detect only spatial density dependence. Two of the exceptions confounded time and space during data analysis (Stimson, 1990; Caselle, 1999). Only Victor (1986b) provided a truly temporal (albeit indirect) analysis.

A. Recruitment

Table 1 summarizes studies that examined whether recruitment is density dependent. Although there have been many publications on recruitment, few report data allowing tests for density dependence. Density dependence is not falsified simply by demonstrating high variation in recruitment in time and space, which is characteristic of virtually all reef fishes examined to date (reviews by Doherty and Williams, 1988; Doherty, 1991; Jones, 1991; Williams, 1991; Booth and Brosnan, 1995; Caley *et al.,* 1996; see also Dixon *et al.,* 1999).

1. Observational Studies

We found five nonexperimental studies that explicitly examined whether recruitment was density dependent (Table 1). Seven species from four families were studied in the Pacific and western Atlantic/Caribbean. In all cases, the total recruitment rate was reported (Section II,C,2). Four different patterns were evident: (1) density dependence (Stimson, 1990; Tupper and Hunte, 1994), (2) density independence (Tupper and Hunte, 1994; Forrester, 1999), (3) inverse density dependence (Booth, 1992), and even (4) a unimodal relationship between the total recruitment rate and local population size (Schmitt and Holbrook, 1996). Note that Victor's (1983b, 1986b) extensive study of the bluehead wrasse (*Thalassoma bifasciatum*) in the Caribbean is also suggestive of density-independent recruitment, but the data are presented in such a way that we were unable to extract an explicit statistical analysis. On conversion to per-capita rates, recruitment appeared to be density dependent (i.e., pseudo-density dependent) in every case (Section II,C,2).

2. Experimental Studies

We found six studies that tested for density-dependent recruitment experimentally, all of which reported total recruitment rates (Table 1). These studies examined six species from four families in the Pacific and western Atlantic/Caribbean. Again, a variety of patterns emerged: (1) density dependence (Stimson, 1990; Forrester, 1995; Steele *et al.,* 1998), (2) density independence (Forrester, 1999; Webster, in preparation), and (3) inverse density dependence (Sweatman, 1985b; Booth, 1992). Note that Stimson's (1990) experiment was unreplicated. Again, conversions to per-capita rates resulted in recruitment that was pseudo-density dependent in every case (Section II,C,2).

3. Interpretation and Conclusions

Overall, given the few studies available, the wide taxonomic and geographical coverage of those studies, and the broad variety of relationships documented, we could detect no general patterns in recruitment. Such results reflect the highly variable nature of recruitment in reef fishes.

TABLE 1 Observational and Experimental Studies of the Local Total Recruitment Rate[a]

Family and species	Location	Total recruitment rate	Sampling frequency	Study duration	Reference
Observational studies					
Chaetodontidae	Hawaii				
Chaetodon miliaris	Oahu	DD	1 mo	>6 yr	Stimson (1990)
Pomacentridae	Caribbean				
Stegastes partitus	Barbados	DI	<2 wk	1 yr	Tupper and Hunte (1994)
	Hawaii				
Dascyllus albisella	Oahu	IDD	5 days	14 mo	Booth (1992)
	French Polynesia				
Dascyllus trimaculatus	Moorea	Unimodal	1 day	6 days	Schmitt and Holbrook (1996)
Dascyllus trimaculatus	Moorea	Unimodal	<2 mo	<2 mo	Schmitt and Holbrook (1996)
Labridae	Caribbean				
Halichoeres garnoti	Barbados	DD	<2 wk	1 yr	Tupper and Hunte (1994)
Thalassoma bifasciatum	Barbados	DI	<2 wk	1 yr	Tupper and Hunte (1994)
Gobiidae	Bahamas				
Coryphopterus glaucofraenum	Lee Stocking Island	DI	1 day	7 days	Forrester (1999)
	Caribbean				
Coryphopterus glaucofraenum	Guana Island, BVI[b]	DI	1 day	12 days	Forrester (1999)
Experimental studies					
Grammatidae	Bahamas				
Gramma loreto	Lee Stocking Island	DI	1 day	2 mo	Webster (in preparation)
Chaetodontidae	Hawaii				
Chaetodon miliaris	Oahu	DD	<1 mo	4 mo	Stimson (1990)
Pomacentridae	Great Barrier Reef				
Dascyllus aruanus	Lizard Island	IDD	<1 wk	2 mo	Sweatman (1985b)
Dascyllus reticulatus	Lizard Island	IDD	<1 wk	2 mo	Sweatman (1985b)
	Hawaii				
Dascyllus albisella	Oahu	IDD	1–3 days	~6 mo	Booth (1992)
Gobiidae	Caribbean				
Coryphopterus glaucofraenum	Guana Island, BVI	DD	2.5 mo	2.5 mo	Forrester (1995)
Coryphopterus glaucofraenum	Guana Island, BVI	DI	1 day	14 days	Forrester (1999)
	Bahamas				
Coryphopterus glaucofraenum	Lee Stocking Island	DD	2–4 wk	2–4 wk	Steele *et al.* (1998)

[a]These studies determined whether the local total recruitment rate was density independent (DI), density dependent (DD), inversely density dependent (IDD), or unimodal. Note that the local per-capita recruitment rate was "pseudo-density dependent" in every case (see Fig. 3). Studies with shorter sampling frequencies more accurately estimated actual patterns of settlement.
[b]BVI, British Virgin Islands.

Figure 3 illustrates four conversions from total to per-capita recruitment rates—two from observational studies and two from field experiments. Notice that this conversion shifts the rich variety of recruitment patterns to uniform apparent density dependence (i.e., a negative relationship between recruitment and local population size). These examples demonstrate how such pseudo-density dependence is a purely mathematical phenomenon rather than a biological mechanism (Section II,C,2). However, per-capita recruitment can nonetheless be used to examine equilibrial dynamics of a local population studied in isolation (see Section III,C).

B. Mortality

Compared to recruitment, far more explicit data have been published regarding postsettlement mortality

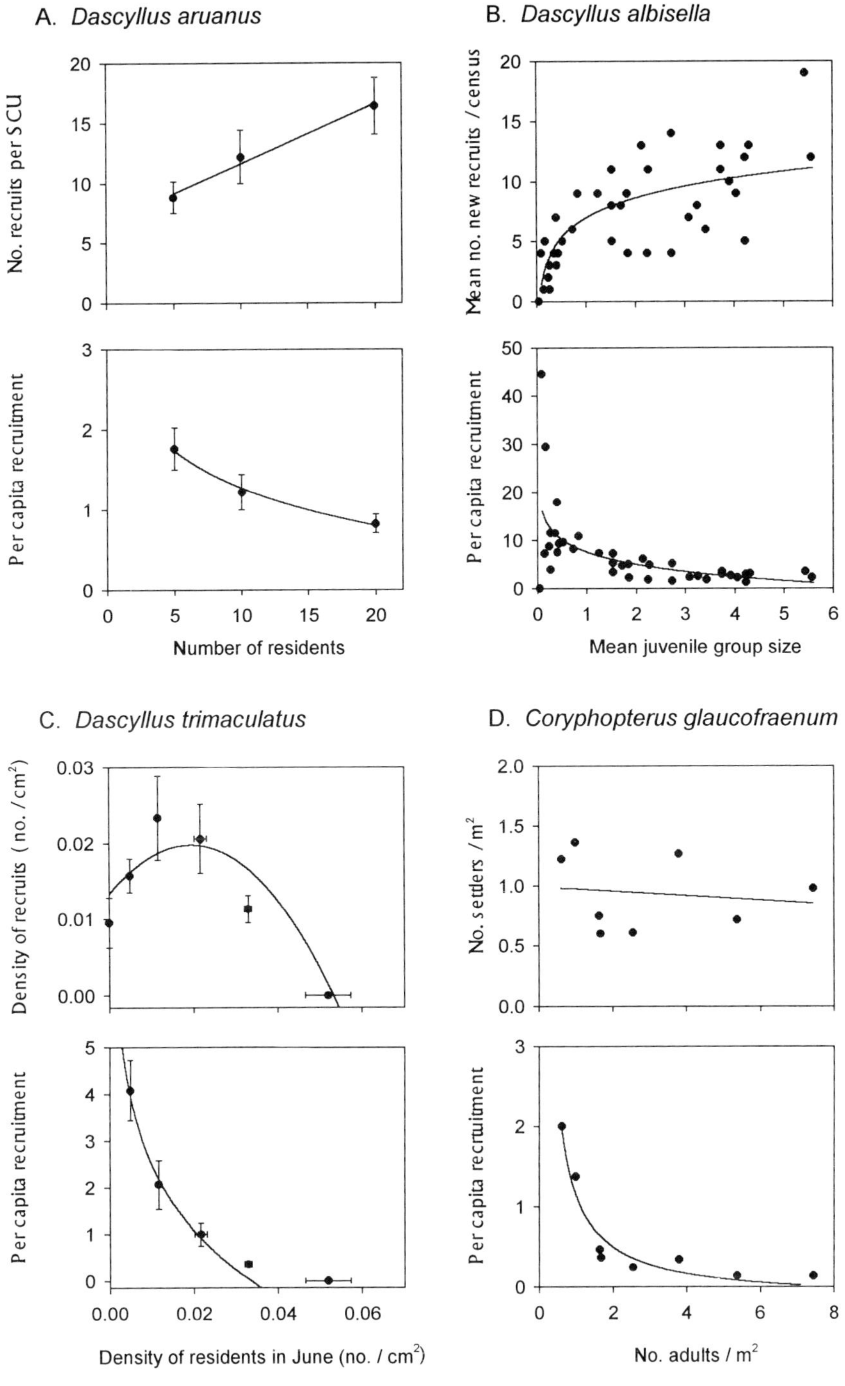

FIGURE 3 Example conversions of total recruitment (upper plots) to per-capita recruitment (lower plots) from studies of (A) a damselfish in the Great Barrier Reef [experimental data from Sweatman (1985b)], (B) a damselfish in Hawaii [observational data from Booth (1992)], (C) a damselfish in French Polynesia [observational data from Schmitt and Holbrook (1996)], and (D) a goby in the British Virgin Islands [experimental data from Forrester (1999)]. Despite large variation in the shape of the total recruitment curve, per-capita recruitment appears to be density dependent (i.e., "pseudo-density dependent") in every case (A; $r^2 = 0.45$, $P = 0.006$; B, $r^2 = 0.34$, $P = 0.0001$; C, $r^2 = 0.51$, $P < 0.0001$; D, $r^2 = 0.74$, $P = 0.006$).

(Table 2). Due to difficulties of interpretation, we did not include studies in which species were pooled for analysis (e.g., Caley, 1995a; Connell, 1997a). Also inappropriate were studies (reporting density-dependent mortality) that followed members of only a single cohort through time, thus confounding density and age along a survivorship curve (e.g., Hunte and Côté, 1989). Given that type 3 (hyperbolic) postsettlement survivorship curves are common in reef fishes (reviews by Hixon, 1991; Caley, 1998), including such data would artificially inflate the reported occurrence of density dependence (Caley *et al.*, 1996).

Either from the original publication or directly from the author, we were able to extract data from most studies listed in Table 2 to calculate the instantaneous daily per-capita mortality rate as a function of local

TABLE 2 Observational and Experimental Studies of the Local Per-Capita Mortality Rate[a]

Family and species	Location[b]	Per-capita mortality rate: Overall	DI component (y intercept)	DD component (slope)	Initial age of fish	Study duration	Reference[c]
Observational studies							
Serranidae	Indian Ocean						
Epinephelus merra	Reunion Island	DD	-4.2×10^{-2}	3.9×10^{-2}**	4–5 days	7 wk	1
Pomacentridae	Bahamas						
Chromis cyanea	Lee Stocking Island 1994	DI	2.1×10^{-2}**	5.0×10^{-3}	<1 wk	1 mo	2
Chromis cyanea	Lee Stocking Island 1995	DD	-4.2×10^{-2}	3.3×10^{-2}**	<1 wk	1 mo	2
Chromis cyanea	Lee Stocking Island 1996	DD	8.3×10^{-3}	1.5×10^{-2}*	<1 wk	1 mo	2
	Hawaii						
Dascyllus albisella	Oahu	IDD	—	—	1–5 days	8 mo	3
	French Polynesia						
Dascyllus aruanus	Moorea	DI/DD	1.4×10^{-2}*	7.6×10^{-5}**	1 day	2 wk	4
Dascyllus aruanus	Moorea	DI	1.3×10^{-2}**	-2.1×10^{-5}	2–30 days	2 wk	4
Dascyllus falvicaudus	Moorea	DI/DD	2.0×10^{-2}***	1.8×10^{-4}***	1 day	2 wk	4
Dascyllus falvicaudus	Moorea	di	1.5×10^{-2}	-3.7×10^{-5}	2–30 days	2 wk	4
Dascyllus trimaculatus	Moorea	DD	—	—	1 day	2 wk	5
Dascyllus trimaculatus	Moorea	DD	1.1×10^{-2}	4.7×10^{-4}***	1 day	2 wk	4
Dascyllus trimaculatus	Moorea	di	2.4×10^{-3}	1.0×10^{-4}	2–30 days	2 wk	4
	Great Barrier Reef						
Pomacentrus molluccensis	Lizard Island	DD	2.9×10^{-4}	6.7×10^{-3}*	<1 mo	1 yr	6
Pomacentrus moluccensis	Southern GBR	DI	5.7×10^{-4}**	1.4×10^{-4}	2–5 mo	9 yr	7, 8
Pomacentrus wardi	Southern GBR	di	3.1×10^{-4}	6.3×10^{-4}	2–5 mo	9 yr	8
	Caribbean						
Stegastes partitus	Barbados	DD	—	—	<2 wk	1 yr	9
Labridae	Caribbean						
Halichoeres garnoti	Barbados	DI	—	—	<2 wk	1 yr	9
Thalassoma bifasciatum	Panama	DI	—	—	1 day	3 days	10
Thalassoma bifasciatum	Panama	DI	—	—	4 days	31 days	10
Thalassoma bifasciatum	Barbados	DD	—	—	Adults	4–5 mo	11
Thalassoma bifasciatum	Barbados	DD	—	—	<2 wk	1 yr	9
Thalassoma bifasciatum	St. Croix, USVI	DD	—	—	1 mo	4–6 mo	12
	French Polynesia						
Thalassoma hardwicke	Moorea	DD	—	—	1 day	90 days	13
Acanthuridae	Caribbean						
Acanthurus bahianus SH	Panama	di	7.3×10^{-4}	2.5×10^{-3}	1 mo	8 yr	14
Acanthurus bahianus AH	Panama	DD	6.7×10^{-4}	2.9×10^{-3}**	1 mo	8 yr	14
Acanthurus chirurgus SH	Panama	DI/DD	1.6×10^{-3}*	3.0×10^{-3}**	1 mo	8 yr	14
Acanthurus chirurgus AH	Panama	DI	1.9×10^{-3}*	2.1×10^{-3}	1 mo	8 yr	14
Acanthurus coeruleus SH	Panama	di	7.6×10^{-6}	1.0×10^{-2}	1 mo	8 yr	14
Acanthurus coeruleus AH	Panama	DD	-5.7×10^{-4}	2.0×10^{-2}**	1 mo	8 yr	14

Experimental studies							
Grammatidae	Bahamas						
Gramma loreto	Lee Stocking Island	DD	—	—	All ages	2 mo	15
Pomacentridae	Great Barrier Reef						
Acanthochromis polyacanthus	One Tree Island	DD	—	—	<2 days	<1 mo	16
Dascyllus aruanus	One Tree Island	DD	—	—	Recent settlers	7 mo	17
Dascyllus aruanus	One Tree Island	DD	—	—	Recent settlers	1 yr	18
Dascyllus aruanus	One Tree Island	DD	—	—	3–12 days	10 mo	19
Pomacentrus ambionensis	One Tree Island	DI	—	—	Recent settlers	1 yr	20
Pomacentrus ambionensis	One Tree Island	DI	—	—	1 yr	2 yr	20
Pomacentrus ambionensis	One Tree Island	IDD	—	—	Recent settlers	1 yr	18
Pomacentrus ambionensis	Lizard Island	DI/DD	1.7×10^{-3}**	1.6×10^{-4}*	Recent settlers	17 mo	21
Pomacentrus flavicauda	One Tree Island	DI	—	—	2–5 wk	1 yr	22
Pomacentrus wardi	One Tree Island	di	2.4×10^{-3}	3.1×10^{-5}	2–5 wk	1 yr	22
	Hawaii						
Dascyllus albisella	Oahu	IDD	—	—	<1 day	35 days	3
	French Polynesia						
Dascyllus trimaculatus	Moorea	DD	-6.7×10^{-4}	9.8×10^{-5}***	≤2 days	2 wk	4
Dascyllus trimaculatus	Moorea	DI/DD	9.0×10^{-3}*	8.2×10^{-5}***	≤2 days	2 wk	4
	Bahamas						
Chromis cyanea	Lee Stocking Island	DD	8.1×10^{-3}	1.1×10^{-2}**	Recent settlers	1 mo	23
Stegastes partitus	Lee Stocking Island	DD	1.9×10^{-3}	1.6×10^{-2}*	<1 wk	54 days	24
Labridae	Bahamas						
Halichoeres garnoti	Lee Stocking Island	DD	1.2×10^{-3}	2.5×10^{-2}***	<1 wk	3 wk	25
	Caribbean						
Thalasoma bifasciatum	St. Croix, USVI	DD	—	—	1–3 days	1 day	12
Thalasoma bifasciatum	St. Croix, USVI	DI	—	—	3–6 days	1 mo	12
Gobiidae	Caribbean						
Coryphopterus glaucofraenum	Guana Island, BVI	DI/DD	8.4×10^{-3}***	1.6×10^{-3}***	Adults	2.5 mo	26
	Bahamas						
Coryphopterus glaucofraenum	Lee Stocking Island	DD	—	—	Adults	30 days	27
Coryphopterus glaucofraenum	Lee Stocking Island	DD	—	—	Adults	55–64 days	27

[a]These studies determined whether the overall local per-capita mortality rate was density independent (DI), density dependent (DD), both (DI/DD), or inversely density dependent (IDD). Determination was made by testing whether the *y* intercept (DI component) and positive slope (DD component) of the instantaneous daily per-capita mortality curve (see Fig. 1) were significantly different from zero by least-squares linear regression (*P <0.05; **P <0.01; ***P <0.001). Cases in which neither parameter was significant (di) indicate a low level of density-independent mortality. In cases in which the original data were unavailable, the author's published interpretation is listed without *y* intercept and slope values. Note that the observational study of Acanthuridae (reference 14) examined density on the basis of both a settlement habitat (SH) and adult habitat (AH), which provided different conclusions.

[b]USVI, United States Virgin Islands; BVI, British Virgin Islands.

[c]1, Letourneur *et al.* (1998); 2, Hixon and Carr (in preparation); 3, Booth (1995); 4, Schmitt and Holbrook (1999b); 5, Schmitt and Holbrook (1996); 6, Beukers and Jones (1997); 7, Doherty and Fowler (1994a); 8, Doherty and Fowler (1994b); 9, Tupper and Hunte (1994); 10, Victor (1986b); 11, Hunte von Herbing and Hunte (1991); 12, Caselle (1999); 13, Shima (1999a); 14, Robertson (1988a); 15, Webster (in preparation); 16, Connell (1998a); 17, Jones (1987a); 18, Jones (1988b); 19, Forrester (1990); 20, Jones (1987b); 21, Hixon and Jones (in preparation); 22, Doherty (1982); 23, Hixon and Carr (1997); 24, Carr *et al.* (in preparation); 25, M. A. Hixon *et al.* (previously unpublished); 26, Forrester (1995); 27, Forrester and Steele (2000).

population density. This conversion in turn allowed us to test via standard least-squares linear regression whether mortality had a statistically significant density-independent component (i.e., positive y intercept) and density-dependent component (i.e., positive slope), because both can determine the overall mortality rate (Fig. 1). [Data were approximately linear for all data sets except that of Schmitt and Holbrook (1999b).] In cases in which we could neither extract nor obtain the required data, we relied on the author's conclusions.

1. Observational Studies

We found 14 studies that provided explicit nonexperimental tests for density dependence in postsettlement mortality (Table 2). These studies examined 15 species representing four families, and were conducted at various sites in the Pacific, the Caribbean/western Atlantic, and the Indian Ocean. Excluding three surgeonfishes (see next paragraph), per-capita mortality was always density dependent (with or without an additional density-independent component) in 3 species (a grouper, a damselfish, and a wrasse), sometimes density dependent and sometimes density independent in 6 species (those which were the subject of multiple studies or were examined over different time periods), exclusively density independent in 2 species (a damselfish and a wrasse), and inversely density dependent in 1 species of damselfish. Including surgeonfishes, the 12 species that showed density-dependent mortality at some time and place represented four major families of reef fishes and all three major oceans. In virtually all cases, density dependence was an early postsettlement phenomenon. The damselfish (*Pomacentrus wardi*) and the wrasse (*Halichoeres garnoti*) that showed density-independent mortality exclusively were studied in the southern Great Barrier Reef and Barbados, respectively. The single case of inverse density dependence was apparently due to antipredatory benefits of schooling in the highly social Hawaiian damselfish *Dascyllus albisella* (Booth, 1995).

Robertson's (1988a) study of Caribbean surgeonfishes defies easy categorization and shows how methodological variation can alter conclusions regarding density dependence. He forthrightly reported fish densities in terms of both "settler habitat" and "adult habitat." If one considers only settler habitat, then two surgeonfish species suffered density-independent mortality exclusively and the third species experienced a combination of density-independent and density-dependent mortality. However, if densities are calculated using adult habitat (which we believe is of ultimate demographic importance), then only one species suffered density-independent mortality and two species experienced density dependence.

In the one case in which different age classes were compared, only day-old new settlers of three species of humbug damselfish (*Dascyllus* spp.) in Moorea, French Polynesia, suffered density-dependent mortality (Schmitt and Holbrook, 1999b). Mortality of older juveniles was density independent (see also Section III,B,2), but older fishes were studied in the absence of larger juveniles and adults.

Two species were the subjects of multiple studies, allowing regional comparisons. The wrasse *Thalassoma bifasciatum* was studied off the Caribbean coast of Panama (Victor, 1986b), off Barbados (Hunte von Herbing and Hunte, 1991; Tupper and Hunte, 1994), and in the United States Virgin Islands (Caselle, 1999). Victor reported density-independent mortality, whereas the other three studies detected density dependence. The reason for this discrepancy is unknown, given that recruitment densities were similar among regions. There were differences in methods: Victor's analysis was temporal (patterns on one reef through time) and of relatively indirect and low resolution, those in Barbados were spatial (comparing multiple reefs), and Caselle's analysis combined time and space.

The other regional comparison involved studies of the damselfish *Pomacentrus moluccensis* at both the southern (Doherty and Fowler, 1994a,b) and northern (Beukers and Jones, 1997) Great Barrier Reef (GBR). The former study reported density-independent postsettlement mortality, whereas the latter detected density dependence (Fig. 4). Although both studies employed a spatial design using similar sample units (patch reefs), the methods differed in geographic coverage and temporal scale. Doherty and Fowler censused patches over an entire archipelago annually for 9 years and compared seven islands, whereas Beukers and Jones censused patches within a single location twice (1 year apart) and compared eight patches. Also, the initial age of new recruits was up to 5 months postsettlement in the former study (Doherty and Fowler, 1994a,b), but less than 2 months in the latter (J. S. Beukers, personal communication). Finally, the initial density of new recruits ranged from 3 to 40 fishes per 100 m^2 in Doherty and Fowler's (1994a,b) study, and from 25 to nearly 100 per 100 m^2 in Beukers and Jones' (1997) study.

What explains the difference in the mortality rates in the southern and northern Great Barrier Reef? There are three major possibilities. First, the different spatial resolutions of the two studies could have led to different conclusions despite no real difference, i.e., density dependence may be detectable only at the local scale of patch reefs within islands rather than at the regional scale of patch reefs pooled by island [see Hassell *et al.*

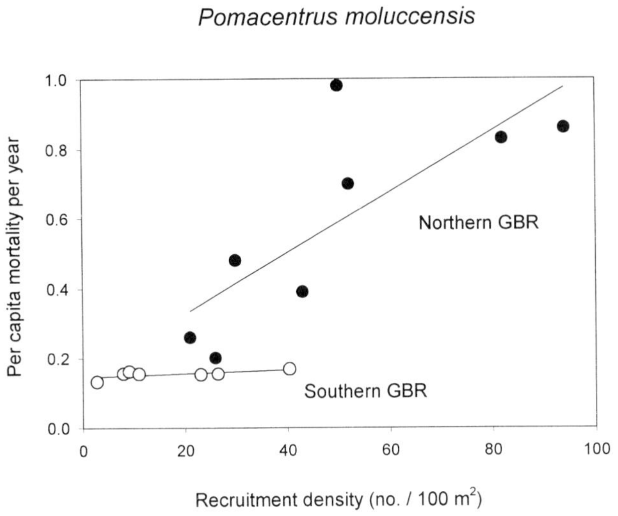

FIGURE 4 Comparison of annual per-capita mortality of the damselfish *Pomacentrus moluccensis* between the southern Great Barrier Reef (GBR) [data from Doherty and Fowler (1994a)] and the northern GBR [data from Beukers and Jones (1997)]. Mortality rates were roughly estimated from the data of Doherty and Fowler (1994a) as follows: 1 − [(average "annual recruitment density" × 9 years)/"adjusted catch"], which gave the overall per-capita mortality rate for a combination of 1- to 9-year-olds, so this value was divided by the average age of 5 years to estimate annual mortality.

(1987), for a similar comparison of insects on leaves vs. entire plants]. Second, density dependence may occur only at higher recruitment densities characteristic of the northern GBR (see Sweatman, 1985b; Caley, 1995b). Finally, Doherty and Fowler may have missed early postsettlement density dependence by starting their observations with older fishes. The argument that the density of older fishes in Doherty and Fowler's (1994a,b) study was highly correlated with the density of new settlers (or even larval supply), based on a study by Williams *et al.* (1994), is flawed by two serious problems in observational and analytical design of the latter study. First, multiple 40-day census intervals without removals of new recruits means that juveniles counted as settlers were up to several months old, allowing plenty of time for postsettlement mortality to obscure true patterns of settlement. Sampling settlement accurately often requires at least daily censuses to ensure that new settlers are counted before they die (Booth, 1991; personal observations). Indeed, crest net samples have shown that most nocturnally settling larvae entering the lagoon in Moorea, French Polynesia, disappear before they can be counted as new recruits (Dufour *et al.*, 1996; Doherty *et al.*, in preparation). Second, the conclusion by Williams *et al.* (1994) of a tight relationship between settler density and older recruit density was based on correlations drawn between "net total gain" of recruits and "end of season count," which as defined could be (and were in many cases) exactly the same data!

2. Experimental Studies

We found 16 experimental studies that tested explicitly whether postsettlement mortality was density dependent (Table 2). These studies examined 13 species from four families, and were conducted at various locations in the Pacific Ocean and the Caribbean/western Atlantic region. Seven species representing all four families always showed density-dependent per-capita mortality (with or without an additional density-independent component) (e.g., Fig. 5), three species (a damselfish, a wrasse, and a goby) showed density-dependent mortality at some times and density-independent mortality at other times, and two damselfishes showed density-independent mortality exclusively. The thirteenth species showed inverse density dependence, as did another damselfish in one case (see next paragraph), apparently due to antipredatory benefits of sociality.

The three species that exhibited density dependence at some times and density independence at other times experienced density-dependent mortality only as very recent recruits. This pattern suggests that local regulation due to mortality may be an early postsettlement phenomenon. One of these species, the damselfish *Pomacentrus amboinensis,* also exhibited slightly inverse density dependence at the southern Great Barrier Reef (Jones, 1988b), as well as density-independent mortality at the same site (Jones, 1987b), whereas mortality at the northern GBR showed density dependence (Hixon and Jones, in preparation). The experimental designs among sites were comparable, so the difference appeared to be related to a much greater settlement rate at the northern site (Caley, 1995b), perhaps similar to the situation with *P. moluccensis* (Section III,B,1).

Both damselfish species that exhibited density-independent mortality exclusively (then named *P. flavicauda* and *P. wardi*) were studied as older juveniles at One Tree Lagoon in the southern Great Barrier Reef (Doherty, 1982). The results reviewed above suggest that this study may have missed early postsettlement density dependence by focusing on older fish. Additionally, One Tree Lagoon receives relatively low settlement compared to more equatorial locations in the northern GBR (Sweatman, 1985b; Caley, 1995b; see also Hughes *et al.*, 1999), suggesting that density dependence required higher initial densities. This hypothesis is bolstered by the fact that two

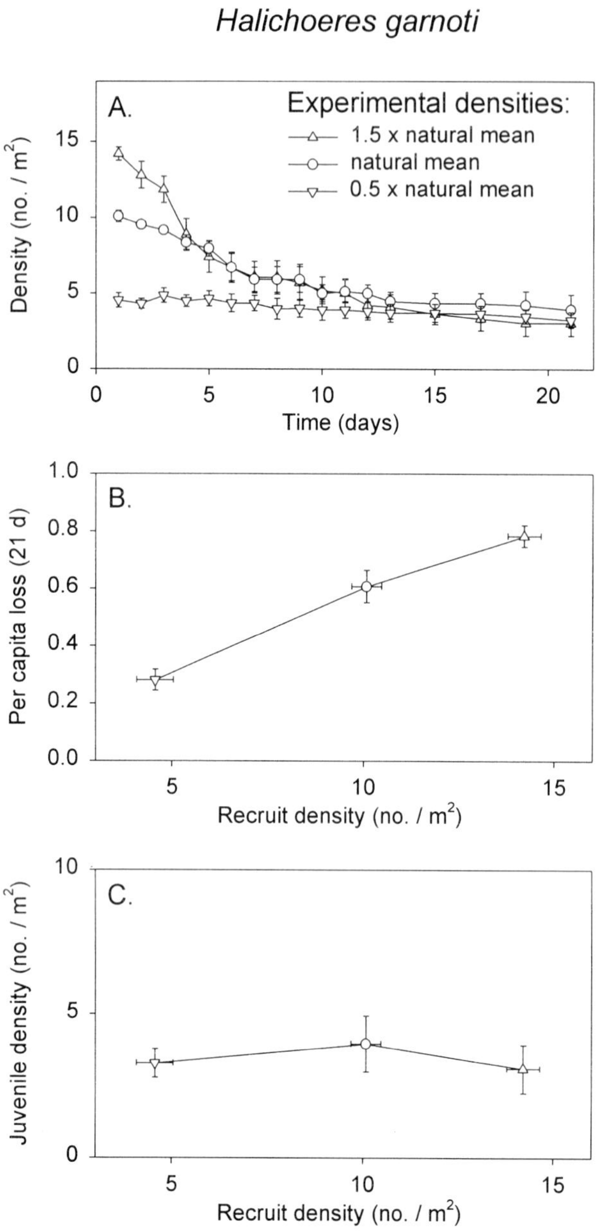

FIGURE 5 Example of a field experiment demonstrating density-dependent mortality. Numbers of new settlers of the wrasse *Halichoeres garnoti* were adjusted on small patch reefs in the Bahamas to mean natural densities, 150% natural, and 50% natural ($n = 5$ reefs each; results given as mean ± SEM). Results are equivalently presented as (A) raw survivorship curves, (B) per-capita loss, and (C) abundance of 21-day-old juveniles as a function of the initial abundance of new settlers. Note that the per-capita presentation shows density dependence most forthrightly. Fishes were not tagged in this study, so some loss may have been due to emigration [data from M. A. Hixon, M. H. Carr, and A. T. Kaltenberg (previously unpublished)].

congeneric species (*P. moluccensis* and *P. amboinensis*) exhibited density-dependent mortality at Lizard Island in the northern GBR (Beukers and Jones, 1997; Hixon and Jones, in preparation).

3. Meta-analysis and Conclusions

The studies summarized in Table 2 provide sufficient data for a simple meta-analysis, or analysis of analyses (Gurevitch and Hedges, 1993; Osenberg *et al.*, 1999). The goal was to determine whether density-dependent mortality was common across studies and species. We measured the strength of density dependence as the correlation (r) between local population density and instantaneous daily per-capita mortality rate for 24 separate data sets extracted from 14 studies of 15 species (Table 2). "Effect size" was tested as Fisher's z-transformation of r in a random-effects model using the MetaWin 2.0 software package (Rosenberg *et al.*, 2000). We ran the analysis twice, once including only the "settler habitat" data from Robertson (1988a) and once including only the "adult habitat" data. Weighted mean correlations from the two meta-analyses were 0.77 (95% CI: 0.63–0.86) and 0.79 (95% CI: 0.65–0.87), respectively. There was no evidence of publication bias or other violation of assumptions.

The high values of the mean correlations from the meta-analyses suggest that postsettlement density-dependent mortality is widespread among coral reef fishes. This conclusion is likely to be robust, given the wide range of species and sampling/experimental methods used in the 14 studies. Because we were able to extract sufficient information from only 13 of 27 studies listed in Table 2, more detailed meta-analyses will certainly be possible if and when additional studies are collated into a central archive (see Section V,D).

C. Recruitment and Mortality Combined

To our knowledge, only four published studies have simultaneously examined whether both recruitment and mortality are density dependent (Tables 1 and 2) (Booth, 1992, 1995; Tupper and Hunte, 1994; Forrester, 1995; Schmitt and Holbrook, 1996). We plot the combined results from two of these studies in Fig. 6, both as total rates and as per-capita rates. In theory, if these curves remained constant through time, then populations would stabilize at the equilibrium density where the recruitment and mortality curves intersect (see Fig. 1). However, such constancy is not apparent in reef fish populations, nor should it be expected (see Section V,C). Importantly, these demographic "snapshots" are consistent with the idea

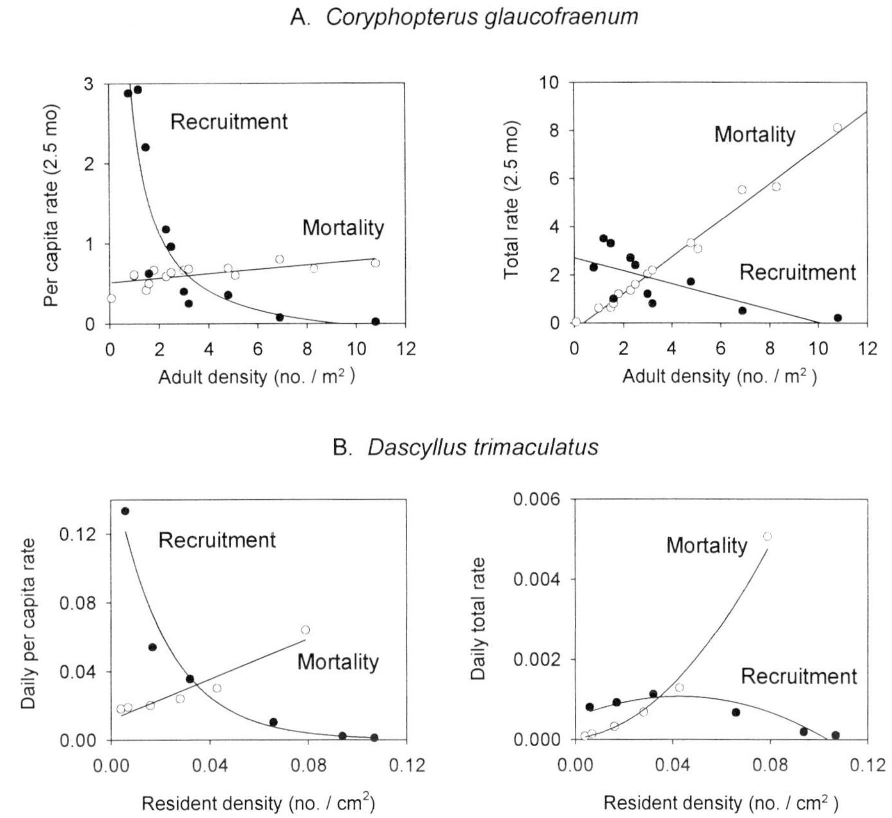

FIGURE 6 Comparisons of recruitment and mortality curves, presented equivalently as per-capita rates (left plots) and as total rates (right plots), from studies of (A) a goby in the British Virgin Islands [experimental data from Forrester (1995)] and (B) a damselfish in French Polynesia [observational data from Schmitt and Holbrook (1996)]. If these demographic curves were static and postsettlement movement was negligible, each population would come to an equilibrium at the density where the recruitment and mortality curves cross. In any case, the data suggest regulated local populations.

that local population fluctuations are bounded due to density-dependent processes.

IV. Mechanisms Causing Density Dependence

Few studies have provided data on the processes responsible for density-dependent recruitment and mortality in coral reef fishes, so we not only review available data but also offer speculation.

A. Recruitment

The studies reviewed in Section III,A indicated that the total recruitment rate can be density dependent, density independent, inversely density dependent, or a combination of these in the form of a unimodal relationship between recruitment and local population density. Explicit data on underlying mechanisms are lacking. Density dependence in the total recruitment rate indicates that increasing densities of resident fishes inhibit subsequent establishment of new recruits, presumably due to intraspecific competition and/or cannibalism. Density independence suggests no effect of resident fishes on recruitment. Inverse density dependence indicates facilitation whereby settling fishes are attracted to conspecifics and/or enjoy enhanced survival in larger groups (Sweatman, 1985b). The unimodal pattern reported for humbug damselfishes (genus *Dascyllus*) by Schmitt and Holbrook (1996) suggests that facilitation operates at lower population densities, whereas inhibition occurs at higher densities, apparently as a function of increasing intraspecific competition. Booth (1995) documented and modeled demographic cost–benefit trade-offs in *Dascyllus*. Survival of recruits was enhanced in larger groups, but growth was reduced.

B. Mortality

Predation is the predominant source of mortality in coral reef fishes, especially shortly after settlement [reviewed through 1990 by Hixon (1991); subsequently, see references in Table 2] (see also Shpigel and Fishelson, 1991; Caley, 1993, 1995a,b; Hixon and Beets, 1993; Carr and Hixon, 1995; Caley and St. John, 1996; Connell, 1996, 1997a; Beets, 1997; Craig *et al.*, 1997; Eggleston *et al.*, 1997; Nemeth, 1998). Seven of the studies listed in Table 2 provide data directly implicating predation as the source of density-dependent mortality, as well as indicating the role of habitat complexity and the synergistic effects of various groups of predators and interference competitors. These studies are interrelated conceptually, so we simply present them chronologically.

1. Empirical Evidence

Forrester (1990) determined experimentally that juveniles of the damselfish *Dascyllus aruanus* suffered density-dependent mortality (and growth) over 10 months in One Tree Lagoon, Great Barrier Reef. As part of the experiment, planktonic food was supplemented artificially, causing enhanced growth rates relative to controls. Forrester concluded that, at higher local population densities, competition for food inhibited growth, which translated to density-dependent mortality, possibly via predation. This conclusion is in accordance with what has been called the "growth-mortality hypothesis" or "stage-duration hypothesis" from fisheries biology, which predicts that increasing competition for food will increase the time required for small fishes to grow to a size relatively invulnerable to predation, thereby increasing overall mortality (Ricker and Foerster, 1948; Shepherd and Cushing, 1980; Chambers and Leggett, 1987; Houde, 1987; review by Sogard, 1997).

Beukers and Jones (1997) observed density-dependent mortality of new recruits of the damselfish *Pomacentrus moluccensis* over 1 year at Lizard Island, Great Barrier Reef (see Section III,B,1). They also observed that the mortality rate of new recruits was directly correlated with the density of piscivores, and inversely correlated with the percent cover of structurally complex corals. They determined in laboratory experiments that three common piscivores (*Pseudochromis fuscus*, *Cephalopholis boenak*, and *Thalassoma lunare*) were more successful at capturing these prey when they inhabited structurally simple corals (*Acropora nobilis*) rather than complex ones (*Acropora nasuta* and *Pocillopora damicornis*). In a field experiment, they cross-factored predators (present vs. absent) and corals (simple vs. complex) with one level of initial prey density. After 2 months, reefs of high complexity with predators supported the same number of prey as reefs of low or high complexity without predators. Damselfish abundance was significantly lower on the low-complexity reefs with predators. These results implicate predation as the source of density-dependent mortality, and indicate the importance of habitat complexity in mediating predation (review by Hixon, 1991).

Hixon and Carr (1997) were the first to demonstrate unequivocally that predation causes density-dependent mortality in marine fishes. They ran a factorial experiment on new recruits of the damselfish *Chromis cyanea* at Lee Stocking Island, Bahamas, using 32 patch reefs in a regression design: eight levels of initial population size (within the natural range) × resident predators (serranid groupers, etc.) present vs. absent × transient predators (carangid jacks, etc.) present vs. absent. Predator treatments were effected using large enclosures surrounding individual reefs, which produced no substantial artifacts (see also Steele, 1996; Connell, 1997a). After 1 month, mortality of new recruits was density dependent only in the natural control treatment in which both sources of predation were present (Fig. 7A). In the absence of all predators, mortality was low and density independent. There was a trend toward density dependence in the treatments with only one source of predation, but high variance precluded statistical significance.

Behavioral observations via automated time-lapse video samples showed that these results were due to "synergistic predation" involving the combined effects of resident and transient piscivores. On reefs with no transient predators, recruits sheltered high in the water column, where they fed on zooplankton. (Resident predators did not enter the water column, presumably because they were prey of still larger piscivores, such as sharks and barracuda, as indicated by observations of such predation and tooth scars on resident predators.) Conversely, on reefs with no resident predators, *Chromis* recruits simply sheltered in the reef at the approach of schools of transient predators. In both circumstances, predation was highly variable in time and space due to the vicissitudes of prey vigilance and predator stealth. Only in the presence of both suites of predators were the prey particularly vulnerable—sandwiched between resident predators attacking from below and transient predators attacking from above. There was evidence that the mechanism of density dependence involved transient jacks (*Caranx ruber*) exhibiting an "aggregative response" (sensu Hassell and May, 1974), congregating at reefs with many prey and ignoring reefs with few prey. Hixon (1998) had documented

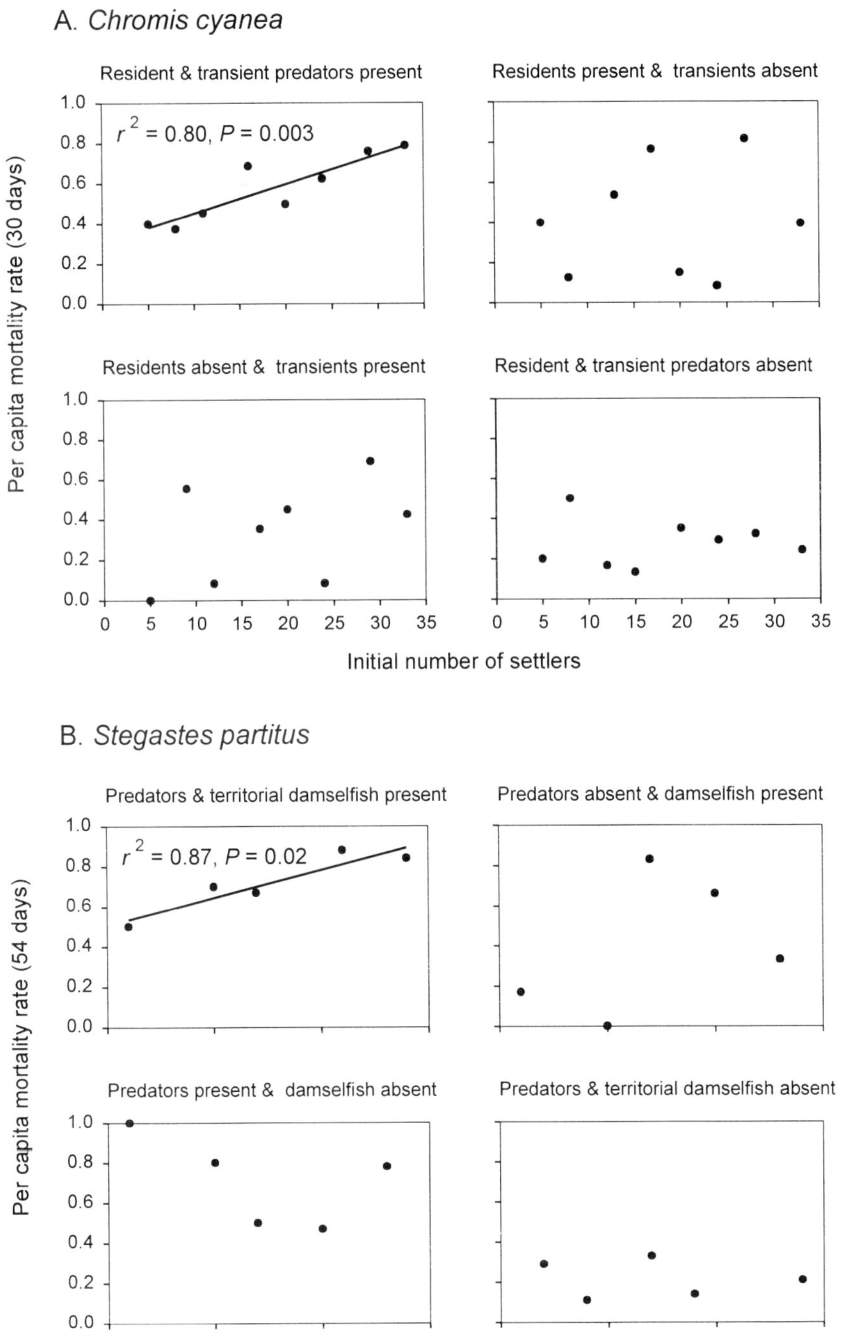

FIGURE 7 Factorial field experiments in the Bahamas illustrating that synergistic interactions cause density-dependent mortality in new settlers of damselfishes. Each point represents one patch reef. (A) For nonterritorial schooling *Chromis cyanea,* significant density dependence occurred only in the natural control treatment, where both resident and transient predators were present. Mortality was low and density independent in the complete absence of predation [data from Hixon and Carr (1997)]. (B) For individually territorial *Stegastes partitus,* significant density dependence occurred only in the natural control treatment, where both predators and larger territorial damselfishes were present. Mortality was low and density independent in the complete absence of these strong interactors [data from Carr *et al.* (in preparation)].

a temporal aggregative response involving the same predator and prey species in St. Thomas, United States Virgin Islands (see also Hixon and Beets, 1993).

Steele *et al.* (1998), using the same experimental system in the Bahamas as Hixon and Carr (1997), found that per-capita recruitment of the goby *Coryphopterus glaucofraenum* was density dependent in the first of 2 years (the experiments running 32 and 15 days, respectively). Interestingly, this pattern appeared to occur despite predation rather than because of it. In fact, there was a significant positive effect of resident predators on goby recruitment (and no effects of transient predators). The authors speculated that this pattern may have been an indirect effect: resident piscivores may have displaced or consumed smaller generalized predators (such as wrasses) that attacked gobies. They concluded that interference competition with adult conspecifics was the predominant mechanism causing density-dependent recruitment in this goby. In the same experiments, Forrester and Steele (2000) reported that mortality was apparently density dependent regardless of the presence or absence of predators. Besides the possibility that the cages did not exclude very small predators (such as wrasses and stomatopods), it is possible that interference competition caused gobies to emigrate from predator-exclusion cages, where they became vulnerable to predation.

Connell (1998a) also used cages to exclude larger predators from newly hatched schools of *Acanthochromis polyacanthus,* one of the few species of coral reef teleosts that do not undergo larval dispersal. Working in One Tree Lagoon in the Great Barrier Reef, he found that the difference in hatchling survival between caged and uncaged plots was the same as the difference between reef habitats where piscivores were naturally low and high, respectively. In a second experiment, per-capita mortality during the first 2 weeks posthatching was density dependent in uncaged plots but not caged plots, demonstrating that predation was the source of density dependence.

Hixon and Jones (in preparation) manipulated predators by removal at Lizard Island on the Great Barrier Reef. On a set of 48 patch reefs in the sandy central lagoon, they cross-factored recruitment density of the damselfish *Pomacentrus amboinensis* (three natural levels) and resident predators (present vs. absent), with eight replicates of each treatment (two-way ANOVA design). In this case, per-capita mortality after 17 months was density dependent regardless of the presence or absence of resident predators. It appeared that the highly complex coral that dominated these reefs (*Porites cylindrica*) inhibited piscivores (see also Beukers and Jones, 1997). Because density-dependent growth was evident, the ultimate cause of density-dependent mortality appeared to be exploitative competition for planktonic food, with predation on slower growing fishes being the proximate mechanism. These findings are reminiscent of the conclusions of Forrester (1990).

Finally, Carr *et al.* (in preparation) experimentally examined the effects of both predators (residents and transients combined) and interference competitors (territorial damselfish) on new recruits of the damselfish *Stegastes partitus* at Lee Stocking Island, Bahamas. This damselfish is loosely territorial and interacts aggressively with several congeners (Robertson, 1996). Using the same patch reefs and cages as Hixon and Carr (1997), Carr *et al.* (in preparation) used a regression design subjecting five natural density levels of new recruits to two cross-factored treatments: all predators present or absent × all territorial damselfishes present or absent. After 54 days, the results were similar to those of Hixon and Carr (1997) in that density-dependent mortality occurred only in the natural control treatment where all predators and territorial damselfish were present (Fig. 7B). Unlike the results of Hixon and Carr (1997), the mechanism underlying Carr *et al.'s* (in preparation) results appeared to be a synergistic interaction of predators and interference competitors (rather than a synergism between two kinds of predators). Again, behavioral field observations provided insight. In the absence of larger territorial damselfishes but in the presence of predators, new recruits of *S. partitus* vigilantly remained close to small shelter holes in the reef, and mortality was occasional and density independent. In the presence of larger interference competitors and the absence of predators, the larger fish territorially harassed the new recruits, but again, mortality was density independent. Only when both territorial damselfishes and predators were present was mortality density dependent, apparently because aggressive harassment by territorial fishes rendered new recruits more vulnerable to density-dependent predation (due to expulsion from shelter, increased conspicuousness, and/or decreased vigilance).

2. Mechanistic Issues

Predation has been demonstrated or implicated in most studies to date as the proximate or ultimate mechanism causing postsettlement density-dependent mortality (see also Rice *et al.,* 1997). Limited evidence suggests the importance of an aggregative response of transient predators, congregating at sites where new recruits are abundant and ignoring sites where new recruits are uncommon. Note that such spatial density dependence may not necessarily translate to the temporal density dependence required for population regulation (Stewart-Oaten and Murdoch, 1990). Predator aggregation may simply redistribute mortality among patches without changing its overall level (Harrison and Cappuccino, 1995). Of importance is the spatial range of the predator population relative to that of the prey (Hanski, 1990). The detection of density-dependent predation at scales larger than the movement patterns of transient piscivores would reflect a type 3 functional response of those predators and thus approximate temporal density dependence (Myers and Rothman, 1995). A type 3 functional response occurs if individual predators switch to alternative prey when any given prey species is scarce (for details, see Murdoch and Oaten, 1975; Murdoch and Bence, 1987; Sinclair and Pech, 1996). In reef fishes, each piscivore would disproportionately consume new recruits of a particular species when those prey are abundant, and disproportionately ignore that prey species when it is rare. Additional, mechanistic studies of piscivory in reef fishes are sorely needed.

Few studies have adequately examined the extent to which estimates of mortality are confounded by emigration. There are two ways to separate these processes: (1) study highly reef-attached species on highly isolated reefs, or (2) tag fish and search surrounding habitats for emigrants. Of the 27 studies listed in Table 2, only 6 both examined reef-attached species and reported reef isolation distances of at least 25 m (Doherty, 1982; Hixon and Carr, 1997, in preparation; Forrester and Steele, 2000; Carr *et al.*, in preparation; Hixon and Jones, in preparation), and only 5 used tagged fish (Jones, 1987b; Forrester, 1990, 1995; Booth, 1995; Webster, in preparation). Increased effort at understanding the demographic consequences of emigration is clearly needed.

Competition can potentially be the ultimate cause of demographic density-dependent mortality. Because habitat complexity can mediate the effects of piscivory on coral reef fishes (e.g., Beukers and Jones, 1997; Hixon and Jones, in preparation), predation may induce competition for structural shelter if such prey refuges are limiting (Hixon, 1991). Regarding competition for food, aggressive interference appears to be the ultimate source of density-dependent mortality in highly social species, such as planktivorous damselfishes of the genus *Dascyllus* (e.g., Forrester, 1990), and territorial species, such as more benthivorous damselfishes of the genus *Stegastes* (Carr *et al.*, in preparation). Density-dependent growth (review by Jones, 1991) presumably leads to density-dependent mortality as smaller fishes are consumed by predators (Hixon and Jones, in preparation). Proximally, aggressive harassment of smaller fishes by larger fishes may increase the vulnerability of small fishes to density-dependent predation (Carr *et al.*, in preparation).

V. Discussion and Synthesis

A. Empirical Summary

This literature review suggests two major patterns. First, observational and experimental studies of 10 species from five families show that the total local recruitment rate varies in a broad variety of patterns as a function of local population size (Table 1; Section III,A). Tentatively, we conclude that recruitment not a consistent source of demographic density dependence. However, in every study, conversion from total to per-capita recruitment resulted in apparent or pseudo-density dependence (Section II,C,2), which can provide a nonmechanistic kind of local population equilibrium unrelated to the regulation of the entire metapopulation (Bence and Nisbet, 1989; Hughes, 1984, 1990; Caley *et al.*, 1996) (Section III,C).

Second, data on 20 species from six families show that postsettlement mortality is often density dependent, especially shortly after settlement (Table 2; Section III,B), and caused largely by predation (Section IV,B). Overall, 17 of 20 species experienced density-dependent per-capita mortality at some time and place. In most studies that reported density-independent mortality, older juvenile fishes were studied, suggesting that regulation via mortality is an early postsettlement phenomenon.

The prevalence of density-dependent mortality has two ramifications. First, local populations may be an important source of regulation for the entire metapopulation, suggesting the importance of conserving the mechanisms causing local density dependence. Second, the more restrictive version of the recruitment limitation hypothesis, which predicts that postsettlement mortality will be density independent (Section II,C,2), has been falsified in most cases. However, evidence that some local populations are recruitment limited nonetheless exists (Section III,B). Thus, recruitment limitation remains a viable hypothesis, provided that it is defined explicitly (Caley *et al.*, 1996; Hixon, 1996a, 1998).

B. Are Reef Populations Model Systems for Marine Fishes?

Coral reef fishes support major fisheries (Russ, 1991; Polunin and Roberts, 1996) (see Chapters 16–18, this volume), are the object of conservation efforts in their own right (Roberts and Hawkins, 1999; Hawkins *et al.*, 2000), and are ideal subjects for both observational and experimental population studies (Sale, 1991b). Yet, are they relevant to understanding population dynamics in other demersal marine fishes, especially fishery species? Certainly, there are major differences between tropical reef species and temperate demersal species in terms of phylogeny, species diversity, the complexity of habitats and communities, and a variety of other parameters. Yet, we are more impressed by two major similarities.

First, despite earlier generalizations to the contrary (e.g., Shepherd and Cushing, 1990), density-dependent mortality (especially early during the postsettlement phase) is becoming increasingly well documented in temperate fishes, both in fishery species (e.g., Lockwood, 1980; Cook and Armstrong, 1986; van der

Veer, 1986; van der Veer and Bergman, 1987; Sundby *et al.*, 1989; Fogarty *et al.*, 1991; Laevastu and Bax, 1991; Beverton and Iles, 1992; Myers and Cadigan, 1993a,b; Bailey, 1994; Rijnsdorp, 1994; Myers *et al.*, 1995a; Ralston and Howard, 1995; Bailey *et al.*, 1996; Bjørnstad *et al.*, 1999) and in nonfishery species (e.g., Jones, 1984; Carr, 1994; Tupper and Boutilier, 1995b; Sano, 1997; Steele, 1997a,b, 1998; Planes *et al.*, 1998c; Anderson, 2001; see also reviews by Jones, 1988a; Ebeling and Hixon, 1991).

Second, predation is coming under increasing scrutiny as the cause of density dependence in temperate fishes, both in fishery species (e.g., Sissenwine, 1984; van der Veer and Bergman, 1987; Fogarty *et al.*, 1991; Laevastu and Bax, 1991; Bailey, 1994) and in nonfishery species (e.g., Tupper and Boutilier, 1995b; Steele, 1997b, 1998; Anderson, 2001). These similarities suggest that concepts and hypotheses derived from mechanistic studies of coral reef fishes are worthy of consideration by temperate fisheries biologists and fish ecologists for understanding sources of population regulation.

C. Conceptual Synthesis

What generalities, perhaps relevant to other demersal marine species, can be derived from this review regarding population dynamics and regulation of coral-reef fishes? Importantly, we are not currently in a position to evaluate what regulates entire metapopulations (see Section V,D). The substantial evidence to date focuses on local recruitment and postsettlement mortality, which do interact in a way suggesting local regulation (Section III,C), so we synthesize empirical generalities only at the local scale with a simple model.

Consider an open local population whose dynamics involve only recruitment (input) and mortality (loss). Regarding input, given that any pattern of total recruitment seems to be possible (Table 1; Section III,A), assume (to be conservative) that the total recruitment rate is generally density independent. From the discussion in Section II,C,2, we know that conversion to a per-capita rate results in apparent or pseudo-density dependence (i.e., a hyperbolic relationship between per-capita recruitment and local population size). It is also well-documented that recruitment is variable at the spatial and temporal scales thus far examined (reviews by Doherty and Williams, 1988; Doherty, 1991; Jones, 1991; Williams, 1991; Booth and Brosnan, 1995; Caley *et al.*, 1996), so temporal variation in the recruitment rate must be part of any synthesis.

Regarding loss, given that local mortality is density dependent at some time and place for the great majority of species studied to date (Table 2; Section III,B), assume that per-capita mortality is density dependent, and for simplicity, assume a linear function. Assume further that mortality, like recruitment, is also variable through time. This assumption is based on the observation that density-dependent mortality in reef fishes appears to be caused mostly by predation (see Section IV,B), and that recruitment of predatory reef fishes is as variable as that experienced by prey species (e.g., Shenker *et al.*, 1993; see also Levin, 1998).

The above assumptions give the following per-capita model for an open local population (n):

$$\frac{1}{n}\frac{dn}{dt} = \frac{R(t)}{n} - [D(t)n + I(t)],$$

where $R(t)$ is the total density-independent recruitment rate (dividing by n makes this rate pseudo-density dependent). The quantity $D(t)n + I(t)$ describes the per-capita mortality rate, where $D(t)n$ is the density-dependent component [$D(t)$ defining the slope] and $I(t)$ is the density-independent component (y intercept; cf. Fig. 1). Both the recruitment rate and the mortality rate vary through time and thus are functions of t (Fig. 8A). Converting to total rates gives the following equation (Fig. 8B):

$$\frac{dn}{dt} = R(t) - [D(t)n + I(t)]n.$$

An equilibrial population size can be calculated from either equation, but will be of little use due to the unpredictable nature of variance in recruitment and mortality, both driven by external processes. Any theoretical equilibrium point will vary through time as the positions of the recruitment and mortality curves shift in slope and position (Fig. 8). Thus, local population size will range between some lower and upper limits, and the projected equilibrial point at which the curves cross will appear through time as a "cloud of points" (sensu Wolda, 1989). The key point is that this variation in abundance is nonetheless regulated (i.e., bounded above zero) via density-dependent mortality, but only as long as recruitment does not go to zero for the remaining life-span of the longest lived individual [cf. Chesson's (1983, 1984) "storage effect"]. (Of course, due to metapopulation structure at the global scale, a local population can go extinct and start again with a new pulse of recruitment.) Depending on time lags, the population may track but never reach the ever-changing equilibrium point, such that population fluctuations may appear stochastic, i.e., regulated populations need not be constant in size and may appear to fluctuate randomly as a stationary stochastic process (Turchin, 1990, 2001).

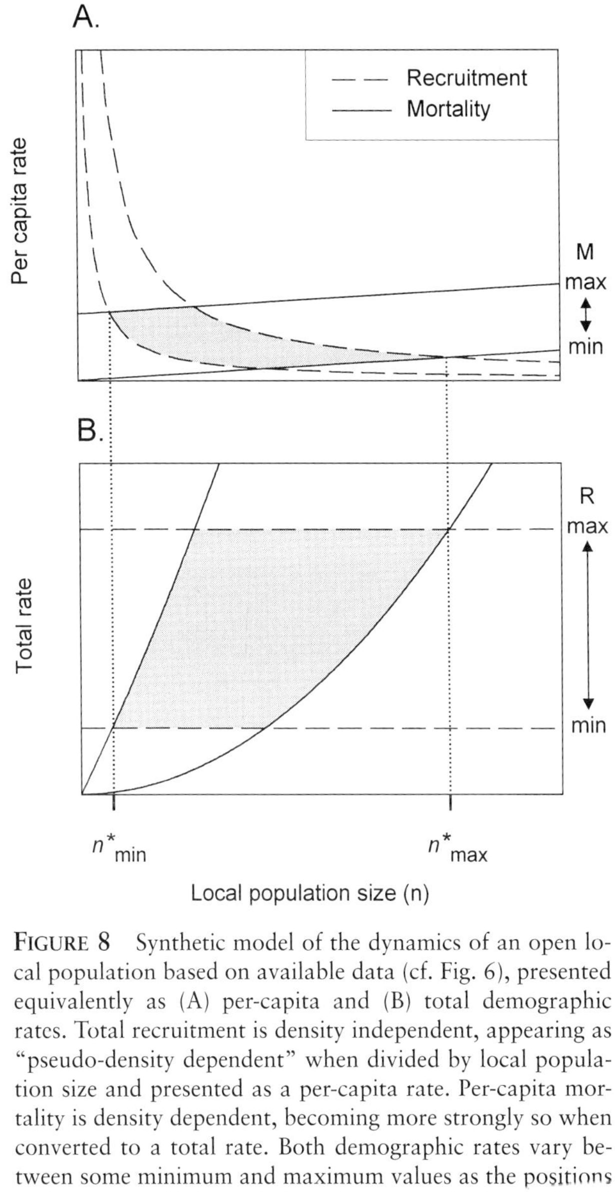

FIGURE 8 Synthetic model of the dynamics of an open local population based on available data (cf. Fig. 6), presented equivalently as (A) per-capita and (B) total demographic rates. Total recruitment is density independent, appearing as "pseudo-density dependent" when divided by local population size and presented as a per-capita rate. Per-capita mortality is density dependent, becoming more strongly so when converted to a total rate. Both demographic rates vary between some minimum and maximum values as the positions of the curves shift through time. Consequently, the equilibrial intersection of the curves varies throughout the shaded area, and local population size varies between n^*_{min} and n^*_{max}, perhaps never reaching equilibrium. Thus, temporal variability leads to dynamics that may be substantial and appear random, even though the local population is regulated.

Note that this simple model does not include an Allee effect, whereby population growth becomes inversely density dependent at low population sizes, pushing the population toward extinction (review by Courchamp *et al.*, 1999). The focus here is a local population sufficiently large to preclude this effect. Overall, we simply wish to stress that reef fish population dynamics cannot be explained by variation in a single process (see also Hixon, 1991, 1998; Jones, 1991; Roberts, 1996; Chesson, 1998b).

D. Future Directions

Coral reef fish ecology is a new discipline, having emerged in the 1960s but not gaining wide recognition until the 1970s. In "The Ecology of Fishes on Coral Reefs" (Sale, 1991a), several major hypotheses relevant to reef fish population dynamics were reviewed (Doherty, 1991; Hixon, 1991; Sale, 1991b; Williams, 1991), yet none of them were then corroborated by substantial data (see also Mapstone and Fowler, 1988; Sale, 1990). The new millennium is an exciting time in reef fish ecology because detailed observational and experimental studies are becoming more the norm than the exception, so definitive hypothesis testing is on the rise. With an eye on the future, we offer several suggestions for continued study of reef fish populations.

1. Explore Mechanisms Underlying Demographic Patterns

Testing for and detecting density dependence buried in a field of stochastic demographic variation is certainly a challenge in and of itself. However, for practical aspects of managing fisheries and conserving fish populations, understanding the sources and causes of population dynamics in general, and density dependence in particular, is crucial. We believe that the most useful empirical approach combines observational and experimental exploration of multiple processes at multiple spatial and temporal scales (see Hixon, 1998). Nested spatial designs, such as those reported by Hughes *et al.* (1999) for corals, would be especially valuable to determine the scale at which density dependence operates (see Ray and Hastings, 1996; Sale, 1998; Chesson, 1998b). Additional exploration of the density-independent component of postsettlement mortality is also needed (e.g., Shulman and Ogden, 1987; Levin, 1998), as is greater knowledge of apparently special cases of inversely density-dependent mortality (e.g., Booth, 1995). Especially welcome would be more studies examining the relative contributions of density-dependent and density-independent factors in driving population dynamics (e.g., Schmitt and Holbrook, 1999b; Schmitt *et al.*, 1999; Shima, 1999b).

Currently, there are three major approaches to examine the relative contribution of density-dependent vs. density-independent factors. The simplest way to compare the two components of mortality is to determine the slope and *y* intercept of the per-capita mortality rate plotted as a function of initial population size. Assuming the slope is positive, the *y*-intercept divided by the overall mortality rate at any given population size is the proportion of mortality that is density independent (Fig. 1). A more sophisticated approach is to fit a Shepherd (1982) stock-recruitment curve to a recruit-adult (or recruit-juvenile) plot and interpret the components of the descriptive equation (Schmitt *et al.*, 1999). This method has the advantage of providing the relative contribution of recruitment and both components of mortality to local population size. The disadvantage is that this approach is data intensive. The third approach is to construct detailed Leslie-matrix-type demographic models and examine the relative contribution of proportional variation in various parameters via elasticity analysis (Heppell *et al.*, 2000).

Although requiring much time and effort to implement, studies of temporal density dependence (preferably over multiple generations) are also needed. Nearly all studies to date investigated and detected only spatial density dependence, which may not necessarily translate to the temporal density dependence required for population regulation (see Section IV,B,2). Fortunately, otolith microchemistry (Thorrold and Hare, this volume), elastomer injections and microtags (Buckley *et al.*, 1994; Beukers *et al.*, 1995), and other recent advances in methodology now allow the entire lives of fishes to be tracked in unprecedented detail.

When and where possible (admittedly a major challenge), we encourage multifactorial field experiments for the strongest possible inference on causation (see Connell, 1975). Given that postsettlement density dependence is often induced by mortality caused directly or indirectly by predation (see Section IV,B), we suggest an experimental design in which a natural range of initial population sizes is monitored in the presence and absence of predators (Caley *et al.*, 1996; Hixon, 1998). Assuming that (as the data in Section III,B indicate), first, predators in fact increase prey mortality rates, and, second, mortality is not inversely density dependent, the four possible outcomes of this experimental design provide insight on the relative importance of predation and competition in causing density-dependent mortality (Fig. 9). For example, Hixon and Carr's (1997) experiment (Fig. 7A; Section IV,B,1) produced results illustrated in Fig. 9C, demonstrating that predation was the sole cause of early postsettlement density dependence in *Chromis cyanea* in the Bahamas. Carr *et al.'s* (in preparation) experiment gave the same results for *Stegastes partitus* in the Bahamas (Fig. 7B), but with the important difference that interference competitors were necessary for the predation effect to occur. Finally, Forrester and Steele's (2000) experiment produced results illustrated in Fig. 9D, indicating several possible sources of density dependence for *Coryphopterus glaucofraenum* in the Bahamas.

Accurate interpretation of experimental outcomes requires detailed knowledge of the study system, including the natural history and behavior of the species

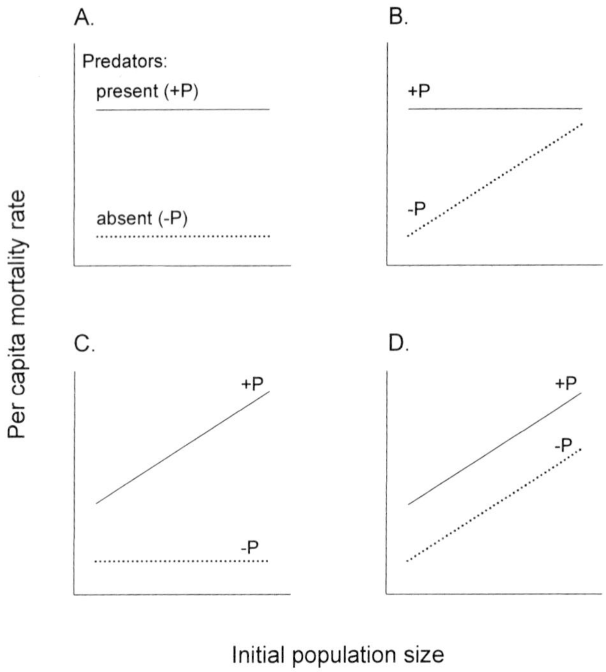

FIGURE 9 Alternative outcomes of a factorial experiment designed to test for both the presence of and the source of density-dependent mortality. (A) Mortality is density independent regardless of the presence or absence of predators that consume the study species, which indicates recruitment limitation (sensu Doherty, 1998). (B) Mortality is density independent in the presence of predators, but density dependent when predators are removed, indicating that predation normally precludes density dependence caused by competition. (C) Mortality is density dependent in the presence of predators, but density independent when predators are removed, indicating that predation alone induces density dependence. (D) Mortality is density dependent regardless of the presence or absence of predators. This fourth outcome could be due to competition causing density dependence regardless of predators, or because competition causes density dependence in the absence of predation whereas predators cause density dependence when they are present.

involved. Experiments alone cannot tell the whole story. Innovative observational methods, including ultrasonic tagging (e.g., Holland *et al.*, 1996) and automated video monitoring (e.g., Carr and Hixon, 1995; Holbrook and Schmitt, 1997), provide means of sampling fish behavior in the absence of divers. Such approaches will help to identify the behavioral mechanisms driving and regulating population dynamics, including the roles of predation (Section IV,B), competition (both within and between species), and postsettlement movement (Section II,C,2). Carefully executed laboratory studies can also provide a useful supplement for field data (e.g., Anderson, 2001).

2. Create a Centralized Demographic Data Base

Reviewing the literature for this chapter was tedious because there was extreme variation in the way demographic data were presented. Particularly difficult to interpret were indirect indices of recruitment and mortality. Recent strides in interpreting fisheries data, including meta-analyses, have been facilitated by creation of a centralized archive of stock-recruitment data by Myers *et al.* (1995b). Centrally collating and updating demographic data on reef fishes in standardized units and format would greatly facilitate exploration for general patterns.

3. Develop Empirically Based Demographic Models

There is an immense literature regarding population models based on field data [for entry, see Cappuccino and Price (1995)]. Such models allow one to explore the relative importance of different processes in driving and regulating population dynamics, as well as the ability to project population trends. Unfortunately, most models assume closed populations and so are not applicable to many local populations of reef fishes. However, the past decade has seen the development of practical approaches for open populations, including both size- and age-structured models (see Chapter 3, this volume, on the importance of this distinction). Earlier models of coral reef fishes were simple simulations focusing on system-specific questions (e.g., Shulman and Ogden, 1987; Booth, 1995). A spatially explicit analytical model is developed in Chapter 13 of this volume. Another approach is Leslie matrix modeling modified for open populations (e.g., Hughes, 1984, 1990). Pfister (1996) applied such a model to temperate tide pool fishes and demonstrated how one can evaluate the relative importance of recruitment vs. mortality in driving changes in local population size via sensitivity analysis. More appropriate for comparing factors measured in different currencies is elasticity analysis (Heppell *et al.*, 2000). Greater use of empirically based models, especially regarding the dynamics of metapopulations, will help to increase conceptual generalization and develop more explicit hypotheses (Gaines and Lafferty, 1995).

4. Expand Taxonomic Coverage and Demographic Scale of Study

Damselfishes (family Pomacentridae) are disproportionately represented in the studies listed in Tables 1 and 2. Pomacentrids are certainly superb study species, but we believe that the time has come to expand the taxonomic coverage of demographic studies, and especially to include important food fishes (see Chapter 17, this volume). Greater exploration of multispecies density dependence is needed. To truly understand population dynamics, one must examine all demographic rates and their components simultaneously. In particular, greater focus on the demographic consequences of variable growth and fecundity is needed. Ultimately, integrated studies of the entire life cycle, including the pelagic larval phase, would provide a complete picture of population dynamics (Hixon, 1998). Although theory suggests that a metapopulation can persist only if some local populations are regulated (e.g., Hanski *et al.*, 1996), the spatiotemporal pattern of density dependence required for global persistence remains unknown. Although the logistic challenges are immense, combined demographic studies of both the presettlement and postsettlement stages are needed for insight on the dynamics and regulation of metapopulations (Section II,C,1).

Studies of coral reef fishes have contributed much to basic ecology, including concepts later adopted by terrestrial ecologists, such as competitive lotteries (Sale, 1977), recruitment limitation (Doherty, 1981), the storage effect (Chesson, 1983, 1984), diffuse predation (Hixon, 1991), and synergistic interactions (Hixon and Carr, 1997). We believe that, with increased emphasis on understanding population dynamics, reef fish ecologists also have much to contribute conceptually to applied problems in fisheries management and conservation of demersal fishes worldwide.

Acknowledgments

Many thanks to colleagues who generously shared their raw data for reanalysis, especially D. J. Booth, G. E. Forrester, S. J. Holbrook, R. J. Schmitt, and H. P. A. Sweatman. Constructive reviews by T. W. Anderson, P. R. Armsworth, B. A. Byrne, P. L. Chesson, S. J. Holbrook, R. A. Myers, K. L. Overholtzer, P. F. Sale, R. J. Schmitt, and P. Turchin greatly improved the manuscript. Special thanks to P. R. Armsworth, P. L. Chesson, H. P. Possingham, and P. Turchin for help in clarifying our theoretical musings. Our sincere apologies to authors whose relevant work we either overlooked or misinterpreted. We acknowledge support for our research cited herein provided by the National Science Foundation (OCE-92-17163, OCE-96-17483, and OCE-00-93976 to MAH and a Predoctoral Fellowship to MSW), by NOAA's National Undersea Research Program (CMRC-92-46, 93-12, 94-15, 95-3042, and 97-3109 to MAH), and by the Caribbean Marine Research Center. MAH dedicates this chapter to his children, Chelsea Renee and Sean Wolf, and MSW is grateful to T. L. Freidenburg for continued support.

CHAPTER 15

Variable Replenishment and the Dynamics of Reef Fish Populations

Peter J. Doherty
Australian Institute of Marine Science
Cape Ferguson, Townsville
Queensland 4810, Australia

I. Introduction: Why Study Recruitment?

In late 1995, four dozen marine ecologists retreated to an Australian rainforest to review the state of coral reef fish ecology (Jones *et al.*, 1998). Selected contributions were published as a special issue of the *Australian Journal of Ecology* (Caley, 1998) and stand as a midterm review between this book and the last one (Sale, 1991a). American and Australian granting agencies were persuaded to fund this bilateral meeting largely on arguments that the "recruitment bandwagon" (Sale, 1990) needed a major wheel alignment. At the center of the debate was a suggestion that "recruitment limitation" is inconsistent with population regulation (Hixon, 1998), but in this chapter it will be shown that fish populations can display both properties. This conclusion is not original but it obviously needs restatement. Like many others, I believe that it is the needless polarization of positions rather than the bandwagon that has run its course.

There is ample evidence that populations of coral reef fishes can experience variable replenishment (Doherty and Williams, 1988; Doherty, 1991) and that these variations can have lasting impacts on patterns of demography and abundance (Section V). However, to suggest that recruitment can be a major determinant of spatial and temporal pattern in fish populations is not to deny that these populations can be regulated through different mechanisms operating at other times and places (Sale and Tolimieri, 2000). All populations have ultimate limits on abundance set by finite resources, and coral reef fishes are no exception. Populations of siganids in Guam (Kami and Ikehara, 1976), acanthurids in Minicoy Atoll (Pillai *et al.*, 1983) and Johnston Atoll (Sancho *et al.*, 1997), balistids in Panama (Robertson, 1988b), and serranids in Moorea (Dufour *et al.*, 1996) and La Reunion (Letourneur *et al.*, 1998) have all shown conspicuous mass mortalities following exceptional replenishment. There is no disputing that fish stocks are regulated at places and times by density-dependent processes.

Modern theoretical models of reef fish populations tend to be based on bounded stochastic dynamics (Chesson, 1998a; Armsworth, 2002), with variable replenishment of populations providing a strong source of change when populations are not near their upper limits. Under this density-vague scenario (Strong, 1986), population dynamics will feature both density-dependent and density-independent fluctuations, and the relative importance of each varies with context. When replenishment is drawn from strongly skewed probability distributions, density dependence will be most important at the extremes of density and its primary effect may be to dampen exceptional cohorts (Armsworth, 2002). Between such episodes of regulation, there may be little evidence of density dependence and density-independent fluctuations can be dominant.

A couple of examples show the value of monitoring replenishment. In 1992, the Great Barrier Reef Marine Park Authority (GBRMPA) halted all fishing on Bramble Reef because of community concerns about the state of fish stocks at that popular destination. A consultant was hired to monitor the recovery of coral trout (*Plectropomus leopardus*), which is the primary target of the line fishery (Ayling and Ayling, 1992, 1994, 1997). In 1996, GBRMPA reopened Bramble Reef to fishing in response to community demand and because coral trout had become five times more numerous than before the closure; this was a very satisfactory outcome for management. What the fishing public never heard was that densities of this grouper were higher also on control reefs that had remained open to fishing because of strong replenishment of all populations during the 1992 spawning season. Without such knowledge, the closure of Bramble Reef would impart a misleading impression of the resilience of coral trout populations when rested from fishing pressure. Russ and Alcala (1996a) showed similar effects on the accumulation of biomass within marine reserves in the Philippines. These are just two examples among many that show that recruitment matters at scales relevant to human interactions with tropical fish stocks.

II. Background

Marine ecologists have long debated the relative importance of external forcing (via variable replenishment) versus internal regulation (via density-dependent change in demographic variables) as determinants of population abundance. This debate has been widespread because so many marine organisms have complex life cycles (*sensu* Roughgarden *et al.*, 1988), which involve multiple life history stages and/or more than one habitat. Our examples will be the benthic (bottom-dwelling) fishes of coral reefs that spawn propagules (gametes or larvae) into the ocean, where they have a pelagic (water column) existence before returning to the reef to become the next generation of benthic juveniles and adults.

Benthic life starts with the settlement of pelagic juveniles, although this transition is more or less abrupt for different species (Kaufman *et al.*, 1992; Finn and Kingsford, 1996; McCormick and Makey, 1997). The abundance of presettlement fish is often called *larval supply*, even though few fish remain in a larval state by this time (Balon, 1999). Settlement is generally nocturnal, presumably to reduce predation risk on settlers (but see Connell, 1998b; Danilowicz and Sale, 1999). It is also patchy in space and time (Doherty, 1991). This combination of factors means that direct observations of settlement are rare without cumbersome technology (Holbrook and Schmitt, 1997). Typically, divers count newly settled fish in recruitment surveys, often with an intention to record an index of settlement and/or larval supply (Doherty, 1987a; Masterson *et al.*, 1997).

Spatial context is also important. Coral reefs vary in geomorphology from the isolated oceanic pinnacles of French Polynesia to the complex archipelagoes of the Great Barrier Reef (GBR) and continental fringing barrier reefs stretching hundreds of kilometers (mesoamerican system). At finer resolution, all of these environments are fragmented into complex habitat mosaics. Some fish respond to very fine grain in their environment and may pass their entire benthic existence within a single coral head or solitary anemone (Munday and Jones, 1998). In contrast, large predators may roam over many square kilometers (Samoilys, 1997b; Zeller, 1997). Between these extremes, there is plenty of room for movement among local patches of habitat (Frederick, 1997; Lewis, 1997). These differences in lifestyle limit the utility of physical descriptors of habitat to define populations, so that standardized comparisons require focus on functional units.

Regardless of spatial scale, a *local population* is defined here to be an isolated group of individuals that does not exchange postsettlement stages (juveniles and adults) with others. Eliminating migration from the equation means that abundance within these benthic groups is simply the balance between external supply and postsettlement mortality (Talbot *et al.*, 1978). Local populations provide a logical scale to examine the sensitivity of demographic variables to changes in density, including growth (Doherty, 1982) and maturation (Forrester, 1990). The management of fish resources often requires tracking population dynamics on larger scales (Chesson, 1998b).

The proportion of settlement to a population that is sourced from spawnings by its own members is *self-recruitment* and the proportion derived from spawners belonging to another population *is one measure of connectivity*. Because benthic reef fishes are sedentary compared with their dispersive offspring (Sale, 1978b), local populations defined by sufficiently small habitat patches will have negligible self-recruitment and high connectivity. In this condition, local populations are open rather than closed with respect to direct feedback loops between generations (i.e., local settlement is independent of local spawning). At larger size, local populations may be partially closed, with a proportion of supply sourced from self-recruitment. Complete closure will be rare at the scale of a single local population, with the most likely case being very isolated reefs in oceanic settings, as proved by the existence of endemic species (Victor and Wellington, 2000).

An ensemble of local populations connected by larval exchange is a *metapopulation*. A metapopulation that includes all local populations with significant interconnectivity encompasses the *unit stock* of management. This population will often occupy less geographic space than a *genetic stock*, because panmixis (genetic cohesion) can be maintained by very weak connectivities (Shulman, 1998).

Within a metapopulation, a local population in which the birth rate exceeds the death rate is a *source* (Pulliam, 1988; Chesson, 1998b). There are other definitions around based on weaker conditions, but theoreticians have shown that sources with positive growth are essential to the persistence of metapopulations (Chesson, 1998b; Armsworth, 2002). Such sources must have significant levels of self-recruitment although offspring do not have to return to the natal population in the next generation. They may be exchanged through a network of connected populations with the proviso that local reproduction will cause long-term growth of a source population in the absence of regulation (Armsworth, 2002). Conversely, a local population that depends on propagules from other populations (exogenous supply) for its persistence is a *sink* [see Danilowicz (1997), for an example]. Naturally, the only propagules counted are those that survive the pelagic phase to settle somewhere.

The interaction of geography, physics, and biology (Cowen *et al.*, 2000) (see also Chapter 8, this volume) will determine levels of self-recruitment, connectivity, and source–sink relationships within a metapopulation (James *et al.*, submitted). Temporal variability in the oceanographic domain (Wing *et al.*, 1995b; Masterson *et al.*, 1997; Robertson *et al.*, 1999) and nonlinear responses of larval fish to their noisy physical environment (Dixon *et al.*, 1999) have been implicated in the erratic replenishment of open marine populations.

III. Paradigm Evolution

The scientific literature on the population dynamics of coral reef fishes covers little more than 30 years, and each decade has seen a significant shift or modification of the dominant paradigm. In the first phase, the high species diversity of coral reef fishes was accepted to be a logical consequence of a long history of coevolution among species structured by strong resource competition (Ehrlich, 1975). As a corollary, communities were expected to have stable composition and populations were expected to have equilibrium dynamics, i.e., saturated densities that tracked the carrying capacities of benthic habitats. Shelter and/or food were assumed to be limiting resources and larval supply was assumed to be limitless compared with the capacity of saturated benthic populations to absorb new individuals (Sale, 1980).

The second phase was characterized by falsification of the equilibrium model and a shift toward nonequilibrial alternatives. Two major themes to emerge from this period of empiricism were a focus on replenishment [starting with Sale (1977)] and a focus on predation [starting with Talbot *et al.* (1978)] as major drivers of change within populations. Until recently, these factors have been explored largely within their own space (Doherty, 1991; Hixon, 1991) with tacit acknowledgment that real populations are affected by multiple factors operating over a wide variety of spatial and temporal scales (Sale, 1991b, 1998).

The current phase can be characterized as development of theory for multifactorial models (Jones, 1991), and that is the mode of this chapter. To meet the aspirations of the title of this chapter, however, progress will be traced toward a new synthesis through historical perspectives on recruitment. Balancing views are provided in other chapters in this volume (see Chapters 10 and 14).

Doherty (1980) found that local populations of two damselfishes could not be at carrying capacity because they absorbed variable replenishment without detectable compensatory mortality, and settlement was a primary determinant of their total abundance. This is an almost complete, yet not sufficient, description of the corpus of ideas that became labeled as the "recruitment limitation hypothesis."

Doherty (1981) showed that average recruitment provided an adequate and sufficient explanation for the different densities observed between two groups of local populations (same species, same habitat) within one coral reef lagoon. There was a critical caveat implicit in that demonstration. The test was made deliberately between group averages to factor out deterministic and stochastic differences among individual reefs in settlement and/or mortality. In other words, recruitment limitation makes prediction about differences among places and times due to larval supply when settlement stimuli and predation risks are equal. The caveat does not result in a tautology because prediction will fail once the first population is saturated, but it does mean that recruitment limitation is not a universal theory. It is silent about population outcomes between different habitats (Wellington, 1992; Beukers and Jones, 1997; Gutierrez, 1998; Elliott and Marsical, 2001), outcomes among places with different abundance of settlement habitat (Tolimieri, 1998a; Holbrook *et al.*, 2000), outcomes after serious environmental disturbance (R.D. Clarke, 1996; Lewis, 1998), and outcomes between places with different regional species pools (Caley, 1995b). This has not deterred authors from using these inappropriate contexts to attack the concept.

Victor (1986b) distinguished between "primary" and "secondary" states of recruitment limitation. To précis his distinction, populations are in the first state when settlement does not exceed adult losses (even with no losses of juveniles, the breeding pool should shrink or remain the same), whereas populations are in the second state when settlement would exceed adult losses except for the mortality of juveniles. The corollary is that there is a third state, when settlement exceeds adult losses despite juvenile mortality. In this event, the breeding pool will expand unless restrained by density-dependent processes, which clearly must apply at some level. It is clear, therefore, that recruitment limitation refers to one end of a continuum of population responses to variable replenishment (Forrester, 1998; Levin, 1998).

IV. A Quantitative Synthesis Incorporating Recruitment Limitation

Schmitt *et al.* (1999) recognized the continuum of population responses to variable replenishment by partitioning the relative contributions made by three processes (larval supply, density-independent and density-dependent mortality) to the local abundance of a damselfish dependent on anemone hosts in Moorea Lagoon. They started with a three-parameter model, loosely based on a generalized stock-recruitment function from fisheries science (Shepherd, 1982), in order to describe the relationship between settlement and subsequent abundance within a cohort:

$$A = aS/[1 + (a/b)S^d], \qquad (1)$$

where A is the density of older stages (adults) produced from settler density (S), and a, b, and d are fitted parameters derived from empirical investigations of the relationship between A and S. The shape of the function is controlled by d, which can produce linear ($d = 0$), decelerating ($0 < d < 1$), asymptotic ($d = 1$), and dome-shaped ($d > 1$) solutions (Shepherd, 1982). Based on evidence that high settlement of the damselfish results in asymptotic densities, they reduced the general model to one with two parameters:

$$A = aS/[1 + (a/b)S], \qquad (2)$$

where a is survivorship measured at low densities (estimating density-independent mortality) and b is the maximum density achieved by a cohort when larval supply is limitless (estimating carrying capacity).

The model has very simple properties. Without any density-independent mortality ($a = 1$), Eq. (2) reduces a cohort by half ($0.5S$) when $S = b$. Final densities converge rapidly with the asymptote (b) when $S > 10b$ and converge rapidly with settlement (S) when $S < 0.1b$ (Fig. 1). Adding density-independent mortality to the equation alters the absolute but not the relative outcomes at low levels of settlement (Fig. 1). It has negligible effect on outcomes at input densities where $S > b$.

Schmitt *et al.* (1999) suggested two ways to calculate the relative contribution of the three processes (S, a, b) to the density observed from a given level

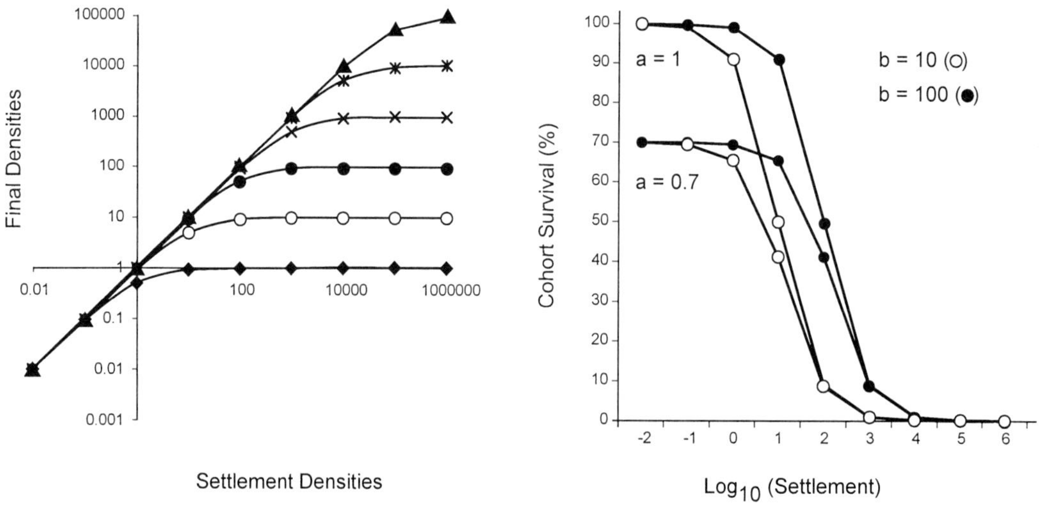

FIGURE 1 Left: Settler–recruit relationships predicted by Eq. (2) with perfect survival ($a = 1$). Right: Cohort dynamics with density-independent mortality of zero ($a = 1$) and 30% ($a = 0.7$). Common symbols identify the equivalent trajectories in both plots.

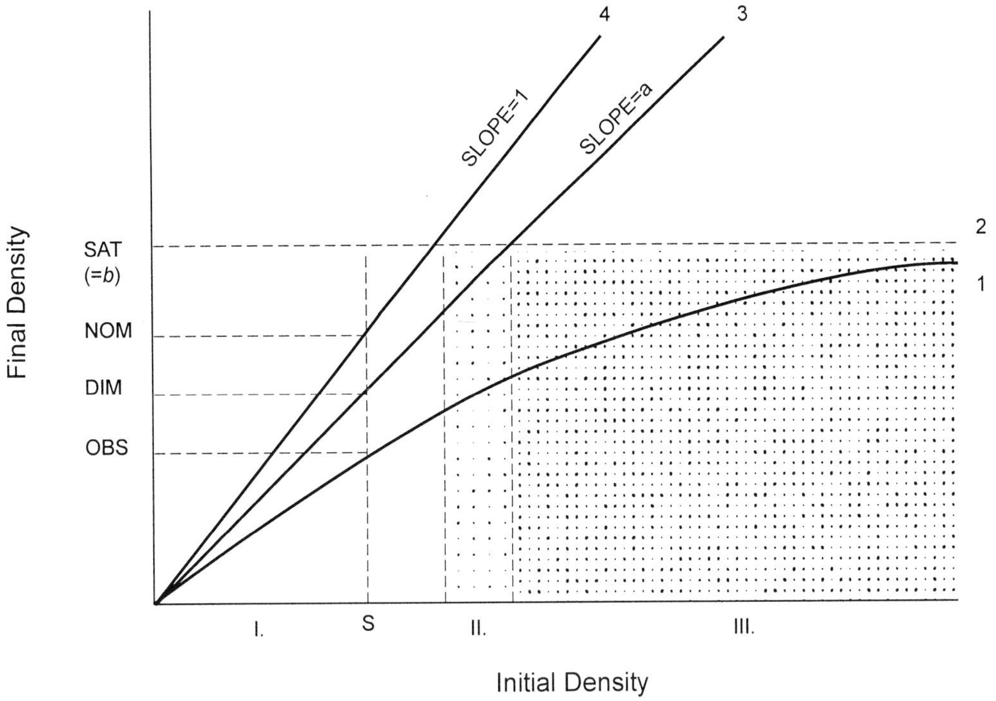

FIGURE 2 Cohort analysis model. Line 1 is the observed settler–recruit curve. Line 2 is the saturation (SAT) density. Line 3 is the theoretical trajectory with only density-independent mortality (DIM). Line 4 is the theoretical trajectory with perfect survival (no mortality, NOM). Zones delimited by shades of stippling correspond to primary recruitment limitation (I), secondary recruitment limitation (II), and resource limitation (III). Labeled terms on the y axis identify values used in the text to calculate the relative contributions of the three processes to an observed outcome (OBS) from a given settlement density, S. Redrawn from Shima (1999b).

of settlement. One way was to compare the observed density with one expected when a specific constraint was removed from Eq. (2). Thus, limitation by settlement was calculated as the difference between observed densities and those predicted if settlement had been infinite. They also provided elasticity analyses to determine the proportional change in output that should result from infinitesimal changes in each of the three variables.

Shima (1999b) provided a graphical explanation of the first method (Fig. 2). Assuming that b (the saturation density) is known, the quantity b-observed (OBS) can be partitioned into the three components. The contribution of larval supply is estimated by the difference between b and the theoretical density expected with no mortality (Fig. 2, NOM, $a = 1$). The contribution of density-independent mortality (DIM) is estimated by the difference between NOM and the theoretical density expected with constant application of the mortality observed at low settlement (DIM, $0 < a < 1$). The contribution of density-dependent mortality is estimated by the difference between DIM and OBS. All components can be standardized by dividing them by their sum, providing an estimate of their relative importance at that level of settlement.

According to Eq. (2), all three processes contribute to population density at all levels of settlement. However, the relative importance of the three factors depends on the level of settlement. Assuming that b defines the saturation state for a given cohort, primary recruitment limitation (when $a = 1$) is most likely in the region $0 < S \leq b$. Similarly, secondary limitation (when $0 < a < 1$) can extend the influence of supply into the region $b < S < b/a$. At higher levels of settlement, compensatory mortality should be the dominant process. The relative contributions of the three processes to abundance at low densities will be determined by the curvature of the settler–recruit (S–R) relationship (Shima, 1999b).

Schmitt *et al.* (1999) derived such a relationship 6 months after transplanting newly settled *Dascyllus trimaculatus* into anemones (Fig. 3). Although Eq. (2) explained a very high proportion of the variance, they noted that "deviations between the data and model

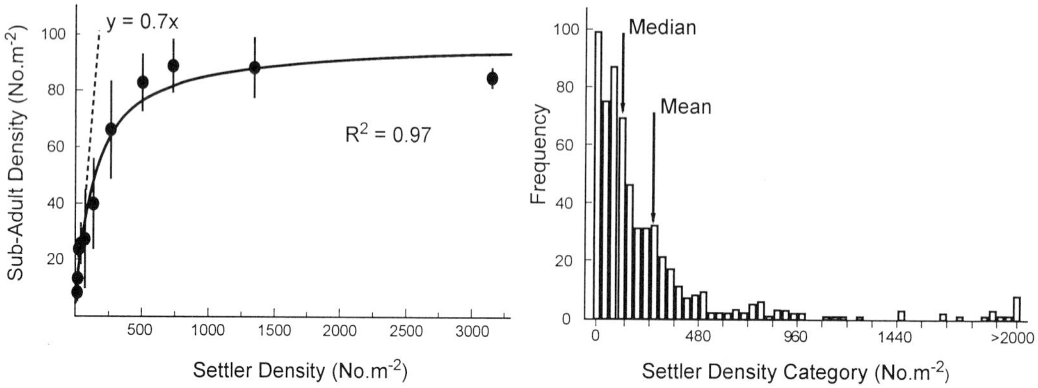

FIGURE 3 Left: Settler–recruit curve for *Dascyllus trimaculatus* over 180 days. Dotted line is the trajectory based on density-independent mortality. Right: Distribution of natural settlement densities from six cohorts and 10 locations. Redrawn from Schmitt *et al.* (1999).

suggest the true asymptote was slightly lower and approached somewhat more quickly than indicated by the model." In other words, the empirical relationship might require a more rectangular function than that estimated by Eq. (2) (Section V,F).

The fitted model estimated density-independent mortality $(1-a)$ as 0.30 and the saturation density (b) at 97.9 subadults m^{-2} (Schmitt *et al.*, 1999). Density-independent mortality was also estimated as 0.36 from observations on natural settlers (Schmitt and Holbrook, 1999a), increasing confidence in the fitted values. Using the method of limitation, the transition between dominance by supply to dominance by density-dependent mortality was estimated at 141 recruits m^{-2}. Up to this value (b/a), final densities increased as a linear function of settlement (Fig. 3).

Given their key conclusion that the dominant process is determined by how close the system is to saturation, Schmitt *et al.* (1999) examined the implications for observed settlement, which averaged 241 recruits m^{-2} across six cohorts and 10 sites. Adopting the mean input density as their reference point, they concluded that removing density dependence would have a larger effect than would making supply limitless and a much larger effect than would eliminating density-independent mortality. Similarly, the elasticity analysis indicated that density dependence was the strongest constraint on numbers. They also recognized that these analyses could be misleading because they ignored variability among individual habitat units.

Settlement on individual anemones varied from 0 to 4630 recruits m^{-2} (Fig. 3). Consequently, outcomes on single units ranged from complete limitation by settlement to complete limitation by density-dependent mortality. Although the mean settlement rate (241 recruits m^{-2}) exceeded the threshold density (141 recruit m^{-2}) above which density dependence became the dominant process, the median settlement rate was only 107 recruits m^{-2}. Consequently 58% of the anemones received settlement less than the threshold density. The authors concluded that "the production of sub-adults was more limited by supply than density-dependence on a majority of anemones." The analysis of elasticities showed that "a small proportionate change in settlement (or survivorship) would yield a 40% larger change in sub-adult density than... the same change in [the asymptote, *b*]."

Schmitt *et al.* (1999) anticipated that density-independent mortality $(1-a)$ and saturation density (b) could vary in space and time, but it was their colleague who demonstrated such effects. Shima (1999b) used the graphical approach of Fig. 2 to estimate the relative importance of the triple processes to the development over 90 days of three cohorts of the six-bar wrasse (*Thalassoma hardwicke*) on small natural patch reefs at eight sites in the northern lagoon of Moorea. After transformation, these relative contributions (determined for each patch reef) were tested with analysis of variance.

All factors (cohorts, zones within cohorts, sites within zones within cohorts) were significant despite strong variation among patch reefs within a treatment (Fig. 4). For all three processes (primary recruitment limitation, density-independent limitation, density-dependent limitation), the largest components of variance were measured among cohorts and among "replicate" patch reefs. Variance components for other spatial scales were small by comparison, albeit dependent on process. Primary recruitment limitation and density dependence varied in similar ways [10% between zones (cohorts) and 4% among sites (zones (cohorts))], whereas density-independent mortality

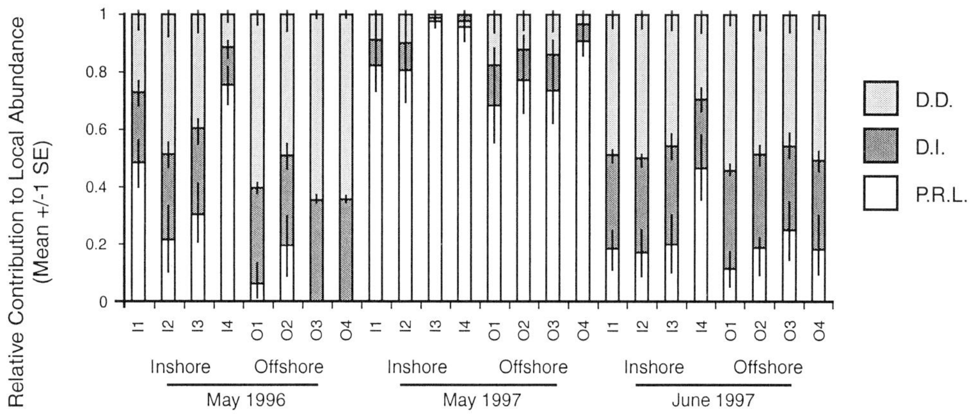

FIGURE 4 Proportional contributions over 90 days of density-dependent mortality (DD), density-independent mortality (DI), and primary recruitment limitation (PRL) to the abundance of three cohorts (May 1996, May 1997, June 1997) of *Thalassoma hardwicke* at eight sites in Moorea Lagoon. Redrawn from Shima (1999b).

explained the least amount of variation at both scales (3% and 1%, respectively).

Shima concluded that "larval supply was the predominate factor shaping patterns of abundance of juvenile wrasse in specific locations (inshore sites . . .) and times (May 1997), but was superceded by post-settlement mortality in other times and locations" and that "the role played by primary recruitment limitation . . . was more variable than both forms of post-settlement mortality [which] appears to be representative of . . . systems in which larval inputs are notoriously variable . . . and density-independent mortality is . . . constant."

The cohort analysis illustrated by these two studies from Moorea falls far short of a mature multifactorial model for the population dynamics of coral reef fishes, particularly because it ignores interactions with other cohorts (Schmitt *et al.*, 1999; Shima, 1999b). Nonetheless, it does give a simple framework and common language for comparing different studies.

V. Case Studies of the Impacts of Variable Recruitment

A key message from the cohort analysis is that the relative importance of pre- and postsettlement influences on local abundance depends on how near or far a system is from its saturation state. Consequently, for this chapter, studies were sought, especially those done in the past 10 years, that position natural populations along the continuum between dominance by supply and dominance by resources. The studies have been aggregated by family because short- and long-lived species ought to display different population dynamics (Warner and Hughes, 1988).

A. Gobies

The small body size of many gobies permits them to exploit fine structure within reefs (Munday and Jones, 1998), with the result that there are many habitat specialists, including a diverse group of obligate coral dwellers (Munday *et al.*, 1997). At sites on the northern GBR, members of the genus *Gobiodon* are highly overdispersed (resulting in an even distribution with each coral supporting just a single mated pair), and vacant habitat space is rare (P. L. Munday, personal communication). This situation, in which it takes just two fishes to saturate a habitat unit, looks like a case of strong numerical regulation. Variable replenishment will have little impact on the composition of the adult pool so long as larval supply exceeds the turnover of adults.

The small body size, sedentary lifestyle, and short life-span of many gobies make them ideal experimental animals. Graham Forrester and Mark Steele have coordinated studies over a number of years on tropical and temperate gobies to test the impact of pre- and post-settlement processes on their abundance (Steele *et al.*, 1998; Forrester and Steele, 2000). For this reason, the scope here is broadened to include their non-tropical examples.

Forrester (1995) discussed the difficulty of detecting weak density effects from observational and/or correlative studies. He argued that experiments are the definitive way to determine the impacts of density on demographic variables, but noted that experiments

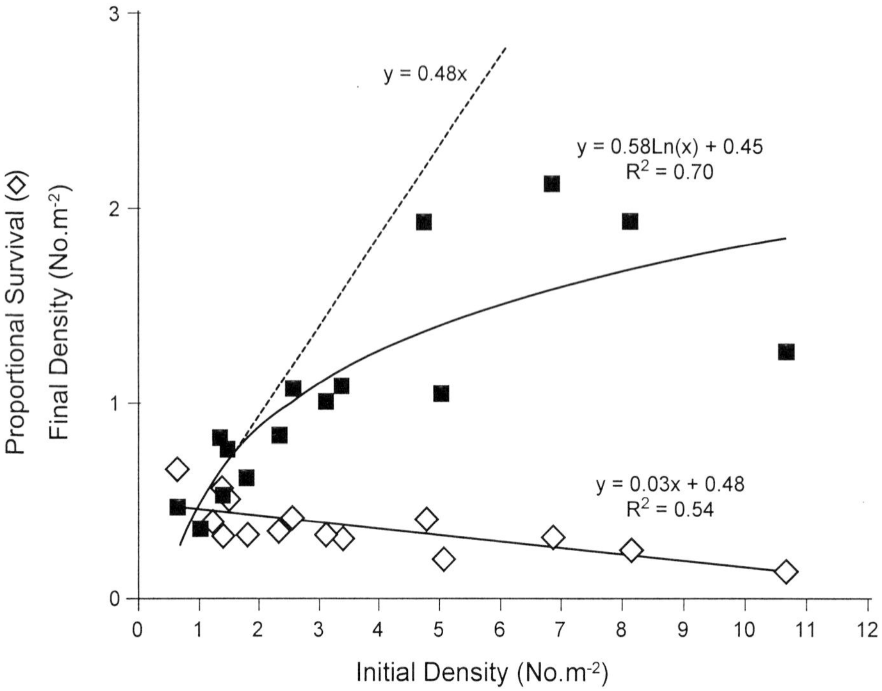

FIGURE 5 Outcomes from transplants of adult *Coryphopterus glaucofraenum*: per-capita mortalities (diamonds), final densities (squares), and theoretical trajectory based on density-independent mortality alone. Redrawn from Forrester (1995).

based on long-lived animals are only able to focus on interactions during parts of the life cycle. Accordingly, he chose a western Atlantic goby (*Coryphopterus glaucofraenum*) that lives for less than a year. Forrester transplanted adult gobies at 0.7–10.7 fishes m^{-2} to artificial reefs of $8.5m^2$. He also monitored survival of marked and unmarked individuals on natural reefs. After 75 days, survival of all three groups showed a common negative linear relationship with stocking density such that the proportion surviving at the highest density was only half that at the lowest initial densities (Fig. 5). Survival at low densities averaged 48% but was highly variable (30–66%), reflecting the impact of stochastic mortality on very small groups, or something else (see below). The relationship between final and stocking densities was linear up to 2 transplants m^{-2}.

Forrester suggested that limited shelter from predators caused the density-dependent survival of *C. glaucofraenum*. Steele (1997a) showed that predators could exert both direct and indirect effects on prey. He found that recruitment of *Lythrypnus dalli* was enhanced by the presence of older conspecifics (1.37–1.46 increase over unstocked reefs) but was halved in the presence of predators (only enhanced by 1.1). From this, he was able to partition the impact of predators on natural populations into a direct effect (from mortality) and a smaller indirect effect from the reduced settlement caused by the lower density of residents. To emphasize the species-specific nature of such effects, Steele found that conspecific densities had no impact on the settlement of another goby, *Coryphopterus nicholsii*. Within predator exclusion cages, both gobies showed only density-independent mortality so that variations in initial density were preserved among replicates. On control sites, such spatial patterns were quickly eliminated.

Steele (1997b) tested the impact of variable replenishment on these gobies by transplanting young recruits at a range of densities to artificial reefs (Fig. 6). Both species showed density-dependent losses on the first day of the experiment, although linear relationships between initial and final densities were preserved until they matured—after day 16 for *Lythrypnus* and day 36 for *Coryphopterus*. By the end of the experiment, both recruit–adult relationships were asymptotic, with saturation densities of 10 and 3 adults m^{-2}, respectively. Steele hypothesized that these gobies become vulnerable to predators during territorial combats and/or sexual displays.

The density-dependent losses observed immediately after the release of transplants is a consistent feature associated with this procedure and, in some

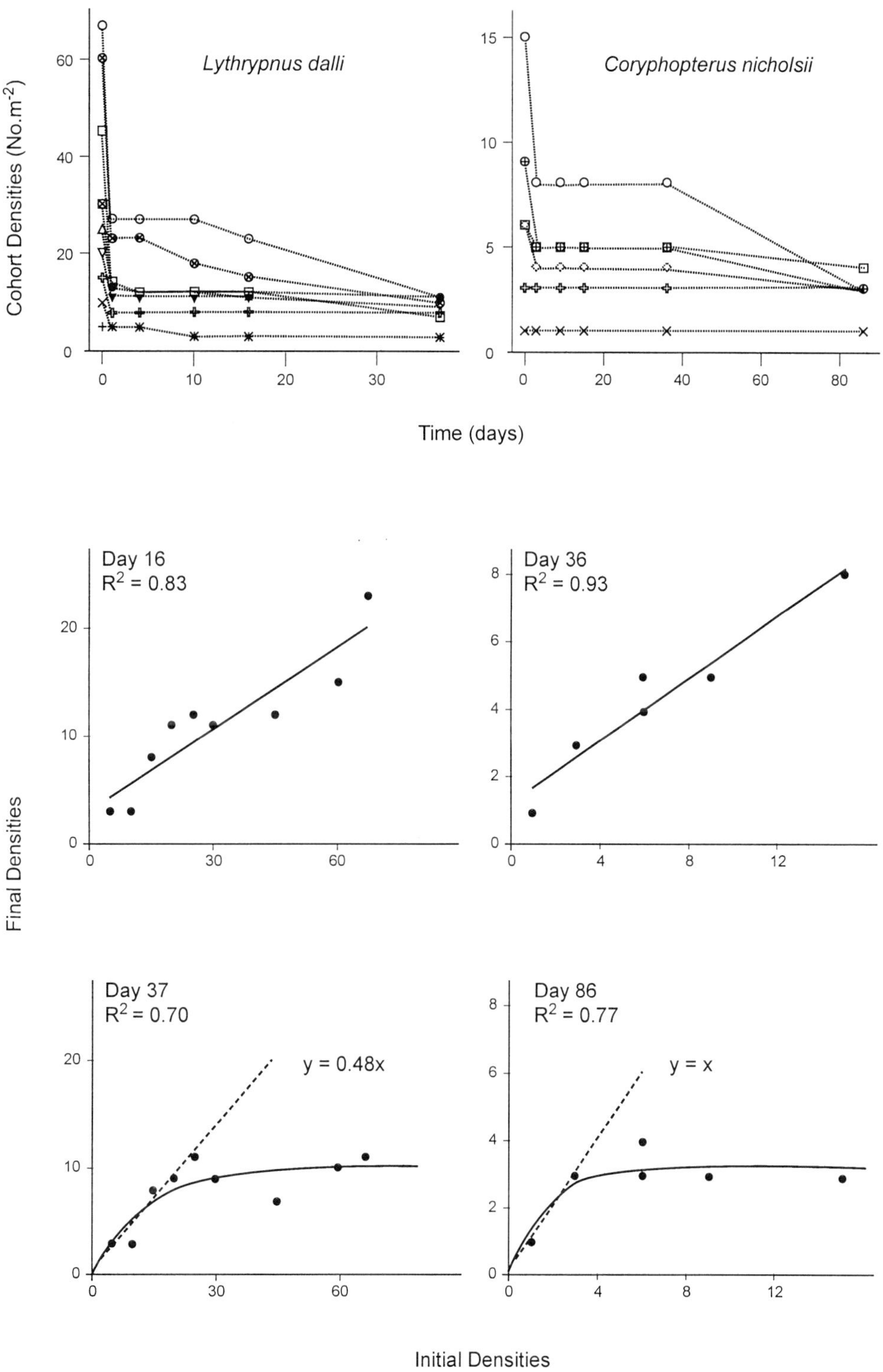

FIGURE 6 Top: Densities of experimental cohorts of two gobies transplanted to artificial reefs. Middle: Settler–recruit relationships prior to sexual maturation. Bottom: Settler–recruit curves after sexual maturation. Dotted lines are theoretical trajectories based on density-independent mortalities. Redrawn from Steele (1997b).

cases, it is clear that emigration is involved (Carr and Hixon, 1995). Doherty (1982) argued that it is an artifact and chose to base similar experiments on fish that had been acclimatized for a few days to a range of densities. Although others (Jones, 1987a, 1988b, 1990) have stocked reefs repeatedly until densities stabilize, Steele (1997b) argued that all such strategies "may bias against detecting density-dependent mortality... because those fish not tolerant to high densities disappeared before the experiment was initiated." Rather than debate this novel hypothesis, I prefer to highlight how choice affects the analysis.

The high survivorship rate of *Coryphopterus* shows that all mortality before day 40 occurred between release and the first census (Fig. 6). For the next 35 days, there was perfect survivorship and Steele attributed no density-independent mortality to this species. Yet the linear relationship that he chose to present for day 36 indicates density-independent loss of roughly 40% between initial and final densities. There is no way of knowing which value (0 or 40%) is correct, but we can be certain that absolute settlement is unknowable for most natural populations (Booth, 1991).

Obtaining values for S, a, and A from Fig. 6, Eq. (2) predicts saturation densities (b) of ~13 recruits m^{-2} for *Lythrypnus* and ~4 recruits m^{-2} for *Coryphopterus*. When the results expected from density-independent mortality alone (aS) are projected onto the same space, the settler–adult relationships are linear up to input densities of 15 and 3 recruits m^{-2}, respectively. This matches Steele's qualitative assessment that the density of adults is limited by supply when settlement is <19 (for *Lythrypnus*) and <3 recruits m^{-2} (for *Coryphopterus*). Using these benchmarks, he observed that "such low densities are commonly encountered at certain sites and occasionally in most areas during certain years but at other places and times recruit densities exceed the levels above which additional recruits did not produce additional adults."

Sano (1997) provides an example from gobies of spatial and temporal change in the importance of density dependence. He monitored the abundance of three consecutive year-classes of an annual goby, *Sagamia geneionema,* on the same eight sand patches (52–180 m^2). Postsettlement mortality over a 6-month period produced relatively constant densities of adults (0.34 adults m^{-2}) in 2 of the 3 years despite two- to threefold differences in recruitment. In 1993, however, mortality was spatially uniform, resulting in final densities that scaled directly to settlement and produced densities up to 1.3 adults m^{-2}. Sano speculated that the different outcomes were due to differential abundance of predators among years, but there is a possibility that settlement was more continuous in the anomalous year, which simply highlights the weakness of correlative studies to reveal causal mechanisms.

B. Wrasses

The laboratory rat for fish ecologists working in the tropical western Atlantic has been a common wrasse, *Thalassoma bifasciatum*. It is a short-lived species that can mature within 3 months of settlement and most of its adult populations consist of individuals that have settled during the previous year (Warner and Hughes, 1988).

Victor (1983b, 1986b) suggested that populations of *T. bifasciatum* on the Atlantic coast of Panama are recruitment limited. Because the emphasis here is new evidence, his conclusions and the nature of the supporting evidence are only summarized. Victor (1983b) observed that *T. bifasciatum* settled in sporadic pulses that did not reflect either the daily spawning pattern of this species or the mortality of residents (i.e., settlement was unrelated to changes in local population size). Back-calculating age from otolith records, he showed that the age structure of one adult population retained the recruitment signal monitored on an adjacent reef during the previous year, which would not be expected if the sacrificed population had been near saturation. He concluded that "if recruitment rates are not sufficient to enable the population to reach the carrying capacity of the habitat, then the population is limited by recruitment" and that "rates of recruitment apparently determine a large part of the population dynamics of the bluehead wrasse."

Warner and Hughes (1988) used life history and demographic data from *T. bifasciatum* to make three points about recruitment limitation and settler–recruit relationships. Their first point was that local populations can maintain constant numbers without density dependence when recruitment and mortality rates are both constant, because per-capita settlement declines with increasing population size. [Caley *et al.* (1996) called recruitment limitation density-dependent for this reason, but Sale and Tolimieri (2000) introduced "density-dependent determinism" to make clear that this relationship with density was free of feedback.] Warner and Hughes' second point was that population size and age structure will be closely correlated to variations in recruitment if there is no density dependence, but that such correlations will be difficult to detect if mortality rates are low or if the animals are long-lived. [This is due to the storage effect (Warner and Chesson, 1985), whereby populations accumulate the effects of many past recruitment events.] Warner and Hughes' final point was that even moderate variability in the rate of postsettlement mortality would corrode settler–recruit relationships, such that an obscured relationship between the two variables does not indicate the absence of recruitment limitation. They concluded that it is imperative to know the mean and variance of postsettlement mortality in populations.

Caselle (1999) took up this challenge by measuring the mortalities of nine cohorts of *T. bifasciatum* at St. Croix, United States Virgin Islands. In 1991, she ran the same experiment four times with initial densities of 0.04–0.87 fish m^{-2}. In 1992, she ran another five trials, with higher densities of 0.15–1.43 fish m^{-2}. Each of these trials followed the same general protocol. Newly settled (1- to 3-day old) wrasses were collected and transplanted at dusk onto 11 natural reefs (15–50 m^2). All reefs had similar adult densities (~0.24 adults m^{-2}) and the densities of resident juveniles were manipulated before each trial to a constant 0.25 juveniles m^{-2} in order to isolate intracohort effects. Experimental reefs were inoculated with transplanted juveniles for up to 3 days and then followed for around 30 days. The large number of repeated trials allowed rotation of density treatments among replicate reefs so that the overall design was very robust. Initial and final densities were log-transformed to stabilize their variances and then tested by linear regression on pooled data. On a double-log plot, a slope of unity indicates a linear relationship whereas slopes of <1 indicate curvilinear relationships.

In the four trials of 1991, densities established on the last day of transplanting were used as the baselines. Measured against these inputs, there was no significant effect of density on per-capita mortality over 30 days and stocking densities explained 64% of the variance in final densities. The slope of the double-log plot (cohorts pooled) was not different from unity, consistent with an absence of density dependence. Mortality rates averaged 44% and were most variable at the lowest initial densities.

In the five trials of 1992, Caselle monitored instantaneous mortalities (losses in the first 24 hours) as well as losses over 30 days. Like Steele (Section V,A), she observed density-dependent losses during the first day after transplantation. However, these 24-hour rates were extremely variable and density explained only 14% of the variance in per-capita mortality, which supports the view that these losses may include a component due to emigration. In 1992, as in the previous year, stocking density explained virtually none of the variance in per-capita mortality over the 30-day experimental period. Initial density explained 62% of the variation in final density. The slope of the double-log plot (cohorts pooled) was not statistically different from unity, indicating an absence of density dependence despite the higher initial densities. Average mortality for this set of trials was 42%.

In addition to her manipulations, Caselle measured per-capita mortality over 30 days for nine cohorts settling at eight locations around St. Croix. The average mortality of these undisturbed fishes was slightly lower (0.35 versus 0.44) than that of transplants; rates were also less variable for higher initial densities (> 0.5 recruits m^{-2}). At the lowest densities, Caselle reported per-capita mortalities of between 0 and 85%. The equivalent double-log plot, however, showed that a third of the observations from settlement of < 0.5 recruits m^{-2} produced final densities greater than the initial ones. The most likely source of these invalid results (McGuiness and Davis, 1989) is unmeasured settlement. Because of their position at the low end of the density spectrum, inclusion of these values levers the slope of the log-log linear regression downward, which means that density dependence was overestimated at this scale. Sensitivity analysis on the recovered data suggests that suppressing invalid outcomes would change the slope on the log-log plot from 0.64 to 0.75 but would have minor effect on the untransformed relationship between initial and final densities (Fig. 7a). Consequently, the major conclusion that there was some density-dependence is sustained.

Caselle nominated 1 recruit m^{-2} as the benchmark to separate supply and density effects. This input was exceeded in 6 (19%) of 32 monthly site counts in 1991 and 11 (20%) of 54 in 1992. Although only a minority of places and times exceeded the threshold, the distribution of settlement resulted in 2004 (55%) and 3635 (60%) juveniles being exposed to densities greater than or equal to the benchmark. In addition, there was spatial consistency to this pattern. Over 2 years of monitoring, five sites never received recruitment levels above the threshold density; one site received higher recruitment just twice, and two sites on opposite corners of the island received higher recruitment more than half of the time. Caselle (1997) showed that differences among sites resulted from larval supply and not from habitat selection.

The consistent differences in recruitment among sites (Caselle and Warner, 1996) resulted in a positive relationship between cumulative inputs and population densities in consecutive annual assessments (Caselle, 1999). Recruitment explained half of the variance in adults, but the settler–adult relationship departed from linear when cumulative inputs exceeded 5–10 recruits m^{-2} yr^{-1} (Fig. 7b).

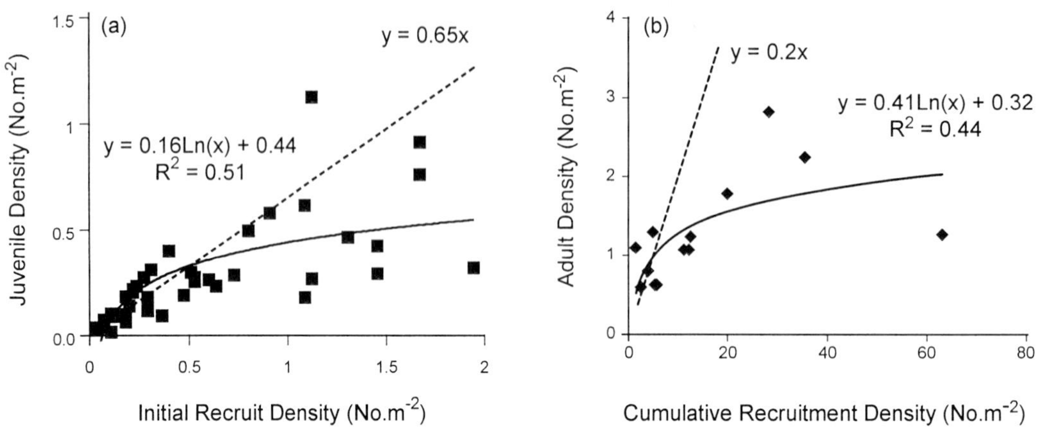

FIGURE 7 (a) Final densities after 30 days of nine transplanted cohorts of newly settled *Thalassoma bifasciatum*. The dotted line is the theoretical trajectory based on density-independent mortality. (b) Final densities of two year-classes at eight sites around St. Croix compared with cumulative monthly settlements. The dotted line is the theoretical trajectory based on density-independent mortality calculated from a combination of literature values [data from Caselle (1999)].

In another ocean, Shima (1999a,b) studied the replenishment of an Indo-Pacific congener, *Thalassoma hardwicke,* to natural patch reefs in the fringing lagoon of Moorea. This study was introduced in Section IV to show that the determinants of abundance need not be fixed across time and space (Fig. 4). Shima (2001) analyzed the fractional losses over 90 days of three cohorts of *T. hardwicke* settling on 96 natural patch reefs. To achieve equal precision in his estimates of per-capita mortality, he pooled individuals from different reefs containing similar densities and allocated them to density categories that contained similar numbers of initial recruits. Reefs receiving a single recruit were listed as a separate category. In addition, he transplanted 300 new recruits (<5 days postsettlement age) to 58 natural patch reefs (mean size 3.5 m^2) in five density treatments: solitary settler and 1, 2, 3, and 5 recruits m^{-2}, which represented a natural span of postsettlement densities. The solitary settler treatment consisted of 21 fish transplanted at initial densities of 0.3 fish m^{-2}. Per-capita mortality in this experiment was estimated by the fraction of transplants lost over 2 weeks. To isolate effects of density and group size, a further transplant experiment was done on 150 reefs (five sites, 30 replicates) with two crossed treatments (0.25 or 0.5 fishes m^{-2}; 1 or 2 fishes per reef). All reefs were cleared of residents prior to inoculating them with transplants (<14 days old), which were then monitored for 7 days.

Despite different intensities of settlement among the three cohorts, mortality rates could be pooled into a single relationship showing increasing per-capita losses with higher initial densities (Fig. 8a). In each cohort, the lowest turnover was observed from solitary settlers ($\sim$25% loss). Transplants extended the range of initial densities and provided qualitatively similar results (Fig. 8a), although the mortality of solitary transplants (61%) was much higher (after 2 weeks) than those of solitary settlers (after 90 days), raising further questions about transplantation. Regardless, both studies produced congruent estimates of the settler–recruit relationship (Fig. 8b).

Using Eq. (2), Shima (2001) estimated saturation densities of 0.32–0.53 juveniles m^{-2} from the three cohorts. Using average outcomes from the low-density treatments of his experiment yields an estimate of density-independent mortality. A line with this slope ($a = 0.37$) intersects the mean saturation density ($b = 0.42$ juveniles m^{-2}) at an initial density of $\sim$1 recruit m^{-2} (Fig. 8b) to indicate the upper boundary of limitation by supply (Schmitt *et al.,* 1999). The distribution of settlement densities (Fig. 8c) suggests that density-dependent mortality occurred on a minority of the patch reefs, although the proportions varied substantially among years (Fig. 4).

Holding density constant, Shima found higher losses (35%) from reefs with two recruits instead of one (25%). This confirms a numerical effect on mortality at very low densities, which is obvious in the ascending limb of Fig. 8a. He suggested that territorial contests between individuals increase the risk of predation, which contrasts with some examples from the next section showing that a benefit is derived from living with conspecifics. For *T. harwicke,* this numerical effect involved only fish from the same cohort because removal of older cohorts did not alter the outcomes.

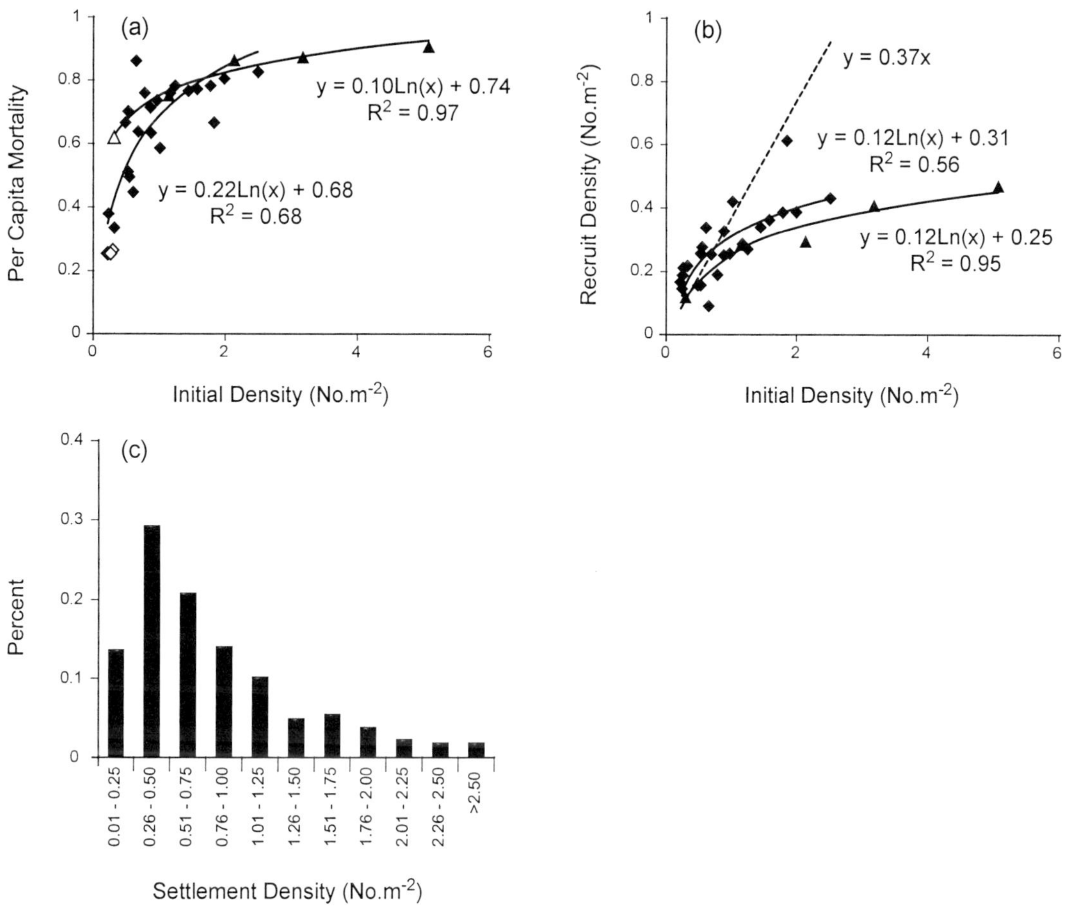

FIGURE 8 (a) Losses of transplanted (triangles) and natural (diamonds) settlers of *Thalassoma hardwicke* over 14 and 90 days, respectively. Open symbols identify treatments with solitary individuals. (b) Settler–recruit curves for the two cohorts. The dotted line is the theoretical trajectory based on density-independent mortality alone. (c) Distribution of settlement densities from three cohorts on eight sites in Moorea Lagoon [data from Shima (2001)].

C. Sex-Changing Damselfishes

Dascyllus aruanus, the banded humbug, is found throughout the Indo-Pacific and has become a favorite with experimentalists in that part of the world. These fish are protogynous hermaphrodites that live in small to large social groups that shelter within live branching coral colonies. Pelagic juveniles settle preferentially into corals occupied by resident conspecifics (Sweatman, 1985b), responding to chemical cues during nocturnal settlement (Sweatman, 1988). Once established, their aggressive nature is capable of reducing the settlement and/or survival of other settlers seeking shelter in the same colonies (Jones, 1987b).

Forrester (1990) transplanted *D. aruanus* to small coral colonies (0.125 m^2) at three locations in One Tree Lagoon (southern GBR) to examine interactions among cohort densities, adult presence, and food supply. Although restricted by just three levels of density (16–112 recruits m^{-2}), he found that per-capita mortality increased with density, albeit with less effect in the presence of adults (Fig. 9) and slight differences among locations. All factors had small impact on density so that survivors remaining in the habitat units at 10 months were strongly related to stocking densities. In contrast, the number of recruits that matured during the period of the experiment was enhanced by the absence of adults and the supplementation of diet but was independent of stocking density (Fig. 9). This study confirmed earlier demonstrations that crowding has a stronger impact on growth and reproduction than it does on total abundance (Doherty, 1982).

Booth (1992) showed that conspecifics also facilitate settlement of *Dascyllus albisella* in Hawaii and that settlers select larger groups (although this effect may be no more than the greater efficiency of a stronger

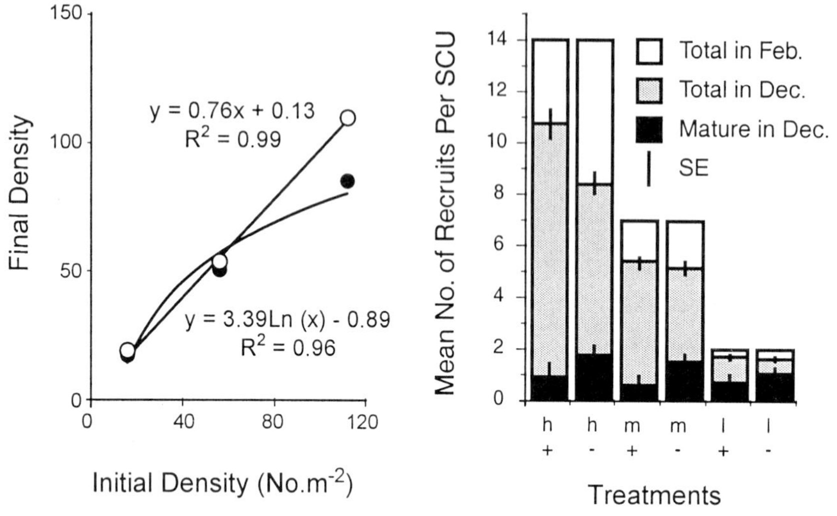

FIGURE 9 Left: Densities 10 months after *Dascyllus aruanus* were transplanted to isolated coral heads with (open symbols) and without (filled symbols) conspecifics. Right: Survival and maturation rates in the three density treatments (high, medium, low) crossed with presence/absence (+/−) of adults. Redrawn from Forrester (1990).

chemical odor). He also established that there are both costs and benefits to individual fitness flowing from these choices (Booth, 1995). Survival is enhanced in larger groups, especially for small fishes, but the formation of dominance hierarchies results in growth stunting, especially for fishes of low social status. Because growth rates determine the onset of maturity, fishes settling into smaller groups should mature sooner if they survive. Booth found that the balance of risks produced different outcomes for different year-classes.

Dascyllus aruanus, Dascyllus flavicaudus, and *Dascyllus trimaculatus* have been the focus of multiple published studies from Moorea (Holbrook and Schmitt, 1997; Schmitt and Holbrook, 1996, 1999a,b; Schmitt *et al.,* 1999). *Dascyllus trimaculatus* settles on anemones; the others prefer live branching corals. Schmitt and Holbrook (1999b) monitored settlement of the three species to empty habitats transplanted to mid-lagoon sites at nine locations around the island. At two locations, they added additional habitat to inshore and offshore sites to examine patterns of settlement across the lagoon. Observations on cohorts settling in August 1996 and July 1997 revealed consistent differences at both spatial scales. Settlement of *D. trimaculatus* was heaviest at northern locations, and the others showed the opposite pattern. *Dascyllus trimaculatus* settled uniformly across the lagoon, whereas *D. aruanus* settled mainly on inshore corals and *D. flavicaudus* preferred corals in the outer lagoon. A competition experiment showed that *D. flavicaudus* had a greater effect on the population growth of *D. aruanus* than vice versa.

Schmitt and Holbrook (1999a) measured mortalities of the next cohort to settle in these corals and anemones. Just prior to that settlement pulse, juveniles were removed from half of the habitats, so that the mortality of the newly settled cohort could be estimated in the presence and absence of the fish that had settled 14 days earlier. During the two weeks of this experiment, losses across species averaged 41% of the new and 15% of the old fishes. Losses from the six groups (three species by two cohorts) declined exponentially with increasing size. Losses of new settlers were 1.6–1.8 times greater in the presence of older conspecifics. For all species, per-capita mortality of the new cohort increased with initial density, whereas per-capita mortality of the older cohort was both lower and without pattern. When recovered data for the older cohorts were replotted on log-log scales (not shown), all of the slopes for the settler–recruit relationship were close to unity (*D. trimaculatus,* 1.04; *D. flavicaudus,* 1.00; *D. aruanus,* 0.93), which confirms that mortality of these fish was density independent. In the untransformed data (Fig. 10), density dependence of newly settled fish is obvious only for *D. aruanus,* but the double-log plots (not shown) suggested that all relationships were curvilinear (*D. trimaculatus,* 0.82; *D. flavicaudus,* 0.79; *D. aruanus,* 0.44).

Experiments with additional anemones confirmed that the mortality of newly settled *D. trimaculatus* was influenced by the density of older conspecifics and that the presence of such fish amplified any density dependence (not always detected) within the younger cohort. Experiments with additional corals showed that postsettlement losses in both species were influenced by the density of older *D. flavicaudus* and the density of newly settled *D. aruanus.* A parsimonious explanation of these mixed results is that these two identities were simply the most abundant of the four possible categories (Schmitt and Holbrook, 1999a). Nonetheless, the outcomes of this experiment illustrate the potential for complex (interspecific, intercohort, and

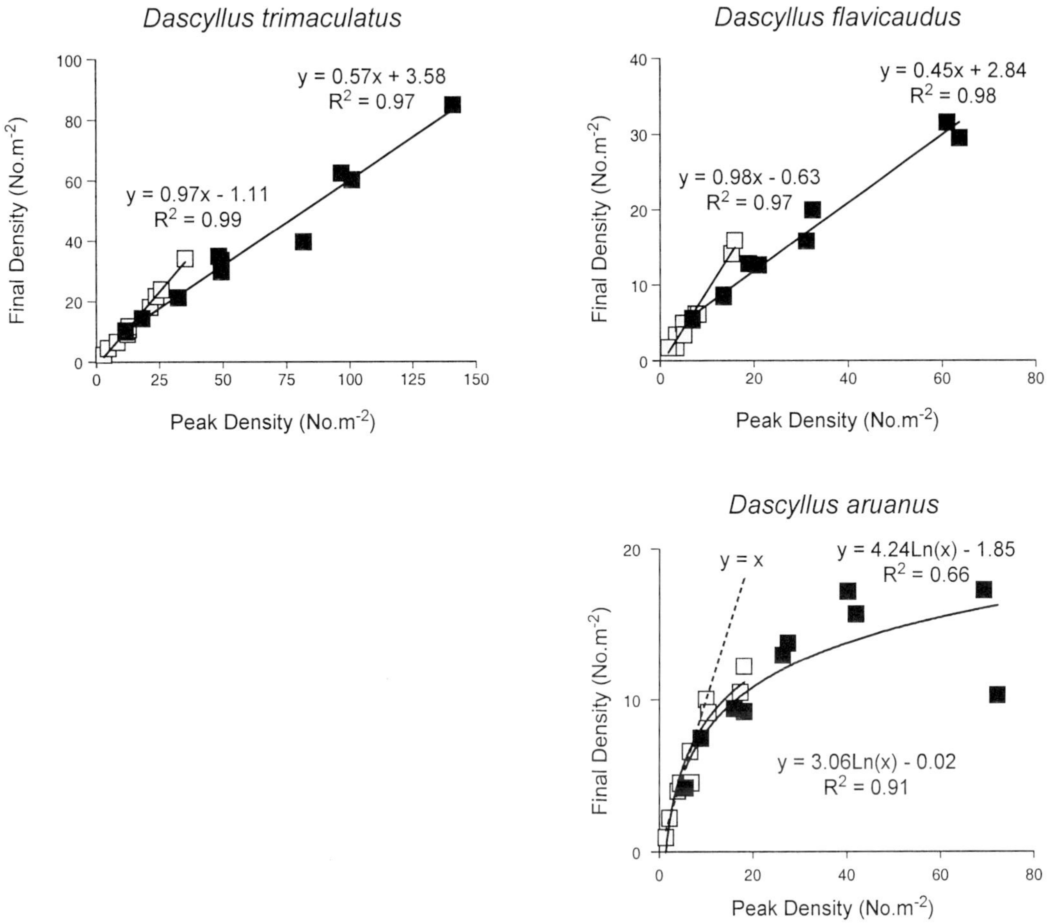

FIGURE 10 Settler–recruit relationships over 14 days in new settlers (dark symbols) and the previous cohort (light symbols) of three damselfishes from Moorea Lagoon. The dotted line is a theoretical trajectory based on perfect survival. Redrawn from Schmitt and Holbrook (1999a).

intracohort) interactions within these very small local populations.

Averaged across species, postsettlement mortality reduced spatial variance in new settlers by 87% among sites around the island and a similar amount among replicates within sites. Even though half of the reduction was attributed to density-dependent processes, linear regressions of settlement and final densities after 14 days were highly significant at both spatial scales. In fact, despite the evidence of density-dependent mortality after settlement, a nonlinear model did not provide a better fit than the simple linear alternative, except for *D. aruanus* (Fig. 10). Settlement explained more variance among sites than within, but mainly because of the greater range of input densities measured across the larger scale. At 14 days, primary recruitment limitation was the dominant process affecting abundance among sites (Table 1).

TABLE 1 Relative Contributions of the Three Processes to Cohort Formation for Three Damselfishes in Moorea Lagoon[a]

Process	*Dascyllus flavicaudus*	*Dascyllus aruanus*	*Dascyllus trimaculatus*
Primary recruitment limitation	79	66	69
Density-independent mortality	12	20	18
Density-dependent mortality	9	14	13

[a] Data from Schmitt and Holbrook (1999a).

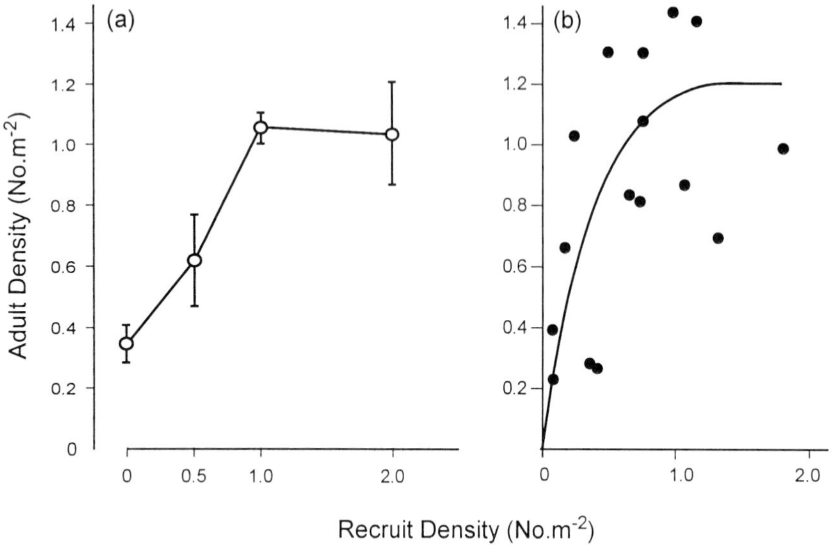

FIGURE 11 (a) Final average densities of *Pomacentrus amboinensis* on 16 patch reefs after 3 years of absorbing fixed recruitment levels. (b) Matching data on 16 control reefs where recruitment density was the average of four year-classes. Redrawn from Jones (1990).

Dascyllus aruanus has been the subject of numerous experiments also by Australian researchers. Jones (1991) summarized the main ones for those interested in detailed comparisons with the more recent work from French Polynesia. On the GBR, the banded humbug has limited spatial overlap with congeneric species. It is more likely to interact with confamilials, especially *Pomacentrus amboinensis,* which is a sex-changing damselfish found in small harems near reef/sand borders. Because the latter is common on small patch reefs, it has become a favored model for fishes lacking the habitat specialization of *Dascyllus* and *Amphiprion* (Jones, 1987a,b, 1988b, 1990, 1997; Kerrigan, 1996, 1997; Meekan *et al.,* 1993).

Jones (1987a) showed that *P. amboinensis* had no effect on the recruitment of other fishes. He also found no effects of density on the mortality of juveniles transplanted to small artificial reefs (1–2 m^2) at densities of 2–6 recruits m^{-2}, although he recorded slight positive effects of density on survival in an experiment using smaller reefs and densities of 4–12 recruits m^{-2} (Jones, 1988b). In a third trial, Jones (1987b) transplanted 0+ fishes at 2–6 recruits m^{-2} and 1+ fishes at 1.5–5 subadults m^{-2} and recaptured them at four intervals to follow their growth over the same year. The design crossed density treatments with the presence and absence of a pair of conspecific adults. At the end of a year, there were no consistent effects of stocking density and/or conspecific presence on the survivorship of either cohort (0+, 1+), although there were substantial effects on growth and maturation. Adults checked the growth of the largest, most dominant subadults. In their absence, some juveniles from both cohorts reached a mature size within the year but the numbers of ripe fishes did not reflect their initial densities. As with Forrester's demonstration with *D. aruanus,* Jones concluded that size-based dominance hierarchies can control recruitment to the breeding pool.

Jones (1990) tested this model with populations living on natural patch reefs (mean size 7.8 m^2). Surveys of 32 reefs at the end of the 1984 recruitment season showed average densities of 0.2 recruits m^{-2}, 0.7 subadults m^{-2}, and 0.8 adults m^{-2}. For the next 3 years, half of the reefs were monitored for natural changes. Replenishment on the remaining half was manipulated by redistributing (0+) fish among them at four fixed densities (0, 0.5, 1, and 2 recruits m^{-2} yr^{-1}) at the conclusion of four consecutive recruitment seasons. By the time of the last restocking, there was a nonlinear relationship between treatment and adult abundance (Fig. 11a). There was a direct relationship between input and abundance up to 1 recruit m^{-2} yr^{-1} but no enhancement from higher levels of recruitment. Adult abundance halved on the treatment starved of recruitment (zero recruits m^{-2} yr^{-1}), showing that these populations have limited storage (*sensu* Warner and Chesson, 1985). Average annual replenishment of >0.5 recruits m^{-2} yr^{-1} was required to balance turnover in these populations.

Superficially, results from the unmanipulated reefs were consistent with the experimental outcomes and seem better able to define the shape of the nonlinear relationship (Fig. 11b). Closer inspection raises some interesting issues. Jones (1990) shows stable adult densities on control reefs over 3 years despite large interannual variability in four recruitment events. The last of these events coincided with the end of the experiment and therefore cannot have had any influence over adult abundance. This year-class (1.8 recruits m^{-2}) was threefold stronger than the average of the previous three (0.46 recruits m^{-2}), which means that the shape of Fig. 11b is altered radically by a decision to pool or not to pool this last recruitment year into the independent variable. Regardless, it serves to remind that the relationships in this study were mostly correlative rather than causal. A more intriguing question is why the adult numbers on control reefs increased during the trial (0.75 to 0.85 adults m^{-2}), when supplied by a mean input of 0.46 recruits m^{-2} yr^{-1}, whereas the numbers on experimental reefs halved after being supplied with 0.5 recruits m^{-2} yr^{-1}. Given random assignment of treatments, both groups should have had equal storage (Warner and Chesson, 1985), which again hints at a possible transplant effect.

Jones (1990) estimated that the four recruitment episodes resulted in varying proportions (0, 25, 6, and 50%) of the patch reefs receiving inputs above the threshold (1 recruit m^{-2} yr^{-1}) that defined excess settlement. When allowance is made for the spatial dispersion of fish, 75% of the regional population (living on half of the reefs) may have experienced overcrowding during the best year. In contrast, it is clear that recruitment levels were below replacement for many reefs during the two poorest years. Such simple analysis, however, does not account for potential recruitment of adults from a pool of stunted subadults, which may moderate the impacts of variable replenishment on total egg production.

Jones (1997) transplanted corals to locations near the center and perimeter of One Tree Lagoon (southern GBR) to compare spatial intensity of settlement with measures of postsettlement fitness. Two fish (*Pomacentrus amboinensis* and *D. aruanus*) settled most densely at peripheral sites despite evidence that the latter survived better at central ones and performance of the former was unaffected by location. Jones suggested that this disconnect arises because both species settle in suitable habitat at the first opportunity after entering the lagoon and that the lower replenishment of central sites reflects depleted larval supply. He detected similar recruitment shadows at smaller scales through lower settlement of *P. amboinensis* on reefs downstream of other reefs. Both outcomes show that larval supply is finite rather than limitless.

D. Other Damselfishes

Booth *et al.* (2000) described recruitment of 23 species of damselfishes in One Tree Lagoon, confirming the recruitment-shadow hypothesis (Section V,C). Ault and Johnson (1998b) cited stochastic supply to an adjacent coral reef (Heron) as one factor that weakens relationships between recruitment and microhabitat availability. Tolimieri (1995) found that a Caribbean damselfish (*Stegastes planifrons*) selects *Montastrea* corals at small spatial scales, but that the abundance of this coral did not predict the recruitment of the fish at larger scales.

Doherty and Fowler (1994a) described the first large-scale empirical test of recruitment limitation after monitoring all visible recruitment for 9 years to large patch reefs (~90m^{-2}) within seven adjacent lagoon systems. For one damselfish, *Pomacentrus moluccensis*, this monitoring revealed both stochastic and deterministic influences (Fig. 12). Treating year-classes as replicates, different lagoons received sustained inputs at three statistically significant levels (low, medium, and high). In this regard, the sampling design looks like a scaled-up natural version of the experiment done by Jones (1990). Just before the tenth season, whole assemblages were recovered from each lagoon. From these collections, 150–200 *P. moluccensis* were aged using validated protocols (Fowler and Doherty, 1992). For each reef, this allowed the standing densities of nine year-classes of this fish to be compared with their initial densities. A worked example is provided from Lady Musgrave Reef (Fig. 13), which received most replenishment.

Average recruitment was an excellent predictor of the gross harvest of *P. moluccensis* from each reef. For six reefs, receiving initial densities of 0.03–0.27 recruits m^{-2} yr^{-1}, there was a strong linear relationship between average input and average catch. The exception was Lady Musgrave, which with highest recruitment (0.4 recruits m^{-2} yr^{-1}) yielded the same catch as Fitzroy (0.27 recruits m^{-2} yr^{-1}). Even including this outlier, mean recruitment predicted 84% of the variance in total catch among lagoons, a statistic that improved after age was brought into the equation.

Warner and Hughes (1988) showed that it would be difficult to demonstrate linkages between recruitment and abundance in populations with high storage, even if they were recruitment limited. Knowledge of the age structure avoids this problem. In the case of *P. moluccensis*, all populations contained fishes aged 10 years or older that recruited before the start of monitoring. In two populations, more than a third were teenagers. Removing them from the analysis made a small improvement in the fit of the linear regression to the overall data (r^2 = 89%) but a large improvement to the fit for the subset of six reefs. After adjustment, recruitment history explained 99% of the variance in adult numbers on these reefs.

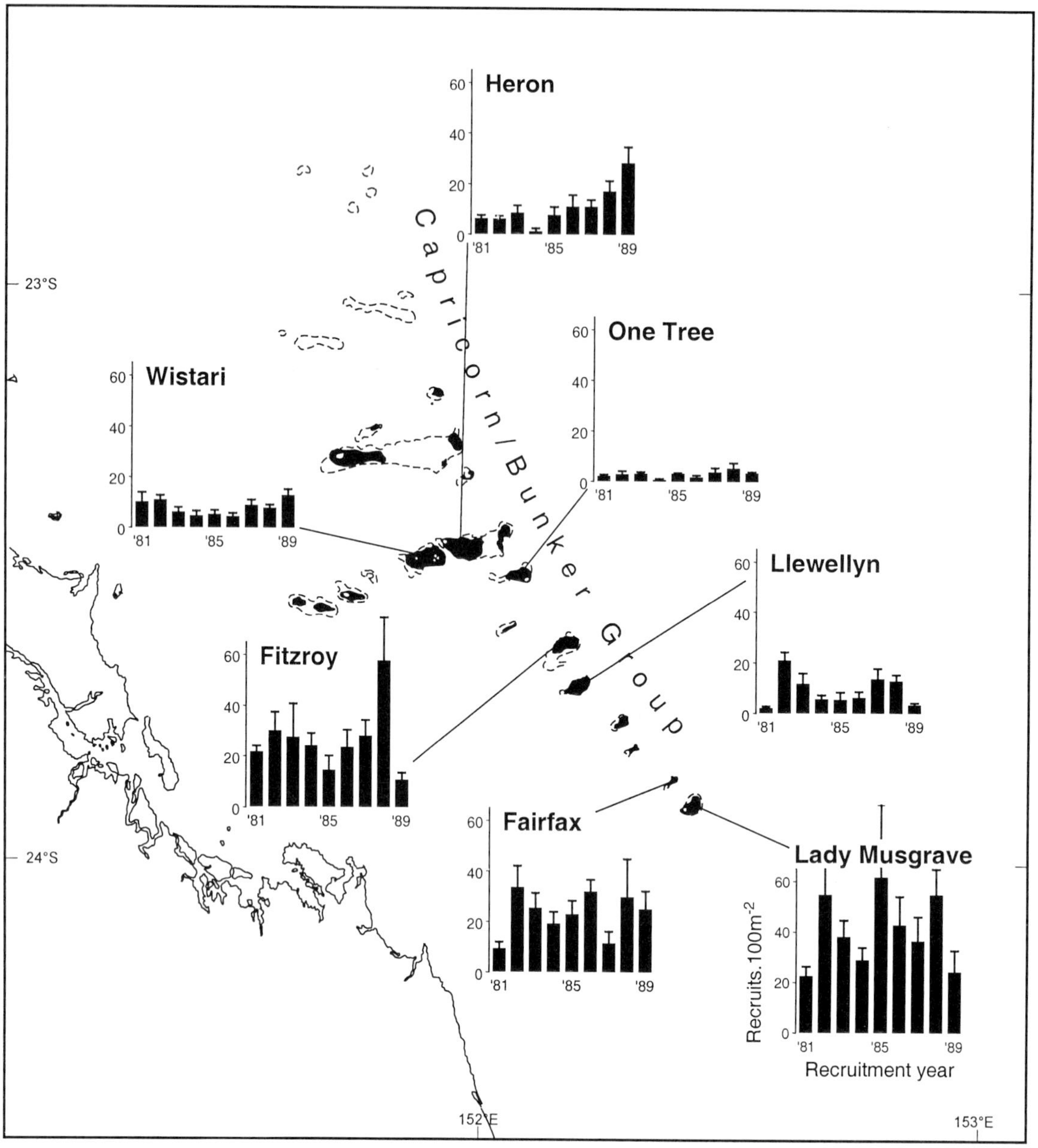

FIGURE 12 Recruitment densities in April of nine year-classes of *Pomacentrus moluccensis* counted on 10 large patch reefs within seven lagoons from the southern GBR. Error bars show standard errors. Redrawn from Doherty and Fowler (1994a).

Doherty and Fowler (1994b) extended the same analysis to a congeneric species, *Pomacentrus wardi,* which settled in an almost complementary pattern among the seven lagoons. Demographic profiles for the two species were developed from two reefs: Lady Musgrave (ranked 1 by recruitment and abundance of *P. moluccensis,* 5 by *P. wardi*) and Wistari (ranked 1 by *P. wardi,* 6 by *P. moluccensis*). On Lady Musgrave, both populations contained relatively few old fishes that pre-dated the monitoring (2% *P. moluccensis,* 8% *P. wardi*), whereas on Wistari, the same age-classes contributed 30 and 18%, respectively. This shows that longevity was controlled by differences among reefs (most likely predation) independent of conspecific density. On Wistari Reef, where turnover of both species was lower, the age structures resolved in 1989 showed significant correlations with the historical records of recruitment ($r = 0.79$, *P. moluccensis*; $r = 0.69$, *P. wardi*). On Lady Musgrave Reef, correlations were limited to the four or five youngest

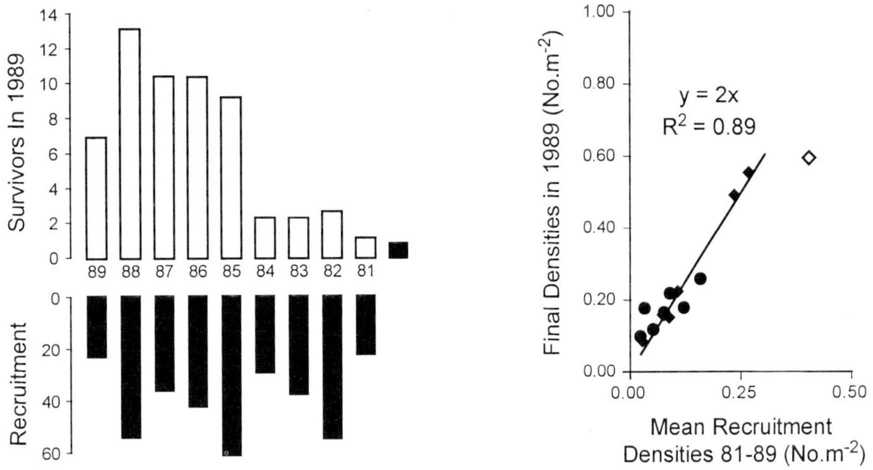

FIGURE 13 Left: Age structure of 154 *Pomacentrus moluccensis* collected from Lady Musgrave Reef in October 1989 plotted against the initial densities of the originating year-classes. Right: Average densities of *P. moluccensis* (diamonds) and *P. wardi* (circles) aged 1–9 years recovered from the seven lagoons plotted against initial year-class strength. The solid line is a regression constrained to pass through the origin. The open symbol shows the outcome for *P. moluccensis* from Lady Musgrave Reef, which was excluded. Redrawn from Doherty and Fowler (1994b).

age-classes because of an obvious depletion of older cohorts. Average recruitment was an effective predictor of the abundance of *P. wardi* among reefs and the recruit–adult values for this species clustered on the same line developed for *P. moluccensis* (Fig. 13). Inclusion of this second data set, however, did not clarify the status of *P. moluccensis* on Lady Musgrave because the highest average recruitment density of *P. wardi* was just 0.16 recruits m^{-2} yr^{-1}.

At the Kuranda workshop (Jones *et al.*, 1998), a lively discussion was had about the correct interpretation to place on the recruit–adult relationship from Doherty and Fowler (1994a). Great importance was attached to the result from Lady Musgrave, which deviates statistically from the linear relationship established from the other six reefs. Qualitatively, the pattern looks a lot like the nonlinear response obtained at smaller scales by Jones (1990). In our defense, Fowler and I decided at the time of publishing that it was premature to conclude anything about the exact shape of the relationship at the upper end of the density spectrum and implemented the most parsimonious model. We did state that there was an urgent need to repeat our experiment in regions receiving higher levels of recruitment (Doherty and Fowler, 1994a).

In light of the attention given to the status of the Lady Musgrave populations (Forrester, 1998; Armsworth, 2002), I have resurrected unpublished analyses where we looked for evidence of compensatory mortality on this reef. Per-capita survival of year-classes over the 9 years (Fig. 14a) showed no relationship to initial year-class strength, despite twofold differences in the latter (Fig. 13), but did suggest an ontogenetic pattern with a rate change obvious after the fourth year. This rate change introduces an inflection into the relationship between cumulative recruitment and stored abundance, which is linear for the four youngest age groups (Fig. 14a). If crowding affected adult and not juvenile mortality (as in Jones, 1987b, 1990), the trajectory for *P. moluccensis* from Lady Musgrave could be another example of density-dependent mortality among adult fishes consistent with examples from Forrester (1995) and Steele (1997b).

The only difficulty with drawing this conclusion is that several other populations among the seven showed the same pattern independent of density. For example, the population on Llewellyn Reef also displayed a nonlinear relationship between cumulative input and stored abundance, even though it received only a quarter of the replenishment received by Lady Musgrave populations (Fig. 14b). When both relationships are standardized, their response trajectories are identical despite being derived from different age structures and abundance levels (Fig. 14c). This suggests that sexual activity in *P. moluccensis* is accompanied by increased mortality but with no evidence that the effect is dependent upon absolute abundance.

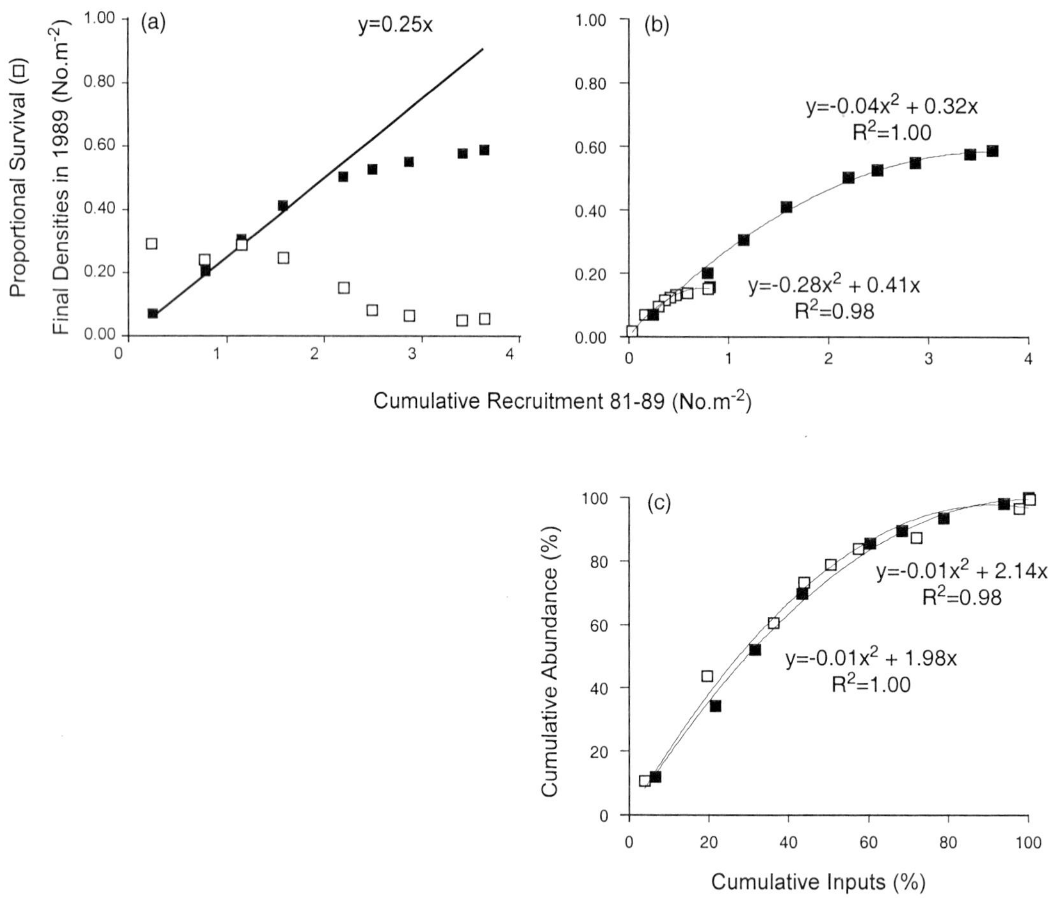

FIGURE 14 (a) Cumulative final densities of individual year-classes (starting with the most recent) of *Pomacentrus moluccensis* recovered from Lady Musgrave Reef plotted against cumulative year-class strength (filled symbols). Per-capita mortalities of the nine year-classes are superimposed (open symbols) but not plotted against the *x* axis. (b) Cumulative plots for Lady Musgrave (filled symbols) and Llewellyn (open symbols) reefs. (c) Cumulative plots for Lady Musgrave (filled symbols) and Llewellyn (open symbols) reefs standardized to relative abundance.

Participants in the Kuranda workshop also debated what it means to base initial densities of year-classes on juveniles that would have been 3–4 months old (Hixon, 1998) when counted in April of each year (Sale *et al.*, 1984). Williams *et al.* (1994) showed that end-of-season surveys do capture the densities measured at shorter census intervals, but neither study is able to account for the rapid mortality changes that can occur immediately after settlement (Doherty and Sale, 1985; Victor, 1986b; Booth, 1991; Schmitt and Holbrook, 1999a).

Doherty and Fowler (1994a) relied on previous longitudinal studies (mostly from One Tree Lagoon), which failed to find density-dependent mortality even when recruitment was monitored daily (Doherty, 1980, 1983a). Above all, they assumed that early losses were similar among the seven reef systems. Because there is no way to know retrospectively whether these assumptions were valid, one response would be to suspend judgement on whether the annual counts of recruitment were unbiased reflections of the relative abundance of larval supply among the reefs. However, studies from the northern GBR have been able to trace relative abundance patterns from supply through settlement to recruitment at the age of 3 months (Milicich *et al.*, 1992; Meekan *et al.*, 1993). Because these studies were from a location with much higher rates of settlement (Caley, 1995b) (see Chapter 14, this volume), it is not unreasonable to extrapolate similar linkages to the southern reefs.

Although the outcomes in Doherty and Fowler (1994a,b) cannot be linked for certain to settlement density, they provide an unambiguous demonstration that the size of a year-class measured within months

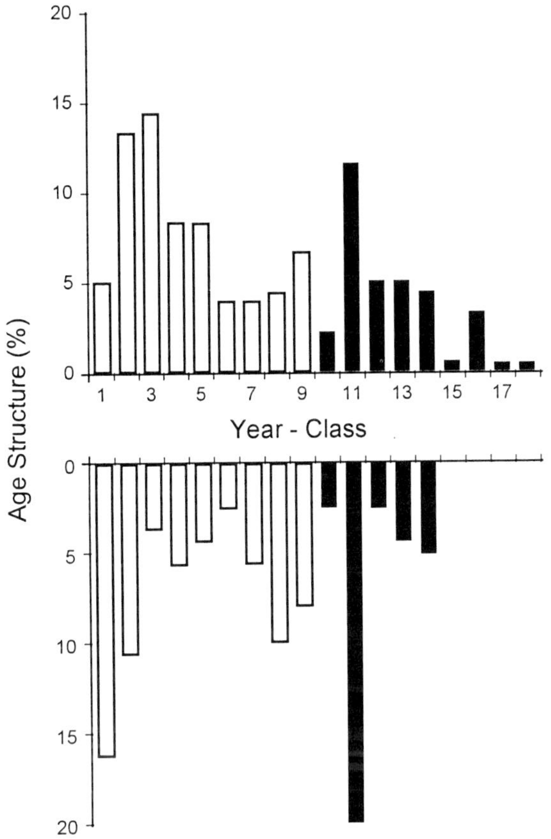

FIGURE 15 Age-structures of *Pomacentrus moluccensis* recovered from One Tree (upper) and Wistari (lower) reefs in 1989 showing a dominant year-class of 11-year-old fishes on both reefs that settled in 1979. Redrawn from Doherty and Fowler (1994a).

of settlement can have lasting consequences for abundance in a species that lives up to 20 years. On two reefs, where a third or more of the fishes were teenagers, *P. moluccensis* that settled 11 years before the harvest date yielded 12–20% of the recovered fishes (Fig. 15). This spatial coherence of an old dominant cohort shows that the processes responsible for year-class formation can operate over scales greater than a single reef system. Meekan *et al.* (2001) found a correlation between regional oceanographic conditions (El Niño) and the formation of dominant year-classes in populations of damselfishes from the Galapagos Islands.

E. Other Taxa

Spatial patterns in contemporary abundance and recruitment are ambiguous when both are correlated with the distribution of some habitat variable (e.g., Tolimieri, 1998a). In contrast, spatial variations in recruitment among sites of the same character, or temporal variations among fixed sites in the absence of disturbance, are explained most parsimoniously by variable larval supply, especially when the former are manifest over large spatial scales (Doherty, 1987a). Tolimieri *et al.* (1998) monitored recruitment of 14 reef fishes over 4 years at 10 sites in the waters around three Caribbean islands. Six species had ambiguous patterns, where variation among sites accounted for most of the pattern and ranking was preserved over times. Eight species were temporally variable, with much of the variance due to interactions between sites and years.

On the GBR, strong year-classes are often synchronous over broad spatial scales. Fowler *et al.* (1992) showed that a strong recruitment year-class of a butterflyfish was present across neighboring reefs (within 100 km) and also detectable among reefs separated by >800 km on the GBR. Russ *et al.* (1996) found a dominant year-class in the age structures of a grouper (*Plectropomus leopardus*) on coral reefs about 10 km apart in the central GBR. In the first collection, fishes belonging to this cohort were already 6 years old yet still contributed more than half of the numbers in unfished stocks (Fig. 16). Over the next 3 years, the strength of the cohort dwindled slowly as its members aged but survivors still made up 36% of the population at 9 years of age (Fig. 16). Newman *et al.* (1996), sampling in the same general area, found variable age structures for two species of snapper (Family: Lutjanidae) taken from interreef habitats which showed that these fish also store the effects of good recruitment years in their populations.

Because dominant cohorts by definition stand out from the background, a pertinent question is "What is the long-term probability distribution of recruitment densities observed at a particular scale?" With this in mind, the review of original data is closed here with unpublished results covering a wide range of species and situations from the GBR. The take-home message from these frequency distributions is the strong skew toward very low recruitment densities.

The first of these unpublished data sets represents 17 years of monitoring by David Williams and Susan English of fish recruitment to backreef sites on five coral reefs in the central GBR. In each year, recruits were counted once at the end of summer (Williams *et al.*, 1994) on 10 contiguous plots (each 25 m^2) at each of three sites with similar exposure in depths of 2–7 m. The length of the time series yielded ~20,000 densities after removing species with features (cryptic, schooling, excessively mobile) that encourage imprecise counts. The distribution of these densities is highly skewed, with 90% under 0.5 recruit m^{-2} (Fig. 17).

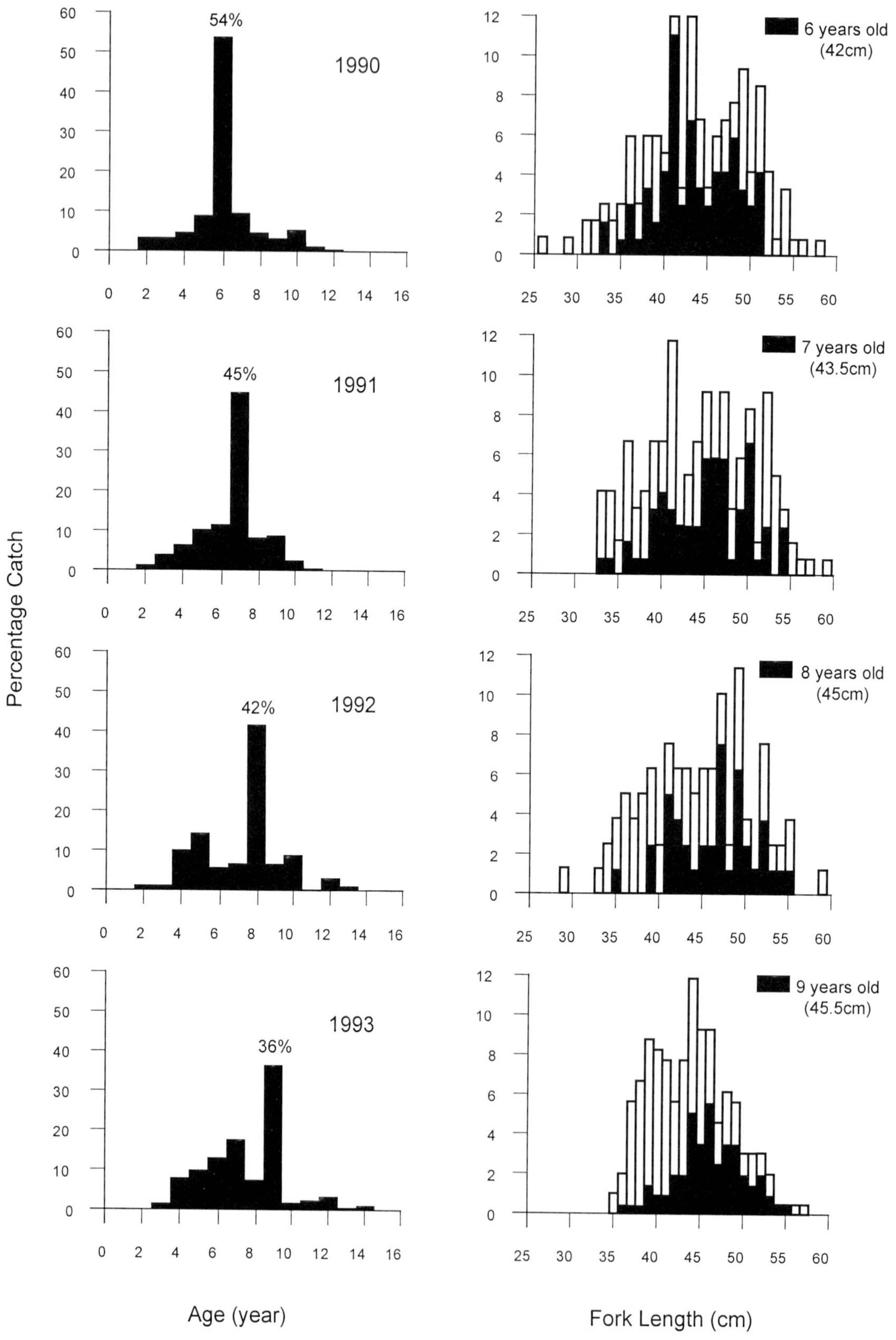

FIGURE 16 Left: Longitudinal samples of age-structures of coral trout, *Plectropomus leopardus,* showing persistence of a dominant year-class from 1984. Right: Size frequency distributions for the same samples with the dominant cohort shown by the dark bars. Redrawn from Russ *et al.* (1996).

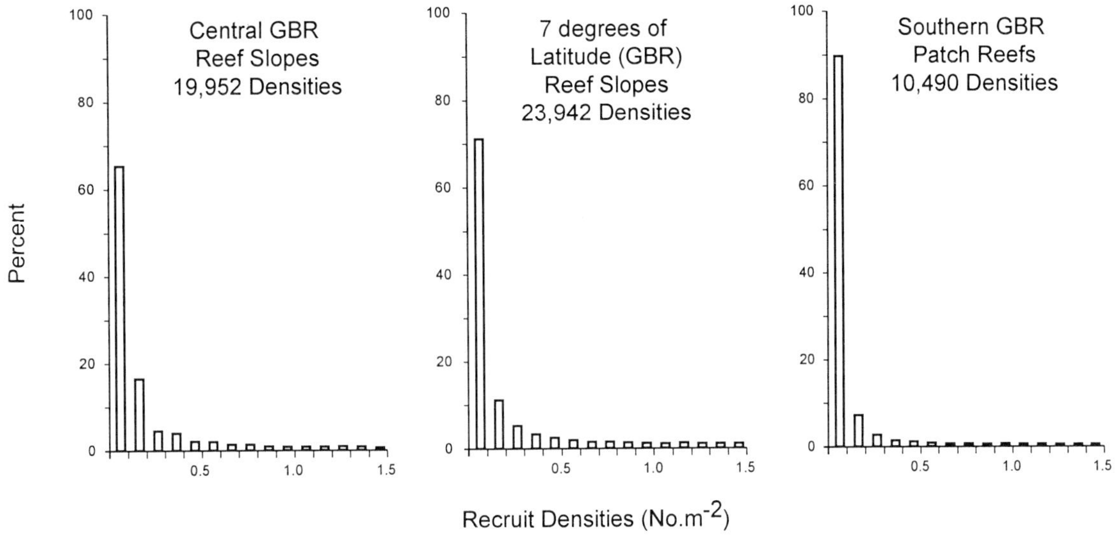

FIGURE 17 Frequency distributions of recruit densities in April counted in three independent studies of the central and southern GBR (see text for details).

The second data set was collected independent of the first, providing useful cross-validation. Doherty, Williams, and Sale led teams that counted fish recruitment on midshelf reefs from five regions of the southern and central GBR spread across 7° of latitude (>800 km). In each region, five reefs were selected on the basis of similarity in attributes such as coral cover and exposure. On each reef, recruitment was counted once at the ends of three consecutive summers (1985–1987) on three sites using strip transects of 40 m^2 laid parallel to the reef crest at depths of 6–9 m. Although three reefs were shared with the Townsville study, the methodology and locations were different. Nonetheless, the distribution of ~24,000 recruitment densities was very similar to the previous study, with 90% under 0.5 recruits m^{-2} (Fig. 17).

The final data set was collected from annual exhaustive counts of 10 patch reefs averaging 90 m^2 (Doherty and Fowler, 1994a) from the lagoons of seven coral reefs from the southern GBR (Fig. 12). These surveys provided >10,000 densities from 13 year-classes. The distribution of these densities was skewed even more strongly than those from the two surveys of reef slopes: >99% of the counts on patch reefs were <0.5 recruits m^{-2} and >95% were <0.2 recruits m^{-2} (Fig. 17).

Although these broad-scale recruitment surveys underestimate settlement (Sale and Ferrell, 1988; Booth and Beretta, 1994), the strong skew toward juvenile densities far below experimental benchmarks (Section V) suggests little scope for density dependence in the vast majority of cases. This would be consistent with the detection of supply signals in the distributions of juvenile densities (Doherty and Williams, 1988; Doherty, 1987a, 1991) and evidence that these variations are preserved with little modification in population structures (Doherty and Fowler, 1994a,b). Even on the GBR, not all reefs are expected to have these properties (Section VI). Unfortunately, there are few comparative large-scale and long-term data from other places (but see Tolimieri *et al.*, 1998; Masterson *et al.*, 1997). Robertson and Kaufman (1998) established that one reason for this is that sampling protocols developed for the GBR are inappropriate for much of the tropical western Atlantic, where recruitment is more continuous and geographically variable. It may be no coincidence that the more sustained replenishment of Caribbean populations has been associated with a greater focus on density-dependent regulation (Chapter 14, this volume).

F. Summary

Schmitt *et al.* (1999) introduced a simple cohort analysis to describe how density at a given stage in the life cycle affects the relationship between present (N_0) and future (N_t) abundance. Although this model is too simplistic for whole populations consisting of mixed age groups (Schmitt *et al.*, 1999; Shima, 1999b), or situations with significant interspecific interactions, it provides a portable analytical framework that can be applied (as I have) to standardize comparisons among species across places and time. The requirement to define N_0 and N_t anchors the comparisons along ontogenetic schedules and relative or absolute time scales.

Schmitt and co-workers concentrated on newly settled coral reef fishes because settlement represents peak abundance for all benthic cohorts and small inexperienced recruits have the highest daily losses (Victor, 1986b; Booth, 1991). For many species, the first week of benthic life is indicated as a window when mortality can rapidly modify the settlement signal. For some species, the window of high mortality closes slowly and only a small proportion of settlers ever reach maturity (Shulman and Ogden, 1987). For others, there may be a second mortality window associated with entry to the breeding pool (Steele, 1997b). In view of these several possibilities, my review showed that we do not have enough information about the linkages between initial and final densities in coral reef fishes. Robust theory will require more empirical data, preferably collected through controlled manipulations once questions have been resolved about possible artifacts from the use of transplants.

To boost the number of relationships, I reconstructed some from studies of mortality rates and admitted Jones' (1990) study, which was not fully causal. A residual problem with the dozen or so qualifying relationships (Table 2) is that they are a heterogeneous group. Results came from natural and experimentally formed cohorts, from pure and mixed cohorts, from different life histories and distant geographies, from different spatial and temporal scales, from different ontogenetic starting points, and from three orders of magnitude difference in initial densities. With virtually all degrees of freedom exhausted, it would be easy to end up with no trends, but not so.

Four relationships were linear. In all cases, this can be attributed to the range of densities examined by these studies. For example, *Dascyllus trimaculatus* showed a linear relationship in an experiment based on initial densities of up to 143 recruits m^{-2} (Schmitt and Holbrook, 1999a), but an asymptotic relationship in another experiment based on initial densities of up to 3160 recruits m^{-2} (Schmitt *et al.*, 1999). Ironically, analysis of the second experiment suggested that the upper limit for dominance by supply was around 140 recruits m^{-2}. The majority of relationships in Table 2 were curvilinear and are consistent with asymptotic limits to abundance.

There was more evidence for density dependence from cohorts of newly settled fish, but older cohorts may produce linear or weakly curvilinear relationships only because typical sampling designs confound lower rates of mortality (Sale and Ferrell, 1988; Victor, 1986b) with a lower range of densities. Some have suggested that age is the main reason why studies based on older juveniles have failed to find evidence of density-dependent mortality, but compensatory mortality was observed after manipulating the densities of older juveniles (Jones, 1990) and sexually mature

TABLE 2 Summary of Input–Output Relationships Analyzed in the Text[a]

Species	Maximum input density	Shape	Asymptotic density (*b*)	Threshold density (%)	Source
Pomacentrus wardi	0.16	Linear	na	na	Fig. 13
Pomacentrus moluccensis	0.3	Linear	na	na	Fig. 13
Pomacentrus amboinensis	1.8	Curve	1.2	75	Fig. 11
Pomacentrus amboinensis	2	Nonlinear	~1	>100	Fig. 11
Thalassoma bifasciatum	2	Curve	0.5	100	Fig. 7a
Thalassoma hardwicke	5	Curve	0.5	200	Fig. 8b
Coryphopteryus glaucofraenum	11	Curve	2	100	Fig. 5
Coryphopteryus nicholsii	15	Curve	3	100	Fig. 6
Thalassoma bifasciatum	60	Curve	2.5	200	Fig. 7b
Dascyllus flavicaudus	63	Linear	na	na	Fig. 10
Lythrypnus dalli	65	Curve	10	150	Fig. 6
Dascyllus aruanus	71	Curve	~20	~75	Fig. 10
Dascyllus aruanus	111	Mixed[b]	na	na	Fig. 9a
Dascyllus trimaculatus	143	Linear	na	na	Fig. 10
Dascyllus trimaculatus	3160	Curve	100	150	Fig. 3a

[a]All densities are fishes m^{-2}. Asymptote estimates the saturation density (not applicable for linear relationships). Threshold identifies the input density that delineates limitation by supply from limitation by resources expressed as a percentage of the saturation density. na, Not applicable.

[b]Shape was dependent on the presence/absence of conspecific adults.

gobies (Forrester, 1995). The model at the center of this review (Fig. 2) assumes that a cohort will remain subject to density-dependent pressures as long as its abundance exceeds a critical threshold.

The fifth column of Table 2 shows that this critical density was consistent across a broad range of examples. For each nonlinear relationship, the initial density was estimated beyond which density compensation was obvious. Where data were adequate, the heuristic was used, which indicates that this threshold should be located around $S = b/a$. Figure 6 provides examples that validate this rule of thumb with two gobies that experienced different rates of density-independent mortality. In *Coryphopterus*, $a = 1$ and the settler–adult relationship flattened abruptly at $S = b$. In *Lythrypnus*, $a = 0.5$ and the relationship flattened abruptly at $S = 2b$. In both cases, this response involved adult fishes, showing that density dependence need not be restricted to settlers. Considering their many differences, the 10 curvilinear relationships in Table 2 showed remarkable consistency, with threshold densities falling within a narrow range (0.75–$2b$). Because the average ($1.25b$) is equivalent to density-independent survival (a) of 0.8, the examples would appear to be biased toward fishes with relatively low turnover. Shulman and Ogden (1987) suggested that populations with turnover of <30% may be more likely to retain the imprint of variable supply. Consequently, there is a need to extend similar analyses to species with high natural mortalities, even though they provide less tractable experimental subjects.

For the examples covered by this review, cohorts were formed by the product of settlement and density-independent mortality ($A = aS$) until $S > b/a$, after which density-dependent mortality set in rapidly with strong compensation, producing asymptotic densities when a sufficient range of densities was explored. Getz (1996) suggested that organisms that produce large numbers of zygotes and exploit resources in a uniform manner could be expected to show little if any density dependence while they remain below a critical density. In his model, which was also based on a generalized form of the Beverton–Holt formulation for stock recruitment in marine fish (cf. Shepherd, 1982), the rate of onset of density dependence was controlled by an abruptness factor (γ).

The third parameter (d) of Eq. (1) changes the abruptness of the transition between dominance by density-independent and density-dependent mortality functions (Fig. 1), but scenario modeling shows that it is not equivalent to γ. Because d introduces a power function into the denominator of the density-dependent term, even small increases in its value ($d > 1$) produce unlikely parabolic relationships between initial and final densities. In conclusion, it seems that the cohort growth curve should be described with a modified term for density dependence that can be tuned against empirical data for the abruptness of the shift between states of limitation. There is every reason to suspect that such flexibility will be required to accommodate the responses of different species.

VI. Discussion

There are compelling theoretical arguments for believing that fish stocks must experience density-dependent regulation at some scale to prevent unbounded random walks to either extinction or overpopulation (Murdoch, 1994). This well-founded embrace of bounded persistence is not inconsistent with recruitment-limited local populations. Density dependence is not required of any demographic sink that depends on external inputs for its persistence (Armsworth, 2002). The debate about regulation has been clouded by issues of focus, definition (Caley *et al.*, 1996; Hixon, 1998), and legitimate issues of scale (Chesson, 1998b; Sale, 1998). Until I found ways to sample larval supply directly (Doherty, 1987b; Milicich and Doherty, 1994; Doherty and McIlwain, 1996), I spent a decade mapping patterns of fish recruitment on the GBR because I was searching for the oceanographic processes responsible for variable replenishment (Doherty, 1983c, 1987a). The best places to look for clues are in recruitment-limited systems because the supply signal has to be filtered from other noises. In surveys and experiments, this means factoring out variables that affect settlement preference and predation risk (e.g., Beukers and Jones, 1997). For some, this singular focus on recruitment patterns has been equated with relegation of all other factors. Hopefully this essay will have reminded readers again that recruitment limitation is simply one extreme in a continuum of responses to variable replenishment.

Primary recruitment limitation can only occur when settlement is low relative to the potential saturation state for an additional cohort. In this condition, abundance can be the simple product of settlement and density-independent mortality (Warner and Hughes, 1988). In this condition, variable replenishment can have lasting effects on spatial patterns of abundance, demographic profiles, and marine harvests (Guillou and Michel, 1993). The regulation debate simply says that this state cannot extend to all places and times.

Correlations and demography cannot discriminate between the presence and absence of weak density dependence (Forrester, 1995). In fact, several authors have suggested that perfect compensation is implausible in natural populations (Holm, 1990), which means that variable replenishment will nearly always leave its mark within benthic populations. For this reason, Forrester (1990) offered "recruitment determination" as a more general term to describe populations affected by varying combinations of density-independent and density-dependent fluctuations.

Schmitt and Holbrook (1999a) found that mortality reduced the spatial variation of cohorts of damselfishes at two different scales. Notwithstanding evidence that this reduction in variance involved a density-dependent component, they still found strong correlations between initial and final densities (i.e., recruitment determination). The correlations were stronger at the larger spatial scale, which they attributed to the different ranges of initial densities at the two scales. Although this was undoubtedly a factor, it may also reflect the way that predators are likely to operate.

Crowding *per se* rarely if ever kills fishes. The clearest evidence of this is the presence inside cages of supply effects that disappear with uncaged controls (Steele, 1997a). Predators will nearly always be the terminal agent of mortality, even though other factors may change their efficiency (Beukers and Jones, 1997; Letourneur *et al.,* 1998). Both contest and scramble competitions promote predation, albeit with slightly different outcomes. When there is an excess supply of hole-dwelling blennies (Bucheim and Hixon, 1992), contest for a limiting discontinuous resource results in marginalized individuals and differential predation ensures a constant maximum population. When there is an excess supply of schooling rabbitfish, which can overgraze their food source, the scramble to secure food can lead to mass starvation (Kami and Ikehara, 1976) and the subsequent predation can cause overcompensation. For sedentary prey that fall between these extremes, the foraging behavior of predators interacting with the spatial dispersion of settlers has the potential to produce density-dependent outcomes linked only indirectly to absolute abundance.

Predation is not an easy process to observe (Sweatman, 1984; Holbrook and Schmitt, 1997) and experimental manipulations are notoriously prone to artifacts (Doherty and Sale, 1985; Steele, 1996; Connell, 1997b). Furthermore, Hixon and Carr (1997) have shown that experiments can fail to detect density dependence because of the importance of synergistic interactions between resident and transient predators. This effect is analogous to the interactions among competitors for limiting shelter that can expose some individuals to greater risk. Prey can seek refuge in the water column to hide from resident benthic predators and they can shelter in the reef matrix away from transient pelagic predators. They are in trouble, however, when both types of predators act in concert. Because residents by definition reside with their targets, this suggests a critical role for the transient predators in a system.

By definition, transient predators forage over spatial scales that encompass many patches of potential prey, providing them with feedback about relative prey density. Any such predator following an optimal foraging strategy should direct its effort to obtain greatest reward. Options include switching targets according to relative abundance and aggregating when prey are underdispersed (clumped) in a patchy environment. Together, these functional and numerical responses have the potential to produce regulation that is driven by patterns of relative abundance.

Pelagic predators will generally forage over larger spatial scales, compared to benthic predators, but both categories should aggregate on patches of prey offering higher rewards (Kingsford, 1992). Pelagic species are very difficult to monitor (Thresher and Gunn, 1986), but Stewart and Jones (2001) found a positive correlation between the abundance of benthic predators and prey among sites within one coral reef (Fig. 18a). The fact that the relationship was not linear may suggest some constraints on the mobility (or supply) of benthic predators or a limit to the numbers that can operate in one site without interference (Rohner and Krebs, 1998). Although the aggregative response resulted in density-dependent per-capita mortality (Fig. 18b), which reduced the variance in prey among the sites, it did not suppress the spatial effect of differential supply (Fig. 18c). An analogous case can be argued for predation reducing but not eliminating a strong year-class in a time series.

More work needs to be done on the home ranges and foraging behavior of transient predators (Zeller, 1998), but most reef ecosystems contain natural barriers to the migration of adult fishes, which will set an upper limit on their activities. For the GBR, this scale corresponds to platform coral reefs. Most of these units measure 1–10 km and they are separated by similar distances. Monitoring data suggest that neighboring platform reefs are replicates on a regional scale (Ninio *et al.,* 2000). Over the long-term, they will share similar larval supply and accumulate similar assemblages of piscivores. At interannual scales, however, these predators are offered a moving feast of choices through the varied settlement of multispecific pulses of prey (Doherty, 1991). Although transient predators can respond to differential abundance within their home ranges, they cannot respond to variations in replenishment at larger scales. All things being equal, the spatial restriction on

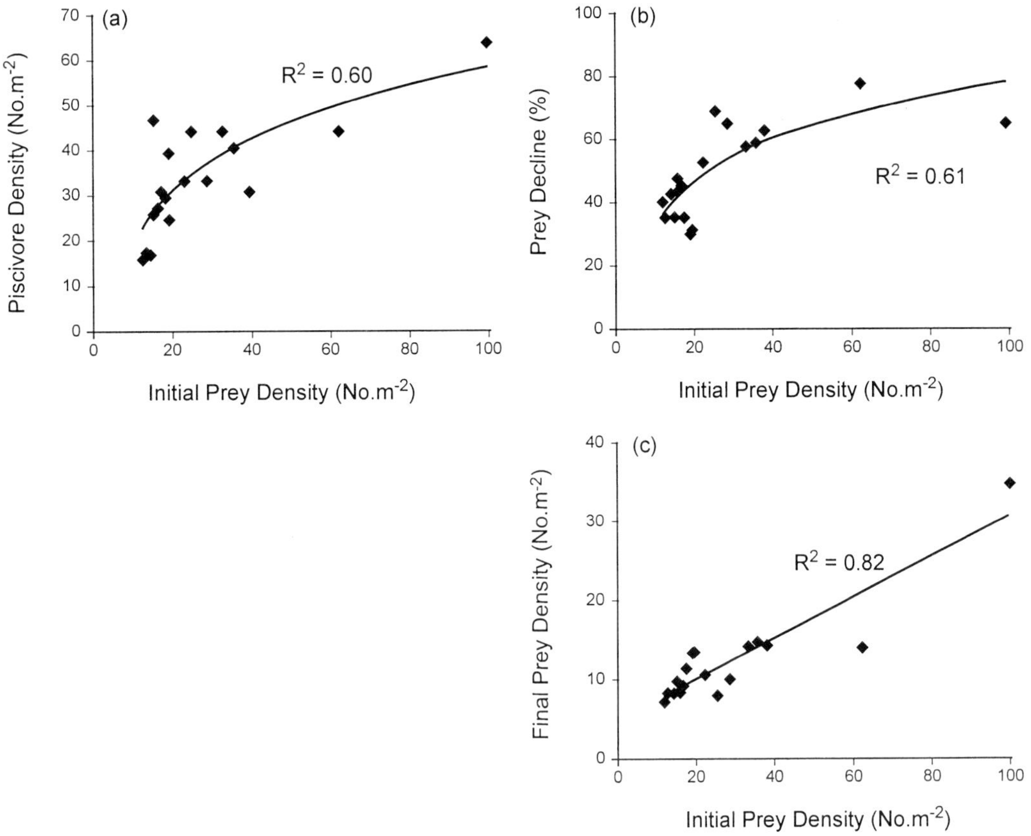

FIGURE 18 (a) Contemporaneous counts of predator and prey densities on patch reefs in February 1996. (b) Per-capita losses of prey over 6 months plotted against their initial densities. (c) Final densities in August 1996 versus their initial abundance. Redrawn from Stewart and Jones (2001).

the scale of the numerical response predicts stronger regulation within rather than among predator neighborhoods.

The tendency for mobile predators to aggregate at sites of high reward within their home ranges should be exaggerated when their prey are more clumped (Connell, 2000). For prey with random settlement, spatial variance in densities will be correlated with year-class strength. For species with greater contagion (Doherty, 1981), rising year-class strength will result in larger proportions of a year-class at risk in fewer high-density patches. Another numerical effect may enhance survival at very low abundance when most settlement involves solitary fish (Shima, 2001). Thus predators provide a responsive mechanism for regulation, but one that operates within constrained spatial scales.

Hixon and Webster (Chapter 14) invoked precedents from insect ecology (Hassell, 1986; Ray and Hastings, 1996) to argue that "studying spatial scales... much greater than that of the local population inhibits one's ability to detect biologically meaningful density dependence." Forrester (1995) agreed that the density dependence measurable in small-scale experiments may not be important, or observable, at larger scales. He proposed that density dependence in reef fishes tends to occur at small scales (among interacting neighbors) but not at large scales, where oceanographic factors may dominate. Chesson (1998b) labeled this disjunction as the scale transition. He showed that spatial variation in population density interacts with density-dependent processes to change the nature of population dynamics with a change in scale—largely as a result of nonlinear averaging (see Schmitt *et al.*, 1999). He argued also that dynamics on larger scales are no less real and therefore must be factored into models for the metapopulation.

Armsworth (2002) provides mathematical proof that metapopulations require density dependence at some (but not all) scales. For example, demographic sinks (those reliant on external inputs for their persistence) need not display any density dependence. Conversely the definition of demographic sources requires that they must be regulated, else abundance could ratchet to impossible levels, although density dependence need only act strongly in some years. A metapopulation may contain many sinks but, for persistence, it must contain an adequate number of self-sustaining sources. Armsworth rejects the notion of a persistent chain of populations passing propagules downstream (Roberts, 1997b). For persistence in a linear system, such as the GBR, some parts of the system must be self-sustaining, although propagules can be exchanged through a network of connected sources over many generations. Invoking mathematical analogy, P. R. Armsworth (personal communication) suggests that we look to isolated populations, which *in extremis* have to be totally self-recruiting and therefore represent the limiting case.

The literature confirms that almost all documented cases of strong density dependence at large spatial scales come from isolated oceanic systems (Section I). In 1998, I witnessed the spectacular aftermath of a strong pulse of settlement of *Ctenochaetus striatus* to Moorea. Everywhere we dived, inshore habitats were partially obscured by clouds of these juvenile surgeonfish. Half remained in a schooling presettlement stage, characterized by bright, transitional colors that I have never seen on the GBR, while equal numbers of drab grey benthic juveniles milled below. Contests and chases between settled individuals were as common as their frenzied feeding and many individuals displayed ragged fins and signs of emaciation. Numbers were so high that it was impossible to know whether there was additional settlement, but there was a clear visual perception of declining densities in both groups over the next 3 weeks, indicative of very high turnover.

In 25 years, I have never seen a similar event on the GBR, but Tahitian spearfishermen confirmed that this event was not atypical for this surgeonfish in French Polynesia (Stosh Thompson, personal communication). They also spoke of dramatic influxes and declines of other food fishes [see Dufour *et al.* (1996) and Letourneur *et al.* (1998) for scientific corroboration]. Because the Tahitian fishermen had no interest in damselfishes, it was not possible to conclude that similar dynamics apply to these species, although the self-recruitment inherent in isolated sources suggests that all species should require intermittent regulation. In Reunion Island (Indian Ocean), Letourneur (1996) cited poor recruitment as the reason for an island-wide decline in fish abundance.

Theory shows that self-recruitment and the associated regulation are just as important for the persistence of metapopulations as for isolated sources (Armsworth, 2002). On the GBR, the first empirical demonstration of self-recruitment for reef fishes came from Lizard Reef (Jones *et al.*, 1999), which lies in a region predicted to have high rates of self-seeding during periods of slow low-frequency longshore currents (Black *et al.*, 1995). Recent hydrodynamic dispersal modeling has shown that Lizard is one of a few dozen sources in this region predicted to have significant self-recruitment in most years (James *et al.*, submitted). When these sources (<10% of all reefs in the model domain) were suppressed, the model predicted the steady collapse of the metapopulation. In contrast, the model predicted little change following the suppression of an equal amount of habitat that was selected at random.

Comparative studies at Lizard Reef and One Tree Reef from the southern GBR have shown that settlement and postsettlement mortality of reef fishes are both higher at Lizard (Caley, 1995b) (see Chapter 14, this volume). This is consistent with the notion that one reef is a source and the other a sink. This review of case studies covering a broad range of situations found little evidence of density-dependent mortality in coral reef fishes at densities below 1 fish m^{-2}. Unpublished data from three surveys of the central and southern GBR, representing >55,000 densities sampled over >250 reef-years, revealed that 95% of counts were <0.5 recruits m^{-2} and 75% were <0.1 recruits m^{-2}. Experimental manipulations of abundance have shown no evidence of compensatory mortality trimming supply to such low densities so soon after settlement. The observed distributions of densities raise the possibility that many reefs in the central and southern GBR are demographic sinks sustained ultimately by reproduction from the northern region (Armsworth, 2002; James *et al.*, submitted). This status would be consistent with their recorded demographics and numerical population dynamics (Doherty and Fowler, 1994a; Russ *et al.*, 1996). In making this last point, I acknowledge yet again (Doherty, 1982) that a singular focus on numbers is a one-dimensional way of looking at regulation.

Fish populations are very tolerant of crowding, with numerical responses detected only at very high (Schmitt *et al.*, 1999) and very low (Shima, 2001) levels of abundance. Between these extremes, social behavior can have a big effect on individual access to essential resources, including food (G. P. Jones, 1986; Forrester, 1990; Webster and Hixon, 2000), with impacts on growth (Doherty, 1982; Jones, 1987b) and maturation (Forrester, 1990). Some populations may be able to store fecundity in the form of stunted subadults that could be recruited into the breeding pool as required to buffer egg production from the impacts of variable settlement (Jones, 1987b, 1990). Consequently, it is very likely that numerical fluctuations in fish populations will be a poor proxy for changes in egg production.

Kerrigan (1997) and McCormick (1998b, 1999a) have raised the debate about regulation to a new level of sophistication with their demonstrations that the density of competitors and egg predators can change the levels of hormones (cortisol, testosterone) in egg-guarding females, with impacts on the size and nutritive status of progeny at hatching. It seems logical that such parental legacies will affect the fitness of offspring (Kerrigan and McCormick, unpublished), either in the pelagic race or in the settlement transition. These and other examples (see Chapter 10, this volume) suggest that models of metapopulation dynamics may need to be translated to a common currency, such as eggs per recruit, as is the case with managed fish stocks (Rothschild, 1986).

Irrespective of the currency in which they are expressed finally, robust models of marine metapopulation dynamics are required to solve practical problems for human society. Millions of persons in the tropics depend on the value extracted from coral reef fish populations, whether for food or trade. Many commentators have seen networks of marine protected areas (MPAs) as the only salvation of degraded habitats and overfished stocks in developing countries (Roberts and Polunin, 1993) (see Chapter 19). Crowder *et al.* (2000) suggest that we need to understand source–sink relationships to site MPAs properly, and they conclude that improper placement has potential to harm fish populations. Roberts (1998, 2000) has made a counterargument that action cannot await such knowledge, and he proposed that "we create source areas when[ever] we create no-take reserves." The definition of a source (Pulliam, 1988; Crowder *et al.,* 2000) suggests that this may be wishful thinking when much of the habitat consists of recruitment-limited sinks.

Theoretically, recruitment-limited sinks will be the first places to decline in response to overfishing, because they have minimal redundancy. Although they would seem to be ideal candidates for stock enhancement (Peterson *et al.,* 1996; Maroz and Fishelson, 1997), the vulnerability of transplanted natural settlers revealed by this review suggests that piscivores will be a formidable obstacle to any sea-ranching in coral reef lagoons (Doherty, 1999). Despite the potential for rapid depletion, populations in recruitment-limited sinks should be resilient and persistent, provided that their sources remain healthy. The corollary is that source reefs will be productive and initially resilient to the effects of fishing; both features that will tend to attract fishing effort. Strong sources should be the last places to decline, but when overfished the whole system could be at risk (James *et al.,* submitted). This sets a clear challenge for scientists and natural resource managers to identify and differentiate those places with robust levels of recruitment from those where weak and erratic replenishment limits abundance.

Acknowledgments

I thank colleagues for permission to present some of our unpublished data and all of the students and volunteers who have assisted us over the past two decades. I thank the participants in the Kuranda workshop for our thoughtful discussions and a week of great fun. I am greatly indebted to Peter Sale for his patience and encouragement of this recalcitrant author, and Paul Armsworth for useful comments on the manuscript. Last, I honor Ern and Jake, who started this journey with me but who had to leave before its end. Now we can all rest in peace.

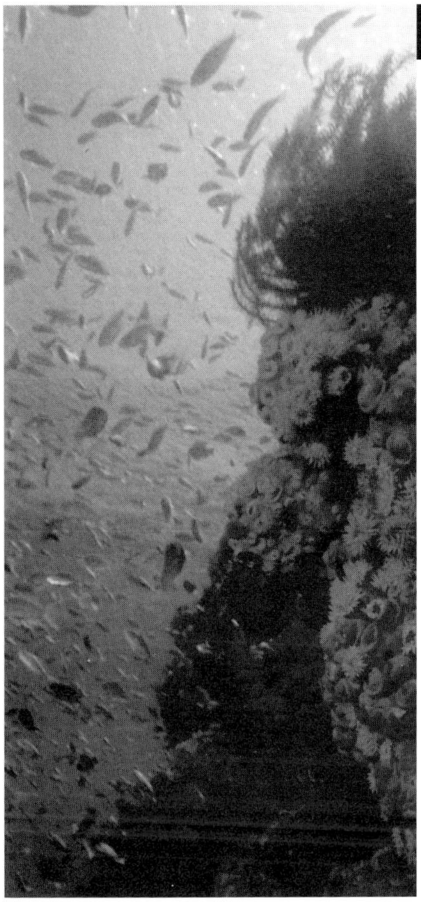

SECTION IV

Management of Coral Reef Fishes

Introduction

In preparing chapters for *The Ecology of Fishes on Coral Reefs* (Sale, 1991a), I selected 1 out of 19 chapters (Russ, 1991) to deal with issues of reef fish management. In the present book, 4 of 19 chapters address that topic. This is so, despite the publication in 1996 of *Reef Fisheries* (Polunin and Roberts, 1996), in which 11 of 14 chapters dealt with aspects of management. Clearly I thought the importance of the topic has increased (some will say there should have been greater representation in 1991).

The fishes that occupy coral reefs are the richest vertebrate communities on Earth, and an important part of the richest biotic community on the planet (if one measures richness at the family level or higher, to avoid all those rain forest beetles!). They are a wonderful system in which to explore ecological principles using powerful, experimental approaches. But they also provide food and other economic benefits for large numbers of people, chiefly in developing countries that need these resource benefits. We have not been doing a satisfactory job of managing them so that these economic benefits, and their persistence as shimmering pools of biodiversity, can be sustained. It would be inappropriate to not include a significant component on management in this book, because improving management should be a priority, and will require all the scientific understanding we can generate.

Chapter 16 is my own mea culpa, an acknowledgment that I now understand that academic ecologists should not remain cloistered, and a serious attempt to identify the critical issues that link science and management. Unless we build the new science that will help answer management questions, reef fish communities, as we know them, will dwindle away. With them will go the great economic values they provide us by being on this planet.

In Chapter 17, Phil Levin and Churchill Grimes review the conservation and management of groupers, attempting to draw lessons from the ecological literature on reef fishes. In a sense, they explore the degree to which it is possible to take all that academic research on "pretty little fishes" and use it to derive new insights about these valuable but vulnerable species. Their chapter demonstrates the substantial information we do have about these important species, and highlights the gaps that need to be filled. I am encouraged by their closing plea for powerful, experimental approaches to test ideas; powerful adaptive management approaches are coming.

Yvonne Sadovy and Amanda Vincent use Chapter 18 to review two important, but non-traditional fisheries for coral reef fishes: the live restaurant trade and the aquarium/curio trade. Though the latter has long been a part of the reef fishery mix, the former exploded in importance during the past decade. In both trades, value increases markedly with rarity, so that fishing effort does not respond conventionally to decline of fish stocks. For this reason, traditional fishery management approaches are poorly equipped to deal with them. Add to that the enormous profits that encourage poaching, and the general lack of effective fishery management in many coral reef areas, and it is clear that these trades represent potentially very serious threats to the continued viability of many reef fish populations and species. This chapter will disturb. I hope it will provoke action.

In Chapter 19, Garry Russ provides us with "yet another review" (his words) of marine reserves as fishery management tools. He critically reviews the many studies of the effects (or not) of reserves, and sets out a clear statement of what is needed to demonstrate that a reserve is functioning to improve the surrounding fishery. His main conclusion will surprise many. It is that, as yet, there is no definitive evidence that reserves augment fisheries in the ways conventional wisdom claims. However, Russ closes his chapter with five arguments why marine reserves should continue to be developed, for fisheries purposes, even while the needed, definitive experiments to justify their use as fishery tools are being done. Managers work in the real world, and on too many reefs there is not the luxury of time to await definitive answers before acting. His closing paragraph both justifies the attention to management given by this book, and calls on the scientific community to contribute to addressing these very real management questions.

CHAPTER 16

The Science We Need to Develop for More Effective Management

Peter F. Sale
Department of Biological Sciences and
Great Lakes Institute for Environmental Research
University of Windsor
Windsor, Ontario, Canada N9B 3P4

I. Introduction

Around the world, reef fish assemblages provide important food resources and sources of income for coastal populations. Worldwide, our use of these living resources, and our other activities, are reducing their capacity to be productive, and even to be sustained. This is so, despite the fact that around the world, coastal states have management agencies that seek to make our use of these important resources sustainable. Annual fishery harvests from areas of coral reefs range from less than 1 to 44 t km^{-2} yr^{-1}, with a mode around 5–10 t km^{-2} yr^{-1} (Dalzell, 1996). [Note: the weight unit used here is the metric ton (t).] Yields of 10–20 t km^{-2} yr^{-1} seem feasible, but higher rates of extraction have not been sustained (Russ, 1991; Birkeland, 1997b). Catch per unit of effort (CPUE) fell 70% in the Samoan fishery, with a marked shift in species composition, in the 15 years following an initial report (Wass, 1982) of catches as high as 44 t km^{-2} yr^{-1}. Comparable declines are widespread (Birkeland, 1997b), despite the efforts of management agencies to ensure sustainable fishing. Less well quantified, but equally important, are the indirect effects of intensive fishing on the nonextractive values of reef fish assemblages for tourism and recreation. The goal here is to look at what has gone wrong, and to suggest steps to improve our management of tropical coastal resources, and reef fish communities in particular.

Although all management involves control of human impacts on natural systems, the thesis here is that environmental and fisheries management protocols have been developed for tropical coastal systems without as much real attention to the structure and dynamics of the ecological systems impacted as there should have been. Reasons for this lack include the chronic shortage of human and other resources in developing countries, and a widespread failure among professionals to realize that most tropical coastal fisheries are fundamentally different from the capital-intensive, large-scale, commercial fisheries of northern seas. In addition, management actions have been developed on the basis of ecological principles for which there may be scant scientific justification (particularly in the context of complex tropical communities), and using an extensive basis of common sense about how ecological systems operate—a body of knowledge that may include many errors.

If political correctness requires that we never point out our problems, finding real solutions may take a lot longer than we can afford. Still, before I risk alienating all readers, I am not advocating that ecological principles are the only type of knowledge that is required to provide effective management. Environmental and resource management decisions involve complex and difficult processes that require economic, sociological, ethical, and a range of other data and perspectives, as well as sound ecology. Nor do I imply, by referring to our failure to manage effectively, that the managers have not done the best they could with the resources and knowledge at their disposal. Above all, I do not imply that I am the first ecologist to recognize this problem—several ecologists have worked long and hard to integrate modern ecological understanding with management practices (for example, Walters, 1986; Lubchenco *et al.*, 1991; Daily *et al.*, 1996; Birkeland, 1997b; Johannes, 1998; Dayton, 1998; Lubchenco, 1998; Steneck, 1998; Dayton *et al.*, 2000). I am repeating their messages one more time.

I believe that we ecologists and our natural science allies have permitted ourselves to be moved away (or have taken ourselves away) from the table at which management policies are developed. Ecological systems are complex entities, and tropical marine systems, particularly on coral reefs, are among the most complex. Ecologists continue to learn new things about their structure and function, and are far from agreed about the fundamental principles defining them (Table 1). The vibrant discussions and the exciting new discoveries along the cutting edge of ecology are not properly conveyed outside "the club," with the result that others operate under the false impression that the functioning of ecological systems has been well understood for some time. For example, Agardy (1997) lists seven commonly accepted myths concerning management of human impacts on marine resources—most are ecological. If management skills are to improve, we must improve the scientific base for them. This means that ecologists must come back to the table.

TABLE 1 Some Not-So-Fundamental Principles of Ecology[a]

- Ecological communities achieve and maintain an equilibrium composition except when unusual, often anthropogenic, disturbances intervene
- In ecological trophic webs, complexity—or degree of connectance—confers stability of structure
- Ecological communities are primarily regulated by "top-down" effects of predators
- Ecological communities are primarily regulated by "bottom-up" effects of resource limitation
- Natural selection ensures that organisms usually are optimally adapted for the environment in which they live
- Invasions by new species, whether naturally or anthropogenically caused, invariably damage the invaded community, and low-diversity communities are more susceptible than others to being invaded
- The Ricker and the Beverton and Holt models adequately account for relationships between population size and rate of recruitment in fishery species
- The theory of island biogeography provides an adequate framework for the design and implementation of protected areas to maintain biodiversity

[a]There are currently substantial differences of opinion among ecologists concerning each of these assertions, yet each is often presented to the public as a well-established principle of ecological science.

II. Why Ecology Is Not Well Integrated into Management

Here I discuss in turn each of the three reasons for the lack of representation of ecologists at the management table:

- Ecologists have tended to explore questions using convenient species and systems rather than those that are economically important.
- Academic ecologists until recently actively avoided "applied" questions in favor of "pure" research.
- Ecology is not a profession with membership limited to those accredited by national or international bodies that certify qualifications.

A. Use of Economically Less Important Species

Academic ecologists for many years have explored questions using systems that were convenient rather than economically important. The ecology of the 1960s was developed through work on songbirds, small mammals, insects (butterflies as often as pest species), intertidal invertebrates, the fishes of ponds and streams, and herbaceous plants. True, there were ecologists working on big game, on commercial fisheries, on crop plants, and on insect pests. But to a very large degree, these individuals were not the ones who developed the underlying principles that drove ecology. There are two reasons for this.

First, working with many of the economically important organisms is difficult. When working with game mammals, it is not easy to replicate well, or even to do field experiments, and a great portion of work on these larger organisms has been descriptive. This difficulty of replication and experimental design shows up even in studies of small rodents (hardly game mammals), which have turned out to be quite mobile and difficult to manipulate. Methodologies using large cages (e.g., 250 m^2) of sophisticated design have been required (Valone and Brown, 1996) in studies spanning several years, simply to explore competitive interactions. Ecologists wishing to work on commercial fishery species have been drawn to fishery management agencies, and to a strong interdependence with the fishery simply for access to data on the fish populations. Fishery-independent sampling is expensive, and opportunities to manipulate the fishery in order to run field experiments have been rare. It is far easier to do innovative work with guppies, sticklebacks, or a host of other small species. The consequence is *not* that good science has not been done on economically important species, but that the science has tended to move in directions compatible with the logistic constraints, and to address features specific to particular species or populations, rather than to tackle general ecological questions.

Second, many of the ecologists who worked on the more "important" species have been directly employed by management agencies. Their freedom to explore new ideas has been constrained by the need to do the work the agency had a legal responsibility to do, and, to a considerable degree, they have worked in an environment that is pragmatic, and problem-solving, rather than one that encourages the building of general scientific principles. Career advancement has derived from solving particular problems, rather than from publishing these solutions in high-profile journals, and extensive "gray" literatures have been developed and used. This has further isolated many agency ecologists from the academic ecologists who published their exciting, but often esoteric, new ideas about economically quite unimportant organisms in the peer-reviewed literature!

I am deliberately overstating here to make my point. In governmental departments of fisheries, entomology, and agriculture, there have always been some individuals who played substantial roles in the development of ecology as a science. W. E. Ricker, R. J. H. Beverton, D. H. Cushing, and J. A. Gulland are four giants of fisheries biology whose work influenced the wider ecological world, and H. G. Andrewartha and L. C. Birch always approached their science from the pragmatic perspective of the argricultural research institute, while addressing fundamental ecological questions. I suggest, however, that such individuals are the exceptions that prove the general rule.

B. Active Disinterest in "Applied" Questions

There has been a long period of active disinterest by the academic ecological community in "applied" questions. We frequently prided ourselves on working on systems that were not economically important, or on systems largely unimpacted by human activities. Indeed, human impact was seen frequently as an "unnatural" disturbance making a system less suitable for ecological study. In academic institutions, well into the 1970s, work that was economically relevant was often disparaged. It was "applied," rather than "basic" (or, more self-servingly, "pure"). Now, with human impacts substantially greater and more widespread, it is difficult to find "unimpacted" systems, and academic ecologists seem much more interested in the systems that our species is impacting severely. This is good, but it is a recent development (Lubchenco *et al.,* 1991), and the major advances in ecological understanding over the past 100 years have come about because of work on economically less important organisms. Ecology did not evolve as a practical science.

C. Lack of Professional Status for Ecologists

Ecologists do not wear rings or insignia, and are not licensed or certified (until recently and only on application). Anyone who seems to be "interested in" natural systems can identify as an ecologist, and in many people's minds, an individual who practices the science of ecology is not different from one who enjoys natural history or from one who espouses a conservationist ethic. The distinction between ecologists and conservationists has not been clear.

Although ecologists have busily been avoiding research that would have ensured we sat at the table where management goals, policies, and practices were developed, the conservationists have been eager to take places there. I suggest the absence of ecologists was not even noticed by others at the table, because conservationists frequently seemed to look, walk, and talk like ecologists. Places at the table seemed to be filled, and an ecological perspective was present. It is not going to be easy to fight our way back in, because ecologists have not been missed. I am not deriding the important role of conservationists. They should be participants at the management table, but they do not replace ecologists. The conservationist ethic is not science, despite having many scientific elements. It is an ethical perspective on the world and our place in it. Although most conservationists have sought ecological evidence to support their arguments, and while many have formal ecological training, there are necessarily delays between discovery and dissemination, and ecological concepts have been evolving rapidly. Conservationists should not be expected to provide a sound ecological basis for management action in addition to providing their important ethical perspective.

To get back to the management table, we ecologists need to demonstrate that our science can help improve management, and we need to become better informed about the management process and the other kinds of expertise that contribute to it. I believe ecologists need to articulate our science more clearly, and to apply it to the development of scientifically valid management practices, and I am encouraged that this is now happening (Doak and Mills, 1994; Bohnsack, 1998; Roberts, 1998; Steneck, 1998; S. N. Murray *et al.*, 1999; Russ and Alcala, 1999; Walters *et al.*, 1999; Daily *et al.*, 2000; Pauly *et al.*, 2000). In becoming more active participants in the debates that lead to management decisions, we will contribute directly to the evolution of management practices, and our exposure to real-world management issues will stimulate our science, leading us to new questions that are important to tackle.

III. Our Present Condition

I will briefly summarize some essential aspects of reef fish ecology. I will then consider current management practice. In so doing, I make a distinction between fisheries management and conservation management, because fishes are subject to both, and these have different goals and sometimes different approaches.

A. Reef Fish Ecology

Coral reef fishes are the shorefishes of tropical coastlines. A few species are restricted to the topographically complex, biogenic habitats of coral reefs, but many of them occur also over seagrass beds, in shallow or deeper off-reef lagoons, and in mangroves and estuaries (Robertson, 1998b; Bellwood, 1998) . In all of these habitats, but especially on the reefs, coral reef fishes are notably abundant and speciose. Species lists for remote locations such as Easter and Ascension islands number less than 100, but most coral reef regions include from 500 to 1000 or more species (Thresher, 1991). Biomass estimates vary considerably among regions, due both to levels of production and of harvest, but range from 26 to 238 t km^{-2} yr^{-1} (Polunin, 1996). This biomass is always distributed among many species, and fisheries therefore tend to include many species. This is in marked contrast to fisheries in temperate waters (where fisheries management evolved), in which single species can dominate the catch.

Reef fish species fill a broad range of consumer roles; they are planktivores, carnivores on demersal invertebrates, piscivores, and grazing and browsing herbivores (see Chapter 2, this volume). In many systems, they are the most important participants in several of these trophic roles (in terms of energy or nutrient transfer). Further, because most species experience at least an order of magnitude change in size during the life-span, trophic roles can change substantially during life of the individual. Such ontogenetic niche shifts mean that individuals use their world in two or more quite different ways during the course of their life-span, with the difference between the pelagic larval and the demersal juvenile and adult phases being the most pronounced.

With very few exceptions, reef fishes have life cycles with a more or less lengthy pelagic larval phase followed by a demersal and usually strongly site-attached adult phase. The larval phase lasts from 10 to 100 days, more or less, and the sedentary phase lasts from one to several years (often to several decades in unfished systems; see Chapter 3, this volume). Until the 1980s, ecological studies of reef fishes focused almost exclusively on the reef-associated juvenile and adult phases, and our knowledge of the very different larval life has lagged seriously (Leis, 1991a) (see Chapters 7 and 8, this volume).

Immense differences exist among reef fish species in expression of sexual roles (gonochores and simultaneous and sequential hermaphrodites all exist), in fecundity, in frequency and periodicity of spawning, in duration of the larval phase, in survivorship during the first critical days in the demersal habitat, and in longevity of settled fishes. Until recently, there has been relatively little attention to these differences (Chapter 8, this volume). In addition, it is now clear that there can be substantial spatial and temporal differences within species in these same life history features (Chapter 10, this volume). In one sense, the many different life history patterns exhibited by reef fishes are alternate (and usually successful?) solutions to the challenges of living in a spatially very patchy environment, bathed by waters that move in complex, nonconstant ways. The diversity of life styles means parenthetically that no single management approach will be optimal for all species!

Early ecological research on reef fishes made use of a rather narrow suite of species. Species that were used were small, easily caught, site attached, and reasonably abundant. As a consequence we know a lot about the commonest damselfishes, somewhat less about the most abundant parrotfishes, wrasses, and butterflyfishes, and next to nothing about the great range of species that each account for less than 1% of all individuals present in a site (Chapter 4, this volume). A focus on common, small, easily caught species has permitted a strongly experimental approach in reef fish ecology—one that has challenged several widely held paradigms of ecology (Sale, 1988). But our understanding may not scale up easily to larger, rarer species. Although damselfishes are attractive animals with which to work, and will continue to yield important ecological insights, they cannot be representative of all reef fish species. It is important that recent attempts to work with logistically more "difficult" groups continue. The demographic differences among species are particularly important in determining the spatial and temporal scales at which individual species live out their lives (Sale, 1998), and we need a reasonable representation of "types" in our body of ecological data.

The pelagic larval phase and the patchy reef environment ensure that reef fish populations function as open systems of separate subpopulations interconnected by the recruitment of juveniles. However, perhaps the major unanswered ecological question at the present time is how open these populations are (see Section IV,B). Ecologists have swung the pendulum. In the 1970s and early 1980s, when the prevailing ecological paradigm was one of closed populations, we emphasized the dispersive abilities of pelagic larvae. There was abundant evidence that fishes spawned in ways that facilitated export of larvae to the open ocean (Johannes, 1978). Larval lives were of substantial duration, and larvae were specialized morphologically, and in other ways, for a pelagic existence (Leis, 1991a). Clearly, larvae dispersed.

During the 1990s, it became apparent that reef fish populations are less than totally open (see Section IV,B,2), and two empirical studies showed the "self-recruitment" of reef fish populations (Jones *et al.*, 1999; Swearer *et al.*, 1999) by demonstrating that a substantial portion of larvae produced appear to be retained within an area measured in tens to hundreds of square kilometers. This area remains large enough to contain a substantial number of local populations of the species concerned (a damselfish and a wrasse, respectively), and it is important that in emphasizing this evidence of control of dispersal, we do not fall into the trap of claiming that larval reef fishes do not disperse. The extent of larval dispersal appears to be much less than the early enthusiasm suggested, but reef fish populations are still open. [Parenthetically, I suggested long ago (Sale, 1970), and based on hardly any data, that surgeonfish larvae would be found better at getting themselves to a reef habitat than if they acted as passive particles!]

At present, we do not know the magnitude of the interconnection among local populations on any spatial scale, and are unable to assess whether it is such that these groups of populations will function as metapopulations. At one extreme (low levels of dispersal among populations), the separate populations will function almost independently. At the other extreme (higher rates of dispersal among populations), they may be so well interconnected that they will function as a single, if subdivided, population. Only at intermediate levels of dispersal will they operate as a metapopulation with the "special" dynamics that characteristically buffer such systems from both local and global extinction (Hanski, 1998, 1999). I anticipate that the range of reproductive patterns, larval durations, and spatial arrays of habitat found among coral reef fishes surely will ensure that some populations form metapopulations, and I will be surprised if this trait turns out to be rare.

Because we have focused study of most reef fish species within the structurally complex reef environment, we are largely ignorant of the ecology of populations living in other habitats, the relative magnitude of reef and nearby nonreef populations, and the role if any of nonreef populations in sustaining populations on reefs. There are many studies of reef fishes in lagoonal, seagrass bed, and estuarine regions (Laegdsqaard and Johnson, 1995; Ley *et al.*, 1999; Lindeman and Snyder, 1999; Ronnback *et al.*, 1999; Kulbicki *et al.*, 2000; Nagelkerken *et al.*, 2000), but I do not think these yet give a clear picture of the relative densities in these different habitats, or of the degree of connection due to animals moving between these habitats. Are mangroves and seagrass beds vital to the success of populations of a broad range of species in an adjacent coral reef habitat? Or are the animals seen in these nonreef habitats destined never to find their way to reefs, and perhaps never to spawn? A full comprehension of the dynamics of reef fish populations will require far better information than we now possess on fishes in nonreef habitats.

B. Management Goals and Approaches

Reef fish populations are affected by management actions undertaken in two separate contexts: fishery management and conservation management. These contexts have different goals and are frequently handled by different individuals and agencies, sometimes with rather little communication between them (Table 2).

1. Fishery Management

Fishery management aims to preserve economic value of a fishery, usually by implementing a set of regulations that will lead to an economically beneficial, but demographically sustainable, harvest of desired species. Ideally (meaning if the logistic growth equation adequately describes a targeted fish population), by adjusting the rate at which fish are taken by the fishery, it should be possible to depress the population to a size (typically half its usual unfished size) at which population growth will be maximal. Fishing at a rate that keeps the population at this size will yield the maximum biomass that that population can sustain without going extinct. Achieving this maximum sustainable yield (MSY) was the elusive holy grail of fisheries management science for many years. More recently, attention has shifted to ensuring a maximum economic yield (MEY) that is usually less than MSY, but

TABLE 2 Coral Reef Fishes Are Affected Both by Fisheries Management and by Conservation Management[a]

Concern	Fisheries management	Conservation management
Goal	Sustainable economic benefit from consumption of fishery resources	Sustainable ecosystems that function to preserve biodiversity and its intrinsic value for the biosphere
Agency	Fisheries department (or equivalent) of local, regional or national government, sometimes acting in cooperation with adjacent agencies in management of fishery stocks that cross jurisdictional borders	Conservation department (or equivalent) of local, regional, or national government, or nongovernmental organization with conservation objectives
Measure of success	Sustained economic value of the catch, or related measures such as sustained employment in the industry or sustained yield (tonnage) of product	Successful mitigation of harmful human impacts on ecosystems or specific endangered species; successful preservation of ecological systems and their biota
Complementarity	Sustainable fishery practices usually minimize deleterious impacts on the ecological system fished beyond the removal of the harvest; use of no-take reserves as a way to control fishing effort also protects the reserve from other fishery impacts	Protected areas established for conservation of representative ecosystems and their biota may harbor fishes that become a protected broodstock to re-supply fished regions beyond the area of protection
Conflicts	A well-managed fishery may still damage aspects of ecosystems that are not essential for the continued production of fishery species	Restriction of extractive activities, including fishing, as a way to promote conservation reduces access of the fishery to fish stocks that might otherwise be available to it

[a]These two kinds of management can operate synergistically, but they do have different objectives and are frequently undertaken by separate management agencies.

maximizes profits rather than total catch. Coincidentally, this buffers the population from inadvertent overfishing in periods of poorer than expected recruitment (Fig. 1). Concepts such as MSY and MEY, which are core to fishery management theory, are based on models of exploitation of single populations. These have no significant competitors for their trophic resources, are not interfered with by habitat damage due to fishing activities, and, in the absence of a fishery, would reach abundances set by the carrying capacity of their environment. These models also assume a stable price structure for the product (otherwise the income curve in Fig. 1 is not fixed).

These core concepts of fisheries management do not appear to translate very well to many real fishes and their fishing fleets (Caddy, 1999; Hilborn *et al.*, 2001), and current fishery science includes many attempts to improve the theory. Coral reef fisheries clearly depart in a number of respects from the underlying assumptions: each fishery targets many species simultaneously rather than one (Dalzell, 1996; Johannes, 1998), few species are targeted by only a single fishery method, and most catching techniques have substantial deleterious effects via discarded (unrecorded) by-catch, and via impacts on aspects of the habitat that are important for the fishes (Turner *et al.*, 1999; McManus *et al.*, 2000). These are not simple relationships between a predator (the industry) and its sole prey species within an unchanging environment, as required by traditional fisheries theory. In addition, if fish populations are naturally regulated by mechanisms other than a limitation of resources (Doherty and Fowler, 1994a; Sale and Tolimieri, 2000;) (see Chapter 10, this volume), there is no basis on which to expect that MEY can be obtained by fixing effort.

Even if traditional theory were to hold in multispecies fisheries such as those on coral reefs, it is not possible to regulate effort to keep yield of all species close to the MEY. Some species will be impacted more severely by a given level of effort, and a change in species composition of the catch and of the community will occur. Such shifts in composition are well documented (Russ, 1991; Jennings and Lock, 1996; Birkeland, 1997b), and may even lead to additional indirect fishing effects on other components of the reef system (McManus *et al.*, 2000).

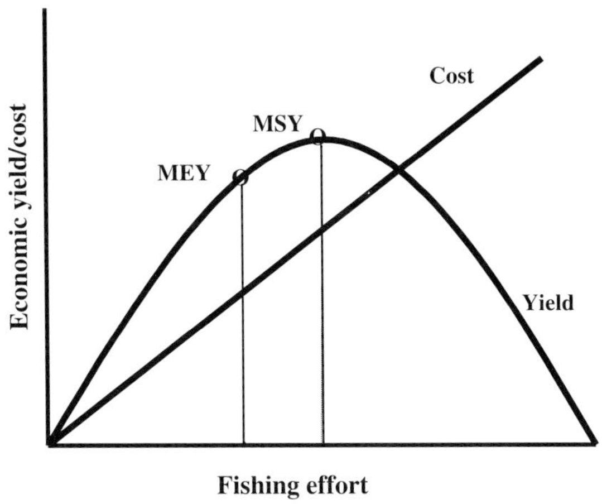

FIGURE 1 Diagram showing how the relationship between cost of fishing and yield from fishing interacts to define the maximum sustainable yield (MSY) and the maximum economic yield (MEY). This model assumes the fishery species has a logistic pattern of population growth, that its demographic parameters, other than fishing mortality, are constant over time, and that the economic value per unit of the catch does not change through time if the species becomes rare. It also assumes that the fishery is for a single species, and that fishing activities have no effects on production other than through the direct removal of biomass to market.

The management approach based on MSY or MEY is not suited to management of fisheries in developing countries. It is a data-intensive approach originally developed for the single-species fisheries of Europe and North America (Caddy, 1999). To implement management of effort it is necessary to know the state of the fish population. This requires sophisticated statistical analyses of catch statistics collected over many years. Although such methods theoretically can be applied to the multigear, multispecies fisheries typical of coral reef environments, they can rarely be applied successfully given the limits of adequate human resources and expertise in fisheries management agencies in the great majority of coral reef regions. Catches are inadequately sampled, data are recorded for large pooled groups of species, rather than for single species, and the people collecting and analyzing the data are often inadequately trained. Adjustment to fishing effort is almost impossible because the inadequate resources make enforcement of regulations difficult, and, in any event, there is little political will to restrict fishing when people have few other sources of employment and need fish for their tables. McManus (1996), Ruddle (1996), Adams (1996), and Russ (Chapter 19, this volume) provide abundant evidence of these administrative difficulties. The result is that, irrespective of claims or intentions or of the efforts of dedicated managers, most reef fisheries are unmanaged or undermanaged. The management activity serves primarily a public relations function, while perhaps deterring gross overexploitation and use of the most egregious examples of ecologically inappropriate fishing methods.

Exceptions to this disappointing picture exist in places where there has been effectively a single predominant fishery in operation, and an adequate investment in management. The predominantly trap fisheries of Jamaica were managed relatively successfully for many years through the 1970s. The trawl and trap finfish fishery and the lobster fishery of Cuba appear to be being managed successfully because of the strongly centralized control of the fishing endeavor and the high levels of compliance by the industry. The snapper/grouper hook-and-line and trawl fisheries of the extensive non-reef habitats of the southeast United States have been successfully managed over many decades, although there is evidence of some overfishing (see Chapter 17, this volume).

The difficulties of employing conventional effort-based fisheries management approaches have encouraged use by fisheries managers of two additional approaches. The first is to enrich habitat structure by means of artificial reefs and fish-attracting devices (FADs) in locations that are not structurally complex naturally, and to fish the resulting aggregations of fishes. The second is to create "no-take" marine protected areas (MPAs), on the understanding that the fishes within these will populate (either through reproduction or emigration) nearby fished regions, enhancing fishery yield. Unfortunately, as yet there is no evidence that artificial reefs or FADs do more than aggregate existing fishes, making them easier to harvest. However, they may enhance recruitment by offering a greater amount of structurally complex habitat for larval fishes to colonize. There is also no evidence that MPAs serve to enhance fishery yields in the region surrounding them to a degree that fully compensates for the loss of fishery access to the area they enclose (Chapter 19, this volume). However, MPAs do enhance fish biomass within their borders, and by so doing they provide a valuable safety net to many fisheries operating on the brink of disaster. As discussed below, MPAs are also a favored technique for conservation management, and can provide conservation benefits irrespective of any fishery management benefit they offer.

2. Conservation Management

Conservation management seeks to regulate human activities to minimize direct and indirect negative impacts on valued sites and/or valued species, with the goal of sustaining existence of specific species or of biodiversity in general. In either case, activities managed include those that might have direct negative impacts on the target, and those that have only indirect effects, and may have these effects at some considerable distance from the location where the activity takes place. Thus conservation of a coastal marine site or species may require that industrial, agricultural, commercial, and recreational activities conducted on land be regulated, along with fishing or other activities conducted on the water. Although conservation management directed at specific valued species, such as green turtles, manatees, or the jewfish, *Epinephelus itajara,* has the longer history, the comments here are restricted to management directed at valued sites because that has been overwhelmingly the more prevalent form of conservation management applied to coral reef regions.

Central to conservation management is the concept of the marine management area (MMA), a specific coastal or open ocean location to which specific management actions are directed. Marine protected areas are one of several types of MMA and MPAs also occur in several forms (Agardy, 1997). Without distinguishing terrestrial and marine areas, the International Union for Conservation of Nature (IUCN, 1994) recognizes six categories of protected areas, ranging from nature reserves and wilderness areas (Category 1) to managed resource protected areas (Category 6), which would include biosphere reserves and other multiple-use MPAs such as the Great Barrier Reef Marine Park. Examples of all six occur in marine environments (Agardy, 1997).

Given the broad range of types of managed areas, and even of MPAs, let alone the fact that some MPAs are managed specifically as no-take zones for fishery purposes (above), generalizations are difficult. In all managed areas, however, there is a formal (usually legal) declaration of the boundaries of each area, and an attempt to manage one or more human impacts on it. [I use "attempt" because of the unfortunately very widespread occurrence of "paper parks" with little real management in place (Kelleher *et al.,* 1995).] Fish populations within MPA boundaries frequently benefit from this management along with other components of the ecosystem, showing greater abundances, individual size, and longevity in MPAs where fishing activities are restricted (Chapter 19, this volume).

Although fisheries management is based (even if not successfully) on a rich body of theory defining how harvested populations grow, conservation management is based far more on compromise and consensus among competing user groups as they reluctantly yield their "right" to use all parts of the commons comprising the reef. Underlying conservation management is a relatively thin base of ecological theory, and the understanding that overexploitation or mismanagement kills reefs.

The theory is thin for two reasons. First, the low level of participation of the ecological community has led to reliance on ecological concepts that are a decade or two out of date. For example, the competitively mediated equilibrium in relative abundances of species that is supposed to exist in unimpacted systems is a good example of a concept that is far less solid than many conservationists appear to recognize. Second, conservation has a longer history in terrestrial systems, and conservation biology is almost entirely terrestrial in focus, so that most theory derives from terrestrial examples. The substantial differences between marine and terrestrial systems mean that this theory does not transfer readily (Agardy, 1997).

The gaps in relevant ecological knowledge extend broadly. For example, there exist many studies documenting the negative impacts of sedimentation on coral survival and growth, but attempts to mitigate anthropogenic sedimentation are driven by rule-of-thumb guides to acceptable levels because the ecological research has not yet led to formulation of tolerance limits. Similarly, artificial reefs are seen as a valuable conservation tool in regions where natural reefs are sparse but a shallow shelf that could support them exists. However, the building of such artificial reefs is usually done without ecological data on how many, how large, how close together, what shapes, what structural complexity, or what building materials will be best. Too often, they are built of whatever is at hand, and may be more useful as garbage disposal mechanisms than as conservation tools. Studies such as that of Lindberg (Frazer and Lindberg, 1994; Lindberg, 1997) on the west Florida shelf are far too rare. Lindberg used the placement of new artificial reefs by the management agency as an opportunity to test explicit hypotheses about size and placement effects in a properly designed long-term ecological project.

The most egregious gaps in knowledge concern the scale, connectivity, and ecosystem dynamics of marine systems, because these could provide sound ecological reasons guiding management decisions such as determining the siting, sizes, and shapes of MPAs (Roberts, 1997b; but see Sale and Cowen, 1998; Bellwood *et al.,* 1998). Instead, these important decisions are made in a strongly political process, pitting those who wish to protect sites from human impacts against those who wish to impose minimal restraints on human activities. Despite several recent, and welcome, calls for the development of integrated networks of MPAs (whether for conservation or for fisheries management), the need to develop the ecological theory necessary to help guide the detailed decisions needed in such an endeavor has scarcely been mentioned (Agardy, 1997; Roberts, 1997b). Indeed some argue that to suggest this subverts the urgent need to increase protection now (Ballantine, 1997). In a strongly worded call for reserve networks, one large group including several highly regarded ecologists (S. N. Murray *et al.*, 1999) noted that "more information about reserve size and . . . spacing" was needed if networks were to meet management objectives. But they suggested vaguely that the best way to gain this information was to implement reserve systems and then study how they function. There are more than 1300 MPAs already in existence, many in reasonable proximity to one another, so I wonder why establishing more is necessary before this neglected research can take place. Groups of influential ecologists should be able to mount a stronger argument in favor of setting this research as a high priority.

We can be a little optimistic given the signs that a small but growing number of ecologists is undertaking research that will help build a science of reef management. Notwithstanding the important caution by Russ (Chapter 19, this volume) about the need for adequate experimental design, major experiments are being done. These are listed in Section IV,D, with the argument for the need to do much more. Along with targeted field studies, there must be a greater integration of these with theoretical studies if a scientific base for conservation management is to be built (Doak and Mills, 1994). The path forward is clear.

IV. The Steps We Need to Take

A. A New Perspective: Coral Reefs Are Not a Commons

When Garrett Hardin (1968) first wrote of the tragedy of the commons, it was widely appreciated that marine environments provided prime examples of this phenomenon. In western culture, the seas have traditionally been unowned, open to exploitation by all. Fishery management has always struggled with the view that the fishery resources are unowned until they are captured, and new regulations to limit entry or to otherwise reduce effort in a fishery are usually hotly contested. In commons situations, when a number of independent harvesters seek economic gain by competing to extract a shared resource, it is in the self-interest of every individual to cheat, or at least to take the maximum yield obtainable within the regulations. An individual who takes less forgoes potential profit with no guarantee of future profits. Indeed this individual is the ultimate altruist, who will lose status among his peers (who measure worth in economic terms) rather than be seen as a leader. Over time, he or she will probably be eliminated by market forces.

Unlike many marine habitats, coral reefs have a rich local topography, and the fishes that occupy them are closely associated with local sites. This offers managers the opportunity to manage fisheries by regulating entry, not to a fishery for a specified product, but to a specific location. The majority of MPAs established in coral reef waters have enhancement of fish stocks as one of their objectives, and are managed to ban all or specific types of fishing within their borders. Examples of limited entry to specific fishing grounds also exist, but are less widespread than "no-take" MPAs, and many jurisdictions continue to attempt to manage fishing effort only on the larger geographic scale of the coastal state. A concerted exposition is badly needed of the arguments that show spatially explicit management of entry to be the best, if not the only useful one, of all currently available approaches for managing reef fisheries, particularly in regions where human and other resources to support management are rare. This educational effort should be mounted by fisheries economists and scientists, including ecologists. It is time to educate the many small local and national fisheries departments with responsibility for managing reef fisheries—and more importantly, the governments to which they report and the local communities that depend on them for food and livelihood—that the old ways that permitted open access do not work. This is true whether or not catch limits or gear restrictions are in place. Wealthy nations may be able to implement an effective management regime using these other approaches in place of spatially explicit management of entry, but this is not a viable option for the majority of nations with coral reefs.

Reliance on spatially explicit control of entry does not preclude addition of measures to regulate local catch, or of measures such as recruitment surveys (which may be more easily implemented for reef species than for species that are less easily censused) to forecast accurately potential local catch. These approaches, however, must be seen as secondary to the establishment of spatially explicit management of entry, and their addition must not deprive this primary mechanism of resources needed for its effective implementation.

Introducing spatially explicit management of reef fisheries to new locations will not be politically easy, but it can be done. Coral reefs and other inshore habitats are not the open sea, and national ownership of coastal waters is clearly recognized. Coastal waters, particularly reef waters, provide many visible geographic features, and shallow depths make installation of marker bouys straightforward. The reefs offer a natural mosaic of discrete patches of habitat that can be managed differently. There is no legal or technical reason to avoid treating the coastal ocean as subtidal land divided into patches, within which certain activities can be permitted and others prohibited. Given that artisanal fisheries are significant in many coral reef regions, regulations governing fishing access will have to be sensitively framed and administered and will be most effective if "ownership" is ceded to local coastal communities or fishery cooperatives. Indeed, the existence of traditional fishing rights and/or fishery cooperatives can facilitate spatially explicit management.

Spatially based management has been done successfully in past centuries by Pacific Island governments with far fewer resources or ecological knowledge than we now possess (Johannes, 1981, 1998). Effective policing of fishing activities will be facilitated if there is community participation and sufficient political will. The introduction of international programs to register fishery product obtained from sustainably managed fisheries and development of the Code of Conduct for Responsible Fisheries (FAO, 1995) should help build this political will. The need, in most regions, to reduce overall catch in order to achieve sustainability will increase the difficulty of introducing spatially explicit management, as will the lack of detailed science concerning spatial scales at which to manage (below).

Given that the data-intensive, single-species fisheries management methods developed in the north temperate regions are not appropriate for multispecies, multigear reef fisheries, that they cannot be applied effectively when management resources are limited, that the fishes are rather sedentary (at least over structurally complex habitat), and that site-based management of fishing effort is a viable alternative approach, fishery management should become based predominantly on the appropriate management of activities at specific locations, and use of management resources should reflect this priority. Fishery management should be done in conjunction with conservation management, which also should focus on regulating activities at defined sites.

B. We Must Define the Spatial Scales

At the present time, the largest gap in our understanding of the ecology of coral reef fishes is to define the spatial scales on which populations operate (Sale, 1998; Normile, 2000). The extent of dispersal of larval fishes and the pattern of interconnection of local populations are vital information for the effective implementation of spatially explicit management (Table 3). The extent and pattern of interconnection may be substantially different in regions of different geomorphology or hydrology, as well as among species of differing life history. Effective management will require that we know the rules governing such variation. Several potentially useful techniques now exist to aid the closing of this knowledge gap, but there is much work to do, and effective interdisciplinary collaborations will be needed to make progress.

TABLE 3 Defining Spatial Scales: Some Reasons Why This Important Task Is Far from Trivial

- Reef fishes commence their pelagic lives as passive (eggs), or nearly passive particles (newly hatched larvae). They end it as more or less accomplished swimmers. We need to know the ontogenetic changes in swimming ability and associated sensory capabilities that take place.
- Ocean currents are temporally variable and are modified in complex ways close to topographically rich reef structures. Reef fish larvae are introduced into the current flow within, or very close to, the boundary layer. We need to know about the spatiotemporal patterns of water movement in close proximity to reefs
- Larval reef fishes, are difficult to sample quantitatively and are frequently uncommon. Their taxonomy is challenging when they are young. They offer few obvious cues to where they have been, and, if kept alive, they have a habit of maturing into juveniles with different behavior and sensory capabilities
- Reef fish species vary substantially in timing of reproduction and in mode of entry to the pelagic zone, in duration of larval life, and in behavioral capabilities as larvae. There will not be one spatial scale that correctly defines all reef fish species

1. "Conventional" In-Water Ecology

Considerable progress has been made in delineating the ecology of late-stage larvae and settled juveniles, but important questions remain. We now have several techniques for collecting late-stage reef fish larvae (Choat *et al.*, 1993) that permit more accurate assessment of the spatial and temporal distributions of larvae just prior to settlement and recruitment. These techniques also provide a way of obtaining living larvae, in good condition, for use in physiological, behavioral, and ecological experiments. Led by several Australian ecologists, we are developing techniques for studying the behavior and sensory capabilities of reef fishes during the pelagic larval phase and at the time of settlement to the reef environment (for example,

Stobutzki and Bellwood, 1994, 1998; Stobutzki, 1998; Leis and Carson-Ewart, 1997, 1998, 2001; Tolimieri *et al.*, 2000) (see Chapters 6 and 8, this volume). Reef fish larvae of many species are surprisingly strong swimmers and may be capable of orienting their swimming toward the reef habitat. Their larval dispersal will not be a simple advection/diffusion process, although movement of the water mass will remain a part of the process. Cowen and Sponaugle and their oceanographer co-workers have shown the way for an effective exploration of larval dispersal in conjunction with a contemporaneous study of the physics of water movement (Cowen and Castro, 1994; Sponaugle and Cowen, 1996a; Cowen *et al.*, 2000). The results are not simple, but they are determinable. Water moves around reefs in highly complex ways, and reef fish larvae sometimes travel with the flow. These small beginnings must be built on to yield an adequate description of how larval dispersal and subsequent settlement act to define the spatial patterns of reef fish species.

At the same time, there is a need for caution in accepting the results of simple behavioral experiments done with larval fish that have either been reared in captivity or captured at sea and held for several hours before being introduced into an experimental chamber or back into the field. Until now the exciting new results have not received the sceptical evaluation they need, given the flexibility of animal behavior and the likelihood of serious artifacts in at least some of the studies reported to date. Studies of the behavior of larval fishes also need to be done with due notice of the physics of the world they face, and of the sensory equipment they use in deciphering it (Chapter 6, this volume). Theoretical studies by Armsworth (2000) are tempering some of the early enthusiasm regarding the capabilities of late-stage larvae to use their own swimming abilities to counteract prevailing current patterns. They are capable, but they are not "Superfishes."

Data on ecology of fishes during their first months on the reef fill an important gap between the larval data and the much more extensive data on ecology of adult reef fishes. This work has confirmed highly variable patterns of settlement and/or recruitment, across years and locations, for a broad range of species and regions (for example, Sale *et al.*, 1984; Fowler *et al.*, 1992; Sponaugle and Cowen, 1996a; Masterson *et al.*, 1997; Tolimieri *et al.*, 1998; reviewed in Doherty and Williams, 1988; Caley *et al.*, 1996). Although there are strong differences over whether this variability is established during larval life, or by processes that act in the hours and days immediately following settlement, the consequence remains that recruitment of new cohorts of fishes to local populations is highly variable in time and space. Metapopulation models of population (and community) dynamics will have to deal with local populations that are constantly modified by an apparently stochastic and temporally decoupled recruitment of new individuals. That is, the additions to one population in one time period may be orders of magnitude different from the additions to nearby populations during the same time period, and a period of rapid growth may be followed by a period of stagnation, or even decline when recruitment at that site fails completely. This variability will make it more difficult to discriminate among sites that differ in the average levels of recruitment they receive, because the data will have to be collected over several recruitment seasons to yield reliable estimates.

We know that some species show precise responses to microhabitat (e.g., Tolimieri, 1998a; Munday, 2000) or to presence of conspecifics or other species (Sweatman, 1988), but we do not know whether these are backed up by equally strong responses to larger scale environmental features in determining where settlement will be most pronounced. Nor do we yet know whether patterns of abundance established at settlement are routinely "evened out" by migration within reef habitat following the first critical weeks. (The numerous experiments using replicate artificial or natural but constructed reeflets cannot address what happens to juveniles that settle to the extensive areas of contiguous reef habitat, but manipulative studies in such habitats are much more difficult.)

A number of manipulative studies that put newly settled juveniles onto replicate microreefs and subject them to various experimental treatments (such as presence or absence of older conspecifics or predators) have demonstrated density-dependent effects on growth (reviewed in Chapter 14, this volume). There are significant differences in interpretation of the importance of these compensatory effects (cf. Chapter 14, and my introductory comments to Section III). It seems unlikely to me that results obtained for one cohort in an experiment will be repeated in detail for other cohorts of the same species at other times and places, yet alone be repeated generally across species.

Although these explorations with early juvenile fishes are vitally important in providing the data that will be necessary for serious metapopulation modeling, the critical question of the scale of metapopulations will not be answered by reef-based ecological and behavioral studies alone. In regions where mangroves, seagrass beds, and "hard bottom" are significant components of the enviroment adjacent to reefs (all continental coral reef regions?), there is a need to extend the SCUBA-based methods that have been successful in studying reef-based populations, to populations of fishes in these other habitats. Do the majority of fishes that settle to mangrove habitats ever move out to a coral reef environment? If so, do they contribute significantly to the abundance of that species on the reef location? Until recently, very little work that had been done in these systems would address such issues—the less tractable environmental conditions precluded other than simple descriptive studies (cf. Section III,A). Again, we can expect to obtain different answers for different species, and in different locations. It is very likely, however, that for some species, metapopulations will contain important nonreef cells, though at present its unclear whether they will function usually as sources or sinks.

2. Other Approaches to Scale Questions

The problem of defining scales of connectivity will not be solved without a multidisciplinary effort combining ecological with other research approaches. There is a need for a much better characterization of the physical dynamics of the larval environment, and for efficient ways to tag larval fish to learn how far and where they disperse. These call, respectively, for participation of physical oceanographers and of molecular biologists.

A number of individuals have attempted to build effective partnerships between reef fish ecologists and physical oceanographers. These are intrinsically difficult partnerships to build because the tradition in physical oceanography has been to keep ships away from reefs, and to avoid the computationally much more complex patterns of water movement close to reefs. Yet these more complex flows are precisely the ones into which fishes dump their eggs or newly hatched larvae. These partnerships have also been difficult to build because many reef fish ecologists lack the mathematics to comprehend the science done by their physical oceanographer colleagues. Time must be taken to learn to speak to one another, to discover that there are exciting questions in physical oceanography that can be undertaken close to reefs, and to build the mathematical strengths of the ecologists. Such partnerships need to be cultivated. A good garden for growing them may be the multidisciplinary marine research facility, where the early courtship stages of conversations over coffee are easily undertaken. Two of the most successful such partnerships have been that involving R. Cowen and K. Lwiza, which originated when both were in the Marine Science Research Center, State University of New York, Stony Brook, and that involving E. Wolanski, L. Bode, K. Black, P. Doherty, and D. Williams, which was nurtured at the Australian Institute of Marine Science and the James Cook University. One serious handicap for such partnerships is that their research proposals, being interdisciplinary, tend to get double peer review, with its potential for double jeopardy. Funding agencies must ensure that efforts to build effective partnerships do not get overzealous treatment at the time of review, while ensuring that weak proposals do not get funded just because they are multidisciplinary. I raise this point here because the biology-physics partnership involves building a bridge across a wider gulf than for partnerships between ecologists and molecular scientists. Chapter 7 of this volume provides an excellent overview of the progress that has been made by biologist-ocean physicist teams in understanding the environment of larvae, and the responses of larvae to it. Armsworth *et al.* (2001) also provide an important study that combines physical oceanography, empirical data, and new behavioral theory on this complex topic.

A broad range of molecular genetic techniques now exist, several of which may provide effective ways to establish relatedness of nearby populations of a species (Schultz and Cowen, 1994) (see also Chapter 9, this volume). Although genetic relatedness does not directly demonstrate degree of interconnection through dispersal, it may be that genetic data will be more easily obtained than direct data on dispersal. The latter are likely to require detailed information on water movement, as well as knowledge of fish behavior during larval life. Thus effort should be made to make genetic relatedness an effective proxy for ecological connection among populations, because it would then be possible to sample genetics in a new location and make decisions that could be used for management.

Among nongenetic approaches, use of otolith microchemistry appears to be a particularly promising tool to define larval dispersal and interconnectedness of populations (Campana, 1999) (Chapter 11, this volume). Usefulness of this approach may be restricted to regions in which there are unusual geochemically defined spatial patterns, but such locations may be more usual than we anticipate. Microchemical approaches may turn out to provide more direct proxies for ecological connectivity than genetics. Otolith microchemistry requires equipment and skills not normally found in the average ecology lab, but workers in several labs in the United States and Australia are beginning to explore use of these methods in coral reef environments (e.g., Swearer *et al.,* 1999, M. J. Kingsford, personal communication).

There may be additional kinds of "tags" on reef fish larvae. A careful study of parasite communities occupying particular species of fishes might provide cues, so long as the parasites invade at or soon after the egg stage (P. Sasal, personal communication). Bacterial floras within the mucus coat of fishes may offer another possibility (B. S. Danilowicz, personal communication). Fish species that brood their eggs, or whose larvae aggregate at sea, may be particularly appropriate subjects for such studies. Regardless of the "tagging" technique used, study of the interrelatedness of larval fishes will be truly effective only if integrated with synoptic ecological work to delineate patterns of settlement and recruitment.

C. Extend the Theory of Open Populations

It is time that reef fish ecologists engaged the interest of theroreticians skilled in metapopulation theory. What are the dynamics of a species, sometimes recruitment limited and existing as a set of x open populations interconnected by larval dispersal? How do such species respond to competition, to predation, or to fishing pressure? At present we can only wave our arms about and guess, because the theoreticians happen to have backgrounds in terrestrial ecology, and have naturally used terrestrial systems as examples to explore the theory. Marine and terrestrial systems appear to differ profoundly in their degree of openness, and the theory being developed may not be appropriate for marine systems. However, as noted earlier, it seems very likely that metapopulation dynamics will apply to some reef fish species at some spatial scales, so we need to encourage exploration of our systems.

The spatial extent, and the number of subpopulations included in a functioning reef fish metapopulation, must vary with geography, with hydrodynamics, with behavioral and sensory capabilities of the larvae, and with the particular reproductive patterns of the particular species. However, even the most reluctant dispersers seem likely to travel kilometers from home, and this still means that the larvae live on a spatial scale two or more orders of magnitude greater than the home ranges of adults. Pinning down, for some selected species, the demographic consequences of actual rates of dispersal among populations, and the variations due to degree of spatial separation, must be top priorities for reef fish ecologists. Clearly, this work will demand effective collaborations between field ecologists and theoreticians, and it will likely develop in parallel with study of the effectiveness of MPAs. Man *et al.* (1995) may provide a useful beginning for delineating reef fish metapopulations. Other theoretical studies are those by Armsworth (2001), on the impacts of fishing on sequential hermaphrodites; by Crowder *et al.* (2000), exploring consequences of placing MPAs in source-sink systems; and by Walters (2000) and Bohnsack *et al.* (in press), dealing with other aspects of MPA siting, size, and number.

Extending the theory of open populations in directions relevant to our needs will help us pose new questions that drive new empirical studies. Perhaps the persistence of the very large numbers of rare species on coral reefs (Chapter 4, this volume) will become much better understood only after this extension of theory.

D. Undertake Deliberately Adaptive Management

Decisions concerning the size, shape, and placement of protected areas, as well as the extent and nature of the protection, will always be subject to a political process. However, a stronger body of theory concerning how MPAs aid the populations occurring within them will improve an otherwise "I say–you say" process. Compromise achieved by each group moving halfway toward the viewpoint of an opposing group does not always result in the best management decision (half the current fishing pressure, or pollution, or other impact is not necessarily the right amount to permit!)

Improved knowledge of the spatial scales on which reef fish species operate, and the degree to which the spatial arrangement of members of a population is a function of hydrographic or geomorphic properties of the region, will both lead to better informed decisions on MPA boundaries. But gaining that knowledge will proceed very slowly if the efforts to gain it are kept independent of the management process. Resources to support large-scale ecological experiments are limited, and permission to do large-scale manipulations simply to answer scientific questions is increasingly difficult to obtain. Without experiments to test hypotheses, science cannot advance quickly.

The process of implementing management decisions invariably manipulates our impacts on ecological systems. Thus there exists the opportunity to use the implementation process deliberately to further scientific understanding, while aiding management of particular locations. This integration of science and management, or "adaptive management," offers the only realistic way of conducting large-scale ecological experimentation (Walters, 1986). Does the size of a marine protected area matter? If so, how small is too small? And does minimal viable size vary if several MPAs are located within a region? Is there any benefit in making an MPA larger than the minimum viable size? These questions could be answered in the process of implementing MPAs. What is needed is a close working relationship between scientists and managers, and a coordinated monitoring plan to measure effects of management actions.

As an ecologist, I suspect that there is a minumum viable size for an MPA in a coral reef region. I suspect that MPAs larger than this may be more beneficial, but I doubt that the value increases continuously. I suspect that minimum viable size and the relation of benefit with size will both vary with the inclusion of additional MPAs nearby. But I do not know the answer to any of these questions for coral reefs, and neither does anyone else.

A management process should begin with the assumption that the management plan can be altered in the future (the Great Barrier Reef Marine Park zoning plan is reviewed on a rotating 5-year cycle). Then it becomes possible deliberately to introduce variation in management among managed sites, undertake quantitative monitoring, and draw conclusions. It was possible over 20 years ago to provide a series of unharvested plots of Amazon rain forest of varying size to determine the effects of plot size on species maintenance and long-term viability in that environment (Bierregaard *et al.*, 2001). It is surely possible now to use a set of MPAs of different sizes on coral reef habitats to see how population density, diversity, and species composition respond in that environment. All that is required is that the management process be driven, at least partly, by a desire to develop a science for management.

Adaptive management can be introduced in many more ways than in sizing of MPAs. How much recreational spearfishing, hook-and-line fishing, and commercial trap fishing can a reef sustain? Is one of these more destructive than the others? Do they interact other than additively? Do any of these rules change if the site is also subject to hordes of underwater photographers? Adaptive management provides the tools to answer these questions, and it is encouraging to see that its use is growing as we seek to better manage coral reefs.

Habitat and fishery effects of closures of areas to trawling are being assessed on the northwest shelf of Australia (Sainsbury, 1991; Sainsbury *et al.*, 1993), and closures to line fishing are being examined in the Great Barrier Reef Marine Park (Mapstone *et al.*, 1997b; Adams *et al.*, 2000). Effects of no-take MPAs on fish and other biota are being monitored in the Florida Keys National Marine Sanctuary (Bohnsack, 1999, also personal communication). Each of these projects uses adaptive management to answer ecological questions requiring large-scale manipulative experiments. The answers improve management decisions at other locations. If we fail to move from "rule of thumb" toward a science of management, we risk ultimate failure because the process of political compromise will steadily compromise toward overexploitation, whether for fishing, for tourism , or for other endeavors.

E. Recognize that Spatially Based Management Is More than MPAs

The marine protected area is a simple concept—a defined portion of the coastal ocean that is protected from one or more human impacts. MPAs exist surrounded by a matrix of differently managed environments. Unless the MPA is very large, the populations of fishes and other organisms that use it will interact strongly with populations lying outside it. Having defined an MPA, there is no reason not to provide some form of explicit management to the matrix within which it resides.

The science of environmental management will be enhanced if we reduce the emphasis on the creation and management of individual marine protected areas in favor of an emphasis on spatially defined regional management that will include MPAs. Such an emphasis recognizes that the task of the management agency is the effective management of the resources, the populations, and the biodiversity of a particular coastal region (Fig. 2). This region may be the entire coastline of a country, such as Belize, in which case its boundaries are politically defined. But it may be the coastline of an ecologically unified portion of a country, such as the Great Barrier Reef, or the coastline of an ecologically significant region encompassing several adjacent countries, such as the mesoamerican reef system of Mexico, Belize, Guatemala, and Honduras. In these cases regional boundaries are defined on a biogeographic basis, and the rationale for using ecological principles in formulating management decisions may be more readily appreciated (though politically even more challenging to achieve).

Regionally based management is not an untried approach. The majority of coastal biosphere reserves are examples, although most are relatively small (Agardy, 1997). The Great Barrier Reef Marine Park, the Florida Keys National Marine Sanctuary, and the new Northwestern Hawaiian Islands Coral Reef Ecosystem Reserve are all examples established at appropriately large scales (Fig. 2). However, that these are all in first-world nations, and that only one is more than a decade old, are signs of how new and how relatively difficult an approach this is. Nor can we yet claim that the opportunities for adaptive management have been fully seized even in these three cases. Yet, the path is set out before us, and it is one worth following.

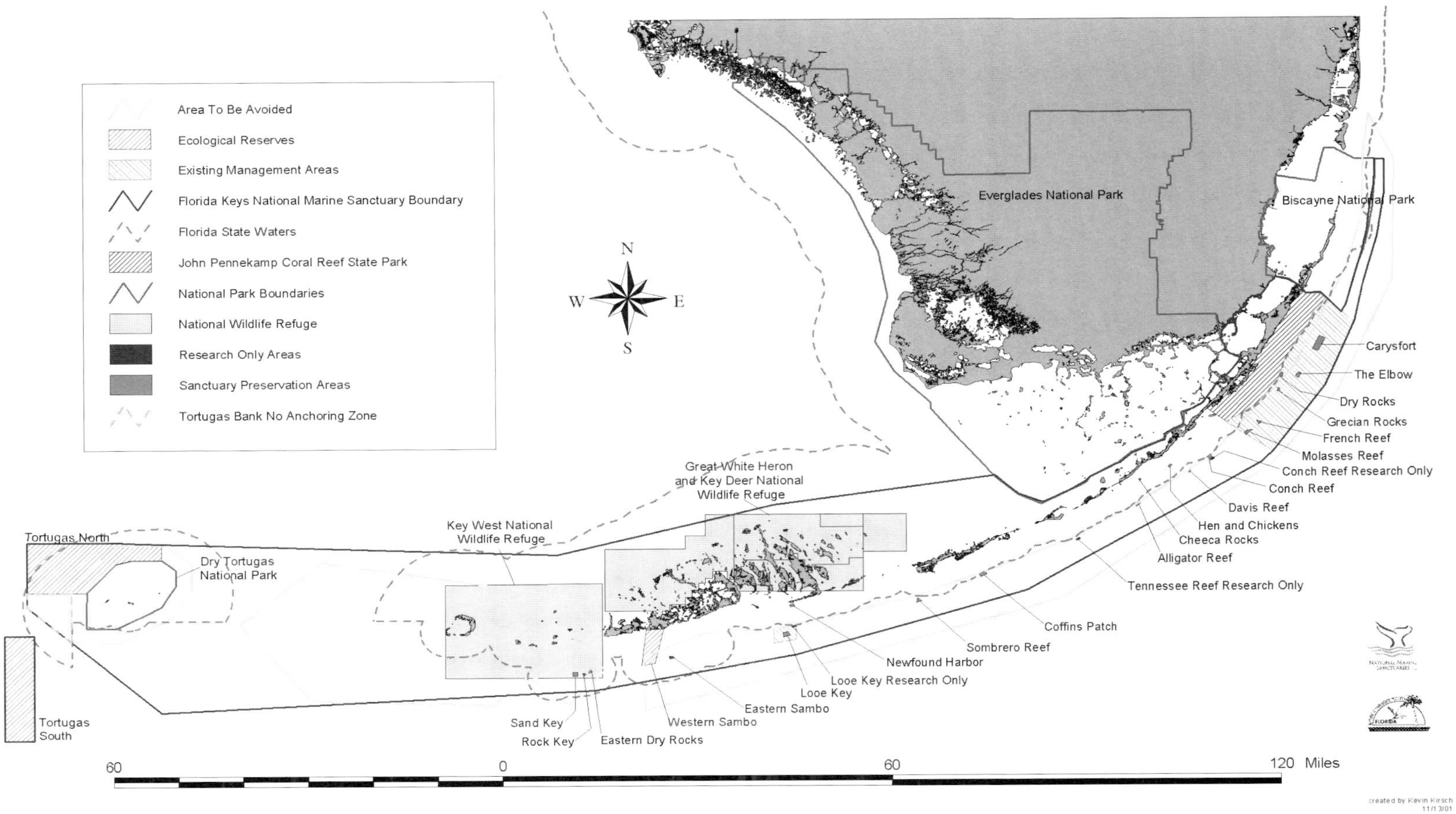

FIGURE 2 Map of the Florida Keys National Marine Sanctuary. The boundaries surround all reefs of the Florida Keys reef tract south to and including the Dry Tortugas. There are nine (mostly much smaller) contiguous, or surrounded managed areas (not all marked on the map). These are the Everglades, Biscayne, and Dry Tortugas national parks, the John Pennekamp Coral Reef State Park, the Card Sound Aquatic Preserve, the Lignumvitae Aquatic Preserve, the National Key Deer Refuge, and the Great White Heron and the Key West national wildlife refuges. Cooperative management of these entities allows for a regionally based management of the entire Florida Keys system from Miami south and west to the Dry Tortugas. This is an appropriate spatial scale for coral reef management. (Map courtesy of Florida Keys National Marine Sanctuary.)

Whether the boundaries of the region of interest are politically or biogeographically defined, the emphasis on management of the region, rather than on declaration of one or more MPAs within the region, is a major advantage. Management of the region need not be all or nothing, but can be varied from place to place. Apart from facilitating the inevitable compromises necessary to make management a reality, this approach opens up the opportunity for different forms of impact control in different parts of the total region. Now the MPA is just one of a suite of site-specific tools that can be used. These other tools will include differential regulation of human activities, as well as use of artificial reefs and other techniques for enhancing recruitment of reef species. In such a zoned managed region, adaptive management is easily implemented—to the ultimate benefit of management science—and the differing management regimes among locations provide a powerful educational tool. By advocating a shift toward regionally based management, we do not limit the activities of conservationists who wish to continue to strive for the special protection of specific sites. However, we do provide a rationale for a broader and, I believe, a scientifically more sound perspective on environmental management.

V. Summary

We currently manage reef fishes either for fishery use or as part of reef conservation. We need to manage for both simultaneously, and in an integrated way. We frequently manage using actions justified by inappropriate science (most fisheries management) or with little scientific justification (most conservation). We need to build a science of environmental management that is appropriate to coral reefs and coral reef fishes. This requires coral reef ecologists and other scientists to direct their efforts to management questions and to articulate their science more effectively, and it requires much closer collaboration of scientists and managers.

To build this science there are certain areas of reef fish ecology that must be expanded, and there are new techniques from other fields that must be brought to bear to answer the overwhelmingly most important question: What is the spatial scale at which reef fish species act out their lives? There are other questions more directly related to specific management techniques that need to be given an underpinning of science.

The spatial scales at which management must be done preclude the use of small manipulative ecological experiments to answer questions about management techniques. To build this science it will be essential for ecologists and managers to work together using adaptive management as a way of deliberately advancing management theory while simultaneously putting new management into place in particular locations.

The prevailing view that coastal environmental management consists of defining marine protected areas at significant sites focuses attention at the wrong spatial scale. Adoption of the view that one manages a particular coastal region, within which one or more MPAs will be established, makes the introduction of adaptive management a much easier process. It is in the self-interest everyone to improve our management of tropical coastal oceans and the reefs they contain, before it is too late, but unless people understand this, those who wish to manage better will ultimately fail as the growing human population increases the pressure to use what is there.

Acknowledgments

This chapter benefited, at successive stages in its development, from critical reviews by Drs. Bruce Hatcher, Stuart Ludsin, and Jake Kritzer. I doubt my revisions have fully satisfied them, but I thank them for forcing the improvements that occurred. My recent experience as a member of the International Network on Water, Environment and Health, of the United Nations University (UNU-INWEH), has given me the opportunity to begin to experience the challenges of building a science for coral reef management.

CHAPTER 17

Reef Fish Ecology and Grouper Conservation and Management

Phillip S. Levin
National Marine Fisheries Service
Northwest Fisheries Science Center
Seattle, Washington 98112

Churchill B. Grimes
National Marine Fisheries Service
Southwest Fisheries Science Center
Santa Cruz Laboratory
Santa Cruz, California 95060

I. Introduction

The management of fishery resources may ideally be defined as articulate intervention, with the central constituent of such intervention being thoughtful consideration of alternative management strategies prior to implementation (Boothroyd, 1978). Informed decisions require reflection on how fish populations are structured, how they function, and how they are likely to respond to intervention (Baskerville, 1997). Although exploited fishes of coral reefs are important resources for both commercial and artisanal fishers, collection of the detailed knowledge necessary to make informed decisions is made difficult by several factors.

First, the fisheries of coral reefs are typically multispecific, with 200–300 fish species harvested in the Indo-Pacific, greater than 100 species in the Caribbean (Munro, 1996), and more than double these numbers consumed in subsistence fisheries in both the Pacific and Atlantic oceans (FAO, 1997). Additionally, the prevalence of artisanal fishers in developing countries results in diffuse fishing effort, with the catch landed at a number of sites, making even the most basic fishery statistics such as catch and effort difficult to obtain (Russ, 1991). Detailed research on exploited reef fishes in many locations is not accorded a high priority by management agencies because of their relatively low economic value (Russ, 1991), despite the fact that the people of many developing island nations utterly depend on reef fishes as sources of protein and employment. Even in developed countries in which capital-intensive commercial and recreational fisheries target high-value reef fishes, much of the research that is performed is descriptive in nature, and thus cannot provide a thorough understanding of how these fish populations will respond to various management schemes (Cushing, 1995).

The harvest of reef fishes has provided about 10% of the commercial yield of fishes worldwide, and this harvest has been critically important socially and economically in many developing countries (Medley *et al.*, 1993). However, the overwhelming pressures of poverty, along with escalating human populations, have resulted in the depletion of reef fish resources to extraordinarily low levels (Munro, 1996). The social, economic, and political complexities of the developing world make management of reef fishes difficult under the best of circumstances, and the task is made more challenging by the dearth of detailed data on exploited species (Polunin *et al.*, 1996).

Unlike their exploited relatives, small-bodied reef fishes have been intensively studied over the past three decades. These fish are easily observed and manipulated and have proved to be excellent models for investigating demographic processes in nature (Sale, 1991b). But can current ecological understanding of small-bodied reef fishes provide the insight needed for "articulate intervention" in reef fisheries? Certainly, the reef fish literature is replete with statements averring that particular data sets are necessary for the management of reef fisheries. However encouraging such statements may be, ecological research on small-bodied reef fishes has been criticized as being irrelevant to the formulation of appropriate models or to the provision of management advice for reef fisheries (Medley *et al.*, 1993; Polunin *et al.*, 1996). Although it is possible that bottom-up ecological processes are more important to populations of larger exploited reef species such as grouper than to populations of their smaller brethren, there is no reason to suspect that the population dynamics of exploited reef fishes differs qualitatively from that of the small-bodied species that have been the object of ecological study (Doherty *et al.*, 1994b; DeMartini, 1998).

This chapter evaluates the degree to which basic reef fish ecology can offer advice to reef fish managers by reviewing key aspects of the life history of groupers (Serranidae: Epinephelinae). Approximately 150 species of groupers occur worldwide, with the genera *Ephinephelus* and *Mycteroperca* most widely distributed. Ranging in size from 25 cm to over 3 m, groupers occur mostly in shallow water, with a few species occurring in depths as great as 300 m. They are important components of most reef fish fisheries, e.g., they comprise 42% of commercial landings (by weight), and in the United States a single grouper species, gag *Mycteroperca microlepis*, accounts for 27% of recreational catches (by weight) (NMFS, 1999). They are highly esteemed as food fishes and are very susceptible to nearly every type of fishing gear (Dalzell, 1996), a combination that is typically catastrophic for fish populations. For example, in the Caribbean, such traditional mainstays as Nassau grouper (*Ephinephelus striatus*), speckled hind (*Ephinephelus drummondhayi*), and Warsaw grouper (*Ephinephelus nigritus*) have practically disappeared from commercial and recreational catches (Huntsman *et al.*, 1999; NMFS, 1999). Additionally, landings of species of more recent importance, such as red hind (*Ephinephelus guttatus*), have declined since the late 1970s (NMFS, 1999). Perhaps the best documented example of the extreme vulnerability of groupers to fishing is the case of snowy grouper (*Ephinephelus neviatus*) in the United States South Atlantic. Epperly and Dodrill (1995) documented that one fisher using a small (15-m) vessel fishing on a previously unknown reef site removed virtually all catchable fish within a single month. Worldwide, the circumstances of grouper are also dire. In many cases, protection for grouper stocks is inadequate or nonexistent, and given their high value, rapid overfishing of groupers is common (Russ, 1991). As with most reef fishes, attempts at stock assessments and management of groupers are relatively recent, and data shortages hamper our ability to make informed management decisions (Huntsman *et al.*, 1999). This chapter is not meant to be an exhaustive review, but focuses on this important exploited group and selected ecological processes in an effort to highlight the potential contribution of the voluminous literature on small-bodied reef fishes to the management of reef fisheries for grouper.

II. Reproductive Ecology

Below we address two key attributes of grouper reproductive ecology that are critical to population dynamics and fisheries management: (1) spatial and temporal concentration of reproductive activities and (2) protogynous hermaphroditism. We then discuss how knowledge gained from ecological research on small-bodied reef fishes may be applied to problems involving grouper stocks. [For more detailed information on grouper reproductive biology, see reviews by Thresher (1984), Shapiro (1987a) and Sadovy (1996).]

A. Spawning in Time and Space

Many reef fishes migrate tens to hundreds of meters to specific spawning sites (Robertson and Hoffman, 1977; Colin and Clavijo, 1978; Robertson, 1983; Warner, 1988a), and numerous groupers similarly travel to specific locations to form spawning aggregations (Sadovy, 1996). In some cases these aggregations can be spectacular. The Nassau grouper, for example, may travel more than 100 km (Colin, 1992) to traditional spawning sites, where up to 30,000 individuals have been reported in a single aggregation (Smith, 1972). Perhaps more typical, red hinds form aggregations of 750–2900 individuals (Shapiro, 1987a; Shapiro *et al.*, 1993) and coral trouts (*Plectropomus leopardus*), scamp (*Myctoperca phenax*), and gags occur in aggregations of about 75 individuals (Samoilys and Squire, 1994; Coleman *et al.*, 1999). Many other species of groupers are known or suspected to aggregate for spawning (e.g., Johannes, 1978; Shapiro, 1987a; Colin, 1992; Sadovy and Domeier, 1994; Sadovy *et al.*, 1994; Coleman *et al.*, 1996; Domier and Colin 1997), and spawning sites are often well known to local fishers (Olson and La Place, 1979; Aguilar-Perera and Aguilar-Davila, 1996; Huntsman *et al.*, 1999; Sadovy and Eklund, 1999). Nonetheless, data documenting details of spawning aggregations or characteristics of spawning sites are restricted to a small number of species (e.g., Colin, 1992; Shapiro *et al.*, 1993; Sadovy *et al.*, 1994; Zeller, 1998).

Groupers tend to spawn in early spring and summer, with a general trend toward later spawning with increasing latitude (Thresher, 1984; Shapiro, 1987a). Although some species [e.g., graysby (*Ephinephelus*

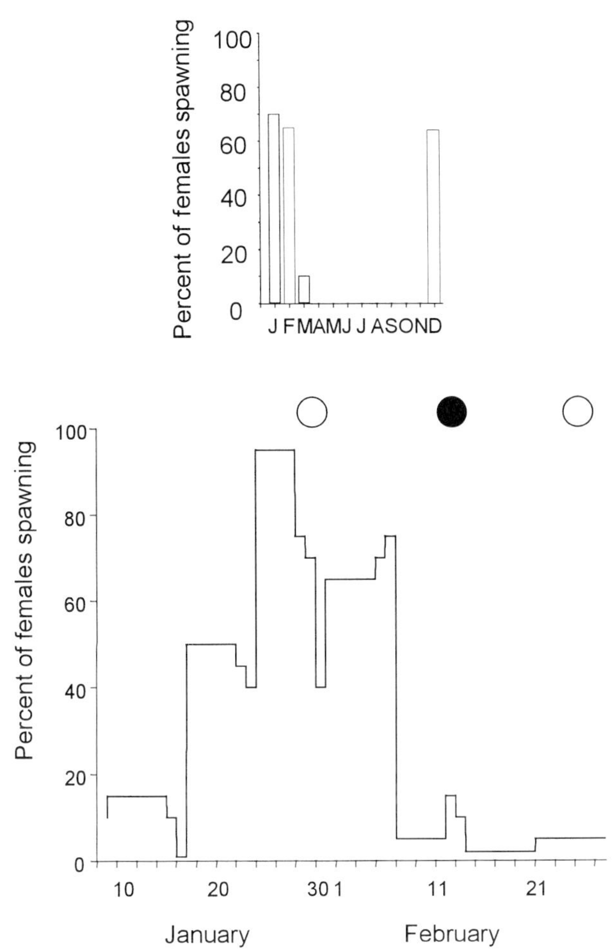

FIGURE 1 Percent of female red hinds (*Epinephelus guttatus*) that were actively spawning, as determined histologically. The top panel shows results from monthly samples taken between 1987 and 1991, and the bottom panel shows results from daily samples taken from spawning aggregations in 1991 and 1992. Data are from Sadovy *et al.* (1994).

cruentatus)] reproduce for extended periods annually, most groupers spawn over a 1- to 5-month interval, with peak spawning periods frequently much more restricted (Shapiro, 1987a; Sadovy, 1996; Collins *et al.*, 1998) (Fig. 1). For example, coral trouts have ripe gonads from September through February (Shapiro, 1987a). However, surveys and otolith back-calculations by Doherty and colleagues (1994b) strongly suggest that the majority of spawning occurs during a single period lasting about 2 weeks. Similarly, red hinds have ripe gonads from December through April (Shapiro, 1987a), but spawning aggregations appear to occur only during a brief period of days to weeks (Shapiro *et al.*, 1993; Sadovy *et al.*, 1994). Female gags in the Gulf of Mexico have hydrated ova from February to April (Collins *et al.*, 1998), but most successful recruits appear to come from only 3 weeks of the spawning period (Koenig and Coleman, 1998) (Fig. 2).

Spawning activity in groupers is often correlated with lunar or semilunar rhythms (Johannes, 1978; Thresher, 1984; Shapiro, 1987a; Sadovy, 1996). As examples, (1) spawning of coral trouts peaks in a 4- to 5-day period around the new moon (Samoilys, 1997a), (2) spawning of red hinds occurs within 7 days of the full moon (Shapiro *et al.*, 1993; Sadovy *et al.*, 1994) (Fig. 1), and (3) reproductive activity of Nassau groupers has been reported to peak around the new moon in the Bahamas (Smith, 1972) and around the full moon in Belize (Johannes, 1978), the Bahamas (Colin, 1992), and Mexico (Aguilar-Perera and Aguilar-Davila, 1996). Additionally, groupers, like many pelagic spawning reef fishes, spawn primarily near dusk (Ukawa *et al.*, 1966; Thresher, 1984; Colin, 1992; Samoilys, 1997a), although actual spawning observations are few. For example, Colin (1992) observed Nassau groupers spawning about 20 minutes prior to sunset, with a peak 10 minutes before to 10 minutes after sunset, after which spawning appeared to cease (Fig. 3).

B. Sex Change, Sex Ratios, and the Effects of Fishing

Groupers are usually considered to be protogynous hermaphrodites (Shapiro, 1987a). Although primary males may occur in some species (Sadovy *et al.*, 1994), juvenile groupers typically mature and function as females. Males are produced when adult females change sex. Thus, sex ratios tend to be female biased, although there is great variation both among and within species (Shapiro, 1987a). Red hinds and gags have highly female-biased sex ratios, though the ratio of females to males varies more than 7-fold and 12-fold, respectively. Other species such as Nassau groupers and yellowfin groupers, (*Myctoperca venenosa*) appear to approach a sex ratio of 1:1.

Sex change in sequential hermaphrodites has been suggested to be age or size dependent (Bullough, 1947; McErlean and Smith, 1964) and behaviorally (socially) mediated (Shapiro, 1987b). Socially mediated sex change occurs either through assessment of size (Ross *et al.*, 1990; Lutnesky, 1994) or sex ratio (Shapiro and Lubbock, 1980). In size-ratio assessment, a female changes sex when she perceives some ratio of smaller to larger individuals within the social group. Sex allocation theory suggests that sex-ratio assessment would determine the rate of sex change (Charnov, 1982). For example, where males are abundant, the proportion

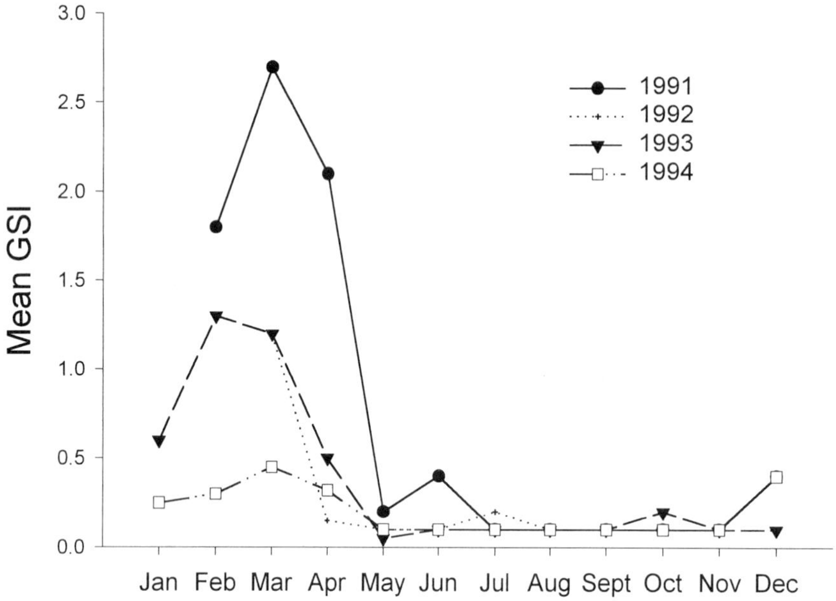

FIGURE 2 Mean gonadosomatic index (GSI) by month and year for gag (*Mycteroperca microlepis*) females from the northeastern Gulf of Mexico. Redrawn from Collins *et al.* (1998).

of females changing sex will be low; conversely, when males are rare, more females will change sex. Although the exact mechanism of sex change has not been determined for any grouper, whatever the mechanism, the consequence is that the largest animals in the population tend to be all one sex.

Sex-ratio or size-ratio assessments can occur only when both sexes are present, so there must be sufficient opportunities in time and space for individuals to evaluate population sex/size ratios. These opportunities will vary greatly among species, depending on their propensity to spawn in aggregations or the propensity for both sexes to co-occur year round. For example, in red hinds (Shapiro, 1987a; Sadovy *et al.*, 1994) and gags (Coleman *et al.*, 1996) males and females co-occur only during the spawning season. Outside the spawning

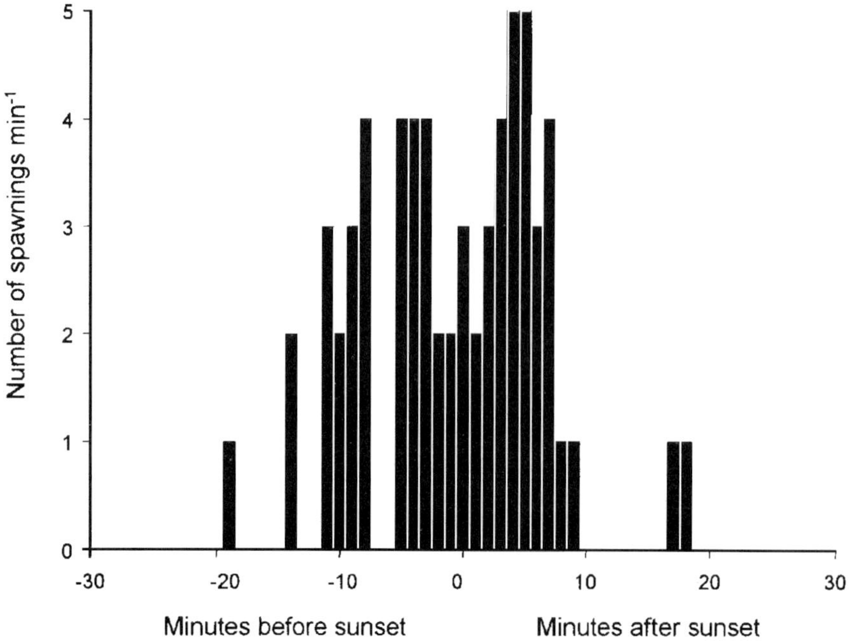

FIGURE 3 Frequency and timing of spawning by Nassau groupers (*Epinephelus striatus*) relative to time of sunset. Data are from Colin (1992).

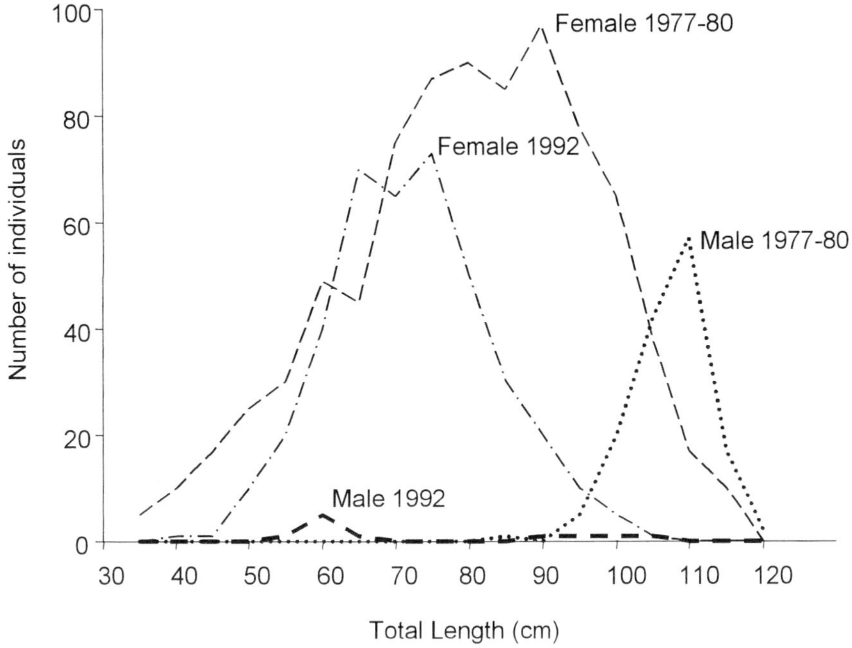

FIGURE 4 Size frequency distributions for gags (*Mycteroperca microlepis*) in the eastern Gulf of Mexico before (1977–1980) and after the onset of intense fishing pressure. Both size and sex ratio of gags appear to have responded to fishing. Data are from Koenig *et al.* (1996).

period, females occur inshore at shallower depths and males remain offshore. Conversely, red grouper (*Ephinephelus morio*) females and males co-occur year round. Thus, the ability of females to assess their potential future reproductive value as a male is limited to a period of days to weeks for species that form spawning aggregations, whereas other groupers may be able to assess sex ratios year round (Shapiro, 1987a).

Exploitation of grouper aggregations clearly has profound implications for their reproductive ecology. If larger and older fishes are more vulnerable to capture—for example, if there is selective fishing for behaviorally dominant males—then the proportion of males in a population will decline, because males are the terminal sex (Fig. 4) (Coleman *et al.*, 1996; Vincent and Sadovy, 1998). Sex-selective fishing has been reported on spawning aggregations of gags and scamps, because males remained high in the water column and took baited hooks more aggressively than did females (Gilmore and Jones, 1992; R. G. Gilmore, personal communication, cited in Coleman *et al.*, 1996).

If sex change is socially mediated, reductions in the proportion of males should result in a compensatory increase in the rate of female sex change to male (Coleman *et al.*, 1996; Vincent and Sadovy, 1998). In studies on small-bodied reef fishes (*Pseudanthias squamipinnis* and bluehead wrasse, *Thalassoma bifasciatum*), removal of males created temporary increases in the female sex ratio that quickly disappeared as females changed sex in compensation (Shapiro, 1980; Warner and Swearer, 1991). Groupers that spawn in aggregations appear unable to compensate for the loss of males. Perhaps the rapid and drastic removal of males from fishing on aggregations disrupts the socially mediated system of sex change such that the compensatory response is overridden (Collins *et al.*, 1987; Hood and Schleider, 1992; Coleman *et al.*, 1996; McGovern *et al.*, 1998). The initiation of sex change in *P. squamipinnis* (a largely unexploited serranid) is heralded by females adopting stereotypical male behaviors before morphological evidence of sex change (Shapiro, 1987a); this may suggest that sex-changing females of exploited serranids would also be behaviorally dominant and more vulnerable to fishing, further disrupting the mechanism of sex change and the breeding system.

Fishing on aggregations need not be size or sex selective to disrupt reproduction if members of the group depend on each other for spawning cues (Coleman *et al.*, 1996) or if there is a threshold size for the spawning aggregation below which courtship declines or ceases (Vincent and Sadovy, 1998). Further, when individuals return to a site repeatedly to spawn (Gilmore and Jones, 1992), removal of older fishes may leave no experienced fishes for new recruits to follow to traditional spawning sites, resulting in disappearance of the aggregation. This phenomenon probably caused the collapse of Nassau grouper fisheries in the Caribbean and southeastern U.S. Atlantic (Olson and LaPlace, 1979; Bannerot *et al.*, 1987; Sadovy, 1994). Even when aggregations remain functional, reduction in the number of males at spawning aggregations could result in sperm limitation (Shapiro *et al.*, 1994b), or disruption of social hierarchies that results in some females going unspawned. For example, Coleman *et al.* (1996) found in gag ovaries unusually high rates of atresia that they attributed to unspawned eggs resulting from lost spawning opportunities.

III. Larval Settlement and Postsettlement Survival and Movement

The rate at which stocks are replenished by the arrival of new settlers from the pelagic realm is obviously of critical importance to fisheries, and fisheries biologists have vigorously debated the causes of spatial and temporal variability in recruitment of fish stocks. Fisheries scientists have long known that fluctuations in the landings of fishes are profoundly affected by variable year-class strength (Rothschild, 1986), and for the last century investigators have argued that variation in year-class strength is largely determined by the end of the larval stage (e.g., Cushing, 1991). This view was further reinforced as workers on small-bodied reef fishes repeatedly documented patterns of adult abundance that were determined at or before the time of larval settlement (e.g., Doherty, 1983a; Victor, 1983b, 1986b; Doherty and Williams, 1988; Doherty and Fowler, 1994b). However, reef fish researchers have also demonstrated the frequent importance of postsettlement mortality on population dynamics (reviews by Hixon, 1991; Jones, 1991; Caley *et al.*, 1996), resulting in a dichotomous controversy regarding the degree to which the dynamics of reef fish populations are driven by processes occurring prior to or after larval settlement (Hixon, 1998).

The generally low densities of grouper larvae and recruits (Koenig and Coleman, 1998; Sadovy and Eklund, 1999), as well as the cryptic nature of new settlers (e.g., Keener *et al.*, 1988; Koenig and Coleman, 1998), make studies of settlement and postsettlement in groupers difficult, if not impossible, for many species. Nonetheless, a number of workers have studied grouper settlement and postsettlement survival, with most of the work focused on Nassau groupers in the Caribbean, coral trouts in the Indo-Pacific, and gags in the southeastern United States. Below we discuss temporal and spatial patterns of settlement, postsettlement mortality, and movements of these species, and we argue that the concentration of reproduction in time and space results in discrete settlement pulses that could result in density-dependent interactions even when adult populations are severely depleted.

A. Temporal Patterns of Settlement

Episodic settlement is characteristic of many grouper species, particularly those that aggregate for spawning. Nassau groupers provide a dramatic example of such mass settlement. As discussed above, Nassau grouper adults are typically widely dispersed and assemble in large spawning aggregations during one or two winter full moons for about 1 week. This short and intense period of spawning is followed by a concentrated recruitment episode. For example, in 1991 in Exuma Sound, Bahamas, Shenker *et al.* (1993) estimated that 86% of the total annual recruitment of Nassau groupers occurred during a single 4-day period. *Epinephelus merra* has also been reported to settle *en masse* with densities as high as 7–20 fishes m^{-2} on Reunion reefs (Letourneur *et al.*, 1998). Gags also settle in a pulse, with about 70% of the annual cohort recruiting to North Carolina estuaries within a 3-week period (Ross and Moser, 1995). Similarly, 76% of the annual gag cohort settled in northwest Florida between mid-April and early May at densities of about 500 fishes ha^{-1} (Koenig and Coleman, 1998).

Nassau groupers, coral trouts, and gags all appear to settle primarily during new moons (Shenker *et al.*, 1993; Doherty *et al.*, 1994b; Colin *et al.*, 1997), but also demonstrate enormous interannual variation in settlement (Shenker *et al.*, 1993; Russ *et al.*, 1996). The near-surface distribution of late-stage larvae at night suggests that wind and storm events could generate such variability (Keener *et al.*, 1988; Kingsford and Finn, 1997). Shenker *et al.* (1993) demonstrated that the episodic settlement of Nassau groupers is a result of the cooccurrence of strong winds that produce cross-shelf flow and transport grouper larvae. In the absence of storms that create cross-shelf flow, settlement of Nassau groupers is extremely low. Similarly, Grimes *et al.* (1997) suggested that changes in the direction of surface Eckman transport related to seasonally shifting winds was a principal mechanism controlling the magnitude and timing of settlement of gags in seagrass meadows along the west Florida shelf. Thus, given adequate larval production, it is the fortuitous occurrence of wind-driven transport at the precise time when grouper larvae are present and competent to settle that drives variability in recruitment to local populations. The critical nature of sporadic weather events has also been suggested for coral trout (Doherty *et al.*, 1994b) other smaller bodied reef fishes (e.g., Milicich *et al.*,

1992; Levin *et al.*, 1997a) and temperate commercial fishes (Myers, 1991).

B. Spatial Patterns of Settlement

Settling reef fishes are frequently distributed nonrandomly with respect to a number of factors, including depth (Doherty, 1982; G. P. Jones, 1986; Shapiro, 1987c), habitat complexity (Roberts and Ormond, 1987), the density of conspecifics (Sweatman, 1985b; Jones, 1987a; Forrester, 1990; Booth, 1992), coral cover or species composition (Jones, 1988b; Booth and Beretta, 1994; Tolimieri, 1995; Gutierrez, 1998), algal cover (Levin, 1991; Carr, 1994; Anderson, 1994; Levin and Hay, 1996), and presence of seagrass (Shulman, 1984; Ross and Moser, 1995). Although most studies have not performed the necessary manipulations to demonstrate conclusively that these patchy distributions of settlers result from habitat preference rather than differential predation or migration, they are highly suggestive that many fishes exhibit habitat selection at settlement (Booth and Wellington, 1998). Additionally, when habitat selection is strong it can have profound effects on demography (e.g., Tupper and Boutilier, 1997; Schmitt and Holbrook, 1999b), although this is not universal (e.g., Tolimieri, 1995; Caselle and Warner, 1996).

Although groupers tend to be cryptic and relatively rare as new settlers, the limited evidence available suggests that some species select specific habitats at settlement. For example, recently settled Nassau groupers reside within coral clumps (*Porites* spp.) covered by masses of the macroalga *Laurencia* spp. (Eggleston, 1995), or in holes adjacent to algal-covered coral (Eggleston *et al.*, 1997). The percent cover of algae explained nearly half of the spatial variation in settlement of the Nassau grouper (Eggleston, 1995). Coral trouts show consistent spatial variability in recruitment, with some sites typically receiving higher settlement than others (e.g., Doherty *et al.*, 1994b; Light and Jones, 1997). Comparison of habitats available versus habitats used suggests that newly settled coral trouts select microhabitats associated with expanses of sand covered with coral rubble (Fig. 5) (Light and Jones, 1997). This microhabitat preference scaled up such that among-reef patterns of settlement were explained by large-scale differences in the availability of sand/rubble habitat (Light and Jones, 1997). Also in Australia, groupers as a group (Connell and Kingsford, 1998) as well as coral trouts specifically (Kingsford, 1992) tend to concentrate on the inner edge rather than the center of the lagoon, and this pattern appears consistent over time. Further, this grouper assemblage responds positively to the cover of live coral and topographic complexity of reefs, thereby creating a patchy distribution both within and among reefs (Connell and Kingsford, 1998). Juvenile gags also respond to habitat attributes at landscape scales. In extensive trawl surveys of gag settlement in North Carolina estuaries, newly settled gags occurred only in large, continuous seagrass meadows (rather than in fragmented beds), and fishes occurred in greater densities in seagrass meadows associated with marshes (mean catch per unit effort = 0.21 fish/tow) than in seagrass meadows away from marshes (mean catch per unit effort = 0.01 fish/tow) (P. S. Levin, unpublished). This landscape-scale pattern appears to be generated by small-scale patterns of habitat use. Gags appear to forage in seagrass meadows at high tide (Ross and Moser,

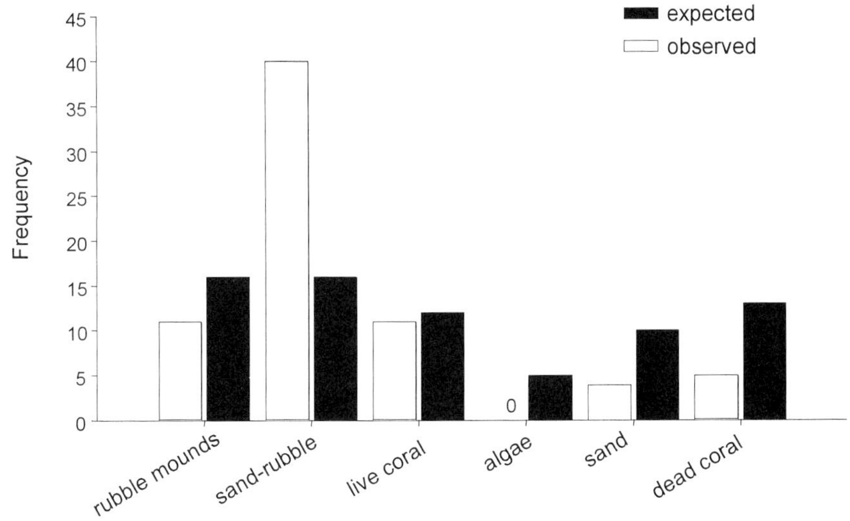

FIGURE 5 Observed and expected frequency of occurrence of newly settled coral trouts (*Plectropomus leopardus*) in habitats within a study area of 50 × 50 m on Green Reef of the Great Barrier Reef. Modified from Light and Jones (1997).

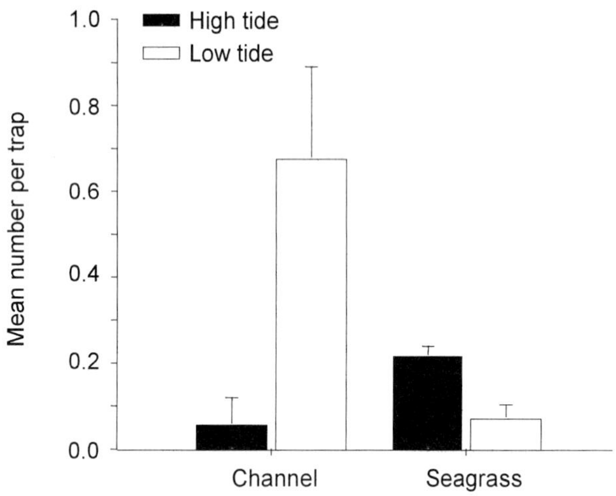

FIGURE 6 The average number of newly settled gags (*Mycteroperca microlepis*) captured at high and low tides in seagrass and channel habitats associated with marsh edges. Gags appear to move between the shallow seagrass habitat and deeper channel habitats with the tide (P. S. Levin, unpublished).

1995), and use deep (2 m) channels that meander through the much shallower marsh–seagrass landscape as a refuge when the tide ebbs (Fig. 6).

C. Postsettlement Movement

Like most reef fishes, newly settled groupers appear to be site attached (e.g., Beets and Hixon, 1994; Eggleston *et al.*, 1997; Koenig and Coleman, 1998; Heinisch and Fable, 1999), although they may have larger home ranges as compared to smaller-bodied reef fishes (DeMartini, 1998). For example, in a mark–recapture study of juvenile gags, Heinisch and Fable (1999) relocated 86% of fishes at their original capture location. Similarly, newly settled Nassau groupers and coral trouts do not stray far from cover and exhibit strong site fidelity (Light and Jones, 1997; Sadovy and Eklund, 1999).

Although groupers may remain initially within small home ranges, they may also undergo distinct ontogenetic movements. Coral trouts, for instance, move from sandrubble settlement sites to higher relief habitats as they age (Light and Jones, 1997). Likewise, gags initially settle in nearshore seagrass beds before moving to offshore reefs 5 to 6 months postsettlement (Ross and Moser, 1995; Hastings, 1997; Koenig and Coleman, 1998). In a particularly thorough examination, Eggleston (1995) demonstrated a series of ontogenetic habitat shifts in juvenile Nassau groupers. Newly settled Nassau groupers [25–35 mm total length (TL)] appear to reside exclusively in macroalgal/*Porites* clumps (described above). When juveniles reach about 60 mm TL, they occupy habitats adjacent to but outside of algal-covered clumps of coral. Finally, larger juveniles (>150 mm TL) generally occur in crevices and holes on natural or patch reefs (see also Beets and Hixon, 1994).

Ontogenetic habitat shifts are not unusual in fishes generally (Werner and Gilliam, 1984) and in reef fishes in particular (Jones, 1991). For example, several reef species settle in seagrass and mangrove habitats before migrating to reef habitats as they mature (Parrish, 1989; Ross and Moser, 1995; Koenig and Coleman, 1998). Presumably, fishes benefit from changes in resources as body size increases (Werner and Gilliam, 1984; Ebenman, 1992), although rigorous demonstration of this is lacking for reef fishes. As body size increases, foraging efficiency is increased (MacKenzie *et al.*, 1990), and susceptibility to starvation (Miller *et al.*, 1988) and predators (Sogard, 1997) declines. Habitat shifts with increasing body size are also associated with shifts in trophic habits (Werner and Gilliam, 1984), and the combination of rapid growth with pronounced habitat shifts produces clear dietary shifts in groupers (Kingsford, 1992; Mullaney and Gale, 1996; Weaver, 1997; Eggleston *et al.*, 1998; Grover *et al.*, 1998; St. John, 1999). When Nassau groupers settle into macroalgal clumps, for example, their diet is dominated by gammaridean amphipods and harpacticoid copepods (Fig. 7) (Grover *et al.*, 1998). However, as they grow and shift from within isolated algal clumps to habitats adjacent to algal clumps, their diets shift to include more amphipods. Small prey, such as harpacticoid copepods, disappear from the diet (Fig. 7). As Nassau groupers continue to grow and move to reefs, their diet shifts further and is dominated by crabs and fish (Heck and Weinstein, 1989; Eggleston *et al.*, 1998).

The close association of small groupers with structurally complex habitats that contain appropriate food resources may simultaneously reduce foraging time and exposure to predators (Marx and Hernkind, 1985; Eggleston, 1995). As fishes grow, their increasing metabolic demands, in concert with a reduction in vulnerability to predators, may result in a shift toward habitats with higher predation risk but greater food reward (Werner and Gilliam, 1984; Holbrook *et al.*, 1990). Additionally, rapidly growing fishes may face a shortage of appropriate shelter, prompting size-related habitat shifts, as has been observed in decapods (e.g., Eggleston *et al.*, 1990; Wahle, 1992).

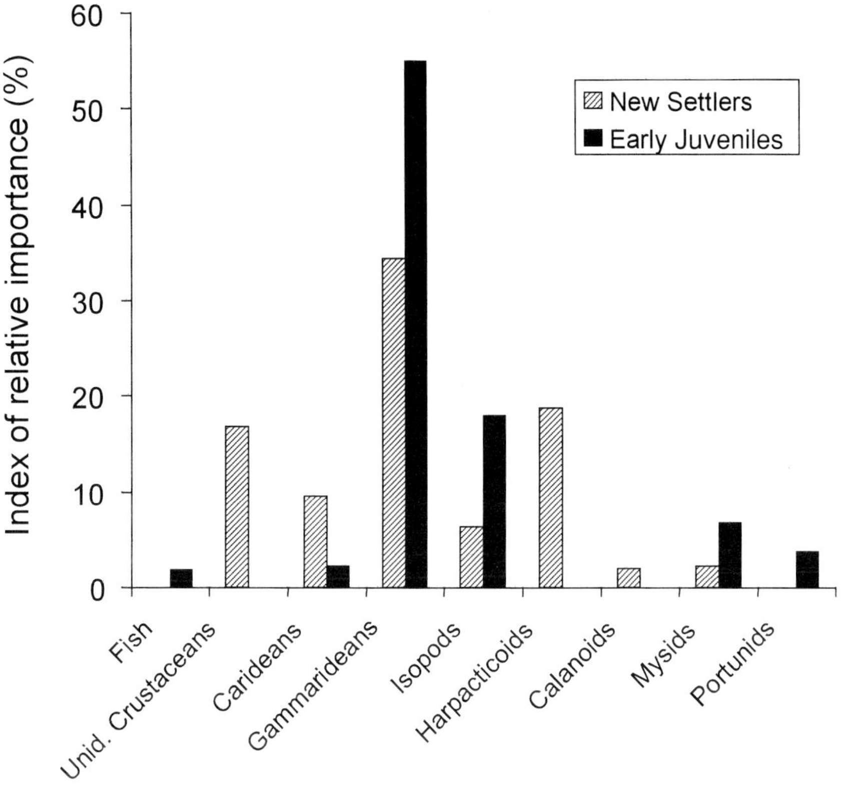

FIGURE 7 The relative importance of prey to newly settled and juvenile Nassau groupers (*Epinephelus striatus*). Newly settled fishes were 22–26 mm standard length and early juveniles were 50–75 mm total length. Modified from Grover *et al.* (1998).

As adults, large-bodied groupers typically have larger home ranges than do their small-bodied counterparts, because home range generally increases with body size (DeMartini, 1998). Even so, with the exception of annual migrations to spawning aggregations, the limited available data suggest that routine movements of adult groupers are somewhat limited. Using ultrasonic telemetry, Zeller (1997, 1998) and Zeller and Russ (1997) determined that home ranges for coral trouts averaged between about 10,000 and 19,000 m^2, with daily fish movements averaging 192 m within home ranges. Further, coral trouts appear to prefer a small number of locations within their home ranges, and they return to the same location each night (Zeller, 1997). Telemetry on gags off west Florida also suggested limited home range: movement among experimental patch reefs with 25-m spacing was extensive, but there was no movement among patch reefs with 225-m spacing (Kellogg *et al.*, 1997). Shapiro *et al.* (1994a) estimated home ranges of red hinds as 100 to 5600 m^2. Nassau groupers also appear to be site attached (Sadovy and Eklund, 1999), with daily movements of 400 m from a home site to which they return at night (Carter *et al.*, 1994).

D. Demographic Importance of Synchronous Reproduction, Concentrated Settlement, and Postsettlement Movement

The synchrony of reproductive activities within short time periods and in specific locations produces large groups of eggs, and later larvae, that experience similar oceanographic conditions and may be advected *en masse* to locations where they later settle (Shenker *et al.*, 1993; Lee *et al.*, 1994; Colin *et al.*, 1997). Similar episodic settlement as a function of synchronous reproductive events may occur in damselfishes (Danilowicz, 1997; but see Danilowicz and Sale, 1999). Thus, even though many grouper stocks are severely depleted and in decline (e.g., Sadovy and Eklund, 1999), the occurrence of dense aggregations of settlers sets the stage for density-dependent mortality.

There are no rigorous experimental data demonstrating density-dependent mortality in juvenile groupers, but the available indirect evidence suggests that this may be an important demographic process that affects recruitment in some species of groupers. After settling in high numbers, some grouper species suffer high early postsettlement mortality. Following an unusually strong settlement pulse, Letourneur *et al.* (1998) reported an average mortality rate of 90% of newly settled *Epinephelus merra,* with the variance around this average explained by initial density of settlers. Similarly, Beets and Hixon (1994) documented an 85% mortality rate following an uncommon pulse of recruitment. A requirement for very specific habitat (e.g., Eggleston, 1995), in combination with unusually high numbers of settlers, may induce density dependence if predation pressure is high and shelter is in short supply (Hixon and Menge, 1991; Hixon and Carr, 1997). Although there certainly are too few data to conclude that density-dependent mortality is an important process affecting the recruitment dynamics of these groupers, the combination of extreme spatial and temporal concentration of settlement, the need for specific settlement habitat that may be in short supply, and high mortality rates strongly suggest that compensatory mortality may occur.

Although density-dependent mortality may be important in some juvenile groupers, different processes are clearly the leading determinants of recruitment success in other species. For example, Ferreira and Russ (1995) documented the settlement of a strong cohort of coral trouts to reefs closed to fishing. Subsequent age-frequency data from these reefs showed this strong year-class dominating the population for several years (Russ *et al.,* 1996). This, in concert with data linking settlement and recruitment of coral trouts with larval supply (Doherty *et al.,* 1994b), suggests that variability in year-class strength of this species is determined by vagaries of larval supply. Similar mechanisms may also affect the dynamics of gag populations. Survival of newly settled gags in seagrass meadows approaches 100% (Koenig and Coleman, 1998), and the abundance of juveniles in seagrass meadows corresponds to subsequent year-class strength in regional fisheries (Johnson and Koenig, 2001), suggesting that year-class strength has already been determined at settlement. Thus, although data directly linking larval supply and settlement in gags are lacking, the available evidence indicates that variability in larval supply may be a major determinant of population size in gags as well.

IV. Applications of Ecological Knowledge in the Management and Conservation of Grouper Fisheries

We now return to our initial question and ask if current ecological understanding can provide the insight needed for "articulate intervention" in reef fisheries? We first address how the reproductive ecology of groupers impacts the implementation of stock assessment models. Second, we explore how knowledge of recruitment dynamics might be usefully applied to grouper conservation and management.

A. Stock Assessment in Grouper Fisheries

Typical assessment of the status of exploited reef fish populations in general, and groupers in particular, has used age-based analytical models to simulate the effects of different levels and strategies of fishing on populations, and predict future yields. Historical catch-at-age data from the fishery are used in virtual population analysis (VPA) as the basis for backward calculation of cohort-specific abundances and fishing mortality rates (Gulland, 1965), which are then reversed to project into the future (Hilborn and Walters, 1992). Dynamic-pool yield models (Thompson and Bell, 1934; Ricker, 1954; Beverton and Holt, 1957) and their derivative spawning stock biomass-per-recruit (SSBR) models (Gabriel *et al.,* 1989; Goodyear, 1993) are used to estimate potential yield-per-recruit and SSBR relative to fishing mortality and size at entry into the fishery. Yield-per-recruit and SSBR are calculated as a summation across discrete age increments, under specific conditions of growth and mortality. SSBR models allow estimation of the spawning potential ratio (SPR), the ratio of present spawning biomass to that of the stock in an unfished condition. Target SPR values that maintain adequate spawning potential for the stocks were initially developed by examination of the SPR and exploitation status for many stocks (Gabriel *et al.,* 1989), then improved later by simulation to maximize the minimum yield over a range of productivity conditions (Clark, 1991, 1993). The results of virtual population analysis and SSBR are used to project a trajectory of future catches that will allow the target SPR to be maintained or will attain the SPR target within a specific recovery time frame, usually determined as a function of the generation time of the species.

Simulated effects of protogyny on yield-per-recruit, SSBR, and other measures of reproductive capability suggest that protogynous species lose reproductive capacity more rapidly with increasing fishing and sustain reproductive failure at lower levels of fishing, as compared to gonochoristic species. Shepard and Idoine (1993), for example, developed a length-based dynamic-pool yield model for the protogynous serranid, black seabass (*Centropristis striata*), that allowed for changes in growth and mortality following sexual transition not accounted for in conventional yield-per-recruit models. Their results suggested that spawning stock biomass-per-recruit was considerably less than estimated by conventional models. Likewise, Huntsman and Schaff (1994) used simulations to investigate the effects of fishing on the reproductive capacity of protogynous graysby, and compared their model's implications for management to those from a conventional yield-per-recruit model. They examined the effect of various compensatory mechanisms on maintaining reproductive capacity. Under conditions of uncompensated protogyny, reproductive capacity declined rapidly with increasing fishing mortality and reproduction failed at lower fishing mortality than it did in gonochoristic stocks. Compensation through conservation of the numerical sex ratio, and through conservation of the cohort male biomass:fecundity ratio, reduced and eliminated effects of protogyny, respectively. All conditions modeled incurred drastic reductions in reproductive capacity even at modest levels of fishing mortality. Bannerot *et al.* (1987) also suggested that as increasing mortality disrupts the ability of males and females to come in contact, the reproductive success of protogynous species declines relative to that of gonochorists. They concluded that management strictly by yield-per-recruit models was risk prone.

Interestingly, Bannerot *et al.* (1987) highlighted the need for several lines of field research critical to better understand the response of protogynous fishes to exploitation: (1) that exploited populations be monitored to determine if increasing proportions of eggs go unfertilized as fishing mortality increases, (2) that changes in population sex ratio with exploitation be followed, (3) that the ability of populations to compensate to avoid sperm limitation be characterized, (4) that the time required for hermaphroditic individuals to change sex and the range of sizes capable of changing sex be determined, and (5) that the extent to which spawning is disrupted by fishing be evaluated. Some of these questions have now been answered, at least to some extent. For example, the proportion of males has declined from ~20 to 3% with increasing exploitation of gag populations over about 30 years in the U.S. South Atlantic and Gulf of Mexico (Collins *et al.*, 1987; Hood and Schleider, 1992; Coleman *et al.*, 1996; McGovern *et al.*, 1998), suggesting that any ability for compensation in the breeding system has been overridden. Direct selective fishing on breeding aggregations for more aggressive males (Gilmore and Jones, 1992; R. G. Gilmore, personal communication, cited in Coleman *et al.*, 1996), and male residence at aggregation sites up to ninefold longer than that of females (Zeller, 1998), which would lead to greater cumulative effects of fishing on males, have been shown for protogynous groupers. Both the direct and indirect selective imposition of fishing mortality on males could certainly disrupt spawning. The time required for sex change is not known for any grouper, although it has been shown to occur in about 2 weeks in the confamilial *Pseudoanthias squamipinnis* (Shapiro, 1987a).

Perhaps a fatal shortcoming of the SSBR approach to simulating effects of fishing on protogynous populations is the assumption that female biomass or egg production is a relevant measure of stock reproductive capacity. A demonstrable effect of fishing on protogynous grouper populations that may influence reproductive capacity is the drastic reduction in the proportion of males in the population (Coleman *et al.*, 1996; McGovern *et al.*, 1998). Assuming that breeding aggregations remain functional, a reduction of males from an unfished condition of 20% to a fished condition of 2% would require males to increase harem size in breeding aggregations up to 10-fold or females would go unmated (Coleman *et al.*, 1999). This scenario also assumes that there is no sperm limitation and that fertilization rates at the new sex ratio remain the same, which seems unlikely. Obviously, if fertilization rates decline, so does the reproductive output of the stock. If a simple decrease in fertilization rate is the response of the stocks to reduction in the proportion of males, then SSBR might be made a useful analytical tool to prevent recruitment overfishing by determining the fertilization rate and adjusting the computed SSBR accordingly. Alternatively, male biomass of the stock might be used as the proxy for stock reproductive potential.

It seems more likely that a reduction in the proportion of males, especially if fishing is directed at spawning aggregations, is likely to disrupt the social organization and function of spawning aggregations (e.g., Coleman *et al.*, 1996; Vincent and Sadovy, 1998), rendering them nonfunctional and eliminating their contribution to the aggregate reproduction of the stock altogether. The critical point is that the functional unit of reproduction is the spawning aggregation, and the traditional SSBR models female biomass or egg production as the measure of stock reproductive capacity. Thus, SPR target values that are intended to assure reproductive integrity of a stock may be irrelevant for that purpose. It seems pointless and irresponsible to continue to apply modeling approaches such as SSBR to assess and manage grouper stocks in light of the available knowledge both from simulations of the response of protogynous fish to exploitation, and from the observed reproductive ecology and behavior of exploited grouper and unexploited model species. These findings indicate that traditional approaches do not accurately simulate the response of protogynous grouper stocks to exploitation. They generally underestimate stock decline from fishing and overestimate stock productivity, and thus lead to overly optimistic assessments of the status of grouper stocks. Clearly, new modeling approaches are called for, especially if traditional management approaches such as catch/effort controls and size restrictions are applied to attain management objectives. These approaches have not been successful for conserving grouper stocks in most instances, owing at least in part to the modeling and assessment problems discussed above [but see Coleman *et al.* (1999) for a more general evaluation of traditional management techniques for groupers]. Because spawning aggregations occur at specific locations year after year, marine harvest reserves, a vigorously debated and increasingly popular fishery management tool (Agardy, 1997), are an obvious management technique that could be effectively used to protect spawning aggregations and preserve the reproductive capacity of grouper stocks.

B. Impacts of Recruitment Dynamics on Grouper Conservation and Management

The development of the concept of recruitment limitation has profoundly affected the direction of reef fish ecology, and clearly has the potential to increase the efficacy with which reef fisheries are managed. If variability in the abundance of fishes reflects variability of their accumulated settlement [i.e., they are recruitment limited (Doherty and Fowler, 1994b)], studies of larval supply or settlement provide a means to accurately forecast year-class strength, thus providing key information that may be used to modify and enhance stock assessment and management approaches currently in use. For species such as coral trouts and gags, in which initial spatial and temporal patterns in the magnitude of settlement are reflected as variability in adult stock size, knowledge of settlement strength or juvenile abundance may permit prediction of the size of the harvestable population several years in advance (Russ *et al.*, 1996; Koenig and Coleman, 1998). Such data would allow the "fine-tuning" of stock assessments to the year-classes of most significance for management (Myers *et al.*, 1997), and would provide resource managers with more detailed information to adapt management strategies to spatial and temporal variability in recruitment stock size. Importantly, data on the location of sources of larvae and sinks for settlers would allow networks of harvest reserves to be created, such that source reefs and nursery habitats would have the capacity to restore depleted stocks and protect stock reproductive capacity most efficiently (Russ and Alcala, 1996b).

On the other hand, if large episodic pulses of settlement, in concert with habitat specificity and high early mortality, produce density dependence as in Nassau groupers and in species with similar recruitment dynamics, then settlement dynamics may still be important, but in a somewhat different fashion. Population sizes of these species may be limited by resource levels, at least at the scale of a single reef in a single year. At this scale stock size should be set by levels of both settlement and resources, with resource levels limiting the number of fishes per year and temporal variability in the number of settlement pulses determining the number of cohorts that are present in the population. We can expand this concept beyond the scale of a single reef to include an entire population. If we assume that some larvae arising from different spawning aggregations settle at different reefs (Danilowicz, 1997), and, as we have argued, that local populations are resource limited, then total population size is a function of the carrying capacity of each reef and the number of reefs that receive settlement pulses. Although such populations would be locally limited by resources, at large scales they would be recruitment limited in the sense that population size would be determined by the number of reefs that receive settlers. If this is true, then from a management perspective it would be critical to maintain as many spawning aggregations as possible rather than many fishes within few aggregations, because the addition of more settlers to a single resource-limited reef will not produce more adults, but adding more reefs does produce large changes in population size (cf. Beets and Friedlander, 1998). This line of reasoning obviously contains many assumptions, assumptions that may, in some cases, be easily tested on small-bodied reef fishes as well as on groupers. The rigorous ecological work on settlement and recruitment of small-bodied reef fishes has been crucial in bringing the importance of recruitment to the fore (e.g., Doherty and Fowler, 1994b). In addition, the similarity in early life histories of small-bodied and exploited reef fishes (DeMartini, 1998) suggests that recruitment studies of small reef fishes can continue to provide insight and direction to work on exploited species.

V. Conclusions and Recommendations for Future Research

Over the past 25 years, studies of coral reef fishes have had a profound impact on ecological understanding and have challenged reef ecologists to use their growing understanding of small-bodied reef fishes to better manage coral reef fisheries (Sale, 1988). In an age when the once mighty Canadian Grand Banks cod fishery is closed and Pacific salmon throughout the western United States are now listed as endangered species, one hopes that the lessons learned from small-bodied reef fishes can now be successfully applied to the management of reef fisheries.

Our review suggests several lines of inquiry that could provide information relevant to improving the conservation and management of exploited groupers. For many, if not most groupers, habitats used by larvae and juveniles are unknown or incompletely known. Such data are critical for mechanistic pre- and postsettlement studies that are required to develop indices for prediction of year-class strength. As is now the norm with small-bodied reef fishes, investigations of both pre- and postsettlement processes must be conducted in an effort to determine what processes determine recruitment success and how these processes vary in time and space. Demonstration that presettlement processes determine variability in year-class strength (e.g., Russ *et al.*, 1996, Koenig *et al.*, 1996) would provide a powerful new tool to those managing reef fish stocks. In other cases in which postsettlement mortality is density dependent, it may still be possible to develop predictive indices if the shape of the survivorship curve is described.

Research to understand the source, connectivity, and sink dynamics of populations is needed to provide critical information for management. For example, if the locations of spawning adults, settlement sites, and recruits and the magnitude and direction of recruitment were known, it would be possible to identify which spawning and settlement habitats were most critical to sustaining populations. Models of ocean circulation and larval transport (e.g., Cowen *et al.*, 2000), studies of otolith microchemistry and microstructure (e.g., Swearer *et al.*, 1999) (see Chapter 11, this volume), and microtagging of larvae and juveniles (e.g., Jones *et al.*, 1999) offer great promise to help solve these problems. This information would allow managers to identify and protect the habitats critical to sustaining populations, for example, by strategically locating marine protected areas. However, if these source-sink dynamics change with time due to factors such as change in ocean climate or spatial changes in reproduction, these problems will be difficult to insoluble.

Although the aforementioned contributions to age-based population modeling (Bannerot *et al.*, 1987, Shepherd and Idoine, 1993; Huntsman and Schaff, 1994) that simulate protogynous populations and their response to fishing are pioneering, there is much that remains to be done to incorporate accurately the reproductive ecology and behavior of hermaphroditic groupers into stock assessment models. For example, it may be worthwhile to explore the possible utility of male-based SSBR models for more realistically representing the male-limited nature of reproductive capacity of protogynous populations. Field studies to determine fertilization rates in fished and unfished spawning aggregations would greatly aid in improving traditional SSBR models. Further study of the mechanism of sex change in exploitable protogynous groupers to identify the cues that initiate sex change, the sizes/ages of transition, and the ability of these breeding systems to compensate for male loss would be valuable. In addition, we encourage field studies on exploited populations and controlled investigations on small-bodied model species to determine the extent to which fishing on spawning aggregations is selective for males and disrupts spawning in aggregations.

Populations of small-bodied reef fishes have been used as living laboratories for investigating processes affecting population dynamics (Sale, 1991b). Although there are obviously differences among species, many manipulable small-bodied fishes share a fundamentally similar early life history with exploited groupers. As such, basic ecological research on reef fishes has had a profound influence on both the hypotheses tested and the approaches used to test these hypotheses in exploited species. Continued basic research on reef fishes is likely to keep bearing applied fruits. Nonetheless, there is also a clear need to move beyond the living laboratories of small fishes into the real world of exploited fishes. Applying the principles of experimental design employed in many areas of fundamental research to fisheries research could help reduce the uncertainty managers presently face. Experiments on exploited fishes are very difficult to conduct; however, management activities provide opportunities to conduct such experiments. Although management actions are rarely designed as "good" experiments, they provide unique opportunities for the resolution of uncertainty. As relatively small, replicated systems, reefs provide excellent opportunities to apply experimental management. Indeed, perhaps the greatest lesson small-bodied reef fishes can provide is that experiments offer the greatest hope of understanding the dynamics of fish populations. As experimental management becomes increasingly more common in reef fisheries, we hope such an approach will be equally productive with exploited reef fishes.

CHAPTER 18

Ecological Issues and the Trades in Live Reef Fishes

Yvonne J. Sadovy
Department of Ecology & Biodiversity
The University of Hong Kong
Hong Kong

Amanda C. J. Vincent
Project Seahorse, Department of Biology
McGill University
Penfield, Montreal, Quebec H3A 1B1
Canada

I. Introduction

The trades in reef fishes marketed alive for table food and aquarium display are relatively new fisheries, with novel problems and promises. These are high value-to-volume fisheries for luxury markets, often focused on species with life histories unsuited to heavy exploitation. Some aquarium fisheries are carefully managed and the trades are well operated, but the dearth of management initiatives in the major source areas of Southeast Asia for both trades means that overfishing and poorly directed selective fishing can threaten exploited populations. In addition, fishing practices employed to capture high-value live reef fishes often generate wasteful mortality of target and nontarget species and damage reef habitats. Yet these fisheries and trades, for all their problems, can offer important livelihoods for local communities, thus potentially promoting marine conservation, by providing a strong incentive for subsistence fishers to maintain fish populations and keep reefs in good condition. Such promise can only be fulfilled, however, if fisheries that supply these trades are operated in a way that does not compromise the long-term future of wild reef fish populations.

Trades for live food fishes and ornamental marine fishes are striking in their similarities, and show marked differences from most conventional (dead food) fisheries (Table 1). They are often based on coral reef fish species not otherwise considered important as food in many source countries and have increased greatly in the past one to three decades (live food and aquarium trades, respectively), as economic and technological changes prompted greater demand for table foods and for marine aquaria. Both trades are directed at luxury export markets and are worth millions of retail dollars (United States) annually and both largely originate in Southeast Asia, where reef fishes are extracted from developing countries to supply wealthy markets in both the East and West. Although many species are traded, relatively few make up the bulk of the trade, particularly in the case of food fishes. Gourmets and aquarium hobbyists both select their live fishes individually, favoring attractive appearance and a narrow range of sizes (plate size for food and appropriate fits for small aquaria). Indeed, rarity may be particularly valued by some customers and seems to spur demand (Johannes and Riepen, 1995). Such preference for certain sizes or species, and the need to maintain and transport fishes alive, have provoked the use of fishing and fish handling methods and practices that arouse concern for reef fish populations and reef habitats. On the positive side, such methods have increasingly prompted the introduction of quality controls, at least within the aquarium trade, and have led to the emergence of nondestructive fishing. A relatively recent change is the increased focus on supplying both trades through culturing, an approach with its own suite of problems and prospects that we shall also examine.

The two trades do differ in important ways (Table 2). Live food fishes are in high demand in Asian restaurants, where consumers can select their dinner directly from a restaurant's display tank. In contrast,

TABLE 1 Similarities between Food Trade and Aquarium Trade Fisheries[a]

- High value*
- Rarity confers value*
- Luxury items*
- Fishes often individually selected (for size, sex, color, etc.)*
- Relatively new fisheries (mostly since WW II)*
- Trades not included in FAO world production figures*
- Immature fishes often targeted*
- Mainly originating in Indonesia and the Philippines, although resource base is shifting*
- Smaller fishes can be more valuable than larger fishes*
- Target species seldom valued as food in source country*
- High hidden mortalities mean trade numbers are underestimates*
- Appearance of fishes matters to consumer*
- Wealthy consumers able to influence nature of consumption*
- Standard fishery models generally do not apply*
- Animal welfare issues in shipping live organisms*
- Trades poorly monitored
- Mainly subsistence fisheries with low opportunity costs
- Concentrated in tropics and coral reef fishes
- Commonly supporting subsistence fisheries
- Can be combined with food fishing
- Many species potentially vulnerable to overfishing
- Some fishing techniques destructive
- Fisheries rarely managed
- Low local accountability, especially by importers
- Most species unstudied
- Predominantly export fisheries
- Demand expected to grow and tied to economics/social events
- Mariculture has small potential role in lowering fishing pressure

[a]Although fisheries that serve the food and aquarium trades have similarities, they both may differ from many conventional fisheries. These aspects are denoted here with an asterisk (*).

TABLE 2 Differences between Food Trade and Aquarium Trade Fisheries

- Food trade extracts particularly high numbers of juveniles for mariculture
- Food trade takes larger species/individuals, aquarium trade takes smaller species/individuals
- Food trade is only in Indo-Pacific; aquarium trade is global in tropics/subtropics
- Food trade has associated human health issues
- Food trade has high mortalities and demand for feed in mariculture
- Aquarium trade is better monitored in international trade
- Aquarium trade outside of Southeast Asia may be monitored/managed
- Aquarium trade may have long-term stability in specific areas; food trade typically has little stability
- Aquarium trade seeks more species than food trade
- Value of tanks and other hardware substantial for aquarium trade, negligible for food trade
- Aquarium trade requires better catch survivorship compared to food trade catch
- Aquarium trade culture based on hatcheries; food trade culture mainly based on wild capture of juveniles
- Major demand for aquarium trade is in the West; demand for food trade is in the East
- Public aquaria can promote education and conservation; food trade groups do not
- Aquarium trade prices generally more stable in the long term

aquarium fishes are primarily destined for North American and European hobbyists, who keep marine fishes for their entertainment, relaxation, and educational value (Kidd and Kidd, 1999). Only the trade in live food fishes needs to consider human health issues. Because they deal in live fishes, both trades have ethical and animal welfare concerns not associated with conventional fisheries, although these are beyond the scope of this chapter. Also not covered, but nonetheless worthy of mention, are the invertebrate components of both trades. In the case of the aquarium trade, in particular, invertebrates, including "live or coral rock" for home tanks, make up the fastest growing sector (Cesar, 1996; International Trade Subgroup, 2000).

Live reef fisheries, like conventional fisheries, act on marine communities in three ways: (1) they exert pressure directly on captured individuals, removing the fishes or returning them in a compromised state (damaged or to the wrong location); (2) they affect the members of populations that are not directly removed, by disrupting their social and spatial structures, and (3) they place pressures on the habitats and communities on which these species depend. The use of cyanide as a fish capture method is largely responsible for particular attention to these live-marketed fisheries, because of the possible direct and indirect effects of this poison on reef biota (Johannes and Riepen, 1995; Barber and Pratt, 1997). However, overfishing and selective fishing are, almost certainly, the predominant threats to reef species of value to live fish markets. In this chapter we focus on the live reef fish trades derived from capture fisheries in Southeast Asia, the major suppliers. We describe these fisheries in terms of species, sources, volumes, values, destinations, capture methods, selectivity, mortality, seasonality, monitoring, trade structure, and policy. We then present theoretical issues, and evaluate what is known of the direct and indirect biological impacts of fishing for live reef fishes. We conclude by discussing management and research issues and the roles of mariculture and stakeholders in contributing to sustainable use of these resources. The dearth of information in the standard scientific literature on these relatively new fisheries means that we are heavily dependent on reports, proceedings, policy documents, and personal communications, often from those within the live trade industries

II. Nature of Live Reef Fish Capture and Trade

The nature of the live food fish trade (FT) and the marine aquarium (ornamental) fish trade (AT) warrants special examination because of the way in which socioeconomic forces and selective fishing act on wild populations. These live fisheries differ considerably from the less selective conventional fisheries that are the focus of most models of fisheries management, and therefore require us to undertake new analyses and devise novel responses to their monitoring and management. We here outline the background on the two trades by turn.

A. Species and Preferences

1. Live Food Trade

Only a handful of species are sought in high volumes for the FT, with particular emphasis on the groupers (Serranidae), wrasses (Labridae), and snappers (Lutjanidae) (Table 3). The humphead (Napoleon or Maori) wrasse (*Cheilinus undulatus*), the high-finned grouper (*Cromileptes altivelis*), and giant grouper (*Epinephelus lanceolatus*) are traded only in small volumes but are particularly valued (Johannes and Riepen, 1995; Lau and Parry-Jones, 1999; Chan, 2001).

Consumers show clear preferences for certain sizes, colors, and species of live food fishes in restaurants and markets, and confer a high value on rarity. Regardless of species, diners prefer fishes of a size that will fit onto a plate to feed one or two people, or else will please larger groups during banquets or for business dinners. Fishes in the size range of 0.6–1.5 kg tend to be most in demand, with highest prices for individuals between 0.8 and 1.2 kg (Johannes and Riepen, 1995), and most individual fishes in trade measure between 30 and 50 cm in total length (Fig. 1). Groupers are favored for their taste, as is the humphead wrasse, and Chinese consumers prefer fishes that are reddish, such as the coral trout, believing that color to be auspicious (Lau and Parry-Jones, 1999). Species, or color forms, that are rare (e.g., the albino coral trout), or that have

TABLE 3 Economic Data for Fishes Used in the Food Trade[a]

Family[b]	Number of species	Prices in Hong Kong (1997) (U.S. $/kg)[c]		Volume for 1997 (mt)
		Wholesale	Retail	
Carangidae	2	—	—	—
Centropomidae and Percichthyidae	1	2.5	—	1346
Haemulidae	2	(6)	(10)	—
Labridae (humphead wrasse)[d]	1	54–97	92–174	2
Labridae (other and Scaridae)	6	11–18	—	10
Lethrinidae	1	2	6	—
Lutjanidae[d]	8	9–18	25	—
Scorpaenidae	2	(8)	(17)	—
Serranidae[d,e]	30	6–31	29–169	5714
Sparidae	3	(8)	(12)	—
Other marine fish[f]	—	6	—	13,994

[a]Sources: Lee and Sadovy, (1998); Lau and Parry-Jones (1999).
[b]Bold type indicates inclusion in trade of 20 or more species per family.
[c]Prices in parentheses are from 1999 (P. S. W. Chan, personal communication).
[d]Higher retail prices are sometimes realized for smaller individuals within a species. For example, small humphead wrasses bring U.S. $174/kg, but larger ones bring U.S. $92/kg; small giant groupers bring U.S. $169/kg, larger ones bring U.S. $63/kg.
[e]Family includes some cultured species (hatchery reared).
[f]Unclassified reef fishes, includes mullids, siganids, scarids and haemulids.

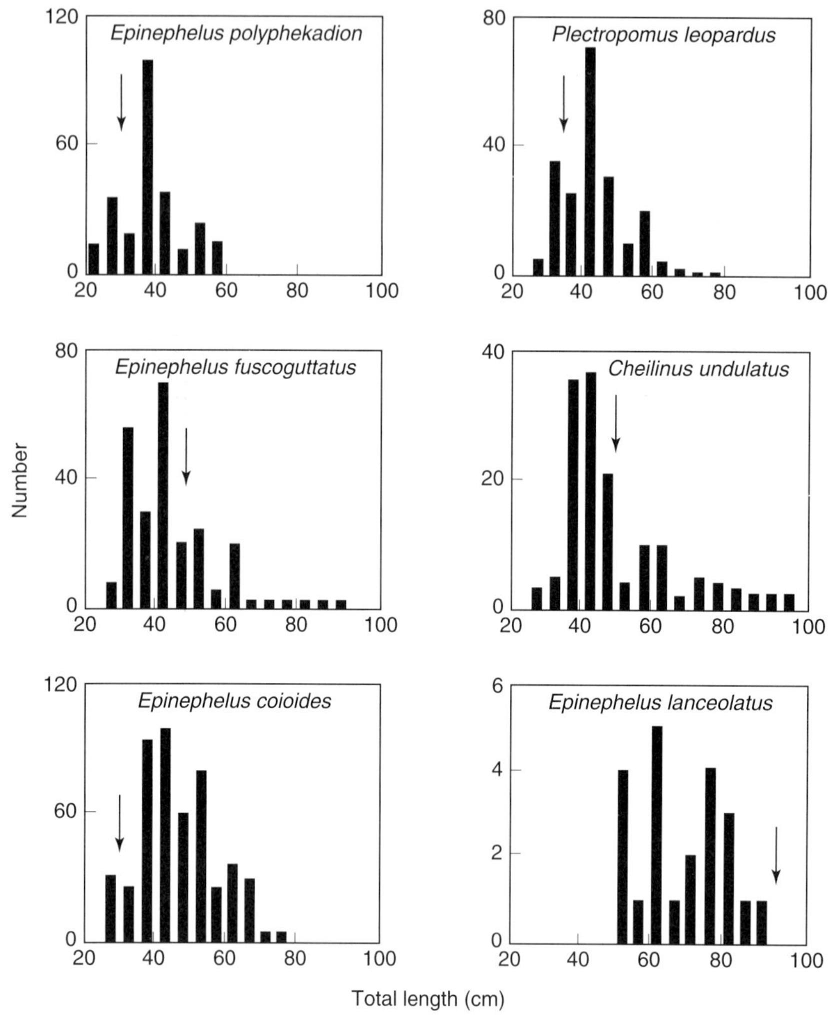

FIGURE 1 Size–frequency distributions of six species frequently sold in the food trade in a major Hong Kong retail market (Lei Yue Mun) between December 1995 and February 1996 (modified from Lee and Sadovy, 1998). Bars indicate approximate length at sexual maturation. Sizes for sale are likely to match consumer preferences for common species, and fish availability for uncommon species, such as the humphead wrasse (*Cheilinus undulatus*) and the giant grouper (*Epinephelus lanceolatus*).

medicinal value (e.g., the giant grouper), fetch particularly high prices (Sadovy, 2001a).

A further preference is for fish likely to be free of ciguatoxins (naturally occurring poisons that bioaccumulate in food chains where certain dinoflagellates associate with the algae eaten by herbivores). Ciguatoxins accumulate in reef fishes in certain areas of the tropics and can cause a condition in humans known as ciguatera, which has on occasion proved fatal (Lewis, 1992). Ciguatoxins, therefore, can drive the preference for small to medium-sized fishes, because larger individuals often carry a higher risk of being ciguatoxic, with consequent danger to human health (Dalzell, 1992).

2. Aquarium Trade

The trade in marine ornamentals embraces a wide diversity of fishes, with over 1000 species recorded in trade globally from over 50 families (Wood, 2001). About 386 species from 79 families are exported from the Philippines, a major supplier of aquarium fishes (Vallejo, 1997) and almost 600 species in 40 families were being sold in the United States in 1999, this country accounting for about 60% of world import trade (Warmolts, 2000) (Table 4). Despite such diversity, most species are from just seven families (angelfishes, Pomacanthidae; wrasses, Labridae; damselfishes, Pomacentridae; surgeonfishes, Acanthuridae; butterfly fishes, Chaetodontidae; gobies, Gobiidae; and the groupers and their relatives, Serranidae), with relatively few species dominating the trade in many areas (Conroy, 1975; Randall, 1987b; Sadovy, 1992; Pyle, 1993; Rajasuriya *et al.*, 1995; Wood, 2001).

Preferences for particular sizes, color forms, or rarity can produce extremes in selective harvesting. Most consumers are keen to purchase animals of about 2 to 10 cm in length, typically finding smaller or larger individuals more difficult to maintain within the constraints of the typical home aquarium (Chan

TABLE 4 Marine Aquarium Trade: 1998–1999 Retailers' Prices for North American Markets[a]

Family[b]	Number of species	Price range (U.S. $/animal)	Family[b]	Number of species	Price range (U.S. $/animal)
Acanthuridae	**30**	8–200	Malacanthidae	7	17–52
Anomalopidae	1		Monacanthidae	12	7–29
Antennariidae	4	11–52	Mullidae	3	11–24
Apogonidae[c]	6	6–37	Ostraciidae	9	
Aulostomidae	1	9–13	**Pomacanthidae**[d]	**58**	13–179
Balistidae[d]	18	9–96	**Pomacentridae**[c]	**48**	2–110
Blenniidae	**20**	5–22	Pseudochromidae[c]	14	10–38
Callionymidae	4	8–14	Scaridae	13	7–37
Carangidae	3		Scorpaenidae	14	10–37
Chaetodontidae	**68**	9–200	**Serranidae**[c]	**48**	7–67
Channichthyidae	1		Siganidae	3	
Cirrhitidae	12	10–56	Syngnathidae	4	7–48
Eels (various)	15	15–89	Tetraodontidae	11	9–150
Enoplosidae	1	22	Triglidae	3	30
Gobiidae[c]	**43**	4–250	Zanclidae	1	22–28
Holocentridae	7	10–19	Flatfish	3	15–17
Labridae	**82**	6–120	Haemulidae	3	
Lutjanidae	5	7–52	Elasmobranchs	18	22–375

[a]Sources: Lee and Sadovy (1998); Lau and Parry-Jones (1999). Prices are from Aquacon (Sunrise, Florida) and are comparable to those listed by Flying Fish Express (Los Angeles, California) and Harspen Marine Fish (Echo Bay, Ontario, Canada).

[b]Bold type indicates inclusion in trade of 20 or more species per family.

[c]Family includes some cultured species (hatchery reared).

[d]Higher retail prices are sometimes realized for smaller individuals within a species. For example, small clown triggerfish, *Balistoides conspicillum*, cost twice as much as large ones.

and Sadovy, 1998; Wood, 2001). Attractive color and physical shape are especially important to hobbyists. Indeed, one factor driving a preference for smaller fishes is the distinctive coloration of the juveniles and subdominants of many species (e.g., damselfishes, angelfishes). Preference for bright color may also explain why male wrasses are collected much more frequently than their drabber conspecific females (J. Tullock, founder of American Marinelife Dealers' Association, September 29, 1999). Male mandarinfishes, *Synchiropus splendidus*, are preferred over females for their particularly attractive dorsal fins and displays (Sadovy *et al.*, 2001). Hobbyists, like fish gourmets, also favor rarity. For example, melanistic forms of anemone fishes such as *Amphiprion percula* and *Amphiprion ocellaris*, the yellow-banded form of *Premnas biaculeatus*, and the gold form of the guineafowl pufferfish, *Arothron meleagris*, are particularly sought, as are hard-to-find species such as the longnose hawkfish (*Oxycirrhites typus*). This proclivity for rarity is revealed by the higher prices quoted for less readily available species on dealers' lists. Unusual hybrids or deep-water species, for example, costing hundreds to thousands of dollars, find a specialist, albeit limited, market (Wood, 2001).

Many dealers in the aquarium trade prefer getting fishes from areas where animals are known to be caught and handled well, because they survive longer, a positive force for promoting better practices. For example, fishes from the Red Sea or central Pacific islands often command higher prices than do those from the Philippines and Indonesia, where fishers may use cyanide and often have poor shipping and handling practices (Section II,D) (Vallejo, 1997; J. Baquero, Ocean Voice International, September 29, 1999).

B. Sources and Destinations

1. Live Food Trade

Although the live food trade has only recently become large, marine fishes have been transported and maintained alive until minutes before cooking since at least the eighteenth century in Asia (Johannes and Riepen, 1995; Akimichi, 1998). Originally, most were caught in the northern sector of the South China Sea, but Indonesia and the Philippines became significant suppliers in the 1980s and the trade expanded greatly in the 1990s. Fishes sold live for food now come from much of the Indo-Pacific (Table 5; Fig. 2) (Johannes and Riepen, 1995; Chan, 2001). Trial shipments by sea have begun from the Seychelles (Bentley and Aumeeruddy, 1999) and source countries lie far into the Pacific. New, less exploited populations are continually being sought, with foreign-based businesses often setting up joint ventures in source countries (P. S. W. Chan, Chairman of the Hong Kong Chamber of Seafood Merchants, 11 November 1999). The demand for live reef fishes is spreading. A domestic trade in several reef fish species (such as the California sheephead wrasse, *Semicossyphus pulcher*) began off the coast of California in the United States in 1988, mainly for the local Chinese communities, and experimental shipments of live reef fishes for food have just begun out of Fiji to the United States west coast (Tegner and Dayton, 1997; McKee-Lewis, 1997; Y.J.S. personal observation).

A small proportion of fishes in the food trade is purportedly "cultured," but this sector also depends on wild populations, because most of these "cultured" fishes are simply wild-caught as juveniles and then grown out in floating net cages or shallow ponds until they attain a marketable size. Only Taiwan regularly exports a significant quantity of cultured food fishes that derive from broodstock (i.e., full-cycle culture including hatchery rearing), although many researchers are working to increase hatchery success and the number of species reared (Lau and Parry-Jones, 1999; Cesar *et al.*, 2000).

Fishes are culturally and gastronomically important in Asia, signifying wealth and abundance, so being able to serve expensive live fishes to your family or guests carries considerable social status. The market for live reef fishes for food is largely within Southeast Asia. About 60% of the trade goes to Hong Kong, with the rest destined mainly for Taiwan, Singapore, and Japan, and to meet a growing demand from Chinese communities in Southeast Asia, Australia, and elsewhere (Johannes and Riepen, 1995; Indrawan, 1999; Chan, 2001). Of all the reef fishes imported live to Hong Kong, over 50% may be reexported to China, a consumer market expected to expand rapidly in the future (Chan, 2001). There is also growing interest from North American companies to become involved in the lucrative Asian market by importing fishes from the central Pacific.

TABLE 5 Source and Destination Countries in the Food Trades[a]

Source countries[b]	Destination countries[b]
Australia	Australia[d]
Bangladesh	**China**[c]
Cambodia[c]	Fiji[d]
Fiji	**Hong Kong**[c]
Indonesia[c]	Indonesia[d]
Kiribati	Japan
Malaysia[c]	Malaysia[d]
Maldives	Philippines[d]
Marshall Islands	Singapore
Myanmar	**Taiwan**[c]
(Palau)	*United States of America*[d]
Papua-New Guinea	
Philippines[c]	
(Seychelles)	
Solomon Islands	
Sri Lanka	
Thailand[c]	
Vietnam[c]	

[a]Sources: Perino, 1990; Lau and Parry-Jones (1999); International Trade Subgroup (2000); Wood (2001); R. van der Elst, Oceanographic Research Institute (April 20, 1999); K. Davenport (August 25, 2000); M. Rolon, Caribbean Fishery Management Council (personal communication, November 24, 2000); A. Perry, Project Seahorse (January 21, 2000); N. Dulvy, University of Newcastle (November 7, 2000); see also references in Section II,B,2.

[b]Obvious transshipment points and nonsource countries appearing on some consignment lists (e.g., Singapore, Hong Kong, and Denmark) were not included. Food trade into the western United States is starting with experimental shipments from Fiji. Those countries that operate the highest levels of trade are indicated by bold type; those that operate a domestic market only for local fishes, with or without import, are indicated by italic type; those in which trade has been or is suspended are enclosed in parentheses.

[c]Suppliers of cultured/grown-out fishes or fish fry exports for/from mariculture. Note that hatchery rearing for food fishes occurs at commercial levels only in Taiwan.

[d]Destinations are Chinese communities.

2. Aquarium Trade

Ornamental marine fishes have been traded since the 1930s, but demand from the West increased significantly in the 1970s, as capacity to keep marine aquaria improved (Wood, 2001). About 80 countries now supply marine aquarium fishes globally (Table 6; Fig. 3). As for the food trade, the great majority of all ornamental fishes are caught in Indonesia and the Philippines, although significant trade occurs from

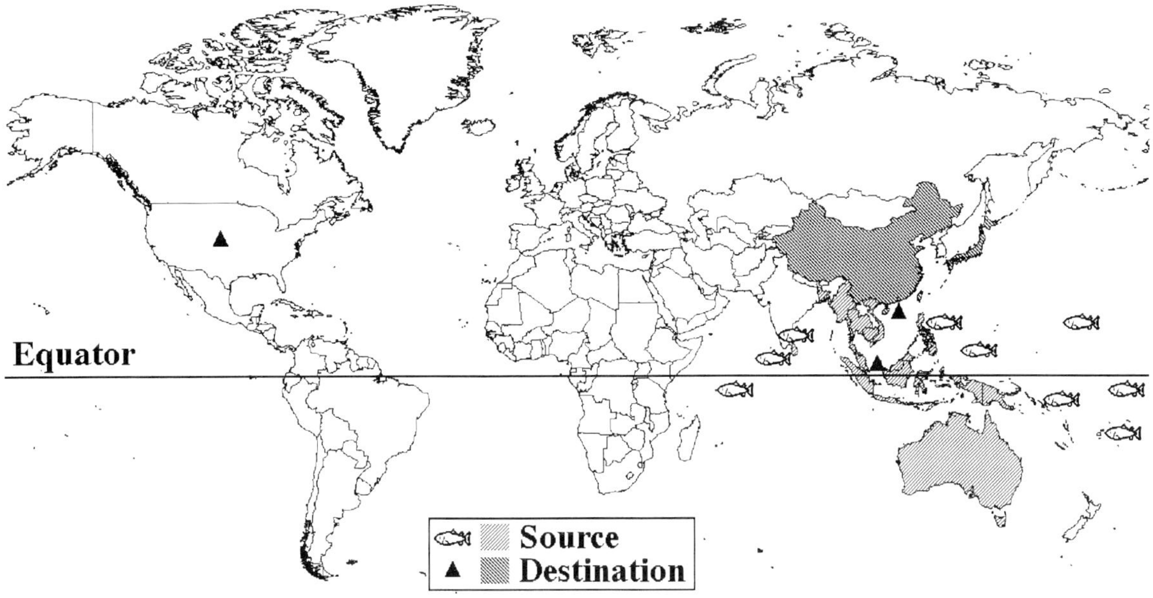

FIGURE 2 Map showing source and destination countries for international trade in live reef fish for food. Domestic use and negligible levels of trade are not shown. See Table 5 for countries and additional details on trade. The black triangle in the United States reflects a recent food trade development.

Australia, Hawaii, Florida, and the greater Caribbean (Pyle, 1993; Davenport, 1996; Barber and Pratt, 1997; Chapman and Fitz-Coy, 1997; Wood, 2001).

Virtually all marine ornamental fishes are wild-caught, with a few percent at most being hatchery reared (Warmolts, 2000; Wood, 2001). Yet many aquarium aficionados declare an interest in, or concern for, environmental issues associated with the wild capture of aquarium fishes. They often look favorably on captive rearing and indicate that they are prepared to pay higher prices if the culture (as opposed to wild) source can be assured; about 20 species of aquarium fishes, including several anemone fishes, dottybacks (Pseudochromidae), and some gobies can be reliably hatchery reared at commercial levels (J. Tullock, August 12, 1999). Although their colors are often not as vibrant as wild-caught fish, an important market-acceptance issue, they are often easier to maintain in captivity.

Globally, the hobby trade consumes vastly more ornamental fishes than do public facilities, with 99% of imported marine fishes and invertebrates purchased by hobbyists (Warmolts, 2000). It is a highly popular hobby in Europe and the United States. In Britain, marine fish keepers number around 100,000, and there are about 500,000–700,000 keepers in the United States, with approximately 800,000 in the European Union (P. Holthus, Marine Aquarium Council, 2000; R. Sankey, founder of Tropical Marine Centre, UK, July 4, 2000; K. Davenport, Ornamental Aquatic Trade Association, August 25, 2000 and December 7, 2000). The major demand for aquarium fishes for hobbyists and public aquaria comes from the United States and western Europe, although the Asian market is rapidly expanding. Keeping ornamental fishes is one of the most popular hobbies in North America and 4–10% of fishes traded are marine (Biffar, 1997; Chapman and Fitz-Coy, 1997). Europe has about 80 public aquaria with at least 5 new major aquaria under construction, and 10–20 more in the planning stage. In the United States, the American Zoo and Aquarium Association (AZA) has 34 member institutions with some aquarium component, and these received 39 million visitors in 1998 (AZA, 2000). In Japan alone, there are now more than 250 large public aquaria, and 12 major public and private aquaria exist in China (thanks to an explosion of aquarium construction there since 1997), with smaller ones in most Chinese provinces (H. Hall, Zoological Society of London, February 26, 2000; Liu Min, Xiamen Underwater World Aquarium, China, January 31, 2000). It is worrisome that recent poor economic performance has resulted in poorly planned closures and abandonments of some of the Chinese aquaria (South China Morning Post, April 10, 2000). In Australia, the major public aquaria host a total of more than 3 million visitors annually [Queensland Fisheries Management Authority (QFMA), 1999].

TABLE 6 Source and Destination Countries in the Aquarium Trade[a]

Source countries[b]	Source countries[b]	Destination countries[b]
American Samoa	Mexico	Australia
Argentina	(Mozambique)	**Belgium**
Australia	Myanmar	Canada
Bahrain	Netherlands Antilles	China
Barbados	New Zealand	Denmark
Bangladesh	Nigeria	**France**
Belize	Norway	**Germany**
Brazil	Oman	Guyana
Brunei Darussalem	Palau	Hong Kong
Cambodia	**Philippines**	**Italy**
Cape Verde	Papua New Guinea	**Japan**
Cocos Keeling Island	Poland	*Malaysia*
Cook Islands	Portugal	Mexico
Columbia	Puerto Rico, U.S.A.	**Netherlands**
Costa Rica	Qatar	Pakistan
Cyprus	Republic of Korea	Singapore
(Djibouti)	Samoa	*South Africa*
Dominican Republic	Saudi Arabia	Soviet block[c]
(Egypt)	Seychelles	Taiwan
(Eritrea)	Solomon Islands	**United Kingdom**[d]
Fiji	Spain	**United States of America**[d]
Finland	Sri Lanka	
Ghana	Sweden	
Greece	Switzerland	
Guadaloupe	Syria	
Guam, U.S.A.	Taiwan	
Guyana	Tanzania	
Haiti	Thailand	
Hungary	Tonga	
India	Trinidad	
Indonesia	United Arab Emirates	
Japan	**United States of America** (especially Hawaii and Florida)	
Kiribati	Vanuatu	
Kenya	Venezuela	
Kuwait	Vietnam	
Line Islands	Yemen	
Maldives	Zaire	
Marshall Islands		
Mauritius		

[a]Sources: Perino, 1990; Lau and Parry-Jones (1999); International Trade Subgroup, (2000); Wood, (2001); R. van der Elst, Oceanographic Research Institute, (April 20, 1999); K. Davenport, (August 25, 2000); M. Rolon, Caribbean Fishery Management Council (November 24, 2000); A. Perry, Project Seahorse, (January 21, 2000); N. Dulvy, University of Newcastle, (November 7, 2000); See also references in Section II,B,2.

[b]Obvious transshipment points and nonsource countries appearing on some consignment lists (e.g., Singapore, Hong Kong and Denmark) were not included. Those countries that operate the highest levels of trade are indicated by bold type; those that operate a domestic market only for local fishes, with or without import, are indicated by italic type, those in which trade has been or is suspended are enclosed in parentheses.

[c]Countries included within the former Soviet block are unspecified.

[d]Suppliers of cultured fishes.

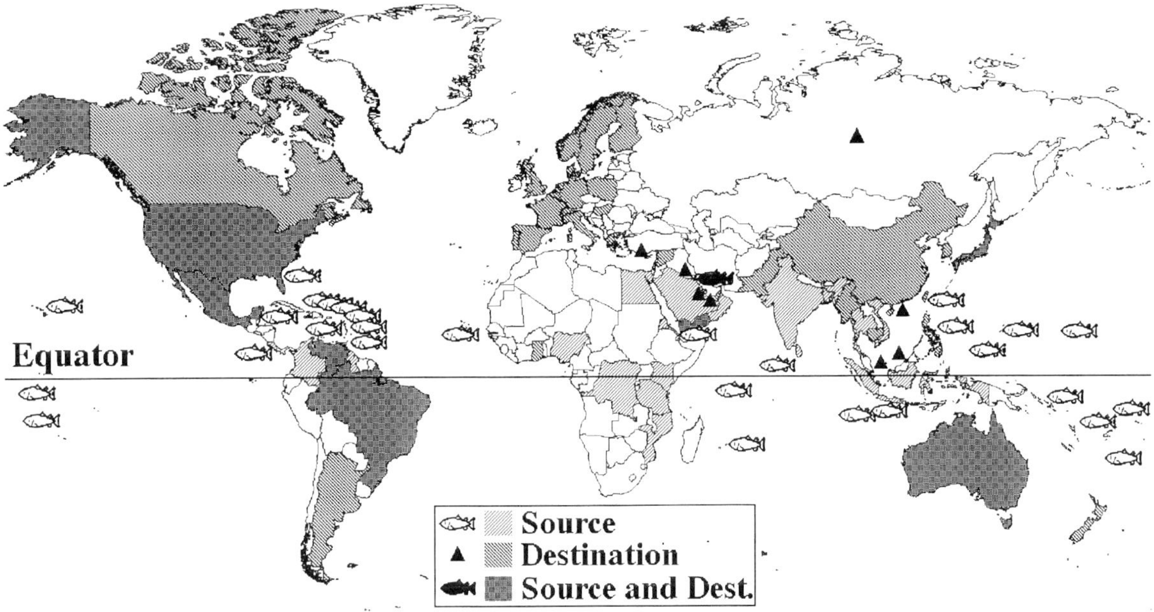

FIGURE 3 Map showing source and destination countries for international trade in live reef fishes for marine aquaria. Domestic use and negligible levels of trade are not shown. See Table 6 for countries and additional details on trade. The black triangle in the former Soviet block indicates that the specific countries involved could not be determined.

C. Trade Volumes and Values

1. Monitoring (Food and Aquarium Trades)

The quality of data on levels of capture and fishing effort in countries of origin (sources) is generally poor, or unreliable, and that on international trade in live reef fishes is variable. High rates of mortality under some conditions of capture, holding, and transport, prior to export (e.g., Johannes and Lam, 1999), mean that export and import records underestimate, sometimes markedly, the number of animals actually removed. Capture mortalities may or may not add to pressures on wild populations of reef fishes, depending on whether these dead fishes would have been numerous enough to reduce demand on wild populations (Section II,D,2). On the import side, for the food trades, imports are typically not inspected and data collection is invariably incomplete. Even the major food trade importer, Hong Kong, does not require locally registered vessels, key importers, to declare live imported fishes, relying on informal and unstandardized data collection methods to gauge this sector of the trade (Sadovy, 1998). Mortality and capture rates are rarely recorded in the aquarium trades, although imports to the United States and Europe must at least pass through regulated designated ports of entry such that data on numbers of consignments, species, boxes, and fishes, as well as source/consignment country, should be available (K. Davenport, August 25, 2000); in reality, however, data remain confused and inconclusive for the aquarium trades too.

2. Live Food Trade

Although trade in live marine fishes has long been practiced in Asia, the large food trade that we see today emerged only within the past decade, with Hong Kong the major demand center. Fishers started visiting reefs within a few hundred kilometers of Hong Kong in the late 1960s, bringing back small catches of exotic reef fishes. In the 1970s and 1980s, many fishing boats and businesses moved further afield and grew, with the number of importing and exporting countries expanding. Hong Kong was importing about 1000–2000 metric tonnes (1 mt = 1000 kg) per annum in the late 1980s and early 1990s, which had increased to 30,000 to 35,000 mt by the late 1990s (Johannes and Riepen, 1995; Lau and Parry-Jones, 1999; Chan, 2001). If Hong Kong imports still account for about 60% of the total import trade in the region, as it did in the mid-1990s (Johannes and Riepen, 1995), the global annual volume of live reef fishes traded for food is currently in the order of 50,000 mt. At a mean of approximately 1 kg per fish, about 50 million fishes of just a few species are being sold annually. However, approximately double that number is caught if we factor in the likely mortality incurred between capture and consumer (Section II,D,2) bringing the estimated total to 100,000 mt. This trade is clearly significant, with the estimated volume comparable to the Food and Agriculture Organization (FAO) estimate in the mid-1990s of 98,000 mt for conventional, dead, grouper landings for all Asian economies combined (FAO, 1996).

Live reef fishes are very valuable when marketed for food. The average wholesale live price was U.S. $17–22 per kg in the late 1990s, whereas dead fishes fetched less than one-tenth of that value in Hong Kong over the same period [Hong Kong Special Administrative Region Agriculture and Fisheries Department (HKSAR-AFCD), unpublished data; Chan, 2001]. Total wholesale value of the trade was about U.S. $500 million in Hong Kong alone in 1998, giving a global trade worth about U.S. $830 million wholesale, again, assuming that Hong Kong accounts for 60% of the world trade. Prices range markedly depending on the species, body size, time of year, local festivals, volume available on the market, and the general state of the economy, because eating live reef fishes is a luxury and an important status symbol (Johannes and Riepen, 1995; Lau and Parry-Jones, 1999) (Table 3). Prices can reach extremes, with one massive giant grouper fetching U.S. $10,000 [South China Morning Post, 1996 (04/14/96)] and humphead wrasse at times exceeding U.S. $150/kg at retail. Cultured fishes in the food trade are typically up to 60% cheaper at retail compared with wild-caught conspecifics (Li, 1999), because of the perceived intrinsic value of wild-caught animals, even though tastes are evidently indistinguishable by all but the most discerning gourmet (OmniTrak Group, 1997; W. W. Chan, 2000; Sadovy, 2001a).

3. Aquarium Trade

The fishery for marine ornamental fishes is thought to have begun in the 1930s in Sri Lanka, with considerable expansion in the 1970s and trade somewhat static in the 1990s (Wood, 2001). Exports began from the Philippines, still a key source country, in the 1950s (McAllister *et al.*, 1999). Despite the poor quality of available trade data, estimates suggest that from 15 to 36 million marine fishes are traded annually (Wheeler, 1996; Wood, 2001; K. Davenport, August 25, 2000); actual captures are much higher when mortalities are factored in (see Section II,D,2). Such numbers are not unlikely. In 1995, for example, the small Australian aquarium trade on the Great Barrier Reef had about 60 licensed fishers who reported about 250,000 fishes (about 4000 fishes each) while the ~50 aquarium trade fishers in west Hawaii reported 422,823 fishes (over 8000 fishes each), annually. The Philippines alone had an estimated 2500 fish collectors who, even at Australian collection rates, could take at least 10 million fish a year, before mortality is factored in (Pajaro, 1993; QFMA, 1999; Tissot, 1999).

The marine aquarium trade includes some very high-value species, with retail prices varying from a few to thousands of dollars (Table 4). This trade may well exceed the freshwater ornamental fish trade in unit value. Although making up <10% of the entire ornamental fish trade in volume (marine and freshwater species combined), marine species comprise 10–20% of the value (Biffar, 1997; Chapman and Fitz-Coy, 1997; Young, 1997; Warmolts, 2000). In general, estimates of trade are rather unreliable and calculations of the annual global import value of the aquarium trade in marine fishes vary greatly, from U.S. $24–40 million both in 1985 (Wood, 1985) and in the late 1990s (Wood, 2001) for fishes plus invertebrates, to U.S. $250 million in the late 1980s (Barratt and Medley, 1990). Other estimates place the global retail value of the trade in the 1990s at anywhere between U.S. $90 and $300 million (Wheeler, 1996; Biffar, 1997; Warmolts, 2000). In contrast to the food trade, aquarium products (tanks, hardware, chemicals, etc.) form a very important part of the trade and are worth far more than the fishes. For example, in the United States alone, aquarium products were valued to be worth at least U.S. $300 million a year at retail (Baquero, 1999).

Prices vary with consumer preference (e.g., fish sex and size), species availability and ease of shipping, as well as market forces and source (Section II,A,2) (Perino, 1990; J. Tullock, September 29, 1999; L. Squire Jr., Cairns Marine, June 20, 2000). For example, males of the brightly colored bird wrasse, *Gomphosus varius,* fetch double the price of the duller and more common female of comparable size, and the orange-tailed male blue devil, *Chrysiptera cyanea,* sells at about twice the price of the duller female (J. Tullock, December 6, 1999). Small clown triggers, *Balistoides conspicillum,* are worth twice as much as those that are extra large. Increasing rarity often means higher prices (Wood, 2001). As examples, individual angelfishes of two rare species, the yellow-faced angelfish, *Pomacanthus (Euxiphopops) xanthometapon,* and the blue-girdled angelfish, *Pomacanthus (Euxiphopops) navarchus,* exceptionally fetch hundreds of dollars in the United States market. Deep-reef species are also highly valued. For example, Tinker's butterfly fish, *Chaetodon tinkeri,* can sell for U.S. $1000 per pair, and the swissguard, *Liopropoma rubre,* and blackcap, *Gramma melacara,* basslets retail at about three times that of the similar but shallow-dwelling royal gramma, *Gramma loreto.* The peppermint angel (*Centropyge boylei*) may fetch as much as U.S. $10,000 in some markets because of its scarcity. Whether such species are truly rare or supply is being kept low to keep prices high is not always easy to determine. However, they are likely to be expensive if they are difficult, or dangerous, to obtain, or available from only a few divers. Cultured fishes will find a small market, especially among conservation-minded aquarists. Unlike the case for food fishes, however, cultured aquarium fishes may be twice as expensive as their wild-caught conspecifics. Industry sources suggest that the aquarium trade will remain reliant on wild fishes for some time, given the economic realities of culturing and the relatively few species currently suitable for culture (K. Davenport, August 25, 2000).

D. Capture Methods and Mortality

1. Capture Methods

Live food and aquarium fishes share a number of capture methods, so will be considered together in this section. These methods range from hook and line, traps, and nets, to artificial reefs and chemicals, such as quinaldine, bleach, tobacco, formalin, and cyanide. Often some form of air supply, either self-contained underwater breathing apparatus (SCUBA) or hookah (surface-supplied compressed air), is used. Each type of gear and each method of handling and shipping incur mortality for the target species, ranging from negligible to high, and some gears and methods may take considerable by-catch of nontarget species and/or damage habitats (Table 7) (see also Wood, 2001).

Cyanide is a particularly controversial fish capture method associated with the live fish trades. Because of its infamy as a fishing gear—it is highly toxic—it has attracted particular interest as an unusual and relatively new problem in management of both the food and the aquarium trades and merits a more detailed discussion. Typically, cyanide tablets are dissolved in a bottle of seawater and then squirted at the target fish, stunning it by affecting the oxygen uptake mechanism of the fish's gills. The fish is then easy to retrieve by hand and it is removed to clean water for a rapid recovery. Certain species are more readily, or are exclusively, caught with cyanide in some areas, such as the humphead wrasse for food, and secretive, inaccessible, or prime value angelfishes in the aquarium trade (Vallejo, 1997; Wood, 2001; J. Tullock, January 12, 2000).

Despite the considerable attention recently focused on the use of cyanide as a capture method (Johannes and Riepen, 1995; Barber and Pratt, 1997), little research has yet been done on the long-term and nontarget effects of this or other fishing methods on reef communities (but see Section IV,C). Controlled studies have shown that cyanide can adversely affect fishes and corals. Indeed, traders of marine ornamentals prefer those advertised as "net-caught" (the veracity of which can be difficult to confirm) because of their better survivorship postcapture (Section II,D,2) (van Poollen and Obara, 1984; Wood, 1985; Rubec, 1988; Johannes and Riepen, 1995; Jones and Steven, 1997; Vallejo, 1997; Bentley, 1999). Research indicates that cyanide use is widespread in Southeast Asia, the area central to both live fish trades, despite being illegal in most countries (Johannes and Riepen, 1995; International Trade Subgroup, 2000). In Indonesia, 90% of live fish vessels in eastern islands had cyanide on board in the mid-1990s, and the large-scale poison fishery in Indonesia was calculated to produce a net quantifiable loss to the country of U.S. $46 million over a 4-year period (Cesar *et al.*, 1997). In the Philippines, approximately 70% of ornamental fishes and significant numbers of food fishes are caught with cyanide (Hingco and Rivera, 1991; Barber and Pratt, 1997; McManus *et al.*, 1997; Bentley, 1999). In 1996, random analyses by cyanide detection laboratories set up by the International Marinelife Alliance (IMA) in the Philippines found that 19% of food and aquarium fishes, collectively, tested positive for cyanide shortly after capture (Barber and Pratt, 1997). This figure is likely to be an underestimate because cyanide levels rapidly become undetectable following capture (e.g., Barber and Pratt, 1997; McAllister *et al.*, 1999). Only the Philippines has active nongovernmental organization programs to eliminate the use of cyanide (e.g., Haribon Foundation and IMA). However, the proposed introduction of a certification system for the aquarium trade being developed by the Marine Aquarium Council (MAC) should stimulate wider interest and concern in capture methods and develop the means whereby good capture practices can be encouraged and rewarded (Holthus, 1999). It may also help preclude the spread of cyanide fishing into the Pacific islands where, along with other prime aquarium trade collection areas such as Australia, Sri Lanka, and the United States, it is not yet found.

2. Mortality

Estimates indicate that cumulative fish mortality from capture to consumer can be high and can produce significant wastage in both the food and aquarium trades, but that mortality levels can be lessened with training and experience. Mortality levels in both trades range from a few percent to 80% or more for cyanide-caught fishes and/or where poor capture, handling, and maintenance practices produce stressed animals, although the source of mortality is not always clear (Hall and Bellwood, 1995; Hanawa *et al.*, 1998). Mortality depends on the capture method, the experience of the fishers or mariculturists, the expertise of the

TABLE 7 Capture Methods and Impacts

Gear type	Fishery	Function and selectivity	Mortality at time of capture	By-catch	Source
Hand operation	Aquarium	Selective: microspear, slurp gun, iron bar; hand gathering, sometimes after cyaniding	Not recorded	None known	Vincent, (1996); QFMA (1999); Y. Sadovy (personal observation)
Hook and line	Aquarium and live food	Somewhat selective: juvenile/adult, barbed and barbless hooks	Variable, from 4 to 60%; depends on experience; problems with foul-hooking, and degassing if taken from deeper water	Nontarget species taken; some sold frozen locally, exported, or consumed	Johannes and Riepen (1995); Bentley (1999)
Trap	Live food	Nonselective: "bubu" traps for adults, small traps for juveniles	Variable, from 10 to 20%; depends on handling; less if bait is cyanide laced to calm fish	Nontarget species taken; some sold or consumed; some damage to coral	Bentley (1999); Johannes and Ogburn (1999); Erdmann and Pet-Soede (1998)
Nets	Aquarium and live food	Somewhat selective: hand nets for small aquarium fishes, and humphead wrasses at night; barrier/stop/gill/drop nets; bag/fyke nets take post larvae	Variable, but appears to be relatively low at 1–2% except mortality can be high for postlarval fishes taken by bag nets for mariculture grow-out	Can be high if bag nets used, otherwise not high	van Poollen and Obara (1984); Johannes and Ogburn (1999); Sadovy (2000); L. Squire, Jr. (June 6, 2000)
Artificial reef	Live food	Nonselective: stones, twigs, branches	Unknown	Can be high: one example 99% of catch	Mous *et al.*, (1999); Johannes and Ogburn (1999)
Chemical	Aquarium and live food	Can be selective: cyanide dissolved in water and squirted, or laced in bait; also quinaldine, formalin, clove oil, detergent, pesticides, tobacco, bleach, *Derris* sp.	Cyanide: can be high at 10–60%, often difficult to separate from effects of other impacts because effects often delayed by months, but immediate mortalities can be as much as 25%. Other chemicals: effects unknown	Cyanide: Nontarget species that do not escape effects; habitat damage inferred because high doses can kill coral experimentally. Effects of other chemicals unknown	van Poollen and Obara (1984); Johannes and Riepen (1995); Barber and Pratt (1997); Bentley (1999); Jones and Steven (1997); Sadovy (1992); Pyle (1993); Perino (1990)

traders, the quality of the facilities and care (water quality, feed, stocking densities), as well as the biological characteristics of individual species (e.g., van Poollen and Obara, 1984; Johannes and Riepen, 1995; Bentley, 1999). Training in appropriate handling can reduce such mortalities. In food trade operations, capture and in-transit mortalities may also be high, likely on the order of 50%, overall (Johannes and Riepen, 1995; Bentley, 1999; Johannes and Lam, 1999). In the aquarium trade, maintenance conditions following capture may also be poor. For example, in the Philippines mortality can reach 20% within days of capture because fishes may be held, until collection by middlemen, in small plastic bags with infrequent water changes (Baquero, 1995; J. Baquero, September 29, 1999). This situation is in marked contrast to areas, such as Hawaii or Australia, where the aquarium trade collector is often also the exporter, businesses are small, and experienced operators and good practice minimize overall mortality levels.

High mortality in the aquarium trade can directly increase pressures on wild populations if shops or hobbyists then purchase replacement animals. A survey by the IMA in the United States showed that overall mortality in 300 aquarium shops ranged from 30 to 60%, with only about 20% mortality in shops with careful practices (J. Tullock, October 1, 1999). Much depends on the source and prior treatment of the fishes (M. Lidster, Aquatic Technologies, Inc., January 20, 2000). With poor management, 75% of even hardy species, such as the flame angelfish, *Centropyge loriculus,* and other dwarf angelfishes, died within 6 weeks of retail sale (Pyle, 1993; M. Lidster, January 20, 2000). Yet some species appear to be generally unsuitable to aquaria, dying even with best practices: a survey of marinelife dealers by the American Marinelife Dealers' Association (AMDA) indicated that about 10% of species sold as ornamental fishes are almost impossible to maintain, and another 30% need advanced skill (Fig. 4). Members of the Ornamental Aquatic Trade Association (OATA) in the United Kingdom have also long avoided trading in species listed as impossible to keep, such as obligate coral feeders (R. Sankey, personal communication, July 4, 2000; K. Davenport, August 25, 2000).

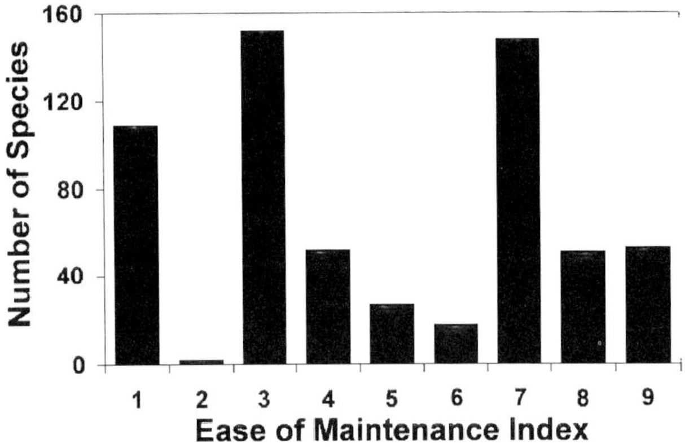

FIGURE 4 Classifications of ease of maintenance, from 1 to 9 (where 1 is easy to keep, and 7 and above require expert knowledge), for 612 marine ornamental species imported into the United States. Classifications were adapted from the American Marinelife Dealers' Association Ecolist, which classifies species for the benefit of retailers and hobbyists. The higher the index, the harder the species is to maintain in captivity.

E. Trade Structure

In sheer volume, most fishes destined for these trades are caught in Southeast Asia by subsistence fishers, usually men but also women and children (e.g., Akimichi, 1998; A.C.J.V. personal observation). Primary buyers commonly have crude fish-holding facilities (perhaps oxygen and a few net cages to hold aquarium fishes or to grow food fishes to market size), and typically sell to secondary buyers/consolidators through one or more middlemen (e.g., Perino, 1990). In such a system, exporters may even promote destructive fishing techniques, either directly by supplying gear or chemicals, or indirectly by urging unrealistically large catches in a short period of time. They have reportedly, for example, been highly influential in the expansion of cyanide use combined with hookah and SCUBA dive gear (Johannes and Riepen, 1995; Barber and Pratt, 1997). In addition, buyers' high seasonal demand for small juveniles (destined for mariculture grow-out and often referred to as "fry" or "fingerlings") prompts the capture of hundreds of millions of grouper fry in the Philippines, Indonesia, and Thailand, using gears that often cause high mortalities (Sadovy, 2000). A lack of accountability in many of these export fisheries makes management particularly difficult.

In the Pacific islands and elsewhere outside of Southeast Asia, such as the Caribbean, Hawaii and Florida, practices may be better, at least in the aquarium trade. The trade structure may be simpler, with collectors often being the exporters, and businesses are typically small. In Australia, 90% of aquarium trade collectors are wholesalers (Couchman and Beumer, 1992). Fishers have often chosen their livelihood and are unencumbered by many of the problems found in the trades in Southeast Asia. This also means that they are better able to respond to management initiatives and are well placed to participate actively in protecting the resources on which they depend.

F. Summary

The live reef fish trades for food and ornamentals are highly valuable international trade operations that rely on a range of species from a relatively small number of families. Such trade supplies luxury markets around the world and may involve damaging fishing techniques. Captures are largely unrecorded and estimates of trade volume often unreliable and incomplete. The international nature of these trades, the many countries and species involved, the lack of reliable capture data, the incomplete monitoring, and the high mortalities known to occur in some cases mean that any estimates of trade volume are generally poor indicators of actual production (i.e., number captured). Clearly, guidelines and practices for capturing and shipping live animals need to be implemented and refined to demand greater responsibility and accountability at all levels of trade (e.g., Olin, 1997; Rollin, 1997). We do, however, already know enough to consider probable impacts of the fisheries and trades in live reef fishes.

III. Theoretical Considerations

Live reef fisheries can be examined within the same conceptual framework as conventional fisheries, but a combination of features contributes to their overall vulnerability: (1) opportunity costs (see below) to the fishers are low, (2) the value of individual fishes often increases with rarity, and (3) the fisheries are highly selective for certain sizes and/or species (also see Table 1). Incorporating these three factors into a classical fisheries approach—first the two economic elements and then the biological issue of selectivity—shows that there is a very real risk for live reef fisheries to overexploit and potentially threaten targeted fish populations. We will subsequently evaluate the evidence for such biological impacts.

A. Economic Issues

1. Background

As a fishery develops, according to the classic Schaefer yield curve (Schaefer, 1954), an increase in fishing effort produces an increased biomass yield (Fig. 5a). The increasing catch will stimulate more capital investment, or more people to enter the fishery.

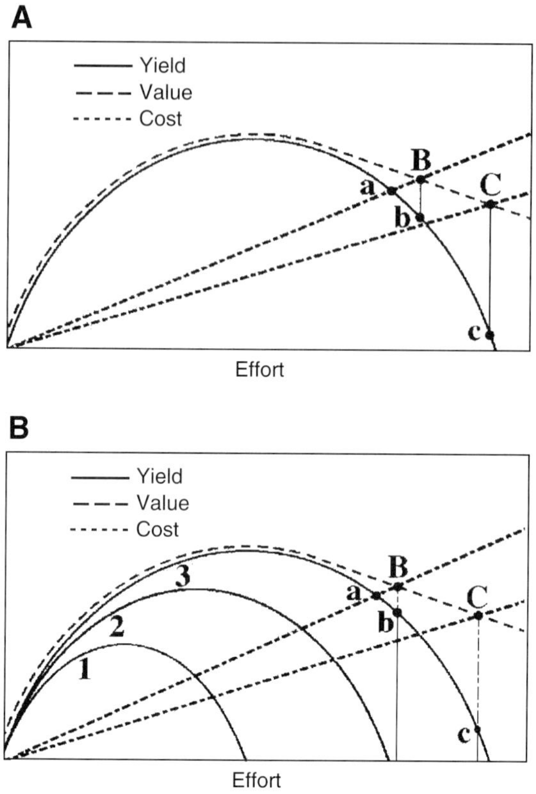

FIGURE 5 (a) Curve of biomass yield(—)with increasing effort based on the Schaefer yield curve. Fishing is no longer economically viable where the cost line (------) intersects the total value curve (–––) Usually the total value curve is assumed to track the biomass yield curve closely, and we would expect the upper cost line to intersect the total value curve at effort point a, where fishing would no longer be economically viable. In fisheries where value per fish increases with rarity, however, the total value curve declines at a slower rate than the yield curve and diverges from it. Consequently, the upper cost line intersects the total value curve at point B, which maps onto effort point b as the economic equilibrium point (to the right of point a along the yield curve). In fisheries where costs are pushed down (e.g., low opportunity costs), then the new cost line might, for example, intersect the total value curve at point C, which maps onto effort point c, where the fishery is severely overexploited. (b) Curves of biomass yield(—), total value (– – –), and cost lines (------) with increasing effort, as in (a). In this case, however, the yield/effort curves of three species (1, 2, 3) are represented. The effort at which fishing stops will depend on which species is driving the system. A fishery based on species 3, with the upper cost line, would undoubtedly overfish all species, but effectively eliminate species 1 and 2.

Fishing effort will further increase and catch per unit of effort (CPUE) will start to decline.

If effort continues to increase, then recruitment and growth cannot keep pace, the number of adults will decline, biological overfishing will occur, CPUE and overall yield will decline, and the population will eventually collapse. Fisheries operating on the descending, right-hand, limb of the curve are in decline. In theory, it will no longer be economically viable to continue fishing when CPUE has declined greatly (unless the fishery is subsidized) and the fishery will stop before the population collapses (although collapse may still follow later because of latent problems arising from overfishing). However, biomass yield and value will become decoupled and diverge if (1) other economic opportunities are poor or few and (2) the economic benefit per animal increases as the target species becomes increasingly rare. The point at which fishing stops once biological overfishing occurs, therefore, depends largely on the economics of the fishery.

2. Low Opportunity Cost

For a fishery to be economically viable, benefits must outweigh costs. The economic breakeven point is where the cost–benefit line (for example B or C in Fig. 5a) crosses the biomass yield curve and costs equal benefits. The position of this cost–benefit line in a given fishery depends, in part, on the economic return to the fisher. Theoretically, fishers will continue fishing only while this occupation brings greater rewards than are offered by working elsewhere.

Opportunity costs (the rewards to be gained from working or investing elsewhere) are typically low in live reef fisheries because (1) fishers have few other income-earning opportunities and (2) more unusually for small-scale reef fisheries, and regardless of the lack of other opportunities, the products are very valuable. The subsistence fishers who supply the greater volume of the live reef trades in Southeast Asia generally lack land, capital, or labor skills that they can market. Indeed, fishing is commonly the occupation of last resort or the only option (Pauly, 1997). With so few opportunities for work outside the fishery or in other fisheries (due to declining resource bases), low levels of high-value catch still return a profit even if the costs of fishing are high, and the potential for extreme biological overfishing is great. Indebtedness to buyers, consequent financial entanglements, and/or investment in skill development and fishing equipment all make it difficult for fishers to extricate themselves from the fishery, or to readily make a better living elsewhere. Such relationship with a particular buyer often traps the fisher into effective bondage that dictates his or her fishing practices (Akimichi, 1998; Bentley, 1999; J. Baquero, February 21, 1999; M. Pajaro, National Coordinator, Project Seahorse/Haribon Foundation, July 5, 2000). Fishers also keep catching live reef fishes because the high values per kilogram and per fish keep returns high compared with possible earnings in other sectors, or in other forms of fishing with the same scale of capital investment (Bentley, 1999). In 1996, for example, live food fishers in Indonesia earned U.S. $150–500 per month, 3 to 30 times the average monthly salary of artisanal fishers and 1 to 3 times that of a university lecturer (Erdmann and Pet-Soede, 1998). A cyanide fisher's net return is believed to greatly exceed that of an average fisher in the Philippines (Añonuevo, 1989).

3. Increasing Value with Increasing Rarity

If, as in the live reef fisheries, the value per unit of catch increases with rarity, then continued fishing will remain profitable even at high effort and low catch (i.e., low CPUE on the descending limb of the Schaefer yield curve). As a particular target fish declines in availability, its increase in value uncouples value from biomass yield, thereby maintaining a profitable fishery (see revenue curve, Fig. 5a, dashed line). This provides an incentive to continue fishing (to points B and C) even as catches decline. This could allow the fishery to persist down to very low population levels (e.g., point c in Fig. 5a). Although this phenomenon is known from other fisheries, it is more extreme in the live fish trades. One example of particularly high worth in the food trade is the rare and highly esteemed giant grouper (Section II,C,2); by selling at up to U.S. $10,000 per fish, this species provides tremendous incentive for impoverished fishers to pursue individual fishes even at very low densities. Similarly, rare angelfishes can fetch hundreds or thousands of dollars (U.S.) in the aquarium trade (Section II,C,3) and an official protected-species status may yield particularly high prices to the few licensed to trade in them (e.g., seadragons, *Phycodurus eques* and *Phyllopteryx taeniolatus;* L. Squire, Jr., June 20, 2000). An important distinction, however, must be made here between true biological rarity and imposed scarcity due to market forces (Section II,C,3). The general issue of rarity and risk of extinction in marine fishes will be addressed in Section IV.

4. Exacerbating Conditions

The potential threats posed by low opportunity costs and increasing value with rarity become more marked if the population suffers a depensatory response when its numbers decline, or if the fishery takes many species at once (e.g., Sissenwine *et al.,* 1988).

For example, the population may collapse when male biomass becomes too low in a heavily fished protogynous species, even if female biomass remains high, or if density becomes too low to find a mate or avoid predation (Knowlton, 1992; Huntsman *et al.*, 1995). In multispecies fisheries, on the other hand, such as the aquarium fishery, pressure will be driven largely by the most available species, or by that with highest yield for effort (species 3 in Fig. 5b). As long as it makes economic sense for fishers to pursue the high-return species, they will also take low-return species as encountered, even if these alone could no longer sustain a fishery (i.e., species 1 and 2 with lower yield per effort curves in Fig. 5b). Although many of these conditions are common to more conventional fisheries (Clark, 1973), live food fish fisheries represent extreme cases of low opportunity costs and increasing value with rarity. Such factors can drive the number of individuals in targeted populations to extremely low levels, and set the stage for extirpation through a combination of environmental, demographic effects and/or genetic stochasticity. Threats to wild populations could be further exacerbated by size selectivity, the third distinguishing feature of these fisheries, which we now explore.

B. Biological Issues

1. High Selectivity

The live reef fisheries tend to concentrate on juveniles (for grow-out as live food, plate-sized in large species, or for aquarium display), on intermediate-size fish (because they fit plates or home aquaria), and on breeders (because they are easy to catch in spawning aggregations, or are attractive). These preferences are in contrast to the general tendency for conventional fisheries to target the largest individuals available [Plan Development Team (PDT), 1990], and need to be evaluated for possible risks to exploited populations. It is also possible that the high degree of selectivity in the live reef fish trades confers certain refuges by size, sex, or color, and that shifting preferences may rotate pressure off certain species or fishing locations.

Focusing attention on smaller animals may or may not be costly in conservation terms. Management practices and legislation usually seek to avoid catching fishes that still have high growth potential (primarily juveniles), as a means of avoiding growth overfishing, but management is also beginning to consider protecting the largest and most fecund individuals. If size/age-specific mortality is very high in young (small) animals, as is assumed for marine fishes, and if the young are taken from high-mortality cohorts, then the effects of a fishery may be inconsequential (Roberts, 1996). Removal of fishes from older, lower (natural) mortality cohorts will have a much more significant impact on the population, however (Sadovy and Pet, 1998). The difficulty, of course, lies in determining size- or age-specific mortalities, especially in the early postsettlement phase when these are hypothesized to be highest. Research completed to date suggests that natural mortality levels of several reef fishes drop substantially within a few weeks or months postsettlement, indicating that, thereafter, most individuals are likely to survive to adulthood (Doherty and Williams, 1988; Sale and Ferrell, 1988; Doherty *et al.*, 1994b; Koenig and Colin, 1999). Juveniles are particularly targeted in the food trade in the case of larger species (e.g., humphead wrasse and brown-marbled grouper), when it is the juveniles that are plate-sized (Fig. 1), and as fingerlings for mariculture grow-out for a range of species. In the aquarium trade juveniles are preferred when they are more attractive or easy to maintain in captivity. In Hong Kong, for example, 56% of 12,652 fishes ready for retail sale were estimated to be juveniles (Chan and Sadovy, 1998).

The effects of targeting intermediate-sized individuals or breeders, particularly those in spawning aggregations, depend on species-level, size-specific, fecundity and survival, rates of removal, and the effects of fishing. If fecundity is relatively much higher in larger (older) animals, as is usual in marine fishes, then selecting medium-sized individuals may reduce pressure on the population, especially on larger individuals, unless more fishes need be taken in compensation. Certainly, older animals commonly carry a range of benefits in conservation terms because they are known breeders, may be more heterozygous (Smith *et al.*, 1991), and clearly have survival capabilities under current conditions (which may or may not be heritable). If, on the other hand, a high proportion of breeding fishes is disrupted or removed from spawning aggregations, the reproductive output of targeted populations may rapidly be compromised (Section IV,A,2).

C. Summary

The general tendency, or need, for fishers to stay in the live reef fish trades and the enhancement of value with rarity have the potential to push biological overfishing far beyond levels predicted in more conventional fisheries. The effects of the preferences of the live reef fisheries and their selectivity for juveniles, medium-sized and breeding fishes are difficult to gauge but could be problematic if they compromise spawning stock. The possible consequences of juvenile-focused fisheries on the persistence of local populations have yet to receive serious attention. What evidence do we have, therefore, as to the nature and extent of impact of the food and aquarium trades on reef fish communities and their habitat?

IV. Biological Impacts

A population's response to exploitation will depend on its biological characteristics and how it is affected by larger disruption of its community and ecosystem (e.g., Russ, 1991; Roberts, 1995; Jennings *et al.*, 1999). Many exploited reef fishes have biological traits that increase risk of endangerment in fishes, including life history stages with specialized needs, complex life histories, highly structured social behavior, low reproductive rates, long life, slow growth and low natural mortality, limited larval dispersal or geographic range, site fidelity, migration, and spawning aggregations (Angemeier, 1995; Coleman *et al.*, 1996; Jennings *et al.*, 1999; Roberts and Hawkins, 1999; Sadovy, 1996, 1999; Hawkins *et al.*, 2000). Even smaller reef species, such as damselfishes and surgeonfishes, may live for 20 and 70 years, respectively (Doherty and Fowler, 1994a; Choat and Axe, 1996), and larger reef species can also attain several decades. Certain life history characteristics, such as spawning aggregations, represent significant bottlenecks for many commercially exploited reef fishes, and hermaphroditic sexual patterns may produce heavily biased sex ratios under exploitation, with the possibility of reproductive failure (Huntsman *et al.*, 1995; Vincent and Sadovy, 1998).

Exploitation must be carefully judged in the case of species exhibiting high degrees of endemism, such as the Banggai cardinal fish, *Pterapogon kauderni,* and the resplendent pygmy angelfish, *Centropyge resplendens,* or those with highly specific habitat needs, such as the seadragons, or those that are naturally rare. It is important, however, to distinguish true rarity from market forces in evaluating the significance of high prices in the aquarium trade, as was illustrated in the case of desirable, but inaccessible, deep-water species (see Section II,A,2). For example, certain species, such as the racoon butterfly fish, *Chaetodon lunula,* and the regal angelfish, *Pygoplites diacanthus,* are rare off Sri Lanka but are evidently "abundant" in the Maldives (Wood, 2001). On the other hand, certain deep-water fishes may not be naturally rare but difficult to obtain and therefore are infrequently available. Rarity, therefore, must be evaluated both in relation to conservation concern and also in relation to market supply to assess the true vulnerability of species that are actually or apparently rare in trade.

Conservation assessments of extinction risk in marine fishes have developed particularly during the middle to late 1990s and vary in their focus on specific life history characteristics. Assessments for inclusion on the World Conservation Union (IUCN) Red List of Threatened Animals, for example, are made on the basis of (1) limited area of occurrence (high endemicity); (2) limited area of occupancy (the area within a species' geographic range, on which it depends for particular activities—for example, when spawning); (3) few populations; (4) few animals (rarity); (5) low densities; and (6) significantly and persistently declining population size (Baillie and Groombridge, 1996; Hudson and Mace, 1996). Six species (*Centropyge resplendens, Lachnolaimus maximus, Epinephelus lanceolatus, Chromis sanctaehelenae, Balistes vetula,* and *Cheilinus undulatus*) and many sea horses and their relatives (family Syngnathidae) traded for live food or aquarium display are now included in the IUCN Red List of Threatened Animals as "vulnerable." These species are considered to be at risk because of their natural rarity, and/or from a range of human-induced threats, including the live fish trade (Baillie and Groombridge, 1996). None of the species traded live is listed in the appendices of the international agreement, the Convention on International Trade in Endangered Species of Wild Fauna and Flora (CITES).

A more fisheries-theory-oriented approach to assessing extinction risk in marine fishes, based on a productivity index and population decline rates, has also been developed. This approach uses *r* (intrinsic rate of increase), the von Bertalanffy growth coefficient, age at sexual maturity, maximum age, fecundity, and expert opinions on the species in question (Musick, 1999b). This approach assumes that the high fecundity often associated with pelagic spawning (i.e., the typical reproductive mode in reef fishes) confers an intrinsic higher resilience. Yet, we have no *a priori* confirmation that high rates of egg production confer resilience to exploitation. Indeed, a number of fished species with high fecundity failed to recover as anticipated from fisheries-induced collapses, providing clear warnings that high-fecundity species are not immune to potentially threatening depletions (Hutchings, 2000; Sadovy, 2001b). Indeed, the fecund sciaenid, the totoaba, *Totoaba macdonaldi,* was listed on CITES Appendix I in 1975.

We here review the evidence that fisheries and practices in the food and aquarium trades may directly or indirectly pose conservation threats to target populations, nontarget populations, and reef communities in general. Meticulous analyses are largely impossible with the current *ad hoc* data collection on live reef fisheries, and we are forced to infer what we can from a mixture of quantitative and anecdotal information. Nonetheless, these patchy sources, together, give sufficient indication of trouble to alert us to pending conservation and management crises and to highlight areas for further research.

A. Direct Effects on Target Species

1. Background

The following section reports on evidence for quantitative changes in populations of fishes directly exploited for the food and aquarium trades. Clear studies of population persistence remain few and limited and are essentially restricted to changing numbers and sizes of fishes, without an exploration of the reasons for such changes or an understanding of cause and effect. In order to understand the full impact of fishing, however, we need to consider the impacts of selective mortality on specific size classes, color phases or morphs, and social structure in target populations and under experimentally controlled situations. For example, the faithful monogamy of some sea horse species means that individuals are slow to repair after one partner has disappeared, and their site fidelity means that depleted areas will not rapidly be replenished (Vincent and Sadler, 1995). As another example, the aggregations of sex-changing groupers may be a critical opportunity for the fishes to assess population sex ratios, thus determining whether or not to change sex. Fishing such aggregations, especially if one or other sex is differentially removed, could result in altered sex change rates and overly skewed sex ratio biases (Shapiro *et al.*, 1993; Koenig *et al.*, 1996; Johannes *et al.*, 1999).

2. Live Food Trade

Fishing pressure on many species valued in the food trade is intense. Of particular concern is the growing practice of targeting spawning aggregations of groupers; for many species, virtually all reproduction seems to occur in such brief, easily located, gatherings (Bohnsack, 1989; Domeier and Colin, 1997; Johannes, 1998). Much of the recent live food fishing activity in the Solomon Islands is based on aggregation fishing (Johannes and Lam, 1999), with aggregations increasingly targeted in eastern Indonesia (Pet *et al.*, 1999). In Pohnpei, Micronesia, research indicates that, over a 7-day aggregation period of camouflage grouper in 1997, about 4000 fishes were removed for the local fishery from one site, representing about one-third of all fishes assembled (Rhodes, 1999). In the Solomon Islands, the 4–12 mt of groupers (about 4000 to 12,000 fishes) stripped from one spawning aggregation during one fishing expedition was probably a significant percentage of animals using that site at the time (Johannes and Lam, 1999). These findings are by no means unique, because heavy pressure on grouper aggregations in conventional (dead food) fisheries, leading to declines or extirpation within a few years, is well documented in the western Atlantic (e.g., Bohnsack, 1989; Sadovy and Eklund, 1999).

a. Changes in Number or Size Populations of fishes valued in the food trade are definitely being depleted, to the extent that businesses are forced to move around the Indo-Pacific and ever further from Hong Kong, in search of new sources of fishes (Richards, 1993; Johannes and Riepen, 1995; A. J. Smith, 1997; Sadovy, 2001a) (Fig. 6). Fishing pressure by one grouper export business in Palau was so acute over a 3-year period in the late 1980s that it virtually eliminated a spawning aggregation (Johannes *et al.*, 1999). The practice of targeting spawning aggregations to meet high demand is doubtless a contributor to these declines, although demand is the key factor. In Indonesia, fishers, exporters, and importers have all expressed concerns over rapid declines in live fish catches (Erdmann and Pet Soede, 1998; Bentley, 1999; Indrawan, 1999). Underwater observers noted that key target species have become extremely rare, preferred fishes have become harder to find in trade, and prices paid to fishers for live coral trouts and humphead wrasses almost doubled between 1995 and 1997, very possibly the result of declining supply. Indeed, many businesses have moved on from Indonesia largely because of declining catch rates, with an industry spokesman acknowledging in the mid-1990s that Indonesian populations of target species would soon be depleted (Bentley, 1999). In Indonesia and Malaysia, rapidly declining catch rates of high-finned groupers, the humphead wrasses, and some *Epinephelus* species have forced fishers to search progressively further from home ports to obtain sufficient fishes (Johannes and Riepen, 1995; Pet-Soede and Erdmann, 1998; Bentley, 1999; Indrawan, 1999). A decline in cyanide use in Indonesia occurred following declines in target species in the shallow-water areas where cyanide can be used (P. S. W. Chan, 2000).

Declines in numbers of humphead wrasse have been particularly alarming in some areas, prompting protective measures in Indonesia, the Maldives, and Palawan (Philippines) (Donaldson and Sadovy, 2001). The humphead wrasse is now uncommon in western Sumatra, Indonesia, with catches per trip declining from 50–70 kg per month in the early 1990s to 10–50 kg per month by the end of the decade (Bentley, 1999). This species also disappeared from divers' favorite

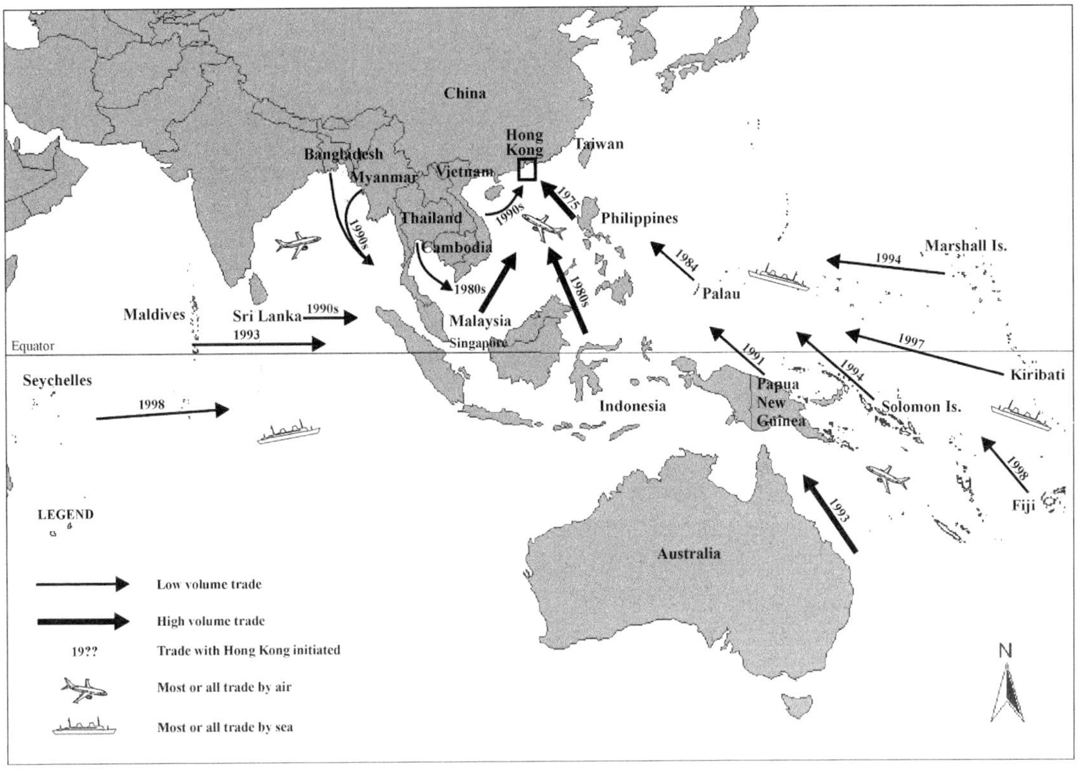

FIGURE 6 Map of the geographic spread of the Hong Kong-based food trade. Years over the arrows indicate approximate time that food trade with the indicated country began. Thick arrows indicate heavier trade than thin arrows. Symbols represent major transport mode (by air or by sea).

reefs shortly after a food trade was established in the Maldives, the resultant outcry leading to an export ban (Shakeel, 1994; Shakeel and Ahmed, 1997). Rapid depletions of humphead wrasse in Palawan after introduction of the food trade likewise led to a regional ban in exports of this species in the mid-1990s. This species is not compromised everywhere however. On the Great Barrier Reef, where it is managed, catches have remained relatively stable at 10–15 t yr^{-1} for the past 10 years and individuals are not rare in visual surveys (M. Samoilys, Fisheries, Queensland Department of Primary Industries, May 10, 2000).

The taking of juveniles, or of fishes below market size, for mariculture grow-out is gathering momentum in Southeast Asia, as mariculture gains popularity and buyers or exporters find it increasingly difficult to obtain a regular supply of market-size adults. However, such grow-out operations may be further contributing to overexploitation of source populations. In Thailand, the Philippines and elsewhere, the export of hundreds of millions of juveniles annually is possibly a causal factor in the reduction in numbers of juveniles and adults taken in recent years (Bentley, 1999; Sadovy, 2000). The weight of evidence from interviews, reports, and anecdotal accounts strongly indicates that declines are greater where fishing pressure is higher, either because of the pressure on juveniles, or on the adults, or both, and/or degradation of nearshore habitats on which fish depend. Although declines have not been noted everywhere that juveniles are harvested, there is a need to look more closely at this rapidly expanding capture fishery (Sadovy and Pet, 1998; Johannes and Ogburn, 1999).

There is some evidence of reductions in fish size for exploited species following the introduction of the food trade, and no evidence of populations in which fish size was unaffected by exploitation. Fishers in Indonesia and Malaysia report that the groupers they catch have become steadily smaller since the early 1990s. For example, in Indonesia size declines have been noted coincident with fishing pressure in the humphead wrasse. In the-mid 1990s, fishes weighing 45 kg were readily available whereas 25-kg fishes are now rare (Bentley, 1999). Most humpheads imported into Hong Kong now weigh less than 5 kg and are sexually immature (Fig. 1). Fishes are now smaller at retail, perhaps because of (1) declines in size availability and/or (2) demand for smaller, cheaper, fishes in the depressed economic environment of Asia, and/or (3) consumer preferences for smaller fishes due to concerns over ciguatera in larger individuals (Section II,A,1).

b. Estimating Sustainability Support for the view that overfishing may be severely compromising wild populations of fishes in the food trade comes from calculations of annual sustainable production for groupers in Southeast Asia and the Maldives. Assuming a reef area of 68,100 km^2(Spalding and Grenfell, 1997) and average (optimistic) sustainable grouper production of 1 mt km^{-2} yr^{-1} for reefs of fair quality and under medium fishing pressure [based on figures selected from McAllister (1988) and Dalzell (1996)], then an upper level for sustainable production for the region is calculated at about 68,000 mt yr^{-1}. For 1999, the estimated actual production was in the order of 100,000 mt [calculated from the total trade, i.e., sales, of 50,000 mt and 50% mortality prior to sale (Section II,D,2)]. This level of production substantially exceeds the estimated sustainable production for the region (K. Warren-Rhodes *et al.*, unpublished manuscript; Marine ecosystem appropriation by the live reef fish trade). In the Maldives, estimated sustainable grouper harvest of 1800 mt yr^{-1} was surpassed in 1995 (Shakeel and Ahmed, 1997), prompting calls for a halt on expanding the food trade.

3. Aquarium Trade

Few changes in populations of ornamental fishes have been well documented and there are no estimates of sustainability. Most evaluations of direct impacts of the aquarium trade on reef fish populations come from visual censuses of fish densities, calculations of potential yield from modeling, estimated exports from customs records, or observations by experienced biologists and commercial fishers, often without quantitative validation. Nonetheless, a pattern may be emerging and indiscriminate exploitation of particular ornamental fishes certainly raises concerns, especially if one considers the fate of populations.

a. Changes in Number or Size The sole systematic research study on the effects of the aquarium trade, comparing control and impact sites, found significant numeric declines in several reef fishes in Hawaii (Tissot and Hallacher, 1999). Underwater surveys were used to compare the effect of collectors at collection (impact) sites with prohibited sites as controls at each of two reef areas. The abundance of several species targeted for the aquarium trade showed significant declines over the 2-year study period at impact sites, whereas little change was noted for reef fishes not targeted for the aquarium trade. Care is needed, however, when evaluating the results of such short-term studies, given that temporal effects (such as yearly fluctuations in recruitment of common species found at these sites, e.g., the yellow tang, *Zebrasoma flavescens*) can be substantial (J. E. Randall, Bernice Bishop Museum, January 31, 2000).

Anecdotal evidence raises similar concerns for other fished populations and points to slow recoveries after collection. Although inconclusive, the sum weight of implications of decline cannot readily be dismissed. In Indonesia, intensive collecting of aquarium fishes allegedly depleted populations of some species in the Seribu islands and around eastern Java (Soegiarto and Polunin, 1982). In the Philippines, biologists who undertook underwater surveys that revealed unusually low numbers of butterfly fishes, angelfishes, and triggerfishes on reefs argued that these findings reflected overfishing for the aquarium trade (Albaladejo and Corpuz, 1981). Exporters from the Philippines perceived declines in catches of valuable species and were concerned over loss of variety (Vallejo, 1997). In the Maldives, 27 of the 65 ornamental fish species examined in the mid-1980s (including butterfly fishes, damselfishes, and cleaner wrasses) were being collected at levels high enough to provoke concern, and a further 12 species were thought to be exploited at or near the calculated potential yield (Edwards and Shepherd, 1992). Surveys in Curaçao and Hawaii have suggested that collecting exceeded sustainable levels and may have been a contributing factor in the collapse of the Hawaiian fishery (Kruijf, 1978; Pfeffer and Tribble, 1985). A coral head in the Bahamas, where fishes were experimentally removed using rotenone, required between 4 and 9 months to reestablish precollection population equilibrium (Smith, 1973).

Certain species may be particularly vulnerable to overfishing in the aquarium trade. A combination of heavy exploitation and destructive fishing methods has led to concern about the fate of the Banggai cardinal fish found only in the Banggai Islands off Sulawesi in Indonesia (Allen and Steene, 1995). Indeed, this species was first noted in 1920 and not again until 1994. It is now the focus of serious conservation efforts as well as encouraging captive breeding programs (Allen, 2000; K. Hunter, Oceans, Reefs and Aquariums, Inc., January 17, 2000). Sea horses, too, are threatened by overexploitation because of their status as aquarium fishes, their use in traditional medicines, and their appeal as curios. For example, fishers and buyers in eastern Java (where sea horses are taken almost solely as ornamental fishes) were agreed that sea horse numbers had declined by perhaps 15–50% over the 5 years from 1990 to 1995 (Vincent, 1996). Concerns have been expressed by fishery participants that some populations of scribbled angelfishes, *Chaetodontoplus duboulayi,* and western yellowtail angelfishes, *Chaetodontoplus personifer,* may be depleted in Australia (QFMA, 1999), with depletions also suggested in the Philippines for emperor angelfishes, *Pomacanthus imperator* (Rubec, 1987). In Mexico, there is concern over the heavily collected clarion angelfishes, *Holacanthus clarionensis,* after visual census surveys suggested declines of 95% in fished areas (cited in Wood, 2001).

Several studies outside Southeast Asia detected no numeric impact of the aquarium trade on fished populations. For example, Perino reports no noticeable decline in fish numbers in Fiji after 6 years of collecting activity (Perino, 1990). Other studies found that removals of cleaner wrasses had no negative effects on source populations (e.g., Grutter, 1996). In Fiji, fishing impacts were thought to be minimal because collecting was confined to a low percentage of available reef habitat (Lewis, 1988), and export figures indicated no declines in CPUE in the Cook Islands (Bertram, 1996). Underwater observations in Hawaii suggested that populations of a surgeonfish (tang) were not affected by collecting (Taylor and Nolan, 1978), and that the pygmy angelfish, *Centropyge potteri,* was still "abundant" despite heavy collecting activity (Randall, 1987b). Randall argued that the impact of the aquarium trade is likely to be negligible on common species with pelagic eggs (but see introduction to Section IV for counterargument).

Other findings in regions distant from Southeast Asia suggested no effects of fishing or were ambiguous, and help illustrate the difficulties of assessing such impacts. One 2.5-year study in the 1970s found that the five most heavily collected species of Hawaiian reef fishes were equally numerous on reefs where collection did and did not occur, and might even be more numerous on the former (Nolan, 1978). However, the uncontrolled nature of the design prevented clear conclusions (Tissot and Hallacher, 1999). In Australia, a study comparing densities of anemone fishes and anemone colonies between protected and unprotected reefs found no significant differences between and within reefs over five sites (QFMA, 1999). Underwater visual censuses elsewhere have yielded varying results but have generally lacked replication. For example, lower densities of butterfly fishes were noted on collected versus uncollected areas in Australia, whereas there were no differences in densities noted in 48 species between collected and uncollected sites in Kenya (Samoilys, 1988; Samoilys and Green, 1990).

B. Effects on Nontarget Reef Fish Species

Certain reef species taken as part of the operations of the live reef fish trades are not sold live and may be considered nontarget species. For the food trade, these may be taken either to provide bait or to feed groupers during grow-out mariculture, often to the frustration of local people who might value the same species for human consumption (Johannes and Riepen, 1995; Bentley, 1999). Nontarget species may also be caught accidentally, because many of the fishing methods for live reef fishes are indiscriminate (Section II,D,1). One method that has attracted considerable attention is cyanide, which, when widely applied to a reef to catch groupers or humphead wrasses, likely kills most of the organisms that cannot rapidly escape its effects (Heming and Thurston, 1984; Eisler, 1991). Some gears used to catch juveniles for food trade mariculture, including artificial reefs (fish attraction devices), may also trap large numbers of nontarget species, which are unlikely to survive collection because they are harvested by nets and beached such that they suffocate if not rapidly returned to water. A 12-month study in Indonesia, for example, using an artificial reef of branches and rocks, recorded 200 target grouper juveniles and 15,000 nontarget fishes, with few of the nontarget individuals surviving collection (Mous *et al.,* 1999). Fyke (bag) nets, widely used to capture juveniles for mariculture, also take large volumes of by-catch under certain conditions (Sadovy, 2000).

C. Effects on the Reef Ecosystem

The effects of fishing on reef ecosystems have been reviewed elsewhere (see Russ, 1991; Jennings and Lock, 1996). Clearly, the impacts of selective removals or destructive fishing extend far beyond simple extractions of target species at a particular time and place and can have severe economic and biological impacts (e.g., Jennings and Lock, 1996; Wilkinson, 1996; Cesar *et al.,* 1997). Most studies consider only the impact of conventional fisheries but their findings are directly relevant to this chapter because conventional fisheries remove some of the same species (e.g., surgeonfishes, snappers, groupers, wrasses) as do the food and aquarium trades, although not necessarily in the same way. For example, removals of the red-lined triggerfish, *Balistapus undulatus,* a major predator of sea urchins, from an East African conventional (i.e., dead food) fishery resulted in proliferation of sea urchins and required a recovery period of decades (McClanahan, 2000).

Fishing for the live fish trades often physically damages corals. As specific examples, corals are broken off to camouflage the weighted "bubus" (large fish traps) that are often used to catch groupers in Indonesia, and corals may be ripped apart as fishers pursue recently cyanided fish (Bertram, 1996; Erdmann and Pet-Soede, 1998; Pet-Soede and Erdmann, 1998). On the other

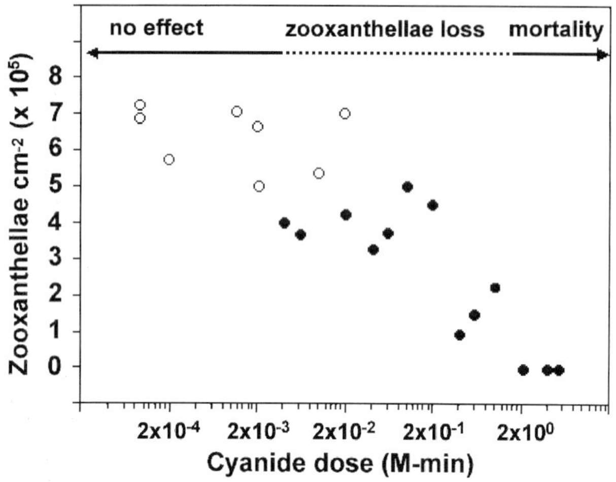

FIGURE 7 Relationship between cyanide dose (molar concentration per minute) and zooxanthellae density (100,000 zooxanthellae cm^{-2}) in colonies of *Pocillopora damicornis* 12 days after exposure. Each point represents a mean of five corals, including controls. Filled symbols represent significant differences in algal densities relative to handling controls (ANOVA, $P < 0.05$) Reproduced with permission from CSIRO Publishing; from Jones and Steven (1997).

hand, fishing gears such as hook and line, small (stop) nets, and hand nets evidently cause relatively little physical damage, although quantitative data are lacking in both cases. Likewise, the dangers of cyanide fishing have been much discussed, but there is surprisingly little information on the real effects of this fishing method (see Mous *et al.*, 2000). It does appear clear, however, that sodium cyanide can stress, and sometimes kill, corals under experimental conditions (using concentrations and dosages probable while fishing), by impairing photosynthesis of zooxanthellae, causing corals to release their symbiotic zooxanthellae (Jones and Hoegh-Guldberg, 1999) (Fig. 7).

The loss of physical structure induced by live reef fisheries and the selective removal of certain species can cause ecological change, but results are not consistent. On the one hand, long-term, large-scale destruction of live coral cover may produce replacement by benthic, turfing algae (Moran, 1986). On the other hand, a controlled study in Hawaii showed no increase in the abundance of macroalgae in exploited areas despite the finding that herbivore (algal predator) abundance was reduced by aquarium collecting (Dart, 1972; Jennings and Lock, 1996; Tissot and Hallacher, 1999). In principle, extensive loss of or damage to coral cover and complexity significantly reduces the diversity and abundance of fishes on the reef (e.g., Carpenter *et al.*, 1981; Russ and Alcala, 1989), although such direct effects of collection for the aquarium or food trade have yet to be conclusively demonstrated.

Some species may actually do rather well under disturbed conditions, although the long-term effects are little understood. Certain wrasses appear in greater numbers when the benthic substratum is damaged and the percentage cover of coral rubble increases (Russ and Alcala, 1989). As another example, it has been feared that the removal of "cleaner" fishes for the aquarium trade would result in higher parasite loads on the fishes they groom, but no such effects have yet been found (e.g., Grutter, 1996). Effects on the reef ecosystem resulting from loss of predators or other reef fish components, however, are unclear and probably complex (e.g., Jennings and Lock, 1996). As an example, two guilds of predatory fishes of aquarium trade type affect different stages of the life history of a sea urchin—small-mouthed wrasses and gobies control small urchins and larger mouthed predators control larger urchins—so differential removal of the fishes might produce trophic cascade effects of unknown consequence (Sala *et al.*, 1998). Clearly, substantial alterations to reef ecosystems can have marked and unexpected consequences.

One worrisome aspect of these international trades is the potential for species to be introduced into waters well outside their usual geographic range. In the food trade, exotic species that are imported to be sold or grown-out in Hong Kong waters (e.g., brown-marbled groupers) sometimes escape and are found in open waters near mariculture zones (Sadovy and Cornish, 2000). Hundreds of thousands of grouper juveniles from all over Southeast Asia are being released into Taiwan waters for restocking and public relations exercises (T. Yang, Fish Breeding Association, Taiwan, August 31, 2000); in the aquarium trade, hobbyists or public aquaria may dispose of specimens by releasing them into alien waters. Although we encountered no confirmed reports that any such releases had yet become established, other introduced marine species certainly have persisted (Randall, 1987a; Bohnsack, 1996), potentially accompanied by the introduction and spread of associated diseases (Courtenay and Stauffer, 1990; Biffar, 1997).

D. Summary

The accumulated evidence, from anecdotal to quantitative, indicates that live reef fish trades can—and sometimes do—provoke severe overfishing of certain target species, exacting a heavy toll on reef habitats. Most studies to date, however, are of insufficient duration and with uncontrolled design such that cause and effect have been difficult to establish. Given the generally poor documentation of actual capture levels and trends, the impacts may be worse than we yet realize, or not as extreme as they appear. Nonetheless, we do have legitimate conservation concerns about population declines in number, and/or reductions in size, such as for the humphead wrasse and the Banggai cardinal fish (among others). We can also infer impacts of unknown severity and consequence on nontarget species, reefs, and reef ecosystems that require further scrutiny. The destructive and wasteful nature of some capture and culture practices add to the pressures, as does the focus on selecting certain species or individuals. Available information on declines in size or numbers in Southeast Asia (although not necessarily elsewhere) is consistent with the growing realization that coral reef fisheries cannot withstand commercial export fisheries without careful controls, and might even lead to biological endangerment. Long-term, well-designed studies on more vulnerable, important, and representative species in trade are needed to gauge the impacts of removals and habitat damage.

V. Management Issues

A. Context for Management and Conservation

Long-term persistence of healthy populations of reef fishes will best be assured through a combination of biological and fisheries research, resource management, informed aquaculture development, stakeholder involvement, and legislative support. Maintenance of viable populations is key to the retention of species, the maintenance of long-term healthy ecosystem function and stability, and the development of sustainable use (Ehrlich and Daily, 1993; Polunin *et al.*, 1996). Integrated management initiatives will thus need to ensure that populations remain numerically strong, sex and size cohorts stay balanced, behavioral and color morphs are safeguarded, and genetic diversity retained.

The recognition that people, especially those living on the coast in developing countries, will always eat fishes means that conservationists and fisheries scientists need to seek common ground, and look for holistic solutions. Certainly there is great danger in reducing catches of one species or in one region, if that leads to increased fishing effort on another species or in another region. Importers in the food trade, for example, have already shown a tendency to deplete an area then move to another region, in a sequential exhaustion of resources that makes their management particularly difficult and their accountability a challenge. Indeed, the advent of large and mobile fish carrier vessels has led to regional raids akin to gold rushes (Bentley, 1999). The aquarium trade differs somewhat in being more closely tied to coastal communities or transport facilities, resulting in concentration of effort close to major transport hubs. Nonetheless, the ability of businesses based in importer countries to seek supplies from a broad range of sources worldwide reduces accountability and dependence on resources in any one region.

In our planning, we should recognize that economic extinction in live fish trades need not precede biological extirpations or extinction (Section III,A,3). Fishers may well continue to exploit live-traded fishes until the resources are exhausted: this dearth of final economic constraints is also found in conventional fisheries, which are maintained well beyond sustainable levels by subsidies of somewhere between U.S. $14 and 54 billion per annum (FAO, 1993; Milazzo, 1998). However, the real dearth of economic constraints, or livelihood options, for fishers engaged in the live reef fish trades makes management all the more difficult, and the trades, in some areas, are being subsidized at the expense of future reef production (Cesar *et al.*, 1997).

B. Biological and Fisheries Research

The large number of species, multiple sources of fishes, and wide range of gears involved in reef fisheries make it essential but difficult to identify research priorities. We certainly need to understand three key areas: (1) the biology of species and populations, (2) the intensity and effects of fishing and fishing methods on target and nontarget species, and (3) the rates and causes of age-specific natural mortalities.

The selective nature of live reef fisheries makes it essential that we understand the importance of different cohorts and classes (color and sex) of fishes within a population. Studies of basic behavior, ecology, longevity, and demography of the fishes exploited for the food and aquarium trades would help build urgently needed life history tables and thus allow population assessments. It would be particularly valuable to understand reproductive and foraging behavior of targeted species, adult mobility and larval recruitment patterns, interconnectedness among populations, and how differential removal of one sex or life history stage or color morph might affect populations. As a specific example, we know very little of the natural mortality schedules during, and shortly after, settlement and prior to sexual maturation. Yet a better understanding of mortality schedules for early life history stages would be enormously useful in managing new artisanal fisheries that capture and culture settlement-stage larvae, or when relying on juvenile capture for grow-out, whether for AT or FT (Dufour and Galzin, 1993; Sadovy and Pet, 1998; Bell *et al.*, 1999).

Research priorities for the food trade would include basic information on population sizes, age and size of maturity, growth/survivorship of juveniles, and determinants of sex change (many species in this trade are hermaphroditic). Additionally, given that most of the larger species in the food trade form spawning aggregations, we urgently need to discover what drives the formation and collapse of these groupings, ascertain the importance of individual aggregations for recruitment, and examine the effects of fishing. In the year 2000, concerns about overexploitation of spawning aggregations led to the formation of the Society for the Conservation of Reef Fish Aggregations (SCRFA) to promote management, awareness, and research of this vulnerable period in the life history of larger reef species.

Similarly, most species in the aquarium trade have never been studied in any systematic fashion in the wild. We need to resolve taxonomic uncertainties, document areas of occupancy (see Section IV), and determine the independence or interdependence of populations (metapopulations). In the vast array of fishes traded for the aquarium trade, it may be best first to focus on species that are most commonly traded and on those whose life history traits (to the extent that they are understood) suggest greater vulnerability to overexploitation; this would include species that show low replacement rates, high longevity, parental care, highly structured or rigid social and spatial behavior, specialized habitat needs, restricted geographic range, and/or rarity.

Very few existing studies of exploitation for the food and aquarium trades are rigorous over sufficiently large temporal and spatial scales. It is essential to undertake a suite of controlled long-term studies of fished and unfished populations and associated community structures under different exploitation regimes; relevant models already exist for traditional reef fisheries (e.g., Jennings and Polunin, 1996). Ideally, full investigation would require (1) quantitative comparisons both before and after a given phase of exploitation or a particular management initiative, and (2) that multiple control and impact sites, or populations, be evaluated [beyond BACI (Before/After Control-Impact-type design) (Underwood, 1992)]. Control sites and populations should be well removed from spillover effects of impact sites and populations, in order to be confident that effects are due to fishing activity as opposed to natural variations or other human activities. In looking at possible depletions, such studies should particularly address the effects of repeated removals of particular sizes or classes of fishes, and the responses by other, nontarget, species in the community. Of particular interest would be careful long-term BACI analyses of the effects of different types of nondestructive fishing gear and methods on corals and on both target and nontarget coral reef fish populations.

Good documentation and analysis of live reef fisheries should be a priority. Currently, those few that are monitored typically yield little information about resource status or fishing effort. Improved fishery logbooks, with clear indication of fishing effort, trade records, and documentation of consumption patterns could aid in predicting new pressures and threats to wild populations, particularly if they identified species and classes of individuals sought as well as origins, trade routes, and destinations. Record-keeping will be most valuable where it is periodically validated through independent monitoring. In the case of Hong Kong's imports for the food trade, for example, where key species are only partially monitored, size data are lacking, and sources go unnoted for most sea imports, the International Marinelife Alliance is working to improve import data (F. McGilvray, IMA, November 9, 2000). More encouragingly, Australia monitors exports of sea horses and their relatives by species and destination (under a permit system), and is encouraging fishers to record their take by size and sex.

To be really useful, results of biological, fisheries, and trade research should be disseminated widely. Relevant new and useful data bases are emerging, including REEFBASE and FISHBASE managed by the International Center for Living Aquatic Resources Management (ICLARM) in Malaysia. Most recently (in 2000), the World Conservation Monitoring Center (WCMC) in Cambridge, UK, began developing the Marine Ornamentals Information System (MOIS) in close collaboration with the Marine Aquarium Council. This system will be based on importers' and exporters' records and should help generate information for an aquarium industry-wide self-monitoring system.

C. Resource Management

The very limited formal management of the food trade, and the only slightly better management of the aquarium trade, in many countries is of considerable concern. (Kim and Case, 1996; International Trade Subgroup, 2000). The few regulations of the food trade generally involve export bans or limits, restricted entry to foreign fishing vessels, or fish size limits (Sections IV,A,2a and V,F). In Australia, for example, the valuable coral trout fishery is managed by imposing minimum capture sizes, although enforcement of this measure is not always effective (e.g., Johannes *et al.*, 1999).

Spawning aggregations have been protected for years in several Pacific locations by local communities, such as in Palau, a measure now being adopted more broadly, such as in Pohnpei, as the food trade grows (Rhodes, 1999). Although the value of aggregation protection as a sole management measure is impossible to predict, all protected aggregations thus far have persisted. Other measures for the food trade variously under consideration include the banning of hookah or compressed air (necessary for cyanide fishing), vessel monitoring systems (VMSs) to locate foreign vessel activities in local waters, and the development of more sensitive cyanide detection kits (e.g., Johannes and Riepen, 1995. In the aquarium trade, the United States has imposed licensing, size limits, minimum and maximum sizes, restrictions on quinaldine concentrations, and daily quotas and fishing seasons [e.g., Dufour, 1997; Wood, 2001; Marine Fisheries Commission Regulations (Florida) Chapter 68B-42-Marine Life]. Countries in the Indo-Pacific have developed aquarium trade management measures, including the designation of fishing zones, collection prohibitions, quotas by species on captures or exports, limits to total trade volume, restrictions on size or class of fishes, gear restrictions, and limited entry for fishers (Graham, 1996; Dufour, 1997; Wood, 2001; QFMA, 1999).

Although tropical, multispecies fisheries such as those for the food and aquarium trades will always present challenges to data collection, some management principles are biologically obvious and should be undertaken despite a dearth or paucity of data (Johannes, 1998). For example, recognition of a particular suite of vulnerable life history characteristics in a species (see Section IV), even in the absence of solid data, should prompt close attention (Jennings *et al.*, 1999). And certain actions, such as reducing exploitation and disruption of spawning animals or those providing parental care, seem self-evident, and are indeed becoming more common. Prohibiting the capture or export of vulnerable species (as for the humphead wrasse) is another apparently obvious approach, although the worry is that such action might lead to a cascade effect, transferring pressure to other vulnerable resources, if such action is not adopted universally (see Section V,A), or encourage illegal trading.

The live reef fish trades are unusual in often favoring smaller or intermediate-size fish and in their quest for variety. Slot-size restrictions, in which neither the smallest nor the largest individuals are taken, should help protect the particularly small individuals of aquarium trade species that are unlikely to survive the collection process, and the large fish that are often important spawners. Such controls have already been incorporated into Florida's management plan for the aquarium trade, although a lack of biological knowledge means that restrictions tend to be based on trade convenience rather than on biological principles [e.g., Marine Fisheries Commission Regulations (Florida) Chapter 68B-42-Marine Life]. Also effective in reducing undue pressure, in this case on aquarium trade species, are industry agreements to adopt voluntary guidelines and to reduce trade in unsuitable species (Section II,D,2)

The current dearth of biological understanding suggests that live reef fish exploitation might benefit greatly from adaptive management, in which new findings are constantly integrated into fisheries planning, allowing catch levels to be varied in response to new information. It may, however, be difficult to employ such a responsive system in communities with little history of, or capacity for, formal assessment or management, or where operations are moved from one area to another in serial depletions (as in the food trade). Moreover, the costs of effective management in some situations may well exceed the value of the trade (Johannes and Lam, 1999; Johannes *et al.*, 1999). Another approach is precautionary management: for example, the 1972 Fisheries Act of the Solomon Islands provides for impact assessments in response to new proposals to exploit fish (Oreihaka, 1999).

Marine protected areas (MPAs) should be useful tools for managing the live reef fisheries. Indeed MPAs are increasingly being explored as generic management tools for a wide variety of marine organisms, and especially for reef-associated species (PDT, 1990; Roberts and Polunin, 1993; Russ and Alcala, 1994; Bohnsack, 1996), including those traded live. Most notably, Hawaii recently designated no-take MPAs covering 35% of the Kona coast of Hawaii Island to help manage its aquarium trade, following conflicts between recreational divers and aquarium trade fishers (Tissot and Hallacher, 1999). Such zones can act as insurance policies against poor or uninformed management, can help protect pools of biodiversity, and can act as sources from which individuals emigrate to surrounding regions. MPAs established on the abused reefs of the Philippines produced significant increases in the number of species, number of fishes per species, and size of fishes in a very short period of time (e.g., Roberts and Polunin, 1993; Russ and Alcala, 1994). Indeed, site-based management, including protection of aggregations, is an important focus of The Nature Conservancy (TNC) in both the Indo-Pacific and the Atlantic (J. Pet, TNC, November 24, 2000).

D. Mariculture as a Tool

In theory, mariculture could reduce pressure on wild fish populations, and thus on nontarget species and habitats. In reality, most mariculture remains rather problematic in both technical and socioeconomic terms (Rosamond *et al.*, 2000). Attempts to close life cycles repeatedly (full-cycle culture)—with spawning, rearing, and mating in closed systems—are proving technically challenging for all but a very few species. This is disappointing because reliable production of juveniles and market-size fish year-round, of a wide range of species, could potentially help meet growing market requirements, and take pressure off wild stocks. Culturing fishes for the food trade could also reduce the risk of ciguatera (natural fish poisoning) as a human health concern. In addition, aquarium trade fishes that have been cultured *ex situ* (especially for several generations) are likely to fare better in captivity, surviving longer and thus reducing demand for replacements, as well as be acceptable to conservation-conscientious buyers.

As currently practiced, and given market forces, commercially viable culture for the food and aquarium trades is likely to be able to satisfy only a small proportion of total demand and is not expected to become a major substitute in either trade. In the food trade, Taiwan is a notable exception, having established commercially viable full-cycle culture for two species (Cesar *et al.*, 2000). Elsewhere, food trade mariculture is usually based on repeated capture of wild juveniles and breeders and international trade of the juveniles for grow-out to market size and sale. The effects of heavy fishing pressure on juveniles, especially the capture of massive numbers of those long past the early (presumably high-mortality) postsettlement phase, could certainly have a significant impact on wild populations. Indeed, the typical grow-out approach is unlikely to be ecologically sustainable: coastal habitats are damaged as facilities are created; international trade means that alien species or genotypes may be introduced to new regions, with high chances of escapes; effluent (including antibiotics) is discharged from intensive farms in coastal areas; and other fish are taken to feed high-trophic-level carnivorous cultured species. In the aquarium trade, fishes are usually cultured on a much smaller scale, often by aquarium traders, in pilot projects, facing severe technical limitations and postlarval culture is being developed. In addition, the high price of some fishes cultured for aquaria, compared to those imported from the Philippines or Indonesia, undermines their economic viability, and the demand for a wide diversity of species in the aquarium trade, most of which are difficult to raise, means that culture may never be able significantly to replace wild sources.

Aquaculture research may well eventually make simple low-technology aquaculture feasible for many species, but such enterprises should be evaluated for their real conservation and human benefits before they are embraced unconditionally. Mariculture will reduce pressure on wild resources only if (1) it is practiced sustainably, (2) if the cultured fish can meet all market needs, (3) if fishers earn more from fish farming than from wild capture, and/or (4) wild-caught animals can be distinguished from captive-reared animals. The first and second conditions are unlikely to be met because aquaculture business interests tend to create new market demand (even more than can be met by aquaculture) in promoting their products. The third condition is also problematic because mariculture seldom employs many local fishers, other than for the capture of juveniles for grow-out, and may indeed take away their livelihood if juvenile production is concentrated in big business hatcheries (Sadovy, 2000). Ensuring that captive-reared fishes are traded in place of wild fishes will require research and development of appropriate labeling protocols acceptable to all players.

E. Stakeholder Involvement

The lack of livelihood alternatives means that the impoverished fishers at the heart of these trades continue catching live reef fishes long after numbers have been depleted. Successful management of these fisheries will depend to a considerable extent on integrated programs in fishing communities, including education and awareness building, and on well-conceived regulations or legislation controlling live fish capture and export. Fishers can prove good allies in the long-term management of their reef resources because the depletion of reef fish populations in general, the degradation of many marine environments, and the lack of other income-earning options mean that they usually value long-term availability of remaining marketable commodities. The capacity of local people to organize and take leadership, and their confidence to confront abuses directly, will be enhanced by dissemination of biological information, so they can devise fisheries management approaches. Access to trade information, to allow development of cooperative action with their neighbors and negotiate with buyers for fair prices, is also important (e.g., Johannes and Riepen, 1995; Bentley, 1999). On the other hand, stakeholders must seek a full understanding of the real costs of getting fishes to consumers as well as an appreciation of whether the resources are more valuable as export commodities or alive on reefs, for tourism perhaps (Spurgeon, 1992; Cesar, 1996; Birkeland, 1999).

Fishers need to control their own stake in the fishery and ensure that resources are managed well. It often takes longer to implement effective community-based management than it does for overfishing to occur, especially when aggregation sites are being targeted. In the western province of the Solomon Islands, for example, some communities are aware that targeting grouper aggregations has depleted fish numbers to the point of aggregation collapse, yet they persisted in this approach because groupers were not traditionally valued as a local food fish (A. J. Smith, The Nature Conservancy, March 29, 2000). In some areas, fishers have adopted cyanide fishing for the live trade to the point where it is becoming a "traditional" fishing method (e.g., Galvez *et al.,* 1989; Barber and Pratt, 1997). Training fishers in the use of less destructive gear, such as hook and line for the food trade and hand nets for the aquarium trade, has limited long-term impact unless it is woven into a comprehensive package of community engagement and does not further contribute to unsustainable levels of fishing.

Penalties for violations need also to be reasonable, because punishment perceived as being overly severe is seldom imposed (Galvez, 1988), whereas punishment too light is not taken seriously. Detecting violations (poaching by live fish transport vessels in the food trade, for example) requires new approaches, such as the recent proposal for foreign vessels to carry tracking devices (VMSs) (as is done in the Pacific tuna industry) and as proposed in all Southern Pacific Forum Fisheries Agency members (Alee and Eke, 1999). There is also a need to inform and educate lawmakers and those who preside over court cases to ensure that violations are taken seriously.

Market forces may be important in helping to reform the live reef fish trades: it is most encouraging that a recent consumer survey in Hong Kong indicated that 80% of food trade fish consumers, and especially younger people, would be willing to change their consumption to protect vulnerable or declining species (W. W. Chan, 2000). Consumers can exert considerable pressure for improvement, for example, by (1) accepting the need to pay real costs of best ecological management practices, research, and species and habitat restoration, (2) buying cultured fish instead of wild-caught animals (but see Section V,D with respect to ecological concerns for cultured fish), (3) eschewing food or aquarium trade fishes considered to be vulnerable or aquarium species documented as difficult to keep in captivity, and (4) demanding improved husbandry of food trade and aquarium species to limit waste through mortality and reduce repeat purchases. In the food trade, consumer concerns are a foundation for modifying eating habits. In the aquarium trade, public aquaria have the potential to play a great role in informing and educating a wide range of stakeholders, with their access to millions of receptive visitors, many of them consumers of live food fishes and/or aquarium hobbyists.

Formal certification for quality and sustainability (sometimes referred to as "ecolabeling") could offer consumers the opportunity to reduce the ecological and socioeconomic costs of their demand for live food and ornamental fishes, thereby motivating traders to improve trade practices (Wood, 1992) and discouraging businesses not committed in the long term. Such certification might succeed particularly well for live reef fishes because they are purchased as luxury items by people who can afford to exercise choice. The Marine Stewardship Council (MSC) and the Marine Aquarium Council are developing international criteria for sustainability for food fishes and aquarium fishes, respectively (Holthus, 1999). Both are now independent entities, after being launched as joint ventures between nongovernmental organizations and industry (World Wildlife Fund and Unilever for the MSC, and a range of nongovernmental organizations with the aquarium industry for MAC). A code of conduct has yet to be developed for the food trades. MAC is assisted by aquarium industry associations that promote responsible trade, such as AMDA, OATA, and OATA–World Wide and Ornamental Fish International (OFI). Both MSC and MAC are in the early stages of certification, and still need to build stronger support in source countries and public awareness among consumers. Globally, however, market assessments indicate a strong demand for the certified aquarium trade being developed by MAC (Holthus, 1999).

F. Trade Legislation

Thoughtful trade controls and legislative policies on the live reef fish trades are needed to complement biological research, fisheries management, community-based programs, and market pressures. A number of nations are devising appropriate precautionary policies, even in the face of current ignorance. For example, the Seychelles Fishing Authority has repeatedly refused requests for expansion of the food trades, considering it unlikely that the currently estimated maximum sustainable yield of 100 mt could support more than the single current operation, and noting that international experience suggests that risks of reef damage can be greater than potential benefits (Bentley and Aumeeruddy, 1999). Likewise, concerns over the difficulty of controlling cyanide use has led Malaysia to prohibit any export of marine fishes or invertebrates for aquarium purposes (Biffar, 1997). In Mozambique, the Ministry of Agriculture and Fisheries ordered a 2-year moratorium in 1999 on capture, export, and trade in aquarium fishes because of the current lack of management and laws to avoid damage to and depletions of reef resources. As well, Puerto Rico is considering banning exports for the aquarium trade because of concerns about such a fishery. We should, however, be mindful that very high value per fish can lead to a lack of respect for management regimes (Birkeland, 1997a), especially where there is no accountability, as in the food trades, in which businesses simply move on after depleting an area. Demand centers need also to bear their responsibility for unsustainable and destructive practices: it is encouraging, therefore, that an executive order (No. 13089) for the protection of coral reefs charges the United States to examine its role in international trade and protection of coral reef species to ensure that it is not contributing to destructive practices (International Trade Subgroup, 2000). Broader controls could be achieved through measures such as the Lacey Act in the United States that bans import of fishes or wildlife obtained in violation of any foreign law (Lathrop and Hourigan, 1998) and permits prosecution of importers and retailers who handle such animals.

Certain restrictions will be most appropriate when undertaken in a regional context, with support from other jurisdictions. For example, in Southeast Asia, intergovernmental organizations such as the Asia–Pacific Economic Cooperation (APEC), which includes major live fish exporting and importing economies, could play a key role in fostering a proactive approach, could act to ensure compliance with export controls in cyanide-free certifications, or could promote import controls on fishes banned from trade (Sadovy, 1998; International Trade Subgroup, 2000). On a global scale, the Food and Agriculture Organization (FAO) has developed a Code of Conduct for Responsible Fishing that bears on many of the issues in the aquarium and food trades and that could be widely adopted by participant economies (FAO, 1995). It is worrisome, however, that the FAO undertakes no systematic monitoring of either live fish trade. It is thus a matter for concern that a major intergovernmental agreement on species conservation, the Convention on International Trade in Endangered Species of Wild Fauna and Flora, tends to consider matters affecting commercial fisheries to be the domain of the FAO. If it so chooses, CITES could offer a useful approach to managing fishes found in the live trade by requiring evidence that international trade is not proving detrimental to species that have been brought under its jurisdiction (i.e., species listed in Appendix II of CITES).

G. Summary

Biological research is urgently required to respond to the immediate need for management decisions. Exploitation proceeds even in the face of biological ignorance, forcing managers to act on available information. It is up to biologists and nongovernmental organizations to work with managers to ascertain what research is necessary, to provide essential information, to rely on precedents set elsewhere, and to make these accessible and applicable to local contexts. In the absence of adequate knowledge of a species, biologists can draw inferences from similar species and situations elsewhere, even while warning about such extrapolations (e.g., Johannes, 1998). In making decisions about live fish fisheries, we can draw from early ecological studies of small reef species (e.g., Sale, 1991a) and information from conventional (dead fish) reef fisheries. Qualitative and anecdotal information from fishers and traders can also provide guidance. The solutions to attaining sustainable live reef fisheries will not be found only in alternatives, such as mariculture, at least not in the short term. Rather, they will emerge through an understanding of patterns and limits of resource use and the active engagement of stakeholders at all levels, but especially fishers and consumers.

VI. Conclusions

The fisheries for live food and display animals are harbingers of things to come: an emerging genre of such "extraordinary fisheries" is increasingly producing high-value items for sale to growing international markets. Similar sorts of nonconventional fisheries are those targeting species for traditional medicines (e.g., pipefishes and sea moths), research and education (e.g., dogfishes and eels), high-value gourmet fish parts (e.g., flying fish roe or shark fins), and curiosities (e.g., pufferfishes and sea horses). Managing these fisheries and trades in a sustainable fashion will pose new challenges. The target species are largely unstudied, their life history characteristics tend to make them ill suited for heavy exploitation, the fisheries are not well served by existing management models, juveniles are heavily targeted, the trades effectively go unrecorded, and the commodities are commonly nonessentials. Moreover, by ignoring the full costs to the environment and future resources, governments and consumers are in effect heavily subsidizing these trades.

The poor documentation of—and disorganized response to—problems in the live reef fish trades cannot hide the fact that management is clearly necessary and ignores their considerable economic worth. These fisheries are already exacting their toll on some wild populations, with demand projected to continue increasing rapidly. The expectation is that consumption of food fishes and interest in ornamentals will grow particularly markedly in Asian countries, where rapid economic growth (recent crises notwithstanding) means that many consumers have more disposable income than ever before. Solutions will depend on cooperation at all levels of society and government. The international nature of the trade requires commitment from both exporting and importing economies, as well as intergovernmental organizations, to ensure the monitoring and compliance that support management, such as through certification.

Well-managed live reef fish fisheries offer an excellent chance for high-value and relatively low-volume trades that could employ many fishers without damaging wild populations, with the income providing strong local incentives to care for marine resources. Similarly, well-considered and cautious mariculture has the potential to help support local people, but all mariculture operations and proposals should be very carefully scrutinized for ecological integrity as well as economic viability and social implications. Currently, mariculture proposals are often viewed as intrinsically positive without consideration of potential impacts. Mariculture, like the import of exotic fishes, could threaten local fish fauna, because of the potential for releases and escapes, as well as the spread of disease. Indeed, the IUCN actively discourages releases unless they are part of a well-conceived and long-term international program. Moreover, the polluting and resource-expensive (for feed) demands of the larger carnivorous species being cultured pose a range of problems that have yet to be addressed. Mariculture is not now, and is unlikely to become, an alternative to management in these trades, or indeed, in reef fisheries in general.

Successful management of live fish fisheries for sustainability will require a creative and interdisciplinary mix of management measures, well supported by government, local fishing communities, traders, and consumers. Biologists and managers must make information available and widely accessible. Ecosystem-level approaches, such as MPAs, offer considerable hope because they help conserve habitat and biodiversity simultaneously. Preemptively established no-take zones will become increasingly critical as the live fish trades, and associated abusive fishing practices, spread into new regions. One difficulty in managing live reef fish fisheries is that neither the food nor the aquarium species are vital for food in many source countries. The nonessential nature of these luxury products makes it more difficult to push for effective stewardship of the resource and more likely that populations will be heavily overfished. In the food trade, as in so many other fisheries, it will often be more profitable to "strip-mine" the fish quickly (and move on) and put the money in the bank than to manage the fishes cautiously to harness their intrinsic rate of increase in the longer term (Clark, 1973). On the other hand, the luxury nature of the fishes means that consumers have some flexibility in their purchases, and can exert considerable market pressure on suppliers to conform to agreed ecological standards, perhaps under a certification scheme. This is particularly promising in the case of the aquarium trade. Such decisions will depend on access to information, as will fishers' decisions on exploitation and sale of their resources. The challenge is to create the market incentives to promote the management practices necessary to maximize the benefits of, minimize the negative impacts for, and reward good practice in these trades.

The live reef fish trades raise ethical issues well beyond those of conventional fisheries. The high levels of associated indiscriminate mortalities and stresses may or may not add to our concerns about the stability of wild populations, but they certainly raise more general questions about animal welfare and rights that stakeholders will wish to address. An underlying but common assumption is that marine fishes are not wildlife, and thus not worthy of the same conservation action as terrestrial species (Safina, 1995). Such attitudes influence almost every aspect of their management and mean, for example, that (unlike terrestrial native species) they were explicitly excluded from the Wildlife Protection Act in Australia until sea horses and their relatives (traded for aquarium pets, traditional medicines, and curios) were listed in January 1998. Hong Kong still exempts live food fishes from the definition of "food," thereby largely avoiding health checks and monitoring. The devotion of aquarium aficionados to their hobby and the capacity of public aquaria to educate an interested public may help to give marine fishes in general their rightful place in the pantheon of conservation concerns, but this is far from being fully realized.

The big question that source countries need to ask is whether indiscriminate trade in live fishes is worth the ecological risks and how best to benefit from continued trade. The direct loss of populations (through capture and associated disruption), the damage to other species, and the degradation of reefs from illegal capture methods are all very costly in terms of future fishing income and alternative income-earning opportunities, such as tourism (e.g., Bryant *et al.*, 1998). Certainly the Seychelles are concerned that the long-term costs of the food trade outweigh short-term gains, and Malaysia, and Mozambique, among others, no longer permit exports for one or both of the trades for similar reasons. Yet, although a total export ban on live fishes would solve almost all the problems in these trades, it is possible that well-managed live reef fish fisheries—based, where possible, on well-enforced adaptive management and with the appropriate checks and balances—offer the best chance to stimulate interest in protecting the communities and coastal habitats on which a stunning variety of reef fishes depend.

Acknowledgments

We are extremely grateful to many people who have provided advice, information, and support. We especially acknowledge Dale Marsden for tireless research in gathering details, preparing many of the figures, and reviewing the manuscript, and Susan Innis for her valuable contribution to the review of the aquarium trade. Rachel Wong also provided excellent technical assistance; P. Holthus, B. Tissot, and E. Wood granted us access to unpublished manuscripts that have been particularly useful; J. Baquero, M. Berica-Rasotto, K. Davenport, J. Dawes, P. Holthus, B. Johannes, P. Mous, T. Pitcher, M. Samoilys, R. Sankey, L. Squire Jr., J. Tullock, and K. Warren-Rhodes provided valuable information and insightful comments and have furnished data. Information on various aspects of both trades was also provided by G. Allen, C. Bull, P. Burgess, P. S. W. Chan, C. Chu, T. Donaldson, M.V. Erdmann, T. Graham, H. Hall, J. Hawkins, S. Larkin, P. Lau, M. Lidster, L. Min, L. Perino, A. Perry, J. E. Randall, C. Raymakers, J. Resor, C. Roberts, J. Schulz, A. J. Smith, D. Turgeon, L. Ty, and R. van der Elst; M. Legault prepared one of the maps.

CHAPTER 19

Yet Another Review of Marine Reserves as Reef Fishery Management Tools

Garry R. Russ
School of Marine Biology and Aquaculture
James Cook University
Townsville, Queensland 4811, Australia

I. Introduction

The term "marine reserves" is defined in this chapter to simply mean "no-fishing" areas in the marine environment, that is, areas permanently closed to fishing. The goal in this chapter is to review some of the major issues of marine reserves as reef fisheries management tools. Other benefits of marine reserves, including maintenance of biodiversity and ecosystem structure and enhanced tourism, have been reviewed many times elsewhere (Table 1). Marine reserves are not being advocated here as the only, or even the optimum, method of reef fisheries management. In fact, it is vital to stress from the outset that other forms of fisheries and habitat management should be encouraged and attempted in areas open to fishing. In addition, in many developing nations, managing fisheries on coral reefs requires that the number of fishers be reduced considerably (Munro and Williams, 1985; Russ, 1991; Munro, 1996). Without such measures as finding alternative livelihoods for fishers and reducing rates of human population growth, most attempts at managing reef fisheries will probably be futile in many developing nations.

In the past decade the topic of marine reserves as potential fisheries management tools has produced a burgeoning literature (Tables 1 and 2). There are at least two major reasons for this. First, scientists working in developing nations and/or on the management of coral reef fisheries (e.g., A. C. Alcala, J. A. Bohnsack, G. E. Davis, T. R. McClanahan, N. V. C. Polunin, and C. M. Roberts) have realized that there are probably few other viable management alternatives. In such situations it is almost socially immoral to try to impose fishing effort or catch restrictions on subsistence and artisanal fishers. You cannot tell a fisher in a developing nation that they must throw a fish back into the ocean because it is too small, or that they must catch only four fish per day when they have eight family members to feed. Second, traditional fisheries management (effort, catch controls) has generally failed to prevent massive overfishing globally. The dismal state of most of 20 stocks of cod in the North Atlantic, exploited by the highly developed nations of Canada, the United States, and Europe, are good examples of this (Myers *et al.*, 1996). Marine reserves are now seen as an insurance policy against such management failures, something Jim Bohnsack was advocating for reef fisheries a decade ago [Plan Development Team (PDT), 1990].

But there is more to the popularity of marine reserves as fisheries management tools than indicated above. The topic has an almost seductive quality about it. It offers conservation and sustainable exploitation. It offers, simultaneously, what in the past would have been seen on coral reefs as almost conflicting objectives—maintaining the aesthetic qualities of reefs for tourism while also providing sustainable fisheries. The topic has attracted marine ecologists and (some!) fisheries scientists and managers, conservationists, social scientists, environmental managers, policy makers, economists, and mathematical modelers, to name just a few (see reviews and modeling studies in Tables 1 and 2). In a sense, the topic is popular because it proverbially offers us a chance to have our fishes and eat them too!

Despite this popularity, or perhaps because of it, the current literature on marine reserves as fisheries management tools is dominated by papers reviewing or modeling what marine reserves *could* do as fisheries management tools. There seems to be a remarkable

TABLE 1 Selected List of Major Reviews of Marine Reserves since 1990: Reviews Having a Strong Emphasis on the Role of Marine Reserves in Reef Fisheries Management

Authors (year)	Title and reference
PDT (1990)	The potential of marine fisheries reserves for reef fisheries management in the U.S. South Atlantic. *NOAA Tech. Memo.* **NMFS-SEFC 261,** 1–40.
Dixon and Sherman (1990)	"Economics of Protected Areas: A New Look at Benefits and Costs." Island Press, Washington, DC.
Polunin (1990)	Marine regulated areas: Expanded approach for the tropics. *Resour. Manage. Optim.* 7, 283–299.
Roberts and Polunin (1991)	Are marine reserves effective in management of reef fisheries? *Rev. Fish Biol. Fish.* **1,** 65–91.
Fiske (1992)	Sociocultural aspects of establishing marine protected areas. *Ocean Coast. Manage.* **17,** 25–46.
Jones *et al.* (1993)	Marine reserves: Do they work? *2nd Int. Temp. Reef Symp.* NIWA Marine Wellington, NZ.
Bohnsack (1993)	Marine reserves: They enhance fisheries, reduce conflicts, and protect resources. *Oceanus* **36,** 63–71.
Carr and Reed (1993)	Conceptual issues relevant to marine harvest refuges: Examples from temperate reef fishes. *Can. J. Fish. Aquat. Sci.* **50,** 2019–2028.
Dixon (1993)	Economic benefits of marine protected areas. *Oceanus* **36,** 35–40.
Dugan and Davis (1993)	Applications of marine refugia to coastal marine fisheries. *Can. J. Fish. Aquat. Sci.* **50,** 2029–2042.
Roberts and Polunin (1993)	Marine reserves: Simple solutions to managing complex fisheries? *Ambio* **22,** 363–368.
Agardy (1994)	Advances in marine conservation: The role of marine protected areas. *Trends Ecol. Evol.* **7,** 267–270.
McNeill (1994)	The selection and design of marine protected areas: Australia as a case study. *Biodiv. Conserv.* **3,** 586–605.
Rowley (1994)	Case studies and reviews: Marine reserves in fisheries management. *Aquat. Conserv. Mar. Freshwater Ecosyst.* **5,** 233–254.
Roberts *et al.* (1995)	Review of the use of marine fishery reserves in the U.S. SE Atlantic. *NOAA Tech. Mem.* **NMFS-SEFSC-376.**
Shackell and Willison (1995)	Marine protected areas and sustainable fisheries. *Science and Management of Protected Areas Association,* Wolfville, Nova Scotia.
Alder (1996)	Have tropical marine protected areas worked? An initial analysis of their success. *Coast. Manage.* **24,** 97–114.
Bohnsack (1996)	Maintenance and recovery of fishery productivity. *In* "Tropical Reef Fisheries" (N. V. C. Polunin and C. M. Roberts, eds.), pp. 283–313. Chapman & Hall, London.
Bohnsack and Ault (1996)	Management strategies to conserve marine biodiversity. *Oceanography* **9,** 73–82.
Clark (1996)	Marine reserves and the precautionary management of fisheries. *Ecol. Appl.* **6,** 369–370.
Farrow (1996)	Marine protected areas: Emerging economics. *Mar. Policy* **20,** 439–446.
Agardy (1997)	"Marine Protected Areas and Ocean Conservation." Landes Publishers, Austin Texas/Academic Press, San Diego, CA.
Ballantine (1997)	"No take" marine reserve networks support fisheries. *2nd World Fish. Congr.,* pp. 702–706.
Roberts (1997a)	Ecological advice for the global fisheries crisis. *Trends Ecol. Evol.* **12,** 35–38.
Schmidt (1997)	"No take" zones spark fisheries debate. *Science* **277,** 489–491.
Allison *et al.* (1998)	Marine reserves are necessary but not sufficient for marine conservation. *Ecol. Appl.* **8,** s79–s92.
Bohnsack (1998)	Application of marine reserves to reef fisheries management. *Aust. J. Ecol.* **23,** 298–304.

(*continues*)

TABLE 1 *(continued)*

Authors (year)	Title and reference
Guennette *et al.* (1998)	Marine reserves: From Beverton and Holt to the present. *Rev. Fish. Biol. Fish.* **8**, 1–21.
S. J. Hall (1998)	Closed areas for fisheries management—The case consolidates. *Trends Ecol. Evol.* **13**, 297–298.
Lauck *et al.* (1998)	Implementing the precautionary principle in fisheries management through marine reserves. *Ecol. Appl.* **8**, s72–s78
Roberts (1998)	Sources, sinks, and the design of marine reserve networks. *Fisheries* **17**, 16–19.
Environmental Biology of Fishes (1999)	Fish behavior and the design of coral reef reserves. *Environ. Biol. Fishes* **55**, 53–98. (4 papers).
Fogarty (1999)	Essential habitat, marine reserves, and fishery management. *Trends Ecol. Evol.* **14**, 133–134.
S. N. Murray *et al.* (1999)	No-take reserve networks: Sustaining fishery populations and marine ecosystems. *Fisheries* **24**, 11–25.
Dayton *et al.* (2000)	Marine protected areas: Parks, baselines, and fishery enhancement. *Bull. Mar. Sci.* **66**, 617–634.
Palumbi (2000)	The ecology of marine protected areas. *In* "Marine Ecology: The New Synthesis" (M. Bertness, ed.). Sinauer Press, Sunderland, MA.
Ecological Applications (2002)	Special Issue on Marine Reserves. *Ecol. Appl.* (in press).

paucity of good empirical studies telling us what they *can* do as fisheries management tools. In a fisheries context, one can define seven basic expectations of marine reserves. Five concern effects on the fishes inside the reserves, and the other two relate to the fishery continuing outside the reserve. To be effective as fisheries management tools, marine reserves should display net export of fish biomass that more than compensates for the loss of fishing area required to set up the reserves. The bipartite life cycle of most reef fishes (an adult benthic phase capable of dispersing distances of tens to thousands of meters and a larval phase capable of dispersing distances of tens to hundreds of kilometers) means that there are two potential ways by which marine reserves could display net export of fish biomass. Each will most likely act at very different spatial scales. The seven expected effects, summarized graphically in Fig. 1, are as follows:

Effects inside reserves

1. Significantly lower fishing mortality (F) than in fished areas, or even $F = 0$.
2. Significantly higher density of target species.
3. Significantly higher mean size/age of target species.
4. Significantly higher biomass of target species.
5. Significantly higher production of propagules (eggs/larvae) of target species per unit area.

Effects outside reserves—fisheries enhancement

6. Effects 1–4 above result in net export of adult (postsettlement) fishes (the "spillover effect"). This may occur by density-dependent movements (e.g., space limitation and territorial interactions) or may simply arise because higher densities of larger than average fishes occur in reserves and these fishes flux randomly across the unfished–fished boundary.
7. Effects 1–5 above result in net export of eggs/larvae (the "recruitment effect"). The result is an enhanced supply of recruits to fished areas.

Note that for marine reserves to be successful as fisheries management tools, the fisheries enhancement effects, particularly the "recruitment effect," are vitally important. Their effects, in the medium to long term, should more than compensate for the loss of fishing area required to set up the reserves. However, even if they do not fully compensate for this loss of fishing area, they will still act as a reserve of spawning stock from which to start a recovery in the event that other management measures fail.

In reviewing the literature on each of the seven topics above, the focus is on reef fisheries particularly coral reef fisheries and reef fishes. The main conclusions of the review are that the empirical evidence for effects 1–5 listed above is still surprisingly equivocal, with a lack of any well-designed, definitive experiments

TABLE 2 Selected List since 1990 of Major Papers Modeling Potential Impacts of Marine Reserves on Reef Fisheries

Authors (year)	Title and reference
Spillover/SSB[a]	
Polacheck (1990)	Year around closed areas as a management tool. *Nat. Res. Mod.* **4**, 327–354.
Russ *et al.* (1992)	Marine reserves and fisheries management on coral reefs with preliminary modeling of the effects on yield per recruit. *Proc. 7th Int. Coral Reef Symp.*, Vol. 2, pp. 978–985.
DeMartini (1993)	Modeling the potential of fishery reserves for managing Pacific coral reef fishes. *Fish. Bull.* **91**, 414–427.
Attwood and Bennett (1995)	Modeling the effect of marine reserves on the recreational shore-fishery on the south-western cape, South Africa. *S. Afr. J. Mar. Sci.* **16**, 227–240.
Recruitment	
Quinn *et al.* (1993)	Harvest refugia in marine invertebrate fisheries: Models and applications to the red sea urchin, *Strongylocentros franciscanus*. *Am. Zool.* **33**, 537–550.
Man *et al.* (1995)	Roles of marine reserves in recruitment to reef fisheries: A metapopulation model. *Biol. Conserv.* **71**, 197–204.
Holland and Brazee (1996)	Marine reserves for fisheries management. *Mar. Res. Econ.* **11**, 157–171.
Roberts (1997b)	Connectivity and management of Caribbean coral reefs. *Science* **278**, 1454–1457.
Wolanski *et al.* (1997)	Directional swimming of fish larvae determines connectivity of fish populations on the Great Barrier Reef. *Naturwissenschaften* **84**, 262–268.
Hannesson (1998)	Marine reserves: What would they accomplish? *Mar. Res. Econ.* **13**, 159–170.
Lauck *et al.* (1998)	Implementing the precautionary principle in fisheries management through marine reserves. *Ecol. Appl.* **8**, s72–s78.
Guennette and Pitcher (1999)	An age structured model showing the benefits of marine reserves in controlling exploitation. *Fish. Res.* **39**, 295–303.
Sladek-Nowlis and Roberts (1999)	Fisheries benefits and optimal design of marine reserves. *Fish. Bull.* **97**, 604–616.
Hastings and Botsford (1999)	Equivalence in yield from marine reserves and traditional fisheries management. *Science* **284**, 1537–1538.
Cowen *et al.* (2000)	Connectivity of marine populations: Open or closed? *Science* **287**, 857–859.
Tuck and Possingham (2000)	Marine protected areas for spatially structured exploited stocks. *Mar. Ecol. Prog. Ser.* **192**, 89–101.

[a]SSB, Spawning stock biomass.

carried out at appropriate scales of time and space. The evidence for the "spillover effect" is even more equivocal, for the same reasons. The evidence for the "recruitment effect" is virtually nonexistent, due to the incredible design, scale, and logistic difficulties of actually measuring net larval export from reserves. The questions regarding fisheries enhancement by net larval export, like all large-scale spatial and temporal questions in marine ecology, will probably not be answered unequivocally by empirical experiments alone. This will probably require a modeling approach (see Table 2), armed with the best possible empirical data derived from large temporal- and spatial-scale management experiments.

Despite these conclusions, it is vitally important to stress that marine reserves may be one of the few viable fisheries management options available to the reef fisheries of developing nations. Marine reserves should not be rejected, nor their implementation even be delayed, because the definitive experiments have not shown us the optimal design of reserves to ensure recruitment supply to fisheries. Marine reserves as fisheries management tools should be viewed for what they are—a healthy dose of the precautionary principle. They are reserves for and of the target fish stocks. They are our insurance policy against future fisheries management failures.

II. Effects inside Reserves

A. Effects of Marine Reserves on Fishing Mortality of Target Species

A basic expectation of a marine reserve is that it will reduce fishing mortality (F) to levels significantly lower than those in fished areas, perhaps even achieve $F = 0$ in the reserve. Unfortunately this has been a somewhat

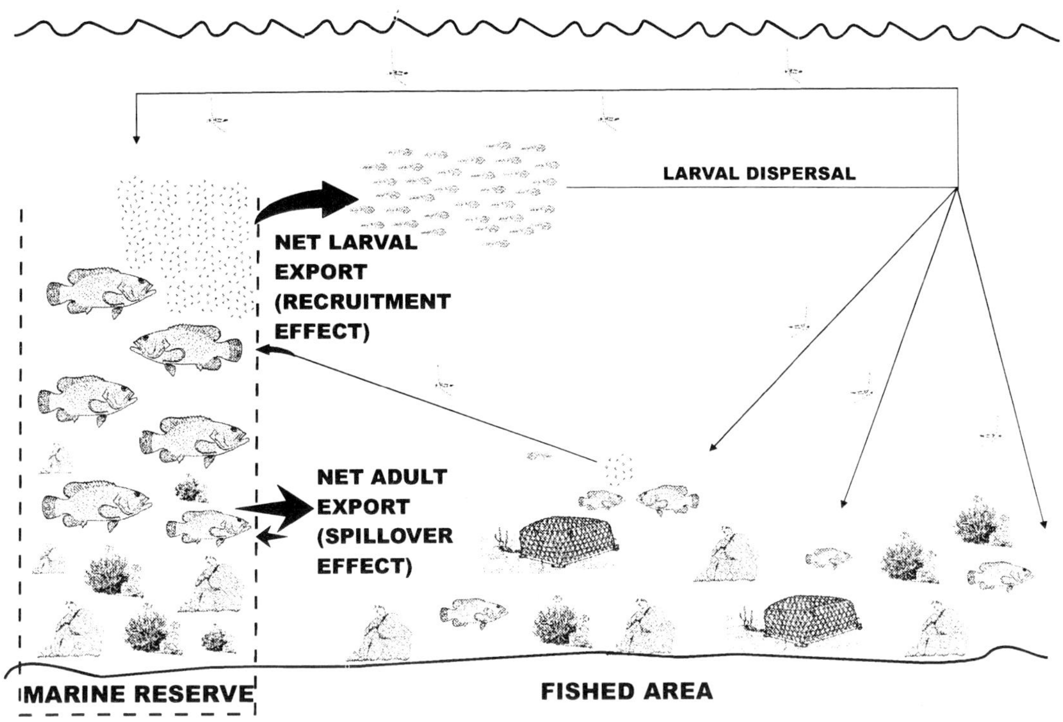

FIGURE 1 A visual representation of the seven basic expectations of marine reserves as fisheries management tools. Inside reserves, one expects significantly lower fishing mortality and significantly higher density, mean size/age, biomass, and production of propagules per unit area of target species, as compared to fished areas. Reserves are expected to become net exporters of postsettlement fishes (spillover effect) and propagules (recruitment effect). Fisheries outside reserves may be enhanced by these two mechanisms.

naive, even forlorn, hope for many marine reserves. Poaching is often the rule, rather than the exception. Kelleher *et al.* (1995), for example, pointed out that of 1306 marine protected areas (marine reserves) worldwide, the effectiveness of management was known for only 29% (383) of them. Furthermore, of these 383, almost 70% had only moderate to low management effectiveness. Given this, it is rather surprising that virtually no studies of the effects of marine reserves on attributes such as density, average size, and biomass of target species (see Section II,B) actually measure *F* inside the reserve(s) studied and in fished areas over time. The implicit assumption is that *F* is lower in the reserve, by some unspecified amount. This is equivalent to trying to measure the effects of predation on intertidal gastropods on rocky shores, and having some unknown amount of predation occurring inside predator exclusion cages. This is a rather unsatisfactory situation, experimentally, because a relatively short period of poaching can remove density and biomass gained over relatively long periods of protection. For example, Russ and Alcala (1996a) showed that 1–2 years of uncontrolled fishing removed density and biomass of target species accumulated over 5- to 10-year periods of protection at Sumilon Island, Philippines. It is acknowledged, however, that measuring *F* inside reserves would require some form of tagging or destructive sampling (e.g., to determine age structure), both often difficult to justify, or would require almost constant surveillance of poaching activity.

To emphasize the point that simply legislating a coral reef as a marine reserve does not necessarily result in $F = 0$, consider the largest and best known marine park in the world, the 350,000 km^2 Great Barrier Reef Marine Park (GBRMP) in Australia. This was certainly included in the 30% of marine protected areas considered by Kelleher *et al.* (1995) to have good management effectiveness. Approximately 5% of the area of this park consists of "no-fishing" zones, which include the so-called green reefs. However, studies by Ferreira and Russ (1995) and Adams *et al.* (2000), working on coral trout, the main target species of the GBR commercial and recreational line fishery, were unable to detect significant differences in the age frequency distributions between green and fished ("blue") reefs in three of four regions of the GBRMP. These results are of particular significance, because the reefs had been zoned "no fishing" for up to

8 to 12 years. Furthermore, Mapstone *et al.* (1997a) used visual census to compare 10 green reefs with 14 blue outer-shelf reefs in the Cairns section of the GBRMP and concluded that, after 8 years of zoning, there was virtually no difference in the density of coral trout between green and blue reefs. This lack of an effect of zoning is unlikely to be due to a low fishing intensity on this species on blue reefs (Mapstone *et al.*, 1996a), and movement of adult coral trout between coral reefs is known to be minimal (Davies, 1995). The most likely reason for the lack of effect of zoning is poaching on the green reefs, something that is, anecdotally at least, known to occur. Most reefs of the GBR are 50 to 100 km off the coast, and often hundreds of kilometers from the nearest towns. Aerial surveillance is logistically difficult and, until recently, enforcement of zoning regulations has been minimal. Thus, the shining example of management success, the largest marine park in the world, in a developed nation and sitting on one of the most sparsely populated coastlines in the world, has probably been rather unsuccessful in reducing F substantially on coral reefs nominally "closed" to fishing.

Some studies have been successful at measuring mortality rates of target species on reefs nominally closed to fishing. Buxton (1993) used age-based catch curves and an empirical estimate of natural mortality to estimate that the exponential rate of fishing mortality (F) on two species of sparid was 5.8 and 4.8 times higher in fished locations than in the adjacent Tsitsikamma National Park in South Africa. This marine reserve had been protected for 25 years at the time of sampling. Russ *et al.* (1998) were able to estimate the exponential rate of total mortality (Z) of a strong cohort of coral trout over a 4-year period on two reefs nominally closed to fishing in the Townsville region of the GBRMP. This region had a strong contrast in age frequency distributions between green and blue reefs of the GBRMP (Adams *et al.*, 2000). The estimate of Z on the green reefs, based on validated ages of fishes in the cohort, was two to three times less than for estimates on the reefs open to fishing (Russ *et al.*, 1998).

B. Effects of Marine Reserves on Density, Mean Size/Age, and Biomass of Target Species

At first sight, the evidence that marine reserves increase the density, average size, and biomass of target species on reefs significantly appears considerable (Tables 3 and 4). In reviews of these subjects Roberts and Polunin (1991), Dugan and Davis (1993), Rowley (1994), Bohnsack (1996, 1998), and Roberts (1997a), for example, all suggested that the evidence for enhanced abundance of target species in marine reserves was good. The majority of this evidence involves comparisons of abundance of target species at the one time at sites with and without marine reserve protection. Such comparisons are invariably confounded by habitat, history, and/or larval supply differences between reserve and fished locations, a point stressed nearly a decade ago by Roberts and Polunin (1991) and Dugan and Davis (1993). Table 3 is a summary of such spatial comparisons. In this table the focus is on the effects on "large predatory reef fishes" as the target group of the fisheries. Not included are the many excellent studies of invertebrates or intertidal organisms. Dugan and Davis (1993) provide the best summary of this work. The main reasons for restricting the table to predatory reef fishes were to maintain consistency with the theme of this book, to keep the table to a reasonable size, and to concentrate on a group of fishes likely to provide the best possible indication of the effects of marine reserves. Periods of marine reserve protection range from 3 to 36 years, although in many cases this protection has probably been ineffective (see Section II,A). Almost all studies reported significantly larger mean size of fishes inside reserves compared to fished locations. However, the effects of the reserves on density (and biomass, where data was available), expressed as the ratio of reserve density/fished density, range from 0.63 to 5.2. A plot of reserve density/fished density against years of nominal reserve protection suggests that, on average, reserves have a 2.5 times higher density than fished locations, based purely on spatial comparisons (Fig. 2). However, Fig. 2 also suggests that there is no evidence of a significant relationship between a higher reserve/fished location density ratio and years of reserve protection. This is true, even when data from Kenya, the Great Barrier Reef, California, and the Red Sea, areas having marine reserves with long periods of nominal protection but very low ratios of reserve/fished density, are excluded from the plot. The only change to the relationship is that reserves have, on average, a density three times higher than fished locations.

The spatial comparison evidence is clearly not telling us much about rates of increase of density (and biomass) over time (Table 3; Fig. 2). In fact, it never will tell us anything unequivocal about such things. Nor will it unequivocally demonstrate the causes of density differences between reserves and fished areas. For the past decade the data in Fig. 3 have been shown to students in an introductory marine ecology course, and it does not matter how long or hard these data are studied: it cannot be concluded that the reserve at Sumilon Island

TABLE 3 Summary of Spatial Comparisons of Density/Biomass/Mean Size of Large Predatory Reef Fishes between Marine Reserves and Fished Locations

Author (year)	Marine reserve[a]	Years of protection	No. of fished locations	Target fish	Reserve density/fished density[b]	Reserve biomass/fished biomass[b]	>Mean size?
Craik (1981)	Heron Island, GBR, AUS	5	7 GBR reefs	*Plectropomus leopardus*	3.8 (s)	—	Yes
Bohnsack (1982)	Key Largo, Florida	21	1	Large predators	3.9 (s)	—	Yes
Bell (1983)	Banyuls-Cerbere, FR	3	1	*Diplodus* (two species)	2.6 (s)	—	Yes
Russ (1985)	Sumilon, PH	9	1	Serranidae	5.2 (s)	7.6 (s)	Yes
Ayling and Ayling (1986)	Heron Island, GBR, AUS	12	6	*Plectropomus leopardus*	1.5 (ns)	—	Yes
Ayling and Ayling (1986)	Two green reefs	5	6	*Plectropomus leopardus*	1.2 (ns)	—	Yes
Ayling and Ayling (1986)	Two green reefs	3	6	*Plectropomus leopardus*	0.9 (ns)	—	No
McCormick and Choat (1987)	Leigh, NZ	11	1	*Cheilodactylus spectabilis*	2.3 (s)	—	Yes
Samoilys (1988)	Malindi, Watamu, KE	18	10	Large predators	0.63 (ns)	0.91 (ns)	No
Buxton and Smale (1989)	Tsitsikamma, SA	21	1	Sparidae	4.2 (s)	—	Yes
McClanahan and Shafir (1990)	Malindi, Watamu, KE	20	4	Labridae	1.3 (ns)	—	Not stated
Roberts and Polunin (1992)	Ras Mohammed, Red Sea	15	3	Serranidae	1.11 (ns)	1.4 (ns)	Yes
Polunin and Roberts (1993)	Saba, NA	4	6	Haemulidae (shallow)	4.0 (s)	6.0 (s)	Yes
Polunin and Roberts (1993)	Hol Chan, Belize	4	3	Lutjanidae (shallow)	2.1 (s)	3.3 (s)	Yes
Jennings *et al.* (1996)	Cousin Island, SEY	26	1	Large predators	—	3.5 (s)	Not stated
Jennings *et al.* (1996)	St. Anne, SEY	21	1	Large predators	—	1.2 (ns)	Not stated
Rakitin and Kramer (1996)	Barbados Marine Reserve	11	5	Trappable species	2.0 (ns)	—	Yes
Mapstone *et al.* (1997a)	10 GBR green outer-shelf reefs, AUS	8	14	*Plectropomus leopardus*	0.98 (ns)	—	Yes
Chapman and Kramer (1999)	Barbados Marine Reserve	15	2	*Haemulon* spp.	3.2 (ns)	—	Yes
Paddack and Estes (2000)	Point Lobos, CA	36	2	*Sebastes atrovirens*	1.3 (ns)	2.2 (s)	Yes
Paddack and Estes (2000)	Hopkins Marine Area, CA	12	2	*Sebastes atrovirens*	1.4 (ns)	2.88 (s)	Yes
Paddack and Estes (2000)	Big Creek, CA	2	2	*Sebastes atrovirens*	1.2 (ns)	1.0 (ns)	No

[a]Abbreviations: AUS, Australia; CA, California; GBR, Great Barrier Reef; KE, Kenya; FR, France; NA, Netherlands Antilles; NZ, New Zealand; SA, South Africa; SEY, Seychelles; PH, Philippines.

[b]s, Significant difference; ns, not significant.

TABLE 4 Summary of Studies with Temporal Monitoring of Density/Biomass of Large Predatory Reef Fishes in Marine Reserves and Fished Locations

Author (year)	Marine reserve[a]	Target	Years of protection at monitoring	Duration of monitoring (years)	Final density/ initial density (DR)[b]	DR/duration of monitoring reserve
White (1988)	Apo, PH	Serranidae	2, 3, 4	2	7.6	3.8
White (1988)	Balicasag, PH	Serranidae	0, 0, 1	2	3.9	2.0
White (1988)	Pamilacan, PH	Serranidae	0, 1	1	8.4	8.4
Clark *et al.* (1989)	Looe Key, Florida	Haemulidae	0, 1, 2	2	5.4	2.7
Cole *et al.* (1990)[c]	Leigh, NZ	*Cheilodactylus spectabilis*	1, 2, 3, 4, 7, 11	10	3.1	0.3
Bennett and Attwood (1991)	De Hoop, SA	*Coracinus capensis*	0, 2.5, 5	5	5.1[d]	1.0
Christie *et al.* (1994)	San Salvador, PH	Target species	0, 0.5, 1, 1.5, 2	2	2.9	1.5
Roberts (1994)	Saba, NA	Large Predators	4, 6	2	1.0	0.5
McClanahan and Kaunda-Arara (1996)	Malindi, Watamu, KE	Lethrinidae	20, 24, 25, 26	6	2.2[e]	0.4
McClanahan and Kaunda-Arara (1996)	Mombasa Marine National Preserve, KE	Lethrinidae	0, 1, 2, 3	3	13.5[e]	4.5
Russ and Alcala (1996a)	Apo, PH	Large predators	1, 3, 6, 8, 9, 10, 11	10	7.1	0.7
Russ and Alcala (1996a)	Sumilon, PH	Large predators	−2, −1.5, −1, 2, 4, 5, 9	11	4.5	0.4
Russ and Alcala (1996a)	Sumilon nonreserve, PH	Large predators	0, 0, 2, 4, 5	8	2.6	0.3
Edgar and Barrett (1997)	Maria Island, Tasmania	*Penicipelta vittiger*	1, 2	1	1.3	1.3

[a]Abbreviations: KE, Kenya; NA, Netherlands Antilles; NZ, New Zealand; PH, Philippines; SA, South Africa.
[b]DR, Density ratio.
[c]Data of McCormick and Choat (1987), used at 11 years of protection.
[d]Catch rate data.
[e]Biomass, not density used.

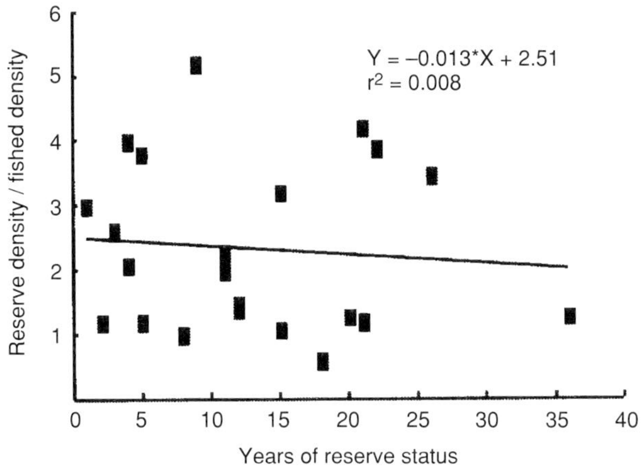

FIGURE 2 Reserve density/fished density of large predatory reef fishes versus years of reserve status. Data from 16 studies in Table 3. All studies made spatial comparisons of density at one time between reserve and fished areas.

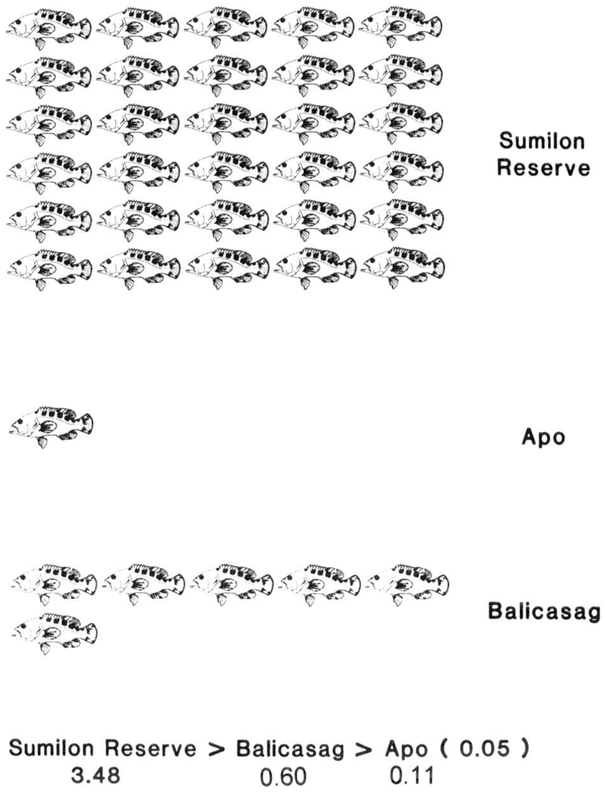

FIGURE 3 Mean standing crop (kg wet weight/1000 m^2) of epinepheline serranids at three locations in the Philippines in December 1983. The Sumilon Reserve had been protected fairly effectively from fishing for 9 years. The Apo and Balicasag locations had been essentially open to fishing indefinitely. You cannot conclude from such data that the reserve caused the difference in standing crop between locations. There were no baseline data available before reserve establishment, and no temporal monitoring before and after reserve establishment at reserve and control locations. Data from Russ (1985).

caused the difference in biomass between the three locations. This was pointed out when the data were first published (Russ, 1985). Without data on what the locations were like before the reserve was set up, and temporal monitoring to show that biomass increased over time in the reserve but not the fished locations, the difference could have existed for reasons other than reserve protection. Thus, to demonstrate effects of marine reserves on density/biomass of target species unequivocally, one needs studies that incorporate both spatial and temporal comparisons, preferably with several sampling times at all locations *before* reserves are established.

Spatial and temporal comparisons of density (biomass) of large predatory reef fishes at reserve and fished locations are summarized in Table 4. This table includes data from nine studies, indicating a greater availability of studies incorporating spatial and temporal comparisons than were available to earlier reviews (e.g., Roberts and Polunin, 1991; Dugan and Davis, 1993). Most of the studies in Table 4 have been of relatively short duration. Changes in density *inside* the reserves over time, expressed as the ratio of final density/initial density, range from 1.3 (in 1 year) to 13.5 (in 3 years). When final density/initial density is divided by duration of reserve monitoring, a pattern does emerge. Rapid rates of increase have invariably been reported by short-term studies (Table 4). The most spectacular of these are an 8.4-fold increase in density of serranids at Pamilacan reserve over 1 year (White, 1988) and a 13.5-fold increase in biomass of lethrinids in the Mombasa National Park over 3 years (McClanahan and Kaunda-Arara, 1996). Such data may result from very successful recruitment soon after closure of a reserve (see Russ and Alcala, 1996a; Russ *et al.*, 1996). The five studies of reasonable duration (6–11 years at Leigh, Malindi, Sumilon, Apo, and Sumilon Nonreserve) all reported relatively low figures for final density/initial density divided by duration of reserve monitoring. The mean increase in density for these five reserves was by a factor of 3.9 over a mean period of monitoring of 9 years.

The low rates of mean density increase for the longer term monitoring studies are not due to density reaching an asymptote in the reserves. This does not appear to have happened in any of the studies in Table 4, which have assessed monitoring over 6 years or greater. For example, significant positive linear correlations of mean density of large predators against years of reserve protection were observed at both Sumilon and Apo

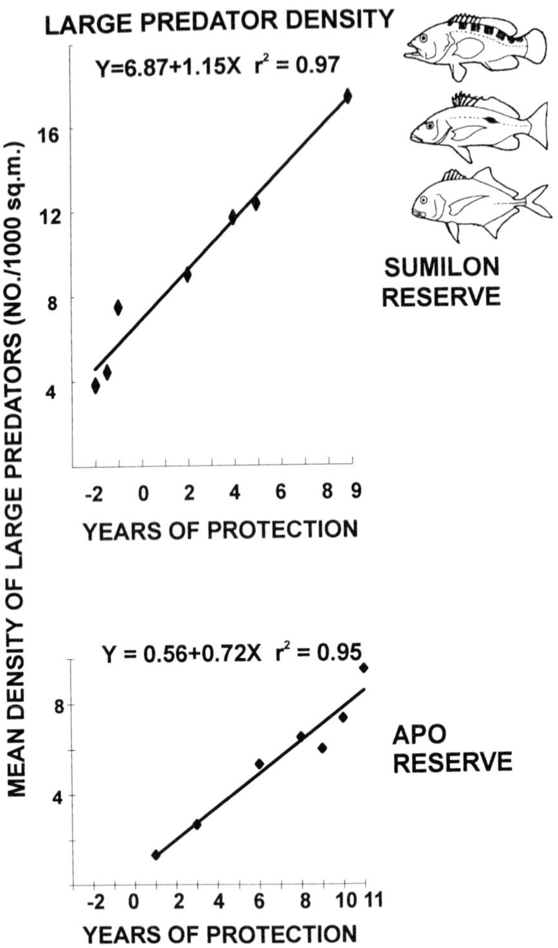

FIGURE 4 Relationship between mean density of large predatory reef fishes [families Serranidae (Epinephelinae), Lutjanidae, Lethrinidae, and Carangidae as a group] and years of marine reserve protection at Sumilon and Apo islands, Philippines. Negative years of protection indicates years open to fishing. Reprinted with permission from Russ and Alcala (1996a).

reserves over a decade (Fig. 4). An asymptote in density over time was not observed. Monitoring of these reserves (and fished locations) now spans 15 years, still with no clear evidence of an asymptote in density of large predatory reef fishes at the Apo reserve (G. R. Russ, unpublished data). Over a 15-year period of monitoring (from 1 to 16 years of reserve protection), density of large predatory fishes at Apo reserve has increased by a factor of 8.4. The pattern of increase of mean biomass against years of reserve protection during the first decade of monitoring at Sumilon and Apo was more curvilinear than that of mean density, particularly at Sumilon, where a slow increase was observed in the first 3 to 5 years (reflecting delayed recruitment and a natural delay to the period of maximum individual weight growth), followed by an increasing rate over the next 4 years (Russ and Alcala, 1996a). The potential generality of a curvilinear pattern of buildup of biomass of large predatory fish following marine reserve protection was tested with a simple age-structured computer simulation model (Russ and Alcala, 1996a). In 36% of 300 simulations, the pattern of buildup of biomass was curvilinear, suggesting that the empirical results from Sumilon and Apo reserves may have some generality. A curvilinear pattern occurred when recruitment was relatively low during the first 2 to 4 years after closure to fishing. Eventually, of course, an asymptote is expected. No empirical data appear to be available, as yet, to demonstrate this.

Table 5 provides a summary of the sampling designs of the 25 studies reviewed in Tables 3 and 4. The conclusions drawn from Table 5 are almost identical to those reached by Jones *et al.* (1993), almost a decade ago. The majority (64%) of studies involved spatial comparisons only of reserve and fished locations. Furthermore, only 24% involved spatial comparisons of more than two reserve and fished locations. Just 16% involved spatial and temporal comparisons, with no data collected before the reserves were established. Of these, Cole *et al.* (1990), Russ and Alcala (1996a), and Edgar and Barrett (1997) collected data within the first year of protection of some reserves. Another 16% had spatial and temporal comparisons with before data (White, 1988; Clark *et al.*, 1989; Christie *et al.*, 1994; McClanahan and Kaunda-Arara, 1996), but often the "before" data spanned a relatively short period. One study had "before/after" data inside but not outside a reserve (Bennett and Attwood, 1991). Another (Russ and Alcala, 1996a) collected convincing before/after data at a "reserve" (Sumilon Nonreserve!) and a fished site (Apo Nonreserve) over the period 1983–1991 (Fig. 5) somewhat by default when fishing was banned around the whole of Sumilon Island from 1987 to 1991. Thus, extensive "before" data is still very rare in studies of the effects of marine reserves.

Another common design fault is the lack of nested sites within reserve and fished locations. Only about half of the studies reviewed did this (Table 5), and thus could have confounded "reserve" effects with small-scale spatial variability. Most studies measured benthic habitat variables, and many stratified sampling by depth, at reserve and fished locations. However, only 20% of all studies made a formal (statistical) attempt to partition out such confounding effects. The best attempts at this so far are the studies of McCormick and Choat (1987) and Chapman and Kramer (1999). Finally, no study measured fishing mortality at reserve and fished locations, especially over time (see Section II, A). Other studies sometimes

TABLE 5 Summary of Sampling Designs of 25 Studies Used to Detect Effects of Marine Reserves on Density/Biomass/Mean Size of Large Predatory Reef Fishes

Study design	Number of studies
Spatial comparisons only	16 (64%)
Spatial comparisons with >2 reserves and fished locations	6 (24%)
Spatial and temporal comparisons without BEFORE data	4 (16%)
Spatial and temporal comparisons with BEFORE data	4 (16%)
Nested sites within both reserve and fished locations	14 (56%)
Habitat variables measured	21 (84%)
Habitat variables factored out of reserve/fished comparison	5 (20%)
Fishing mortality measured at reserve and fished sites	0

compare density/biomass of reef fishes along presumed gradients of fishing pressure (reviewed by Jennings and Lock, 1996; Jennings and Kaiser, 1998).

Do marine reserves increase density, average size/age, and biomass of target species significantly? Current evidence supporting these contentions (Tables 3 and 4) appears to be more equivocal than has been suggested (e.g., Bohnsack, 1996; Roberts, 1997a). The empirical data consisting of spatial comparisons of reserve and fished locations suggest that density is, at best, on average three times higher inside reserves. The few long-term data on temporal change in density of such fishes inside reserves suggest a mean increase in density of 3.9 over a mean period of 9 years. Given that almost all studies report larger mean sizes of fishes inside reserves, the differences for biomass should be greater.

The major reason for the somewhat equivocal nature of the data is that, as yet, no definitive experiment has been performed that includes all of the following protocols:

1. Data on all variates of interest collected before and after reserves are established, in sufficiently replicated pairs of reserve and fished locations [i.e., A "Before–After Control-Impact-Pair" (BACIP) design]. Such designs are the generally accepted method of unequivocally measuring environmental impacts in marine ecology (Jones *et al.*, 1993; Underwood, 1994).
2. Factoring out the confounding effects of habitat and larval supply differences between reserve and fished sites [e.g. by multiple regression techniques; see Chapman and Kramer (1999)].
3. Including replicate sites nested inside reserve and fished locations. Many studies are pseudoreplicated. That is, they have replicated their basic sampling unit (e.g., transects), but have sampled only one site per reserve and control location. Thus, significant differences may reflect variation among sites within reserve and control locations, rather than between reserves and controls.
4. Measuring F in reserve and fished locations before and after reserves are established and over time, to ensure that F is reduced significantly in reserves.
5. Carrying out the experiment on a time scale of 5–20 years, because such time scales are probably the appropriate ones for most reef fishes (e.g., Russ and Alcala, 1996a).
6. Replicating such an experiment regionally to ensure the generality of the result.

The studies currently underway that come closest to these rather stringent requirements are those of Hatcher *et al.* (1995) on coral reefs of St. Lucia, Edgar and Barrett (1997) on temperate reefs in Tasmania, Mapstone *et al.* (1996b) on the Great Barrier Reef, G. P. Jones and colleagues in Papua-New Guinea (G. P. Jones, personal communication), and Russ and Alcala in three new marine reserves in the Philippines. An experimental design to measure buildup of biomass of target species adequately in marine reserves, and spillover of biomass from the reserves, is presented in Section III,B.

Despite the lack of such a comprehensive experiment, there is no doubt that some studies, involving temporal monitoring over 6 to 10 years, have produced convincing evidence that marine reserves can increase variates such as density, mean size/age, and biomass (e.g., Cole *et al.*, 1990; McClanahan and Kaunda-Arara, 1996; Russ and Alcala, 1996a, 1998a,b). The most important message from such studies is that the time scales of continuous protection from fishing required to ensure sufficient biomass buildup for most coral reef fish will, in many cases, probably be measured in decades.

C. Effects of Marine Reserves on Production of Eggs/Larvae of Target Species per Unit Area

If marine reserves increase the density, mean size, and thus biomass per unit area of target species of reef fishes, a logical expectation is that the reproductive output per unit area of target species will not just increase, but increase substantially. This is so because fecundity of reef fishes increases exponentially with body size

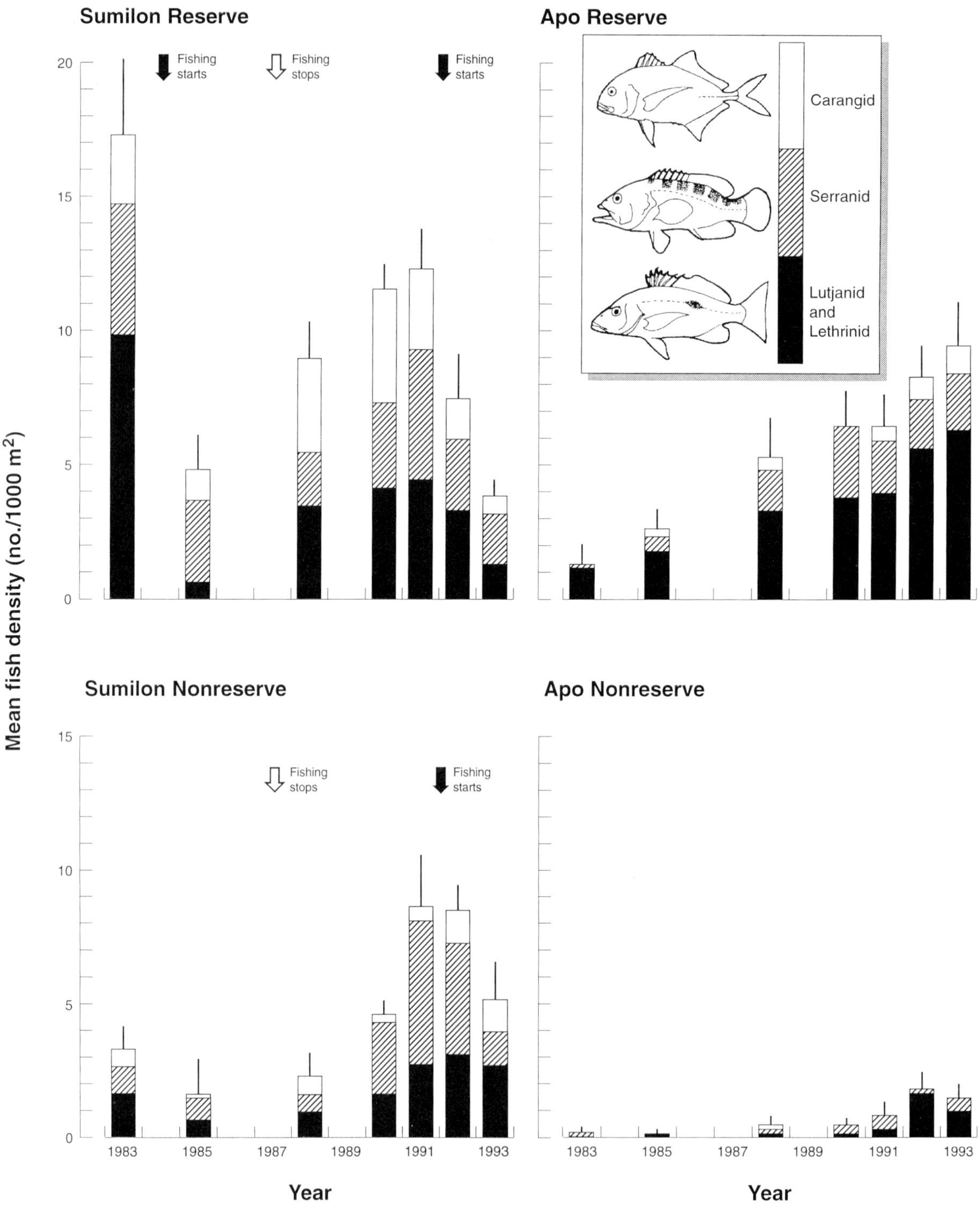

FIGURE 5 Mean (±SE) density (number/1000 m^2) of large predatory reef fishes [families Serranidae (Epinephelinae), Lutjanidae, Lethrinidae, and Carangidae as a group] at four sites at seven sampling times from 1983 to 1993. Black arrows indicate when fishing began and the open arrows indicate when fishing stopped at the two Sumilon sites. Fishing stopped at Apo Reserve in 1982 and the site remained unfished throughout the study. The Apo nonreserve area was open to fishing throughout the study. Reprinted with permission from Russ and Alcala (1996a).

(Thresher, 1984). This point was made more than adequately by the early reviews of Bohnsack and colleagues (PDT, 1990) and Roberts and Polunin (1991). Their arguments were based on evidence of greater biomass of target reef fishes inside reserves and reproductive biology of reef fishes, derived from the literature. Readers are referred to Table 3 of Roberts and Polunin (1991) and the very memorable Figure 6 in the Plan Development Team Publication (PDT, 1990), depicting that a 61-cm-long female red snapper had a fecundity equal to that of 212 42-cm long females. This calculation did not take into account potential differences in the frequency of spawning, but it was suggested that bigger fishes may spawn more frequently (PDT, 1990). In addition, a number of modeling studies (e.g., Polacheck, 1990; DeMartini, 1993) demonstrated a long time ago that marine reserves increase spawning stock biomass per recruit for fishes of moderate vagility.

I will again restrict the discussion to reef fishes, however, the best empirical evidence in support of greater reproductive output per unit area inside reserves comes from studies of invertebrates (e.g., lobsters, abalone, urchins). In one of the most comprehensive reviews available, all examples of the effects of marine reserves on reproductive output were studies of invertebrates (Dugan and Davis, 1993). Davis and Dodrill (1989), for example, suggested that the lobsters in the Dry Tortugas marine reserve may have had a greater total reproductive output than much of the fished areas in the lower Florida Keys. Probably the best data on potential reproductive output of reef fishes from marine reserves also come from the Florida Keys. Bohnsack and colleagues have data on the spawning potential ratio (potential fecundity of a recruit in a fished compared with an unfished stock) for 16 species of groupers, 13 species of snappers, and 5 species of grunts in the Florida region (Bohnsack and Ault, 1996; Ault *et al.*, 1997; Bohnsack, 1998). Most species of groupers had spawning potentials less than 20% of that expected for unfished populations, for example. Presumably such ratios will be monitored in the region over time to test if the ratios increase as spawning stock builds up over time in the "no-fishing" sections of the Florida Keys National Marine Sanctuary (Bohnsack and Ault, 1996).

Surprisingly few empirical studies attempt to estimate reproductive output per unit area of reef fishes inside and outside reserves. Only one of the 25 studies reviewed in Tables 3 and 4 estimated this, for example. Paddack and Estes (2000) used density, size structure, and length–fecundity relationships of rockfishes (*Sebastes* spp.) to estimate that the reproductive potential (eggs produced per unit area) was 2.9 and 2.7 times higher inside than outside the Hopkins and Point Lobos marine reserves (36 and 12 years of protection, respectively), but was almost identical inside and outside the Big Creek reserve (2 years of protection). The main impediments to collection of similar data for other marine reserves are the paucity of good length–fecundity relationships for most reef fishes, and a poor understanding of the frequency of spawning of many reef fishes.

As concluded for studies of the effects of marine reserves on density, size structure, and biomass of reef fishes, no empirical studies capable of unequivocally testing effects on reproductive output exist. An experimental design to do so is given in Section III,B.

III. Effects outside Reserves—Fisheries Enhancement

A. The Spillover Effect

Spillover was defined in Section I as net export of adult (postsettlement) fishes from reserves to fished areas. The term "emigration of adults from reserves," or similar terms that mean movement of fishes from reserves to surrounding areas, are still used commonly in the literature to mean spillover (e.g., Rowley, 1994; Bohnsack, 1996; Roberts, 1997a). The distinction between "emigration" and "net export" is not subtle. The latter term requires that you measure both emigration and immigration. For example, most of the major reviews of marine reserves cite the paper of Attwood and Bennett (1994) as some of the best available evidence for spillover. The paper is still one of the most extensive tagging studies in the marine reserve literature, with 11,022 fishes tagged and 1008 tag recoveries. Emigration of fishes from the reserve was demonstrated unequivocally. However, fishes were tagged only in the reserve. Immigration of fishes into the reserve could not be measured and, thus, net export could not be demonstrated. The point is that a clear idea of what spillover is clarifies the experimental design required to measure it.

Again, the discussion of spillover here is restricted to studies of reef fishes. However, some of the best evidence for spillover involves invertebrates, particularly lobsters, prawns, and crabs (see Dugan and Davis, 1993; Bohnsack, 1996), and reef fish ecologists are encouraged to consult this excellent literature base. For example, Yamasaki and Kuwahara (1990), studying snow crabs in Japan, demonstrated enhanced catch rates in a fished area surrounding a reserve after

5 years of reserve protection *and* demonstrated movement from the unfished to the fished area.

Studies of spillover of reef fishes are of four main types. First, there are those that model spillover (see Table 2). Second, some studies have monitored catch and catch per unit effort (CPUE) of fisheries adjacent to reserves over time. Third, a few studies have measured gradients of either density or CPUE across or near reserve boundaries. Fourth, many studies have measured movement patterns of fishes near and across reserve boundaries by tagging fish.

Several studies have modeled the potential effects of movements of adult reef fishes from marine reserve to fished areas on yield per recruit (YPR) of target species (Polacheck, 1990; Russ *et al.*, 1992; DeMartini, 1993; Attwood and Bennett, 1995). Yield per recruit is a function of reserve size, fishing mortality rate (F) outside the reserve, and life history characteristics of the target species, including movement rates. Most studies concluded that reserves contribute little, if anything, to YPR, except when F is very high. Even in the latter case, gains are very moderate.

Two studies have monitored catch and catch per unit effort of fisheries adjacent to reserves over time. Alcala and Russ (1990), studying coral reef fishes at Sumilon Island in the Philippines, demonstrated a significant decline in catch rates and total catch after a marine reserve (with almost 10 years of protection) was heavily fished (Russ and Alcala, 1989) (Fig. 6), suggesting that migration of fishes from the reserve to the nonreserve area during protection was enhancing fisheries yield. The study lacked data on movement patterns of fishes and spatial replication, and remains suggestive but inconclusive. McClanahan and Kaunda-Arara (1996) monitored catch and CPUE close to (1.5 km) and far away from (6 km) the Mombasa marine park in Kenya, before and after the establishment of the park. The park removed 65% of the fishing area used by the closest fish-landing site, and about 65% of fishers quit that landing site. Although CPUE increased by 110% at the landing site closest to the reserve 2 years after establishment of the reserve, total catch was still 35% lower than before the reserve was in place. Note that a neutral reserve effect would suggest a 65% lower total catch. Further monitoring presumably will demonstrate if total catch does approach previous levels from the area, providing further evidence for export from the reserve.

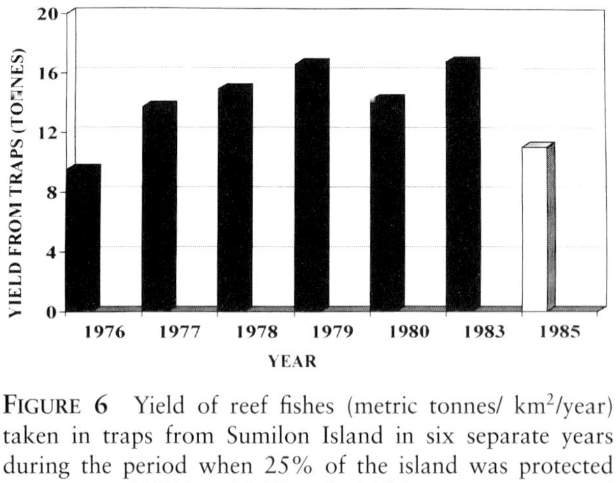

FIGURE 6 Yield of reef fishes (metric tonnes/ km^2/year) taken in traps from Sumilon Island in six separate years during the period when 25% of the island was protected as a reserve (1974 to 1984) and yield from traps measured 18 months after protection broke down (1985/1986). The yield from traps (approximately 45% of the total yield) from the whole island in 1985/1986 was significantly less than the average yield from the nonreserve area (75% of the reef) measured over six separate years during the period of protection (one-sample t test, $t_5 = 3.05$, $p < 0.05$). Redrawn with permission from Alcala and Russ (1990).

At least two studies have measured gradients of either density or CPUE across or near reserve boundaries. Russ and Alcala (1996b) reported some circumstantial evidence, derived from underwater visual census monitoring of densities of large predatory coral reef fishes inside and adjacent to a small marine reserve, at Apo Island in the central Philippines over a 10-year period (1983–1993) (Fig. 7). They proposed a simple set of models of change in the ratio of fish densities inside and outside marine reserves that one might expect over time and showed that the data collected for the Apo reserve was consistent with spillover occurring. They concluded, however, that the spillover effect may be slow to develop (up to 8 years), and may be important on scales of a few hundred meters only. Rakitin and Kramer (1996) found that the density of trappable species of reef fishes was higher in the Barbados marine reserve (Table 3). They also showed that trap catches decreased gradually with distance from the center of this reserve. For example, trap catch rates 100–500 m from the reserve boundary were up to twice those 1.5 km from the boundary. However, such a gradient was less evident in visual census data.

A large number of studies have measured movement patterns of fishes near and across reserve boundaries by tagging fishes. Movement patterns of reef fishes were reviewed very well by Roberts and Polunin (1991). Bryant *et al.* (1989) and Rutherford *et al.* (1989) used tag–recapture studies to demonstrate that several species of reef fishes moved out of the Everglades National Park in Florida Bay seasonally, and that *Lutjanus griseus* moved, on average, 18.3 km, with the movement apparently related to an ontogenetic shift in habitat preference. Buxton and Allen (1989) tagged 464 individuals of two species of sparid in the Tsitsikamma Coastal National Park in South Africa. Of the 12 individuals recaptured, all were within 2 km of their

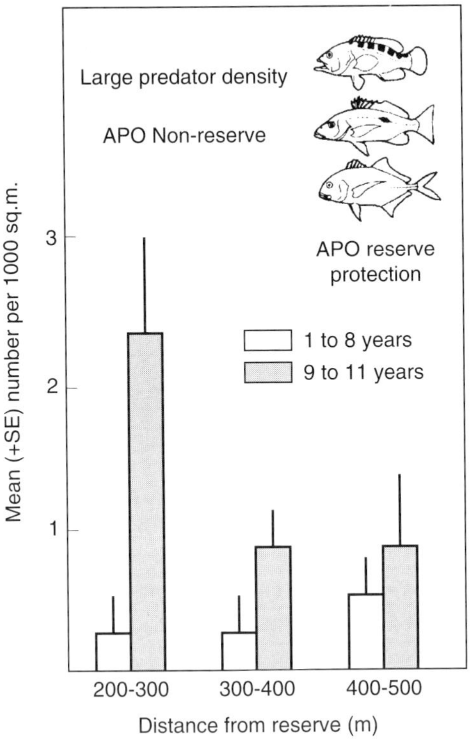

FIGURE 7 Mean (+SE) density of large predatory fish [Serranidae (Epinephelinae), Lutjanidae, Lethrinidae, and Carangidae as a group] at different distances from the Apo Island (Philippines) reserve boundary during the first 8 years of reserve protection (four sampling periods pooled) and the last 3 years of reserve protection (three sampling periods pooled). Reprinted with permission from Russ and Alcala (1996b).

release point, and none had moved out of the park. The study by Attwood and Bennett (1994) was discussed above. Beinssen (1989) showed that 29% of tagged individuals of the serranid *Plectropomus leopardus* moved distances of up to 500m in 3 weeks, with movements recorded from an unfished to a fished area of Heron Island reef on the Great Barrier Reef, Australia. Davies (1995) investigated the movement of the same species studied by Beinssen between and within five coral reefs of the Great Barrier Reef. Davies concluded that less than 1% of the tagged population of 4627 fishes moved between reefs, but that movement within reefs was considerable. Twenty six percent of tagged fishes moved distances within reefs of 2–2.5 km; 7%, distances of 2.5–5 km; and 2%, distances of 5–7.5 km over a period of 22 months. Most of the movements greater than 1 km were likely to be to spawning sites (Zeller, 1998).

Zeller and Russ (1998), also working on coral trout on the GBR, and using freeze branding and visual recapture, were unable to detect any movement of fishes across reserve/fished boundaries at Lizard Island. This was probably because both tagging and resighting effort were distributed evenly through the study area, rather than concentrated near the boundaries. The same study used ultrasonic telemetry to show that coral trout with territories near reserve boundaries could cross the boundary up to 15 times per month. Holland *et al.* (1993), working in Hawaii, used conventional tag–recapture and ultrasonic telemetry to show that the mullid (*Mulloides flavolineatus*) demonstrated high site fidelity, but exhibited crepuscular movements away from daytime schooling sites to nighttime feeding grounds up to 600 m away. The only movements out of a 1.37 km^2 no-fishing conservation zone were related to summer spawning migrations. Holland *et al.* (1996), using similar techniques at the same marine reserve in Hawaii, showed that *Caranx melampygus* had strong site fidelity, with 75% of tagged fishes recaptured within 0.5 km of the release point. Kramer and Chapman (1999) and Chapman and Kramer (2000) used conventional tagging techniques, traps, and visual resighting to study movement patterns of coral reef fishes near the boundary of the Barbados Marine Reserve. Most species were strongly site attached, with the majority of recaptures and resightings occurring at the site of tagging. Standardized recapture ranges for traps were 0–116 m, and for visual re-sightings, 0–44 m. Maximum recorded distances of movement were 50–150 m for most reef fish families, with some acanthurids moving 200–500 m. A total of 1443 individual reef fishes in 35 families were tagged. Despite this considerable level of effort, no tagged fishes were detected to move from the reserve to the nonreserve area, and only three movements were detected from the nonreserve to the reserve.

Virtually all of these tag–recapture studies confirm that reef fishes, particularly large predatory reef fishes, are certainly likely to move over distances of a few hundred meters to a kilometer or more, and that such fish may move considerably larger distances during ontogenetic shifts in habitat preference, or during spawning migrations. My interpretation of these data is in agreement with that of Roberts and Polunin (1991), who concluded that considerable potential exists for movement of adult fishes across reserve boundaries, but that any significant enhancement of fisheries by net export of adult biomass would be expected to be over distances of less than 1 km. Measurable benefits of spillover (net export of postsettlement reef fish) are likely to operate on scales of about 100–500 m for most medium-sized coral reef fishes.

Despite the impressive data on fish movements, some in and near marine reserves, few studies have measured movement patterns of reef fishes along gradients encompassing several reserves, their boundaries, and adjacent fished areas. One such study is that of B. G. Hatcher and colleagues at the Soufriere marine management area in St. Lucia, Caribbean (Hatcher *et al.*, 1995; Corless *et al.*, 1996). This group collected movement data along a gradient encompassing two reserve and fished areas, and measured movement patterns of reef fishes during the first year of reserve protection. Their initial conclusions (Corless *et al.*, 1996) were that the marine reserves had no significant effect on fish movement, but they acknowledged the short duration of reserve protection. They also demonstrated strong site fidelity of most reef fishes, with mean distance moved per family ranging from 10 to 260 m. In a similar study of two reserve/fished boundaries and two control areas at Lizard Island, GBR, G. R. Russ, D. C. Zeller, and C. Hatcher (unpublished data) found that 69% of 1188 recaptured predatory reef fishes had not moved from their 50 × 30-m area of tagging, and that only 8% moved distances of 100 m or more. The only reserve boundary with a clear difference in density (~1.5 times higher in the reserve) had a 3:1 bias in movement from the reserve to the fished area. Net movement was from high to low density in three of four areas.

Do marine reserves enhance fisheries in areas outside the reserves by spillover? This requires a demonstration of net export of fish biomass from reserves to fished areas that, to a reasonable extent, compensates for the loss of area required to set up the reserve. The evidence in support of spillover is even more equivocal than the evidence for significant biomass buildup reported in Section II,B. If it occurs, it will likely require a long period of reserve protection to develop, and probably act on scales of hundreds of meters to a few kilometers, depending on species. As yet, there are no experiments that include all six points listed in the experimental design in Section II,B and also incorporate two additional points: (7) measurement of movement patterns across a gradient of reserve/fished sites before and after the reserves have been established and (8) measurement of catch and catch per unit effort (CPUE) close to and far away from reserves, before and after they are established.

B. An Experimental Design to Measure the Effects of Marine Reserves on Biomass of Target Species and Spillover

The following experimental design incorporates all eight points listed in Sections II,B and III,A, and could be used as a general guide to measuring the effects of marine reserves on density, mean size/age, biomass, reproductive output per unit area, and spillover of target species. The first requirement (Point 1) is a Before–After Control-Impact-Pairs (BACIP) design (Fig. 8). Such designs are generally considered superior for detecting environmental impacts (Underwood, 1994). In such a design, you have control-impact (CI) pairs (in this case fished and reserve locations) monitored through time, starting before the reserves are established (at least two or three estimates of all variates of interest should be taken before the reserves are protected, including estimates made at both spawning and nonspawning seasons) and continuing after they are established. Statistical power of such designs increases with the number of CI pairs ($n = 3$ shown in Fig. 8) and the number of sampling times, and, of course, depends on the strength of the treatment effect. The latter is the effect of reduction in fishing mortality (F) in the reserve on the variate of interest.

Control-impact-pair locations should ideally be selected so that the benthic habitats (and perhaps the larval supply characteristics—although this is very difficult to measure) are as similar as possible. Such confounding characteristics should be measured and monitored, and "factored out" where possible (Point 2) with multiple regression techniques [see Chapman and Kramer (1999) for a good example of factoring-out effects of benthic habitat]. In addition, replicate sampling units should be monitored at replicate sites, nested within reserve and control locations (Point 3). Replicate nested sites within reserve and fished locations are labeled S1 to S12 in Fig. 8.

Some attempt should be made to measure fishing mortality (F) inside and outside all of the reserve-fished locations (Point 4). Again this is very difficult to do, because it usually requires some form of tagging or destructive sampling for collection of age frequency distributions. Reasonable estimates of fishing effort (E) inside and outside the reserves could well suffice, particularly if some independent estimate of the catchability coefficient (q) were available.

The experiment probably needs to run successfully over a time period of 5 to 20 years (Point 5), depending on the life history characteristics of the target species (particularly longevity, age at first reproduction, and recruitment rates). Ideally, the experiment should be duplicated regionally (Point 6), to ensure the generality of the result.

The BACIP design presented so far needs to be extended to examine gradients of fish biomass and fish yield across reserve and fished locations. This requires that a large transect be set up that traverses all reserve and control locations, somewhat like that shown in Fig. 9. Such a transect would be divided into smaller sections, e.g., 50, 100, or 500 m in length, depending

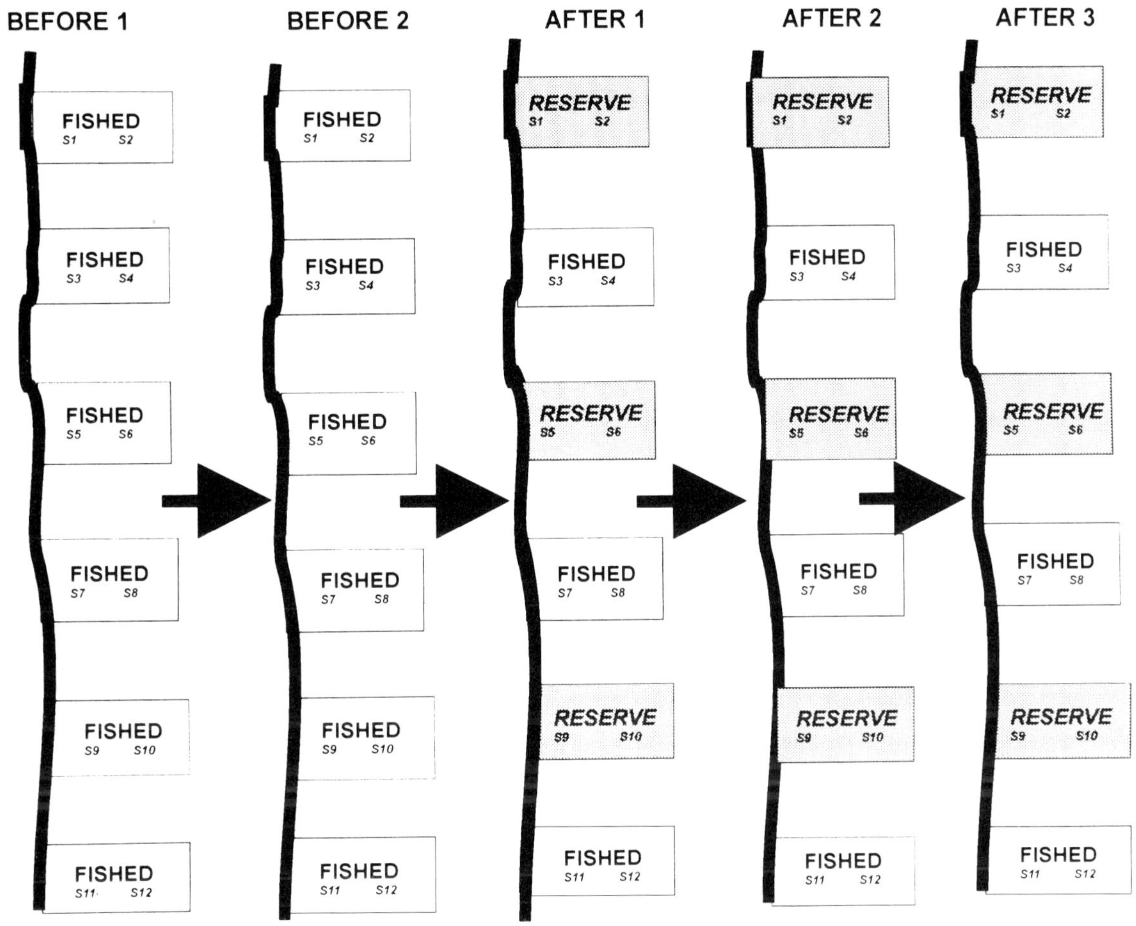

FIGURE 8 A Before–After Control-Impact-Pairs (BACIP) design to test the effects of marine reserves on variates such as fish density and biomass inside reserves, and local yields adjacent to the reserves. The design has control-impact pairs (in this case fished and reserve locations) monitored through time, starting before the reserves are established (at least two or three estimates of all variates of interest should be taken before the reserves are protected) and continuing after they are established. The statistical power of such designs increases with the number of control-impact pairs ($n = 3$ shown) and the number of sampling times, and, of course, depends on the strength of the treatment effect. The treatment effect is how strongly the reduction of fishing mortality (F) affects the variate of interest. S1 to S12 are sampling sites nested in reserve and fished locations. Modified from Hatcher *et al.* (1995).

on the target species and size of reserves. It is then assumed that the investigators can externally tag individual fishes, such that the tag color/position/type, etc. uniquely identifies the section of the transect in which the fish was first tagged (Fig. 9). These fishes can then be recaptured over time, either by underwater visual census (UVC) or by sampling with standard fishing gears. Initial tagging and subsequent recapture effort should be allocated as evenly as possible across all portions of the transect shown in Fig. 9. Such a design could then measure dispersion of postsettlement fishes, before and after establishment of the reserves, to test the idea of net export of biomass from reserves (Point 7). This extension of the design is important, because it would allow explanation of the mechanism of enhanced local yield adjacent to marine reserve boundaries (measured close to and far from reserve boundaries before and after reserves are established—Point 8), should such enhancement be detected.

Figures 10–12 provide some theoretical "expected" results of such an experiment. If the reserves result in a significant increase in biomass of the target species (e.g., measured by UVC), one would expect a change over time in the spatial distribution of biomass somewhat like that in Fig. 10. Biomass would gradually build up inside the reserves, but also show a tendency to be higher in those areas open to fishing but closest to the reserve boundaries (Fig. 10). In addition, one may expect

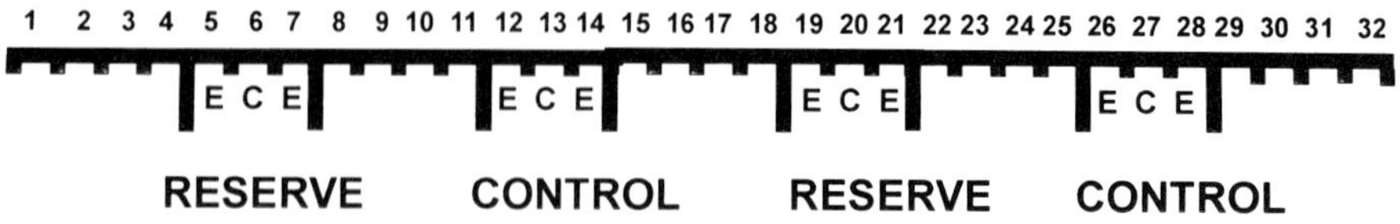

FIGURE 9 A transect across two reserve and two control locations shown in Fig. 8. The transect is divided into 32 sections (e.g., of 50, 100, or 500 m). It is assumed that the investigators can externally tag individual fishes such that the tag color/position/type uniquely identifies the section in which a fish was first tagged. Initial tagging and subsequent recapture effort should be allocated as evenly as possible across all portions of the transect. Such a design could then measure dispersion of postsettlement fishes, before and after establishment of the reserves. C, Center; E, edge.

a gradual increase in yield over time in those areas closest to the reserve boundaries (Fig. 11). If fishes tagged inside reserve locations gradually display a greater tendency over time to disperse from reserves than from fished (control) locations (Fig. 12), this provides a mechanism to explain enhanced local yield adjacent to marine reserve boundaries. It is acknowledged that the effects of this mechanism could be dampened by any behavioral response of fishes to fishing pressure outside the reserve, causing them to migrate into the reserve.

Nobody (including the author, despite 16 years of monitoring marine reserves!) has carried out such an idealistic experiment. Scientists should not be discouraged from attempting such a design for fear that they cannot hope to incorporate *all* aspects of it. Inclusion of *any* aspects of it will considerably improve the quality of scientific information we currently have on the effects of marine reserves on biomass buildup and spillover (see Table 5). The most crucial points are probably the collection of "before" data, long-term monitoring, and attempting to factor out potentially confounding habitat effects.

Finally, a brief comment on the significance of spillover. Establishment of marine reserves, particularly in the developing world, almost invariably requires the strong support of the local people living in and adjacent to the area concerned (e.g., Russ and Alcala, 1999). In many cases promises of "stock-wide" (i.e., recruitment effect—Section III,C) rather than local (i.e., spillover effect—Section III,A) benefits can appear rather nebulous. It is a common assumption, not just of local peoples, but often of the community-based managers encouraging establishment of marine reserves, that fish populations will increase in the reserve and that this will have direct benefits to fished areas nearby. The reasoning is that both adults and their young will move out from the reserve. Convincing a village or island community that a marine reserve may benefit the reef fisheries 10 to 100 km downstream (because that is where the larvae are likely to go—Section III,C) will inevitably be difficult. Thus, any demonstration that marine reserves may affect nearby fisheries in a positive manner, even if the effect is only minor and on a very small spatial scale, may be critical to the successful establishment of community-based marine reserves. Nevertheless, there is little doubt that the primary benefit of marine reserves is protection of a critical minimum spawning stock biomass and subsequent enhancement of recruitment at broad spatial scales via net export of eggs/larvae—the "recruitment effect."

C. The Recruitment Effect

There is general agreement that the major benefit of marine reserves as fisheries management tools will be maintenance of a critical minimum spawning stock

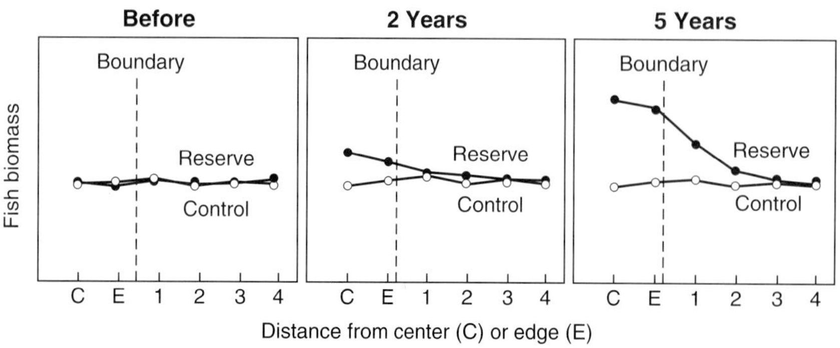

FIGURE 10 Expected change over time in the spatial distribution of biomass if reserves result in a significant decrease in fishing mortality. Biomass would gradually build up inside the reserves, but also show a tendency to be higher in those areas open to fishing but closest to the reserve boundaries. C, Center; E, edge. Numbers represent sections of a transect similar to that shown in Fig. 9.

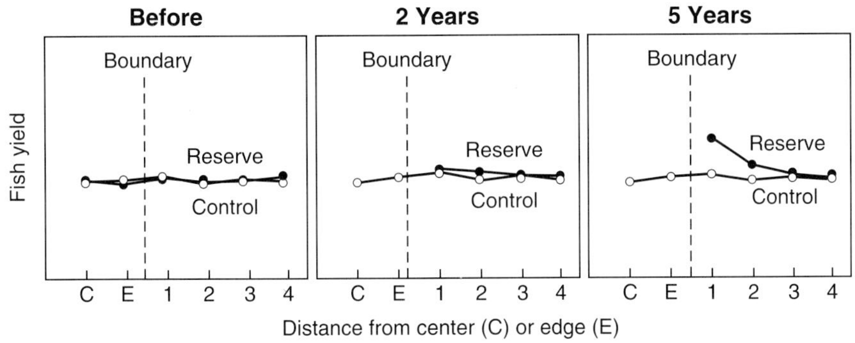

FIGURE 11 Expected change over time in yield if reserves result in a significant decrease in fishing mortality. Yield should gradually increase in those areas closest to the reserve boundaries. C, Center; E, edge. Numbers represent sections of a transect similar to that shown in Fig. 9.

biomass inside reserves, and subsequent maintenance or even enhancement of recruitment to fished areas by larval dispersal. Such an expectation is perfectly reasonable, provided that there is a greater production of propagules per unit area in reserves compared to fished areas, the scale of dispersal of larvae is much larger than the spatial scale of the individual reserves, and that reserves are located in areas from which larvae are likely to disperse (i.e., in "source" areas). However, for a reserve—or in the case of the recruitment effect, a spatial network of reserves—to be an effective fisheries management tool, they should be net exporters of eggs/larvae of target species (Fig. 1). The evidence for populations of target species in marine reserves replenishing target species outside the reserves by the process of net export of eggs/larvae (the "recruitment effect") is essentially nonexistent. In fact, Dayton *et al.* (1995) suggest that the issue of "where larvae go" is one of the most important, yet difficult, problems in marine ecology. The evidence for the recruitment effect has been reviewed by PDT (1990), Roberts and Polunin (1991), Carr and Reed (1993), Dugan and Davis (1993), Rowley (1994), Roberts (1997b, 1998), Allison *et al.* (1998), and S. N. Murray *et al.* (1999). The reason why evidence is lacking, despite such a process being a perfectly reasonable expectation, is that an unequivocal demonstration of the recruitment effect would require an experiment that includes all eight points listed in the experimental design to measure spillover (Section III,B), replacing the requirement to measure net movement patterns of postsettlement fishes between reserve and fished areas with measuring net movement patterns of eggs/larvae. This difference results in enormous logistic difficulties, in that the scale of larval movement is so much greater (potentially tens to hundreds of kilometers, compared with hundreds to thousands of meters), and larvae are difficult to sample, tag, and track from source to site of settlement.

The best reviews of the topic of marine reserves affecting recruitment to fished populations are probably those of Carr and Reed (1993) and Allison *et al.* (1998). The topic is a complex interface of oceanography, larval biology, and computer modeling. Chapter 7 in this volume and Boehlert (1996) review larval dispersal and survival of tropical reef fishes. The aim here is to provide a brief overview of some of the issues involved in the "recruitment effect" of marine reserves. The strategy here is to discuss the major issues from spawning to subsequent settlement of coral reef fishes. In the context of a network of marine reserves, where and how many propagules are produced, where do they go, and what are their survival rates along the way? In

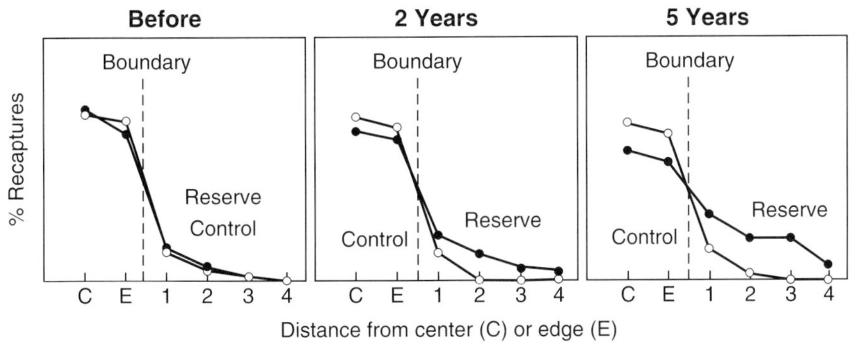

FIGURE 12 Expected pattern of dispersion of reef fishes over time if reserves result in a significant decrease in fishing mortality. Fishes may tend to show a greater propensity to disperse from the reserves than from the control areas over time. C, Center; E, edge. Numbers represent sections of a transect similar to that shown in Fig. 9.

the jargon of the field, what are the sources, sinks, and levels of connectivity?

The evidence for greater production per unit area of propagules inside than outside reserves was presented in Section II,C. The next question concerns where the larvae go. Allison *et al.* (1998) described four models of population replenishment, related to distance of propagule dispersal (short, long) and number of local propagule sources (single, multiple). Most coral reef fishes appear to fit best into the category of multiple sources and long dispersal. Johannes (1978) argued that there is a strong selective advantage to most coral reef fishes to get their propagules off the reef as quickly as possible, to avoid high levels of egg and larval predation. He also argued that many reef fishes spawn at times and places that enhance the chances of larvae returning to the originating reef system. Although there are some examples of retention of larvae of coral reef fishes in lagoons (Leis, 1994), the duration of larval life of most coral reef fishes [9–100 days, with a strong mode between 20 and 30 days (Brothers and Thresher, 1985)] suggests that a substantial number of larvae would be dispersed broadly. Potential dispersal distances of tens to hundreds of kilometers are often cited (Doherty, 1991; Leis, 1991a). Such large potential dispersal distances do not necessarily mean that some level of larval retention is not occurring at such scales. Sinclair (1988) argued that the stock structure of Atlantic herring was maintained largely by breeding populations spawning at times and places that enhanced the chances of retention of propagules in "larval retention areas." Such areas can be as small as 800 km^2. Cowen and Castro (1994) produced evidence of physical mechanisms maintaining larvae of coral reef fishes in the vicinity of Barbados Island. Nevertheless, the basic assumption of most studies of dispersal of coral reef fishes is that a substantial number of larvae are broadly dispersed.

In the context of the dispersal versus self-recruitment issue, it is important to note that, for a network of marine reserves to be useful as a fisheries management tool, it must be both a net exporter of propagules and a largely self-sustaining system. This implies that a reserve network must display a significant amount of both self-recruitment and dispersal of larvae to fished areas. If the network of reserves is not capable of sustaining itself, it will not be an effective bet-hedge against recruitment overfishing in fished areas. In other words, if in overfishing the fished component of the stock we also undermine the reserves of spawning stock in the network, the network is not much of an insurance policy against overfishing.

Early modeling studies of potential dispersal of coral reef organisms (e.g., Williams *et al.*, 1984) used field point measurements (current meters) and progressive vector plots to predict the direction and strength of surface layer, mainstream currents. This type of information, combined with larval duration, was used to predict dispersal direction and distance. Two examples of such models applied to the question of larval dispersal from marine reserves are those of McManus (1994) in the Spratly Islands and Roberts (1997b) in the Caribbean. However, the limitations of combining mainstream flow patterns with information on larval duration to describe adequately dispersal of fish larvae have long been recognized (e.g., Sinclair, 1988; see also Cowen *et al.*, 2000). A major limitation was that flow and dispersal was modeled in two dimensions. Some of the first attempts to actually model currents and larval dispersal in reef systems more realistically, rather than extrapolating from point sources, included depth-integrating approximations (e.g., Dight *et al.*, 1994). Such models went beyond surface flow and generally performed well, except in the narrow boundary layers at the sea surface and near the bottom. Three-dimensional models of flow and larval dispersal in reef systems are now available (e.g. Black, 1994), which may perform better than depth-integrated models if larvae enter boundary layers. However, until recently, all larval dispersal models of coral reef organisms assumed larvae were passive, neutrally buoyant particles. In other words, most early models largely ignored larval biology.

Considerable advances have occurred recently in our knowledge of the behavioral and sensory abilities of larvae of coral reef fishes. Rather than being passive particles, larvae can control their depth distribution, even early in larval life, and thus can have some effect on the degree of their dispersal in surface or subsurface currents (Leis, 1991a). In the late stages of larval life, coral reef fish larvae can potentially swim tens of kilometers (Stobutzki and Bellwood, 1994, 1997; Leis *et al.*, 1996) and pelagic juveniles can orient and swim toward reefs (Leis *et al.*, 1996; Stobutzki and Bellwood, 1998; Armsworth, 2000). Furthermore, the swimming and behavioral capabilities of these larvae are size related, so that growth and mortality schedules during the pelagic phase will likely influence dispersal success considerably.

An example of the new generation of dispersal models is that of Wolanski *et al.* (1997), a computer program modeling the three-dimensional patterns of water movement on Australia's Great Barrier Reef. The realism of models should always be tested against field observations. This study provides one of the best examples that models must be based on the best possible empirical data. The study had considerable computing power at its disposal. This was required to model adequately water movement on this large spatial scale in three-dimensions, and over highly complex topography. The model could describe current patterns in and around reefs adequately. However, as long as the larvae were treated as passive particles, the model could not predict empirically observed patterns of distribution and abundance of larval reef fishes around reefs. Only when models included directional swimming of reef fish larvae, at realistic speeds, could the model reproduce field results. Thus a major modeling project, worth many millions of dollars, could not attain adequate realism until inclusion of larval swimming speed data collected in a piece of apparatus that cost less than $250 (D. R. Bellwood, personal communication). A little investment in better empirical data can go a long way!

An approach that could improve our empirical data on larval dispersal is the "tagging" of larvae. Jones *et al.* (1999) were able to mark the otoliths of prehatched larvae of *Pomacentrus amboinensis* with tetracyline at Lizard Island, GBR, and showed that propagules spawned on the reef could return to that reef as settlers. Otolith microchemistry provides a possible means of detecting, at least in general terms, the area of origin of a settled larvae (e.g., Thorrold *et al.*, 1999) (Chapter 11, this volume). Swearer *et al.* (1999) used trace element concentrations in otoliths to show that recruitment to an island population of a widely distributed coral reef fish may often result from local retention. Novel means of larval tagging are likely to be an important aspect of investigations of the "recruitment effect" in the near future.

In the context of larval tags, the current focus of the larval connectivity issue is self-recruitment of reefs (e.g., Jones *et al.*, 1999; Swearer *et al.*, 1999; Cowen *et al.*, 2000). As interesting and exciting as this is, self-recruitment is almost certainly *not* the critical question regarding whether marine reserve networks will be useful fisheries management tools for coral reef systems. The majority of the world's coral reefs are not isolates. In extensive reef systems, such as the Great Barrier Reef, current modeling in the northern section of the GBR suggests that the mean percentage of self-recruitment of individual reefs is likely to be relatively modest (M. James and L. Mason, personal communication). The critical measure is net larval export from a reserve, or from a network of reserves (see Fig. 1).

Direct measurements of average larval dispersal are needed to understand connectivity in marine reserve networks. However, such measurements, by means of natural or artificial tags, as described above, are extremely difficult to obtain. Palumbi (2000, in press) has argued that genetic patterns of isolation by distance have the potential to provide direct measurement of larval dispersal distance. Genetic isolation by distance results in close populations being more genetically similar to each other than distant ones. Use of appropriate genetic markers, sampled over long-distance transects, could help set appropriate geographic scales on which marine reserve networks will function adequately. Palumbi (in press) calibrated available data on larval dispersal regimes of fish and invertebrates with computer simulations incorporating genetic isolation by distance. He suggested average dispersal distances of 4 to 30 km.

There are probably three geographic regions where our knowledge of the sources and fates of coral reef fish larvae from marine reserves is likely to increase considerably in the near future: (1) the Great Barrier Reef Marine Park, being studied intensively by the Australian Institute of Marine Science (E. Wolanski and colleagues), James Cook University (M. James, L. Bode, L. Mason, and colleagues) and the Reef Co-Operative Research Center (P. Doherty and colleagues); (2) the Florida region, particularly around the area of the Florida Keys National Marine Sanctuary (see Bohnsack and Ault, 1996, and oceanographic and modeling references therein); and (3) the Caribbean region, building on the work of Roberts (1997b) and Cowen *et al.* (2000) (see also Chapter 7, this volume). Although these detailed studies of larval dispersal will provide considerable insight into where to place reserves, more marine reserve-specific models may be required as a guide to how much area needs to be encompassed by reserves to provide sufficient recruitment to ensure sustainable fisheries. Actual percentages of reef regions that are closed to fishing include 5% (Great Barrier Reef), 6% (Florida Keys), and 20% (Bermuda) (see Bohnsack, 1996). Modeling studies now suggest that to enhance recruitment and provide a sufficient hedge against recruitment failure, up to 40–50% of regions may need to be reserves (Lauck *et al.*, 1998; Sladek-Nowlis and Roberts, 1999), although considerable variation probably exists around such estimates. If such estimates are correct, closures of that magnitude will require a considerable change in societal attitudes, probably over several generations!

The difficulties of carrying out an empirical field test of the recruitment effect on an appropriate spatial scale, over a period of perhaps 10–20 years, are so large, that an experiment demonstrating net larval export unequivocally from a network of marine reserves may be almost impossible. The appropriate scales of space and time for such an experiment are so large, that the potential for confounding (coral reef scientists would be well aware of large-scale phenomena such as El Niños, global warming, coral bleaching, crown-of-thorns starfish outbreaks, as relatively recent examples), and thus of equivocal results, are considerable. The solution to this will probably be a cooperative, large-scale modeling approach [e.g., (1) the combined empirical and modeling efforts described above for the Great Barrier Reef in Australia (AIMS, JCU, and CRC) and (2) the National Center for Ecological Analysis and Synthesis (NCEAS) in the United States, through the research group of J. Lubchenco], with models that are based on the best possible empirical data (e.g., oceanography, larval biology), backed up with large-scale management experiments, often of imperfect design [the active adaptive management approach of Hilborn and Walters (1992)]. Ultimately, the efficacy of marine reserve networks as net exporters of recruits of target species, and their ability to maintain recruitment supply to fisheries, will be demonstrated, or not, by their *application* at appropriate scales of space and time. The proof of the pudding will be in the eating.

IV. Conclusions

Do marine reserves increase density, average size/age, biomass, and reproductive output per unit area of target species significantly? Current evidence supporting these contentions appears to be more equivocal than other recent reviews have suggested. The major reason for this is that, as yet, no definitive experiment has been performed that includes the first six points given in Section III,B.

Do marine reserves enhance fisheries in areas outside them by spillover? This requires a demonstration of net export of fish biomass from reserves to fished areas, which, to a reasonable extent, compensates in the medium to long term for the loss of area required to set up the reserve. The evidence in support of spillover is even more equivocal than the evidence for significant biomass buildup reported in Section II,B. As yet, there are no experiments that include all eight points listed in the experimental design in Section III,B.

Do marine reserves enhance fisheries by net export of recruits to fished areas? The empirical evidence for populations of target species in marine reserves replenishing target species outside the reserves by the process of net export of eggs/larvae (the "recruitment effect") is essentially nonexistent. The solution to this will probably be a large-scale, cooperative modeling approach, with models that are based on the best possible empirical data (e.g., oceanography, larval biology), backed up with large-scale management experiments. Ultimately, the efficacy of marine reserve networks as net exporters of recruits of target species, and their ability to maintain recruitment supply to fisheries, will be demonstrated, or not, by their application at appropriate scales of space and time.

There is a plethora of reviews on what marine reserves *could* do as fisheries management tools (see Tables 1 and 2). Yet, there is a distinct paucity of empirical studies demonstrating what they *can* do. Until we begin carrying out experiments that follow the guidelines in Section III,B, the cupboard containing the solid empirical data base on what reserves can do as fisheries management tools may still be fairly bare when somebody else reviews this topic in 10 or 20 years.

The preceding paragraphs outlining the lack of data and the difficulty of collecting unequivocal data almost read like an argument against the use of marine reserves as fisheries management tools! The experiments and modeling exercises discussed in this chapter are very tall orders. The lack of solid empirical knowledge about what marine reserves can achieve as fisheries management tools has sometimes been cited by fisheries managers as a reason not to use such "novel" approaches to fisheries management. Some scientists take the scientific moral high ground and refuse to endorse the use of marine reserves as fisheries management tools until there is certainty that they will be effective, and until reserves are of optimum size and placement to ensure maximum fisheries benefit. I believe such positions are foolish, because they ignore some very simple facts:

1. Conventional fisheries management techniques (effort/catch controls) have, by themselves, generally failed to control fishing effort, and have often failed to prevent recruitment overfishing.
2. Techniques such as effort and catch controls are often difficult to administer in the coral reef fisheries of many developing nations.
3. Marine reserves may be one of the few viable fisheries management options available to the reef fisheries of developing nations. Unlike conventional effort/catch controls, they appear to have a chance of being accepted by subsistence and artisanal fishing communities, often because they offer a variety of other benefits to the community, such as income from tourism.
4. Marine reserves simplify management of multispecies fisheries. For example, they provide a refuge for species with life histories that make them more susceptible to intense, relatively nonselective fishing.
5. Marine reserves as fisheries management tools should be viewed for what they are—a healthy dose of the precautionary principle. They are reserves for and of the target fish stocks. They are not being

advocated as the only, nor even the optimum, method of fisheries management. They are simply our insurance policy against future fisheries management failures and against overfishing.

Marine reserves should not be rejected, nor their implementation be delayed, because the definitive experiments have not shown us the optimal design of reserves to ensure recruitment supply to fisheries. In most developing nations the fishes on coral reefs are so overfished that there simply is no time to wait.

Acknowledgments

I thank Paul Armsworth, Lance Bode, Rick Nemeth, Peter Sale, and Dirk Zeller for useful comments on the manuscript.

Bibliography

Abbott, J. C., and Dill, L. M. (1985). Patterns of aggressive attacks in juvenile steelhead trout (*Salmo gairdneri*). *Can. J. Fish. Aquat. Sci.* **42,** 1702–1706.

Acosta, A., and Appeldoorn, R. S. (1992). Estimation of growth, mortality and yield per recruit for *Lutjanus synagris* (Linnaeus) in Puerto Rico. *Bull. Mar. Sci.* **50,** 282–291.

Adams, C. E., and Huntingford, F. A. (1996). What is a successful fish? Determinants of competitive success in Arctic char (*Salvelinus alpinus*) in different social contexts. *Can. J. Fish. Aquat. Sci.* **53,** 2446–2450.

Adams, C. G. (1981). An outline of Tertiary palaeogeography. *In* "Chance, Change and Challenge: The Evolving Earth" (P. H. Greenwood and R. L. M. Cocks, eds.), pp. 221–235. Cambridge University Press, Cambridge, UK.

Adams, C. G., and Ager, D. V. (1967). "Aspects of Tethyan Biogeography," No. 7. Systematics Association, London.

Adams, C. G., Gentry, A. W., and Whybrow, P. J. (1983). Dating the terminal Tethyan event. *Utrecht. Micropalaeontol. Bull.* **30,** 273–298.

Adams, S., Mapstone, B. P., Russ, G. R., and Davies, C. R. (2000). Geographic variation in the sex ratio, sex specific size, and age structure of *Plectropomus leopardus* (Serranidae) populations between reefs open and closed to fishing on the Great Barrier Reef. *Can. J. Fish. Aquat. Sci.* **57,** 1448–1458.

Adams, T. J. H. (1996). Modern institutional framework for reef fisheries management. *In* "Reef Fisheries" (N. V. C. Polunin and C. M. Roberts, eds.), pp. 337–360. Chapman & Hall, London.

Aerts, P., and Verraes, W. (1974). Theoretical analysis of a planar four-bar linkage in the teleostean skull. The use of mathematics in biomechanics. *Ann. Soc. R. Zool. Belg.* **114,** 273–290.

Agardy, T. S. (1994). Advances in marine conservation: The role of marine protected areas. *Trends Ecol. Evol.* **9,** 267–270.

Agardy, T. S. (1997). "Marine Protected Areas and Ocean Conservation." Academic Press, Austin, TX.

Aguilar-Perera, A., and Aguilar-Davila, W. (1996). A spawning aggregation of Nassau grouper *Epinephelus striatus* (Pisces: Serranidae) in the Mexican Caribbean. *Environ. Biol. Fishes* **45,** 351–361.

Ahrenholz, D. W. (2000). Periodicity of growth increment formation in otoliths of juvenile gray snapper (*Lutjanus griseus*) and lane snapper (*Lutjanus synagris*). *J. Elisha Mitchell Sci. Soc.* **116,** 251–259.

Akimichi, T. (1998). Live grouper trade and ethno-networks in Southeast Asia. *In* "Proceedings of the APEC Workshop on the Impacts of Destructive Fishing Practices on the Marine Environment, December, 1997, Hong Kong," pp. 240–248. Agriculture and Fisheries Department, Hong Kong, China.

Albaladejo, V. D., and Corpuz, V. T. (1981). A market study of the aquarium fish industry of the Philippines: An assessment of the growth and mechanics of the trade. *Proc. 4th Int. Coral Reef Symp.* Vol. 1, pp. 75–81.

Alcala, A. C., and Russ, G. R. (1990). A direct test of the effects of protective management on abundance and yield of tropical marine resources. *J. Cons. Int. Explor. Mer* **47,** 40–47.

Aldenhoven, J. M. (1986). Different reproductive strategies in a sex-changing coral reef fish *Centropyge bicolor* (Pomacanthidae). *Aust. J. Mar. Freshwater Res.* **37,** 353–360.

Alder, J. (1996). Have tropical marine protected areas worked? An initial analysis of their success. *Coastal Manage.* **24,** 97–114

Alexander, R. M. (1967). "Functional Design in Fishes." Hutchinson, London.

Alee, I., and Eke, S. (1999). Live reef fish trade in the Republic of Marshall Islands. *In* "Proceedings of the first Asia-Pacific Seminar/Workshop on the Live Reef Fish Trade, August 11–14, 1998," pp. 12–14. International Marinelife Alliance, Manila, Philippines.

Alfaro, M., and Westneat, M. W. (1999). Motor patterns of herbivorous feeding: Electromyographic analysis of biting in the parrotfishes *Cetoscarus bicolor* and *Scarus iserti*. *Brain, Behav. Evol.* **54,** 205–222.

Al-Hussaini, A. H. (1947). The feeding habits and morphology of the alimentary tract of some teleosts living in the neighborhood of the Marine Biological Station, Ghardaqa, Red Sea. *Mar. Biol. Sta. Ghardaqa, Red Sea., Publ.* **5,** 1–61.

Ali, M., and Wootton, R. J. (1999). Effect of variable food levels on reproductive performance of breeding female three-spined sticklebacks. *J. Fish Biol.* **55,** 1040–1053.

Alldridge, A. L., and Hamner, W. M. (1980). Recurring aggregation of zooplankton by a tidal current. *Estuarine Coastal Mar. Sci.* **10,** 31–37.

Allen, G. R. (1972). "The Anemonefishes: Their Classification and Biology." TFH Publications, Neptune City, NJ.

Allen, G. R. (1975). "Damselfishes of the South Seas." TFH Publications, Neptune City, NJ.

Allen, G. R. (1979). "Butterfly and Angelfishes of the World 2." Mergus, Hans A. Baensch, Melle.

Allen, G. R. (1985). "Snappers of the World: FAO Species Catalogue," Vol. 6. FAO, Rome.

Allen, G. R. (1991). "Damselfishes of the World." Mergus, Hans A. Baensch, Melle.

Allen, G. R. (1994). "Damselfishes of the World." Mergus Hans A. Baensch, Melle.

Allen, G. R. (1999). *Altrichthys,* a new genus of damselfish (Pomacentridae) from Philippine Seas with description of a new species. *Rev. fr. Aquariol.* **26,** 23–28.

Allen, G. R. (2000). Threatened fishes of the world: *Pterapogon kauderni* Koumans (1933) (Apogonidae). *Environ. Biol. Fishes* **57,** 142.

Allen, G. R., and Steene, R. (1988). "Fishes of Christmas Island and Indian Ocean." Christmas Island Natural History Association, Christmas Island.

Allen, G. R., and Steene, R. C. (1995). Notes on the ecology and behaviour of the Indonesian cardinalfish (Apogonidae) *Pterapogon kauderni* Koumans. *Rev. fr. Aquariol.* **22,** 7–9.

Allen, G. R., Hoese, D. F., Paxton, J. R., Randall, J. E., Russell, B. C., Stark, W. A., II, Talbot, F. H., and Whitley, G. P. (1976). Annotated checklist of the fishes of Lord Howe Island. *Rec. Aus. Mus.* **30,** 365–454.

Allen, G. R., Steene, R., and Allen, M. A. (1998). "Guide to Angelfishes and Butterflyfishes." Odyssey Publishing/ Tropical Reef Research, Australia.

Allendorf, F. W., and Phelps, S. R. (1981). Use of allelic frequencies to describe population structure. *Can. J. Fish. Aquat. Sci.* **38,** 1507–1514.

Allison, G. W., Lubchenco, J., and Carr, M. H. (1998). Marine reserves are necessary but not sufficient for marine conservation. *Ecol. Appl.* **8,** s79–s92.

Allman, R. J., and Grimes, C. B. (1998). Growth and mortality of little tunny (*Euthynnus alletteratus*) larvae off the Mississippi River plume and Panama City, Florida. *Bull. Mar. Sci.* **62,** 189–197.

Almeida, A. J. (1996). Structure and spatial variability of the rocky fish fauna in the protected marine "Reserva natural de Berlenga" (Portugal). *Arq. Mus. Bocage* [2] **35,** 633–642.

American Zoo and Aquarium Association (AZA) (2000). "The AZA Membership Directory." AZA, Silver Spring, MD.

Anderson, C. S. (1995). Calculating size-dependent relative survival from samples taken before and after selection. *In* "Recent Developments in Fish Otolith Research" (D. H. Secor, J. M. Dean, and S. E. Campana, eds.), pp. 455–466. University of South Carolina Press, Columbia.

Anderson, G. R. V., Ehrlich, A. H., Ehrlich, P. R., Roughgarden, J. D., Russell, B. C., and Talbot, F. H. (1981). The community structure of coral reef fishes. *Am. Nat.* **117,** 476–495.

Anderson, J. T. (1988). A review of size dependent survival during pre-recruit stages of fishes in relation to recruitment. *J. Northwest Atl. Fish. Sci.* **8,** 55–66.

Anderson, T. W. (1994). Role of macroalgal structure in the distribution and abundance of a temperate reef fish. *Mar. Ecol. Prog. Ser.* **113,** 279–290.

Anderson, T. W. (2001). Predator responses, prey refuges, and density-dependent mortality of a marine fish. *Ecology* **82,** 245–257.

Anderson, T. W., and Sabado, B. D. (1995). Correspondence between food availability and growth of a planktivorous temperate reef fish. *J. Exp. Mar. Biol. Ecol.* **189,** 65–76.

Andrade, C. A. P., and Albuquerque, F. M. M. (1995). Fish assemblages associated with bottom habitats on the south coast of Madeira. *Bol. Mus. Mun. Funchal, Suppl.* **4,** 9–20.

Andrew, N. L., and Mapstone, B. D. (1987). Sampling and the description of spatial pattern in marine ecology. *Oceanogr. Mar. Biol.* **25,** 39–89.

Andrewartha, H. G., and Birch, L. C. (1954). "The Distribution and Abundance of Animals." University of Chicago Press, Chicago.

Andrews, A. H., Cailliet, G. M., and Coale, K. H. (1999). Age and growth of the Pacific grenadier (*Coryphaenoides acrolepis*) with age estimate validation using an improved radiometric ageing technique. *Can. J. Fish. Aquat. Sci.* **56,** 1339–1350.

Andrews, J. C. (1983). Water masses, nutrient levels and seasonal drift on the outer central Queensland shelf (Great Barrier Reef). *Aust. J. Mar. Freshwater Res.* **34,** 821–834.

Andrews, J. C., and Gentien, P. (1982). Upwelling as a source of nutrients for the Great Barrier Reef ecosystems: A solution to Darwin's question? *Mar. Ecol. Prog. Ser.* **8,** 257–269.

Andrews, J. C., and Pickard, G. L. (1990). The physical oceanography of coral-reef systems. *In* "Coral Reefs. Ecosystems of the World" (Z. Dubinsky, ed.), Vol. 25, pp. 11–48. Elsevier, Amsterdam.

Angemeier, P. (1995). Ecological attributes of extinction-prone species: Loss of freshwater fishes of Virginia. *Conserv. Biol.* **9,** 143–158.

Anholt, B. R., and Werner, E. E. (1995). Interaction between food availability and predation mortality mediated by adaptive behavior. *Ecology* **76,** 2230–2234.

Anker, G. C. (1974). Morphology and kinematics of the stickleback, *Gasterosteus aculeatus. Trans. Zool. Soc. London* **32,** 311–416.

Anneville, O., Cury, P., Page, C. L., and Treuil, J. P. (1998). Modelling fish spatial dynamics and local density-dependence relationships: Detection of patterns at a global scale. *Aquat. Living Resour.* **11,** 305–314.

Añonuveo, C. (1989). The economics of municipal fisheries: The case of the Lingayen Gulf. *ICLARM Conf. Proc.* **17,** 141–155.

Appeldoorn, R. S., Hensley, D. A., Shapiro, D. Y., Kioroglou, S., and Anderson, B. G. (1994). Egg dispersal in a Caribbean coral reef fish, *Thalassoma bifasciatum*. II. Dispersal off the reef platform. *Bull. Mar. Sci.* **54,** 271–280.

Arai, N., Sakamoto, W., and Maeda, K. (1996). Correlation between ambient temperature and strontium-calcium concentration ratios in otoliths of red sea bream *Pagrus major. Fish. Sci.* **62,** 652–653.

Arai, T., Otake, T., Limbong, D., and Tsukamoto, K. (1999). Early life history and recruitment of the tropical eel *Anguilla bicolor pacifica,* as revealed by otolith microstructure and microchemistry. *Mar. Biol.* **133,** 319–326.

Arambourg, C. (1927). Les poissons fossiles d'Oran. *Matér. Carte Geo. Algér., 1er Sér. Paléontol.* **6,** 154–157.

Arita, H. T., Robinson, J. G., and Redford, K. H. (1990). Rarity in Neotropical forest mammals and its ecological correlates. *Conserv. Biol.* **4,** 181–192.

Armsworth, P. R. (2000). Modelling the swimming response of late stage larval reef fish to different stimuli. *Mar. Ecol.: Prog. Ser.* **195,** 231–247.

Armsworth, P. R. (2001). Effects of fishing on a protogynous hermaphrodite. *Can. J. Fish. Aquat. Sci.* **58,** 568–578.

Armsworth, P. R. (2002) Recruitment limitation, population regulation and larval connectivity in reef fish metapopulations. *Ecology* (in press).

Armsworth, P. R., James, M. K., and Bode, L. (2001). When to press on or turn back: Dispersal strategies for reef fish larvae. *Am. Nat.* **157,** 434–450.

Arvedlund, M., and Nielson, L. E. (1996). Do anemonefish *Amphiprion ocellaris* (Pisces: Pomacentridae) imprint themselves to their host sea anemone *Heteractis magnifica* (Anthozoa: Actinidae)? *Ethology* **102,** 197–211.

Arvedlund, M., McCormick, M. I., Fautin, D. G., and Bildsøe, M. (1999). Host recognition and possible imprinting in the anemonefish *Amphiprion melanopus* (Pisces: Pomacentridae). *Mar. Ecol.: Prog. Ser.* **188,** 207–218.

Atema, J. (1971). Structures and functions of the sense of taste in the catfish *(Ictalurus natalis). Brain, Behav. Evol.* **4,** 273–294.

Atema, J. (1980). Chemical signals and feeding behavior in fishes. *In* "Fish Behavior and its Use in the Capture and Culture of Fishes" (J. E. Bardach, J. J. Magnuson, R. C. May, and J. M. Reinhart, eds.), pp. 57–94. ICLARM, Manila.

Atema, J. (1988). Distribution of chemical stimuli. *In* "Sensory Biology of Aquatic Animals" (J. Atema, R. R. Fay, A. N. Popper, and W. N. Tavolga, eds.), pp. 29–56. Springer-Verlag, New York.

Atema, J., Holland, K., and Ikehara, W. (1980). Olfactory response of yellowfin tuna (*Thunnus albacares*) to prey odors: Chemical search image. *J. Chem. Ecol.* **6,** 457–465.

Atkinson, D. (1994). Temperature and organism size—a biological law for ectotherms? *Adv. Ecol. Res.* **25,** 1–58.

Attwood, C. G., and Bennett, B. A. (1994). Variation in dispersal of galjoen (*Coracinus capensis*) from a marine reserve. *Can. J. Fish. Aquat. Sci.* **51,** 1247–1257.

Attwood, C. G., and Bennett, B. A. (1995). Modelling the effect of marine reserves on the recreational shore-fishery on the south-western cape, South Africa. *S. Afr. J. Mar. Sci.* **16,** 227–240.

Ault, J. S, Bohnsack, J. A. and Meester, G. A. (1997). Florida Keys National Marine Sanctuary: Retrospective (1979–1995) reef fish assessment and a case for protected marine areas. *In* "Developing and Sustaining World Fisheries Resources. 2nd World Fisheries Congress" (D. A. Hancock, D. C. Smith, A. Grant, and J. P. Beumer, eds.), pp. 415–425, CSIRO Publishing, Collingwood, Australia.

Ault, T. R., and Johnson, C. R. (1998a). Spatially and temporally predictable fish communities on coral reefs. *Ecol. Monogr.* **68,** 25–50.

Ault, T. R., and Johnson, C. R. (1998b). Relationships between habitat and recruitment of three species of damselfish (Pomacentridae) at Heron Reef, Great Barrier Reef. *J. Exp. Mar. Biol. Ecol.* **223,** 145–166.

Avise, J. C. (1992). Molecular population structure and biogeographic history of regional fauna: A case history with lessons for conservation biology. *Oikos* **63,** 2–76.

Avise, J. C. (1994). "Molecular Markers, Natural History and Evolution." Chapman & Hall, New York.

Avise, J. C., and Shapiro, D. Y. (1986). Evaluating kinship of newly settled juveniles within social groups of the coral reef fish *Anthias squamipinnis. Evolution (Lawrence, Kans.)* **40,** 1051–1059.

Avise, J. C., Arnold, J., Ball, R. M., Jr., Bermingham, E., Lamb, T., Neigel, J. E., Reeb, C. A., and Saunders, N. C. (1987). Intraspecific phylogeography: The mitochondrial DNA bridge between population genetics and systematics. *Annu. Rev. Ecol. Syst.* **18,** 489–522.

Ayling, A. L. (1992). "Long Term Surveys of *Acanthaster,* Coral Trout and Chaetodontids on Three Reefs off Townsville: 1983–1989," Res. Publ. No. 15. Great Barrier Reef Marine Park Authority, Townsville, Australia.

Ayling, A. M., and Ayling, A. L. (1986). "A Biological Survey of Selected Reefs in the Capricorn Section of the Great Barrier Reef Marine Park," Report. Great Barrier Reef Marine Park Authority, Townsville, Australia.

Ayling, A. M., and Ayling, A. L. (1992). "Bramble Reef Replenishment Area—Baseline Survey," Report. Great Barrier Reef Marine Park Authority, Townsville, Australia.

Ayling, A. M., and Ayling, A. L. (1994). "Bramble Reef Replenishment Area—1993 and 1994 Surveys," Report. Great Barrier Reef Marine Park Authority, Townsville, Australia.

Ayling, A. M., and Ayling, A. L. (1997). "Bramble Reef Replenishment Area: Pre- and Post-opening Surveys," Report. Great Barrier Reef Marine Park Authority, Townsville, Australia.

Azevedo, J. M. N. (1995). Food web of the Azorean shallow water marine ichthyological communities: A guild approach. *Bol. Mus. Mun. Funchal, Suppl.* **4,** 29–53.

Bagarinao, T., and Hunter, J. R. (1983). The visual feeding threshold and action spectrum of northern anchovy (*Engraulis mordax*) larvae. *Calif. Coop. Oceanic Fish. Invest. Rep.* **24,** 245–254.

Bailey, K. M. (1994). Predation on juvenile flatfish and recruitment variability. *Neth. J. Sea Res.* **32,** 175–189.

Bailey, K. M., and Houde, E. D. (1989). Predation on eggs and larvae of marine fishes and the recruitment problem. *Adv. Mar. Biol.* **25**, 1–83.

Bailey, K. M., Brodeur, R. D., and Hallowed, A. B. (1996). Cohort survival patterns of walleye pollock, *Theragra chalcogramma,* in Shelikof Strait, Alaska: A critical factor analysis. *Fish. Oceanogr.* **5**, 179–188.

Baillie, J., and Groombridge, B. (1996). "1996 IUCN Red List of Threatened Animals." IUCN, Gland, Switzerland.

Baird, T. A. (1988). Female and male territoriality and mating system of the sand tilefish, *Malacanthus plumeri. Environ. Biol. Fishes* **22**, 101–116.

Bakun, A. (1986). Local retention of planktonic early life stages in tropical reef/bank demersal systems: The role of vertically-structured hydrodynamic processes. *In* "IOC/FAO Workshop on Recruitment in Tropical Coastal Demersal Communities" (D. Pauly and A. Yanez-Arancibia, eds.), IOC Workshop Rep. 44, Suppl., pp. 15–32. UNESCO, Paris.

Ballantine, W. (1997). No take marine reserve networks support fisheries. *In* "Developing and Sustaining World Fisheries Resources" (D. A. Hancock, D. C. Smith, A. Grant, and J. P. Beumer, eds.), pp. 702–706. CSIRO Publishing, Collingwood, Australia.

Ballantyne, J. S., Mercure, F., Gerrits, M. F., Van Der Kraak, G., McKinley, S., Martens, D. W., Hinch, S. G., and Diewert, R. E. (1996). Plasma nonesterified fatty acid profiles in male and female sockeye salmon, *Oncorhynchus nerka,* during spawning migration. *Can. J. Fish. Aquat. Sci.* **53**, 1418–1426.

Ballintijn, C. M., van den Burg, A., and Egberink, B. P. (1972). An electromyographic study of the adductor mandibulae complex of a free swimming carp (*Cyprinus carpio*) during feeding. *J. Exp. Biol.* **57**, 261–283.

Balon, E. K. (1999). Alternative ways to become a juvenile or a definitive phenotype (and on some persisting linguistic offenses). *Environ. Biol. Fishes* **56**, 17–38.

Banner, A. (1967). Evidence of sensitivity to acoustic displacements in the lemon shark, *Negaprion brevirostris (Poey). In* "Lateral Line Detectors" (P. Cahn, ed.), pp. 265–273. Indiana University Press, Bloomington.

Banner, A. (1972). Use of sounds in predation by young lemon sharks, *Negaprion brevirostris (Poey). Bull. Mar. Sci.* **22**, 251–283.

Bannerot, S. P., Fox, W. W., Jr., and Powers, J. E. (1987). Reproductive strategies and the management of groupers and snappers in the Gulf of Mexico and Caribbean. *In* "Tropical Snappers and Groupers: Biology and Fisheries Management" (J. J. Polovina and S. Ralston, eds.), pp. 561–606. Westview Press, Boulder, CO.

Baquero, J. (1995). The stressful journey of ornamental marine fish. *Sea Wind* **9**, 19–21.

Baquero, J. (1999). "Marine Ornamentals Trade: Quality and Sustainability for the Pacific Region," Report. South Pacific Forum Secretariat, Suva, Fiji.

Barber, C. V., and Pratt, V. R. (1997). "Sullied Seas: Strategies for combating cyanide fishing in Southeast Asia and Beyond." World Resources Institute and International Marinelife Alliance, Manila, Philippines/Washington, D.C.

Bardach, J. E., and Villars, T. (1974). The chemical senses of fishes. *In* "Chemoreception in Marine Organisms" (P. T. Grant and A. M. Mackie, eds.), pp. 49–104. Academic Press, New York.

Bardach, J. E., Winn, H. E., and Menzel, D. W. (1959). The role of the senses in the feeding of the nocturnal reef predators, *Gymnothorax moringa* and *G. vicinus. Copeia,* pp. 133–139.

Bardach, J. E., Todd, J. H., and Crickmer, R. (1967). Orientation by taste in fish of the genus *Ictalurus. Science* **155**, 1276–1278.

Barel, C. D. N. (1983). Towards a constructional morphology of the cichlid fishes (Teleostei, Perciformes). *Neth. J. Zool.* **33**, 357–424.

Barel, C. D. N., Vermeulen, J. W., and Berkoudt, H. (1977). Kinematicher Transmissionknoeffizient und Vierstangensystem als Kunkitionparameter und Formmodel fur mandibulare Depressionsapparate bei Teleostiern. *Anat. Anz.* **142**, 21–31.

Barkman, R. C., and Bengtson, D. A. (1987). The record of daily growth in otoliths of Atlantic silversides, *Menidia menidia,* from field and laboratory. *J. Fish Biol.* **31**, 683–695.

Barlow, G. W. (1981). Patterns of parental investment, dispersal and size among coral-reef fishes. *Environ. Biol. Fishes* **6**, 65–85.

Barlow, G. W. (1984). Patterns of monogamy among teleost fishes. *Arch. Fischerei wiss.* **35**, 75–123.

Barnes, D. J., and Chalker, B. E. (1990). Calcification and photosynthesis in reef building corals and algae. *In* "Coral Reefs. Ecosystems of the World." (Z. Dubinsky, ed.), Vol. 25, pp. 109–31. Elsevier, Amsterdam.

Barnett, C. (1977). Chemical recognition of the mother by the young of the cichlid fish, *Cichlasoma citrinellum. J. Chem. Ecol.* **3**, 461–466.

Barnwell, F. H. (1979). Variation the form of the tide and some problems it poses for biological timing systems. *In* "Biological Rhythms in the Marine Environment" (P. J. DeCoursey, ed.), pp. 161–187. University of South Carolina, Columbia.

Barratt, L., and Medley, P. (1990). Managing multi-species ornamental reef fisheries. *Prog. Underwater Sci.* **15**, 55–72.

Barron, E. J., and Peterson, W. H. (1989). Model simulation of the Cretaceous ocean circulation. *Science* **244**, 684–686.

Bartels, P. J. (1984). Extra-territorial movements of a perennially territorial damselfish, *Eupomacentrus dorsopunicans. Behaviour* **91**, 312–321.

Barton, E. D., Aristegui, J., Tett, P., Canton, M., Garcia-Braun, J., Hernandez-Leon, S., Nykjaer, L., Almeida, C., Almunia, J., Ballesteros, S., Basterretxea, G., Escanez, J., Garcia-Weill, L., Hernandez-Guerra, A., Lopez-Laatzen, F., Molina, R., Montero, M. F., Navarro-Perez, E.,

Rodriguez, J. M., van Lenning, K., Velez, H., and Wild, K. (1998). The transition zone of the Canary Current upwelling region. *Prog. Oceanogr.* **41**, 455–504.

Baskerville, G. L. (1997). Advocacy, science, policy and life in the real world. *Conserv. Ecol.* **1**, 9. Available at URL: *http://www.consecol.org/vol1/iss1/art9*

Bass, A. H. (1986). Electric organs revisited: Evolution of a vertebrate communication and orientation organ. *In* "Electroreception" (T. H. Bullock and W. Heiligenberg, eds.), pp. 13–70. Wiley, New York.

Bass, A. H. (1990). Sounds from the intertidal zone: Vocalizing fish. *BioScience* **40**, 249–258.

Bass, A. H. (1993). From brains to behavior: Hormonal cascades and alternative mating tactics in teleost fishes. *Rev. Fish Biol. Fish.* **3**, 139–145.

Batchelor, G. K. (1967). "An Introduction to Fluid Dynamics." Cambridge University Press, Cambridge, UK.

Bath, G. E., Thorrold, S. R., Jones, C. M., Campana, S. E., McLaren, J. W., and Lam, L. W. H. (2000). Strontium and barium uptake in aragonitic otoliths of marine fish. *Geochim. Cosmochim. Acta* **64**, 1705–1714.

Bazin, C., Capy, P., Higuet, D., and Langin, T. (1997). Séquences d'ADN mobiles et évolution du génome. *Pour Sci., Hors. Sér. Janvier* **97**, 106–109.

Beck, J. W., Edwards, R. L., Ito, E., Taylor, F. W., Recy, J., Rougerie, F., Joannot, P., and Henin, C. (1992). Sea-surface temperature from coral skeletal strontium/calcium ratios. *Science* **257**, 644–647.

Beets, J. (1997). Effects of predatory fish on the recruitment and abundance of Caribbean coral reef fishes. *Mar. Ecol.: Prog. Ser.* **148**, 11–21.

Beets, J., and Friedlander, A. (1998). Evaluation of a conservation strategy: A spawning aggregation closure for red hind, *Epinephelus guttatus,* in the U.S. Virgin Islands. *Environ. Biol. Fishes* **55**, 91–98.

Beets, J., and Hixon, M. A. (1994). Distribution, persistence, and growth of groupers (Pisces: Serranidae) on artificial and natural patch reefs in the Virgin Islands. *Bull. Mar. Sci.* **55**, 470–483.

Beinssen, K. (1989). Heron reef demersal reef fish movement study. Unpublished report to the Department of Conservation, Parks and Wildlife, Rockhampton, Queensland, Australia.

Belcher, A. M., Wu, X. H., Christensen, R. J., Hansma, P. K., Stucky, G. D., and Morse, D. E. (1996). Control of crystal phase switching and orientation by soluble mollusc-shell proteins. *Nature (London)* **381**, 56–58.

Bell, J. D. (1983). Effects of depth and marine reserve fishing restrictions on the structure of a rocky reef fish assemblage in the northwestern Mediterranean Sea. *J. Appl. Ecol.* **20**, 357–369.

Bell, J. D., and Harmelin-Vivien, M. L. (1983). Fish fauna of French Mediterranean *Posidonia oceanica* seagrass meadows. II: Feeding habits. *Tethys* **11**, 1–14.

Bell, J. D., Harmelin-Vivien, M. L., and Galzin, R. (1985). Large scale spatial variation in abundance of butterflyfishes (Chaetodontidae) on Polynesian reefs. *Proc. 5th Int. Coral Reef Congr.,* Vol. 5, pp. 421–426.

Bell, J., Doherty, P., and Hair, C. (1999). The capture and culture of postlarval coral reef fish: Potential for new artisanal fisheries. *Secretariat of the Pacific Community Live Reef Fish Inf. Bull.* **6**, 31–34.

Bell, L. J., Moyer, J. T., and Numachi, K. (1982). Morphological and genetic variation in Japanese populations of the anemone fish *Amphiprion clarkii. Mar. Biol.* **72**, 99–108.

Bell, M. A., and Andrews, C. A. (1997). Evolutionary consequences of postglacial colonization of freshwater by primitively anadromous fishes. *In* "Evolutionary Ecology of Freshwater Animals" (B., Streit, T., Stadler, and C., Lively, eds.), pp. 323–363. Birkhaeuser, Basel.

Bellwood, D. R. (1988a). Seasonal changes in the size and composition of the fish yield from reefs around Apo Island, Central Philippines, with notes on methods of yield estimation. *J. Fish Biol.* **32**, 881–893.

Bellwood, D. R. (1988b). Ontogenetic changes in the diet of early post-settlement *Scarus* species. *J. Fish Biol.* **33**, 213–219.

Bellwood, D. R. (1990). A new fossil fish *Phyllopharyngodon longipinnins* gen. et sp. nov. (family Labridae) from the Eocene, Monte Bolca, Italy. *Stud. Ric. Giacimenti Terziari Bolca, Mus. Civ. Stor. Nat. Verona* **6**, 149–160.

Bellwood, D. R. (1994). A phylogenetic study of the parrotfishes family Scaridae (Pisces: Labroidei), with a revision of genera. *Rec. Aust. Mus.*, Suppl. **20**, 1–86.

Bellwood, D. R. (1995a). Direct estimate of bioerosion by two parrotfish species *Chlorurus gibbus* and *Chlorurus sordidus* on the Great Barrier Reef. *Mar. Biol.* **121**, 419–429.

Bellwood, D. R. (1995b). Carbonate transport and intrareefal patterns of bioerosion and sediment release by parrotfishes (family Scaridae) on the Great Barrier Reef. *Mar. Ecol.: Prog. Ser.* **117**, 127–136.

Bellwood, D. R. (1996a). The Eocene fishes of Monte Bolca: The earliest coral reef fish assemblage. *Coral Reefs* **15**, 11–19.

Bellwood, D. R. (1996b). Production and reworking of sediment by parrotfishes (family Scaridae), on the Great Barrier Reef, Australia. *Mar. Biol.* **125**, 795–800.

Bellwood, D. R. (1997). Reef fish biogeography: Habitat associations, fossils and phylogenies. *Proc. 8th Int. Coral Reef Symp.*, Panama, Vol. 1, pp. 379–384.

Bellwood, D. R. (1998). What are reef fishes? Comment on the report by R. Robertson: Do coral reef fish faunas have a distinctive taxonomic structure? *Coral Reefs* **17**, 187–189.

Bellwood, D. R. (1999). Fossil pharyngognath fishes from Monte Bolca, Italy, with a description of a new pomacentrid genus and species. *Stud. Ric. Giacimenti Terziari Bolca, Misc. Palaeontol. Mus. Civ. Stor. Nat. Verona* **8**, 207–217.

Bellwood, D. R., and Choat, J. H. (1989). A description of the juvenile phase colour patterns of 24 parrotfish species (family Scaridae) from the Great Barrier Reef, Australia. *Rec. Aust. Mus.* **41**, 1–41.

Bellwood, D. R., and Choat, J. H. (1990). A functional analysis of grazing in parrotfishes (family Scaridae): The ecological implications. *Environ. Biol. Fishes* **28**, 189–214.

Bellwood, D. R., and Fisher, R. (2001). Relative swimming speeds in reef fish larvae. *Mar. Ecol.: Prog. Ser.* **211**, 299–303.

Bellwood, D. R., and Hughes, T. P. (2001). Regional-scale assembly rules and biodiversity of coral reefs. *Science* **292**, 1532–1534.

Bellwood, D. R., and Schultz, O. (1991). A review of the fossil record of the parrotfishes (Labroidei: Scaridae) with a description of a new *Calotomus* species from the Middle Miocene (Badenian) of Austria. *Ann. Naturhist. Mus. Wien* **92**, 55–71.

Bellwood, D. R., and Sorbini, L. (1996). A review of the fossil record of the Pomacentridae (Pisces: Labroidei) with a description of a new genus and species from the Eocene of Monte Bolca, Italy. *Zool. J. Linn. Soc.* **117**, 159–174.

Bellwood, D. R., and Wainwright, P.C. (2001). Swimming ability in labrid fishes: Implications for habitat use and cross-shelf distribution on the Great Barrier Reef. *Coral Reefs* **20**, 139–150.

Bellwood, D. R., Leis, J. M., and Stobutzki, I. C. (1998). Fishery and reef management. *Science* **279**, 2021.

Bence, J. R., and Nisbet, R. M. (1989). Space-limited recruitment in open systems: The importance of time delays. *Ecology* **70**, 1434–1441.

Bennett, B. A., and Attwood, C. G. (1991). Evidence for recovery of a surf zone fish assemblage following the establishment of a marine reserve on the southern coast of South Africa. *Mar. Ecol.: Prog. Ser.* **75**, 173–181.

Bennett, J. T., Boehlert, G. W., and Turekian, K. K. (1982). Confirmation of longevity in *Sebastes diploproa* (Pisces: Scorpaenidae) from $^{210}Pb/^{226}Ra$ measurements in otoliths. *Mar. Biol.* **71**, 209–215.

Bennett, M. V. L. (1971a). Electric organs. *Fish Physiol.* **5**, 347–491.

Bennett, M. V. L. (1971b). Electroreception. *Fish Physiol.* **5**, 493–574.

Benoit, H. P., Pepin, P., and Brown, J. A. (2000). Patterns of metamorphic age and length in marine fishes, from individuals to taxa. *Can. J. Fish. Aquat. Sci.* **57**, 856–869.

Bentley, N. (1999). "Fishing for Solutions: Can the Live Trade in Wild Groupers and Wrasses from Southeast Asia be Managed?" TRAFFIC Southeast Asia, TRAFFIC, Petaling Jaya, Malaysia.

Bentley, N., and Aumeeruddy, R. (1999). The live reef fishery in the Seychelles. *Secretariat of the Pacific Community Live Reef Fish Inf. Bull.* **6**, 5–7.

Bentzen, P., Taggart, C. T., Ryzzante, D. E., and Cook, D. (1996). Microsatellite polymorphism and the population structure of Atlantic cod (*Gadus morhua*) in the northwest Atlantic. *Can. J. Fish. Aquat. Sci.* **53**, 2706–2721.

Benzie, J. A. H. (1994). Patterns of gene flow in the Great Barrier Reef and the Coral Sea. *In* "Genetics and Evolution of Aquatic Organisms" (A. R. Beaumont, ed.), pp. 67–79. Chapman & Hall, London.

Benzie J. A. H., and Williams, S. T. (1997). Genetic structure of giant clam (*Tridacna maxima*) populations in the west Pacific is not consistent with dispersal by present-day ocean currents. *Evolution (Lawrence, Kans.)* **51**, 768–783.

Bergenius, M. A. J., Meekan, M. G., Robertson, R., Mc Cormick, M. I. (2002). Larval growth predicts the recruitment success of a coral reef fish. *Oecologia* (in press).

Bernardi, G., and Bernardi, G. (1986). Compositional constraints and genome evolution. *J. Mol. Evol.* **24**, 1–11.

Bernardi, G., Robertson, D. R., Clifton, K. E., and Azzurro, E. (2000). Molecular systematics, zoogeography, and evolutionary ecology of the Atlantic parrotfish genus *Sparisoma*. *Mol. Phyl. Evol.* **15**, 292–300.

Bertram, D. F., and Leggett, W. C. (1994). Predation risk during the early life history periods of fishes: Separating the effects of size and age. *Mar. Ecol.: Prog. Ser.* **109**, 105–114.

Bertram, D. F., Chambers, R. C., and Leggett, W. C. (1993). Negative correlations between larval and juvenile growth rates in winter flounder: Implications of compensatory growth for variation in size-and-age. *Mar. Ecol.: Prog. Ser.* **96**, 209–215.

Bertram, I. (1996). The aquarium fishery in the Cook Islands. Is there a need for management? *Secretariat of the Pacific Community Live Reef Fish Inf. Bull.* **1**, 10–12.

Beukers, J. S., and Jones, G. P. (1997). Habitat complexity modifies the impact of piscivores on a coral reef fish population. *Oecologia* **114**, 50–59.

Beukers, J. S., Jones, G. P., and Buckley, R. M. (1995). Use of implant microtags for studies on populations of small reef fishes. *Mar. Ecol.: Prog. Ser.* **125**, 61–66.

Beverton, R. J. H. (1992). Patterns of reproductive strategy parameters in some marine teleost fishes. *J. Fish Biol.* **41**(Suppl. B), 137–160.

Beverton, R. J. H., and Holt, S. J. (1957). On the dynamics of exploited fish populations. *Fish. Invest. Minist. Agric. Fish. Food (GB)* **2**, 1–533.

Beverton, R. J. H., and Iles, T. C. (1992). Mortality rates of 0-group plaice (*Pleuronectes platessa* L.), dab (*Limanda limanda* L.) and turbot (*Scophthalmus maximus* L.) in European waters. III. Density-dependence of mortality rates of 0-group plaice and some demographic implications. *Neth. J. Sea Res.* **29**, 61–79.

Bierregaard, R. O., Jr., Lovejoy, T. E., Gascon, G. C., and Wilson, E. O. (2001). "Lessons from Amazonia: The Ecology and Conservation of a Fragmented Forest." Yale University Press, New Haven, CT.

Biffar, M. (1997). The worldwide trade in ornamental fish: Current status, trends and problems. *Bull. Eur. Assoc. Fish Pathol.* **17**, 201–204.

Billard, R., Bry, C., and Gillet, C. (1981). Stress, environment and reproduction in teleost fish. *In* "Stress and Fish" (A. D. Pickering, ed.), pp. 185–208. Academic Press, London.

Birkeland, C. (1997a). Symbiosis, fisheries and economic development on coral reefs. *Trends Ecol. Evol.* **12**, 364–367.

Birkeland, C. (1997b). Implications for resource management. *In* "The Life and Death of Coral Reefs" (C. Birkeland, ed.), pp. 411–435. Chapman & Hall, New York.

Birkeland, C. (1999). Live reef fish trade in the Carolines and Marianas. *In* "Proceedings of the first Asia-Pacific Seminar/Workshop on the Live Reef Fish Trade, August 11–14, 1998," pp. 25–27. International Marinelife Alliance, Manila.

Bisazza, A. and Marconato, A. (1988). Female mate choice, male-male competition and parental care in the river bullhead, *Cottus gobio* (Pisces, Cottidae). *Anim. Behav.* **36**, 1352–1360.

Bjørnstad, O. N., Fromentin, J.-M., Stenseth, N. C., and Gjøsæter, J. (1999). A new test for density dependent survival: The case of coastal cod populations. *Ecology* **80**, 1278–1288.

Black, K. P. (1988). The relationship of reef hydrodynamics to variations in numbers of planktonic larvae on and around coral reefs. *Proc. 6th Int. Coral Reef Symp.*, Vol. 2, pp. 125–130.

Black, K. P. (1994). Developments in our knowledge of dispersal on the Great Barrier Reef. *In* "The Bio-physics of Larval Dispersal" (P. W. Sammarco and M. L. Heron, eds.), pp. 159–192. American Geophysical Union, Washington, DC.

Black, K. P. (1995). Response to (E. Wolanski) 'Facts and numerical artefacts in modelling the dispersal of Crown-of-thorns starfish larvae in the Great Barrier Reef.' *Mar. Freshwater Res.* **46**, 883–887.

Black, K. P. (1998). Are model behaviors actively or passivly determining outcomes of biological hypothesis testing? *Abstr., 3rd Int. Larval Biol. Meet.*, Melbourne, *1998*, p. 17.

Black, K. P., and Gay, S. L. (1987). Eddy formation in unsteady flows. *J. Geophys. Res.* **92**, 9514–9522.

Black, K. P., and Moran, P. J. (1991). Influence of hydrodynamics on the passive dispersal and initial recruitment of larvae of *Acanthaster planci* (Echinodermata:Asteroidea) on the Great Barrier Reef. *Mar. Ecol.: Prog. Ser.* **69**, 55–65.

Black, K. P., Moran, P. J., and Hammond, L. S. (1991). Numerical models show coral reefs can be self-seeding. *Mar. Ecol.: Prog. Ser.* **74**, 1–11.

Black, K. P., Moran, P. J., Burrage, D., and De'ath, G. (1995). Association of low-frequency currents and crown-of-thorns starfish outbreaks. *Mar. Ecol.: Prog. Ser.* **125**, 185–194.

Blackburn, T. M., and Gaston, K. J. (1997). A critical assessment of the form of the interspecific relationship between abundance and body size in animals. *J. Anim. Ecol.* **66**, 233–249.

Blackburn, T. M., and Lawton, J. H. (1994). Population abundance and body size in animal assemblages. *Philos. Trans. R. Soc. London, Ser. B* **343**, 33–39.

Blackburn, T. M., Brown, V. K., Doube, B. M., Greenwood, J. J. D., Lawton, J. H., and Stork, N. E. (1993). The relationship between abundance and body size in natural animal assemblages. *J. Anim. Ecol.* **62**, 519–528.

Blanc, F., and Bonhomme, F. (1987). Polymorphisme gènètique des populations naturelles de mollusques diintèrit aquacole. *Proc. World Symp. Sel., Hybrid., Gene. Eng. Aquacul.*, Bordeaux, *1986*, pp. 59–78.

Blaxter, J. H. S. (1968). Visual thresholds and spectral sensitivity of herring larvae. *J. Exp. Biol.* **48**, 39–53.

Blaxter, J. H. S. (1969). Visual thresholds and spectral sensitivity of flatfish larvae. *J. Exp. Biol.* **51**, 221–230.

Blaxter, J. H. S. (1986). Development of sense organs and behavior of teleost larvae with special reference to feeding and predator avoidance. *Trans. Am. Fish. Soc.* **115**, 98–114.

Blaxter, J. H. S. (1988a). Sensory performance, behavior, and ecology of fish. *In* "Sensory Biology of Aquatic Animals" (J. Atema, R. R. Fay, A. N. Popper, and W. N. Tavolga, eds.), pp. 203–232. Springer-Verlag, New York.

Blaxter, J. H. S. (1988b). Pattern and variety in development. *Fish Physiol.* **11A**, 1–58.

Blaxter, J. H. S. (1991). Sensory systems and behavior of larval fish. *In* "Marine Biology: Its Accomplishment and Future Prospect" (J. Mauchline and T. Nemoto, eds.), pp. 15–38. Elsevier, Amsterdam.

Blaxter, J. H. S., and Batty, R. S. (1985). The development of startle responses in herring larvae. *J. Mar. Biol. Assoc. U.K.* **66**, 737–750.

Blaxter, J. H. S., and Fuiman, L. A. (1989). Function of the free neuromasts of marine teleost larvae. *In* "The Mechanosensory Lateral Line: Neurobiology and Evolution" (S. Coombs, P. Görner, and H. Münz, eds.), pp. 481–499. Springer-Verlag, New York.

Blaxter, J. H. S., and Fuiman, L. A. (1990). The role of the sensory systems of herring larvae in evading predatory fishes. *J. Mar. Biol. Assoc. U.K.* **70**, 413–427.

Bleck, R., and Chassignet, E. P. (1994). Simulating the oceanic circulation with isopycnic-coordinate models. *In* "The Oceans: Physical-chemical Dynamics and Human Impact" (S. K. Majumdar, E. W. Miller, G. S. Forbes, R. F. Schmalz, and A. A. Panah, eds.), pp. 17–39. Pennsylvania Academy of Science, Pittsburgh.

Bleckmann, H. (1986). Role of the lateral line in fish behavior. *In* "The Behavior of Teleost Fishes" (T. J. Pitcher, ed.), 1st ed., pp. 177–202. The Johns Hopkins Univ. Press, Baltimore.

Bleckmann, H. (1993). Role of the lateral line in fish behaviour. *In* "The Behaviour of Teleost Fishes" (T. J. Pitcher, ed.), 2nd ed., pp. 201–246. Chapman & Hall, New York.

Blot, J. (1980). La faune ichthyologique des gisements du Monte Bolca (Province de Vérone, Italie). Catalogue systématique présentant l'état actuel des recherches concernant cette faune. *Bull. Mus. Natl. Hist. Nat., Ser. 4* **2**(Sect. C, No. 4), 339–396.

Blot, J., and Tyler, J. C. (1990). New genera and species of fossil surgeon fishes and their relatives (Acanthuroidei, Teleostei) from the Eocene of Monte Bolca, Italy, with application of the Blot formula to both fossil and recent forms. *Studi e-Ric. Giacimenti Terziari Bolca, Mus. Civ. Stor. Nat., Verona* **6**, 13–92.

Blum, S. D. (1988). Osteology and phylgeny of the Chaetodontidae (Pisces: Perciformes). Ph.D. Dissertation, University of Hawaii, Honolulu.

Blum, S. D. (1989). Biogeography of the Chaetodontidae: An analysis of allopatry among closely related species. *Environ. Biol. Fishes* **25**, 9–31.

Board, P. A. (1956). The feeding mechanism of the fish *Sparisoma cretense*. *Proc. Zool. Soc. London* **127**, 59–78.

Boden, B. P. (1952). Natural conservation of insular plankton. *Nature (Lodon)* **169**, 697–699.

Boehlert, G. W. (1996). Larval dispersal and survival in tropical reef fishes. *In* "Reef Fisheries" (N. V. C. Polunin and C. M. Roberts, eds.), pp. 61–84. Chapman & Hall, London.

Boehlert, G. W., and Mundy, B. C. (1988). Roles of behavioral and physical factors in larval and juvenile fish recruitment to estuarine nursery areas. *Am. Fish. Soc. Symp.* **3**, 51–67.

Boehlert, G. W., and Mundy, B. C. (1993). Ichthyoplankton assemblages at seamounts and oceanic islands. *Bull. Mar. Sci.* **53**, 336–361.

Boehlert, G. W., and Mundy, B. C. (1994). Vertical and onshore-offshore distributional patterns of tuna larvae in relation to physical habitat features. *Mar. Ecol.: Prog. Ser.* **107**, 1–13.

Boehlert, G. W., and Mundy, B. C. (1996). Ichthyoplankton vertical distributions near Oahu, Hawai'i, 1985–1986: Data report. *NOAA Tech. Memo. NMFS-SWFSC* **NMFS-SWFSC 235**, 1–148.

Boehlert, G. W., Watson, W., and Sun, L. C. (1992). Horizontal and vertical distributions of larval fishes around an isolated oceanic island in the tropical Pacific. *Deep-Sea Res.* **39**, 439–466.

Böhlke, J. E., and Chaplin, C. C. G. (1968). "Fishes of the Bahamas and Adjacent Tropical Waters." Livingston Publ. Wynnewood, PA.

Böhlke, J. E., and Chaplin, C. C. G. (1993). "Fishes of the Bahamas and Adjacent Tropical Waters," 2nd ed. University of Texas Press, Austin.

Bohnsack, J. A. (1982). Effects of piscivorous predator removal on coral reef fish community structure. *In* "Gutshop 81: Fish Food Habit Studies" (G. M. Cailliet and C. A. Simstead, eds.), Washington Sea Grant Publication, pp. 258–267. University of Washington, Seattle.

Bohnsack, J. A. (1989). "Protection of Grouper Spawning Aggregations." Coast. Resour. Div. Contrib. 88-89-06. National Oceanic and Atmospheric Administration, Washington, DC.

Bohnsack, J. A. (1993). Marine reserves: They enhance fisheries, reduce conflicts, and protect resources. *Oceanus* **36**, 63–71.

Bohnsack, J. A. (1996). Maintenance and recovery of reef fish productivity. *In* "Reef Fisheries" (N. V. C. Polunin & C. M. Roberts, eds.), pp. 283–313. Chapman & Hall, London.

Bohnsack, J. A. (1998). Application of marine reserves to reef fisheries management. *Aust. J. Ecol.* **23**, 298–304.

Bohnsack, J. A. (1999). Incorporating no-take marine reserves into precautionary management and stock assessment. *NOAA Tech. Memo NMFS-F/SPO-40,* **NMFS-F/SPO-40,** 8–16.

Bohnsack, J. A., and Ault, J. S. (1996). Management strategies to conserve marine biodiversity. *Oceanography* **9**, 73–82.

Bohnsack, J. A., and Bannerot, S. P. (1986). A stationary visual census technique for quantitatively assessing community structure of coral reef fishes. *NOAA Tech. Rep. NMFS* **41**, 1–15.

Bohnsack, J. A., Causey, B., Crosby, M., Griffis, R., Hixon, M., Hourigan, T., Koltes, K., Maragos, J., Simons, A., and Tilmant, J. (in press). Rationale for minimum 20–30% no-take reef protection. *Proc. 9th Int. Coral Reef Symp., Bali* (in press).

Bone, Q., Marshall, N. B., and Blaxter, J. H. S. (1995). "Biology of Fishes." Chapman & Hall, London.

Boonstar, R., Krebs, C. J., and Stenseth, N. C. (1998). Population cycles in small animals: The problem of explaining the low phase. *Ecology* **79**, 1479–1488.

Boorman, S. A., and Levitt, P. R. (1973). Group selection on the boundary of a stable population. *Theor. Popul. Biol.* **4**, 85–128.

Booth, D. J. (1991). The effect of sampling frequency on estimates of recruitment of the domino damselfish *Dascyllus albisella* Gill. *J. Exp. Mar. Biol. Ecol.* **145**, 149–159.

Booth, D. J. (1992). Larval settlement patterns and preferences by domino damselfish, *Dascyllus albisella* Gill. *J. Exp. Mar. Biol. Ecol.* **155**, 85–104.

Booth, D. J. (1995). Juvenile groups in a coral-reef damselfish: Density-dependent effects on individual fitness and population demography. *Ecology* **76**, 91–106.

Booth, D. J., and Beretta, G. A. (1994). Seasonal recruitment and habitat associations in a guild of pomacentrid reef fishes at two sites around St. Thomas, USVI. *Coral Reefs* **13**, 81–89.

Booth, D. J., and Brosnan, D. M. (1995). The role of recruitment dynamics in rocky shore and coral reef fish communities. *Adv. Ecol. Res.* **26**, 309–385.

Booth, D. J., and Hixon, M. A. (1999). Food ration and condition affect early survival of the coral reef damselfish, *Stegastes partitus*. *Oecologia* **121**, 364–368.

Booth, D. J., and Wellington, G. (1998). Settlement preferences in coral-reef fishes: Effects on patterns of adult and juvenile distributions, individual fitness and population structure. *Aust. J. Ecol.* **23**, 274–279.

Booth, D. J., Kingsford, M. J., Doherty, P. J., and Beretta, G. A. (2000). Recruitment of damselfishes in the One Tree Island lagoon: Persistent interannual spatial patterns. *Mar. Ecol. Prog. Ser.* **202**, 219–230.

Boothroyd, H. (1978). "Articulate Intervention." Taylor & Francis, London.

Borden, W. C. (1998). Phylogeny of the unicornfishes (*Naso,* Acanthuridae) based on soft anatomy. *Copeia,* pp. 104–113.

Bortone, S. A., Van Tassel, J., Brito, A., Falcón, J. M., and Bundrick, C. M. (1991). A visual assesment of the inshore fishes and fishery resources off El Hierro, Canary Islands: A baseline survey. *Sci. Mar.* **55**, 529–541.

Bouchon-Navaro, Y. (1980). Quantitative distribution of the Chaetodontidae on a fringing reef of the Jordanian coast (Gulf of Aqaba, Red Sea). Tethys 9, 247–251.

Bouchon-Navaro, Y., and Bouchon, C. (1989). Correlations between chaetodontid fishes and coral communities of the Gulf of Aqaba (Red Sea). *Environ. Biol. Fishes* **25**, 47–60.

Bouchon-Navaro, Y., Bouchon, C., and Harmelin-Vivien, M. (1985). Impact of coral degradation on the chaetodontid fish assemblage (Moorea, French Polynesia). *Proc. 5th Int. Coral Reef Congr.*, Vol. 5, pp. 427–432.

Boucot, A. J. (1983). Does evolution occur in an ecologic vacuum? *J. Paleontol.* **57**, 1–30.

Bowen, B. W., Clark, A. M., Abreu-Grobois, F. A., Chaves, A., Reichart, H. A., and Ferl, J. A. (1998). Global phylogeography of the ridley sea turtles (*Lepidochelys* spp.) as inferred from mitochondrial DNA sequences. *Genetica* **101**, 179–189.

Bowmaker, J. K. (1990). Visual pigments of fishes. *In* "The Visual System of Fish" (R. H. Douglas and M. B. A. Djamgoz, eds.), pp. 81–107. Chapman & Hall, New York.

Bowmaker, J. K. (1995). The visual pigments of fish. *Prog. Retinal Eye Res.* **15**, 1–31.

Bowman, M. J., Stansfield, K. L., Fauria, S. J., and Wilson, T. C. (1994). Coastal ocean circulation near Barbados, West Indies, Spring 1990 and 1991. *J. Geophys. Res.* **99**, 16, 131–16, 142.

Bradbury, J. W., and Vehrencamp, S. L. (1977). Social organization and foraging in emballonurid bats. III: Mating systems. *Behav. Ecol. Sociobiol.* **2**, 1–17.

Brander, K. M. (1995). The effect of temperature on growth of Atlantic cod (*Gadus morhua* L.). *ICES J. Mar. Sci.* **52**, 1–10.

Brantley, R. K., and Bass, A. H. (1994). Alternative male spawning tactics and acoustic signals in the plainfin midshipman fish *Porichthys notatus Girard* (Teleostei, Batrachoididae). *Ethology* **96**, 213–232.

Breitburg, D. L. (1989). Demersal schooling prior to settlement by larvae of the naked goby. *Environ. Biol. Fishes* **26**, 97–103.

Breitburg, D. L. (1991). Settlement patterns and presettlement behavior of the naked goby, *Gobiosoma bosci*, a temperate oyster reef fish. *Mar. Biol.* **109**, 213–221.

Breitburg, D. L., Palmer, M. A., and Loher, T. (1995). Larval distributions and the spatial patterns of settlement of an oyster reef fish: Responses to flow and structure. *Mar. Ecol.: Prog. Ser.* **125**, 45–60.

Bretsky, P. W., and Lorenz, D. M., (1970). An essay on genetic adaptative strategies and mass extinction. *Geol. Soc. Am. Bull.* **81**, 2449–2456.

Brey, T., Rumohr, H., and Ankar, S. (1988). Energy content of macrobenthic invertebrates: General conversion factors from weight to energy. *J. Exp. Mar. Biol. Ecol.* **117**, 271–278.

Briggs, J. C. (1961). The east pacific barrier and the distribution of marine shorefishes. *Evolution (Lawrence, Kans.)* **15**, 545–554.

Briggs, J. C. (1974). "Marine Zoogeography." McGraw-Hill, New York.

Briggs, J. C. (1989). The historic biogeography of India: Isolation or contact? *Syst. Zool.* **38**, 322–332.

Briggs, J. C. (1996). "Global Biogeography," Dev. Paleonto. Stratigr., No. **14**. Elsevier, Amsterdam.

Briggs, J. C. (1999). Coincident biogeographic patterns: Indo-west Pacific Ocean. *Evolution (Lawrence, Kans.)* **53**; 326–335.

Brogan, M. W. (1994a). Two methods of sampling fish larvae over reefs: A comparison from the Gulf of California. *Mar. Biol.* **118**, 33–44.

Brogan, M. W. (1994b). Distribution and retention of larval fishes near reefs in the Gulf of California. *Mar. Ecol.: Prog. Ser.* **115**, 1–13.

Brooks, D. R., and McLennan, D. A. (1991). "Phylogeny, Ecology and Behavior: A Research Program in Comparative Biology." University of Chicago Press, Chicago.

Brothers, E. B., and Thresher, R. E. (1985). Pelagic duration, dispersal, and the distribution of Indo-pacific coral-reef fishes. *In:* "The Ecology of Coral Reefs" (M. L. Reaka, ed.), pp. 53–69. NOAA symposia series for undersea research, **3(1)**. Nat. Oceanographic Atmospheric Admin., United States Department of Commerce, Washington, D.C.

Brothers, E. B., Williams, D. McB., and Sale, P. F. (1983). Length of larval life in twelve families of fishes at "One Tree. Lagoon," Great Barrier Reef, Australia. *Mar. Biol.* **76**, 319–324.

Browman, H. I., Novales-Flamarique, I., and Hawryshyn, C. W. (1994). Ultraviolet photoreception contributes to prey search behavior in two species of zooplanktivorous fishes. *J. Exp. Biol.* **186**, 187–198.

Brown, A. H. D. (1979). Enzyme polymorphism in plant populations. *Theor. Popul. Biol.* **15**, 1–42.

Brown, G. E., Chivers, D. P., and Smith, R. J. F. (1995a). Fathead minnows avoid conspecific and heterospecific alarm pheromones in the faeces of northern pike. *J. Fish Biol.* **47**, 387–393.

Brown, G. E., Chivers, D. P., and Smith, R. J. F. (1995b). Localized defecation by pike: A response to labelling by cyprinid alarm pheromones? *Behav. Ecol. Sociobiol.* **36**, 105–110.

Brown, J. A., Pepin, P., Methven, D. A., and Somerton, D. C. (1989). The feeding, growth and behavior of juvenile cod, *Gadus morhua* L., in cold environments. *J. Fish Biol.* **35**, 373–380.

Brown, J. H. (1984). On the relationship between abundance and distributions of species. *Am. Nat.* **124**, 255–279.

Brown, J. H. (1995). "Macroecology." Chicago University Press, Chicago.

Brown, J. H., and Maurer, B. A. (1989). Macroecology: The division of food and space among species on continents. *Science* **243**, 1145–1150.

Brown, J. H., and Nicoletto, P. F. (1991). Spatial scaling of species composition: Body masses of North American land mammals. *Am. Nat.* **138,** 1478–1512.

Brown, J. H., Mehlman, D. W., and Stevens, G. C. (1995). Spatial variation in abundance. *Ecology* **76,** 2028–2043.

Brown, J. H., Stevens, G. C., and Kaufman, D. M. (1996). The geographic range: Size, shape, boundaries, and internal structure. *Annu. Rev. Ecol. Syst.* **27,** 597–623.

Bruckmoser, P. (1973). Beziehungen zwischen Struktur und Funktion in der Evolution des Telencephalon. *Verh. Dtsch. Zool. Ges.* **66,** 219–229.

Bruggemann, J. H., Begeman, J., Bosma, E. M., Verburg, P., and Breeman, A. M. (1994a). Foraging by the stoplight-parrotfish *Sparisoma viride.* 2. Intake and assimilation of food, protein and energy. *Mar. Ecol.: Prog. Ser.* **106,** 57–71.

Bruggemann, J. H., Kuyper, M. W. M. and Breeman, A. M. (1994b). Comparative-analysis of foraging and habitat use by the sympatric caribeean parrotfish *Scarus vetula* and *Sparisoma viride* (Scaridae). *Mar. Ecol.: Prog. Ser.* **112,** 51–66.

Bruggemann, J. H., van Oppen, M. J. H., and Breeman, A. M. (1994c). Foraging by the stoplight parrotfish *Sparisoma viride.* 1. Food selection in different socially determined habitats. *Mar. Ecol.: Prog. Ser.* **106,** 51–55.

Bruggemann, J. H., van Kessel, A. M., van Rooij, J. M., and Breeman, A. M. (1996). Bioerosion and sediment ingestion by the Caribbean parrotfish *Scarus vetula* and *Sparisoma viride*: Implications of fish size, feeding mode and habitat use. *Mar. Ecol.: Prog. Ser.* **134,** 59–71.

Bryan, P. G. (1975). Food habits, functional digestive morphology, and assimilation efficiency of rabbitfish, *Siganus spinus* (Pisces, Siganidae) on Guam. *Pac. Sci.* **29,** 269–277.

Bryan, P. G., and Madraisau, B. B. (1977). Larval rearing and development of *Siganus lineatus* (Pisces: Siganidae) from hatching through metamorphosis. *Aquaculture* **10,** 243–252.

Bryant, D., Burke, L., McManus, J., and Spalding, M. (1998). "Reefs at Risk: A Map-based Indicator of Potential Threats to the World's Coral Reef." World Resources Institute, Washington, DC.

Bryant, H. E., Dewey, M. R., Funicelli, N. A., Meineke, D. A., and Mengel, J. (1989). Movement of five selected species of fish in the Everglades National Park. *Bull. Mar. Sci.* **44,** 515.

Buchheim, J. R., and Hixon, M. A. (1992). Competition for shelter holes in the coral-reef fish *Acanthemblemaria spinosa* Metzelaar. *J. Exp. Mar. Biol. Ecol.* **164,** 45–54.

Buckley, R. M., West, J. E., and Doty, D. C. (1994). Internal micro-tag systems for marking juvenile reef fishes. *Bull. Mar. Sci.* **55,** 848–857.

Budd, A. F., Johnson, K. G., and Stemann, T. A. (1994). Plio-Pleistocene extinctions and the origin of the modern Caribbean coral reef fauna. *In* "Global Aspects of Coral Reefs: Health, Hazards and History" (R. N. Ginsburg, ed.), pp. 7–13. University of Miami Publ., Miami.

Budd, A. F., Johnson, K. G., and Stemann, T. A. (1996). Plio-Pleistocene turnover and extinctions in the Caribbean reef-coral fauna. *In* "Evolution and Environment in Tropical America" (J. B. C. Jackson, A. F. Budd, and A. G. Coates, eds.), pp. 168–204. University of Chicago Press, Chicago.

Buijse, A. D., and Houthuijzen, R. P. (1992). Piscivory, growth, and size selective mortality of age 0 pikeperch (*Stizostedion lucioperca*). *Can. J. Fish. Aquat. Sci.* **49,** 894–902.

Bullock, A. E., and Monod, T. (1997). Cephalic myology of two parrotfishes (Teleostei: Scaridae). *Cybium* **21,** 173–199.

Bullock, L. H., and Murphy, M. D. (1994). Aspects of the life history of the yellowmouth grouper, *Mycteroperca interstitialis,* in the eastern gulf of Mexico. *Bull. Mar. Sci.* **55,** 30–45.

Bullock, L. H., Murphy, M. D., Godcharles, M. F., and Mitchell, M. E. (1992). Age, growth and reproduction of jewfish *Epinephelus itajara* in the eastern Gulf of Mexico. *Fish. Bull.* **90,** 243–249.

Bullock, T. H. (1973). Seeing the world through a new sense: Electroreception in fish. *Am. Sci.* **61,** 316–325.

Bullock, T. H., and Heiligenberg, W. (1986). "Electroreception." Wiley, New York.

Bullough, W. S. (1947). Hermaphroditism in the lower vertebrates. *Nature (London)* **266,** 828–830.

Burrage, D., Hughes, R., Bode, L., and Williams, D. (1994). Dynamic features and transports of the Coral Sea circulation. *In* "Recent Advances in Marine Science and Technology" (O. Bellwood, H. Choat, and N. Saxena, eds.), pp. 95–105. PACON International and James Cook University, Townsville, Australia.

Burt, A., Kramer, D. L., Nakatsuru, K., and Spry, C. (1988). The tempo of reproduction in *Hyphessobrycon pulchripinnis* (Characidae), with a discussion of the biology of "multiple spawning" in fishes. *Environ. Biol. Fishes* **22,** 15–27.

Burton, R. S. (1983). Protein polymorphism and genetic differentiation in marine invertebrate populations. *Mar. Biol. Lett.* **4,** 193–206.

Buxton, C. D. (1993). Life-history changes in exploited reef fishes on the east coast of South Africa. *Environ. Biol. Fishes* **36,** 47–63.

Buxton, C. D., and Allen, J. A. (1989). Mark and recapture studies of two reef sparids in the Tsitsikamma Coastal National Park. *Koedoe* **32,** 39–45.

Buxton, C. D., and Smale, M. J. (1989). Abundance and distribution patterns of three temperate marine fish (Teleostei: Sparidae) in exploited and unexploited areas off the southern cape coast. *J. Appl. Ecol.* **26,** 441–451.

Buzas, M. A., and Culver, S. J. (1991). Species diversity and dispersal of benthic foraminifera. *BioScience* **41,** 483–489.

Bystrom, P., Persson, L., and Wahlstrom, E. (1998). Competing predators and prey: Juvenile bottlenecks in whole-lake experiments. *Ecology* **79,** 2153–2167.

Caddy, J. F. (1999). Fisheries management in the twenty-first century: Will new paradigms apply? *Rev. Fish Biol. Fish.* **9**, 1–43.

Cahn, P. H., Shaw, L. A., and Atz, E. H. (1968). Lateral line histology as related to the development of schooling in the atherinid fish *Menidia*. *Bull. Mar. Sci.* **18**, 660–670.

Caley, M. J. (1993). Predation, recruitment and the dynamics of communities of coral-reef fishes. *Mar. Biol.* **117**, 33–43.

Caley, M. J. (1995a). Community dynamics of tropical reef fishes: Local patterns between latitudes. *Mar. Ecol.: Prog. Ser.* **129**, 1–18.

Caley, M. J. (1995b). Reef-fish community structure and dynamics: An interaction between local and larger-scale processes? *Mar. Ecol.: Prog. Ser.* **129**, 19–29.

Caley, M. J. (1997). Local endemism and the relationship between local and regional diversity. *Oikos* **79**, 612–615.

Caley, M. J. (1998). Age-specific mortality rates in reef fishes: Evidence and implications. *Aust. J. Ecol.* **23**, 241–245.

Caley, M. J., and Schluter, D. (1997). The relationship between local and regional diversity. *Ecology* **78**, 70–80.

Caley, M. J., and St. John, J. (1996). Refuge availability structures assemblages of tropical reef fishes. *J. Anim. Ecol.* **65**, 414–428.

Caley, M. J., Carr, M. H., Hixon, M. A., Hughes, T. P., Jones, G. P., and Menge, B. A. (1996). Recruitment and the local dynamics of open marine populations. *Annu. Rev. Ecol. Syst.* **27**, 477–500.

Calow, L. J., and Alexander, R. M. (1973). A mechanical analysis of a hind leg of a frog (*Rana temporaria*). *J. Zool.* **171**, 293–321.

Campana, S. E. (1983). Feeding periodicity and the production of daily growth increments in otoliths of steelhead trout (*Salmo gairdneri*) and starry flounder (*Platichthys stellatus*). *Can. J. Zool.* **61**, 1591–1597.

Campana, S. E. (1990). How reliable are growth back-calculations based on otoliths? *Can. J. Fish. Aquat. Sci.* **47**, 2219–2227.

Campana, S. E. (1992). Measurement and interpretation of the microstructure of fish otoliths. *Can. Fish. Aquat. Sci., Spec. Publ.* **117**, 59–71.

Campana, S. E. (1996). Year-class strength and growth rate in young Atlantic cod *Gadus morhua*. *Mar. Ecol.: Prog. Ser.* **135**, 21–26.

Campana, S. E. (1997). Use of radiocarbon from nuclear fallout as a dated marker in the otoliths of haddock, *Melanogrammus aeglefinus*. *Mar. Ecol.: Prog. Ser.* **150**, 49–56.

Campana, S. E. (1999). Chemistry and composition of fish otoliths: Pathways, mechanisms, and applications. *Mar. Ecol.: Prog. Ser.* **188**, 263–297.

Campana, S. E., and Casselman, J. M. (1993). Stock discrimination using otolith shape analysis. *Can. J. Fish. Aquat. Sci.* **50**, 1062–1083.

Campana, S. E., and Jones, C. M. (1992). Analysis of otolith microstructure data. *Can. Fish. Aquat. Sci., Spec. Publ.* **117**, 73–100.

Campana, S. E., and Jones, C. M. (1998). Radiocarbon from nuclear testing applied to age validation of black drum, *Pogonias chromis*. *Fish. Bull.* **96**, 185–192.

Campana, S. E., and Moksness, E. (1991). Accuracy and precision of age and hatch date estimates from otolith microstructure examination. *ICES J. Mar. Sci.* **48**, 303–316.

Campana, S. E., and Neilson, J. D. (1985). Microstructure of fish otoliths. *Can. J. Fish. Aquat. Sci.* **42**, 1014–1032.

Campana, S. E., and Thorrold, S. R. (2001). Otoliths, increments, and elements: Keys to a comprehensive understanding of fish populations? *Can. J. Fish. Aquat. Sci.* **58**, 30–38.

Campana, S. E., Gagne, J. A., and Munro, J. (1987). Otolith microstructure of larval herring (*Clupea harengus*): Image or reality? *Can. J. Fish. Aquat. Sci.* **44**, 1922–1929.

Campana, S. E., Zwanenburg, K. C. T., and Smith, J. N. (1990). $^{210}Pb/^{226}Ra$ determination of longevity in redfish. *Can. J. Fish. Aquat. Sci.* **47**, 163–165.

Campana, S. E., Oxenford, H. A., and Smith, J. N. (1993). Radiochemical determination of longevity in flyingfish *Hirundichthys affinis* using $^{228}Th/^{228}Ra$. *Mar. Ecol.: Prog. Ser.* **100**, 211–219.

Campana, S. E., Thorrold, S. R., Jones, C. M., Günther, D., Tubrett, M., Longerich, H., Jackson, S., Halden, N. M., Kalish, J. M., Piccoli, P., de Pontual, H., Troadec, H., Panfili, J., Secor, D. H., Severin, K. P., Sie, S. H., Thresher, R., Teesdale, W. J., and Campbell, J. L. (1997). Comparison of accuracy, precision and sensitivity in elemental assays of fish otoliths using the electron microprobe, PIXE and laser ablation ICPMS. *Can. J. Fish. Aquat. Sci.* **54**, 2068–2079.

Campbell, P. M., Pottinger, T. G., and Sumpter, J. P. (1992). Stress reduces the quality of gametes produced by rainbow trout. *Biol. Rep.* **47**, 1140–1150.

Campbell, P. M., Pottinger, T. G., and Sumpter, J. P. (1994). Preliminary evidence that chronic confinement stress reduces the quality of gametes produced by brown and rainbow trout. *Aquaculture* **120**, 151–169.

Cappo, M., Eden, P., Newman, S. J., and Robertson, S. (2000). A new approach to validation of periodicity and timing of opaque zone formation in the otoliths of 11 species of *Lutjanus* from the central Great Barrier Reef. *Fish. Bull.* **98**, 474–488.

Cappuccino, N., and Price, P. W. (1995). "Population Dynamics: New Approaches and Synthesis." Academic Press, San Diego, CA.

Caprio, J. (1988). Peripheral filters and chemoreceptor cells in fishes. *In* "Sensory Biology of Aquatic Animals" (J. Atema, R. R. Fay, A. N. Popper, and W. N. Tavolga, eds.), pp. 313–338. Springer-Verlag, New York.

Carlton, J. T., Geller, J. B., Reaka-Kudla, M. L., and Norse, E. A. (1999). Historical extinctions in the sea. *Annu. Rev. Ecol. Syst.* **30**, 515–538.

Carpenter, K. E. (1987). A taxonomic revision of the Indo-Pacific fish family Caesionidae (Perciformes: Lutjanoidea), with descriptions of five new species. *Indo-Pac. Fish.* **15**, 1–56.

Carpenter, K. E. (1990). A phylogenetic analysis of the Caesionidae (Perciformes: Lutjanoidea). *Copeia,* pp. 692–717.

Carpenter, K. E. (1993). Optimal cladistic and quantitative evolutionary classifications as illustrated by fusilier fishes (Teleostei: Caesionidae). *Syst. Biol.* **42,** 142–154.

Carpenter, K. E., Miclat, R. I., Albaladejo, V. D., and Corpuz, V. T. (1981). The influence of substrate structure on the local abundance and diversity of Philippine reef fishes. *Proc. 4th Int. Coral Reef Symp.,* Vol. 2, pp. 497–502.

Carpenter, R. C. (1986). Partitioning herbivory and its effects on coral reef algal communities. *Ecol. Monogr.* **56,** 345–363.

Carr, M. H. (1994). Effects of macroalgal dynamics on recruitment of a temperate reef fish. *Ecology* **75,** 1320–1333.

Carr, M. H., and Hixon, M. A. (1995). Predation effects on early post-settlement survivorship of coral-reef fishes. *Mar. Ecol.: Prog. Ser.* **124,** 31–42.

Carr, M. H., and Reed, D. C. (1993). Conceptual issues relevant to marine harvest refuges: Examples from temperate reef fishes. *Can. J. Fish. Aquat. Sci.* **50,** 2019–2028.

Carr, W. E. S. (1976). Chemoreception and feeding behavior in the pigfish, *Orthopristis chrysopterus:* Characterization and identifcation of stimulatory substances in a shrimp extract. *Comp. Biochem. Physiol. A* **55A,** 153–157.

Carr, W. E. S., and Chaney, T. B. (1976). Chemical stimulation of feeding behavior in the pinfish, *Lagodon rhomboides*: Characterization and identification of stimulatory substances extracted from shrimp. *Comp. Biochem. Physiol. A* **54A,** 437–441.

Carr, W. E. S., Gondeck, A. R., and Delaney, R. L. (1977). Chemical stimulation of feeding behavior in the pinfish, *Lagodon rhomboides*: A new approach to an old problem. *Comp. Biochem. Physiol. A* **54A,** 161–166.

Carragher, J. F., and Pankhurst, N. W. (1991). Stress and reproduction in a commercially important marine fish, *Pagrus auratus* (Sparidae). *In* "Reproductive Physiology of Fish 1991" (A. P. Scott, J. P. Sumpter, D. E. Kime, and M. S. Rolfe, eds.), Fish Symp. 91, pp. 253–255. Sheffield.

Carroll, R. L. (1988). "Vertebrate Paleontology and Evolution." Freeman, New York.

Carter, J. (1989). Grouper sex in Belize. *Nat. Hist.* **10,** 61–68.

Carter, J. G., Marrow, G. J., and Pryor, C. (1994). Aspects of the ecology and reproduction of Nassau grouper, *Epinephelus striatus,* off the coast of Belize, Central America. *Proc. Annu. Gulf. Carrib. Fish. Inst.* **43,** 65–111.

Carvalho, G. R. (1993). Evolutionary aspects of fish distribution: Genetic variability and adaptation. *J. Fish Biol.* **43**(Suppl. A), 53–73.

Carvalho, G. R., and Hauser, L. (1998). Advances in the molecular analysis of fish population structure. *Ital. J. Zool.* **65,** 21–33.

Caselle J. E. (1997). Small-scale spatial variation in early life history characteristics of a coral reef fish: Implications for dispersal hypotheses. *Proc. 8th Int. Coral Reef Symp.,* Vol. 2, pp. 1161–1166.

Caselle, J. E. (1999). Early post-settlement mortality in a coral reef fish and its effect on local population size. *Ecol. Monogr.* **69,** 177–195.

Caselle, J. E., and Warner, R. R. (1996). Variability in recruitment of coral reef fishes: The importance of habitat at two spatial scales. *Ecology* **77,** 2488–2504.

Castro, L. R. (1995). Ichthyoplankton distribution around Barbados: Patterns and processes conducive to retention. Ph.D. Dissertation, SUNY, Stony Brook, NY.

Cato, D. H. (1978). Marine biological choruses observed in tropical waters near Australia. *J. Acoust. Soc. Am.* **64,** 736–743.

Cato, D. H. (1980). Some unusual sounds of apparent biological origin responsible for sustained noise in the Timor Sea. *J. Acoust. Soc. Am.* **68,** 1056–1060.

Cato, D. H. (1992). The biological contribution to the ambient noise in waters near Australia. *Acoust. Aust.* **20,** 76–80.

Cesar, H. (1996). "Economic Analysis of Indonesian Coral Reefs," World Bank Environ. Dep. Pap., Environ. Econ. Ser. Agriculture Operations, Country Department III, East Asia and Pacific Region and Environment Department, World Bank, Washington, DC.

Cesar, H., Lundin, C. G., Bettencourt, S., and Dixon, J. (1997). Indonesian coral reefs—an economic analysis of a precious but threatened resource. *Ambio* **26,** 345–350.

Cesar, H., Warren, K., Sadovy. Y., Lau, P., Meijer, S., and van Ierland, E. (2000). Marine market transformation of the live reef fish food trade in Southeast Asia. *In* "Collected Essays on the Economics of Coral Reefs" (H. Cesar, ed.), pp. 137–157. CORDIO, Kalmar University, Kalmar, Sweden.

Cha, S. S., McGowan, M. F., and Richards, W. J. (1994). Vertical distribution of fish larvae off the Florida Keys, 26 May–5 June 1989. *Bull. Mar. Sci.* **54,** 828–842.

Chabernet, P., Ralambondrainy, H., Amanieu, M., Faure, G., and Galzin, R. (1997). Relationships between coral reef substrata and fish. *Coral Reefs* **16,** 93–102.

Chakraborty, S. K., and Vidyasagar, K. D. (1996). Growth, mortality and stock assessment of two perches, moontail bull's eye *Priacanthus hamrur* (Perciformes/Priacanthidae) and thornycheek grouper *Epinephelus diacanthus* (Perciformes/Serranidae) from Bombay waters. *Indian J. Mar. Sci.* **25,** 312–315.

Chambers, R. C., and Leggett, W. C. (1987). Size and age at metamorphosis in marine fishes: An analysis of laboratory-reared winter flounder (*Pseudopleuronectes americanus*) with a review of variation in other species. *Can. J. Fish. Aquat. Sci.* **44,** 1936–1947.

Chambers, R. C., and Leggett, W. C. (1996). Maternal influences on variation in egg sizes in temperate marine fishes. *Am. Zool.* **36,** 180–196.

Chambers, R. C., and Miller, T. J. (1995). Evaluating fish growth by means of otolith increment analysis: special properties of individual-level longitudinal data. *In* "Recent Developments in Fish Otolith Research" (D. H.

Secor, J. M. Dean, and S. E. Campana, eds.), pp. 155–175. University of South Carolina Press, Columbia.

Champalbert, G., Macquart-Moulin, C., Partiti, G., and Chiki, D. (1991). Ontogenetic variations in the phototaxis of larval and juvenile sole *Solea solea L. J. Exp. Mar. Biol. Ecol.* **149**, 207–255.

Chan, P. S. W. (2000). Current status of the live reef fish trade based in Hong Kong. *Secretariat of the Pacific Community Live Reef Fish Inf. Bull.* **7**, 8–9.

Chan, P. S. W. (2001). Wholesale and retail marketing aspects of Hong Kong's live seafood business. *In* "Marketing and Shipping Live Aquatic Products: Proceedings of the Second International Conference and Exhibition, November 1999, Seattle, WA" (B. C. Paust and A. A. Rice, eds.), AK-SG-01-03, pp. 201–206. University of Alaska Sea Grant, Fairbanks.

Chan, T. T. C., and Sadovy, Y. (1998). Profile of the marine aquarium fish trade in Hong Kong. *Aquar. Sci. Conserv.* **2**, 1–17.

Chan, W. W. (2000). "An Integrated Attitude Survey on Live Reef Food Fish Consumption in Hong Kong." World Wide Fund for Nature, Hong Kong.

Chao, N. L. (1986). A synopsis on zoogeography of the Sciaenidae. *In* "Indo-Pacific Fish Biology" (T. Uyeno, R. Arai, T. Taniuchi, and K. Matsuura, eds.), Proc. 2nd Int. Conf. Indo-Pac. Fishes, pp. 570–589. Ichthyological Society of Japan, Tokyo.

Chapman, F. A., and Fitz-Coy, S. A. (1997). United States of America trade in ornamental fish. *J. World Aquacult. Soc.* **28**, 1–10.

Chapman, M. R., and Kramer, D. L. (1999). Gradients in coral reef fish density and size across the Barbados Marine Reserve boundary: Effects of reserve protection and habitat characteristics. *Mar. Ecol.: Prog. Ser.* **181**, 81–96.

Chapman, M. R., and Kramer, D. L. (2000). Movements of fishes within and among fringing coral reefs in Barbados. *Environ. Biol. Fishes* **57**, 11–24.

Charnov, E. L. (1982). "The Theory of Sex Allocation." Princeton University Press, Princeton, NJ.

Chatfield, C. (1979). "The Analysis of Time Series: An Introduction." Chapman & Hall, London.

Chesson, P. L. (1981). Models for spatially distributed populations: The effect of within-patch variability. *Theor. Popul. Biol.* **19**, 288–323.

Chesson, P. L. (1983). Coexistence of competitors in a stochastic environment: The storage effect. *Lect. Notes Biomath.* **52**, 188–198.

Chesson, P. L. (1984). The storage effect in stochastic population models. *Lect. Notes Biomath.* **54**, 76–89.

Chesson, P. L. (1996). Matters of scale in the dynamics of populations and communities. *In* "Frontiers in Population Biology" (R. B. Floyd and A. W. Sheppard, eds.), pp. 353–368. CSIRO, Melbourne.

Chesson, P. L. (1998a). Spatial scales in the study of reef fishes: A theoretical perspective. *Aust. J. Ecol.* **23**, 209–215.

Chesson, P. L. (1998b). Recruitment limitation: A theoretical perspective. *Aust. J. Ecol.* **23**, 234–240.

Choat, J. H. (1991). The biology of herbivorous fishes on coral reefs. *In* "The Ecology of Fishes on Coral Reefs" (P. F. Sale, ed.), pp. 120–155. Academic Press, San Diego, CA.

Choat, J. H., and Axe, L. (1996). Growth and longevity in acanthurid fishes; an analysis of otolith increments. *Mar. Ecol.: Prog. Ser.* **134**, 15–26.

Choat, J. H., and Bellwood, D. R. (1985). Interactions amongst herbivorous fishes on a coral reef: Influence of spatial variation. *Mar. Biol.* **89**, 221–234.

Choat, J. H., and Bellwood, D. R. (1991). Reef fishes: Their history and evolution. *In* "The Ecology of Fishes on Coral Reefs" (P. F. Sale, ed.), pp. 39–66. Academic Press, San Diego, CA.

Choat, J. H., and Clements, K. D. (1998). Vertebrate herbivores in marine and terrestrial environments: A nutritional ecology perspective. *Annu. Rev. Ecol. Syst.* **29**, 375–403.

Choat, J. H., Ayling, A. M., and Schiel, D. R. (1988). Temporal and spatial variation in an island fish fauna. *J. Exp. Mar. Biol. Ecol.* **121**, 31–42.

Choat, J. H., Doherty, P. J., Kerrigan, B. A., and Leis, J. M. (1993). A comparison of towed nets, purse seines and light aggregation devices for sampling larvae and pelagic juveniles of coral reef fishes. *Fish. Bull.* **91**, 195–209.

Choat, J. H., Axe, L., and Lou, D. C. (1996). Growth and longevity in fishes of the family Scaridae. *Mar. Ecol.: Prog. Ser.* **145**, 33–41.

Christie, P., White, A.T., and Buhat, D. (1994). Community-based coral reef management on San Salvador Island, the Philippines. *Soc. Nat. Resour.* **7**, 103–117.

Chu, F. L., and Ozkizilcik, S. (1995). Lipid and fatty acid composition of striped bass (*Morone saxatilis*) larvae during development. *Comp. Biochem. Physiol. B* **111**, 665–674.

Cione, A. L., Azpelicueta, M. M., and Bellwood, D. R. (1994). An oplegnathid fish from the Eocene of Antarctica. *Paleontology* **37**, 931–940.

Clark, C. W. (1973). The economics of overexploitation. *Science* **181**, 630–634.

Clark, C. W. (1996). Marine reserves and the precautionary management of fisheries. *Ecol. Appl.* **6**, 369–370.

Clark, J. R., Causey, B., and Bohnsack, J. A. (1989). Benefits of coral reef protection: Looe Key Reef, Florida. *Proc. 6th Symp. Coastal Ocean Manag.*, Vol. 4, pp. 3076–3086.

Clark, M. M., and Galef, B. G. (1995). Prenatal influences on reproductive life history strategies. *Trends Ecol. Evol.* **10**, 151–153.

Clark, M. M., and Galef, B. G. (1998). Perinatal influences on the reproductive behavior of adult rodents. *In* "Maternal Effects as Adaptations" (T. A. Mousseau and C. W. Fox, eds.), pp. 261–271. Oxford University Press, New York.

Clark, W. G. (1991). Groundfish exploitation rates based on life history parameters. *Can. J. Fish. Aquat. Sci.* **48**, 734–750.

Clark, W. G. (1993). "The Effect of Recruitment Variability on the Choice of a Target Level of Spawning Biomass per Recruit, pp. 233–246. Management of Exploited Fish, Alaska Sea Grant, Fairbanks.

Clarke, A. (1996). The influence of climate change on the distribution and evolution of organisms. *In* "Animals and Temperature: Phenotypic and Evolutionary Adaptation" (I. A. Johnston and A. F. Bennett, eds.), pp. 377–407. Cambridge University Press, Cambridge, UK.

Clarke, R. D. (1989). Population fluctuation, competition and microhabitat distribution of two species of tube blennies, *Acanthemblemaria* (Teleostei: Chaenopsidae). *Bull. Mar. Sci.* **44**, 1174–1185.

Clarke, R. D. (1992). Effects of microhabitat and metabolic rate on food intake, growth and fecundity of two competing coral reef fishes. *Coral Reefs* **11**, 199–205.

Clarke, R. D. (1996). Population shifts in two competing fish species on a degrading coral reef. *Mar. Ecol.: Prog. Ser.* **137**, 51–58.

Clarke, T. A. (1991). Larvae of nearshore fishes in oceanic waters near Oahu, Hawaii. *NOAA Tech. Rep. NMFS* **101**, 1–19.

Clarke, T. A. (1995). Larvae of near-shore fishes in oceanic waters of the central equatorial Pacific. *Pac. Sci.* **49**, 134–142.

Clements, K. D., and Bellwood, D. R. (1988). A comparison of the feeding mechanisms of two herbivorous labroid fishes, *Odax pullus* and *Scarus rubroviolaceus*. *Aust. J. Mar. Freshwater Res.* **39**, 87–107.

Clements, K. D., and Choat, J. H. (1995). Fermentation in tropical marine herbivorous fishes. *Physiol. Zool.* **68**, 355–378.

Clements, K. D., Sutton, D. C., and Choat, J. H. (1989). Occurrence and characteristics of unusual protistan symbionts from surgeonfishes (Acanthuridae) of the Great Barrier Reef, Australia. *Mar. Biol.* **102**, 403–412.

Clifton, K. E. (1990). The costs and benefits of territory sharing for the Caribbean coral reef fish, *Scarus iserti*. *Behav. Ecol. Sociobiol.* **26**, 139–147.

Clifton, K. E. (1995). Asynchronous food availability on neighboring Caribbean coral reefs determines seasonal patterns of growth and reproduction for the herbivorous parrotfish *Scarus iserti*. *Mar. Ecol.: Prog. Ser.* **116**, 39–46.

Clifton, K. E., and Motta, P. J. (1998). Feeding morphology, diet, and ecomorphological relationships among five Caribbean labrids (Teleostei, Labridae). *Copeia*, pp. 953–966.

Clifton, K. E., and Robertson, D. R. (1993). Risks of alternative mating tactics. *Nature (London)* **366**, 520.

Coates, A. G. (1973). Cretaceous Tethyan coral-rudist biogeography related to the evolution of the Atlantic Ocean. *In* "Organisms and Continents Through Time" (N. F. Hughes, ed.), Spec. Pap. Paleonto., No. 12, pp. 169–174. Paleontological Association, London.

Coates, A. G., and Jackson, J. B. C. (1985). Morphological themes in the evolution of clonal and aclonal marine invertebrates. *In* "Population Biology and Evolution of Clonal Organisms" (J. B. C. Jackson, L. W. Buss, and R. E. Cook, eds.), pp. 67–106. Yale University Press, New Haven, CT.

Coates, A. G., and Obando, J. A. (1996). The geologic evolution of the Central American isthmus. *In* "Evolution and Environment in Tropical America" (J. B. C. Jackson, A. F. Budd, and A. G. Coates, eds.), pp. 21–56. University of Chicago Press, Chicago.

Coates, A. G., Jackson, J. B. C., Collins, L. S., Cronin, T. M., Dowsett, H. J., Bybell, L. M., Jung, P. J., and Obando, J. A. (1992). Closure of the Isthmus of Panama: The nearshore record of Costa Rica and western Panama. *Geol. Soc. Am. Bull.* **104**, 814–828.

Cobb, J. S., Wang, D., Campbell, D. B., and Rooney, P. (1989). Speed and direction of swimming by post-larvae of the American lobster. *Trans. Am. Fish. Soc.* **118**, 82–86.

Cole, K. S., and Sadovy, Y. (1995). Evaluating the use of spawning success to estimate reproductive success in a Caribbean reef fish. *J. Fish. Biol.* **47**, 181–191.

Cole, K. S., and Shapiro, D. Y. (1995). Social facilitation and sensory mediation of adult sex change in a cryptic, benthic marine goby. *J. Exp. Mar. Biol. Ecol.* **186**, 65–75.

Cole, L. C. (1954). The population consequences of life history phenomena. *Q. Rev. Biol.* **29**, 103–137.

Cole, R. G., Ayling, A. M., and Creese, R. G. (1990). Effects of marine reserve protection at Goat Island, northern New Zealand. *N. Z. J. Mar. Freshwater Res.* **24**, 197–210.

Coleman, F. C., Koenig, C. C., and Collins, L. A. (1996). Reproductive styles of shallow-water groupers (Pisces: Serranidae) in the eastern Gulf of Mexico and the consequences of fishing spawning aggregations. *Environ. Biol. Fishes* **47**, 129–141.

Coleman, F. C., Koenig, C. C., Eklund, A., and Grimes, C. B. (1999). Management and conservation of temperate reef fishes in the grouper-snapper complex of the Southeastern United States. *In* "Life in the Slow Lane: Ecology and Conservation of Long-Lived Marine Animals" (J. A. Musick, ed.), pp. 233–242, American Fisheries Society, Bethesda, MD.

Colin, P. L. (1992). Reproduction of the Nassau grouper, *Epinephelus striatus* (Pisces: Serranidae) and its relationship to environment conditions. *Environ. Biol. Fishes* **34**, 357–377.

Colin, P. L., and Bell, L. J. (1991). Aspects of the spawning of labrid and scarid fishes (Pisces: Labroidei) at Enewetak Atoll, Marshall Islands with notes on other families. *Environ. Biol. Fishes* **31**, 229–260.

Colin, P. L., and Clavijo, I. E. (1978). Mass spawning by the spotted goatfish, *Pseudupeneus maculates* (Black) (Pisces: Mullidae). *Bull. Mar Sci.* **28**, 780–782.

Colin, P. L., and Clavijo, I. E. (1988). Spawning activity of fishes producing pelagic eggs on a shelf edge coral reef, Southwestern Puerto Rico. *Bull. Mar. Sci.* **43**, 249–279.

Colin, P. L., Laroche, W. A., and Brothers, E. B. (1997). Ingress and settlement in the Nassau grouper, *Epinephelus striatus* (Pices: Serranidae), with relationship to spawning occurrence. *Bull. Mar. Sci.* **60**, 656–667.

Collin, S. P. (1997). Specializations of the teleost visual system: Adaptive diversity from shallow-water to deep-sea. *Acta Physiol. Scand.* **161**(Suppl. 638), 5–24.

Collin, S. P. (1999). Behavioral ecology and retinal cell topography. *In* "Adaptive Mechanisms in the Ecology of Vision" (S. N. Archer, M. B. A. Djamgoz, E. Loew, J. C.

Partridge, and S. Vallerga, eds.), pp. 509–535. Kluwer Academic Publishers, Dordrecht, The Netherlands.

Collin, S. P., and Pettigrew, J. D. (1988a). Retinal topography in reef teleosts. I. Some species with well developed areae but poorly developed streaks. *Brain, Behav. Evol.* **31**, 269–282.

Collin, S. P., and Pettigrew, J. D. (1988b). Retinal topography in reef teleosts. II. Some species with prominent horizontal streaks and high density areae. *Brain, Behav. Evol.* **31**, 283–295.

Collin, S. P., and Pettigrew, J. D. (1989). Quantitative comparison of the limits on visual spatial resolution set by the ganglion cell layer in twelve species of reef teleosts. *Brain, Behav. Evol.* **34**, 184–192.

Collins, L. A., Johnson, A. G., Koenig, C. C., and Baker, M. S., Jr. (1998). Reproductive patterns, sex ratio and fecundity in gag, a protogynous grouper (Serranidae: *Mycteroperca microlepis*) from the northeastern Gulf of Mexico. *Fish. Bull.* **96**, 415–427.

Collins, M. R., Waltz, C. W., Roumillat, and Stubbs, D. L. (1987). Contribution to the life history and reproductive biology of gag, *Mycteroperca microlepis* (Serranidae) in the South Atlantic Bight. *Fish. Bull.* **85**, 648–653.

Colson, D. J., Patek, S. N., Brainerd, E. L., and Lewis, S. M. (1998). Sound production during feeding in *Hippocampus* seahorses (Syngnathidae). *Environ. Biol. Fishes* **51**, 221–229.

Connell, J. H. (1975). Some mechanisms producing structure in natural communities: A model and evidence from field experiments. *In* "Ecology and Evolution of Communities" (M. L. Cody and J. A. Diamond, eds.), pp. 460–490. Belknap-Harvard University Press, Cambridge, MA.

Connell, S. D. (1996). Variations in mortality of a coral-reef fish: Links with predator abundance. *Mar. Biol.* **126**, 347–352.

Connell, S. D. (1997a). The relationship between large predatory fish and recruitment and mortality of juvenile coral reef-fish on artificial reefs. *J. Exp. Mar. Biol. Ecol.* **209**, 261–278.

Connell, S. D. (1997b). Exclusion of predatory fish on a coral reef: The anticipation, pre-emption and evaluation of some caging artefacts. *J. Exp. Mar. Biol. Ecol.* **213**, 181–198.

Connell, S. D. (1998a). Effects of predators on growth, mortality and abundance of a juvenile reef-fish: Evidence from manipulations of predator and prey abundance. *Mar. Ecol.: Prog. Ser.* **169**, 251–261.

Connell, S. D. (1998b). Patterns of piscivory by resident predatory reef fish at One Tree Reef, Great Barrier Reef. *Mar. Freshwater Res.* **49**, 25–30.

Connell, S. D. (2000). Is there safety-in-numbers for prey? *Oikos* **88**, 527–532.

Connell, S. D., and Jones, G. P. (1991). The influence of habitat complexity on postrecruitment processes in a temperate reef fish population. *J. Exp. Mar. Biol. Ecol.* **151**, 271–294.

Connell, S. D., and Kingsford, M. J. (1998). Spatial, temporal and habitat-related variation in the abundance of large predatory fish at One Tree Reef, Australia. *Coral Reefs* **17**, 49–57.

Conroy, D. A. (1975). "An Evaluation of the Present State of World Trade in Ornamental Fish." FAO, Rome, and Inland Water Fishery Development Programme, Bogota, Columbia.

Constantz, G. D. (1985). Allopaternal care in the tesselated darter, *Etheostoma olmstedi* (Pisces: Percidae). *Environ. Biol. Fishes* **14**, 175–183.

Cook, R. M., and Armstrong, D. W. (1986). Stock-related effects in the recruitment of North Sea haddock and whiting. *J. Cons. Int. Explor. Mer* **42**, 272–280.

Coombs, S., and Popper, A. N. (1979). Hearing differences among Hawaiian squirrelfish (family: Holocentridae) related to differences in the peripheral auditory system. *J. Comp. Physiol. A* **132**, 203–207.

Cordes, J. F. (2000). Application of genetic markers to provide species identification and define stock structure: Analysis of selected marine fishes of the Mid-Atlantic Bight. Ph.D. Thesis, College of William and Mary, Williamsburg, VA.

Corless, M., Hatcher, B. G., Hunte, W., and Scott, S. (1996). Assessing the potential for fish migration from marine reserves to adjacent fished areas in the Soufriere Marine Management Area, St. Lucia. *Proc. Annu. Gulf Caribb. Fish. Inst.*, pp. 82–92.

Corwin, J. T. (1981a). Audition in elasmobranchs. *In* "Hearing and Sound Communication in Fishes" (W. N. Tavolga, A. N. Popper, and R. R. Fay, eds.), pp. 81–102. Springer-Verlag, New York.

Corwin, J. T. (1981b). Peripheral auditory physiology in the lemon shark: Evidence of parallel otolithic and non-otolithic sound detection. *J. Comp. Physiol. A* **142**, 379–390.

Corwin, J. T. (1981c). Postembryonic production and aging of inner ear hair cells in sharks. *J. Comp. Neurol.* **201**, 541–553.

Corwin, J. T. (1983). Postembryonic growth of the macula neglecta auditory detector in the ray, *Raja clavata*: Continual increases in hair cell number, neural convergence, and physiological sensitivity. *J. Comp. Neurol.* **217**, 345–356.

Corwin, J. T. (1989). Functional anatomy of the auditory system in sharks and rays. *J. Exp. Zool., Suppl.* **2**, 62–74.

Côté, I. M., and Hunte, W. (1989). Male and female mate choice in the redlip blenny: Why bigger is better. *Anim. Behav.* 38, 78–88.

Côté, I. M., and Hunte, W. (1993). Female redlip blennies prefer older males. *Anim. Behav.* **46**, 203–205.

Cotgreave, P. (1993). The relationship between body size and population abundance in animals. *Trends Ecol. Evol.* **8**, 244–248.

Couchman, D., and Beumer, J. P. (1992). "The Commercial Fishery for the Collection of Marine Aquarium Fishes in Queensland: Status and Management Plan." Fisheries Division, Department of Primary Industries, Brisbane, Queensland.

Coughlin, D. J. (1994). Suction prey capture by clownfish larvae (*Amphiprion perideraion*). *Copeia*, pp. 242–246.

Courchamp, F., Clutton-Brock, T., and Grenfell, B. (1999). Inverse density dependence and the Allee effect. *Trends Ecol. Evol.* **14**, 405–410.

Courtenay, W. R., Jr., and Stauffer, J. R., Jr. (1990). The introduced fish problem and the aquarium fish industry. *J. World Aquacult. Soc.* **21**, 145–159.

Cowen, R. K. (1985). Large scale pattern of recruitment by the labrid, *Semicossyphus pulcher*: Causes and implications. *J. Mar. Res.* **43**, 719–742.

Cowen, R. K. (1991). Variation in the planktonic larval duration of the temperate wrasse *Semicossyphus pulcher*. *Mar. Ecol.: Prog. Ser.* **69**, 9–15.

Cowen, R. K., and Castro, L. R. (1994). Relation of coral reef fish larval distributions to island scale circulation around Barbados, West Indies. *Bull. Mar. Sci.* **54**, 228–244.

Cowen, R. K., and Sponaugle, S. (1997). Relationships between early life history traits and recruitment among coral reef fishes. *In* "Early Life History and Recruitment in Fish Populations" (R. C. Chambers and E. A. Trippel, eds.), pp. 423–449. Chapman & Hall, London.

Cowen, R. K., Hare, J. A., and Fahay, M. P. (1993). Beyond hydrography: Can physical processes explain larval fish assemblages within the middle Atlantic bight? *Bull. Mar. Sci.* **53**, 567–587.

Cowen, R. K., Lwiza, K. M. M., Sponagle, S., Limouzy-Paris, C. B., and Olson, D. B. (2000). Connectivity of marine populations: Open or closed? *Science* **287**, 857–859.

Cox, E. F. (1994). Resource use by corallivorous butterflyfishes (Family Chaetodontidae) in Hawaii. *Bull. Mar. Sci.* **54**, 535–545.

Crabtree, R. (1995). Relationship between lunar phase and spawning activity of tarpon, *Megalops atlanticus,* with notes on the distribution of larvae. *Bull. Mar. Sci.* **56**, 895–899.

Crabtree, R. E., and Bullock, L. H. (1998). Age, growth, and reproduction of black grouper, *Mycteroperca bonaci,* in Florida waters. *Fish. Bull.* **96**, 735–753.

Cracraft, J. (1985). Biological diversification and its causes. *Ann. Mo. Bot. Gard.* **72**, 794–822.

Craig, P. C. (1999). The von Bertalanffy growth curve: When a good fit is not enough. *Naga* **22**, 28–30.

Craig, P. C., Choat, J. H., Axe, L. M., and Saucerman, S. (1997). Population biology and harvest of the coral reef surgeonfish *Acanthurus lineatus* in American Samoa. *Fish. Bull.* **95**, 680–693.

Craik, G. J. S. (1981). Underwater survey of coral trout *Plectropomus leopardus* (Serranidae) populations in the Capricornia Section of the Great Barrier Reef Marine Park. *Proc. 4th Int. Coral Reef Symp.,* Vol. 1, pp. 53–58.

Crame, J. A., and Clarke, A. (1997). The historical component of marine taxonomic diversity gradients. *In* "Marine Biodiversity: Patterns and Processes" (R. F. G. Ormond, J. D. Gage, and M. V. Angel, eds.), pp. 258–273. Cambridge University Press, Cambridge, UK.

Criales, M. M., and Lee, T. N. (1995). Larval distribution and transport of penaeiod shrimps during the presence of the Tortugas Gyre in May-June 1991. *Fish. Bull.* **93**, 471–482.

Crisp, D. J. (1978). Genetic consequences of different reproductive strategies in marine invertebrates. *In* "Marine Organisms: Genetics, Ecology and Evolution" (B. Battaglia and J. A. Beadmore, eds.), pp. 257–273. Plenum Press, New York.

Croizat, L., Nelson, G., and Rosen, D. E. (1974). Centres of origin and related concepts. *Syst. Zool.* **23**, 265–287.

Cronin, T. M., and Dowsett, H. J. (1996). Biotic and oceanographic response to the Pliocene closing of the Central American Isthmus. *In* "Evolution and Environment in Tropical America" (J. B. C. Jackson, A. F. Budd, and A. G. Coates, eds.), pp. 76–104. University of Chicago Press, Chicago.

Crowder, L. B., Lyman, S. J., Figueira, W. F., and Priddy, J. (2000). Source-sink population dynamics and the problem of siting marine reserves. *Bull. Mar. Sci.* **66**, 799–820.

Cummings, W. C., Brahy, B. D., and Herrnkind, W. F. (1964). The occurrence of underwater sounds of biological origin off the west coast of Bimini, Bahamas. *In* "Marine Bio-acoustics" (W. N. Tavolga, ed.), pp. 27–43. Pergamon Press, New York.

Cummins, K. W., and Wuycheck, J. C. (1971). Caloric equivalents for investigations in ecological energetics. *Int. Assoc. Theoret. Appl. Limnol. Commun.* **18**, 1–158.

Currie, D. J. (1993). What shape is the relationship between body size and population density? *Oikos* **66**, 353–358.

Cushing, D. H. (1972). The production cycle and the numbers of marine fish. *Symp. Zool. Soc. London* **29**, 213–232.

Cushing, D. H. (1975). "Marine Ecology and Fisheries." Cambridge University Press, Cambridge, UK.

Cushing, D. H. (1976). Biology of fishes in the pelagic community. *In* "The Ecology of the Seas" (D. H. Cushing, D. Henry, and J. J. Walsh, eds.), pp. 317–340. Blackwell Scientific Publications, Oxford.

Cushing, D. H. (1991). The study of stock and recruitment. *In* "Fish Population Dynamics" (J. A. Gulland, ed.), pp. 105–128. Wiley, Chichester.

Cushing, D. H. (1995). "Population Production and Regulation in the Sea: A Fisheries Perspective." Cambridge University Press, Cambridge, UK.

Cushing, D. H. (1996). "Towards a Science of Recruitment in Fish Populations." Ecology Institute, Oldendorf.

Cutwa, M. M., and Turingan, R. G. (2000). Intralocality variation in feeding biomechanics and prey use in *Archosargus probatocephalus* (Teleostei, Sparidae), with implications for the ecomorphology of fishes. *Environ. Biol. Fishes* **59**, 191–198.

Daily, G. C., Ehrlich, P. R., and Alberti, M. (1996). Managing earth's life support systems: The game, the players, and getting everyone to play. *Ecol. Appl.* **6**, 19–21.

Daily, G. C., Soderqvist, T., Aniyar, S., Arrow, K., Dasgupta, P., Ehrlich, P. R., Folke, C., Jansson, A., Jansson, B. O., Kautsky, N., Levin, S., Lubchenco, J., Maler, K. G., Simpson, D., Starrett, D., Tilman, D., and Walker, B. (2000). Ecology—The value of nature and the nature of value. *Science* **289**, 395–396.

Dall, W. (1991). Zoogeography of the Penaeidae. *Mem. Queensl. Mus.* **31**, 39–50.

Dalzell, P. (1992). Ciguatera fish poisoning and fisheries development in the South Pacific region. *Bull. Soc. Patho. Exot. Ses. Fil.* **85**, 435–444.

Dalzell, P. (1996). Catch rates, selectivity and yields of reef fishing. *In* "Reef Fisheries" (N. V. C. Polunin and C. M. Roberts, eds.), pp. 161–192. Chapman & Hall, London.

Dalzell, P., and Adams, T. J. H. (1997). Sustainability and management of reef fisheries in the Pacific islands. *Proc. 8th Int. Coral Reef Symp.* Vol. 2, pp. 2027–2032.

Damuth, J. (1981). Population density and body size in mammals. *Nature (London)* **290**, 699–700.

Dandonneau, Y., Dugas, F., Fourmanoir, P., Magnier, Y., Rougerie, F., and Debenay, J. P. (1983). Courantologie et climatologie. *In* "Atlas de la Nouvelle Caledonie," Plate 8. ORSTOM, Paris.

Danilowicz, B. S. (1995). The role of temperature in spawning of the damselfish *Dascyllus albisella*. *Bull. Mar. Sci.* **57**, 624–636.

Danilowicz, B. S. (1996). Choice of coral species by naïve and field-caught damselfish. *Copeia,* pp. 735–739.

Danilowicz, B. S. (1997). A potential mechanism for episodic recruitment of a coral reef fish. *Ecology* **78**, 1415–1423.

Danilowicz, B. S., and Sale, P. F. (1999). Relative intensity of predation on the French grunt, *Haemulon flavolineatum,* during diurnal, dusk, and nocturnal periods on a coral reef. *Mar. Biol.* **133**, 337–343.

Dart, J. K. G. (1972). Echinoids, algal lawn and coral recolonization. *Nature (London)* **239**, 50–51.

Davenport, K. E. (1996). Characteristics of the current international trade in ornamental fish, with special reference to the European Union. *Rev. Sci. Tech. Of. Int. Epizoot.* **15**, 435–443.

Davidson, J., and Andrewartha, H. G. (1948). The influence of rainfall, evaporation and atmospheric temperature on fluctuations in the size of a natural population of *Thrips imaginis* (Thysanoptera). *J. Anim. Ecol.* **17**, 200–222.

Davies, C. R. (1995). Patterns of movement of three species of coral reef fish on the Great Barrier Reef. Unpublished Ph.D. Thesis, James Cook University of North Queensland, Australia.

Davies, P. J. (1988). Evolution of the Great Barrier Reef—reductionist dream or expansionist vision. *Proc. 6th Int. Coral Reef Symp.* Vol. 1, pp. 9–17.

Davis, C. S., Gallager, S. M., Marra, M., and Stewart, W. K. (1996). Rapid visualization of plankton abundance and taxonomic composition using the Video Plankton Recorder. *Deep-Sea Res.* **43**, 1947–1970.

Davis, E. A., and Hopkins, C. D. (1988). Behavioral analysis of electric signal localization in the electric fish, *Gymnotus carapo* (Gymnotiformes). *Anim. Behav.* **36**, 1658–1671.

Davis, G. E., and Dodrill, J. W. (1989). Recreational fishery and population dynamics of spiny lobster, *Panulirus argus,* in Florida Bay, Everglades National Park, 1977–1980. *Bull. Mar. Sci.* **44**, 78–88.

Davis, T. L. O., and West, G. J. (1993). Maturation, reproductive seasonality, fecundity, and spawning frequency in *Lutjanus vittus* (Quoy and Gaimard) from the North West shelf of Australia. *Fish. Bull.* **91**, 224–236.

Dawson, E. Y. (1966). "Marine Botany," Holt, Rinehart & Winston, New York.

Dayton, P. K. (1998). Ecology—Reversal of the burden of proof in fisheries management. *Science* **279**, 821–822.

Dayton, P. K., Thrush, S. F., Agardy, T. S., and Hofman, R. J. (1995). Environmental effects of marine fishing. *Aqua. Conserv.* **5**, 205–232.

Dayton, P. K., Sala, E., Tegner, M. J., and Thrush, S. F. (2000). Marine protected areas: Parks, baselines, and fishery enhancement. *Bull. Mar. Sci.* **66**, 617–634.

D'Croz, L., and Robertson, D. R. (1997). Coastal oceanographic conditions affecting coral reefs on both sides of the isthmus of Panama. *Proc. 8th Int. Coral Reef Symp.* Vol. 2, pp. 2053–2058.

DeFerrari, H. A., and Green, W. C. (1968). Ambient noise studies—Bimini, Bahamas—1967–1968. *Inst. Mar. Sci., Univ. Miami Tech. Rep.* **ML 69066**, 1–62.

Degens, E. T., Deuser, W. G., and Haedrich, R. L. (1969). Molecular structure and composition of fish otoliths. *Mar. Biol.* **2**, 105–113.

DeMartini, E. E. (1987). Paternal defence, cannibalism and polygamy; factors influencing the reproductive success of painted greenling (Pisces, Hexagrammidae). *Anim. Behav.* **35**, 1145–1158.

DeMartini, E. E. (1991). Annual variations in fecundity, egg size and the gonadal and somatic conditions of queenfish *Seriphus politus* (Sciaenidae). *Fish. Bull.* **89**, 9–18.

DeMartini, E. E. (1993). Modelling the potential of fishery reserves for managing Pacific coral reef fishes. *Fish. Bull.* **91**, 414–427.

DeMartini, E. E. (1998). How might recruitment research on coral-reef fishes help manage tropical reef fisheries. *Aust. J. Ecol.* **23**, 305–310.

DeMartini, E. E., Roberts, D. A., and Anderson, T. W. (1989). Contrasting patterns of fish density and abundance at an artificial rock reef and cobble-bottom kelp forest. *Bull. Mar. Sci.* **44**, 881–892.

Dempsey, C. H. (1978). Chemical stimuli as a factor in feeding and intraspecific behavior of herring larvae. *J. Mar. Biol. Assoc. U.K.* **58**, 739–747.

Dempster, T., Gibbs, M. T., Rissik, D., and Suthers, I. M. (1997). Beyond hydrography: Daily ichthyoplankton variability and short term oceanographic events on the Sydney continental shelf. *Cont. Shelf Res.* **17**, 1461–1481.

den Boer, P. J. (1968). Spreading the risk and stabilization of animal numbers. *Acta Biotheor.* **18**, 165–194.

den Boer, P. J., and Reddingius, J. (1996). "Regulation and Stabilization Paradigms in Population Ecology." Chapman & Hall, London.

Denman, K. L., and Gargett, A. E. (1983). Time and space scales of vertical mixing and advection of phytoplankton in the upper ocean. *Limnol. Oceanogr.* **28**, 801–815.

Denman, K. L., and Powell, T. M. (1984). Effects of physical processes on planktonic ecosystems in the coastal ocean. *Oceanogr. Mar. Biol.* **22**, 33–37.

Denton, E. J. (1990). Light and vision at depths greater than 200 metres. *In* "Light and Life in the Sea" (P. J. Herring, A. K. Campbell, M. Whitfield, and L. Maddock, eds.), pp. 127–148. Cambridge University Press, Cambridge, UK.

Denton, E. J., and Blaxter, J. H. S. (1976). The mechanical relationship between the clupeid swimbladder, inner ear, and lateral line. *J. Mar. Biol. Assoc. U.K.* **56**, 787–807.

Denton, E. J., and Gray, J. A. B. (1988). Mechanical factors in the excitation of the lateral line of fishes. *In* "Sensory Biology of Aquatic Animals" (J. Atema, R. R. Fay, A. N. Popper, and W. N. Tavolga, eds.), pp. 595–617. Springer-Verlag, New York.

Diebel, C. E., Proksch, R., Green, C. R., Neilson, P., and Walker, M. M. (2000). Magnetite defines a vertebrate magnetoreceptor. *Nature (Lond.)* **406**, 299–302.

Dight, I. J., James, M. K., and Bode, L. (1994). Modelling the larval dispersal of *Acanthaster planci* II. Patterns of reef connectivity. *Coral Reefs* **9**, 125–134.

Dilley, F. C. (1973). Larger foraminefera and seas through time. *In* "Organisms and Continents Through Time." (N. F. Hughes, ed.), Spe. Pap. Paleontol., No. 12, pp. 155–168. Paleontological Association, London.

DiMichele, L. and Power, D. A. (1982). Ldh-B genotype specific hatching times of *Fundulus heteroclitus* embryos. *Nature (London)* **296**, 563–564.

Di Rienzo, A., Peterson, A. C., Garza, J. C., Valdes, A. M., and Slatkin, M. (1994). Mutational processes of simple sequence repeat loci in human populations. *Proc. Natl. Acad. Sci. U.S.A.* **91**, 3166–3170.

Dixon, J. A. (1993). Economic benefits of marine protected areas. *Oceanus* **36**, 35–40.

Dixon, J. A. and Sherman, P. B. (1990). "Economics of Protected Areas: A New Look at Benefits and Costs." Island Press, Washington, DC.

Dixon, P. A., Milicich, M. J., and Sugihara, G. (1999). Episodic fluctuations in larval supply. *Science* **283**, 1528–1530.

Doak, D. F., and Mills, L. S. (1994). A useful role for theory in conservation. *Ecology* **74**, 615–626.

Dobzhansky, T. (1951). "Genetics and the Origin of Species," 3rd ed. Columbia University Press, New York.

Doherty, P. J. (1980). Biological and physical constraints on the populations of two sympatric territorial damelfishes on the southern Great Barrier Reef. Ph.D. Thesis, University of Sydney.

Doherty, P. J. (1981). Coral reef fishes: Recruitment-limited assemblages? *Proc. 4th Int. Coral Reef Symp.*, Vol. 2, pp. 465–470.

Doherty, P. J. (1982). Some effects of density on the juveniles of two species of tropical, territorial damselfishes. *J. Exp. Mar. Biol. Ecol.* **65**, 249–261.

Doherty, P. J. (1983a). Tropical territorial damselfishes: Is density limited by aggression or recruitment? *Ecology* **64**, 176–190.

Doherty, P. J. (1983b). Diel, lunar and seasonal rhythms in the reproduction of two tropical damselfishes: *Pomacentrus flavicauda* and *P. wardi*. *Mar. Biol.* **75**, 215–224.

Doherty, P. J. (1983c). Recruitment surveys of coral reef fishes as tools for science and management. *In* "Proceedings of the Inaugural Great Barrier Reef Conference, Townsville, August 28–September 2, 1983" (J. T. Baker, R. M. Carter, P. W. Sammarco, and K. P. Stark, eds.), pp. 191–196. James Cook University Press, Townsville, Australia.

Doherty, P. J. (1987a). The replenishment of populations of coral reef fishes: Problems of variation manifest at multiple scales. *Bull. Mar. Sci.* **41**, 411–422.

Doherty, P. J. (1987b). Light traps: Selective but useful devices for quantifying the distributions and abundances of larval fishes. *Bull. Mar. Sci.* **41**, 423–431.

Doherty, P. J. (1991). Spatial and temporal patterns in recruitment. *In* "The Ecology of Fishes on Coral Reefs" (P. F. Sale, ed.), pp. 261–293. Academic Press, San Diego, CA.

Doherty, P. J. (1998). Recruitment limitation: Definitions, predictions and tests. *In* "Reefish '95: Recruitment and Population Dynamics of Coral Reef Fishes" (G. P. Jones, P. J. Doherty, B. D. Mapstone, and L. Howlett, eds.), pp. 129–131. CRC Reef Research Center, Townsville, Australia.

Doherty, P. J. (1999). Recruitment-limitation is the theoretical basis for stock enhancement in marine populations. *In* "Stock Enhancement and Sea Ranching" (B. R. Howell, E. Moksness, and T. Svasand, eds.), pp. 9–21. Fishing News Books, Blackwell.

Doherty, P. J., and Carleton, J. H. (1997). The distribution and abundance of pelagic juvenile fish near Grub Reef, Central Great Barrier Reef. *Proc. 8th Int. Coral Reef Symp.*, Vol. 2, pp. 1155–1160.

Doherty, P. J., and Fowler, A. J. (1994a). An empirical test of recruitment limitation in a coral reef fish. *Science* **263**, 935–939.

Doherty, P. J., and Fowler, A. J. (1994b). Demographic consequences of variable recruitment to coral reef fish populations: A congeneric comparison of two damselfishes. *Bull. Mar. Sci.* **54**, 297–313.

Doherty, P. J., and McIlwain, J. (1996). Monitoring larval fluxes through the surf zones of Australian coral reefs. *Mar. Freshwater Res.* **47**, 383–390.

Doherty P. J., and Sale, P. F. (1985). Predation on juvenile coral reef fishes: An exclusion experiment. *Coral Reefs* **4**, 225–234.

Doherty, P. J., and Williams, D. McB. (1988). The replenishment of coral reef fish populations. *Oceanogr. Mar. Biol.* **26**, 487–551.

Doherty P. J., Mather P., and Planes S. (1994a). *Acanthochromis polyacanthus,* a fish lacking larval dispersal, has genetically differentiated populations at local and regional scales on the Great Barrier Reef. *Mar. Biol.* **121**, 11–21.

Doherty, P. J., Fowler, A., Samoilys, M. A., and Harris, D. A. (1994b). Monitoring the replenishment of coral trout (Pisces: Serranidae) populations. *Bull. Mar. Sci.* **54**, 343–355.

Doherty, P. J., Planes, S., and Mather, P. (1995). Gene flow and larval duration in seven species of fish from the Great Barrier Reef. *Ecology* **76**, 2373–2391.

Doherty, P. J., Kingsford, M. J., Booth, M. J., and Carleton, J. H. (1996). Habitat selection before settlement by *Pomacentrus coelestus*. *Mar. Freshwater Res.* **47**, 391–399.

Domeier, M. L. (1994). Speciation in the serranid fish *Hypoplectrus*. *Bull. Mar. Sci.* **54**, 103–141.

Domeier, M. L., and Colin, P. L. (1997). Tropical reef fish spawning aggregations: Defined and reviewed. *Bull. Mar. Sci.* **60**, 698–726.

Donaldson, T. J. (1986). Distribution and species richness patterns of the Indo-West Pacific Cirrhitidae: Support for Woodland's hypothesis. *In* "Indo-Pacific Fish Biology" (T. Uyeno, R. Arai, T. Taniuchi, and K. Matsuura, eds.) Proc. 2nd Int. Conf. Indo-Pac. Fishes, pp. 623–628. Ichthyological Society of Japan, Tokyo.

Donaldson, T. J., and Sadovy, Y. (2001). Threatened fishes of the world: *Cheilinus undulatus* Rüppell, 1835 (Labridae). *Environ. Biol. Fishes* **62**, 428.

Donat, J. R., and Bruland, K. W. (1995). Trace elements in the oceans. *In* "Trace Elements in Natural Water" (B. Salbu and E. Steinnes, eds.), pp. 247–281. CRC Press, Boca Raton, FL.

Done, T. J. (1998). Science for management of the Great Barrier Reef. *Nat. Resour.* **34**, 16–29.

Done, T. J., and Reichelt, R. E. (1998). Integrated coastal zone and fisheries ecosystem management: generic goals and performance indices. *Ecol. Appl.* **8**, S110–S118.

Doty, M. S., and Oguri, M. (1956). The island mass effect. *J. Conserv.* **22**, 33–37.

Douglas, R. H., and Hawryshyn, C. W. (1990). Behavioral studies of fish vision: An analysis of visual capabilities. *In* "The Visual System of Fish" (R. H. Douglas and M. B. A. Djamgoz, eds.), pp. 373–418. Chapman & Hall, London.

Dove, S. G., Gillanders, B. M., and Kingsford, M. J. (1996). An investigation of chronological differences in the deposition of trace metals in the otoliths of two temperate reef fishes. *J. Exp. Mar. Biol. Ecol.* **205**, 15–33.

Døving, K. B., and Knutsen, J. A. (1993). Chemokinesis in marine fish larvae. *In* "Physiological and Biochemical Aspects of Fish Development" (B. T. Walther and H. J. Fyhn, eds.), pp. 139–145. University of Bergen, Norway.

Døving, K .B., Dubois-Dauphin, M., Holley, A., and Jourdan, F. (1977). Functional anatomy in the olfactory organs of fish and the ciliary mechanisms of water transport. *Acta Zool. (Stockholm)* **58**, 245–255.

Døving, K. B., Marstol, M., Andersen, J. R., and Knutsen, J. A. (1994). Experimental evidence of chemokinesis in newly-hatched cod larvae (*Gadus morhua* L). *Mar. Biol.* **120**, 351–358.

Dower, J. F., Miller, T. J., and Leggett, W. C. (1997). The role of microscale turbulence in the feeding ecology of larval fish. *Adv. Mar. Biol.* **31**, 170–220.

Dower, J. F., Pepin, P., and Leggett, W. C. (1998). Enhanced gut fullness and an apparent shift in size selectivity by radiated shanny (*Ulvaria subbifurcata*) larvae in response to increased turbulence. *Can. J. Fish. Aquat. Sci.* **55**, 128–142.

Druce, B. E., and Kingsford, M. J. (1995). An experimental investigation on the fishes associated with drifting objects in coastal waters of temperate Australia. *Bull. Mar. Sci.* **57**, 376–390.

Drucker, E. G., and Jensen, J. S. (1991). Functional analysis of a specialized prey processing behavior: Winnowing by surfperches (Teleostei: Embiotocidae). *J. Morphol.* **210**, 268–287.

Druffel, E. M., and Linick, T. W. (1978). Radiocarbon in annual coral rings of Florida. *Geophys. Res. Lett.* **5**, 913–916.

Dufour, V. (1991). Variations d'abondance des larves de poissons en milieu recifal: Effet de la lumière sur la colonisation larvaire. *C. R. seances Acad. Sci., Ser. 3* **313**, 187–194.

Dufour, V. (1997). Pacific Island countries and the aquarium fish market. *Secretariat of the Pacific Community Live Reef Fish Inf. Bull.* **2**, 6–11.

Dufour, V., and Galzin, R. (1993). Colonization patterns of reef fish larvae to the lagoon at Moorea Island, French Polynesia. *Mar. Ecol.: Prog. Ser.* **102**, 143–152.

Dufour, V., Riclet, E., and Lo-Yat, A. (1996). Colonization of reef fishes at Moorea Island, French Polynesia: Temporal and spatial variation of the larval flux. *Mar. Freshwater Res.* **47**, 413–422.

Dufour, V., Pierre, C., and Rancher, J. (1998). Stable isotopes in fish otoliths discriminate between lagoonal and oceanic residents of Taiaro Atoll (Tuamoto Archipelago, French Polynesia). *Coral Reefs* **17**, 23–28.

Dugan, J. E., and Davis G. E. (1993). Applications of marine refugia to coastal fisheries management. *Can. J. Fish. Aquat. Sci.* **50**, 2029–2042.

Dunning, D. J., Ross, Q. E., Geoghegan, P., Reichle, J. J., Menezes, J. K., and Watson, J. K. (1992). Alewives avoid high-frequency sound. *North Am. J. Fish. Manage.* **12**, 407–416.

Durie, C. J., and Turingan, R. G. (2001). Relationship between durophagy and feeding biomechanics in gray triggerfish, *Balistes capriscus*: Intraspecific variation in ecological morphology. *Fla. Sci.* **64**, 20–28.

Duston, J., and Saunders, R. L. (1999). Effect of winter food deprivation on growth and sexual maturity of Atlantic salmon (*Salmo salar*) in seawater. *Can. J. Fish. Aquat. Sci.* **56**, 201–207.

Duthie, G. G. (1987). Observations of poor swimming performance among hatchery-reared rainbow trout, *Salmo gairdneri*. *Environ. Biol. Fishes* **18**, 309–311.

Dutkiewicz, S., Griffa, A., and Olson, D. B. (1993). Particle diffusion in a meandering jet. *J. Geophys. Res.* **98**, 16,487–16,500.

Ebeling, A. W., and Hixon, M. A. (1991). Tropical and temperate reef fishes: Comparison of community structure. *In* "The Ecology of Fishes on Coral Reefs" (P. F. Sale, ed.), pp. 509–563. Academic Press, San Diego, CA.

Ebeling, A. W., Larson, R. J., Alevizon, W. S., and Bray, R. N. (1980). Annual variability of reef-fish assemblages in kelp forests off Santa Barbara, California. *Fish. Bull.* **78,** 361–377.

Ebenman, B. (1992). Evolution in organisms that change their niches during the life cycle. *Am. Nat.* **139,** 990–1021.

Ebenman, B., and Persson, L. (1988). "Size Structured Populations: Ecology and Evolution." Springer-Verlag, Berlin.

Eckman, J. (1996). Closing the larval loop: Linking larval ecology to the population dynamics of marine benthic invertebrates. *J. Exp. Mar. Biol. Ecol.* **200,** 207–237.

Ecological Applications (2002). Special issue. Theory of marine reserves. *Ecol. Appl.* (in press).

Edgar, G. J., and Barrett, N. S. (1997). Short term monitoring of biotic change in Tasmanian marine reserves. *J. Exp. Mar. Biol. Ecol.* **213,** 261–279.

Edinger, E. N., and Risk, M. J. (1994). Oligocene-Miocene extinction and geographic restriction of Caribbean corals: Role of turbidity, temperature, and nutrients. *Palaios* **9,** 576–598.

Edmonds, J. S., and Fletcher, W. J. (1997). Stock discrimination of pilchards *Sarinops sagax* by stable isotope ratio analysis of otolith carbonate. *Mar. Ecol.: Prog. Ser.* **152,** 241–247.

Edmonds, J. S., Steckis, R. A., Moran, M. J., Caputi, N., and Morita, M. (1999). Stock discrimination of pink snapper and tailor from Western Australia by analysis of stable isotope and strontium/calcium ratios in otolith carbonate. *J. Fish Biol.* **55,** 243–259.

Edwards, A. J., and Shepherd, A. D. (1992). Environmental implications of aquarium-fish collection in the Maldives, with proposals for regulation. *Environ. Conserv.* **19,** 61–72.

Eggleston, D. B. (1995). Recruitment of Nassau grouper *Epinephelus striatus*: Post-settlement abundance, microhabtiat features, and ontogenetic habitat shifts. *Mar. Ecol.: Prog. Ser.* **124,** 9–22.

Eggleston, D. B. (1997). Predator and shelter-size effects on coral reef fish and spiny lobster prey. *Mar. Ecol.: Prog. Ser.* **149,** 43–59.

Eggleston, D. B., Lipcius, R. N., Miller, D. L., and Coba-Cetina, L. (1990). Shelter scaling regulates survival of juvenile Caribbean spiny lobster *Panulirus argus. Mar. Ecol.: Prog. Ser.* **62,** 79–88.

Eggleston, D. B., Lipcius, R. N., and Grover, J. J. (1997). Predator and shelter-size effects on coral reef fish and spiny lobster prey. *Mar. Ecol.: Prog. Ser.* **149,** 43–59.

Eggleston, D. B., Grover, J. J., and Lipcius, R. N. (1998). Ontogenetic diet shifts in Nassau grouper: Trophic linkages and predatory impact. *Bull. Mar. Sci.* **63,** 111–126.

Ehrlich P. R. (1975). The population biology of coral reef fishes. *Annu. Rev. Ecol. Syst.* **6,** 211–247.

Ehrlich, P. R., and Daily, G. C. (1993). Population extinction and saving biodiversity. *Ambio* **22,** 64–68.

Eisler, R. (1991). "Cyanide Hazards to Fish, Wildlife and Invertebrates: A Synoptic Review." U.S. Department of the Interior, Fish and Wildlife Service, Washington, DC.

Ekman, S. (1953). "Zoogeography of the Sea." Sidgwick & Jackson, London.

Elliot, N. G. (1996). Allozyme and mitochomdrial DNA analysis of the tropical saddle-tail perch, *Lutjanus malabaricus* (Schneider), from Australian waters. *Mar. Freshwater Res.* **47,** 869–876.

Elliott, J. K., and Marsical R. N. (2001). Coexistence of nine anemonefish species: Differential host and habitat utilization, size and recruitment. *Mar. Biol.* **138,** 23–36.

Elliott, J. K., Elliott, J. M., and Mariscal, R. N. (1995). Host selection, location, and association behaviors of anemonefishes in field settlement experiments. *Mar. Biol.* **122,** 377–389.

Ellis, T., and Gibson, R. N. (1997). Predation of 0-group cod: Handling times and size-selection. *Mar. Ecol.: Prog. Ser.* **149,** 83–90.

Elton, C. S. (1958). "The Ecology of Invasion by Animals and Plants." Methuen, London.

Emery, A. R. (1972). Eddy formation from an oceanic island: Ecological effects. *Caribb. J. Sci.* **12,** 121–124.

Emery, A. R. (1973). Comparative ecology and functional osteology of fourteen species of damselfish (Pisces: Pomacentridae) at Alligator Reef, Florida Keys. *Bull. Mar. Sci.* **23,** 649–770.

Emlen, S. T., and Oring, L. W. (1977). Ecology, sexual selection, and the evolution of mating systems. *Science* **197,** 215–223.

Emslie, M. J., and Jones, G. P. (2001) Patterns of embryo mortality in a demersally spawning coral reef fish and the role of predatory fishes. *Environ. Biol. Fishes* **60,** 363–373.

Enger, P. S. (1976). On the orientation of hair cells in the labyrinth of perch (*Perca fluviatilus*). *In* "Sound Reception in Fish" (A. Schuijf and A. D. Hawkins, eds.), pp. 49–61. Elsevier Scientific Publishing, New York.

Enger, P. S., Kalmijn, A. J., and Sand, O. (1989). Behavioral investigations on the functions of the lateral line and inner ear in predation. *In* "The Mechanosensory Lateral Line: Neurobiology and Evolution" (S. Coombs, P. Görner, and H. Münz, eds.), pp. 575–587. Springer-Verlag, New York.

Enger, P. S., Karlsen, H. E., Knudsen, F. R., and Sand, O. (1993). Detection and reaction of fish to infrasound. *ICES Mar. Sci. Symp.* **196,** 108–112.

Environmental Biology of Fishes. (1999). Fish behavior and the design of coral reef reserves. *Environ. Biol. Fishes* **55,** 53–98 (4 papers).

Epifanio, C. E. (1988). Dispersal strategies of two species of swimming crab on the continental shelf adjacent to Delaware Bay. *Mar. Ecol.: Prog. Ser.* **49,** 243–248.

Epperly, S. P., and Dodrill, J. W. (1995). Catch rates of snowy grouper, *Epinephelus niveatus,* on the deep reefs of Onslow Bay, Southeastern U.S.A. *Bull. Mar. Sci.* **56,** 450–461.

Erdmann, M. V., and Pet-Soede, L. (1998). An overview of destructive fishing practices in Indonesia. *Proc. APEC Workshop Impacts Destruct. Fish. Pract. Mar. Enviro.,* December, 1997, Hong Kong, pp. 25–34. Agriculture and Fisheries Department, Hong Kong, China.

Eschmeyer, W. N. (1990). "Catalog of the Genera of Recent Fishes." California Academy of Sciences, San Fransisco.

Estes, J. A., Jameson, R. J., and Rhode, E. B. (1982). Activity and prey selection in the sea otter: Influence of population status on community structure. *Am. Nat.* **120,** 242–258.

Estes, J. A., Duggins, D. O., and Rathbun, G. B. (1989). The ecology of extinctions in kelp forest communities. *Conserv. Biol.* **3,** 252–264.

Evans, B. I., and Fernald, R. D. (1990). Metamorphosis and fish vision. *J. Neurobiol.* **21,** 1037–1052.

Evans, B. I., Harosi, F. I., and Fernald, R. D. (1993). Photoreceptor spectral absorbency in larval and adult winter flounder (*Pseudopleuronectes americanus*). *Vision Neurosci.* **10,** 1065–1071.

Fahay, M. P. (1983). Guide to the early stages of marine fishes occurring in the western north Atlantic Ocean, Cape Hatteras to the southern Scotian shelf. *J. Northwest Atl. Fish. Sci.* **4,** 1–423.

Falini, G., Albeck, S., Weiner, S., and Addadi, L. (1996). Control of aragonite or calcite polymorphism by mollusk shell macromolecules. *Science* **271,** 67–69.

FAO (1993). "Marine Fisheries and the Law of the Sea: A Decade of Change," Special chapter (revised) of The State of World Food and Agriculture 1992, FAO Fish. Circ. No. 853. Food and Agriculture Organization of the United Nations, Rome.

FAO (1995). "Code of Conduct for Responsible Fisheries." Adopted at the Twenty-eighth Session of the FAO Conference, October 1995.

FAO (1996). "Fishery Statistics Catches and Landings," Vol. 78. Fisheries Department, Food and Agriculture Organization of the United Nations, Rome.

FAO (1997). "Review of the State of World Fishery Resources: Marine Fisheries," FAO Fish. Circ. No. 920. Fisheries Department, Food and Agriculture Organization of the United Nations, Rome.

Farias, I. P., Orti, G., Sampaio, I., Schneider, H., and Meyer, A. (1999). Mitochondrial DNA phylogeny of the family Cichlidae: Monophyly and fast molecular evolution of the Neotropical assemblage. *J. Mol. Evol.* **48,** 703–711.

Farmer, M. W., and Berg, C. J., Jr. (1990). Circulation around islands, gene flow, and fisheries management. *Proc. Annu. Gulf Caribb. Fish. Inst.* **39,** 318–330.

Farrell, J., and Campana, S. E. (1996). Regulation of calcium and strontium deposition on the otoliths of juvenile Tilapia, *Oreochromis niloticus*. *Comp. Biochem. Physiol.* **115A,** 103–109.

Farrow, S. (1996). Marine protected areas: Emerging economics. *Mar. Policy* **20,** 439–446.

Fautin, D. G. (1986). Why do anemonefishes only inhabit some host actinians? *Environ. Biol. Fishes.* **15,** 171–180.

Fautin, D. G. (1992). Anemonefish recruitment: The roles of order and chance. *Symbiosis* **14,** 143–160.

Fautin, D. G., and Allen, G. R. (1992). "Field Guide to Anemonefishes and Their Host Sea Anemones." Western Australian Museum, Perth.

Fay, R. R. (1984). The goldfish codes the axis of particle motion in three dimensions. *Science* **225,** 951–954.

Fay, R. R. (1988). "Hearing in Vertebrates. A Psychophysics Databook." Hill-Fay Assoc., Winnetka, IL.

Fay, R. R., and Popper, A. N. (1980). Structure and function in teleost auditory systems. *In* "Comparative Studies of Hearing in Vertebrates" (A. N. Popper and R. R. Fay, eds.), pp. 3–42. Springer-Verlag, New York.

Fenical, W., and Pawlik, J. R. (1991). Defensive properties of secondary metabolites from the Caribbean gorgonian coral *Erythropodium caribaerum*. *Mar. Ecol.: Prog. Ser.* **75,** 1–8.

Fenton, G. E., and Short, S. A. (1992). Fish age validation by radiometric analysis of otoliths. *Aust. J. Mar. Freshwater Res.* **43,** 913–922.

Fenton, G. E., and Short, S. A. (1994). Radiometric analysis of blue grenadier, *Macruronus novaezelandiae,* otolith cores. *Fish. Bull.* **93,** 391–396.

Fenton, G. E., Ritz, D. A., and Short, S. A. (1990). $^{210}Pb/^{226}Ra$ disequilibria in otoliths of blue grenadier, *Macruronus novaezelandiae*; Problems associated with radiometric ageing. *Aust. J. Mar. Freshwater Res.* **41,** 467–473.

Fenton, G. E., Short, S. A., and Ritz, D. A. (1991). Age determination of orange roughy, *Hoplostethus atlanticus* (Pisces: Trachichthyidae) using ^{210}Pb:^{226}Ra disequilibria. *Mar. Biol.* **109,** 197–202.

Fernald, R. D. (1988). Aquatic adaptations in fish eyes. *In* "Sensory Biology of Aquatic Animals" (J. Atema, R. R. Fay, A. N. Popper, and W. N. Tavolga, eds.), pp. 435–466. Springer-Verlag, New York.

Fernald, R. D. (1990). The optical system of fishes. *In* "The Visual System of Fish" (R. H. Douglas and M. B. A. Djamgoz, eds.), pp. 45–61. Chapman & Hall, London.

Ferreira, B. P., and Russ, G. R. (1992). Age, growth and mortality of the inshore coral trout, *Plectropomus maculatus* (Pisces: Serranidae) from the central Great Barrier Reef, Australia. *Aust. J. Mar. Freshwater Res.* **43,** 1301–1312.

Ferreira, B. P., and Russ, G. R. (1994). Age validation and estimation of growth rate of the coral trout, *Plectropomus leopardus,* (Lacepede 1802) from Lizard Island, Northern Great Barrier Reef. *Fish. Bull.* **92,** 46–57.

Ferreira, B. P., and Russ, G. R. (1995). Population structure of the leopard coral grouper *Plectropomus leopardus,* on fished and unfished reefs off Townsville, central Great Barrier Reef, Australia. *Fish. Bull.* **93,** 629–642.

Ferry-Graham, L. A., and Lauder, G. V. (2001). Prey capture performance in fishes: Digital particle image velocimetry for detecting suction generation. *J. Morphol.* **248,** 99–119.

Ferry-Graham, L. A., Wainwright, P. C., and Bellwood, D. R. (2001a). Prey capture in long-jawed butterflyfishes (Chaetodontidae): The functional basis of novel feeding habits. *J. Exp. Mar. Biol. Ecol.* **256,** 167–184.

Ferry-Graham, L. A., Wainwright, P. C., Hulsey, C. D., and Bellwood, D. R. (2001b). Evolution and mechanics of long jaws in butterflyfishes (Family Chaetodontidae). *J. Morphol.* **248,** 120–143.

Findley, J. S., and Findley, M. T. (1985). A search for pattern in butterfly fish communities. *Am. Nat.* **126,** 800–816.

Findley, J. S., and Findley, M. T. (1989). Circumtropical patterns in butterflyfish communities. *Environ. Biol. Fishes* **25,** 1–3.

Fine, M. L., Winn, H. E., and Olla, B. (1977). Communication in fishes. *In* "How Animals Communicate" (T. A. Sebeok, ed.), pp. 472–518. Indiana University Press, Bloomington.

Finger, T. E. (1988). Organization of chemosensory systems within the brains of bony fishes. *In* "Sensory Biology of Aquatic Animals" (J. Atema, R. R. Fay, A. N. Popper, and W. N. Tavolga, eds.), pp. 339–363. Springer-Verlag, New York.

Finger, T. E. (1997). Evolution of taste and solitary chemoreceptor cell systems. *Brain, Behav. Evol.* **50,** 234–243.

Finn, M. D., and Kingsford, M. J. (1996). Two-phase recruitment of apogonids (Pisces) on the Great Barrier Reef. *Mar. Freshwater Res.* **47,** 423–432.

Fischer, E. A. (1984). Egg trading in the chalk bass, *Serranus tortugarum,* a simultaneous hermaphrodite. *Z. Tierpsychol.* **66,** 143–151.

Fischer, E. A., and Petersen, C. W. (1987). The evolution of sexual patterns in the seabasses. *BioScience* **37,** 482–489.

Fischer, W., Bauchot, M. L., and Schneider, V. (1987). "Fiches FAO d'identification des espèces pour les besoins de la pêche. Méditerranée et Mer Noire, Zone de pêche 37," Vol. II, pp. 761–1530. FAO, Rome.

Fish, J. F. (1972). The effect of sound playback on the toadfish. *In* "Behavior of Marine Animals: Current Perspectives in Research...", Vol. 2 (H. E. Winn and B. L. Olla, eds.), pp. 386–434. Plenum Press, New York.

Fish, J. F., and Cummings, W. C. (1972). A 50–dB increase in sustained ambient noise from fish (*Cynoscion xanthulus*). *J. Acoust. Soc. Am.* **52,** 1246–1270.

Fish, M. P. (1964). Biological sources of sustained ambient noise. *In* "Marine Bio-acoustics" (W. N. Tavolga, ed.), pp. 175–194. Pergamon Press, New York.

Fishelson, L. (1977). Sociobiology of the feeding behavior of coral fish along the coral reef of the Gulf of Elat (=Gulf of Aqaba), Red Sea. *Isr. J. Zool.* **26,** 114–134.

Fishelson, L. (1995). Comparative morphology and cytology of the olfactory organs in moray eels with remarks on their foraging behavior. *Anat. Rec.* **243,** 403–412.

Fishelson, L. (1997). Olfaction and visual detection of food and relevant morphometric characters in some species of moray eels (Muraenidae). *Isr. J. Zool.* **43,** 367–375.

Fishelson, L., and Baranes, A. (1997). Ontogenesis and cytomorphology of the nasal olfactory organs in the Oman shark, *Iago omanensis* (Triakidae), in the Gulf of Aqaba, Red Sea. *Anat. Rec.* **249,** 404–421.

Fisher, R., Bellwood, D. R., and Job, S. D. (2000). The development of swimming abilities in reef fish larvae. *Mar. Ecol.: Prog. Ser.* **202,** 163–173.

Fiske, S. J. (1992). Sociocultural aspects of establishing marine protected areas. *Ocean Coastal Manage.* **17,** 25–46.

Fitch, W. T. S., and Shapiro, D. Y. (1990). Spatial dispersion and nonmigratory spawning in the bluehead wrasse (*Thalassoma bifasciatum*). *Ethology* **85,** 199–211.

Fleminger, A. (1986). The Pleistocene equatorial barrier between the Indian and Pacific Oceans and a likely cause for Wallace's line. *UNESCO Tech. Pap. Mar. Sci.* **49,** 84–97.

Flock, Å. (1967). Transducing mechanisms in the lateral line canal organ receptors. *Cold Spring Harbor Symp. Quant. Biol.* **30,** 133–144.

Flock, Å. (1971). The lateral line organ mechanoreceptors. *Fish Physiol.* **5,** 241–263.

Floeter, S. R., and Gasparini, J. L. (2000). The southwestern Atlantic reef fish fauna: Composition and zoogeographic patterns. *J. Fish Biol.* **56,** 1099–1114.

Fogarty, M. J. (1993). Recruitment in randomly varying environments. *ICES J. Mar. Sci.* **50,** 247–260.

Fogarty, M. J. (1999). Essential habitat, marine reserves, and fishery management. *Trends Ecol. Evol.* **14,** 133–134.

Fogarty, M. J., Sissenwine, M. P., and Cohen, E. B. (1991). Recruitment variability and the dynamics of exploited marine populations. *Trends Ecol. Evol.* **6,** 241–246.

Foote, M. (1996). Ecological controls on the evolutionary recovery of post-Paleozoic crinoids. *Science* **274,** 1492–1495.

Forrester, G. E. (1990). Factors influencing the juvenile demography of a coral reef fish. *Ecology* **71,** 1666–1681.

Forrester, G. E. (1991). Social rank, individual size and group composition as determinants of food consumption by humbug damselfish, *Dascyllus aruanus. Anim. Behav.* **42,** 710–711.

Forrester, G. E. (1995). Strong density-dependent survival and recruitment regulate the abundance of a coral reef fish. *Oecologia* **103,** 275–282.

Forrester, G. E. (1998). Definitions and tests of recruitment limitation. *In* "ReeFish '95: Recruitment and Population Dynamics of Coral Reef Fishes" (G. P. Jones, P. J. Doherty, B. D. Mapstone, and L. Howlett, eds.), pp. 135–139. CRC Reef Research Center, Townsville, Australia.

Forrester, G. E. (1999). The influence of adult density on larval settlement in a coral reef fish, *Coryphopterus glaucofraenum. Coral Reefs* **18,** 85–89.

Forrester, G. E., and Steele, M. A. (2000). Variation in the presence and cause of density-dependent mortality in three species of reef fishes. *Ecology* **81,** 2416–2427.

Forsgren, E., Karlsson, A., and Kvarnemo, C. (1996). Female sand gobies gain direct benefits by choosing males with eggs in their nests. *Behav. Ecol. Sociobiol.* **39,** 91–96.

Fortier, L., and Quiñonez-Velazquez, C. (1998). Dependence of survival on growth in larval pollock *Pollachius virens* and haddock *Melanogrammus aeglefinus*: A field study based on individual hatchdates. *Mar. Ecol.: Prog. Ser.* **174,** 1–12.

Forward, R. B., Burke, J. S., Rittschof, D., and Welch, J. M. (1996). Photoresponses of larval Atlantic menhaden (*Brevoortia tyrannus* Latrobe) in offshore and estuarine waters: implications for transport. *J. Exp. Mar. Biol. Ecol.* **199,** 123–135.

Foster, S. A. (1985). Group foraging by a coral reef fish: A mechanism for gaining access to defended resources. *Anim. Behav.* **33**,782–792.

Foster, S. A. (1987). Acquisition of a defended resource: A benefit of group foraging for the neotropical wrasse, *Thalassoma lucasanum*. *Environ. Biol. Fishes* **19**, 215–222.

Fowler, A. J. (1989). Description, interpretation and use of the microstructure of otoliths from juvenile butterflyfishes (family Chaetodontidae). *Mar. Biol.* **102**, 167–181.

Fowler, A. J. (1990a). Validation of annual increments in the otoliths of a small, tropical coral reef fish. *Mar. Ecol.: Prog. Ser.* **64**, 25–38.

Fowler, A. J. (1990b). Spatial and temporal patterns of distribution and abundance of chaetodontid fishes at One Tree Reef, southern GBR. *Mar. Ecol.: Prog. Ser.* **64**, 39–53.

Fowler, A. J. (1995). Annulus formation in the otoliths of coral reef fish: A review. *In* "Recent Developments in Fish Otolith Research" (D. H. Secor, J. M. Dean, and S. E. Campana, eds.), pp. 45–63. University of South Carolina Press, Columbia.

Fowler, A. J., and Doherty, P. J. (1992). Validation of annual growth increments in the otoliths of two species of damselfish from the southern Great Barrier Reef. *Aust. J. Mar. Freshwater Res.* **43**, 1057–1068.

Fowler, A. J., and Short, D. A. (1996). Temporal variation in the early life-history characteristics of the King George Whiting (*Sillaginodes punctata*) from analysis of otolith microstructure. *Mar. Freshwater Res.* **47**, 809–818.

Fowler, A. J., Doherty, P. J., and Williams, D. McB. (1992). Multi-scale analysis of recruitment of a coral reef fish on the Great Barrier Reef. *Mar. Ecol.: Prog. Ser.* **82**, 131–141.

Fowler, A. J., Campana, S. E., Jones, C. M., and Thorrold, S. R. (1995). Experimental assessment of the effect of temperature and salinity on elemental composition of otoliths using solution-based ICPMS. *Can. J. Fish. Aquat. Sci.* **52**, 1421–1430.

Fox, L. R., and Morrow, P. A. (1981). Specialization: Species property or local phenomenom. *Science* **211**, 887–893.

Francis, M. P. (1994). Growth of juvenile snapper, *Pagrus auratus*. *N. Z. J. Mar. Freshwater Res.* **28**, 201–218.

Francis, R. C., and Barlow, G. W. (1993). Social control of primary sex differentiation in the Midas Cichlid. *Proc. Nat. Acad. Sci. U.S.A.* **90**, 10673–10675.

Francis, R. I. C. C. (1990). Back-calculation of fish length: A critical review. *J. Fish. Biol.* **36**, 883–902.

Francis, R. I. C. C., Williams, M. W., Pryce, A. C., Pollard, S., and Scott, S. G. (1993). Uncoupling of otolith and somatic growth in *Pagrus auratus* (Sparidae). *Fish. Bull.* **91**, 159–164.

Fraser, D. F., and Gilliam, J. F. (1992). Nonlethal impacts of predator invasion: Facultative suppression of growth and reproduction. *Ecology* **73**, 959–970.

Fratantoni, P. S., Lee, T. N., Podesta, G., and Muller-Karger, F. (1998). The influence of loop current perturbations on the formation and evolution of Tortugas eddies in the southern Straits of Florida. *J. Geophys. Res.* **103**, 24,759–24,779.

Frazer, T. K., and Lindberg, W. J. (1994). Refuge spacing similarly affects reef-associated species from three phyla. *Bull. Mar. Sci.* **55**, 388–400.

Frazer, T. K., Lindberg, W. J., and Stanton, G. R. (1991). Predation on sand dollars by grey triggerfish, *Balistes capriscus*, in the northeastern Gulf of Mexico. *Bull. Mar. Sci.* **48**, 159–164.

Frederick, J. L. (1997). Post-settlement movement of coral reef fishes and bias in survival estimates. *Mar. Ecol.: Prog. Ser.* **150**, 65–74.

Freeman, B., and Smith, D. (1990). Variation of density dependence with spatial scale in the leaf mining fly *Liriomyza commelinae* (Diptera: Agromyzidae). *Ecol. Entomol.* **15**, 265–274.

Fricke, H. W. (1971). Fische als feinde tropischer seeigel. *Mar. Biol.* **9**, 328–338.

Fricke, H. W. (1975). Losen einfacher probleme bei einem fisch (freiswasser an *Balistes fuscus*). *Z. Tierpsychol.* **38**, 18–33.

Fricke, H. W. (1983). Social control of sex: Field experiments with the anemonefish *Amphiprion bicinctus*. *Z. Tierpsychol.* **61**, 71–77.

Friedland, K. D., Reddin, D. G., Shimizu, N., Haas, R. E., and Youngson, A. F. (1998). Strontium:calcium ratios in Atlantic salmon (*Salmo salar*) otoliths and observations on growth and maturity. *Can. J. Fish. Aquat. Sci.* **55**, 1158–1168.

Friel, J. P., and Wainwright, P. C. (1997). A model system of structural duplication: Homologies of the adductor mandibulae muscles in tetraodontiform fishes. *Syst. Biol.* **46**, 441–463.

Friel, J. P., and Wainwright, P. C. (1998). Evolution of motor pattern in Tetraodontiform fishes: Does muscle duplication lead to functional diversification? *Brain, Behav. Evol.* **53**, 159–170.

Friel, J. P., and Wainwright, P. C. (1999). Evolution of complexity in motor patterns and jaw musculature of tetraodontiform fishes. *J. Exp. Biol.* **202**, 867–880.

Frith, C. A., Leis, J. M., and Goldman, B. (1986). Currents in the Lizard Island Region of the Great Barrier Reef Lagoon and their relevance to potential movements of larvae. *Coral Reefs* **5**, 81–92.

Fuiman, L. A. (1997). What can flatfish ontogenies tell us about pelagic and benthic lifestyles? *J. Sea Res.* **37**, 257.

Fuiman, L. A., and Batty, R. S. (1997). What a drag it is getting cold: Partitioning the physical and physiological effects of temperature on fish swimming. *J. Exp. Biol.* **200**, 1745–1755.

Fuiman, L. A., and Delbos, B. C. (1998). Developmental changes in visual sensitivity of red drum, *Sciaenops ocellatus*. *Copeia,* pp. 936–943.

Fuiman, L. A., and Higgs, D. M. (1997). Ontogeny, growth, and the recruitment process. *In* "Early Life History and Recruitment in Fish Populations" (R. C. Chambers and E. A. Trippel, eds.), pp. 225–249. Chapman & Hall, New York.

Fuiman, L. A., Smith, M. E., and Malley, V. N. (1999). Ontogeny of routine swimming speed and startle responses in red drum, with a comparison of responses to acoustic and visual stimuli. *J. Fish Biol.* **55**(Suppl. A), 215–226.

Fulton, C. J., Bellwood, D. R., and Wainwright, P. C. (2001). The relationship between swimming ability and habitat use in wrasses (Family Labridae). *Mar. Biol.* **20**, 139–150.

Furukawa, K., and Wolanski, E. (1998). Shallow-water frictional effects in island wakes. *Estuarine Coastal Shelf Sci.* **46**, 599–608.

Gabriel, W. L., Sissenwine, M. P., and Overholz, W. J. (1989). Analysis of spawning stock biomass per recruit: An example of Georges Bank haddock. *North Am. J. Fish. Manage.* **9**, 383–391.

Gaines, S. D., and Lafferty, K. D. (1995). Modeling the dynamics of marine species: The importance of incorporating larval dispersal. *In* "Ecology of Marine Invertebrate Larvae" (L. McEdward, ed.), pp. 389–412. CRC Press, New York.

Galis, F., and Drucker, E. G. (1996). Pharyngeal biting mechanics in centrarchid and cichlid fishes: Insights into a key evolutionary innovation. *J. Evol. Biol.* **9**, 641–670.

Gallahar, N. K., and Kingsford, M. J. (1996). Factors influencing Sr/Ca ratios in otoliths of *Girella elevata:* An experimental investigation. *J. Fish Biol.* **48**, 174–186.

Gallego, A., and Heath, M. R. (1994). The development of schooling behavior in Atlantic herring *Clupea harengus*. *J. Fish Biol.* **45**, 569–588.

Gallego, A., Heath, M. R., McKenzie, E., and Cargill, L. H. (1996). Environmentally induced short-term variability in the growth rates of larval herring. *Mar. Ecol. Prog. Ser.* **137**, 11–23.

Galvez, R. (1988). Blast fishing: A Philippine case study. *Trop. Coastal. Area Manage., 1988.*

Galvez, R. E., Hingco, T. G., Bautista, C., and Tungpalan, M. T. (1989). Sociocultural dynamics of blast fishing and sodium cyanide fishing in two fishing villages in the Lingayen Gulf Area. *In* "Towards Sustainable Development of the Coastal Resources of Lingayen Gulf" (G. Silvestre, E. Miclat, and T.-E. Chua, eds.), pp. 43–62. Council for Aquatic and Marine Research and Development and the International Center for Living Aquatic Resources Management, Manila, Philippines.

Galzin, R. (1985). Ecologie des poissons récifaux de Polynésie française. Variations spatio-temporelles des peuplements; Dynamique de population de trois espèces dominantes des lagons nord de Moorea; Evaluation de la production ichtyologique d'un secteur récifo-lagonaire. Ph.D. Thesis, Université de Sciences et Techniques du Languedoc.

Galzin, R. (1987). Structure of fish communities of French Polynesian coral reefs. I: Spatial scales. *Mar. Ecol.: Prog. Ser.* **41**, 129–136.

Galzin, R., Planes S., Adjeroud, M., Chauvet, C., Doherty, P. J., and Poupin, J. (1998). Objectives and background to the 1994 Franco-Australian expedition to Taiaro Atoll (Tuamotu Archipelago, French Polynesia). *Coral Reefs* **17**, 15–21.

Garcia Rubies, A. (1997). Estudi ecològic de les poblacions de peixos litorals sobre substrat rocòs a la Mediterrània occidental: Effectes de la fondària, el substrat, l'estacionalitat i la protecció. Ph.D. Thesis, University of Barcelona.

Gardner, T. R. (1997). Commercial breeding of the dottybacks. *Seascope* **14**, 1–2.

Gaston, K. J. (1994a). "Rarity." Chapman & Hall, London.

Gaston, K. J. (1994b). Measuring geographic range sizes. *Ecography*. **17**, 198–205.

Gaston, K. J. (1996). The multiple forms of the interspecific abundance-distribution relationship. *Oikos* **76**, 211–220.

Gaston, K. J. (1998). Rarity as double jeopardy. *Nature (London)* **394**, 229–230.

Gaston, K. J. (1999). Why Rapoport's rule does not generalize. *Oikos* **84**, 309–312.

Gaston, K. J., and Blackburn, T. M. (1996). Range size-body size relationships: Evidence of scale dependence. *Oikos* **75**, 479–485.

Gaston, K. J., and Lawton, J. H. (1988a). Patterns in the distribution and abundance of insect populations. *Nature (London)* **331**, 709–712.

Gaston, K. J., and Lawton, J. H. (1988b). Effects of scale and habitat on the relationship between regional distribution and local abundance. *Oikos* **58**, 329–335.

Gaston, K. J., and Lawton, J. H. (1990a). The population ecology of rare species. *J. Fish Biol.* **37**, 97–104.

Gaston, K. J., and Lawton, J. H. (1990b). Effects of scale and habitat on the relationship between regional distribution and local abundance. *Oikos* **58**, 329–335.

Gaston, K. J., Blackburn, T. M., and Lawton, J. H. (1997). Interspecific abundance-range size relationships: An appraisal of mechanisms. *J. Anim. Ecol.* **66**, 579–601.

Gauldie, R. W. (1996a). Effects of temperature and vaterite replacement on the chemistry of metal ions in the otoliths of *Oncorhynchus tshawytscha*. *Can. J. Fish. Aquat. Sci.* **53**, 2015–2026.

Gauldie, R. W. (1996b). Biological factors controlling the carbon isotope record in fish otoliths: Principles and evidence. *Comp. Biochem. Physiol.* **115B**, 201–208.

Gauldie, R. W., Sharma, S. K., and Volk, E. (1997). Micro-Raman spectral study of vaterite and aragonite otoliths of the coho salmon, *Oncorhynchus kisutch*. *Comp. Biochem. Physiol.* **118A**, 753–757.

Gay, S. L., and Andrews, J. C. (1994). The effects of recruitment strategies on coral larvae settlement distributions at Helix Reef. *In* "The Bio-physics of Marine Larval Dispersal" (P. W. Sammarco and M. L. Heron, eds.), pp. 73–88. American Geophysical Union, Washington, DC.

Geffen, A. J. (1982). Otolith ring deposition in relation to growth rate in herring (*Clupea harengus*) and turbot (*Scopthalmus maximus*) larvae. *Mar. Biol.* **71**, 317–326.

Geffen, A. J. (1992). Validation of otolith increment deposition rate. *Can. Fish. Aquat. Sci., Spec. Publ.* **117**, 101–113.

Genin, A., and Boehlert, G. W. (1985). Dynamics of temperature and chlorophyll structures above a seamount: An oceanic experiment. *J. Mar. Res.* **43**, 907–924.

Gerking, S. D. (1994). "Feeding Ecology of Fish." Academic Press, San Diego, CA.

Getz, W. M. (1996). A hypothesis regarding the abruptness of density dependence and the growth rate of populations. *Ecology* 77, 2014–2026.

Gibb, A. C. (1995). Kinematics of prey capture in a flatfish, *Pleuronichthys verticalis*. *J. Exp. Biol.* **198,** 1171–1183.

Gibb, A. C. (1996). The kinematics of prey capture in *Xystreurys liolepis:* Do all flatfish feed asymmetrically? *J. Exp. Biol.* **199,** 2269–2283.

Gibson, R. M., and Höglund, J. (1992). Copying and sexual selection. *Trends Ecol. Evol.* **7,** 229–232.

Gibson, R. N. (1972). The vertical distribution and feeding relationship of intertidal fish on the Atlantic coast of France. *J. Anim. Ecol.* **41,** 189–207.

Gilbert, P. S., Lee, T. N., and Podesta, G. P. (1996). Transport of anomalous low-salinity waters from the Mississippi River flood of 1993 to the Straits of Florida. *Cont. Shelf Res.* **16,** 1065.

Gilbert, P. W., Hodgson, E. S., and Mathewson, R. F. (1964). Electroencephalograms of sharks. *Science* **145,** 949–951.

Gill, D. E. (1978). The metapopulation ecology of the red-spotted newt, *Notophthalmus viridescens* (Rafinesque). *Ecol. Monogr.* **48,** 145–166.

Gillanders, B. M., and Kingsford, M. J. (1996). Elements in otoliths may elucidate the contribution of estuarine recruitment to sustaining coast reef populations of a temperate reef fish. *Mar. Ecol.: Prog. Ser.* **141,** 13–20.

Gillanders, B. M., and Kingsford, M. J. (2000). Elemental fingerprints of otoliths of fish may distinguish estuarine nursery habitats. *Mar. Ecol.: Prog. Ser.* **201,** 273–286.

Gilmartin, M., and Relevante, N. (1974). The 'island mass' effect on the phytoplankton and primary production of the Hawaiian Islands. *J. Exp. Mar. Biol. Ecol.* **16,** 181–204.

Gilmore, R. G. (1997). Fishes of the Indian River Lagoon and adjacent waters, Florida. *Bull. Fla. State Mus. Biol. Sci.* **22,** 101–147.

Gilmore, R. G., and Jones, R. S. (1992). Color variation and associated behavior in the Epinepheline groupers, *Myceroperca microlepis* (Goode and Bam) and *M. phenax* Jordan & Swain. *Bull. Mar. Sci.* **51,** 84–103.

Gladfelter, W. B., Ogden, J. C., and Gladfelter, B. H. (1980). Similarity and diversity among coral reef fish communities: A comparison between tropical western Atlantic (Virgin Islands) and tropical central Pacific (Marshall Islands). *Ecology* **61,** 1156–1168.

Gladstone, W. (1987a). The courtship and spawning behaviors of *Canthigaster valentini* (Tetraodontidae). *Environ. Biol. Fishes* **20,** 255–261.

Gladstone, W. (1987b). Role of female territoriality in social and mating systems of *Canthigaster valentini* (Pisces: Tetraodontidae): Evidence from field experiments. *Mar. Biol.* **96,** 185–191.

Glazier, D. S. (1980). Ecological shifts and the evolution of geographically restricted species of North American *Peromyscus*. *J. Biogeogr.* **7,** 63–83.

Gluckmann, I., and Vandewalle, P. (1998). Morpho-functional analysis of the feeding apparatus in four Pomacentridae species: *Dascyllus aruanus, Chromis retrofasciata, Chrysiptera biocellata* and *C. unimaculata*. *Ital. J. Zool.* **65,** 421–424.

Gobalet, K. W. (1989). Morphology of the parrotfish pharyngeal jaw apparatus. *Am. Zool.* **29,** 319–331.

Godwin, J. (1995). Phylogenetic and habitat influences on mating system structure in the humbug damselfishes (*Dascyllus,* Pomacentridae). *Bull. Mar. Sci.* **57,** 637–652.

Goldschmid, A., and Kotrschal, K. (1985). Morphological and functional adaptations in different feeding types of blennies (Perciformes, Teleostei). *In* "Functional Morphology in Vertebrates" (H. R. Duncker and G. Fleischer, eds.), pp. 241–244. Fischer, New York.

Goldschmid, A., Kotrschal, K., and Wirtz, P. (1984). Food and gut length of 14 Adriatic blenniid fish (Blenniidae; Percomorpha; Teleostei). *Zool. Anz.* **213,** 145–150.

Goldstein D. B., Linares, A. R., Cavalli-Sforza, L. L., and Feldman, M. W. (1995). An evaluation of genetic distances for use with microsatellite loci. *Genetics* **139,** 463–471.

Gomon, M. F. (1997). Relationships of fishes of the labrid tribe Hypsigenyini. *Bull. Mar. Sci.* **60,** 789–871.

Goodman, D. (1987). Considerations of stochastic demography in the design and management of reserves. *Nat. Res. Mod.* **1,** 205–234.

Goodrich, J. S., Sanderson, S. L., Batjakas, I. E., and Kaufman, L. S. (2000). Branchial arches of supension feeding *Oreochromis esculentus:* Sieve or sticky filter? *J. Fish Biol.* **56,** 858–875.

Goodyear, C. P. (1993). Spawning stock biomass per recruit in fisheries management: Foundation and current use. *Can. Fish. Aquat. Sci. Spec. Publ.* **120,** 67–81.

Gordon, H. B., and Hughes, R. L. (1981). A study of rotating baroclinic nonlinear flow around an island. *J. Phys. Oceanogr.* **11,** 1011–1014.

Gosline, W. A. (1971). "Functional Morphology and Classification of Teleostean Fishes." University of Hawaii Press, Honolulu.

Gosline, W. A. (1984). Structure, function and ecology of the goatfishes (Family Mullidae). *Pac. Sci.* **38,** 312–323.

Gould, S. J., and Vrba, E. S. (1982). Exaptation—a missing term in the science of form. *Paleobiology* **8,** 4–15.

Govoni, J. J. (1993). Flux of larval fishes across frontal boundaries—examples from the Mississippi River plume front and the western Gulf-Stream front in winter. *Bull. Mar. Sci.* **53,** 538–566.

Govoni, J. J., and Grimes, C. B. (1992). The surface accumulation of larval fishes by hydrodynamic convergence within the Mississippi river plume front. *Cont. Shelf Res.* **12,** 1265–1276.

Govoni, J. J., Hoss, D. E., and Colby, D. (1989). The spatial distribution of larval fishes about the Mississippi River Plume. *Limnol. Oceanogr.* **34,** 178–187.

Graham, T. (1996). Managing Palau's aquarium fishery. *Secretariat of the Pacific Community Live Reef Fish Inf. Bull.* **1,** 13–18.

Graham, W. M., Field, J. G., and Potts, D. C. (1992). Persistent "upwelling shadows" and their influence on zooplankton distributions. *Mar. Biol.* **114,** 561–570.

Graves, J. E., McDowell, J. R., and Jones, M. L. (1992). A genetic analysis of weakfish (*Cynoscion regalis*) stock structure along the mid-Atlantic coast. *Fish. Bull.* **90**, 469–475.

Gray, C. A. (1998). Diel changes in the vertical distributions of larval fishes in unstratified coastal waters off southeastern Australia. *J. Plankton Res.* **20**, 1539–1552.

Green, A. L. (1993). Damselfish territories: Focal sites for studies of the early life history of labroid fishes. *In* "Proc. 7th Int. Coral Reef Symp." **Vol. 1**, pp. 601–605.

Green, B., and McCormick, M. I. (1999). Influence of larval feeding history on the body condition of *Amphiprion melanopus*. *J. Fish Biol.* **55**, 1273–1289.

Greene, C. H. (1983). Selective predation in freshwater zooplankton communities. *Int. Rev. Gesamten Hydrobiol.* **68**, 297–315.

Greenfield, D. W. (1968). The zoogeography of *Myripristis* (Pisces: Holocentridae). *Syst. Zool.* **17**, 76–87.

Grimes, C. B., and Kingsford, M. J. (1996). How do riverine plumes of different sizes influence fish larvae: Do they enhance recruitment? *Mar. Freshwater Res.* **47**, 191–208.

Grimes, C. B., Fitzhugh, G., Koenig C. C., and Coleman, F. C. (1997). Factors influencing the timing and magnitude of settlement of reef fish recruiting to seagrass meadows along the West Florida Shelf. *In* "Florida Big Bend Coastal Research Workshop. Toward a Scientific Basis for Ecosystem Management" (W. J. Lindberg, ed.), Proc. Meet. May 7–8, 1997, Steinhatchee, Florida, pp. 71–72.

Grobecker, D. B. (1983). The "lie-in-wait" feeding mode of a cryptic teleost, *Synanceia verrucosa*. *Environ. Biol. Fishes* **8**, 191–202.

Grobecker, D. B., and Pietsch, T. W. (1979). High-speed cinematographic evidence for ultrafast feeding in antennariid anglerfishes. *Science* **205**, 1161–1162.

Grobecker, D. B., and Pietsch, T. W. (1987). "Frogfishes of the World: Systematics, Zoogeography, and Behavioral Ecology." Stanford University Press, Stanford, CA.

Gronell, A. M. (1989). Visiting behavior by females of the sexually dichromatic damselfish, *Chrysiptera cyanea* (Teleostei: Pomacentridae): A probable method of assessing male quality. *Ethology* **81**, 89–122.

Grover, J. J., Eggleston, D. B., and Shenker, J. M. (1998). Transition from pelagic to demersal phase in early-juvenile Nassau grouper, *Epinephelus striatus:* Pigmentation, squamation and ontogeny of diet. *Bull. Mar. Sci.* **62**, 97–113.

Grubich, J. R. (2000). Crushing motor patterns in drum (Teleostei: Sciaenidae): Functional novelties associated with molluscivory. *J. Exp. Biol.* **203**, 3161–3176.

Grubich, J. R., and Wainwright, P. C. (1997). Motor basis of feeding performance in largemouth bass (*Micropterus salmoides*). *J. Exp. Zool.* **277**, 1–13.

Grütter, A. S. (1996). Cleaning symbioses: Proximate and adaptive explanations. *BioScience* **46**, 517–521.

Guenette, S., and Pitcher, T. J. (1999). An age structured model showing the benefits of marine reserves in controlling exploitation. *Fish. Res.* **39**, 295–303.

Guenette, S., Lauck, T., and Clark, C. (1998). Marine reserves: From Beverton and Holt to the present. *Rev. Fish. Biol. Fish.* **8**, 1–21.

Guiasu, R. C., and Winterbottom, R. (1998). Yellow juvenile color pattern, diet switching and the phylogeny of the surgeonfish *Zebrasoma* (Percomorpha, Acanthuridae). *Bull. Mar. Sci.* **63**, 277–294.

Guilderson, T. P., and Schrag, D. P. (1998). Abrupt shift in subsurface temperatures in the eastern tropical Pacific associated with recent changes in El Niño. *Science* **281**, 240–243.

Guilderson, T. P., Fairbanks, R. G., and Rubenstone, J. L. (1994). Tropical temperature variations since 20,000 years ago: Modulating interhemispheric climate change. *Science* **263**, 663–665.

Guilderson, T. P., Schrag, D. P., Kashgarian, M., and Southon, J. (1998). Radiocarbon variability in the western equatorial Pacific inferred from a high-resolution coral record from Nauru Island. *J. Geophys. Res.* **103**, 24641–24650.

Guillou, M., and Michel, C. (1993). Influence of recruitment variability on the marketable stock of the sea-urchin (*Sphaerechinus granularis*) in the Glenan archipelago (South-Brittany). *Oceanol. Acta.* **16**, 423–430.

Gulland, J. A. (1965). Estimation of mortality rates. Annex to Arctic Fisheries Working Group Report. *ICES, C.M.*, Doc. No. 3, mimeo.

Gunn, J. S., Harrowfield, I. R., Proctor, C. H., and Thresher, R. E. (1992). Electron probe microanalysis of fish otoliths—evaluation of techniques for studying age and stock discrimination. *J. Exp. Mar. Biol. Ecol.* **158**, 1–36.

Gurevitch, J., and Hedges, L. V. (1993). Meta-analysis: Combining the results of independent experiments. *In* "Design and Analysis of Ecological Experiments" (S. M. Scheiner and J. G. Gurevitch, eds.), pp. 378–398. Chapman & Hall, New York.

Gurney, W. S. C., and Nisbet, R. M. (1985). Fluctuation periodicity, generation separation, and the expression of larval competition. *Theor. Popul. Biol.* **28**, 150–180.

Gutierrez, L. (1998). Habitat selection by recruits establishes local patterns of adult distribution in two species of damselfishes: *Stegastes dorsopunicans* and *S. planifrons*. *Oecologia* **115**, 268–277.

Gyllensten, U. (1985). The genetic structure of fish: Differences in the intraspecific distribution of biochemical genetic variation between marine, anadromous, and freshwater species. *J. Fish Biol.* **26**, 691–699.

Ha, S. J. (1973). Aspects of sound communication in the damselfish *Eupomacentrus partitus*. Unpublished Ph.D. Dissertation, University of Miami, Miami, FL.

Haldane, J. B. S. (1953). Animal populations and their regulation. *New Biol.* **15**, 9–24.

Hall, K. C., and Bellwood, D. R. (1995). Histological effects of cyanide, stress and starvation on the intestinal mucosa of *Pomacentrus coelestis*, a marine aquarium fish species. *J. Fish Biol.* **47**, 438–454.

Hall, R. (1998). The plate tectonics of Cenozoic SE Asia and the distribution of land and sea. *In* "Biogeography

and Geological Evolution of SE Asia" (R. Hall, and J. D. Holloway, eds.), pp. 99–132. Backhuys Publishers, Leiden.
Hall, S. J. (1998). Closed areas for fisheries management—the case consolidates. *Trends Ecol. Evol.* **13,** 297–298.
Hallam, A. (1973). Distributional patterns in contemporary terrestrial and marine animals. *In* "Organisms and Continents Through Time" (N. F. Hughes, ed.), Spec. Pape. Paleonto., No. 12, pp. 93–105. Paleontological Association, London.
Halliday, A. N., Lee, D.-C., Christensen, J. N., Rehkämper, M., Yi, W., Luo, X., Hall, C. M., Ballentine, C. J., Pettke, T., and Stirling, C. (1998). Applications of multiple collector-ICPMS to cosmochemistry, geochemistry, and paleoceanography. *Geochim. Cosmochim. Acta* **62,** 919–940.
Hamner, W. M., and Hauri, I. R. (1977). Fine-scale surface currents in the Whitsunday Islands, Queensland, Australia: Effect of tide and topography. *Aust. J. Mar. Freshwater Res.* **28,** 333–359.
Hamner, W. M., and Hauri, I. R. (1981). Effects of island mass: Water flow and plankton pattern around a reef in the Great Barrier Reef lagoon, Australia. *Limnol. Oceanogr.* **31,** 171–177.
Hamner, W. M., Jones, M. S., Carleton, J. H., Hauri, I. R., and Williams, D.McB. (1988). Zooplankton, planktivorous fish, and water currents on a windward reef face: Great Barrier Reef, Australia. *Bull. Mar. Sci.* **42,** 459–479.
Hamrick, J. L., Linhart, Y. B., and Mitton, J. B. (1979). Relationships between life history characteristics and electrophoretically detectable genetic variation in plants. *Annu. Rev. Ecol. Syst.* **10,** 173–200.
Hanawa, M., Harris, L., Graham, M., Farrell, A. P., and Bendell-Young, L. I. (1998). Effects of cyanide exposure on *Dascyllus aruanus,* a tropical marine fish species: Lethality, anaesthesia and physiological effects. *Aquar. Sci. Conserv.* **2,** 21–34.
Hannesson, R. (1998). Marine reserves: What would they accomplish? *Mar. Res. Econ.* **13,** 159–170.
Hanski, I. (1982). Communities of bumblebees: Testing the core-satellite species hypothesis. *Ann. Zool. Fenn.* **19,** 65–73.
Hanski, I. (1990). Density dependence, regulation and variability in animal populations. *Philos. Trans. R. Soc. London, Ser. B* **330,** 141–150.
Hanski, I. (1996). Metapopulation ecology. *In* "Population Dynamics in Ecological Space and Time" (O. E. Rhodes, R. K. Chesser, and M. H. Smith, eds.), pp. 13–44. University of Chicago Press, Chicago.
Hanski, I. (1998). Metapopulation dynamics. *Nature (London)* **396,** 41–49.
Hanski, I. (1999). "Metapopulation Ecology." Oxford University Press, Oxford.
Hanski, I., and Gilpin, M. E. (1997). "Metapopulation Biology: Ecology, Genetics, and Evolution." Academic Press, San Diego, CA.
Hanski, I., Kouki, J., and Halkka, A. (1993). Three explanations of the positive relationship between distribution and abundance of species. *In* "Species Diversity in Ecological Communities" (R. E. Ricklefs and D. Schluter, eds.), pp. 108–116. University of Chicago Press, Chicago.
Hanski, I., Foley, P., and Hassell, M. (1996). Random walks in a metapopulation: How much density dependence is necessary for long-term persistence. *J. Anim. Ecol.* **65,** 274–282.
Hanson, P. J., and Zdanowicz, V. S. (1999). Elemental composition of otoliths from Atlantic croaker along an estuarine pollution gradient. *J. Fish Biol.* **54,** 656–668.
Hara, T. J. (1971). Chemoreception. *Fish Physiol.* **5,** 79–120.
Hara, T. J. (1993). The role of olfaction in fish behavior. *In* "The Behavior of Teleost Fishes" (T. J. Pitcher, ed.), pp. 171–199. Chapman & Hall, New York.
Harden Jones, F. R. (1968). "Fish Migration." Arnold, London.
Hardin, G. (1968). The tragedy of the commons. *Science* **162,** 1243–1248.
Hare, J. A., and Cowen, R. K. (1991). Expatriation of *Xyrichtys novacula* (Pisces: Labridae) larvae: Evidence of rapid cross-slope exchange. *J. Mar. Res.* **49,** 801–823.
Hare, J. A., and Cowen, R. K. (1995). Effect of age, growth rate, and ontogeny on the otolith size—fish size relationship in bluefish, *Pomatomus saltatrix,* and the implications for back-calculation of size in fish early life history stages. *Can. J. Fish. Aquat. Sci.* **52,** 1909–1922.
Hare, J. A., and Cowen, R. K. (1996). Transport mechanisms of larval and pelagic juvenile bluefish (*Pomatomus saltatrix*) from South Atlantic Bight spawning grounds to mid-Atlantic Bight nursery habitats. *Limnol. Oceanogr.* **41,** 1264–1280.
Hare, J. A., and Cowen, R. K. (1997). Size, growth, development, and survival of the planktonic larvae of *Pomatomus saltatrix* (Pisces: Pomatomidae). *Ecology* **78,** 2415–2431.
Hare, J. A., Cowen, R. K., Zehr, J. P., Juanes, F., and Day, K. H. (1994). Biological and oceanographic insights from larval labrid (Pisces: Labridae) identification using mtDNA sequences. *Mar. Biol.* **118,** 17–24.
Harmelin, J. G. (1987). Structure et variabilité de l'ichtyofaune d'une zone rocheuse protégée en Méditerranée (Parc National de Port-Cros, France). *P.S.Z.N.I.: Mar. Ecol.* **8,** 263–284.
Harmelin, J. G., Bachet, F., and Garcia, F. (1995). Mediterranean marine reserves: Fish indices as tests of protection efficiency. *P.S.Z.N.I.: Mar. Ecol.* **16,** 233–250.
Harmelin-Vivien, M. L. (1979). Ichtyofaune des récifs coralliens de Tuléar (Madagascar): Ecologie et relations trophiques. Ph.D. Thesis, Vol. 2, pp. 165–258. Université Aix-Marseille.
Harmelin-Vivien, M. L. (1981). Trophic relationships of reef fishes in Tuléar (Madagascar). *Oceanol. Acta* **4,** 365–374.
Harmelin-Vivien, M. L. (1983). Etude comparative de l'ichtyofaune des herbiers de phanérogames marines en milieux tropical et tempéré. *Rev. Ecol. (Terre Vie)* **38,** 179–210.

Harmelin-Vivien, M. L. (1989). Reef fish community structure: An Indo-Pacific comparison. *Ecol. Stud.* **69**, 21–60.

Harmelin-Vivien, M. L., and Bouchon-Navarro, Y. (1983). Feeding diets and significance of coral feeding among chaetodontid fishes in Moorea (French Polynesia). *Coral Reefs* **2**, 119–127.

Harper, E. M. (1991). The evolution of the cemented habit in bivalved molluscs. *Palaeontology* **34**, 455–460.

Harper, J. L. (1981). The meanings of rarity. *In* "The Biological Aspects of Rare Plant Conservation" (H. Synge ed.), pp. 190–203. Wiley, New York.

Harris, G. G. (1964). Considerations on the physics of sound production by fishes. *In* "Marine Bio-acoustics" (W. N. Tavolga, ed.), pp. 233–247. Pergamon Press, New York.

Harrison, S. (1991). Local extinction in a metapopulation context: An empirical evaluation. *Biol. J. Linn. Soc.* **42**, 73–88.

Harrison, S., and Cappuccino, N. (1995). Using density-manipulation experiments to study population regulation. *In* "Population Dynamics: New Approaches and Synthesis" (N. Cappuccino and P. W. Price, eds.), pp. 131–147. Academic Press, San Diego, CA.

Harrison, S., and Taylor, A. D. (1997). Empirical evidence for metapopulation dynamics. *In* "Metapopulation Biology: Ecology, Genetics, and Evolution" (I. Hanski and M. E. Gilpin, eds.), pp. 27–42. Academic Press, San Diego, CA.

Hart, A. M., and Russ, G. R. (1996). Response of herbivorous fishes to crown-of-thorns starfish *Acanthaster planci* outbreaks. III. Age, growth, mortality and maturity indices of *Acanthurus nigrofuscus*. *Mar. Ecol.: Prog. Ser.* **136**, 25–35.

Harvey, P. H., Read, A. F., and Nee, S. (1995). Why ecologists need to be phylogenetically challenged. *J. Ecol.* **83**, 535–536.

Hasler, A. D. (1957). The sense organs: Olfactory and gustatory senses of fishes. *In* "The Physiology of Fishes" (M. E. Brown, ed.), Vol. 2, pp. 187–209. Academic Press, New York.

Hassan E. S. (1989). Hydrodynamic imaging of the surroundings by the lateral line of the blind cave fish *Anoptichthys jordani*. *In* "The Mechanosensory Lateral Line: Neurobiology and Evolution" (S. Coombs, P. Görner, and H. Münz, eds.), pp. 217–228. Springer-Verlag, New York.

Hassell, M. P. (1986). Detecting density dependence. *Trends Ecol. Evol.* **1**, 90–93.

Hassell, M. P. (1987). Detecting regulation in patchily distributed animal populations. *J. Anim. Ecol.* **56**, 705–713.

Hassell, M. P., and May, R. M. (1974). Aggregation of predators and insect parasites and its effect on stability. *J. Anim. Ecol.* **43**, 567–594.

Hassell, M. P., Southwood, T. R. E., and Reader, P. M. (1987). The dynamics of the viburnum whitefly (*Aleurotrachelus jelinekii*): A case study of population regulation. *J. Anim. Ecol.* **56**, 283–300.

Hassell, M. P., Latto, J., and May, R. M. (1989). Seeing the wood for the trees: Detecting density dependence from existing life-table studies. *J. Anim. Ecol.* **58**, 883–892.

Hassell, M. P., May, R. M., Pacala, S. W., and Chesson, P. L. (1991). The persistence of host-parasitoid associations in patchy environments. I. A general criterion. *Am. Nat.* **138**, 568–583.

Hastings, A., and Botsford, L. W. (1999). Equivalence in yield from marine reserves and traditional fisheries management. *Science* **284**, 1537–1538.

Hastings, P. A. (1990). Phylogenetic relationships of tube blennies of the genus *Acanthemblemaria* (Pisces: Blennioidei). *Bull. Mar. Sci.* **47**, 725–738.

Hastings, P. A., and Springer, V. G. (1994). Review of *Stathmonotus,* with redefinition and phylogenetic analysis of the Chaenopsidae (Teleostei: Blennioidei). *Smithson. Contrib. Zool.* **558**, 1–48.

Hastings, R. W. (1997). The origin and seasonality of the fish fauna on a new jetty in the northeastern Gulf of Mexico. *Bull. Fla. State Mus. Biol. Sci.* **24**, 1–122.

Hatcher, B. G., Corless, M., Goodridge, R., Oxenford, H., and Scott, S. (1995). Testing mechanisms by which marine protected areas export fish to adjacent habitats: The Soufriere experiment in reef fisheries sustainability (SEFRS). *Proc. Annu. Gulf Caribb. Fish. Inst.* **48**, 101–118.

Haury, L. R., McGowen, J. A., and Wiebe, P. H. (1978). Patterns and processes in the time-space scales of plankton distributions. *In* "Spatial Pattern in Plankton Communities" (J. M. Steele, ed.), pp. 277–327. Plenum Press, New York.

Hawkins, A. D. (1973). The sensitivity of fish to sounds. *Oceanogr. Mar. Biol.* **11**, 291–340.

Hawkins, A. D. (1993). Underwater sound and fish behavior. *In* "The Behavior of Teleost Fishes" (T. J. Pitcher, ed.), pp. 127–168. Chapman & Hall, New York.

Hawkins, A. D., and Horner, K. (1981). Directional characteristics of primary auditory neurons from the codfish ear. *In* "Hearing and Sound Communication in Fishes" (W. N. Tavolga, A. N. Popper, and R. R. Fay, eds.), pp. 311–328. Springer-Verlag, New York.

Hawkins, A. D., and Myrberg, A. A., Jr. (1983). Hearing and sound communication underwater. *In*. "Bioacoustics: A Comparative Approach" (B. Lewis, ed.), pp. 347–405. Academic Press, London.

Hawkins, A. J. S. (1996). Temperature adaptation and genetic polymorphism in aquatic animals. *In*. "Animals and Temperature: Phenotypic and Evolutionary Adaptation" (I. A. Johnston and A. F. Bennett, eds.), pp. 103–125. Cambridge University Press, Cambridge, UK.

Hawkins, J. P., Roberts, C. M., and Clark, V. (2000). The threatened status of restricted range coral reef fish species. *Anim. Conserv.* **3**, 81–88.

Hay, M. E. (1994). Species as 'noise' in community ecology: Do seaweeds block our view of the kelp forest? *Trends Ecol. Evol.* **9**, 414–416.

Hay, M. E. (1996). Marine chemical ecology: What's known and what's next? *J. Exp. Mar. Biol. Ecol.* **200**, 103–134.

He, Ji. X. (2001). Age and size at first reproduction of fishes: Predictive models based only on growth trajectories. *Ecology* **82**, 784–791.

Heads, P. A., and Lawton, J. H. (1983). Studies on the natural enemy complex of the holly leaf-miner: The effects of scale on the detection of aggregative responses and the implications for biological control. *Oikos* **40**, 267–276.

Heath, M. R. (1992). Field investigations of the early life stages of marine fish. *Adv. Mar. Biol.* **28**, 1–174.

Heck, K. L., and Weinstein, M. P. (1989). Feeding habits of juvenile reef fishes associated with Panamanian seagrass meadows. *Bull. Mar. Sci.* **45**, 629–636.

Hedderson, T. A. (1992). Rarity at range limits; dispersal capacity and habitat relationships of extraneous moss species in a boreal Canadian National Park. *Biol. Conserv.* **59**, 113–120.

Hedgecock, D. (1986). Is gene flow from larval dispersal important in the adaptation and evolution of marine invertebrates? *Bull. Mar. Sci.* **39**, 550–564.

Hedgecock, D. (1994). Temporal and spatial genetic structure of marine animal populations in the California Current. *CalCOFI* **35**, 73–81.

Hedgecock, D., Chow, V., and Waples, R. E. (1992). Effective population numbers of shellfish broodstocks estimated from temporal variance in allelic frequencies. *Aquaculture* **108**, 215–232.

Heemstra, P. C., and Randall, J. E. (1993). "Groupers of the World (family Serranidae subfamily Epinephalinae)," FAO Species Catalogue, Vol. 16. FAO, Rome.

Heinisch, B. V., and Fable, W. A. (1999). Movement of Gag, *Mycteroperca microlepis,* in and from St. Andrew Bay, Florida. *Bull. Mar. Sci.* **64**, 501–508.

Helfman, G. S., Collette, B. B., and Facey, D. E. (1997). "The Diversity of Fishes." Blackwell Science, Oxford.

Heming, T. A., and Thurston, R. V. (1984). Physiological and toxic effects of cyanide to fishes: A review and recent advances. *Conf. Cyanide Environ.* Tucson, AZ, *1984*.

Hensley, D. A., Appeldoorn, R. S., Shaprio, D. Y., Ray, M., and Turingan, R. G. (1994). Egg dispersal in a Caribbean coral reef fish, *Thalassoma bifasciatum.* I. Dispersal over the reef platform. *Bull. Mar. Sci.* **54**, 256–270.

Heppell, S., Pfister, C. A., and Kroon, D. D. (2000). Elasticity analysis in population biology: Methods and applications. *Ecology* **81**, 605–606.

Hernamen, V., Munday, P. L., and Schlappy, M. L. (2000). Validation of otolith growth increment periodicity in tropical gobies. *Mar. Biol.* **137**, 715–726.

Hernandez, F. J., and Lindquist, D. G. (1999). A comparison of two light-trap designs for sampling larval and presettlement juvenile fish above a reef in Onslow Bay, North Carolina. *Bull. Mar. Sci.* **64**, 173–184.

Hernandez, L. P., and Motta, P. J. (1997). Trophic consequences of differential performance: Ontogeny of oral jaw-crushing performance in the sheepshead, *Archosargus probatocephalus.* (Teleostei, Sparidae). *J. Zool.* **243**, 737–756.

Hernandez-Leon, S. (1988). Gradients of mesozooplankton biomass and ETS activity in the wind-shear area as evidence of an island mass effect in the Canary Island waters. *J. Plankton Res.* **10**, 1141–1154.

Hernandez-Leon, S. (1991). Accumulation of mesoplankton in a wake area as a conservative mechanism of the island-mass effect. *Mar. Biol.* **109**, 141–147.

Hess, H. C. (1993). Male mouthbrooding in jawfishes (Opistognathidae): Constraints on polygyny. *Bull. Mar. Sci.* **52**, 806–818.

Heywood, K. J., Barton, E. D., and Simpson, J. H. (1990). The effects of flow distribution by an oceanic island. *J. Mar. Res.* **48**, 55–73.

Heywood, K. J., Scrope-Howe, S., and Barton, E. D. (1991). Estimation of zooplankton biomass from shipborne ADCP backscatter. *Deep-Sea Res.* **38**, 677–691.

Hiatt, R. W., and Strasburg, D. W. (1960). Ecological relationships of fish fauna on coral reefs of the Marshall Islands. *Ecol. Monogr.* **30**, 65–127.

Higgs, D. M., and Fuiman, L. A. (1996). Ontogeny of visual and mechanosensory structure and functions in Atlantic menhaden. *J. Exp. Biol.* **199**, 2619–2629.

Higgs, D. M., and Fuiman, L. A. (1998a). Associations between sensory development and ecology in three species of clupeoid fish. *Copeia,* pp. 133–144.

Higgs, D. M., and Fuiman, L. A. (1998b). Associations between behavioral ontogeny and habitat change in clupeoid larvae. *J. Mar. Biol. Assoc. U. K.* **78**, 1281–1294.

Hilborn, R., and Walters, C. J. (1992). "Quantitative Fisheries Stock Assessment. Choice, Dynamics and Uncertainty." Chapman & Hall, New York.

Hilborn, R., Maguire, J. J., Parma, A. M., and Rosenberg, A. A. (2001). The Precautionary Approach and risk management: Can they increase the probability of successes in fishery management? *Can. J. Fish. Aquat. Sci.* **58**, 99–107.

Hingco, T. G., and Rivera, R. (1991). Aquarium fish industry in the Philippines: Toward development or destruction? *ICLARM Conf. Proc.* **22**, 249–253.

Hislop, J. R. G., Robb, A. P., and Gauld, J. A. (1978). Observations on effects of feeding level on growth and reproduction in haddock, *Melanogrammus aeglefinus* (L.) in captivity. *J. Fish Biol.* **13**, 85–89.

Hitchcock, G. L., Arnold, W. S., and Cowen, R. K. (in review). Role of diffusion in the dispersion of larval *Mercenaria mercenaria* (Linne) in the Indian River lagoon, Florida, USA. (In review).

Hixon, M. A. (1991). Predation as a process structuring coral reef fish communities. *In* "The Ecology of Fishes on Coral Reefs" (P. F. Sale, ed.), pp. 475–508. Academic Press, San Diego, CA.

Hixon, M. A. (1996a). What is "recruitment limitation" anyway? *Reef Encounter* **19**, 5–6.

Hixon, M. A. (1996b). Effects of reef fishes on corals and algae. *In* "Life and Death on Coral Reefs" (C. Birkeland, ed.), pp. 230–248. Chapman & Hall, London.

Hixon, M. A. (1998). Population dynamics of coral-reef fishes: Controversial concepts and hypotheses. *Aust. J. Ecol.* **23**, 192–201.

Hixon, M. A., and Beets, J. P. (1989). Shelter characteristics and Caribbean fish assemblages: Experiments with artificial reefs. *Bull. Mar. Sci.* **44**, 666–680.

Hixon, M. A., and Beets, J. P. (1993). Predation, prey refuges, and the structure of coral-reef fish assemblages. *Ecol. Monogr.* **63**, 77–101.

Hixon M. A., and Carr, M. H. (1997). Synergistic predation, density dependence, and population regulation in marine fish. *Science* **277**, 946–949.

Hixon, M. A., and Menge, B. A. (1991). Species diversity: Prey refuges modify the interactive effects of predation and competition. *Theor. Popul. Biol.* **39**, 178–200.

Hjort, J. (1914). Fluctuation in the great fisheries of northern Europe reviewed in the light of biological research. *Rapp. P.-V. Reun., Cons. Perm. Int. Explor. Mer.* **20**, 1–228.

Hobson, E. S. (1963). Feeding behavior in three species of sharks. *Pac. Sci.* **17**, 171–194.

Hobson, E. S. (1972). Activity of Hawaiian reef fishes during the evening and morning transitions between daylight and darkness. *Fish. Bull.* **70**, 715–740.

Hobson, E. S. (1974). Feeding relationships of teleostean fishes on coral reefs in Kona, Hawaii. *Fish. Bull.* **72**, 915–1031.

Hobson, E. S. (1991). Trophic relationships of fishes specialized to feed on zooplankters above coral reefs. *In* "The Ecology of Fishes on Coral Reefs" (P. F Sale, ed.), pp. 69–95. Academic Press, San Diego, CA.

Hobson, E. S. (1994). Ecological relations in the evolution of acanthopterygian fishes in warm-temperate communities of the northern Pacific. *Environ. Biol. Fishes* **40**, 49–90.

Hobson, E. S., and Chess, J. R. (1978). Trophic relationships among fishes and plankton in the lagoon at Enewetak Atoll, Marshall Islands. *Fish. Bull.* **76**, 133–153.

Hobson, E. S., and Chess, J. R. (1986). Relationships among fishes and their prey in a nearshore sand community off southern California. *Environ. Biol. Fishes* **17**, 201–226.

Hodgson, J. G. (1986). Commonness and rarity in plants with special reference to the Sheffield flora. Part I: The identity, distribution and habitat characteristics of the common and rare species. *Biol. Conserv.* **36**, 199–252.

Hoffman, S. G. (1983). Sex-related foraging behavior in sequentially hermaphroditic hogfishes (*Bodianus* spp.). *Ecology* **64**, 798–808.

Hoffman, S. G. (1985). Effects of size and sex on the social organization of reef-associated hogfishes, *Bodianus* spp. *Environ. Biol. Fishes* **14**, 185–197.

Hofmann, E. E., and Lascara, C. M. (1998). An overview of interdisciplinary modeling for marine ecosystems. *In* "The Sea, The Global Coastal Ocean: Processes and Methods" (A. R. Robinson and K. H. Brink, eds.), Vol. 10, pp. 507–540. Wiley, New York.

Hogg, N. G. (1972). Steady flow past an island with application to Bermuda. *Geophys. Fluid Dyn.* **4**, 55–81.

Hogg, N. G. (1980). Effects of bottom topography on ocean currents. *GARP Pub. Ser.* **23**, 6659–6682.

Hogg, N. G., Katz, E. J., and Sanford, T. B. (1978). Eddies, islands, and mixing. *J. Geophys. Res.* **83**, 2921–2938.

Högstedt, G. (1983). Adaptation unto death: Function of fear screams. *Am. Nat.* **121**, 562–570.

Holbrook, S. J., and Schmitt, R. J. (1988). Effects of predation risk on foraging behavior: mechanisms altering patch choice. *J. Exp. Mar. Biol. Ecol.* **121**, 151–63.

Holbrook, S. J., and Schmitt, R. J (1997). Settlement patterns and process in a coral reef damselfish: *In situ* nocturnal observations using infared video. *Proc. 8th Int. Coral Reef Symp*, Vol. 2, pp. 1143–1148.

Holbrook, S. J., and Schmitt, R. J. (1999). *In situ* nocturnal observations of reef fishes using infared video. *In* "Proceedings of the 5th Indo-Pacific Fish Conference, Nouméa, 1997" (B. Séret and J.-Y. Sire, eds), pp. 805–812. Société Française d'Ichtyologie & Institut de Recherche pour le Développement, Paris.

Holbrook, S. J., Carr, M. H., Schmitt, R. J., and Coyer, J. A. (1990). Effect of giant kelp on local abundance of reef fishes: The importance of ontogenetic resource requirements. *Bull. Mar. Sci.* **47**, 104–114.

Holbrook, S. J., Forrester, G. E., and Schmitt, R. J. (2000). Spatial patterns in abundance of a damselfish reflect availability of suitable habitat. *Oecologia* **122**, 109–120.

Holland, D. S., and Brazee, R. J. (1996). Marine reserves for fisheries management. *Mar. Res. Econ.* **11**, 157–171.

Holland, K. N. (1975). A behavioral and electrophysiological investigation of the taste function of the barbels of an Hawaiian goatfish. Unpublished MSc Thesis, University of Hawaii, Honolulu.

Holland, K. N. (1978). Chemosensory orientation to food by a Hawaiian goatfish (*Parupeneus prophyreus,* Mullidae). *J. Chem. Ecol.* **4**, 173–186.

Holland, K. N., Peterson, J. D., and Wetherbee, B. M. (1993). Movements, distribution and growth rates of the white goatfish *Mulloides flavolineatus* in a fisheries conservation zone. *Bull. Mar. Sci.* **52**, 982–992.

Holland, K. N., Lowe, C. G., and Wetherbee, B. M. (1996). Movements and dispersal patterns of blue trevally (*Caranx melampygus*) in a fisheries conservation zone. *Fish. Res.* **25**, 279–292.

Holldobler, B., and Wilson, E. O. (1990). "The Ants." Harvard University Press, Cambridge, MA.

Holling, C. S. (1959). The components of predation as revealed by a study of small-mammal predation of the European pine sawfly. *Can. Entomol.* **91**, 293–320.

Holm, E. R. (1990). Effects of density-dependent mortality on the relationship between recruitment and larval settlement. *Mar. Ecol.: Prog. Ser.* **60**, 141–146.

Holt, R. D. (1997). Rarity and evolution: Some theoretical considerations. *In* "The Biology of Rarity" (W. E. Kunin and K. J. Gaston, eds.), pp. 210–234. Chapman & Hall, London.

Holthus, P. (1999). Conservation, culture and certification: The future of marine ornamentals. *Proc. First Int. Conf. Marine Ornamentals: Collection, Culture and Conservation.* Kailua-Kona, Hawaii, Nov. 16–19, 1999. Abstracts, p. 5.

Holyoak, M. (1993). The frequency of detection of density dependence in insect orders. *Ecol. Entomol.* **18**, 339–347.

Hood, P. B., and Johnson, A. K. (1999). Age, growth, mortality, and reproduction of vermilion snapper, *Rhomboplites aurorubens,* from the eastern Gulf of Mexico. *Fish. Bull.* **97,** 828–841.

Hood, P. B., and Schleider, R. A. (1992). Age, growth and reproduction of gag *Mycteroperca microlepis* (Pisces:Serranidae) in the eastern Gulf of Mexico. *Bull. Mar. Sci.* **51,** 337–352.

Hopkins, C. D. (1972). Sex differences in electric signaling in an electric fish. *Science* **176,** 1035–1037.

Hopkins, C. D. (1988). Social communication in the aquatic environment. *In* "Sensory Biology of Aquatic Animals" (J. Atema, R. R. Fay, A. N. Popper, and W. N. Tavolga, eds.), pp. 233–268. Springer-Verlag, New York.

Hopkins, C. D., Shieh, K-T., McBride, D. W., Jr., and Winslow, M. (1997). A quantitative analysis of passive electrolocation behavior in electric fish. *Brain, Behav. Evol.* **50**(Suppl. 1), 32–59.

Hopper, K. R., Powell, J. E., and King, E. G. (1991). Spatial density dependence in parasitism of *Heliothis virescens* (Lepidoptera, Noctuidae) by *Microplitis croceipes* (Hymenoptera, Braconidae) in the field. *Environ. Entomol.* **20,** 292–302.

Horn, M. H. (1989). Biology of marine herbivorous fishes. *Oceanogr. Mar. Biol.* **27,** 167–272.

Horn, M. H., and Gibson, R. N. (1990). Effects of temperature on the food processing of three species of seaweed-eating fishes from European coastal waters. *J. Fish Biol.* **37,** 237–247.

Houde, E. D. (1987). Fish early life dynamics and recruitment variability. *Am. Fish. Soc. Symp. Ser.* **2,** 17–29.

Houde, E. D. (1989). Comparative growth, mortality and energetics of marine fish larvae: Temperature and implied latitudinal effects. *Fish Bull.* **87,** 471–495.

Houde, E. D. (1997). Patterns and consequences of selective processes in teleost early life histories. *In* "Early Life History Traits and Recruitment among Coral Reef Fishes" (R. C. Chambers and E. A. Trippel, eds.), pp. 173–196. Chapman & Hall, London.

Houde, E. D., and Zastrow, C. E. (1993). Ecosystem- and taxon-specific dynamic and energetics properties of larval fish assemblages. *Bull. Mar. Sci.* **53,** 290–335.

Houghton, R. W., and Visbeck, M. (1998). Upwelling and convergence in the Middle Atlantic Bight shelfbreak front. *Geophys. Res. Lett.* **25,** 2765–2768.

Hourigan, T. F. (1989). Environmental determinants of butterflyfish social systems. *Environ. Biol. Fishes* **25,** 61–78.

Hourigan, T. F., and Reese, E. S. (1987). Mid-ocean isolation and the evolution of Hawaiian reef fishes. *Trends Ecol. Evol.* **2,** 187–191.

Huckins, C. J. F. (1997). Functional linkages among morphology, feeding performance, diet, and competitive ability in molluscivorous sunfish. *Ecology* **78,** 2401–2414.

Hudson, E. J., and Mace, G. M. (1996). "Marine Fish and the IUCN Red List of Threatened Animals," Report of the Workshop held in Collaboration with the WWF and IUCN at the Zoological Society of London.

Hughes, T. P. (1984). Population dynamics based on individual size rather than age: A general model with a coral reef example. *Am. Nat.* **123,** 778–795.

Hughes, T. P. (1990). Recruitment limitation, mortality, and population regulation in open systems: A case study. *Ecology* **71,** 12–20.

Hughes, T. P. (1994). Catastrophies, phase shifts, and large-scale degradation of a Caribbean coral reef. *Science* **265,** 1547–1551.

Hughes, T. P., Baird, A. H., Dinsdale, E. A., Moltschaniwskyj, N. A., Pratchett, M. S., Tanner, J. E., and Willis, B. L. (1999). Patterns of recruitment and abundance of corals along the Great Barrier Reef. *Nature (London)* **397,** 59–63.

Humphreys, C. J., and Parenti, L. R. (1986). "Cladistic Biogeography," Oxford Monogr. Biogeogr. No. 2. Oxford University Press, Oxford.

Hunt von Herbing, I., and Boutilier, R. G. (1996). Activity and metabolism of larval Atlantic cod (*Gadus morhua*) from Scotian Shelf and Newfoundland source populations. *Mar. Biol.* **124,** 607–617.

Hunt von Herbing, I., and Hunte, W. (1991). Spawning and recruitment of the bluehead wrasse *Thalassoma bifasciatum* in Barbados, West Indies. *Mar. Ecol.: Prog. Ser.* **72,** 49–58.

Hunt von Herbing, I., Boutilier, R. G., Miyake, T., and Hall, B. K. (1996). Effects of temperature on morphological landmarks critical to growth and survival in larval Atlantic Cod *(Gadus morhua). Mar. Biol.* **124,** 593–606.

Hunte, W., and Côté, I. M. (1989). Recruitment in the redlip blenny *Ophioblennius atlanticus:* Is space limiting? *Coral Reefs* **8,** 45–50.

Hunter, J. R. (1981). Feeding ecology and predation of marine fish larvae. *In* "Marine Fish Larvae: Morphology, Ecology, and Relation to Fisheries" (R. Lasker, ed.), pp. 34–77. Washington Sea Grant Program, University of Washington, Seattle.

Huntsman, G. R., and Schaaf, W. E. (1994). Simulations of the impact of fishing on reproduction of a protogynous grouper, the graysby. *North Am. J. Fish. Manage.* **14,** 41–52.

Huntsman, G. R., Vaughan, D. S., and Potts, J. (1995). "Trends in Population Status of the Red Porgy, *Pagrus pagrus,* in the Atlantic Ocean off North Carolina and South Caroline, USA, 1972–1992." South Atlantic Fishery Management Council, Charleston, SC.

Huntsman, G. R., Potts, J., Mays, R. W., and Vaughan, D. S. (1999). Groupers (Serranidae, Epinephelinae): Endangered apex predators of reef communities. *In* "Life in the Slow Lane. Ecology and Conservation of Long-Lived Marine Animals" (J. A. Musick ed.), pp. 217–232. American Fisheries Society, Bethesda, MD. December, 1997, Hong Kong, pp. 25–34. Agriculture and Fisheries Department, Hong Kong, China.

Huston, M. A. (1979). A general hypothesis of species diversity. *Am. Nat.* **113,** 39–59.

Hutchings, J. A. (2000). Collapse and recovery of marine fishes. *Nature (London)* **406,** 882–885.

Hutchings, P. A. (1986). Biological destruction on coral reefs—a review. *Coral Reefs* **4**, 239–252.

Hutchins, J. B., and Pearce, A. F. (1994). Influence of the Leeuwin current on the recruitment of tropical reef fishes at Rottnest Island, Western Australia. *Bull. Mar. Sci.* **54**, 245–255.

Hwang, P. P., Wu, S. M., Lin, J. H., and Wu, L. S. (1992). Cortisol content of eggs and larvae of teleosts. *Gen. Comp. Endocrinol.* **86**, 189–196.

Iacumin, P., Bianucci, G., and Longinelli, A. (1992). Oxygen and carbon isotopic composition of fish otoliths. *Mar. Biol.* **113**, 537–542.

Ibara, R. M., Penny, L. T., Ebeling, A. W., van Dyhuiszen, G., and Cailliet, G. (1983). The mating call of the plainfin midshipman fish, *Porichthys notatius*. *In* "Predators and Prey in Fishes" (D. L. G. Noakes, D. G. Lindquist, G. S. Helfman, and J. A. Ward, eds.), pp. 205–212. Junk Publ., The Hague.

Idler, D. R., Fagerlund, U. H. M., and Mayoh, H. (1956). Olfactory perception in migrating salmon. 1. L-serine, a salmon repellent in mammalian skin. *J. Gen. Physiol.* **39**, 889–892.

IGOSS (1998). The Climate Library of the International Research Institute for Climatic Predictions/Lamont Doherty Earth Observatory. Available at URL: *http://ingrid.Igdo.columbia.edu/SOURCES/.IGOSS/*

Indrawan, M. (1999). Live reef food fish trade in the Banggai Islands (Sulawesi, Indonesia): A case study. *Secretariat of the Pacific Community Live Reef Fish Inf. Bull.* **6**, 7–14.

International Trade Subgroup (2000). "International Trade in Coral and Coral Reef Species: The Role of the United States." Draft Report of the Trade Subgroup of the International Working Group to the U.S. Coral Reef Task Force, Washington, DC.

Ishioka, H., Hatakeyama, Y., and Sakaguchi, S. (1987). Development of the startle response to sound stimulation in the red sea bream, *Pagrus major* Temminck et Schlegel. *Bull. Nansei Reg. Fish. Res. Lab.* **21**, 17–23.

Isles, T. D., and Sinclair, M. (1982). Atlantic herring: Stock discreteness and abundance. *Science* **215**, 627–633.

IUCN (1994). "Guidelines for Protected Area Management Categories." International Union for the Conservation of Nature and Natural Resources, Gland, Switzerland.

Ivany, L. C., Patterson, W. P., and Lohmann, K. C. (2000). Cooler winters as a possible cause of mass extinctions at the Eocene/Oligocene boundary. *Nature (London)* **407**, 887–890.

Jablonski, D. (1986). Larval ecology and macroevolution in marine invertebrates. *Bull. Mar. Sci.* **39**, 565–587.

Jablonski, D. (1987). Heritability at the species level: Analysis of geographic range of Cretaceous molluscs. *Science* **238**, 360–363.

Jablonski, D. (1993). The tropics as a source of evolutionary novelty through geological time. *Nature (London)* **364**, 142–144.

Jackson, D. A., and Harvey, H. H. (1989). Biogeographic associations in fish assemblages: Local vs. regional processes. *Ecology* **70**, 1472–1484.

Jackson, J. B. C. (1991). Adaptation and diversity of reef corals. *Science* **41**, 475–482.

Jackson, J. B. C. (1994). Constancy and change of life in the sea. *Philos. Trans. R. Soc. London, Ser. B* **344**, 55–60.

Jackson, J. B. C. (1997). Reefs since Columbus. *Coral Reefs* **16**, S23–S32.

Jackson, J. B. C., and Hughes, T. P. (1985). Adaptive strategies of coral-reef invertebrates. *Am. Sci.* **73**, 265–274.

Jackson, J. B. C., and McKinney, F. K. (1990). Ecological processes and progressive macroevolution of marine clonal benthos. *In* "Causes of Evolution: A Paleontological Perspective" (R. M. Ross and W. D. Allmon, eds.), pp. 173–209. University of Chicago Press, Chicago.

Jackson, J. B. C., Jung, P., Coates, A. G., and Collins, L. S. (1993). Diversity and extinction of tropical American mollusks and emergence of the Isthmus of Panama. *Science* **260**, 1624–1626.

Jackson, J. B. C., Budd, A. F., and Coates, A. G., eds. (1996). "Evolution and Environment in Tropical America." University of Chicago Press, Chicago.

Jackson, L. L., Baedecker, P. A., Fries, T. L., and Lamothe, P. J. (1993). Geological and inorganic materials. *Anal. Chem.* **65**, 12–28.

Jaffe, J. S., Reuss, E., McGehee, D., and Chandran, G. (1995). FTV: Sonar for tracking macrozooplankton in three dimensions. *Deep-Sea Res.* 1. **42**, 1495–1512.

James, M. K., Armsworth, P. R., Mason, L. B., and Bode, L. (in review). The structure of reef fish metapopulations: Modeling larval dispersal and retention patterns (submitted for publication).

Jamieson, I. (1996). Do female fish prefer to spawn in nests with eggs for reasons of mate choice copying or egg survival? *Am. Nat.* **145**, 824–832.

Janssen, J. (1990). Localization of substrate vibrations by the mottled sculpin *(Cottus bairdi)*. *Copeia*, pp. 349–355.

Janssen, J., Jones, W. R., Whang, A., and Oshel, P. E. (1995). Use of the lateral-line in particulate feeding in the dark by juvenile alewife *(Alosa pseudoharengus)*. *Can. J. Fish. Aquat. Sci.* **52**, 358–363.

Jarrige, F., Rotschi, H., Krause, G., and Rual, P. (1975). Courants dans le lagon de Noumea (Nouvelle Caledonie). ORSTOM and Horace Lamb Institute of Oceanography, Flinders University of South Australia, Australia.

Jaume, D., and Boxshall, G. A. (1996). The persistence of an ancient marine fauna in Mediterranean waters: New evidence from misophrioid copepods living in anchihaline caves. *J. Nat. Hist.* **30**, 1583–1595.

Jenkins, G. P., and May, H. M. A. (1994). Variation in settlement and larval duration of King George whiting, *Sillaginodes punctata* (Sillaginidae), in Swan Bay, Victoria, Australia. *Bull. Mar. Sci.* **54**, 281–296.

Jenkins, T. M., Diehl, S., Kratz, K. W., and Cooper, S. D. (1999). Effects of population density on individual growth of brown trout in streams. *Ecology* **80**, 941–956.

Jennings, S., and Kaiser M. J. (1998). The effects of fishing on marine ecosystems. *Adv. Mar. Biol.* **34,** 201–351.

Jennings, S., and Lock, J. M. (1996). Population and ecosystem effects of reef fishing. *In* "Reef Fisheries" (N. V. C. Polunin and C. M. Roberts, eds.), pp. 193–218. Chapman & Hall, London.

Jennings, S., and Polunin N. V. C. (1996). Impacts of fishing on tropical reef ecosystems. *Ambio* **25,** 44–49.

Jennings, S., Marshall, S. S., and Polunin, N. V. C (1996). Seychelles' marine protected areas: Comparative structure and status of reef fish communities. *Biol. Conserv.* **75,** 201–209.

Jennings, S., Reynolds, J. D., and Polunin, N. V. C. (1999). Predicting the vulnerability of tropical reef fishes to exploitation with phylogenies and life histories. *Conserv. Biol.* **13,** 1466–1475.

Jernavall, J., Hunter, J. P., and Fortelius, M. (1996). Molar tooth diversity, disparity, and ecology in Cenozoic ungulate radiations. *Science* **274,** 1489–1492.

Job, S. D., and Bellwood, D. R. (1996). Visual acuity and feeding in larval *Premnas biaculeatus*. *J. Fish Biol.* **48,** 952–963.

Job, S. D., and Bellwood, D. R. (2000). Light sensitivity in larval fishes: Implications for vertical zonation in the pelagic zone. *Limnol. Oceanogr.* **45,** 362–371.

Johannes, R. E. (1978). Reproductive strategies of coastal marine fishes in the tropics. *Environ. Biol. Fishes* **3,** 65–84.

Johannes, R. E. (1981). "Words of the Lagoon. Fishing and Marine Lore in the Palau District of Micronesia." University of California Press, Los Angeles.

Johannes, R. E. (1998). The case for data-less marine resource management: Examples from tropical nearshore finfisheries. *Trends Ecol. Evol.* **13,** 243–246.

Johannes, R. E., and Lam, M. (1999). The live reef food fish trade in the Solomon Islands. *Secretariat of the Pacific Community Live Reef Fish Inf. Bull.* **5,** 8–15.

Johannes, R. E., and Ogburn, N. J. (1999). Collecting grouper seed for aquaculture in the Philippines. *Secretariat of the Pacific Community Live Reef Fish Inf. Bull.* **6,** 35–48.

Johannes, R. E., and Reipen, M. (1995). "Environmental, Economic, and Social Implications of the Live Reef Fish Trade in Asia and the Western Pacific." Report to The Nature Conservancy and the Forum Fisheries Agency.

Johannes, R. E., Squire, L., Graham, T., Sadovy, Y., and Renguul, H. (1999). "Spawning Aggregations of Groupers (Serranidae) in Palau," Mar. Conserv. Res. Ser. Publ. No. 1. The Nature Conservancy, Honolulu, Hawaii.

Johnson, A. G., and Koenig, C. C. (in press). Age and size structure of the fishery and juvenile abundance of gag, *Mycteroperca microlepis,* from the northeastern Gulf of Mexico. *Proc. Annu. Gulf Carib. Fish. Inst.* **47,** (in press).

Johnson, C. N. (1998). Rarity in the tropics: Latitudinal gradients in distribution and abundance in Australian mammals. *J. Anim. Ecol.* **67,** 689–698.

Johnson, D. G. (1981). The limits and relationships of the Lutjanidae and associated families. *Bull. Scripps Inst. Oceanogr.* **24,** 1–114.

Johnson, G. D. (1993). Percomorph phylogeny: Progress and problems. *Bull. Mar. Sci.* **52,** 3–28.

Johnson, G. D., and Patterson, C. (1993). Percomorph phylogeny: A survey of acanthomorphs and a new proposal. *Bull. Mar. Sci.* **52,** 554–626.

Johnson, J. H. (1961). "Limestone-Building Algae and Algal Limestones." Johnson Publishing Company, Boulder, CO.

Johnson, K. G., Budd, A. F., and Stemann, T. A. (1995). Extinction selectivity and ecology of Neogene Caribbean reef corals. *Paleobiology* **21,** 52–73.

Johnson, M. (2000). A reevaluation of density-dependent population cycles in open systems. *Am. Nat.* **155,** 36–45.

Johnson, M. S., Hebbert, D. R., and Moran, M. J., (1993). Genetic analysis of populations of North-western Australian fish species. *Aust. J. Mar. Freshwater Res.* **44,** 673–685.

Johnson, M. S., Watts, R. J., and Black, R. (1994). High levels of genetic subdivision in peripherally isolated populations of the atherinid fish *Craterocephalus capreoli* in the Houtman Abrohlos Islands, Western Australia. *Mar. Biol.* **119,** 179–184.

Jokiel, P., and Martinelli, F. J. (1992). The vortex model of coral reef biogeography. *J. Biogeogr.* **19,** 449–458.

Jolicoeur, P. (1975). Linear regression in fishery research: Some comments. *J. Fish Res. Board Can.* **32,** 1491–1494.

Jones, C. M. (1986). Determining age of larval fish with the otolith increment technique. *Fish. Bull.* **84,** 91–104.

Jones, C. M. (2000). Fitting growth curves to retrospective size-at-age data. *Fish. Res.* **46,** 123–130.

Jones, C. M., and Wells, B. (1998). Age, growth, and mortality of Black Drum, *Pogonias cromis,* in the Chesapeake Bay region. *Fish. Bull.* **96,** 451–461.

Jones, G. P. (1984). Population ecology of the temperate reef fish *Pseudolabrus celidotus* Block & Schneider (Pisces: Labridae). II. Factors influencing adult density. *J. Exp. Mar. Biol. Ecol.* **75,** 277–303.

Jones, G. P. (1986). Food availability affects growth in a coral reef fish. *Oecologia* **70,** 136–139.

Jones, G. P. (1987a). Some interactions between residents and recruits in two coral reef fishes. *J. Exp. Mar. Biol. Ecol.* **114,** 169–182.

Jones, G. P. (1987b). Competitive interactions among adults and juveniles in a coral reef fish. *Ecology* **68,** 1534–1547.

Jones, G. P. (1988a). Ecology of rocky reef fish of north-eastern New Zealand: A review. *Mar. Freshwater Res.* **22,** 445–462.

Jones, G. P. (1988b). Experimental evaluation of the effects of habitat structure and competitive interactions on the juveniles of two coral reef fishes. *J. Exp. Mar. Biol. Ecol.* **123,** 115–126.

Jones, G. P. (1990). The importance of recruitment to the dynamics of a coral reef fish population. *Ecology* **71,** 1691–1698.

Jones, G. P. (1991). Post-recruitment processes in the ecology of coral reef fish populations: A multifactorial perspective.

In "The Ecology of Fishes on Coral Reefs" (P. F. Sale, ed.), pp. 294–328. Academic Press, San Diego, CA.

Jones, G. P. (1997). Relationships between recruitment and postrecruitment processes in lagoonal populations of two coral reef fishes. *J. Exp. Mar. Biol. Ecol.* **213,** 231–246.

Jones, G. P., and Kaly, U. L. (1995). Conservation of rare, threatened and endemic marine species in Australia. *In* "The State of the Marine Environment Report for Australia Tech. Annexe 1, (L. P. Zann and P. Kailola, eds.), Ocean Rescue 2000, pp. 183–191. Department of the Environment, Sport and Territories, Canberra.

Jones, G. P., and Milicich, M. (1995). Dispersal of coral reef fish larvae: Application of otolith marking techniques. *Int. Larval Fish Conf.,* Sydney, Australia, 1995, p. 61.

Jones, G. P., and Milicich, M. J. (1997). Dispersal of reef fish larvae: Application of an otolith marking technique. *Abstr., 8th Int. Coral Reef Symp.,* p. 101.

Jones, G. P., and Syms, C. (1998). Disturbance, habitat structure and the ecology of fishes on coral reefs. *Aust. J. Ecol.* **23,** 287–297.

Jones, G. P., Cole, R. C., and Battershill, C. N. (1993). Marine reserves: do they work? *In* "The Ecology of Temperate Reefs" (C. N. Battershill, D. R. Schiel, G. P. Jones, R. G. Creese, and A. B. MacDiarmid eds.), Proc. 2nd Int. Temp. Reef. Symp., pp. 29–45. NIWA Publications, Wellington.

Jones G. P., Doherty P. J., Mapstone B. D. and Howlett L. (1998). "ReeFish '95: Recruitment and Population Dynamics of Coral Reef Fishes." CRC Reef Research Centre, Townsville, Australia.

Jones, G. P., Milicich, M. I., Emslie, M. J., and Lunow, C. (1999). Self-recruitment in a coral reef fish population. *Nature (London)* **402,** 802–804.

Jones, R. J., and Hoegh-Guldberg, O. (1999). Effects of cyanide on coral photosynthesis: Implications for identifying the cause of coral bleaching and for assessing the environmental effects of cyanide fishing. *Mar. Ecol.: Prog. Ser.* **177,** 83–91.

Jones, R. J., and Steven, A. L. (1997). Effects of cyanide on corals in relation to cyanide fishing on reefs. *Mar. Freshwater Res.* **48,** 517–522.

Jones, R. S. (1968). Ecological relationships in Hawaiian and Johnston Island Acanthuridae (surgeonfishes). *Micronesica* **4,** 309–361.

Jónsson, L. (1980). Chemical stimuli: Role in the behavior of fishes. *In* "Environmental Physiology of Fishes" (M. A. Ali, ed.), pp. 353–367. Plenum Press, New York.

Jouvenel, J. Y. (1997). Ichtyofaune de la côte rocheuse des Albères (Méditerranée N.-O., France). Ph.D. Thesis, EPHE, Perpignan.

Juanes, F., and Conover, D. O. (1994). Piscivory and prey size selection in young-of-the-year bluefish: Predator preference or size-dependent capture success? *Mar. Ecol.: Prog. Ser.* **114,** 59–69.

Juliano, S. A. (1983). Body size, disperal ability, and range size in North American species of *Brachinus* (Coleoptera: Carabidae). *Coleopt. Bull.* **37,** 232–238.

Kalish, J. M. (1989). Otolith microchemistry: Validation of the effects of physiology, age and environment on otolith composition. *J. Exp. Mar. Biol. Ecol.* **132,** 151–178.

Kalish, J. M. (1991a). Determinants of otolith chemistry: Seasonal variation in the composition of blood plasma, endolymp and otoliths of bearded rock cod *Pseudophycis barbatus*. *Mar. Ecol.: Prog. Ser.* **74,** 137–159.

Kalish, J. M. (1991b). ^{13}C and ^{18}O isotopic disequilibria in fish otoliths: Metabolic and kinetic effects. *Mar. Ecol.: Prog. Ser.* **75,** 191–203.

Kalish. J. M. (1991c). Oxygen and carbon stable isotopes in the otoliths of wild and laboratory-reared Australian salmon (*Arripis trutta*). *Mar. Biol.* **110,** 37–47.

Kalish, J. M. (1993). Pre- and post-bomb radiocarbon in fish otoliths. *Earth Planet. Sci. Lett.* **114,** 549–554.

Kalish, J. M. (1995). Application of the bomb radiocarbon chronometer to the validation of redfish *Centroberyx affinis* age. *Can. J. Fish. Aquat. Sci.* **52,** 1399–1405.

Kalish, J. M., Beamish, R. J., Brothers, E. B., Casselman, J. M., Francis, R. I. C. C., Mosegaard, H., Panfili, J., Prince, E. D., Thresher, R. E., Wilson, C. A., and Wright, P. J. (1995). Glossary of otolith studies. *In* "Recent Developments in Fish Otolith Research" (D. H. Secor, J. M. Dean, and S. E. Campana, eds.), pp. 723–729. University of South Carolina Press, Columbia.

Kalish, J. M., Johnston, J. M., Gunn, J. S., and Clear, N. P. (1996). Use of the bomb radiocarbon chronometer to determine age of southern bluefin tuna *Thunnus maccoyii*. *Mar. Ecol.: Prog. Ser.* **143,** 1–8.

Kalish, J. M., Johnston, J. M., Smith, D. C., Morrison, A. K., and Robertson, S. G. (1997). Use of the bomb radiocarbon chronometer for age validation in the blue grenadier *Macruronus novaezelandiae*. *Mar. Biol.* **128,** 557–563.

Kalmijn, A. J. (1971). The electric sense of sharks and rays. *J. Exp. Biol.* **55,** 371–383.

Kalmijn, A. J. (1974). The detection of electric fields from inanimate and animate sources other than electric organs. *In* "Handbook of Sensory Physiology" (A. Fessard, ed.), Vol. 3, Part 3, pp. 147–200. Springer-Verlag, New York.

Kalmijn, A. J. (1977). The electric and magnetic sense of sharks, skates, and rays. *Oceanus* **20,** 45–52.

Kalmijn, A. J. (1978). Electrical and magnetic sensory world of sharks, skates, and rays. *In* "Sensory Biology of Sharks, Skates, and Rays" (E. S. Hodgson and R. F. Mathewson, eds.), pp. 507–528. U.S. Govt. Printing Office, Washington, DC.

Kalmijn, A. J. (1981). Biophysics of geomagnetic field detection. *IEEE Trans. Magn.* **17,** 1113–1124.

Kalmijn, A. J. (1982). Electric and magnetic field detection in elasmobranch fishes. *Science* **218,** 915–918.

Kalmijn, A. J. (1984). Theory of electromagnetic orientation: a further analysis. *In* "Comparative Physiology of Sensory Systems" (L. Boils, R. D. Keynes, and S. H. P. Maddrell, eds.), pp. 525–560. Cambridge University Press, New York.

Kalmijn, A. J. (1988a). Hydrodynamic and acoustic field detection. *In* "Sensory Biology of Aquatic Animals"

(J. Atema, R. R. Fay, A. N. Popper, and W. N. Tavolga, eds.), pp. 83–130. Springer-Verlag, New York.

Kalmijn, A. J. (1988b). Detection of weak electric fields. *In* "Sensory Biology of Aquatic Animals" (J. Atema, R. R. Fay, A. N. Popper, and W. N. Tavolga, eds.), pp. 151–186. Springer-Verlag, New York.

Kalmijn, A. J. (1989). Functional evolution of lateral line and inner ear systems. *In* "The Mechanosensory Lateral Line: Neurobiology and Evolution" (S. Coombs, P. Görner, and H. Münz, eds.), pp. 187–215. Springer-Verlag, New York.

Kalmijn, A. J. (1997). Electric and near-field acoustic detection, a comparative study. *Acta Physiol. Scand.* **161**(Suppl.), 25–38.

Kalmijn, A. J., Kolba, C. A., and Kalmijn, V. (1976a). Orientation of catfish (*Ictalurus nebulosa*) in strictly uniform electric fields. I. Sensitivity of response. *Biol. Bull. (Woods Hole, Mass.)* **151**, 415.

Kalmijn, V., Kolba, C. A., and Kalmijn, A. J. (1976b). Orientation of catfish (*Ictalurus nebulosa*) in strictly uniform electric fields. II. Spatial discrimination. *Biol. Bull. (Woods Hole, Mass.)* **151**, 415–416.

Kami, H. T., and Ikehara, I. I. (1976). Notes on the annual juvenile siganid harvest in Guam. *Micronesica* **12**, 323–325.

Kanashiro, K. (1998). Morphology, and changes of distribution and food habits with growth, of late larvae and juveniles of black-spot tuskfish, *Choerodon schoenleinii* (Labridae), settled on seagrass beds of Okinawa Island, the Ryukyus. *Nippon Suison Gakkaishi* **64**, 427–434.

Kara, M. H., and Derbal, F. (1995). Morphometry, growth and mortality of the grouper *Epinephelus marginatus* (Serranidae) from the eastern coasts of Algeria. *Cah. Biol. Mar.* **36**, 229–237.

Kareiva, P. (1987). Habitat fragmentation and the stability of predator-prey interactions. *Nature (London)* **326**, 380–388.

Kareiva, P., and Anderson, M. (1986) Spatial aspects of species interactions: The wedding of models and experiments. *In* "Community Ecology: Lecture Notes in Biomathematics" (A. Hastings, ed.), pp. 35–50. Springer-Verlag, Berlin.

Karlsen, H. E. (1992a). The inner ear is responsible for detection of infrasound in the perch (*Perca fluviatilus*). *J. Exp. Biol.* **171**, 163–172.

Karlsen, H. E. (1992b). Infrasound sensitivity in the plaice (*Pleuronectes platessa*). *J. Exp. Biol.* **171**, 173–187.

Kastelle, C. R., Kimura, D. K., Nevissi, A E., and Gunderson, D. R. (1994). Using Pb^{210}/Ra^{226} disequilibria for sablefish, *Anoplopoma fimbria,* age validation. *Fish. Bull.* **92**, 292–301.

Katsuki, Y., and Yanagisawa, K. (1982). Chemoreception in the lateral line organ. *In* "Chemoreception in Fishes" (T. J. Hara, ed.), pp. 227–242. Am. Elesvier, New York.

Katsuki, Y., Yanagisawa, K., Tester, A. L., and Kendall, J. I. (1969). Shark pit organs: Response to chemicals. *Science* **163**, 405–407.

Kauffman, E. G., and Fagerstrom, J. A. (1993). The Phanerozoic evolution of reef diversity. *In* "Species Diversity in Ecological Communities, Historical and Geological Perspectives" (R. E. Ricklefs and D. Schluter, eds.), pp. 315–329. University of Chicago Press, Chicago.

Kauffman, E. G., and Sohl, N. F. (1974). Structure and evolution of Antillean Cretaceous rudist frameworks. *Verh. Naturforsch. Ges. Basel* **84**, 399–467.

Kaufman, L. S., and Liem, K. F. (1982). Fishes of the suborder Labroidei (Pisces: Perciformes): Phylogeny, ecology, and evolutionary significance. *Breviora* **472**, 1–19.

Kaufman, L. S., Ebersole, J., Beets, J., and McIvor, C. C. (1992). A key phase in the recruitment dynamics of coral reef fishes: Post-settlement transition. *Environ. Biol. Fishes* **34**, 109–118.

Kavanagh, K. D. (2000). Larval brooding in the marine damselfish *Acanthochromis polyacanthus* (Pomacentridae) is correlated with highly divergent morphology, ontogeny and life history traits. *Bull. Mar. Sci.* **66**, 321–337.

Kawakami, Y., Mochioka, N., Morishita, K., Tajima, T., Nakagawa, H., Toh, H., and Nakazono, A. (1998). Factors influencing otolith strontium/calcium ratios in *Anguilla japonica* elvers. *Environ. Biol. Fishes* **52**, 299–303.

Kawase, H., Mochioka, N., and Nakazono, A. (1993). Otolith increment formation and planktonic larval duration of a temperate damselfish, *Chromis notatus*. *Jpn. J. Ichthyol.* **40**, 377–380.

Kay, E. A. (1990). Cypraeidae of the Indo-Pacific: Cenozoic fossil history and biogeography. *Bull. Mar. Sci.* **47**, 23–43.

Kay, E. A., and Palumbi, S. R. (1987). Endemism and evolution in Hawaiian marinee invertebrates. *Trends Ecol. Evol.* **2**, 183–187.

Keast, A., and Webb, D. (1966). Mouth and body form relative to feeding ecology in the fish of a small lake, Lake Opinicon, Ontario. *J. Fish. Res. Board Can.* **25**, 1133–1144.

Keefe, M., and Able, K. W. (1993). Patterns of metamorphosis in summer flounder, *Paralichthys dentatus*. *J. Fish Biol.* **42**, 713–728.

Keener, P., Johnson, G. D. Stender, B. W., Brothers, E. B., and Beatty, H. R. (1988). Ingress of postlarval gag, *Mycteroperca microlepis* (Pisces: Serranidae) through a South Carolina barrier island inlet. *Bull. Mar. Sci.* **42**, 376–396.

Keenleyside, M. H. A. (1972). The behavior of *Abudefduf zonatus* (Pisces: Pomacentridae) at Heron Island Great Barrier Reef. *Anim. Behav.* **20**, 763–775.

Kelleher, G., Bleakley, C., and Wells, S. (1995). "A Global Representative System of Marine Protected Areas," Vols. 1–4. The Great Barrier Reef Marine Park Authority, The World Bank, The World Conservation Union. Environment Department, The World Bank, Washington, DC.

Keller, G., Zenker, C. E., and Stokes, S. M. (1989). Late Neogene history of the Pacific-Caribbean gateway. *J. South Am. Earth Sci.* **2**, 73–108.

Kellogg, L., Kiel, B., Loftin, J., Hale, J., Heck, A., and Lindberg, W. (1997). Spatial patterns of grouper (Pisces:

Serranidae) occupying experimental patch reefs. *In* "Florida Big Bend Coastal Research Workshop: Toward a Scientific Basis for Ecosystem Management." (W. J. Lindberg, ed.), Proc. Meet., Steinhatchee, FL, May 7–8, 1997, pp. 84–86. Florida Sea Grant, University of Florida, Gainesville, Florida.

Kelly, J. C., and Nelson, D. R. (1975). Hearing thresholds of the horn shark, *Heterodontus francisci*. *J. Acoust. Soc. Am.* **58**, 905–909.

Kelly, P. S., Lwiza, K. M. M., Cowen, R. K., and Goni, G. (2000). Low-salinity lenses at Barbados, West Indies: Their origin, frequency and variability. *J. Geophys. Res.* **105**, 19,699–19,708.

Kennedy, B. D., Folt, C. L., Blum, J. D., and Chamberlain, C. P. (1997). Natural isotope markers in salmon. *Nature (London)* **387**, 766–767.

Kennedy, B. P., Blum, J. D., Folt, C. L., and Nislow, K. H. (2000). Using natural strontium isotopic signatures as fish markers: Methodology and application. *Can. J. Fish. Aquat. Sci.* **57**, 2280–2292.

Kenyon, T. N. (1994). The significance of sound interception to males of the bicolor damselfish, *Pomacentrus partitus*, during courtship. *Environ. Biol. Fishes* **40**, 391–405.

Kenyon, T. N. (1995). Ontogenetic changes in the auditory sensitivity of the bicolor damselfish, *Pomacentrus partitus*. Unpublished Ph.D. Dissertation, University of Miami, Miami, FL.

Kenyon, T. N. (1996). Ontogenetic changes in the auditory sensitivity of damselfishes (Pomacentridae). *J. Comp. Physiol. A* **179**, 553–561.

Kerrigan, B. A. (1994). Post-settlement growth and body condition in relation to food availability in a juvenile tropical reef fish. *Mar. Ecol.: Prog. Ser.* **111**, 7–15.

Kerrigan, B. A. (1995). Variability in condition and morphological traits of two tropical reef fish (Pomacentridae): Implications for recruitment success. Unpublished Ph.D. Thesis, James Cook University, Townsville, Australia.

Kerrigan, B. A. (1996). Temporal patterns in size and condition at settlement in two tropical reef fishes (Pomacentridae: *Pomacentrus amboinensis* and *P. nagasakiensis*). *Mar. Ecol.: Prog. Ser.* **135**, 27–41.

Kerrigan, B. A. (1997). Variability in larval development of a tropical reef fish (Pomacentridae: *Pomacentrus amboinensis*): The parental legacy. *Mar. Biol.* **127**, 395–402.

Khoda, M. (1988). Diurnal periodicity of spawning activity of permanently territorial damselfishes (Teleostei: Pomacentridae). *Environ. Biol. Fishes* **21**, 91–100.

Khoury, C. (1987). Ichtyofaune des herbiers de posidonies du parc national de Port-Cros: composition, éthologie alimentaire et rôle dans le réseau trophique. Thèse 3ème cycle, Océanographie, Université Aix-Marseille 2.

Kidd, A. H., and Kidd, R. M. (1999). Benefits, problems and characteristics of home aquarium owners. *Psychol. Rep.* **84**, 998–1004.

Kiene, W. E. (1988). A model of bioerosion on the Great Barrier Reef. *Proc. 6th Int. Coral Reef Symp.*, Vol. 3, pp. 449–454.

Kiflawi, M., Mazeroll, A. I., and Goulet, D. (1998). Does mass spawning enhance fertilization in coral reef fish? A case study of the brown surgeonfish. *Mar. Ecol.: Prog. Ser.* **172**, 107–114.

Kim, S. J., and Case, S. D. (1996). "Analysis of Laws Addressing Destructive Fishing Practices in the Asia/Pacific Region: Sodium Cyanide/Live Reef Fish Trade." The Nature Conservancy Asia/Pacific Program, Honolulu, Hawaii.

Kim, S. T., and O'Neil, J. R. (1997). Equilibrium and nonequilibrium oxygen isotope effects in synthetic carbonates. *Geochim. Cosmochim. Acta* **61**, 3461–3475.

Kimura, D. K. (1980). Likelihood methods for the von Bertalanffy growth curve. *Fish. Bull.* **77**, 765–774.

Kimura, M. (1983). "The Neutral Theory of Molecular Evolution". Cambridge University Press, Cambridge, UK.

Kimura, M. (1991). The neutral theory of molecular evolution: A review of recent evidence. *Jpn. J. Genet.* **66**, 367–386.

Kimura, M., and Weiss, G. H. (1964). The stepping stone model of population structure and the decrease of genetic correlation with distance. *Genetics* **49**, 561–571.

Kimura, S., and Kiriyama, T. (1993). Development of eggs, larvae and juveniles of the labrid fish, *Halichoeres poecilopterus*, reared in the laboratory. *Jpn. J. Ichthyol.* **39**, 371–378.

Kimura, S., Nakayama, Y., and Kiriyama, T. (1998). Comparison of laboratory-reared eggs, embryos and larvae of five labrid fishes. *Environ. Biol. Fishes* **52**, 187–202.

Kingsford, M. J. (1990). Linear oceanographic features: A focus for research on recruitment processes. *Aust. J. Ecol.* **15**, 391–401.

Kingsford, M. J. (1992). Spatial and temporal variation in predation on reef fishes by coral trout (*Plectropomus leopardus*, Serranidae). *Coral Reefs* **11**, 193–198.

Kingsford, M. J. (1998). The relative importance of feeding, predation and transport processes on the survival of presettlement fish. *In* "ReeFish '95: Recruitment and Population Dynamics of Coral Reef Fishes" (G. P. Jones, P. J. Doherty, B. D. Mapstone, and L. Howlett, eds.), pp. 115–122. CRC Reef Research Centre, Townsville, Australia.

Kingsford, M. J., and Choat, J. H. (1986). The influence of surface slicks on the distribution and onshore movement of small fish. *Mar. Biol.* **91**, 161–171.

Kingsford, M. J., and Finn, M. (1997). The influence of phase of the moon and physical processes on the input of presettlement fishes to coral reefs. *J. Fish. Biol.* **51**, 176–205.

Kingsford, M. J., and MacDiarmid, A. B. (1988). Interrelations between planktivorous reef fish and zooplankton in temperate waters. *Mar. Ecol.: Prog. Ser.* **48**, 103–117.

Kingsford, M. J., and Milicich, M. J. (1987). Presettlement phase of *Parika scaber* (Pisces: Monacanthidae): A temperate fish. *Mar. Ecol.: Prog. Ser.* **36**, 65–79.

Kingsford, M. J., and Tricklebank, K. A. (1991). Ontogeny and behaviour of *Aldrichetta forsteri* (Teleostei: Mugilidae). *Copeia*, pp. 9–16.

Kingsford, M. J., Wolanski, E., and Choat, J. H. (1991). Influence of tidally induced fronts and Langmuir circulation on distribution and movement of pre-settlement fishes around a coral reef. *Mar. Biol.* **109,** 167–180.

Kirkpatrick M., and Slatkin, M. (1993). Searching for evolutionary patterns in the shape of a phylogenetic tree. *Evolution (Lawrence, Kans.)* **47,** 1171–1181.

Kishiro, T. K., and Nakazono, A. (1991). Seasonal patterns of larval settlement and daily otolith increments in the temperate wrasse *Halichoeres tenuispinis*. *Nippon Suisan Gakkaishi* **57,** 409–415.

Kjesbu, O. S., Klungsoyr, J., Kryvi, H., Witthames, P. R., and Greer Walker, M. (1991). Fecundity, atresia and egg size of captive Atlantic cod (*Gadus morhua*) in relation to proximate body composition. *Can. J. Fish. Aquat. Sci.* **48,** 2333–2343.

Kjesbu, O. S., Witthames, P. R., Solemdal, P., and Greer Walker, M. (1998). Temporal variations in the fecundity of Arcto-Norwegian cod (*Gadus morhua*) in response to natural changes in food and temperature. *J. Sea Res.* **40,** 303–321.

Klassen, G. J. (1995). Phylogeny and biogeography of the Ostraciinae (Tetraodontiformes: Ostraciidae). *Bull. Mar. Sci.* **57,** 393–441.

Kleerekoper, H. (1969). "Olfaction in Fishes." Indiana University Press, Bloomington.

Klimley, A. P. (1993). Highly directional swimming by scalloped hammerhead sharks *Sphyrna lewini,* and subsurface irradiance, temperature, bathymetry and goemagnetic field. *Mar. Biol.* **117,** 1–22.

Knapp, R. A. (1993). The influence of egg survivorship on the subsequent nest fidelity of female bicolor damselfish, *Stegastes partitus*. *Anim. Behav.* **46,** 111–121.

Knapp, R. A., and Kovach, J. T. (1991). Courtship as an honest indicator of male parental quality in the bicolor damselfish, *Stegastes partitus*. *Behav. Ecol.* **2,** 295–300.

Knowlton, N. (1992). Thresholds and multiple stable states in coral reef community dynamics. *Am. Zool.* **32,** 674–682.

Knowlton, N. (1993). Sibling species in the sea. *Annu. Rev. Ecol. Syst.* **24,** 189–216.

Knudsen, E. I. (1974). Behavioral thresholds to electric signals in high frequency electric fields. *J. Comp. Physiol.* **91,** 333–353.

Knutsen, J. A. (1992). Feeding behavior of North Sea turbot (*Scophthalmus maximus*) and Dover sole (*Solea solea*) larvae elicited by chemical stimuli. *Mar. Biol.* **113,** 543–548.

Koebele, B. P. (1985). Growth and the size hierarchy effect: An experimental assessment of three proposed mechanisms; activity differences, disproportional food acquisition, physiological stress. *Environ. Biol. Fishes* **12,** 181–188.

Koehl, M. A. R. (1996). When does morphology matter? *Annu. Rev. Ecol. Syst.* **27,** 501–542.

Koenig, C. C., and Coleman, F. C. (1998). Absolute abundance and survival of juvenile gags in sea grass beds of the northeastern Gulf of Mexico. *Trans. Am. Fish. Soc.* **127,** 44–55.

Koenig, C. C., and Colin, P. L. (1999). Distribution, abundance and survival of juvenile gag grouper, *Mycteroperca microlepis* (Pisces: Serranidae), in seagrass beds of the northeastern Gulf of Mexico. *Proc. Annu. Gulf Caribb. Fish. Inst.* **45,** 37–54.

Koenig, C. C., Coleman, F. C., Collins, L. A., Sadovy, Y., and Colin, P. L. (1996). Reproduction in gag (*Mycteroperca microlepis*) (Pisces: Serranidae) in the eastern Gulf of Mexico and the consequences of fishing spawning aggregations. In: "Biology, Fisheries, and Culture of Tropical Groupers and Snappers." (F. Arregui'n-Sanchez, J. L. Munro, M. C. Balgos, and D. Pauly, eds.) *ICLARM Conf. Proc.* **48,** 307–323.

Kohn, A. J. (1985). Evolutionary ecology of *Conus* on Indo-Pacific coral reefs. *Proc. 5th Int. Coral Reef Congr.* Vol. 4, pp. 139–144.

Kohn, A. J. (1997). Why are coral reef communities so diverse? *In* "Marine Biodiversity: Patterns and Processes" (R. F. G. Ormond, J. D. Gage, and M. V. Angel, eds.), pp. 205–215. Cambridge University Press, Cambridge, UK.

Koslow J. A., Bell, J., Virtue, P., and Smith, D.C. (1995). Fecundity and its variability in orange roughy: Effects of population density, condition, egg size, and senescence. *J. Fish. Biol.* **47,** 1063–1080.

Kotrschal, K. (1988). A catalogue of skulls and jaws of eastern tropical Pacific blennioid fishes (Blennioidei: Teleostei): A proposed evolutionary sequence of morphological change. *Z. Zool. Syst. Evol.* **26,** 442–466.

Kotrschal, K. (1989a). Trophic ecomorphology in eastern Pacific blennioid fishes: Character transformation of oral jaws and associated change of their biological roles. *Environ. Biol. Fishes.* **24,** 199–218.

Kotrschal, K. (1989b). On the biological role of form: Evolutionary trophic ecomorphology in 34 Gulf of California Mexico blennioid fishes. *In* "Trends in Vertebrate Morphology" (H. Splechtna and H. Hilgers, eds.), pp. 517–518. Fischer, New York.

Kozloski, J., and Crawford, J. D. (1998). Functional neuroanatomy of auditory pathways in the sound-producing fish *Pollimyrus*. *J. Comp. Neurol.* **401,** 227–252.

Kraak, S. B. M. (1996). Female preference and filial cannibalism in *Aidablennius sphynx* (Teleostei, Blenniidae); a combined field and laboratory study. *Behav. Processes.* **36,** 85–98.

Kraak, S. B. M., and Videler, J. J. (1991). Mate choice in *Aidablennius sphynx* (Teleostei, Blenniidae); females prefer nests containing more eggs. *Behaviour* **119,** 243–266.

Kramer, D. L., and Chapman, M. R. (1999). Implications of fish home range size and relocation for marine reserve function. *Environ. Biol. Fishes* **55,** 65–79.

Krebs, C. J. (1985). "Ecology—The Experimental Analysis of Distribution and Abundance." Harper & Row, New York.

Krebs, C. J. (1995). Two paradigms of population regulation. *Wildl. Res.* **22,** 1–10.

Kritzler, H., and Wood, L. (1961). Provisional audiogram for the shark, *Carcharhinus leucas*. *Science* **133,** 1480–1482.

Kroese, A. B. A., and Schellart, N. A. M. (1992) Velocity- and acceleration-sensitive units in the trunk lateral line of the trout. *J. Physiol. (London)* **68**, 2212–2221.

Kruijf, H. A. M. (1978). Report on the export of fishes and invertebrates for the aquarium trade from Curaçao 1972–1977. *Carmabi Coll. Pap. Stinapa Doc. Ser. 1.* **13**, 1–24.

Kühme, W. (1963). Chemisch ausgelöste Brutflege und Schwarmverhalten bei *Hemichromis bimaculatus* (Pisces). *Z. Tierpsychol.* **20**, 688–704.

Kuiter, R. H. (1996). "Guide to Sea Fishes of Australia." New Holland Publishers, Sydney.

Kulbicki, M., Labrosse, P., and Letourneur, Y. (2000). Fish stock assessment of the northern New Caledonian lagoons: 2. Stocks of lagoon bottom and reef-associated fishes. *Aquat. Living Resour.* **13**, 77–90.

Kumpf, H. E. (1964). Use of underwater television in bio-acoustic research. *In* "Marine Bio-acoustics" (W. N. Tavolga, ed.), pp. 45–57. Pergamon Press, New York.

Kunin, W. E., and Gaston, K. J. (1993). The biology of rarity: Patterns, causes and consequences. *Trends Ecol. Evol.* **8**, 298–301.

Kunin, W. E., and Gaston, K. J. (1997). "The Biology of Rarity: Causes and Consequences of Rare-common Differences." Chapman & Hall, London.

Kuwamura, T. (1984). Social structure of the protogynous fish *Labroides dimidiatus*. *Publ. Seto Mar. Biol. Lab.* **29**, 117–177.

Kuwamura, T., and Nakashima, Y. (1998). New aspects of sex change among reef fishes: Recent studies in Japan. *Environ. Biol. Fishes* **52**, 125–135.

Kuwamura, T., Yogo, Y., and Nakashima, Y. (1993). Size-assortative monogamy and parental egg care in a coral goby *Paragobiodon echinocephalus*. *Ethology* **95**, 65–75.

Kuwamura, T., Nakashima, Y., and Yogo, Y. (1996). Plasticity in size and age at maturity in a monogamous fish: Effect of host coral size and frequency dependence. *Behav. Ecol. Sociobiol.* **38**, 365–370.

Kvenseth, A. M., Pittman, K., and Helvik, J. V. (1996). Eye development in Atlantic halibut (*Hippoglossus hippoglossus*): Differentiation and development of the retina from early yolk sac stages through metamorphosis. *Can. J. Fish. Aquat. Sci.* **53**, 2524–2532.

Kwang-Tse, S., Wilson, W., Winslow, M., McBride, D., and Hopkins, C. D. (1996). Short-range orientation in electric fish: An experimental study of passive electrolocation. *J. Exp. Biol.* **199**, 2383–2393.

Labelle, M., and Nursall, J. R. (1985). Some aspects of the early life history of the redlip blenny, *Ophioblennius atlanticus* (Teleostei: Blenniidae). *Copeia*, pp. 39–49.

Lack, D. (1954). "The Natural Regulation of Animal Numbers." Oxford University Press, Oxford.

Lacson, J. M. (1992). Minimal genetic variation among samples of six species of coral reef fishes collected at La Parguera, Puerto Rico, and Discovery Bay, Jamaica. *Mar. Biol.* **112**, 327–331.

Lacson, J. M. (1994). Fixed allele frequecy differences among Palauan and Okinawan populations of the damselfishes *Chrysiptera cyanea* and *Pomacentrus coelestis*. *Mar. Biol.* **118**, 359–365.

Lacson, J. M., and Clark, S. (1995). Genetic divergence of Maldivian and Micronesian demes of the damselfishes *Stegastes nigricans, Chrysiptera biocellata, C. glauca* and *C. leucopoma* (Pomacentridae). *Mar. Biol.* **121**, 585–590.

Lacson, J. M., and Morizot, D. C. (1991). Temporal genetic variation in subpopulations of bicolor damselfish (*Stegastes partitus*) inhabiting coral reefs in the Florida Keys. *Mar. Biol.* **110**, 353–357.

Lacson, J. M., Riccardi, V. M., Calhoun, S. W., and Morizot, D. C. (1989). Genetic differentiation of bicolor damselfish (*Eupomacentrus partitus*) populations in Florida Keys. *Mar. Biol.* **103**, 445–451.

Ladd, H. S. (1960). Origin of the Pacific island molluscan fauna. *Am. J. Sci.* **258**, 137–150.

Ladich, F. (1997). Agonistic behaviour and significance of sounds in vocalizing fish. *Mar. Freshwater Behav. Physiol.* **29**, 87–108.

Laegdsgaard, P., and Johnson, C. R. (1995). Mangrove habitats as nurseries: Unique assemblages of juvenile fish in subtropical mangroves in eastern Australia. *Mar. Ecol.: Prog. Ser.* **126**, 67–81.

Laevastu, T., and Bax, N. J. (1991). Predation controlled recruitment in the Bering Sea fish ecosystem. *ICES Mar. Sci. Symp.* **193**, 147–152.

Laine, P., and Rajasilta, M. (1999). The hatching success of Baltic herring eggs and its relation to female condition. *J. Exp. Mar. Biol. Ecol.* **237**, 61–73.

Landman, N. H., Druffel, E. R. M., Cochran, J. K., Donahue, D. J., and Jull, A. J. T. (1988). Bomb-produced radiocarbon in the shell of the chambered nautilus: Rate of growth and age at maturity. *Earth Planet. Sci. Lett.* **89**, 28–34.

Lannoo, M. J., and Lannoo, S. J. (1996). Development of the electrosensory lateral line lobe in the channel catfish, *Ictalurus punctatus*, with reference to the onset of swimming and feeding behaviors. *Mar. Freshwater Behav. Physiol.* **28**, 45–53.

Lara, M. R. (1999). Sensory development in settlement-stage larvae of Caribbean labrids and scarids—a comparative study with implications for ecomorphology and life history strategies. Unpublished Ph.D. Dissertation, College of William and Mary, Williamsburg, VA.

Lara, M. R. (2001). Morphology of the eye and visual acuities in the settlement—intervals of some coral reef fishes (Labridae, Scaridae). *Environ. Biol. Fishes* **62**, 365–378.

Lathrop, C., and Hourigan, T. (1998). The United States and the live reef fish trade: Trade-based solutions to cyanide fishing in the Asia-Pacific region. *Proc. APEC Workshop Impacts Destruct. Fish. Pract. Mar. Environ.*, Hong Kong, *1997*, pp. 190–199.

Lau, P. P. F., and Parry-Jones, R. (1999). "The Hong Kong Trade in Live Reef Fish for Food." TRAFFIC, Hong Kong.

Lauck, T., Clark, C. W., Mangel, M., and Munro, G. R. (1998). Implementing the precautionary principle in

fisheries management through marine reserves. *Ecol. Appl.* **8**, s72–s78.
Lauder, G. V. (1980). The suction feeding mechanism in sunfishes (*Lepomis*): An experimental analysis. *J. Exp. Biol.* **88**, 49–72.
Lauder, G. V. (1982). Patterns of evolution in the feeding mechanism of actinopterygian fishes. *Am. Zool.* **22**, 275–285.
Lauder, G. V. (1983a). Prey capture hydrodynamics in fishes: Experimental tests of two models. *J. Exp. Biol.* **104**, 1–13.
Lauder, G. V. (1983b). Functional and morphological bases of trophic specialization in sunfishes (Teleostei: Centrarchidae). *J. Morphol.* **178**, 1–21.
Lauder, G. V. (1983c). Functional design and evolution of the pharyngeal jaw apparatus in euteleostean fishes. *Zool. J. Linn. Soc.* **77**, 1–38.
Lauder, G. V. (1985). Aquatic feeding in lower vertebrates. *In* "Functional Vertebrate Morphology" (M. Hildebrand, D. Bramble, K. Liem, and D. Wake, eds.), pp. 210–229. Harvard University Press, Cambridge, MA.
Lauder, G. V. (1990). Functional morphology and systematics: Studying functional patterns in an historical context. *Annu. Rev. Ecol. Syst.* **21**, 317–340.
Laverack, M. S. (1988). The diversity of chemoreceptors. *In* "Sensory Biology of Aquatic Animals" (J. Atema, R. R. Fay, A. N. Popper, and W. N. Tavolga, eds.), pp. 287–312. Springer-Verlag, New York.
Lavery, S., Moritz, C., and Fielder, D. R. (1995). Changing patterns of population structure and gene flow at different spatial scales in the coconut crab *(Birgus latro). Heredity* **74**, 531–541.
Lavery, S., Moritz, C., and Fielder, D. R. (1996). Indo-Pacific population structure and evolutionary history of the coconut crab *Birgus latro*. *Mol. Ecol.* **5**, 557–570.
Lavin, P. A., and McPhail, J. D. (1986). Adaptive divergence of trophic phenotypes among freshwater populations of the threespine stickleback (*Gasterosteus aculeatus*). *Can. J. Fish. Aquat. Sci.* **43**, 2455–2463.
Lawrence, J. M. (1957). Estimated size of various forage fishes largemouth bass can swallow. *Proc. Southeast. Assoc. Game Fish Comm.* **11**, 220–226.
Laws, E. A., and Archie, J. W. (1981). Appropriate use of regression analysis in marine biology. *Mar. Biol.* **65**, 13–16.
Lawton, J. H. (1989). What is the relationship between population density and body size in animals? *Oikos* **55**, 429–434.
Lawton, J. H. (1990). Species richness and population dynamics of animal assemblages: Patterns in body size, abundance, space. *Philos. Trans. R. Soc. London, Ser. B* **330**, 283–291.
Lawton, J. H. (1999). Are there general laws in ecology? *Oikos* **84**, 177–192.
Lea, D. W., Shen, G. T., and Boyle, E. A. (1989). Coralline barium records temporal variability in equatorial Pacific upwelling. *Nature (London)* **340**, 373–376.
Lea, D. W., Mashiotta, T. A., and Spero, H. J. (1999). Controls on magnesium and strontium uptake in planktonic foraminifera determined by live culturing. *Geochim. Cosmochim. Acta* **63**, 2369–2379.
LeBorgne, R., Dandonneau, Y., and Lemasson, L. (1985). The problem of the island mass effect on chlorophyll and zooplankton standing crops around Mare (Loyalty Islands) and New Caledonia. *Bull Mar. Sci.* **37**, 450–459.
Leder, J. J., Swart, P. K., Szmant, A., and Dodge, R. E. (1996). The origin of variations in the isotopic record of scleractinian corals: I. Oxygen. *Geochim. Cosmochim. Acta* **60**, 2857–2870.
Lee, C., and Sadovy, Y. (1998). A taste for live fish: Hong Kong's live reef fish market. *NAGA, ICLARM Q.*, **1998**, pp. 38–42.
Lee, T. N., and Mayer, D. A. (1977). Low-frequency current variability and spin-off Southeast Florida. *J. Mar. Res.* **35**, 193–220.
Lee, T. N., and Williams, E. (1999). Mean distribution and seasonal variability of coastal currents and temperature in the Florida Keys with implications for larval recruitment. *Bull. Mar. Sci.* **64**, 35–64.
Lee, T. N., Yoder, J. A., and Atkinson, L. P. (1991). Gulf Stream frontal eddy influence on productivity of the southeast U.S. continental shelf. *J. Geophys. Res.* **96**, 22,191–22,205.
Lee, T. N., Rooth, C., Williams, E., McGowan, M., Szmant, A. F., and Clarke, M. E. (1992). Influence of Florida Current, gyres and wind-driven circulation on transport of larvae and recruitment in the Florida Keys coral reefs. *Cont. Shelf. Res.* **12**, 971–1002.
Lee, T. N., Clarke, M. E., Williams, E, Szmant, A. F., and Berger, T. (1994). Evolution of the Tortugas gyre and its influence on recruitment in the Florida keys. *Bull. Mar. Sci.* **64**, 621–646
LeFervre, J. (1986). Aspects of the biology of frontal zones. *Adv. Mar. Biol.* **23**, 163–299.
Leggett, W. C., and Deblois, E. (1994). Recruitment in marine fishes—is it regulated by starvation and predation in the egg and larval stages. *Neth. J. Sea Res.* **32**, 119–134.
Legler, J. M., and Georges, A. (1993). Biogeography and phylogeny of the Chelonia. *In* "Fauna of Australia" (C. J. Glasby, G. J. B. Ross, and P. L. Beesley, eds.), Vol. 2A, pp. 129–138. Australian Government Publishing Service, Canberra.
Leichter, J. J., Wing, S. R., Miller, S. L., and Denny, M. W. (1996). Pulsed delivery of subthermocline water to Conch Reef (Florida Keys) by intertidal bores. *Limnol. Oceanogr.* **41**, 1490–1501.
Leichter, J. J., Shellenbarger, G., Genovese, S. J., and Wing, S. R. (1998). Breaking internal waves on a Florida (USA) coral reef: A plankton pump at work? *Mar. Ecol. Prog. Ser.* **166**, 83–97.
Leis, J. M. (1978). Systematics and zoogeography of the porcupine fishes (*Diodon,* Diodontidae, Tetraodontiformes), with comments on egg and larval development. *Fish. Bull.* **76**, 535–567.
Leis, J. M. (1986a). Vertical and horizontal distribution of

fish larvae near coral reefs at Lizard Island, Great Barrier Reef. *Mar. Biol.* **90**, 505–516.

Leis, J. M. (1986b). Epibenthic schooling by larvae of the clupeid fish *Spratelloides gracilis*. *Jpn. J. Ichthyol.* **33**, 67–69.

Leis, J. M. (1989). Larval biology of butterflyfishes (Pisces, Chaetodontidae): What do we really know? *Environ. Biol. Fishes* **25**, 87–100.

Leis, J. M. (1991a). The pelagic stage of reef fishes: The larval biology of coral reef fishes. *In* "The Ecology of Fishes on Coral Reefs" (P. F. Sale, ed.), pp. 183–230. Academic Press, San Diego, CA.

Leis, J. M. (1991b). Vertical distribution of fish larvae in the Great Barrier Reef Lagoon, Australia. *Mar. Biol.* **109**, 157–166.

Leis, J. M. (1993). Larval fish assemblages near Indo-Pacific coral reefs. *Bull. Mar. Sci.* **53**, 362–392.

Leis, J. M. (1994). Coral Sea atoll lagoons—closed nurseries for the larvae of a few coral reef fishes. *Bull. Mar. Sci.* **54**, 206–227.

Leis, J. M., and Carson-Ewart, B. M. (1997). *In situ* swimming speeds of the late pelagic larvae of some Indo-Pacific coral-reef fishes. *Mar. Ecol.: Prog. Ser.* **159**, 165–174.

Leis, J. M., and Carson-Ewart, B. M. (1998). Complex behaviour by coral-reef fish larvae in open-water and near-reef pelagic environments. *Environ. Biol. Fishes* **53**, 259–266.

Leis, J. M., and Carson-Ewart, B. M. (1999). *In situ* swimming and settlement behaviour of larvae of an Indo-Pacific coral-reef fish, the Coral Trout (Pisces, Serranidae, *Plectropomus leopardus*). *Mar. Biol.* **134**, 51–64.

Leis, J. M., and Carson-Ewart, B. M., eds. (2000). "The Larvae of Indo-Pacific Coastal Fishes. An Identification Guide to Marine Fish Larvae," Fauna Malesiana Handbooks 2. Brill, Leiden.

Leis, J. M., and Carson-Ewart, B. M. (2001). Behaviour of pelagic larvae of four coral-reef fish species in the ocean and an atoll lagoon. *Coral Reefs* **19**, 247–257.

Leis, J. M., and Rennis, D. S. (1983). "The Larvae of Indo-Pacific Coral Reef Fishes." University of Hawaii Press, Honolulu.

Leis, J. M., and Stobutzki, I. C. (1999). Swimming performance of late pelagic larvae of coral-reef fishes: *In situ* and laboratory-based measurements. *In* "Proceedings of the 5th Indo-Pacific Fish Conference, Nouméa, 1997" (B. Séret and J.-Y. Sire, eds.), pp. 575–583. Société Française d'Ichtyologie & Institut de Recherche pour le Développement, Paris.

Leis, J. M., Goldman, B., and Reader, S. E. (1989). Epibenthic fish larvae in the Great Barrier Reef Lagoon near Lizard Island, Australia. *Jpn. J. Ichthyol.* **35**, 428–433.

Leis, J. M., Sweatman, H. P. A., and Reader, S. E. (1996). What the pelagic stages of coral reef fishes are doing out in blue water: Daytime field observations of larval behavioural capabilities. *Mar. Freshwater Res.* **47**, 401–411.

Leis, J. M., Trnski, T., Doherty, P. J., and Dufour, V. (1998). Replenishment of fish populations in the enclosed lagoon of Taiaro Atoll: Evidence from eggs and larvae. *Coral Reefs* **17**, 1–8.

Leis, J. M., Carson-Ewart, B. M., and Webley, J. (in press). Settlement behaviour of coral-reef fish larvae at subsurface artificial-reef moorings. *Mar. Freshwater Res.* **53**, (in press).

Lessios, H. A., Allen, G. R., Wellington, G. M., and Bermingham, E. (1995). Genetic and morphological evidence that the Eastern Pacific damselfish *Abudefduf declivifrons* is distinct from *A. concolor* (Pomacentridae). *Copeia*, pp. 277–288.

Letourneur, Y. (1996). Dynamics of fish communities of Reunion fringing reefs, Indian Ocean. 2. Patterns of temporal fluctuations. *J. Exp. Mar. Biol. Ecol.* **195**, 31–52.

Letourneur, Y., Chabanet, P., Vigliola, L., and Hamelin-Vivien, M. (1998). Mass settlement and post-settlement mortality of *Epinephelus merra* (Pisces: Serranidae) on Reunion coral reefs. *J. Mar. Biol. Assoc. U.K.* **78**, 307–319.

Levin, P. S. (1991). Effects of microhabitat on recruitment variation in a temperate reef fish. *Mar. Ecol.: Prog. Ser.* **75**, 183–189.

Levin, P. S. (1996). Recruitment in a temperate demersal fish—does larval supply matter. *Limnol. Oceanogr.* **41**, 672–679.

Levin, P. S. (1998). The significance of variable and density-independent post-recruitment mortality in local populations of reef fishes. *Aust. J. Ecol.* **23**, 246–251.

Levin, P. S., and Hay, M. E. (1996). Responses of temperate reef fishes to alterations in algal structure and species composition. *Mar. Ecol. Prog. Ser.* **134**, 37–47.

Levin, P. S., Chiasson, W, and Green, J. M. (1997a). Geographic differences in recruitment and population structure of a temperate reef fish. *Mar. Ecol.: Prog. Ser.* **161**, 23–35.

Levin, P. S., Petrik, R., and Malone, J. (1997b). Interactive effects of habitat selection, food supply and predation on recruitment of an estuarine fish. *Oecologia* **112**, 55–63.

Levins, R. (1968). "Evolution in Changing Environments." Princeton University Press, Princeton, NJ.

Levins, R. (1969). Some demographic and genetic consequences of environmental heterogeneity for biological control. *Bull. Entomol. Soc. Am.* **15**, 237–240.

Levins, R. (1970). Extinction. *In* "Some Mathematical Questions in Biology. Lectures on Mathematics in the Life Sciences"(Dans M. Gertenhaber, eds.), Vol. 2, pp. 77–101. Am. Math. Soc., Providence, RI.

Levitan, D. R., and Sewell, M. A. (1998). Fertilization success in free-spawning marine invertebrates: Review of the evidence and fisheries implications. *Can. Fish. Aquat. Sci., Spec. Publ.* **125**, 159–164.

Lewis, A. D. (1988). A coral reef fishery for aquarium fish—the Fiji experience. *Secretariat of the Pacific Community Workshop Inshore Fish. Resour.*, Noumea, New Caledonia, *1988*.

Lewis, A. R. (1997). Recruitment and post-recruit immigration affect the local population size of coral reef fishes. *Coral Reefs* **16,** 139–149.

Lewis, A. R. (1998). Effects of experimental coral disturbance on the population dynamics of fishes on large patch reefs. *J. Exp. Mar. Biol. Ecol.* **230,** 91–110.

Lewis, R. J. (1992). Socioeconomic impacts and management of ciguatera in the Pacific. *Bull. Soc. Pathol. Exot.* **85,** 427–434.

Lewis, S. M., and Wainwright, P. C. (1985). Herbivore abundance and grazing intensity on a Caribbean coral reef. *J. Exp. Mar. Biol. Ecol.* **87,** 215–228.

Lewontin R. C. (1974). "The Genetic Basis of Evolutionary Change." Columbia University Press, New York.

Ley, J. A., McIvor, C. C., and Montaque, C. L. (1999). Fishes in mangrove prop-root habitats of northeastern Florida Bay: Distinct assemblages across an estuarine gradient. *Estuarine Coastal Shelf Sci.* **48,** 701–723.

Li, K. T., Wetterer, J. K., and Hairston, N. G. (1985). Fish size, visual resolution, and prey selectivity. *Ecology* **66,** 1729–1735.

Li, L. W. H. (1999). Live marine fish trade and mariculture in Hong Kong. *Aquacult. Coral Fishes Sustainable Reef Fish.* Sabah, Malaysia, *1996,* pp. 224–234.

Lidicker, W. Z., Jr. (1978). Regulation of numbers in small mammal populations—historical reflections and a synthesis. *In* "Populations of Small Mammals under Natural Conditions" (D. Snyder, ed.), pp. 122–141. University of Pittsburgh Press, Pittsburgh.

Liem, K. F. (1970). Comparative functional anatomy of the Nandidae (Pisces: Teleostei). *Fieldiana, Zool.* **56,** 7–166.

Liem, K. F. (1980). Acquisition of energy by teleosts: Adaptive mechanisms and evolutionary patterns. *In* "Environmental Physiology of Fishes" (M. A. Ali, ed.), pp. 299–334. Plenum Press, New York.

Liem, K. F. (1986). The pharyngeal jaw apparatus of the Embiotocidae (Teleostei): A functional and evolutionary perspective. *Copeia,* pp. 311–323.

Liem, K. F., and Sanderson, S. L. (1986). The pharyngeal jaw apparatus of labrid fishes: A functional morphological perspective. *J. Morphol.* **187,** 143–158.

Light, P. (1995). The early life history of coral trout (*Plectropomus leopardus*) at Green Island and Arlington Reef, Australia. Unpublished Ph.D. thesis, James Cook University, Townsville, Australia.

Light, P. R., and Jones, G. P. (1997). Habitat preference in newly settled coral trout (*Plectropomus leopardus,* Serranidae). *Coral Reefs* **16,** 117–126.

Lima, S. L. (1998). Stress and decision-making under the risk of predation: Recent developments from behavioral, reproductive and ecological perspectives. *Adv. Stud. Behav.* **27,** 215–290.

Limouzy-Paris, C. B., Graber, H. C., Jones, D. L., Ropke, A. W., and Richards, W. J. (1997). Translocation of larval coral reef fishes via sub-mesoscale spin-off eddies from the Florida Current. *Bull. Mar. Sci.* **60,** 966–983.

Lindberg, W. J. (1997). Can science resolve the attraction-production issue? *Fisheries* **22,** 10–13.

Lindeman, K. C., and Snyder, D. B. (1999). Nearshore hard-bottom fishes of southeast Florida and effects of habitat burial caused by dredging. *Fish. Bull.* **97,** 508–525.

Lindeman, K. C., Lee, T. N., Wilson, W. D., Claro, R., and Ault, J. S. (2001). Transport of larvae originating in southwest Cuba and the Dry Tortugas: Evidence for partial retention in grunts and snappers. *Proc. Annu. Gulf Caribb. Fish. Inst.* **52,** 732–747.

Lindström, J., Kokko, H., and Lindén, H. (1999). Density dependence and the response surface methodology. *Oikos* **85,** 40–52.

Lissman, H. W. (1958). On the function and evolution of electric organs in fish. *J. Exp. Biol.* **35,** 156–191.

Lissman, H. W., and Machin, K. E. (1958). The mechanism of object location in *Gymnarchus niloticus* and similar fish. *J. Exp. Biol.* **35,** 451–486.

Lobel, P. S. (1989). Ocean variability and the spawning season of Hawaiian reef fishes. *Environ. Biol. Fishes* **24,** 161–171.

Lobel, P. S. (1992). Sounds produced by spawning fishes. *Environ. Biol. Fishes* **33,** 351–358.

Lobel, P. S. (1996). Spawning sound of the trunkfish, *Ostracion meleagris* (Ostraciidae). *Biol. Bull. (Woods Hole, Mass.)* **191,** 308–309.

Lobel, P. S., and Mann, D. A. (1995). Spawning sounds of the damselfish, *Dascyllus albisella* (Pomacentridae) and relationship to male size. *BioAcoustics* **6,** 187–198.

Lobel, P. S., and Robinson, A. R. (1986). Transport and entrapment of fish larvae by ocean mesoscale eddies and currents in Hawaiian waters. *Deep-Sea Res.* **33,** 483–500.

Lobel, P. S., and Robinson, A. R. (1988). Larval fishes and zooplankton in a cyclonic eddy in Hawaiian waters. *J. Plankton Res.* **10,** 1209–1223.

Lockwood, S. J. (1980). Density-dependent mortality in 0-group plaice (*Pleuronectes platessa* L.) populations. *J. Cons. Int. Explor. Mer* **39,** 148–153.

Loew, E. R., and McFarland, W. N. (1990). The underwater visual environment. *In* "The Visual System of Fish" (R. H. Douglas and M. B. A. Djamgoz, eds.), pp. 1–43. Chapman & Hall, London.

Loew, E. R., McFarland, W. N., Mills, E. L., and Hunter, D. (1993). A chromatic action spectrum for planktonic predation by juvenile yellow perch, *Perca flavescens. Can. J. Zool.* **71,** 1552–1557.

Loew, E. R., McAlary, F. A., and McFarland, W. N. (1996). Ultraviolet sensitivity in the larvae of two species of marine atherinid fishes. *In* "Zooplankton: Sensory Ecology and Physiology" (P. H. Lenz, D. K. Hartline, J. E. Percel, and D. L. Macmillan, eds.), pp. 195–209. Gordon & Breach, Sydney.

Logan, A. (1977). Reef-dwelling articulate brachiopods from Grand Cayman, B. W. I. *Proc. 3rd Int. Coral Reef Symp.,* Vol. 1, pp. 87–93.

Lombarte, A., and Popper, A. N. (1994). Quantitative analyses of postembryonic hair cell addition in the otolithic end organs of the inner ear of the European hake, *Merluccius*

merluccius (Gadiformes, Teleostei). *J. Comp. Neurol.* **345**, 419–428.

Long, D. J. (1992). An Eocene wrasse (Perciformes: Labridae) from Seymour Island. *Antarct. Sci.* **4**, 235–237.

Longhurst, A. R., and Pauly, D. (1987). "Ecology of Tropical Oceans." Academic Press, Orlando, FL.

López-Albors, O., Gil, F., Ramírez-Zarzosa, G., Vázquez, J. M., Latorre, R., García-Alcázar, A., Arencibia, A., and Moreno, F. (1998). Muscle development in gilthead sea bream (*Sparus aurata*, L.) and sea bass (*Dicentrarchus labrax*, L.): Further histochemical and ultrastructural aspects. *Anat. Histol. Embryol.* **27**, 223–229.

Losey, G. S., Jr. (1969). Sexual pheromone in some fishes of the genus *Hypsoblennius* Gill. *Science* **163**, 181–183.

Losey, G. S., Jr., Cronin, T. W., Goldsmith, T. H., Hyde, D., Marshall, N. J., and McFarland, W. N. (1999). The UV visual world of fishes: A review. *J. Fish Biol.* **54**, 921–943.

Lou, D. C. (1993). Growth in juvenile *Scarus rivulatus* and *Ctenochaetus binotatus*: A comparison of families Scaridae and Acanthuridae. *J. Fish. Biol.* **42**, 15–23.

Loubens, G. (1978). Biologie de quelque espèces de poisson du lagon Neo-Caledonien. I. Determination de l'age (otolithmetrie). *Cah. ORSTOM, Ser. Oceanogr.* **16**, 263–283.

Love, D. P., and Proudfoot, D. A. (1946). Underwater noise due to marine life. *J. Acoust. Soc. Am.* **18**, 446–449.

Lowe-McConnell, R. H. (1987). "Ecological Studies in Tropical Fish Communities." Cambridge University Press, Cambridge, UK.

Lu, Z., and Popper, A. N. (1998). Morphological polarizations of sensory hair cells in the three otolithic organs of a telelost fish: Fluorescent imaging of ciliary bundles. *Hear. Res.* **126**, 47–57.

Lubchenco, J. (1978). Plant species diversity in a marine intertidal community: Importance of herbivore food preference and algal competitive abilities. *Am. Nat.* **112**, 23–39.

Lubchenco, J. (1998). Entering the century of the environment: The need for a new social contract for science. *Science* **279**, 491–497.

Lubchenco, J., Olson, A. M., Brubaker, L. B., Carpenter, S. R., Holland, M. M., Hubbell, S. P., Levin, S. A., MacMahon, J. A., Matson, P. A., Melillo, J. M., Mooney, H. A., Peterson, C. H., Pulliam, H. R., Real, L. A., Regal, P. J., and Risser, P. G. (1991). The sustainable biosphere initiative: An ecological research agenda. *Ecology* **72**, 371–412.

Luh, H. K., and Mok, H. K. (1986). Sound production in the domino damselfish, *Dascyllus trimaculatus* (Pomacentridae) under laboratory conditions. *Jpn. J. Ichthyol.* **33**, 70–74.

Lundberg, J. G. (1993). African-South American freshwater fish clades and continental drift: problems with a paradigm. *In* "The African-South America Connection" (W. B. George and R. Lavocat, eds.), Oxford Monogr. Biogeog 7, pp. 156–199. Clarendon Press, Oxford.

Lutnesky, M. M. F. (1994). Density dependent protogynous sex change in territorial haremic fishes: Models and evidence. *Behav. Ecol.* **5**, 375–383.

Lythgoe, J. N., Muntz, W. R. A., Partridge, J. C., Shand, J., and Williams, D. McB. (1994). The ecology of the visual pigments of snappers (Lutjanidae) on the Great Barrier Reef. *J. Comp. Physiol. A* **174**, 461–467.

Ma, Y., Kjesbu, O. S., and Jorgensen, T. (1998). Effects of ration on the maturation and fecundity in captive Atlantic herring *(Clupea harengus)*. *Can. J. Fish. Aquat. Sci.* **55**, 900–908.

MacArthur, R. H. (1968). The theory of the niche. *In* "Population Biology and Evolution" (R. C. Lewontin, ed.), pp. 159–176. Syracuse University Press, Syracuse, NY.

MacArthur, R. H. (1972). "Geographical Ecology: Patterns in the Distribution of Species." Harper & Row, New York.

MacGinitie, G. E. (1939). The natural history of the blind goby, *Typhlogobius californiensis* Steindachner. *The Aquarium, Philadelphia* **21**, 489–505.

Mackas, D. L., Denman, K. L., and Abbott, M. R. (1985). Plankton patchiness: Biology in the physical vernacular. *Bull. Mar. Sci.* **37**, 652–674.

MacKenzie, B. R. (2000). Turbulence, larval fish ecology and fisheries recruitment: A review of field studies. *Oceanol. Acta* **23**, 357–375.

MacKenzie, B. R., Leggett, W. C., and Peters, R. H. (1990). Estimating larval fish ingestion rates: Can laboratory derived values be reliably extrapolated to the wild? *Mar. Ecol.: Prog. Ser.* **67**, 209–225.

MacKenzie, B. R., Miller, T. J., Cyr, S., and Leggett, W. C. (1994). Evidence for a dome-shaped relationship between turbulence and larval ingestion rates. *Limnol. Oceanogr.* **39**, 1790–1799.

Magnhagen, C. (1993). Conflicting demands in gobies: When to eat, reproduce and avoid predators. *Mar. Behav. Physiol.* **23**, 79–90.

Magurran, A. E. (1988). "Ecological Diversity and its Measurement." Croom Helm, London.

Maisey, J. G. (1991). "Santana Fossils: An Illustrated Atlas." TFH Publications, Neptune City, NY.

Maisey, J. G. (1994). Predator-prey relationships and trophic level reconstruction in a fossil fish community. *Environ. Biol. Fishes* **40**, 1–22.

MalÉcot, G. (1968). "The Mathematics of Heredity." Freeman, San Francisco.

Man, A., Law, R., and Polunin, N. V. C. (1995). Roles of marine reserves in recruitment to reef fisheries: A metapopulation model. *Biol. Conserv.* **71**, 197–204.

Manickchand-Heileman, S. C., and Phillip, D. A. T. (2000). Age and growth of the yellowedge grouper, *Epinephelus flavolimbatus* and the yellowmouth grouper, *Mycteroperca interstitialis* off Trinidad and Tobago. *Fish. Bull.* **98**, 290–298.

Mann, D. A., Lu, Z., and Popper, A. N. (1997). A clupeid fish can detect ultrasound. *Nature (London)* **389**, 341.

Mann, S., Sparks, N. H. C., Walker, M. M., and Krischvink, J. L. (1988). Ultrastructure, morphology and organization of biogenic magnitite from sockeye salmon, *Oncorhynchus nerka:* Implications for magnetoreception. *J. Exp. Biol.* **140**, 35–49.

Manooch, C. S., III (1987). Age and growth of Snappers and Groupers. *In* "Tropical Snappers and Groupers: Biology and Fisheries Management." Westview Press, Boulder, Co.
Mapstone, B. D., and Fowler, A. J. (1988). Recruitment and the structure of assemblages of fish on coral reefs. *Trends Ecol. Evol.* **3**, 72–77.
Mapstone, B. D., McKinlay, J. P., and Davies, C. R. (1996a). "A Description of the Commercial Reef Line Fishery Logbook Data held by The Queensland Fisheries Management Authority." CRC Reef Research Centre, Townsville, Australia.
Mapstone, B. D., Campbell, R. A., and Smith, A. D. M. (1996b). "Design of Experimental Investigations of the Effects of Line and Spear Fishing on the Great Barrier Reef," Reef Res. Cent. Tech. Rep. No. 7. CRC Reef Research Centre, Townsville, Australia.
Mapstone, B. D., Ayling, A. M., and Choat, J. H. (1997a). "A Visual Survey of Demersal Biota in the Cairns Section of the Great Barrier Reef Marine Park," A report to the Great Barrier Reef Marine Park Authority, Townsville, Australia.
Mapstone, B. D., Davies, C. R., and Robertson, J. W. (1997b). The effects of line fishing on the Great Barrier Reef: Available evidence and future directions. *Proc. Great Barrier Reef: Sci., Use Manage. Natl. Conf.*, Townsville, 1996, Vol. 1, pp. 178–192.
Marchand, F., and Boisclair, D. (1998). Influences of fish density on the energy allocation pattern of juvenile brook trout (*Salvelinus fontinalis*). *Can. J. Fish. Aquat. Sci.* **55**, 796–805.
Marconato, A. (1985). Reproductive behaviour and mate choice in *Cottus gobio* (Pisces, Cottidae). *Monit. Zool. Ital.* [N.S.] **19**, 60–61.
Marconato, A., and Shapiro, D. Y. (1996). Sperm allocation, sperm production and fertilization rates in the bucktooth parrotfish. *Anim. Behav.* **52**, 971–980.
Marconato, A., Tessari, V., and Marin, G. (1995). The mating system of *Xyrichthys novacula:* Sperm economy and fertilization success. *J. Fish Biol.* **47**, 292–301.
Marconato, A., Shapiro, D. Y., Petersen, C. W., Warner, R. R., and Yoshikawa, T. (1997). Methodological analysis of fertilization rate in the bluehead wrasse, *Thalassoma bifasciatum:* Pair vs. group spawns. *Mar. Ecol.: Prog. Ser.* **161**, 61–70.
Margulies, D. (1993). Assessment of the nutritional condition of larval and early juvenile tuna and Spanish mackerel (Pisces, Scombridae) in the Panama Bight. *Mar. Biol.* **115**, 317–330.
Margulies, D. (1997). Development of the visual system and inferred performance capabilities of larval and early juvenile scombrids. *Mar. Freshwater Behav. Physiol.* **30**, 75–98.
Markert, J. A., Arnegard, M. E., Danley, P. D., and Kocher, T. D. (1999). Biogeography and population genetics of the Lake Malawi cichlid *Melanochromis auratus:* Habitat transience, philopatry and speciation. *Mol. Ecol.* **8**, 1013–1026.
Markle, D. F., Harris, P. M., and Toole, C. L. (1992). Metamorphosis and an overview of early-life-history stages in Dover sole *Microstomus pacificus. Fish. Bull.* **90**, 285–301.
Maroz, A., and Fishelson, L. (1997). Juvenile production of *Amphiprion bicinctus* (Pomacentridae, Teleostei) and rehabilitation of impoverished habitats. *Mar. Ecol.: Prog. Ser.* **151**, 295–297.
Marquet, P. A., Navarrete, S. A., and Castilla, J. C. (1990). Scaling population density to body size in rocky intertidal communities. *Science* **250**, 1125–1127.
Marshall, C. T., and Frank, K. T. (1999a). The effect of interannual variation in growth and condition on haddock recruitment. *Can. J. Fish. Aquat. Sci.* **56**, 347–355.
Marshall, C. T., and Frank, K. T. (1999b). Implications for density dependent juvenile growth for compensatory recruitment regulation of haddock. *Can. J. Fish. Aquat. Sci.* **56**, 356–363.
Marshall, S. L., and Parker, S. S. (1982). Pattern identification in the microstructure of sockeye salmon (*Oncorhynchus nerka*) otoliths. *Can. J. Fish. Aquat. Sci.* **39**, 542–547.
Martin, A. P., Naylor, G. J. P., and Palumbi, S. R. (1992). Rates of mitochondrial DNA evolution in sharks are slow compared to mammals. *Nature (London)* **357**, 153–155.
Martins, E. P. (2000). Adaptation and the comparative method. *Trends. Ecol. Evol.* **15**, 296–299.
Marx, J. M., and Herrnkind, W. F. (1985). Macroalgae (Rhodophyta: *Laurencia* spp.) as habitat for young spiny lobsters, *Panulirus arugus. Bull. Mar. Sci.* **36**, 423–431.
Masterson, C. F., Danilowicz, B. S., and Sale, P. F. (1997). Yearly and inter-island variation in the recruitment dynamics of the bluehead wrasse *(Thalassoma bifasciatum). J. Exp. Mar. Biol. Ecol.* **214**, 149–166.
Masuda, R., and Tsukamoto, K. (1996). Morphological development in relation to phototaxis and rheotaxis in the striped jack, *Pseudocaranx dentex. Mar. Freshwater Behav. Physiol.* **28**, 75–90.
Masuda, R., and Tsukamoto, K. (1998). The ontogeny of schooling behavior in the striped jack. *J. Fish Biol.* **52**, 483–493.
Masuma, S., Tezuka, N., and Teruya, K. (1993). Embryonic and morphological development of larval and juvenile coral trout, *Plectropomas leopardus. Jpn. J. Ichthyol.* **40**, 333–342.
Matheson, R. H. I., and Huntsman, G. R. (1984). Growth, mortality and yield-per-recruit models for speckled hind (*Epinephelus drummondhayi*) and snowy grouper (*Epinephelus niveatus*) from the U.S.A south Atlantic bight. *Trans. Am. Fish. Soc.* **113**, 607–616.
Mathewson, R. F., and Hodgson, E. S. (1972). Klinotaxis and rheotaxis in orientation of sharks toward chemical stimuli. *Comp. Biochem. Physiol.* **42A**, 79–84.
Mathis, A., and Smith, R. J. F. (1992). Avoidance of areas marked with a chemical alarm substance by fathead minnows (*Pimephales promelas*) in a natural habitat. *Can. J. Zool.* **70**, 1473–1476.
Mathis, A., Chivers, D. P., and Smith, R. J. F. (1995). Chemical alarm signals: Predator deterrents or predator attractants? *Am. Nat.* **145**, 994–1005.

Matsuoka, M. (1998). Development of the lateral muscle in the Japanese sardine *Sardinops melanostictus*. *Fish. Sci.* **64,** 83–88.
Maul, G. A. (1977). The annual cycle of the Gulf Loop Current. Part I: Observations during a one year time series. *J. Mar. Res.* **35,** 29–47.
Maul, G. A., and Vukovich, F. (1993). The relationship between variations in the Gulf of Mexico Loop Current and Straits of Florida volume transport. *J. Phys. Oceanogr.* **23,** 785–796.
May, R. M. (1974). General introduction. *In* "Ecological Stability" (M. B. Usher and M. H. Williamson, eds.), pp. 1–14. Chapman & Hall, London.
May, R. M. (1975). Patterns of species abundance and diversity. *In* "Ecology and Evolution of Communities" (M. L. Cody and J. M. Diamond, eds.), pp. 81–120. Belknap Press of Harvard University Press, Cambridge, MA.
Maynard Smith, J. (1977). Parental investment—a prospective analysis. *Anim. Behav.* **25,** 1–9.
Mayr, E. (1954). Change of genetic environment and evolution. *In* "Evolution as a Process" (J. Huxley, A. Hardy, and E. Ford, eds.), pp. 157–180. Allen & Unwin, London.
McAllister, D. E. (1988). Environmental, economic and social costs of coral reef destruction in the Philippines. *Galaxea* **7,** 161–178.
McAllister, D. E., Schueler, F. W., Roberts, C. M., and Hawkins, J. P. (1994). Mapping and GIS analysis of the global distribution of coral reef fishes on an equal-area grid. *In* "Advances in Mapping the Biodiversity of Nature" (R. Miller, ed.), pp. 155–175. Chapman & Hall, London.
McAllister, D. E., Caho, N. L., and Shih, C. T. (1999). Cyanide fisheries: Where did they start? *Secretariat of the Pacific Community Live Reef Fish Inf. Bull.* **5,** 18–21.
McArdle, B. H., Gaston, K. J., and Lawton, J. H. (1990). Variation in the size of animal populations: Patterns, problems and artefacts. *J. Anim. Ecol.* **59,** 439–454.
McBride, R. S., and Able, K. W. (1998). Ecology and fate of butterflyfishes, *Chaetodon* spp., in the temperate, Western North Atlantic. *Bull. Mar. Sci.* **63,** 401–416.
McCauley, R. D., and Cato, D. H. (1998). Evening fish choruses near coral reef systems in the Great Barrier Reef, Australia. *Proc. 16th Int. Congr. Acoust.,* Seattle, WA, Vol. 2, pp. 1029–1030.
McClanahan, T. R. (1994). Kenyan coral reef lagoon fish: Effects of fishing, substrate complexity, and sea urchins. *Coral Reefs* **13,** 231–241.
McClanahan, T. R. (2000). Recovery of a coral reef keystone predator, *Balistapus undulatus,* in East African marine parks. *Biol. Conserv.* **94,** 191–198.
McClanahan, T. R., and Kaunda-Arara, B. (1996). Fishery recovery in a coral-reef marine park and its effect on the adjacent fishery. *Conserv. Biol.* **10,** 1187–1199.
McClanahan, T. R., and Shafir, S. H. (1990). Causes and consequences of sea urchin abundance and diversity in Kenyan coral reef lagoons. *Oecologia* **83,** 362–370.
McCleave, J. D., Kleckner, R. C., and Castonguay, M. (1987). Reproductive sympatry of American and European eels and implications for migration and taxonomy. *Am. Fish. Soc. Symp.* **1,** 286–297.
McCleave, J. D., Brickley, P. J., O'Brien, K. M., Kistner, D. A., Wong M. W., Gallagher, M., and Watson, S. M. (1998). Do leptocephali of the European eel swim to reach continental waters? Status of the question. *J. Mar. Biol. Assoc. U.K.* **78,** 285–306.
McCormick, M. I. (1993). Development and changes at settlement in the barbel structure of the reef fish, *Upeneus tragula* (Mullidae). *Environ. Biol. Fishes* **37,** 269–282.
McCormick, M. I. (1994). Variability in age and size at settlement of the tropical goatfish *Upeneus tragula* (Mullidae) in the northern Great Barrier Reef. *Mar. Ecol.: Prog. Ser.* **103,** 1–15.
McCormick, M. I. (1995). Fish feeding on mobile benthic invertebrates: Influence of spatial variability in habitat associations. *Mar. Biol.* **121,** 627–637.
McCormick, M. I. (1998a). Condition and growth of reef fish at settlement: Is it important? *Aust. J. Ecol.* **23,** 258–264.
McCormick, M. I. (1998b). Behaviorally induced maternal stress in a fish influences progeny quality by a hormonal mechanism. *Ecology* **79,** 1873–1883.
McCormick, M. I. (1998c). Ontogeny of diet shifts by a micro-carnivorous fish, *Cheilodactylus spectabilis:* Relationship between feeding mechanics, microhabitat selection and growth. *Mar. Biol.* **132,** 9–20.
McCormick, M. I. (1999a). Experimental test of the effect of maternal hormones on larval quality of a coral reef fish. *Oecologia* **118,** 412–422.
McCormick, M. I. (1999b). Delayed metamorphosis of a tropical reef fish (*Acanthurus triostegus*): A field experiment. *Mar. Ecol.: Prog. Ser.* **176,** 25–38.
McCormick, M. I., and Choat, J. H. (1987). Estimating total abundance of a large temperate reef fish using visual strip-transects. *Mar. Biol.* **96,** 469–478.
McCormick, M. I., and Kerrigan, B. A. (1996). Predation and its influence on the condition of a newly settled tropical demersal fish. *Mar. Freshwater Res.* **47,** 557–562.
McCormick, M. I., and Makey, L. J. (1997). Post-settlement transition in coral reef fishes: Overlooked complexity in niche shifts. *Mar. Ecol.: Prog. Ser.* **153,** 247–257.
McCormick, M. I., and Makey, L. J. (2002). Comparative study of metamorphosis in tropical reef fishes. *Mar. Biol.* (in press).
McCormick, M. I., and Milicich, M. J. (1993). Late pelagic-stage goatfishes: Distribution patterns and inferences on schooling behaviour. *J. Exp. Mar. Biol. Ecol.* **174,** 15–42.
McCormick, M. I., and Molony, B. W. (1992). Effects of feeding history on the growth characteristics of a reef fish at settlement. *Mar. Biol.* **114,** 165–173.
McCormick, M. I., and Molony, B. W. (1993). Quality of the reef fish *Upeneus tragula* (Mullidae) at settlement: Is size a good indicator of condition? *Mar. Ecol.: Prog. Ser.* **98,** 45–54.
McCormick, M. I., and Molony, B. W. (1995). Influence of water temperature during the larval stage on size, age and

body condition of a tropical reef fish at settlement. *Mar. Ecol.: Prog. Ser.* **118**, 59–68.

McCormick, M. I., and Shand, J. (1992). Metamorphosis of the visual and barbel sensory systems at settlement in the reef fish *Upeneus tragula* (Family Mullidae). *Proc. 7th Int. Coral Reef Symp.*, Vol. 1, pp. 616–623.

McCoy, E. D., and Mushinsky, H. R. (1992). Rarity of organisms in the Sand Pine scrub habitat of Florida. *Conserv. Biol.* **6**, 537–547.

McDonald, D. J., Juritz, J. M., Cowling, R. M., and Knottenbelt, W. J. (1995). Modelling the biological aspects of local endemism in South African fynbos. *Plant Syst. Evol.* **195**, 137–147.

McErlean, A. J., and Smith, C. L. (1964). The age of sexual succession in the protogynous hermaphrodite *Mycteroperca microlepis*. *Trans. Am. Fish. Soc.* **93**, 301–302.

McFarland, W. N. (1986). Light in the sea—correlations with behaviors of fishes and invertebrates. *Am. Zool.* **26**, 389–401.

McFarland, W. N. (1991). The visual world of coral reef fishes. *In* "The Ecology of Fishes on Coral Reefs" (P. F. Sale, ed.), pp. 16–38. Academic Press, San Diego.

McFarland, W. N., and Loew, E. R. (1994). Ultraviolet visual pigments in marine fishes of the family Pomacentridae. *Vision Res.* **34**, 1393–1396.

McFarland, W. N., and Munz, F. W. (1975a). Part II. The photic environment of clear tropical seas during the day. *Vision Res.* **15**, 1063–1070.

McFarland, W. N., and Munz, F. W. (1975b). Part III. The evolution of photopic visual pigments in fishes. *Vision Res.* **15**, 1071–1080.

McFarland, W. N., Brothers, E. B., Ogden, J. C., Shulman, M. J., Bermingham, E. L., and Kotchian-Prentiss, N. M. (1985). Recruitment patterns in young French grunts, *Haemulon flavolineatum* (Family Haemulidae), at St. Croix, Virgin Islands. *Fish. Bull.* **83**, 413–426.

McFarlane, G. A., and Beamish, R. J. (1995). Validation of the otolith cross-section method of age determination for Sablefish (*Anoplopoma fimbria*) using oxytetracycline. *In* "Recent Developments in Fish Otolith Research" (D. H. Secor, J. M. Dean, and S. E. Campana, eds.), pp 319–330. University of South Carolina Press, Columbia.

McGovern, J. C., Wyanski, D. M., Pashuk, O., Manooch, C. S., III, and Sedberry, G. R. (1998). Changes in the sex ratio and size at maturity of gag, *Mycteroperca microlepis*, from the Atlantic coast of the southeastern United States during 1976–1995. *Fish. Bull.* **96**, 797–807.

McGuiness K. A., and Davis A. R. (1989). Analysis and interpretation of the recruit-settler relationship. *J. Exp. Mar. Biol. Ecol.* **134**, 197–202.

McIntosh, R. P. (1987). Pluralism in ecology. *Annu. Rev. Ecol. Syst.* **18**, 321–341.

McKee-Lewis, K. K. (1997). Rapid changes and growth of California's live finfish fishery. *In* "Marketing and Shipping Aquatic Products," pp. 116–120. Northeast Regional Agricultural Engineering Service, Ithaca, NY.

McKinnon, A. D., and Thorrold, S. R. (1993). Zooplankton community structure and copepod egg production in coastal waters of the central Great Barrier Reef lagoon. *J. Plankton Res.* **15**, 1387–1411.

McManus, J. W. (1985). Marine speciation, tectonics and sea-level changes in southeast Asia. *Proc. 5th Int. Coral Reef Congr.*, Vol. 4, pp. 133–138.

McManus, J. W. (1994). The Spratly Islands: A marine park? *Ambio* **23**, 181–186.

McManus, J. W. (1996). Social and economic aspects of reef fisheries and their management. *In* "Reef Fisheries" (N. V. C. Polunin and C. M. Roberts, eds.), pp 249–281. Chapman & Hall, London.

McManus, J. W., and Menez, L. A. B. (1997). The proposed international Spratly Island Marine Park: Ecological considerations. *Proc. 8th Int. Coral Reef Symp.* Vol. 2, pp. 1943–1948.

McManus, J. W., Nanola, C. L. J., Jr., Reyes, R. B. Jr., and Kesner, K. N. (1992). "Resource Ecology of the Bolinao Coral Reef System," ICLARM Study Rev. 22. ICLARM, Philippines.

McManus, J. W., Reyes, R. B. Jr., and Nanola, C. L. J., Jr. (1997). Effects of some destructive fishing methods on coral cover and potential rates of recovery. *Environ. Manage.* **21**, 69–78.

McManus, J. W., Menez, L. A. B., Kesner, K. N., Reyes, R. B. Jr., Vergara, S. G., and Ablan, M. C. (2000). Coral reef fishing and coral-algal phase shifts: Implications for global reef status. *ICES J. Mar. Sci.* **57**, 572–578.

McMillan, W. O., and Palumbi, S. R. (1995). Concordant evolutionary patterns among Indo-West Pacific butterflyfishes. *Proc. R. Soc. London, Ser. B* **260**, 229–236.

McMillan, W. O., and Palumbi, S. R. (1997). Rapid rate of control-region evolution in Pacific butterflyflishes (Chaetodontidae). *J. Mol. Evol.* **45**, 473–484.

McMillan, W. O., Weigt, L. A., and Palumbi, S. R. (1999). Color pattern evolution, assortative mating, and genetic differentiation in brightly colored butterflyfishes (Chaetodontidae). *Evolution (Lawrence, Kans.)* **53**, 247–260.

McNeill, S. E. (1994). The selection and design of marine protected areas: Australia as a case study. *Biodivers. Conserv.* **3**, 586–605.

McShea, D. W. (1994). Mechanisms of large-scale evolutionary trends. *Evolution (Lawrence, Kans.)* **48**, 1747–1763.

Medley, P. A., Guadian, G., and Wells, S. (1993). Coral reef fisheries stock assessment. *Rev. Fish Biol. Fish.* **3**, 242–285.

Meekan, M. G. (1992). Limitations to the back-calculation of recruitment patterns from otoliths. *Proc. 7th Int. Coral Reef Symp.*, Vol. 1, pp. 624–628.

Meekan, M. G., and Choat, J. H. (1997). Latitudinal variation in abundance of herbivorous fishes: A comparison of temperate and tropical reefs. *Mar. Biol.* **128**, 373–383.

Meekan, M. G., and Fortier, L. (1996). Selection for fast growth during the larval life of Atlantic cod *Gadus*

morhua on the Scotian Shelf. *Mar. Ecol.: Prog. Ser.* **137**, 25–37.
Meekan, M. G., Milicich, M. J., and Doherty, P. J. (1993). Larval production drives temporal patterns of larval supply and recruitment of a coral reef damselfish. *Mar. Ecol.: Prog. Ser.* **93**, 217–225.
Meekan, M. G., Dodson, J. J., and Ryan, D. A. J. (1998). Otolith and fish size relationships, measurement error, and size-selective mortality during the early life of Atlantic salmon *(Salmo salar). Can. J. Fish. Aquat. Sci.* **55**, 1663–1673.
Meekan, M. G., Wellington, G. M., and Axe, L. (1999). El-Niño-Southern Oscillation events produce checks in the otoliths of coral reef fishes in the Galápagos Archipelago. *Bull. Mar. Sci.* **64**, 383–390.
Meekan, M. G., White, L., and Doherty, P. J. (2000). Recapture experiments show the low sampling efficiency of light-traps. *Bull. Mar. Sci.* **67**, 875–885.
Meekan, M. G., Ackerman, J. L., and Wellington, G. M. (2001). Demography and age structures of coral reef damselfishes in the tropical eastern Pacific Ocean. *Mar. Ecol.: Prog. Ser.* **212**, 223–232.
Meesters, E. H., Noordeloos, M., and Bak, R. P. M. (1994). Damage and regeneration: Links to growth in the reef-building coral *Montastrea annularis. Mar. Ecol.: Prog. Ser.* **122**, 119–128.
Meng, L. (1993). Sustainable swimming speeds of striped bass larvae. *Trans. Am. Fish. Soc.* **122**, 702–708.
Mesa, M. G., Poe, T. P., Gadomski, D. M., and Petersen, J. H. (1994). Are all prey created equal? A review and synthesis of differential predation on prey in substandard condition. *J. Fish Biol.* **45**, 81–96.
Metcalfe, N. B. (1986). Interspecific variation in competitive ability and food intake in salmonids: Consequences for energy budgets and growth rates. *J. Fish Biol.* **28**, 525–531.
Metcalfe, N. B. (1993). Behavioral causes and consequences of life-history variation in fish. *Mar. Behav. Physiol.* **23**, 205–217.
Metcalfe, N. B., Huntingford, F. A., Thorpe, J. E., and Adams, C. E. (1990). The effects of social status on life-history variation in juvenile salmon. *Can. J. Zool.* **68**, 2630–2636.
Meyer, D. L., and Macurda, D. B., Jr. (1977). Adaptive radiation of the cromatulid crinoids. *Paleobiology* **3**, 74–82.
Miagawa, K. (1989). Experimental analysis of the symbiosis between anemonefishes and sea anemones. *Ethology* **80**, 19–46.
Milazzo, M. (1998). "Subsidies in World Fisheries: A Reexamination." Tech. Pap. No. 406. World Bank, Washington, DC.
Milicich, M. J. (1994). Dynamic coupling of reef fish replenishment and oceanographic processes. *Mar. Ecol.: Prog. Ser.* **110**, 135–144.
Milicich, M. J., and Doherty, P. J. (1994). Larval supply of coral reef fish populations: Magnitude and synchrony of replenishment to Lizard Island, Great Barrier Reef. *Mar. Ecol.: Prog. Ser.* **110**, 121–134.
Milicich, M. J., Meekan, M. G., and Doherty, P. J. (1992). Larval supply: A good predictor of recruitment of three species of reef fish (Pomacentridae). *Mar. Ecol. Prog. Ser.* **86**, 153–166.
Miller, A. I. (1997). Dissecting global diversity patterns: Examples from the Ordovician radiation. *Annu. Rev. Ecol. Syst.* **28**, 85–104.
Miller, T. J., Crowder, L. B., Rice, J. A., and Marschall, E. A. (1988). Larval size and recruitment mechanisms in fishes: Toward a conceptual framework. *Can. J. Fish. Aquat. Sci.* **45**, 1657–1670.
Millwood, G. E., and Turner, A. (1995). Trace elements in estuaries. *In* "Trace Elements in Natural Waters" (B. Salbu and E. Steinnes, eds.), pp. 223–245. CRC Press, Boca Raton, FL.
Milton, D. A., Short, S. A., O'Neill, M. F., and Blaber, S. J. M. (1995). Ageing of three species of tropical snapper (Lutjanidae) from the Gulf of Carpentaria, Australia, using radiometry and otolith ring counts. *Fish. Bull.* **93**, 103–115.
Mittelbach, G. G. (1984). Predation and resource partitioning in two sunfishes. *Ecology* **65**, 499–513.
Mittelbach, G. G., and Chesson, P. L. (1987). Predation risk: Indirect effects on fish populations. *In* "Predation: Direct and Indirect Impacts on Aquatic Communities" (W. C. Kerfoot and A. Sih, eds.), pp. 315–330. University of New England Press, Hanover, NH.
Mittelbach, G. G., and Osenberg, C. W. (1994). Using foraging theory to study trophic interactions. *In* "Theory and Application in Studies of Fish Feeding Ecology" (D. J. Stouder, K. L. Fresh, and B. Feller, eds.), pp. 45–59. University of South Carolina Press, Columbia.
Mitton, J. B., and Lewis, W. M., Jr. (1989). Relationships between genetic variability and life history features of bony fishes. *Evolution (Lawrence, Kans.)*, **43**, 1712–1723.
Mohler, M. H. (1984). Aspects of sexual selection, female fecundity and functions of male-sound on females of the bicolor damselfish, *Pomacentrus partitus*. Unpublished Ph.D. Dissertation, University of Basel, Switzerland.
Mok, H.-K. (1977). Gut patterns of the Acanthuridae and Zanclidae. *Jpn. J. Ichthyol.* **23**, 214–219.
Mok, H.-K. (1981). Sound production in the naked goby, *Gobiosoma bosci* (Pisces, Gobiidae). *In* "Hearing and Sound Communication in Fishes" (W. N. Tavolga, A. N. Popper, and R. R. Fay, eds.), pp. 447–456. Springer-Verlag, New York.
Monod, T. (1951). Notes sur le squelette viscéral des Scaridae. *Bull. Soc. Hist. Nat. Toulouse* **86**, 191–194.
Montgomery, J. C. (1984). Noise cancellation in the electrosensory system of the thornback ray: Common mode rejection of input produced by the animal's own ventilatory movement. *J. Comp. Physiol. A* **155**, 103–111.
Montgomery, J. C. (1989). Lateral line detection of planktonic prey. *In* "The Mechanosensory Lateral Line: Neurobiology and Evolution" (S. Coombs, P. Görner, and H. Münz, eds.), pp. 561–574. Springer-Verlag, New York.
Montgomery, J. C., and Saunders, A. J. (1985). Functional-morphology of the piper *Hyporhamphus ihi* with

reference to the role of the lateral line in feeding. *Proc. R. Soc. London, Ser. B* **224**, 197–208.
Montgomery, J. C., Coombs, S., and Halstead, M. (1995). Biology of the mechanosensory lateral line in fishes. *Rev. Fish Biol. Fish.* **5**, 399–416.
Montgomery, J. C., Baker, C. F., and Carton, A. G. (1997). The lateral line can mediate rheotaxis in fish. *Nature (London)* **389**, 960–963.
Montgomery, W. L., and Pollak, P. E. (1988). *Epulopiscium fishelsoni* N. G., N. Sp., a protist of uncertain taxonomic affinities from the gut of an herbivorous reef fish. *J. Protozool.* **35**, 565–569.
Mooi, R. D. 1993. Phylogeny of the Plesiopidae (Pisces: Perciformes) with evidence for the inclusion of the Acanthoclinidae. *Bull. Mar. Sci.* **52**, 284–326.
Morales-Nin, B., Torres, G. J., Lombarte, A., and Recasens, L. (1998). Otolith growth and age estimation in the European Hake. *J. Fish Biol.* **53**, 1155–1168.
Moran, P. J. (1986). The *Acanthaster* phenomenon. *Oceanogr. Mar. Biol.* **24**, 379–480.
Moreno, C. A., and Jara, H. F. (1984). Ecological studies on fish fauna associated with *Macrocystis pyrifera* belts in the south of Fueguian Islands, Chile. *Mar. Ecol.: Prog. Ser.* **15**, 99–107.
Morgan, M. J., Wilson, C. E., and Crim, L. W. (1999). The effect of stress on reproduction in Atlantic cod. *J. Fish Biol.* **54**, 477–488.
Morgan, S. G. (1995). The timing of larval release. *In* "Ecology of Marine Invertebrate Larvae" (L. McEdward, ed.), pp. 157–191. CRC Press, Boca Raton, FL.
Morgan, S. G., and Christy, J. H. (1994). Plasticity, constraint and optimality in reproductive timing. *Ecology* **75**, 2185–2203.
Morris, A. V., Roberts, C. M., Hawkins, J. P. (2000). The threatened status of groupers (Epinephelinae). *Biodivers. Conserv.* **9**, 919–942.
Moser, H. G., ed. (1996). "The Early Stages of Fishes in the California Current Region," Atlas No. 33. California Cooperative Oceanic Fishery Investigations, San Diego.
Motta, P. J. (1982). Functional morphology of the head of the inertial suction feeding butterflyfish, *Chaetodon miliaris* (Perciformes, Chaetodontidae). *J. Morphol.* **174**, 283–312.
Motta, P. J. (1984a). Mechanics and function of jaw protrusion in teleost fishes: A review. *Copeia*, pp. 1–18.
Motta, P. J. (1984b). Tooth attachment, replacement, and growth in the butterflyfish *Chaetodon miliaris* (Chaetodontidae, Perciformes). *Can J. Zool.* **62**, 183–189.
Motta, P. J. (1985). Functional morphology of the head of Hawaiian and Mid-Pacific butterflyfishes (Perciformes, Chaetodontidae). *Environ. Biol. Fishes.* **13**, 253–276.
Motta, P. J. (1987). A quantitative analysis of ferric iron in butterflyfish teeth (Chaetodontidae, Perciformes) and the relationship to feeding ecology. *Can. J. Zool.* **65**, 106–112.
Motta, P. J. (1988). Functional morphology of the feeding apparatus of ten species of Pacific butterflyfishes (Perciformes, Chaetodontidae): An ecomorphological approach. *Environ. Biol. Fishes* **22**, 39–67.
Motta, P. J. (1989). Dentition patterns among Pacific and western Atlantic butterfly fishes (Perciformes, Chaetodontidae): Relationship to feeding ecology and evolutionary history. *Environ. Biol. Fishes* **25**, 159–170.
Motta, P. J., and Kotrshal, K. M. (1992). Correlative, experimental, and comparative experimental approaches in ecomorphology. *Neth. J. Zool.* **42**, 400–415.
Motta, P. J., and Wilga, C. D. (1999). Anatomy of the feeding apparatus of the nurse shark, *Ginglymostoma cirratum. J. Morphol.* **241**, 33–60.
Moulton, L. L. (1977). An ecological analysis of fishes inhabiting the rocky nearshore regions of northern Puget Sound. Ph.D. Thesis, University of Washington, Seattle.
Moura, R. L., Martins Rodrigues, M. C., Franchini-Filho, R. B., and Sazima, I. (1999). Unexpected richness of reef corals near the southern Amazon River mouth. *Coral Reefs* **18**, 170.
Mous, P. J., Halim, A., and Pet, J. S. (1999). "Harvest Characteristics of Gango, a Method to Capture Fingerling Groupers from Mangrove Areas in West Flores, Indonesia." A Report from the TNC/YPAN Komodo Marine Conservation Project, The Nature Conservancy.
Mous, P. J., Pet-Soede, L., Erdmann, M., Cesar, H. S. J., Sadovy, Y., and Pet, J. S. (2000). Cyanide fishing on Indonesian coral reefs for the live food fish market—What is the problem? *Secretariat of the Pacific Community Live Reef Fish Inf. Bull.* **7**, 20–26.
Mousseau, T. A., and Fox, C. W. (1998). "Maternal Effects as Adaptations." Oxford University Press, New York.
Moyer, J. T. (1979). Mating strategies and reproductive behavior of ostraciid fishes at Miyake-jima, Japan. *Jpn. J. Ichthyol.* **26**, 48–160.
Moyer, J. T. (1987). Quantitative observations of predation during spawning rushes of the labrid fish *Thalassoma cupido* at Miyake-jima, Japan. *Jpn. J. Ichthyol.* **34**, 76–81.
Moyer, J. T. (1989). Reef channels as spawning sites for fishes on the Shiraho coral reef, Ishigaki Island, Japan. *Jpn. J. Ichthyol.* **36**, 371–375.
Moyer, J. T. (1991). "Comparative Mating Strategies of Labrid Fishes," Monogr. No. 1. Watanabe Ichthyological Institute, Tokyo.
Mugiya, Y., and Yoshida, M. (1995). Effects of calcium antagonists and other metabolic modulators on in vitro calcium deposition in the rainbow trout, *Oncorhynchus mykiss. Fish. Sci.* **61**, 1026–1030.
Mukai H. (1992). Biogeography of the tropical sea-grasses in the western Pacific. *Aust. J. Mar. Freshwater Res.* **44**, 1–17.
Mulcahy, S. A., Killingley, J. S., Phleger, C. F., and Berger, W. H. (1979). Isotopic composition of otoliths from a benthopelagic fish, *Coryphaenoides acrolepis,* Macrouridae: Gadiformes. *Oceanol. Acta* **2**, 423–427.
Mullaney, M. D., Jr., and Gale, L. D. (1996). Ecomorphological relationships in ontogeny: Anatomy and diet in gag, *Mycteroperca microlepis* (Pisces: Serranidae). *Copeia,* pp. 167–180.

Muller, M. (1987). Optimization principals applied to the mechanism of neurocranium levation and mouth bottom depression in bony fishes (Halecostmoi). *J. Theor. Biol.* **126,** 343–368.

Muller, M. (1989). A quantitative theory of expected volume changes of the mouth during feeding in teleost fishes. *J. Zool.* **217,** 639–661.

Muller, M. (1996). A novel classification of planar four-bar linkages and its application to the mechanical analysis of animal systems. *Philos. Trans. R. Soc. London, Ser. B* **351,** 689–720.

Muller, M., and Osse, J. W. M. (1984). Hydrodynamics of suction feeding fish. *Trans. Zool. Soc. London* **37,** 51–135.

Muller, M., Osse, J. W. M., and Verhagen, J. H. G. (1982). A quantitative hydrodynamical model of suction feeding in fish. *J. Theor. Biol.* **95,** 49–79.

Müller-Karger, F., McClain, C. R., Fisher, T. R., Esaias, W. E., and Varela, R. (1989). Pigment distribution in the Caribbean Sea: Observations from space. *Prog. Oceanogr.* **23,** 23–64.

Mulligan, T. J., and Leaman, B. M. (1992). Length-at-age analysis: Can you get what you see? *Can. J. Fish. Aquat. Sci.* **49,** 632–643.

Mullin, M. M. (1993). "Webs and Scales: Physical and Ecological Processes in Marine Fish Recruitment," Washington Sea Grant. University of Washington Press, Seattle.

Mulvey, M., Keller, G., and Meffe, G. K. (1994). Single- and multiple-locus genotypes and life-history responses of *Gambusia holbrooki* reared at two temperatures. *Evolution (Lawrence, Kans.)* **48,** 1810–1819.

Munday, P. L. (2000). Interactions between habitat use and patterns of abundance in coral-dwelling fishes. *Environ. Biol. Fishes* **58,** 355–369.

Munday, P. L. (2001). Fitness consequences of habitat use and competition among coral-dwelling fishes. *Oecologia* **128,** 585–593.

Munday, P. L. (in press). Does habitat availability determine geographic-scale abundances of coral-dwelling fishes? Coral Reefs (in press).

Munday, P. L., and Jones, G. P. (1998). The ecological implications of small body size among coral-reef fishes. *Oceanogr. Mar. Biol.* **36,** 373–411.

Munday, P. L., Jones, G. P., and Caley, M. J. (1997). Habitat specialization and the distribution and abundance of coral-dwelling gobies. *Mar. Ecol.: Prog. Ser.* **152,** 227–239.

Munday, P. L., Jones G. P., Öhman M. C., and Kaly U. L. (1998). Enhancement of recruitment to coral reefs using light-attractors. *Bull. Mar. Sci.* **63,** 581–588.

Munday, P. L., Jones, G. P., and Caley, M. J. (2001). Interspecific competition and coexistence in a guild of coral-dwelling gobies. *Ecology,* **82,** 2177–2189.

Munro, J. L. (1983). Coral reef fish and fisheries of the Caribbean Sea. *In* "Caribbean Coral Reef Fishery Resources" (J. L. Munro, ed.), ICLARM Stud. Rev. 7, pp. 1–9. ICLARM, Manila.

Munro, J. L. (1996). The scope of tropical reef fisheries and their management. *In* "Reef Fisheries" (N. V. C. Polunin and C. M. Roberts, eds.), pp. 1–14. Chapman & Hall, London.

Munro, J. L., and Polunin, N. V. C. (1997). A decade of progress in coral reef fisheries research: 1986–1995. *Proc. 8th Int. Coral Reef Symp.,* Vol. 2, pp. 2003–2008.

Munro, J. L., and Williams, D. McB. (1985). Assessment and management of coral reef fisheries: Biological environmental and socio-economic aspects. *Proc. 5th Int. Coral Reef Congr.,* Vol. 4, pp. 545–581.

Munro, J. L., Jones, A. C., and Dimitriou, D. (1968). Abundance and distribution of the larvae of the pink shrimp (Panaeus duorarum) on the Tortugas Shelf of Florida, August 1962–October 1964. *Fish. Bull.* **67,** 165–181.

Munro, J. L., Gaut, V. C., Thompson, R., and Reeson, P. H. (1973). The spawning seasons of Caribbean reef fishes. *J. Fish Biol.* **5,** 69–78.

Muntz, W. R. A. (1990). Stimulus, environment and vision in fishes. *In* "The Visual System of Fish" (R. H. Douglas and M. B. A. Djamgoz, eds.), pp. 491–511. Chapman & Hall, New York.

Munz, F. W., and McFarland, W. N. (1973). The significance of spectral position in the rhodopsins of tropical marine fishes. *Vision Res.* **13,** 1829–1874.

Münz, H. (1989). Functional organization of the lateral line periphery. *In* "The Mechanosensory Lateral Line: Neurobiology and Evolution" (S. Coombs, P. Görner, and H. Münz, eds.), pp. 285–297. Springer-Verlag, New York.

Murdoch, W. W. (1970). Population regulation and population inertia. *Ecology* **51,** 497–502.

Murdoch, W. W. (1994). Population regulation in theory and practice. *Ecology* **75,** 271–287.

Murdoch, W. W., and Bence, J. (1987). General predators and unstable prey populations. *In* "Predation: Direct and Indirect Effects on Aquatic Communities" (W. Kerfoot, and A. Sih, eds.), pp. 17–30. University Press of New England, Hanover, NH.

Murdoch, W. W., and Oaten, A. (1975). Predation and population stability. *Adv. Ecol. Res.* **9,** 1–132.

Murdoch, W. W., and Oaten, A. (1989). Aggregation by parasitoids and predators: Effects on equilibrium and stability. *Am. Nat.* **134,** 288–310.

Murdoch, W. W., and Walde, S. J. (1989). Analysis of insect population dynamics. *In* "Towards a More Exact Ecology" (P. J. Grubb and J. B. Whittaker, eds.), pp. 113–140. Blackwell, Oxford.

Murphy, W. J., and Collier, G. E. (1997). A molecular phylogeny for aplocheiloid fishes (Atherinomorpha, Cyprinodontiformes): The role of vicariance and the origins of annualism. *Mol. Biol. Evol.* **14,** 790–799.

Murray, A. M. (2001). The oldest fossil cichlids (Teleostei: Perciformes): Indication of a 45 million-year-old species flock. *Proc. R. Soc. Lond. B.* **268,** 679–684.

Murray, B. G., Jr. (1999). Can the population regulation controversy be buried and forgotten? *Oikos* **84,** 148–152.

Murray, B. R., Rice, B. L., Keith, D. A., Myerscough, P. J.,

Howell, P. J., Floyd, A. G., Mills, K., and Westoby, M. (1999). Species in the tail of rank-abundance curves. *Ecology* **80**, 1806–1816.

Murray, S. N., Ambrose, R. F., Bohnsack, J. A., Botsford, L. W., Carr, M. E., Davis, G. E., Dayton, P. K., Gotshall, D., Gunderson, D. R., Hixon, M. A., Lubchenco, J., Mangel, M., MacCall, A., McArdle, D., Ogden, J. C., Roughgarden, J., Starr, R. M., Tegner, M. J., and Yoklavich, M. M. (1999). No-take reserve networks: Sustaining fishery populations and marine ecosystems. *Fisheries* **24**(11), 11–25.

Murray, S .P., Hsu, S: A., Roberts, H. H., Owens, E. H., and Crout, R. L. (1982). Physical processes and sedimentation on a broad, shallow bank. *Estuarine Coastal Shelf Sci.* **14**, 135–157.

Musick, J. A. (1999a). Ecology and conservation of long-lived marine animals. *In* "Life in the Slow Lane" (J. A. Musick, ed.), pp. 1–10. American Fisheries Society, Baltimore, MD.

Musick, J. A. (1999b). Criteria to define extinction risk in marine fishes. *Fisheries* **24**, 6–12.

Myers, R. (1986). "Classical and Modern Regression with Applications." Duxbury Press, Boston.

Myers, R. A. (1991). Recruitment variability and range of 3 fish species. *NAFO Sci. Counc. Stud.* **16**, 21–24.

Myers, R. A. (1997). Recruitment variation in fish populations assessed using meta-analysis. *In* "Early Life History and Recruitment in Fish Populations" (R. C. Chambers and E. A. Trippel, eds.), pp. 451–467. Chapman & Hall, London.

Myers, R. A., and Barrowman, N. J. (1996). Is fish recruitment related to spawner abundance? *Fish. Bull.* **94**, 707–724.

Myers, R. A., and Cadigan, N. G. (1993a). Density-dependent juvenile mortality in marine demersal fish. *Can. J. Fish. Aquat. Sci.* **50**, 1576–1590.

Myers, R. A., and Cadigan, N. G. (1993b). Is juvenile natural mortality in marine demersal fish variable? *Can. J. Fish. Aquat. Sci.* **50**, 1591–1598.

Myers, J. H., and Rothman, L. D. (1995). Field experiments to study regulation of fluctuating populations. *In* "Population Dynamics: New Approaches and Synthesis" (N. Cappuccino and P. W. Price, eds.), pp. 229–250. Academic Press, San Diego, CA.

Myers, R. A., Barrowman, N. J., Hutchings, J. A., and Rosenberg, A. A. (1995a). Population dynamics of exploited fish stocks at low population levels. *Science* **269**, 1106–1108.

Myers, R. A., Bridson, J., and Barrowman, N. J. (1995b). Summary of worldwide stock and recruitment data. *Can. Tech. Rep. Fish. Aquat. Sci.* **2024.**

Myers, R. A., Hutchings, J. A., and Barrowman, N. J. (1996). Hypotheses for the decline of cod in the North Atlantic. *Mar. Ecol.: Prog. Ser.* **138**, 293–308.

Myers, R. A., Hutchings, J. A., and Barrowman, N. J. (1997). Why do fish stocks collapse? The example of cod in eastern Canada. *Ecol. Appl.* **7**, 91–106.

Myers, R. F. (1989). "Micronesian Reef Fishes." Coral Graphics, Guam.

Myers, R. F. (1999) "Micronesian Reef Fishes," 3rd ed. Coral Graphics, Guam.

Mylonas, C. C., Sullivan, C. V., and Hinshaw, J. M. (1994). Thyroid hormones in brown trout (*Salmo trutta*) reproduction and early development. *Fish Physiol. Biochem.* **13**, 485–493.

Myrberg, A. A., Jr. (1966). Parental recognition of young in cichlid fishes. *Anim. Behav.* **14**, 565–571.

Myrberg, A. A., Jr. (1972). Ethology of the bicolor damselfish, *Eupomacentrus partitus* (Pisces: Pomacentridae). A comparative analysis of laboratory and field behavior. *Anim. Behav. Monogr.* **5**, 199–283.

Myrberg, A. A., Jr. (1975). The role of chemical and visual stimuli in the preferential discrimination of young by the cichlid fish, *Cichlasoma nigrofasciatum* (Gunther). *Z. Tierpsychol.* **37**, 274–296.

Myrberg, A. A., Jr. (1978). Underwater sound—its effects on the behavior of sharks. *In* "Sensory Biology of the Sharks, Skates and Rays" (T. Hodgson and R. Mathewson, eds.), pp. 391–417. U.S. Govt. Printing Office, Washington, DC.

Myrberg, A. A., Jr. (1980). Ocean noise and the behavior of marine animals. *In* "Advanced Measurements in Ocean Sampling" (F. Diemer, F. J. Vernberg, and D. Z. Mirkes, eds.), pp. 461–491. University of South Carolina Press, Columbia.

Myrberg, A. A., Jr. (1981). Sound communication and interception in fishes. *In* "Hearing and Sound Communication in Fishes" (W. N. Tavolga, A. N. Popper, and R. R. Fay, eds.), pp. 395–425. Springer-Verlag, New York.

Myrberg, A. A., Jr. (1997a). Underwater sound: Its relevance to behavioral functions among fishes and marine mammals. *Mar. Freshwater Behav. Physiol.* **29**, 3–21.

Myrberg, A. A., Jr. (1997b). Sound production by a coral reef fish (*Pomacentrus partitus*): Evidence for a vocal, territorial "keep-out" signal. *Bull. Mar. Sci.* **60**, 1017–1025.

Myrberg, A. A., Jr., and Riggio, R. J. (1985). Acoustically mediated individual recognition by a coral reef fish (*Pomacentrus partitus*). *Anim. Behav.* **33**, 411–416.

Myrberg, A. A., Jr., and Spires, J. Y. (1972). Sound discrimination by the bicolor damselfish, *Eupomacentrus partitus* (Poey) (Pisces: Pomacentridae). *J. Exp. Biol.* **57**, 727–735.

Myrberg, A. A., Jr., and Spires, J. Y. (1980). Hearing in damselfishes: An analysis of signal detection among closely related species. *J. Comp. Physiol. A* **140**, 135–144.

Myrberg, A. A., Jr., Banner, A., and Richard, J. D. (1969). Shark attraction using a video-acoustic system. *Mar. Biol.* **2**, 264–276.

Myrberg, A. A., Jr., Gordon, C. R., and Klimley, A. P. (1976). Attraction of free-ranging sharks by low frequency sound, with comments on its biological significance. *In* "Sound Reception in Fish" (A. Schuijf and A. D. Hawkins, eds.), pp. 205–228. Am. Elsevier, New York.

Myrberg, A. A., Jr., Gordon, C. R., and Klimley, A. P. (1978). Rapid withdrawal from a sound source by open-ocean sharks. *J. Acoust. Soc. Am.* **64**, 1289–1297.

Myrberg, A. A., Jr., Mohler, M., and Catala, J. (1986). Sound production by males of a coral reef fish (*Pomacentrus partitus*): Its significance to females. *Anim. Behav.* **34,** 923–933.

Myrberg, A. A., Jr., Ha, S. J., and Shamblott, M. J. (1993). The sounds of bicolor damselfish (*Pomacentrus partitus*): Predictors of body size and a spectral basis for individual recognition and assessment. *J. Acoust. Soc. Am.* **94,** 3067–3070.

Nagelkerken, I., vanderVelde, G., Gorissen, M. W., Meijer, G. J., vantHof, T., and denHartog, C. (2000). Importance of mangroves, seagrass beds and the shallow coral reef as a nursery for important coral reef fishes, using a visual census technique. *Estuarine Coastal Shelf Sci.* **51,** 31–44.

Nakashima, Y., Kuwamua, T., and Yogo, Y. (1995). Why be a both-ways sex changer? *Ethology* **101,** 301–307.

National Research Council (1999). "Sustaining Marine Fisheries." National Academy Press, Washington, DC.

Naval Ordinance Laboratory (1942). "Measurement of Background Noise in the Water at Cape Henry, Virginia due to Surf and Marine Life," N.O.L. Rep. No. 594. NOL.

Nei, M. (1987). "Molecular evolutionary genetics." Columbia Univ. Press, New York.

Nei, M., and Graur, D. (1984). Extent of protein polymorphism and the neutral mutation theory. *Evol. Biol.* **17,** 73–118.

Neill, W. H., Miller, J. M., van der Veer, H. W., and Winemiller, K. O. (1994). Ecophysiology of marine fish recruitment: A conceptual framework for understanding interannual variability. *Neth. J. Sea Res.* **3,** 135–152.

Neilson, J. D. (1992). Sources of error in otolith microstructure examination. *Can. Fish. Aquat. Sci., Spec. Publ.* **117,** 115–125.

Neilson, J. D., and Perry, R. I. (1990). Diel vertical migrations of marine fishes: An obligate or facultative process? *Adv. Mar. Bio.* **26,** 115–168.

Nelson, D. R. (1967). Hearing thresholds, frequency discrimination, and acoustic orientation in the lemon shark, *Negaprion brevirostris* (Poey). *Bull. Mar. Sci.* **17,** 741–768.

Nelson, D. R., and Johnson, R. H. (1972). Acoustic attraction of Pacific reef sharks: Effect of pulse intermittency and variability. *Comp. Biochem. Physiol.* **42A,** 85–95.

Nelson, D. R., Johnson, R. H., and Waldrop, L. G. (1969). Responses in Bahamian sharks and groupers to low-frequency, pulsed sounds. *Bull. South. Calif. Acad. Sci.* **68,** 131–137.

Nelson, J. S. (1994). "Fishes of the World," 3rd ed. J. Wiley, New York.

Nelson, K., and Hedgecock, D. (1980). Enzyme polymorphism and adaptative strategy in the decapod Crustacea. *Am. Nat.* **116,** 238–280.

Nemeth, D. H. (1997). Modulation of attack behavior and its effect on feeding performance in a trophic generalist fish, *Hexagrammos decagrammus*. *J. Exp. Biol.* **200,** 2155–2164.

Nemeth, R. S. (1998). The effects of natural variation in substrate architecture on the survival of juvenile bicolor damselfish. *Environ. Biol. Fishes* **53,** 129–141.

Nemtzov, S., and Clark, E. (1994). Intraspecific egg predation by male razorfishes (Labridae) during broadcast spawning: Filial cannibalism or intra-pair parasitism? *Bull. Mar. Sci.* **55,** 133–141.

Nestler, J. M., Ploskey, G. R., Pickens, J., Menzes, J., and Schilt, C. (1992). Responses of blueback herring to high-frequency sound and implications for reducing entrainment at hydropower dams. *North Am. J. Fish. Manage.* **12,** 667–683.

Nevo, E. (1978). Genetic variation in natural population: Patterns and theory. *Theor. Popul. Biol.* **13,** 121–177.

Newell, N. B. (1972). The evolution of reefs. *Sci. Am.* **226,** 54–65.

Newman, S. J., Williams, D. McB., and Russ, G. R. (1996). Age validation, growth, and mortality rates of the tropical snappers (Pisces: Lutjanidae) *Lutjanus adetii* (Castelnau, 1873) and *L. quinquelineatus* (Bloch, 1790) from the central Great Barrier Reef. *Fish. Bull.* **94,** 313–329.

Newman, S. J., Cappo, M., and Williams, D. McB. (2000a). Age, growth, mortality rates and corresponding yield estimates using otoliths of the tropical snappers, *Lutjanus erythropterus, L. malabaricus* and *L. sebae,* from the central Great Barrier Reef. *Fish. Res.* **48,** 1–14.

Newman, S. J., Cappo, M., and Williams, D. McB. (2000b). Age, growth and mortality of the stripey, *Lutjanus carponotatus* (Richardson) and the brown-stripe snapper, *L. vitta* (Quoy and Gaimard) from the central Great Barrier Reef, Australia. *Fish. Res.* **48,** 263–275.

Newman, S. J., Steckis, R. A., Edmonds, J. S., and Lloyd, J. (2000c). Stock structure of the goldband snapper *Pristopomoides multidens* (Pisces: Lutjanidae) from the waters of northern and western Australia by stable isotope ratio analysis of sagittal otolith carbonate. *Mar. Ecol.: Prog. Ser.* **198,** 239–247.

Nichida, M., and Lucas, J. S. (1988). Genetic differences between geographic populations of the crown-of-thorns starfish throughout the Pacific region. *Mar. Biol.* **98,** 359–368.

Nicholson, A. J. (1954). An outline of the dynamics of animal populations. *Aust. J. Zool.* **2,** 9–65.

Nicholson, A. J. (1957). The self adjustment of populations to change. *Cold Spring Harbor Symp. Quant. Biol.* **22,** 153–172.

Nicholson, A. J., and Bailey, V. A. (1935). The balance of animal populations. *Proc. Zool. Soc. London* **3,** 551–598.

Nicol, J. A. C. (1989). "The Eyes of Fishes." Clarendon Press, Oxford.

Ninio, R., Meekan, M. G., Done, T. J., and Sweatman, H. P. A. (2000). Temporal patterns in coral assemblages on the Great Barrier Reef from local to large spatial scales. *Mar. Ecol.: Prog. Ser.* **194,** 65–74.

Nisbet, R. M., and Bence, J. R. (1989). Alternative dynamic regimes for canopy forming kelp: A variant on density-vague population regulation. *Am. Nat.* **134,** 377–408.

Nisbet, R. M., and Onyiah, L. C. (1994). Population dynamic consequences of competition within and between age classes. *J. Math. Biol.* **32,** 329–344.

Nisbet, R. M., Murdoch, W. W., and Stewart-Oaten, A. (1996). Consequences for adult fish stocks of human-induced mortality on immatures. *In* "Detecting Ecological Impacts: Concepts and Applications in Coastal Habitats" (R. J. Schmitt and C. W. Osenberg, eds.), pp. 257–277. Academic Press, San Diego, CA.

NMFS. (1999). "Our Living Oceans. Report on the Status of U.S. Living Marine Resources, 1999," NOAA Tech. Memo. NMFS-F/SPO-4. U.S. Department of Commerce, Washington, DC.

Nolan, R. S. (1978). Hawaii tropical reef fish study. *In* "Papers and Comments on Tropical Reef Fish," Working Pap. No. 34, pp.27–34. Sea Grant College Program, University of Hawaii, Honolulu.

Nolf, D. (1985). Otolithi Piscium. *In* "Handbook of Paleoichthyology" (H. P. Schultze, ed.), Vol. 10, pp. 1–145. Fischer, Stuttgart.

Normile, D. (2000). Reef migrations, bleaching effects stir the air in Bali. *Science* **290,** 1282–1283.

Norris, K. S. (1963). The functions of temperature in the ecology of the percoid fish *Girella nigricans* (Ayres). *Ecol. Monogr.* **33,** 23–62.

Northcutt, R. G. (1989). The phylogenetic distribution and innervation of craniate mechanoreceptive lateral lines. *In* "The Mechanosensory Lateral Line: Neurobiology and Evolution" (S. Coombs, P. Görner, and H. Münz, eds.), pp. 481–499. Springer-Verlag, New York.

Norton, S. F. (1991). Capture success and diet of cottid fishes: The role of predator morphology and attack kinematics. *Ecology* **72,** 1807–1819.

Norton, S. F. (1995). A functional approach to ecomorphological pattersn of feeding in cottid fishes. *Environ. Biol. Fishes* **44,** 61–78.

Norton, S. F., and Brainerd, E. L. (1993). Convergence in the feeding mechanics of ecomorphologically similar species in the Centrarchidae and Cichlidae. *J. Exp. Biol.* **176,** 11–29.

Nunney, L. N., and Campbell, K. A. (1993). Assessing minimum viable population size: Demography meets population genetics. *Trends Ecol. Evol.* **8,** 234–239.

Nursall, J. R. (1996). Distribution and ecology of pycnodont fishes. *In* "Mesozoic Fishes—Systematics and Palaeoecology." (G. Arratia and G. Viohl, eds.), pp. 115–124. Verlag F. Pfeil, Munchen.

Nursall, J. R., and Turner, L. J. (1985). The liver supports metamorphosis in the Caribbean reef blenniid, *Ophioblennius atlanticus*. *Proc. 5th Int. Coral Reef Congr.* Vol. 5, pp. 457–462.

Oakley, N. S., and Elliott, J. A. (1982). Dissipation within the surface mixed layer. *J. Phys. Oceanogr.* **12,** 171–185.

O'Brien, W. J. (1979). The predator-prey interaction of planktivorous fish and zooplankton. *Am. Sci.* **67,** 572–581.

O'Connell, M., and Wright, J. M. (1997). Microsatellite DNA in fishes. *Rev. Fish Biol. Fish.* **7,** 331–363.

O'Connell, M., Dillon, M. C., Wright, J. M., Bentzen, P., Merkouris, S., and Seeb, J. (1998). Genetic structuring among Alaskan Pacific herring populations identified using microsatellite variation. *J. Fish Biol.* **53,** 150–153.

Oda, D. K., and Parrish, J. D. (1981). Ecology of commercial snappers and groupers introduced to Hawaiian reefs. *Proc. 4th Int. Coral Reef Symp.* Vol. 1, pp. 59–67.

Ogden, J. C., and Quinn, T. P. (1984). Migration in coral reef fishes: Ecological significance and orientation mechanisms. *In* "Mechanisms of Migration in Fishes" (J. D. McCleave, G. P. Arnold, J. J. Dodson, and W. H. Neill, eds.), pp. 293–308. Plenum Press, New York.

Öhman, M. C., Munday, P. L., Jones, G. P., and Caley, M. J. (1998a). Settlement strategies and distribution patterns of coral-reef fishes. *J. Exp. Mar. Biol. Ecol.* **225,** 219–238.

Öhman, M. C., Rajasuriya, A. and Svensson, S. (1998b). The use of butterflyfishes (Chaetodontidae) as bio-indicators of habitat structure and human disturbance. *Ambio* **27,** 708–716.

Okubo, A. (1994). The role of diffusion and related physical processes in dispersal and recruitment of marine populations. *In* "The Bio-physics of Marine Larval Dispersal" (P. W. Sammarco and M. L. Heron, eds.). American Geophysical Union, Washington, DC.

Olafsson, E. B., Peterson, C. H., and Ambrose, W. G., Jr. (1994). Does recruitment limitation structure populations and communities of macro-invertebrates in marine soft sediments: The relative significance of pre- and post-settlement processes. *Oceanogr. Mar. Biol.* **32,** 65–109.

Olin, P. G. (1997). The ornamental fish trade: Status, trends, and shipping techniques. *Proc. 1st Conf. Market. Shipp. Aquat. Prod.,* Seattle, WA, *1996,* pp. 141–144.

Olivar, M. P., and Sabatés, A. (1997). Vertical distribution of fish larvae in the North-west Mediterranean Sea in spring. *Mar. Biol.* **129,** 289–300.

Olivieri, I., Couvet, D., and Gouyon, P. H., (1990). The genetics of transient populations: research at the metapopulation level. *Trends Ecol. Evol.* **5,** 207–210.

Olla, B. L., Davis, M. W., and Ryer, C. H. (1998). Understanding how the hatchery environment represses or promotes the development of behavioral survival skills. *Bull. Mar. Sci.* **62,** 531–550.

Olsen, D. A., and La Place, J. A. (1979). A study of a Virgin Islands grouper fishery based on a breeding aggregation. *Proc. Annu. Gulf Caribb. Fish. Inst.* **31,** 130–144.

Olson, D. B., and Backus, R. H. (1985). The concentrating of organisms at fronts: A cold-water fish and a warm-core Gulf Stream ring. *J. Mar. Res.* **43,** 113–137.

Olson, D. B., Hitchcock, G. L., Mariano, A. J., Ashjian, C. J., Peng, G., Nero, R. W., and Podesta, G. P. (1994). Life on the edge: Marine life and fronts. *Oceanography* **7,** 52–60.

Olson, M. H. (1996). Ontogenetic niche shifts in largemouth bass: Variability and consequences for first-year growth. *Ecology* **77,** 179–190.

Olson, M. H., Mittelbach, G. G., and Osenberg, C. W. (1995). Competition between predator and prey: Resource-based mechanisms and implications for stage-structured dynamics. *Ecology* **76,** 1758–1771.

OmniTrak Group (1997). "Summary on the Taste Test Between the Mariculture and the Wild-caught Malabar Grouper." The Nature Conservancy.

Oreihaka, E. (1999). Live reef fish trade in the Solomon Islands. *Proc. 1st Asia-Pac. Semin./Workshop Live Reef Fish Trade, 1998,* pp. 15–18.

O'Reilly, P., and Wright, J. M. (1995). The evolving technology of DNA fingerprinting and its application to fisheries and aquaculture. *J. Fish Biol.* **47,** 29–55.

Orians, G. H. (1969). On the evolution of mating systems in birds and mammals. *Am. Nat.* **103,** 589–603.

Ormond, R. F. G., and Roberts, C. M. (1997). The biodiversity of coral reef fishes. *In* "Marine Biodiversity: Patterns and Processes" (R. F. G. Ormond, J. D. Gage, and M. V. Angel, eds.), pp. 216–257. Cambridge University Press, Cambridge, UK.

Ormond, R. F. G., Roberts, J. M., and Jan, R. Q. (1996). Behavioral differences in microhabitat use by damselfishes (Pomacentridae): Implications for reef fish biodiversity. *J. Exp. Mar. Biol. Ecol.* **202,** 85–95.

Osenberg, C. W., and Mittelbach, G. G. (1989). Effects of body size on the predator-prey interaction between pumpkinseed sunfish and gastropods. *Ecol. Monogr.* **59,** 405–432.

Osenberg, C. W., and Mittelbach, G. C. (1996). The relative importance of resource limitation and predator limitation in food chains. *In* "Food Webs: Integration of Patterns and Dynamics" (G. A. Polis and K. O. Winemiller, eds.), pp. 134–148. Chapman & Hall, New York.

Osenberg, C. W., Sarnelle, O., Cooper, S. D., and Holt, R. D. (1999). Resolving ecological questions through meta-analysis: Goals, metrics, and models. *Ecology* **80,** 1105–1117.

Owen, R. W. (1980). Eddies of the California Current system: Physical and ecological characteristics. *In* "The California Islands: Proceedings of a Multidisciplinary Symposium" (D. M. Power, ed.), pp. 237–263. Santa Barbara Museum of Natural History, Santa Barbara, CA.

Paddack, M. J., and Estes, J. A. (2000). Implications of a marine reserve effect on reef fish populations in central California kelp forests. *Ecol. Appl.* **10,** 855–887.

Pagel, M. (1997). Inferring evolutionary processes from phylogenies. *Zool. Sci.* **26,** 331–348.

Paine, M. D. (1990). Life history tactics of darters (Percidae: Etheostomatiini) and their relationship with body size, reproductive behaviour, latitude and rarity. *J. Fish Biol.* **37,** 473–488.

Pajaro, M. (1993). Alternatives to sodium cyanide use in aquarium fish collection: A community-based approach. *Sea Wind* **6,** 9–11.

Palumbi, S. R. (1996). What can molecular genetics contribute to marine biogeography? An urchin's tale. *J. Exp. Mar. Biol. Ecol.* **203,** 75–92.

Palumbi, S. R. (1997). Molecular biogeography of the Pacific. *Coral Reefs* **16** (Suppl.), S47–S52.

Palumbi, S. R. (2000). The ecology of marine protected areas. *In* "Marine Ecology: The New Synthesis" (M. Bertness, ed.), pp. 509–530. Sinauer Press, Sunderland, MA.

Palumbi, S. R. (in press). Population genetics, demographic connectivity and the design of marine reserves. *Ecol. Appl.* (in press).

Palumbi, S. R., and Metz, E. (1991). Strong reproductive isolation between closely related tropical sea urchin (genus *Echinometra*). *Mol. Biol. Evol.* **8,** 227–239.

Palumbi, S. R., Grabowsky, G., Duda, T., Geyer, L., and Tachino, N. (1997). Speciation and population genetic structure in tropical Pacific sea urchins. *Evolution (Lawrence, Kans.)* **51,** 1506–1517.

Pandian, T. J., and Vivekanandan, E. (1985). Energetics of feeding and digestion. *In* "Fish Energetics–New Perspectives" (P. Tyler and P. Calow, eds.), pp. 99–124. Croom Helm, London.

Pandolfi J. M. (1992). Successive isolation rather than evolutionary centers for the origination of the Indo-Pacific reef corals. *J. Biogeogr.* **19,** 593–609.

Panella, G. (1971). Fish otoliths: Daily growth layers and periodic patterns. *Science* **173,** 1124–1127.

Pankhurst, N. W. (1989). The relationship of ocular morphology to feeding modes and activity periods in shallow marine teleosts from New Zealand. *Environ. Biol. Fishes* **26,** 201–211.

Pankhurst, N. W., and Van der Kraak, G. (1997). Effects of stress on reproduction and growth of fish. *In* "Fish Stress and Health in Aquaculture" (G. K. Iwana, A. D. Pickering, J. P. Sumpter, and C. B. Schreck, eds.), pp. 73–93. Cambridge University Press, Cambridge, UK.

Pankhurst, P. M. (1994). Age related changes in the visual acuity of larvae of New Zealand snapper *Pagrus auratus. J. Mar. Biol. Assoc. U.K.* **74,** 79–90.

Pankhurst, P. M., and Butler, P. E. (1996). Development of the sensory organs in the greenback flounder *Rhombosolea tapirina. Mar. Freshwater Behav. Physiol.* **28,** 55–73.

Pankhurst, P. M., and Hilder, P. E. (1998). Effect of light intensity on feeding of striped trumpeter *Latris lineata* larvae. *Mar. Freshwater Res.* **49,** 363–368.

Pankhurst, P. M., Pankhurst, N. W., and Montgomery, J. C. (1993). Comparison of behavioral and morphological measures of visual acuity during ontogeny in a teleost fish, *Forsterygion varium,* Tripterygiidae (Forster, 1801). *Brain, Behav. Evol.* **42,** 178–188.

Paperno, R., Targett, T. E., and Grecay, P. A. (1997). Daily growth increments in otoliths of juvenile weakfish, *Cynoscion regalis*: Experimental assessment of changes in increment width with changes in feeding rate, growth rate, and condition factor. *Fish. Bull.* **95,** 521–529.

Parenti, L. R. (1991). Ocean basins and the biogeography of freshwater fishes. *Aust. Syst. Bot.* **4,** 137–149.

Parker, P. G., Snow, A. A., Schug, M. D., Booton, G. C., and Fuerst, P. A. (1998). What molecules can tell us about populations: Choosing and using a molecular marker. *Ecology* **79,** 361–382.

Parnell, K. E. (1988). The hydrodynamics of the fringing reef bays in the Great Barrier Reef marine Park. *Proc. 6th Int. Coral Reef Symp.* Vol. 2, pp. 503–508.

Parrish, J. D. (1989). Fish communities of interacting shallow-water habitat in tropical oceanic regions. *Mar. Ecol.: Prog. Ser.* **58**, 143–160.

Parrish, J. D., Callahan, M. W., and Norris, J. D. (1985). Fish trophic relationships that structure reef communities. *Proc. 5th Int. Coral Reef Congr.* Vol. 4, pp. 73–78.

Partridge, B. L., and Pitcher, T. J. (1980). The sensory basis of fish schools—relative roles of lateral line and vision. *J. Comp. Physiol.* **135**, 315–325.

Parvulescu, A. (1964). Problems in propagation and processing. *In* "Marine Bio-acoustics" (W. N. Tavolga, ed.), pp. 87–100. Pergamon Press, New York.

Pascual, M., and Levin, S. A. (1999). Spatial scaling in a benthic population model with density-dependent disturbance. *Theor. Popul. Biol.* **56**, 106–122.

Patterson, C. (1981). Significance of fossils in determining evolutionary relationships. *Annu. Rev. Ecol. Syst.* **12**, 195–223.

Patterson, C. (1993). Osteichthyes: Teleostei. *In* "The Fossil Record 2" (M. J. Benton, ed.), pp. 621–656. Chapman & Hall, London.

Patterson, H. M., Thorrold, S. R., and Shenker, J. M. (1999). Analysis of otolith chemistry in Nassua grouper (*Epinephelus striatus*) from the Bahamas and Belize using solution-based ICP-MS. *Coral Reefs* **18**, 171–178.

Patterson, W. P., Smith, G. R., and Lohmann, K. C. (1993). Continental paleothermometry and seasonality using the isotopic composition of aragonitic otoliths of freshwater fishes. *Geophys. Monogr.* **78**, 191–202.

Pattiaratchi, C., James, A., and Collins, M. (1986). Island wakes and headland eddies: A comparison between remotely sensed data and laboratory experiments. *J. Geophys. Res.* **92**, 783–794.

Patzert, W. C. (1970). Eddies in Hawaiian waters. *Hawaii Inst. Geophys. [Rep.] HIG* **HIG-69-8.**

Paulay, G. (1990). Effects of late Cenozoic sea-level fluctuations on the bivalve faunas of tropical oceanic islands. *Paleobiology* **16**, 415–434.

Paulay, G. (1996). Dynamic clams: Changes in the bivalve faunas of Pacific islands as a result of sea-level fluctuations. *Am. Malacol. Bull.* **12**, 45–57.

Pauly, D. (1994). "On the Sex of Fish and the Gender of Scientists: Collected Essays in Fisheries Science." Chapman & Hall, London.

Pauly, D. (1997). Small-scale fisheries in the tropics: Marginality, marginalization, and some implications for fisheries management. *In* "Global Trends: Fisheries Management" (E. K. Pikitch, D. D. Huppert, and M. P. Sissenwine, eds.), Am. Fish. Soc. Symp. 20, pp. 1–10. Am. Fish. Soc., Bethesda, MD.

Pauly, D. (1998). Tropical fishes: Patterns and propensities. *J. Fish Biol.* **53**(Suppl. A), 1–17.

Pauly, D., Christensen, V., and Walters, C. (2000). Ecopath, Ecosim, and Ecospace as tools for evaluating ecosystem impact of fisheries. *ICES J. Mar. Sci.* **57**, 697–706.

Pawlik, J. R., Chanas, B., Toonen R. J., and Fenical, W. (1995). Defenses of Caribbean sponges against predatory reef fish. I. Chemical deterrency. *Mar. Ecol.: Prog. Ser.*, **127**, 183–194.

Payan, P., Borelli, G., Boeuf, G., and Mayer-Gostan, N. (1998). Relationship between otolith and somatic growth: Consequence of starvation on acid-base balance in plasma and endolymph in the rainbow trout. *Fish Physiol. Biochem.* **19**, 35–41.

Pearre, S. (1979). Problems of detection and interpretation of vertical migration. *J. Plankton Res.* **1**, 29–44.

Pechenik, J. A. (1990). Delayed metamorphosis by larvae of benthic marine invertebrates: does it occur? Is there a price to pay? *Ophelia* **32**, 63–94.

Pechenik, J. A., Wendt, D. E., and Jarrett, J. N. (1998). Metamorphosis is not a new beginning. Larval experience influences juvenile performance. *BioScience* **48**, 901–910.

Pereira, D. L., Bingham, C., Spangler, G. R., Conner, D. J., and Cunningham, P. K. (1995). Construction of a 110–year biochronology from sagittae of freshwater drum (*Aplodinotus grunniens*). *In* "Recent Developments in Fish Otolith Research" (D. H. Secor, J. M. Dean, and S. E. Campana, eds.), pp. 177–196. University of South Carolina Press, Columbia.

Perino, L. (1990). "Assessment of the Feasibility of Establishing an Aquarium Fish Industry in Papua New Guinea," FFA Rep. 90/30. South Pacific Forum Fisheries Agency.

Pet, J., Squire, L., Subagyo, C., and Mulyadi, A. (1999). "Grouper and Napoleon Wrasse Spawning Aggregation Sites: Komodo National Park. Monitoring Report 1998–1999," Unpublished report. The Nature Conservancy, Jakarta Selantan, Indonesia.

Peters, R. H. (1983). "The Ecological Implications of Body Size." Cambridge University Press, Cambridge, UK.

Petersen, C. W. (1987). Reproductive behavior and gender allocation in *Serranus fasciatus*, a hermaphroditic reef fish. *Anim. Behav.* **35**, 1601–1614.

Petersen, C. W. (1989). Females prefer mating males in the carmine triplefin, *Axoclinus carminalis*, a paternal brood guarder. *Environ. Biol. Fishes* **26**, 213–221.

Petersen, C. W. (1990a). The relationships among population density, individual size, mating tactics, and reproductive success in a hermaphroditic fish, *Serranus fasciatus*. *Behaviour* **113**, 57–80.

Petersen, C. W. (1990b). Variation in reproductive success and gonadal allocation in the simultaneous hermaphrodite, *Serranus fasciatus*. *Oecologia* **83**, 62–67.

Petersen, C. W. (1991a). Variation in fertilization rate in the tropical reef fish, *Halichoeres bivattatus:* Correlates and implications. *Biol. Bull (Woods Hole, Mass.)* **181**, 232–237.

Petersen, C. W. (1991b). Sex allocation in hermaphroditic sea basses. *Am. Nat.* **138**, 650–667.

Petersen, C. W. (1995). Male mating success and female choice in permanently territorial damselfishes. *Bull. Mar. Sci.* **57**, 690–704.

Petersen, C. W., and Fischer, E. A. (1986). Mating system of the hermaphroditic coral reef fish, *Serranus baldwini*. *Behav. Ecol. Sociobiol.* **19**, 171–178.

Petersen, C. W., and Fischer, E. A. (1996). Intraspecific variation in sex allocation in a simultaneous hermaphrodite: The effect of individual size. *Evolution (Lawrence, Kans.)* **50**, 636–645.

Petersen, C. W., and Hess, H. C. (1991). The adaptive significance of spawning synchronization in the Caribbean damselfish *Stegastes dorsopunicans* (Poey). *J. Exp. Mar. Biol. Ecol.* **151**, 155–167.

Petersen, C. W., and Levitan, D. L. (2001). The Allee effect: A barrier to recovery by exploited species. *In* "Conservation of Exploited Species." (J. D. Reynolds, G. Mace, K. Redford, and J. Robinson, eds.), pp. 281–300. Cambridge University Press, Cambridge, UK.

Petersen, C. W., and Marchetti, K. (1989). Filial cannibalism in the Cortez damselfish *Stegastes rectifraenum*. *Evolution (Lawrence, Kans.)* **43**, 158–168.

Petersen, C. W., and Warner, R. R. (1998). Sperm competition in fishes. *In* "Sperm Competition and Sexual Selection." (T. R. Birkhead and A. P. Moller, eds.), pp. 435–463. Academic Press, London.

Petersen, C. W., Warner, R. R., Cohen, S., Hess, H. C., and Sewell, A. T. (1992). Variation in pelagic fertilization success: Implications for mate choice and spatial patterns of mating. *Ecology* **73**, 391–401.

Petersen, C. W., Warner, R. R., Shapiro, D. Y., and Marconato, A. (2001). Components of fertilization success in the bluehead wrasse, *Thalassoma bifasciatum*. *Behav. Ecol.* **12**, 237–245.

Peterson, B. J., and Fry, B. (1987). Stable isotopes in ecosystem studies. *Annu. Rev. Ecol. Syst.* **18**, 293–320.

Peterson, C. H., Summerson, H. C., and Luettich, R. A. (1996). Response of bay scallops to spawner transplants: A test of recruitment limitation. *Mar. Ecol.: Prog. Ser.* **132**, 93–107.

Pet-Soede, L., and Erdmann, M. (1998). An overview and comparison of destructive fishing practices in Indonesia. *Secretariat of the Pacific Community Live Reef Fish Inf. Bull.* **4**, 28–36.

Pettigrew, J. D., Collin, S. P., and Ott, M. (1999). Convergence of specialized behavior, eye movements and visual optics in the sandlance (Teleostei) and the chameleon (Reptilia). *Curr. Biol.* **9**, 421–424.

Petuch, E. J. (1982). Geographical heterochrony: Contemporaneous coexistence of Neogene and Recent molluscan faunas in the Americas. *Palaeogeogr. Palaeoclimatol. Palaeoecol.* **37**, 277–312.

Pfeffer, R. A., and Tribble, G. W. (1985). Hurricane effects on an aquarium fish fishery in the Hawaiian Islands. *Proc. 5th Int. Coral Reef Congr.*, Vol. 3, pp. 331–336.

Pfeiffer, W. (1960). Über die Schreckreaktionen bei Fischen und die Herkunft des Schreckstoffes. *Z. Vergl. Physiol.* **43**, 578–614.

Pfeiffer, W. (1977). The distribution of fright reaction and alarm substance cells in fishes. *Copeia*, pp. 653–665.

Pfeiler, E., and Luna, A. (1984). Changes in biochemical composition and energy utilization during metamorposis of leptocephalous larvae of the bonefish *(Albula)*. *Environ. Biol. Fishes* **10**, 243–251.

Pfeiler, E., Lindley, V. A., and Elser, J. J. (1998). Elemental (C, N and P) analysis of metamorphosing bonefish (*Albula* sp.) leptocephali: Relationship to catabolism of endogenous organic compounds, tissue remodeling, and feeding ecology. *Mar. Biol.* **132**, 21–28.

Pfister, C. A. (1996). The role and importance of recruitment variability to a guild of tide pool fishes. *Ecology* **77**, 1928–1941.

Philander, S. G. (1992). El Niño. *Oceanus* **35**, 56–61.

Pianka, E. R. (1966). Latitudinal gradients in species diversity: A review of concepts. *Am. Nat.* **100**, 33–46.

Pianka, E. R. (1978). "Evolutionary Ecology," 2nd ed. Harper & Row, NewYork.

Piggott, C. L (1964). Ambient sea noise at low frequencies in shallow water of the Scotian shelf. *J. Acoust. Soc. Am.* **36**, 2152–2163.

Pillai, C. S. G., Mohan, M., and Kunhikoya, K. K. (1983). On an unusual massive recruitment of the reef fish *Ctenochaetus strigosus* (Bennet) (Perciformes: Acanthuridae) to the Minicoy Atoll and its significance. *Indian J. Fish.* **30**, 261–268.

Pilling, G. M., Millner, R. S., Easy, M. W., Mees, C. C., Rathacharen, S. and Azemia, R. (2000). Variation of annual growth increments in the otoliths of the lethrinid *Lethrinus mahsena* and the lutjanid *Aprion. virescens* from sites in the tropical Indian Ocean with notes on the nature of growth increments in *Pristipomoides filamentosus*. *Fish. Bull.* **98**, 600–611.

Pineda, J. (1991). Predictable upwelling and the shoreward transport of planktonic larvae by intertidal bores. *Science* **253**, 548–551.

Pineda, J. (1994). Internal tidal bores in the nearshore: Warm-water fronts, seaward gravity currents and the onshore transport of neustonic larvae. *J. Mar. Res.* **52**, 427–458.

Pineda, J. (1995). An internal tidal bore regime at nearshore stations along western U.S.A.: Predictable upwelling within the lunar cycle. *Cont. Shelf Res.* **15**, 1023–1041.

Pineda, J. (1999). Circulation and larval distribution in internal tidal bore warm fronts. *Limnol. Oceanogr.* **44**, 1400–1414.

Pingree, R. D., and Maddock, L. (1980). The effects of bottom friction and the earth's rotation on an island's wake. *J. Mar. Biol. Assoc. U.K.* **60**, 499–508.

Pitcher, C. R. (1988). Validation of a technique for reconstructing daily patterns in the recruitment of coral reef damselfish. *Coral Reefs* **7**, 105–111.

Plan Development Team (PDT) (1990). The potential of marine fishery reserves for reef management in the U. S. Southern Atlantic (J. A. Bohnsack, ed.). *NOAA Tech. Memo.* **NMFS-SEFC-261**, 1–40.

Planes, S. (1993). Generic differentiation in relation to restricted larval dispersal of the convict surgeon fish,

Acanthurus triostegus in French Polynesia. *Mar. Ecol.: Prog. Ser.* **98,** 237–246.

Planes, S. (1998). Genetic diversity and dispersal capabilities in Marine Fish. *Evol. Biol.* **30,** 253–298.

Planes, S., and Doherty, P. J. (1997a). Genetic and color interactions at a contact zone of *Acanthochromis polyacanthus:* A marine fish lacking pelagic larvae. *Evolution (Lawrence, Kans.)* **51,** 1232–1243.

Planes, S., and Doherty, P. J. (1997b). Genetic relationships of the color morphs of *Acanthochromis polyacanthus* (Pomacentridae) in the northern Great Barrier Reef. *Mar. Biol.* **130,** 109–117.

Planes, S., and Lecaillon, G. (1998). Consequence of the founder effect in the genetic structure of introduced island coral reef fish populations. *Biol. J. Linn. Soc.* **63,** 537–552.

Planes, S., and Lecaillon, G. (2002). Intra-cohort genetics variability of *Naso unicornis,* a coral reef fish. *J. Fish Biol.* (in press).

Planes, S., Galzin, R., and Bonhomme, F. (1993). Genetic structure of *Dascyllus aruanus* populations in French Polynesia. *Mar. Biol.* **117,** 665–674.

Planes, S., Borsa, P., Galzin, R., and Bonhomme, F. (1994). Genetic structure and gene flow in the manini (convict surgeonfish, *Acanthurus triostegus*) in the south-central Pacific. *In* "Genetics and Evolution of Aquatic Organisms" (A. R. Beaumont, ed.), pp. 113–121. Chapman & Hall, London.

Planes, S., Bonhomme, F., and Galzin, R. (1996). A genetic metapopulation model for reef fishes in oceanic islands: The case of the surgeonfish, *Acanthurus triostegus. J. Evol. Biol.* **9,** 103–117.

Planes, S., Parroni, M., and Chauvet, C. (1998a). Genetic analysis of population structure in relation with larval duration in three species of coral reef fishes from New Caledonia. *Mar. Biol.* **130,** 361–368.

Planes, S., Romans, P., and Lecomte-Finiger, R. (1998b). Genetic evidence of closed life-cycles for some coral reef fishes within Taiaro Lagoon (Tuamotu Archipelago, French Polynesia).*Coral Reefs* **17,** 9–14.

Planes, S., Jouvenel, J. Y., and Lenfant, P. (1998c). Density dependence in post-recruitment processes of juvenile sparids in the littoral of the Mediterranean Sea. *Oikos* **83,** 293–300.

Platt, C. (1988). Equilibrium in the vertebrates: Signals, senses, and steering underwater. *In* "Sensory Biology of Aquatic Animals" (J. Atema, R. R. Fay, A. N. Popper, and W. N. Tavolga, eds.), pp. 783–809. Springer-Verlag, New York.

Platt, C., and Popper, A. N. (1984). Variation in lengths of ciliary bundles on hair cells along the macula of the sacculus in two species of teleost fishes. *Scanning Electron Microsc.* **4,** 1915–1924.

Polacheck, T. (1990). Year round closed areas as a management tool. *Nat. Res. Mod.* **4,** 327–354.

Poling, K. R., and Fuiman, L. A. (1998). Sensory development and its relation to habitat change in three species of sciaenids. *Brain, Behav. Evol.* **52,** 270–284.

Poling, K. R., and Fuiman, L. A. (1999). Behavioral specialization in developing sciaenids and its relationship to morphology and habitat. *Environ. Biol. Fishes* **54,** 119–133.

Polovina, J. J., and Ralston, S. (1987). "Tropical Snappers and Groupers: Biology and Fisheries Management." Westview Press, Boulder, CO.

Polovina, J. J., Kleiber, P., and Kobayashi, D. R. (1999). Application of TOPEX-POSEIDON satellite imagery to simulate transport dynamics of larvae of spiny lobster, *Panulirus marginatus,* in the Northwestern Hawaiian Islands, 1993–1996. *Fish. Bull.* **97,** 132–143.

Polunin, N. V. C. (1990). Marine regulated areas: An expanded approach for the tropics. *Resour. Manage. Optim.* **7,** 283–299.

Polunin, N. V. C. (1996). Trophodynamics of reef fisheries productivity. *In* "Reef Fisheries" (N. V. C. Polunin and C. M. Roberts, eds.), pp. 113–135. Chapman & Hall, London.

Polunin, N. V. C., and Roberts, C. M. (1993). Greater biomass and value of target coral-reef fishes in two small Caribbean marine reserves. *Mar. Ecol.: Prog. Ser.* **100,** 167–176.

Polunin, N. V. C., and Roberts, C. M. (1996). "Reef Fisheries." Chapman & Hall, London.

Polunin, N. V. C., Harmelin-Vivien, M. L., and Galzin, R. (1995). Contrasts in food processing by five herbivorous coral-reef fishes. *J. Fish Biol.* **47,** 455–465.

Polunin, N. V. C., Roberts, C. M., and Pauly, D. (1996). Developments in tropical reef fisheries science and management. *In* "Reef Fisheries" (N. V. C. Polunin and C. M. Roberts, eds.), pp. 361–377. Chapman & Hall, London.

Popper, A. N. (1977). A scanning electron microscopic study of the sacculus and lagena in the ears of fifteen species of teleost fishes. *J. Morphol.* **153,** 397–417.

Popper, A. N. (1983). Organization of the inner ear and auditory processing. *In* "Fish Neurobiology and Behavior" (R. G. Northcutt and R. E. Davies, eds.), Vol. 1, pp. 125–178. University of Michigan Press, Ann Arbor.

Popper, A. N., and Carlson, T. J. (1998). Application of sound and other stimuli to control fish behavior. *Trans. Am. Fish. Soc.* **127,** 673–707.

Popper, A. N., and Coombs, S. (1980a). Acoustic detection by fishes. *In* "Environmental Physiology of Fishes" (M. A. Ali, ed.), pp. 403–430. Plenum Press, New York.

Popper, A. N., and Coombs, S. (1980b). Auditory mechanisms in teleost fishes. *Am. Sci.* **68,** 429–440.

Popper, A. N., and Fay, R. R. (1993). Sound detection and processing by fish: Critical review and major research questions. *Brain, Behav. Evol.* **50,** 213–221.

Popper, A. N., and Hoxter, B. (1981). The fine structure of the sacculus and lagena of a teleost fish. *Hear. Res.* **5,** 245–263.

Popper, A. N., and Hoxter, B. (1984). Growth of a fish ear: 1. Quantitative analysis of hair cell and ganglion cell proliferation. *Hear. Res.* **15,** 133–142.

Porch, C. E. (1998). A numerical study of larval fish

retention along the southeast Florida coast. *Ecol. Model.* **109,** 35–59.

Possingham, H. P., Tuljapurkar, S., Roughgarden, J., and Wilks, M. (1994). Population cycling in space-limited organisms subject to density-dependent predation. *Am. Nat.* **143,** 563–582.

Post, J. R., and Evans, D. O. (1989). Experimental evidence of size-dependent predation mortality in juvenile yellow perch. *Can. J. Zool.* **67,** 521–523.

Post, J. R., Johannes, M. H. R., and McQueen, D. J. (1997). Evidence of density-dependent cohort splitting in age-0 yellow perch (*Perca flavescens*): Potential behavioural mechanisms and population-level consequences. *Can. J. Fish. Aquat. Sci.* **54,** 867–875.

Postner, M., and Kramer, B. (1995). Electrosensory thresholds in larvae of the weakly electric fish *Pollimyrus isidori* (Mormyridae, Teleostei). *J. Exp. Biol.* **198,** 783–791.

Potts, D. C. (1985). Sea-level fluctuations and speciation in Scleractinia. *Proc. 5th Int. Coral Reef Congr.* Vol. 4, pp. 127–132.

Potts, J. C., and Manooch, C. S., III (1995). Age and growth of red hind and rock hind collected from North Carolina through the Dry Tortugas, Florida. *Bull. Mar. Sci.* **56,** 784–794.

Potts, J. C., Manooch, C. S., III, and Vaughan, D. S. (1998). Age and growth of vermilion snapper from the southeastern United States. *Trans. Am. Fish. Soc.* **127,** 787–795.

Powers, M. K., and Raymond, P. A. (1990). Development of the visual system. *In* "The Visual System of Fish" (R. H. Douglas and M. B. A. Djamgoz, eds.), pp. 419–442. Chapman & Hall, London.

Powers, M. K., Bassi, C. J., Rone, L. A., and Raymond, P. A. (1988). Visual detection by the rod system in goldfish of different sizes. *Vision Res.* **28,** 211–221.

Powles, H. (1975). Abundance, seasonality, distribution, and aspects of the ecology of some larval fishes off Barbados. Ph.D Dissertation, McGill University, Montreal, Quebee, Canada.

Pratchett, M. S. (1995). Spatial distribution, abundance and diet selection of butterflyfish (Pisces: Chaetodontidae). Honors Dissertation, James Cook University, Townsville, Australia.

Prejs, A. (1987). Risk of predation and feeding rate in tropical freshwater fishes: Field evidence. *Oecologia* **63,** 6–12.

Pressey, R. L., Humphries, C. J., Margules, C. R., Vane-Wright, R. I., and Williams, P. H. (1993). Beyond opportunism: Key principles for systematic reserve selection. *Trends Ecol. Evol.* **8,** 124–128.

Preston, F. W. (1948). The commonness and rarity of species. *Ecology* **29,** 254–283.

Price, T. (1998). Maternal and paternal effects in birds: Effects on offspring fitness. *In* "Maternal Effects as Adaptations" (T. A. Mousseau and C. W. Fox, eds.), pp. 202–226. Oxford University Press, New York.

Proctor, C. H., and Thresher, R. E. (1998). Effects of specimen handling and otolith preparation on concentration of elements in fish otoliths. *Mar. Biol.* **131,** 681–694.

Pulliam, H. R. (1988). Sources, sinks and population regulation. *Am. Nat.* **132,** 652–661.

Purcell, S. W., and Bellwood, D. R. (1993). A functional analysis of food procurement in two surgeonfish species, *Acanthurus nigrofuscus* and *Ctenochaetus striatus* (Acanthuridae). *Environ. Biol. Fishes* **37,** 139–159.

Pyle, R. L. (1993). Marine aquarium fish. *In* "Nearshore Marine Resources of the South Pacific" (A. Wright and L. Hill, eds.), pp. 135–176. Forum Fisheries Agency, Honiara, Solomon Islands.

Pyle, R. L., and Chave, E. H. (1994). First record of the chaetodontid genus *Prognathodes* from the Hawaiian Islands. *Pac. Sci.* **48,** 90–93.

Queensland Fisheries Management Authority (QFMA) (1999). "Queensland Marine Aquarium Fish and Collecting Fisheries," QFMA Discus. Pap. No. 10. QFMA, Queensland, Australia.

Quéro, J. C., Hureau, J. C., Karrer, C., Post,V., and Saldanha, L. (1990). "Check-list of the Fishes of the Eastern Tropical Atlantic." JNICT-Portugal, UNESCO.

Quinn, J. F., Wing, S. R., and Botsford, L. W. (1993). Harvest refugia in marine invertebrate fisheries—models and applications to the Red Sea urchin, *Strongylocentrotus franciscanus*. *Am. Zool.* **33,** 537–550.

Quinn, T. J., and Deriso, R. B. (1999). "Quantitative Fish Dynamics." Oxford University Press, Oxford.

Quinn, T. P., and Brannon, E. L. (1982). The use of celestial and magnetic cues by orienting sockeye salmon smolts. *J. Comp. Physiol.* **147,** 547–552.

Quinn, T. P., Merrill, R. T., and Brannon, E. L. (1981). Magnetic field orientation in sockeye salmon. *J. Exp. Zool.* **217,** 137–142.

Rabinowitz, D. (1978). Abundance and diaspore weight in rare and common prairie grasses. *Oecologia* **37,** 213–219.

Rabinowitz, D. (1981). Seven forms of rarity. *In* "The Biological Aspects of Rare Plant Conservation" (H. Synge, ed.), pp. 205–217. Wiley, New York.

Radtke, R. L. (1985). Recruitment patterns resolved from structural and chemical components of juvenile *Dascyllus albisella* otoliths. *Proc. 5th Int. Coral Reef Congr.*, Vol. 5, pp. 397–401.

Radtke, R. L., and Kinzie, R. A. III. (1996). Evidence of a marine larval stage in endemic Hawaiian stream gobies from isolated high-elevation locations. *Trans. Am. Fish. Soc.* **125,** 613–621.

Radtke, R. L., and Schafer, D. J. (1992). Environmental sensitivity of fish otolith chemistry. *Aust. J. Mar. Freshwater Res.* **43,** 935–951.

Radtke, R. L., Kinzie, R. A., III, and Folsom, S. D. (1988). Age at recruitment of Hawaiian freshwater gobies. *Environ. Biol. Fishes* **23,** 205–213.

Radtke, R. L., Townsend, D. W., Folsom, S. D., and Morrison, M. A. (1990). Strontium:calcium concentration ratios in otoliths of herring larvae as indicators of environmental histories. *Environ. Biol. Fishes* **27,** 51–61.

Radtke, R. L., Showers, W., Moksness, E., and Lenz, P. (1996). Environmental information stored in otoliths: Insights from stable isotopes. *Mar. Biol.* **127**, 161–170.

Radtke, R. L., Townsend, D. W., Kinzie, R. A., III, and Fey, D. (1999). Two-dimensional X-ray mapping of otoliths, a high-resolution technique for determining amphidromy in the tropical goby *Lentis concolor* (Gill). *J. Exp. Mar. Biol. Ecol.* **238**, 21–27.

Rajasuriya, A., Ranjith, M. W., De Silva, N., and Öhman, M. C. (1995). Coral reefs of Sri Lanka: Human disturbance and management issues. *Ambio* **24**, 428–437.

Rakitin, A., and Kramer, D. L. (1996). Effect of a marine reserve on the distribution of coral reef fishes in Barbados. *Mar. Ecol.: Prog. Ser.* **131**, 97–113.

Ralston, K. R., and Wainwright, P. C. (1997). Functional consequences of trophic specialization in pufferfishes. *Funct. Ecol.* **11**, 43–52.

Ralston, S., and Howard, D. F. (1995). On the development of year-class strength and cohort variability in two northern California rockfishes. *Fish. Bull.* **93**, 710–720.

Ramírez-Zarzosa, G., Gil, F., Vázquez, J. M., Arencibia, A., Latorre, R., López-Albors, O., Ortega, A., and Moreno, F. (1998). The post-larval development of lateral musculature in gilthead sea bream *Sparus aurata* (L.) and sea bass *Dicentrarchus labrax* (L.). *Anat. Histol. Embryol.* **27**, 21–29.

Rand, D. M. (1994). Thermal habit, metabolic rate and the evolution of mitochondrial DNA. *Trends Ecol. Evol.* **9**, 125–131.

Randall, J. E. (1961). A contribution to the biology of the convict surgeonfish of the Hawaiian Islands, *Acanthurus triostegus sandvicensis*. *Pac. Sci.* **15**, 215–272.

Randall, J. E. (1965). Food habits of the Nassau grouper *(Epinephalus striatus)*. *Proc. Assoc. Island Mar. Lab.* **6**, 13–16.

Randall, J. E. (1967). Food habits of reef fishes of the West Indies. *Stud. Trop. Oceanogr.* **5**, 665–847.

Randall, J. E. (1968). "Caribbean Reef Fishes." T. F. H. Publications, Neptune City, NJ.

Randall, J. E. (1983). "Caribbean Reef Fishes." T. F. H. Publications, Hong Kong.

Randall, J. E. (1987a). Introduction of marine fishes to the Hawaiian islands. *Bull. Mar. Sci.* **41**, 490–502.

Randall, J. E. (1987b). Collecting reef fishes for Aquaria. *In* "Human Impacts on Coral Reefs: Facts and Recommendations" (B. Salvat, ed.), pp. 29–38. Antenne Museum, EPHE, French Polynesia.

Randall, J. E. (1998). Zoogeography of shore fishes of the Indo-Pacific region. *Zool. Stud.* **37**, 227–268.

Randall, J. E., and Randall, H. A. (1963). The spawning and early devolopment of the Atlantic parrot fish *Sparisoma rubripinne*, with notes on other scarid and labrid fishes. *Zoologica (N.Y.)* **48**, 49–59.

Randall, J. E., Allan, G. R., and Steene, R. C. (1990). "Fishes of the Great Barrier Reef and Coral Sea." Crawford House, Bathurst, Australia.

Randall, J. E., Allen, G. R., and Steene, R. C. (1997). "Fishes of the Great Barrier Reef and Coral Sea," rev. ed. Crawford House, Bathurst, Australia.

Rapoport, E. H. (1994). Remarks on marine and continental biogeography: An areographical viewpoint. *Philos. Trans. R. Soc. London, Ser. B* **343**, 71–78.

Ray, C., and Hastings, A. (1996). Density dependence: Are we searching at the wrong spatial scale? *J. Anim. Ecol.* **65**, 556–566.

Ray, G. C., and Grassle, J. F. (1991). Marine biological diversity. *BioScience* **41**, 453–457.

Reaka, M. L. (1980). Geographic range, life history patterns, and body size in a guild of coral-dwelling mantis shrimps. *Evolution (Lawrence, Kans.)* **34**, 1019–1030.

Real, L. (1977). The kinetic of functional response. *Am. Nat.* **111**, 289–300.

Reed, S. J. B. (1990). Recent developments in geochemical microanalysis. *Chem. Geol.* **83**, 1–9.

Reid, D. G., Rumbak, E., and Thomas, R. H. (1996). DNA, morphology and fossils: Phylogeny and evolutionary rates of the gastropod genus *Littorina*. *Philos. Trans. R. Soc. London, Ser. B* **351**, 877–895.

Reid, W. V. (1998). Biodiversity hotspots. *Trends Ecol. Evol.* **13**, 275–280.

Reinhardt, U. G. (1999). Predation risk breaks size-dependent dominance in juvenile coho salmon (*Oncorhynchus kisutch*) and provides growth opportunities for risk prone individuals. *Can. J. Fish. Aquat. Sci.* **56**, 1206–1212.

Reutter, K. (1978). Taste organ in the bullhead (Teleostei). *Adv. Anat. Embryol. Cell Biol.* **55**, 1–98.

Reutter, K. (1982). Taste organ in the barbel of the bullhead. *In* "Chemoreception in Fishes" (T. J. Hara, ed.), pp. 77–91. Am. Elsevier, New York.

Reutter, K., and Witt, M. (1993). Morphology of vertebrate taste organs and their nerve supply. *In* "Mechanisms of Taste Transduction" (S. A. Simon and S. D. Roper, eds.), pp. 29–82. CRC Press, Boca Raton, FL.

Reynolds, J. D., and Côté, I. M. (1995). Direct selection on mate choice: Female redlip blennies pay more for better mates. *Behav. Ecol.* **6**, 175–181.

Reyns, N., and Sponaugle, S. (1999). Patterns and processes of brachyuran crab settlement to Caribbean coral reefs. *Mar. Ecol.: Prog. Ser.* **185**, 155–170.

Rhodes, K. L. (1999). Grouper aggregation protection in proactive Pohnpei. *Secretariat of the Pacific Community Live Reef Fish Inf. Bull.* **6**, 14–15.

Rice, J. A., Miller, T. J., Rose, K. A., Crowder, L. B., Marschall, E. A., Trebitz, A. S., and DeAngelis, D. L. (1993). Growth rate variation and larval survival: Inferences from an individual-based size-dependence predation model. *Can. J. Fish. Aquat. Sci.* **50**, 133–142.

Rice, J. A., Crowder, L. B., and Marschall, E. A. (1997). Predation on juvenile fishes: Dynamic interactions between size structured predators and prey. *In* "Early Life History and Recruitment in Fish Populations" (R. C. Chambers and E. A. Trippel, eds.), pp. 333–356. Chapman & Hall, New York.

Richard, B. A., and Wainwright, P. C. (1995). Scaling the

feeding mechanism of largemouth bass (*Micropterus salmoides*): I. Kinematics of prey capture. *J. Exp. Biol.* **198**, 419–433.

Richards, A. (1993). Live reef fish export fisheries in Papua New Guinea: Current status and future prospects. *Forum Fish. Agency Rep.* **93/10**, 1–14.

Richards, W. J., ed. (in preparation). Preliminary guide to the early life history stages of fishes of the western central Atlantic. URL: *http://www4.cookman.edu/NOAA*

Richards, W. J., and Lindeman, K. C. (1987). Recruitment dynamics of reef fishes: Planktonic processes, settlement and demersal ecologies, and fishery analysis. *Bull. Mar. Sci.* **41**, 392–410.

Richards, W. J., McGowan, M. F., Leming, T., Lamkin, J. T., and Kelley, S. (1993). Larval fish assemblages at the Loop Current boundary in the Gulf of Mexico. *Bull. Mar. Sci.* **53**, 475–537.

Ricker, W. E. (1954). Stock and recruitment. *J. Fish. Res. Board Can.* **11**, 559–623.

Ricker, W. E. (1973). Linear regressions in fishery research. *J. Fish. Res. Board Can.* **30**, 409–434.

Ricker, W. E. (1975). Computation and interpretation of biological statistics of fish populations. *Bull. Fish. Res. Board Can.* **191**, 1–382.

Ricker, W. E., and Foerster, R. E. (1948). Computation of fish production. *Bull. Bingham Oceanogr. Collect.* **11**, 173–211.

Rickleffs, R. E. (1987). Community diversity: Relative roles of local and regional processes. *Science* **235**, 167–171.

Rickleffs, R. E., and Schluter, D. eds. (1993a). "Species Diversity in Ecological Communities, Historical and Geological Perspectives." University of Chicago Press, Chicago.

Rickleffs, R. E., and Schluter, D. (1993b). Species diversity: Regional and historical influences. *In* "Species Diversity and Ecological Communities" (R. E. Ricklefs and D. Schluter, eds.), pp. 350–363. University of Chicago Press, Chicago.

Ridley, M., and Rechten, C. (1981). Female sticklebacks prefer to spawn with males whose nests contain eggs. *Behaviour* **76**, 152–161.

Rijnsdorp, A. D. (1994). Population-regulating processes during the adult phase in flatfish. *Neth. J. Sea Res.* **32**, 207–223.

Riley, C. M., and Holt, G. J. (1993). Gut contents of larval fishes from light trap and plankton net collections at Enmedio Reef near Veracruz, Mexico. *Rev. Biol. Trop., Suppl.* **41**, 53–57.

Rissik, D., Suthers, I. M., and Taggart, C. T. (1997). Enhanced zooplankton abundance in the lee of an isolated reef in the South Coral Sea: The role of flow disturbance. *J. Plankton Res.* **19**, 1347–1368.

Roberts, C. M. (1993). Coral reefs: Health, hazards and history. *Trends Ecol. Evol.* **8**, 425–427.

Roberts, C. M. (1994). Rapid build-up of fish biomass in a Caribbean marine reserve. *Conserv. Biol.* **9**, 815–826.

Roberts, C. M. (1995). Effects of fishing on the ecosystem structure of coral reefs. *Conserv. Biol.* **9**, 988–995.

Roberts, C. M. (1996). Settlement and beyond: population regulation and community structure of reef fishes. *In* "Reef Fisheries" (N. V. C. Polunin and C. M. Roberts, eds.), pp. 85–112. Chapman & Hall, UK.

Roberts, C. M. (1997a). Ecological advice for the global fisheries crisis. *Trends Ecol. Evol.* **12**, 35–38.

Roberts, C. M. (1997b). Connectivity and management of Caribbean coral reefs. *Science* **278**, 1454–1457.

Roberts, C. M. (1998). Sources, sinks, and the design of marine reserve networks. *Fisheries* **23**(7), 16–19.

Roberts, C. M. (2000). Selecting marine reserve locations: Optimality versus opportunism. *Bull. Mar. Sci.* **66**, 581–592.

Roberts, C. M., and Hawkins, J. P. (1999). Extinction risk in the sea. *Trends Ecol. Evol.* **14**, 241–246.

Roberts, C. M., and Ormond, R. F. G. (1987). Habitat complexity and coral reef fish diversity and abundance on Red Sea fringing reefs. *Mar. Ecol.: Prog. Ser.* **41**, 1–8.

Roberts, C. M., and Polunin, N. V. C. (1991). Are marine reserves effective in management of reef fisheries? *Rev. Fish Biol. Fish.* **1**, 65–91.

Roberts, C. M., and Polunin, N. V. C. (1992). Effects of marine reserve protection on northern Red Sea fish populations. *Proc. 7th Int. Coral Reef Symp.*, Vol. 2, pp. 969–977.

Roberts, C. M., and Polunin, N. V. C. (1993). Marine reserves: Simple solutions to managing complex fisheries? *Ambio* **22**, 363–368.

Roberts, C. M., Ballantine, W. J., Buxton, C. D., Dayton, P. K., Crowder, L. B., Milon, W., Orbach, M. K. Pauly, D. and Trexler, J. (1995). Review of the use of marine fishery reserves in the US SE Atlantic. *NOAA Tech. Mem.* **NMFS-SEFSC-376**.

Roberts, C. M., Hawkins, J. P., McAllister, D. E., and Schueler, F. W. (in press). Global distribution and conservation of reef fish biodiversity. *In* "Coral Reef Status Report" (D. E. McAllister, ed.). IUCN Species Survival Commission. Washington, DC.

Roberts, H. H., and Murray, S. P. (1983). Controls on reef development and the terrigenous-carbonate interface on a shallow shelf, Nicaragua (Central America). *Coral Reefs* **2**, 71–80.

Roberts, H. H., Wilson, P. A., and Lugo-Fernandez, A. (1992). Biological and geologic responses to physical processes: Examples from modern reef systems of the Caribbean-Atlantic region. *Cont. Shelf Res.* **12**, 809–834.

Roberts, W. M., Howard, J., and Hudspeth, A. J. (1988). Hair cell: Transduction, tuning and transmission in the inner ear. *Annu. Rev. Cell Biol.* **4**, 63–92.

Robertson, D. R. (1972). Social control of sex-reversal in a coral-reef fish. *Science* **177**, 1007–1009.

Robertson, D. R. (1973). Field observations on the reproductive behavior of the labrid fish, *Acanthochromis polyacanthus*. *Z. Tierpsychol.* **32**, 319–324.

Robertson, D. R. (1983). On the spawning behavior and spawning cycles of eight surgeon fishes (Acanthuridae) from the Indo-Pacific. *Environ. Biol. Fishes* **9**, 193–223.

Robertson, D. R. (1988a). Abundances of surgeonfishes on patch-reefs in Caribbean Panamá: Due to settlement, or post-settlement events? *Mar. Biol.* **97,** 495–501.

Robertson D. R. (1988b). Extreme variation in settlement of the Caribbean triggerfish *Balistes vetula* in Panama. *Copeia,* pp. 698–703.

Robertson, D. R. (1990). Differences in the seasonalities of spawning and recruitment of small neotropical reef fishes. *J. Exp. Mar. Biol. Ecol.* **144,** 49–62.

Robertson, D. R. (1991a). The role of adult biology in the timing of spawning of tropical reef fishes. *In* "The Ecology of Fishes on Coral Reefs" (P. F. Sale, ed.), pp. 356–386. Academic Press, San Diego, CA.

Robertson, D. R. (1991b). Increases in surgeonfish populations after mass mortality of the sea urchin *Diadema antillarum* in Panama indicate food limitation. *Mar. Biol.* **111,** 437–444.

Robertson, D. R. (1992). Patterns of lunar settlement and recruitment in Caribbean reef fishes at Panama. *Mar. Biol.* **114,** 527–537.

Robertson, D. R. (1996). Interspecific competition controls abundance and habitat use of territorial Caribbean damselfishes. *Ecology* **77,** 885–899.

Robertson, D. R. (1998a). Implications of body size for interspecific interactions and assemblage organization among coral reef fishes. *Aust. J. Ecol.* **23,** 252–257.

Robertson, D. R. (1998b). Do coral reef fish faunas have a distinctive taxonomic structure? *Coral Reefs* **17,** 179–186.

Robertson, D. R. (2001). Population maintenance among tropical reef fishes: Inferences from small-island endemics. *Proc. Natl. Acad. Sci. U.S.* **98,** 5667–5670.

Robertson, D. R., and Gaines, S. D. (1986). Interference competition structures habitat use in a local assemblage of coral reef surgeonfish. *Ecology* **67,** 1372–1382.

Robertson, D. R., and Hoffman, S. G. (1977). The role of female mate choice and predation in the mating systems of some tropical labroid fishes. *Z. Tierpsychol.* **45,** 298–320.

Robertson, D. R., and Justines, G. (1982). Protogynous hermaphroditism and gonochorism in four Caribbean reef gobies. *Environ. Biol. Fishes* **7,** 137–142.

Robertson, D. R., and Kaufman, K. (1998). Assessing early recruitment dynamics and its demographic consequences among tropical reef fishes: Accommodating variation in recruitment seasonality and longevity. *Aust. J. Ecol.* **23,** 226–233.

Robertson, D. R., and Warner, R. R. (1978). Sexual patterns in the labroid fishes of the western Caribbean, II: The parrotfishes (Scaridae). *Smithson. Contrib. Zool.* **255,** 1–26.

Robertson, D. R., Sweatman, H. P. A., Fletcher, E. A., and Cleland, M. G. (1976). Schooling as a mechanism for circumventing the territoriality of competitors. *Ecology* **57,** 1208–1220.

Robertson, D. R., Green, D. G., and Victor, B. C. (1988). Temporal coupling of production and recruitment of larvae of a Caribbean reef fish. *Ecology* **69,** 370–381.

Robertson, D. R., Petersen, C. W., and Brawn, J. D. (1990). Lunar reproductive cycles of benthic-brooding reeffishes: Reflections of larval biology or adult biology. *Ecol. Monogr.* **60,** 311–329.

Robertson, D. R., Schober, U. M., and Brawn, J. D. (1993). Comparative variation in spawning output and juvenile recruitment of some Caribbean reef fishes. *Mar. Ecol.: Prog. Ser.* **94,** 105–113.

Robertson, D. R., Swearer S. E., Kaufmann, K., and Brothers, E. B. (1999). Settlement vs. environmental dynamics in a pelagic-spawning reef fish at Caribbean Panama. *Ecol. Monogr.* **69,** 195–218.

Robinson, A. R., ed. (1983). "Eddies in Marine Sciences." Springer-Verlag, Berlin.

Robinson, B. W., and Schluter, D. (1999). Natural selection and the evolution of adaptive genetic variation in northern freshwater fishes. *In* "Adaptive Genetic Variation in the Wild" (T. A. Mousseau, B. Sinervo, and J. A. Endler, eds.), pp. 65–94. Oxford University Press, New York.

Robinson, B. W., and Wilson, D. S. (1994). Character release and displacement in fishes: A negelected literature. *Am. Nat.* **144,** 596–627.

Robinson, B. W., Wilson, D. S. and Margosian, A. S. (2000). A pluralistic analysis of character release in pumpkinseed sunfish (*Lepomis gibbosus*). *Ecology* **81,** 2799–2812.

Rocha-Olivares, A. (1998). Age, growth, mortality, and population characteristics of the Pacific red snapper, *Lutjanus peru,* off the southeast cost of Baja California, Mexico. *Fish. Bull.* **96,** 562–574.

Roden, G. I. (1987). Effect of seamounts and seamount chains on ocean circulation and thermohaline structure. *In* "Seamounts, Islands and Atolls" (B. H. Keating, P. Fryer, R. Batiza, and G. W. Boehlert, eds.), Geophys. Monogr. 43, pp. 335–354. American Geophysical Union, Washington, DC.

Roff, D. A. (1984). The evolution of life history parameters in teleosts. *Can. J. Fish. Aquat. Sci.* **41,** 989–1000.

Rogers, J. S., Hare, J. A., and Lindquist, D. G. (2001). The otolith record of age, growth and ontogeny in larval and pelagic juvenile *Stephanolepis hispidus* (Pisces: Monacanthidae). *Mar. Biol.* **138,** 945–953.

Rogers, P. H., and Cox, M. (1988). Underwater sound as a biological stimulus. *In* "Sensory Biology of Aquatic Animals" (J. Atema, R. R. Fay, A. N. Popper, and W. N. Tavolga, eds.), pp. 131–149. Springer-Verlag, New York.

Rogers, P. H., Popper, A. N., Hastings, M. C., and Saidel, W. M. (1988). Processing of acoustic signals in the auditory system of bony fish. *J. Acoust. Soc. Am.* **83,** 388–349.

Rögl, F., and Steininger, F. F. (1983). Neogene Paratethys, Mediterranean and Indo-Pacific seaways: Implications for the paleobiogeography of marine and terrestrial biotas. *Geol. J., Spec. Issue* **11,** 171–200.

Rognes, K. (1973). Head skeleton and jaw mechanism in Labrinae (Teleostei: Labridae) from Norwegian waters. *Arb. Univ. Bergen, Mat. Naturvitensk.* Ser. **4,** 1–149.

Rohde, K. (1978). Latitudinal gradients in species diversity and their causes. I. A review of the hypotheses explaining the gradients. *Zentralbl. Biol.* **97,** 393–403.

Rohde, K., Heap, M., and Heap, D. (1993). Rapoport's rule

does not apply to marine teleosts and cannot explain latitudinal gradients in species richness. *Am. Nat.* **142**, 1–16.

Rohner, C., and Krebs, C. J. (1998). Response of great horned owls to experimental "hot spots" of snowshoe hare density. *Auk* **115**, 694–705.

Rollin, B. E. (1997). Animal production and the new social ethic for animals. *Proc. 1st Conf. Market. Shipp. Aquat. Prod.*, Seattle, WA., *1996.*

Ronnback, P., Troell, M., Kantsky, N., and Primavera, J. H. (1999). Distribution pattern of shrimps and fish along *Avicennia* and *Rhizophora* microhabitats in the Pagbilae mangroves, Philippines. *Estuarine Coastal Shelf Sci.* **48**, 223–234.

Rosa, R. S., and Moura, R. L. (1997). Visual assessment of reef fish community structure in the Atol das Rocas biological reserve, off Northeastern Brazil. *Proc. 8th Int. Coral Reef Symp.*, Vol. 1, pp. 983–986.

Rosamond, L. N., Goldburg, R. J., Primavera, J. H., Kautsky, N., Beveridge, M. C. M., Clay, J., Folke, C., Lubchenco, J., Mooney, H., and Troell, M. (2000). Effect of aquaculture on world fish supplies. *Nature (London)* **405**, 1017–1024.

Rosen, B. R. (1984). Reef coral biogeography and climate through the late Cainozoic: Just islands in the sun or a critical pattern of islands. *Geol. J., Spec. Issue* **11**, 201–262.

Rosen B. R. (1988). Progress, problems and patterns in the biogeography of reef corals and other tropical marine organisms. *Helgol. Meer. Sunters.* **42**, 269–301.

Rosen, B. R., and Smith, A. B. (1988). Tectonics from fossils? Analysis of reef-coral and sea-urchin distributions from late Cretaceous to Recent, using a new method. *Spec. Publ.—Geol. Soc. London* **37**, 275–306.

Rosenberg, M. S., Adams, D. C., and Gurevitch, J. (2000). "Meta Win: Statistical Software for Meta-analysis." Sinauer Associ., Sunderland, MA.

Rosenthal, Y., Sherrell, R. M., and Field, P. (1999). A novel method for precise determination of element:calcium ratios in calcareous samples using sector field inductively coupled plasma mass spectrometry. *Anal. Chem.* **71**, 3248–3253.

Rosenzweig, M. L., and Lomolino, M. V. (1997). Who gets the short bits of the broken stick? *In* "The Biology of Rarity" (W. E. Kunin and K. J. Gaston, eds.), pp. 63–90. Chapman & Hall, London.

Ross, Q. E., Dunning, D. J., Thorne, R., Menezes, J. K., Tiller, G. W., and Watson, J. K. (1993). Response of alewives to high-frequency sound at a power plant intake on Lake Ontario. *North Am. J. Fish. Manage.* **13**, 291–303.

Ross, R. M. (1990). The evolution of sex-change mechanisms in fishes. *Environ. Biol. Fishes* **29**, 81–93.

Ross, R. M., Hourigan, T. F., Lutnesky, M. M. F., and Singh, I. (1990). Multiple simultaneous sex change in social groups of a coral reef fish. *Copeia,* pp. 427–433.

Ross, S. W., and Moser, M. L. (1995). Life history of juvenile gag, *Mycteroperca microlepis,* in North Carolina estuaries. *Bull. Mar. Sci.* **56**, 222–237.

Rothman, L. D., and Darling, D. C. (1990). Parasitoids of the goldenrod gall moth: Effects of scale on spatial density dependence. *Oecologia* **83**, 1–6.

Rothschild, B. J. (1986). "Dynamics of Marine Fish Populations." Harvard University Press, Cambridge, MA.

Rothschild, B. J., and Osborn, T. R. (1988). Small-scale turbulence and plankton contact rates. *J. Plankton Res.* **10**, 465–474.

Rougerie, F., and Wauthy, B. (1986). L' océan qui nous entoure. *In* "Encyclopédie de la Polynésie" (C. Gleizal, ed.), Vol. 1, pp. 73–88. Multipress, Papeete, Polynésie Française.

Roughgarden, J., Iwasa, Y., and Baxter, C. (1985). Demographic theory for an open marine population with space-limited recruitment. *Ecology* **66**, 54–67.

Roughgarden, J., Gaines, S. D., and Pacala, S. W. (1987). Supply side ecology: The role of physical transport processes. *In* "Organization of Communities" (J. H. R Gee and P. S. Giller, eds.), pp. 491–518. Blackwell, London.

Roughgarden, J., Gaines, S., and Possingham, H. (1988). Recruitment dynamics in complex life cycles. *Science* **241**, 1460–1466.

Rowley, R. J. (1994). Marine reserves and fisheries management. *Aquat. Conserv.* **4**, 233–254.

Royama, T. (1977). Population persistence and density dependence. *Ecol. Monogr.* **47**, 1–35.

Royama, T. (1992). "Analytical Population Dynamics." Chapman & Hall, London.

Rubec, P. J. (1987). The effects of sodium cyanide on coral reefs and marine fish in the Philippines. *In* "The First Asian Fisheries Forum" (J. L. Maclean, L. B. Dizon, and L. V. Hosillos, eds.), pp. 297–302. Asian Fisheries Society, Manila.

Rubec, P. J. (1988). The need for conservation and management of Philippine coral reefs. *Environ. Biol. Fishes* **23**, 141–154.

Ruddle, K. (1996). Geography and human ecology of reef fisheries. *In* "Reef Fisheries" (N. V. C. Polunin and C. M. Roberts, eds.), pp. 137–160. Chapman & Hall, London.

Rundle, H. D., Nagel, L., Boughman, J. W., and Schluter, D. (2000). Natural selection and parallel speciation in sympatric sticklebacks. *Science* **287**, 306–308.

Russ, G. R. (1984). Distribution and abundance of herbivorous grazing fishes in the central Great Barrier Reef. II. Patterns of zonation of mid-shelf and outershelf reefs. *Mar. Ecol.: Prog. Ser.* **20**, 35–44.

Russ, G. R. (1985). Effects of protective management on coral reef fishes in the central Philippines. *Proc. 5th Int. Coral Reef Congr.*, Vol. 4, pp. 219–224.

Russ, G. R. (1991). Coral reef fisheries: effects and yields. *In* "The Ecology of Fishes on Coral Reefs" (P. F. Sale, ed.), pp. 601–635. Academic Press, San Diego, CA.

Russ, G. R., and Alcala, A. C. (1989). Effects of intense fishing pressure on an assemblage of coral reef fishes. *Mar. Ecol.: Prog. Ser.* **56**, 13–27.

Russ, G. R., and Alcala, A. C. (1994). Marine reserves: They enhance fisheries, reduce conflicts and protect resources. *Naga, ICLARM Q.* **17**, 4–7.

Russ, G. R., and Alcala, A. C. (1996a). Marine reserves: Rates and patterns of recovery and decline in abundance of large predatory fish. *Ecol. Appl.* **6**, 947–961.

Russ, G. R., and Alcala, A. C. (1996b). Do marine reserves export adult fish biomass? Evidence from Apo Island, central Philippines. *Mar. Ecol.: Prog. Ser.* **132**, 1–9.

Russ, G. R., and Alcala, A. C. (1998a). Natural fishing experiments in marine reserves 1983–1993: Community and trophic responses. *Coral Reefs* **17**, 383–397.

Russ, G. R., and Alcala, A. C. (1998b). Natural fishing experiments in marine reserves 1983–1993: Roles of life history and fishing intensity in family responses. *Coral Reefs* **17**, 399–416.

Russ, G. R., and Alcala, A. C. (1999). Management histories of Sumilon and Apo marine reserves, Philippines, and their influence on national marine resource policy. *Coral Reefs* **18**, 307–319.

Russ, G. R., Alcala, A. C., and Cabanban, A. S. (1992). Marine reserves and fisheries management on coral reefs with preliminary modeling of the effects on yield per recruit. *Proc. 7th Int. Coral Reef Symp.*, Vol. 2, pp. 988–995.

Russ, G. R., Lou, D. C., and Ferreira, B. P. (1996). Temporal tracking of a strong cohort in the population of a coral reef fish, the coral trout, *Plectropomus leopardus* (Serranidae: Epinephelinae), in the central Great Barrier Reef, Australia. *Can. J. Fish. Aquat. Sci.* **53**, 2745–2751.

Russ, G. R., Lou, D. C., Higgs, J., and Ferreira, B. P. (1998). Mortality rate of a cohort of the coral trout, *Plectropomus leopardus,* in no fishing zones of the Great Barrier Reef Marine Park. *Mar. Freshwater Res.* **49**, 507–511.

Russell, B. C. (1983). "Annotated Checklist of the Coral Reef Fishes in the Capricorn-Bunker Group Great Barrier Reef Australia." Great Barrier Reef Marine Park Authority, Townsville, Australia.

Russell, F. S. (1976). "The Eggs and Planktonic Stages of British Marine Fishes." Academic Press, London.

Rutherford, E. S., and Houde, E. D. (1995). The influence of temperature on cohort-specific growth, survival, and recruitment of striped bass, *Morone saxatilis,* larvae in Chesapeake Bay. *Fish. Bull.* **93**, 315–332.

Rutherford, E. S., Tilmant, J. T., Thue, E. B., and Schmidt, T. W. (1989). Fishery harvest and population dynamics of gray snapper, *Lutjanus griseus,* in Florida Bay and adjacent waters. *Bull. Mar. Sci.* **44**, 139–154.

Ruzzante, D. E., Taggart, C. T., and Cook, D. (1996). Spatial and temporal variation in the genetic composition of a larval cod (*Gadus morhua*) aggregation: Cohort contribution and genetic stability. *Can. J. Fish. Aquat. Sci.* **53**, 2695–2705.

Sabatés, A., and Maso, M. (1992). Unusual larval fish distribution pattern in a coastal zone of the western Mediterranean. *Limnol. Oceanogr.* **37**, 1252–1260.

Sabatés, A., and Olivar, M. (1996). Variation of larval fish distributions associated with variability in the location of a shelf-slope front. *Mar. Ecol.: Prog. Ser.* **135**, 11–20.

Sadovy, Y. (1992). A preliminary assessment of the marine aquarium export trade in Puerto Rico. *Proc. 7th Int. Coral Reef Symp.*, pp. 1014–1022.

Sadovy, Y. (1993). The Nassau Grouper, endangered or just unlucky? *Reef Encounter,* July, pp. 10–12.

Sadovy, Y. (1994). Grouper stocks in the western central Atlantic: The need for management and management needs. *Proc. Annu. Gulf Caribb. Fish. Inst.* **43**, 43–64.

Sadovy, Y. (1996). Reproduction of reef fishery species. *In* "Reef Fisheries" (N. V. C. Polunin and C. M. Roberts, eds.), pp. 15–59. Chapman & Hall, London.

Sadovy, Y. (1998). The live reef fish trade: A role for importers in combating destructive fishing practices—example of Hong Kong, China. *Proc. APEC Workshop Impacts of Destruct. Fish. Practi. Mar. Environ.* Hong Kong, 1997, pp. 200–207.

Sadovy, Y. (1999). The case of the disappearing grouper, *Epinephelus striatus,* the Nassau grouper in the Caribbean and western Atlantic. *Proc. Annu. Gulf Caribb. Fish. Inst.* **45**, 5–22.

Sadovy, Y. (2000). "Regional Survey of Fry/Fingerling Supply and Current Practices for Grouper Mariculture: Evaluating Current Status and Long-Term Prospects for Grouper Mariculture in South East Asia," Final report. Collaborative APEC Grouper Research and Development Network (FWG 01/99).

Sadovy, Y. (2001a). The live reef fish trade in Hong Kong: Problems and prospects. *In* "Marketing and Shipping Live Aquatic Products: Proceedings of the Second International Conference and Exhibition, November 1999, Seattle, WA" (B. C. Paust and A. A. Rice, eds.), pp. 183–192. University of Alaska Sea Grant, AK-SG-01-03, Fairbanks.

Sadovy, Y. (2001b). The threat of fishing to highly fecund fishes. *J. Fish Biol.* **59(Suppl. A)**, 90–108.

Sadovy, Y., and Cornish, A. S. (2000). "Reef Fishes of Hong Kong." Hong Kong University Press, Hong Kong.

Sadovy, Y., and Domeier, M. L. (1994). Aggregation and spawning in the tiger grouper, *Mycteroperca tigris* (Pisces: Serranidae). *Copeia,* pp. 511–516.

Sadovy, Y., and Eklund, A. M. (1999). "Synopsis of Biological Data on the Nassau Grouper, *Epinephelus striatus* (Bloch, 1792) and Jewfish, *E. itajara* (Lichtenstein, 1822)," NOAA Tech. Rep., NMFS 146. NOAA, U.S. Department of Commerce, Washington, DC.

Sadovy, Y., and Pet, J. (1998). Wild collection of juveniles for grouper mariculture: Just another capture fishery? *Secretariat of the Pacific Community Live Reef Fish Inf. Bull.* **4**, 36–39.

Sadovy, Y., and Severin, K. P. (1992). Trace elements in biogenic aragonite: Correlation of body growth rate and strontium levels in the otoliths of the white grunt, *Haemulon plumieri* (Pisces: Haemulidae). *Bull. Mar. Sci.* **50**, 237–257.

Sadovy, Y., and Severin, K. P. (1994). Elemental patterns in red hind (*Epinephelus guttatus*) otoliths from Bermuda and Puerto Rico reflect growth rate, not temperature. *Can. J. Fish. Aquat. Sci.* **51**, 133–141.

Sadovy, Y., and Shapiro, D. Y. (1987). Criteria for the diagnosis of hermaphroditism in fishes. *Copeia,* pp. 136–156.

Sadovy, Y., Figuerola, M., and Román, A. (1992). Age and growth of red hind *Epinephelus guttatus* in Puerto Rico and St. Thomas. *Fish. Bull.* **90,** 516–528.

Sadovy, Y., Rosario, A., and Román, A. (1994). Reproduction in an aggregating grouper, the red hind, *Epinephelus guttatus. Environ. Biol. Fishes* **41,** 269–286.

Sadovy, Y., Mitcheson, G., and Rasotto, M. B. (2001). Early development of the mandarinfish, *Synchiropus splendidus* (Callionymidae), with notes on its fishery and potential for culture. *Aquar. Sci. Conserv.* **3,** 253–263.

Safina, C. (1995). The world's imperiled fish. *Sci. Am.* **273**(5), 46–53.

Sainsbury, K. J. (1991). Application of an experimental management approach to management of a tropical multispecies fishery with highly uncertain dynamics. *ICES Mar. Sci. Symp.* **193,** 301–320.

Sainsbury, K. J., Campbell, R. A., and Whitelaw, A. W. (1993). Effects of trawling on the marine habitat on the Northwest Shelf of Australia and implications for sustainable fisheries management. *In* "Sustainable fisheries through sustaining fish habitat" (D. A. Hancock, ed.), *Aust. Soc. Fish. Biol. Workshop Bureau of Resource Science, Proceedings* **17,** 137–145.

Sala, E., and Ballesteros, E. (1997). Partitioning of space and food resources by three fish of the genus *Diplodus* (Sparidae) in a Mediterranean rocky infralittoral ecosystem. *Mar. Ecol.: Prog. Ser.* **152,** 273–283.

Sala, E., Boudouresque, C. F., and Harmelin-Vivien, M. (1998). Fishing, trophic cascades, and the structure of algal assemblages: Evaluation of an old but untested paradigm. *Oikos* **83,** 425–439.

Sale, P. F. (1970). Distribution of larval Acanthuridae off Hawaii. *Copeia,* pp. 765–766.

Sale, P. F. (1976). The effect of territorial adult pomacentrid fishes on the recruitment and survival of juveniles on patches of coral rubble. *J. Exp. Mar. Biol. Ecol.* **24,** 297–306.

Sale, P. F. (1977). Maintenance of high diversity in coral reef fish communities. *Am. Nat.* **111,** 337–359.

Sale, P. F. (1978a). Coexistence of coral reef fishes—a lottery for living space. *Environ. Biol. Fishes* **3,** 85–102.

Sale, P. F. (1978b). Reef fish and other vertebrates: A comparison of social structure. *In* "Contrasts in Behavior" (E. S. Reese and F. J. Lighter, eds.), pp. 313–346. Wiley-Interscience, New York.

Sale, P. F. (1980). The ecology of fishes on coral reefs. *Oceanogr. Mar. Biol.* **18,** 367–421.

Sale, P. F. (1988). What coral reefs can teach us about ecology. *Proc. 6th Int. Coral Reef Symp.,* Vol. 1, pp. 19–27.

Sale, P. F. (1990). Recruitment of marine species: Is the bandwagon rolling in the right direction? *Trends Ecol. Evol.* **5,** 25–27.

Sale, P. F., ed. (1991a). "The Ecology of Fishes on Coral Reefs." Academic Press, San Diego, CA.

Sale, P. F. (1991b). Reef fish communities: Open nonequilibrial systems. *In* "The Ecology Fishes on Coral Reefs" (P. F. Sale, ed.), pp. 564–598. Academic Press, San Diego, CA.

Sale, P. F. (1998). Appropriate scales for studies of reef-fish ecology. *Aust. J. Ecol.* **23,** 202–208.

Sale, P. F., and Cowen, R. K. (1998). Not quite so simple: Reply to Roberts. *Science* **279,** 2021.

Sale, P. F., and Ferrell, D. J. (1988). Early survivorship of juvenile coral reef fishes. *Coral Reefs* **7,** 117–124.

Sale, P. F., and Tolimieri, N. (2000). Density-dependence at some time and place? *Oecologia* **124,** 166–171.

Sale, P. F., Doherty, P. J., Eckert, G. J., Douglas, W. A., and Ferrell, D. J. (1984). Large scale spatial and temporal variation in recruitment to fish populations on coral reefs. *Oecologia* **64,** 191–198.

Sale, P. F., Guy, J. A., and Steel, W. J. (1994). Ecological structure of assemblages of coral reef fishes on isolated patch reefs. *Oecologia* **98,** 83–99.

Salvat, B. (1971). Biogéographie malacologique de la Polynesie à la lumière des recentes recherches sur l'histoire géologique des iles hautes et des atolls de cette region. *Atti Soc. Itol. Sci. Nat. Mus. Civ. Ster. Nat. Milano* **112,** 330–334.

Sammarco, P. W. (1994). Larval dispersal and recruitment processes in Great Barrier Reef corals: Analysis and synthesis. *In* "The Bio-physics of Marine Larval Dispersal" (P. W. Sammarco and M. L. Heron, eds.), pp. 35–72. American Geophysical Union, Washington, DC.

Samoilys, M. A. (1988). Abundance and species richness of coral reef fish on the Kenyan coast: The effects of protective management and fishing. *Proc. 6th Int. Coral Reef Symp.,* Vol. 2, pp. 261–266.

Samoilys, M. A. (1997a). Periodicity of spawning aggregations of coral trout *Plectropomus leopardus* (Pisces: Serranidae) on the northern Great Barrier Reef. *Mar. Ecol.: Prog. Ser.* **160,** 149–159.

Samoilys M. A. (1997b). Movement in a large predatory fish: Coral trout, *Plectropomus leopardus* (Pisces: Serranidae), on Heron Reef, Australia. *Coral Reefs* **16,** 151–158.

Samoilys, M. A, and Green, A. (1990). "The Effects of Collecting Coral Reef Fish for the Aquarium Trade: A Review, and a Preliminary Study on the Northern Great Barrier Reef," Internal Report. Department of Primary Industries, Queensland.

Samoilys, M. A., and Squire, L. C. (1994). Preliminary observations on the spawning behavior of coral trout, *Plectropomus leopardus* (Pisces: Serranidae), on the Great Barrier Reef. *Bull. Mar. Sci.* **54,** 332–342.

Sancho, G., Ma, D., and Lobel, P. S. (1997). Behavioral observations of an upcurrent reef colonization event by larval surgeonfish *Ctenochaetus strigosus* (Acanthuridae). *Mar. Ecol.: Prog. Ser.* **153,** 311–315.

Sancho, G., Petersen, C. W., and Lobel, P. S. (2000a). Predator-prey relations at a spawning aggregation site of coral reef fishes. *Mar. Ecol.: Prog. Ser.* **203,** 275–288.

Sancho, G., Solow, A. R., and Lobel, P. S. (2000b).

Environmental influences on the diel timing of spawning in coral reef fishes. *Mar. Ecol.: Prog. Ser.* **206**, 193–212.

Sand, O. (1974). Directional sensitivity of microphonic potentials from the perch ear. *J. Exp. Biol.* **60**, 881–899.

Sand, O., and Karlsen, H. E. (1986). Detection of infrasound by the Atlantic cod. *J. Exp. Biol.* **125**, 197–204.

Sander, F. (1981). A preliminary assessment of the main causative mechanisms of the "island mass" effect of Barbados. *Mar. Biol.* **64**, 199–205.

Sanderson, S. L. (1988). Variation in neuromuscular activity during prey capture by trophic specialists and generalists (Pisces: Labridae). *Brain, Behav. Evol.* **32**, 257–268.

Sanderson, S. L. (1990). Versatility and specialization in labrid fishes: Ecomorphological implications. *Oecologia* **84**, 272–279.

Sanderson, S. L. (1991). Functional stereotypy and feeding performance correlated in a trophic specialist. *Funct. Ecol.* **5**, 795–803.

Sanderson, S. L., and Wassersug, R. (1993). Convergent and alternative designs for vertebrate suspension feeding. *In* "The Skull" (J. Hanken and B. K. Hall, eds.), Vol. 3, pp. 37–112. University of Chicago Press, Chicago.

Sanderson, S. L., Chestnutt, C. R., and Lobel, P. S. (1996). Evidence for ram suspension feeding by the piscivore, *Seriola dumerli* (Carangidae). *Environ. Biol. Fishes* **46**, 365–373.

Sano, M. (1989). Feeding habits of Japanese butterflyfishes (Chaetodontidae). *Environ. Biol. Fishes* **25**, 195–203.

Sano, M. (1997). Temporal variation in density dependence: Recruitment and postrecruitment demography of a temperate zone goby. *J. Exp. Mar. Biol. Ecol.* **214**, 67–84.

Sanvicente-Añorve, L., Flores-Coto, C., and Sanchez-Velasco, L. (1998). Spatial and seasonal patterns of larval fish assemblages in the southern Gulf of Mexico. *Bull. Mar. Sci.* **62**, 17–30.

Sargent, R. C. (1988). Paternal care and egg survival both increase with clutch size in the fathead minnow, *Pimephales promelas*. *Behav. Ecol. Sociobiol.* **23**, 33–37.

Sarkar, S. P. (1960). A study of the functional morphology of the head of an Indian pufferfish, *Sphoeroides oblongus* (Bloch). Ph.D. Dissertation, University of Leiden, The Netherlands.

Schaefer, M. B. (1954). Some aspects of the dynamics of populations important to the management of commercial marine fisheries. *Bull. Int-Am. Trop. Tuna Comm.* **1**, 27–56.

Schaeffer, B., and Rosen, D. E. (1961). Major adaptive levels in the evolution of the actinopterygian feeding mechanism. *Am. Zool.* **1**, 187–204.

Schellart, N. A. M., and Popper, A. N. (1992). Functional aspects of the evolution of the auditory system of actinopterygian fish. *In* "The Evolutionary Biology of Hearing" (D. B. Webster, R. R. Fay, and A. N. Popper, eds.), pp. 295–322. Springer-Verlag, New York.

Scheltema, R. S. (1986). On dispersal and planktonic larvae of benthic invertebrates: An eclectic overview and summary of problems. *Bull. Mar. Sci.* **39**, 290–322.

Schluger, J., and Hopkins, C. D. (1987). Electric fish approach stationary signal sources by following electric current lines. *J. Exp. Biol.* **130**, 359–367.

Schluter, D. (1996) Ecological causes of adaptive radiation. *Am. Nat.* **148**, S40–S64.

Schluter, D., and McPhail, J. D. (1993). Character displacement and replicate adaptive radiation. *Trends Ecol. Evol.* **8**, 197–200.

Schmale, M. C. (1981). Sexual selection and reproductive success in males of the bicolor damselfish, *Eupomacentrus partitus* (Pisces: Pomacentridae). *Anim. Behav.* **29**, 1171–1184.

Schmidt, K. F. (1997). "No take" zones spark fisheries debate. *Science* **277**, 489–491.

Schmitt, E., and Kunz, Y. W. (1989). Retinal morphogenesis in the rainbow trout, *Salmo gairdneri*. *Brain, Behav. Evol.* **34**, 48–64.

Schmitt, R. J., and Holbrook, S. J. (1996). Local-scale patterns of larval settlement in a planktivorous damselfish: Do they predict recruitment? *Mar. Freshwater Res.* **47**, 449–463.

Schmitt, R. J., and Holbrook, S. L. (1999a). Mortality of juvenile damselfish: Implications for assessing processes that determine abundance. *Ecology* **80**, 35–50.

Schmitt, R. J., and Holbrook, S. J. (1999b). Settlement and recruitment of three damselfish species: Larval delivery and competition for shelter space. *Oecologia* **118**, 76–86.

Schmitt, R. J., and Holbrook, S. J. (2000). Habitat-limited recruitment of coral reef damselfish. *Ecology,* **81**, 3479–3494.

Schmitt, R. J., Holbrook, S. J., and Osenberg, C. W. (1999). Quantifying the effects of multiple processes on local abundance: A cohort approach for open populations. *Ecol. Lett.* **2**, 294–303.

Schmitt, T. M., Hay, M. E., and Lindquist, N. (1995). Constraints on chemically mediated coevolution: Multiple functions for seaweed secondary metabolites. *Ecology* **76**, 107–123.

Schneider, H. (1961). Neue Ergebnisse der Lautforschung bei Fischen. *Naturwissenschaften* **48**, 513–518.

Schneider, H. (1964a). Physiologische und morphologische Untersuchungen zur Bioakustik der Tigerfische (Pisces, Theraponidae). *Z. Vergl. Physiol.* **47**, 493–558.

Schneider, H. (1964b). Bioakustische Untersuchungen an Anemonenfischen der Gattung *Amphiprion* (Pisces). *Z. Morphol. Oekol. Tiere* **53**, 453–474.

Schoener, T. W. (1987). The geographical distribution of rarity. *Oecologia* **74**, 161–173.

Schopf, T. J. M. (1980). "Paleoceanography." Harvard University Press, Cambridge, MA.

Schreck, C. B., Fitzpatrick, M. S., Feist, G. W., and Yeoh, C. G. (1991). Steroids: Developmental continuum between mother and offspring. *In* "Reproductive Physiology of Fish" (A. P. Scott, J. P. Sumpter, D. E. Kinne, and M. S. Rolfe, eds.), Proc. 4th Int. Symp. Reprod. Physiol. Fish, pp. 256–258. University of East Anglia, Norwich, UK.

Schreck, C. B., Olla, B. L., and Davis, M. W. (1997).

Behavioral responses to stress. *In* "Fish Stress and Health in Aquaculture" (G. K. Iwama, A. D. Pickering, J. P. Sumpter, and C. B. Schreck, eds.), pp. 145–170. Cambridge University Press, New York.

Schuijf, A. (1981). Models of acoustic localization. *In* "Hearing and Sound Communication in Fishes" (W. N. Tavolga, A. N. Popper, and R. R. Fay, eds.), pp. 267–307. Springer-Verlag, New York.

Schuijf, A., and Buwalda, R. J. A. (1980). Underwater localization—a major problem in fish acoustics. *In* "Comparative Studies of Hearing in Vertebrates" (A. N. Popper and R. R. Fay, eds.), pp. 43–77. Springer-Verlag, New York.

Schuijf, A., and Siemelink, M. E. (1974). The ability of cod (*Gadus morhua*) to orient towards a sound source. *Experientia* **30,** 773–774.

Schuijf, A., Baretta, J. W., and Wildschut, J. T. (1972). A field investigation on the discrimination of sound direction in *Labrus berggylta* (Pisces: Perciformes). *Neth. J. Zool.* **22,** 81–104.

Schultz, E. T., and Cowen, R. K. (1994). Recruitment of coral-reef fishes to Bermuda: Local retention or long-distance transport? *Mar. Ecol.: Prog. Ser.* **109,** 15–28.

Schultz, E. T., and Warner, R. R. (1989). Phenotypic plasticity in life-history traits of female *Thalassoma bifasciatum* (Pisces: Labridae): 1. Manipulations of social structure in tests for adaptive shifts of life-history allocations. *Evolution (Lawrence, Kans.)* **43,** 1497–1506.

Schultz, E. T., and Warner, R. R. (1991). Phenotypic plasticity in life-history traits of female *Thalassoma bifasciatum* (Pisces: Labridae). 2. Correlation of fecundity and growth rate in comparative studies. *Environ. Biol. Fishes* **30,** 333–344.

Schultz, O. (1993). Der Nachweis von *Scorpaena* s. s. (Pisces: Teleostei) im Badenien von St. Margarethen, Burgenland, Osterreich. Revision von *Scorpaena prior* Heckel in Heckel & Kner, 1861. *Ann. Naturhist. Mus. Wien* **95,** 127–177.

Schutz, F. (1956). Vergleichende Untersuchungen über die Schreckreaktion bei Fischen und deren Verbreitung. *Z. Vergl. Physiol.* **38,** 84–135.

Schwarcz, H. P., Gao, Y., Campana, S., Browne, D., Knyf, M., and Brand, U. (1998). Stable carbon isotope variations in otoliths of Atlantic cod (*Gadus morhua*). *Can. J. Fish. Aquat. Sci.* **55,** 1798–1806.

Scott, P. J. B. (1990). Chronic pollution recorded in coral skeletons in Hong Kong. *J. Exp. Mar. Biol. Ecol.* **139,** 51–64.

Searcy, S. P., and Sponaugle, S. (2000). Variable larval growth in a coral reef fish. *Mar. Ecol.: Prog. Ser.* **206,** 213–226.

Searcy, S. P., and Sponaugle, S. (2001). Early life history traits and survival in two coral reef fishes. *Ecology* **82,** 2452–2470.

Secor, D. H., and Dean, J. M. (1989). Comparison of otolith-based back-calculation methods to determine individual growth histories of larval striped bass, *Morone saxatilis*. *Can. J. Fish. Aquat. Sci.* **49,** 1439–1454.

Secor, D. H., Dean, J. M., and Laban, E. H. (1992). Otolith removal and preparation for microstructural examination. *Can. Fish. Aquat. Sci., Spec. Publ.* **117,** 19–57.

Secor, D. H., Dean, J. M., and Campana, S. E., eds. (1995a). "Recent Developments in Fish Otolith Research." University of South Carolina Press, Columbia.

Secor, D. H., Houde, E. D., and Monteleone, D. M. (1995b). A mark-release experiment on larval striped bass *Morone saxatilis* in a Chesapeake Bay tributary. *ICES J. Mar. Sci.* **52,** 87–101.

Selander, R. K., and Kaufman, D. W. (1973). Genetic variability and strategies of adaptation in animals. *Proc. Natl. Acad. Sci. U.S.A* **70,** 1875–1877.

Sette, O. E. (1955). Consideration of midocean fish production as related to oceanic circulatory systems. *J. Mar. Res.* **14,** 398–414.

Shackell, N. L., and Willison, J. H. M. (1995). "Marine Protected Areas and Sustainable Fisheries." Science and Management of Protected Areas Association, Wolfville, Nova Scotia.

Shakeel, H. (1994). "Study of Grouper Fishery and Live Grouper Holding Operations in the Maldives." Manu. Inshore Fisheries Research Unit, Marine Research Section, Ministry of Fisheries and Agriculture, Malé, Maldives.

Shakeel, H., and Ahmed, H. (1997). Exploitation of reef resources, grouper and other food fishes in the Maldives. *Secretariat of the Pacific Community Live Reef Fish Inf. Bull.* **2,** 14–20.

Shaklee, J. B. (1984). Genetic variation and population structure in the damselfish, *Stegastes fasciolatus,* throughout the Hawaiian archipelago. *Copeia,* pp. 629–640.

Shaklee, J. B., and Samollow, P. B. (1984). Genetic variation and population structure in a deep-water snapper, *Pristipomoides filamentosus,* in the Hawaiian archipelago. *Fish. Bull.* **82,** 703–713.

Shaklee, J. B., Tamaru, C. S., and Waples, R. S. (1982). Speciation and evolution of marine species studied by the electrophoretic analysis of proteins. *Pac. Sci.* **36,** 141–157.

Shand, J. (1993). Changes in the spectral absorption of cone visual pigments during the settlement of the goatfish *Upeneus tragula* and the loss of red sensitivity as a benthic existence begins. *J. Comp. Physiol.* **173,** 115–121.

Shand, J. (1994a). Changes in retinal structure during development and settlement of the goatfish *Upeneus tragula*. *Brain, Behav. Evol.* **43,** 51–60.

Shand, J. (1994b). Changes in the visual system of teleost fishes during growth and settlement: An ecological perspective. Unpublished Ph.D. Dissertation. James Cook University, Townsville, Australia.

Shand, J. (1997). Ontogenetic changes in retinal structure and visual acuity: A comparative study of coral-reef teleosts with differing post-settlement lifestyles. *Environ. Biol. Fishes* **49,** 307–322.

Shand, J., and Lythgoe, J. N. (1987). Light induced changes in corneal iridescence in fish. *Vision Res.* **276,** 303–305.

Shand, J., Partridge, J. C., Archer, S. N., Potts, G. W., and Lythgoe, J. N. (1988). Spectral absorbance changes in

the violet blue sensitive cones of the juvenile pollack, *Pollachius pollachius*. *J. Comp. Physiol.* **163**, 699–703.
Shand, J., Døving, K. B., and Collin, S. P. (1999). Optics of the developing fish eye: comparisons of Matthiessen's ratio and the focal length of the lens in the black bream *Acanthopagrus butcheri* (Sparidae, Teleostei). *Vision Res.* **39**, 1071–1078.
Shanks, A. L. (1983). Surface slicks associated with tidally forced internal waves may transport pelagic larvae of benthic invertebrates and fishes shoreward. *Mar. Ecol.: Prog. Ser.* **13**, 311–315.
Shanks, A. L. (1986). Tidal periodicity in the daily settlement on intertidal barnacle larvae and a hypothesized mechanism for the cross-shelf transport of cyprids. *Biol. Bull. (Woods Hole, Mass.)* **170**, 429–440.
Shanks, A. L. (1995). Orientated swimming by Megalopae of several eastern North Pacific crab species and its potential role in their onshore migration *J. Exp. Mar. Biol. Ecol.* **186**, 1–16.
Shapiro, D. Y. (1979). Social behavior, group structure, and the control of sex reversal in hermaphroditic fish. *Adv. Study Behav.* **10**, 43–102.
Shapiro, D. Y. (1980). Serial female changes after simultaneous removal of males from social groups of coral reef fish. *Science* **209**, 1136–1137.
Shapiro, D. Y. (1983). On the possibility of kin groups in coral reef fishes. *In* "Ecology of Deep and Shallow Coral Reefs" (M. L. Reaka, ed.) pp. 39–45. NOAA symposia series for undersea research, **Vol. 1.** Nat. Oceanographic Atmospheric Admin., United States Department of Commerce, Washington DC.
Shapiro, D. Y. (1987a). Reproduction in groupers. *In* "Tropical Snappers and Groupers: Biology and Fisheries Management" (J. J. Polovina and S. Ralston, eds.), pp. 295–327. Westview Press, Boulder, CO.
Shapiro, D. Y. (1987b). Differentiation and evolution of sex change in fishes. *BioScience* **37**, 490–497.
Shapiro, D. Y. (1987c). Inferring larval recruitment strategies from the distributional ecology of settled individuals of a coral reef fish. *Bull. Mar. Sci.* **41**, 289–295.
Shapiro, D. Y. (1989). Sex change as an alternative life-history style. *In* "Alternative Life-History Styles of Animals" (M. N. Bruton, ed.), pp. 177–195. Kluwer Academic Publishers, Dordrecht, The Netherlands.
Shapiro, D. Y. (1991). Intraspecific variability in social systems of coral reef fishes. *In* "The Ecology of Fishes on Coral Reefs" (P. F. Sale, ed.), pp. 331–355. Academic Press, San Diego, CA.
Shapiro, D. Y., and Giraldeau, L. A. (1996). Mating tactics in external fertilizers when sperm is limited. *Behav. Ecol.* **7**, 19–23.
Shapiro, D. Y., and Lubbock, R. (1980). Group sex ratio and sex reversal. *J. Theor. Biol.* **82**, 411–426.
Shapiro, D. Y., Hensley, D. A., and Appeldoorn, R. S. (1988). Pelagic spawning and egg transport in coral-reef fishes: A skeptical overview. *Environ. Biol. Fishes* **22**, 3–14.
Shapiro, D. Y., Sadovy, Y., and McGehee, A. (1993). Size composition, and spatial structure of the annual spawning of the red hind, *Epinephelus guttatus* (Pisces: Serranidae). *Copeia*, pp. 367–374.
Shapiro, D. Y, Garcia-Moliner, G., and Sadovy, Y. (1994a). Social system of an inshore stock of the red hind grouper, *Epinephelus guttatus* (Pisces: Serranidae*). Environ. Biol. Fishes* **41**, 415–422.
Shapiro, D. Y., Marconato, A., and Yoshikawa, T. (1994b). Sperm economy in a coral reef fish, *Thalassoma bifasciatum. Ecology* **75**, 1334–1344.
Shen, C. C., Lee, T., Chen, C. Y., Wang, C. H., Dai, C. F., and Li, L. A. (1996). The calibration of D[Sr/Ca] versus sea surface temperature relationship for *Porites* corals. *Geochim. Cosmochim. Acta* **60**, 3849–3858.
Shenk, T. M., White, G. C., and Burnham, K. P. (1998). Sampling-variance effects on detecting density dependence from temporal trends in natural populations. *Ecol. Monogr.* **68**, 445–463.
Shenker, J. M., Maddox, E. D., Wishinski, R., Pearl, A., Thorrold, S. R., and Smith, N. (1993). Onshore transport of settlement-stage Nassau grouper *Epinephelus striatus* and other fishes in Exuma Sound, Bahamas. *Mar. Ecol.: Prog. Ser.* **98**, 31–43.
Shepherd, G. R., and Idoine, J. S. (1993). Length based analysis of yield and spawning biomass per recruit for black sea bass *Centropristis striata,* a protogynous hermaphrodite. *Fish. Bull.* **91**, 328–337.
Shepherd, J. G. (1982). A versatile new stock-recruitment relationship for fisheries, and the construction of sustainable yield curves. *J. Cons. Int. Explor. Mer.* **40**, 67–75.
Shepherd, J. G., and Cushing, D. H. (1980). A mechanism for density-dependent survival of larval fish as the basis of a stock-recruitment relationship. *J. Cons. Int. Explor. Mer* **39**, 160–167.
Shepherd, J. G., and Cushing, D. H. (1990). Regulation in fish populations: Myth or mirage? *Philos. Trans. R. Soc. London, Ser. B* **330**, 151–164.
Shieh, K.-T., Wilson, W., Winslow, M., McBride, D. W., Jr., and Hopkins, C. D. (1996). Short-range orientation in electric fish: An experimental study of passive electrolocation. *J. Exp. Biol.* **199**, 2383–2393.
Shima, J. S. (1999a). An evaluation of processes that influence variability in abundance of a coral reef fish. Ph.D. Dissertation, University of California at Santa Barbara.
Shima, J. S. (1999b). Variability in relative importance of determinants of reef fish recruitment. *Ecol. Lett.* **2**, 304–310.
Shima J. S. (2001). Regulation of local populations of a coral reef fish via joint effects of density- and number-dependent mortality. *Oecologia* **126**, 58–65.
Shpigel, M., and Fishelson, L. (1991). Experimental removal of piscivorous groupers of the genus *Cephalopholis* (Serranidae) from coral habitats in the Gulf of Aqaba (Red Sea). *Environ. Biol. Fishes* **31**, 131–138.
Shulman, M. J. (1984). Resource limitation and recruitment patterns in coral reef fish assemblages. *J. Exp. Mar. Biol. Ecol.* **74**, 85–109.
Shulman, M. J. (1998). What can population genetics tell us

about dispersal and biogeographic history of coral-reef fishes? *Aust. J. Ecol.* **23**, 216–225.

Shulman, M. J., and Bermingham, E. (1995). Early life histories, ocean currents, and the population genetics of Caribbean reef fishes. *Evolution (Lawrence, Kans.)* **49**, 897–910.

Shulman, M. J., and Ogden, J. C. (1987). What controls tropical reef fish populations: Recruitment or benthic mortality? An example in the Caribbean reef fish *Haemulon flavolineatum*. *Mar. Ecol.: Prog. Ser.* **39**, 233–242.

Sibbing, F. A., Osse, J. W. M., and Terlow, A. (1986). Food handling in the carp (*Cyprinus carpio*): Its movement patterns, mechanisms and limitations. *J. Zool.* **210**, 161–203.

Sikkel, P. C. (1989). Egg presence and developmental stage influence spawning site choice by female garibaldi. *Anim. Behav.* **38**, 447–456.

Simberloff, D. (1988). The contribution of population and community biology to conservation science. *Annu. Rev. Ecol. Syst.* **19**, 473–511.

Simpson, A. L. (1992). Differences in body size and lipid reserves between maturing and no-maturing Atlantic salmon parr, *Salmo salar*. *Can. J. Zool.* **70**, 1737–1742.

Simpson, J. H., Tett, P. B., Argote-Espinoza, M. L., Edwards, A., Jones, K. J., and Savidge, G. (1982). Mixing and phytoplankton growth around an island in a stratified sea. *Cont. Shelf Res.* **1**, 15–31.

Sinclair, A. R. E. (1989). Population regulation in animals. *In* "Ecological Concepts: The Contribution of Ecology to an Understanding of the Natural World" (J. M. Cherrett, ed.), pp. 197–241. Blackwell, Oxford.

Sinclair, A. R. E., and Pech, R. P. (1996). Density dependence, stochasticity, compensation and predator regulation. *Oikos* **75**, 164–173.

Sinclair, D. J., Kinsley, L. P. J., and McCullogh, M. T. (1998). High resolution analysis of trace elements in corals by laser ablation ICP-MS. *Geochim. Cosmochim. Acta* **62**, 1889–1901.

Sinclair, M. (1988). "Marine Populations: An Essay on Population Regulation and Speciation," Washington Sea Grant Program. University of Washington Press, Seattle.

Sisneros, J. A., Tricas, T. C., and Luer, C. A. (1998). Response properties and biological function of the skate electrosensory system during ontogeny. *J. Comp. Physiol.* **183**, 87–99.

Sissenwine, M. P. (1984). Why do fish populations vary? *In* "Exploitation of Marine Communities" (R. M. May, ed.), pp. 59–94. Springer-Verlag, New York.

Sissenwine, M. P., Fogarty, M. J., and Overholz, W. J. (1988). Some fisheries management implications of recruitment variability. *In* "Fish Population Dynamics" (J. A. Gulland, ed.), 2nd ed. pp. 129–152. Wiley, London.

Skelton, P. W. (1988). The trans-Pacific spread of equatorial shallow-marine benthos in the Cretaceous. *Spec. Publ.—Geol. Soc. London* **37**, 247–253.

Sladek Nowlis, J., and Roberts, C. M. (1997). You can have your fish and eat it too: Theoretical approaches to marine reserve design. *Proc. 8th Int. Coral Reef Symp.*, Vol. 2, pp. 1907–1910.

Sladek Nowlis, J., and Roberts, C. M. (1999). Fisheries benefits and optimal design of marine reserves. *Fish. Bull.* **97**, 604–616.

Slatkin, M. (1995). A measure of population subdivision based on microsatellite allele frequencies. *Genetics* **139**, 457–462.

Smith, A. G., Hurley, A. M., and Briden, J. C., (1981). "Phanerozoic Paleocontinental World Maps." Cambridge Earth Science Series, Cambridge University Press, Cambridge, UK.

Smith, A. J. (1997). Live reef fisheries activities in the Republic of the Marshall Islands. *Secretariat of the Pacific Community Live Reef Fish Inf. Bull.* **3**, 14–15.

Smith, C. L. (1972). A spawning aggregation of Nassau grouper, *Epinephelus striatus* (Bloch). *Trans. Am. Fish. Soc.* **101**, 257–261.

Smith, C. L. (1973). Small rotenone stations: A tool for studying coral reef fish communities. *Am. Mus. Novit.* **2512**, 1–21.

Smith, C. L., and Tyler, J. C. (1975). Succession and stability in fish communities of the dome-shaped patch reefs in the West Indies. *Am. Mus. Novit.* **2572**, 1–18.

Smith, D. C., Fenton, G. E., Robertson, S. G., and Short, S. A. (1995). Age determination and growth of Orange Roughy (*Hoplostethus atlanticus*): A comparison of annulus counts with radiometric ageing. *Can. J. Fish. Aquat. Sci.* **52**, 391–401.

Smith, D. G. (1981). "The Cambridge Encyclopedia of Earth Sciences." Cambridge University Press, Cambridge, UK.

Smith, F. E. (1961). Density dependence in the Australian thrips. *Ecology* **42**, 403–407.

Smith, P. J. (1979). Esterase gene frequencies and temperature relationships in the New Zealand snapper *Chrysophrys auratus*. *Mar. Biol.* **53**, 305–310.

Smith, P. J., and Fujio, Y. (1982). Genetic variation in marine teleosts: High variability in habitat specialists and low variability in habitat generalists. *Mar. Biol.* **69**, 7–20.

Smith, P. J., Jamieson, A., and Birley, A. (1990). Electrophoretic studies and the stock concept in marine teleosts. *J. Cons. Int. Explor. Mer.* **47**, 231–245.

Smith, P. J., Francis, R. C. I. I., and McVeagh, M. (1991). Loss of genetic diversity due to fishing pressure. *Fish. Res.* **10**, 309–316.

Smith, R. J. F. (1992). Alarm signals in fishes. *Rev. Fish Biol. Fish.* **2**, 33–63.

Smith, R. J. F. (1997). Avoiding and deterring predators. *In* "Behavioral Ecology of Teleost Fishes" (J. J. Godin, ed.), pp. 163–190. Oxford University Press, New York.

Smith, R. J. F., and Lemly, A. D. (1986). Survival of fathead minnows after injury by predators and its possible role in the evolution of alarm signals. *Environ. Biol. Fishes* **15**, 147–149.

Smith, T. B., and Skulason, S. (1996). Evolutionary significance of resource polymorphism in fishes, amphibians, and birds. *Annu. Rev. Ecol. Syst.* **27**, 111–133.

Smith, T. B., Wayne, R. K., Girman, D. J., and Bruford, M. W. (1997). A role for ecotones in generating rainforest diversity. *Science* **276**, 1855–1857.

Smith, T. D. (1994). "Scaling Fisheries: The Science of Measuring the Effects of Fishing, 1855–1955." Cambridge University Press, Cambridge, England.

Snedecor, G. W, and Cochran, W. G. (1980). Statistical Methods, 7th ed. Iowa State University Press, Ames.

Soegiarto, A., and Polunin, N. V. C. (1982). "The Marine Environment of Indonesia." Report for the Government of Indonesia under Sponsorship of IUCN and WWF.

Sogard, S. M. (1997). Size-selective mortality in the juvenile stage of teleost fishes: A review. *Bull. Mar. Sci.* **60**, 1129–1157.

Sokolowski, B. H. A., and Popper, A. N. (1988). Transmission electron microscopic study of the saccule in the embryonic, larval, and adult toadfish, *Opsanus tau. J. Morphol.* **198**, 49–69.

Song, Y. H., and Haidvogel, D. (1994). A semiimplicit ocean circulation model using a generalized topography-following coordinate system. *J. Comput. Phys.* **115**, 228–244.

Sorbini, L. (1979). Segnalazione di un Plettognato Cretacico *Plectrocretacicus nov. gen. Boll. Mus. Civ. St. Nat., Verona* **4**, 1–4.

Sorbini, L., and Bannikov, A. F. (1991). The Cretaceous fishes of Nardo. II. An enigmatic spiny-rayed fish. *Bull. Soc. Paleontol. Ital.* **30**, 239–249.

Soulé, M. (1972). Phenetics of natural populations. III. Variation in insular populations of a lizard. *Am. Nat.* **106**, 429–446.

Soulé, M. (1976). Allozyme variation: Its determination in space and time. *In* "Molecular Evolution" (F. J. Ayala, ed.), pp. 60–77, Sinauer, Boston.

Soulé, M. E., and Wilcox, B. A. (1980). "Conservation Biology: An Evolutionary-Ecological Approach." Sinauer Asso. Sunderland, MA.

Spalding, M., and Grenfell, A. (1997). New estimates of global and regional coral reef areas. *Coral Reefs* **16**, 225–230.

Spanier, E. (1979). Aspects of species recognition by sound in four species of damselfish, genus *Eupomacentrus* (Pisces: Pomacentridae). *Z. Tierpsychol.* **51**, 301–316.

Sponaugle, S., and Cowen, R. K. (1994). Larval duration and recruitment patterns of two Caribbean gobies (Gobiidae): Contrasting early life histories in demersal spawners. *Mar. Biol.* **120**, 133–143.

Sponaugle, S., and Cowen, R. K. (1996a). Nearshore patterns of coral reef fish larval supply to Barbados, West Indies. *Mar. Ecol.: Prog. Ser.* **133**, 13–28.

Sponaugle, S., and Cowen, R. K. (1996b). Larval supply and patterns of recruitment for two Caribbean fishes, *Stegastes partitus* and *Acanthurus bahianus*. *Mar. Freshwater Res.* **47**, 433–447.

Sponaugle, S., and Cowen, R. K. (1997). Early life history traits and recruitment patterns of Caribbean wrasses (Labridae). *Ecol. Monogr.* **67**, 177–202.

Springer, V. G. (1982). Pacific plate biogeography, with special reference to shorefishes. *Smithson. Contrib. Zool.* **367**, 1–182.

Springer, V. G. (1988). The Indo-Pacific blenniid fish genus *Ecsenius*. *Smithson. Contrib. Zool.* **465**, 1–134.

Springer, V. G., and Williams, J. T. (1990). Widely distributed Pacific plate endemics and lowered sea level. *Bull. Mar. Sci.* **47**, 631–640.

Springer, V. G., and Williams, J. T. (1994). The Indo-West Pacific blenniid fish genus *Istiblennius* reappraised: A revision of *Istiblennius, Blenniella,* and *Paralticus,* new genus. *Smithson. Contrib. Zool.* **565**, 1–193.

Spurgeon, J. P. G. (1992). The economic valuation of coral reefs. *Mar. Pollut. Bull.* **24**, 529–536.

Srinivasan, M., Jones, G. P., and Caley, M. J. (1999). Experimental evaluation of the roles of habitat selection and interspecific competition in determining patterns of host use in two anemonefishes. *Mar. Ecol.: Prog. Ser.* **186**, 283–292.

St. John, J. (1999). Ontogenetic changes in the diet of the coral reef grouper *Plectropomus leopardus* (Serranidae): Patterns in taxa, size and habitat of prey. *Mar. Ecol.: Prog. Ser.* **180**, 233–246.

St. Mary, C. M. (1993). Novel sexual patterns in two simultaneously hermaphroditic gobies, *Lythrypnus dalli* and *Lythrypnus zebra*. *Copeia,* pp. 1062–1072.

Stadler, J. H. (1999). Species-recognition and sex-discrimination by males of the notchtongue goby, *Bathygobius curacao*. Unpublished Ph.D. Dissertation, University of Miami, Miami, FL.

Stanley, S. M. (1974). What has happened to the articulate brachiopods? *Geol. Soc. Am., Abstr. Program* **6**, 966–967.

Stanley, S. M. (1977). Trends, rates and patterns of evolution in the Bivalvia. *In* "Patterns of Evolution, as Illustrated by the Fossil Record" (A. Hallam, ed.), pp. 209–250. Elsevier, Amsterdam.

Stanley, S. M. (1984). Temperature and biotic chrises in the marine realm. *Geology* **12**, 205–208.

Stansfield, K. L., Bowman, M. J., Fauria, S. J., and Wilson, T. C. (1995). Water mass and coastal current variability near Barbados, West Indies. *J. Geophys. Res.* **100**, 24,819–24,830.

Stearns, D. E., Holt, G. J., Forward, R. B., Jr., and Pickering, P. L. (1994). Ontogeny of phototactic behavior in red drum larvae (Sciaenidae: *Sciaenops ocellatus*). *Mar. Ecol.: Prog. Ser.* **104**, 1–11.

Stearns, S. C. (1992). "The Evolution of Life Histories." Oxford University Press, New York.

Steele, M. A. (1996). Effects of predators on reef fishes: Separating cage artifacts from effects of predation. *J. Exp. Mar. Biol. Ecol.* **198**, 249–267.

Steele, M. A. (1997a). The relative importance of processes affecting recruitment of two temperate reef fishes. *Ecology* **78**, 129–145.

Steele, M. A. (1997b). Population regulation by post-settlement mortality in two temperate reef fishes. *Oecologia* **112**, 64–74.

Steele, M. A. (1998). The relative importance of predation and competition in two reef fishes. *Oecologia* **115**, 222–232.

Steele, M. A. (1999). Effects of shelter and predators on reef fishes. *J. Exp. Mar. Biol. Ecol.* **233**, 65–79.

Steele, M. A., Forrester, G. E., and Almany, G. R. (1998). Influences of predators and conspecifics on recruitment of a tropical and a temperate reef fish. *Mar. Ecol.: Prog. Ser.* **172**, 115–125.

Stein, R. A., Goodman, C. G., and Marschall, E. A. (1984). Using time and energetic measures of cost in estimating prey value for fish predators. *Ecology* **65**, 702–715.

Steinberg, J. C., Cummings, W. C., Brahy, B. D., and McBain (Spires), J. Y. (1965). Further bio-acoustic studies off the west coast of North Bimini, Bahamas. *Bull. Mar. Sci.* **15**, 942–963.

Steneck, R. S. (1983). Escalating herbivory and resulting adaptive trends in calcareous algal crusts. *Palaeobiology* **9**, 44–61.

Steneck, R. S. (1988). Herbivory on coral reefs; a synthesis. *Proc. 6th Int. Coral Reef Symp.*, Vol. 1, pp. 37–49.

Steneck, R. S. (1997). Crustose corallines, other algal functional groups, herbivores and sediments: Complex interactions along reef productivity gradients. *Proc. 8th Int. Coral Reef Symp.*, Vol. 1, pp. 695–700.

Steneck, R. S. (1998). Human influences on coastal ecosystems: Does overfishing create trophic cascades? *Trends Ecol. Evol.* **13**, 429–430.

Steneck, R. S., and Dethier, M. N. (1994). A functional group approach to the structure of algal-dominated communities. *Oikos* **69**, 476–498.

Stephens, P. A., and Sutherland, W. J. (1999). Consequences of the Allee effect for behaviour, ecology, and conservation. *Trends Ecol. Evol.* **14**, 401–405.

Stephens, P. A., Sutherland, W. J., and Freckleton, R. P. (1999). What is the Allee effect? *Oikos* **87**, 185–190.

Stepien, C. A., Randall, J. E., and Rosenblatt, R. H. (1994). Genetic and morphological divergence of a circumtropical complex of goatfishes: *Mulloidichthys vanicolensis, M. dentatus,* and *M. martinicus*. *Pac. Sci.* **48**, 44–56.

Stevens, G. C. (1989). The latitudinal gradient in geographical range: How so many species coexist in the tropics. *Am. Nat.* **133**, 240–256.

Stevens, G. C. (1996). Extending Rapoport's rule to Pacific marine fishes. *J. Biogeogr.* **23**, 139–148.

Stewart, B. D. (1998). Interactions between piscivorous coral reef fishes and their prey. Ph.D. Dissertation, James Cook University, Townsville, Australia.

Stewart, B. D., and Jones, G. P. (2001). Associations between the abundance of piscivorous fishes and their prey on coral reefs: Implications for prey-fish mortality. *Mar. Biol.* **138**, 383–397.

Stewart, B. D., Fenton, G. E., Smith, D. C., and Short, S. A. (1995). Validation of otolith-increment age estimates for a deepwater fish species, the warty oreo *Allocyttus verrucosus,* by radiometric analysis. *Mar. Biol.* **123**, 29–38.

Stewart-Oaten, A., and Murdoch, W. W. (1990). Temporal consequences of spatial density dependence. *J. Anim. Ecol.* **59**, 1027–1045.

Stiassny, M. L. (1991). Phylogenetic intrarelationships of the family Cichlidae: An overview. *In* "Cichlid Fishes Behavior, Ecology and Evolution" (M. H. A. Keenleyside, ed.), pp. 1–35. Chapman & Hall, London.

Stiassny, M. L., and Jensen, J. S. (1987). Labroid intrarelationships revisted: Morphological complexity, key innovations, and the study of comparative diversity. *Bull. Mus. Comp. Zool.* **151**, 269–319.

Stiling, P., Throckmorton, A., Silvanima, J., and Strong, D. R. (1991). Does spatial scale affect the incidence of density dependence—a field test with insect parasitoids. *Ecology* **72**, 2143–2154.

Stimson, J. S. (1990). Density dependent recruitment in the reef fish *Chaetodon miliaris*. *Environ. Biol. Fishes* **29**, 1–13.

Stobutzki, I. C. (1998). Interspecific variation in sustained swimming ability of late pelagic stage reef fish from two families (Pomacentridae and Chaetodontidae). *Coral Reefs* **17**, 111–119.

Stobutzki, I. C., and Bellwood, D. R. (1994). An analysis of the sustained swimming abilities of pre- and post-settlement coral reef fishes. *J. Exp. Mar. Biol. Ecol.* **175**, 275–286.

Stobutzki, I. C., and Bellwood, D. R. (1997). Sustained swimming abilities of the late pelagic stages of coral reef fishes. *Mar. Ecol.: Prog. Ser.* **149**, 35–41.

Stobutzki, I. C., and Bellwood, D. R. (1998). Nocturnal orientation to reefs by late pelagic stage coral reef fishes. *Coral Reefs* **17**, 103–110.

Stoddart, D. R. (1992). Biogeography of the Tropical Pacific. *Pac. Sci.* **46**, 276–293.

Stoner, A. W., Mehta, N., and Lee, T. N. (1997). Recruitment of *Strombus* veligers to the Florida reef tract: Relation to hydrographic events. *J. Shellfish Res.* **16**, 1–6.

Strathmann, R. R. (1974). The spread of sibling larvae of sedentary marine invertebrates. *Am. Nat.* **108**, 29–44.

Strathmann, R. R. (1982). Selection for retention or export of larvae in estuaries. *In* "Estuarine Comparisons" (V. S. Kennedy, ed.), pp. 521–536. Academic Press, New York.

Strathmann, R. R. (1986). What controls the type of larval development? Summary statement from the evolution session. *Bull. Mar. Sci.* **39**, 616–622.

Strathmann, R. R., and Eernisse, D. J. (1994). What molecular phylogenies tell us about the evolution of larval forms. *Am. Zool.* **34**, 502–512.

Streelman, J. T., and Karl, S. A. (1997). Reconstructing labroid evolution with single-copy nuclear DNA. *Proc. R. Soc. London, Ser. B* **264**, 1011–1020.

Strong, D. R. (1986). Density-vague population change. *Trends. Ecol. Evol.* **1**, 39–42.

Strong, D. R., Jr., Simberloff, D., Abele, L. G., and Thistle, A. B. (1984). "Ecological Communities: Conceptual Issues and the Evidence." Princeton University Press, Princeton, NJ.

Sturges, W. (1992). An estimate of the spectrum of Loop

Current variability from gappy data. *J. Phys. Oceanogr.* **22,** 1245–1256.

Suga, S., Wada, K., Taki, Y., and Ogawa, M. (1989). Iron concentration in teeth of Tetraodontiform fishes and its phylogenetic significance. *J. Dent. Res.* **68,** 1115–1123.

Sugihara, G. (1980). Minimal community structure: an explanation of species abundance patterns. *Am. Nat.* **116,** 770–787.

Summers, A. P., Darouian, K. F. Richmond, A. M., and Brainerd, E. L. (1998). Kinematics of aquatic and terrestrial prey capture in *Terrapene carolina,* with implications for the evolution of feeding in cryptodire turtles. *J. Exp. Zool.* **281,** 280–287.

Sundby, S., Bjorke, H., Soldal, A. V., and Olsen, S. (1989). Mortality rates during the early life stages and yearclass strength of Arctic cod (*Gadus morhua* L). *Rapp. P.-V. Reun.—Cons. Int. Explor. Mer* **191,** 351–358.

Suthers, I. M. (1996). Spatial variability of recent otolith growth and RNA indices in pelagic juvenile *Diaphus kapalae* (Myctophidae): An effect of flow disturbance near an island? *Mar. Freshwater Res.* **47,** 273–282.

Suthers, I. M. (1998). Bigger? Fatter? Or is faster growth better? Considerations on condition in larval and juvenile coral-reef fish. *Aust. J. Ecol.* **23,** 265–273.

Swart, P. K., Leder, J. J., Szmant, A. M., and Dodge, R. E. (1996). The origin of variations in the isotopic record of scleractinian corals II. Carbon. *Geochim. Cosmochim. Acta* **60,** 2871–2885.

Swearer, S. E., Caselle, J. E., Lea, D. W., and Warner, R. R. (1999). Larval retention and recruitment in an island population of a coral-reef fish. *Nature (London)* **402,** 799–802.

Sweatman, H. P. A. (1984). A field study of the predatory behavior and feeding rate of a piscivorous coral reef fish, the lizardfish *Synodus englemani. Copeia,* pp. 187–194.

Sweatman, H. P. A. (1985a). The timing of settlement by larval *Dascyllus aruanus*: Some consequences for larval habitat selection. *Proc. 5th Int. Coral Reef Congr.,* Vol. 5, pp. 367–371.

Sweatman, H. P. A. (1985b). The influence of adults of some coral reef fishes on larval recruitment. *Ecol. Monogr.* **55,** 469–485.

Sweatman, H. P. A. (1988). Field evidence that settling coral reef fish larvae detect resident fishes using dissolved chemical cues. *J. Exp. Mar. Biol. Ecol.* **124,** 163–174.

Sweatman, H. P. A. (1993). Tropical snapper (Lutjanidae) that is piscivorous at settlement. *Copeia,* pp. 1137–1139.

Sweatman, H. P. A., and St. John, J. (1990). Effects of selective settlement and of aggression by residents on distribution of young recruits of two tropical damselfishes. *Mar. Biol.* **105,** 247–252.

Symonds, G., Black, K. P., and Young, I. R. (1995). Wave-driven flow over shallow reefs. *J. Geophys. Res.—Oceans* **100,** 2639–2648.

Syms, C., and Jones, G. P. (2000). Disturbance, habitat structure, and the dynamics of a coral-reef fish community. *Ecology* **81,** 2714–2729.

Szedlmayer, S. T. (1998). Comparison of growth rate and formation of otolith increments in age-0 red snapper. *J. Fish Biol.* **53,** 58–65.

Talbot, F. H., Russell, B. C., and Anderson, G. R. V. (1978). Coral reef fish communities: unstable, high diversity systems? *Ecol. Monogr.* **48,** 425–440.

Tamarin, R. H. (1983). Animal population regulation through behavioral interactions. *In* "Advances in the Study of Mammalian Behavior" (J. F. Eisenberg and D. G. Kleiman, eds.), Am. Soc. Mammol. Spe. Publ. No. 7, pp. 698–720. Allen Press, Manhattan, KS.

Tang, K. L., Berendzen, P. B., Wiley, E. O., Morrissey, J. F., Winterbottom, R., and Johnson, G. D. (1999). The phylogenetic relationships of the suborder Acanthuroidei (Teleostei : Perciformes) based on molecular and morphological evidence. *Mol. Phylog. Evol.* **11,** 415–425.

Tannenbaum, R., Fishelson, L., and Harpaz, S. (1992). Prey odors as chemical stimulants of feeding behaviour in the Red Sea moray eel *Siderea grisea. In* "Chemical Signals in Vertebrates" (R. L. Doty and D. Müller-Schwarze, eds.), pp. 383–387. Plenum Press, New York.

Tarr, H. L. A. (1969). Contrast between fish and warm blooded vertebrates in enzyme systems of intermediary metabolism. *In* "Fish in Research" (O. W. Neuhaus and J. E. Halver, eds.), pp. 155–174. Academic Press, New York.

Tavolga, W. N. (1956). Visual, chemical and sound stimuli in sex-discriminatory behavior in the gobiid fish *Bathygobius soporator. Zoologica (N.Y.)* **41,** 49–64.

Tavolga, W. N. (1958a). The significance of underwater sounds produced by males of the gobiid fish, *Bathygobius soporator. Physiol. Zool.* **31,** 259–271.

Tavolga, W. N. (1958b). Underwater sounds produced by males of the blenniid fish, *Chasmodes bosquianus. Ecology* **39,** 759–760.

Tavolga, W. N. (1960). Sound production and underwater communication in fishes. *In* "Animal Sounds and Communication" (W. E. Lanyon and W. N. Tavolga, eds.), pp. 93–136. A. I. B. S., Washington, DC.

Tavolga, W. N. (1974). Signal/noise ratio and the critical band in fishes. *J. Acoust. Soc. Am.* **55,** 1323–1333.

Tavolga, W. N., and Wodinsky, J. (1963). Auditory capacities in fishes. *Bull. Am. Mus. Nat. Hist.* **126,** 179–239.

Tavolga, W. N., and Wodinsky, J. (1965). Auditory capacities in fishes: Threshold variability in the blue-striped grunt, *Haemulon sciurus. Anim. Behav.* **13,** 301–311.

Taylor, L., and Nolan, R. (1978). "Papers and Comments on Tropical Reef Fish," Working Pap. No. 34. Sea Grant College Program, University of Hawaii, Honolulu.

Tchernavin, V. V. (1953). "The Feeding Mechanism of a Deep Sea Fish, *Chauliodus sloani* (Schneider)." British Museum of Natural History, London.

Tedman, R. A. (1980a). Comparative study of the cranial morphology of the labrids *Choerodon venustus* and *Labroides dimidiatus* and the scarid *Scarus fasciatus* (Pisces: Perciformes). I. Head skeleton. *Aust. J. Mar. Freshwater Res.* **31,** 337–349.

Tedman, R. A. (1980b). Comparative study of the cranial morphology of the labrids *Choerodon venustus* and *Labroides dimidiatus* and the scarid *Scarus fasciatus* (Pisces: Perciformes). II. Cranial myology and feeding mechanisms. *Aust. J. Mar. Freshwater Res.* **31**, 351–372.

Tegner, M. J., and Dayton, P. K. (1997). The live-fish fishery of California. *Secretariat of the Pacific Community Live Reef Fish Inf. Bull.* **2**, 25–26.

Teichmann, H. (1959). Über die Leistung des Geruchssinnes beim Aal *(Anguilla anguilla L.). Z. Vergl. Physiol.* **42**, 206–254.

Terborgh, J. W. (1985). The vertical component of plant species diversity in temperate and tropical forests. *Am. Nat.* **126**, 760–776.

Tessier, N., Bernatchez, L., Presa, P., and Angers, B. (1995). Gene disversity analysis of mitochondrial DNA, microsatellites and allozymes in landlocked Atlantic salmon. *J. Fish Biol.* **47**, 156–163.

Tester, A. L. (1963). The role of olfaction in shark predation. *Pac. Sci.* **17**, 145–170.

Teyke, T. (1989). Learning and remembering the environment in the blind cave fish *Anoptichthys jordani. J. Comp. Physiol. A* **164**, 655–662.

Thompson, W. F., and Bell, F. H. (1934). Biological statistics of the Pacific halibut fishery. Effects of changes in intensity upon total yield and yield per unit gear. *Rep. Int. Halibut Comm.*, pp. 1–49.

Thorpe, J. E. (1986). Age at first maturity in Atlantic salmon, *Salmo salar*: Freshwater period influences and conflicts with smolting. *Can. Fish. Aquat. Sci., Spec. Publ.* **89**, 7–14.

Thorpe, J. E. (1989). Developmental variation in salmonid populations. *J. Fish Biol.* **35**, 295–303.

Thorpe, J. E., Talbot, C., Miles, M. S., and Keay, D. S. (1990). Control of maturation in cultured Atlantic salmon, *Salmo salar,* in pumped seawater tanks, by restricting food intake. *Aquaculture* **86**, 315–326.

Thorrold, S. R. (1993). Post-larval and juvenile scombrids captured in light traps: Preliminary results from the central Great Barrier Reef lagoon. *Bull. Mar. Sci.* **52**, 631–641.

Thorrold, S. R., and McKinnon, A. D. (1995). Response of larval fish assemblages to a riverine plume in coastal waters of the central Great Barrier Reef Lagoon. *Limnol. Oceanogr.* **40**, 177–181.

Thorrold, S. R., and Milicich, M. J. (1990). Comparison of larval duration and pre- and post-settlement growth in two species of damselfish, *Chromis atripectoralis* and *Pomacentrus coelestis* (Pisces: pomacentridae), from the Great Barrier Reef. *Mar. Biol.* **105**, 375–384.

Thorrold, S. R., and Shuttleworth, S. (2000). In situ analysis of trace elements and isotope ratios in fish otoliths using laser ablation sector field inductively coupled plasma mass spectrometry. *Can. J. Fish. Aquat. Sci.* **57**, 1232–1242.

Thorrold, S. R., and Williams, D. McB. (1989). Analysis of otolith microstructure to determine growth histories in larval cohorts of a tropical herring *(Herklotsichthys castelnaui). Can. J. Fish. Aquat. Sci.* **46**, 1615–1624.

Thorrold, S. R., Shenker, J. M., Mojica, R., Maddox, E. D., and Wishinski, E. (1994a). Temporal patterns in the larval supply of summer-recruitment reef fishes to Lee Stocking Island, Bahamas. *Mar. Ecol.: Prog. Ser.* **112**, 75–86.

Thorrold, S. R., Shenker, J. M., Wishinksi, E., Mojica, R., and Maddox, E. D. (1994b). Larval supply of shorefishes to nursery habitats around Lee Stocking Island, Bahamas. I. Small-scale distribution patterns. *Mar. Biol.* **118**, 555–566.

Thorrold, S. R., Shenker, J. M., Maddox, E. D., Mojica, R., and Wishinksi, E. (1994c). Larval supply of shorefishes to nursery habitats around Lee Stocking Island, Bahamas. II. Lunar and oceanographic influences. *Mar. Biol.* **118**, 567–578.

Thorrold, S. R., Jones, C. M., and Campana, S. E. (1997a). Response of otolith chemistry to environmental variations experienced by larval and juvenile Atlantic croaker *(Micropogonias undulatus). Limnol. Oceanogr.* **42**, 102–111.

Thorrold, S. R., Campana, S. E., Jones, C. M., and Swart, P. K. (1997b). Factors determining $\delta^{13}C$ and $\delta^{18}O$ fractionation in aragonitic otoliths of marine fish. *Geochim. Cosmochim. Acta* **61**, 2909–2919.

Thorrold, S. R., Jones, C. M., Swart, P. K., and Targett, T. E. (1998). Accurate classification of juvenile weakfish *Cynoscion regalis* to estuarine nursery habitats based on chemical signatures in otoliths. *Mar. Ecol.: Prog. Ser.* **173**, 253–265.

Thorrold, S. R., Campana, S. E., Jones, C. M., McLaren, J. W., and Lam, J. W. H. (1999). Trace element signatures in otoliths record natal river of juvenile American shad *(Alosa sapidissima). Limnol. Oceanogr.* **43**, 1826–1835.

Thorrold, S. R., Latkoczy, C., Swart, P. K., and Jones, C. M. (2001). Natal homing in a marine fish metapopulation. *Science* **291**, 297–299.

Thorson, G. (1950). Reproduction and larval development of Danish marine bottom invertebrates, with special reference to the planktonic larvae in the Sound (Oresund). *Medd. Komm. (Danm. Fisk.-og) Havunders., Ser. Plankton* **4**, 1–529.

Thresher, R. E. (1980). Clustering: Non-agonistic group contact in territorial reef fishes, with special reference to the Sea of Cortez damselfish, *Eupomacentrus rectifraenum. Bull. Mar. Sci.* **30**, 252–260.

Thresher, R. E. (1983a). Environmental correlates of the distribution of planktivorous fishes in the One Tree Reef Lagoon. *Mar. Ecol.: Prog. Ser.* **10**, 137–145.

Thresher, R. E. (1983b). Habitat effects on reproductive success in the coral reef fish, *Acanthochromis polyacanthus* (Pomacentridae). *Ecology* **64**, 1184–1199.

Thresher, R. E. (1984). "Reproduction in Reef Fishes." TFH Publication, Neptune City, NJ.

Thresher, R. E. (1985). Distribution, abundance, and reproductive success in the coral reef fish *Acanthochromis polyacanthus. Ecology* **66**, 1139–1150.

Thresher, R. E. (1991). Geographic variability in the ecology of coral reef fishes: evidence, evolution and possible implications. *In* "The Ecology of Fishes on Coral Reefs" (P. F. Sale, ed.), pp. 401–436. Academic Press, London.

Thresher, R. E., and Brothers, E. B. (1985). Reproductive ecology and biogeography of Indo-West Pacific angelfishes (Pisces: Pomacanthidae). *Evolution. (Lawrence, Kans.)* **39,** 878–887.

Thresher, R. E., and Gunn, J. S. (1986). Comparative analysis of visual census techniques for highly mobile, reef-associated piscivores (Carangidae). *Environ. Biol. Fishes* **17,** 93–116.

Thresher, R. E., Colin, P. L., and Bell, L. J. (1989). Planktonic duration, distribution and population structure of western and central Pacific damselfishes (Pomacentridae). *Copeia,* pp. 420–434.

Tintori, A. (1995). Biomechanical fragmentation in shell-beds from the Late Triassic of the Lombardian Basin (Northern Italy). Preliminary report. *Riv. Ital. Paleontol. Stratigr.* **101,** 371–380.

Tissot, B. N. (1999). Adaptive management of aquarium fish collecting in Hawaii. *Secretariat of the Pacific Community Live Reef Fish Inf. Bull.* **6,** 16–19.

Tissot, B. N., and Hallacher, L. E. (1999). "Impacts of Aquarium Collectors on Coral Reef Fishes in Kona, Hawaii," Tech. Rep. Department of Land and Natural Resources. Division of Aquatic Resources, Honolulu, HI.

Todd, J. H. (1971). The chemical languages of fishes. *Sci. Am.* **224,** 98–108.

Tolimieri, N. (1995). Effects of microhabitat characteristics on the settlement and recruitment of a coral reef fish at two spatial scales. *Oecologia* **102,** 52–63

Tolimieri, N. (1998a). The relationship among microhabitat characteristics, recruitment and adult abundance in the stoplight parrotfish, *Sparisoma viride,* at three spatial scales. *Bull. Mar. Sci.* **62,** 253–268.

Tolimieri, N. (1998b). Effects of substrata, resident conspecifics and damselfish on the settlement and recruitment of the stoplight parrotfish, *Sparisoma viride. Environ. Biol. Fishes* **53,** 393–404.

Tolimieri, N., Sale, P. F., Nemeth, R. S., and Gestring, K. B. (1998). Replenishment of populations of Caribbean reef fishes: Are spatial patterns of recruitment consistent through time? *J. Exp. Mar. Biol. Ecol.* **230,** 55–71.

Tolimieri, N., Jeffs, A., and Montgomery, J. (2000). Ambient sound as a cue for navigation by the pelagic larvae of reef fishes. *Mar. Ecol.: Prog. Ser.* **207,** 219–224.

Torricelli, P., Lugli, M., and Pavan, G. (1990). Analysis of sounds produced by male *Padogobius martensi* (Pisces, Gobiidae). *BioAcoustics* **2,** 261–275.

Townsend, D. W., Radtke, R. L., Malone, D. P., and Wallinga, J. P. (1995). Use of otolith strontium:calcium ratios for hindcasting larval cod distributions relative to water masses on Georges Bank. *Mar. Ecol.: Prog. Ser.* **119,** 37–44.

Tracey, D. M., and Horn, P. L. (1999). Background and review of ageing orange roughy (*Hoplostethus atlanticus,* Trachichthyidae) from New Zealand and elsewhere. *N. Z. J. Mar. Freshwater* **33,** 67–86.

Tremblay, M. J., Loder, J. W., Werner, F. E., Naimie, C. E., Page, F. H., and Sinclair, M. M. (1994). Drift of sea scallop larvae *Placopecten magellanicus* on Georges Bank: A model study of the roles of mean advection, larval behavior and larval origin. *Deep-Sea Res.* **41,** 7–49.

Tricas, T. C., and New, J. G. (1998). Sensitivity and response dynamics of elasmobranch electrosensory primary afferent neurons to near threshold fields. *J. Comp. Physiol.* **182,** 89–101.

Tricas, T. C., Michael, S. W., and Sisneros, J. A. (1995). Electrosensory optimization to conspecific phasic signals for mating. *Neurosci. Lett.* **202,** 129–132.

Trippel, E. A., Kjesbu, O. S., and Solemdal, P. (1997). Effects of adult age and size structure on reproductive output in marine fishes. *In* "Early Life History and Recruitment in Fish Populations," (R. C. Chambers and E. A. Trippel, eds.), pp. 31–62. Chapman & Hall, London.

Trivers, R. L. (1972). Parental investment and sexual selection. *In* "Sexual Selection and the Descent of Man, 1871–1971" (B. Campbell, ed.), pp. 136–179. Aldine Press, Chicago.

Tsuda, R. T., and Bryan, P. G. (1973). Food preferences of juvenile *Siganus rostratus* and *S. spinus* in Guam. *Copeia,* pp. 604–606.

Tsukamoto, K. (1992). Discovery of the spawning area for the Japanese eel. *Nature (London)* **356,** 789–791.

Tsukamoto K., Umezawa A., Tabeta O., Mochioka N., and Kajihara T. (1989). Age and birth date of *Anguilla japonica* (leptocephali) collected in Western North Pacific in September 1986. *Nippon Suisan Gakkaishi* **55,** 1023–1028.

Tsukamoto, Y., and Okiyama, M. (1997). Metamorphosis of the Pacific tarpon, *Megalops cyprinoides* (Elopiformes, Megalopidae) with remarks on development patterns in the Elopomorpha. *Bull. Mar. Sci.* **60,** 23–36.

Tuck, G. N., and Possingham, H. P. (2000). Marine protected areas for spatially structured exploited stocks. *Mar. Ecol.: Prog. Ser.* **192,** 89–101.

Tucker, J. W. (1998). "Marine Fish Culture." Kluwer Academic Publications, Boston.

Tupper, M., and Boutilier, R. G. (1995a). Size and priority at settlement determine growth and competitive success of newly settled Atlantic cod. *Mar. Ecol.: Prog. Ser.* **118,** 295–300.

Tupper, M., and Boutilier, R. G. (1995b). Effects of conspecific density on settlement, growth and post-settlement survival of a temperate reef fish. *J. Exp. Mar. Biol. Ecol.* **191,** 209–222.

Tupper, M., and Boutilier, R. G. (1997). Effects of habitat on settlement, growth, predation risk and survival of a temperate reef fish. *Mar. Ecol.: Prog. Ser.* **151,** 225–236.

Tupper, M., and Hunte, W. (1994). Recruitment dynamics of coral reef fishes in Barbados. *Mar. Ecol.: Prog. Ser.* **108,** 225–235.

Turchin, P. (1990). Rarity of density dependence or population regulation with lags? *Nature (London)* **344,** 660–663.

Turchin, P. (1995). Population regulation: Old arguments and a new synthesis. *In* "Population Dynamics" (N. Cappuccino and P. Price, eds.), pp. 19–40. Academic Press, San Diego, CA.

Turchin, P. (1999). Population regulation: A synthetic view. *Oikos* **84,** 153–159.

Turchin, P. (2002). "Complex Population Dynamics: A Theoretical/Empirical Synthesis." Princeton University Press, Princeton, NJ.

Turingan, R. G. (1994). Ecomorphological relationships among Caribbean tetraodontiform fishes. *J. Zool.* **233,** 493–521.

Turingan, R. G., and Wainwright, P. C. (1993). Morphological and functional bases of durophagy in the queen triggerfish, *Balistes vetula* (Pisces, Tetraodontiformes). *J. Morphol.* **215,** 101–118.

Turingan, R. G., Wainwright, P. C., and Hensley, D. A. (1995). Interpopulation variation in prey use and feeding biomechanics in Caribbean triggerfishes. *Oecologia* **102,** 296–304.

Turner, D. R., Whitfield, M., and Dickson, A. G. (1981). The equilibrium speciation of dissolved components in freshwater and seawater at 25°C and 1 atm pressure. *Geochim. Cosmochim. Acta* **45,** 855–881.

Turner, S. J., Thrush, S. F., Hewitt, J. E., Cummings, V. J., and Funnell, G. (1999). Fishing impacts and the degradation or loss of habitat structure. *Fish. Manage. Ecol.* **6,** 401–420.

Tyler, J. C. (1980). Osteology, phylogeny and higher classification of the fishes of the order Plectognathi (Tetraodontiformes). *NOAA tech. Rep. NMFS* **434,** 1–422.

Tyler, J. C. (1997). The Miocene fish *Marosichthys,* a putative tetraodontiform, actually a perciform surgeon fish (Acanthuridae) related to the recent *Naso. Beaufortia* **47,** 1–10.

Tyler, J. C., Johnson, G. D., Nakamura, I., and Collette, B. B. (1989). Morphology of *Luvarus imperialis* (Luvaridae), with a phylogenetic analysis of the Acanthuroidei (Pisces). *Smithson. Contrib. Zool.* **485,** 1–78.

Tyler, J. C., Johnson, G. D., Brothers, E. B., Tyler, D. M., and Smith, C. L. (1993). Comparative early life history of Western Atlantic squirrelfishes (Holocentridae): Age and settlement of *Rhynchichthys meeki,* and juvenile stages. *Bull. Mar. Sci.* **53,** 1126–1150.

Tyler, W. A., III, and Stanton, F. G. (1995). Potential influence of food abundance on spawning patterns in a damselfish, *Abudefduf abdominalis. Bull. Mar. Sci.* **57,** 610–623.

Tzeng, W. N. (1996). Effects of salinity and ontogenetic movements on strontium:calcium ratios in the otoliths of the Japanese eel, *Anguilla japonica. J. Exp. Mar. Biol. Ecol.* **199,** 111–122.

Ukawa, M., Higuchi, M., and Mito, S. (1966). Spawning habitats and early life history of a serranid fish, *Epinephelus akaara* (Temminck et Schlegel). *J. J. Ichthyol.* **13,** 156–161.

Underwood, A. J. (1992). Beyond BACI: The detection of environmental impacts on populations in the real, but variable, world. *J. Exp. Mar. Biol. Ecol.* **161,** 145–178.

Underwood, A. J. (1994). On beyond BACI: Sampling designs that might reliably detect environmental disturbances. *Ecol. Appli.* **4,** 3–15.

Underwood, A. J., and Fairweather, P. G. (1989). Supply-side ecology and benthic marine assemblages. *Trends Ecol. Evol.* **4,** 16–20.

Unger, L. M., and Sargent, R. C. (1988). Allopaternal care in the fathead minnow, *Pimephales promelas*:Females prefer males with eggs. *Behav. Ecol. Sociobiol.* **23,** 27–32.

Urick, R. J. (1975). "Principles of Underwater Sound." McGraw-Hill, New York.

Valdes, A. M., Slatkin, M., and Freimer, N. B. (1993). Allele frequencies at microsatellite loci: The stepwise mutation model revisited. *Genetics* **105,** 437–460.

Valentine, J. W., and Ayala, F. J. (1975). Genetic variation in *Frieleia halli,* a deep-sea brachiopod. *Deep-Sea Res.* **22,** 37–44.

Valiela, I. (1995). "Marine Ecological Processes." Springer, New York.

Valkenburgh, B. V. (1994). Ecomorphological analysis of fossil vertebrates and their palaeocommunities. *In* "Ecological Morphology—Integrative Organismal Biology" (P. C. Wainwright and S. M. Reilly, eds.), pp. 140–166. University of Chicago Press, Chicago.

Vallejo, B. V. (1997). Survey and review of the Philippine marine aquarium fish industry. *Sea Wind* **11,** 2–16.

Valone, T. J., and Brown, J. H. (1996). Desert rodents: Long-term responses to natural changes and experimental manipulations. *In* "Long-Term Studies of Vertebrate Communities" (M. L. Cody and J. A. Smallwood, eds.), pp. 555–583. Academic Press, San Diego, CA.

van Bergeijk, W. A. (1964). Directional and nondirectional hearing in fish. *In* "Marine Bio-acoustics" (W. N. Tavolga, ed.), pp. 281–298. Pergamon Press, New York.

Van Damme, R., and Aerts, P. (1997). Kinematics and functional morphology of aquatic feeding in Australian red necked turtles. *J. Morphol.* **233,** 113–125.

van den Berghe, E. P., Wernerus, F., and Warner, R. R. (1989). Female choice and the mating cost of peripheral males. *Anim. Behav.* **38,** 875–884.

van der Klaauw, C. J. (1948). Ecological studies and reviews. IV. Ecological morphology. *Bibl. Biotheor.* **4,** 27–111.

van der Veer, H. W. (1986). Immigration, settlement, and density-dependent mortality of a larval and early postlarval 0–group plaice (*Pleuronectes platessa*) population in the western Wadden Sea. *Mar. Ecol.: Prog. Ser.* **29,** 223–236.

van der Veer, H. W., and Bergman, M. J. N. (1987). Predation by crustaceans on a newly settled 0–group plaice *Pleuronectes platessa* population in the western Wadden Sea. *Mar. Ecol.: Prog. Ser.* **35,** 203–215.

Vandewalle, P., Havard, M, Claes, G., and De Vree, F. (1922). Mouvements des machoires pharyngiennes pendant la prise de nourriture chez la *Serranus scriba* (Linne, 1758) (Pisces: Serranidae). *Can J. Zool.* **70,** 145–160.

van Hasselt, M. J. F. M. (1978). A kinematic model for the jaw movements in some Labrinae (Piseces, Perciformes). *Neth. J. Zool.* **28,** 548–558.

van Hasselt, M. J. F. M. (1979a). Morphology and movements of jaw apparatus of some Labrinae (Pisces: Perciformes). *Neth J. Zool.* **29**, 52–108.

van Hasselt, M. J. F. M. (1979b). A kinematic model for the rapid opening and closing of the mouth in some Labrinae (Pisces, Perciformes). *Neth. J. Zool.* **29**, 352–401.

van Hasselt, M. J. F. M. (1980). A kinematic model for upper jaw protrusion in some Labrinae (Pisces, Perciformes). *Neth. J. Zool.* **30**, 82–128.

van Poollen, H. W., and Obara, A. M. (1984). "Hawaii's Marine Aquarium Fish Industry Profile. Studies on Marine Economics No. 3," UNIHI-SEAGRANT-ME-84-03, University of Hawaii Sea Grant College Program; Ocean Resources Office Contribution No. 14, Department of Planning and Economic Development, State of Hawaii, Honolulu.

van Rooij, J. M., Bruggemann, J. H., Videler, J. J., and Breeman, A. M. (1995). Plastic growth of the herbivorous reef fish *Sparisoma viride:* Field evidence for a trade-off between growth and reproduction. *Mar. Ecol.: Prog. Ser.* **122**, 93–105.

Van Valen, L. (1965). Morphological variation and width of ecological niche. *Am. Nat.* **94**, 377–390.

Vari, R. P. (1978). The terapon perches (Percoidei, Teraponidae). A cladistic analysis and taxonomic revision. *Bull. Am. Mus. Nat. Hist.* **159**, 175–340.

Varley, G. C., Gradwell, G. R., and Hassell, M. P. (1973). "Insect Population Ecology: An Analytical Approach." Blackwell Scientific, Oxford.

Veevers, J. J., and Ettriem, S. L. (1988). Reconstruction of Antarctica and Australia at breakup (95 ± 5 Ma) and before rifting (160 Ma). *Aust. J. Earth Sci.* **35**, 355–362.

Verlaque, M. (1987). Contribution à l'étude du phytobenthos d'un écosystème photophile thermophile marin en Méditerranée occidentale. Ph.D. Thesis, Universitè Aix-Marseille 2, France.

Vermeij, G. J. (1977). The Mesozoic marine revolution: Evidence from snails, predators and grazers. *Paleobiology* **3**, 245–258.

Vermeij, G. J. (1987). The dispersal barrier in the tropical Pacific: Implications for molluscan speciation and extinction. *Evolution (Lawrence, Kans.)* **41**, 1046–1058.

Vermeij, G. J. (1997). Strait answers from a twisted isthmus. *Paleobiology* **23**, 263–269.

Vermeij, G. J., and Petuch, E. J. (1986). Differential extinction in tropical American mollusks: Endemism, architecture, and the Panama land bridge. *Malacologia* **27**, 29–41.

Veron, J. E. N. (1995). "Corals in Space and Time. The Biogeography and Evolution of the Scleractinia." University of New South Wales Press, Sydney.

Victor, B. C. (1982). Daily growth increments and recruitment in two coral-reef wrasses, *Thalassoma bifasciatum* and *Halichoeres bivittatus*. *Mar. Biol.* **71**, 203–208.

Victor, B. C. (1983a). Settlement and larval metamorphosis produce distinct marks on the otoliths of the slippery dick, *Halichoeres bivittatus*. *In*: "The Ecology of Deep and Shallow Coral Reefs" (M. L. Reaka, ed.) pp. 47–67. NOAA symposia series for undersea research, **Vol. 1.** Nat. Oceanographic Atmospheric Admin., United States Department of Commerce, Washington D.C.

Victor, B. C. (1983b). Recruitment and population dynamics of a coral reef fish. *Science* **219**, 419–420.

Victor, B. C. (1986a). Duration of the planktonic larval stage of one hundred species of Pacific and Atlantic wrasses (family Labridae). *Mar. Biol.* **90**, 317–326.

Victor, B. C. (1986b). Larval settlement and juvenile mortality in a recruitment-limited coral reef fish population. *Ecol. Monogr.* **56**, 145–160.

Victor, B. C. (1986c). Delayed metamorphosis with reduced larval growth in a coral reef fish *(Thalassoma bifasciatum)*. *Can. J. Fish. Aquat. Sci.* **43**, 1208–1213.

Victor, B. C. (1987). The mating system of the Caribbean rosy razorfish *Xyrichtys martinicensis*. *Bull. Mar. Sci.* **40**, 152–160.

Victor, B. C. (1991). Settlement strategies and biogeography of reef fishes. *In* "The Ecology of Fishes on Coral Reefs" (P. F. Sale, ed.), pp. 231–260. Academic Press, San Diego, CA.

Victor, B. C., and Wellington, G. M. (2000). Endemism and the pelagic larval duration of reef fishes in the eastern Pacific Ocean. *Mar. Ecol.: Prog. Ser.* **205**, 241–248.

Vijayan, M. M., Mommsen, T. P., Glémet, H. C., and Moon, T. W. (1996). Metabolic effects of cortisol treatment in a marine teleost, the sea raven. *J. Exp. Biol.* **199**, 1509–1514.

Viladiu, C., Vandewalle, P., Osse, J. W. M., and Casinos, A. (1999). Suction feeding strategies of two species of Mediterranean Serranidae (*Serranus cabrilla* and *Serranus scriba*). *Neth. J. Zool.* **49**, 81–95.

Vincent, A. C. J. (1996). "The International Trade in Seahorses." Traffic International, Cambridge, U.K.

Vincent, A. C. J., and Sadler, L. M. (1995). Faithful pair bonds in wild seahorse, *Hippocampus whitei*. *Anim. Behav.* **50**, 1557–1569.

Vincent, A. C. J., and Sadovy, Y. (1998). Reproductive ecology in the conservation and management of fishes. *In* "Behavioral Ecology and Conservation Biology" (T. Caro, ed.), pp. 209–245 Oxford University Press, New York.

Viohl, G. (1990). Piscivorous fishes of the Solnhofen lithographic limestone. *In* "Evolutionary Palaeobiology of Behaviour and Coevolution" (A. J. Boucot, ed.), pp. 287–303. Elsevier, Amsterdam.

Vischer, H. A., Lannoo, M. J., and Heiligenberg, W. (1989). Development of the electrosensory nervous system in *Eigenmannia* (Gymnotiformes): I. The peripheral nervous system. *J. Comp. Neurol.* **290**, 16–40.

Vogel, S. (1994). "Life in Moving Fluids," 2nd ed. Princeton University Press, Princeton, NJ.

Wahle, R. A. (1992). Body-size dependent anti-predator mechanisms of the American lobster. *Oikos* **65**, 52–60.

Wainwright, P. C. (1987). Biomechanical limits to ecological performance: Mollusc crushing by the Caribbean hogfish, *Lachnolaimus maximus* (Labridae). *J. Zool.* **213**, 283–298.

Wainwright, P. C. (1988). Morphology and ecology: The functional basis of feeding constraints in Caribbean labrid fishes. *Ecology* **69**, 635–645.

Wainwright, P. C. (1989a). Prey processing in haemulid fishes: Patterns of variation in pharyngeal jaw muscle activity. *J. Exp. Biol.* **141**, 359–376.

Wainwright, P. C. (1989b). Functional morphology of the pharyngeal jaws in perciform fishes: An experimental analysis of the Haemulidae. *J. Morphol.* **200**, 231–245.

Wainwright, P. C. (1991). Ecomorphology: Experimental functional anatomy for ecological problems. *Am. Zool.* **31**, 680–693.

Wainwright, P. C. (1996). Ecological explanation through functional morphology: The feeding biology of sunfishes. *Ecology* **77**, 1336–1343.

Wainwright, P. C., and Friel, J. P. (2000). The effect of prey type on motor pattern variance in tetraodontiform fishes. *J. Exp. Zool.* **286**, 563–571.

Wainwright, P. C., and Reilly, S. M. (1994). "Ecological Morphology: Integrative Organismal Biology." University of Chicago Press, Chicago.

Wainwright, P. C., and Richard, B. A. (1995). Predicting patterns of prey use from morphology with fishes. *Environ. Biol. Fishes* **44**, 97–113.

Wainwright, P. C., and Shaw, S. S. (1999). Morphological basis of kinematic diversity in feeding sunfishes. *J. Exp. Biol.* **202**, 3101–3110.

Wainwright, P. C., and Turingan, R. G. (1993). Coupled versus uncoupled functional systems: Motor plasticity in the queen triggerfish *Balistes vetula*. *J. Exp. Biol.* **180**, 209–227.

Wainwright, P. C., and Turingan, R. G. (1997). Evolution of pufferfish inflation behavior. *Evolution (Lawrence, Kans.)* **51**, 506–518.

Wainwright, P. C., Westneat, M. W., and Bellwood, D. R. (2000). Linking feeding behavior and jaw mechanics in fishes. *In* "Biomechanics in Behavior" (P. Domenici and R. Blake, eds.), pp. 207–221. BIOS, Oxford.

Wainwright, P. C., Bellwood, D. R., and Westneat, M. W. Ecomorphology of locomotion in labrid fishes. *Environ. Biol. Fishes* (in press).

Walde, S. J. (1995). Internal dynamics and metapopulations: Experimental tests with predator-prey systems. *In* "Population Dynamics: New Approaches and Synthesis" (N. Cappuccino and P. W. Price, eds.), pp. 173–193. Academic Press, San Diego, CA.

Walker, J. A., and Westneat, M. W. (2000). Mechanical performance of aquatic rowing and flying. *Philos. Trans. R. Soc. London, Ser. B* **267**, 1875–1881.

Walker, M. M. (1984). Learned magnetic field discrimination in yellowfin tuna, *Thunnus albacares*. *J. Comp. Physiol. A* **155**, 673–679.

Walker, M. M., Diebel, C. E., Haugh, C. V., Pankhurst, P. M., Montgomery, J. C., and Green, C. R. (1997). Structure and function of the vertebrate magnetic sense. *Nature (London)* **390**, 371–376.

Walker, M. M., Diebel, C. E., and Green, C. R. (2000). Structure, function, and use of the magnetic sense in animals. *J. Appl. Physics* **87**, 4653–4658.

Walker, S. D. (1991). Reproducing the Atlantic Spadefish at the Tulsa zoo. *Seascope* **8**, 1–3.

Wallace, C. C. (1997). The Indo-Pacific centre of coral diversity re-examined at species level. *Proc. 8th Int. Coral Reef Symp.*, Vol. 1, pp. 365–370.

Wallace, C. C., Pandolfi, J. M., Ross, R., and Cronin, T. M. (1989). Indo-Pacific coral biogeography: A case study from the *Acropora selago* group. *Aust. Syst. Bot.—Aust. Biogeogr.* **4**, 199–210.

Walters, C. (1986). "Adaptive Management of Renewable Resources." Macmillan, New York.

Walters, C. (2000). Impacts of dispersal, ecological interactions, and fishing effort dynamics on efficacy of marine protected areas: How large should protected areas be? *Bull. Mar. Sci.* **66**, 745–757.

Walters, C., Pauly, D., and Christensen, V. (1999). Ecospace: Prediction of mesoscale spatial patterns in trophic relationships of exploited ecosystems, with emphasis on the impacts of marine protected areas. *EcoSystems* **2**, 539–554.

Wang, J. D. (1990). Numerical modeling of bay circulation *In* "The Sea" (B. LeMéhauté and D. Hanes, eds.), Vol. 9, Part B, pp. 1033–1066. Wiley, New York.

Ward, R. D., Woodwark, M., and Skibinski, D. O. F. (1994). A comparison of genetic diversity levels in marine, freshwater and anadromous fishes. *J. Fish Biol.* **44**, 213–232.

Warmolts, D. I. (2000). Marine medicinals in the aquarium trade. *In* "The Workshop on the Management and Culture of Marine Species used in Traditional Medicines" (M.-A. Moreau, M. J. Hall, and A. C. J. Vincent, eds.), pp. 107–109, Project Seahorse, Montreal, Canada.

Warner, R. R. (1975). The adaptive significance of sequential hermaphroditism in animals. *Am. Nat.* **109**, 61–84.

Warner, R. R. (1980). The coevolution of behavioral and life history characterisitics. *In* "Sociobiology: Beyond Nature-Nurture?" (G. W. Barlow and J. Silverberg, eds.), pp. 151–188. Westview, Boulder, CO.

Warner, R. R. (1984a). Deferred reproduction as a response to geared selection in a coral reef fish: A test of life history consequences. *Evolution (Lawrence, Kans.)* **38**, 148–162.

Warner, R. R. (1984b). Mating systems and hermaphroditism in coral reef fishes. *Am. Sci.* **72**, 128–136.

Warner, R. R. (1986). The environmental correlates of female infidelity in a coral reef fish. *In* "Indo-Pacific Fish Biology" (T. Uneyo, R. Arai, T. Taniuchi, and K. Matsuura, eds.), Proc. 2nd Int. Conf. Indo-Pac. Fishes, pp. 803–810. Ichthyological Society of Japan, Tokyo.

Warner, R. R. (1987). Female choice of sites versus males in a coral reef fish, *Thalassoma bifasciatum*. *Anim. Behav.* **35**, 1470–1478.

Warner, R. R. (1988a). Traditionality of mating-site preferences in a coral-reef fish. *Nature (London)* **335**, 719–721.

Warner, R. R. (1988b). Sex change and the size advantage model. *Trends Ecol. Evol.* **3**, 133–136.

Warner, R. R. (1988c). Sex change in fishes: Hypotheses, evidence, and objections. *Environ. Biol. Fishes* **22,** 81–90.

Warner, R. R. (1990a). Male versus female influences on mating-site determination in a coral reef fish. *Anim. Behav.* **39,** 540–548.

Warner, R. R. (1990b). Resource assessment versus traditionality in mating site determination. *Am. Nat.* **135,** 205–217.

Warner, R. R. (1991). The use of phenotypic plasticity in coral reef fishes as tests of theory in evolutionary ecology. *In* "The Ecology of Fishes on Coral Reefs" (P. F. Sale, ed.), pp. 387–398. Academic Press, San Diego, CA.

Warner, R. R. (1995). Large mating aggregations and daily long-distance spawning migrations in the bluehead wrasse, *Thalassoma bifasciatum. Environ. Biol. Fishes* **44,** 337–345.

Warner, R. R. (1997a). Evolutionary ecology: How to reconcile pelagic dispersal with local adaptation. *Coral Reefs* **16,** s115–s120.

Warner, R. R. (1997b). Sperm allocation in coral reef fishes. *BioScience* **47,** 561–564.

Warner, R. R. (1998). The role of extreme iteroparity and risk-avoidance in the evolution of mating systems. *J. Fish. Biol.* **53**(Supp. A), 82–93.

Warner, R. R., and Chesson, P. L. (1985). Coexistence mediated by recruitment fluctuations: A field guide to the storage effect. *Am. Nat.* **125,** 769–787.

Warner, R. R., and Dill, L. (2000). Courtship displays and coloration as indicators of safety rather than of male quality: The safety assurance hypothesis. *Behav. Ecol.* **11,** 444–451.

Warner, R. R., and Hoffman, S. G. (1980). Population density and the economics of territorial defense in a coral reef fish. *Ecology* **61,** 772–780.

Warner, R. R., and Hughes, T. P. (1988). The population dynamics of reef fishes. *Proc. 6th Int. Coral Reef Symp.*, Vol. 1, pp. 149–155.

Warner, R. R., and Robertson, D. R. (1978). Sexual patterns in the labroid fishes of the western Caribbean. I. The wrasses (Labridae). *Smithson. Contrib. Zool.* **254,** 1–27.

Warner, R. R., and Schultz, E. T. (1992). Sexual selection and male characteristics in the bluehead wrasse, *Thalassoma bifasciatum*: Mating site acquisition, mating site defense, and female choice. *Evolution (Lawrence, Kans.)* **46,** 1421–1442.

Warner, R. R., and Swearer, S. E. (1991). Social control of sex change in the bluehead wrasse, *Thalassoma bifasciatum* (Pisces: Labridae). *Biol. Bull. (Woods Hole, Mass.)* **181,** 199–204.

Warner, R. R., Shapiro, D. Y., Marconato, A., and Petersen, C. W. (1995). Sexual conflict: males with highest mating success convey the lowest fertilization benefits to females. *Proc. R. Soc. London, Ser. B* **262,** 135–139.

Warner, R. R., Swearer, S. E. and Caselle, J. E. (2000). Larval accumulation and retention: Implications for the design of marine reserves and essential fish habitat. *Bull. Mar. Sci.* **66,** 821–830.

Wass, R. C. (1982). The shoreline fishery of American Samoa: Past and present. *In* "Marine and Coastal Processes in the Pacific: Ecological Aspects of Coastal Zone Management" (J. L. Munro, ed.), pp. 51–83. UNESCO-ROSTSEA, Jakarta.

Watson, W. (1996). Chaetodontidae: Butterflyfishes. *In* "The Early Stages of Fishes in the California Current Region," Atlas No. 33 (H. G. Moser, ed.), pp. 1050–1053. California Cooperative Oceanic Fishery Investigations, San Diego.

Watson, W., and Leis, J. M. (1974). Ichthyoplankton of Kaneohe Bay, Hawaii: A oneyear study of fish eggs and larvae. *Univ. Hawaii Sea Grant Program Tech. Rep.* **75–01,** 1–178.

Weaver, D. C. (1997). Feeding ecology and morphology of three sea basses (Pisces: Serranidae) in the Northeastern Gulf of Mexico. *In* "Florida Big Bend Coastal Research Workshop. Toward a Scientific Basis for Ecosystem Management" (W. J. Lindberg, ed.), pp. 81–83.

Webb, J. F. (1998). Laterophysic connection: A unique link between the swimbladder and the lateral line system in *Chaetodon* (Perciformes: Chaetodontidae). *Copeia,* pp. 1032–1036.

Webb, P. W., and Weihs, D. (1986). Functional locomotor morphology of early life history stages of fishes. *Trans. Am. Fish. Soc.* **115,** 115–127.

Webster, M. S., and Hixon, M. A., (2000). Mechanisms and individual consequences of intraspecific competition in a coral-reef fish. *Mar. Ecol.: Prog. Ser.* **196,** 187–194.

Weidman, C. R., and Millner, R. (2000). High-resolution stable isotope records from North Atlantic cod. *Fish. Res.* **46,** 327–342.

Weinberg, J. R., Caswell, H., and Whitlatch, R. B. (1986). Demographic importance of ecological interactions: How much do statistics tell us? *Mar. Biol.* **93,** 305–310.

Weir, B. S., and Cockerham, C. C. (1984). Estimating F-statistics for the analysis of population structure. *Evolution (Lawrence, Kans.)* **38,** 822–831.

Welch, J. M., Forward, R. B., and Howd, P. A. (1999). Behavioral responses of blue crab *Callinectes sapidus* postlarvae to turbulence: Implications for selective tidal stream transport. *Mar. Ecol.: Prog. Ser.* **179,** 135–143.

Wellington, G. M. (1992). Habitat selection and juvenile persistence control the distribution of two closely-related Caribbean damselfishes. *Oecologia* **90,** 500–508.

Wellington, G. M., and Victor, B. C. (1988). Variation in components of reproductive success in an undersaturated population of coral-reef damselfish: A field perspective. *Am. Nat.* **131,** 588–601.

Wellington, G. M., and Victor, B. C. (1989a). Planktonic larval duration of one hundred species of Pacific and Atlantic damselfishes (Pomacentridae). *Mar. Biol.* **101,** 557–567.

Wellington, G. M., and Victor, B. C. (1989b). Regional differences in duration of the planktonic larval stage of reef fishes in the eastern Pacific Ocean. *Mar. Biol.* **113,** 491–498.

Werner, E. E. (1977). Species packing and niche complementarity in three sunfishes. *Am. Nat.* **111,** 553–578.

Werner, E. E., and Anholt, B. R. (1993). Ecological consequences of the tradeoff between growth and mortality rates mediated by foraging activity. *Am. Nat.* **142,** 242–272.

Werner, E. E., and Gilliam, J. F. (1984). The ontogenetic niche and species interactions in size-structured populations. *Annu. Rev. Ecol. Syst.* **15,** 393–425.

Werner, E. E. and Hall, D. J. (1974). Optimal foraging and size selection of prey by the bluegill sunfish (*Lepomis macrochirus*). *Ecology* **55,** 1042–1052.

Werner, E. E., Gilliam, J. F., Hall, D. J., and Mittelbach, G. (1983). An experimental test of the effects of predation risk on habitat use in fish. *Ecology* **64,** 1540–1548.

Werner, F. E., Page, F. H., Lynch, D. R., Loder, J. W., Lough, R. G., Perry, R. I., Greenberg, D. A., and Sinclair, M. M. (1993). Influences of mean advection and simple behavior on the distribution of cod and haddock early life stages on Georges Bank. *Fish. Oceanogr.* **22,** 43–64.

Werner, F. E., Perry, R. I., Lough, R. G., and Naimie, C. E. (1996). Trophodynamic and advective influences on Georges Bank larval cod and haddock. *Deep-Sea Res.* **43,** 1,793–1,822.

Werner, F. E., Quinlin, J. A., Blanton J. O., and Luettich, R. A. (1997). The role of hydrodynamics in explaining variability in fish populations. *J. Sea Res.* **37,** 135–212.

West, I. F., and Gauldie, R. W. (1994). Determination of fish age using ^{210}Pb:^{226}Ra disequilibrium methods. *Can. J. Fish. Aquat. Sci.* **51,** 2333–2340.

Westneat, M. W. (1990). Feeding mechanics of teleost fishes (Labridae: Perciformes): A test of four-bar linkage models. *J. Morphol.* **205,** 269–295.

Westneat, M. W. (1991). Linkage biomechanics and evolution of the jaw protrusion mechnism of the sling-jaw wrasse, *Epibulus insidiator*. *J. Exp. Biol.* **159,** 165–184.

Westneat, M. W. (1994). Transmission force and velocity in the feeding mechanisms of labrid fishes (Teleostei, Perciformes). *Zoomorphology* **114,** 103–118.

Westneat, M. W. (1995). Feeding, function, and phylogeny: Analysis of historical biomechanics in labrid fishes using comparative methods. *Syst. Biol.* **44,** 361–383.

Westneat, M. W. (1996). Functional morphology of aquatic flight in fishes: Kinematics, electromyography, and mechanical modelling of labriform locomotion. *Am. Zool.* **36,** 582–598.

Westneat, M. W., and Wainwright, P. C. (1989). The feeding mechanism of the sling-jaw wrasse, *Epibulus insidiator* (Labridae; Teleostei): Evolution of a novel functional system. *J. Morphol.* **202,** 129–150.

Wheeler, J. A. (1996). "The Marine Aquarium Trade: A Tool for Coral Reef Conservation," Sustainable Development and Conservation Biology Program. University of Maryland, College Park.

White, A. T. (1988). The effect of community-managed marine reserves in the Philippines on their associated coral reef fish populations. *Asian Fish. Sci.* **2,** 27–41.

White, B. N. (1986). The isthmian link, antitropicality and American biogeography: Dstributional history of the Atherinopsinae (Pisces: Atherinidae). *Syst. Zool.* **35,** 176–194.

Wieser, W., and Kaufmann, R. (1998). A note on interactions between temperature, viscosity, body size, and swimming energetics in fish larvae. *J. Exp. Biol.* **201,** 1369–1372.

Wilga, C. D., and Motta, P. J. (2000). Durophagy in sharks: Feeding mechanics of the hammerhead *Sphyrna tiburo*. *J. Exp. Biol.* **203,** 2781–2796.

Wilga, C. D., Wainwright, P. C., and Motta, P. J. (2000). Evolution of jaw mechanics in vertebrates: Insights from Chondrichthys. *Biol. J. Linn. Soc.* **71,** 165–185.

Wilkinson, C. R. (1996). Global change and coral reefs: Impacts on reefs, economies and human cultures. *Global Change Biol.* **2,** 547–558.

Williams, D. McB. (1980). Dynamics of the pomacentrid community on small patch reefs in One Tree Lagoon (Great Barrier Reef). *Bull. Mar. Sci.* **30,** 159–170.

Williams, D. McB. (1982). Patterns in the distribution of fish communities across the central Great Barrier Reef. *Coral Reefs* **1,** 35–43.

Williams, D. McB. (1986). Temporal variation in the structure of reef slope fish communities (central Great Barrier Reef): Short-term effects of *Acanthaster planci* infestation. *Mar. Ecol.: Prog. Ser.* **28,** 157–164.

Williams, D. McB. (1991). Patterns and process in the distribution of coral reef fishes. *In* "The Ecology of Fishes on Coral Reefs" (P. F. Sale, ed.), pp. 437–474. Academic Press, San Diego, CA.

Williams, D. McB., and English, S. (1992). Distribution of fish larvae around a coral reef: Direct detection of a mesoscale, multi-specific patch? *Cont. Shelf Res.* **12,** 923–937.

Williams, D. McB., and Hatcher, A. I. (1983). Structure of fish communities on outer slopes of inshore, mid-shelf and outer shelf reefs of the Great Barrier Reef. *Mar. Ecol.: Prog. Ser.* **10,** 239–250.

Williams, D. McB., Wolanski, E., and Andrews, J. C. (1984). Transport mechanisms and the potential movement of planktonic larvae in the central region of the Great Barrier reef. *Coral Reefs* **3,** 229–236.

Williams, D. McB., English, S., and Milicich, M. J. (1994). Annual recruitment surveys of coral reef fishes are good indicators of patterns of settlement. *Bull. Mar. Sci.* **54,** 314–331.

Williams, P. C., Gibbons, D., Margules, C., Rebelo, A., Humphries, C., and Pressey, R. A. (1996a). Comparison of richness hotspots, rarity hotspots, and complementarity areas for conserving diversity of British birds. *Conserv. Biol.* **10,** 155–174.

Williams, P. C., Brown, J. A., Gotceitas, V., and Pepin, P. (1996b). Developmental changes in escape response performance of five species of marine larval fish. *Can. J. Fish. Aquat. Sci.* **53,** 1246–1253.

Willis, B. L., Babcock, R. C., Harrison, P. L., and Wallace, C. C. (1997). Hybridization and breeding incompatibilities within the mating systems of mass spawning reef corals. *Proc. 8th Int. Coral Reef Symp.,* Vol. 1, pp. 81–90.

Wilson, A. B., Noack-Kunnmann, K., and Meyer, A. (2000). Incipient speciation in sympatric Nicaraguan crater lake cichlid fishes: Sexual selection versus ecological diversification. *Proc. R. Soc. London, Ser. B* **267**, 2133–2141.

Wilson, B. R., and Allen, G. R. (1987). Major components and distribution of marine fauna. *In* "Fauna of Australia 1A" (G. R. Dyne and D. W. Walton, eds.), pp. 43–68. Australian Government Publishing Service, Canberra.

Wilson, D. T. (2000). Patterns and processes of coral reef fish larval supply to the San Blas Archipelago, Caribbean Panama. Unpublished Ph.D. Thesis, James Cook University, Townsville, Australia.

Wilson, D. T., and McCormick, M. I. (1997). Spatial and temporal validation of settlement marks in the otoliths of tropical reef fishes. *Mar. Ecol.: Prog. Ser.* **153**, 259–271.

Wilson, D. T., and McCormick, M. I. (1999). Microstructure of settlement-marks in the otoliths of tropical reef fishes. *Mar. Biol.* **134**, 29–41.

Wilson, M. E. J., and Rosen, B. R. (1998). Implications of paucity of corals in the Paleogene of SE Asia: Plate tectonics or Centre of Origin? *In* "Biogeography and Geological Evolution of SE Asia" (R. Hall and J. D. Holloway, eds.), pp. 165–195. Backhuys Publishers, Leiden.

Wilson, S., and Bellwood, D. R. (1997). Cryptic dietary components of territorial damselfishes (Pomacentridae, Labroidei). *Mar. Ecol.: Prog. Ser.* **153**, 299–310.

Wing, S. R., Largier, J. L., Botsford, L. W., and Quinn, J. F. (1995a). Settlement and transport of benthic invertebrates in an intermittent upwelling region. *Limnol. Oceanogr.* **40**, 316–329.

Wing, S. R., Botsford, L. W., Largier, J. L., and Morgan, L. E. (1995b). Spatial structure of relaxation events and crab settlement in the northern California upwelling system. *Mar. Ecol.: Prog. Ser.* **128**, 199–211.

Winn, H. E. (1964). The biological significance of fish sounds. *In* "Marine Bio-acoustics" (W. N. Tavolga, ed.), pp. 213–231. Pergamon Press, New York.

Winn, H. E. (1972). Acoustic discrimination by the toadfish with comments on signal systems. *In* "Behavior of Marine Animals. Current Perspectives in Research" (H. E. Winn and B. L. Olla, eds.), Vol. 2, pp. 361–385. Plenum Press, New York.

Winterbottom, R. (1974). The familial phylogeny of the Tetraodontiformes (Acanthopterygii: Pisces) as evidenced by their comparative myology. *Smithson. Contrib. Zool.* **155**, 1–201.

Winterbottom, R. (1986). Revision and vicariance biogeography of the subfamily Congrogadinae (Pisces: Perciformes: Pseudochromidae). *Indo-Pac. Fishes* **9**, 1–34.

Winterbottom, R., and McLennan, D. A. (1993). Cladogram versatility: Evolution and biogeography of Acanthuroid fishes. *Evolution (Lawrence, Kans.)* **47**, 1557–1571.

Winterbottom, R., Reist, J. D., and Goodchild, C. D. (1984). Geographic variation in *Congrogadus subductens* (Teleostei, Perciformes, Congrogadidae). *Can. J. Zool.* **62**, 1605–1617.

Wolanski, E. (1993). Facts and numerical artefacts in modelling the dispersal of Crown-of-thorns starfish larvae in the Great Barrier Reef. *Aust. J. Mar. Freshwater Res.* **44**, 427–436.

Wolanski, E. (1994). "Physical Oceanographic Processes of the Great Barrier Reef." CRC Press, Boca Raton, FL.

Wolanski, E., and Delesalle, B. (1995). Upwelling by internal waves, Tahiti, French Polynesia. *Cont. Shelf Res.* **15**, 357–368.

Wolanski, E., and Hamner, W. M. (1988). Topographically controlled fronts in the ocean and their biological significance. *Science* **241**, 177–181.

Wolanski, E., and Sarenski, J. (1997). Larvae dispersion in coral reefs and mangroves. *Am. Sci.* **85**, 236–243.

Wolanski, E., and Van Seden, D. (1983). Mixing of Burdekin River flood waters in the Great Barrier Reef. *Aust. J. Mar. Freshwater Res.* **34**, 23–47.

Wolanski, E., Imberger, J., and Heron, M. L. (1984a). Island wakes in shallow coastal waters. *J. Geophys. Res.* **89**, 10,553–10,569.

Wolanski, E., Pickard, G. L., and Jupp, D. L. B. (1984b). River plumes, coral reefs and mixing in the Gulf of Papua and northern Great Barrier Reef. *Estuarine Coastal Shelf Sci.* **18**, 291–314.

Wolanski, E., Burrage, D., and King, B. (1989). Trapping and dispersion of coral eggs around Bowden Reef, Great Barrier Reef, following mass coral spawning. *Cont. Shelf Res.* **9**, 479–496.

Wolanski, E., Doherty, P., and Carleton, J. (1997). Directional swimming of fish larvae determines connectivity of fish populations on the Great Barrier Reef. *Naturwissenschaften* **84**, 262–268.

Wolanski, E., King, B., and Spagnol, S. (1999). The implications of oceanographic chaos for coastal management. *In* "Environmental Science: Perspectives on Integrated Coastal Zone Management" (W. Salomons, R. K. Turner, L. Drude de Lacerda, and S. Ramachandran, eds.), pp. 129–141. Springer-Verlag, Berlin.

Wolda, H. (1989). The equilibrium concept and density dependence tests: What does it all mean? *Oecologia* **81**, 430–432.

Wolda, H. (1995). The demise of the population regulation controversy? *Res. Popul. Ecol.* **37**, 91–93.

Wolda, H., and Dennis, B. (1993). Density dependence tests, are they? *Oecologia* **95**, 581–591.

Wolff, J. O. (1997). Population regulation in mammals: An evolutionary perspective. *J. Anim. Ecol.* **66**, 1–13.

Wood, E. (1985). "Exploitation of Coral Reef Fishes for the Aquarium Trade." Marine Conservation Society, Ross-on-Wye, UK.

Wood, E. (1992). "Trade in Tropical Marine Fish and Invertebrates for Aquaria: Proposed Guidelines and Labelling Scheme." Marine Conservation Society, Ross-on-Wye, UK.

Wood, E. M. (2001). "Collection of Coral Reef Fish for Aquaria: Global Trade, Conservation Issues and Management Strategies." Marine Conservation Society, Ross-on-Wye, UK.

Wood, R. (1993). Nutrients, predation and the history of reefs. *Palaios* **8**, 526–543.

Wood, R. (1995). The changing biology of reef-building. *Palaios* **10**, 517–529.

Wood, R. (1999). "Reef Evolution." Oxford University Press, Oxford.

Woodland, D. J. (1983). Zoogeography of the Siganidae (Pisces): An interpretation of distribution and richness patterns. *Bull. Mar. Sci.* **33**, 713–717.

Woodland, D. J. (1986). Wallace's line and the distribution of inshore fishes. *In* "Indo-Pacific Fish Biology" (T. Uyeno, R. Arai, T. Taniuchi, and K. Matsuura, eds.), Proc. 2nd Int. Conf. Indo-Pac. Fishes, pp. 453–469. Ichthyological Society of Japan, Tokyo.

Woodland, D. J. (1990). Revision of the fish family Siganidae with descriptions of two new species and comments on distribution and biology. *Indo-Pac. Fish.* **19**, 1–136.

Wooninck, L., Strassman, J. E., Queller, D. C., Fleischer, R., and Warner R. R. (1998). Characterization of hypervariable microsatellite markers in the bluehead wrasse, *Thalassoma bifasciatum*. *Mol. Ecol.* **7**, 1613–1614.

Wootton, R. J. (1998). "The Ecology of Teleost Fishes." Kluwer, Dordrecht, The Netherlands.

Workman, P. L., and Niswander, J. D. (1970). Population studies on southwestern Indian tribes. II. Local genetic differentiation in Papago. *Am. J. Hum. Genet.* **22**, 24–29.

Worthington, D. G., Doherty, P. J., and Fowler, A. J. (1995). Variation in relationship between otolith weight and age—Implications for the estimation of age of two tropical damselfish (*Pomacentrus moluccensis* and *P. wardi*). *Can. J. Fish. Aquat. Sci.* **52**, 233–242.

Wright, S. (1943). Isolation by distance. *Genetics* **28**, 114–138.

Wright, S. (1946). Isolation by distance under diverse systems of mating. *Genetics* **31**, 39–59.

Wright, S. (1978). "Evolution and the Genetics of Populations," Vol. 4. University of Chicago Press, Chicago.

Wust, G. (1964). "Stratification and Circulation in the Antillean-Caribbean Basins." Columbia University Press, New York.

Wyanski, D. M., White, D. B., and Barans, C. A. (2000). Growth, population age structure, and aspects of the reproductive biology of the snowy grouper, *Epinephelus niveatus*, off North Carolina and South Carolina. *Fish. Bull.* **98**, 199–218.

Yager, D. D., and Hopkins, C. D. (1993). Directional characteristics of tuberous electroreceptors in the weakly electric fish, *Hypopomus* (Gymnotiformes). *J. Comp. Physiol. A* **173**, 401–414.

Yamamori, K., Nakamura, N., Matsui, T., and Hara, T. J. (1988). Gustatory responses to tetrodotoxin and saxitoxin in fish: A possible mechanism for avoiding marine toxins. *Can. J. Fish. Aquat. Sci.* **45**, 2182–2186.

Yamano, H., Kayanne, H., Yonekura, N., Nakamura, H., and Kudo, K. (1998). Water circulation in a fringing reef located in a monsoon area: Kabira Reef, Ishigaki Island, Southwest Japan. *Coral Reefs* **17**, 89–99.

Yamaoka, K. (1978). Pharyngeal jaw structure in labrid fish. *Publ. Seto Mar. Biol. Lab.* **24**, 409–426.

Yamasaki, A., and Kuwahara, A. (1990). Preserved area to effect recovery of overfished Zuwai crab stocks off Kyoto Prefecture. *In* "Proceedings of the International Symposium on King and Tanner Crabs," Alaska Sea Grant College Program, pp. 575–578. University of Alaska, Fairbanks.

Yeung, C. (1996). Transport and retention of lobster phyllosoma larvae in the Florida Keys. Ph.D. Dissertation, University of Miami, Coral Gables, FL.

Yin, M. C., and Blaxter, J. H. S. (1987). Escape speeds of marine fish larvae during early development and starvation. *Mar. Biol.* **96**, 459–468.

Young, I. R., Black, K. P., and Heron, M. L. (1994). Circulation in the Ribbon Reef region of the Great Barrier Reef. *Cont. Shelf Res.* **14**, 117–142.

Young, L. G. L. (1997). Sustainability issues in the trade for wild and cultured aquarium species. *Market. Shipp. Aquat. Prod.*, Seattle, WA, *1996*, pp. 145–151.

Youson, J. H. (1988). First metamorphosis. *Fish Physiol.* **11**, 135–196.

Zakon, H. (1988). The electroreceptors: Diversity in structure and function. *In* "Sensory Biology of Aquatic Animals" (J. Atema, R. R. Fay, A. N. Popper, and W. N. Tavolga, eds.), pp. 813–850. Springer-Verlag, New York.

Zander, C. D. (1982). Feeding ecology of littoral gobiid and blenniid fish of the Banyuls area (Mediterranean Sea). I. Main food and trophic dimension of niche and ecotope. *Vie Milieu* **32**, 1–10.

Zeiske, E., Theison, B., and Breuker, H. (1992). Structure, development and evolutionary aspects of the peripheral olfactory system. *In* "Fish Chemoreception" (T. J. Hara, ed.), pp. 13–39. Chapman & Hall, New York.

Zeller, D. C. (1997). Home range and activity patterns of the coral trout *Plectropomus leopardus* (Serranidae). *Mar. Ecol.: Prog. Ser.* **154**, 65–77.

Zeller, D. C. (1998). Spawning aggregations: Patterns of movement of coral trout *Plectropomus leopardus* (Serranidae) as determined by ultrasonic telemetry. *Mar. Ecol.: Prog. Ser.* **162**, 253–263.

Zeller, D. C., and Russ, G. R. (1998). Marine reserves: Patterns of adult movement of the coral trout *(Plectropomus leopardus)* (Serranidae). *Can. J. Fish. Aquat. Sci.* **55**, 917–924.

Zhang, G., and Takemura, A. (1989). Acoustical behavior of brown goby, *Bathygobius fuscus*. *Bull. Fac. Fish., Nagasaki Univ.* **66**, 21–30.

Zottoli, S. J., and Danielson, P. D. (1989). Lateral line afferent and efferent systems of the goldfish with special reference to the Mauthner cell. *In:* "The Mechanosensory Lateral Line: Neurobiology and Evolution" (S. Coombs, P. Görner, and H. Münz, eds.), pp. 461–480. Springer-Verlag, New York.

Taxonomic Index

Subject Index

M

Q

R

S

T

U

V